AF668607

KHANNA'S

MECHANICAL
CONVENTIONAL TYPE

For All Types of Examinations & Interviews

U.P.S.C. Engineering Services Examination, I.A.S. Exams. Graduate Aptitude Test in Engineering (GATE) Conducted by I.I.T's for admission to M. Tech Courses, Objective Tests for Public Undertaking and Other Organizations, M.Tech Entrance Exams., Interviews and for Quick Review of the subject.

OTHER BOOKS BY THE SAME AUTHOR

- **Strength of Materials**
  ISBN : 978-93-81068-80-9
  Pages (Approx.) : 600

- **Fluid Mechanics**
  ISBN : 978-93-81068-65-6
  Pages (Approx.) : 680

- **Machine Design**
  ISBN : 978-93-82609-57-5
  Pages (Approx.) : 832

- **Machine Design Data Book**
  ISBN : 978-93-82609-51-3
  Pages (Approx.) : 300

- **Strength of Materials-I**
  ISBN : 978-93-81068-61-8
  Pages (Approx.) : 400

- **Fluid Machinery (Hydraulic Machines)**
  ISBN : 978-93-82609-28-5
  Pages (Approx.) : 500

- **Khanna's Mechanical Objective Type**
  ISBN : 978-93-82609-60-5
  Pages (Approx.) : 1050

- **Khanna's Mechanical Conventional Type**
  ISBN : 978-93-82609-61-2
  Pages (Approx.) : 1050

KHANNA'S

MECHANICAL CONVENTIONAL TYPE

For All Types of Examinations & Interviews

U.P.S.C. Engineering Services Examination, I.A.S. Exams. Graduate Aptitude Test in Engineering (GATE) Conducted by I.I.T's for admission to M. Tech Courses, Objective Tests for Public Undertaking and Other Organizations, M.Tech Entrance Exams., Interviews and for Quick Review of the subject.

Dr. Sadhu Singh

Formerly, **Professor & Head**, Mechanical Engineering Department
& **Dean,** Faculty of Engineering & Technology
Govind Ballabh Pant University of Agriculture & Technology,
Pantnagar - 263 145 (Uttrakhand)

Formerly, **Director** (Colleges),
Punjab Technical University, Jalandhar

KHANNA BOOK PUBLISHING CO. (P) LTD.

Publisher of Engineering and Computer Books

4C/4344, Ansari Road, Darya Ganj, New Delhi-110002
Phone: 011-23244447-48 **Mobile:** +91-99109 09320
E-mail: contact@khannabooks.com,
sales@khannapublishers.com
Website: www.khannabooks.com, www.khannapublishers.com

KHANNA'S MECHANICAL CONVENTIONAL TYPE

Dr. Sadhu Singh

Copyright © **Khanna Book Publishing Co. (P) Ltd.**

This book is sold subject to the condition that it shall not, by way of trade or otherwise, be lent, resold, hired out, or otherwise circulated without the publisher's prior consent in any form of binding or cover other than that in which it is published and without a similar condition including this condition being imposed on the subsequent purchaser and without limiting the rights under copyright reserved above, no part of this publication may be reproduced, stored in or introduced into retrieval system, or transmitted any form or by any means (electronic, mechanical, photocopying, recording or otherwise), without the prior written permission of both the copyright owner and the above mentioned publisher of this book.

ISBN: 978-93-82609-61-2

First Edition : **2013**
Second Edition : **2015**

Published by:
Khanna Book Publishing Co. (P) Ltd.
4C/4344, Ansari Road, Darya Ganj, New Delhi-110 002
Phone: 011-2324 44 47-48 **Mobile :** +91 9910909320
E-mail: contact@khannabooks.com
Website: www.khannabooks.com

Cover Design & Layout by
Ajay Arts, New Delhi

Printed in India by:
S.P.S. Printers & Binders, Delhi

PREFACE

Indian Engineering Services Examination (IESE) is conducted every year by Union Public Service Commission. All the central Govt. jobs in Engineering are filled through this examination.

An attempt has been made by the author to present as correct Solutions to the Questions as possible to the previous Question Papers from 1992 to 2014.

It is expected that the method of presenting the solutions shall be liked by the candidates while preparing for the examination. Further suggestions for the improvement of presenting the solutions shall be welcome by the author.

The author is extremely thankful to the publishers for bringing out this books in a short span of time. Special thanks are due to Mr. Mukul Seth and Mr. Puneet Khanna for their interest in checking the proofs of the manuscript so precisely. They have been otherwise very helpful during the preparation of the manuscript. Thanks are also due to my family members for their cooperation and help. Special thanks are due to my grand children Kanupreet and Amitoej for keeping my mood cheerful during odd hours of the day.

Dr. Sadhu Singh

CONTENTS

MECHANICAL ENGINEERING PAPER I—1992

(Conventional)

Q.1. **(*a*)** 3 kg of air at 1.50 bar pressure and 87°C temperature at condition 1 is compressed polytropically to condition 2 at pressure 7.50 bar, index of compression being 1.2. It is then cooled at constant pressure to condition 3 and then finally heated at constant temperature to its original condition 1. Find the net work done and heat transferred.

Sol.

Given: $p_1 = 1.5$ bar, $p_2 = p_3 = 7.5$ bar, $T_1 = 273 + 87 = 360$ K,

The $p - v$ diagram is shown in Fig. 1.

Process 1 – 2:

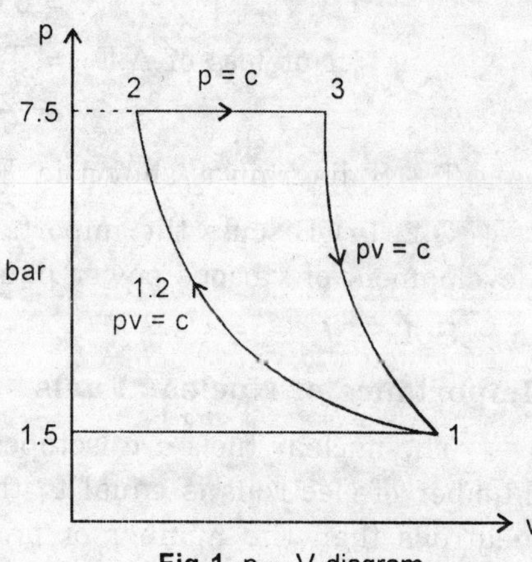

Fig 1. p – V diagram

$$\frac{T_2}{T_1} = \left(\frac{p_2}{p_1}\right)^{\frac{n-1}{n}}$$

$$T_2 = 360\left(\frac{7.5}{1.5}\right)^{\frac{1.2-1}{1.2}}$$

$$= 470.76 \text{ K}$$

Work done, $$W_{1-2} = \frac{mR(T_1 - T_2)}{n-1}$$

$$= \frac{3 \times 0.287(360 - 470.76)}{1.2 - 1}$$

$$= -476.82 \; kJ$$

Process 2 – 3:

Work done, $$W_{2-3} = mR(T_3 - T_2) = mR(T_1 - T_2) \qquad (\because T_3 = T_1)$$

$$= 3 \times 0.287\,(360 - 470.76)$$

$$= -95.36 \text{ kJ}$$

Proces 3 – 1:

Work done, $$W_{3-1} = mRT_3 \ln(p_3/p_1)$$

$$= 3 \times 0.287 \times 360 \ln(7.5/1.5) = 448.86 \text{ kJ}$$

Net work done, $$W_{net} = W_{1-2} + W_{2-3} + W_{3-1}$$

According to first law of thermodynamies.

$$\oint dQ = \oint dW$$

$\therefore$ Heat transferred $= -73.32 \; kJ$

(*b*) 500 kJ of heat is removed from a constant temperature heat reservoir maintained at 835K. The heat is received by a system at constant temperature of 720K. The temperature of the surroundings, the lowest available temperature, is 280K. Illustrate the problem by T-S diagram and calculate the net loss of available energy as a result of this irreversible heat transfer.

Sol.

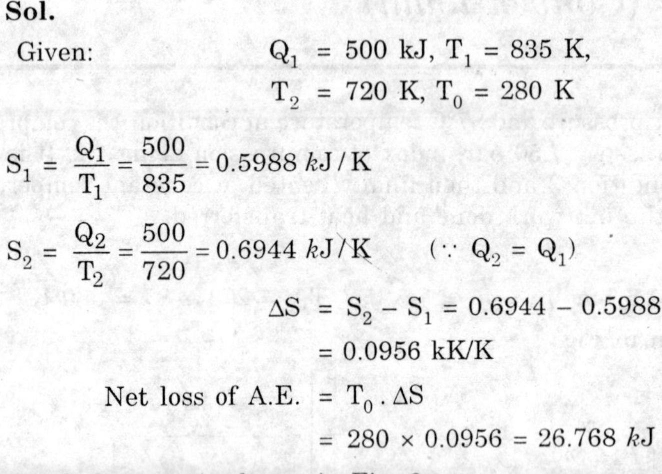

Given: $Q_1 = 500$ kJ, $T_1 = 835$ K,

$T_2 = 720$ K, $T_0 = 280$ K

$$S_1 = \frac{Q_1}{T_1} = \frac{500}{835} = 0.5988\ kJ/K$$

$$S_2 = \frac{Q_2}{T_2} = \frac{500}{720} = 0.6944\ kJ/K \qquad (\because Q_2 = Q_1)$$

$$\Delta S = S_2 - S_1 = 0.6944 - 0.5988$$

$$= 0.0956 \text{ kK/K}$$

$$\text{Net loss of A.E.} = T_0 \cdot \Delta S$$

$$= 280 \times 0.0956 = 26.768\ kJ$$

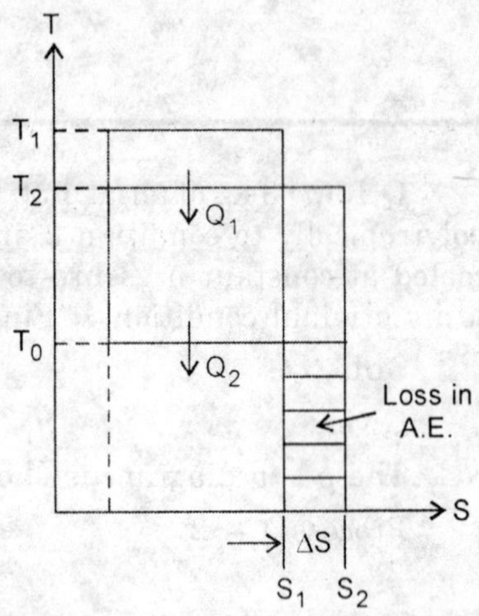

Fig 2. T – S Diagram

T – S diagram is shown in Fig. 2.

Q.2. (*a*) Discuss the importance of nuclear fuels and their properties required in the development of various power plants.

Sol.

Importance of Nuclear Fuels

The nuclear fuels are isotopes of Uramium, Thorium and Plutonium. In any atom, the number of electrons is equal to the number of protons. Atoms having different number of neutrons than the number of protons are known as isotopes. Only U^{235} is the naturally occuring isotope. Other isotopes can be formed from Th^{232} and Th^{238} by the absorption of one neutron. The isotopes have the same chemical properties and have the same atomic number. But the nuclear properties of each of the isotopes are different because of the different number of neutrons in the nucleus.

During a nuclear reaction, the change in the mass of the particle represents the release or absoption of energy. If the total mass of the particle after the reaction is reduced, the process releases the energy of the order of MeV (Mega electron Volt, $1\text{eV} = 1.602 \times 10^{-19}$ J). Nuclear reaction takes place either by fission or fusion. When a nuecleus splits into two equal masses, the transformation is known as fission. This reaction is suited only to heavy nucleus such as U^{233}, U^{235}, Pu^{239}, etc. This reaction always releases the energy. The number of neutrons keeps on multiplying rapidly during fission. This is called the chain reaction. Nuclear fussion is the process of combining or fusing two lighter nuclei into a stable and heavier nuclei. In this case also, large amount of energy is released due to decrease in mass of the combined nuclei. The apparatus in which nuclear fission is produced in the form of a controlled self-sustaining chain reaction is called the nuclear reactor. Nuclear fuels are assembed in a specific shape, called the fuel elements.

Nuclear fuels have great importance in view of the depleting natural resources for powr generation. The fuel consumption is very small as compared to conventional fuels, therefore, cost of fuel transportation is very less. Nuclear fuels produce valuable fissible material.

Required Properties of Nuclear Fuels

The nuclear fuel elements should have the following properties:

1. High tensile strength at high temperature to prevent buckling and to bear thermal stresses.
2. High thermal conductivity for better heat transfer.
3. Better machinability and higher ductility to reduce machining cost.
4. High corrosion resistance against coolant.
5. High radiation stability to avoid distortion and buckling.

(*b*) A simple jet carburettor has to supply 5 kg of air per minute. The air is at a pressure of 1.013 bar and at a temperature of 27°C. Calculate the throat diameter of the choke for air flow velocity of 90 m/s. Take velocity coefficient to be 0.8. Assume isentropic flow. Assume the flow to be compressible.

Sol.

Given: $\dot{m}_a = 5$ kg/min, $p_1 = 1.013$ bar, $T_1 = 273 + 27 = 300$ K,

$c_2 = 90$ m/s, $C_c = 0.8$

Fig. 3 show a simple jet carburettor. Applying Bernoulli's equation between entrance (1) and throat (2), we have

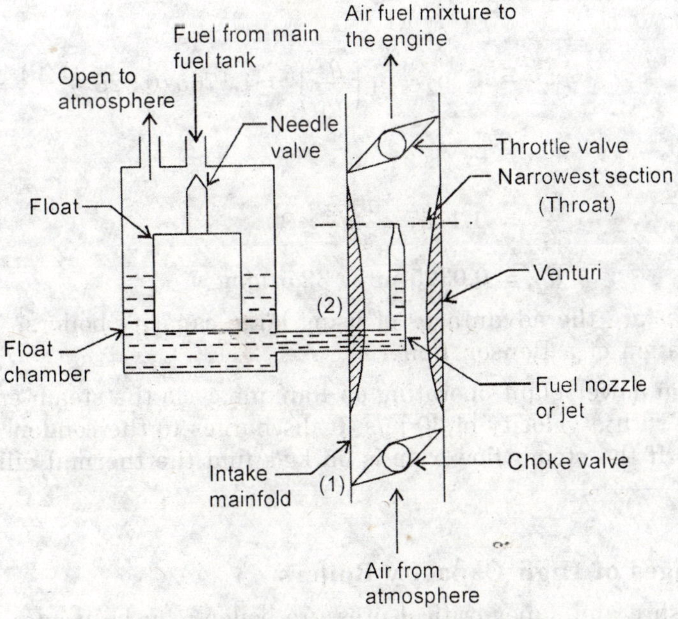

Fig 3. A simple jet carburettor.

$$h_1 + \frac{c_1^2}{2} = h_2 + \frac{c_2^2}{2}$$

Taking $c_1 = 0$, we get

$$c_2 = \sqrt{2(h_1 - h_2)}$$

Actual throat velocity,

$$c_{2a} = C_c c_2 = C_c\sqrt{2c_p\,(T_1 - T_2)} = C_c\sqrt{2c_p\,T_1\left(1 - \frac{T_2}{T_1}\right)}$$

Now

$$\frac{T_2}{T_1} = \left(\frac{p_2}{p_1}\right)^{\frac{\gamma-1}{\cdot\gamma}}$$

$\therefore$

$$90 = 0.8\sqrt{2 \times 1.005 \times 10^3 \times 300\left(1 - \frac{T_2}{T_1}\right)}$$

$$\frac{T_2}{T_1} = 0.979$$

$$\frac{p_2}{p_1} = (0.979)^{\frac{1.4}{0.4}} = (0.979)^{3.5} = 0.9284$$

$$p_2 = 1.013 \times 0.9284 = 0.9405 \text{ bar}$$

Density of air at entrance,

$$\rho_1 = \frac{p_1}{RT_1} = \frac{1.013 \times 10^2}{0.287 \times 300} = 1.1765\ kg/m^3$$

For

$$pv^\gamma = C,\ \rho_2 = \rho_1\left(\frac{p_2}{p_1}\right)^{\frac{1}{\gamma}} = 1.1765\,(0.9284)^{1/1.4} = 1.1157\ kg/m^3$$

At the throat,

$$\dot{m}_a = \rho_2 A_2 c_2$$

$$\frac{5}{60} = 1.1157 \times \frac{\pi}{4} d_2^2 \times 90$$

$$d_2 = 0.0325 \text{ m or } 32.5 \text{ mm}$$

Q.3. (*a*) What are the advantages of using large capacity boilers? With neat sketches describe the operation of a Benson Boiler.

(*b*) In a steam power plant operating on Rankine cycle the steam enters the turbine at 70 bar and 550°C with a velocity of 30 m/s. It discharges to the condenser at 0.20 bar with a velocity 90 m/s. If the steam flow rate is 35 kg/s find the thermal efficiency and the net power produced.

Sol.

(*a*) Advantages of High Capacity Boilers

1. High pressure and super critical pressure boilers can be used.
2. Cost of drums is eliminated.
3. Circulating pumps and down comers are eliminated.

4. The boiler can be located outdoors, being self-contained.

5. The boiler is comparatively lighter in weight.

6. They have high specific outputs and higher factor of sefety.

7. The boiler can be put into service in a short time.

Benson Boiler:

The Benson boiler is illustrated in Fig. 4.

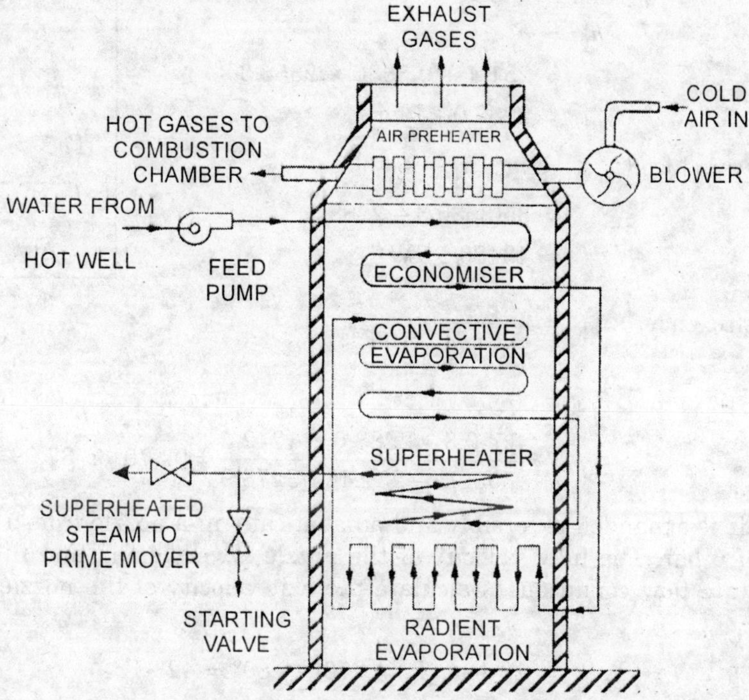

Fig 3. Benson boiler.

The water is passed through the economizer into the radiant evaporator where the major portion of the water is converted into steam. The remaining water is evaporated in the final evaporator absorbing the heat from hot gases by convection. The supercritical pressure steam (at 225 at.) is then passed through the superheater.

Steam upto a temperature of 650°C, can be raised by this boiler with a steam flow of 150 tonnes/hour.

(b) Given: $p_1 = 70$ bar, $t_1 = 550°C$, $c_1 = 30$ m/s,

$p_2 = 0.20$ bar, $c_2 = 90$ m/s,

$\dot{m}_a = 35$ kg/s

The simple Rankine cycle and T – s diagram, is shown in Fig. 4.

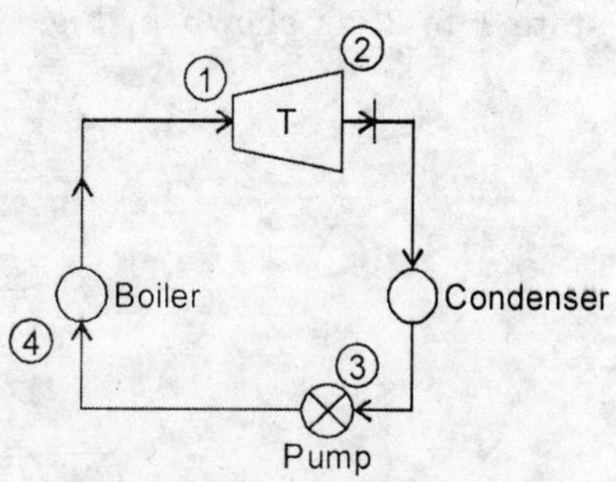

(a) Simple Rankine cycle.

From steam tables, we have

At $\quad p_1 = 70 \text{ bar and } t_1 = 550°\text{C}, \; h_1 = 3530.3 \text{ kJ/kg}$

$\quad s_1 = 6.9435 \text{ kJ/kh.K}$

At $\quad p_2 = 0.2 \text{ bar}, \; h_{f2} = 251.4 \text{ kJ/kh}, \; h_{fg2} = 2358.3 \text{ kJ/kg}$

$\quad s_{f2} = 0.832 \text{ kJ/kg.K}, \; s_{fg2} = 7.077 \text{ kg/kg.K}$

Now $\quad s_1 = s_2 = s_{f2} + x_2 \, s_{fg2}$

$$6.9435 = 0.832 + x_2 \times 7.077$$

$$x_2 = 0.8636$$

$$h_2 = h_{f2} + x_2 \, h_{fg2}$$

$$= 251.4 + 0.8636 \times 2358.3$$

$$= 2288.0 \text{ kJ/kg}$$

Net power produced $= \dot{m}_s \,(h_1 - h_2)$

$$= 35(3530.3 - 2288.0)$$

$$= 43480.5 \text{ kW}$$

(b) T – s diagram

Fig 4.

Thermol efficiency, $\eta_{th} = \dfrac{h_1 - h_2}{h_1 - h_{f3}}$

Now $\quad h_{f3} = h_{f2}$

$$\eta_{th} = \frac{3530.3 - 2288.0}{3530.3 - 251.4} = \frac{1242.3}{3278.9} = 0.379 \text{ or } 37.9\%$$

Q.4. (*a*) Air is expanded reversibly and adiabatically in a nozzle from 13 bar and 150°C to a pressure of 6 bar. The inlet velocity of the nozzle is very small and the process occurs under steady state flow conditions. Calculate the exit velocity of the nozzle.

Sol.

Given: $\quad p_1 = 13 \text{ bar}, \; T_1 = 273 + 180° = 423 \text{ K},$

$\quad p_2 = 6 \text{ bar}, \; c_1 = 0$

SFEE for unit mass flow rate is :

$$\left(h_1 + \frac{1}{2}c_1^2 + gz_1\right) + q = \left(h_2 + \frac{1}{2}c_2^2 + gz_2\right) + w$$

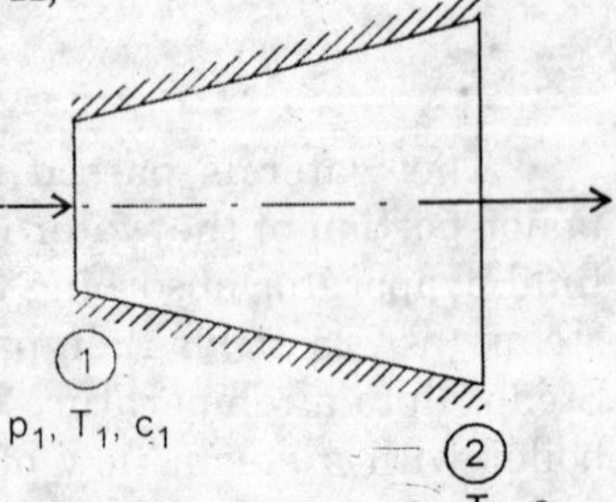

Fig 5.

Since the air expands reversibly and adiabatically from state 1 to 2, as shown in Fig. 5, we have

$$q = 0, \quad w = 0, \quad \text{and} \quad z_1 = z_2$$

$$\therefore \quad h_1 = h_2 + \frac{1}{2}c_2^2$$

$$c_2 = \sqrt{2(h_1 - h_2)} = \sqrt{2c_p(T_1 - T_2)}$$

Now $\quad \dfrac{T_2}{T_1} = \left(\dfrac{p_2}{p_1}\right)^{\frac{\gamma - 1}{\gamma}}$

$$T_2 = 423\left(\frac{6}{13}\right)^{\frac{1.4-1}{1.4}} = 339.156 \text{ K}$$

Exit velocity of the nozzle,

$$c_2 = \sqrt{2 \times 1.005 \times 10^3 \,(423 - 339.156)} = 410.52 \; m/s$$

(*b*) A gas turbine unit receives air at I bar and 300K and compresses it adiabatically to 6.20 bar. The compressor efficiency is 88%. The fuel has a heating value of 44186 kJ/kg and the fuel air ratio is 0.017 kg fuel/kg of air. The turbine internal efficiency is 90%. Calculate the work of turbine and compressor per kg of air compressed and thermal efficiency. For products of combussion c_p = 1.147 kJ/kg K, γ = 1.333.

Sol.

Given: p_1 = 1 bar, T_1 = 300 K, p_2 = 6.20 bar, pv^γ = const.,

η_c = 0.88, CV = 44186 kJ/kg, F/A = 0.099 kg fuel/kg of air,

η_t = 0.90, c_{pg} = 1.147 kJ/kg.K γ_g = 1.333

The T – s diagram for gas turbine is shown in Fig. 6.

Process 1 – 2 (compressor):

$$\frac{T_2}{T_1} = \left(\frac{p_2}{p_1}\right)^{\frac{\gamma-1}{\gamma}}$$

$$T_2 = 300\left(\frac{6.20}{1.0}\right)^{\frac{1.4-1}{1.4}} = 505.26 \text{ K}$$

$$\eta_c = \frac{T_2 - T_1}{T_2' - T_1}$$

$$0.88 = \frac{505.26 - 300}{T_2' - 300}$$

$$T_2' = 533.26 \text{ K}$$

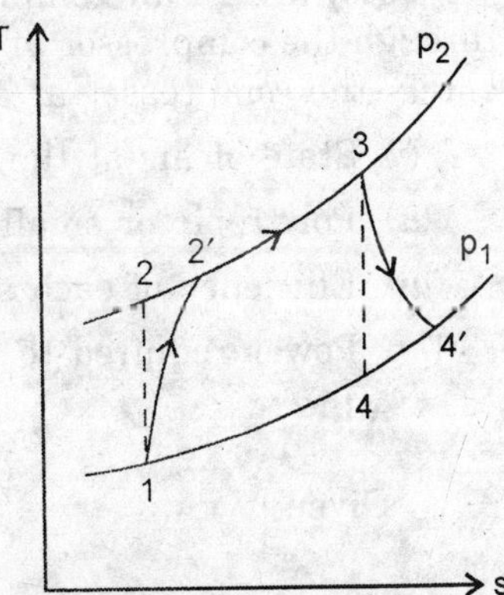

Fig 6. T – S diagram for gas turbine

Process 2′ – 3 (combustor) :

$$(m_a + m_f)\, e_{pg} \times T_3 - m_a\, c_{pa} \times T_2' = m_f \times \text{CV}$$

or

$$\left(1 + \frac{m_f}{m_a}\right) c_{pg} \times T_3 - c_{pa} \times T_2' = \frac{m_f}{m_a} \times \text{CV}$$

$$(1 + 0.017) \times 1.147 \times T_3 - 1.005 \times 533.26 = 0.017 \times 44186$$

$$T_3 = 1103.38 \text{ K}$$

Process 3 – 4 (Turbine) :

$$\frac{T_4}{T_3} = \left(\frac{p_4}{p_3}\right)^{\frac{\gamma_g-1}{\gamma_g}} = \left(\frac{p_1}{p_2}\right)^{\frac{\gamma_g-1}{\gamma_g}}$$

$$T_4 = 1103.38\left(\frac{4.0}{6.2}\right)^{\frac{0.333}{1.333}} = 699.48 \text{ K}$$

$$\eta_t = \frac{T_3 - T_4'}{T_3 - T_4}$$

$$0.9 = \frac{1103.38 - T_4'}{1103.38 - 699.48}$$

$$T_4' = 739.87 \text{ k}$$

Compressor work, $w_c = m_a c_{pa} (T_2' - T_1) = 1 \times 1.005 (533.26 - 300) = 234.43$ kJ/kg

Turbine work, $w_t = \left(1 + \frac{m_f}{m_a}\right) c_{pg} (T_3 - T_4')$

$$= (1 + 0.017) \times 1.147 (1103.38 - 739.87) = 424.03 \text{ kJ/kg}$$

Net work, $w_{net} = w_t - w_c = 424.03 - 234.43 = 189.6$ kJ/kg

Thermol efficiency, $\eta_{th} = \dfrac{w_{net}}{\dfrac{m_f}{m_a} \times CV} = \dfrac{189.6}{0.017 \times 44186} = 0.2524$ or 25.24 %

Q.5. (*a*) An air compressor has eight stages of equal pressure ratio 1.35. The flow rate through the compressor and its overall efficiency are 50 kg/s and 82% respectively. If the air enters the compressor at a pressure of 1.0 bar and temperature of 313K, determine :

(*i*) State of air at the exit of the compressor

(*ii*) Polytropic or small stage efficiency

(*iii*) Efficiency of each stage

(*iv*) Power required to drive the compressor assuming the overall efficiency as 90%.

Sol.

Given: $z = 8, \quad \dfrac{p_z}{p_{z-1}} = 1.35, \quad \dot{m}_a = 50 \text{ kg/s}, \quad \eta_{overall} = 0.82,$

$$p_1 = 1.0 \text{ bar}, \quad T_1 = 313 \text{ K}$$

(*i*)
$$\frac{p_{z+1}}{p_1} = \left(\frac{p_z}{p_{z-1}}\right)^z$$

$$\frac{p_9}{p_1} = (1.35)^8$$

$$p_9 = 1 \times 11.03 = 11.03 \text{ bar}$$

$$\frac{T_9}{T_1} = \left(\frac{p_9}{p_1}\right)^{\frac{\gamma - 1}{\gamma}}$$

$$T_9 = 313 (11.03)^{\frac{0.4}{1.4}} = 621.47 \text{ K}$$

$$\eta_{overall} = \frac{T_9 - T_1}{T_9' - T_1}$$

$$T_9' = T_1 + \frac{T_9 - T_1}{\eta_{overall}}$$

$$T_9' = 313 + \frac{621.47 - 313}{0.82} = 689.18 \text{ K}$$

State of air at exit of compressor is : p_9 = 11.03 bar and T_9' = 689.18 K

(*ii*) Polytropic or small stage efficiency,

$$\eta_p = \frac{\ln(p_9/p_1)^{\frac{\gamma-1}{\gamma}}}{\ln(T_9'/T_9)} = \frac{\ln(11.03/1.0)^{0.4/1.4}}{\ln(689.18/313)}$$

$$= \frac{0.6928}{0.7893} = 0.878 \text{ or } 87.8\%$$

(*iii*) Efficiency of each stage, η_p = 87.8%

(*iv*) Power required $= \dfrac{\dot{m}_a\, c_p\, (T_9' - T_1)}{\eta_{overall}} = \dfrac{50 \times 1.005\,(689.18 - 313)}{0.9} = 21{,}003.4$ kW

(***b***) One stage of an impulse turbine consists of a converging nozzle ring and one ring of moving blades. The nozzles are inclined at 22° to the blades whose tip angles are both 35°. If the velocity of steam at exit from the nozzle is 660 m/s, find the blade speed so that the steam shall pass on without shock. Find the diagram efficiency neglecting losses if the blades are run at this speed.

Sol.

Given: $\alpha_1 = 22°$, $\beta_1 = \beta_2 = 35°$, $v_{a1} = 660$ m/s

The velocity triangles for impulse turbine are shown in Fig. 7.

Maximum blade efficiency,

$$(\eta_b)_{max} = \frac{\cos^2 \alpha_1}{2}(1 + KC)$$

For $\beta_1 = \beta_2$, $C = 1$ and for no friction on the blades, K = 1.

$$\therefore \quad (\eta_b)_{max} = \cos^2 \alpha_1 = \cos^2 22°$$

$$= 0.8597 \text{ or } 85.97\%$$

Also speed ratio, $\rho = \dfrac{u}{v_{a1}} = \dfrac{\cos \alpha_1}{2}$

Blade speed, $u = 660 \times \dfrac{\cos 22°}{2}$

$= 305.97$ m/s.

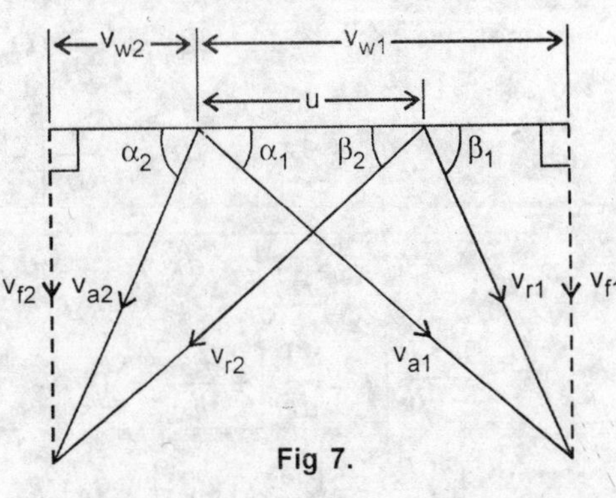

Fig 7.

Q.6. (***a***) Show that temperature distribution in a sphere of radius R, having a surface temperature of T_w, in which heat is generated at a uniform rate of q_g W/m^3 is given by:

$$[T(r) - T_w] = \frac{q_g R^2}{6k}(1 - r^2/R^2)$$

where, k is the thermal conductivity of the sphere material.

Sol. Sphere with Uniform Heat Generation

Consider one dimensional radial conduction of heat, under steady state conditions, through a sphere having uniform heat generation.

Let

R = outside radius of sphere,

k = thermal conductivity (uniform),

$\dot{q}_g$ = uniform heat generation per unit voluem, per unit time within the solid,

t_w = temperature of the outside surface (wall) of the sphere, and

t_a = ambient temperature.

Consider an element at radius r and thickness, dr as shown in Fig. 8.

Heat conducted in at radius r,

$$Q_r = -kA\frac{dT}{dr} = -k \times 4\pi r^2 . \frac{dT}{dr}$$

Heat generated in the element,

$$Q_g = \dot{q}_g \times A \times dr = \dot{q}_g + 4\pi r^2 \times dr$$

Heat conducted *out* at radius $(r + dr)$

$$Q_{(r+dr)} = Q_r + \frac{d}{dr}(Q_r)\,dr$$

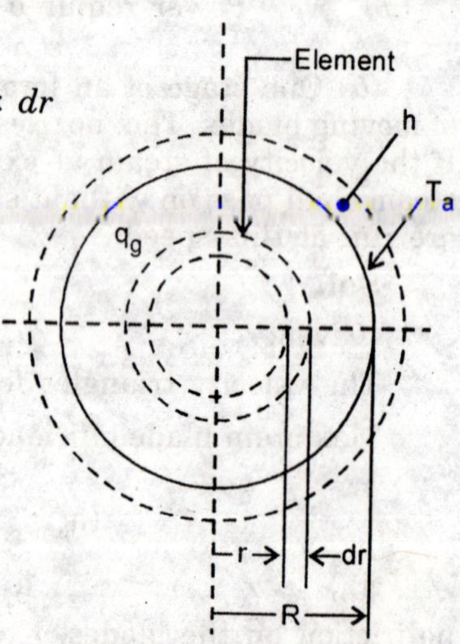

Fig 8. Sphere with uniform heat generation

Under steady state conditions, we have

$$Q_r + Q_g = Q_{(r+dr)} = Q_r + \frac{d}{dr}(Q_r).dr$$

or $$Q_g = \frac{d}{dr}(Q_r)\,dr$$

or $$q_g \times 4\pi r^2 \times dr = \frac{d}{dr}\left[-4\pi k\, r^2 . \frac{dT}{dr}\right] dr$$

or $$q_g \times 4\pi\, r^2 \times dr = -4\pi k \frac{d}{dr}\left[r^2 . \frac{dT}{dr}\right] dr$$

or $$\frac{1}{r^2}\frac{d}{dr}\left[r^2 . \frac{dT}{dr}\right] + \frac{\dot{q}_g}{k} = 0$$

or $$\frac{1}{r^2}\left[r^2 . \frac{d^2T}{dr^2} + 2r . \frac{dT}{dr}\right] + \frac{\dot{q}_g}{k} = 0$$

or $$\frac{d^2T}{dr^2} + \frac{2}{r} \times \frac{dT}{dr} + \frac{\dot{q}_g}{k} = 0$$

or $$r\frac{d^2T}{dr^2} + 2\frac{dT}{dr} + \frac{\dot{q}_g r}{k} = 0$$ (multiplying both sides by r)

or $$r\frac{d^2T}{dr^2}+\frac{dT}{dr}+\frac{dT}{dr}+\frac{q_g r}{k} = 0$$

or $$\frac{dT}{dr}+\left(r\frac{dT}{dr}\right)+\frac{dT}{dr}+\frac{\dot{q}_g r}{k} = 0$$

Integrating both sides, we have

$$r\frac{dT}{dr}+T+\frac{\dot{q}_g}{k}.\frac{r^2}{2} = C_1 \quad ...(ii)$$

or $$\frac{d}{dr}(rT)+\frac{\dot{q}_g}{k}.\frac{r^2}{2} = C_1$$

Integrating again, we have

$$rT+\frac{\dot{q}_g}{k}.\frac{R^3}{6} = C_1 r + C_2 \quad ...(iii)$$

(where C_1, C_2 = constants of integration).

At the centre of sphere, r = = 0 $\therefore$ $C_2 = 0$ [From eqn. (*iii*)]

Applying boundary condition, at r = R, T = T_w, to eqn. (*iii*), we have

$$RT_w+\frac{\dot{q}_g}{k}.\frac{R^3}{6} = C_1 R \quad (\because C_2 = 0)$$

or $$C_1 = T_w+\frac{\dot{q}_g}{6k}R^2$$

By substituting the values of C_1 and C_2 in eqn. (*iii*), we have the temperature distribution as:

$$rT+\frac{\dot{q}_g}{k}.\frac{r^3}{6} = \left[T_w+\frac{\dot{q}_g}{6k}R^2\right]r$$

or $$T+\frac{\dot{q}_g}{6k}.r^2 = T_w+\frac{\dot{q}_g}{6k}R^2$$

or $$T = T_w+\frac{\dot{q}_g}{6k}(R^2-r^2)$$

or $$T(r)-T_w = \frac{\dot{q}_g}{6k}R^2\left(1-\frac{r^2}{R^2}\right)$$

Hence proved.

(b) A chemical having specific heat of 3.3 kJ/kg·K flowing at the rate of 20,000 kg/h entres a parallel flow heat exchanger at 120°C. The flow rate of cooling water is 50,000 kg/h with an inlet temperature of 20°C. The heat transfer area is 10m^2 and the overall heat transfer coefficient is 1050 W/m^2·K. Find the outlet temperatures of water and chemical and effectiveness of the heat exchanger. Take for water specific heat = 4.186 kJ/kg·K.

Sol.

Given: c_{ph} = 3.3 kJ/kg. K, $\dot{m}_a$ = 20,000 kg/h, t_{hi} = 120°,

$\dot{m}_c = 50{,}000$ kg/h, $t_{ei} = 20°C$, $A = 10\ m^2$, $U = 10502\ W/m^2.K$,

$c_{pc} = 4.186$ kJ/kg.K; t_{oh}, t_{oc}, $\varepsilon = ?$

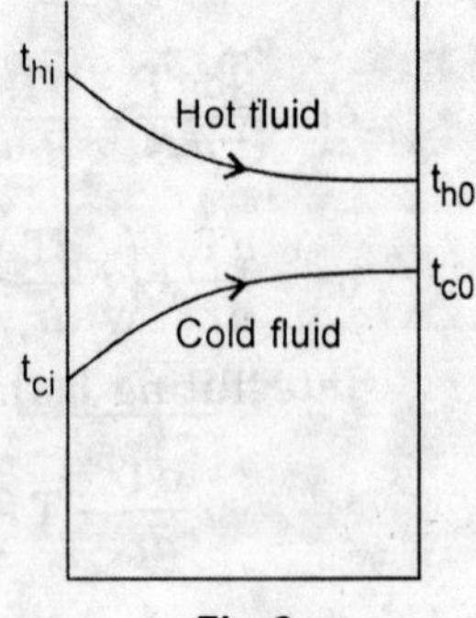

Fig 9.

Heat capacities:

$$C_h = \dot{m}_h \times c_{ph} = 20{,}000 \times 3.3$$
$$= 66{,}000 \text{ kJ/h.K}$$
$$C_c = \dot{m}_c \times c_{pc} = 50{,}000 \times 4.186$$
$$= 209{,}300 \text{ kJ/h.K}$$

We find that $C_h < C_c$. $C_{min} = \dfrac{66{,}000 \times 10^3}{3600} = 18{,}333.3\ J/s.K$

For heat balance,

$$\text{Heat lost by hot fluid} = \text{Heat gained by cold fluid}$$
$$66{,}000\ (120 - t_{ho}) = 209{,}300\ (t_{co} - 20)$$

or
$$120 - t_{ho} = 3.171\ (t_{co} - 20) \qquad ...(1)$$

Now
$$NTU = \frac{UA}{C_{min}} = \frac{1050 \times 10}{18{,}333.3} = 0.5727$$

Effectiveness of parallel flow heat exchanger,

$$\varepsilon = \frac{1 - \exp(-NTU)\left[1 + \dfrac{C_{min}}{C_{max}}\right]}{1 + \dfrac{C_{min}}{C_{max}}}$$

Let
$$R = \frac{C_{min}}{C_{max}} = \frac{66{,}000}{209{,}300} = 0.31534$$

$$\varepsilon = \frac{1 - \exp(-0.5727) \times 1.31534}{1 + 0.31534}$$

$$= \frac{0.25815}{1.31534} = 0.1962$$

Also
$$\varepsilon = \frac{C_h\ (t_{hi} - t_{ho})}{C_{min}\ (t_{hi} - t_{ci})}$$

$$0.1962 = \frac{120 - t_{ho}}{120 - 20}$$

$$t_{ho} = 100.38°C$$

From Eq. (1), we have

$$120 - 100.38 = 3.171\ (t_{co} - 20)$$
$$t_{co} = 26.19°C$$

Q.7. (*a*) A refrigeration machine is required to produce ice at 0°C from water at 20°C. The machine has a condenser temperature of 298 K while the evaporator temperature is 268 K. The relative efficiency of the machine is 50% and 6 kg of Freon-12 refrigerant is circulated through the system per minute. The refrigerant enters the compressor with a dryness fraction of 0.6. Specific heat of water is 4.187 kJ/kg K and the latent heat of ice is 335 kJ/kg. Calculate the amount of ice produced in 24 hours. The table of properties of Freon-12 is given below :

Temperature K	Liquid heat kJ/kg	Latent heat kJ/kg	Entropy of liquid kJ/kg K
298	59.7	138.0	0.2232
268	31.4	154.0	0.1251

Sol.

Given: $x_1 = 0.6$, $T_1 = 268$ K(– 5°C), $T_2 = 298$ K(25°C), $\dot{m}_h = 6$ kg/min,

$\eta_{rel} = 0.50$, $c_{pw} = 4.187$ kJ/kg.K, $(h_{fg})_{ice} = 335$ kJ/kg

The T-s diagram is shown in Fig. 10.

$$h_1 = h_{f1} + x_1\, h_{fg1} = 31.4 + 0.6 \times 154$$
$$= 123.8 \text{ kJ/kg.K,}$$

$$s_1 = s_{f1} + x_1\, s_{fg1} = s_{f1} + x_1 \frac{h_{fg2}}{T_1}$$
$$= 0.1251 + \frac{0.6 \times 154}{268}$$
$$= 0.4698 \text{ kJ/kg·K}$$

$$s_1 = s_2 = s_{f2} + \frac{x_2\, h_{fg2}}{T_2}$$

$$0.4698 = 0.2232 + \frac{x_2 \times 138}{298}$$

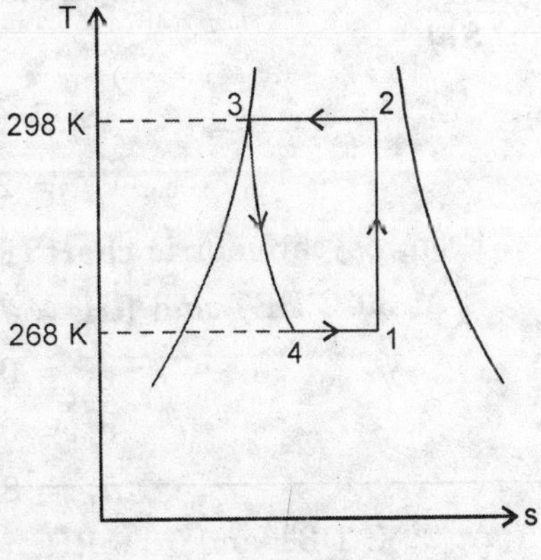

$$x_2 = 0.5325$$
$$h_2 = h_{f2} + x_2\, h_{fg2} = 59.7 + 0.5325 \times 138 = 133.185 \text{ kJ/kg}$$
$$h_3 = h_4 = h_{f2} = 59.7 \text{ kJ/kg}$$

Fig 10.

Theoretical COP $= \dfrac{h_1 - h_4}{h_2 - h_1} = \dfrac{123.8 - 59.7}{133.185 - 123.8} = \dfrac{64.1}{9.385} = 6.83$

$$(COP)_{act} = \eta_{rel} \times (COP)_{th} = 0.5 \times 6.83 = 3.415$$

Heat extracted from 1 kg of water at 20°C to form ice at 0°C.

$$= m_w \times \left[c_{pw} \times \Delta t + (h_{fg})_{ice}\right]$$
$$= 1 \times [4.187 \times 20 + 335] = 418.74 \text{ kJ/kg}$$

Let W = ice formed in 24 hours in tons.

Heat taken out from water per minute, $\dot{q} = \dfrac{1000 \times W \times 418.74}{24 \times 60} = 290.8 \text{ W kJ/min}$

Work done per minute $\dot{w} = \dot{m}_r (h_2 - h_1) = 6 \times 9.385 = 56.31 \text{ kJ/min}$

$$(COP)_{act} = \frac{\dot{q}}{\dot{w}}$$

$$3.415 = \frac{290.8 \text{ W}}{56.31}$$

$$W = 0.661 \text{ tons}$$

(*b*) A small auditorium is required to be maintained at 22°C dry bulb temperature and 70% relative humidity. The ambient conditions are at 30°C dry bulb tempeture and 75% relative humidity. The amount of free air circulated is 200 m³/min. The required conditions are achieved by first cooling coil having apparatus dew point of 14°C and then by heating with the help of psychrometric chart find :

(*i*) The cooling capacity of cooling coil in tons of refreigeration and its bypass factor.

(*ii*) The amount of water vapour removed by the cooling coil in kg/h.

(*iii*) Assuming the by pass factor as 0.2, determine the capacity of the heating coil in kW and its surface temperature.

Sol.

Given: $t_{d2} = 22°C,\ (RH)_2 = 70\%,\ t_{d1} = 30°C,\ (RH)_1 = 75\%,$

$\dot{V}_a = 200 \text{ m}^3/\text{min},\ t_{ADP} = 14°C,\ BPF = 0.2$

From psychrometric chart (Fig. 11), we have

At 30°C DBT and RH = 75%,

$$v_1 = 0.886 \text{ m}^3/\text{kg}$$
$$v_4 = 0.838 \text{ m}^3/\text{kg}$$
$$h_1 = 82 \text{ kJ/kg}$$

At 22°C DBT and 70% RH,

$$h_2 = 51 \text{ kJ/kg}$$
$$h_4 = 47 \text{ kJ/kg}$$
$$t_4 = 16.5°C$$
$$w_1 = 20.5 \text{ gm/kg d.a.}$$
$$w_2 = 11.8 \text{ gm/kg d.a.}$$

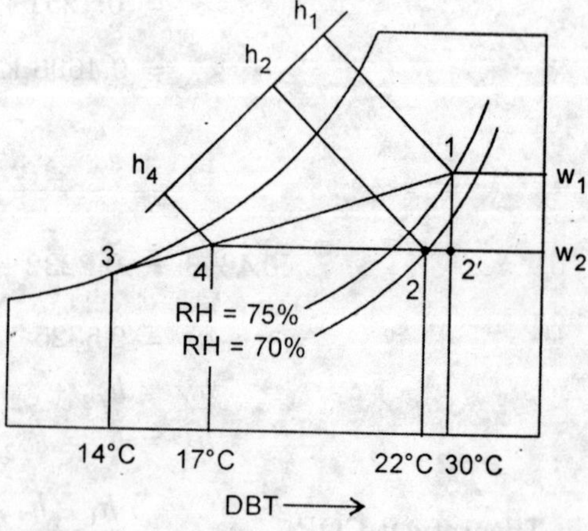

Fig 11. Psychrometric chart

(*i*) Cooling capacity of coil

$$= \frac{\dot{V}_a}{v_1} \times \frac{h_1 - h_4}{211}$$

$$= \frac{200}{0.886} \times \frac{(82 - 47)}{211} = 37.47 \text{ ton}$$

(*ii*) Amount of water vapour removed by cooling coil,

$$\dot{m}_w = \frac{\dot{V}_a}{v_1}(w_1 - w_2)$$

$$= \frac{200}{0.886}(20.5 - 11.8) \times 60 \times 10^{-3} = 117.8 \text{ kg/hr.}$$

(*iii*)

$$\text{BPF} = \frac{t_{d2} - t_{d2}'}{t_{d4} - t_{d2}'}$$

$$0.2 = \frac{22 - t_{d2}}{17 - t_{d2}'}$$

$$3.4 - 0.2\, t_{d2}' = 22 - t_{d2}'$$

$$t_{d2'} = \frac{18.6}{0.8} = 23.25°\text{C}$$

Capacity of heating coil

$$= \frac{\dot{V}_a}{v_4}(h_2 - h_4)$$

$$= \frac{200}{0.838}(51 - 47) \times \frac{1}{60} = 15.91 \text{ kW}$$

Q.8. (*a*) The velocity profile for laminar flow in the boundary layer of a flat plate is given by :

$$\frac{u}{U} = \sin\left(\frac{\pi}{2}\frac{y}{\delta}\right)$$

where, u is the velocity of fluid in the boundary layer at a vertical distance y from the plate surface and U is the free stream velocity. Prove that the boundary layer thickness δ may be given by the expression

$$\delta = \frac{4.795\, x}{(\text{Re}_x)^{1/2}}$$

where, Re_x is the Reynolds number of the fluid at a distance x from the leading edge of the plate.

Sol.

Given :

$$\frac{u}{U} = \sin\left(\frac{\pi}{2}\frac{y}{\delta}\right)$$

$$\frac{\tau_0}{\rho} = \frac{\mu}{\rho}\left.\frac{du}{dy}\right|_{y=0} = U^2 \frac{d}{dx}\int_0^\delta \frac{u}{U}\left(1 - \frac{u}{U}\right) dy \qquad [\mu = \text{dynamic viscosity}]$$

Now

$$\int_0^\delta \frac{u}{U}\left(1 - \frac{u}{U}\right) dy = \int_0^\delta \sin\frac{\pi}{2}\left(\frac{y}{\delta}\right)\left[1 - \sin\frac{\pi}{2}\left(\frac{y}{\delta}\right)\right] dy = \int_0^\delta \sin\left(\frac{\pi}{2}.\frac{y}{\delta}\right) dy - \int_0^\delta \sin^2\left(\frac{\pi}{2}.\frac{y}{\delta}\right) dy$$

$$= \left| -\frac{2\delta}{\pi}\cos\left(\frac{\pi}{2}.\frac{y}{\delta}\right) \right|_0^\delta - \int_0^\delta \frac{1}{2}\left[1-\cos\left(\frac{\pi y}{\delta}\right)\right] dy$$

$$= \frac{2\delta}{\pi} - \frac{1}{2}\left| y - \frac{\delta}{\pi}\sin\left(\frac{\pi y}{\delta}\right)\right|_0^\delta = \frac{2\delta}{\pi} - \frac{1}{2}[\delta - 0] = \frac{2\delta}{\pi} - \frac{\delta}{2}$$

$$\tau_0 = \mu \left.\frac{du}{dy}\right|_{y=0} = \mu \frac{d}{dy}\left[U \sin\left(\frac{\pi}{2}\frac{y}{\delta}\right)\right]_{y=0}$$

$$= \mu U \times \frac{\pi}{2\delta}\left[\cos\left(\frac{\pi}{2}.\frac{y}{\delta}\right)\right]_{y=0} = \frac{\mu U \pi}{2\delta}$$

$$\therefore \quad \frac{\mu U \pi}{2\delta\rho} = U^2 \frac{d}{dx}(0.1366\,\delta)$$

or

$$\frac{\mu\pi}{2\delta\rho} = U \times 0.1366 \times \frac{d\delta}{dx}$$

$$\frac{\pi\mu}{0.2732\,U\rho} = \delta\frac{d\delta}{dx} = \frac{1}{2}\frac{d}{dx}(\delta)^2$$

Integrating both sides, we get

$$\delta^2 = \int \frac{2\pi\mu}{0.2732\,U\rho}\,dx = \frac{\pi}{0.1366}.\frac{\mu}{U\rho}.x + C$$

where C = constant of integration.

At $x = 0, \quad \delta = 0$

$\therefore \quad C = 0$

$$\therefore \quad \delta^2 = \frac{\pi}{0.1366}.\frac{\mu x}{U\rho}$$

$$\frac{\delta^2}{x^2} = \frac{23\,\mu x}{U\rho . x^2} = \frac{23}{Re_x} \qquad \left[\because Re_x = \frac{U x \rho}{\mu}\right]$$

$$\frac{\delta}{x} = \frac{4.795}{(Re_x)^{1/2}}$$

Hence proved.

(b) A swimming pool 12 m long and 7 m wide holds water to a depth of 2 m. If the water is discharged through an opening of area 0.2 m² at the bottom of the pool, find the time required to empty the tank. Take coefficient of discharge for the opening as 0.6.

Sol.

Given : l = 12 m, b = 7 m, h = 2 m, A = 0.2 m², C_d = 0.6

Velocity of water flow, $v = C_d\sqrt{2gh}$

Volume of water flowing out during a short interval of time Δt,

$$\Delta V = Av.\ \Delta t$$

$$= 0.2 \times 0.6\sqrt{2gh} \times \Delta t$$

$$= 0.12\sqrt{2gh}\,.\Delta t$$

Fig 12.

Let Δh = fall in depth of water in the tank in time Δt (Fig 12)

Then $\qquad \Delta h = -\ lb \times \Delta h = -\ 12 \times 7 \times \Delta h$

$$= -\ 84 \times \Delta h$$

$\therefore \qquad 0.12\sqrt{2gh}\,.\Delta t = -\ 84 \times \Delta h$

or $\qquad \dfrac{\Delta h}{\Delta t} = -\dfrac{0.12}{84}\sqrt{2gh} = -\dfrac{0.12 \times \sqrt{2 \times 9.81}}{84} \times \sqrt{h} = -\ 6.328 \times 10^{-3} \times \sqrt{h}$

As $\Delta t \to 0$, $\qquad \dfrac{\Delta h}{\Delta t} = \dfrac{dh}{dt}$

or $\qquad \displaystyle\int_2^0 \frac{dh}{h^{1/2}} = -\ 6.328 \times 10^{-3} \int_0^t dt$

$$\left| 2h^{1/2} \right|_2^0 = -\ 6.328 \times 10^{-3} \times t = -\ 2\sqrt{2}$$

$$t = \frac{2\sqrt{2}}{6.328 \times 10^{-3}} = 446.9\ s$$

(*c*) Give the range of specific speed values of the Kaplan, Francis turbines and Pelton wheels. What factors decide whether Kaplan, Francies or a Pelton wheel type turbine would be used in a hydro-electric project?

Sol.

Specific speed of a turbine,

$$N_s = \frac{N\sqrt{P}}{H^{5/4}}$$

Type of turbine	Specific speed range S.I. Units	Head	Flow Rate
1. Pelton (single jet) Multiple jets.	8.5 – 30 30 – 50	More than 400m (High)	Low
2. Francis	50 – 340	Upto 370 m (Medium)	Medium
3. Kaplan	250 – 1050	Upto 60 m (Low)	High

The selection of turbine can be made based on the basis of head available and flow rate.

MECHANICAL ENGINEERING
PAPER I—1993

(Conventional)

Time allowed : 3 hours *Maximum marks : 200*

Q.1. (*a*) 0.5 kg of air (ideal gas) executes a Carnot power cycle having a thermal efficiency of 50 percent. The heat transfer to the air during the isothermal expansion is 40 kJ. At the beginning of the isothermal expansion the pressure is 7 bar and the volume is 0.12 m^3. Determine the maximum and minimum temperatures for the cycle, in K, the volume at the end of isothermal expansion, in m^3, and the work and heat transfer for each of the four processes, in kJ. For air c_v = 0.721 and c_p = 1.008 kJ/kg.K.

Sol.

Given: m = 0.5 kg, η_{th} = 0.5, Q_s = 40 kJ, c_p = 1.008 kJ/kg.K, c_v = 0.721 kJ/kg·K,

p_2 = 7 bar, V_2 = 0.12 m^3

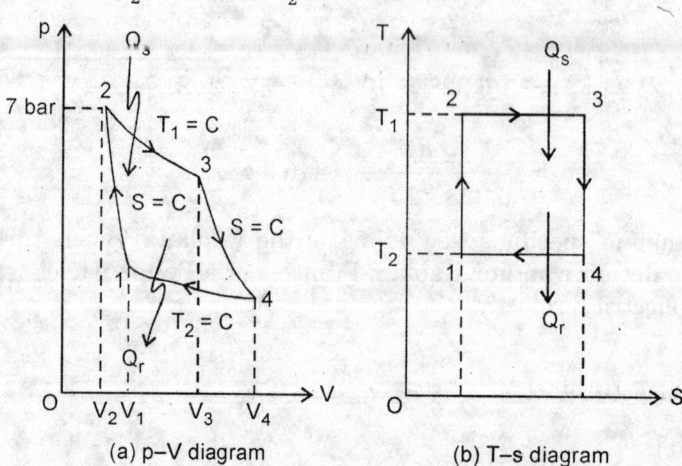

(a) p–V diagram (b) T–s diagram

Fig 1. Carnot cycle

Thermal efficiency of Carnot cycle (Fig. 1),

$$\eta_{th} = 1 - \frac{T_2}{T_1}$$

$$0.5 = 1 - \frac{T_2}{T_1}$$

or $$\frac{T_2}{T_1} = 0.5$$

or $$T_1 = 2T_2$$

For the isothermal expansion process 2 – 3, we have

$$p_2 V_2 = mR\, T_1 \qquad (\because T_2 = T_1]$$

$$7 \times 10^2 \times 0.12 = 0.5 \times (1.008 - 0.721) \times T_2 \qquad [\because R = c_p - c_v]$$

$$T_1 = 585.36 \text{ K or } 312.36°C$$

$$T_2 = 292.68 \text{ K or } 19.68°C$$

During isothermal expansion process 2–3, heat transferred to air,

$$Q_s = mR\,T_1 \ln\left(\frac{V_3}{V_2}\right)$$

$$R = c_p - c_v = 1.008 - 0.721 = 0.287 \text{ kJ/kg·K}$$

$$40 = 0.5 \times 0.287 \times 585.36 \ln\left(\frac{V_3}{0.12}\right)$$

$$V_3 = 0.197 \text{ m}^3$$

	Process	Work transfer W	Heat transfer Q
1.	Reversible adiabatic: 1 – 2	$W_{1-2} = mc_v\,(T_1 - T_2)$ $= 0.5 \times 0.721(585.36 - 292.68)$ $= 105.5$ kJ	$Q_{1-2} = 0$
2.	Isothermal expansion: 2 – 3	$W_{2-3} = mR\,T_1 \ln\left(\frac{V_3}{V_2}\right) = 40$ kJ	$Q_{2-3} = 40$ kJ
3.	Reversible adiabatic: 3 – 4	$W_{3-4} = -\,mc_V\,(T_4 - T_3)$ $= -\,0.5 \times 0.721\,(292.68 - 585.36)$ $= 105.5$ kJ	$Q_{3-4} = 0$
4.	Isothermal compression: 4 – 1	$W_{4-1} = mR\,T_2 \ln\left(\frac{V_1}{V_4}\right)$ $= -\,40$ kJ	$Q_{4-1} = -\,40$ kJ

(*b*) Steam flows through an adiabatic steady flow turbine. The enthalpy at entrance is 4142 kJ/kg and at exit 2585 kJ/kg. The values of flow availability of steam at entrance and exit are 1787 kJ/kg and 140 kJ/kg, respectively. If the dead state temperature T_v is 300 K, determine, per kg of steam, the actual work, the maximum possible work for the given change of state of steam, and the change in entropy of steam. Neglect changes in kinetic and potential energy.

Sol.

Given : $h_1 = 4142$ kJ/kg, $h_2 = 2585$ kJ/kg, $(AE)_1 = 1787$ kJ/kg, $(AE)_2 = 140$ kJ/kg

$T_0 = 300$ K, $W_{act} = ?$, $W_{max} = ?$, $\Delta s = ?$

Actual work done, $W_{act} = h_1 - h_2 = 4142 - 2585 = 1557$ kJ/kg

Maximum work done, $W_{max} = (AE)_1 - (AE)_2 = 1787 - 140 = 1647$ kJ/kg

Decrease in available energy $= T_0 \cdot \Delta s$

$\therefore \qquad 1647 = 300 \times \Delta s$

Change in enterpy, $\Delta s = 5.49$ kJ/kg. K.

Q.2. **(a)** A compressed air bottle of 0.3 m^3 volume contains air at 35 bar and 40°C. This air is used to drive a turbogenerator supplying power to a device which consumes 5W. Calculate the time for which the device can be operated if the actual output of the turbogenerator is 60 percent of the maximum theoretical output. The ambient pressure to which bottle pressure has fallen is 1 bar. For air, $(c_v/c_v) = 1.4$.

Sol.

Given : $V_1 = 0.05$ m^3, $p_1 = 35$ bar, $T_1 = 273 + 40 = 313$ K,

$p_2 = 1$ bar, $P = 5$ W, $\gamma = 1.4$, $\eta = 0.6$

Final temperature, $$T_2 = T_1\left(\frac{p_2}{p_1}\right)^{\frac{\gamma-1}{\gamma}} = 313\left(\frac{1}{35}\right)^{\frac{0.4}{1.4}} = 113.34 \text{ K}$$

Initial mass, $$m_1 = \frac{p_1V_1}{RT_1} = \frac{35\times10^2\times0.3}{0.287\times313} = 11.68 \text{ kg}$$

Final mass, $$m_2 = \frac{p_2V_2}{RT_2} = \frac{1\times10^2\times0.3}{0.287\times113.34} = 0.922 \text{ kg}$$

Work available, $$\begin{aligned} W &= c_v\,(m_1 T_1 - m_2 T_2) - c_p\,(m_1 - m_2)\,T_2 \\ &= 0.718\,(11.68 \times 313 - 0.922 \times 113.34) \\ &\quad - 1.005\,(11.68 - 0.922) \times 113.34 \\ &= 2549.86 - 1225.41 = 1324.45 \text{ kJ} \end{aligned}$$

Input to turbogenerator $$= \frac{P}{\eta} = \frac{5}{0.6} = 8.33 \text{ kJ/s}$$

Time duration for which turbo-generator can run $$= \frac{1324.45}{8.33} = 159\text{s}$$

(b) An insulated vessel is divided into two compartments connected by a valve, one compartment contains steam at 10 bar, 500°C, and the other is evacuated. The valve is opened and the steam is allowed to fill the entire volume, achieving a final pressure of 1 bar. Determine the final temperature, in °C, the percentage of the vessel volume initially occupied by steam, and the amount of entropy produced, in kJ/kg. K.

Sol.

Properties of steam at 10 bar and 500°C from steam tables are:

$$h_1 = 3478.5 \text{ kJ/kg}$$
$$s_1 = 7.762 \text{ kJ/kg.K}$$
$$v_{s1} = 0.3540 \text{ m}^3\text{/kg}$$

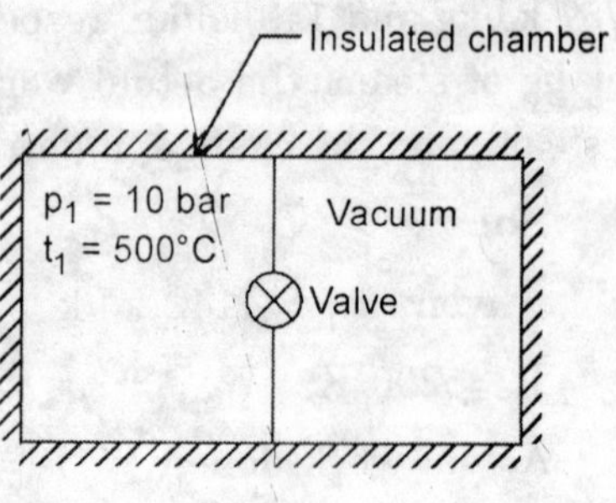

Fig 2.

When the valve is opened, throttling process occurs. Hence,

$$h_2 = h_1 = 3478.5 \text{ kJ/kg}$$

Enthalpy of saturated steam at 1 bar = 2675.5 kJ/kg

Therefore, steam remains superheated after throttling process.

At $p_2 = 1$ bar:

At 400°C, $h_3 = 3278.2$ kJ/kg, $s_3 = 8.544$ kJ/kg.K, $v_{s3} = 3.103$ m^3/kg

At 500°C, $h_4 = 3488.1$ kJ/kg, $s_4 = 8.834$ kJ/kg.K, $v_{s4} = 3.565$ m³/kg

By linear interpolation, we have

$$h_2 = h_3 + \left[\frac{h_4 - h_3}{t_4 - t_3}\right] \times (t_2 - t_3)$$

$$3478.5 = 3278.2 + \left[\frac{3488.1 - 3278.2}{500 - 400}\right] \times (t_2 - 400)$$

Final temperature, $t_2 = 495.43°\text{C}$

Final entropy, $$s_2 = s_3 + \left[\frac{s_4 - s_3}{t_4 - t_3}\right] \times (t_2 - t_3)$$

$$= 8.544 + \left[\frac{8.834 - 8.544}{500 - 400}\right] \times (495.43 - 400)$$

$$= 8.821 \text{ kJ/kg.K}$$

Change in entropy, $\Delta s = s_2 - s_1 = 8.821 - 7.762 = 1.059$ kJ/kg.K

Final specific volume,

$$v_{s2} = v_{s3} + \left[\frac{v_{s4} - v_{s3}}{t_4 - t_3}\right] \times (t_2 - t_3)$$

$$= 3.103 + \left[\frac{3.565 - 3.103}{500 - 400}\right] \times (495.43 - 400) = 3.544 \text{ m}^3/\text{kg}$$

Percentage volume occupied by steam = $\frac{v_{s1}}{v_{s2}} = \frac{0.3540}{3.544} = 0.10$ or 10%

Q.3. (*a*) In an I.C. engine operating on the dual cycle (limited pressure cycle), the temperature of the working fluid (air) at the beginning of compression is 27°C. The ratio of the maximum and minimum pressures of the cycle is 70 and the compression ratio is 15. The amounts of heat added at constant volume and at constant pressure are equal. Compute the air standard thermal efficiency of the cycle. State three main reasons why the actual thermal efficiency is different from the theoretical value.

Sol.

Given: $T_1 = 273 + 27 = 300$ K, $\frac{p_3}{p_1} = 70$,

$\frac{V_1}{V_2} = 15$, $Q_{sv} = Q_{sp}$

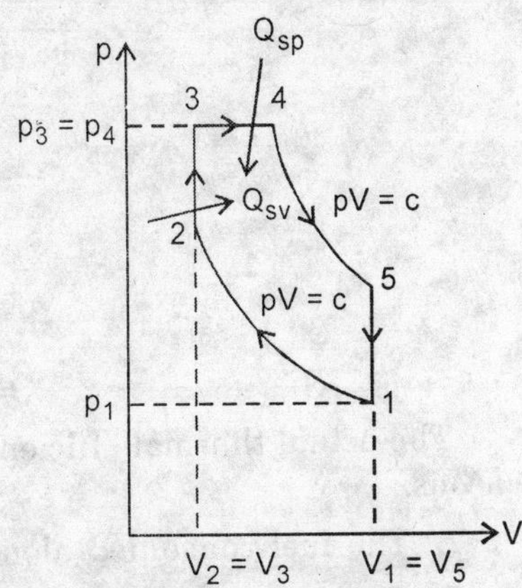

Fig 3. Dual cycle

The dual cycle is shown in Fig 3.

Process 1 – 2 :

$$\frac{T_2}{T_1} = \left(\frac{V_1}{V_2}\right)^{\gamma - 1}$$

$$T_2 = 300\ (15)^{0.4} = 886.25 \text{ K}$$

$$\frac{p_2}{p_1} = \left(\frac{V_1}{V_2}\right)^{\gamma} = (15)^{1.4} = 44.3$$

$$p_2 = 44.3\ p_1$$

Process 2 – 3 :

$$\frac{T_3}{T_2} = \frac{p_3}{p_2}$$

$$T_3 = T_2 \frac{p_3}{p_2} = T_1 \left(\frac{p_3}{44.3\, p_1}\right) = 886.25\left(\frac{70}{44.3}\right) = 1400.4 \text{ K}$$

Now

$$Q_{2-3} = Q_{3-4}$$

$$c_v(T_3 - T_2) = c_p(T_4 - T_3)$$

$$T_4 = T_3 + \frac{1}{\gamma}(T_3 - T_2)$$

$$= 1400.4 + \frac{1}{1.4}(1400.4 - 886.25) = 1767.65 \text{ K}$$

Also

$$\frac{V_4}{V_3} = \frac{T_4}{T_3}$$

$$V_4 = V_3 \times \frac{1767.65}{1400.4} = 1.2622\, V_3$$

$$= 1.2622 \times \frac{V_1}{15} = 0.0841\, V_1$$

Now

$$V_5 = V_1$$

Process 4 – 5 :

$$\frac{T_4}{T_5} = \left(\frac{V_5}{V_4}\right)^{\gamma - 1} = \left(\frac{V_1}{0.0841\, V_1}\right)^{0.4} = 2.692$$

$$T_5 = \frac{1767.65}{2.692} = 656.6 \text{ K}$$

Air standard thermal efficiency,

$$\eta_{th} = 1 - = \frac{Q_r}{Q_s} = 1 - \frac{c_v(T_5 - T_1)}{c_v(T_3 - T_2) + c_p(T_4 - T_3)}$$

$$= 1 - \frac{T_5 - T_1}{(T_3 - T_2) + \gamma(T_4 - T_3)}$$

$$= 1 - \frac{656.6 - 300}{(1400.4 - 886.25) + 1.4(1767.65 - 1400.4)}$$

$$= 0.6532 \text{ or } 65.32\%$$

The actual thermal efficiency is different from the theoretical value due to the following reasons:

1. The fuel consumed along with air is not considered.
2. The fuel combustion phenomenon and associated problems are ignored.
3. The effect of variable specific heat and heat losses have not been accounted for.

Q.3. (*b*) The following data refer to a steam turbine power plant employing one stage of regenerative feed heating :

State of steam entering H.P. stage : 10 MPa, 600°C

State of steam entering L.P. stage : 2 MPa, 400°C

The correct amount of steam is bled for feed heating at exit from the H.P. stage. Calculate the mass of steam bled per kg of steam passing through the H.P. stage and the amount of heat supplied in the boiler per second for an output of 10 MW. Neglect pump work.

Sol.

H.P. stage : $p_1 = 10$ MPa, $t_1 = 600$°C

L.P. stage : $p_2 = 2$ MPa, $t_2 = 400$°C

Condenser : $p_3 = 0.01$ MPa, $x_3 = 0.9$

$W_{out} = 10$ MW

From steam tables, we have (Fig. 4)

$$h_1 = 3625.3 \text{ kJ/kg}$$

$$h_2 = 3247.6 \text{ kJ/kg}$$

$$h_3 = h_{f3} + x_3 h_{fg3}$$

$$= 191.83 + 0.9 \times 2392.8 = 2345.35 \text{ kJ/kg}$$

Heat balance for the heater :

$$(1 - m_1)(h_5 - h_4) = m_1(h_2 - h_5)$$

From steam tables,

$$h_5 = 908.8 \text{ kJ/kg at } p_2 = 2 \text{ MPa}$$

$$h_4 = 191.83 \text{ kJ/kg at } p_3 = 0.01 \text{ MPa}$$

$$\therefore (1 - m_1)(908.8 - 191.83) = m_1(3247.6 - 908.8)$$

$$(1 - m_1) \times 716.97 = 2338.8\, m_1$$

$$m_1 = \frac{716.97}{3055.77} = 0.235 \text{ kg/kg of steam}$$

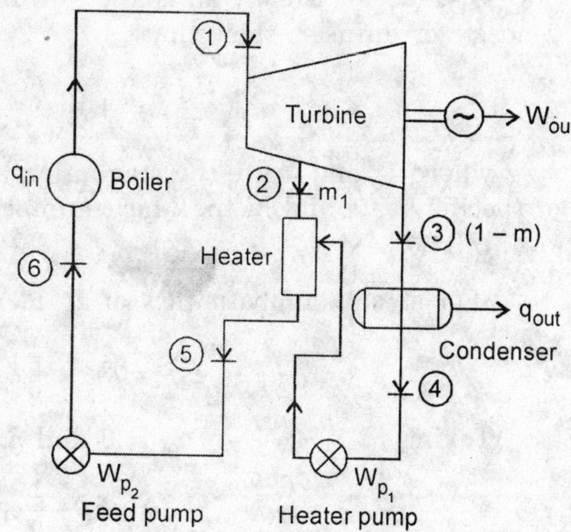

(a) Regenerative feed water heating cycle.

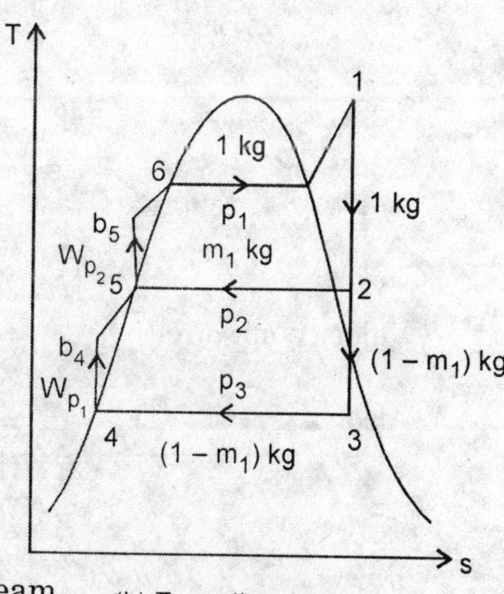

(b) T – s diagram

Fig 4.

Heat supplied in boiler,

$$q_{in} = h_1 - h_5 = 3625.3 - 908.8$$

$$= 2716.5 \text{ kJ/kg}$$

Work output from boiler,

$$W_{out} = (h_1 - h_2) + (1 - m_1)(h_2 - h_3)$$

$$= (3625.3 - 3247.6) + (1 - 0.235)(3247.6 - 2345.35)$$

$$= 1067.92 \text{ kJ/kg}$$

Cycle efficiency, $\eta = \frac{W_{out}}{q_{in}} = \frac{1067.92}{2716.5} = 0.3931$ or 39.31%

Heat supplied in the boiler for 10 MW output = $\frac{10,000}{0.3931} = 25,437$ kW

Q.4. (*a*) For steady adiabatic flow of an ideal gas with constant specific heats through a nozzle or diffuser, show that :

$$\frac{T_0}{T} = 1 + \left(\frac{k-1}{2}\right) M^2$$

where T_0 and T are the stagnation and static temperatures, respectively, k is the ratio of specific heats and M the Mach number.

Sol.

For steady adiabatic flow of an ideal gas,

$$h_1 + \frac{1}{2}c_1^2 = h_2 + \frac{1}{2}c_2^2$$

Taking $c_2 = 0$ and $h_2 = h_0$, we have

$$h_0 = h_1 + \frac{1}{2}c_1^2$$

or

$$c_p T_0 - c_p T_1 = \frac{1}{2}c_1^2$$

$$T_0 = T_1 + \frac{c_1^2}{2c_p}$$

$$\therefore \quad \frac{T_0}{T_1} = 1 + \frac{c_1^2}{2c_p T_1}$$

Velocity of sound, $c_s = \sqrt{kRT}$

Also $c_p - c_v = R$ and $\frac{c_p}{c_v} = k$

$$\therefore \quad c_p - \frac{c_p}{k} = R$$

$$c_p = \frac{kR}{k-1}$$

$$\therefore \quad \frac{T_0}{T} = 1 + \frac{c^2}{2c_p T} = 1 + \frac{c^2(k-1)}{2kRT} = 1 + \frac{(k-1)}{2}\left(\frac{c^2}{kRT}\right) = 1 + \left(\frac{k-1}{2}\right)\left(\frac{c}{c_s}\right)^2$$

Mach number, $M = \frac{c}{c_s}$

$$\therefore \quad \frac{T_0}{T} = 1 + \left(\frac{k-1}{2}\right) M^2$$

Hence proved.

(*b*) In a turbo jet engine, air enters the diffuser at 0.8 bar, 240 K, with a velocity of 1000 km/hr. The pressure ratio across the compressor is 8. The turbine inlet temperature is 1200 K and the pressure at nozzle exit is 0.8 bar. The turbine work just equals the compressor work input. The diffuser, compressor turbine and nozzle processes are isentropic and there is no pressure drop for flow through the combustor. Determine the pressure at exit from the diffuser, the compressor and the turbine, and also the velocity at the nozzle exit. Show the various process on a termperature-entropy diagram. For air, c_p = 1.001 kJ/kg. K and c_p/c_v = 1.4.

Sol.

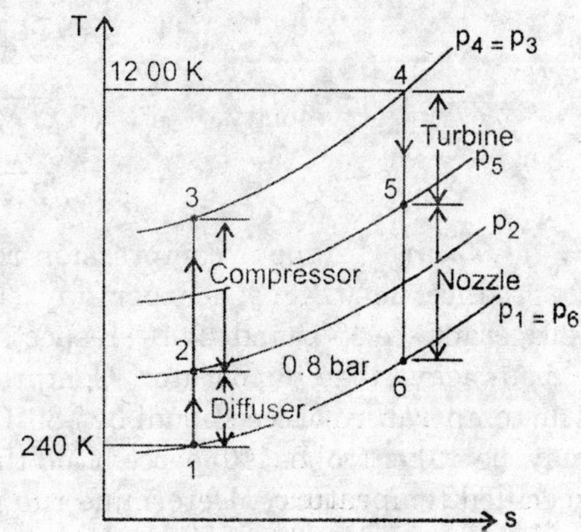

Fig 5. Turbo-jet engine T–S diagram

Given : p_1 = 0.8 bar, T_1 = 240 K,

c_1 = 1000 km/h or 277.78 m/s,

$\frac{p_3}{p_2} = 8$, T_4 = 1200 K, $w_t = w_c$,

c_{pa} = 1.001 kJ/kg·K, γ_a = 1.4

From Fig. 5, we have

$$p_3 = p_4, \quad p_1 = p_6.$$

Assume $$c_{pg} = c_{pa} \text{ and } \gamma_g = \gamma_a$$

Diffuser :

$$T_2 = T_1 + \frac{c_1^2}{2c_{pa}}$$

$$= 240 + \frac{(277.78)^2}{2 \times 1001} = 278.5 \text{ K}$$

$$p_2 = p_1 \left(\frac{T_2}{T_1}\right)^{\frac{\gamma}{\gamma - 1}} = 0.8 \left(\frac{278.5}{2402}\right)^{\frac{1.4}{0.4}} = 1.347 \text{ bar}$$

Compressure :

$$p_3 = 8 \times 1.34 = 10.778 \text{ bar}$$

$$T_3 = T_2 \left(\frac{p_3}{p_2}\right)^{\frac{\gamma}{\gamma - 1}} = 278.5\,(8)^{\frac{0.4}{1.4}} = 504.5 \text{ K}$$

$$p_4 = p_3 = 10.778 \text{ bar}$$

Turbine :

$$w_c = c_p(T_3 - T_2) = 1.001\,(504.5 - 278.5) = 226.226 \text{ kJ/kg}$$

$$w_t = c_p(T_4 - T_5) = 1.001\,(1200 - T_5)$$

Now $$w_t = w_c$$

$$\therefore \quad 1.001\,(1200 - T_5) = 226.226$$

$$T_5 = 974 \text{ K}$$

$$\frac{p_5}{p_4} = \left(\frac{T_5}{T_4}\right)^{\frac{\gamma}{\gamma - 1}}$$

$$p_5 = 10.778\left(\frac{974}{1200}\right)^{\frac{1.4}{0.4}} = 5.19 \text{ bar}$$

Nozzle : $$\frac{T_6}{T_5} = \left(\frac{p_6}{p_5}\right)^{\frac{\gamma-1}{\gamma}} = \left(\frac{p_1}{p_2}\right)^{\frac{\gamma-1}{\gamma}}$$

$$T_6 = 974\left(\frac{0.8}{5.19}\right)^{\frac{0.4}{1.4}} = 570.87 \text{ K}$$

Nozzle jet velocity $$c_6 = \sqrt{2\,c_p(T_5 - T_6)}$$

$$= \sqrt{2 \times 1001\,(974 - 570.87)} = 896.4 \text{ m/s}$$

Q.5. (*a*) A vapour compression refrigeration system using refrigerant – 12 is employed to produce 8640 kg of ice per day. The condensing and evaporating temperatures of the refrigerant are 48°C and –20°C, respectively. Saturated liquid leaves the condenser and saturated vapour leaves the evaporator. Compression is isentropic. Water at 35°C is used to form ice. The temperature of ice shoudl be –80°C. Heat flows into the brine tank from the surroundings may be taken to be 10 percent of the total heat removed from water to form ice at the specified temperature. Determine the power to form ice to drive the compressor, in kW. Take specific heat of ice=2.26 kJ/kg. K, latent heat of ice = 334.72 kJ/kg and specific heat of water = 4.187 kJ/kg K. The following data for refrigerant 12 may be used:

Temperature °C	Pressure bar	Enthalpy, kJ/kg		Entropy, kJ/kg. K	
		liquid	vapour	liquid	vapour
48	11.64	82.83	205.83	0.2973	0.6802
– 20	1.51	17.82	178.74	0.0731	0.7087

Specific heat of vapour = 0.82 kJ/kg. K.

Sol.

Given: m_{ice} = 8640 kg/day, T_1 = 273 – 20 = 253 K,

T_2' = 273 + 48 = 321 K,

t_{wi} = 35°C, t_{ice} = – 80°C,

$(c_p)_{ice}$ = 2.26 kJ/kg·K

(h_{fg}) = 334.72 kJ/kg, c_{pw} = 4.187 kJ/kg·K

c_{ps} = 0.82 kJ/kg. K.

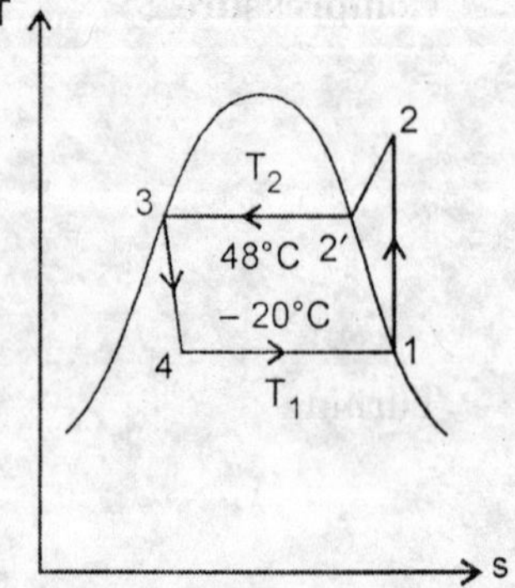

Fig 6. Vapour compression refrigeration system.

Total heat to be removed from water at 35°C to convert it into ice at –80°C

$= m_{ice}\ [c_{pw}\ (t_{wi} - 0) + (h_{fg})_{ice} + (c_p)_{ice}\ [0 - (-8)]]$

$= 8640\ [4.187 \times 35 + 334.72 + 2.26 \times 8]$

= 43,14,340.8 kJ/24 hours

Heat infiltration into brine tank from surroundings = 0.1 × 43,14,340.8

= 4,31,434.08 kJ/24 hours

Total heat of be removed = 43,14,340.8 + 4,31,434.08 = 47,45,774.88 kJ/24 hours

$$= \frac{47,45,774.88}{24 \times 3600} = 54.92 \text{ kJ/s}$$

Now h_1 = 178.74 kJ/kg, s_{g1} = 0.7087 kJ/kg. K (See Fig. 6)

Let temperature of vapour after compression = T_2

Now
$$s_{g1} = s_{g2} + c_{ps} \ln\left(\frac{T_2}{T_2'}\right)$$

$$0.7087 = 0.6802 + 0.82 \ln\left(\frac{T_2}{321}\right)$$

$$T_2 = 332.35 \text{ K}$$

$$h_2 = h_2' + c_{ps} (T_2 - T_2') = 205.83 + 0.82 (332.35 - 321)$$

$$= 215.14 \text{ kJ/kg}$$

$$h_4 = h_{3f} = 82.83 \text{ kJ/kg}$$

Heat removed in evaporator = $h_2 - h_4$ = 178.74 – 82.83 = 95.91 kJ/kg

Flow rate of refrigerant required,

$$\dot{m} = \frac{\text{Total heat of be removed}}{\text{Heat removed in evaporator}} = \frac{54.92}{95.91} = 0.5726 \text{ kg/s}$$

Power required to drive the compressor = $m(h_2 - h_1)$

= 0.5726 (215.14 – 178.74) = 20.843 kW

(***b***) Outside air at 35°C, 60% RH is passed over a cooling coil having an apparatus dew point of 10°C and 0.06 bypass factor,. The cooled air is then supplied to a room which is to be maintained at 22°C dbt and 60% RH. Estimate the sensible and latent heat loads of the room and the heat removed in the cooling coil in kW.

Sol.

Given : t_{d1} = 35°C, $(RH)_1$ = 60%, t_{d4} = 10°C,

t_{d2} = 22°C, x = 0.06, $\dot{m}_a$ = 4500 kg/h

$$\text{BPF} = x = \frac{t_{d3} - t_{d4}}{t_{d1} - t_{d4}}$$

$$0.06 = \frac{t_{d3} - 10}{35 - 10}$$

$$t_{d3} = 11.5°\text{C}$$

From psychrometric chart, we have (Fig. 7)

$$v_{s1} = 0.91 \text{ m}^3/\text{kg}$$

$$\dot{m}_a = \frac{4500}{3600} = 1.25 \text{ kg/s}$$

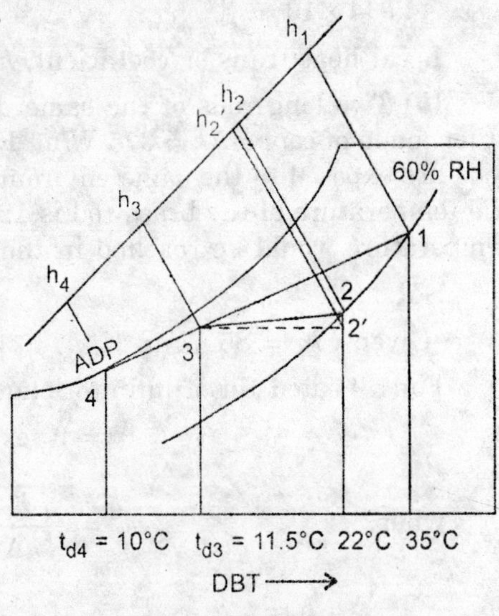

Fig 7.

$$h_3 = 35.5 \text{ kJ/kg}$$
$$h_2 = 48.5 \text{ kJ/kg}$$
$$h_2' = 44.5 \text{ kJ/kg}$$

Sensible heat, $\text{RSH} = \dot{m}_a(h_2' - h_3) = 1.25\,(44.5 - 33.5) = 13.75 \text{ kW}$

$$\text{RLH} = \dot{m}_a(h_2 - h_2') = 1.25\,(48.5 - 44.5) = 5.0 \text{ kW}.$$

Q.6. (*a*) Air flows over a heated flat plate at a velocity of 50 m/s. The local skin friction coefficient at a point on the plate is 0.004. Estimate the local heat transfer coefficient at this point. The following property data for air are given :

Density $= 0.88 \text{ kg/m}^3$;

Viscosity $= 2.286 \times 10^{-5}$ kg/m.s ;

Specific heat, $c_p = 1.001$ kJ/kg. K.

Thermal conductivity, $k = 0.035$ W/m. K.

use $\text{St Pr}^{2/3} = \dfrac{c_f}{2}$.

Sol.

Given : $v = 50$ m/s, $c_f = 0.004$, $\rho = 0.88 \text{ kg/m}^3$, $\mu = 2.286 \times 10^{-5}$ kg/m.s,

$c_p = 1.001$ kJ/kg. K, $k = 0.035$ W/m.K, $\text{St.Pr} = c_f/2$

Prandtl number, $$Pr = \frac{\mu c_p}{k} = \frac{2.286 \times 10^{-5} \times 1.001 \times 10^5}{0.035} = 65.38$$

Stanton number, $$\text{St} = \frac{h}{\rho c_p v} = \frac{h}{0.88 \times 1.001 \times 50 \times 10^3} = \frac{h}{44.044 \times 10^3}$$

$$\text{St.Pr}^{2/3} = \frac{c_f}{2}$$

$$\frac{h}{44.044 \times 10^3} \times (65.38)^{2/3} = \frac{0.004}{2}$$

Local heat transfer coefficient, $h = 5.428 \text{ W/m}^2$. K.

(*b*) Two long rods of the same diameter, one made of brass ($k = 85$ W/m. K) and the other made of copper ($k = 375$ W/m. K), have one of their ends inserted into a furance. Both rods are exposed to the same environment. At a section 10.5 cm away from the furance end, the temperature of the brass rod is 120°C. At what distance from the furance end, the same temperature would be reached in the copper rod? Derive the formula used.

Sol.

Given : $k_b = 85$ W/m. K, $k_c = 375$ W/m. K, $x_b = 10.5$ cm, $t_b = 120°\text{C} = t_c$, $x_c = ?$

For a heated fin of infinite length, temperature at a distance x is,

$$\theta = \theta_0 \exp(-mx)$$

where $$m = \sqrt{\frac{h}{kA}}, \ \theta = t - t_\infty, \text{ and } \theta_0 = t_0 - t_\infty$$

$$t = \text{furnace temperature}$$

$$t_\infty = \text{ambient temperature}$$

$$t_0 = \text{temperature at the point in the fin under consideration.}$$

Now $$m_b\, x_b = m_c\, x_c$$

or $$\sqrt{\frac{h}{k_b\, A_b}} \times x_b = \sqrt{\frac{h}{k_c\, A_c}} \times x_c$$

Here $$A_b = A_c$$

$\therefore$ $$\frac{x_x}{x_b} = \sqrt{\frac{k_c}{k_b}}$$

Length of copper nod, $x_c = 10.5\sqrt{\dfrac{375}{85}} = 22$ cm

Q.7. (*a*) Show that in a Pelton wheel, where the buckets deflect the water through an angle (180° – *a*), the hydraulic efficiency of the wheel is given by :

$$\eta_h = \frac{2u\,(V - u)\,(1 + \cos\alpha)}{V^2}$$

where V is the velocity of the jet, and u the velocity of the wheel at the pitch radius.

Sol.

The velocity diagrams for a Pelton wheel are shown in Fig. 8.

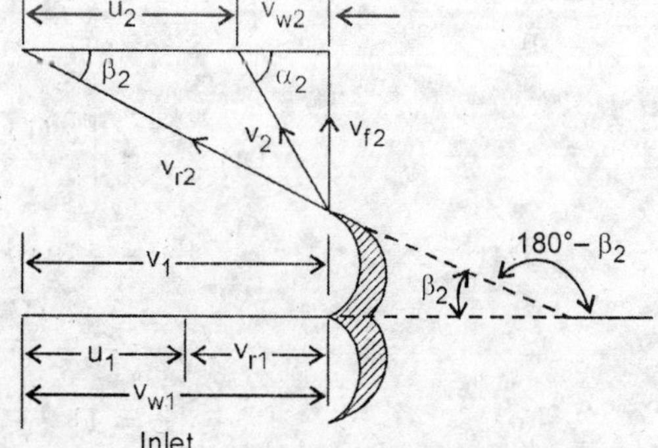

Fig 8. Velocity diagrams for Pelton turbine

An inlet : $v_{r1} = v_1 - u_1 = v_1 - u$

($\because u_1 = u_2 = u$)

$$v_{w1} = v_1,\ \alpha_1 = 0 \text{ and}$$

$$\beta_1 = 0$$

At oulet :

Let K = blade friction coefficient.

Then $$v_{r2} = K\, v_{r1}$$

$$v_{w2} = v_{r2} \cos\beta_2 - u_2$$

$$= v_{r2} \cos\beta_2 - u$$

Force exerted by the jet of water in the direction of its motion,

$$F = \rho A v_1\,(v_{w1} + v_{w2})$$

where ρ = mass density of water, and A = area of jet

Work done by the jet on the runner per second = F × u

$$= \gamma A v_1\,(v_{w1} + v_{w2}) \times u$$

Work done per second per unit mass of water striking the buckets

$$= \frac{\rho A v_1\,(v_{w1} + v_{w2})}{\rho A v_1} = u(v_{w1} + v_{w2})$$

Energy supplied to jet in the form of K.E. per second

$$= \frac{1}{2} m v_1^2 = \frac{1}{2} \rho A v_1 \times v_1^2 = \frac{1}{2} \rho A v_1^3$$

Hydraulic efficiency, $\eta_h = \dfrac{\text{Work done per second}}{\text{K.E. of jet per second}} = \dfrac{\rho A v_1 (v_{w1} + v_{w2}) \times u}{\frac{1}{2} \rho A v_1^3}$

$$= \frac{2 (v_{w1} + v_{w2}) \times u}{v_1^2}$$

Now $v_{w1} = v_1,\ v_{r1} = v_1 - u$

$v_{w2} = v_{r2} \cos \beta_2 - u = \text{K}\, v_{r1} \cos \beta_2 - u = \text{K}\,(v_1 - u) \cos \beta_2 - u$

$\therefore \quad \eta_h = \dfrac{2[v_1 + \text{K}(v_1 - u) \cos \beta_2 - u] \times u}{v_1^2} = \dfrac{2(v_1 - u)(1 + \text{K} \cos \beta_2) \times u}{v_1^2}$

For hydraulic efficiency to be maximum for given value of v_1,

$$\frac{d}{du}(\eta_h) = 0$$

$$\left[\frac{2(1 + \text{K} \cos \beta_2)}{v_1^2}\right] \times \frac{d}{du}(u v_1 - u^2) = 0$$

or $v_1 - 2u = 0$

or $u = \dfrac{v_1}{2}$

$\therefore \quad (\eta_h)_{max} = \left[\dfrac{2(1 + \text{K} \cos \beta_2)}{v_1^2}\right] \times \left(\dfrac{v_1^2}{2} - \dfrac{v_1^2}{4}\right) = \dfrac{1 + \text{K} \cos \beta_2}{2}$

Here $\beta_2 = \propto$ and $v_1 = \text{V}$

$\therefore \quad \eta_h = \dfrac{2u(\text{V} - u)(1 + \text{K} \cos \alpha)}{\text{V}^2}$

For $\text{K} = 1$

$$\eta_h = \frac{2u(\text{V} - u)(1 + \cos \alpha)}{\text{V}^2}$$

Hence proved.

(*b*) A 1/5 scale model of a centrifugal pump absorbs 20 kW when pumping against a test head of 8 m at its best speed of 400 rpm. If the actual pump works against 32 m head, find the speed and power required for the actual pump. Determine also the quantities of water discharged by the two pumps.

Sol.

Given : $\text{P}_m = 20$ kW, $\dfrac{\text{D}_m}{\text{D}_p} = \dfrac{1}{5}$, $\text{H}_m = 8$ m, $\text{N}_m = 400$ spm, $\text{H}_p = 32$ m

For a centrifugal pump,

(i) $$\frac{\sqrt{H}}{DN} = \text{const}$$

$$\therefore \quad \left(\frac{\sqrt{H}}{DN}\right)_m = \left(\frac{\sqrt{H}}{DN}\right)_p$$

$$N_p = N_m \left(\frac{D_m}{D_p}\right)\sqrt{\frac{H_p}{H_m}} = 400 \times \frac{1}{5} + \sqrt{\frac{32}{8}} = 160 \text{ rpm}$$

(ii) $$\frac{P}{D^5 N^3} = \text{const.}$$

$$\therefore \quad P_p = P_m \left(\frac{D_p}{D_m}\right)^5 \left(\frac{N_p}{N_m}\right)^3$$

$$= 20\,(5)^5 \left(\frac{160}{400}\right)^3 = 4000 \text{ kW}$$

(iii) $$\frac{Q}{ND^3} = \text{const.}$$

$$\frac{Q_p}{Q_m} = \left(\frac{N_p}{N_m}\right)\left(\frac{D_p}{D_m}\right)^3$$

$$= \left(\frac{160}{400}\right)(5)^3 = 50$$

Now $$P = \frac{\gamma QH}{10^3} \text{ kW}$$

$$Q_m = \frac{20 \times 10^3}{9.81 \times 10^3 \times 8} = 0.2548 \text{ m}^3/\text{s}$$

$$Q_p = 50 \times 0.2548 = 12.742 \text{ m}^3/\text{s}$$

(c) A single acting reciprocating pump operating at 120 rpm has a piston diameter of 200 mm and stroke of 300 mm. The suction and delivery heads are 4 m and 20 m, respectively. If the efficiency of both suction and delivery strokes is 75 percent, determine the power required by the pump.

Sol.

Given : N = 120 rpm, D = 200 mm, L = 300 mm, H_s = 4m, H_d = 20 m, η = 0.75

Discharge, $$Q = \frac{ALN}{60} = \frac{\frac{\pi}{4}(0.2)^2 \times 0.3 \times 120}{60} = 18.85 \times 10^{-3} \text{ m}^3/\text{s}$$

Power, $$P = \frac{\gamma QH}{\eta \times 10^3} = \frac{9.81 \times 10^3 \times 18.85 \times 10^{-3} \times (4+20)}{0.75 \times 10^3} = 5.917 \text{ kW}.$$

Q.8. (*a*) Determine the adiabatic flame temperature for the combustion of carbon monoxide (CO) with 150 percent theoretical amount of oxygen to form CO_2. The reactants enter the steady flow reactor a 25°C, 1 atm and the products (CO_2 and excess O_2) leave at 1 atm. At 1 atm, 25°C, the enthalpies of formation of O_2, CO and CO_2 are, respectively, 0 kJ/kg. mole, – 110, 418 kJ/kg. mole and – 393, 137 kJ/kg mole. The constant pressure specific heats of CO_2 and O_2 at 1 atm may be assumed to be 56.43 and 36.5 kJ/kg. mole. K. respectively.

Sol.

Combustion equation :

$$2CO + \frac{3}{2}O_2 \longrightarrow 2CO_2 + \frac{1}{2}O_2$$

For adiabatic process, total enthalpy of reactants = Total enthalpy of products.

$$\sum H_r = \sum H_p$$

or

$$\sum_i (n_i\ h_i)_r = \sum_i (n_i\ h_i)_p = \sum_i n_i\ [h_f{}^\circ + \Delta h]_r = \sum_i n_i\ [h_f^\circ + \Delta h]_p$$

where n = number of moles

h = molar specific heat, h_f = enthalply of formation.

Enthalpy of reactants,

$$\sum H_r = 2 \times (-110418) + 0 = -220836 \text{ kJ/kg. mol}$$

Enthalpy of products,

$$\sum H_p = [2 \times (-393137) + 56.43\ (T - 298)] + \left[\frac{1}{2} \times 36.5(T - 298)\right]$$

$$= -786274 + 74.68\ (T - 298) \text{ kJ/kg. mol}$$

$$\therefore \quad -220836 = -786274 + 74.68\ (T - 298)$$

$$T = 7869.49 \text{ K or } 7596.48°C.$$

(*b*) The following data were obtained from a single glas cover flat plate solar heat collector : Mean plate temperature = 70°C; Ambient and sky temperature = 10°C; Back insulation thickness = 5 cm; Insulation thermal conductivity = 0.05 W/m. °C; Coefficient of heat transfer by convection from plate to cover = 3 W/m^2. °C; Equivalent coefficient of heat transfer by radiation from plate to cover = 6 W/m^2 °C; Coefficient of convective heat transfer from cover to ambient air = 25 W/m^2. °C and Equivalent coefficient for radiant heat transfer from cover to sky = 5 W/m^2 °C. Compute the total heat loss per m^2 collector area. Assume that resistance to heat flow by convection at the back surface of the insulation is negligible.

Sol.

Absorber plate surface temperature, T_s = 273 + 70 = 343 K

Ambient temperature, T_∞ = 273 + 10 = 283 K

Sky temperature, T_{sky} = 273 + 10 = 283 K

Back insulation thickness, x_b = 5 cm

Insulation thermal comductivity, k = 0.05 W/m. °C

Convective heat transfer coefficient from plate to cover,

$$h_{ca} = 3 \text{ W/m}^2.\ °\text{C}$$

Equivalent radiation heat transfer coefficient from plate to cover.

$$h_{ra} = 6 \text{ W/m}^2.\ °\text{C}$$

Convective heat transfer coefficient from cover to ambient air,

$$h_{cc} = 25 \text{ W/m}^2.\ °\text{C}$$

Equivalent radiation heat transfer coefficient from cover to sky,

$$h_{rc} = 5 \text{ W/m}^2.\ °\text{C}$$

The control volume for the solar collector is shown in Fig. 9.

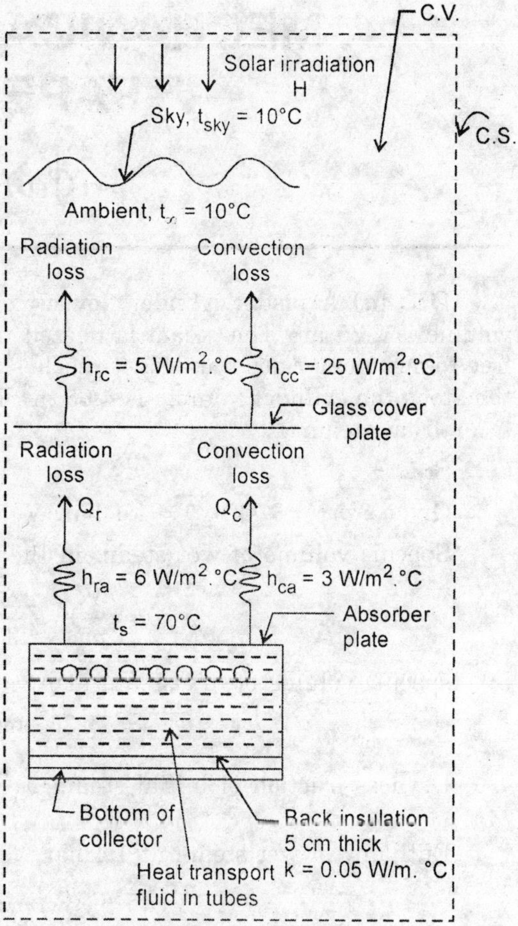

Fig 9.

Let T_c = temperature of cover plate.

For heat balance of cover plate,

Heat absorbed from absorbe plate

= Heat lost to ambient air and sky.

$$Q_{ca} + Q_{ra} = Q_{cc} + Q_{rc}$$

$$h_{ca}\, A(T_s - T_c) + h_{ra}\, A(T_s - T_c)$$

$$= h_{cc}\, A(T_c - T_\infty) + h_{rc}\, A(T_c - T_{sky})$$

For $\qquad A = 1 \text{ m}^2$

$$3(343 - T_c) + 6(343 - T_c) + \frac{0.05(343 - 283)}{5 \times 10^{-2}}$$

$$= 25(T_c - 283) + 5(T_c - 283)$$

$$9(343 - T_c) = 30(T_c - 283)$$

$$3087 - 9T_c = 30T_c - 8490$$

$$39\, T_c = 11577$$

$$T_c = 296.84 \text{ K}$$

Total heat loss per m collector area = 25(296.84 – 283) + 5(296.84 – 283)

$$= 415.4 \text{ W/m}^2$$

Heat lost by conduction through insulation = $kA.\dfrac{\Delta T}{\Delta x}$

$$= \frac{0.05 \times 1 \times (70 - 10)}{0.05} = 60 \text{ W/m}^2$$

Total heat loss = 415.4 + 60 = 475.4 W/m^2

MECHANICAL ENGINEERING
PAPER I—1994

(Conventional)

Q.1. (*a*) A piston-cylinder device contains 3 kg of wet steam at 1.4 bar. The initial volume is 2.25 m^3. The steam is heated until its temperature reaches 400°C. The piston is free to move up or down unless it reaches the stops at the top. When the piston is up against the stops the cylinder volume is 4.65 m^3. Determine the amounts of work and heat transfer to or from steam.

Sol.

Given : m_s = 3 kg, p_1 = 1.4 bar, V_1 = 2.25 m^3, t_2 = 400°C, V_2 = 4.65 m^3

Specific volume of wet steam in the cylinder at p_1 = 1.4 bar,

$$v_{s1} = \frac{V_1}{m_s} = \frac{2.25}{3} = 0.75 \text{ m}^3/\text{kg}$$

specific volume of dry steam at p_1 = 1.4 bar, from steam tables,

$$v_{g1} = 1.2363 \text{ m}^3/\text{kg}$$

Dryness fraction of initial steam, $x_1 = \dfrac{v_{s1}}{v_{g1}} = \dfrac{0.75}{1.2363} = 0.607$

Enthalpy of wet steam at 1.4 bar,

$$H_1 = m_s(h_{f1} + x_1 h_{fg1}) = 3\,(458.4 + 0.607 \times 2231.9) = 5439.5 \text{ kJ}$$

Specific volume of steam after heating, $v_{s2} = \dfrac{V_2}{m_s} = \dfrac{4.65}{3} = 1.55 \text{ m}^3/\text{kg}$

From superheated steam tables, at 400°C, and v_{s2} = 1.55 m^3/kg, we have

$$p_2 = 2 \text{ bar}$$

$$t_s = 120.2\text{°C}$$

Degree of superheat, $(t_{sup} - t_s)$ = 400 – 120.2 = 279.8°C

Enthalpy of steam at 2 bar and 400°C, H_2 = 3 × 3276.7 = 9830.1 kJ

Heat added, $\Delta H = H_2 - H_1$ = 9830.1 – 5439.4 = 4390.6 kJ

Internal energy of wet steam at 1.4 bar,

$$u_1 = u_{f1} + x_1 u_{fg1} = 457.84 + 0.607 \times 2059.34 = 1707.86 \text{ kJ/kg}$$

Internal energy of steam at 2 bar and 400°C,

$$u_2 = h_{sup} - p_2 v_{s2} = 3276.7 - 2 \times 10^2 \times 1.55 = 2966.7 \text{ kJ/kg}$$

Change in internal energy, $\Delta U = m_s(u_2 - u_1)$ = 3(2966.7 – 1707.86) = 3776.52 kJ

Work done = $\Delta H - \Delta U$ = 4390.6 – 3776.52 = 614.08 kJ

(*b*) A evacuated bottle of 0.5 m^3 volume is slowly filled from atmospheric air at 1.0135 bars until the pressure inside the bottle also becomes 1.0135 bar. Due to heat transfer, the temeperature of air inside the bottle after filling is equal to the atmospheric air temperature. Determine the amount of heat transfer.

Sol.

The bottle filling process is shown in Fig. 1.

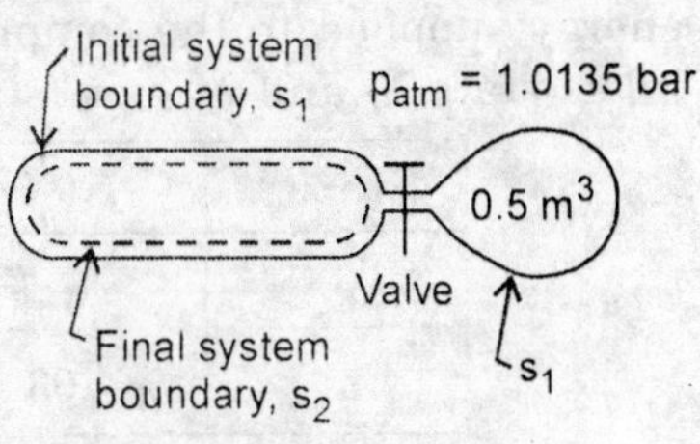

Fig 1.

Work done, $$W = \int_{\text{bottle}} p.dV + \int_{\substack{\text{free–air}\\ \text{boundary}}} p.dV$$

$$= 0 + \int_{0.5}^{0} 1.0135 \times 10^5 \times dV$$

$$= 1.0135 \times 10^5 \, (0 - 0.5)$$

$$= -\,50.675 \text{ kJ}$$

Heat transfer, $$Q = W = -\,50.675 \text{ kJ}$$

Q.2. (***a***) Two blocks of metal, each of mass M and specific heat C, initially at absolute temperature T_1 and T_2 respectively, are brough to the same final temperature by means of a reversible process. Derive an expression for the amount of work obtained during the process in terms of M, C, T_1 and T_2.

Sol.

Let the sink be heated from T to T + dT and the source be cooled from T′ to T′ – *d*T′.

Heat interaction from sink, $dQ_2 = MC[(T + dT) - T] = MC.dT$

Heat interaction from source, $dQ_1 = MC[(T' + dT') - T'] = -\,MC.dT'$

For a reversible heat engine, $\oint \frac{dQ}{T} = 0$

$$\therefore \quad \frac{dQ_1}{dQ_2} = \frac{T'}{T}$$

or $$\frac{dT'}{T'} + \frac{dT}{T} = 0$$

Integrating from initial to final state, we get

$$\ln\frac{T}{T_1} + \ln\frac{T}{T_2} = 0$$

or $$T^2 = T_1 \, T_2$$

or $$T = \sqrt{T_1 \, T_2}$$

Work interaction, $$dW = dQ_1 - dQ_2 = -\,MC(dT' + dT)$$

$$W = -\int MC(dT' - dT) = -\,MC \int_{T_1}^{T} dT' - MC \int_{T_2}^{T} dT$$

$$= -MC(T - T_1) - MC(T - T_2)$$

$$= 2MC\left[\frac{T_1 + T_2}{2} - T\right].$$

(*b*) In a residential heat system the refrigerant flow rate is 412.2 kg/hr and the electric energy supplied to the compressor is 25,480 kJ/hr. The properties of the refrigerant at the inlet (state 1) and at exit (state 2) of the compressor are as given below :

State	***t* (°C)**	***p* (bar)**	***h* (kJ/kg)**	***s* (kJ/kg K)**
1	– 1	2.6	226.35	0.8636
2	93	15.0	286.40	0.9222

Determine the amount of available energy destroyed in the compressor in kJ/hr. All heat transfer, if it occurs, takes place at T_0. Assume T_0 = 273 K, p_0 = 1 atm.

Sol.

Given: $\dot{m}_2$ = 412.2 kg/h, $\dot{Q} = -$ 25,480 kJ/h, T_0 = 273 K, p_0 = 1 atm,

T_1 = 273 – 1 = 272 K, p_1 = 2.6 bar, h_1 = 226.35 kJ/kg,

s_1 = 0.8636 kJ/kg. K,

T_2 = 273 + 93 = 366 K, p_2 = 15.0 bar, h_2 = 286.40 kJ/kg,

s_2 = 0.9222 kJ/kg. K

Minimum work of compression, $W_{min} = (h_1 - T_0 s_1) - (h_2 - T_0 s_2)$

$$= (226.35 - 273 \times 0.8636) - (286.40 - 273 \times 0.9222)$$

$$= -44.0522 \text{ kJ/kg}$$

or $\quad W_{min} = \dot{m}_r W_{min} = 412.2 \times (-44.0522) = -18158.317$ kJ/hr

Amount of available energy destroyed = $\dot{Q} - W_{min}$

$$= 25480 - 18158.7 = 7321.68 \text{ kJ/hr}$$

Q.3. (*a*) In an air-standard Brayton cycle the minimum and maximum temperatures are 300 K and 1200 K, respectively. The pressure ratio is that which maximizes the net work developed by the cyecle per unit mass of air flow. Calculate the compressor and turbine work, each in kJ/kg air, and the thermal efficiency of the cycle.

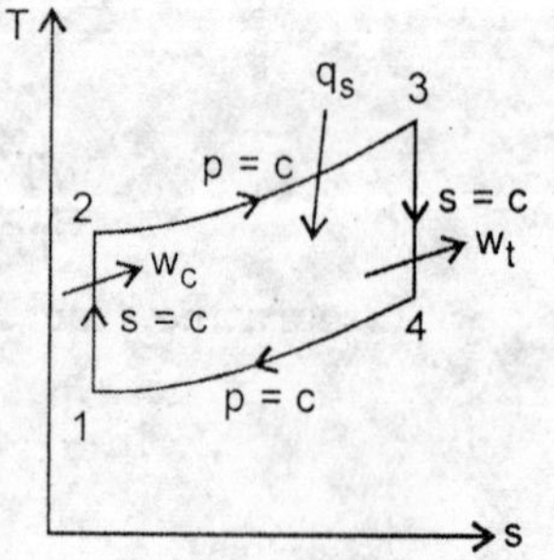

Sol. Given : T_1 = 300 K, T_3 = 1200 K

The Brayton cycle on T – s plot is shown in Fig. 2.

Fig 2. Brayton cycle

For maximum net work developed by the Brayton cycle, pressure, ratio,

$$r_p = \left[\frac{T_3}{T_1}\right]^{\frac{\gamma}{2(\gamma-1)}} = \left[\frac{1200}{300}\right]^{\frac{1.4}{2(1.4-1)}} = 11.314$$

$$T_2 = T_1 (r_p)^{\frac{\gamma-1}{\gamma}} = 300\,(11.314)^{\frac{0.4}{1.4}} = 600 \text{ K}$$

Compressor work, $w_c = c_p (T_2 - T_1) = 1.005 (600 - 300) = 301.5$ kJ/kg

Turbine work, $w_t = c_p (T_3 - T_4) = 1.005 (1200 - 600) = 603$ kJ/kg

Thermal efficiency, $\eta_{th} = \dfrac{w_t - w_c}{q_s} = \dfrac{603 - 301.5}{1.005(1200 - 600)} = \dfrac{301.5}{603} = 0.50$ or 50%.

(***b***) Octane, C_8H_{18}, is burned with dry air. The molar analysis of the dry products is CO_2 = 10%, CO = 0.4%; O_2 = 3% and N_2 = 86.6%. Determine the air fuel ratio on a mass basis and the dew point temperature of the products if the products are at 1 atm pressure. Dry air contains 3.76 moles of nitrogen per mole of oxygen.

Sol.

Constituent	*Percentage by volume* (a)	*Molecular weight* (b)	*Proportional mass* (c) = (a) × (b)	*Percentage by mass* (d) = $\dfrac{(c)}{\Sigma(c)}$
CO_2	10.0	44	440.0	14.805
CO	0.4	28	11.2	0.377
O_2	3.0	32	96.0	3.230
N_2	86.6	28	2424.8	81.588
			Σ(*c*) = 2972.0	Σ(*d*) = 100.00

Mass of C per kg of dry flue gas

$= \text{Mass of C from } CO_2 + \text{Mass of C from CO}$

$= \dfrac{12}{44} \times 0.14805 + \dfrac{12}{28} \times 0.00377 = 0.042 \text{ kg}$

Percentage of C in fuel $= \dfrac{96}{114} \times 100 = 84.21\%$

Mass of dry flue gases formed per kg of fuel $= \dfrac{84.21}{100 \times 0.042} = 20.05$ kg

Mass of air supplied per kg of fuel $= \dfrac{NC}{33(CO_2 + CO)} = \dfrac{86.6 \times 84.21}{33(10 + 0.4)} = 21.25$ kg

Air/Fuel ratio = 21.25 : 1

Dew point temperature of products is the saturation temperature at the partial presure of water vapour.

$$C_8H_{18} + 12.5O_2 + 12.5\,(3.76)\,N_2 \longrightarrow 8CO_2 + 9H_2O + 47N_2$$

12.5 moles of O_2 are required to burn 1 mole C_8H_{18} and 8 moles of H_2O are produced. Stoichiometric mole fraction of water vapour (x_{H_2O}) formed,

$$x_{H_2O} = \frac{n_{H_2O}}{n_{CO_2} + n_{CO} + n_{O_2} + n_{N_2}} = \frac{9}{8 + 0 + 0 + 47} = \frac{9}{55} = 0.1636$$

$$p_{H_2O} = x_{H_2O} \text{ atm} = 0.1636 \text{ bar}$$

The saturation temperature of water vapour corresponding to their partial pressure of 0.1636 bar is, t = 54°.

Q.4. (*a*) A single stage steam turbine is supplied with steam at 5 bar, 200°C at the rate of 50 kg/min. It exhausts into a condenser at a pressure of 0.2 bar. The blade speed is 400 m/s. The nozzles are inclined at an angle of 20° to the plane of the wheel and the outlet blade angle is 30°. Neglecting friction losses, determine the blade efficiency, the stage efficiency and the power developed by the turbine.

Sol.

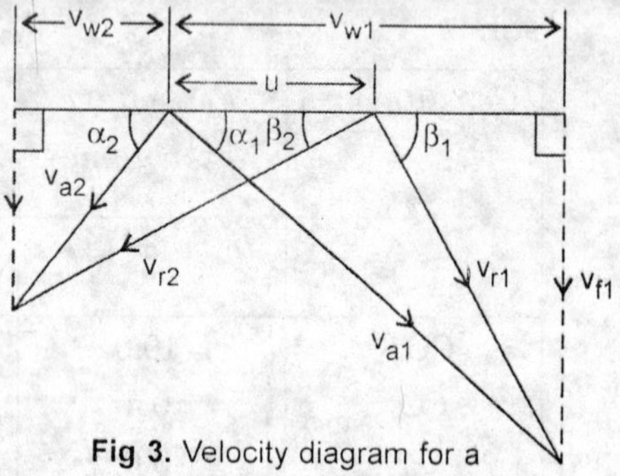

Fig 3. Velocity diagram for a single stage impulse steam turbine.

Given : $\dot{m}$ = 50 kg/min, u = 400 m/s

$$\alpha_1 = 20°, \ \beta_2 = 30°,$$

$$v_{r1} = v_{r2} \text{ as K} = 1$$

At 5 bar, 200°C, $h_1 = 2855.4$ kJ/kg,

$$s_1 = 7.059 \text{ kJ/kg. K}$$

At 0.2 bar, $s_{f2} = 0.832$ kJ/kg. K,

$$s_{fg2} = 7.077 \text{ kJ/kg. K}$$

$$s_2 = s_1 = s_{f2} + x_2 \, s_{fg2}$$

$$7.059 = 0.832 + 7.077 \, x_2$$

$$x_2 = 0.88$$

$$h_2 = h_{f2} + x_2 \, h_{fg2} = 251.4 + 0.88 \times 2358.3 = 2325.7 \text{ kJ/kg}$$

Enthalpy drop, $\Delta h = h_1 - h_2 = 2855.4 - 2325.7 = 528.7$ kJ/kg

$$v_{a1} = \sqrt{2 \times \Delta h} = \sqrt{2 \times 528.7x \, 10^3} = 1028.3 \text{ m/s}$$

$$v_{w1} = v_{a1} \cos \alpha_1 = 1028.3 \cos 20° = 965.3 \text{ m/s}$$

$$v_{f1} = v_{a1} \sin \alpha_1 = 1028.3 \sin 20° = 351.7 \text{ m/s}$$

$$\tan \beta_1 = \frac{v_{f1}}{v_{w1} - u} = \frac{351.7}{966.3 - 400} = \frac{351.7}{566.3} = 0.621$$

$$\beta_1 = 31.84°$$

$$v_{r1} = \frac{v_{f1}}{\sin \beta_1} = \frac{351.7}{\sin 31.84°} = 665.6 \text{ 3m/s}$$

$$v_{r2} = v_{r1} = 665.6 \text{ m/s}$$

$$v_{w2} = v_{r2} \cos \beta_2 - u = 665.6 \cos 30° - 400 = 177. \text{ m/s}$$

$$v_w = v_{w1} + v_{w2} = 965.3 + 177.3 = 1143.6 \text{ m/s}$$

Power developed, $$\text{P} = \frac{\dot{m} u \, v_w}{10^3} = \frac{50}{60} \times 400 \times \frac{1143.6}{10^3} = 381.2 \text{ kW}$$

Blade efficiency, $$\eta_b = \frac{2uv_w}{v_{a1}^2} = \frac{2 \times 400 \times 1143.6}{(1028.3)^2} = 0.8652 \text{ or } 86.52\%$$

(*b*) What is a shock wave? Write the continuity, momentum and energy equations relating points 1 and 2, just upstream and just downstream, respectively, of a normal shock in a constant area duct. State whether the following properties of the flow stream increase, decrease remain unchanged across a normal shock: total pressure, total enegy, static temperature, density and velocity.

Sol.

Shock Wave. An actual wave through which the flow experiences sudden changes in density, pressure and temperature is called a shock wave.

Normal Shock. It is almost perpendicular to the flow in which abrupt change from supersonic conditions to subsonic ones take place through a shock wave. The Mach number and **velocity decrease across a normal shock. There occurs a sudden rise in pressure, density, temperture and entropy. The parameters of normal shock are shown in Fig. 4.**

1. Continutity Equation :

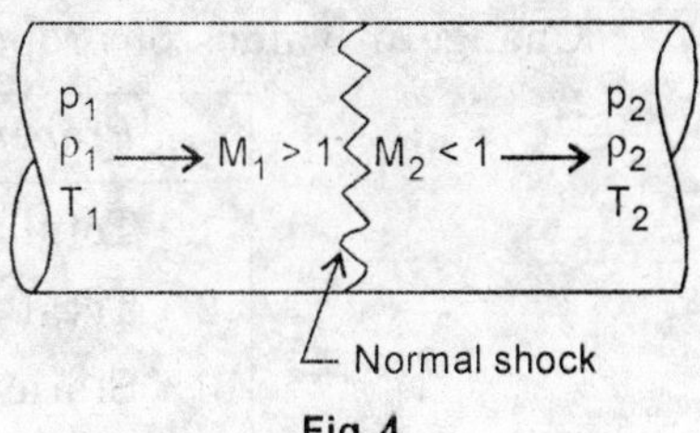

Fig 4.

$$\rho_1 A_1 v_1 = \rho_2 A_2 v_2$$

For a constant area duct,

$$A_1 = A_2$$

$$\therefore \quad \frac{\rho_2}{\rho_1} = \frac{v_1}{v_2}$$

$$v_1 > c \text{ and } v_2 < c$$

$$\rho_2 > \rho_1$$

$$\therefore \quad \rho_1 v_1 = \rho_2 v_2$$

$$c\rho_1 \frac{v_1}{c} = c\rho_2 \frac{v_2}{c}$$

$$\sqrt{\frac{\gamma p_1}{\rho_1}} \times \rho_1 M_1 = \sqrt{\frac{\gamma p_2}{\rho_2}} \times \rho_2 M_2$$

$$\sqrt{\gamma p_1} \cdot \sqrt{\rho_1}\, M_1 = \sqrt{\gamma p_2} \cdot \sqrt{\rho_2}\, M_2$$

$$\rho = \frac{p}{RT}$$

$$\sqrt{p_1} \cdot \sqrt{\frac{p_1}{RT_1}}\, M_1 = \sqrt{p_2} \cdot \sqrt{\frac{p_2}{RT_2}}\, M_2$$

$$\frac{p_1 M_1}{\sqrt{T_1}} = \frac{p_2 M_2}{\sqrt{T_2}}$$

2. Momentum Equation :

$$p_1 + \rho_1 v_1^2 = p_2 + \rho_2 v_2^2$$

$$\frac{p_2}{p_1} = \frac{1 + \gamma M_1^2}{1 + \gamma M_2^2}$$

As $\quad M_1 > 1$ and $M_2 < 1$

$$\therefore \quad p_2 > p_1$$

3. Energy Equation :

$$h_1 + \frac{v_1^2}{2} = h_2 + \frac{v_2^2}{2}$$

$$\frac{T_2}{T_1} = \frac{1 + \frac{\gamma - 1}{2} M_1^2}{1 + \frac{\gamma - 1}{2} M_2^2}$$

As $M_1 > 1$ and $M_2 < 1$

$\therefore$ $T_2 > T_1$

$$\frac{p_2}{p_1} = \frac{(\gamma + 1) M_1^2}{(\gamma - 1) M_2^2 + 2}$$

Change in values of properties of flow stream across a normal shock.

Property	*Change*
1. Total pressure	Constant
2. Total energy	Constant
3. Static temperature	No Change
4. Density	Increases
5. Velocity	Decreases

Q.5. (*a*) A cascade refrigeration system of 100 tons (350 kW) capacity uses ammonia and carbon dioxide. The evaporating and condensing temperature of CO_2 are — 40°C and 5°C, respectively. The evaporating temperature of NH_3 is – 7°C. The power supplied to the ammonia compressor is 96.5 kW. In the CO_2 circuit, the liquid leaving the condenser is saturated, the vapor leaving the evaporator is dry and saturated, and compression is isentropic. Calculate the mass flow rate of CO_2 and the C.O.P. of the system. The following data for CO_2 may be used :

Temperature, °C	Pressure, bars	Enthalpy, kJ/kg		Entropy, kJ/kgK	
		Liquid	Vapor	Liquid	Vapor
– 40	10.05	332.7	652.8	3.8531	5.2262
5	39.70	431.0	649.8	4.2231	5.0097

Specific heat of superheated vapor,

$$c_p = 0.85 \text{ kJ/kg K.}$$

Sol.

The schematic diagram of a two-stage cascade refrigerating system using CO_2 and NH_3 is shown in Fig. 5(*a*).

The corresponding $p - h$ diagram of the system is shown in Fig. 5(*b*). The cycle 1-2-3-4 is for CO2 unit whereas the cycle 5-6-7-8 is for NH_3 unit. Both these cyelces are superimposed in the $p - h$ diagram.

CO_2 - unit :

Refrigerating effect = $h_1 - h_4$

$= h_1 - h_{f3}$

$= 652.8 - 431.0 = 221.8$ kJ/kg

Total refrigerating effect of plant

= 100 tons = 350 kW

Mass flow rate of CO_2,

$$\dot{m}_{CO_2} = \frac{350}{221.8}$$

= 1.578 kg/s or 5680 kg/hr

Now $s_1 = s_2$

$$= s_{g2} + c_p \ln \frac{T_2}{(273+5)}$$

$$5.2262 = 5.0097 + 0.85 \ln\left(\frac{T_2}{278}\right)$$

$T_2 = 358.6$ K

$\therefore h_2 = h_2' + c_p (T_2 - T_2')$

$= 649.8 + 0.85 (358.6 - 278)$

= 718.31 kJ/kg

Work done in CO_2 system,

$$W_{CO_2} = \dot{m}_{CO_2} \times w_{CO_2}$$

$= 1.578 \times 65.51$

= 103.37 kJ/s

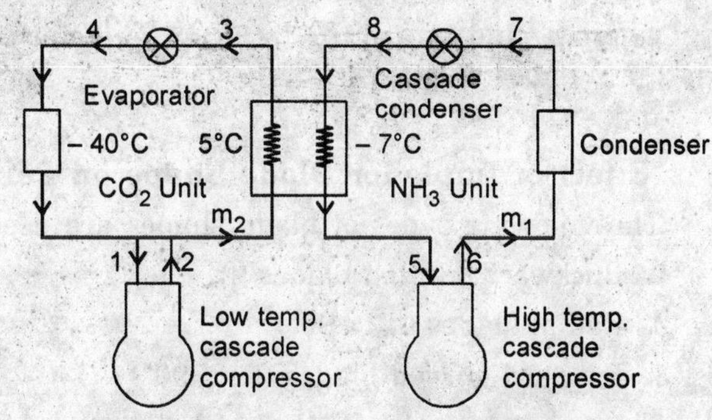

(a) Schematic diagram of a two stage cascade refrigerating system.

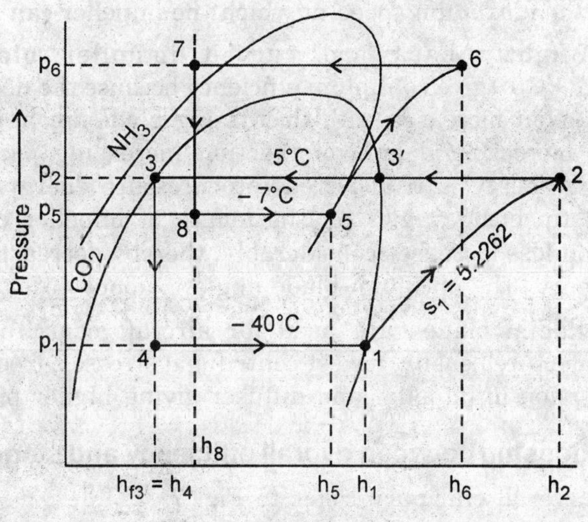

(b) p-h diagram.

Fig 5.

$$\text{COP of system} = \frac{\text{Total refrigerating effect}}{W_{CO_2} + W_{NH_3}}$$

$$(W_{NH_3} = 96.5 \text{ kW})$$

$$= \frac{350}{103.37 + 96.5} = \frac{350}{199.87} = 1.75$$

(b) Sketch cuves showing the variation of pressure ratio (or head) versus volume flow rate for the three types of blades generally used in centrifugal compressors. Indicate on each curve the stable range of operation. Show that stage efficiency η_s and overall efficiency η_0 of an axial flow compressor are related by

$$\eta_0 = \frac{(p_2/p_1)^{\frac{\gamma-1}{\gamma}} - 1}{\left[\dfrac{(p_2/p_1)^{\frac{\gamma-1}{\gamma}} - 1}{\eta_s} + 1\right]^n - 1}$$

where p_1 and p_2 are the total pressures at compressor inlet and exit, respectively, and n is the number of stages.

Sol.

Effect of Impeller Blade Shape on Performance

The following types of blade shapes are used for impellers of centrifugal compressors:

1. Backward–curved blades ($\beta_2 < 90°$)
2. Radial blades ($\beta_2 = 90°$)
3. Forward curved blades ($\beta_2 > 90°$)

The characteristics of these blades are shown in Fig. 6. Centrifugal effects on the curved blades create bending moment and produce increased stresses which limit the maximum speed at which the impeller can run.

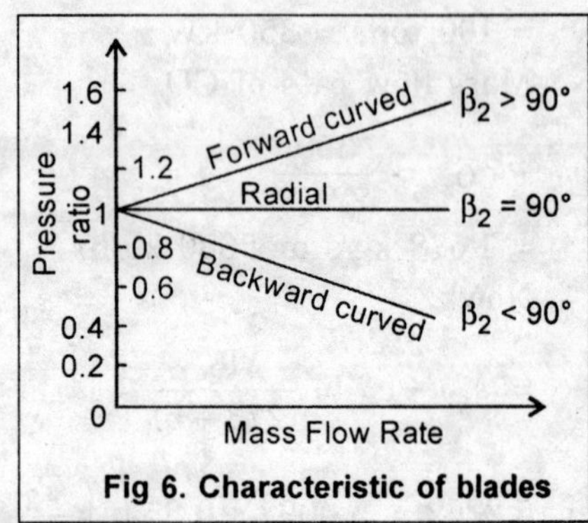

Fig 6. Characteristic of blades

Slightly backward curved imepller blades (*i.e.*, $\beta_2 < 90°$) gives optimum efficiency because the degree of reaction increases with decreasing blade tip angles. With increasing degree of reaction increasing part of kinetic energy is transferred into pressure energy within the impeller. By selecting very small tip blade angles β_2, the length of impeller channel increases and consequently the friction loss increases considerably, thereby decreasing the beneficial effect and reducing the efficiency. The optimum blade angle is about 45°.

Radial blades are used for aircraft centrifugal compressors because they can be manufactured easily, lowest centrifugal stresses, free from bending stresses and equal energy conversion in impeller and diffuser giving higher pressure ratio and efficiency.

Relationship between overall efficiency and stage efficiency of Axial Flow Compressor

Overall efficiency, refer to Fig. 7.

$$\eta_0 = \frac{T_2 - T_1}{T_2' - T_1} = \frac{\left(\frac{T_2}{T_1}\right) - 1}{\left(\frac{T_2'}{T_1}\right) - 1} = \frac{\left(\frac{p_2}{p_1}\right)^{\frac{\gamma-1}{\gamma}} - 1}{\left(\frac{T_2'}{T_1}\right) - 1}$$

Stage efficiency,

$$\eta_s = \frac{T_a - T_1}{T_a' - T_1} = \frac{(T_a / T_1) - 1}{(T_a' / T_1) - 1}$$

$$= \frac{(p_a / p_1)^{\frac{\gamma-1}{\gamma}} - 1}{(T_a'/T_1) - 1}$$

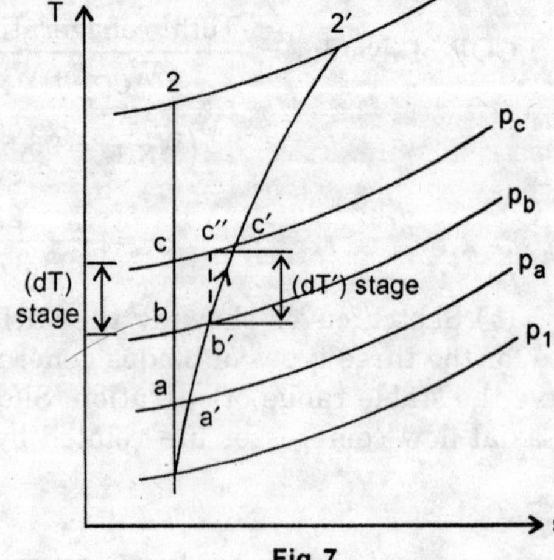

Fig 7.

or $$(T_a'/T_1) = \frac{(p_a / p_1)^{\frac{\gamma-1}{\gamma}} - 1}{\eta_s} + 1$$

Now $$\frac{p_a}{p_1} = \frac{p_b}{p_a} = ... = \left(\frac{p_2}{p_1}\right)^{\frac{1}{n}}$$

$$\frac{T_a'}{T_1} = \frac{(p_2/p_1)^{\frac{\gamma-1}{n\gamma}} - 1}{\eta_s} + 1$$

For the n stages,

$$\frac{T_a'}{T_1} \times \frac{T_b'}{T_a'} \times .... \times \frac{T_2'}{T_{(n-1)}'} = \left[\frac{\left(\frac{p_2}{p_1}\right)^{\frac{\gamma-1}{n\gamma}} - 1}{\eta_s} + 1\right]^n$$

or $$\frac{T_2'}{T_1} = \left[\frac{\left(\frac{p_2}{p_1}\right)^{\frac{\gamma-1}{n\gamma}} - 1}{\eta_s} + 1\right]^n$$

$$\therefore \quad \eta_0 = \frac{(p_2/p_1)^{\frac{\gamma-1}{\gamma}} - 1}{\left[\frac{\left(\frac{p_2}{p_1}\right)^{\frac{\gamma-1}{n\gamma}} - 1}{\eta_s} + 1\right]^n - 1}$$

Hence proved.

Q.6. (*a*) A cylindrical tank with hemispherical ends is used to store liquid oxygen at –183°C. The diameter of the tank is 1.5 m and the total length is 8 m. The tank is covered with a 10 cm thick layer of insulation. Determine the thermal conductivity of the insulation so that the boil-off rate does not exceed 10.8 kg/hr. The latent heat of vaporization of liquid oxygen is 214 kJ/kg. Assume that the outer surface temperature of the insulation is 27°C and that the thermal resistance of the wall of the tank is negligible.

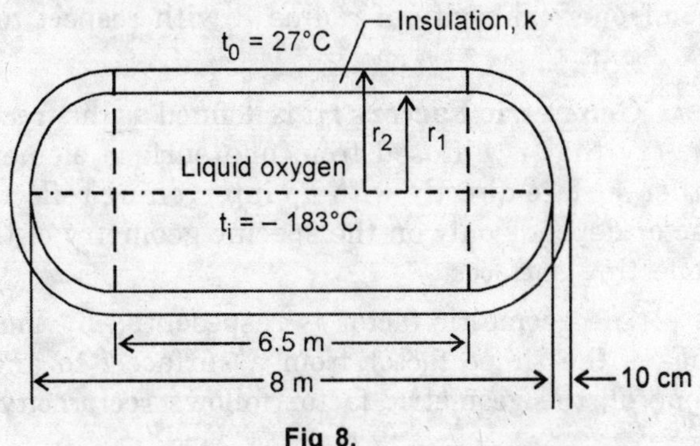

Fig 8.

Sol.

Given : d_1 = 1.5 m, or r_1 = 0.75 m,
L = 8 m, r_2 = 0.75 + 0.10
= 0.85 m,
t_i = – 183°C, t_0 = 27°C, l = L – 2 × 0.75
= 6.5 m
Boil-off rate of liquid oxygen
= 10.8 kg/hr,

$$(h_{fg})_{O_2} = 214 \text{ kJ/kg}$$

Cylindrical section of tank :

$$Q_{cyl} = \frac{2\pi kl\,(t_0 - t_i)}{\ln\left(\frac{r_2}{r_1}\right)}$$

or $$t_0 - t_i = \frac{\ln(r_2/r_1)}{2\pi kl} \times Q_{cyl} \quad ...(1)$$

Hemispherical ends :

$$Q_{hemi} = \frac{4\pi r_1 r_2\, k(t_0 - t_i)}{r_2 - r_1}$$

$$t_0 - t_i = \left(\frac{r_2 - r_1}{4\pi r_1 r_2 k}\right) \times Q_{hemi} \quad ...(2)$$

Equating Eqs. (1) and (2), we get

$$\frac{\ln(r_2/r_1)}{2\pi kl} \times Q_{cyl} = \left(\frac{r_2 - r_1}{4\pi r_1 r_2 k}\right) \times Q_{hemi}$$

$$Q_{hemi} = \frac{2 r_1 r_2 \ln(r_2/r_1)}{(r_2 - r_1)l} \times Q_{cyl}$$

Also $$Q_{cyl} + Q_{hemi} = Q_{boil} = 214 \times 10.8 = 2311.2 \text{ kJ/hr}$$

$$Q_{cyl} + \frac{2 \times 0.75 \times 0.85 \ln(0.85/0.75)}{(0.85 - 0.75) \times 6.5} \times Q_{cyl} = 2311.2$$

$$Q_{cyl}\,(1 + 0.2455) = 2311.2$$

$$Q_{cyl} = 1856.4 \text{ kJ/hr or } 515.66 \text{ g/s}$$

From Eq.(1), we get

Thermal conductivity of insulation,

$$k = \frac{\ln(0.85/0.75)}{2\pi \times 6.5\,(27 + 183)} \times 515.66 = 7.525 \times 10^{-3} \text{ W/m. K}.$$

Q.6. **(*b*)** Explain the meaning of the term 'geometric factor' in relation to heat exchange by radiation. Derive expression for the geometric factor F_{11} for the inside surface of a black hemispherical cavity of radius R with respect to itself.

Sol.

Geometric Factors : It is defined as the fraction of the radiative energy that is diffused from one surface element and strikes the other surface directly with no intervening reflection. The geometric factor depends only on the specific geometry of the emitter and the collecting surfaces.

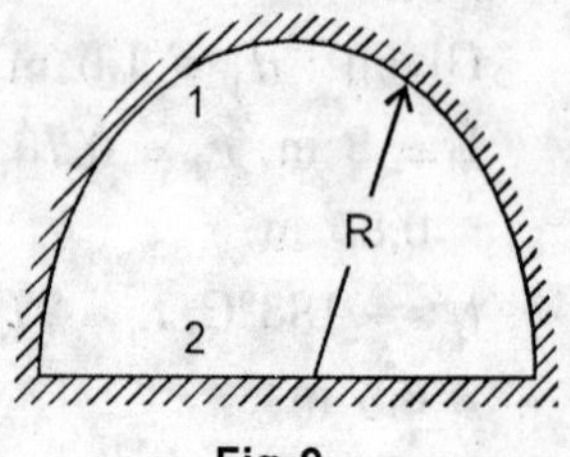

Fig 9.

The geometric factor is respesented by the symbol F_{ij}, which means the shape factor from a surface i to another surface j. In general, the geometric factor follows reciprocity theorem, i.e.

$$A_i\, F_{ij} = A_j\, F_{ji}$$

The geometric factor F_{12} of surface A_1 to surface A_2 is,

$$F_{12} = \frac{\text{Direct radiation from surface 1 incident upon surface 2}}{\text{Total radiation from emitting surface}}$$

Consider a black hemispherical cavity of radius R shown in Fig. 9. Geometric factor F_{11} for the inside surface w.r.t. itself is,

$$F_{11} = 1 - \frac{A_2}{A_1} = 1 - \frac{\pi R^2}{2\pi R^2} = 1 - 0.5 = 0.5$$

Q.7. (*a*) In a double pipe that heat exchanger, $M_h c_h = 0.5\ M_c c_c$, M denotes mass flow rate (kg/s), C denotes specific heat (J/kg K), and suffix *h* and suffix *c* denote hot and cold fluid, respectively. The hot and cold fluids enter the heat exchanger at temperature T_{h1} and T_{c1}, respectively. Deduce an expression, in terms of T_{h1}, T_{c1} and T_{h2}, for for the ratio of the area of a counter flow exchanger to that of a parallel flow exchanger which will give the same hot fluid outlet temperature T_{h2}. What is ratio if T_{h1} = 150°C, T_{c1} = 30°C and T_{h2} = 90°C?

Sol.

Given : $M_h c_h = 0.5\ M_c c_c$, T_{h1} = 30°C, T_{h2} = 90°C, A_{cf}/A_{pf} = ? for same T_{h2}.

$C_h = M_h c_h$ and $C_c = M_c c_c$. Refer to Fig. 10.

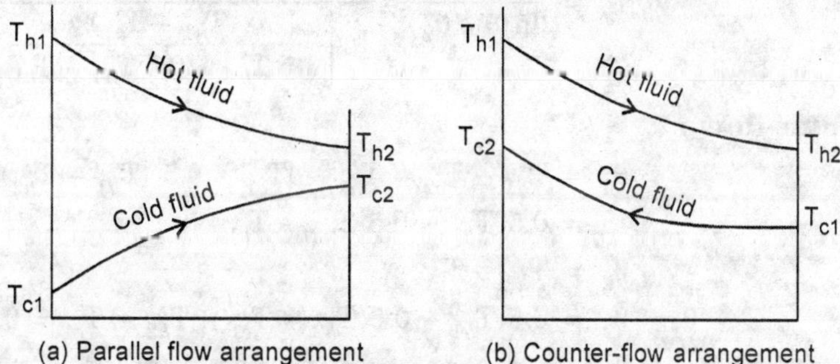

(a) Parallel flow arrangement (b) Counter-flow arrangement

Fig. 10

The effectiveness of the heat exchanger ε is given by

$$\therefore \quad \varepsilon = \frac{C_h(T_{h1} - T_{h2})}{C_{min}(T_{h1} - T_{c1})} = \frac{C_c(T_{c2} - T_{c1})}{C_{min}(T_{h1} - T_{c1})}$$

In this case $C_{min} = C_h$ and $C_h = 0.5\ C_c$ and $C_c = C_{max} = 2C_h$

$$\therefore \quad \varepsilon = \frac{C_h(T_{h1} - T_{h2})}{C_{min}(T_{h1} - T_{c1})} = \frac{2C_h}{C_h}\left(\frac{T_{c2} - T_{c1}}{T_{h1} - T_{c1}}\right)$$

or
$$\varepsilon = \left(\frac{T_{h1} - T_{h2}}{T_{h1} - T_{c1}}\right) = 2\left(\frac{T_{c2} - T_{c1}}{T_{h1} - T_{c1}}\right) \quad ...(1)$$

This *equation is true for counter flow as well as parallel flow.*

Let A_{pf} = area of parallel flow that exchanger, and

A_{cf} = area of counter-flow heat exchanger.

As $(T_{h2})_{pf} = (T_{h2})_{cf}$ therefore, the *heat lost by hot fluid in both the cases is same.*

$\therefore \quad Q = UA_{pf}(\theta_m)_{cf}$, since U is not dependent on the direction of flow.

$$\therefore \quad \frac{A_{cf}}{A_{pf}} = \frac{(\theta_m)_{pf}}{(\theta_m)_{cf}}$$

From eqn. (1), we get T_{c2} in terms of T_{h1}, T_{h2} and T_{c1} is

$$T_{c2} = T_{c1} + 0.5(T_{h1} - T_{h2})$$

(*a*) **Parallel flow :**

$$\theta_1 = T_{h1} - T_{c1}$$

$$\theta_2 = T_{h2} - T_{c2} = T_{h2} - [T_{c1} + 0.5\,(T_{h1} - T_{h2})]$$
$$= T_{h2} - T_{c1} - 0.5T_{h1} + 0.5T_{h2}$$
$$= 1.5\,T_{h2} - 0.5\,T_{h1} - T_{c1}$$

$$\therefore \quad \theta_1 - \theta_2 = T_{h1} - T_{c1} - 1.5\,T_{h2} + 0.5\,T_{h1} + T_{c1}$$
$$= 1.5\,(T_{h1} - T_{h2})$$

Thus,
$$(\theta_m)_{pf} = \frac{\theta_1 - \theta_2}{\ln(\theta_1/\theta_2)} = \frac{1.5\,(T_{h1} - T_{h2})}{\ln\left[\dfrac{(T_{h1} - T_{c1})}{1.5\,T_{h2} - 0.5\,T_{h1} - T_{c1}}\right]}.$$

(*b*) **Counter-flow :**

$$\theta_1 = T_{h1} - T_{c2} = T_{h1} - [T_{c1} + 0.5\,(T_{h1} - T_{h2})]$$
$$= 0.5\,T_{h1} + 0.5\,T_{h2} - T_{c1}$$

$$\theta_2 = T_{h2} - T_{c1}$$

$$\therefore \quad \theta_1 - \theta_2 = 0.5\,T_{h1} - 0.5\,T_{h2} - T_{c1} - T_{h2} + T_{c1} = 0.5\,(T_{h1} - T_{h2})$$

Thus,
$$(\theta_m)_{cf} = \frac{\theta_1 - \theta_2}{\ln(\theta_1/\theta_2)} = \frac{0.5\,(T_{h1} - T_{h2})}{\ln\left[\dfrac{0.5\,(T_{h1} + T_{h2}) - T_{c1}}{(T_{h2} - T_{c1})}\right]}$$

Substituting the values of (*i*) and (*ii*) in eqn. (2), we get

$$\frac{A_{cf}}{A_{pf}} = \frac{1.5\,(T_{h1} - T_{h2})}{\ln\left[\dfrac{(T_{h1} - T_{c1})}{1.5\,T_{h1} - 0.5\,T_{h1} - T_{c1}}\right]} \times \frac{\ln\left[\dfrac{0.5\,(T_{h1} - T_{h2}) - T_{c1}}{(T_{h2} - T_{c1})}\right]}{0.5\,(T_{h1} - T_{h2})}$$

$$= 3\left[\frac{\ln\left\{\dfrac{0.5\,(T_{h1} + T_{h2}) - T_{c1}}{(T_{h2} - T_{c1})}\right\}}{\ln\left\{\dfrac{(T_{h1} - T_{c1})}{1.5\,T_{h2} - 0.5\,T_{h1} - T_{c1}}\right\}}\right]$$

The given data is : T_{h1} : 150°C, T_{c1} = 30°C and T_{h2} = 90°C

$$\therefore \quad \frac{A_{cf}}{A_{pf}} = 3\left[\frac{\ln\left\{\dfrac{0.5\,(150+90)-30}{(90-30)}\right\}}{\ln\left\{\dfrac{(150-30)}{1.5\times 90-0.5\times 150-30}\right\}}\right] = 3\left[\frac{\ln(1.5)}{\ln(4)}\right] = \mathbf{0.877}$$

(b) A spark-ignition engine, designed to run on octane (C_8H_{18}) fuel, is operated on methane (CH_4). Estimate the ratio of the power output of the engine with methane fuel to that with octane. In both cases the fuel-air ratio is stoichiometric, the mixture is supplied to the engine at the same conditions, the engine runs at the same speed, and has the same volumetric and thermal efficiencies. The heating value of methane is 50,150 kJ/kg while that of octane is 44,880 kJ/kg.

Sol.

Indicated power, $\quad IP = \dfrac{p_{im}\,LAN}{60\times 10^3 \times n}$ kW

where $\quad p_{im}$ = indicated mean effective pressure, Pa

n = number of strokes per revolution

L = stroke length, m

A = area of piston, m^2

N = speed, rpm

Now $\quad p_{im} = \dfrac{\text{Net work}}{\text{Stroke length volume}} = \dfrac{Q_{in}\times\eta_{it}}{V_s} \propto$ LCV of fuel.

$$\therefore \quad \frac{P_{octane}}{P_{methane}} = \frac{(LCV)_{octane}}{(LCV)_{methane}} = \frac{44{,}880}{50{,}150} = 0.895$$

Q.8. (a) An inward flow reaction turbine, operating under 30 m head, develops 4000 kW while running at 300 RPM. The overall efficiency of the water turbine is 0.85; the hydraulic efficiency is 0.9; and the radial velocity of flow at inlet is 7 m/s. The inlet guide vane angle at full gate opening is 30°. Calculate the diameter and width of the runner at inlet. Blade thickness coefficient is 5%.

Sol.

Given : $\quad H = 30\text{m},\ P = 400\text{ kW},\ N = 300\text{ rpm},\ \eta_0 = 0.85,\ \eta_{hyd} = 0.90,$

$v_{f1} = 7\text{ m/s},\ \alpha_1 = 30°,\ K_t = 0.95$

Overall efficiency, $\quad \eta_0 = \dfrac{P}{\gamma QH}$

$$0.85 = \frac{400\times 10^3}{9.81\times 10^3\times Q\times 30}$$

$$Q = 15.99\ m^3/s$$

Hydraulic efficiency, $\eta_{hyd} = \dfrac{u_1 v_{w1}}{gH}$

$v_{f1}/v_{w1} = \tan\alpha_1 \quad$ (Fig. 11)

$$v_{w1} = \frac{7}{\tan 30°} = 12.12\text{ m/s}$$

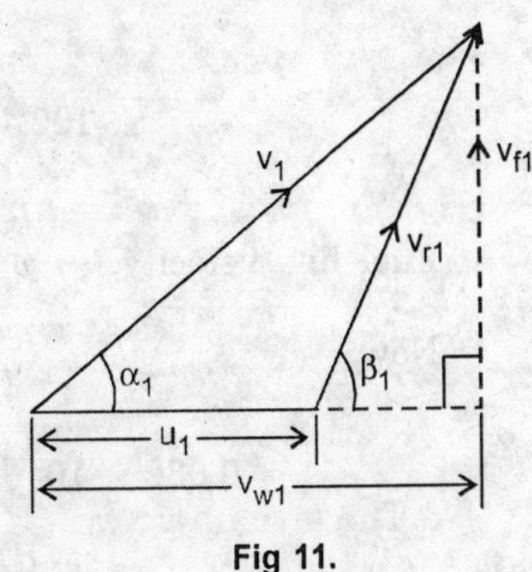

Fig 11.

$$u_1 = \frac{\pi D_1 N}{60} = \frac{\pi D_1 \times 300}{60} = 15.7 \times D_1 \text{ m/s}$$

$$0.90 = \frac{15.7 D_1 \times 12.12}{9.81 \times 30}$$

$$D_1 = 1.392 \text{ m}$$

Now $Q = K_t \times \pi D_1 B_1 v_{f1}$

$$B_1 = \frac{15.99}{0.95 \times \pi \times 1.392 \times 7} = 0.522 \text{ m}$$

(*b*) The plunger area of a fuel injection pump is 1.6 cm² and the plunger velocity is 0.02 cm/s. The pump run at 750 RPM. The density of the fuel is 850 kg/m³ and its viscosity is 0.12 g/cm.s. If the nozzle orifice area is 0.005 cm² and the orifice coefficient of discharge is 0.9, determine the fuel line pressure. The cylinder pressure is 30 bar. Neglect compressibility of the fuel. Determine also the critical fuel line velocity and critical fuel line diameter.

Sol.

Given : $A_p = 1.6 \text{ cm}^2$, $v_p = 0.02$ cm/s, N = 750 rpm, $\rho_f = 850 \text{ kg/m}^3$, $u = 12$ g/cm.s. $A_f = 0.005 \text{ cm}^2$, $C_{df} = 0.9$, $p_i = ?$ $p_a = 30$ bar

Volume of fuel injected per second, $Q_f = A_p\, v_p = 1.6 \times 0.02 = 0.032 \text{ cm}^3/\text{s}$

Also $$Q_f = A_f\, v_f \times \left(\frac{\theta}{360} \times \frac{1}{n}\right) \times \left(\frac{n}{60}\right) \times C_{df}$$

where θ = crank angle during which fuel is supplied

= 15° (assumed)

n = number of injections per minute

$$= \frac{N}{2} \text{ for four stroke engine} = \frac{750}{2} = 375$$

v_f = velocity of fuel passing through the nozzle orifice

$$0.032 = 0.005 \times v_f \times \left(\frac{15}{350} \times \frac{1}{375}\right) \times \left(\frac{375}{60}\right) \times 0.9$$

$$v_f = 102.4 \text{ m/s}$$

Also $$v_f = \sqrt{\frac{2(p_i - p_a)}{\rho_f}}$$

$$102.4 = \sqrt{\frac{2(p_i - 30) \times 10^5}{850}}$$

$$p_i = 74.56 \text{ bar}$$

Fuel line velocity, $v_f = 102.4$ m/s

Now $$Q_f = \frac{\pi}{4} d_f^2 \times v_f$$

$$0.032 \times 10^{-6} = \frac{\pi}{4} d_f^2 \times 102.4$$

$$d_f = 0.02 \times 10^{-3} \text{ m or } 0.02 \text{ mm}$$

MECHANICAL ENGINEERING
PAPER I—1995

(Conventional)

Time allowed : Three hours *Maximum marks : 200*

Q.1. **(*a*)** A tank containing 40 kg of water, initially at 40°C, has one inlet and one exit with equal mass flow rates. Water enters the tank at 40°C at the rate of 200 kg/hr. Energy is removed from the water at the rate of 8 kW by means of a cooling coil immersed in the water. The water is well mixed by a mechnaical stirrer so that the water temeprature in the tank is uniform. The power input to the water from the stirrer is 0.3 kW. Derive an expression for the variation of water temperature in the tank with time. The specific heat of water is 4.2 kJ/kg. K. Neglect kinetic and potential energy effects.

Sol. Given: m_t = 40 kg, t_{it} = 40°C, t_{iw} = 40°C, $\dot{m}_w$ = 200 kg/h, $\dot{Q}$ = – 8 kW,

$\dot{W}$ = + 0.3 kW, c_{pw} = 4.2 kJ/kg. K

Let t = temperature of water in the tank at any time, °C

For 1 kg of water, internal energy,

$$u = c_{pw}\,(t - t_{iw})$$

and
$$\dot{Q} = c_{pw} \times \frac{d}{d\tau}[m_w \times (t - t_{iw})]$$

$$= c_{pw}\,(t - t_{iw})\,\frac{d\,m_w}{d\tau} + c_{pw}\,.m_w\,.\frac{dt}{d\tau}$$

where $\dfrac{dt}{d\tau}$ = variation of water temperature with time

Now $\dfrac{d\,m_w}{d\tau}$ = 0 for steady flow

$$\therefore \quad -8 + 0.3 = 4.2 \times \frac{200}{3600} \times \frac{dt}{d\tau}$$

$$\therefore \quad \frac{dt}{d\tau} = \frac{-7.7 \times 3600}{4.2 \times 200} = -33°\,C/s$$

– ve sign indicates that temperature decreases with time.

(*b*) A closed system executes a reversible cycle 1-2-3-4-5-6-1 consisting of six processes. During processess 1-2 and 3-4 the system receives 1000 kJ and 800 kJ of heat, respectively, at constant temperature of 500 K and 400 K, respectively. Processess 2-3 and 4-5 are adiabatic expansions in which the system temperature is reduced from 500 K to 400 K and from 400 K to 300 K, respectively. During process 5-6 the system rejects heat at a constant temperature of 300 K. Process 6-1 is an adiabatic compression process. Determine the work done by the system during the cycle and the thermal efficiency of the cycle.

Sol.

Given : $Q_{1-2} = +\ 1000$ kJ, $Q_{3-4} = +\ 800$ kJ, $T_1 = 500$ K, $T_3 = 400$ K, $T_5 = 300$ K,

$$Q_{5-6} = -\text{ve}$$

The cycle is shown in Fig. 1.

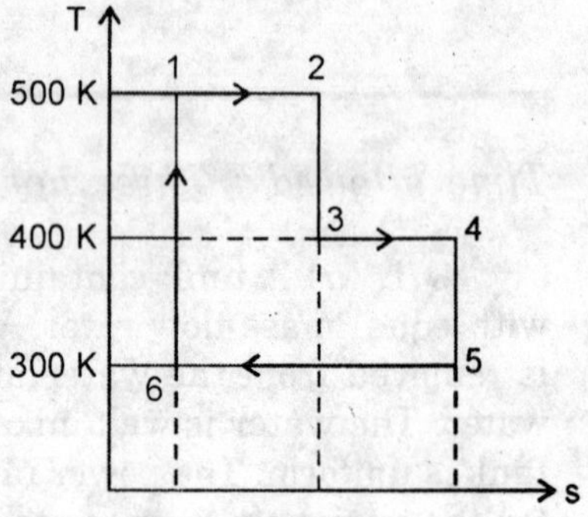

Fig 1.

Total heat added, $\quad Q_a = Q_{1-2} + Q_{3-4}$

$\qquad = 1000 + 800 = 1800$ kJ

Work done, $\quad W_{1-2} = T_1(s_2 - s_1)$

$\qquad 100 = 500\ (s_2 - s_1)$

or $\qquad s_2 - s_1 = 2$ kJ/kg. K

$\qquad W_{3-4} = T_3(s_4 - s_3)$

$$s_4 - s_3 = \frac{800}{400} = 2 \text{ kJ/kg. K}$$

Heat rejected, $\quad Q_r = T_5(s_5 - s_6) = T_5(s_4 - s_1) = 300(2 + 2) = 1200$ kJ

Total work done, $W_{cycle} = T_1(s_2 - s_1) + T_3\ (s_4 - s_3)$

$\qquad = 500 \times 2 + 400 \times 2 = 1800$ kJ

Net work done, $\quad W_{net} = W_{cycle} - Q_r = 1800 - 1200 = 600$ kJ

Thermal efficiency of cycle, $\eta_{th} = \dfrac{W_{net}}{Q_a} = \dfrac{600}{1800} = 0.333$ or 33.3%.

Q.2. (*a*) A rigid vessel is divided into two parts A and B by means of a frictionless, perfectly conducting piston. Intially, part A contains 0.4 m^3 of air (ideal gas) at 10 barspressure and part B constains 0.4 m^3 of wet steam at 10 bar. Heat is now added to both parts until all the water in part B is evaporated. At this condition the pressure in part B is 15 bar. Determine the initial quality of steam in part B and the total amount of heat added to both parts.

Sol. Refer to Fig. 2.

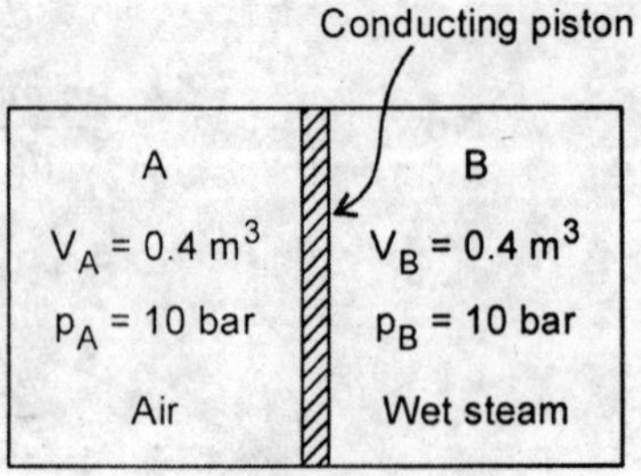

After heat addition, $p_{Bf} = 15$ bar, $(x_f)_B = 1.0$

Let $(x_i)_B$ = initial dryness fraction of steam.

Final pressure in part A, $p_{Af} = 15$ bar

Heat addition in part A at constant volume,

Fig 2.

$$Q_A = V_A\ (p_{Af} - p_A)$$

$$= 0.4\ (15 - 10) \times 10^2$$

$$= 200 \text{ kJ}$$

For part B, total heat = Total heat of steam at 10 bar + Q_A

= Total heat of dry and saturated steam at 15 bar

$$(h_f)_B + (x_i)_B \times (h_{fg})_B + Q_A = (h_g)_B$$

$$736 + (x_i)_B \times 2015 + 200 = 2790$$

$$(x_i)_B = 0.907$$

(*b*) A parallel shaft gearbox receives 800 kW through the high speed shaft but, owing to friction, delivers 784 kW through the low speed shaft. Heat transfer takes place from the outer surface of the gearbox to the environment at a rate given by $hA(T_b - T_0)$ where $h = 0.2$ kW/m^2. K, A = 2 m^2 and T_0, the environment temperature, equals 288 K. The outer surface temperature of the gearbox, T_b, is uniform. Determine the irreversibility rate for the gearbox in kW assuming steady state operation.

Sol.

Given: $W_i = 800$ kW, $W_0 = 784$ kW, $Q = hA(T_b - T_0)$, $h = 0.2$ kW/m².K,

$$A = 2\text{ m}^2,\ T_0 = 288\text{ K}$$

$$W_i - W_0 = hA(T_b - T_0)$$

$$800 - 784 = 0.2 \times 2\,(T_b - 288)$$

$$T_b = \frac{16}{0.4} + 288 = 328\text{ K}$$

Inreversibility, $I = W_{max} - W$

For steady state operation,

$$I = T_0\,[(\Delta s)_{system} + (\Delta s)_{surr}] = 800 - 784 = 16\text{ kW}$$

Q.3. (*a*) Atmospheric air at 40°C dry bulb temperature and 55% relative humidity is passed over a cooling coil at the rate of 600 m³/min. At exit from the coil the air is saturated and its temperature is 8°C. The condensate also leaves at 8°C. Determine the amount of condensate leaving the coil per minute and the refrigeration required in kW. The specific heat of superheated steam may be assumed to be 1.88 kJ/kg. K. Solve the problem without the use of Psychrometic chart.

Sol.

Given : $t_{d1} = 40°C$, $\phi_1 = 55\%$, $V_1 = 600$ m³/min, $t_{d2} = 8°C$, $\phi_2 = 100$ %

Saturation pressure corresponding to 40°C, from steam tables,

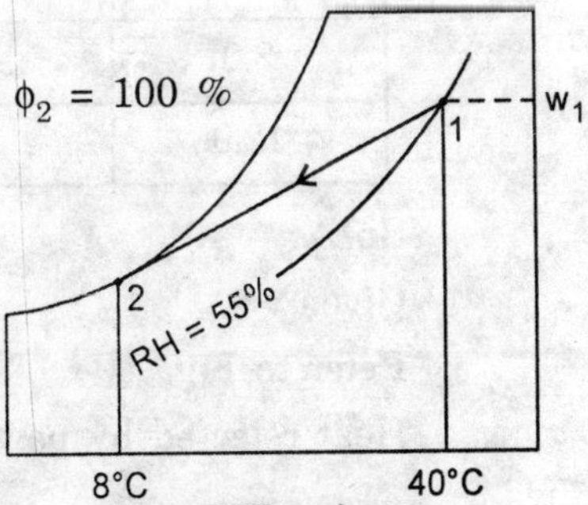

$$p_{s1} = 0.07384\text{ bar}$$

$$p_{v1} = \phi_1\, p_{s1} = 0.55 \times 0.7384 = 0.04061\text{ bar}$$

Humidity ratio,

$$w_1 = \frac{0.622\, p_{v1}}{p_b - p_{v1}} = \frac{0.622 \times 0.04061}{1.01325 - 0.04061} = 0.02597\text{ kg/kg of dry air}$$

At 8°C, $p_{s2} = 0.010724$ bar

$$w_2 = \frac{0.622 \times 0.010724}{1.01325 - 0.010724} = 0.00665\text{ kg/kg of dry air}$$

$$h_1 = 1.005\, t_{d1} + w_1\,(2500 + 1.88\text{ kg})$$

$$= 1.005 \times 40 + 0.2597\,(2500 + 1.88 \times 40) = 107.08\text{ kJ/kg}$$

$$h_2 = 1.005 \times 8 + 0.00665\,(2500 + 1.88 \times 8)$$

$$= 24.76\text{ kJ/kg}$$

Fig 3.

Mass of air, $\dot{m}_a = \dfrac{p_{a1} \dot{V}_1}{R_a T_1}$

Now $p_{a1} = p_b - p_{v1} = 1.01325 - 0.04061 = 0.97264$ bar

$$\dot{m}_a = \frac{0.97264 \times 10^2 \times 600}{0.287 \times 313} = 649.65 \text{ kg/min}$$

Mass of water condensed $= \dot{m}_a (w_1 - w_2)$

$= 649.65\ (0.02597 - 0.00665) = 12.64$ kg/min

Capacity of cooling coil $= \dot{m}_a (h_1 - h_2)$

$$= \frac{649.65}{60} (107.08 - 24.76) = 891.32 \text{ kW}$$

(b) A vapor compression heat pump is driven by a power cycle having a thermal efficiency of 25%. For the heat pump, refrigerant 12 is compressed from saturated vapor at 2.0 bar to the condenser pressure of 12 bar. The isentropic efficiency of the compressor is 80%. Saturated liquid enters the expansion valve at 12 bar. For the power cycle, 80% of the heat rejected by it is transferred to the heated space which has a total heating requirement of 500 kJ/min. Determine the power input to the heat pump compressor. The following data for refrigerant-12 may be used.

Pressure, bar	Temperature, °C	Enthalpy, kJ/kg		Entropy, kJ/kgK	
		Liquid	Vapor	Liquid	Vapor
2.0	– 12.53	24.57	182.07	0.0992	0.7035
12.0	49.31	84.21	206.24	0.3015	0.6799

Sol.

Given : $\eta_{th} = 0.25$, $h_c = 0.8$, $h_s = 0.8\ h_r$, $h_s = 500$ kJ/min,

Refer to Fig. 4 $T_1 = 273 - 12.53 = 260.47$ K, $T_2'' = 273 + 49.31 = 322.31$ K

Heat rejected by power cycle,

$$h_r = \frac{h_s}{0.8} = \frac{500}{0.8} = 625 \text{ kJ/min}$$

For isentropic compression of refrigerant,

$$s_1 = s_2,\ h_{g1} = h_{g2} + c_{ps} \ln \frac{T_2}{T_2''}$$

$$0.7035 = 0.6799 + 0.7 \ln \frac{T_2}{322.31}$$

$$T_2 = 333.4 \text{ K}$$

Compressor efficiency, $\eta_c = \dfrac{T_2 - T_1}{T_2' - T_1}$

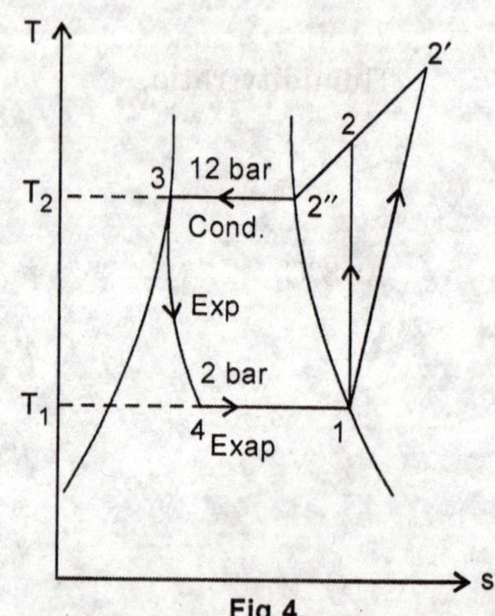

Fig 4.

$$0.8 = \frac{333.4 - 260.47}{T_2' - 260.47}$$

$$T_2' = 351.6 \text{ K}$$

Enthalpy of superheated vapour at 12 bar,

$$h_2' = h_2'' + c_{ps}(T_2' - T_2'')$$

$$= 206.24 + 0.7\,(352.6 - 322.31) = 226.74 \text{ kJ/kg}$$

Heat rejected by cycle,

$$q_r = h_2' - h_{f3} = 226.74 - 84.21 = 142.53 \text{ kJ/kg}$$

Mass flow rate of refrigerant, $\dot{m} = \dfrac{h_r}{q_r} = \dfrac{625}{142.53} = 4.386$ kg/min

Compressor work, $W_c = \dot{m}\,(h_2' - h_1) = 4.386\,(226.74 - 182.07)$

$$= 195.92 \text{ kJ/min or } 3.265 \text{ kW}$$

Q.4. (*a*) In a Brayton cycle gas turbine power plant the minimum and maximum temperatures of the cycle are 300 K and 1200 K. The compression is carried out in two stages of equal pressure ratio with intercooling of the working fluid to the minimum temperature of the cycle after the first stage of compression. The entire expansion is carried out in one stage only. The isentropic efficiency of both compressors is 0.85 and that of the turbine is 0.9. Determine the overall pressure ratio that would give the maximum net work per kg working fluid. Derive the expression you use. $\gamma = 1.4$.

Sol.

Given : $T_1 = T_3 = 300$ K,

$T_5 = 1200$ K,

$$\frac{p_2}{p_1} = \frac{p_3}{p_2}, \eta_{c1} = \eta_{c2} = 0.85,$$

$$\eta_t = 0.90,$$

$$\gamma = 1.4$$

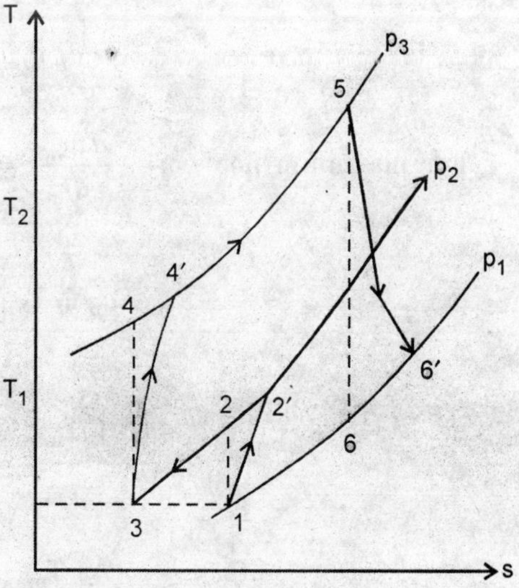

Fig 5.

Compressor 1:

$$\frac{T_2}{T_1} = \left(\frac{p_2}{p_1}\right)^{\frac{\gamma-1}{\gamma}} = \left[\sqrt{r_p}\right]^y,$$

$$y = \frac{\gamma - 1}{\gamma} = r_p^{(y/2)}$$

$$\eta_{c1} = \frac{T_2 - T_1}{T_2' - T_1}$$

$$T_2' - T_1 = \frac{T_2 - T_1}{\eta_{c1}} = \frac{T_1}{\eta_{c1}}\left[\frac{T_2}{T_1} - 1\right] = \frac{T_1}{\eta_{c1}}\left[r_p^{(y/2)} - 1\right]$$

Compressor 2:

$$\frac{T_4}{T_3} = \left(\frac{p_3}{p_2}\right)^y = r_p^{(y/2)} = \frac{T_4}{T_1}$$

$$\eta_{c2} = \frac{T_4 - T_3}{T_4' - T_3}$$

$$T_4' - T_1 = \frac{T_4 - T_1}{\eta_{c2}} = \frac{T_1}{\eta_{c2}}\left[\frac{T_1}{T_1} - 1\right] = \frac{T_4}{\eta_{c2}}\left[r_p^{(y/2)} - 1\right]$$

Compressor work, $w_c = w_{c1} + w_{c2} = c_p\,[(T_2' - T_1) + (T_4' - T_1)]$

$$= \frac{c_p T_1}{\eta_c}\left[(r_p^{(y/2)} - 1) + r_p^{(y/2)} - 1\right] = \frac{2c_p T_1}{\eta_c}\left[r_p^{(y/2)} - 1\right]$$

Turbine :

$$\frac{T_5}{T_6} = \left(\frac{p_3}{p_1}\right)^{\frac{\gamma-1}{\gamma}} = y_{r_p}$$

$$\eta_t = \frac{T_5 - T_6'}{T_5 - T_6}$$

$$T_5 - T_6' = \eta_t (T_5 - T_6) = \eta_t T_5\left(1 - \frac{T_6}{T_5}\right) = \eta_t T_5\left[1 - {}^{-y}_{-y_{r_p}}\right]$$

Turbine work, $w_t = c_p\,[T_5 - T_6'] = c_p \eta_t T_5\left[1 - \frac{-y}{-y_{r_p}} - 1\right]$

Net work done, $w_{net} = w_t - w_c$

$$= c_p \eta_t T_5\left[1 - {}^{-y}_{r_p}\right] - \frac{2c_p T_1}{\eta_c}\left[{}^{(y/2)}_{r_p} - 1\right]$$

For maximum net work, $\frac{dw_{net}}{dr_p} = 0$

or
$$c_p \eta_t T_5\left[y\, r_p^{(-y-1)}\right] - \frac{2c_p T_1}{\eta_c}\left[\frac{y}{2} \times r_p^{\left(\frac{y}{2}-1\right)}\right] = 0$$

or
$$\eta_t T_{max} \times r_p^{-(y+1)} - \frac{T_{min}}{\eta_c} \times r_p^{\left(\frac{y}{2}-1\right)} = 0$$

$$\frac{r_p^{\left(\frac{y}{2}-1\right)}}{r_p^{-(y+1)}} = \left(\frac{T_{max}}{T_{min}}\right) \times \eta_c\, \eta_t$$

or
$$r_p^{\left(\frac{3y}{2}\right)} = \eta_c\, \eta_t\, \frac{T_{max}}{T_{min}}$$

Now
$$\frac{3y}{2} = \frac{3}{2}\left(\frac{\gamma-1}{\gamma}\right) = \frac{3(\gamma-1)}{2\gamma}$$

$$\therefore \qquad (r_p)_{opt} = \left[\eta_c \, \eta_t \left(\frac{T_{max}}{T_{min}}\right)\right]^{\frac{2\gamma}{3(\gamma-1)}}$$

$$= \left[0.85 \times 0.90 \times \frac{120}{300}\right]^{\frac{2 \times 1.4}{3(1.4-1)}} = (3.06)^{2.333} = 13.594$$

(*b*) Air enters a constant-area combustion chamber at a Mach number of 0.25 with a total pressure of 5.5 bar. Combustion gas leaves the chamber with a Mach number of 0.35. Neglecting friction and the mass of the fuel, determine the drop in total pressure of air in the combustion chamber. Assume $c_p/c_v = 1.35$.

Sol.

Given : $M_1 = 0.25,\ p_{01} = 5.5$ bar, $M_2 = 0.35,\ \frac{c_p}{c_v} = 1.35 = k$

Now
$$\frac{p_{02}}{p_{01}} = \left[\frac{1 + kM_1^2}{1 + kM_2^2}\right]\left[\frac{1 + \frac{k-1}{2} M_2^2}{1 + \frac{k-1}{2} M_1^2}\right]^{k/(k-1)}$$

$$= \left[\frac{1 + 1.35\,(0.25)^2}{1 + 1.35\,(0.35)^2}\right]\left[\frac{1 + \left(\frac{1.35-1}{2}\right)(0.35)^2}{1 + \left(\frac{1.35-1}{2}\right)(0.25)^2}\right]^{\left(\frac{1.35}{1.35-1}\right)}$$

$$= 0.9305 \times 0.920 = 0.856$$

$$p_{02} = 5.5 \times 0.856 = 4.7 \text{ bar}$$

$$\text{Drop in total pressure} = p_{01} - p_{02} = 5.5 - 4.7 = 0.8 \text{ bar}$$

Q.5. Steam enters the first stage of a turbinte at 100 bar, 500°C and expands isentropically to 10 bar. It is then reheated to 500°C and expanded in the second stage to the condenser pressure of 0.1 bars. Steam is bled from the first stage at 20 bar and fed to a closed feed water heater. Feed water leaves the closed heater at 100 bar, 200°C (enthalpy = 856.8 kJ/kg) while the condensate exits as saturated liquid at 20 bar. This condensate is supplied to the open heater into which steam is bled at 4 bar pressure. Saturated liquid at 4 bar exits from the open heater and enters the closed heater. The condensate from condenser enters the open heater. The net output of the turbine is 50 MW. Assuming the turbine and pump processes to be isentropic, determine the mass of steam bled at each feed water heater per kg of steam entering the first stage, the mass of steam entering the first stage per second, and the thermal efficiency.

Sol.

Given : $h_9 = 856.8$ kJ/kg, $W_{net} = 50$ MW

$h_5 = (h_f)_{0.1}$ bar $= 191.8$ kJ/kg

$h_6 = (h_f)_4$ bar $= 191.8$ kJ/kg

$h_7 = (h_f)_{10}$ bar $= 604.7$ kJ/kg

$h_8 = (h_f)_{20}$ bar $= 762.6$ kJ/kg

$h_9 = 856.8$ kJ/kg

$h_1 = 3374.6$ kJ/kg (100 bar, 500°C)

$$h_2' = 2935 \text{ kJ/kg (20 bar)}$$
$$h_2 = 2775 \text{ kJ/kg (10 bar, } x = 1.0)$$
$$h_3 = 3500 \text{ kJ/kg (10 bar, 500°C)}$$
$$h_3' = 3180 \text{ kJ/kg (4 bar)}$$
$$h_4 = 2450 \text{ kJ/kg}$$

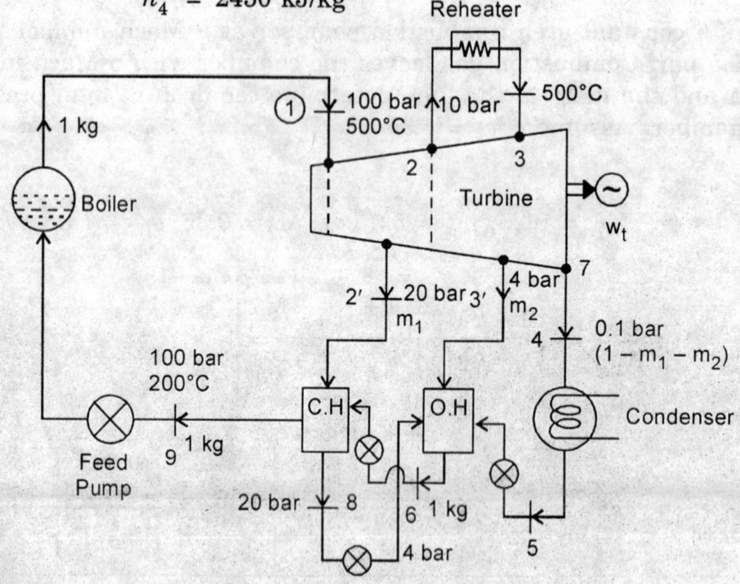

(a) Schematic diagram

Closed Heater :

$$m_1 \times h_2' + 1 \times h_6 = 1 \times h_9 + m_1 \times h_8$$
$$m_1 + 2935 + 1 \times 604.7 = 856.8 + m_1 \times 908.6$$
$$2026.4\, m_1 = 252.1$$
$$m_1 = 0.1244 \text{ kg/kg of steam}$$

Open Heater

$$m_2 \times h_3' + (1 - m_1 - m_2)\, h_5 + m_1 \times h_8 = 1 \times h_6$$
$$m_2 \times 3180 + (1 - 0.1244 - m_2) \times 191.8 + 0.1244 \times 908.6 = 604.7$$
$$m_2 = 0.1083 \text{ kg/kg of steam}$$

(b) T-s diagram

Fig 6.

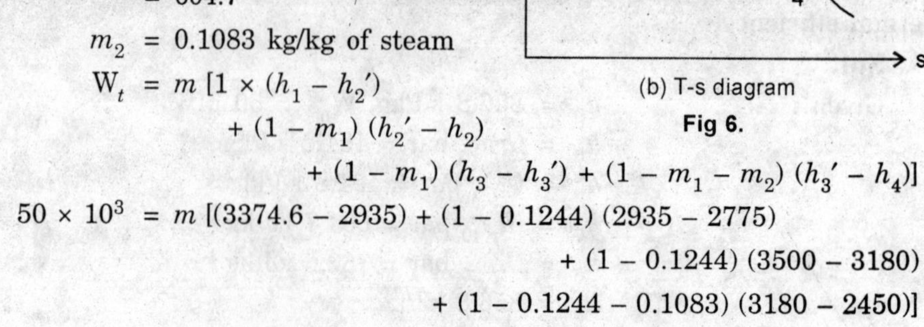

Now $$W_t = m\,[1 \times (h_1 - h_2') + (1 - m_1)(h_2' - h_2) + (1 - m_1)(h_3 - h_3') + (1 - m_1 - m_2)(h_3' - h_4)]$$

$$50 \times 10^3 = m\,[(3374.6 - 2935) + (1 - 0.1244)(2935 - 2775) + (1 - 0.1244)(3500 - 3180) + (1 - 0.1244 - 0.1083)(3180 - 2450)]$$

$$= m \times (439.6 + 140.09 + 280.19 + 560.13)$$

$$= m \times 1420.01$$

$$m = 35.2 \text{ kg/s}$$

Thermal efficiency, $\eta_{th} = \dfrac{W_{net}}{m[(h_1 - h_9) + (1 - m_1)(h_3 - h_2)]}$

$$= \frac{50 \times 10^3}{35.2[(3374.6 - 856.8) + (1 - 0.1244)(3500 - 2775)]}$$

$$= \frac{50 \times 10^3}{110971.87} = 0.4506 \text{ or } 45.06\%$$

Q.6. (*a*) A chemical reaction is being carried out at constant pressure in a packed bed between two coaxial cylinders with radii of 1.2 cm and 1.8 cm. The entire inner wall is at a uniform temperature of 550°C and there is almost no heat transfer from this surface. The reaction releases heat at uniform rate of 550 kW/m^3 throughout the ractor. The effective thermal conductivity of the packed bed is 0.55 W/m. K. Determine the temperature of the outer wall.

Sol.

Governing equation for heat conduction with internal heat generation is, (Fig. 7)

$$\frac{d}{dr}\left(r.\frac{dt}{dr}\right) + \frac{q_g}{k}.r = 0$$

or
$$\frac{d^2t}{dr^2} + \frac{1}{r}\frac{dt}{dr} + \frac{q_g}{k} = 0$$

Integrating, we have

$$r.\frac{dt}{dr} + \frac{r^2}{2}.\frac{q_g}{k} = C_1$$

or
$$\frac{dt}{dr} + \frac{r}{2}.\frac{q_g}{k} = \frac{C_1}{r} \quad ...(1)$$

Integrating again, we have

$$t + \frac{r^2}{4}.\frac{q_g}{k} = C_1 \ln r + C_2 \quad ...(2)$$

where C_1, C_2 = Constants of integration.

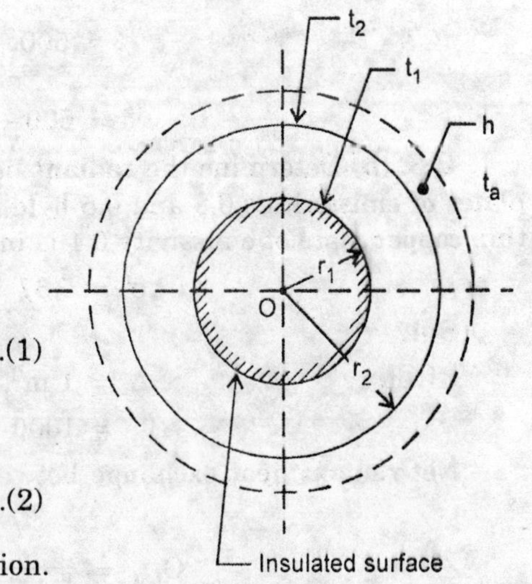

Fig 7.

Boundary conditions are :

(*i*) At $r = r_1, \dfrac{dt}{dr} = 0$

From Eq. (1), we get

$$0 + \frac{r_1}{2}\frac{q_g}{k} = \frac{C_1}{r_1}$$

$$C_1 = \frac{r_1^2}{2}.\frac{q_g}{k}$$

(*ii*) At $r = r_1,\ t = t_1$

From Eq. (2), we get

$$t_1 + \frac{r_1^2}{4}.\frac{q_g}{k} = \frac{r_1^2}{2}.\frac{q_g}{k}.\frac{r_1^2}{4}.\ln r_1 + C_2$$

$$C_2 = t_1 - \frac{r_1^2}{2}.\frac{q_g}{k}.\ln r_1 + \frac{r_1^2}{4}.\frac{q_g}{k}$$

$$= t_1 + \frac{q_g\, r_1^2}{2k}\left(\frac{1}{2} - \ln r_1\right)$$

$$\therefore \quad t = t_1 + \frac{q_g\, r_1^2}{2k}\left(\frac{1}{2} - \ln r_1\right) + \frac{q_g\, r_1^2}{2k} \times \ln r - \frac{q_g\, r^2}{4k}$$

$$= t_1 + \frac{q_g\, r_1^2}{4k}\left[2\ln\left(\frac{r}{r_1}\right) - \left(\frac{r}{r_1}\right)^2 + 1\right]$$

$$\therefore \quad t = 500 + \frac{550 \times 10^3 \times (1.2 \times 10^{-2})^2}{4 \times 0.55}\left[2\ln\left(\frac{1.8}{1.2}\right) - \left(\frac{1.8}{1.2}\right)^2 + 1\right]$$

$$= 500 - 15.8 = 484.2°\text{C}$$

T₁ = 1000 K — Copper plate — T₃ = 500 K

$\varepsilon_1 = 0.8$ $\varepsilon_1 = 0.1$ $\varepsilon_1 = 0.5$

Fig 8.

Q.6. (***b***) Determine the radiant heat exchange in W/m² between two large parallel steel plates of emissivities 0.8 and 0.5 held at temperature of 1000 K and 500 K respectively, if a thin copper plate of emissivity 0.1 is introduced as a radiation shield between the two plates.

$$\sigma = 5.67 \times 10^{-8}\ \text{W/m}^2.\ \text{K}^4.$$

Sol.

Given : $A = 1\ \text{m}^2,\ \sigma = 5.67 \times 10^{-8}\ \text{W/m}^2.\ \text{K}^4,\ \varepsilon_1 = 0.8,\ \varepsilon_2 = 0.1,\ \varepsilon_3 = 0.5,$

$T_1 = 1000\ \text{K},\ T_2 = 500\ \text{K}$

Net radiant heat exchange between two parallel steel plates,

$$Q_{net} = \frac{A\sigma\,(T_1^4 - T_2^4)}{\left[\frac{1}{\varepsilon_1} + \frac{1}{\varepsilon_2} - 1\right] + \left[\frac{1}{\varepsilon_3} + \frac{1}{\varepsilon_2} - 1\right]}$$

$$= \frac{1 \times 5.67\left[\left(\frac{1000}{100}\right)^4 - \left(\frac{500}{100}\right)^4\right]}{\left[\frac{1}{0.8} + \frac{1}{0.1} - 1\right] + \left[\frac{1}{0.5} + \frac{1}{0.1} - 1\right]}$$

$$= \frac{53156.25}{21.25} = 2501.47\ \text{Watts/m}^2.$$

Q.7. Derive an expression for the effectiveness of a parallel flow heat exchanger in terms of the number of transfer units, NTU, and the capacity ratio C_{min}/C_{max}.

In a parallel-flow double-pipe heat exchanger water flows through the inner pipe and is heated from 20°C to 70°C. Oil flowing through the annulus is cooled from 200° to 100°C. It is desired to cool the oil to a lower exit temperature by increasing the length of the heat exchanger. Determine the minimum temperature to which the oil may be cooled.

Sol.

The parallel flow heat exchanger is shown in Fig. 9.

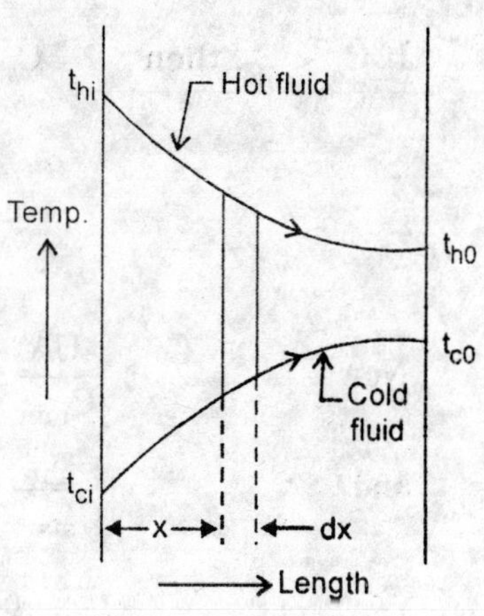

Fig 9.

Heat exchange through an elementary area dA is,

$$dQ = U.dA\,(t_h - t_c)$$
$$= (m_h\, c_h)(-dt_h) = (m_c\, c_c)(dt_c)$$
$$= C_h(-dt_h) = C_c(dt_c)$$

or
$$dt_h = -\frac{dQ}{C_h} \text{ and } dt_c = \frac{dQ}{C_c}$$

∴
$$d(t_h - t_c) = -dQ\left(\frac{1}{C_h} + \frac{1}{C_c}\right)$$

or
$$\frac{d(t_h - t_c)}{(t_h - t_c)} = -U.dA\left(\frac{1}{C_h} + \frac{1}{C_c}\right)$$

Integrating, we get

$$\int_{(t_{hi} - t_{ci})}^{(t_{h0} - t_{c0})} \frac{d(t_h - t_c)}{(t_h - t_c)} = -U\left(\frac{1}{C_h} + \frac{1}{C_c}\right)\int_0^A dA$$

$$\ln\left[\frac{t_{h0} - t_{c0}}{t_{hi} - t_{ci}}\right] = -UA\left(\frac{1}{C_h} + \frac{1}{C_c}\right)$$

Effectiveness,
$$\varepsilon = \frac{C_h(t_{hi} - t_{h0})}{C_{min}(t_{hi} - t_{ci})} = \frac{C_c(t_{c0} - t_{ci})}{C_{min}(t_{hi} - t_{ci})}$$

∴
$$t_{h0} = t_{hi} - \frac{C_{min}}{C_h}(t_{hi} - t_{ci})\,\varepsilon$$

and
$$t_{c0} = t_{hi} + \frac{C_{min}}{C_c}(t_{hi} - t_{ci})\,\varepsilon$$

∴
$$(t_{h0} - t_{c0}) = (t_{hi} - t_{ci}) - C_{min}\,(t_{hi} - t_{ci})\,\varepsilon\left[\frac{1}{C_h} + \frac{1}{C_c}\right]$$
$$= (t_{hi} - t_{ci})\left[1 - \varepsilon\, C_{min}\left(\frac{1}{C_h} + \frac{1}{C_c}\right)\right]$$

Substituting this value in Eq. (1), we get

$$\ln\left[1 - \varepsilon\, C_{min}\left(\frac{1}{C_h} + \frac{1}{C_c}\right)\right] = -UA\left(\frac{1}{C_h} + \frac{1}{C_c}\right)$$

$$\text{or } 1 - \varepsilon\, C_{min}\left(\frac{1}{C_h} + \frac{1}{C_c}\right) = \exp\left[-UA\left(\frac{1}{C_h} + \frac{1}{C_c}\right)\right]$$

$$\varepsilon = \frac{1 - \exp\left[-UA\left(\frac{1}{C_h} + \frac{1}{C_c}\right)\right]}{C_{min}\left(\frac{1}{C_h} + \frac{1}{C_c}\right)} = \frac{1 - \exp\left[-\frac{UA}{C_h}\left(1 + \frac{C_h}{C_c}\right)\right]}{\frac{C_{min}}{C_h}\left[1 - \frac{C_h}{C_c}\right]}$$

If $C_h < C_c$, then $C_h = C_{min}$ and $C_c = C_{max}$

$$\varepsilon = \frac{1 - \exp\left[-\frac{UA}{C_{min}}\left(1 + \frac{C_{min}}{C_{max}}\right)\right]}{\frac{C_{min}}{C_{max}}\left[1 + \frac{C_{min}}{C_{max}}\right]}$$

Now $\frac{UA}{C_{min}}$ = NTU, number of transfer units

and $\frac{C_{min}}{C_{max}}$ = C, capacity ratio

$$\therefore \quad \varepsilon = \frac{1 - \exp[-NTU\,(1 + C)]}{(1 + C)}$$

Given data : $t_{hi} = 200°C,\ t_{h0} = 100°C,\ t_{ci} = 20°C,\ t_{c0} = 70°C$

Let t = lowest temperature to which oil may be cooled, °C

Then $m_h\, c_h\, (200 - t) = m_c\, c_c\, (t - 20)$

$$200 - t = \frac{m_c\, c_c}{m_h\, c_h}(t - 20) = 2\,(t - 20)$$

$$3t = 240$$

$$t = 80°C$$

Q.8. (*a*) Calculate the diameter and speed of the runner of a Kaplan turbine developing 6000 kW under an effective head of 5 m. Overall efficiency of the turbine is 90%. The diameter of the boss is 0.4 times the external diameter of the runner. The turbine speed ratio is 2.0 and flow ratio 0.6. What is the specific speed of the turbine?

Sol.

Given : $P = 6000$ kW, $H = 5$ m, $\eta_0 = 0.9$, $d_b = 0.4 d_0$,

speed ratio = 2, flow ratio = 0.6

$$\text{Speed ratio} = \frac{u}{\sqrt{2gH}} = 2$$

$$\text{Flow ratio} = \frac{v_f}{\sqrt{2gH}} = 0.6$$

$$\therefore \quad u = 2 \times \sqrt{2 \times 9.81 \times 5} = 19.81 \text{ m/s}$$

$$v_f = 0.6 \times \sqrt{2 \times 9.81 \times 5} = 5.94 \text{ m/s}$$

Power developed, $$P = \frac{\gamma QH}{10^3} \times \eta_0$$

$$Q = \frac{6000 \times 10^3}{9.81 \times 10^3 \times 5 \times 0.9} = 135.9 \text{ m}^3/\text{s}$$

Also $$Q = \frac{\pi}{4}(d_0^2 - d_b^2)\, v_f$$

$$135.9 = \frac{\pi}{4}\left[1 - (0.4)^2\right] d_0^2 \times 5.94$$

$$d_0 = 5.89 \text{ m}$$

$$u = \frac{\pi\, d_0 N}{60}$$

$$19.81 = \frac{\pi \times 5.89 \times N}{60}$$

$$N = 64.2 \text{ rpm}$$

Specific speed, $$N_s = \frac{N\sqrt{P}}{H^{5/4}} = \frac{64.2\sqrt{6000}}{5^{5/4}} = 665\,.$$

(*b*) When a fluid flows over a flat plate, the velocity profile within the boundary layer may be assumed to be

$$v_x - v_0\left[\left(\frac{3}{2}\right)\left(\frac{y}{\delta}\right) - \frac{1}{2}\left(\frac{y}{\delta}\right)^3\right] \text{ for } y < \delta$$

and $$v_x = v_0 \text{ for } y \geq \delta$$

where v_0 is a constant and the boundary layer thickness δ is a function of x given by

$$\delta = 5\left(\frac{\mu x}{\rho v_0}\right)^{1/2}$$

Here μ and ρ denote the viscosity and density of the fluid, respectively. Derive an expression for the variation of v_y across the boundary layer.

Sol.

$$v_x = v_0\left[\frac{3}{2}\left(\frac{y}{\delta}\right) - \frac{1}{2}\left(\frac{y}{\delta}\right)^3\right] \text{ for } y \leq \delta$$

$$= v_0 \text{ for } y > \delta$$

and $$\delta = 5\left(\frac{\mu x}{\rho v_0}\right)^{1/2}$$

The Navier-Stokes equation for steady flow is,

$$v_x \frac{\partial v_x}{\partial x} + v_y \frac{\partial v_x}{\partial y} = \nu \frac{\partial^2 v_x}{\partial y^2}$$

Now

$$v_x = v_0\left[\left(\frac{3}{2}\right)\left(\frac{y}{\delta}\right) - \frac{1}{2}\left(\frac{y}{\delta}\right)^3\right] \text{ for } y \le \delta$$

$$\frac{\partial v_x}{\partial x} = 0$$

$$\frac{\partial v_x}{\partial y} = v_0\left[\left(\frac{3}{2}\right)\left(\frac{1}{\delta}\right) - \frac{3}{2}\left(\frac{y^2}{\delta^3}\right)\right]$$

$$= \frac{3}{2}v_0\left[\frac{1}{\delta} - \frac{y^2}{\delta^3}\right]$$

$$\frac{\partial^2 v_x}{\partial y^2} = \frac{3}{2}v_0\left[\frac{1}{\delta} - \frac{2y}{\delta^3}\right] = -\frac{3v_0 y}{\delta^3}$$

$$v_y \times \frac{3}{2}v_0\left[\frac{1}{\delta} - \frac{y^2}{\delta^3}\right] = \nu\left(-\frac{3v_0 y}{\delta^3}\right)$$

$$\frac{3}{2}v_0\, v_y\,(\delta^2 - y^2) = -\,3\nu\, v_0 y$$

$$v_y = \frac{2\nu y}{y^2 - \delta^2} = \frac{2\nu y}{y^2 - \dfrac{25\nu x}{v_0}} = \frac{2\nu v_0 y}{v_0 y^2 - 25\,\nu x}.$$

MECHANICAL ENGINEERING
PAPER I—1996

(Conventional)

Time allowed : Three hours *Maximum marks : 200*

Q.1. (*a*) 0.14 m^3 of steam at 20 bar 250˚C is expanded reversibly and polytropically to 2 bar. Find the final temperature, work done, heat ransferred and change of entropy, if index of expansion is 1.25.

Sol.

Given: $V_s = 0.14$ m^3, $p_1 = 20$ bar,

$$T_1 = 273 + 250 = 523 \text{ K},$$

$$p_2 = 2 \text{ bar}, \ n = 1.25$$

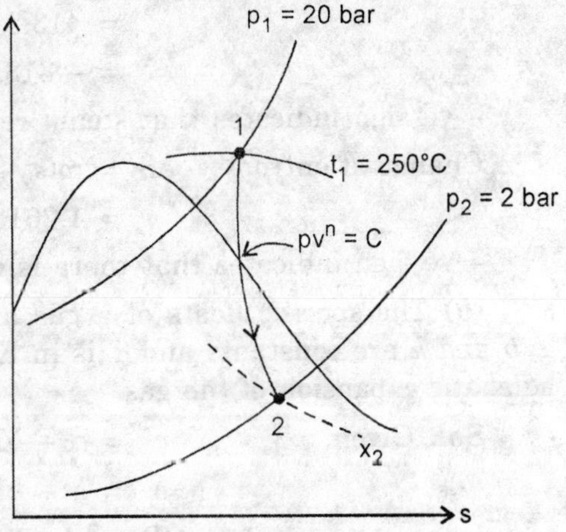

For superheated steam at 20 bar and 250°C, from steam tables, we have (Fig. 1)

$$h_1 = 2902.5 \text{ kJ/kg}$$

$$s_1 = 6.545 \text{ kJ/kg. K}$$

$$v_1 = v_{s1} = 0.1110 \text{ m}^3\text{/kg}$$

$$u_1 = 2680.5 \text{ kJ/kg}$$

Fig 1.

For the polytropic process, we have

$$p_1 v_1^n = p_2 v_2^n$$

$$v_2 = v_1 \left(\frac{p_1}{p_2}\right)^{\frac{1}{n}} = 0.1110 \left(\frac{20}{2}\right)^{\frac{1}{1.25}}$$

$$= 0.7004 \text{ m}^3\text{/kg}$$

At 2 bar, $v_{g2} = 0.886$ m^3/kg and $v_{f2} = 0.001061$ m^3/kg, which is negligible.

$$\therefore \quad x_2 = \frac{v_2}{v_{g2}} = \frac{0.7004}{0.886} = 0.79$$

From Mollier diagram at $x_2 = 0.79$ and $p_2 = 2$ bar,

$s_2 = 5.9$ kJ/kg. K and saturation temperature, $t_2 = 120.2$°C

$$T_2 = 273 + 120.2 = 393.2 \text{ K}$$

At $p_2 = 2$ bar, $h_{f2} = 504.7$ kJ/kg, $h_{g2} = 2706.7$ kJ

$$h_{fg2} = 2201.9 \text{ kJ/kg}$$

$$h_2 = h_{f2} + x_2 h_{fg2} = 504.7 + 0.79 \times 2201.9 = 2244.2 \text{ kJ/kg}$$

$$u_{f2} = h_{f2} - p_2v_2 = 504.7 - 2 \times 10^2 \times 0.001061 = 504.48 \text{ k/kg}$$

$$u_{g2} = h_{g2} - p_2v_{g2} = 2706.7 - 2 \times 10^2 \times 0.886 = 2529.5 \text{ kJ/kg}$$

$$u_2 = (1 - x)\, u_{f2} + xu_{g2} = (1 - 0.79) \times 504.48 + 0.79 \times 2529.5$$

$$= 2104.24 \text{ kJ/kg}$$

Mass of steam, $m = \dfrac{V_s}{v_{s1}} = \dfrac{0.14}{0.1110} = 1.261 \text{ kg}$

$$\text{Work done} = \frac{m(p_1v_{s1} - p_2v_{s2})}{n-1} = \frac{1.261\,(20 \times 0.1110 - 2 \times 0.7004) \times 10^2}{0.25}$$

$$= 413.2 \text{ kJ}$$

Heat transferred, $Q = W + (U_2 - U_1)$

$$= 413.2 + 1.261\,(2104.24 - 2680.5)$$

$$= -\ 313.464 \text{ kJ}$$

– ve sign indicates that steam rejects heat.

Change of entropy, $\Delta S = m(s_2 - s_1)$

$$= 1.261\,(5.9 - 6.545) = -\ 0.8133 \text{ kJ/K}$$

– ve sign indicates that there is decrease of entropy.

(*b*) The specific heats of a gas are of the form $c_p = a + k\mathrm{T}$ and $c_v = b + k\mathrm{T}$, where a, b and k are constants and T is in K. Derive the formula $\mathrm{T}^b\, V^{a-b}\, e^{k\mathrm{T}}$ = constant, for the adiabatic expansion of the gas.

Sol. Given :

$$c_p = a + k\mathrm{T}$$

$$c_v = b + k\mathrm{T}$$

$$\mathrm{R} = c_p - c_v = a - b$$

Now $d\mathrm{Q} = d\mathrm{W} + d\mathrm{U}$

For adiabatic expansion of gas, $d\mathrm{Q} = 0$

$\therefore$ $d\mathrm{Q} = p.d\mathrm{V} + c_v.d\mathrm{T} = 0$

or

$$\frac{\mathrm{Q}}{\mathrm{T}} = \frac{p.d\mathrm{V}}{\mathrm{V}} + c_v \frac{d\mathrm{T}}{\mathrm{T}} = 0$$

$$= \frac{\mathrm{R}.d\mathrm{V}}{\mathrm{V}} + c_v.\frac{d\mathrm{T}}{\mathrm{T}} = 0 \qquad (\because p\mathrm{V} = \mathrm{RT})$$

$$= \int (a-b)\frac{d\mathrm{V}}{\mathrm{V}} + \int (b + k\mathrm{T})\frac{d\mathrm{T}}{\mathrm{T}} = 0$$

Integrating, we get

$(a - b) \ln \mathrm{V} + b \ln \mathrm{T} + k\mathrm{T} = \ln$ (const.)

or $\ln \mathrm{V}^{(a-b)} + \ln \mathrm{T}^b + \ln (e^{k\mathrm{T}}) = \ln$ (const.)

or $\mathrm{T}_b\, \mathrm{V}^{a-b}\, e^{k\mathrm{T}}$ = const.

Hence proved.

(c) A closed system contains 0.5 kg of air. It expands from 2 bar, 60°C to 1 bar, 40°C. During expansion it receives 2 kJ of heat from a reservoir at 100°C. Assuming atmospheric conditions to be at 0.95 bar and 30°C, calculate (*i*) the maximum work, (*ii*) work done on atmospheric, and (*iii*) change in availability.

Sol.

Given :

$$m = 0.5 \text{ kg}, p_1 = 2 \text{ bar}, T_1 = 273 + 60 = 333 \text{ K}, p_2 = 1 \text{ bar},$$
$$T_0 = 273 + 40 = 313 \text{ K},$$
$$Q_R = + 2 \text{ kJ}, T_R = 273 + 100 = 373 \text{ K}, p_0 = 0.95 \text{ bar},$$
$$T_0 = 273 + 30 = 303 \text{ K}$$

(*i*) Maximum work,
$$W_{max} = (U_1 - U_2) + T_0(S_1 - S_2) + Q_R\left(1 - \frac{T_0}{T_R}\right)$$
$$= m\left[c_v(T_1 - T_2) + T_0\left(c_p \ln\frac{T_2}{T_1} - R\ln\frac{p_2}{p_1}\right)\right] + Q_R\left(1 - \frac{T_0}{T_R}\right)$$
$$= 0.5\left[0.718\,(333 - 313) + 303\left(1.005 \ln\frac{313}{333} - 0.287 \ln\frac{1}{2}\right) + 2\left(1 - \frac{303}{313}\right)\right]$$
$$= 28.26 \text{ kJ}$$

(*ii*) Work done on atmosphere,

$$W_{atm} = p_0 . dV$$

Now
$$pV = mRT$$
$$dV = V_1 - V_2 = mR\left[\frac{T_1}{p_1} - \frac{T_2}{p_2}\right]$$
$$= 0.5 \times 0.281\left[\frac{333}{200} - \frac{313}{100}\right] = - 0.21023 \text{ m}^3$$
$$W_{atm} = 0.95 \times 10^2 \times (- 0.21023)$$
$$= - 19.972 \text{ kJ}$$

(*iii*) Change in availability
$$= W_{max} + W_{atm}$$
$$= 28.26 - 19.972 = 8.288 \text{ kJ}$$

Q.2. (*a*) Prove that for the same compression ratio the Otto cycle is more efficient than the Diesel cycle.

Sol.

Let r_c, r_e = Compression and expansion ratios of both cycles respectively.

Air standard efficiency of Otto cycle, $\eta_0 = 1 - \dfrac{1}{r_c^{\gamma - 1}}$

Air standard efficiency of Diesel cycle, $\eta_d = 1 - \dfrac{1}{r_c^{\gamma - 1}}\left[\dfrac{(r_c / r_e)^{\gamma} - 1}{(r_c / r_e) - 1}\right]$

Let $r_e = r_c - \Delta r$

where Δr = a small quantity

$$\frac{r_c}{r_e} = \frac{r_c}{r_e\left(1 - \frac{\Delta r}{r_c}\right)} = \left[1 - \frac{\Delta r}{r_c}\right]^{-1}$$

$$= 1 + \frac{\Delta r}{r_c} - \frac{1(-1-1)}{2!}\left(-\frac{\Delta r}{r_c}\right)^2 + \ldots\ldots$$

$$= 1 + \frac{\Delta r}{r_c} - \left(\frac{\Delta r}{r_c}\right)^2 + \left(\frac{\Delta r}{r_c}\right)^3 + \ldots\ldots$$

$$\left(\frac{r_c}{r_e}\right)^\gamma = \frac{r_c^\gamma}{r_c^r\left(1 - \frac{\Delta r}{r_c}\right)^\gamma} = \left(1 - \frac{\Delta r}{r_c}\right)^{-\gamma}$$

$$= 1 + \gamma\left(\frac{\Delta r}{r_c}\right) + \frac{\gamma(\gamma+1)}{2!}\left(\frac{\Delta r}{r_c}\right)^2 + \frac{\gamma(\gamma+1)(r+2)}{3!}\left(\frac{\Delta r}{r_c}\right)^3 + \ldots\ldots$$

$$\therefore \quad \eta_d = 1 - \frac{1}{r_r^{\gamma-1}}\left[\frac{\frac{\Delta r}{r_c} + \frac{(\gamma+1)}{2!}\left(\frac{\Delta r}{r_c}\right)^2 + \ldots\ldots}{\frac{\Delta r}{r_c} + \left(\frac{\Delta r}{r_c}\right)^2 + \ldots\ldots}\right]$$

$$= 1 - \frac{1}{r_r^{\gamma-1}} \times \left[\frac{1 + \frac{(\gamma+1)}{2!}\left(\frac{\Delta r}{r_c}\right) + \ldots\ldots}{1 + \frac{\Delta r}{r_c} + \ldots\ldots}\right]$$

The ratio within the brackets is more than 1.

$\therefore \quad \eta_d < \eta_0$

or $\quad \eta_0 > \eta_d$

(b) **Sketch and explain the working, of a simple carburettor. Also explain how the slow running and compensating jets work.**

Sol.

CARBURETION

The process of preparing the air-fuel mixture for an I.C. engine is called carburetion. The carburettor is a device which atomises the fuel and mixes it with air.

The Samiple Carburettor

The details of a simple carburettor are shown in Fig. 2. It mainly consists of a float chamber, fuel discharge nozzle and a metering orifice, a venturi, a throttle valve and a choke. A float and a needle valve system maintains a constant level of gasoline in the float chamber. If the amount of fuel in the float chamber falls below the designed level, the float goes down, thereby opening the fuel supply valve and admitting fuel. When the designed level has been reached, the float closes the fuel supply valve thus stopping additional fuel flow from the supply system. The float chamber is vented either to the amosphere or to the upsetream side of the venturi.

During section stroke air is drawn through the venturi (or choke tube). As the air passes through the venturi the velocity increases reaching a maximum at the venturi throat and pressure decreases to minimum. From the float chamber, the fuel is fed to a discharge jet, the tip of which is located in the throat of the venturi, fuel is discharged into the air stream. To avoid overflow of fuel through the jet, the level of the liquid in the float chamber is maintained at a level slightly below the tip of the discharge jet.

The throttle valve controls the amount of charge delivered to the cylinder, which is situated after the venturi tube. As the throttle is closed less air flows through the venturi tube and less is the quantity of air-fuel mixture delivered to the cylinder and hence power output is reduced. The reverse takes place when the throttle is opened.

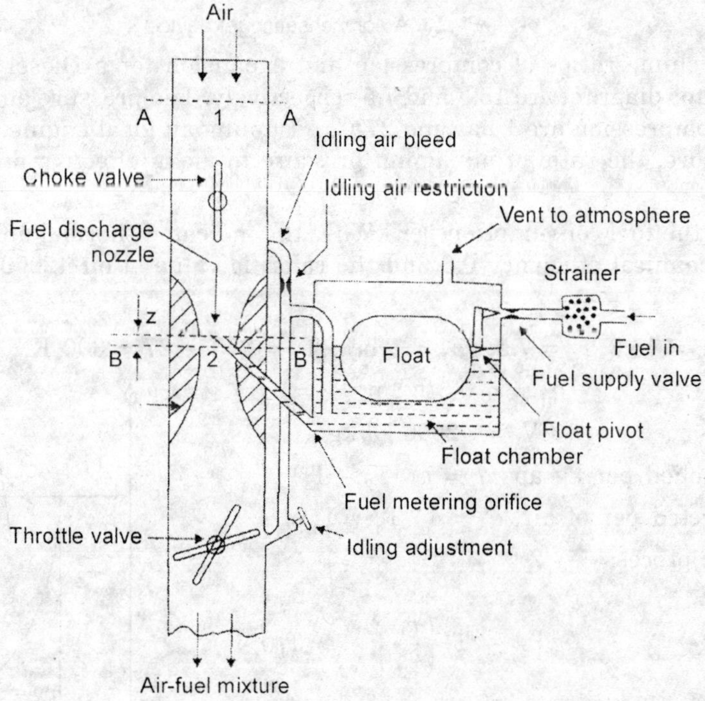

Fig. 2. The Simple Carburettor

Compensating Jet

The function of the compensating jet device is to make the mixture leaner as the throttle opens progressively. As shown in Fig. 3, the compensating jet is connected to the compensating well which is vented to the atmosphere. The compensating well is supplied with fuel from the main float chamber through a restricting orifice. With the increase in air flow rate, there is decrease of fuel level in the compensating well, with the result that fuel supply through the compensating jet decreases. The compensating jet thus progressively makes the mixture leaner as the main jet progressively makes the mixture richer.

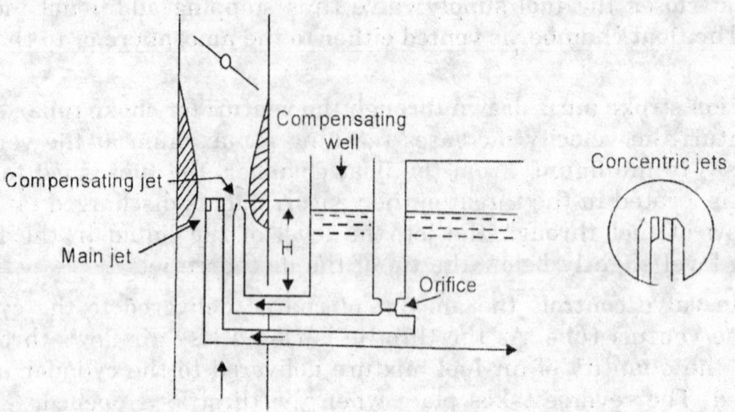

Fig. 3. A compensating jet device

(*c*) The volume ratios of compression and expansion for a Diesel engine as measured from an indicator diagram are 15.3 and 7.5 respectively. The pressure and temperature at the beginning of compression are 1 bar and 27°C. Assuming an ideal engine, determine the mean effective pressure, the ratio of maximum pressure to mean effective pressure and the cycle efficiency.

Also find the fuel consumption per kWh if the indicated thermal efficiency is 0.5 of ideal efficiency, mechanical efficiency 0.8 and the calorific value of oil 42000 kJ/kg.

Sol.

Given : $r_c = 15.3$, $r_e = 7.5$, $rp_1 = 1$ bar, $T_1 = 273 + 27 = 300$ K,

$$(\eta_{th})_i = 0.5\ \eta_a,\ \eta_{mech} = 0.8,$$

$$CV = 42000 \text{ kJ/kg}$$

Heat supplied per kg air, $q_s = c_p\,(T_3 - T_2)$

Heat rejected per of air, $q_r = c_v\,(T_4 - T_1)$

Adiabatic process 1 – 2 :

$$T_2 = T_1\left(\frac{v_1}{v_2}\right)^{\gamma - 1} = T_1(r_c)^{\gamma - 1}$$

$$= 300\,(15.3)^{0.4} = 893.3 \text{ K}$$

$$p_2 = p_1\left(\frac{v_1}{v_2}\right)^{\gamma} = 1\,(15.3)^{1.4}$$

$$= 45.56 \text{ bar}$$

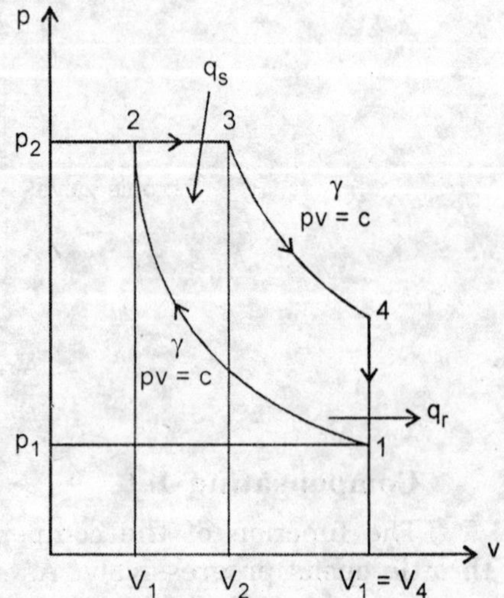

Fig 4.

Isobaric process 2 – 3 :

$$T_3 = T_2\left(\frac{v_3}{v_2}\right) = T_2\left(\frac{v_3}{v_1} \times \frac{v_1}{v_2}\right) = T_2\left(\frac{r_c}{r_e}\right) = 893.3\left(\frac{15.3}{7.5}\right) = 1822.3 \text{ K}$$

Adiabatic process 3 – 4 :

$$\frac{T_3}{T_4} = \left(\frac{v_3}{v_4}\right)^{\gamma - 1} = (7.5)^{0.4} = 2.231$$

$$T_4 = 1822.3/2.239 = 813.95 \text{ K}$$

Mass of air, $m = \dfrac{pV_1}{RT_1} = \dfrac{10^2 \times 15.3}{0.287 \times 300} = 17.77$ kg

Work done,
$$\begin{aligned} W &= m\,(q_s - q_r) \\ &= 17.77\ [1.005\,(1822.33 - 893.3) - 0.718\,(813.95 - 300)] \\ &= 10033.5 \text{ kJ} \end{aligned}$$

Mean effective pressure, $p_m = \dfrac{\text{Work done}}{\text{Swept volume}} = \dfrac{10033.5}{15.3 - 1} = 701.64 \text{ kN/m}^2$ or 7.0164 bar

$$\frac{\text{Maximum pressure, } p_{max}}{p_m} = \frac{p_2}{p_m} = \frac{45.56}{7.0164} = 6.493$$

Cycle efficiency, $\eta_a = \dfrac{\text{Work done}}{\text{Heat supplied}} = \dfrac{10033.5}{17.77 \times 1.005\,(1822.33 - 893.3)} = 0.6047$

Indicated thermal efficiency, $(\eta_{th}) = 0.5\,\eta_a = 0.5 \times 60.47 = 30.24$ %

Brake thermal efficiency, $(\eta_{th}) = \eta_{mech} \times (\eta_{th})_i = 0.8 \times 30.24 = 24.2$ %

Heat supplied per kWh $= \dfrac{3600}{(\eta_{th})_B} = \dfrac{3600}{0.242} = 14876$ kJ/kWh

Fuel consumption $= \dfrac{14876}{42000} = 0.354$ kg/kWh

Q.3. (*a*) **What do you understand by nuclear fission? What are the essential requirements to cause nuclear fission? Draw a neat diagram of a nuclear reactor and explain the function of different components.**

Sol.

Nuclear Fission

When unstable heavy nucleus is bombarded with high energy neutrons, it splits into two fragments more or less of equal mass. This process is known as "*Nuclear Fission*".

The fission fragments formed due to fission are the isotopes which are located in the middle of periodic table.

The binding energy per nucleon for the elements in the middle range of periodic table is more than that of heavy nuclei. Therefore, the nuclear fission is always associated with the release of large amount of energy. The nuclear fission is follwed by the emission of several high energy neutrons per neutron bombarded. This is the essential condition for the self-sustaining fission reaction. The fission reaction is shown in the Fig. 5.

The energy required to break the nucleus must be sufficient to overcome the repulsive forces (nuclear binding energy) between the two fragments. The excitation energy required to split the nucleus is called the *critical energy*.

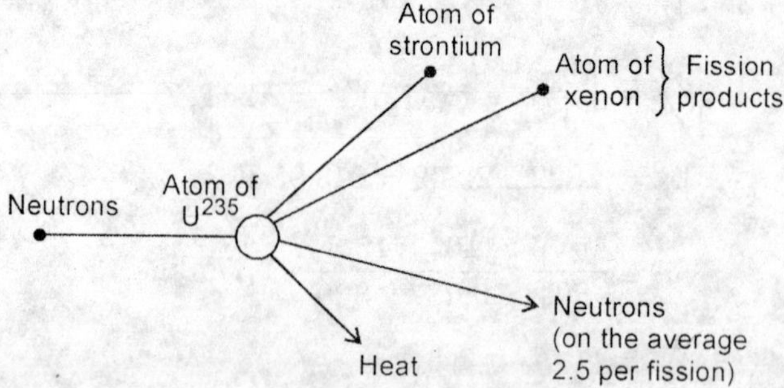

Fig. 5. Fission reaction.

It is obvious that the critical energy is more than neutrons binding energy.

Requirements to sustain the fission processes :

1. The bombarded of neutrons produced must have sufficient energy to cause fission of another nucleus (to overcome binding energy).
2. The number of neutrons produced must be able to increase the rate of fission as certain loss of neutrons by absorbing, and leakage is unavoidable.
3. The fission process must generate energy.
4. The fission process must be controlled.

General Components of Nuclear Reactor :

The nuclear reactor may be regarded as a substitute for the boiler fire box of steam plant and for combustion chamber of a gas turbine plant. The heat produced in the nuclear power plant is by fission where as in steam and gas turbine plant, the heat is produced by combusion. The other cycle of operation and components required are exactly same either as steam plant if steam is generatd by using the heat of fission or a gas turbine plant (closed or open type) if gas is heated by using the heat of fission. The steam or the gas may be used as working fluid in nuclear power plant. The nuclear power plant may be steam driven turbine or gas driven turbine as per the choice of the fluid. The various components of nuclear reactor are shown in Fig. 6.

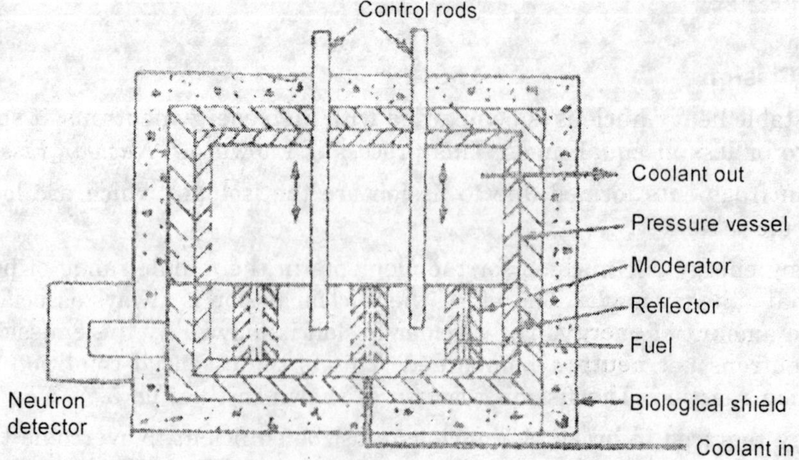

Fig. 6. Components of Nuclear Reactor

1. **Fuel :** The nuclear fuels which are generally used in reactors are $_{92}U^{235}$, $_{94}Pu^{239}$, and $_{92}U^{233}$. Among these three, the $_{92}U^{235}$ is available upto 0.7% in the uranium and the remaining is $_{92}U^{238}$. The other two fuels $_{94}U^{238}$ are formed in the nuclear reactors during fission process from $_{92}U^{238}$ and $_{90}U^{232}$ due to the absorption of netron without fission.

2. **Moderator :** The moderator is a material which reduces the kinetic energy of fast neutron (1 MeV or 13200 km/sec) to slow neutron (0.25 eV or 200 m/sec) and this is done in a fraction of second. The fission chain-reaction in the nuclear reactor is maintained due to slow neutrons when the ordinary uranium is used as fuel. The function of the moderator is to increase the probability of reaction. The slowing down of neutrons is effectively done by the light elements as H_2, D_2, N_2, C and Be as mentioned earlier. THe graphite, heavy water or beryllium can be used as moderator with natural uranium. The ordinary water is used as moderator only with the enriched uranium for the reaction mentioned earlier.

3. **Relfector :** It is always necessary to conserve the neutrons as much as possible in order to reduce the consumption of fissible material and to keep the size of the reactor small. The neutrons which are related in fission process can be absorbed by the fuel itself, moderator, coolant or structural materials. Some neutrons may escape from the core without absorption all will be lost for ever. To reduce the loss of escape, the reactor inner surface is surrounded by a material which reflects the escaping neutrons back into the core.

4. **Coolant :** The main purpose of the coolant in the reactor is to transfer the heat produced inside the reactor. The same heat carried by coolant is used in the heat exchanger for further utilisation in the power generation.

5. **Control Rods :** Following are the functions of control rod.

   (*a*) To start the nuclear chain reaction when the reactor is started from cold.

   (*b*) The chain reaction should be maintained at steady state condition (controlled chain reaction) at the required level.

   (*c*) To shut down the reactor automatically under emergency condition. *i.e.*, a pump circulating the coolant through the reactor fails.

   The control is necessary to prevent the melting of fuel rods, disintegration of coolant and destruction of reactor as the amount of energy released is enormous.

   The control of the chain reaction is possible either by removing the fuel rods or the neutrons continuing the chain reaction from the core of the reactor. It is always easy to get rid of the neutrons. It is generally done by inserting the control rods into the fuel tubes as shown in figure earlier.

6. **Shielding :** The reactor is a source of intense radio activity as mentioned earlier and these radiations are very harmful to the human lifge. The common radiations from the reactor are α-rays, β-rays, γ-rays and fast neutrons. To prevent the effects of these radiation on the human life, it is necessary to absorb them before emitting to the atmosphere. These are effectively absorbed by the concrete and steel. The inner lining of the core is made of 50 to 60 cm thick steel plate and it is further thickened by few meters using concrete. The lining of steel plate absorbs these energies and becomes heated but prevents the adjacent wall of reactor vessel from becoming heated. The thermal shield is cooled by the circulating water.

7. **Reactor vessel :** The reactor vessel encloses the reactor core, reflector and shield. It also provides the entrance and exit passages for directling the flow of coolant. The reactor vessel has to withstand for pressure as high as 200 kgf/cm^2 or above. The holes at the top of the vessel are provided to insert the control rods. The reactor core (i.e. fuel and moderator assembly) is generally placed at the bottom of the vessel.

(b) Following observations were made during the trial of a gas fired boiler of a steam power plant :

Generator output ....50000 kW

Steam conditions .... pressure 50 bar temperature 500°C

Feed water temperature..............80°C

Steam output .... 65 kg/s

Percentage composition ....CH_4 = 96.5
of fuel (natural gas), C_2H_6 = 0.5
by volume (rest incombustibles)

Higher calorific value .... 38700 kJ/m^3, at 1.013
of gas bar and 15°C

Gas consumption ....6.5 m^3/s

Calculate the boiler efficiency and the overall thermal efficiency based on the lower calorific value of the fuel.

Sol.

Combustion equations :

$$CH_4 + 2O_2 \longrightarrow CO_2 + 2H_2O$$

$$C_2H_6 + 2.5O_2 \longrightarrow 2CO_2 + 3H_2O$$

1 mole of CH_4 burns to give 2 moles of water

Therefore, 0.965 moles give 2 × 0.965 × 18 = 34.74 kg of H_2O

Similarly 0.005 moles of C_2H_6 burns to give 3 × 0.005 × 18 = 0.27 kg of H_2O

Assuming steam as a perfect gas, 1 mole of gas at 1.013 bar and OC is 22.41 m^3

$$\therefore \text{ 1 mole of gas at 15°C} = 22.41 \times \frac{288}{273} = 23.64 \text{ m}^3$$

Mass of steam produced,

$$m_s = \frac{(34.74 + 0.27)}{23.64} = 1.481 \text{ kg/m}^3 \text{ of fuel}$$

$$\text{HCV} = \text{LCV + heat in steam}$$

$$\text{HCV} = 38700 - m_s\, h_{fg} \text{ at 15°C}$$

$$= 38700 - 1.481 \times 2465.9 \qquad [h_{fg} \text{ at 15°C} = 2465.9 \text{ kJ/kg}]$$

$$= 35043.07 \text{ kJ/kg}$$

At 50 bar and 500°C, h_1 = 3433.8 kJ/kg

At 80°C, h_{fg} for feed water = 334.98 kJ/kg

$$\text{Enthalpy of steam} = h_1 - h_{fg} = 3433.8 - 334.9 = 3098.9 \text{ kJ/kg}$$

$$\text{Boiler efficiency} = \frac{\text{Heat to water in boiler}}{\text{Heat supplied to boiler}}$$

$$= \frac{m_w (h_1 - h_{fg})}{\dot{V}_f \times \text{HCV}} = \frac{65 \times 3098.9}{6.5 \times 35043.07} = 0.8843 \text{ or } 88.43\ \%$$

$$\text{Overall thermal efficiency} = \frac{\text{Generator output}}{\text{Heat supplied}}$$

$$= \frac{50{,}000}{6.5 \times 35043.07} = 0.2195 \text{ or } 21.95\ \%$$

Q.4. (*a*) Mention the types of nozzles you know. Where are these used?

Sol.

Type of nozzles :

1. **Convergent nozzle :** It is a uniformly decreasing cross-sectional area duct. It is used for accelerating a steadily flowing fluid. It converts internal energy of fluid into kinetic energy.
2. **Divergent nozzle :** It is a uniformly increasing cross-sectional area duct. It is used for decelerating a steadily flowing fluid. It converts kinetic energy into pressure energy.
3. **Convergent divergent nozzle :** It is a combination of above two nozzles.
4. **Steam nozzle :** It is special type convergent nozzle used in steam turbines.

(*b*) From first principles, prove that maximum discharge per unit area in steam nozzle at the throat is given by the expression

$$\frac{\text{m}_{\text{max}}}{A} = \left[n \frac{p_1}{v_1} \left(\frac{2}{n+1} \right)^{\frac{n+1}{n-1}} \right]^{1/2}$$

where p_1 and v_1 are pressure and specific volume at the inlet and n is the index of expansion.

Sol.

Isentropic Flow of Steam Through the Nozzle

Let $pv^n = \text{const.}$

where $n = 1.035 + 0.1\,x$, for wet steam (x = dryness fraction)

$= 1.135$ for saturated steam

$= 1.3$ for superheated steam

As the steam pressure drops while passing through the nozzle, its enthalpy is reduced. This reduction of enthalpy of steam must be equal to the increase in kinetic energy. Hence

$$\frac{c_2^2 - c_1^2}{2} = \left(\frac{n}{n-1} \right) (p_1 v_1 - p_2 v_2)$$

where p_1, p_2 = pressure of steam at entry and exit respectively

If $c_1 << c_2$, then

$$\frac{c_2^2}{2} = \left(\frac{n}{n-1}\right)(p_1v_1 - p_2v_2) = \left(\frac{n}{n-1}\right)p_1v_1\left(1 - \frac{p_2v_2}{p_1v_1}\right)$$

But
$$p_1v_1^n = p_2v_2^n$$

$$\frac{v_2}{v_1} = \left(\frac{p_1}{p_2}\right)^{\frac{1}{n}}$$

$$\frac{c_2^2}{2} = \left(\frac{n}{n-1}\right)p_1v_1\left[1 - \frac{p_2}{p_1}.\left(\frac{p_1}{p_2}\right)^{\frac{1}{n}}\right] = \left(\frac{n}{n-1}\right)p_1v_1\left[1 - \left(\frac{p_2}{p_1}\right)^{\frac{n-1}{n}}\right]$$

$$c_2 = \sqrt{\left(\frac{2n}{n-1}\right)p_1v_1\left[1 - \left(\frac{p_2}{p_1}\right)^{\frac{n-1}{n}}\right]} \quad ...(1)$$

Mass of Steam Discharged

The mass flow of steam per second is flowing with velocity c_2 through a cross-sectional area A_2 and specific volume v_2 is :

$$\dot{m} = \frac{A_2c_2}{v_2}$$

Now
$$v_2 = v_1\left(\frac{p_1}{p_2}\right)^{\frac{1}{n}}$$

Also from eq. (1), we have

$$c_2 = \sqrt{\left(\frac{2n}{n-1}\right)(p_1v_1 - p_2v_2)}$$

$$\therefore \quad \dot{m} = \frac{A_2}{v_1\left(\frac{p_1}{p_2}\right)^{\frac{1}{n}}}\sqrt{\left(\frac{2n}{n-1}\right)(p_1v_1 - p_2v_2)}$$

$$= A_2\left[\left(\frac{2n}{n-1}\right)\frac{p_2}{v_1}\left\{\left(\frac{p_2}{p_1}\right)^{\frac{2}{n}} - \left(\frac{p_2}{p_1}\right)^{\frac{n+1}{n}}\right\}\right]^{\frac{1}{2}} \quad ...(2)$$

Critical Pressure Ratio

From Eq. (2), we have

$$\frac{\dot{m}}{A_2} = \left[\left(\frac{2n}{n-1}\right)\frac{p_2}{v_1}\left\{\left(\frac{p_2}{p_1}\right)^{\frac{2}{n}} - \left(\frac{p_2}{p_1}\right)^{\frac{n+1}{n}}\right\}\right]^{\frac{1}{2}}$$

The mass flow rate per unit area is maximum at the throat which has minimum area of cross-section. $\frac{\dot{m}}{A_2}$ will be maximum when $\left[\left(\frac{p_2}{p_1}\right)^{\frac{2}{n}} - \left(\frac{p_2}{p_1}\right)^{\frac{n+1}{n}}\right]$ is maximum or $$\frac{d}{d\left(\frac{p_2}{p_1}\right)}\left[\left(\frac{p_2}{p_1}\right)^{\frac{2}{n}} - \left(\frac{p_2}{p_1}\right)^{\frac{n+1}{n}}\right] = 0$$

$$\frac{2}{n}\left(\frac{p_2}{p_1}\right)^{\frac{2}{n}-1} - \left(\frac{n+1}{n}\right)\left(\frac{p_2}{p_1}\right)^{\frac{n+1}{n}} = 0$$

or $$\frac{2}{n}\left(\frac{p_2}{p_1}\right)^{\frac{n+1}{n}-1} = \left(\frac{n+1}{n}\right)\left(\frac{p_2}{p_1}\right)^{\frac{1}{n}}$$

or $$\left(\frac{p_2}{p_1}\right)^{2-n} = \left(\frac{n+1}{n}\right)^{n}\left(\frac{p_2}{p_1}\right)$$

or $$\left(\frac{p_2}{p_1}\right)^{1-n} = \left(\frac{n+1}{n}\right)^{n}$$

or $$\frac{p_2}{p_1} = \left(\frac{n+1}{2}\right)^{\frac{n}{1-n}} = \left(\frac{2}{n+1}\right)^{\frac{n}{n-1}} \quad ...(3)$$

Equation (3) gives the critical pressure ratio at which the discharge through the nozzle as maximum.

For saturated steam, $n = 1.135$ and $\frac{p_2}{p_1} = 0.5774$, and for

superheated steam, $n = 1.3$ and $\frac{p_2}{p_1} = 0.5457$

Now $$c_2 = \sqrt{\left(\frac{2n}{n-1}\right)(p_1v_1 - p_2v_2)} = \left[\left(\frac{2n}{n-1}\right)p_2v_2\left\{\left(\frac{p_2}{p_1}\right)^{\frac{1-n}{n}} - 1\right\}\right]^{\frac{1}{2}}$$

Substituting for $\frac{p_2}{p_1}$ from eq. (3), we get

$$c_2 = \left[\left(\frac{2n}{n-1}\right)p_2v_2\left\{\frac{n+1}{2} - 1\right\}\right]^{\frac{1}{2}} = \left[\left(\frac{2n}{n-1}\right)\left(\frac{n-1}{2}\right)p_2v_2\right]^{\frac{1}{2}}$$

$$= \sqrt{n\, p_2v_2} = \sqrt{n\,RT_2} = c_{s2}$$

= Velocity of sound at exit of the nozzle.

Also $$c_2 = \left[\left(\frac{2n}{n-1}\right)p_1v_1\left\{1 - \left(\frac{p_2}{p_1}\right)^{\frac{n-1}{n}}\right\}\right]^{\frac{1}{2}}$$

$$= \sqrt{\left(\frac{2n}{n+1}\right).n\,p_1v_1} = \sqrt{\left(\frac{2n}{n+1}\right).n\,\mathrm{RT}_2}$$

$$= \sqrt{\left(\frac{2n}{n+1}\right)}\,c_{s1} \qquad ...(4)$$

where c_{s1} = Velocity of sound at entry of nozzle.

Maximum Discharge

From Eq. (2), maximum mass flow rate of steam,

$$(\dot{m})_{max} = \frac{A_2}{v_1}\left[\left(\frac{2n}{n-1}\right)p_1v_1\left\{\left(\frac{2}{n+1}\right)^{\frac{2}{n-1}} - \left(\frac{2}{n+1}\right)^{\frac{n+1}{n-1}}\right\}\right]^{\frac{1}{2}}$$

$$= A_2\left[\left(\frac{2n}{n-1}\right)\frac{p_1}{v_1}\left\{\left(\frac{2}{n+1}\right)^{\frac{2}{n-1}} - \left(\frac{2}{n+1}\right)^{\frac{n+1}{n-1}}\right\}\right]^{\frac{1}{2}}$$

$$= A_2\left[\left(\frac{2n}{n-1}\right)\frac{p_1}{v_1}\left(\frac{2}{n+1}\right)^{\frac{n+1}{n-1}}\left\{\left(\frac{2}{n+1}\right)^{\frac{1-n}{n-1}} - 1\right\}\right]^{\frac{1}{2}}$$

$$= A_2\left[\left(\frac{2n}{n-1}\right)\left(\frac{p_1}{v_1}\right)\left(\frac{2}{n+1}\right)^{\frac{n+1}{n-1}}\left\{\left(\frac{2}{n+1}\right)^{-1} - 1\right\}\right]^{\frac{1}{2}}$$

$$= A_2\left[\left(\frac{2n}{n-1}\right)\left(\frac{p_1}{v_1}\right)\left(\frac{2}{n+1}\right)^{\frac{n+1}{n-1}}\left\{\left(\frac{n+1}{2}\right) - 1\right\}\right]^{\frac{1}{2}}$$

$$= A_2\left[\left(\frac{2n}{n-1}\right)\left(\frac{p_1}{v_1}\right)\left(\frac{2}{n+1}\right)^{\frac{n+1}{n-1}}\left(\frac{n-1}{2}\right)\right]^{\frac{1}{2}}$$

$$= A_2\left[\left(\frac{np_1}{v_1}\right)\left(\frac{2}{n+1}\right)^{\frac{n+1}{n-1}}\right]^{\frac{1}{2}}$$

(*c*) At a particular ring of a reaction turbine having 50% degree of reaction, the blade speed is 70 m/s. The mass of steam is 5 kg/s, dry saturated at 1.5 bar. Both fixed and moving blades have inlet and exit angles of 35° and 20° respectively. Determine

(*i*) the power developed in kW by the pair of rings,

(*ii*) the blade height, which is to be one-tenth the mean blade ring diameter,

(*iii*) the heat drop by the pair if the steam expands with and efficiency of 80%.

Sol.

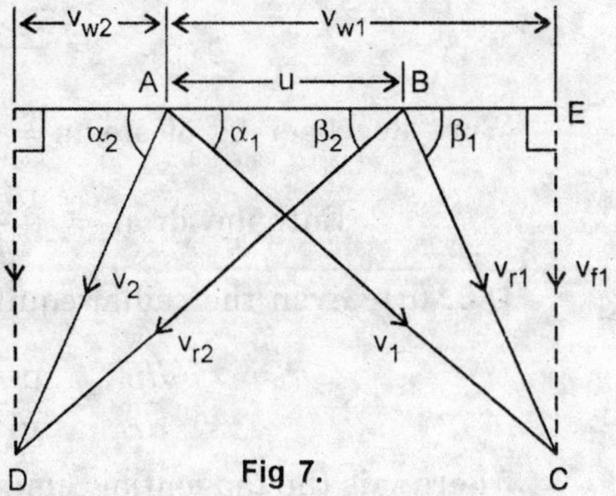

Given : $R_d = 50\%,\ u = 70$ m/s,

$\dot{m} = 5$ kg/s, $p = 1.5$ bar,

$x = 1.0,\ h = 0.1d$

For $R_d = 50\%,\ \alpha_1 = \beta_2 = 20°$

and $\alpha_2 = \beta_1 = 35°,\ v_{f1} = v_{f2}$

Fig 7.

$$v_1 \sin\alpha_1 = v_{r1}\sin\beta_1 = v_{f1}$$

$$v_{r1} = \frac{\sin 20°}{\sin 35°} \times v_1$$

$$= 0.5963\ v_1$$

$$v_{r1}^2 = v_1^2 + u^2 - 2uv_1\cos\alpha_1$$

$$(0.5963\ v_1) = v_1^2 + 70^2 - 2\times 70\times v_1\cos 20°$$

$$0.3556\ v_1^2 = v_1^2 + 4900 - 131.557\ v_1$$

$$0.6444\ v_1^2 - 131.557\ v_1 + 4900 = 0$$

or $v_1^2 - 204.15\text{ v1} + 7603.97 = 0$

$$v_1 = \frac{1}{2}\left[204.15 \pm \sqrt{(204.15)^2 - 4\times 7603.97}\right]$$

$$= \frac{1}{2}(204.15 \pm 106.12) = 155.13 \text{ m/s or } 49.01 \text{ m/s}$$

Take $v_1 = 155.13$ m/s

$$v_{r1} = 0.5963 \times 155.13 = 92.5 \text{ m/s}$$

$$v_{f1} = 155.13 \sin 20° = 53.06 \text{ m/s}$$

$$v_{w2} = v_1\cos\alpha_1 = 155.13\cos 20° = 145.77 \text{ m/s}$$

Now $v_1\sin\alpha_1 = v_2\sin\alpha_2$

$$v_2 = \frac{\sin 20°}{\sin 35°}\times 155.13 = 92.5 \text{ m/s}$$

$$v_{w2} = v_2\cos\alpha_2 = 92.5\cos 35° = 75.77 \text{ m/s}$$

$$v_w = v_{w1} + v_{w2} = 145.77 + 75.77 = 221.54 \text{ m/s}$$

Specific volume of saturated steam at 1.5 bar, from steam tables, is,

$$v_{sg} = 1.1593 \text{ m}^3\text{/kg}$$

Mass flow rate of steam, $\dot{m} = \dfrac{A_f \times v_f}{v_{sg}}$

$$5 = \frac{(\pi dh)\times 53.06}{1.1593} = \frac{\pi \times 0.1d^2 \times 53.06}{1.1593}$$

$$d = 0.59 \text{ m}, \; h = 0.059 \text{ m or } 59 \text{ mm}$$

Power developed, $\quad P = \dot{m}u \times v_w / 10^3$

$$= \frac{5 \times 70 \times 221.54}{10^3} = 77.539 \text{ kW}$$

Work done per kg of steam $= uv_w = 70 \times 221.54 = 15.5078$ kJ/kg

$$\text{Enthalpy drop} = \frac{15.5078}{0.8} = 19.385 \text{ kJ/kg}$$

Q.5. (*a*) Given the radial equilibrium equation in the form

$$\frac{dp}{dr} = \frac{\rho}{g} \frac{v_t^2}{r}$$

where v_t is the tangential component of velocity of a gas element of density ρ at a radius r, the pressure being p, derive the free vortex equation, stating clearly the assumptions made.

Sol.

Assume that the flowing fluid is a perfect gas, then the total energy radius r is given by,

$$h = \frac{1}{2} \frac{c^2}{g} + \frac{c_p T}{g} = \frac{c_a^2}{2g} + \frac{c_w^2}{2g} + \frac{\gamma}{\gamma - 1} \cdot \frac{p}{\rho g} \qquad ...(1)$$

Differentiating Eq. (1) w.r.t. r, we have

$$g . \frac{dh}{dr} = c_a \frac{dc_a}{dr} + c_w \frac{dc_w}{dr} + \left(\frac{\gamma}{\gamma - 1}\right) \left[\frac{1}{\rho} \frac{dp}{dr} - \frac{p}{\rho^2} . \frac{d\rho}{dr}\right] \qquad ...(2)$$

For a small change in pressure dp across the annulus walls, it is assumed that the pressure p and density are related isentropically, i.e.

$$\frac{p}{\rho^\gamma} = \text{const.}$$

Differentiating, we get

$$\frac{1}{\rho^\gamma} . \frac{dp}{dr} + p . \rho^{-\gamma - 1} . (-\gamma) . \frac{d\rho}{dr} = 0$$

or
$$\frac{1}{\rho^r} . \frac{dp}{dr} = r p \rho^{-r-1} . \frac{d\rho}{dr}$$

or
$$\frac{d\rho}{dr} = \frac{\rho}{\gamma p} \frac{dp}{dr} \qquad ...(3)$$

From Eqs. (2) and (3), we have

$$g \frac{dh}{dr} = c_a \frac{dc_a}{dr} + c_w \frac{dc_w}{dr} + \left(\frac{\gamma}{\gamma - 1}\right) \left[\frac{1}{\rho} \frac{dp}{dr} - \frac{p}{\rho^2} \times \frac{\rho}{\gamma p} . \frac{dp}{dr}\right]$$

$$= c_a \frac{dc_a}{dr} + c_w \frac{dc_w}{dr} + \left(\frac{\gamma}{\gamma - 1}\right) \left[\frac{1}{\rho} \frac{dp}{dr}\right] \left(\frac{\gamma - 1}{\gamma}\right)$$

$$= c_a \frac{dc_a}{dr} + c_w \frac{dc_w}{dr} + \frac{1}{\rho} \cdot \frac{dp}{dr}$$

$$= c_a \frac{dc_a}{dr} + c_w \frac{dc_w}{dr} + \frac{c_w^2}{r} \qquad \left[\because v_t = c_w \text{ and } \frac{1}{\rho}\frac{dp}{dr} = \frac{v_t^2}{r}\right] \quad ...(4)$$

Eq. (4) is the basic equation for all vortex flow.

Free vortex flow

1. $\frac{dh}{dr} = 0$

2. Axial velocity is constant at all radii. Thus $\frac{dc_a}{dr} = 0$

Integrating, we get

$$c_w \times r = \text{const}$$

i.e. the which velocity varies inversely as the radius. This is known as the free vortex flow.

(*b*) A centrifugal compressor has a pressure ratio of 4/1 with an isentropic efficiency of 82% when running at 16000 rpm. It takes in air at 17°C. Guide vanes at inlet give the air a pre-whirl of 20°C to the axial direction at all radii and the mean diameter of the eye is 200 mm; the absolute air velocity at inlet is 120 m/s. At exit the blades are radially inclined and the impeller tip diameter is 550 mm. Calculate the slip factor of the compressor.

(*c*) Describe with a neat sketch the working of a turbojet engine.

Sol.

Given : $r_p = 4$, $\eta_{isen} = 0.82$, N= 16000 rpm, $T_1 = 273 + 17 = 290$ K,

prewhirl = 20°, $D_m = 200$ mm, $v_{a1} = 120$ m/s, $\beta_2 = 90°$, $D_t = 550$ mm

Temperature after isentropic compression (Fig. 8),

$$T_2 = T_1 \left(\frac{p_2}{p_1}\right)^{\frac{\gamma-1}{\gamma}} = 290\,(4)^{\frac{0.4}{1.4}} = 430.94 \text{ K}$$

Isentropic temperature rise,

$$(\Delta T)_{isen} = T_2 - T_1$$

$$= 430.94 - 290 = 140.94 \text{ K}$$

Actual temperature rise, $(\Delta T)_{act} = \frac{(\Delta T)_{isen}}{\eta_{isen}}$

$$= \frac{140.94}{0.82} = 171.88 \text{ K}$$

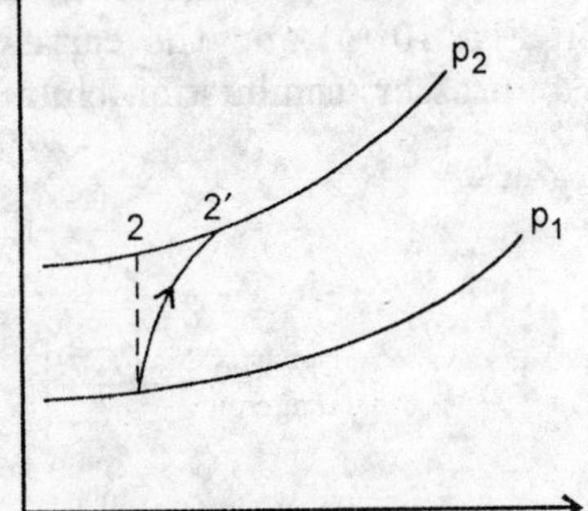

Fig 8. T – S Diagram

Power input per unit air mass flow rate

$$= c_p \times (\Delta T)_{act}$$

$$= 1.005 \times 171.88 = 172.74 \text{ kJ/kg}$$

Now $$u_1 = \frac{\pi D_m N}{60 \times 10^3} = \frac{\pi \times 200 \times 16000}{60 \times 10^3} = 167.58 \text{ m/s}$$

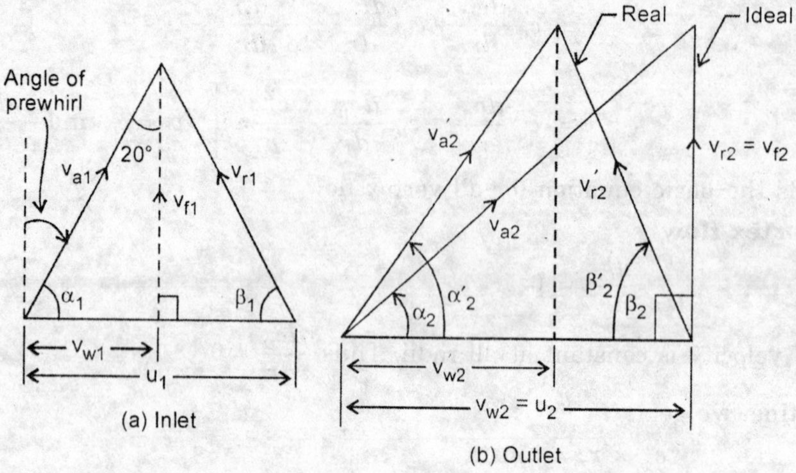

Fig. 9. Velocity triangles

From inlet velocity triangle (Fig. 9),

$$v_{w1} = v_{a1} \sin 20° = 120 \sin 20° = 41.04 \text{ m/s}$$

At impeller exit, $$u_2 = \frac{\pi D_t N}{60 \times 10^3} = \frac{\pi \times 550 \times 16000}{60 \times 10^3} = 460.77 \text{ m/s}$$

For radial discharge, $v_{w2} = u_2 = 460.77$ m/s

With prewhirl, power inut per unit mass flow rate $= u_2 v'_{w2} - u_1 v_{w1} = 172.74 \times 10^3$

$$460.77\, v'_{w2} - 167.58 \times 41.04 = 172.74 \times 10^3$$

$$v'_{w2} = 389.82 \text{ m/s}$$

$$\text{Slip factor, } \phi_s = \frac{v'_{w2}}{v_{w2}} = \frac{389.82}{460.77} = 0.846$$

(*c*) **Turbojet Engine**

The line diagram of the turbojet unit is shown in Fig. 10. (*a*), the schematic arrangments in Fig. 10 (*b*) and the corresponding T-s diagram in Fig. 10 (*c*). It consists of a diffuser, compressor, combustion chamber, turbine and exhaust nozzle.

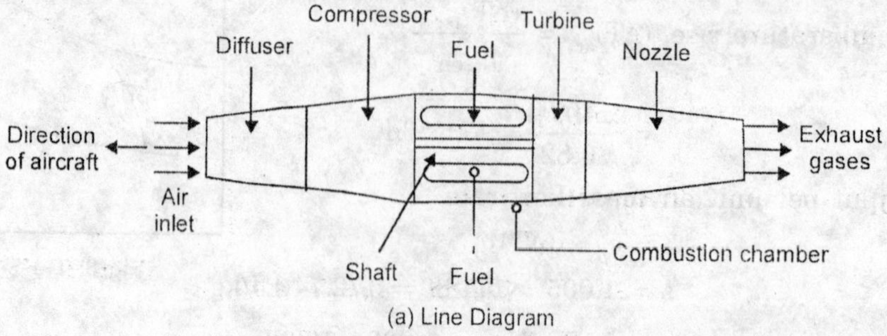

(a) Line Diagram

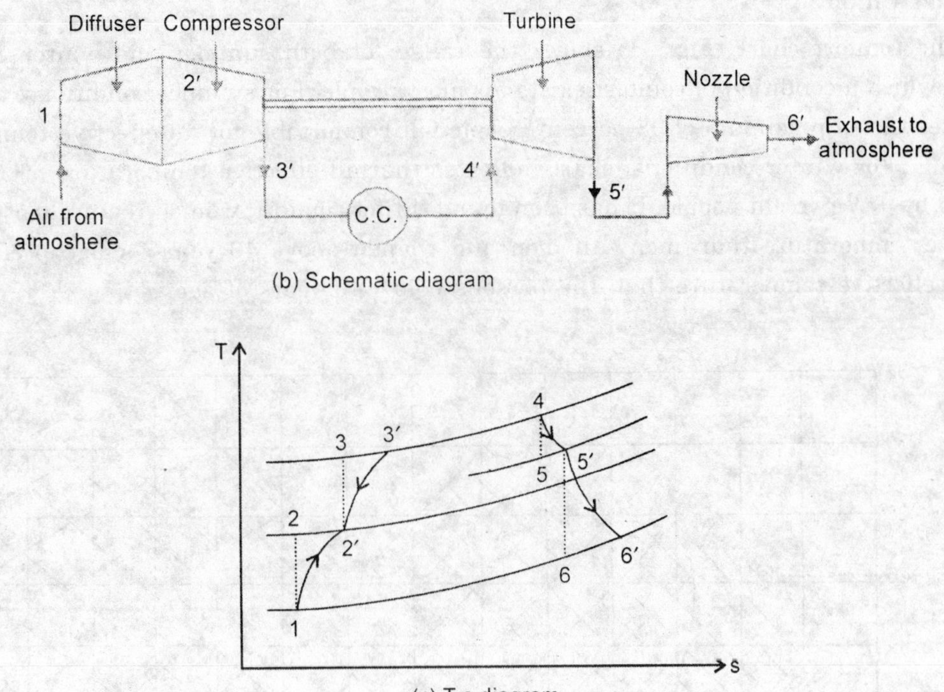

(b) Schematic diagram

(c) T-s diagram

Fig. 10. The trubo-jet engine.

The function of the diffuser is to convert the kinetic energy of the entering air into a static pressure rise which is achieved by the ram effect. After this air enters the axial flow compressor and combustion chamber. The gas turbine produces just sufficient power to drive the compressor by partial expansion in the turbine. The gases coming out from the turbine are expanded in the nozzle and produce a very high velocity jet which gives a forward motion to the aircraft by the jet reaction.

The advantages of a turbo-jet engine are its simpler construction, no vibrations, uninterrupted and smooth operation, lesser weight to power ratio, higher rate of climb, higher propulsion efficiency at higher speeds. However, its disadvantages are : less efficient, shorter life, more noisy, expensive, larger length of take- off required, and low thrust at take-off.

It is suitable for piloted operation of aircraft travelling above 800 km/h.

Q.6. (***a***) What is effective temperature? Sketch a comfort chart and explain its signifance.

Sol.

Effective Temperature

The degree of warmth or cold felt by a human body depends on : (*i*) DBT (*ii*) RH, and (*iii*) Air velocity.

In order to evaluate the combined effect of these feactors, the term effective temperature is employed. It is defined as that index which correlates the combined effects of air temprature relative humidity and air velocity on the human body.

Comfort Chart

The comfort chart (Fig. 11) shows the range for both summer and winter condition within which a condition of comfort exists for most people. For summer conditions, the chart indicates that a maximum of 98 percent people felt comfortable for an effective temperature of 21.6°C. For winter conditions, chart indicates that an effective temperature of 20°C was desired by 97.7 percent people. It has been found tht for comfort, women require 0.5°C higher effective temperature than men. All men and women above 40 years of age prefer 0.5°C higher effective temperature than the persons below 40 years of age.

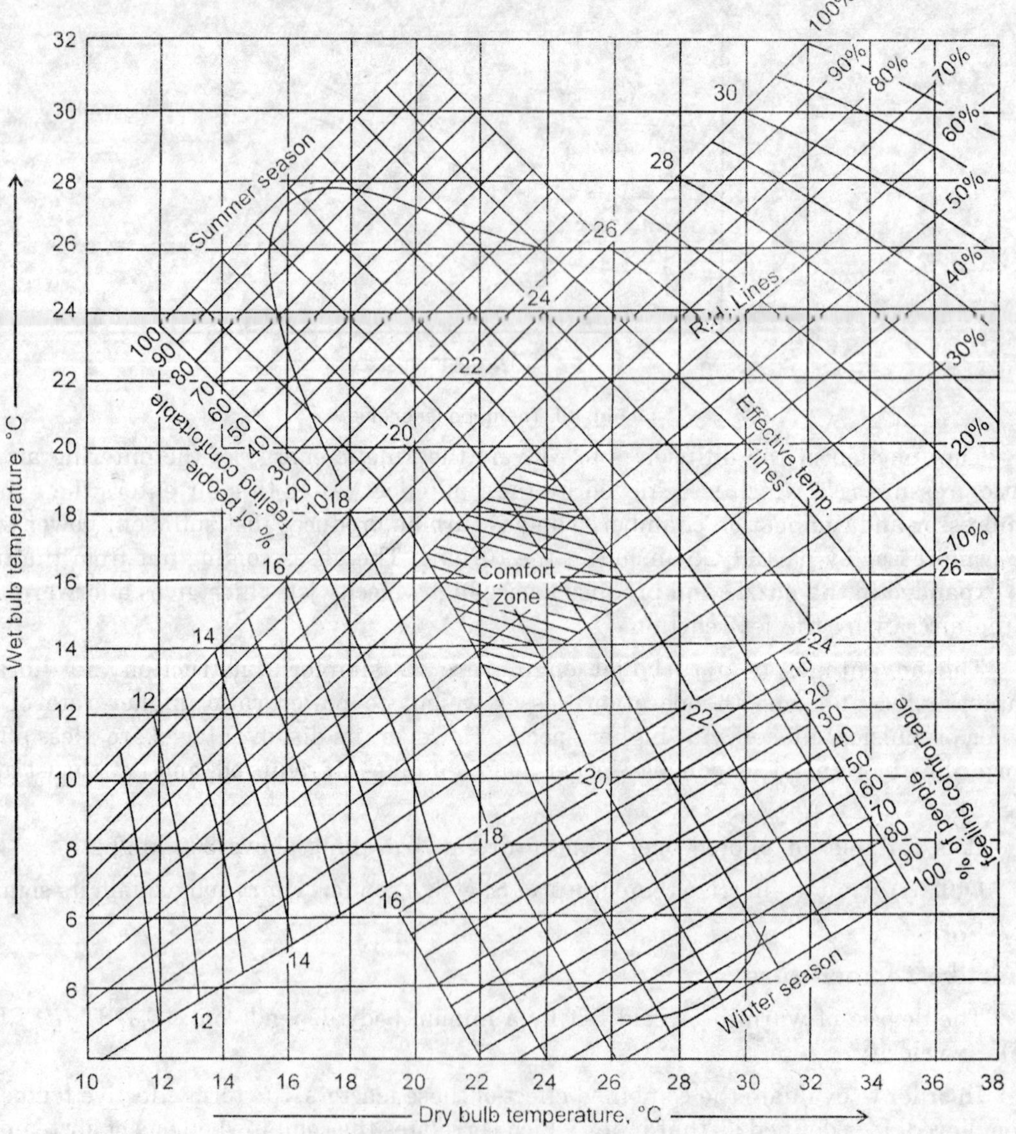

Fig. 11. Comfort chart for still air (air velocities from 5 to 8 m/min).

Q.6. (*b*) The properties of air in a room are: temerature 35°C and relative humidity 70 per cent. Determine (*i*) the partial pressure of air, (*ii*) the humidity ratio, (*iii*) the saturation ratio, (*iv*) dew point temperature, (*v*) density of the air-vapour mixture, (*vi*) the enthalpy of mixture per kg of dry air. Assume room pressure to be 1.0132 bar and the specific heat for the super-heated vapour as 1.863 kJ/kg K. Solve the problem with the help of steam tables.

Sol.

Given : Room conditions: t = 35°C, ϕ = 0.70, p_t = 1.0132 bar,

$$(c_p)_{sup} = 1.863 \text{ kJ/kg. K}$$

From steam tables, saturation pressure at 35°C, p_s = 0.05628 bar

Partial pressure of vapour, $p_v = \phi\, p_s = 0.7 \times 0.5628 = 0.394$ bar

Humidity ratio, w = weight of water vapour/kg of d.a.

$$= \frac{0.622\, p_v}{p_t - p_v} = \frac{0.622 \times 0.0394}{1.0132 - 0.0394} = 0.02517 \text{ kg/kg of d.a.}$$

$$\text{Saturation ratio} = \frac{\text{mass of water vapour/kg of d.a.}}{\text{mass of water vapour/kg of dry air at saturation condition}}$$

$$= \phi\left[\frac{p_t - p_s}{p_t - \phi . p_s}\right] = 0.7\left[\frac{1.0132 - 0.05628}{1.0132 - 0.7 \times 0.05628}\right] = 0.6879$$

Dew point temperature,

DPT = Saturation temperature corresponding to partial pressure of water vapour, p_v

= 28°C

Partial pressure of air,

$$p_a = p_t - p_v = 1.0132 - 0.0394 = 0.9738 \text{ bar}$$

Density of air, $$\rho_a = \frac{p_a}{R_a T} = \frac{0.9738}{0.287 \times (273 + 35)} = 1.102 \text{ kg/m}^3$$

Density of vapour, $$\rho_v = \frac{p_v}{R_v T} = \frac{0.394 \times 10^5}{461 \times 308} = 0.028 \text{ kg/m}^3$$

Density of mixture, $\rho = \rho_a + \rho_v = 1.13$ kg/m^3

Enthalpy of mixture, $h = h_a + h_w$

$$= c_{pa} \times \text{DBT} + w\, [c_w \times \text{DPT} + h_{fg} + (c_p)_{sup}\, (\text{DBT} - \text{DPT})]$$

$$= 1.005 \times 35 + 0.02517\, [4.18 \times 28 + 2450 + 1.863\, (35 - 28)]$$

$$= 100.195 \text{ kJ/kg of d.a.}$$

(*c*) A Freon-12 refrigerator producing a cooling effect of 20 kJ/s operates on a simple cycle with pressurre limits of 1.509 bar and 9.607 bar. The vapour leaves the evaporator dry saturated and there is no undercooling. Determine the power required by the machine.

If the compressor operates at 3000 rpm and has a clearance volume of 3% of stroke volume determine the piston displacement of the compressor. For compressor assume that the expansion follows the law $pv^{1.13}$ = constant.

Given :

Temperature °C	p_s bar	v_g m³/kg	Enthalpy kJ/kg		Entropy kJ/kg K		Specific heat kJ/kg K C_{ps}
			h_f	h_g	s_f	s_g	
– 20	1.509	0.1088	17.8	178.61	0.073	0.7082	
40	9.607		74.53	203.05	0.2716	0.682	0.747

Sol.

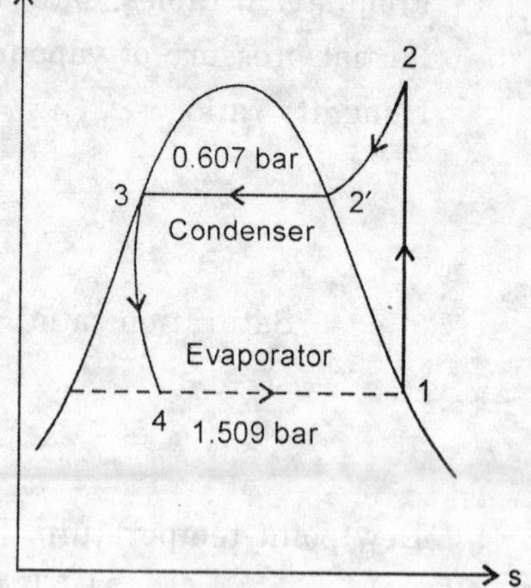

Fig 12.

Refrigeration effect, $R = \dot{m}(h_1 - h_4) = \dot{m}(h_1 - h_{f3})$

$$\dot{m} = \frac{20}{178.61 - 74.53}$$

$$= 0.192 \text{ kg/s}$$

For isentropic process 1 – 2, we have

$$s_2 = s_1$$

$$s'_{g2} + c_{ps} \ln\left(\frac{T_2}{T_2'}\right) = s_{g1}$$

$$0.682 + 0.747 \ln\left(\frac{T_2}{273 + 40}\right) = 0.7082$$

$$T_2 = 324.17 \text{ K}$$

$$h_2 = h_2' + c_{ps} (T_2 - T_2')$$

$$= h'_{g2} + c_{ps} (T_2 - T_2')$$

$$= 203.05 + 0.747 (324.17 - 313) = 211.4 \text{ kJ/kg}$$

Work done $= \dot{m}(h_2 - h_1) = 0.192 (211.4 - 178.61) = 6.296$ kW

Volumetri efficiency of compressor,

$$\eta_{vol} = 1 + c - c\left(\frac{p_2}{p_1}\right)^{1/n}, c = 0.03$$

$$= 1 + 0.03 - 0.03\left(\frac{9.607}{1.509}\right)^{1/1.13} = 0.8756 \text{ or } 87.56\%$$

Volume of refrigerant at the in take condition 1 is :

$$\dot{V}_1 = \dot{m} \times v_{g1} = 0.192 \times 0.1088 = 0.0209 \text{ m}^3/\text{s}$$

Swept volume, $\dot{V}_s = \frac{\dot{v}_1}{\eta_{vol}} = \frac{0.209}{0.8756} = 0.239 \text{ m}^3/\text{s}$

Piston displacement per stroke, $V = \frac{\dot{V}_s \times 60}{N} = \frac{0.0239 \times 60}{3000} = 4.77 \times 10^{-4} \text{ m}^3$

Q.7. (*a*) Derive the equation for temperature distribution in a fin with an insulated tip. Then obtain an expression for heat dissipation by integrating the convective losses along its surface.

Sol.

Heat dissipation from a fin insulated at the tip:

Fig. 13 illustrates a fin of finite length with insulated end (*i.e.* no heat loss from the end of the fin).

The general form of differential equation for a fin is,

$$\frac{d^2\theta}{dx^2} - m^2\theta = 0 \quad ...(1)$$

where $\theta = t - t_a$

$$m = \sqrt{\frac{hP}{kA}}$$

General solution of Eq. (1) is

$$\theta = C_1 e^{mx} + C_2 e - mx \quad ...(2)$$

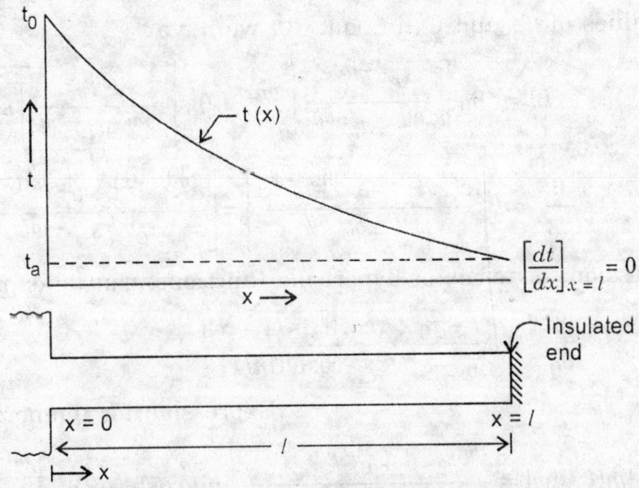

Fig. 13. The fin with insulated end (Case II).

The boundary conditions are:

(*i*) At x = 0, $\theta = \theta_0$

(*ii*) A $x = l$, $\frac{dt}{dx} = 0$

Applying these boundary conditions to eqn. (2), we have

$$C_1 + C_2 = \theta_0$$

Further $t - t_a = C_1 e^{mx} - C_2 e^{-mx}$ ...(*i*)

$$\frac{dt}{dx} = m C_1 e^{ml} - m C_2 e^{-ml}$$

$$\left[\frac{dt}{dx}\right]_{x = l} = m C_1 e^{ml} - m C_2 e^{-ml} = 0$$

$\therefore$ $C_1 e^{ml} - C_2 e^{-ml} = 0$ [As per boundary condition (*ii*)]

Solving eqns. (*i*) and (*ii*), we have

$$C_2 = \theta_0 - C_1 \quad \text{...[From eqn. } (i)]$$

$$C_1 e^{ml} - (\theta_0 - C_1) e^{-ml} = 0$$

or $C_1 e^{ml} - e^{-ml} + C_1 e^{-ml} = 0$

or $$C_1 (e^{ml} + e^{-ml}) = \theta_0 e^{-ml}$$

or $$C_1 = \theta_0 \left[\frac{e^{-ml}}{e^{ml} + e^{-ml}}\right]$$

$\therefore$ $$C_2 = \theta_0 - \left[\theta_0 \left\{\frac{e^{-ml}}{e^{ml} + e^{-ml}}\right\}\right] \quad \text{...[From eqn. } (i)]$$

or $$C_2 = \theta_0 \left[1 - \frac{e^{-ml}}{e^{ml} + e^{-ml}}\right] = \theta_0 \left[\frac{e^{ml}}{e^{ml} + e^{-ml}}\right]$$

Inserting the values of C_1 and C_2 in eqn. (2), we have

$$\theta = \theta_0 \left[\frac{e^{-ml}}{e^{ml} + e^{-ml}}\right] e^{mx} + \theta_0 \left[\frac{e^{ml}}{e^{ml} + e^{-ml}}\right] e^{-mx}$$

or $$\frac{\theta}{\theta_0} = \left[\frac{e^{m(x-1)} + e^{m(l-x)}}{e^{ml} + e^{-ml}}\right] = \left[\frac{e^{m(l-x)} + e^{[-m(l-x)]}}{e^{ml} + e^{-ml}}\right]$$

The above expression, in terms of hyperbolic functions, can be expressed as

$$\frac{\theta}{\theta_0} = \frac{t - t_0}{t_0 - t_a} = \frac{\cosh \{m(1 - x)\}}{\cosh (ml)} \quad \text{...(3)}$$

...Expression for temperature distribution.

$$\left[\because \cosh \{m(l - x)\} = \frac{e^{m(l-x)} + e^{[-m(l-x)]}}{\cosh (ml)}, \text{ and } \cosh (ml) = \frac{e^{ml} + e^{-ml}}{2}\right]$$

The rate of heat flow from the fin is given by:

$$Q_{fin} = -k A_{cs} \left[\frac{dt}{dx}\right]_{x=0}$$

Now, $$t - t_0 = (t_0 - t_a)\left[\frac{\cosh \{m(l - x)\}}{\cosh (ml)}\right] \quad \text{...[From Eq. (3)]}$$

$$\frac{dt}{dx} = (t_0 - t_a)\left[\frac{\sinh \{m(l - x)\}}{\cosh (ml)}\right](-m)$$

$$\left[\because \frac{d}{dx}[\cosh (mx)] = m \sinh (mx)\right]$$

$$\left[\frac{dt}{dx}\right]_{x=0} = -m (t_0 - t_a) \tanh (ml)$$

$$\therefore \qquad Q_{fin} = kA_{cs}\, m\,(t_0 - t_a)\tanh(ml) \text{ (Substituting for } m)$$

$$\text{or} \qquad Q_{fin} = \sqrt{PhkA_{cs}}\,(t_0 - t_a)\tanh(ml).$$

Q.7. (*b*) Define the terms NTU and effectiveness. Derive an expression for effectiveness of a counterflow heat exchanger in terms of NTU and capacity ratio.

Sol.

The counter-flow heat exchanger is shown in Fig. 14.

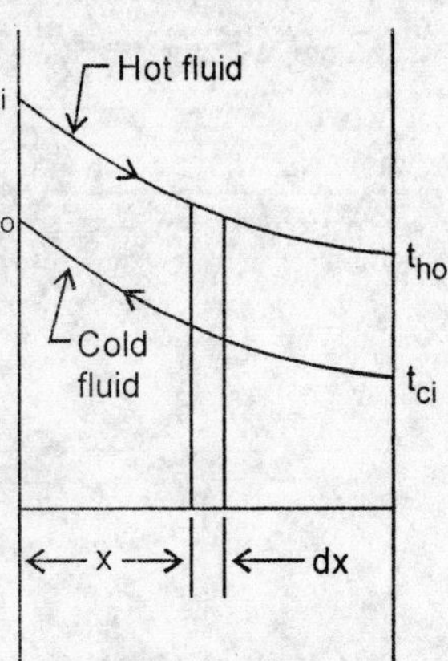

Fig 14. Counter-flow heat exchanger.

Number of Transfer Units, $\text{NTU} = \dfrac{AU}{C_{min}}$

where A = area of heat transfer

U = overall heat transfer coefficient

C_{min} = minimum heat capacity

Effectiveness of a heat exchange is defined as the ratio of actual heat transfer to the maximum possible heat transfer rate.

$$\text{Effectiveness,} \qquad \varepsilon = \frac{C_c(t_{co} - t_{ci})}{C_{min}(t_{hi} - t_{ci})} \text{ or } \frac{C_a(t_{hi} - t_{ho})}{C_{min}(t_{hi} - t_{ci})}$$

Consider an element of heat exchanger dx at a distance x,

$$\theta = t_h - t_c \qquad \text{...(1)}$$

$$d\theta = dt_h - dt_c$$

$$\text{and} \qquad d\theta = -C_h\, dt_h = -C_c\, dt_c = \text{UA}\, d\theta \qquad \text{...(2)}$$

$$\text{where} \qquad C_h = m_h\, c_{ph} \text{ and } C_c = m_c\, c_{pc}$$

$$\therefore \qquad d\theta = -\frac{dQ}{C_h} + \frac{dQ}{C_c}$$

$$= \left(\frac{1}{C_c} - \frac{1}{C_h}\right) U.dA.\theta$$

$$\text{or} \qquad \frac{d\theta}{\theta} = \left(\frac{1}{C_c} - \frac{1}{C_h}\right) U.dA$$

Integrating, we have

$$\int_{t_{hi} - t_{co}}^{t_{h0} - t_{ci}} \frac{d\theta}{\theta} = \left(\frac{1}{C_c} - \frac{1}{C_h}\right) U.\int_0^A \alpha A$$

$$\ln\left[\frac{t_{ho} - t_{ci}}{t_{hi} - t_{co}}\right] = \left(\frac{1}{C_c} - \frac{1}{C_h}\right) UA$$

$$\frac{t_{ho} - t_{ci}}{t_{hi} - t_{co}} = \exp\left[(-\text{UA})\left(\frac{1}{C_c} - \frac{1}{C_h}\right)\right] \qquad \text{...(3)}$$

Using expression for effectiveness, we have

$$t_{ho} = h_{hi} - \frac{C_{min}}{C_h}(t_{hi} - t_{ci})\,\varepsilon$$

$$t_{co} = h_{ci} + \frac{C_{min}}{C_c}(t_{hi} - t_{ci})\,\varepsilon$$

Substituting in Eq. (3), we get

$$\frac{(t_{hi} - t_{ci}) - \dfrac{C_{min}}{C_h}(t_{hi} - t_{ci})\,\varepsilon}{(t_{hi} - t_{ci}) - \dfrac{C_{min}}{C_c}(t_{hi} - t_{ci})^{e}} = \exp\left[(-\,UA)\left(\frac{1}{C_c} - \frac{1}{C_h}\right)\right]$$

or

$$\frac{1 - \dfrac{C_{min}}{C_h}\,\varepsilon}{1 - \dfrac{C_{min}}{C_c}\,\varepsilon} = \exp\left[(-\,UA)\left(\frac{1}{C_c} - \frac{1}{C_h}\right)\right]$$

$$\therefore \qquad \varepsilon = \frac{1 - \exp\left[(-\,UA)\left(\dfrac{1}{C_c} - \dfrac{1}{C_h}\right)\right]}{C_{min}\left[\left(\dfrac{1}{C_c} - \dfrac{1}{C_h}\right)\exp\left\{(-\,UA)\left(\dfrac{1}{C_c} - \dfrac{1}{C_h}\right)\right\}\right]}$$

Let $C_c > C_h$, then $C_{min} = C_c$ and $C_{max} = C_h$

$$\varepsilon = \frac{1 - \exp\left[-\dfrac{UA}{C_{min}}\left(1 - \dfrac{C_{min}}{C_{max}}\right)\right]}{1 - \dfrac{C_{min}}{C_{max}}\exp\left[\left(-\dfrac{UA}{C_{min}}\right)\left(1 - \dfrac{C_{min}}{C_{max}}\right)\right]}$$

$$= \frac{1 - \exp[-\,NTU\,(1 - CR)]}{1 - CR\exp[-\,NTU\,(1 - CR)]}$$

where $\qquad CR = \dfrac{C_{min}}{C_{max}}$ = capacity ratio.

(*c*) Show that the heat emitted per unit area from a black body in the hemisphere above the body surface is given by

$$q_b = \pi I$$

where I is the black body radiation emitted per unit time and solid angle subtended at the source, and per unit area of emitting surface normal to the mean direction in space.

Sol.

Intensity of Radiation

When a surface element emits radiation, all of it will be intercepted by a hemispherical surface placed over the element. The **intensity of radiation** (I) is defined as the *rate of energy leaving a surface in a given direction per unit angle per unit area of the emitting surface with the vertex of the cone at the centre of the sphere.* It is measured by the ratio of the spherical surface enclosed by the cone to the square of the radius of the sphere; its unit is steradian (sr). The solid angle subtended by the complete hemisphere is given by:

$$\frac{2\pi r^2}{r^2} = 2\pi$$

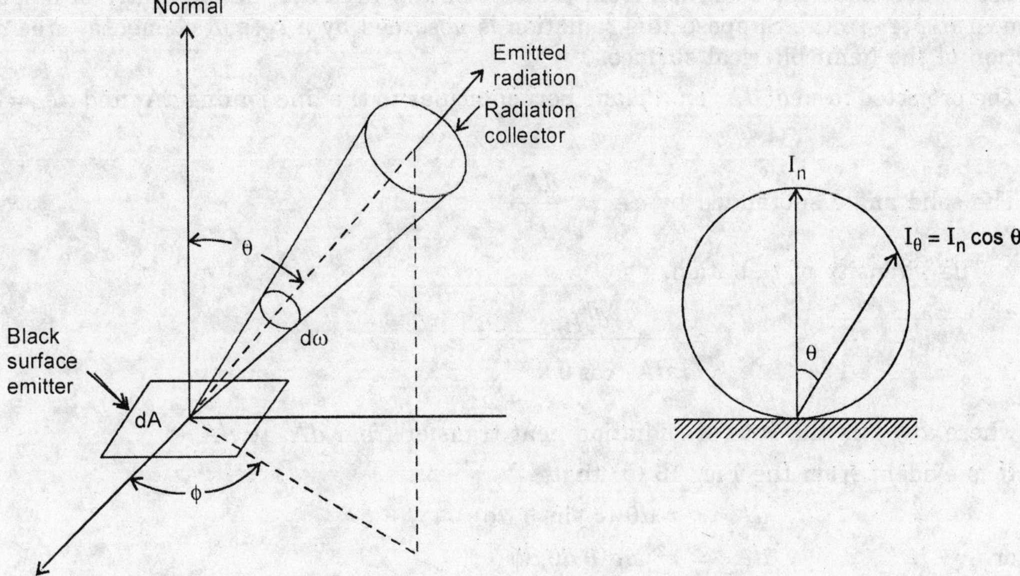

(a) Spacial distribution of radiations emitted from a surface

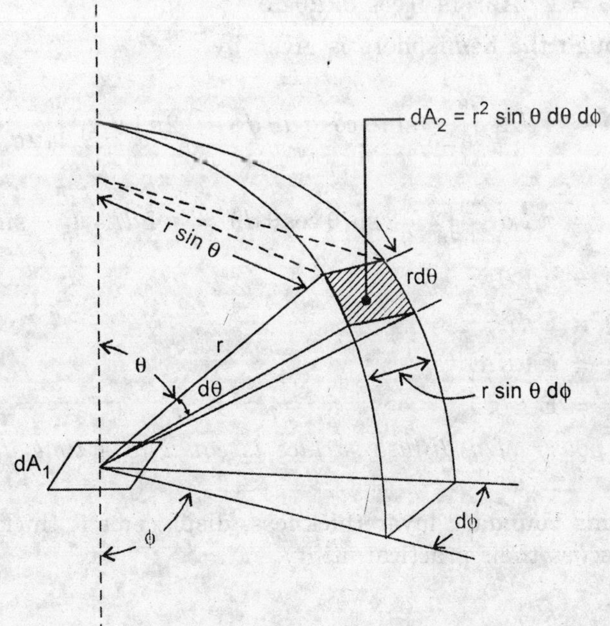

(b) Illustration for evaluating are dA_2

Fig. 15 (*a*) shows a small black surface of area *d*A (emitter) radiation in different directions. A black body radiation collector through which the radiation pass is located at an angular position characterised by *zenith angle* θ towards the surface normal and angle ϕ of a spherical coordinate system. Further the collector subtends a solid angle *d*ω when viewed from a point on the emitter.

Let us now consider radiation from the elementary area dA_1 at the centre of a sphere as shown in Fig. 15 (*b*). Suppose this radiation is absorbed by a second elemental area dA_2, a portion of the hemispherical surface.

The projected area of dA_1 on a plane perpendicular to the line joining dA_1 and $dA_2 = dA_1 \cos\theta$

The solid angle subtended by $dA_2 = \dfrac{dA_2}{r^2}$

∴ The intensity of radiation,

$$I = \frac{dQ_{1-2}}{dA_1 \cos\theta \times \dfrac{dA_2}{r^2}} \quad \text{...(1)}$$

where dQ_{1-2} is the rate of radiation heat transfer from dA_1 to dA_2

It is evident from the Fig. 15 (*b*) that,

$$dA_2 = r\, d\theta\, (r \sin\theta\, d\phi)$$

or

$$dA_2 = r^2 \sin\theta . d\theta . d\phi \quad \text{...(2)}$$

From Eqns. (1) and (2), we obtain

$$dQ_{1-2} = I\, dA_1 . \sin\theta . \cos\theta .\ d\theta . d\phi$$

The total radiation through the hemisphere is given by

$$Q = I\, dA_1 \int_{\theta=0}^{\theta=\frac{\pi}{2}} \sin\theta \cos\theta\, d\theta\, d\phi = 2\pi\, I\, dA_1 \int_{\theta=0}^{\theta=\frac{\pi}{2}} \sin\theta \cos\theta\, d\theta$$

$$= \pi\, I\, dA_1 \int_{\theta=0}^{\frac{\pi}{2}} 2\sin\theta \cos\theta\, d\theta = \pi\, I\, dA_1 \int_{\theta=0}^{\frac{\pi}{2}} \sin 2\theta\, d\theta$$

or $\quad Q = \pi\, I\, dA_1$

Also $\quad Q = E . dA_1$

∴ $\quad E\, dA_1 = \pi\, I\, dA_1$

or

$$q_b = E = \pi\, I \quad \text{...(3)}$$

i.e., The *total emissive power of a diffuse surface is equal to π time its intensity of radiation.*

Q.8. (*a*) Define the terms boundary layer thickness, displacement thickness and the momentum thickness and discuss their practical utility.

Sol.

Boundary Layer Thickness (δ)

The velocity within the boundayr layer increases from zero at the boundary surface to the velocity of the main stream asymptotically. Therefore the thickness of the boundary layer is arbitrarily defined as *that distance from the boundary in which the velocity reaches 99 per cent of the velocity of the free stream* ($u = 0.99U$). *It is denoted by the symbol δ. This definition however gives an approximate value of the boundary layer thickness* and hence δ is generally termed as **nominal thickness** *of the boundary layer.*

The boundary layer thickness for *greater accuracy* is defined in terms of certain mathematical expressions which are the measure of the boundary layer on the flow. The commonly adopted definitions of the boundary layer thickness are :

1. Displacement thickness (δ^*)
2. Momentum thickness (θ)
3. Energy thickness (δ_e).

Displacement Thickness (δ^*)

The *displacement thickness* can be defined as follow:

"It is the distance, measured perpendicular to the boundary, by which the main/free stream is displaced on account of formation of boundary layer".

or

"It is an additional "wall thickness" that would have to be added to compensate for the reduction in flow rate on account of boundary layer formation".

The displacement thickness is denoted by δ^*.

Let fluid of density ρ flow past a stationary plate with velocity U as shown in the Fig. 16. Consider an elementary strip of thickness dy at a distance y from the plate.

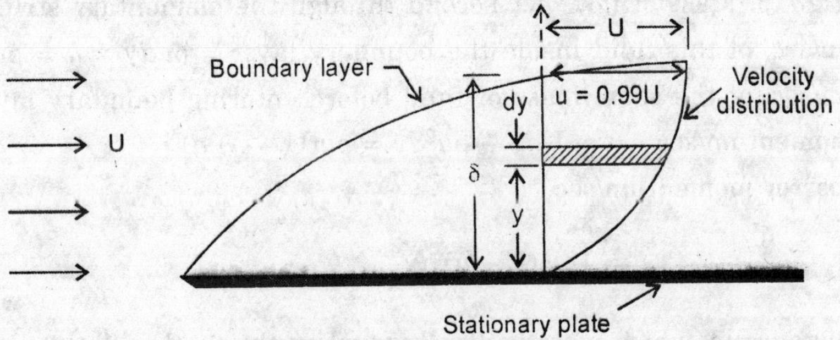

Fig. 16. Displacement thickness.

Assuming *unit width*, the mass flow per second through the elementary strip

$$= \rho u dy$$

Mass flow per second through the elementary strip (unit width) if the plate were not there

$$= \rho \, U.dy \qquad \text{...(ii)}$$

Reduction of mass flow rate through the elementary strip

$$= \rho \, (U - u) \, dy$$

[The difference (U – u) is called **velocity of defect**].

Total reduction of mass flow rate due to introduction of plate

$$= \int_0^\delta \rho \, (U - u) \, dy \qquad \text{...(iii)}$$

(if the fluid is incompressible)

Let the plate is displaced by a distance δ^* and velocity of flow for the distance δ^* is equal to the main/free stream velocity (*i.e.* U). Then, loss of the mass of the fluid/sec. flowing through the distance δ^*

$$= \rho U \delta^* \quad \text{...}(iv)$$

Equating eqns. (*iii*) and (*iv*), we get

$$\pi U \delta^* = \int_0^\delta \rho\,(U - u)\,dy$$

or,
$$\delta^* = \int_0^\delta \left(1 - \frac{u}{U}\right) dy$$

Momentum Thickness (θ)

Momentum Thickness *is defined as the distance through which the total loss of momentum per second be equal to if it were passing a stationary plate.* It is denoted by θ.

It may also be defined as the *distance, measured perpendicular to the bondary of the solid body, by which the boundayr should be displaced to compensate for reduction in momentum of the flowing fluid on account of boundary layer formation.*

Refer to Fig. 16. Mass of flow per second through the elementary strip = $\rho u dy$

Momentum/sec. of this fluid inside the boundary layer = $\rho u dy \times u = \rho u^2 dy$

Momentum/sec. of the same mass of fluid before entering boundary layer = $\rho u U dy$

Loss of momentum/sec. = $\rho u U dy - \rho u^2 dy = \rho u(U - u)\, dy$

∴ Total loss of momentum/sec.

$$= \int_0^\delta \rho u\,(U - u)\,dy$$

Let, θ = Distance by which plate is displaced when the fluid is flowing with a constant velocity U. Then loss of momentum/sec. of fluid flowing through distance θ with a velocity U

Equating eqns. (*i*) and (*ii*), we have

$$\rho \theta U^2 = \int_0^\delta \rho u\,(U - u)\,dy$$

or,
$$\theta = \int_0^\delta \frac{u}{U}\left(1 - \frac{u}{U}\right) dy$$

The momentum thickness is useful in *kinetics*.

(b) Obtain an expression for the specific speed of a hydraulic turbine and explain its significance. Give the range of specific speed values of the Kaplan, Francis turbines and Pelton wheels.

Sol.

Let N_s = specific speed of turbine

D = diameter of turbine runner

N speed of turbine runner, rpm

u = tangential velocity of runner

v = absolute velocity of water

Then $u \propto v \propto \sqrt{2gH}$

$\therefore \quad v \propto \sqrt{2gH} \propto \sqrt{H}$

Now $$u = \frac{\pi DN}{60}$$

$\therefore \quad DN \propto v \propto \sqrt{H}$

or $\quad D \propto \dfrac{\sqrt{H}}{N}$

Let Q = discharge through the turbine

B = width of turbine runner

v_f = velocity of flow

Then $Q = \pi DB\, v_f$

Now $B \,\alpha\, D$

and $v_f \,\alpha\, \sqrt{2gH} \,\alpha\, \sqrt{H}$

$$\therefore \quad Q \,\alpha\, \pi D \times D \times \sqrt{H} \,\alpha\, D^2\sqrt{H} \,\alpha \left(\frac{\sqrt{H}}{N}\right)^2 \sqrt{H} = \frac{H^{3/2}}{N^2}$$

Power produced by turbine, $P = \dfrac{rQH}{10^3} \,\alpha\, QH \,\alpha\, \dfrac{H^{5/2}}{N^2}$

or $$N^2 \,\alpha\, \frac{H^{5/2}}{P}$$

$$N \,\alpha\, \frac{H^{5/4}}{\sqrt{P}}$$

$$= N_s \frac{H^{5/4}}{\sqrt{P}}$$

$$\therefore \quad N_s = \frac{N\sqrt{P}}{H^{5/4}}$$

Range of specific speeds of turbines.

	Type of turbine	Specific speed range (S.I. Units)
1.	Pelton turbine	
	(*i*) Single jet	8.5 – 30
	(*ii*) Multiple jets	30 – 50
2.	Francis turbine	50 – 340
3.	Kaplan turbine	250 – 1025

(c) A centrifugal pump has an impeller with inner and outer diameters of 150 and 250 mm respectively. The pump delivers 0.05 m³/s of water when running at 1500 rev/min. The velocity of flow through the impeller remains constant at 2.5 m/s.

The blades are curved back at an angle of 30° to the tangent at exit and the inlet is axial. The diameters of the suction and delivery pipes are 150 mm and 100 mm respectively. The pressure head at suction is 4 m below and that at delivery is 18 m above atmosphere. The power required to drive the pump is 18 kW.

Find

(*i*) the vane angle at inlet

(*ii*) the overall efficiency, and

(*iii*) the manometric efficiency.

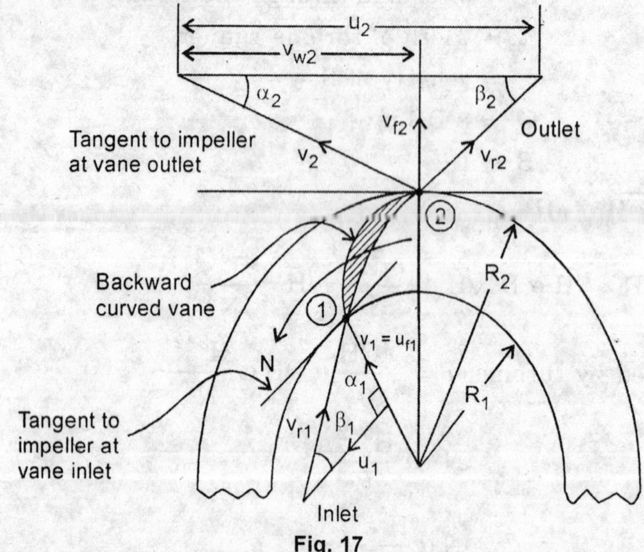

Fig. 17

Sol. Given :

$$D_1 = 150 \text{ mm}, D_2 = \text{ mm},$$

$$Q = 0.05 \text{ m}^3/\text{s}, N = 1500 \text{ rpm},$$

$$v_f = 2.5 \text{ m/s}, d_s = 150 \text{ mm},$$

$$d_d = 100 \text{ mm}, H_s = 4 \text{ m},$$

$$H_d = 18 \text{ m},$$

$$P = 18 \text{ kW}, \beta_2 = 30°, \alpha_1 = 90°$$

$$u_1 = \frac{\pi D_1 N}{10^3 \times 60} = \frac{\pi \times 150 \times 1500}{10^3 \times 60} = 11.78 \text{ m/s}$$

$$u_2 = \frac{11.78 \times 250}{150} = 19.63 \text{ m/s}$$

$$\tan \beta_1 = \frac{v_{f1}}{u_1} = \frac{2.5}{19.78} = 0.21222$$

$$\beta_1 = 11.98°$$

Now $$v_{f1} = v_{f2} = v_f = 2.5 \text{ m/s}$$

Velocity of water in suction pipe,

$$v_s = \frac{4Q}{\pi d_s^2} = \frac{4 \times 0.05}{\pi (0.150)^2} = 2.83 \text{ m/s}$$

Velocity of water in delivery pipe,

$$v_d = \frac{4Q}{\pi d_d^2} = \frac{4 \times 0.5}{\pi (0.1)^2} = 6.37 \text{ m/s}$$

Applying Bernoulli's theorem at inlet (point 1) and outlet (point 2) of impeller, we have

$$\frac{p_2}{\rho} + \frac{v_2^2}{2} = \frac{p_1}{\rho} + \frac{v_1^2}{2} + g.\text{H}_{mano}$$

$$g.\text{H}_{mano} = \frac{p_2 - p_1}{\rho} + \frac{1}{2}(v_2^2 - v_1^2)$$

Now $$\frac{p_1}{\rho g} = \text{Hatm} - 4$$

$$= 10.3 - 4 = 6.3 \text{ m}$$

$$\frac{p_2}{\rho g} = \text{H}_{atm} + 18 = 10.3 + 18 = 28.3 \text{ m}$$

$$\therefore \quad \text{H}_{mano} = (28.3 - 6.3) + \frac{1}{2 \times 9.81}[(6.37)^2 - (2.83)^2] = 23.66 \text{ m}$$

Overall efficiency, $$\eta_0 = \frac{\rho g Q \text{H}_{mano}}{\text{P}}$$

$$= \frac{1000 \times 9.81 \times 0.05 \times 23.66}{18000} = 0.6447 \text{ or } 64.47\%$$

Menometric efficiency,

$$\eta_{mano} = \frac{g\text{H}_{mano}}{u_2 v_{w2}}$$

$$v_{w2} = u_2 - \frac{v_f}{\tan \beta_2} = 19.63 - \frac{2.5}{\tan 30°} = 15.3 \text{ m/s}$$

$$= \frac{9.81 \times 23.66}{19.63 \times 15.3} = 0.7728 \text{ or } 77.28\%$$

MECHANICAL ENGINEERING
PAPER I—1997

(Conventional)

Q.1. ***(a)*** An ideal gas is heated from temperature T_1 to T_2 by keeping its volume constant. The gas is expanded back to its initial temperature according to the law pv^n = constant. If the entropy changes in the two processes are equal, find the value of n in terms of the adiabatic index γ.

Sol.

Change in entropy during constant volume process,

$$(\Delta S)_v = mc_v \ln \frac{T_2}{T_1}$$

Change in entropy during polytropic process (pv^n = const),

$$(\Delta S)_{poly} = \left[\frac{\gamma - n}{n - 1}\right].mc_v \ln \frac{T_2}{T_1}$$

For $(\Delta S)_v = (\Delta S)_{poly}$, we have

$$\frac{\gamma - n}{n - 1} = 1$$

$$n = \frac{\gamma + 1}{2}$$

(b) A cylinder contains one kg of water and steam at a pressure of 3.8 bar and 0.4 dry. Heat is supplied at constant volume until the pressure reaches to 10 bar. The steam is then expanded according to the law pv = constant, until the pressure is 2 bar.

Calculate

(i) Heat transfer during constant volume heating.

(ii) Heat transfer during pv = constant expansion.

(iii) Temperature of steam after the expansion.

Sol.

Given : m = 1 kg, p_1 = 3.8 bar, x_1 = 0.4, p_2 = 10 bar, p_3 = 2 bar

At p_1 = 3.8 bar, v_{g1} = 0.485 m^3/kg

p_2 = 10 bar, v_{g2} = 0.194 m^3/kg

Process 1 – 2 : v = const.

Initial volume of steam, $v_1 = x_1 v_{g1} = 0.4 \times 0.485 = 0.194$ m^3/kg

$$v_2 = v_1$$

$$v_2 = v_1$$

$$x_2\, v_{g2} = v_1$$

$$x_2 \times 0.194 = 0.194$$

$$x_2 = 1$$

∴ Steam at 2 is saturated.

Saturation temperature, t_{s2} = 179.9°C

Internal energy of steam at 1, $u_1 = h_1 - p_1\, x_1\, v_{g1}$

$$h_{f1} = 596.8 \text{ kJ/kg},\ h_{fg1} = 2139.4 \text{ kJ/kg}$$

$$h_1 = h_{f1} + x_1\, h_{fg1} = 596.8 + 0.4 \times 2139.4 = 1452.56 \text{ kJ/kg}$$

$$u_1 = 1452.56 - 3.8 \times 10^2 \times 0.4 \times 0.485 = 1378.84 \text{ kJ/kg}$$

$$u_2 = h_2 - p_2\, x_2\, v_{g2} = h_{g2} - p_2\, v_{g2}$$

$$= 2778.1 - 10 \times 10^2 \times 0.194 = 2584.1 \text{ kJ/kg}$$

Heat supplied during process 1 – 2, $q_{1-2} = u_2 - u_1 = 2584.1 - 1378.84 = 1205.26$ kJ/kg

$$w_{1-2} = 0$$

Process 2 – 3 : pv = const

Now

$$h_2 = 2778.1 \text{ kJ/kg}$$

$$p_3 = 2 \text{ bar},\ t_{s3} = 120.2°\text{C},$$

$$h_3 = h_{g3} = 2706.7 \text{ kJ/kg}$$

$$p_2\, x_2\, v_{g2} = p_3\, x_3\, v_{g3}$$

$$x_3 = \frac{10 \times 0.194}{2 \times 0.886} = 1.095$$

∴ Steam is superheated at 3.

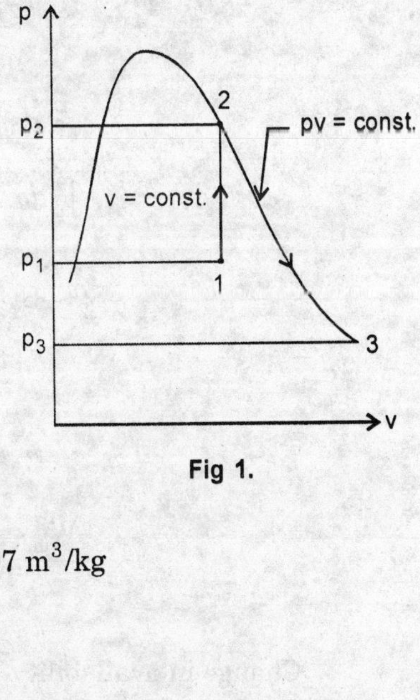

Fig 1.

$$\frac{p_2\, v_{g2}}{T_2} = \frac{p_3\, v_3}{(T_{sup})_3}$$

But

$$p_2\, v_{g2} = p_3\, v_3,\ v_3 = \frac{10 \times 0.194}{2} = 0.97 \text{ m}^3\text{/kg}$$

∴

$$(T_{sup})_3 = T_2$$

$$u_3 - u_2 = (h_3 - p_3\, v_3) - (h_2 - p_2\, v_{g2})$$

$$= (2706.7 - 2 \times 10^2 \times 0.97) - (2778. - 10 \times 10^2 \times 0.194)$$

$$= 2512.7 - 2584.1 = -71.4 \text{ kJ/kg}$$

$$w_{2-3} = p_2 v_2 \ln \frac{v_3}{v_2} = 10 \times 10^2 \times 0.194 \ln \left(\frac{0.97}{0.194}\right) = 312.23 \text{ kJ/kg}$$

$$q_{2-3} = (h_3 - h_2) + p_2 v_2 \ln \frac{v_3}{v_2}$$

$$= (2706.7 - 2778.1) + 312.23 = 240.83 \text{ kJ/kg}$$

(c) An adiabatic cylinder of volume 10 m^3 is divided into two compartments A and B, each of volume 6 m^3 and 4 m^3 respectively, by a thin sliding partion. Initially the compartment A is filled with air at 6 bar and 600 K, whilst there is a vacuum in the compartment B. Suddenly the partition is removed, the fluid in compartment A expands and fills both compartments. Caculate the loss in available energy, Assume atmosphere is at 1 bar and 300 K.

Sol.

Refer to Fig. 2.

A
6 m^3
6 bar, 600 K

B
4 m^3
Vacuum

Fig 2.

Mass of air, $m = \dfrac{p_1 V_1}{RT_1}$

$$= \frac{6 \times 10^2 \times 6}{0.287 \times 600} = 20.9 \text{ kg}$$

For sudden expansion of air,

$$T_2 = T_1 \left(\frac{V_1}{V_2}\right)^{\gamma - 1} = 600 \left(\frac{6}{4}\right)^{0.4}$$

$$= 489.116 \text{ K}$$

$$T.ds = dU + p \, . \, dV$$

$$\int_1^2 ds = \int_1^2 mc_v \frac{dT}{T} + \int_1^2 mR . \frac{dV}{V}$$

$$S_2 - S_1 = mc_v \ln \frac{T_2}{T_1} + mR \ln \frac{V_1}{V_2}$$

$$= 20.9 \left[0.718 \ln \left(\frac{489.116}{600} \right) + 0.287 \ln \left(\frac{10}{6} \right) \right]$$

$$= -2.13 \times 10^{-3} \text{ kJ/K}$$

$$W_{max} = (U_1 - U_2) + T_0(S_1 - S_2) = mc_v(T_1 - T_2) + T_0(S_1 - S_2)$$

$$= 20.9 \times 0.718 \, (600 - 489.116) + 300 \times 2.13 \times 10^{-3}$$

$$= 1664.6 \text{ KJ}$$

Change in availability $= \Phi_1 - \Phi_2 = W_{max} + p_0(V_1 - V_2)$

$$= 1664.6 + 1 \times 10^2(6 - 10) = 1264.6 \text{ kJ}$$

Irreversibility, $I = W_{max} - W_{actual} = -\Delta U - W_{actual} = T_0(S_2 - S_1)$

$$= -300 \times 2.13 \times 10- 3 = -0.639 \text{ kJ}$$

Loss in available energy = 0.639 kJ

Q.2. **(a)** The factors that tend to increase detonation in SI Engine tend to reduce knocking in CI engine. Discuss this statement with reference to the following influencing factors :

(*i*) compression ratio

(*ii*) inlet temperature

(*iii*) self-ignition temperature of fuel

(*iv*) time lag of ignition of fuel, and

(*v*) combustion chamber wall temperature.

Sol.

(*i*) **Compression ratio :** Increase in compression ratio increases the pressure and temperature of the gases at the end of compression stroke which decreases ignition lag of the end gas and increases the tendency of knocking.

(*ii*) **Mass of induced charge :** An increase in mass of induced charge into the cylinder of an engine by throttling or by increasing the amount of supercharging increases both temperature and density of the charge at the time of ignition. This increases the tendency for knocking.

(*iii*) **Inlet temperature of mixture :** Increase in inlet temperature of mixture makes the compression temperature higher thereby, increasing the tendency of knocking.

(*iv*) **Temperature of combustion chamber walls :** Hot spots (like spark plug and exhust value) in combustion chamber promotes knocking.

(*v*) **Retarding spark timing :** Having the spark closer to TDC reduces knocking. However this effects the brake torque and power output of the engine.

(*vi*) **Power output of the engine :** A decrease in the output of the engine reduces the tendency to knock.

Characteristics Tending to Reduce Detonation or knock

S.No.	Characteristics	SI Engines	CI Engines
1.	Ignition temperature of fuel	High	Low
2.	Ignition delay	Long	Short
3.	Compression ratio	Low	High
4.	Inlet temperature	Low	High
5.	Inlet pressure	Low	High
6.	Combustion wall temperature	Low	High
7.	Speed, rpm	High	Low
8.	Cylinder size	Small	Large

(***b***) Derive an expression for air standard efficiency of the following cycle in terms of compression ratio, r, c_v and γ.

(*i*) an isothermal compression, compression ratio, γ

(*ii*) an increase of pressure at constant volume

(*iii*) an adiabatic expansion

Sol.

Refer to Fig. 3 for the p – V diagram.

Process 1 – 2 : Isothermal compression.

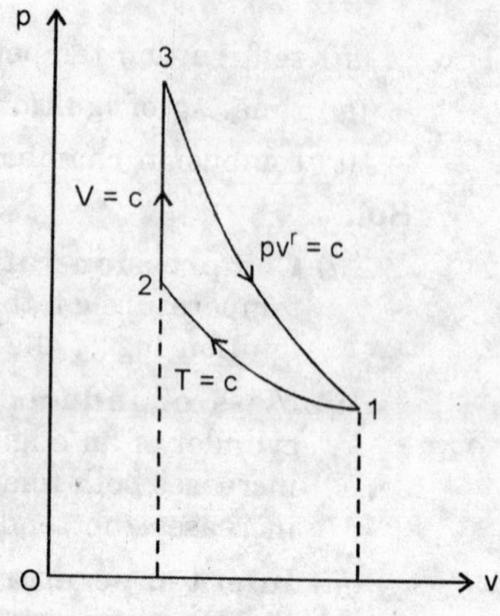

Fig 3.

$$W_{1-2} = Q_{1-2} = mRT_1 \ln \frac{V_1}{V_2}$$

$$= mRT_1 \ln r$$

Process 2 – 3 : Constant volume process.

$$W_{2-3} = \int_2^3 p.dV = 0$$

Now

$$T_2 = T_1$$

$$Q_{2-3} = U_3 - U_2$$

$$= mc_v (T_3 - T_2)$$

$$= mc_v (T_3 - T_1)$$

Process 3 – 1 : Adiabatic expansion.

$$W_{3-1} = mc_v (T_3 - T_1), T_3 = T_1 \left(\frac{V_1}{V_3}\right)^{\gamma-1} = T_1 \left(\frac{V_1}{V_2}\right)^{\gamma-1} = T_1 r^{\gamma-1}$$

$$Q_{3-1} = 0$$

Air standard efficiency, $\eta_a = 1 - \frac{Q_{2-3} + Q_{3-1}}{Q_{1-2}} = 1 - \frac{mc_v (T_3 - T_1)}{mRT_1 \ln r} = 1 - \frac{c_v (T_3 - T_1)}{RT_1 \ln r}$

$$= 1 - \frac{c_v T_1 (r^{\gamma-1} - 1)}{RT_1 \ln r} = 1 - \frac{r^{\gamma-1} - 1}{(\gamma - 1) \ln r}$$

(*c*) During the trial of a single-acting oil engine, cylinder diameter 20 cm, stroke 28 cm, working on the two-stroke cycle and firing every cycle, the following observations were made :

Duration of trial, 1 hour

Total fuel used, 4.22 kg; calorific value, 44670 kJ/kg; proportion of hydrogen in fuel, 15%

Total number of revolutions, 21000

Mean effective pressure, 2.74 bar

Net brake load applied to a drum of 100 cm diameter, 600 N

Total mass of cooling water circulated, 495 kg

Temperature of cooling water : inlet 13°C, oulet 38°C]

Air used, 153 kg

Temperature of air in test room, 20°C and temperature of exhaust gages, 370°C

Assume : $c_{p\text{ gages}}$ = 1.005 kJ/kg K and $c_{p\text{ steam}}$ at atmospheric pressure = 2.093 kJ/kg K.

Calculate the thermal efficiency and draw up the heat balance.

Sol.

Given : d = 20 cm, L = 28 cm, n = 1, t = 1 hour, m_f = 4.22 kg,

CV = 44670 kJ/kg,

$$H_2 = 15\%,\ N = 21000 \text{ rpm},\ p_{mi} = 2.74 \text{ bar},$$
$$d_b = 100 \text{ cm},\ F_b = 600 \text{ N},$$
$$m_w = 495 \text{ kg},\ t_{wi} = 13°C,\ t_{wo} = 38°C,\ m_a = 135 \text{ kg},$$
$$t_a = 20°C,\ t_g = 370°C,$$
$$c_{pg} = 1.005 \text{ kJ/kg. K},\ c_{ps} = 2.093 \text{ kJ/kg. K.}$$

Indicated power developed by single-acting two-stroke cycle oil egine,

$$IP = \frac{p_{mi}\,LAN}{60 \times 10^3 \times n}$$

$$= \frac{2.74 \times 10^5 \times 0.28 \times \frac{\pi}{4}(0.2)^2 \times 21000}{60 \times 10^3 \times 1 \times 60} = 14.06 \text{ kW}$$

Brake power,
$$BP = \frac{2\pi N T_b}{60 + 10^3} = \frac{2\pi N F_b \times (d_b/2)}{60 \times 10^3}$$

$$= \frac{2\pi \times 21000 \times 600 \times 0.5}{60 \times 10^3 \times 60} = 10.9956 \text{ kW}$$

Mechanical efficiency,

$$\eta_{mech} = \frac{BP}{IP} = \frac{10.9956}{14.06} \times 100 = 78.2\%$$

Indicated thermal efficiency,

$$(\eta_{th})_i = \frac{IP}{\left(\frac{m_f}{t}\right) \times CV} = \frac{14.06 \times 3600}{4.22 \times 44670} = 0.2685 \text{ or } 26.85\%$$

Swept volume,
$$V_s = \frac{\pi}{4} d^2 L \times \frac{N}{3600} = \frac{\pi}{4}(0.2)^2 \times 0.28 \times \frac{21000}{3600} = 0.0513 \text{ m}^3/\text{s}$$

Heat supplied,
$$Q_a = \frac{m_f}{3600} \times CV = \frac{4.22}{3600} \times 44670 = 52.363 \text{ kJ/s}$$

Heat input of brake, $Q_b = 10.9956$ kJ/s

Heat lost to cooling water,

$$Q_w = \frac{m_w}{t} \times c_{pw} \times (t_{w0} - t_{wi})$$

$$= \frac{495}{3600} \times 4.18\,(38 - 13) = 14.3685 \text{ kJ/s}$$

Heat carried away by exhaust gases:

Mass of flue gases,
$$\dot{m}_g = \left(\frac{m_a + m_f}{t}\right) = \frac{135 + 4.22}{3600} = 0.03867 \text{ kg/s}$$

Steam in exhaust gases,

$$\dot{m}_s = 9 \times 0.15 \times \frac{4.22}{3600} = 0.00185 \text{ kg/s}$$

Mass of dry exhaust gases

$$= 0.03867 - 0.00158 = 0.03708 \text{ kg/s}$$

Heat in steam and exhaust gases,

$$Q_g = [0.00158 \times 2.093\ (100 - 20) + 0.00158 \times 2256.9] + 0.03708 \times 1.005$$

$$= 3.8304 + 10.0616 = 13.8920 \text{ kJ/s}$$

$$\text{Unaccounted for heat} = Q_a - (Q_b + Q_w + Q_g)$$

$$= 52.363 - (10.9956 + 14.3685 + 13.892)$$

$$= 13.1066 \text{ kJ/s}$$

Heat Balance Sheet

Heat supplied	%	**Heat expenditure**	**kJ/s**	%
Heat supplied by fuel, Q_a = 52.363	100	1. Heat equivalent of BP, Q_b	10.9956	21.00
		2. Heat lost to cooling water, Q_w	14.3685	27.44
		3. Heat lost in steam, Q_s		
		4. Heat lost to exhaust gases, Q_g	10.0616	19.21
		5. Unaccounted heat	13.1066	25.04
				Total = 100%

Q.3. ***(a)*** Sketch and describe the working of a once through boiler. Discuss its special features.

Sol. In a once through boiler, the feed water goes through the economizer, furnance walls, and superheater sections, changing sequentially to saturated water, saturated steam, and super heated steam in one continuous pass. No steam drum is required to separate saturated steam from boiling water and no water circulation takes place. It requires very high purity feed water. It is used for supercritical pressure operation (≈ 221 bar).

Fig. 4 Shows a line diagram of a once-through boiler.

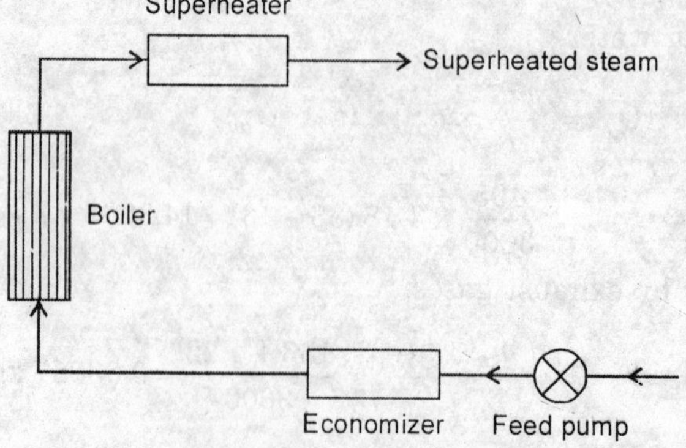

Fig. 4. Once through boiler.

(*b*) Discuss an ideal regenerative cycle for a steam plant. Why is ideal cycle not possible in practice?

Sol.

Regeneration

In the Rankine cycle and reheat cycle, the condensate which is at a fairly low temperature is pumped to the boiler. Thus there is irreversible mixing of the cold condensate with hot boiler water. This results in loss of cycle efficiency. Regeneration is a method to heat the feed water from the hot well of the condenser reversibly by interchange of heat within the system and thus improving the cycle efficiency. The cycle is called regenerative cycle.

In Fig. 5, steam is drawn from the boiler at state point 1 and passes through the turbine and goes to the condenser at state point 2. If no heat is transferred from this steam to the surrounding which include the casing, and the expansion is isentropic, the process is shown by line 1 – 2′ in the T – *s* diagram. But in the regenerative cycle it is assumed that the condensate after being pumped to the boiler pressure at state point 4 passes through the hollow casing surrounding the turbine rotor to the boiler. Thus the steam loses heat to the surrounding water which gets heated. The temperature of steam entering the turbine and temperature of water leaving the turbine are same.

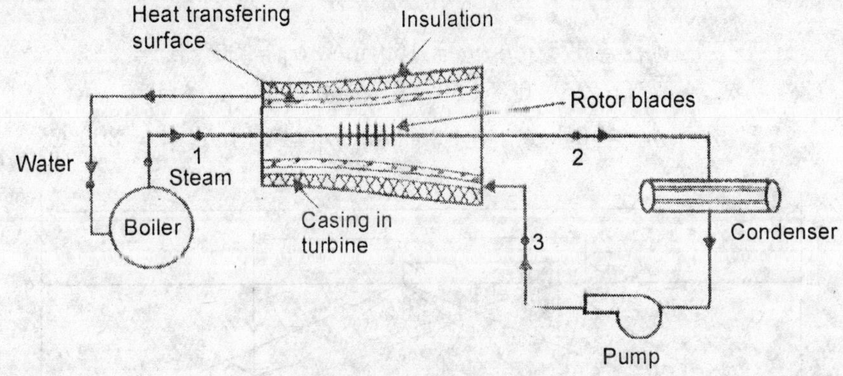

Fig. 5. Regenerative Heating

Let T_1 = saturation temperature of steam at the commencement of expansion at state point 1.

T_3 = temperature of saturated water condensate entering the hollow casing of turbine at state point 3.

Thus one kg of water is gradually heated along the path 3 – 4 and none kg of steam is gradually losing same amount of heat during expansion process 1 – 2. Such a heat exchange within the system is called regenerative heating. Therefore, net heat supplied to the system is in the boiler during process 4 – 1 and net heat rejected from the system is in the condenser during process 2 – 3 (Fig. 6)

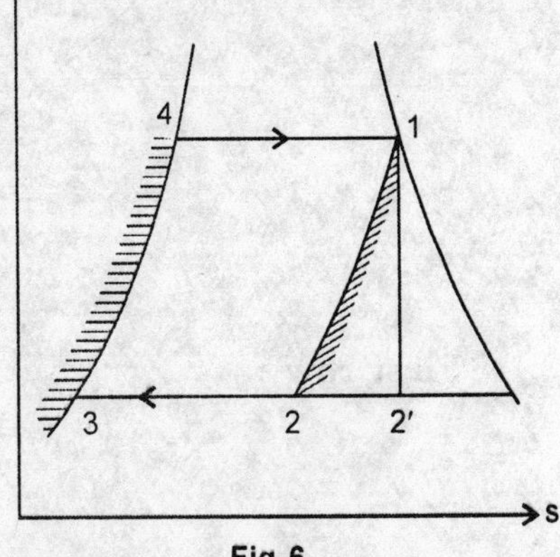

Fig 6.

Heat added, $q_a = T_1 (s_1 - s_4)$

Heat rejected, $q_r = T_2 (s_2 - s_3)$

Net work done,

$$w_{net} = q_a - q_r$$

$$= T_1 (s_1 - s_4) - T_2 (s_2 - s_3)$$

But $$s_1 - s_4 = s_2 - s_3$$

$\therefore$ $$w_{net} = (T_1 - T_2)(s_1 - s_4)$$

Regenerative efficiency,

$$\eta_{reg} = \frac{w_{net}}{q_a} = \frac{(T_1 - T_2)(s_1 - s_4)}{T_1 (s_1 - s_4)} = 1 - \frac{T_2}{T_1}$$

This method of regeneration is not practicable because the steam becomes extremely wet at the later stages of expansion.

(c) In a two-stage velocity compounded impulse turbine the steam leaves the nozzles at a velocity of 675 m/s, when the blade speed is 150 m/s. The nozzle angle is 20°, while the discharge angles are : 25° for the first row blades, 25° for the fixed blades and 30° for the second row blades. There is a 10 per cent drop in velocity during passage of steam through each row of blades. For the steam flow rate of 5 kg/s, the power output of the turbine and the diagram efficiency. Also calculate the length of arc occupied by the nozzles if their height is 25 mm. Steam is supplied to the stage at 12 bar, 300°C.

Sol.

Given : $v_{a1} = 675$ m/s, $u = 150$ m/s, $\alpha_1 = 20°$, $\beta_2 = 25°$,

$\alpha_1' = 25°$, $\beta_2' = 30°$, K = 0.90,

$\dot{m} = 5$ kg/s, $h = 25$ mm, $p_1 = 12$ bar, $t_1 = 300°C$

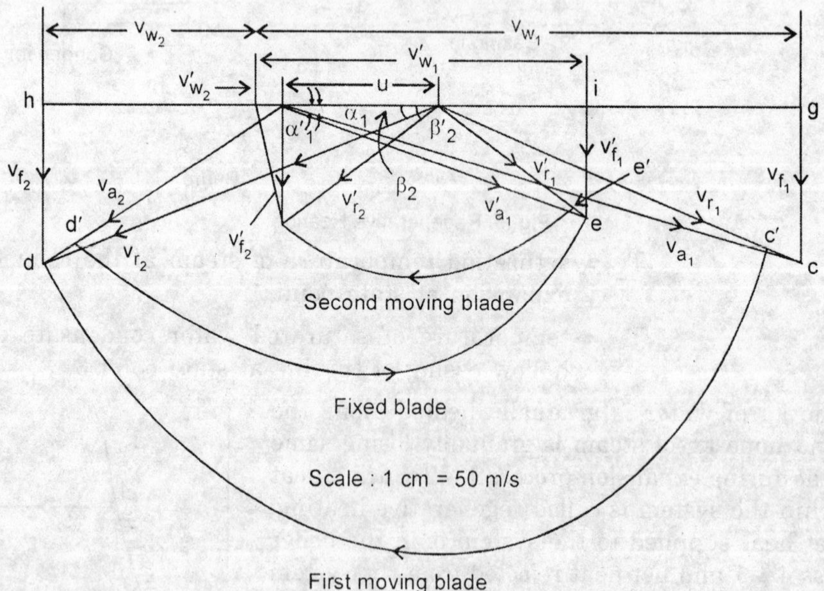

Fig. 5

First row :

$$v_{w1} = v_{a1} \cos \alpha_1 = 675 \cos 20° = 634.29 \text{ m/s}$$

$$v_{f1} = v_{a1} \sin \alpha_1 = 675 \sin 20° = 230.86 \text{ m/s}$$

$$v_{r1} = \sqrt{v_{f1}^2 + (v_{w1} - u)^2}$$

$$= \sqrt{(230.86)^2 + (634.29 - 150)^2} = 536.50 \text{ m/s}$$

$$v_{r2} = \text{K } v_{r1} = 0.90 \times 536.50 = 482.85 \text{ m/s}$$

$$v_{f2} = v_{r2} \sin \beta_2 = 482.85 \sin 25° = 204.06 \text{ m/s}$$

$$v_{w2} = v_{r2} \cos \beta_2 - u = 482.85 \cos 25° - 150 = 287.61 \text{ m/s}$$

$$v_{a2} = \sqrt{v_{f2}^2 + v_{w2}^2} = \sqrt{(204.06)^2 + (287.61)^2} = 352.65 \text{ m/s}$$

$$v_w = v_{w1} + v_{w2} = 634.29 + 287.61 = 921.9 \text{ m/s}$$

Second row :

$$v_{a1}' = 0.90\ v_{a2} = 0.90 \times 352.65$$

$$= 317.38 \text{ m/s}$$

$$v_{w1}' = v_{a1}' \cos \alpha_1' = 317.38 \cos 25° = 287.65 \text{ m/s}$$

$$v_{f1}' = v_{a1}' \sin \alpha_1' = 317.38 \sin 25° = 134.13 \text{ m/s}$$

$$v_{r1}' = [(v_{f1}')^2 + (v_{w1}' - u)^2]^{1/2}$$

$$= [(134.13)^2 + (287.65 - 150)^2]^{1/2} = 192.18 \text{ m/s}$$

$$v_{r2}' = 0.90\ v_{r1}' = 0.90 \times 192.19 = 172.97 \text{ m/s}$$

$$v_{w2}' = u - v_{r2}' \cos \beta_2' = 150 - 172.97 \cos 30° = 0.20 \text{ m/s}$$

$$v_{f2}' = v_{r2}' \sin \beta_2' = 172.97 \sin 30° = 86.48 \text{ m/s}$$

$$v_w' = v_{w1}' - v_{w2}' = 287.65 - 0.20 = 287.45 \text{ m/s}$$

Tangential thrust :

First row, $\quad F_{t1} = \dot{m} v_w = 5 \times 921.9 = 4609.5 \text{ N}$

Second row, $\quad F_{t2} = \dot{m} v_w' = 5 \times 287.45 = 1437.25 \text{ N}$

Power developed, $\quad P = \dfrac{(F_{t1} + F_{t2})u}{10^3} = \dfrac{(4609.5 + 1437.25) \times 150}{10^3} = 907 \text{ kW}$

Diagram efficiency, $\quad \eta_d = \dfrac{2u(v_w + v_w')}{v_{a1}^2} = \dfrac{2 \times 150\,(921.9 + 287.45)}{675^2} = 0.7963 \text{ or } 79.63\%$

At 12 bar 300°C, $\quad v_g = 0.2138 \text{ m}^3/\text{kg},\ h_g = 3045.8 \text{ kJ/kg}$

Area across flow, $\quad A_f = \dfrac{\dot{m} v_g}{v_{f1}} = \dfrac{5 \times 0.2138}{230.86} = 4.6305 \times 10^{-3} \text{ m}^2$

$$(\pi d - l)h = A_f$$

Let $\quad N = 3000 \text{ rpm}$

$$\pi d = \frac{150 \times 60}{3000} = 3 \text{ m}$$

$$(3 - l) \times 25 \times 10^{-3} = 4.6305 \times 10^{-3}$$

$$l = 2.81 \text{ m}$$

Q.4. (*a*) Drive an expression, in terms of initial and final pressures and the volume of air drawn in per stroke, for the minimum work done in the two-stage compression of air. Assume adiabatic compression and perfect interocling. Neglect clearance and show clearly all the steps in the analysis.

Sol.

Two–Stage Compressor

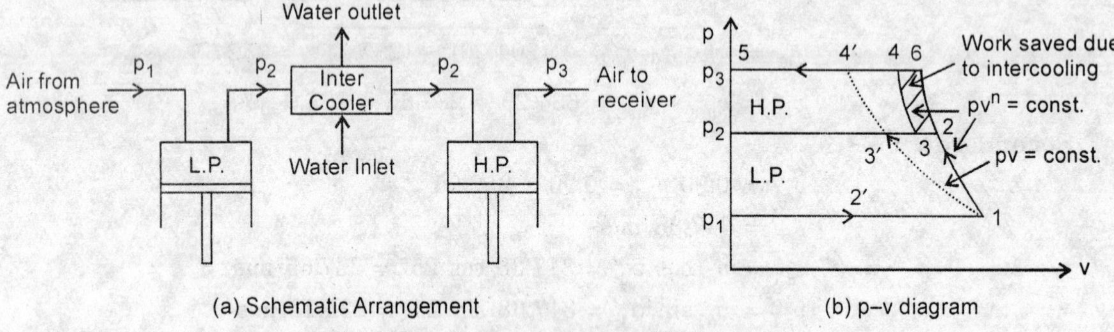

(a) Schematic Arrangement (b) p–v diagram

Fig. 6. Two stage compression with intercooling

The schematic arrangement of two-stage compressor with inter cooler and *p-v* diagram are shown in Fig. 6. The air is first taken into the low pressure (L.P.) cylinder at pressure p_1. After compression to intermediate pressure p_2, the air at condition 2 is passed through the intercooler and leaves it at point 3 where its temperature is reduced from T_2 to T_3. The air may be cooled to point 3′ such that $T_{3'} = T_1$. Finally the air is compressed in high pressure, (H.P.) cylinder from condition 3 to 4 and is discharged to the receiver. Area ′2–3–4–6–2 represent the work saved due to inter cooling.

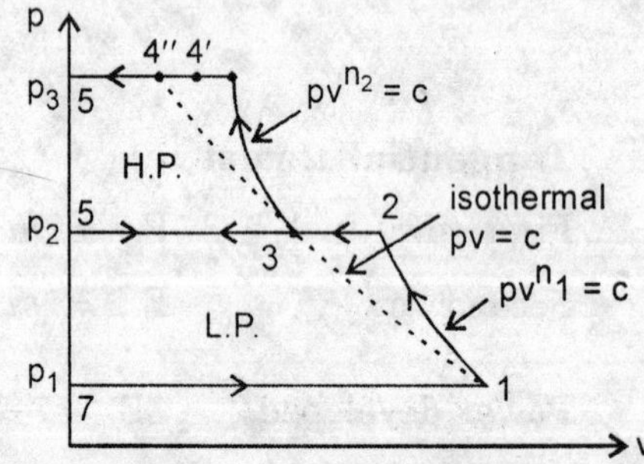

Fig 7. Two stage comression with perfect intercooling

Fig. 7 shows the *p-v* diagram for two–stage compression with perfect inter cooling. Process 1–2 represents the polytropic compression in the L.P. cylinder with pv^{n_1} = const. Process 2–3 represents inter cooling from T_2 to $T_3 = T_1$. Process 3–4 respresents, the polytropic compression in the H.P. cylinder with pv^{n2} = const. Curve 1–3–4′ represents the isothermal compression.

Total work done on the air,

$$W_t = \left(\frac{n_1}{n_1-1}\right)RT_1\left[\left(\frac{p_2}{p_1}\right)^{\frac{n_1-1}{n_1}} - 1\right] + \left(\frac{n_2}{n_2-1}\right)RT_3\left[\left(\frac{p_3}{p_2}\right)^{\frac{n_2-1}{n_2}} - 1\right]$$

Let $$\frac{n_1-1}{n_1} = a \text{ and } \frac{n_2-1}{n_2} = b$$

$$W_t = \frac{RT_1}{a}\left[\left(\frac{p_2}{p_1}\right)^a - 1\right] + \frac{RT_3}{b}\left[\left(\frac{p_3}{p_2}\right)^b - 1\right]$$

For p_1 and p_3 to be constant, intermediate pressure p_2 must be determined for minimum work. Thus

$$\frac{dW_t}{dp_2} = 0$$

$$\frac{T_1}{a}.a\left(\frac{p_2}{p_1}\right)^{a-1}.\frac{1}{p_1}+\frac{T_3}{b}.\left(\frac{p_3}{p_2}\right)^{b-1}.\left(\frac{-p_3}{p_2^2}\right)=0$$

$$T_1\frac{p_2^{a-1}}{p_1^a}-T_3\frac{p_3^b}{p_2^{b+1}} = 0$$

$$T_1\,p_2^{a+b} = T_3\,p_1^a\,p_3^b$$

$$p_2 = \left[\frac{T_3}{T_1}\left(p_1^a\,p_3^b\right)\right]^{\frac{1}{a+b}}$$

For $n_1 = n_2 = n$, $a = b$. Then

$$p_2 = \left[\frac{T_3}{T_1}(p_1\,p_3)^a\right]^{\frac{1}{2a}} = \left[p_1\,p_3\left(\frac{T_3}{T_1}\right)^{\frac{1}{a}}\right]^{\frac{1}{2}}$$

For perfect inter cooling,

$$T_3 = T_1, \text{ and}$$

$$p_2 = \sqrt{p_1\,p_3}$$

$$(W_t)_{min} = \frac{RT_1}{a}\left[\left\{\frac{\sqrt{p_1\,p_3}}{p_1}\right\}^a+\left\{\frac{p_3}{\sqrt{p_1\,p_3}}\right\}^a\quad 2\right]$$

$$= \frac{2RT_1}{a}\left[\left(\frac{p_3}{p_1}\right)^{a/2}-1\right] = \left(\frac{2n}{n-1}\right)RT_1\left[\left(\frac{p_3}{p_1}\right)^{\frac{n-1}{2n}}-1\right]$$

$$= \left(\frac{2n}{n-1}\right)p_1\,v_1\left[\left(\frac{p_3}{p_1}\right)^{\frac{n-1}{2n}}-1\right]$$

In general, with i number of stages, we have

$$\frac{p_2}{p_1} = \frac{p_3}{p_2}=\frac{p_1}{p_3}=.....=\frac{p_{i+1}}{p_i}$$

$$\text{Minimum shaft work, } W_t = \left(\frac{in}{n-1}\right)RT_1\left[\left(\frac{p_{i+1}}{p_i}\right)^{\frac{n-1}{in}}-1\right]$$

The minimum indicated power (I.P.) with imperfect inter cooling is :

$$\text{I.P.} = \left(\frac{n}{n-1}\right)mR\left[2\sqrt{T_1\,T_3}\left(\frac{p_3}{p_1}\right)^{\frac{n-1}{2n}}-(T_1+T_3)\right]\times\frac{1}{10^3}$$

where m = mass of air delivered in kg per second.

(*b*) A jet engine is flying at 300 m/s when the pressure and temperature of the atmosphere are 0.8 bar and 230 K respectively. The compressor pressure ratio is 4/1 and the maximum cycle temperature is 1000 K. Calculate, specific thrust, power produced, propulsive efficiency, overall thermal efficiency and the fuel consumption. Assume : isentropic efficiency of components unity ; nozzle throat area 0.06 m^2; calorific value of fuel 43000 kJ/kg; c_p and γ for the combustion and expansion processes 1.15 kJ/kg K and 1.333.

Sol.

Given :

$$c_1 = 300 \text{ m/s}, \; p_1 = 0.8 \text{ bar},$$
$$T_1 = 230 \text{ K}, \; p_3/p_2 = 4,$$
$$T_4 = 1000 \text{ K},$$
$$\eta_{isen} = 1.0,$$
$$A_t = 0.06 \text{ m}^2,$$
$$CV = 43000 \text{ kJ/kg, 1000 K}$$
$$c_{pa} = 1.005 \text{ kJ/kg. K},$$
$$\gamma_a = 1.4,$$
$$c_{pg} = 1.15 \text{ kJ/kg. K},$$
$$\gamma_g = 1.333$$

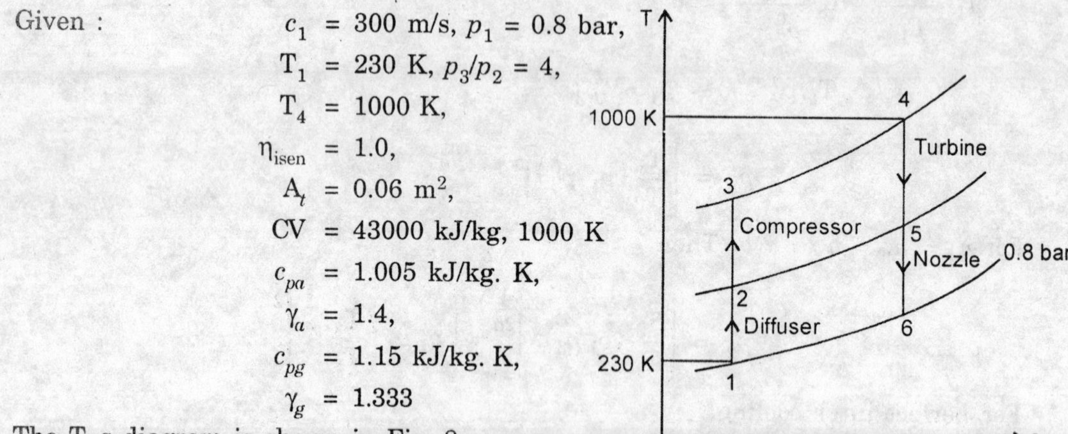

Fig 8.

The T-s diagram is shown in Fig. 8.

Diffuser :

where
$$T_2 = T_1 + \frac{c_1^2}{2c_{pa}} = 230 + \frac{300^3}{2 \times 1.005 \times 10^3} = 274.87 \text{ K}$$

$$p_2 = p_1 \left(\frac{T_2}{T_1}\right)^{\frac{\gamma_a}{\gamma_a - 1}} = 0.8 \left(\frac{274.78}{230}\right)^{\frac{1.4}{0.4}} = 1.49 \text{ bar}$$

Compressor :

$$p_3 = 4p_2 = 4 \times 1.49 = 5.96 \text{ bar}$$

$$T_3 = T_2 \left(\frac{p_3}{p_2}\right)^{\frac{\gamma_a - 1}{\gamma_a}} = 274.78 \, (4)^{\frac{0.4}{1.4}} = 408.32 \text{ K}$$

$$T_4 = 1000 \text{ K}$$

Neglecting mass of fuel as compared to that of air,

Work done by turbine = Work done on compressor

$$c_{pg} (T_4 - T_5) = c_{pa} (T_3 - T_2)$$
$$1.15 \, (1000 - T_5) = 1.005 \, (408.32 - 274.78)$$
$$T_5 = 883.3 \text{ K}$$

Also
$$p_4 = p_3$$

$$\frac{p_4}{p_5} = \left(\frac{T_4}{T_5}\right)^{\frac{\gamma_g}{\gamma_g - 1}}$$

$$p_5 = \frac{5.96}{\left(\frac{1000}{883.3}\right)^{\frac{1.333}{0.333}}} = 3.63 \text{ bar}$$

Let $p_6 = p_1 = 0.8$ bar

$$\frac{T_5}{T_6} = \left(\frac{p_5}{p_6}\right)^{\frac{\gamma_g - 1}{\gamma_g}}$$

$$T_6 = \frac{883.3}{\left(\frac{3.63}{0.8}\right)^{\frac{0.333}{1.333}}} = 605.4 \text{ K}$$

Enthalpy drop in nozzle,

$$\Delta h = c_{pg}(T_5 - T_6) = 1.15\,(883.3 - 605.4) = 319.6 \text{ kJ/kg}$$

Velocity of jet, $c_j = c_6 = \sqrt{2.\Delta h} = \sqrt{2 \times 319.6 \times 10^3} = 799.5$ m/s

Thrust produced, $F = \dot{m}(c_j - c_a) = 1(799.5 - 800) = 499.5$ N/kg of air

Thrust power, $TP = \frac{F\,c_a}{10^3} = \frac{499.5 + 300}{10^3} = 149.85$ kW

$\therefore$ Specific thrust = 149.85 kW/kg of air

Power produced = 149.85 + Turbine output

= 149.85 + 1.15 (1000 – 883.3) = 284.06 kW/kg of air

Propulsive efficiency,

$$\eta_{prop} = \frac{149.85}{284.06} = 0.5275 \text{ or } 52.75\%$$

Overall Thermal efficiency,

$$\eta_{th} = \frac{149.85}{c_{pg}(T_4 - T_3)} = \frac{149.85}{1.15\,(1000 - 408.32)} = 0.2202 \text{ or } 22.02\%$$

(***b***) Considering mass of fuel.

Let m_f = mass of fuel.

Then $m_f \times CV = (m_a + m_f)\,c_{pg} \times T_4 - c_{pa}\,m_a\,T_3$

or $\frac{m_f}{m_a} \times CV = \left(1 + \frac{m_f}{m_a}\right) c_{pg} \times T_4 - c_{pa}.m_a.T_3$

$$\frac{m_f}{m_a}(CV - c_{pg}\,T_4) = c_{pg}\,T_4 - c_{pa}\,T_3$$

$$\frac{m_f}{m_a}(43000 - 1.15 \times 1000) = 1.15 \times 1000 - 1.005 \times 408.32$$

$$\frac{m_f}{m_a} = \frac{739.6384}{41850} = 0.0176 \text{ kg of fuel/kg of air}$$

Now $\quad W_t = W_c$

$$(m_a + m_f)\, c_{pg}\, (T_4 - T_5) = m_a\, c_{pa}\, (T_3 - T_2)$$

$$(1 + 0.0176) \times 1.15\, (1000 - T_5) = 1.005\, (408.32 - 274.78)$$

$$T_5 = 885.3 \text{ K}$$

Also
$$\frac{p_4}{p_5} = \left(\frac{T_4}{T_5}\right)^{\frac{\gamma_g}{\gamma_g - 1}}$$

$$p_5 = \frac{5.96}{\left(\frac{1000}{885.3}\right)^{\frac{1.333}{0.333}}} = 3.66 \text{ bar}$$

$$T_6 = \frac{5.96}{\left(\frac{3.66}{0.8}\right)^{0.333/1.333}} = 604.98 \text{ K}$$

Nozzle :

$$\Delta h = c_{pg}\, (T_5 - T_6) = 1.15\, (885.3 - 604.98) = 322.37 \text{ kJ/kg}$$

$$c_j = c_6 = \sqrt{2.\Delta h} = \sqrt{2 \times 322.37 \times 10^3} = 802.85 \text{ m/s}$$

Specific thrust,
$$F = \left(1 + \frac{m_f}{m_a}\right)(c_j - c_a)$$

$$= (1 + 0.0176)\, (802.95 - 300)$$

$$= 511.8 \text{ N/kg of air}$$

Now
$$R = \left(\frac{\gamma_g - 1}{\gamma_g}\right) c_{pg} = \left(\frac{1.333 - 1}{1.333}\right) \times 1.5 = 0.287 \text{ kJ/kg. K}$$

$$p_6\, v_6 = RT_6$$

$$v_6 = \frac{0.287 \times 604.98}{0.8 \times 10^2} = 2.17 \text{ m}^3\text{/kg of air}$$

Mass flow rate,
$$\dot{m} = \frac{A_f\, c_j}{v_g} = \frac{0.06 \times 802.95}{2.17} = 22.2 \text{ kg/s}$$

Power produced $F \times c_a \times \dot{m}/10^3 = 511.8 \times 300 \times 22.2/10^3 = 3408.6$ kW

Q.5. **(*a*)** A food storage locker requires a refrigeration capacity of 50 kW. It works between a condenser temperature of 35°C and an evaporator temperature of —10°C. The refrigerant is ammonia. It is subcooled by 5°C before entering the expansion valve by the dry saturated vapour leaving the evaporator. Assuming a single cylinder, single-acting compressor operating at 1000 rpm with stroke equal to 1.2 times the bore, determine (*a*) the power required, and (*b*) the cylinder dimensions.

Saturation temperature. C	Pressure bar	Enthalpy. kJ/kg		Entropy. kJ/kg K		Specific volume. m³/kg		Sparific heat kJ/kg K	
		Liquid	Vapour	Liquid	Vapour	Liquid	Vapour	Liquid	Vapour
– 10	2.9157	154.056	1450.22	0.82965	5.7550	–	0.417477	–	2.492
35	13.522	366.072	1488.57	1.56605	5.2086	1.7023	0.0954629	4.556	2.903

Sol.

Refer to Fig. 9.

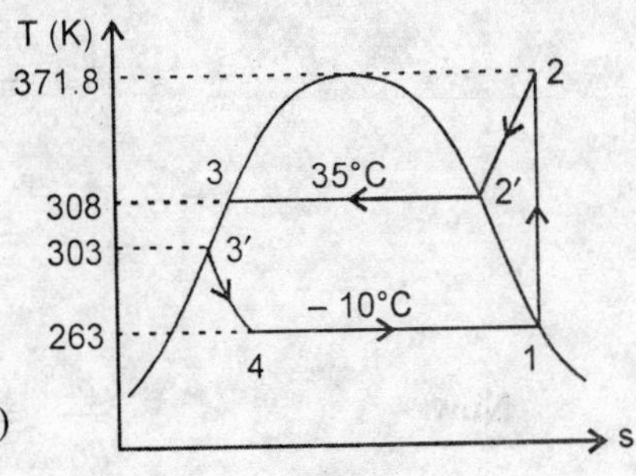

Fig 9.

$$h_1 = 1450.22 \text{ kJ/kg},$$
$$h_2' = 1488.57 \text{ kJ/kg}$$
$$h_{f3} = 366.072 \text{ kJ/kg},$$
$$h_{f3}' = h_4 = h_{f3} - c_{pf}(T_3 - T_3')$$
$$= 366.02 - 4.556\,(308 - 303)$$
$$= 343.29 \text{ kJ/kg}$$

Also
$$s_2 = s_1$$
$$s_2' + c_p \ln \frac{T_2}{T_2'} = 5.7550$$
$$\ln \frac{T_2}{308} = \frac{5.7550 - 5.2086}{2.903} = 0.1882$$
$$T_2 = 371.8 \text{ K}$$
$$h_2 = h_2' + c_p\,(T_2 - T_2') = 1488.57 + 2.903\,(371.8 - 308)$$
$$= 1673.8 \text{ kJ/kg}$$

Mass flow rate of refrigerant,

$$\dot{m} = \frac{50}{h_1 - h_4} = \frac{50}{1450.22 - 343.29} = 0.04517 \text{ kg/s}$$

(*a*) Power required $= \dot{m}\,(h_2 - h_1)$

$$= 0.04517\,(1673.8 - 1450.22) = 10.1 \text{ kW}$$

(*b*) Cylinder dimensions :

$$\dot{m} = \frac{\pi}{4} D^2 L \frac{N}{60} v_1$$
$$D_3 = 0.006888$$
$$D = 0.19 \text{ m},\ L = 0.228 \text{ m}$$

(*b*) What is impulse function? Show that

$$\frac{F}{F^*} = \frac{1 + \gamma M^2}{M\sqrt{\left[2(1+\gamma)\left(1 + \frac{\gamma - 1}{2} M^2\right)\right]}}$$

Sol.

Impulse function, $F = pA + \rho A v^2$

$$= pA + \frac{p}{RT} . Av^2 = pA\left(1 + \frac{\gamma v^2}{c^2}\right), c = \sqrt{\gamma RT}$$

$$= pA\,(1 + \gamma M^2)$$

For M = 1, $F^* = p^* A^* (1 + \gamma)$

$$\frac{F}{F^*} = \frac{p}{p^*} . \frac{A}{A^*}\left[\frac{1+\gamma M^2}{1+\gamma}\right]$$

$$= \frac{p}{p_0} . \frac{p_0}{p^*} . \frac{A}{A^*}\left[\frac{1+\gamma M^2}{1+\gamma}\right]$$

Now $\frac{p}{p_0} = \left(\frac{2}{\gamma+1}\right)^{\frac{\gamma}{\gamma+1}}$

$$\frac{p_0}{p^*} = \left[1 + \frac{\gamma-1}{2} M^2\right]^{\frac{\gamma}{\gamma-1}}$$

$$\frac{A}{A^*} = \frac{1}{M}\left[\frac{2}{\gamma+1} + \left(1 + \frac{\gamma-1}{2} M^2\right)\right]^{\frac{\gamma+1}{2(\gamma-1)}}$$

Then $\frac{F}{F^*} = \frac{1+\gamma M^2}{M\sqrt{2(1+\gamma)\left(1 + \frac{\gamma-1}{2} M^2\right)}}$

(*c*) A supersonic aeroplane flies at a Mach number 1.8 at an altitude of 700 m, where the atmospheric temperature is 10°C. What is the time that elapses, by which the acoustic disturbance reaches an observer on the ground after it is directly overhead ?

Sol.

$$c = \sqrt{\gamma RT} = \sqrt{1.4 \times 0.287 \times 10^3 \times 283} = 337.2 \text{ m/s}$$

$$v = Mc = 1.8 \times 337.2 = 606.97 \text{ m/s}$$

Time taken by sound to reach the ground, $t = \frac{h}{c} = \frac{700}{337.2} = 2.076\ s$

Q.6. (*a*) For the data given below for a Pelton turbine, calculate the friction factor for the buckets, hydraulic efficiency and mechanical efficiency.

Overall efficiency, 0.82; speed ratio, 0.45; coefficient of velocity, 0.92;

jet deflection angle, 160°, windage and bearing friction losses ∝ (rotational speed)^2; volumetric efficiency, 1.

Sol.

Given. $\eta_0 = 0.82,\ K_u = 0.45,\ K_v = 0.98,$

$$\beta_2 = 180° - 160° = 20°, \ \eta_{vol} = 1.0,$$

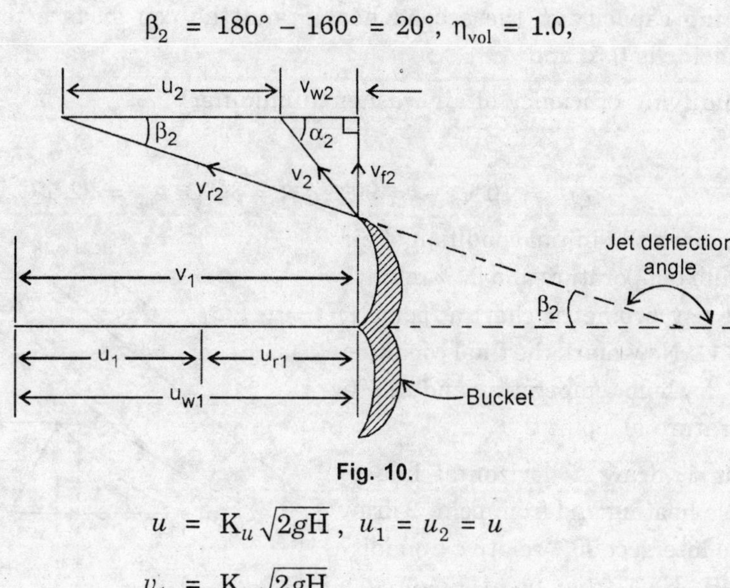

Fig. 10.

$$u = K_u \sqrt{2gH}, \ u_1 = u_2 = u$$

$$v_1 = K_v \sqrt{2gH}$$

Jet efficiency or Head efficiency,

$$\eta_H = 4 K_u (K_v - K_u)$$

$$= 4 \times 0.45 (0.98 - 0.45) = 0.954$$

Hydrantic efficiency, $\eta_h = \eta_H \cdot \eta_{vol} = 0.954 \times 1 = 0.954$ or 95.4 %

Mechanical efficiency, $\eta_{mech} = \dfrac{\eta_0}{\eta_h \cdot \eta_{vol}} = \dfrac{0.82}{0.954 \times 1.0} = 0.8595$ or 85.95%

Let K = friction factor for buckets.

$$v_{r1} = v_1 - u$$

$$v_{r2} = Kv_{r1}$$

$$\eta_h = \frac{2u(v_1 - u)(1 + K\cos\beta_2)}{v_1^2} = \frac{2K_u (K_v - K_u)(1 + K\cos\beta_2)}{K_v^2}$$

$$0.954 = \frac{2 \times 0.45 (0.98 - 0.45)(1 + K\cos 20°)}{(0.98)^2}$$

$$= 0.4967 (1 + 0.9397 K)$$

$$K = 0.98$$

Q.6. (*b*) 300 m³/min of air at 10°C dry bulb temperature and 90% relative humidity is to be heated and humidified to 35°C dry bulb temperature and 22.5°C wet bulb temperature. The required conditions are achieved by heating, humidifying and then again by heating. The relative humidity of the air coming out of the humidifier is 90%. Find

(*i*) the heating capacity of the first heater and the bypass factor if the surface temperature of the coil is 40°C.

(*ii*) the capacity of the humidifier in kg/hour

(*iii*) the heating capacity of the second heater and the coil surface temperature if the bypass factor is 0.5, and

(*iv*) the humidifying efficiency of air washer (humidifier).

Sol.

Given : $t_{d1} = 10°C; \phi_1 = 90\%; t_{d2} = 35°C; t_{w2} = 22.5°C$

First of all, mark the initial condition of air *i.e.* at 10°C dry bulb temperature and 90% relative humidity, on the psychrometric chart at point 1, as shown in Fig. 11. Now mark the final condition of air *i.e.* at 35°C dry bulb temperature and 22.5°C wet bulb temperature at point 2.

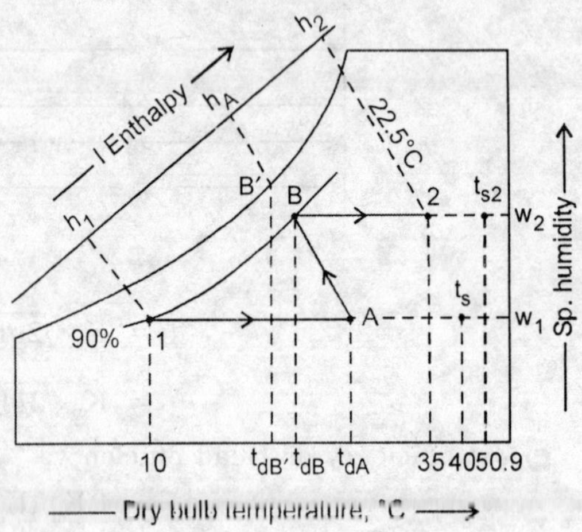

Fig 11.

From point 1, draw a horizontal line to represent sensible heating and from point 2 draw horizontal line to intersect 90% relative humidity curve at point B. Now from point B, draw a constant wet bulb temperature line which intersects the horizontal line drawn through point 1 at point A. The line 1-A represents preheating of air, line AB represents humidification and line B-2 represents reheating to final conditions.

(*i*) Heating capacity of first heater.

From the psychrometric chart, the temperature to which the air should be preheated (corresponding to point A) is

$$t_{dA} = 32.6°\text{ C}$$

$$v_1 = 0.82 \text{ m}^3/\text{kg}$$

$$\dot{m}_a = \frac{300}{60 \times 0.82} = 6.09 \text{ kg/s}$$

From the pyschrometric chart, we find that enthalpy of air at point 1,

$$h_1 = 27.2 \text{ kJ/kg of dry air}$$

Enthalpy of air at point A,

$$h_A = 51 \text{ kJ/kg of dry air}$$

Heating capacity of first heater, $Q_1 = \dot{m}_a (h_A - h_1) = 6.09\ (51 - 27.2) = 144.942$ kW

$$\text{By pass factor} = \frac{t_s - t_{dA}}{t_s - t_1} = \frac{40 - 32.6}{40 - 10} = 0.247$$

(*ii*) Capacity of humidifier.

From the psychrometric chart, we find that specific humidity of entering air,

$$w_1 = 0.0068 \text{ kg/kg of dry air}$$

and specific humidity of leaving air,

$$w_2 = 0.0122 \text{ kg/kg or dry air}$$

Capacity of humidifier = $\dot{m}_a (w_2 - w_1) = 6.09\ (0.0122 - 0.0068) \times 3600 = 118.39$ kg/kr

(*iii*) Heating capacity of second heater.

Enthalpy of air at point 2, $h_2 = 68$ kJ/kg of dry air

Also $t_{dB} = 19.1°C$,

and $h_B = 51$ kJ/kg of dry air

Heating capacity of second heater,

$$Q_2 = \dot{m}_a (h_2 - h_B)$$

$$= 6.09\ (68 - 51) = 103.53 \text{ kW}$$

Now $\dfrac{t_{s2} - t_2}{t_{s2} - t_{dB}} = 0.5$

$$\frac{t_{s2} - 35}{t_{s2} - 19.1} = 0.5$$

$$t_{s2} = 50.9°C$$

(*iv*) Humidifying efficiency of the air washer.

From the psychrometric chart, we find that

$$t_{dB} = 19.1°C \quad \text{and} \quad t_{dB'} = 18°C$$

We know that humidifying efficiency of the air washer,

$$\eta_H = \frac{\text{Actual drop in DBT}}{\text{Ideal drop in DBT}} = \frac{t_{dA} - t_{dB}}{t_{dA} - t_{dB'}}$$

$$= \frac{32.6 - 19.1}{32.6 - 18} = \frac{13.5}{14.6} = 0.924 \text{ or } 92.4\% \textbf{ Ans.}$$

Q.7. (*a*) Derive from the first principles the Euler's equation of motion for steady flow along a strealine. Obtain Bernoulli's equation by its integration. State the assumptions made.

Sol. Euler's Equation for Motion

Consider steady flow of an ideal fluid along the stream tube. Separate out a small element of fluid of cross-sectional area dA and length ds from stream tube as a free body from the moving fluid.

Fig. 12 shows such a small element LM of fluid of cross-section area dA and length ds.

Let, p = Pressure on the element at L,

$p + dp$ = Pressure on the element at M, and

v = Velocity of the fluid element.

ρ = density of fluid

$$\gamma = \rho g = \text{specific weight of fluid.}$$

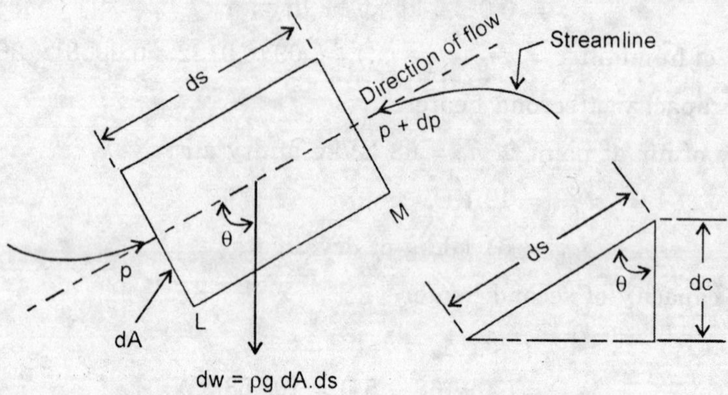

Fig. 12. Forces on a fluid element (Euler's equation).

The external forces tending to accelerate the fluid element in the direction of streamline are as follows:

1. Net pressure force in the direction of flow is,

$$p.dA - (p + dp)dA = -dp \, . \, dA \qquad ...(i)$$

2. Component of the weight of the fluid element in the direction of flow is

$$= -\rho.g.\delta A.ds. \cos\theta$$

$$= -\rho.g.dA.ds\left(\frac{dz}{ds}\right) \qquad \left(\because \cos\theta = \frac{dx}{ds}\right)$$

$$= -\rho.g.dA.dz \qquad ...(ii)$$

Mass of the fluid element

$$= \rho.dA.ds \qquad ...(iii)$$

The acceleration of the fluid element

$$a = \frac{dv}{dt} = \frac{dv}{ds} \times \frac{ds}{dt} = v.\frac{dv}{ds} \qquad ...(iv)$$

Now, according to Newton's second law of motion, Force = mass × acceleration

$$\therefore \quad -dp.dA - \rho.g.dA.dz = \rho.dA.ds \times v.\frac{dv}{ds}$$

Dividing both sides by $\rho.dA$, we get

$$\frac{-dp}{\rho} - g.dz = v.dv$$

or

$$\frac{dp}{\rho} + v.dv + g.dz = 0 \qquad ...(v)$$

This is the required **Euler's equation** for motion, and is in the form of *differential equation*.

Integrating the above Eqn (v), we get

$$\frac{1}{\rho}\int dp + \int v.dv + \int g.dz = \text{constant}$$

$$\frac{p}{\rho} + \frac{v^2}{2} + gz = \text{constant}$$

Dividing by g, we get

$$\frac{p}{\rho g} + \frac{v^2}{2g} + z = \text{constant}$$

or
$$\frac{p}{\gamma} + \frac{v^2}{2g} + z = \text{constant}$$

or, in other words,

$$\frac{p_1}{\gamma} + \frac{v_1^2}{2g} + z_1 = \frac{p_2}{\gamma} + \frac{v_1^2}{2g} + z_2$$

Which proves *Bernoulli's equation*.

Assumptions:

The following *assumptions* are made in the derivation of Bernoulli's equation:

1. The liquid is ideal and incompressibe.
2. The flow is steady and continuous.
3. The flow is along the streamline, *i.e.*, it is one-dimensional.
4. The velocity is uniform over the section and is equal to the mean velocity.
5. The only forces acting on the fluid are the *gravity froces* and the *pressure forces*.

Q.7. (*b*) A cubical tank has sides of 2 m. Upto a height of 0.8 m it is filled with water and the rest space is filled with oil of specific gravity 0.85. For one vertical side calculate (*i*) the total pressure, and (*ii*) the position of the centre of pressure.

Sol.

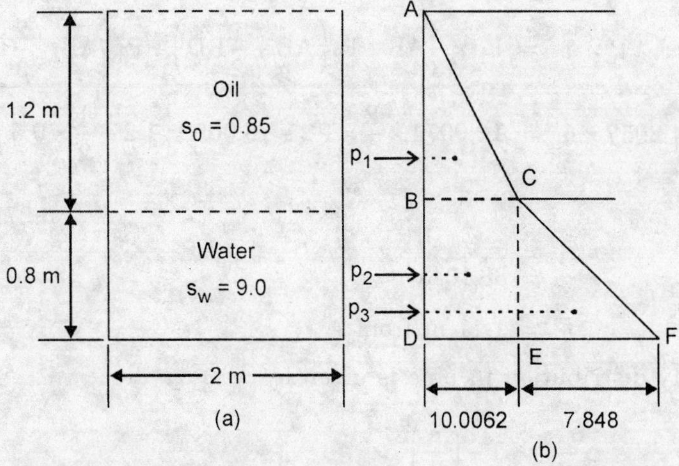

Fig. 13.

Depth of water, $h_w = 0.8$ m

Depth of oil, $h_o = 1.2$ m

Specific gravity of water,

$$s_w = 1.0$$

Specific gravity of oil, $s_o = 0.85$

(*i*) The pressure diagram is shown in Fig. 13(*b*).

Intensity of pressure on top A,

$$p_A = 0$$

Intensity of pressure on BC,

$$p_B = s_o \gamma_w h_o = 0.85 \times 9.81 \times 10^3 \times 1.2 = 10.0062 \text{ kN/m}^2$$

Intensity of pressure on DF,

$$p_D = s_o \gamma_w h_o + s_w \gamma_w h_w = 10.0062 + 1.0 \times 9.81 \times 10^3 \times 0.8$$
$$= 10.0062 + 7.848 = 17.8542 \text{ kN/m}^2$$

Force P_1 = area ΔABC × width of tank

$$= \frac{1}{2} \times AB \times BC \times 2 = \frac{1}{2} \times 1.2 \times 10.0062 \times 2 = 12.0074 \text{ kN}$$

Force P_2 = area of rectangle BCED × width of tank

$$= 0.8 \times 10.0062 \times 2 = 16.0099 \text{ kN}$$

Force P_3 = area of ΔCEB × width of tank

$$= \frac{1}{2} \times 0.8 \times 7.848 \times 2 = 6.2784 \text{ kN}$$

Total pressure, $P = P_1 + P_2 + P_3 = 12.0074 + 16.0091 + 6.2784$

$$= 34.2957 \text{ kN}$$

(*ii*) Centre of pressure :

Taking moments of all the forces about A, we get

$$P \times \bar{h} = P_1 \times \frac{2}{3} AB + P_2 \left(AB + \frac{1}{2} BD \right) + P_3 \left(AB + \frac{2}{3} BD \right)$$

$$34.2957 \times \bar{h} = 12.0074 \times \frac{2}{3} \times 1.2 + 16.0099 \left(1.2 + \frac{1}{2} \times 0.8 \right) + 6.2784 \left(1.2 + \frac{2}{3} \times 0.8 \right)$$

$$= 38.1043$$

$$\bar{h} = 1.11 \text{ m from A}$$

(c) The velocity distribution in pipe is given by

$$\frac{u}{u_s} = 1 - \left(\frac{r}{R} \right)^n$$

where u_s is the maximum velocity at the centre of the pipe, u is the velocity at a distance r from the centre and R is the pipe radius. Obtain an expression for the mean velocity in terms of u_s and n.

Sol.

The velocity distribution in a pipe is, (Fig. 14)

$$\frac{u}{u_s} = 1 - \left(\frac{r}{R}\right)^n$$

Discharge passing through an elementary ring of thickness dh at a radius r of pipe is,

$$dQ = u.dA = u \times 2\pi r\, dr$$

$$= u_s \left[1 - \left(\frac{r}{R}\right)^n\right] \times 2\pi r\, dr$$

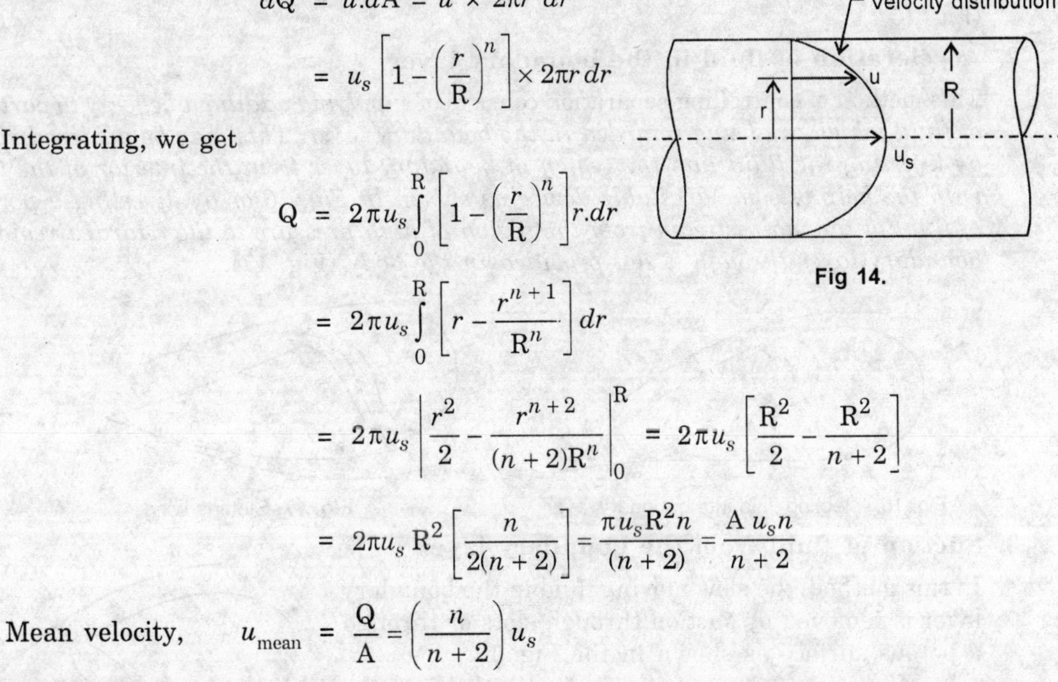

Fig 14.

Integrating, we get

$$Q = 2\pi u_s \int_0^R \left[1 - \left(\frac{r}{R}\right)^n\right] r.dr$$

$$= 2\pi u_s \int_0^R \left[r - \frac{r^{n+1}}{R^n}\right] dr$$

$$= 2\pi u_s \left|\frac{r^2}{2} - \frac{r^{n+2}}{(n+2)R^n}\right|_0^R = 2\pi u_s \left[\frac{R^2}{2} - \frac{R^2}{n+2}\right]$$

$$= 2\pi u_s R^2 \left[\frac{n}{2(n+2)}\right] = \frac{\pi u_s R^2 n}{(n+2)} = \frac{A\, u_s n}{n+2}$$

Mean velocity, $\quad u_{mean} = \frac{Q}{A} = \left(\frac{n}{n+2}\right) u_s$

(d) Explain what is meant by separation of boundary layer. Describe with sketches the methods to control separation.

Sol.

When a solid body is immersed in a flowing fluid, a thin layer of fluid called boundary *layer* is formed, adjacent to the solid body. In this thin layer of fluid, the velocity varies from zero to *free steam velocity* in the direction normal to the solid body. Along the length of the solid body, the thickness of the boundary layer increases. *The fluid layer adjacent to the solid surface has to do work against surface friction at the expense of kinetic energy. This loss of kinetic energy is recovered from the immediate fluid layer in contact with the layer adjacent to the solid surface through momentum exchange process. Thus the velocity of the layer goes on decreasing.* Along the length of solid body, *at a cetain point a stage may come when the boundary layer may not be able to keep sticking to the solid body, if it cannot provide kinetic energy to overcome the resistance offered by the solid body.* In other words, the boundary layer will be *separated from the surface.* This phenomenon is called the **boundary layer separation**. *The point on the body at which the boundary layer is on the verge of separation from the surface is called* **point of separation**.

Methods to control separation:

1. Motion of solid boundary:

By rotating a circular cylinder lying in a stream of fluid, so that the upper side of cylinder where the fluid as well as the cylinder move in the *same direction, the boundary layer does not form.* However on the lower side of cylinder where the *motion is opposite* to that of cylinder *separation would occur* (Fig. 15).

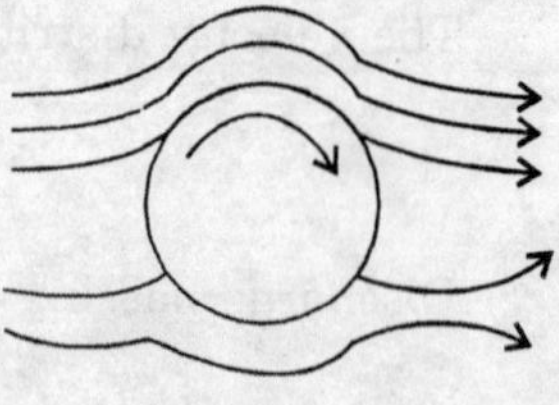

Fig 15.

2. Acceleration of fluid in the boundary layer

This method of controlling separation consists of *supplying additional energy to particles of fluid which are being retarded in the boundary layer. This may be achieved either by injecting the fluid into the region of boundary layer from the interior of the body with the help of some available device as shown in Fig. 16 or by diverting a portion of fluid of the main stream from the region of high pressure to the retarded region of boundary layer through a slot provided in the body* (Fig. 17)

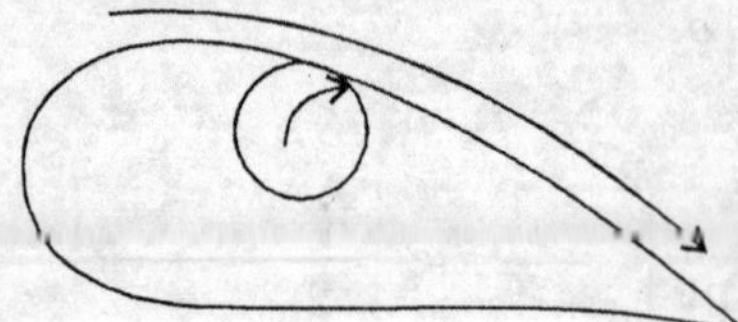

Fig. 16. Injecting fluid into boundary layer.

Fig. 17. Slotting wing.

3. Suction of fluid from the boundary layer

In this method, the slow moving fluid in the boundary layer is removed by suction through slots or through a porous surface as shown in the Fig. 18.

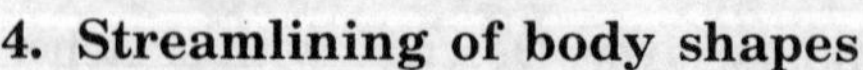

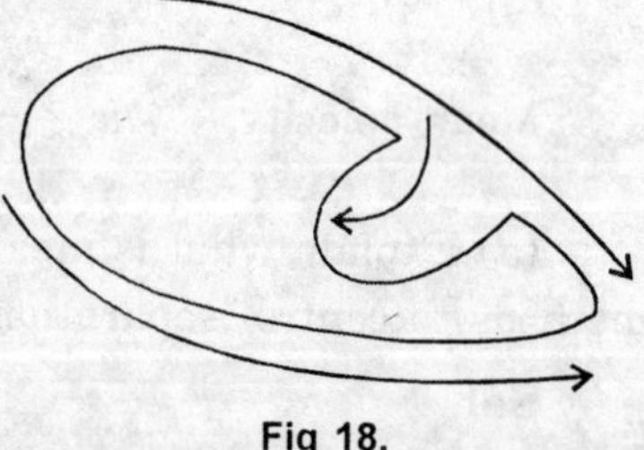

Fig 18.

4. Streamlining of body shapes

By the use of suitably shaped bodies, the point of transition of the boundary layer from laminar to turbulent can be *moved downstream which results in the reduction of the skin friction* drag. Furthermore by *streamlining of body shapes, the separation may be eliminated.*

Q.8. (*a*) During an experiment to measure the temperature of hot gas flowing through a large pipe with thermocouple located centrally in the pipe, the following data were obtained under the steady state conditions:

Velocity of gas in the pipe line, 4 m/s

Temperature of pipe wall, 400°C

Temperature indicated by the thermocouple, 580°C

Diameter of the thermocouple junction, 1.5 mm

Emissivity of the thermocouple, 0.3

Calculate the true gas temperature.

The convective heat transfer coefficient can be calculated from $N_{ud} = 0.5\ \text{Re}_d^{1/2}$

The property values may be used at 400°C and are :

$$k = 49.72 \times 10^{-3} \text{ W/m°C},$$
$$\mu = 32.68 \times 10^{-6} \text{ kg/ms};$$
$$\rho = 0.5224 \text{ kg/m}^3$$

Comment on the methods by which accuracy of the thermocouple reading can be improved.

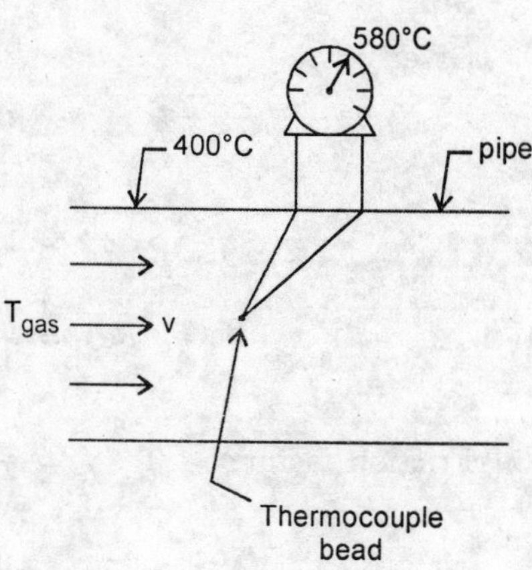

Fig 19.

Sol.

Given : $v = 4$ m/s, $T_w = 273 + 400 = 673$ K,

$$T_c = 272 + 580 = 583 \text{ K},$$
$$d = 1.5 \text{ mm},$$
$$\varepsilon = 0.3, T_g = ?,$$
$$k = 49.72 \times 10^{-3} \text{ W/m.°C},$$
$$\mu = 32.68 \times 10^{-6} \text{ kg/m.s.},$$
$$\rho = 0.5224 \text{ kg/m}^3$$
$$Nu_d = 0.5 \, Re_d^{1/2}$$
$$= 0.5 \, (96.2)^{1/2} = 4.904$$
$$\frac{hd}{k} = Nu_d$$
$$h = \frac{4.904 \times 49.72 \times 10^{-3}}{1.5 \times 10^{-3}} = 162.56 \text{ W/}m^2\text{.°C}$$

Reynolds number, $$Re_d = \frac{vd\rho}{\mu} = \frac{4 \times 1.5 \times 10^{-3} \times 0.5224}{32.68 \times 10^{-6}} = 96.2$$

Heat gained by thermocouple by convection = Heat lost by thermocouple by radiation.

$$Ah(T_g - 583) = A\varepsilon\sigma\,(T_c^4 - T_w^4)$$
$$162.56\,(T_g - 583) = 0.3 \times 5.67\left[\left(\frac{583}{100}\right)^4 - \left(\frac{673}{100}\right)^4\right]$$
$$T_g = 886.7 \text{ K or } 613.9\text{°C}$$

True gas temperature = 613.9°C

(***b***) Air flows through a 25 mm diameter tube with a mean velocity of 30 m/s. The tube wall temperature is 280°C and the air temperature increases from 20°C to 260°C. Using the simple Reynolds analogy calculate, the length of the tube required and the pumping power. For turbulent flow in a tube, take

$$f = \frac{0.046}{Re_d^{0.2}}$$

The properties may be taken at the mean film temperature and from the Tables these are :

$$k = 38.45 \times 10^{-3} \text{ W/m.°C}, \ c_p = 1.0268 \text{ kJ/kg.°C}$$

$$\rho = 0.7306 \text{ kg/m}^3; \ \mu = 26.17 \times 10^{-6} \text{ kg/ms}$$

Sol.

Given : $d = 25$ mm, $v_m = 30$ m/s, $t_w = 280°C$, $t_{ai} = 20°C$, $t_{ao} = 260°C$,

$$f = \frac{0.046}{\text{Re}_d^{0.2}}, \ k = 38.45 \times 10^{-3} \text{ W/m. °C},$$

$$c_p = 1.0268 \text{ kJ/kg. °C}, \ \rho = 0.7306 \text{ kg/m}^3, \ \mu = 26.17 \times 10^{-6} \text{ kg/m.s}$$

$$\text{Re}_d = \frac{v_m \, d\rho}{\mu} = \frac{30 \times 25 \times 10^{-3} \times 0.7306}{26.17 \times 10^{-6}} = 20938$$

Friction factor, $$f = \frac{0.046}{(20938)^{0.2}} = 6.29 \times 10^{-3}$$

$$\text{Pr} = \frac{\mu c_p}{k} = \frac{26.17 \times 10^{-6} \times 1.0268 \times 10^3}{38.45 \times 10^{-3}} = 0.699$$

$$\bar{N}u = \frac{\bar{h}d}{h} = 0.032 \, (\text{Re})^{0.8} \, (\text{P}_r)^{1/3} \text{ for Re} > 2300$$

$$= 0.032 \, (20938)^{0.8} \times (0.699)^{1/3}$$

$$= 81.294$$

$$\bar{h} = \frac{81.294 \times 38.45 \times 10^{-3}}{25 \times 10^{-3}} = 125 \text{ W/m}^2.°\text{C}$$

Heat transferred, $$Q = m. \, c_p \, (t_{ao} - t_{ai})$$

$$= \left(\rho \times \frac{\pi}{4} d^2 \times v_m\right) \times c_p (t_{ao} - t_{ai})$$

$$= 0.7306 \times \frac{\pi}{4} (25 \times 10^{-3})^2 \times 30 \times 1.0268 \times 10^3 \times (260 - 20)$$

$$= 2651.36 \text{ W}$$

Now $$Q = \bar{h} \, A(t_w - t_b)$$

where $$t_b = \frac{20 + 260}{2} = 140°C$$

$$2651.36 = 125 \times \pi \times 25 \times 10^{-3} \times l \times (280 - 140)$$

$$l = 1.929 \text{ m}$$

Frictional heat loss, $$h_f = \frac{f \, l \, v_m^2}{2gd} = \frac{6.29 \times 10^{-3} \times 1.929 \times 30^2}{2 \times 9.81 \times 25 \times 10^{-3}} = 22.26 \text{ m}$$

$$\text{Pumping power} = \rho g \times \frac{\pi}{4} d^2 \times v_m \times h_f$$

$$= 0.7306 \times 9.81 \times \frac{\pi}{4}(25 \times 10^{-3}) \times 30 \times 22.26 = 2.35 \text{ Watts}$$

(*c*) For forced convection heat transfer, when frictional heating in the fluid cannot be neglected, show by dimensional analysis

$$\text{Nu} = f\left[\text{Re}, \text{Pr}, \frac{c^2}{c_p \text{T}}\right].$$

Discuss the physical significance of each term.

Sol.

Heat transfer coefficient in a fully developed forced convection in a tube is a function of the following variables :

$$h = f(\rho, \text{L}, c, \mu, c_p, k, \theta)$$

or $f(h, \rho, \text{L}, c, \mu, c_p, k, \theta) = 0$

Here physical quantities, $m = 8$, and fundamental dimensions, $n = 4$

∴ Number of π-terms $= m - n = 8 - 4 = 4$

∴ $f(\pi_1, \pi_2, \pi_3, \pi_4) = 0$

Variables	**Symbols**	**Dimensions**
Heat transfer coefficient	h	$\text{MT}^{-3}\,\theta^{-1}$
Fluid density	ρ	ML^{-3}
Tube diameter or plate length	D or L	L
Fluid velocity	c	LT^{-1}
Fluid viscosity	μ	$\text{ML}^{-1}\,\text{T}^{-1}$
Specific heat	c_p	$\text{L}^2\,\text{T}^{-2}\,\theta^{-1}$
Thermal conductivity	k	$\text{ML T}^{-3}\,\theta^{-1}$
Temperature	θ	θ

We select h, ρ, L, c as the repeating variables.

Let
$$\pi_1 = (h)^{a1}\,(\rho)^{b1}\,(\text{L})^{c1}\,(c)^{d1}\,\mu$$

$$\text{M}^0\,\text{L}^0\,\text{T}^0\,\theta^0 = [\text{MT}^{-3}\,\theta^{-1}]^{a1}\,[\text{ML}^{-3}]^{b1}\,[\text{L}]^{c1}\,[\text{LT}^{-1}]^{d1}\,[\text{ML}^{-1}\,\text{T}^{-1}]$$

$$= \text{M}^{a1 + b1 + 1}\,\text{L}^{-3b1 + c1 + d1 - 1}\,\text{T}^{-3a1 - d1 - 1}\,\theta^{-a1}$$

$$a_1 + b_1 + 1 = 0$$
$$-3b_1 + c_1 + d_1 - 1 = 0 \qquad a_1$$
$$-3a_1 - d_1 - 1 = 0 \qquad b_1$$
$$-a_1 = 0 \qquad c_1$$
$$d_1$$

∴ $a_1 = 0,\ d_1 = -1,\ b_1 = -1,\ c_1 = 1 - 3 + 1 = -1$

$$\pi_1 = \rho^{-1}\,\text{L}^{-1}\,c^{-1} = \frac{\mu}{\rho \text{LC}}$$

$$\pi_2 = h^{a2}\, \rho^{b2}\, L^{c2}\, c^{d2}\, c_p$$

$$M^0 L^0 T^0 \theta^0 = [MT^{-2}\theta^{-1}]^{a2} [ML^{-3}]^{b2} [L]^{c2} [LT^{-1}]^{d2} [L^2 T^{-2} \theta^{-1}]$$

$$= Ma2 + b2\; L^{-3b2 + c2 + d2 + 2}\; T^{-3a2 - d2}\; \theta^{-a2 - 1}$$

$$\therefore \quad -a_2 - 1 = 0 \quad \text{or} \quad a_2 = -1$$

$$a_2 + b_2 = 0$$

$$b_2 = -a_2 = 1$$

$$-3a_2 - d_2 - 2 = 0$$

$$d_2 = -3a_2 - 2 = 3 - 2 = 1$$

$$-3b_2 + c_2 + d_2 + 2 = 0$$

$$\text{or} \quad -3 + c_2 + 1 + 2 = 0$$

$$c_2 = 0$$

$$\pi_2 = h^{-1}\, \rho^{+1}\, L^0\, c^{+1}\, c_p = \frac{\rho c}{h} c_p$$

Since dimensions of h and $\frac{k}{L}$ are same, hence

$$\pi_2 = \frac{c_p\, \rho c L}{k}$$

$$\pi_3 = h^{a3}\, \rho^{b3}\, L^{c3}\, c^{d3}\, k$$

$$M^0 L^0 T^0 \theta^0 = [MT^{-3}\theta^{-1}]^{a3} [ML^{-3}]^{b3} [L]^{c3} [LT^{-1}]^{d3} [MLT^{-3}\theta^{-1}]$$

$$= [M]^{a3 + b3 + 1}\, [L]^{-3b3 + c3 + d3 + 1}\, [T]^{-3a3 - d3 - 2}\, [\theta]^{-a3 - 1}$$

$$-a_3 - 1 = 0, \quad a_3 = -1$$

$$-3a_2 - d_2 - 3 = 0, \quad d_3 = 3 - b = 0$$

$$a_3 + b_3 + 1 = 0, \quad b_3 = 1 - 1 = 0$$

$$-3b_3 + c_3 + d_3 + 1 = 0, \quad c_3 = 0 - 0 - 1 = -1$$

$$\pi_3 = h^{-1}\, \rho^0\, L^{-1}\, c^0\, k = \frac{k}{hL}$$

$$\pi_3 = f(\pi_1, \pi_2, \pi_4)$$

$$\frac{k}{hL} = f\left(\frac{\mu}{\rho L c}, \frac{c_p\, \rho c L}{k}, \frac{h\theta}{\rho c^3}\right)$$

$$\text{or} \quad \frac{hL}{k} = \oint\left[\left(\frac{\mu}{\rho L c}\right)^x \left(\frac{c_p\, \rho c L}{k}\right)^y \left(\frac{\theta h}{\rho c^3}\right)^z\right]$$

$$\therefore \quad Nu = \oint\left[(Re)^x, (Pr)^y, \frac{c^2}{c_p\, \theta}\right].$$

MECHANICAL ENGINEERING
PAPER I—1998

(Conventional)

Q.1. (*a*) Using the Maxwell relation, derive the following T.ds equation :

$$\text{T}.ds = c_p\, d\text{T} - \text{T}\left(\frac{\partial v}{\partial \text{T}}\right)_p dp \qquad (10)$$

Sol.

Let $\qquad s = f(\text{T}, p)$

Differentiating, we get,

$$ds = \left(\frac{\partial s}{\partial \text{T}}\right)_p .d\text{T} + \left(\frac{\partial s}{\partial \text{T}}\right)_\text{T} .dp$$

Multiplying throughout by T, we have

$$\text{T}.ds = \text{T}.\left(\frac{\partial s}{\partial \text{T}}\right)_p .d\text{T} + \text{T}.\left(\frac{\partial s}{\partial p}\right)_\text{T} .dp$$

From Maxwell's relation, we have

$$\text{T}.\left(\frac{\partial s}{\partial \text{T}}\right)_p = c_p \text{ and } \left(\frac{\partial s}{\partial \text{T}}\right)_\text{T} = -\left(\frac{\partial v}{\partial \text{T}}\right)_p$$

Hence $\qquad \text{T}.ds = c_p.d\text{T} - \text{T}\left(\frac{\partial v}{\partial \text{T}}\right)_p .dp$

(*b*) A compressed air bottle of volume 0.15 m^3 contains air at 40 bar and 27°C. It is used to drive a turbine which exhausts to atmosphere at 1 bar. If the pressure in the bottle is allowed to fall to 2 bar, determine the amount of work that could be delivered by the turbine. (15)

Sol.

Given : $\qquad \text{V}_1 = 0.15 \text{ m}^3, p_1 = 40 \text{ bar},$

$\text{T}_1 = 273 + 27 = 300 \text{ K},$

$p_2 = 1 \text{ bar}, \text{V}_1 = \text{V}_2$

From steady state energy equation, we have

$$m_1\left(u_1 + p_1v_1 + \frac{1}{2}.c_1^2 + gz_1\right) + \text{Q}$$

$$= m_2\left(u_2 + p_2v_2 + \frac{1}{2}.c_2^2 + gz_2\right) + \text{W}$$

Here $m_1 = m_2 = \frac{p_1 V_1}{RT_1} = \frac{40 \times 10^5 \times 0.15}{287 \times 300} = 6.9686$ kg

$z_2 = z_1 = 0,\ c_1 = 0,\ Q = 0$

Work done, $W = m_1 v_1 (p_1 - p_2) = V_1 (p_1 - p_2)$

$= 0.15 \times (40 - 1) \times 10^5 = 585$ kJ

(*c*) A rigid tank contains air at 1.5 bar and 60°C. The pressure of air is raised to 2.5 bar by transfer of heat from a constant temperature reservoir at 400°C. The temperature of surroundings is 27°C. Determine per kg of air, the loss of available energy due to heat transfer. (15)

Sol.

Given : $p_1 = 1.5$ bar, $T_1 = 273 + 60 = 333$ K, $p_2 = 2.5$ bar,

$T_R = 273 + 400 = 673,\ T_0 = 273 + 27 = 300$ K

Assuming constant volume process,

$$T_2 = T_1 . \left(\frac{p_2}{p_1}\right) = 333 \left(\frac{2.5}{1.5}\right) = 555 \text{ K}$$

Heat transferred to rigid tank,

$$Q_R = mc_v (T_2 - T_1)$$

$$= 1 \times 0.714 \times (555 - 333) = 159.396 \text{ kJ/kg}$$

Loss in available energy

$$= T_0 \left[(\Delta s)_{\text{system}} + (\Delta s)_{\text{surrounding}}\right]$$

$$= T_0 \left[c_v \ln\left(\frac{T_2}{T_1}\right) - \frac{Q_R}{T_R}\right]$$

$$= 300 \left[0.718 \ln\left(\frac{555}{333}\right) - \frac{157.396}{673}\right] = 38.978 \text{ kJ/kg}$$

Q.2. (*a*) Discuss the formation of exhaust emission in petrol engine. How to these emissions vary with air-fuel ratio ?

Describe with a neat sketch how a three-way catalytic converter controls pollution in petrol cars? (15)

Sol.

The various exhaust emissions from petrol engine are

1. **Unburnt hydrocarbons (HC).** The causes for the emissions of HC are : Incomplete combustion, crevice volumes and flow in crevices, leakage past the exhaust valve, valve overlap, deposits on wall, and oil on combustion chamber walls.

   The reasons for incomplete combustion are : improper mixing of air and fuel, and flame quenching at the walls of the cylinder. Low load and idle conditions increase HC.

2. **Oxides of Carbon (CO and CO_2).** Carbon monoxide is generated with a rich fuel-air ratio mixture. This happens during starting and accelerating under load. Poor mixing, local rich regions, and imcomplete combustion are the sources of CO emissions.

3. **Oxides of Nitrogen (NO_x).** NO_x are created mostly from nitrogen in the air and fuel blends. In addition to temperature, the formation of NO_x depends on pressure, A/F ratio, combustion duration and location of spark plug.

4. **Oxides of Sulphur (SO_x).** These are generated due to presence of sulphur in the fuel. SO_2 and SO_3 react with water to give rise to H_2SO_4 which causes acid rain.

The variation of emission from a petrol engine with A/F ratio is shown in Fig. 1.

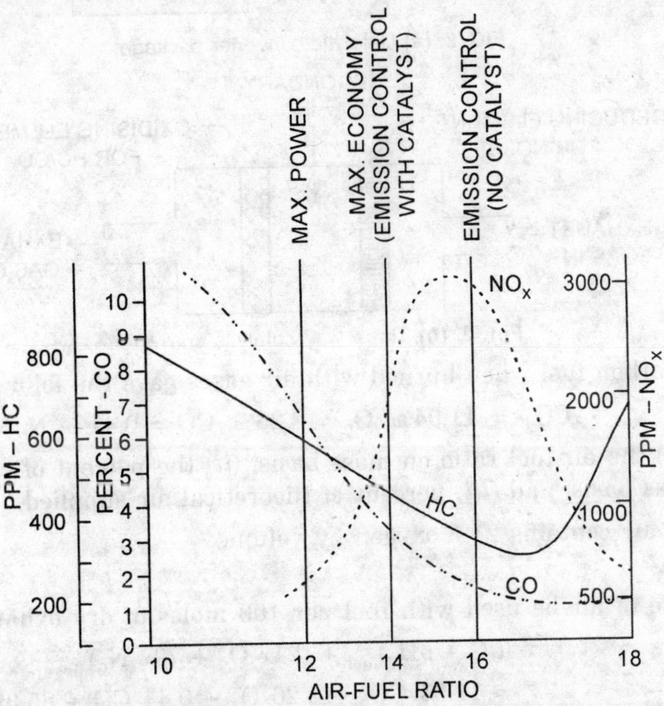

Fig. 1. Variation of emissions from a petrol engine.

Three-way Catalytic Converter. A catalyst is a substance that accelerates a chemical reaction by lowering the energy needed for it to proceed. The catalyst converter reduces the concentration of CO, HC and NO_x in the exhaust. Catalystic converter is usually a stainless steel container mounted along the exhaust pipe of the engine. Inside the container is a porous ceramic (Al_2O_3) structure through which the exhaust gas flows. The ceramic is a single honey-comb structure with many flow passages. The ceramic passages contain small embedded particles of catalytic material like platinum, palladium and rhodium, etc. that promotes the oxidation or reduction reaction in the exhaust gas. Platinum and palladium promote the oxidation of CO and HC while rhodium promotes the reaction of NO_x by reduction process.

The principle of working of a three-way catalytic converter is shown in the following Fig.s 2 (*a*) and (*b*).

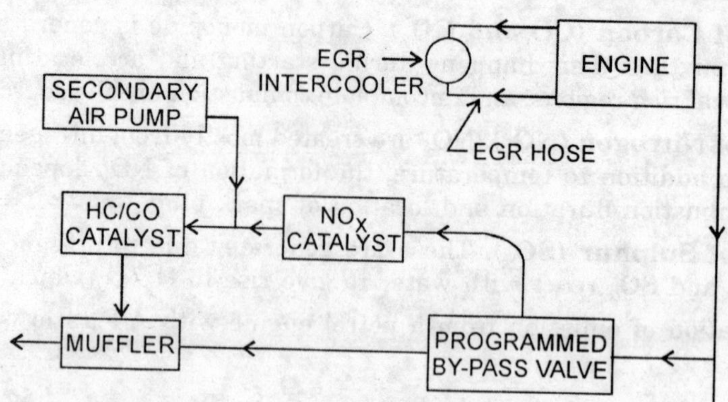

FIG. 2. (*a*) Catalytic converter package.

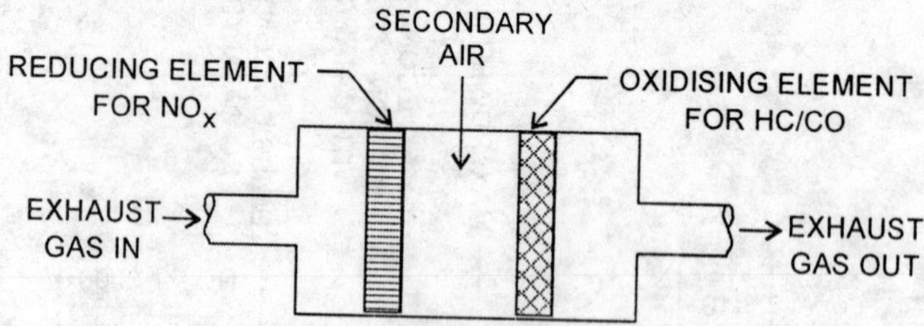

Fig. 2. (*b*) Three way catalytic converter.

(***b***) A hydrocarbon fuel when burned with air gases gave the following Orsat analysis :

$$CO_2 = 11.94\%, \ O_2 = 2.26\%, \ CO = 0.41\%, \ N_2 = 85.39\%.$$

Determine (*i*) the air-fuel ratio on mass basis, (*ii*) the percent of carbon and hydrogen in the fuel on mass basis, and (*iii*) percent of theoretical air supplied.

Assume that air contains 21% oxygen by volume. **(15)**

Sol.

Let $100x$ mole of air be used with fuel per 100 moles of dry exhaust products.

$$(a\text{C} + b\text{H}_2)_{\text{fuel}} + (21x \ \text{O}_2 + 79x \ \text{N}_2)_{\text{air}}$$
$$= 11.94 \ \text{CO}_2 + 2.26 \ \text{O}_2 + 0.41 \ \text{CO} + 85.39 \ \text{N}_2 + d \ \text{H}_2\text{O}$$

By molar balance, we have

Carbon : $a = 11.94 + 0.41 = 12.35$

Hydrogen : $b = d$

Oxygen : $21x = 11.94 + 2.26 + \dfrac{0.41}{2} + \dfrac{d}{2} = 14.405 + 0.5d$

Nitrogen : $79x = 85.39$

or $x = 1.081$

From Eqn. (1), we get,

$$21 \times 1.081 = 14.405 + 0.5d$$
$$d = b = 16.592$$

(*i*) Air-fuel reatio $= \dfrac{100x \times \text{Molecular weight of air}}{a\text{C} + b\text{H}_2}$

$$= \frac{100 \times 1.081 \times 29}{12.35 \times 12 + 16.592 \times 2} = 17.28$$

(*ii*) $\% \text{ Carbon in fuel } = \dfrac{a\text{C} \times 100}{a\text{C} + b\text{H}_2} = \dfrac{12.35 \times 12 + 100}{12.35 \times 12 + 16.592 \times 2} = 81.7\%$

$\% \text{ Hydrogen in fuel } = b\text{H}_2 \times 100/(a\text{C} + b\text{H}_2)$

$$= \frac{16.595 \times 2 \times 100}{12.35 \times 12 + 16.592 \times 2} = 81.7\%$$

(*iii*) % Theoretical air supplied

$$= \frac{100\,(21x\ \text{O}_2 + 97x\ \text{N}_2)}{(a\text{C} + b\text{H}_2) + (21x\ \text{O}_2 + 79x\ \text{N}_2)}$$

$$= \frac{100 \times 1.081 \times (21 \times 32 + 79 \times 28)}{(12.35 \times 12 + 16.592 \times 2) + (21 \times 32 + 79 \times 28) \times 1.081}$$

$= 94.5\%$

(*c*) Draw a schematic flow diagram of a modern steam generator. Discuss the installation of convection and radiation superheaters and their exit temperature response with varying steam flow. (10)

Sol. A modern steam generator is universally used in all modern power plants where pressure are more than 100 bars. The various types of modern steam generator are : La Mont boiler, Benson boilder, Loeffler boiler, Schmidt-Hartmann boiler and Velox boilder. The Benson boilder is illustrated in the following Fig. 3.

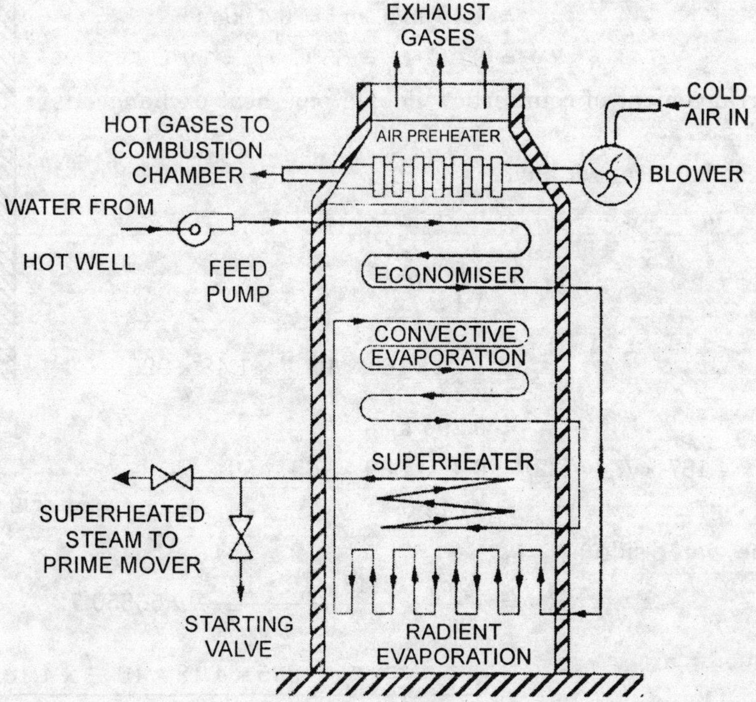

Fig. 3. Benson boiler.

The water is passed through the economizer into the radiant evaporator where the major portion of the water is converted into steam. The remaining water is evaporated in the final evaporator absorbing the heat from hot gases by convection. The supercritical pressure steam (at 225 atm.) is then passed through the superheater. Steam upto a temperature of 650°C, can be raised by this boiler with a steam flow of 150 tonnes/hour.

Q.3. (*a*) In a counterflow double pipe heat exchanger, water flows throug a copper tube (19 mm O.D. and 16 mm I.D.), at a flow rate of 1.48 m/s. The oil flows through the annulus formed by inner copper tube and outer steel tube (30 mm O.D. and 36 mm I.D.). The steel tube is insulated from outside. The oil enters at 0.4 kg/s and is cooled from 65°C to 50°C whereas water enters at 32°C. Neglecting the resistance of the copper tube, calculate the length of the tube required. (25)

Property	Oil	Water
ρ (kg/m^3)	580	995
c_p (kJ/kg. K)	1.89	4.187
k (W/mK)	0.138	0.615
ν (m^2/s)	7.44×10^{-6}	4.18×10^{-7}

Sol.

Given : d_1 = 16 mm, d_2 = 19 mm, d_3 = 26 mm, d_4 = 30 mm

Subscripts :

c, h : for cold fluid (water) and hot fluid (oil), respectively.

i, o : at inlet and oulet respectively.

$$v_c = 1.48 \text{ m/s}, \; m_h = 0.4 \text{ kg/s}$$

$$t_{ci} = 32°\text{C}, \; t_{hi} = 65°\text{C}, \; t_{ho} = 50°\text{C}$$

The arrangement of counterflow double pipe heat exchanger is shown in Fig. 4.

Fig 4. Counter flow double pipe heat exchanger.

For heat balance, we have

$$m_c \, c_{pc} \, (t_{co} - t_{ci}) = m_h \, c_{ph} \, [t_{hi} - t_{ho}]$$

$$m_c = \left(\frac{\pi}{4}\right) d_1^2 . v_c \, \rho$$

$$= \left(\frac{\pi}{4}\right) \times 16^2 \times 10^{-6} \times 1.48 \times 995$$

$$= 0.296 \text{ kg/s}$$

$$0.296 \times 4.187 \times (t_{co} - 32) = 0.4 \times 1.89 \times (65 - 50)$$

$$t_{co} = 41.15°\text{C}$$

For the water side :

$$\text{Re} = \frac{v_c \, d_1}{\nu_c} = \frac{1.48 \times 16 + 10^{-3}}{4.18 \times 10^{-7}} = 56650.72$$

$$\text{Pr} = \frac{\mu \, c_p}{k} = \frac{\rho \nu \, c_p}{k} = \frac{995 \times 4.18 \times 10^{-7} \times 4.18 \times 10^3}{0.615} = 2.83$$

$$\text{Nu} = 0.023 \times (56650.72)^{0.8} \times (2.83)^{0.3} = 199.45$$

$$= \frac{h_c\, d_1}{k_c} = \frac{h_c \times 16 \times 10^{-3}}{0.615}$$

$$h_c = 7666.25 \text{ W/m}^2 \text{ K}$$

For the oil side :

Diameter of passage,

$$d_h = d_3 - d_2 = 26 - 19 = 7 \text{ m}$$

$$\text{Re} = \frac{v_h\, d_d}{\nu_h}$$

$$v_h = \frac{m_h}{\left(\frac{\pi}{4}\right)\left(d_3^2 - d_2^2\right) \times \rho} = \frac{0.4}{\left(\frac{\pi}{4}\right)(26^2 - 19^2)\, 10^{-6} \times 850} = 1.9 \text{ m/s}$$

$$\text{Re} = \frac{1.9 \times 7 \times 10^{-3}}{7.44 \times 10^{-3}} = 1787.6$$

$$\text{Pr} = \frac{\mu\, c_p}{k} = \frac{\rho \nu\, c_p}{k} = \frac{850 \times 7.44 \times 10^{-6} \times 1.89 \times 10^3}{0.138} = 86.61$$

$$\text{Nu} = 0.023 \times (1787.6)^{0.8} \times (86.61)^{0.3} = 35$$

$$= \frac{h_h d_h}{k_h} = \frac{h_c \times 7 \times 10^{-3}}{0.138}$$

$$h_h = 690 \text{ W/m}^2\text{. K}$$

Overall heat transfer co-efficient :

$$\frac{1}{\text{U}_c\text{A}_c} = \frac{1}{\text{U}_h\, \text{A}_h} = \frac{1}{h_c\, \text{A}_c} + \frac{1}{h_{fc}\, \text{A}_c} + \frac{1}{h_h\, \text{A}_h} + \frac{1}{h_{fh}\, \text{A}_h}$$

$$\frac{1}{\text{U}_h} = \left(\frac{\text{A}_h}{\text{A}_c}\right)\left(\frac{1}{h_c} + \frac{1}{h_{fc}}\right) + \left(\frac{1}{h_h} + \frac{1}{h_{fh}}\right)$$

$$\text{U}_h = \frac{1}{\left(\frac{19}{16}\right)\left\{\frac{1}{7666.25} + 0.0005\right\} + \left\{\frac{1}{690} + 0.0008\right\}}$$

$$= 333.564 \text{ W/m}^2\text{. K}$$

$$\Delta t_1 = t_{hi} - t_{co} = 65 - 41.15 = 23.85°\text{C}$$

$$\Delta t_2 = t_{ho} - t_{ci} = 50 - 32 = 18°\text{C}$$

$$\text{LMTD} = \frac{\Delta t_1 - \Delta t_2}{\ln\left(\frac{\Delta t_1}{\Delta t_2}\right)} = \frac{23.85 - 18}{\ln\left(\frac{23.85}{18}\right)} = 20.79°\text{C}$$

$$\text{Q} = \text{UA.LMTD}$$

$$A = \frac{0.4 \times 1.89 \times 10^3 \times 15}{333.564 \times 20.79} = 1.635 \text{ m}^2$$

$$= 2\pi d_2 \text{ L, for two passages}$$

$$L = \frac{1.635}{2\pi \times 19 \times 10^{-3}} = 13.696 \text{ m}$$

(*b*) The following data refers to design of an air-conditioning system of a theatre :

Total seating capacity = 600 persons

Sensible heat per person = 200 kJ/h

Latent heat per person = 130 kJ/h

Solar heat gain through glass, doors and walls = 2,50,000 kJ/h

Outside design condition = 40°C DBT, 27°C WBT

Inside design condition = 24°C DBT, 50% RH

Apparatus dew point temperature of cooling coil = 4°C

50% of return air is circulated and it is mixed with fresh air before entering the cooling coil. Calculate (*i*) the bypass factor of the cooling coil,

(*ii*) the mass flow rate of supply air to the theatre, and

(*iii*) the refrigeration capacity of the coil in tons of refrigeration.

Sol.

$$\text{SHL} = 600 \times 200 + 2{,}50{,}000 = 3{,}70{,}000 \text{ kJ/h}$$

$$\text{LHL} = 600 \times 130 = 78{,}000 \text{ kJ/h}$$

$$\text{THL} = \text{SHL} + \text{LHL} = 448{,}000 \text{ kJ/h}$$

Sensible heat factor,

$$\text{SHF} = \frac{\text{SHL}}{\text{THL}} = \frac{3{,}70{,}000}{4{,}48{,}000} = 0.8259$$

$$\text{ADP} = 4°\text{C}$$

Locate point '*a*' at 40°C DBT and 27°C WBT, point "*b*" at 24°C and 50% RH and point '*c*' on the saturation line of the psychrometric chart. Locate point 1, the mid-point *ab* for 50% return air recirculation. Join 1 and '*c*'. Draw the SHF line to intersect the saturation line at '*d*'. Join *bd* to intersect 1*c* at point 2, as shown in Fig. 5.

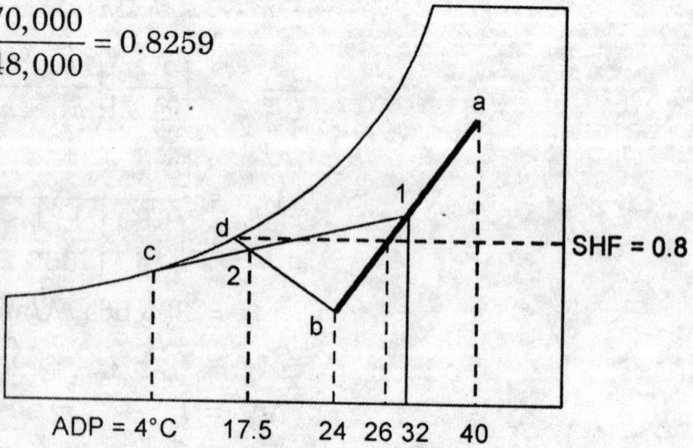

Fig 5. Psychrometric chart.

Then t_1 = 32°C and

t_2 = 17.5°C

(*i*)

$$\text{BPF} = \frac{t_2 - \text{ADP}}{t_1 - \text{ADP}} = \frac{17.5° - 4}{32 - 4} = 0.48$$

(*ii*) Mass flow rate of supply air,

$$m_a = \frac{\text{THL}}{h_1 - h_2} = \frac{4,48,000}{49 - 41} = 56,000 \text{ kg/h}$$

(*iii*) Refrigerating capacity of cooling coil

$$= m_a \times \frac{h_a - h_2}{211 \times 60} = 56,000 \left[\frac{67 - 41}{211 \times 60}\right] = 115 \text{ TR}$$

Q.4. (*a*) A 20 ton vapour compression refrigeration system using Freon-12 operates between evaporator pressure of 1.004 bar and condenser pressure of 13.663 bar. The system uses a subcooling-superheating heat exchanger with 10°C of superheat added. Determine (*i*) the mass flow rate, (*ii*) the COP, (*iii*) the degree of subcooling, and (*iv*) the power required. The refrigerant leaving the condenser is dry saturated liquid and leaving the evaporator is dry saturated vapour. The compression in compressor is isentropic.

Properties for R-12:

Pressure Bar	Temp. C°	Enthalpy kJ/kg		Entropy kJ/kg. K		Specific heat kJ/kg. K	
		Liquid	Vapour	Liquid	Vapour	Liquid	Vapour
1.004	– 30	8.854	174.076	0.0371	0.7165	0.573	0.579
13.663	55	90.201	207.766	0.3194	0.6777	1.074	—

(20)

Sol.

Given :

$$Q = 20 \text{ ton}$$
$$T_1 = 273 - 30 = 243 \text{ K},$$
$$T_2 = T_3 = 273 + 35$$
$$= 328 \text{ K}$$
$$T'_1 = T_1 + 10 = 253 \text{ K}$$

The vapour compression refrigeration cycle is shown in Fig. 6.

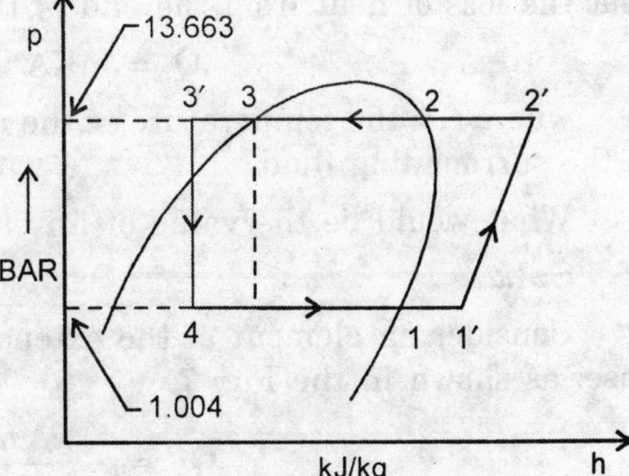

Fig 6.

For constant heat exchange, we have

$$h_1' - h_1 = h_3 - h_3'$$
$$c_{pv} (T_1 - T_1') = C_{pl} (T_3 - T_3')$$
$$0.579 \times 10 = 1.074 (328 - T_3')$$
$$T_3' = 322.61 \text{ K}$$

$$s_1' = s_1 + c_{pv} \ln\left(\frac{T_1'}{T_1}\right) = 0.7165 + 0.579 \ln\left(\frac{253}{243}\right)$$

$$= 0.7398 \text{ kJ'/kg. K}$$

$$= s_2' = s_2 + c_{pv} \ln\left(\frac{T_2'}{T_2}\right)$$

$$T_2' = 365.14 \text{ K}$$

$$h_1' = h_1 + c_{pv} (T_1' - T_1)$$

$$= 174.076 + 0.579 \times 100 = 179.866 \text{ kJ/kg}$$

$$h_2' = h_2 + c_{pv} (T_2' - T_2)$$

$$= 207.766 + 0.579 \times (365.14 - 328)$$

$$= 229.27 \text{ kJ/kg}$$

$$h_3' = h_3 + c_{pl} (T_3' - T_3)$$

$$= 90.201 + 1.074 \times (322.61 - 328) = 84.4 \text{ kJ/kg}$$

$$h_4 = h_3'$$

(*i*) Refrigeration effect,

$$q = h_1' - h_3' = 179.866 - 84.41 = 95.456 \text{ kJ/kg}$$

$$Q = mq, m = \frac{Q}{q} = \frac{20 \times 211}{95.456} = 44.2 \text{ kg/min or } 0.737 \text{ kg/s}$$

(*ii*)
$$\text{COP} = \frac{h_1' - h_3'}{h_2' - h_1'} = \frac{179.866 - 84.41}{229.87 - 179.865} = 1.932$$

(*iii*) Degree of sub-cooling

$$= T_3' - T_3 = 322.61 - 328 = -5.39°\text{C}$$

(*iv*) Power required

$$= m (h_2' - h_1') = 0.737 (229.27 - 179.866) = 36.41 \text{ kW}$$

(*b*) Derive the following expression for heat transfer from an extended surfac assuming that the loss of heat from the end of the extended surface is negligible.

$$Q = mKA (t_e - t_f) \tanh mL$$

where t_e = the temperature at the root of the extended surface, and t_f = the temperature of the surrounding fluid.

What would be the value of fin efficiency in the above case ?

Sol.

Consider an element of the extended surface of thickness dx at a distance x from the base, as shown in the Fig. 7.

$$Q_x = -\frac{kA.dt}{dx}$$

$$Q_{(x+dx)} = Q_x + \left(\frac{dQ_x}{dx}\right).dx$$

$$Q_{conv} = h(P.dx)(t - t_1)$$

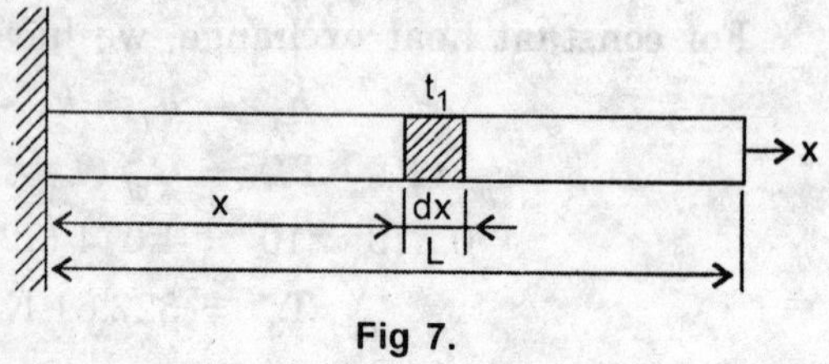

Fig 7.

where
k = thermal conductivity of the extended surface
A = surface area
P = perimeter
t = temperature
h = convective heat transfer co-efficient.

For steady heat flow, we have

$$Q_x = Q_{(x+dx)} + Q_{conv}$$

$$= Q_x + \left(\frac{dQ_x}{dx}\right).dx + Q_{conv}$$

or $$\left(\frac{dQ_x}{dx}\right).dx + Q_{conv} = 0$$

$$\left(\frac{d}{dx}\right)\left(\frac{-kA.dt}{dx}\right).dx + h(P.dx)(t - t_f)$$

or $$-kA\frac{d^2t}{dx^2} + hP(t - t_f) = 0$$

$$\frac{d^2t}{dx^2} - \frac{hP}{kA}(t - t_f) = 0$$

Let $$\theta = t - t_f \text{ and } m^2 = \frac{hP}{kA}$$

then $$\frac{d^2\theta}{dx^2} - m^2\theta = 0$$

The general solution of this equation is,

$$\theta = C_1 \cosh mx + C_2 \sinh mx$$

where C_1 and C_2 are constants.

The boundary conditions are :

At $x = 0 \quad \theta = \theta_0$

And at $x = L, \quad \dfrac{d\theta}{dx} = 0$

Hence $C_1 = \theta_0$

Now, $$\frac{d\theta}{dx} = C_1 m \sinh mx + C_2 \cosh mx$$

$$0 = \theta_0 m \sinh mx + C_2 m \cosh mL$$

$$\theta = \theta_0 [\cosh mx - \tanh mL \sinh mx]$$

$$\frac{\theta}{\theta_0} = \cosh mx - \tanh mL \sinh mx$$

$$= \frac{\cosh m(L - x)}{\cosh mL}$$

$$q_{fin} = -kA\left(\frac{d\theta}{dx}\right) \text{ at } x = 0$$

Now, $$\left.\frac{d\theta}{dx}\right|_{x=0} = -\left(\frac{\theta}{\cosh mL}\right) m \sinh m(L - x)$$

$$q_{fin} = \theta_0 \, kAm \left(\frac{\sinh mL}{\cosh mL} \right)$$

$$= mkA\,(t_a - t_f) \tanh mL$$

Fin efficiency,
$$\eta_{fin} = \frac{q_{fin}}{q_{\max}} = mkA\,(t_a - t_f) \frac{\tanh mL}{[hPL\,(t_a - t_f)]}$$

$$= \left[\frac{hP}{kA}\right]^{0.5} . \frac{kA \tanh mL}{(hPL)} = \left[\frac{kA}{hP}\right]^{0.5} . \frac{(\tanh mL)}{L}$$

$$= \frac{\tanh mL}{mL}$$

(*c*) The filament of a 75 W light bulb may be considered a black body radiating into a black enclosure at 70°C. The filament diameter is 0.10 mm and the length is 5 cm. Considering only radiation, determine the filament temperature.

Sol.

Given :

$$q = 75 \text{ J/s}, \; T_2 = 273 + 30 = 343 \text{ K}$$

$$d = 0.1 \text{ mm}, \; L = 5 \text{ cm}$$

$$q = \sigma \varepsilon A(T_1^4 - T_2^4)$$

$$75 = 5.67 \times 10^{-8} \times 1 \times (\pi \times 0.1 \times 10^{-3} \times 5 \times 10^{-2}) \times (T_1^4 - 343^4)$$

$$T_1^4 = 8.42 \times 10^{13} + 138.413 \times 10^8$$

$$T_1 = 3029 \text{ K or } 2756°\text{C}$$

Q.5. (*a*) A simple impulse turbine has a mean blade speed of 200 m/s. The nozzles are inclined at 20° to the plane of rotation of the blades. The steam velocity from nozzles is 600 m/s. The turbine uses 3500 kg/h of steam. The absolute velocity at exit is along the axis of the turbine. Determine (*i*) the inlet and exit angles of the blades, (*ii*) the power output of the turbine, (*iii*) the diagram efficiency, and (*iv*) the end thrust (per kg steam per second) and its direction. **(20)**

Sol.

Given: $\alpha_1 = 20°$, $\alpha_2 = 90°$,

$$\beta_1 = \beta_2,$$

$$u = 200 \text{ m/s},$$

$$v_{a1} = 600 \text{ m/s},$$

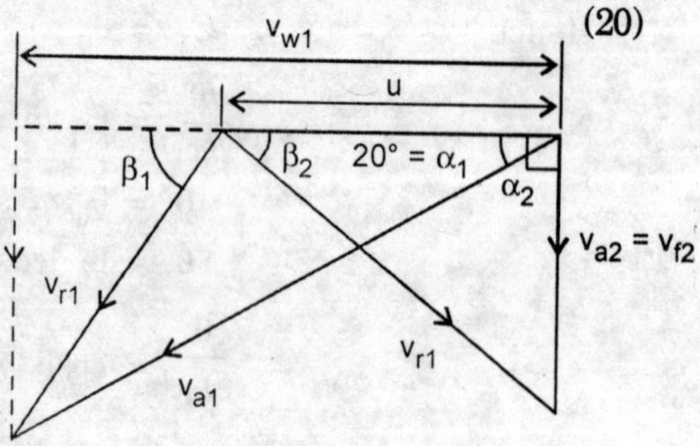

The velocity diagrams are as shown in Fig. 8.

Fig 8.

(*i*) Velocity of flow at inlet

$$v_{f1} = v_{a1} \sin \alpha_1 = 600 \sin 20°$$

$$= 205.2 \text{ m/s}$$

$$\tan \beta_1 = \frac{v_{f1}}{v_{a1} \cos \alpha_1 - u} = \frac{205.2}{600 \cos 20° - 200} = 0.564$$

$$\beta_1 = 29.42° = \beta_2$$

(*ii*) Power developed $= mu \dfrac{(v_{w1} - v_{w2})}{10^3}$

$$= \left(\frac{3500}{3600}\right) \times 200 \times \frac{(600 \cos 20° - 0)}{10^3} = 109.63 \text{ kw}$$

(*iii*) Diagram efficiency $= \dfrac{2v_{w1}\, u}{v_{a1}^2} = \dfrac{2 \times 600 \cos 20° \times 200}{600^2} = 62.65\%$

(*iv*) $v_{f2} = u \tan \beta_2 = 200 \tan 29.42° = 112.78$ m/s

End thrust $= \text{m } (v_{f1} - v_{f2}) = 1 \times (205.2 - 112.78)$

$= 92.42$ N/kg of steam to the left

(*b*) Draw schematic diagram of Pressurized Water Reactor (PWR) and Boiling-Water Reactor (BWR) and discuss their relative merits. (10)

Sol.

Pressurized Water Reactor (PWR). It is light water cooled moderated thermal reactor. It uses both natural and highly enriched fuel. The schematic diagram of PWR is shown below in Fig. 9 (*a*).

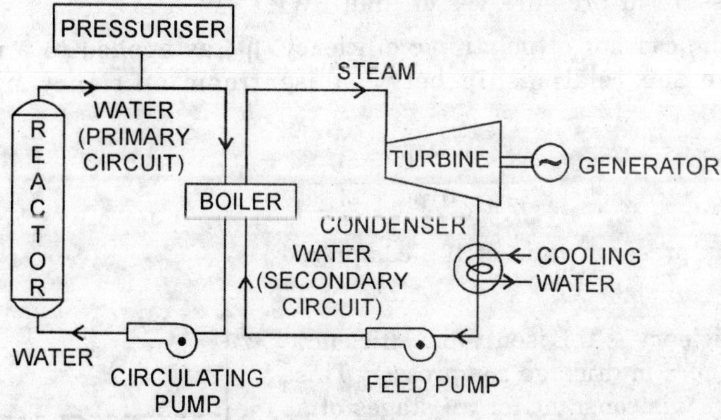

Fig. 9 (*a*) PWR

The principal parts of PWR are : Pressure vessel, reactor thermal shield, fuel elements, control rods, reactor containment and reactor pressuriser.

Merits :

1. Water used in reactor is cheap and easily available.
2. Compact size and power density is high.
3. Fission products remain contained in the reactor.
4. A small number of control rods are required.
5. Complete freedom for inspection and maintenance of turbine, feed heaters and condenser during operation.
6. Less fuel cost per unit of energy produced.

Boiling Water Reactor (BWR). The schematic diagram of BWR is shown below in Fig. 9 (*b*).

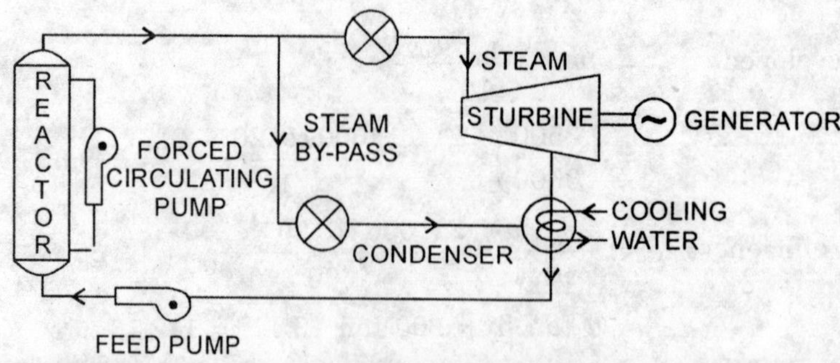

Fig. 9 (*b*) BWR

Merits :

1. Heat exchanger circuit is eliminated resulting in gain in thermal efficiency and cost.
2. Use of lower pressure vessel for the reactor which further reduces cost.
3. Low metal temperature for given output.
4. More efficient cycle than PWR for given containment pressure.
5. Higher outlet temperature of steam than PWR.
6. Lower pressure in pressure vessel than PWR.

(*c*) Explain the concept of polytropic efficiency, η_∞, as applied to a multistage rotary compressor. Drive the relationship between isentropic efficiency η_i and polytropic efficiency η_∞.

$$\eta_i = \frac{r^{(\gamma - 1)/\gamma} - 1}{r^{\{(\gamma - 1)(\gamma\eta_\infty - 1)\}} - 1}$$

where r is the pressure ratio of the compressor. **(10)**

Sol.

Polytropic efficiency is the isentropic efficiency of one small stage of a multistage compressor. This small stage efficiency is constant for all stages of a compressor with infinite number of stages.

The actual and isentropic compression of a single stage compressor is shown in Fig. 10 (*a*) on T-s dsiagram.

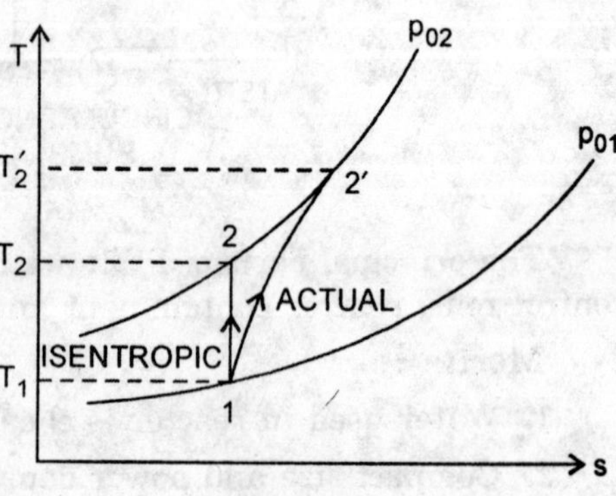

Fig 10 (*a*)

Isentropic efficiency,

$$\eta_i = \frac{\text{Isentropic work}}{\text{Actual work}}$$

$$= \frac{h_{02} - h_{01}}{h_{02}' - h_{01}} = \frac{T_2 - T_1}{T_2' - T_1}$$

$$= \frac{\left(\dfrac{T_2}{T_1} - 1\right)}{\left(\dfrac{T_2'}{T_1} - 1\right)} = \frac{r^{(\gamma - 1)/\gamma} - 1}{r^{(n - 1)/n} - 1}$$

where $r = \frac{p_{02}}{p_{01}}$

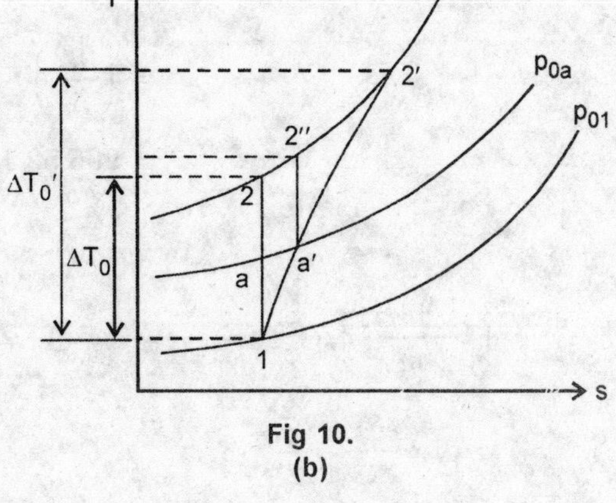

Fig 10. (b)

For a multistage rotary compressor, as shown in Fig. 10 (*b*), actual compression path is 1 – 2′ and isentropic path is 1 – 2.

Polytropic efficiency, $\eta_\infty = \left[\frac{\gamma - 1}{\gamma}\right]\left[\frac{n}{n-1}\right]$

or $$\frac{n}{n-1} = \left[\frac{\gamma - 1}{\gamma\, \eta_\infty}\right]$$

Thus, $$\eta_i = \frac{\left[r^{(\gamma-1)/\gamma} - 1\right]}{\left[r^{\{(\gamma-1)/(\gamma\,\eta_\infty)\}} - 1\right]}$$

Q.6. (*a*) A turbojet unit is flying at a speed of 268 m/s at an altitude where the ambient conditions are 0.20 bar and 220 K. The air enters an ideal diffuser and leaves the combustor at 1350 K and 1 bar. The fuel supplied has a heating value of 43,000 kJ/kg. Assume all compression and expansion process to be isentropic. Determine

(*i*) the air-fuel ratio, (*ii*) the specific thrust, and

(*iii*) the propulsive efficiency.

Sol.

Given. $v_a = 268$ m/s, $p_1 = 0.2$ bar, $T_1 = 202$ K, $T_3 = 1350$ K

$$T_2 = T_1\left(\frac{p_2}{p_1}\right)^{(\gamma-1)/\gamma} = 220\left(\frac{1}{0.2}\right)^{0.4/1.4} = 348.44 \text{ K.}$$

(*i*) For the combustion chamber, we have

$$m_f \times \text{C.V} = c_{pg}\ [(m_a + m_f)\, T_3 - m_a\, T_2]$$

$$\text{C.V} = c_{pg}\left[\left(\frac{m_a}{m_f} + 1\right)T_3 - \left(\frac{m_a}{m_f}\right)T_2\right]$$

$$43{,}000 = 1.102\left[\left(\frac{m_a}{m_f}\right)(1350 - 348.44) + 1350\right]$$

$$\frac{m_a}{m_f} = 37.61$$

or $$\frac{m_f}{m_a} = \frac{1}{37.61} = 0.0265$$

(*ii*) Process 3-4 : As shown in Fig. 11.

$$T_4 = T_3\,(p_4/p_3)^{(n-1)/n}$$

$$= \left(\frac{0.2}{1}\right)^{0.33/1.33}$$

$$= 905.53 \text{ K}$$

For the nozzle :

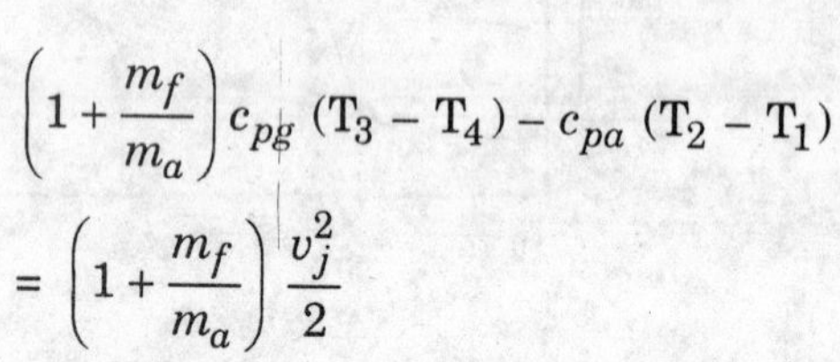

$$\left(1+\frac{m_f}{m_a}\right) c_{pg}\,(T_3 - T_4) - c_{pa}\,(T_2 - T_1) = \left(1+\frac{m_f}{m_a}\right)\frac{v_j^2}{2}$$

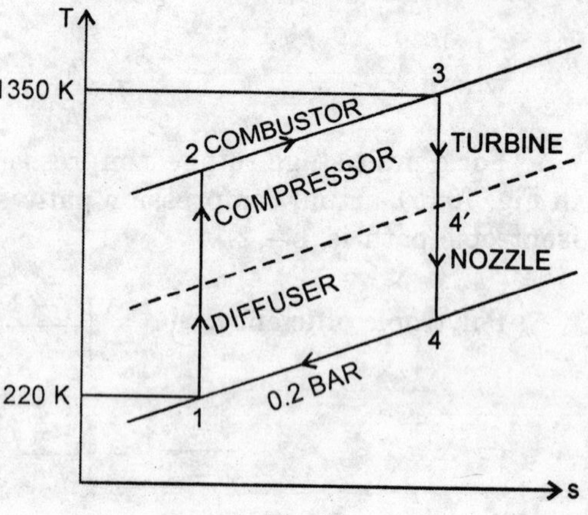

Fig 11.

$$(1 + 0.0265) \times 1.102 \times 10^3 \times (1350 - 905.53) - 1.005 \times 10^3 \times (248.44 - 200) = \frac{(1+0.0265)\,v_j^2}{2}$$

$$v_j = 853.295 \text{ m/s}$$

$$\text{Specific thrust} = \left(1+\frac{m_f}{m_a}\right) v_j - v_a = (1 + 0.0265)\ 853.295 - 268 = 607.7 \text{ N/kg}$$

(*iii*) Propulsive efficiency

$$= \frac{\{(1+m_f/m_a)\,v_j - v_a\}\,v_a}{(1+m_f/m_a)v_j^2/2 - v_a^2/2}$$

$$= \frac{[(1+0.0265\times 853.295 - 268)]\times 268}{(1+0.0265)\times 853.295^2/2 - 268^2/2}$$

$$= 0.4823 \text{ or } 48.23\%$$

Q.6. (*b*) A double-acting single cyclinder reciprocating air compressor has a piston displacement of 0.015 m per revolution, operates at 500 rpm and has 5% clearance. The air is received at 1 bar and delivered at 6 bar. The compression and expansion are polytropic with $n = 1.3$. Determine

(*i*) the volumetric efficiency,

(*ii*) the power required, and

(*iii*) the heat transferred and its direction, during compression and inlet temperature of air at 20°C. (15)

Sol.

Given :

$$v_s = 0.015 \text{ m}^3\text{/rev},\ N = 500 \text{ rpm},$$

$$r_c = \frac{v_c}{v_s} = 0.05$$

$$p_1 = 1 \text{ bar},\ p_2 = 6 \text{ bar},\ n = 1.3,\ T_1 = 273 + 20 = 293 \text{ K}$$

(*i*) Volumetric efficiency,

$$\eta_v = 1 + r_c - r_c \left(\frac{p_2}{p_1}\right)^{1/n}$$

$$= 1 + 0.05 - 0.05\left(\frac{6}{1}\right)^{1/13} = 0.8516 \text{ or } 85.16\%$$

(*ii*) Power required,

$$P = \left[\frac{n}{n-1}\right] p_1 v_a \left[\left(\frac{p_2}{p_1}\right)^{(n-1)/n} - 1\right]$$

$$\eta_v = \frac{v_a}{v_s}, v_a = 0.015 \times 0.8516 \times 500 \times 2 = 12.774 \text{ m}^3/\text{min}$$

$$P = \left(\frac{1.3}{0.3}\right) \times 10^2 \times 12.774 \times \left[\left(\frac{6}{1}\right)^{0.3/1.3} - 1\right] = 2834.54 \text{ kW}$$

(*iii*) Swept volume,

$$V_s = v_s \times 2\text{ N} = 0.015 \times 2 \times 500 = 15 \text{ m}^3/\text{min}$$

Actual air sucked, $V_{sa} = 15 \times 0.8516 = 12.774 \text{ m}^3/\text{min}$

Mass of air sucked,

$$m = \frac{p_1 V_{as}}{RT_1} - \frac{1 \times 10^5 \times 12.744}{287 \times 293} - 15.17 \text{ kg/min}$$

$$T_2 = T_1 \left(\frac{p_2}{p_1}\right)^{(n-1)/n} = 293\left(\frac{6}{1}\right)^{0.3/1.3} = 443 \text{ K}$$

Heat transferred during compression,

$$Q_c = m c_v \left[\frac{\gamma - n}{n-1}\right][T_2 - T_1]$$

$$= 15.19 \times 0.718 \left[\frac{1.4 - 1.3}{1.3 - 1}\right][443 - 293]$$

$$= 545.321 \text{ kJ/min}$$

Q.7. (*a*) The efficiency η of *a* fan depends on th kinematic viscosity ν of the fluid, the angular velocity ω, diameter D of the rotor and the discharge Q. Express η in terms of dimensionless parameter using Buckkingham's π-theorem.

$$\eta = \phi\left[\frac{\omega D^2}{\nu}, \frac{Q}{\omega D^2}\right] \quad (15)$$

Sol.

$$\nu = \frac{\mu}{\rho}$$

Let, $\eta = f(\mu, \rho, \omega, D, Q)$

Sl. No.	Variable	Dimensions
1.	η	$M^{\circ} L^{\circ} T^{\circ}$
2.	μ	$ML^{-1} T^{-1}$
3.	ρ	ML^{-3}
4.	ω	T^{-1}
5.	D	L
6.	Q	$L^3 T^{-1}$

Total number of variables, $n = 6$

Number of fundamental dimensions, $m = 3$

Total numbe of dimensionless groups (or π-terms)

$$= n - m = 6 - 3 = 3$$

Selecting ρ, ω and D as the repeating variables and making use of Buckingham- π theorem, we have

$$\pi_1 = \rho^{a1} \omega^{b1} D^{c1} \eta \quad ...(1)$$

$$\pi_2 = \rho^{a2} \omega^{b2} D^{c2} \mu \quad ...(2)$$

$$\pi_3 = \rho^{a3} \omega^{b3} D^{c3} Q \quad ...(3)$$

Expressing Eqn. (1) dimensionally, we have

$$M^0 L^0 T^0 = [ML^{-3}]^{a1} [T^{-1}]^{b1} [L]^{c1} = M^{a1}.L^{-3a1+c1}.T^{-b1}$$

$$M : 0 = a_1$$

$$L : 0 = -3a_1 + c_1$$

$$T : 0 = -b_1$$

Hence,
$$a_1 = 0, b_1 = 0, c_1 = 0$$

$$\pi_1 = \eta$$

For Eqn. (2), we have

$$M^0 L^0 T^0 = [ML^{-3}]^{a2} [T^{-1}]^{b2} [L]^{c2} (ML^{-1} T^{-1})$$

$$= M^{a_2+1}.L^{-3a_2+c_2-1}.T^{-b_2-1}$$

$$a_2 + 1 = 0, \text{ or } a_2 = -1,$$

$$-3a_2 + c_2 - 1 = 0$$

Thus,
$$c_2 = 1 + 3a_2 = 1 - 3 \times 1 = -2$$

$$-b_2 - 1 = 0, b_2 = -1$$

$$\pi_2 = \rho^{-1} \omega^{-1} D^{-2} \mu = \frac{\mu}{\rho\omega D^2} = \frac{\nu}{\omega D^2}$$

or
$$\pi_2 = \frac{\omega D^2}{\nu}$$

For Eqn. (3), we have

$$M^0 L^0 T^0 = [ML^{-3}]^{a3} [T^{-1}]^{b3} [L]^{c3} (L^3 T^{-1}) = M^{a3}.L^{-3a_3+c_3+2}.T^{-b_3-1}$$

$$a_3 = 0$$

$$-3a_3 + c_3 + 3 = 0, \text{ or } c_3 = -3 + 3 \times 0 = -3$$

$$-b_3 - 1 = 0, \text{ or } b_3 = -1$$

$$\pi_3 = \rho^0\, \omega^{-1}\, D^{-3}\, Q = \frac{Q}{\omega D^3}$$

Thus, $$\eta = \phi\left[\frac{\omega D^2}{\nu}, \frac{Q}{\omega D^3}\right]$$

(b) In an aircraft flying at an altitude where the pressure was 35 kPa and temperature – 38°C, stagnation pressure measured was 65.4 kPa. Calculate the speed of the aircraft.

Sol.

Given : $p = 35$ kPa, T = 273 – 38 = 235 K, $p_0 = 65.4$ kPa

$$T_0 = T\left(\frac{p_0}{p}\right)^{(\gamma-1)/\gamma} = 235\left(\frac{65.4}{35}\right)^{0.4/1.4} = 280.96 \text{ K}$$

$$\frac{T_0}{T} = 1.1956 = 1 + \left[\frac{\gamma - 1}{2}\right] M^2 = 1 + 0.2\, M^2$$

$$M = 0.989$$

Now, $$\frac{v}{v_s} = M$$

$$v_s = [\gamma RT]^{0.5} = [1.4 \times 287 \times 235]^{0.5} = 307.28 \text{ m/s}$$

$$v = 307.28 \times 0.989 = 303.9 \text{ m/s}$$

(c) Discuss the air-fuel ratio requirements of a petrol engine automobile for starting and warm up, idling and low load, normal power range, maximum power range and acceleration. State how these requirements are achieved. (15)

Sol.

Starting and Warm up. During starting a very rich mixture must be supplied, *i.e.*, air-fuel ratio of 3 : 10 to 1.5 : 1 (F/A ratio 0.3 – 0.7) because the engine speed and temperature are low to ignite the fuel. As the engine warms up the mixture ratio should be progressively made leaner to avoid too much evaporated fuel-air ratio. The volatility of the fuel significantly affects the starting and warm up characteristics of the engine.

Idling and low load. During idling (no load) the air supply is restricted by the nearly closed throttle and the suction pressure is very low. This condition of low pressure gives rise to back flow of exhaust gases and air leakage from the various parts of the engine intake system. The problem of dilution by the residual gases become more pronounced at low speeds and idling. This requires that the air-fuel ratio used for idling and low loads should be rich (A/F = 12.5 : 1 or F/A = 0.8 : 1) for smooth engine operation. The richening of mixture increases the probability of contact between fuel and air particles and thus improves combustion.

Normal power (or Cruising) range. In normal power range the prime consideration is usually the fuel economy. Maximum fuel economy occurs at A/F ratio of about 17 : 1 (F/A = 0.06 : 1). Practicle values are about 5% higher (A/F = 16.7 : 1).

Maximum power range. Maximum power is obtained at about 14 : 1 A/F (0.07 : 1 F/A) ratio. Maximum energy is released when slightly excess fuel is introduced so that all the oxygen present in the cylinder is utilized. Mechanical efficiency is maximum at maximum power position.

Acceleration. Acceleration refers to an increase in engine speed or torque resulting from opening the throttle. During acceleration, the liquid fuel lgas behind and temporarily the cylinder receives a lean mixture whereas a rich mixture is needed to produce more instantaneous power for acceleration. To achieve this, additional fuel must be supplied by an acceleration pump.

Q.8. ***(a)*** Explain displacement and momentum boundary layer thickness. Assume that the shear stress varies linearly in a laminar boundary layer such that

$$\tau = \tau_0 \left(1 - \frac{y}{\delta}\right).$$

Calculate the displacement and momentum thickness in terms of δ. **(15)**

Sol.

Displacement thickness, δ*. It may be defined as the distance measured perpendicular to the boundary by which the free stream is displaced on on account the formation of boundary layer. The displacement thickness for two-dimensional boundary layer is shown in Fig. 12.

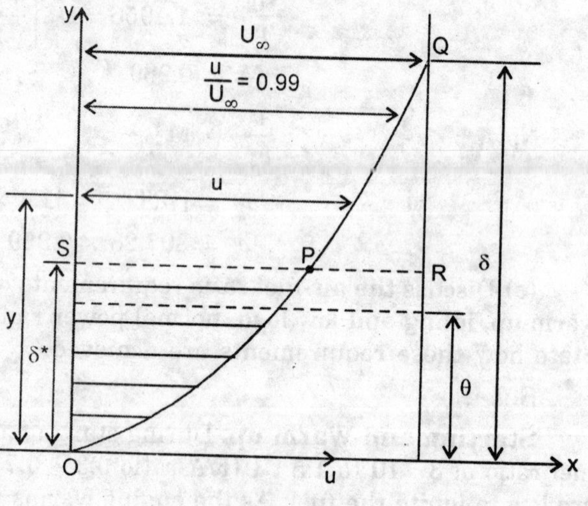

Fig 12.

Let b = Width of the flow section

Total volume rate of flow past a section, considering influence of boundary layer, is

$$Q = b\int_0^\infty u\,dy$$

$$= b\int_0^\delta u\,dy + b\int_\delta^\infty U_\infty\,dy$$

where U_∞ = free stream velocity,

δ = boundary layer thickness.

Without the boundary layer, total volume rate of flow is,

$$Q_i = b\int_0^\infty U_\infty\,dy + b\int_\delta^\infty U_\infty\,dy$$

Reduction in flow due to boundary layer,

$$Q_r = Q_i - Q$$

$$= b\int_0^\infty U_\infty\,dy - b\int_0^\infty u\,dy$$

$$= b\int_0^\infty (U_\infty - u)\,dy$$

Frictionless flow passing through an area $b\delta^* = b\delta^* U_\infty$

Area OSP = Area PQR

$$b\delta^* U_\infty = b\int_0^\infty (U_\infty - u)\,dy$$

or

$$\delta^* = \int_0^\infty \left(1 - \frac{u}{U_\infty}\right) dy$$

For $\tau = \tau_0 \left(1 - \frac{y}{\delta}\right)$

$$\frac{\mu\, du}{dy} = \mu\left(\frac{dU_\infty}{dy}\right)\left(1 - \frac{y}{\delta}\right)$$

$$u = U_\infty \left(1 - \frac{u}{U_\infty}\right) dy$$

$$\delta^d = \int_0^\delta \left(1 - \frac{u}{U_\infty}\right) dy$$

$$= \int_0^\delta \left(\frac{y}{\delta}\right) dy = \left|\frac{y^2}{2\delta}\right|_0^\delta = \frac{\delta}{2}$$

Momentum thickness, θ. It can be defined as the distance corresponding to the momentum defect brought about by an actual profile in relation to the uniform profile for a displacement θ. The momentum transport rate in the boundary layer is

$$M = b\int_0^\delta (\rho u\, dy)\, U_\infty = b\int_0^\delta \rho u^2\, dy$$

Momentum for the same mass flow rate at the free-stream velocity U_∞ is

$$M_i = b\int_0^\delta (\rho u\, dy)\, U_\infty = b\int_0^\delta \rho u\, U_\infty dy$$

Reduction in momentum, $M_r = M_i - M$

$$= b\int_0^\delta \rho(uU_\infty - u^2)\, dy = b\int_0^\delta \rho u\,(U_\infty - u)\, dy$$

Momentum transport through the flow area $b\theta = \rho b\theta U_\infty^2$

$$\rho b\theta U_\infty^2 = \int_0 \rho u\,(U_\infty - u)\, dy$$

$$\theta = \int_0^\delta \left(\frac{u}{U_\infty}\right)\left(1 - \frac{u}{U_\infty}\right) dy$$

$$\theta = \int_0^\delta \left(1 - \frac{y}{\delta}\right)\left(\frac{y}{\delta}\right) dy = \frac{\delta}{6}$$

(*b*) A pipeline, with diameter 0.8 m and length 3000 m connects two open reservoirs of water which have their water surface at elevations of 100 m and 70 m above a datum. In order to increase the rate of flow between the reservoirs by 20% it is decided to lay an additional 0.8 m diameter piepline from the upper reservoir. The second pipeline is to be parallel to the original pipeline and is to be connected to the latter at some suitable point. Determine the point of connection, assuming that the friction factor *f* is 0.04 for each pipeline. Neglect minor losses. (15)

Sol.

Given : $d = 0.8$ m, L = 3000 m, $f = 0.04$

Friction head loss, $$h_f = \frac{f\,Lv^2}{2gd}$$

Now, $$Q = \left(\frac{\pi}{4}\right) d^2 v \text{ and } v = \frac{4Q}{\pi d^2}$$

$$h_f = \frac{f\,L\,Q^2}{12\,d^5}$$

Loss in head = 100 – 70 = 30 m as shown in Fig. 13.

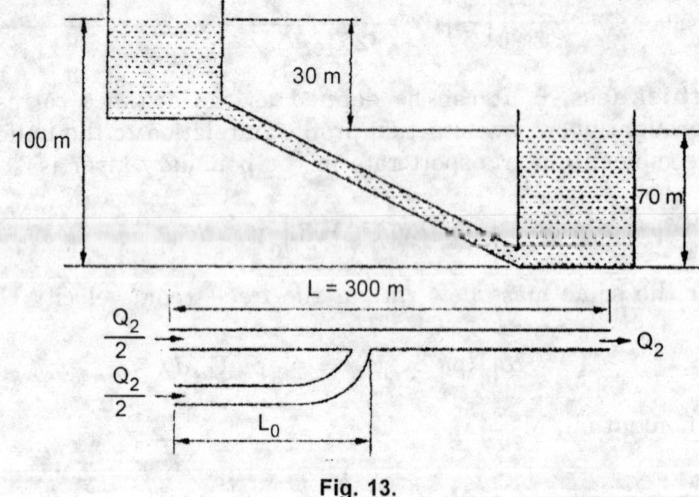

Fig. 13.

With a single piple,

$$30 = \frac{0.04 \times 3000 \times Q_1^2}{12 \times .0.8^5}$$

$$Q_1 = 0.9915 \text{ m}^3/\text{s}$$

Total discharge with additional pipe of same diameter,

$$Q_2 = 1.2\ \ Q_1$$
$$= 1.1898 \text{ m}^3/\text{s}$$

Let the second pipe be connected at a distance L_0 from the higher reservoir. Then,

$$h_f = 2 \times f\,L_0 \frac{\left(\dfrac{Q_2}{2}\right)^2}{(12\,d^5)} + \frac{f(L - L_0)\,Q_2^2}{(12\,d^5)}$$

$$= \left[\frac{f\,Q_2^2}{12\,d^5}\right]\left[\frac{L_0}{2} + (L - L_0)\right]$$

$$\left(\frac{12d^5\, h_f}{f\, Q_2^2}\right) = L - \frac{L_0}{2}$$

$$\frac{12 \times (0.8)^5 \times 30}{0.04 \times (1.1898)^2} = 3000 + \frac{L_0}{2}$$

$$L_0 = 1833.47 \text{ m}$$

(c) A centrifugal pump rotating at 1500 rpm delivers 0.2 m³/s at a head of 15 m. Calculate the specific speed of the pump and the power input. Assume overall efficiency of the pump as 0.68.

If this pump were to operate at 900 rpm, what would be the head, discharge and power required for homologous conditions? Assume that overall efficiency remains uncharged at new rpm.

Sol.

Given : $N = 1500$ rpm, $Q = 0.2$ m³/s, $H = 15$, $\eta_{overall} = 0.68$

$$N_s = \frac{NQ^{1/2}}{H^{3/4}} = \frac{1500 \times (0.2)^{1/2}}{15^{3/4}} = 88.01 \text{ rpm}$$

Power input,

$$P = \frac{\rho g Q H}{\eta_{overall} \times 10^3} \text{ kW}$$

$$= \frac{1000 \times 9.81 \times 0.2 \times 15}{0.68 \times 10^3} = 43.28 \text{ kW}$$

$$H_2 = H_1 \left(\frac{N_2}{N_1}\right)^2 = 15\left(\frac{900}{1500}\right)^2 = 5.4 \text{ m}$$

$$Q_2 = Q_1 \left(\frac{N_2}{N_1}\right) = 0.2\left(\frac{900}{1500}\right) = 0.12 \text{ m}^3/\text{s}$$

$$P_2 = P_1 \left(\frac{N_2}{N_1}\right)^3 = 43.28\left(\frac{900}{1500}\right)^3 = 9.3485 \text{ kW}$$

MECHANICAL ENGINEERING
PAPER I—1999

(Conventional)

Q.1. **(*a*)** Show that the cyclic integral of *d*Q/T for a reversible cycle is equal to zero. (10)

Sol.

Consider a reversible cycle ABCD on *p*-V diagram as shown in Fig. 1. The cycle is sub-divided by a family of reversible adiabatics and reversible isothermals. Every two adjacent adiabatics are connected by two reversible isothermals so that the heat transferred during all isothermal processes are equal to the heat transferred during the original cycle. The cycle can be sub-divided into a large number of Carnot cycles which provide the same amount of work.

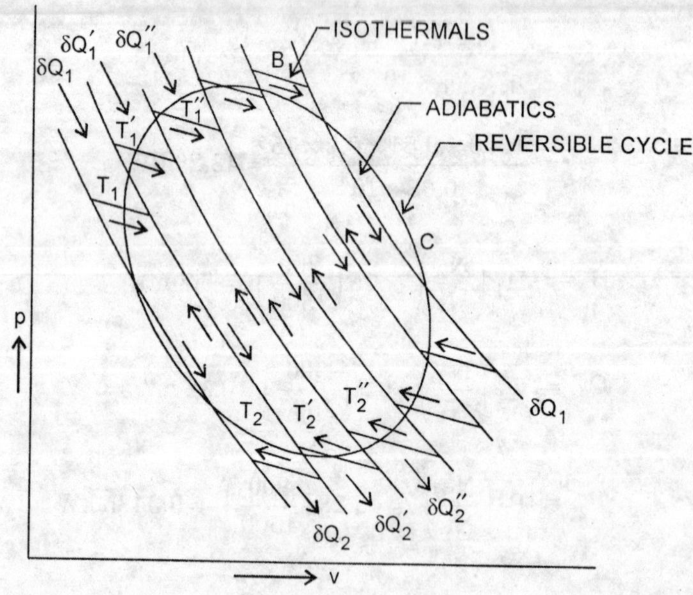

Fig. 1.

For the first small Cannot cycle,

$$\eta_1 = 1 - \frac{\delta Q_2}{\delta Q_1} = 1 - \frac{T_2}{T_1}$$

or $$\frac{T_2}{T_1} = \frac{\delta Q_2}{\delta Q_1}$$

or $$\frac{\delta Q_1}{T_1} = -\frac{\delta Q_2}{T_2}$$

Considering heat added as +ve and heat rejected as –ve, the above equation becomes

$$\frac{\delta Q_1}{T_1} = -\frac{\delta Q_2}{T_2}$$

or $$\frac{\delta Q_1}{T_1} + \frac{\delta Q_2}{T_2} = 0$$

Simiarly for the second cycle,

$$\frac{\delta Q_1'}{T_1'} + \frac{\delta Q_2'}{T_2'} = 0$$

and so on.

For the whole cycle ABCD, we get

$$\left(\frac{\delta Q_1}{T_1} + \frac{\delta Q_2}{T_2}\right) + \left(\frac{\delta Q_1'}{T_1'} + \frac{\delta Q_2'}{T_2'}\right) + \ldots = 0$$

or $$\int_R \frac{\delta q}{T} = 0$$

(*b*) A reversible engine as shown in the Fig. 2 during a cycle of operation draws 5 MJ from the 400 K reservoir and does 840 kJ of work. Find the amount and direction of heat interaction with other reservoirs. (15)

Sol.

Let the heat interaction between the reservoirs be as shown by arrow heads in the Fig. 2.

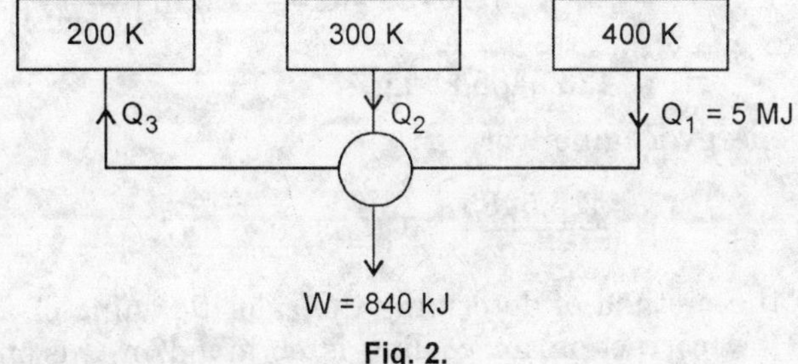

Fig. 2.

For heat balance, we have

$$5000 + Q_2 = 840 + Q_3 \quad \ldots(1)$$

$$Q_3 - Q_2 = 4160$$

Now for a reversible process,

$$\int \frac{dQ}{T} = 0$$

$$\frac{5000}{400} + \frac{Q_2}{300} - \frac{Q_3}{200} = 0$$

$$15{,}000 + 4\,Q_2 - 6\,Q_3 = 0$$

$$Q_2 - 1.5\,Q_3 = -3750 \qquad ...(2)$$

Solving Eqns. (1) and (2), we get

$$Q_2 = -4980 \text{ kJ}, \; Q_3 = -820 \text{ kJ}$$

Negative sign indicates that the direction of heat interaction is opposite to the assumed direction.

(*c*) Exhaust gases leave an internal combustion engine at 800°C and 1 atmosphere, after having done 1050 kJ of work per kg of gas in the engine (c_p = 1.1 kJ/kg. K). The temperature of the surroundings is 30°C.

(*i*) How much available energy per kg of gas is lost by throwing away the exhaust gases?

(*ii*) What is the ratio of the lost available exhaust gas energy to the egine work? (15)

Sol.

Given :

$$T_1 = 273 + 800 = 1073 \text{ K}$$

$$T_0 = 273 + 30 = 303 \text{ K}, \; p_1 = 1 \text{ atm},$$

$$W = 1050 \text{ kJ/kg},$$

$$c_p = 1.005 \text{ kJ/kg K}$$

(*i*) Available energy lost

$$= \int_{T_0}^{T_1} mc_p \left(1 - \frac{T_0}{T}\right) dT$$

$$= mc_p \left| T - T_0 \ln T \right|_{T_0}^{T_1} = mc_p \left[(T_1 - T_0) - T_0 \ln\left(\frac{T_1}{T_0}\right) \right]$$

$$= 1 \times 1.1 \left[(1073 - 303) - 303 \ln\left(\frac{1073}{303}\right) \right]$$

$$= 425.5485 \text{ kJ/kg}$$

(*ii*) Lost available energy/Engine work

$$= \frac{425.5485}{1050} = 0.4053$$

Q.2. (*a*) Describe the method of determing the calorific value of a liquid fuel by means of Bomb calorimeter. How correction for cooling is obtained in this method? (12)

Sol.

The bomb calorimeter (Fig. 3) is used to determine the calorific value of solid and liquid fuels. The arrangement of the calorimeter is shown in the Fig. 3. The bomb is placed in water bath which itself is placed in another container which acts as heat insulator. A thermometer and stirrer are inserted in the outer vessel. A known quantity of liquid fuel is taken in the bomb in such a way that the fuse touches the fuel. The bomb is then charged with oxygen to a pressure of 30 atm. The bomb is placed in water bath and then closed with the cover. The water in the calorimeter is stirred and when the temperature remains steady, the fuel is ignited by passing a current through the fuse wire. The temperature readings are noted with the thermometer at regular intervals and recorded. The maximum temperature reading of the bath is also recorded.

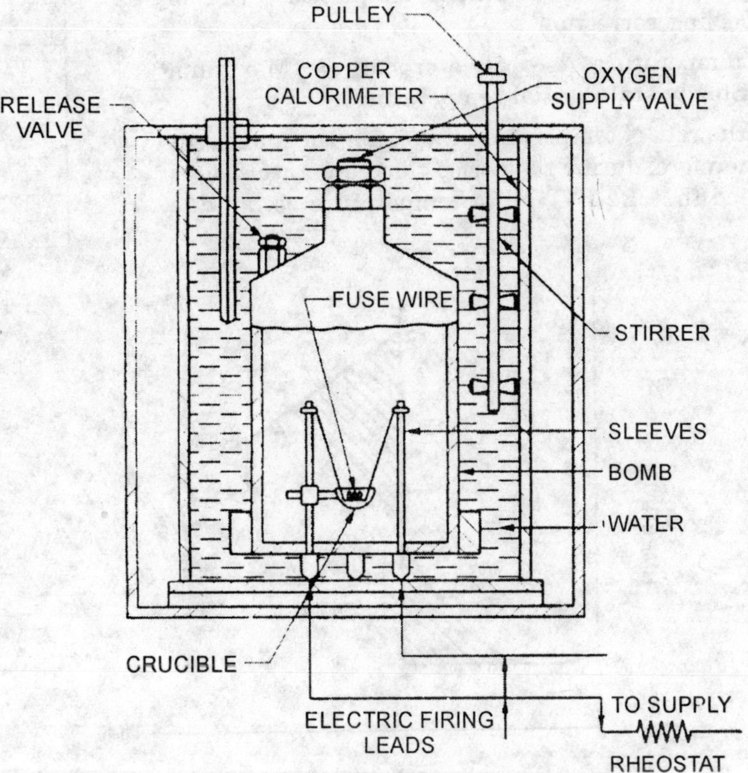

Fig. 3. (*a*) Bomb calorimeter.

For heat balance, we have

Heat given by the fuel due to combustion + Heat given by the combustion of fuse wire = Heat absorbed by water and calorimeter

$$m_f \times \text{HCV} + m_{wf} \times \text{CV} = (m_w + m_e)\,\theta$$

where m_f = mass of fuel

m_{wt} = mass of fuse wire

m_w = mass of water

m_e = water equivalent of calorimeter

θ = corrected temperature rise

$$\text{HCV} = \frac{(m_w + m_e)\,\theta - m_{wf} \times \text{CV}}{m_f}$$

The water equivalent of calorimeter is determined by burning a fuel (like Benzoic acid or Nepthalene) of known quantity and calorific value.

Correction for Cooling. After the maximum temperature is reached, it starts falling slowly. When the temperature fall shows a steady rate, the temperature readings are taken at regular intervals for about 5 minutes. The cooling correction is found by plotting time v's temperature as shown in the Fig. 3 (*b*) and (*c*).

True temperature rise = Apparent temperature rise of water bath + Cooling correction

Apparent temperature rise of water bath = (Maximum temperature – Steady state initial temperature).

Let r°C be the rate of temperature drop after the maximum temperature. Therefore, during the period the temperature rises it can be assumed that 1/2 r°C fall in temperature occurs due to radiation.

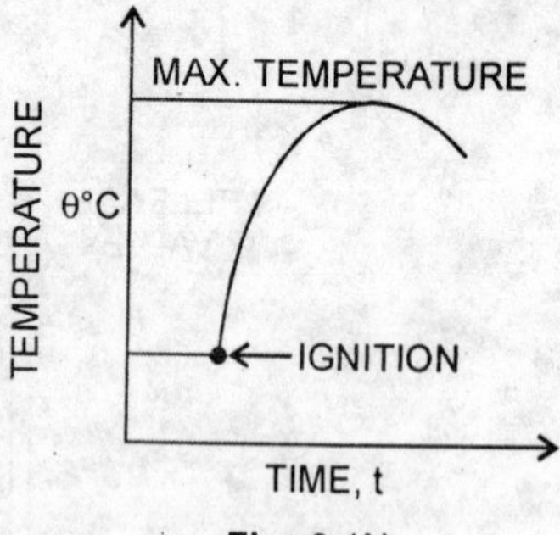

Fig. 3 (*b*).

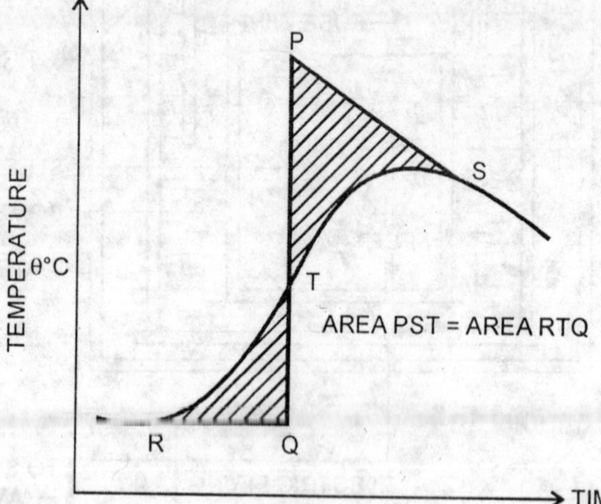

Fig. 3 (*c*).

Let t be the time required for the maximum temperature to be reached. Then temperature fall during temperature rise period due to radiation will be 0.5 rt°C.

Thus, cooling correction = 0.5 rt°C.

(*b*) What is meant by Evaporative cooling used in I.C. engine practice? Explain how the exhaust valves of heavy duty I.C. engines are cooled. (13)

Sol.

Evaporative cooling. It is a process of cooling of I.C. engine because of the evaporation of water in the cylinder jackets into steam. Here, the advantage is taken from the high latent heat of vaporisation of water by allowing it to evaporate in the cylinder jackets. If the steam is formed at a pressure above atmospheric the temperature will be above the normal permissible temperature. Such a cooling system is shown in Fig. 4 (*a*). Evaporative cooling system uses either air-cooled condenser or water - cooled condenser. The cooling circuit is such that the coolant is always liquid but the steam formed is flashed off in a separate vessel. The make-up water so formed is sent back for cooling. This system is used for cooling of industrial stationary engines.

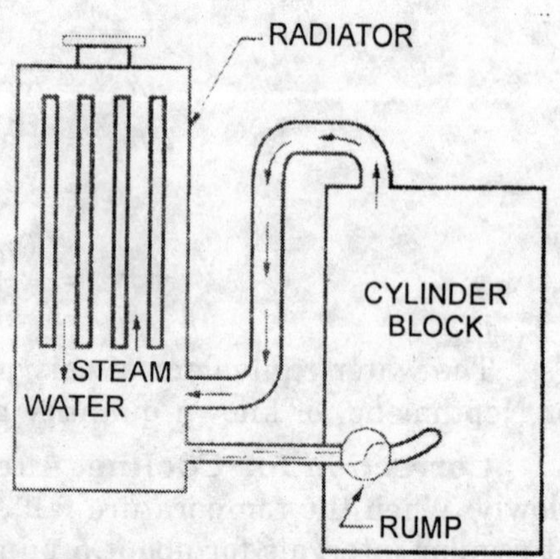

Fig 4 (*a*) Evaporative cooling.

Cooling of Exhaust valves. In the liquid cooled engines the spaces around the exhaust valve seating should be sufficiently large for the coolant to flow. Sometimes, the water circulation provides the jets of water to be directed on the valve seating metal.

The efficient valve cooling requires methods of conducting the heat from the valve head to the cylinder, improved guides and metal, also to the lower part of the valve stem operating under cooler air conditions.

In order to enhance the conductivity of the exhaust valve, the practice of making the valve stem and even the heads hollow is becoming popular. The hollow stem is filled with sodium which melts at 97°C and boils at 880°C. Thus at valve operating temperatures, the valve is filled with conducting material. Such a valve is shown in Fig. 4 (*b*). With a tapered plug seal driven into the hollow stem. The end of the stem is provided with Stellite button, fused on to the seat to prolong the wearing life of stem and end.`

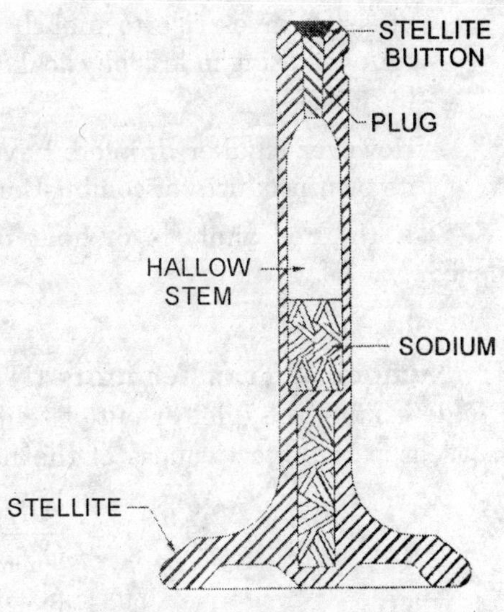

Fig 4 (*b*). Cooling of exhaust valve.

(*c*) Describe four methods for controlling diesel knock, giving advantages and limitations of each method. (15)

Sol.

Diesel knock. If the delay period is long a large amount of fuel will be injected and accumulated in the chamber. The auto-ignition of this large amount of fuel may cause high rate of pressure rise and high maximum pressure which may cause knocking in diesel engines, usually called *diesel knock*. The methods for controlling diesel knock are :

1. **Cetane Number of the Fuel used.** Cetane number is a scale for comparing the ignition delay period of various diesel fuels. A higher Cetane number means a lower delay period, and smoother engine operation, because it reduces the self-ignition temperature. Some compounds called *additives* or *dopes* are used to improve combustion performance of fuels. Various additives used for diesel fuels are : Isopropyl nitrate. Amyl nitrate, Heptyl peroxide and Butyl peroxide. These dopes delay the auto- ignition but they increase NO_x drastically.
2. **Compression ratio.** Increase in compression ratio reduces the delay period as it raises both temperatue and density. At the same time the minimum auto-ignition temperature decreases which reduces the time of reaction when fuel is injected. Thus the delay period decreases. With increase of compression ratio the unused air would be much more decreasing the volumeteric efficiency and power. It further decreases the mechanical efficiency due to increase in weight of reciprocating parts.
3. **Air-fuel ratio.** With decrease in air-fuel ratio (richer mixture) the combustion temperatures are increased and cylinder wall temperatures are increased and hence the delay period decreases. With increase of load air-fuel ratio decreases, operating temperature increases and hence delay period decreases.

4. **Injection pressure.** Lower the injection pressure, the lower the rate of pressure rise during the uncontrolled phase and smoother the running of engine. Lower injection pressure gives rise to fuel droplets of smaller size to obtain largest surface-volume ratio resulting in less physical delay. Large droplets of fuel results in too low subsequent burning.

   However smaller droplets have lesser momentum resulting in partial suffocation by its own products of combustion. It also results in greater uncontrolled pressure rise.

Q.3. (*a*) For analysis of heat exchanger problems, in which cases NTU method is applied?

Sol.

Number of Transfer units (NTU) method. NTU method is used when the problem is to determine the inlet or outlet temperature for a particular heat exchanger. The method is based on the effectiveness of the heat exchanger.

$$\text{NTU} = \frac{\text{UA}}{\text{C}_{\text{min}}}$$

where U = overall heat transfer co-efficient

A = area

C_{min} = minimum heat capacity

$= \min(\dot{m}_h c_{ph}, \dot{m}_c c_{pc})$

c_p = specific heat at constant pressure

m = mass flow rate of the fluid

h, c = subscripts for hot and cold fluid respectively

(*b*) 8,000 kg/h of air at 105°C is cooled by passing it through a counter flow heat exchanger. Find the exit temperature of air if water enters at 15°C and flows at a rate of 7,500 kg/h. U = 145 W/m²K, A = 20 m²

Solve the problem by the NTU method. (20)

Sol.

Given : $\dot{m}_h$ = 8000 kg/h, $\dot{m}_c$ = 7500 kg/h

t_{hi} = 105°C, t_{ci} = 15°C, t_{ho} = ?

Heat capacity of hot fluid (air),

$$\text{C}_h = \dot{m}_h c_{ph} = \frac{8000 \times 1.005}{3600} = 2.233 \text{ kW/K}$$

Heat capacity of cold fluid (water),

$$\text{C}_c = \dot{m}_c c_{pc} = \frac{7500 \times 4.187}{3600} = 8.723 \text{ kW/K}$$

$$\text{C}_h < \text{C}_c$$

Hence $\text{C}_{\text{min}} = \text{C}_h = 2.233$ kW/K and $\text{C}_{\text{max}} = 8.723$ kW/K

Capacity ratio, $$R = \frac{C_{min}}{C_{max}} = \frac{2.233}{8.723} = 0.256$$

$$NTU = \frac{UA}{C_{min}} = \frac{145 \times 20}{2.233 \times 10^3} = 1.2985$$

Effectiveness of heat exchanger,

$$\varepsilon = \frac{1 - \exp\{- NTU(1 - R)\}}{1 - R\exp\{- NTU(1 - R)\}}$$

$$= \frac{1 - \exp\{- 1.2985(1 - 0.256)\}}{1 - 0.256\exp\{- 1.2985(1 - 0.256)\}} = 0.6863$$

Also, $$\varepsilon = \frac{t_{hi} - t_{ho}}{t_{hi} - h_{ci}}$$

$$0.6863 = \frac{105 - t_{ho}}{105 - 15}$$

$$t_{ho} = 43.2°C$$

Exit temperature of air = 43.2°C

(*c*) Air at 200 kPa and 200°C is heated as it flows through a tube with a diameter of 25 mm at a velocity of 10 m/s. Calculate the heat transfer per unit length of tube if a constant heat flux condition is maintained at the wall and wall temperature is 20°C above the air temperature, all along the length of the tube. How much would the bulk temperature increase over a 3 m length of the tube? **(15)**

Sol.

Given : $p_1 = 200$ kPa, $t_i = 200°C$

$$T_i = 273 + 200 = 473 \text{ K}$$

$$L = 3 \text{ m}, d = 25 \text{ mm}$$

$$v = 10 \text{ m/s}, t_w = 200 + 20 = 220°C$$

$$T_w = 273 + 220 = 493 \text{ K}$$

$$\rho = \frac{m}{V} = \frac{p}{RT} = \frac{200 \times 10^3}{287 \times 473} = 1.473 \text{ kg/m}^3$$

For a constant wall heat flux, heat transfer,

$$q = \frac{\rho c_p V\,dt}{d\tau} = \frac{\rho c_p AL\,dt}{d\tau}$$

or $$\left(\frac{q}{L}\right)\int_0^\tau d\tau = \rho c_p A \int_{t_i}^{t_w} dt$$

$$\frac{\tau q}{L} = \rho c_p A(t_w - t_i)$$

Time for 3 m length, $$\tau = \frac{L}{v} = \frac{3}{10} = 0.3 \text{ s}$$

Heat transfer per unit length,

$$\frac{q}{L} = 1.473 \times 1.005 \times 10^3 \times \pi \times 25^2 \times 10^{-6} \times \left[\frac{493 - 473}{4 \times 0.3}\right]$$

$$= 48.445 \text{ W/m}$$

Let t_b be the bulk temperature. Then

$$\frac{3q}{L} = mc_p \text{ A } (t_b - t_i)$$

$$3 \times 48.445 = \frac{1.473 \times \pi \times 25^2 \times 10^{-6} \times 3 \times 1.005 \times 10^3 \, (t_b - 200)}{4}$$

$$t_b = 266.67°\text{C}$$

Q.4. (*a*) An air conditioning system is designed under the following conditions :

Outdoor conditions = 30°C DBT, 75% RH

Required indoor conditions = 22°C DBT, 20% RH

Amount of free air circulated = 3.33 m³/s

Coil dew point temperature = 14°C

The required conditions achieved first by cooling and dehumidification, and then by heating. Estimate

(*i*) The capacity of the cooling coil in tonnes,

(*ii*) The capacity of the heating coil in kW, and

(*iii*) The amount of water vapour removed in kg/s. (20)

Sol.

Given :

$$h_1 = 82 \text{ kJ/kg da},$$
$$h_2 = 52 \text{ kJ/kg } da,$$
$$h_3 = 47 \text{ kJ/kg da},$$
$$h_4 = 40 \text{ kJ/kg da}$$
$$w_1 = 0.020,$$
$$w_2 = w_3 = 0.0115 \text{ kg/kg da},$$
$$v_{s1} = 0.887 \text{ m}^3\text{/kg da (See Fig. 5)}$$

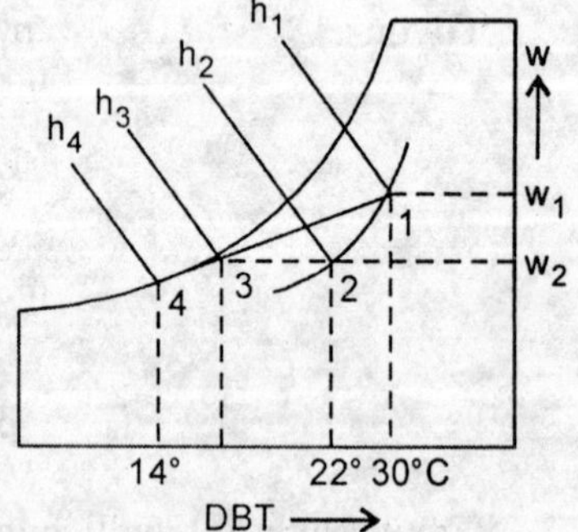

Fig 5.

Dry air flow,

$$G = \frac{\text{Amount of free air circulated}}{v_{s1}}$$

$$= \frac{3.33}{0.887} = 3.754 \text{ kg da/s}$$

(*i*) Capacity of cooling coil

$$= G\,(h_1 - h_3) = \frac{3.754\,(82 - 47)}{3.5167} = 37.362 \text{ tonnes}$$

(*ii*) Capacity of heating coil

$$= G(h_2 - h_3) = 3.753\ (52 - 47) = 18.77 \text{ kW}$$

(*iii*) Rate of water vapour removed

$$= G(w_1 - w_2) = 3.754\ (0.020 - 0.0115) = 0.0319 \text{ kg/s.}$$

(*b*) (*i*) What is the difference between heat pump and refrigeration cycle ?

(*ii*) What is the relation between $(\text{C.O.P.})_{\text{Heat pump}}$ and $(\text{C.O.P.})_{\text{Refrigerator}}$?

(*iii*) What are the differences betwwen Kelvin-Plank and Clausius Statements of Second law ? (20)

Sol. (*i*) **Heat pump cycle.** In the heat pump cycle, heat is taken from a constant temperature source (atmosphere) and then rejected to a body which is to be heated.

Refrigeration cycle. In the refrigeration cycle, heat is taken from a body and then rejected to a constant temperature sink (atmosphere).

(*ii*) $(\text{C.O.P.})_{\text{Heat pump}} = (\text{C.O.P.})_{\text{Refrigerator}} + 1$

(*iii*) **Kelvin-Plank's Statement.** It is impossible to construct an engine which will work in a complete cycle and produce no other effect except the raising of a weight and cooling of heat reservoir.

Clausius statement. It is impossible to construct a heat pump which operating in a cycle will remove heat continuously from a lower temperature reservoir and transfer to a higher temperature reservoir without any amount of external work being done on it.

Both the statements are equivalent in the sense that a violation of either statement means the violation of the other.

Q.5. (*a*) Show that the loss of heat due to friction in pipes can be expressed as

$$h_f = \frac{4f\,Lv^2}{2gd}$$

where f = co-efficient of friction

L = length of pipe

v = velocity of flow through pipe

h_f = loss of head dut to friction. (20)

Sol.

Consider a horizontal pipe of dimater d through which a fluid of viscosity μ and density ρ is flowing with a velocity v, as shown in Fig. 6. Consider a control volume of the fluid bounded by the pipe surface and the two sections at a distance dx apart. The flowing fluid exerts shear stress τ_0 on the wall and the pipe in turn exerts *an* equal and opposite shear stress τ_0 on the fluid. The other forces acting on the control volume are the pressure forces on its end faces. For steady uniform flow, the equation of motion in the flow direction gives.

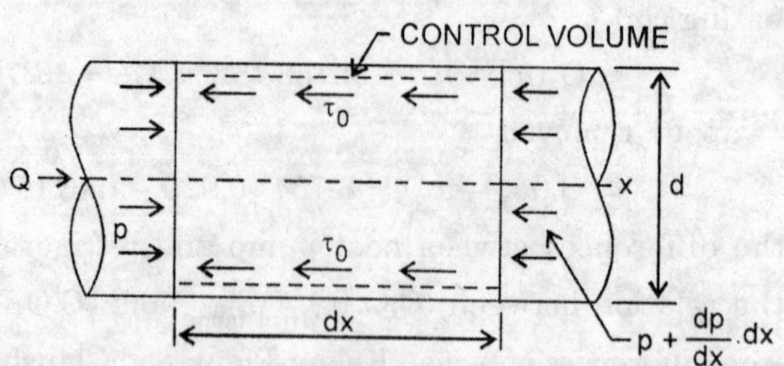

Fig. 6.

$$\left(\frac{\pi}{4}\right)d^2p - \left(\frac{\pi}{4}\right)d^2\left[p + \left(\frac{dp}{dx}\right).dx\right] - \pi d.dx.\tau_0 = 0$$

$$\left(\frac{d}{4}\right).\left(\frac{dp}{dx}\right) + \tau_0 = 0$$

Now $$\frac{dp}{dx} = -\gamma\frac{h_f}{dx}$$

where $$\gamma = \text{specific weight of the fluid}$$

$$\tau_0 = \left(\frac{d}{4}\right).\gamma\frac{h_f}{dx}$$

But $$\tau_0 = 1/2\,.\,f\,\rho\,v^2$$

$$\left(\frac{d}{4}\right).\gamma\frac{h_f}{dx} = \frac{1}{2}.f\,\rho\,v^2$$

$$h_f = \frac{1}{2}.\left(\frac{fv^2}{g}\right)\left(\frac{4}{d}\right).dx$$

For a pipe of length L, head loss

$$h_f = \frac{4\,f\,Lv^2}{2gd}$$

This is know as the Darcy-Weisbach equation.

(*b*) Determine the rate of flow of water through a pipe of diameter 20 cm and length 50 m when one end of the pipe is connected to a tank and other end of the pipe is open to the atmoshpere. The pipe is horizontal and the height of water in the tank is 4 m above the centre of pipe. Consider all minor losses and take f = 0.009.

The loss head (h_i) at the entrance of a pipe may be taken as 0.5 $v^2/2g$. **(20)**

Sol.

Given : d = 20 cm, L = 50 m, h = 4 m, f = 0.009

Total head loss, $$h = \frac{f\,Lv^2}{2gd} + \frac{0.5\,v^2}{2g} + \frac{v^2}{2g}$$ (See Fig. 7)

h

Q, v

Fig. 7

$$4 = \left(\frac{0.009 \times 50}{0.2} + 0.5 + 1\right) \times \left[\frac{v^2}{2 \times 9.81}\right]$$

$$v^2 = 20.928, \; v = 4.5747 \text{ m/s}$$

$$Q = Av = \left(\frac{\pi}{4}\right) \times (0.2)^2 \times 4.5747 = 0.1437 \text{ m}^3\text{/s}$$

Q.6. ***(a)*** With the help of a diagram explain the differences between *(i)* Laminar boundary layer, *(ii)* Turbulent boundary layer, and *(iii)* Laminar sub-layer. (20)

Sol.

(i) **Laminar boundary layer.** In the initial stages of the flow of a fluid over a body, the fluid particles move in an orderly manner in laminas parallel to the solid boundary. Viscous forces slow down the particles near the solid boundary as compared to the fluid motion away from the boundary. The fluid layer where the effects of fluid friction are predominant is called the *boundary layer*. The Reynolds number of flow in the boundary layer is expressed by $Re_x = U_\infty x/\nu$. Laminar boundary layer flow occurs at low Reynolds number. For flow over a flat plate, the nominal laminar boundary layer thickness is

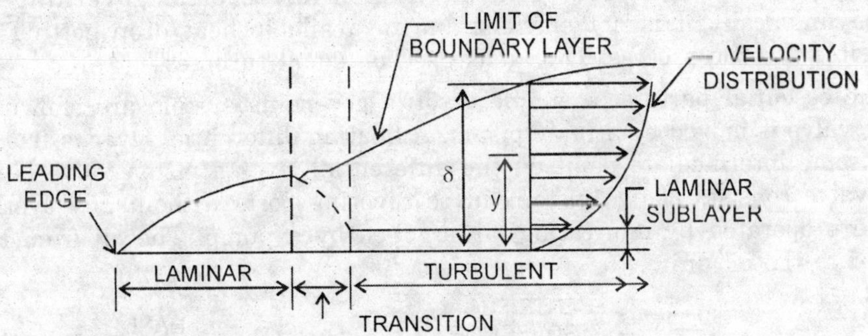

Fig. 8. Boundary layer development over a smooth flat plate.

$$\delta = \frac{5x}{\sqrt{Re_x}}$$

where x = distance from the leading edge of the plate.

The critical value of Re_x for laminar flow may range from 3×10^5 to 6×10^5.

Laminar boundary layer is very stable for $Re_x < 10^5$ for laminar flow. An average value of critcal $Re_x = 5 \times 10^5$.

(ii) **Turbulent boundary layer.** The orderly flow in the laminar boundary layer is not stable and ceases to exist at higher Reynolds number and the flow changes to turbulent motion. Characterized by strong mixing of particles in a direction normal to the boundary. As a consequence, the velocity distribution at a flow section is considerably more uniform in turbulent flow than in laminar flow.

(iii) **Laminar sub-layer.** If the boundary or the wall is very smooth, the turbulent boundary layer has very narrow zone very close to the boundary in which the flow is still laminar. This narrow region exhibiting characteristics of laminar motion, is called the laminar sub-layer. The formation of boundary layer on a flat plate is shown in Fig. 8.

(*b*) Find the displacement thickness and the momentum thickness for the velocity distribution in the boundary layer given by $u/U = y/\delta$ where u is the velocity at a distance y from the plate and $u = U$ at $y = \delta$, where δ = boundary layer thickness.

Sol.

Displacement thickness,

$$\delta^* = \int_0^\delta \left(1 - \frac{u}{U}\right) dy = \int_0^\delta \left(1 - \frac{y}{\delta}\right) dy = -\left| y - \frac{y^2}{2\delta} \right|_0^\delta = \frac{\delta}{2}$$

Momentum thickness,

$$\theta = \int_0^\delta \left(\frac{u}{U}\right)\left(1 - \frac{u}{U}\right) dy = \int_0^\delta \left(\frac{y}{\delta}\right)\left(1 - \frac{y}{\delta}\right) dy = \left| \frac{y^2}{2\delta} - \frac{y^3}{3\delta^2} \right|_0^\delta = \frac{\delta}{6}$$

Q.7. (*a*) What is throttle governing of steam turbine ? (10)

Sol.

The purpose of governing is to maintain the speed of the turbine fairly constant irrespective of the load. In throttle governing of steam turbine, the steam is throttled so as to reduce the steam flow and enthalpy drop whenever there is a reduction of load as compared to design load for maintaining speed. The mechanism of throttle governing is simple but thermodynmmically it is not efficient due to available heat drop getting reduced in the irreversible throttle process. Thus it is used on small turbines.

The essential parts of a simple throttle governing system are shown in Fig. 9. The throttle valve is moved by a relay piston. A floating differential lever is fixed to its one end and at some intermediate point on the differential lever a piston valve is fixed. The pilot piston valve consists of two piston valves covering ports without any overlap. The piston valves are operated by lubricating oil supplied by a pump. The oil from this chamber is returned to the oil drain.

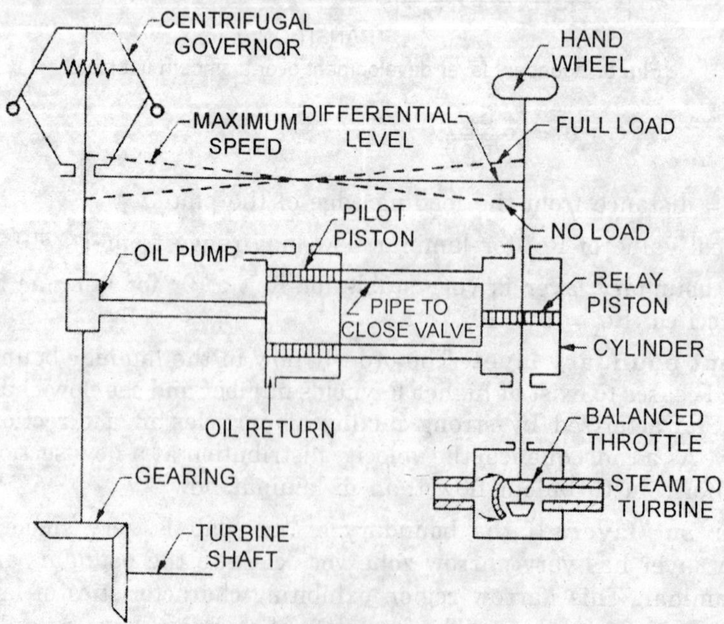

Fig. 9. Throttle governing of steam turbine.

At full load the throttle valve will open to increase the steam flow. If the load is reduced the relay piston will close the throttle valve partially. This will lower the pilot piston spindle and close the ports.

(*b*) Steam at 29.4 N/cm^2 absolute and dry saturated comes out of rotor of a reaction turbine having idential bladings. The velocity of steam entering into the turbine is 100 m/s. The mean blade height is 4 cm and the exit angle of the moving blade is 20°. The blade velocity is 4/3 times of axial flow velocity at the mean radius. If the steam flow rate is 2.5 kg/s, find (*i*) the rotor speed, (*ii*) h.p. develped, (*iii*) the diagram efficiency, (*iv*) the percentage increase in relative velocity in the moving blades, and (*v*) the enthalpy drop of steam in blade passage.

Sol.

Given :

$$p = 29.4 \text{ N/cm}^2,\ h = 4 \text{ cm},$$

$$v_{a1} = 100 \text{ m/s},\ m = 2.5 \text{ kg/s},$$

$$\alpha_1 = \beta_2 = 20°,\ \beta_1 = \alpha_2$$

From velocity triangles, we have (Fig. 10)

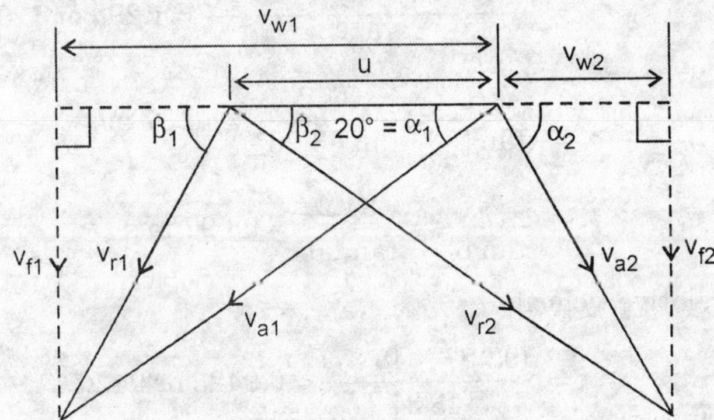

Fig. 10.

$$v_f = v_{f1} = v_{f2} = v_{a1} \sin \alpha_1 = 100 \sin 20° = 34.2 \text{ m/s}$$

$$u = \frac{4}{3} v_f = \frac{4 \times 34.2}{3} = 45.6 \text{ m/s}$$

Specific volume of steam at 29.4 N/cm^2 (from steam tables), v_s = 0.6173 m^3/kg.

$$m \times v_s = \pi\, dh\, v_f$$

$$d = \frac{2.5 \times 0.6173}{\pi \times 4 \times 10^{-2} \times 34.2}$$

$$= 0.359 \text{ m or } 35.9 \text{ cm}$$

(*i*)

$$u = \frac{\pi d N}{60},\ N = \frac{45.6 \times 60}{\pi \times 0.359} = 2426 \text{ rpm}$$

(*ii*) Power developed, $P = mu\,(v_{w1} - v_{w2})$

$$v_{w1} = v_{a1} \cos \alpha_1 = 100 \cos 20° = 93.97 \text{ m/s}$$

$$\tan \beta_1 = \frac{v_{f1}}{v_{w1} - u} = \frac{34.2}{93.97 - 45.6} = 0.707$$

$$\beta_1 = \alpha_2 = 35.26°$$

$$v_{w2} = \frac{v_{f2}}{\tan \alpha_2} = \frac{34.2}{0.707} = 48.37 \text{ m/s}$$

$$P = \frac{2.5 \times 45.6\,(93.97 + 48.37)}{746} = 21.75 \text{ HP}$$

$$v_{a2} = \frac{v_{w2}}{\cos \alpha_2} = \frac{48.37}{\cos 35.26} = 59.24 \text{ m/s.}$$

(*iii*) Diagram efficiency

$$= \frac{2u\,(v_{w1} + v_{w2})}{v_{a1}^2}$$

$$= \frac{2 \times 45.6\,(93.97 + 48.37)}{100^2} = 1.298 \text{ or } 129.8\%$$

(*iv*)

$$v_{r1} = \frac{v_{f1}}{\tan \beta_1} = \frac{34.2}{\tan 35.26°} = 48.37 \text{ m/s}$$

$$v_{r2} = \frac{v_{f2}}{\tan \beta_2} = \frac{34.2}{\tan 20°} = 93.97 \text{ m/s}$$

Increase in relative velocity

$$= \frac{(93.97 - 48.37)}{48.37} = 0.9425 \text{ or } 94.25\%$$

(*v*) Enthalpy drop :

Fixed blades,

$$\Delta h_1 = \frac{m\,(v_{a1}^2 - v_{a2}^2)}{2}$$

$$= \frac{2.5\,(100^2 - 59.24^2)}{2} = 8.113 \text{ kJ}$$

Moving blades,

$$\Delta h_2 = \frac{m\,(v_{r1}^2 - v_{r2}^2)}{2}$$

$$= \frac{2.5\,(93.97^2 - 48.37^2)}{2} = 16.224 \text{ kJ}$$

Q.8. (*a*) Show that the discharge through a reciprocating pump can be expressed as

$$Q = \frac{ALN}{60}$$

where Q = discharge through area of the piston,

A = cross-sectional area of the piston,

L = length of the stroke,

N = rmp of the crank.

Sol.

Cross-sectional area of the piston = A

Volume swept per stroke = AL

Average discharge per second,

$$Q = \frac{ALN}{60} \text{ m}^3/\text{s}$$

(b) **A double acting reciprocating pump, running at 40 rpm, is discharging 1.0 m³ of water per minute. The pump has a stroke of 40 cm. The diameter of the piston is 20 cm. The delivery and suction heads are 20 m and 5 m respectively. Find the slip of the pump and horse power required to drive the pump.**

Sol.

Given : $n = 2$, N = 40 rpm, $Q_a = 1 \text{ m}^3/\text{min}$, L = 40 cm

D = 20 cm, $H_s = 5$ m, $H_d = 20$ m

Theoretical discharge,

$$Q_{th} = nALN = 2 \times \pi \times 0.2^2 \times 0.4 \times \frac{40}{4} = 1.0053 \text{ m}^3/\text{min}$$

$$\text{Slip} = \frac{(Q_{th} - Q_a)}{Q_{th}} = \frac{(1.0053 - 1)}{1.0053} = 0.00528 \text{ or } 0.528\%$$

Power required,

$$P = \frac{\rho g Q(H_s + H_d)}{746}$$

$$= \frac{1000 \times 9.81 \times 1 \times 25}{60 \times 746} = 5.479 \text{ HP}$$

MECHANICAL ENGINEERING
PAPER I—2000

(Conventional)

Q.1. ***(a)*** A reversible engine receives equal quantity of heat from two reservoirs A and B maintained at temperatures T_1 and T_2 respectively. The engine rejects heat to a reservoir C at temperature T_3. In case the thermal efficiency of the above engine is K times the efficiency of reversible engine receiving heat only from reservoir A and rejecting heat to reservoir C and also if the heat supplied by the reservoir A is the same as it supplies in the combined case, show that (20)

$$K = \left(\frac{1}{2}\right)\left[\left\{\frac{T_2 - T_3}{T_1 - T_3}\right\} + \left(\frac{T_2}{T_1}\right)\right] \times \left(\frac{T_1}{2}\right).$$

Sol.

From Fig. 1 (*a*), we have

$$Q_1 = 2Q$$

Carnot efficiency, $\eta_a = 1 - \frac{Q_2}{Q_1} = 1 - \frac{Q_2}{2Q}$ ...(1)

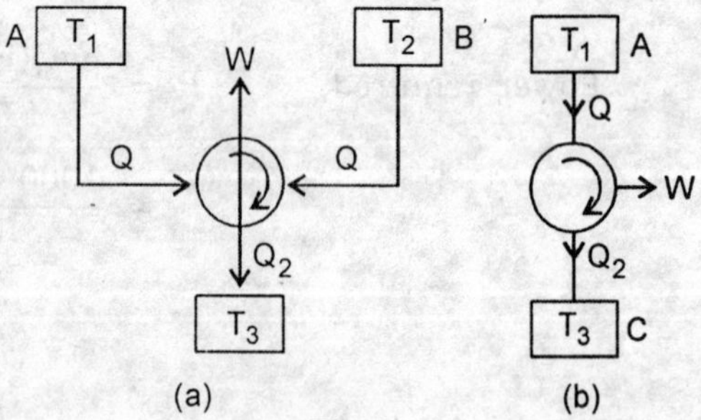

Also $\oint \frac{dQ}{T} = 0$

$$\frac{Q}{T_1} + \frac{Q}{T_2} - \frac{Q_2}{T_3} = 0$$

or $\frac{Q_2}{Q} = \frac{T_3}{T_1} + \frac{T_3}{T_2}$ ...(2)

Fig 1.

Substituting Eqn. (2) in (1), we get

$$\eta_b = 1 - \frac{T_3}{T_1}$$

Also, $\eta_a = K\,\eta_b$

$$1 - \left(\frac{T_3}{2}\right)\left(\frac{1}{T_1} + \frac{1}{T_2}\right) = K\left(1 - \frac{T_3}{T_1}\right)$$

$$K = \frac{1 - \left(\frac{T_3}{2}\right)\left(\frac{1}{T_1} + \frac{1}{T_2}\right)}{1 - \frac{T_3}{T_1}} = \frac{\frac{2T_1T_2 - T_1T_3 - T_2T_3}{2T_1T_2}}{\frac{T_1 - T_3}{T_1}}$$

$$= \frac{T_1(T_2 - T_3) + T_2(T_1 - T_3)}{2T_2(T_1 - T_3)}$$

$$= \left(\frac{1}{2}\right)\left[\left\{\frac{T_2 - T_3}{T_1 - T_3}\right\} + \left(\frac{T_2}{T_1}\right)\right] \times \left(\frac{T_1}{T_2}\right)$$

(*b*) A heat sourc at 627°C transfers heat at the rate of 3000 kJ/min to a system maintained at 287°C. A heat sink is available at 27°C. Assuming these temperatures to remain constant, find

(*i*) Change in entropy of source.

(*ii*) Entropy production accompanying heat transfer.

(*iii*) The original available energy.

(*iv*) The available energy after heat transfer. (20)

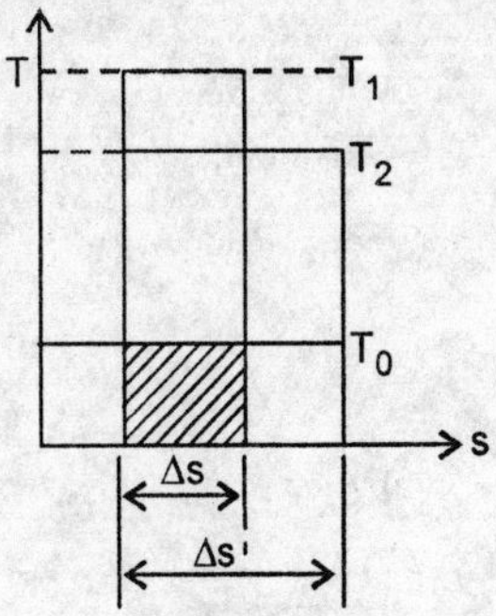

Fig. 2.

Sol.

Given :
$$T_1 = 273 + 627 = 900 \text{ K},$$
$$T_2 = 273 + 287 = 560 \text{ K},$$
$$T_0 = 273 + 27 = 300 \text{ K}$$
$$Q_1 = 3000 \text{ kJ/min}.$$

(*i*) Change in entropy of source,

$$\Delta s = \frac{Q_1}{T_1} = \frac{3000}{900} = 3.333 \text{ kJ/K.min}$$

(*ii*)
$$Q_1 = T_2 \,.\, \Delta s'$$

$$\Delta s' = \frac{3000}{560} = 5.357 \text{ kJ/K.min}$$

(*iii*) Original available energy,

$$A = \left(1 - \frac{T_0}{T_1}\right) Q_1$$

$$= \left(1 - \frac{300}{900}\right) \times 3000 = 2000 \text{ kJ/min}$$

(*iv*) Available energy after heat transfer

$$= A - \text{Unavailable energy due to heat transfer}$$
$$= A - T_0 (\Delta s' - \Delta s)$$
$$= 2000 - 300 (5.357 - 3.333) = 1392.76 \text{ kJ/min}$$

Q.2. (*a*) Calculate the percentage loss in the ideal efficiency of a diesel engine with compression ratio 14, if the fuel cut-off is delayed from 5% to 8%. (10)

Sol.

Given : Compression ratio, $r = 14$, Cut = off ratio, $\rho = 0.05$

Efficiency,
$$\eta_1 = 1 - \left(\frac{1}{r^{\gamma - 1}}\right) \times \left(\frac{1}{\gamma}\right)\left[\frac{\rho^{\gamma} - 1}{\rho - 1}\right]$$

$$= 1 - \left(\frac{1}{14^{1.4-1}}\right) \times \left(\frac{1}{1.4}\right)\left[\frac{0.05^{1.4} - 1}{0.05 - 1}\right] = 0.7423$$

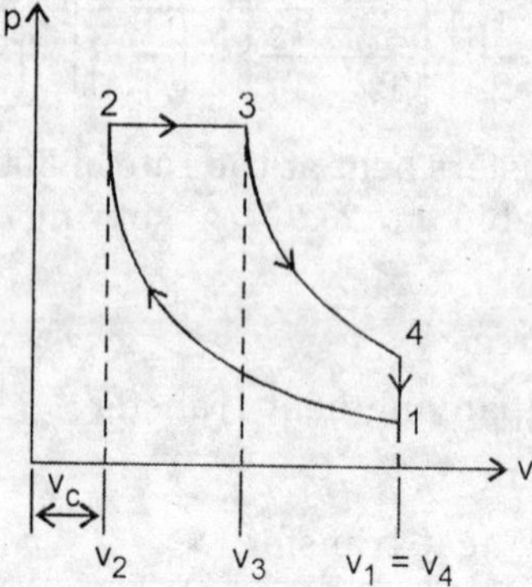

Fig. 3.

For $\rho = 0.08$

$$\eta_2 = 1 - \left(\frac{1}{r^{\gamma-1}}\right) \times \left(\frac{1}{\gamma}\right)\left[\frac{\rho^{\gamma} - 1}{\rho - 1}\right]$$

$$= 1 - \left(\frac{1}{14^{1.4-1}}\right) \times \left(\frac{1}{1.4}\right)\left[\frac{0.08^{1.4} - 1}{0.08 - 1}\right] = 0.7377$$

Change in efficiency

$$= \frac{0.7423 - 0.7377}{0.7423} = 0.006197 \text{ or } 0.6197\%$$

(*b*) Explaing the phenomena of knocking in SI engine. What are the different factors which influence the knocking? Describe the methods to supress it. (15)

Sol.

Phenomenon of Knocking in SI Engine In a SI engine combustion is initiated between the spark plug electrodes and then spreads across the combustible mixture. A definite flame front which separates the fresh mixture from the products of combustion travels from the spark plug to the other end of the combustion chamber. Heat release due to combustion increases the temperature and pressure of the burned part of mixture above those of the unburned mixture. In order to effect pressure equalization the burned part of the mixture will expand and compress the unbured part adiabatically thereby increasing the pressure and temperature of unburnbed part further. If the temperature of unburned mixture exceeds the self-ignition temperature of the fuel and remains at or above this temperature during the ignition lag, spontaneous ignition (or auto-ignition) occurs at various pin- point locations. This phenomenon is called *knocking*.

The phenomenon of knocking may be explained by referring to Fig. 4 (*a*) which shows the cross-section of the combustion chamber with flame advancing from the spark plug location A without knock whereas Fig. 4. (*c*) shows the combustion process with knock. In the normal combustion, the flame travels across the combution chamber from A towards D. The advancing flame front compresses the end charge BB′D farthest from the spark plug, thus raising its temperature. The temperature is also increased due to heat transfer from the hot advancing flame front and preflame oxidation in the end charge. If the temperature of the end charge does not reach its self-ignition temperatuere, the charge would not auto-ignite and flame will advance further and consume the charge BB′D, This is the normal combustion process which is illustrated by means of the pressure-time diagram in Fig. 4 (*b*).

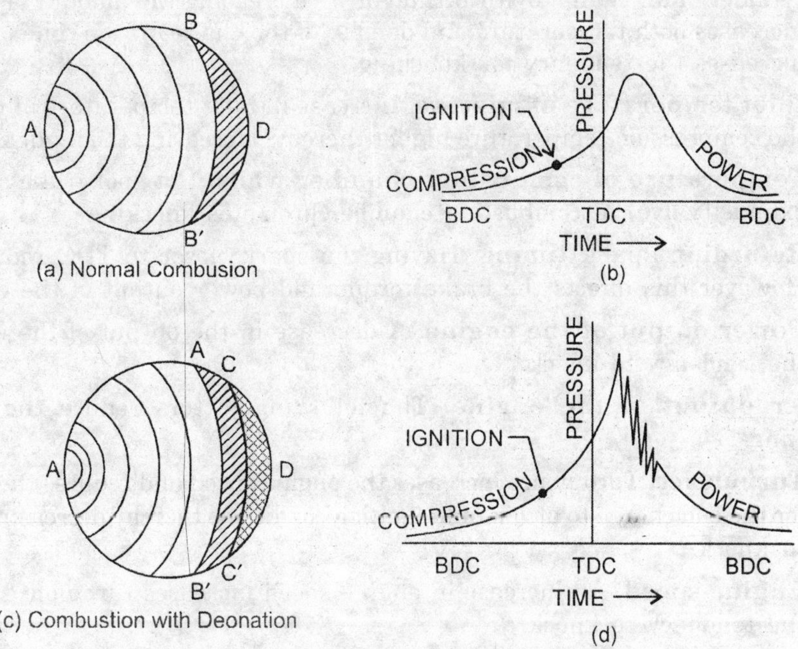

Fig. 4. Normal and Abnormal Combustion

If the end charge BB′D reaches its auto-iginition temperature and remains for some length of time equal to the time of preflame reactions the charge will auto-ignite, leading to knocking combustion. In Fig. 4 (*c*), it is assumed that when the flame has reached the position BB′, the charge ahead of it has reached critical auto-ignition temperature. During the preflame reaction period if the flame front could move from BB′ to only CC′ then the charge ahead of CC′ would auto-ignite. Because of the auto-ignition, another flame front starts travelling in the opposite direction to the main flame front. When the two flame fronts collide, a servere pressure pulse is generated. The gas in the chamber is subjected to compression and rarefaction along the pressure pulse unitl pressure equilibrium is restored. This disturbance can force the walls of the combustion chamber to vibrate at the same frequency as the gas (≈ 5000 Hz). The pressure-time trace of such a situation is shown in Fig. 4 (*d*).

Factor influencing knocking. The factors influencing knokcing are :

1. Density factors

2. Time factors

3. Composition factors

4. Temperature factors

1. Density Factors. Any factor which reduces the density of the charge tends to reduce knocking by providing lower energy release.

(i) **Compression ratio.** Increase in compression ratio increases the pressure and temperature of the gases at the end of compression stroke which decreases the ignition lag of the end gas and increases the tendency of knocking.

(ii) **Mass of induced charge.** An increase in mass of induced charge into the cylinder of an engine by throttling or by increasing the amount of supercharging increases both temperature and density of the charge at the time of ignition. This increases the tendency for knocking.

(iii) **Inlet temperature of mixture.** Increase in inlet temperature of mixture makes the compression temperature higher thereby, increasing the tendency of knocking.

(iv) **Temperature of combustion chamber walls.** Hot spots (like spark plug and exhaust valve) in combustion chamber promotes knocking.

(v) **Retarding spark timing.** Having the spark closer to TDC reduces knocking. However this affects the brake torque and power output of the engine.

(vi) **Power output of the engine.** A decrease in the output of the engine reduces the tendency to knock.

2. **Power output of the engine.** The following factors reduce the possibility of knocking :

(i) **Turbulence.** Turbulence increases the engine speed and reduces the time available for the end charge to attain auto-ignition conditions thereby decreasing the tendency to knock.

(ii) **Engine speed.** An increase in engine speed increases turbulence and decreases the tendency to knock.

(iii) **Flame travel distance.** Shortening the time required for the flame front to traverse the combustion chamber reduces the tendency for knocking. Flame travel distance is governed by combustion chamber size and spark plug position.

(iv) **Engine size.** A large engine (combustion chamber size) has a greater tendency for knocking as the flame requires a longer time to travel across the combustion chamber.

(v) **Combustion chamber shape.** The combustion chamber should be such which promotes turbulence to reduce knocking. Spherical shaped chambers minimizes knocking tendency.

3. **Composition factors :**

(i) **Fuel-Air ratio.** F/A ratio affects the reaction time or ignition delay. When the mixture is nearly 10% richer than stoichiometric (F/A ratio ≈ 0.08) ignition lag of the end gas is minimum and velocity of flame propagation is maximum and knocking tendency is maximum. A rich mixture is effective in decreasing the knocking tendency due to longer delay period and lower temperature of compression.

(ii) **Octane rating of fuel.** Higher the octane number, lesser is the tendency for knocking. Paraffin series have the maximum and aromatic series the minimum tendency to knock. The knocking characteristics of a fuel can be decreased by adding small amounts of additives called *dopes*.

(*iii*) **Humidity of air.** Increasing humidity of atmospheric air decreases the tendency to knock.

4. **Temperature factors.** Increasing the temperature of the unburned mixture by any factor in design or operation will increase the possibility of knocking in SI engine. The temperature of the unburned mixture is increased by the following factors :

   (*i*) **Raising the compression ratio.** For a given engine setting and fuel there will be a critical compression ratio above which knocking occurs. This compression ratio is called the Highest useful compression ratio (HUCR).

   (*ii*) Supercharging.

   (*iii*) Raising the inlet temperature.

   (*iv*) Raising the coolant temperature.

   (*v*) Increasing the load (opening the throttle).

   (*vi*) Raising the temperature of the cylinder and combustion chamber walls.

   (*vii*) Advancing the spark timing.

Methods for Suppressing knocking :

(*i*) **Proper location of spark plug.** A compact combustion chamber with proper central location of the spark plug reduces the path of the flame front travel from the spark plug to the remotest part of the combustion chamber, and thereby, reduces the tendency of knocking. The same result could be obtained by multi-fuel injection system.

(*ii*) **Proper selection of material for piston and cylinder head.** The end charge cooling is much better when piston and cylinder are made of aluminium alloys as they have good thermal conductivity. A better cooling of end charge reduces the tendency for knocking.

(*iii*) **Injecting water into the intake manifold.** Injection of water in the cylinder reduces the temperature of the end gas and increases the delay period, thereby decreasing the tendency to knock.

(*iv*) **Retarding the spark timings.** By retarding the spark timing, the peak pressure are reached only during the power stroke and are of lower magnitude. This reduces knock.

(*v*) **Extremely rich or lean mixture.** By using extremely rich or lean mixture the flame temperature can be kept low thus eliminating considerably the tendency of knock.

(*vi*) **Squish recesses in combution chamber.** The provision of squish recess in combustion chamber cools the last portion of the charge and reduces the tendency of knock.

(*c*) A two stage air compressor takes in air at 1.013 bar and 15°C and delivers at 43.4 bar. The intercooler pressure is 7.56 bar. The intercooling is perfect and the index of compression is 1.3. Calculate the work done in compressing 1 kg of air. If both cylinders have the same stroke and the piston diameters are 9 cm and 3 cm and the volumetric efficiency of the compressor is 90%, will the intercooler pressure be steady or will rise or fall as the compressor continues working? (15)

Sol.

Given : $p_1 = 1.013$ bar, $T_1 = 273 + 15 = 288$ K

$$p_3 = 43.4 \text{ bar}, \; p_2 = 7.56 \text{ bar}, \; n = 1.3, \; m = 1 \text{ kg},$$

$$d_1 = 9 \text{ cm}, \; d_2 = 3 \text{ cm}, \; \eta_v = 0.9$$

Input work per kg for perfect intercooling ($T_3 = T_1$),

$$W = \left[\frac{n}{n-1}\right] RT_1 \left[\left(\frac{p_2}{p_1}\right)^{(n-1)/n} + \left(\frac{p_3}{p_2}\right)^{(n-1)/n} - 2\right]$$

$$= \left(\frac{1.3}{0.3}\right) \times 0.287 \times 288 \left[\left(\frac{7.56}{1.013}\right)^{0.3/1.3} + \left(\frac{43.4}{7.56}\right)^{0.3/1.3} - 2\right]$$

$$= 389.3 \text{ kJ/kg}$$

Compression in LP cylinder :

$$v_2 = v_1 \left(\frac{p_1}{p_2}\right)^{1/n} = \left(\frac{RT_1}{p_1}\right)\left(\frac{p_1}{p_2}\right)^{1/n}$$

$$= \left[\frac{0.287 \times 288}{1.013 \times 10^2}\right]\left(\frac{1.013}{7.56}\right)^{0.3/1.3} = 457.97 \text{ K}$$

For constant pressure in the intercooler, the volume of air leaving the intercooler and entering the HP cylinder,

$$v_2' = \frac{v_2 T_3}{T_2} = 0.17386 \times \frac{288}{457.97} = 0.10933 \text{ m}^3/\text{kg}$$

$$\left(\frac{v_1}{v_2}\right)_{th} = \frac{\left[\dfrac{0.287 \times 288}{1.013 \times 10^2}\right]}{0.17386} = 4.693$$

Ratio of effective cylinder volumes,

$$\frac{v_{1e}}{v_{2e}} = \left(\frac{v_{LP}}{v_{HP}}\right) \times \eta_v$$

$$= \left(\frac{d_1}{d_2}\right)^2 \eta_v = \left(\frac{9}{3}\right)^2 \times 0.9 = 8.1$$

As $\left(\dfrac{v_{1e}}{v_{2e}}\right) > \left(\dfrac{v_1}{v_2}\right)_{th}$,

hence more air is supplied to HP cylinder than it can hold and consequently the pressure in the intercooler will rise.

Q.3. **(*a*)** A steel pipe having 10 cm bore and 12 cm outside diameter carries hot water at 80°C when the surrounding temperature is 15°C. The thermal conductivity of pipe material is 54 W/m.K and inner and outer heat transfer co-efficients are 1 kW/m^2. and 9 W/m^2. K, respectively. Calculate the heat loss per metre length of the pipe and the surface temperatures. Also calculate the heat loss and the surface temperatures when the pipe is covered with a 4 cm thick insulation having thermal conductivity of 0.048 W/m. K with outer surface heat transfer coefficient reduced to 7 W/m^2. K.

Sol.

Given : d_1 = 10 cm, d_2 = 12 cm, t_i = 80°C, t_0 = 15°C, k_1 = 54 W/m.K,
k_1 = 1 kW/m²-K, h_0 = 9 W/m². K, L = 1 m, d_3 = 20 cm, k_2 = 0.048 W/m. K,
h_0' = 7 W/m². K.

(*i*) **Heat loss from the pipe :**

$$q = \frac{t_i - t_0}{\left[\left\{\frac{1}{2\pi L}\right\}\left\{\frac{1}{h_i r_1} + \frac{\ln\left(\frac{r_2}{r_1}\right)}{k_1} + \frac{1}{h_0 r_2}\right\}\right]}$$

$$= \frac{80 - 15}{\left[\left\{\frac{1}{2\pi \times 14}\right\}\left\{\frac{1}{1000 \times 0.05} + \frac{\ln(6/5)}{54} + \frac{1}{(9 \times 0.06)}\right\}\right]}$$

$$= 217.79 \text{ W/m}$$

Also
$$q = \frac{t_i - t_1}{\frac{1}{2\pi L h_i r_1}} = \frac{t_i - t_2}{\left[\ln\left(\frac{r_2}{r_1}\right)\Big/(2\pi k_1 L)\right]}$$

$$217.79 = \frac{80 - t_1}{\left[\frac{1}{2\pi \times 1 \times 10^3 \times 0.005}\right]}$$

$$t_1 = 79.3°C$$

$$217.79 = \frac{79.3 - t_2}{\left[\frac{\ln(6/5)}{2\pi \times 54}\right]}$$

$$t_2 = 79.183°C$$

(*ii*) **Insulated pipe :** r_3 = 6 + 4 = 10 cm

$$q = \frac{t_i - t_0}{\left[\left\{\frac{1}{2\pi L}\right\}\left\{\frac{1}{h_i r_1} + \frac{\ln(r_2/r_1)}{k_1} + \frac{\ln(r_3/r_2)}{k_2} + \frac{1}{h_0' r_3}\right\}\right]}$$

$$= \frac{80 - 15}{\left[\left\{\frac{1}{2\pi \times 1}\right\}\left\{\frac{1}{1000 \times 0.05} + \frac{\ln(6/5)}{54} + \frac{\ln(10/6)}{0.048} + \frac{1}{7 \times 0.1}\right\}\right]}$$

$$= 33.769 \text{ W/m}$$

$$q = \frac{t_i - t_1}{\left[\frac{1}{2\pi L h_i r_1}\right]} = \frac{t_i - t_2}{\left[\frac{\ln(r_2/r_1)}{2\pi k_1 L}\right]} = \frac{t_2 - t_3}{\left[\frac{\ln(r_3/r_2)}{2\pi k_2 L}\right]}$$

$$= 33.769 = \frac{80 - t_i}{\left[\dfrac{1}{2\pi \times 1 \times 10^3 \times 0.05}\right]}$$

$$t_1 = 79.89°\text{C}$$

$$33.769 = \frac{79.89 - t_2}{\left[\dfrac{\ln(6/5)}{2\pi \times 54}\right]}$$

$$t_2 = 79.87°\text{C}$$

$$33.769 = \frac{79.89 - t_3}{\left[\dfrac{\ln(1/6)}{2\pi \times 0.048}\right]}$$

$$t_3 = 22.67°\text{C}$$

(*b*) **Steel balls 12 mm diameter are annealed by heating to 800°C and then slowly cooling to 127°C in air at 50°C. The heat transfer co-efficient for air is 20 W/m^2.K. Calculate the time required for cooling process. The properties of steel are taken as k = 45 W/m.K, ρ = 7830 kg/m^3 and c_p = 600 J/kg.K.** **(20)**

Sol.

Given :

$$d = 12 \text{ mm}, t_1 = 800°\text{C}, t_2 = 127°\text{C}, t_\infty = 50°\text{C}$$

$$h_a = 20 \text{ W/m}^2\text{.K}, k_s = 45 \text{ W/m.K}$$

$$\rho_s = 7830 \text{ kg/m}^3, c_{ps} = 600 \text{ J/kg. K}$$

$$\frac{\theta}{\theta_0} = \frac{t_2 - t_\infty}{t_1 - t_\infty} = \exp(-\text{B}_i \text{F}_0)$$

Characteristic length,

$$\text{L}_c = \frac{\text{V}}{\text{A}} = \frac{\left(\dfrac{4}{3}\right)\pi r^3}{4\pi r^2} = \frac{r}{3} = \frac{6}{3} = 2 \text{ mm} = 2 \times 10^{-3} \text{ m}$$

Thermal disffusivity,

$$\alpha_s = \frac{k_s}{\rho_s c_{ps}} = \frac{45}{7830 \times 600} = 9.578 \times 10^{-6}$$

$$\text{F}_0 = \frac{\alpha\tau}{\text{L}_c^2} = \frac{9.578 \times 10^{-6} \times t}{4 \times 10^{-6}} = 2.3946\ \tau$$

$$\text{B}_i = \frac{h_a \text{L}_c}{k_s} = \frac{20 \times 2 \times 10^{-3}}{45} = 0.889 \times 10^{-3}$$

Since $\text{B}_i < 1$ we use the Lumped capacitance method.

$$\frac{t_2 - t_\infty}{t_1 - t_\infty} = \exp(-\text{B}_i \text{F}_0)$$

$$\frac{127-50}{800-50} = \exp(-0.889 \times 10^{-3} \times 2.3946 \times \tau)$$

$$\tau = 1069.4 \text{ s or } 17.82 \text{ min}$$

Q.4. (*a*) A four cylinder single acting ammonia compressor with cylinder dimensions as 7.5 × 10 cm operates at 600 rpm. Condenser and evaporator pressures are 12 and 2 bar respectively. The vapour from the evaporator to suction of compressor is dry and saturated and there is no undercooling in the condenser. Compression takes place according to law, $pv^{1.2}$ = constant. If clearance is 2% of the stroke, calculate

(*i*) Refrigeration capacity in tons of refrigeration.

(*ii*) Power required to drive the compressor in kW.

(*iii*) Heat rejected in compressor water jacket.

(*iv*) Heat rejected to the condenser in kJ/min. (20)

The properties of ammonia are :

t°C	p kPa	v_s, vapour m3/kg	h_f, liquid kJ/kg	h_g, vapour kJ/kg	s_f, liquid kJ/kg. K	s_g, vapour kJ/kg. K
– 19	200	0.5952	94.9	1421	–	5.6057
31	1200	–	327.6	1469	4.9724	–

Sol.

Given : L = 10 cm, D = 7.5 cm, N = 600 rpm,

$$p_1 = 2 \text{ bar}, p_2 = 12 \text{ bar}$$

$$c = 2\%, n = 1.2,$$

$$i = 4, \gamma = 1.31$$

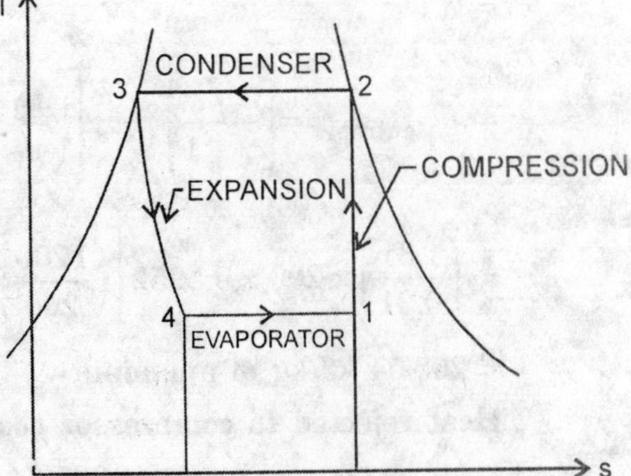

Fig. 5 (a)

Theoretical piston displacement,

$$V_p = \frac{\pi D^2 L N i}{4}$$

$$= \pi \times 7.5^2 \times 10 \times 10^{-6} \times 600$$

$$= 1.06 \text{ m}^3/\text{min}$$

Specific volume of refrigerant at point 1,

$$v_{g1} = 0.5952 \text{ m}^3/\text{kg}$$

Volumetric efficiency of compressor,

$$\eta_{cv} = 1 + c - c\left(\frac{p_2}{p_1}\right)^{1/n}$$

$$= 1 + 0.02 - 0.02\left(\frac{1200}{200}\right)^{1/1.2} = 0.931$$

Swept volume,

$$V_s = V_p\, \eta_{cv} = 1.06 \times 0.931 = 0.987 \text{ m}^3/\text{min}$$

Mass flow rate of refrigerant,

$$\dot{m}_r = \frac{V_s}{v_{g1}} = \frac{0.987}{0.5952} = 1.746 \text{ kg/ min}$$

(*i*) Refrigeration capacity

$$= \dot{m}_r (h_1 - h_{f3}) = 1.746 (1421 - 327.6)$$

$$= \textbf{1909.1 kJ/min or 31.818 kW or 9.048 Ton}$$

Specific work done on the compressor,

$$w_{act} = \left[\frac{n}{n-1}\right] p_1 v_s \left[\left(\frac{p_2}{p_1}\right)^{(n-1)/n} - 1\right]$$

$$= \left(\frac{1.2}{0.2}\right) \times 200 \times 0.5952 \left[\left(\frac{1200}{200}\right)^{0.2/1.2} - 1\right]$$

$$= \textbf{248.56 kJ/kg of ammonia}$$

Enthalpy, $h_2 = h_1 + w$

$= 1421 + 248.56$

$= \textbf{1669.56 kJ/kg}$

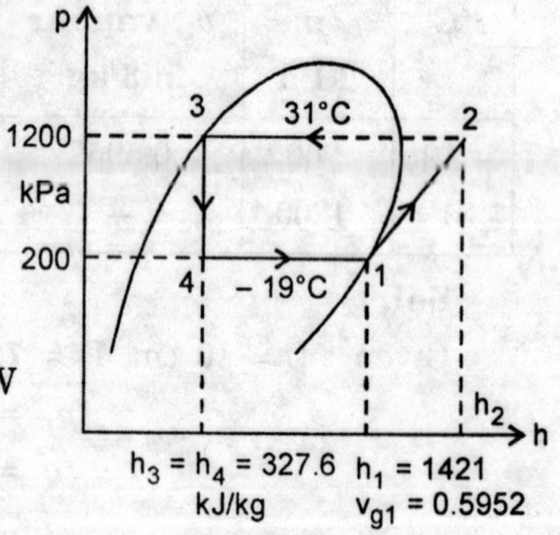

Fig. 5 (b)

(*ii*) Power required to drive the compressor

$$\dot{m}_r w = 1.746 \times 248.56$$

$$= \textbf{433.98 kJ/min or 7.233 kW}$$

(*iii*) Heat rejected to compressor cooling water jacket

$= w_{isentropic} - w_{actual}$

$$= w_{isentropic} = \left[\frac{\gamma}{\gamma - 1}\right] p_1 v_s \left[\left(\frac{p_2}{p_1}\right)^{(n-1)/n} - 1\right]$$

$$= \left(\frac{1.31}{0.31}\right) \times 200 \times 0.5952 \left[\left(\frac{1200}{200}\right)^{0.31/1.31} - 1\right]$$

$= 265.54$ kJ/kg of ammonia

Heat rejected to compressor cooling water jacket

$= 265.64 - 248.56$

$= 17.08$ kJ/kg

$= 17.08 \times 1.746$

$= \textbf{29.82 kJ/min}$

(*iv*) Heat rejected to condenser

$$= \dot{m}_r (h_2 - f_{f3}) = 1.746 (1669.56 - 327.6)$$

$$= \textbf{2343.06 kJ/min or 39.05 kW}$$

(**b**) 100 m³ per minute outdoor air at 43.3°C d.b.t. and 37% relative humidity is mixed with 200 m³ per minute of air at 38.2°C d.b.t. and 24.5° wet bulb temperature. The mixed air is dehumidified first by a cooling coil having bypass factor of 0.32 and apparatus dew point of 15°C and then by a chemical dehumidifier at 30°C dry bulb temperature. Air is then passed over a cooling coil whose surface temperature is 15°C and bypass factor 0.26. Calculate :

(*i*) Capacity of two cooling coils in tons of refrigeration.

(*ii*) Capacity of coil dehumidifier in kg/min.

(*iii*) Capacity of chemical dehumidifier in kg/min.

(*iv*) Total dehumidifying capacity of the system in kg/min.

Sol.

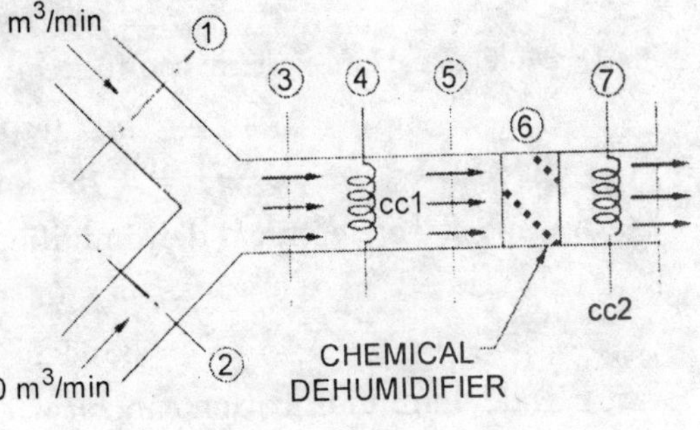

1st stream :

$$\text{Volume flow} = 100 \text{ m}^3/\text{min}$$
$$\text{DBT} = 43.3°\text{C}$$
$$\text{RH} = 37\%$$

Specific volume, $v_{s1} = 0.92 \text{ m}^3/\text{kg d.a}$

2nd stream :

$$\text{Volume flow} = 200 \text{ m}^3/\text{min}$$
$$\text{DBT} = 38.2°\text{C}$$
$$\text{WBT} = 24.5°\text{C}$$

Specific volume, $v_{s2} = 0.90 \text{ m}^3/\text{kg d.a}$

Determination of mixture condition :

From energy balance, we have

$$m_{a1} \cdot c_p \cdot t_1 + m_{a2} \cdot c_p \cdot t_2 = (m_{a1} + m_{a2}) \cdot c_p \cdot t_3$$

$$\left(\frac{100}{0.92}\right) \times 43.3 + \left(\frac{200}{0.90}\right) \times 38.2 = \left(\frac{100}{0.92} + \frac{200}{0.90}\right) t_3$$

$$t_3 = 39.87°\text{C}$$

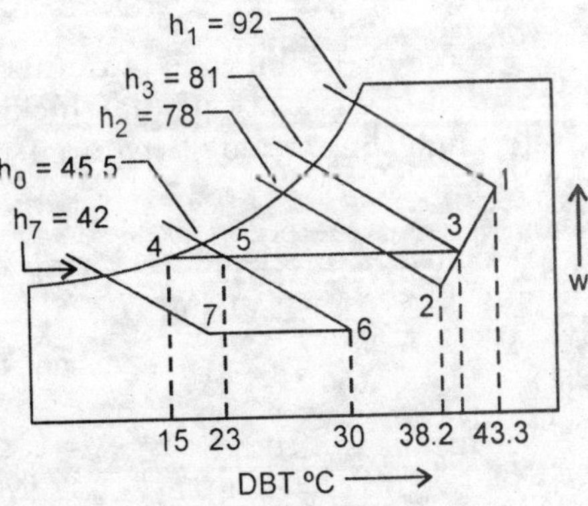

Fig. 6 (b) Psychrometric chart.

Next calculate tmperature t_5 of air leaving cooling coil from by-pass factor of coil.

$$\frac{t_5 - t_4}{t_3 - t_4} = 0.32$$

$$\frac{t_5 - 15}{39.87 - 15} = 0.32$$

$$t_5 = 23°\text{C}$$

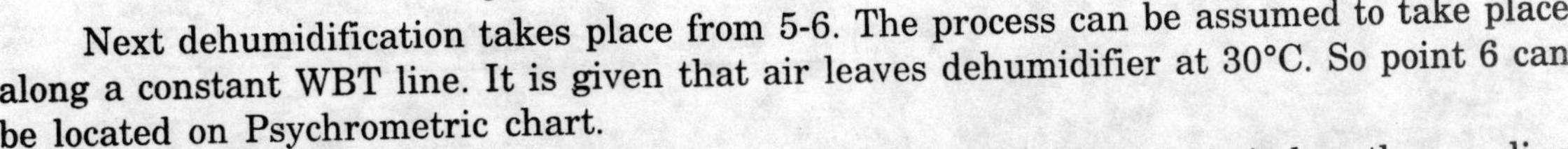

Next dehumidification takes place from 5-6. The process can be assumed to take place along a constant WBT line. It is given that air leaves dehumidifier at 30°C. So point 6 can be located on Psychrometric chart.

Next air is again cooled by a cooling coil at 15°C. The DPT of air is less than cooling coil temperature. Hence process shall be sensible cooling only. Since b.p.f. of second cooling coil is given as 0.26, point 7 can be located on Psychrometric chart, as

$$\frac{t_7 - t_4}{t_7 - t_4} = 0.26$$

$$\frac{t_7 - 15}{30 - 15} = 0.26$$

$$t_7 = 18.9°\text{C} \approx 19°\text{C}$$

(*i*) Total capacity of two cooling coils

$$= m_{a3}\ [(h_3 - h_5) + (h_6 - h_7)]$$

$$= 330.92\ [(81 - 54.5) + (54.5 - 42)]$$

$$= 12909 \text{ kJ/min or } 61.18 \text{ Ton.}$$

(*ii*) Moisture removed in cooling coil

$$= m_{a3}\ (\omega_3 - \omega_5) = 330.92\ (16.0 - 12.5)\ 10^{-3}$$

$$= 1.489 \text{ kg/min.}$$

(*iii*) Capacity of chemical dehumidifier

$$= m_{a3}\ (\omega_5 - \omega_6)$$

$$= 330.92\ (12.5 - 9.2)\ 10^3 = 1.092 \text{ kg/min}$$

(*iv*) Total dehumidification capacity

$$= 1.489 + 1.092 = 2.581 \text{ kg/min}$$

Q.5. (*a*) A solid cone of radius r_0, and vertex angle 2θ is to rotate at an angular velocity ω. An oil of viscosity μ and thickness h fills the gap between the cone and the housing. Determine the torque T required to rotate the cone. **(20)**

Sol.

Consider an elementary ring of the cone between y and $y + dy$ from the apex and of radius r, as shown in Fig. 7.

$$r = y \tan \theta.$$

Thickness of fluid, $h = \dfrac{dy}{\cos \theta}$

Elementary area of fluid,

$$dA = \frac{2\pi r\, dy}{\cos \theta} = \frac{2\pi\, y \tan \theta \times dy}{\cos \theta}$$

Shear stress, $\tau = \dfrac{\mu\, du}{dy} = \dfrac{\mu \omega r}{h} = \dfrac{\mu \omega\, y \tan \theta}{h}$

Elementary force,

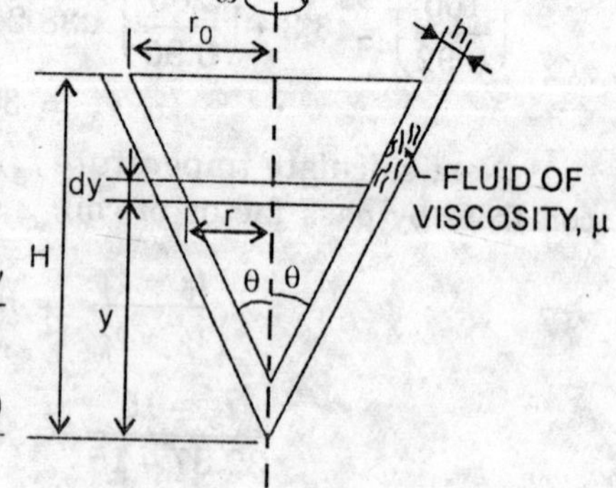

Fig. 7.

$$dF = \tau dA = \left(\frac{\mu\omega\, y \tan \theta}{h}\right) \times \left(\frac{2\pi y \tan \theta \times dy}{\cos \theta}\right)$$

$$= \left(\frac{2\pi\mu \tan^2 \theta}{\cos \theta}\right)\left(\frac{\omega y^2}{h}\right) dy$$

Elementary torque,

$$dT = dF.r = \left(\frac{2\pi\mu \tan^2\theta}{\cos\theta}\right)\left(\frac{wy^2}{h}\right) dy.y\tan\theta$$

$$= \left(\frac{2\pi\mu\omega}{h}\right)\left(\frac{\tan^3\theta}{\cos\theta}\right)\int_0^H y^3\,dy = \left(\frac{2\pi\mu\omega}{h}\right)\left(\frac{\tan^3\theta}{\cos\theta}\right)\left(\frac{H^4}{4}\right)$$

$$= \left(\frac{2\pi\mu\omega}{h}\right)\left(\frac{\tan^3\theta}{\cos\theta}\right)\left(\frac{r_0^4}{(4\tan^4\theta)}\right) = \left[\frac{\pi\mu\omega}{2h}\right]\left(\frac{r_0^4}{\sin\theta}\right)$$

(b) A rectangular container having base area of 1.5 m^2 and length 0.9 m is floating in water with the open end downwards. If the difference in water levels inside and outside the container is 10 cm,

(i) Determine the mass of the container.

(ii) What force will be required to depress the bottom of the container to a depth of 10 m in water, if the trapeed air has constant temperature? (20)

Sol.

Given : $A = 1.5\text{ m}^2,\ l = 0.9\text{ m},\ h = 10\text{ cm},\ p_{atm} = 100\text{ kPa}.$

(i) Let m = Mass of container. Then

$$mg = (p_1 - p_2)\,A = \rho g h A$$

$$m \times 9.81 = 1000 \times 9.81 \times 0.1 \times 1.5$$

$$m - 150\text{ kg}$$

(ii) Let h = Depth of water in the container from the top surface

$$p_1 = \rho g h + p_{atm}$$

Height of air column in teh container,

$$h_a = (h - 10)\ m$$

For constant temperature compression,

$$p_{atm}V_{atm} = p_1V_1$$

$$p_{atm} \times A \times l = p_1 h_a A$$

$$p_1 = p_{atm} \times \frac{0.9}{h_a}$$

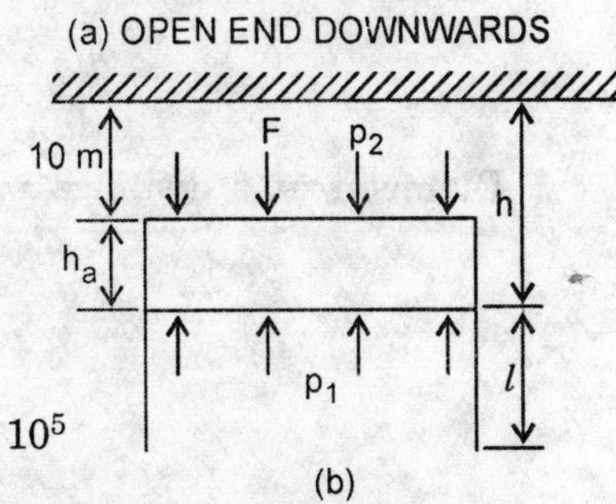

Fig. 8.

From Eqn. (1), we have

$$\frac{10^5 \times 0.9}{h_a} = 1000 \times 9.81 \times (h_a + 10) + 10^5$$

$$0.987\,h_a^2 + 1.981\,h_a - 0.9 = 0$$

$$h_a = \frac{-1.981 + \{1.981^2 + 4 \times 0.981 \times 0.9\}^{0.5}}{2 \times 0.981} = 0.382\text{ m}$$

Force applied, $F = (p_1 - p_2) A -$ weight of container $= \rho g h_a A - mg$

$= 1000 \times 9.81 \times 0.382 \times 1.5 - 150 \times 9.81$

$= 5621.13 - 1471.15 = 4149.98$ N

Q.6. (*a*) A nozzle of diameter 20 mm is fitted to a pipe of diameter 40 mm. Find the force exerted by the nozzle on the water which is flowing through the pipe at the rate of 1.2 m³/min.

Sol.

Given : $d_1 = 40$ mm, $d_2 = 20$ mm, Q = 1.2 m³/min

$$Q = A_1 v_1 = A_2 v_2$$

$$v_1 = \frac{1.2 \times 4}{\pi \times 40^2 \times 10^{-6}} = 954.93 \text{ m/min}$$

$$v_2 = 3819.72 \text{ m/min}$$

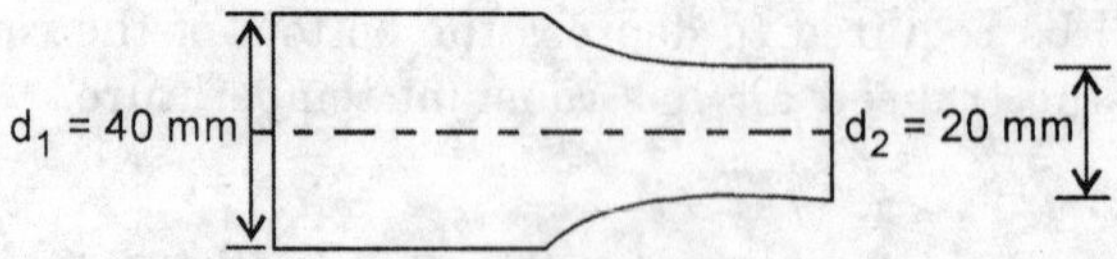

Fig. 9

$$\text{Force} = m(v_2 - v_1) = \left[\frac{1.2 \times 10^3}{60 \times 60}\right] [3818.72 - 954.93]$$

$$= 954.93 \text{ N}$$

(*b*) A 2.0 m long conical diffuser 20 cm in diameter at the upstream end has 80 cm diameter at the downstream end. At a certain instant the discharge through the diffuser is observed to be 200 litres/sec of water and is found to increase uniformly at the rate of 50 litre/sec. Estimate the local, convective and total acceleration at a section 1.5 m from the upstream end. **(20)**

Sol.

Given : $d_1 = 20$ cm, $d_2 = 80$ cm, L = 2 m, Q = 200 lps,

$$\frac{dQ}{dt} = 50 \text{ lps}, \; x = 1.5 \text{ m}$$

Diameter at a distance x cm from the smaller end,

$$d_x = \frac{(d_2 - d_1)\,x}{L} + d_1$$

$$= \frac{(80 - 20)\,x}{200} + 20 = 0.3x + 20 \text{ cm}$$

At $x = 1.5$ m, $d_x = 0.3 \times 150 + 20 = 65$ cm

Area of cross-section,

$$A = \frac{\pi d^2}{4} \; \frac{\pi (0.65)^2}{4} = 0.33183 \text{ m}^2$$

Velocity of flow, $v = \frac{Q}{A} = \frac{200 \times 10^{-3}}{0.33183} = 0.6027$ m/s

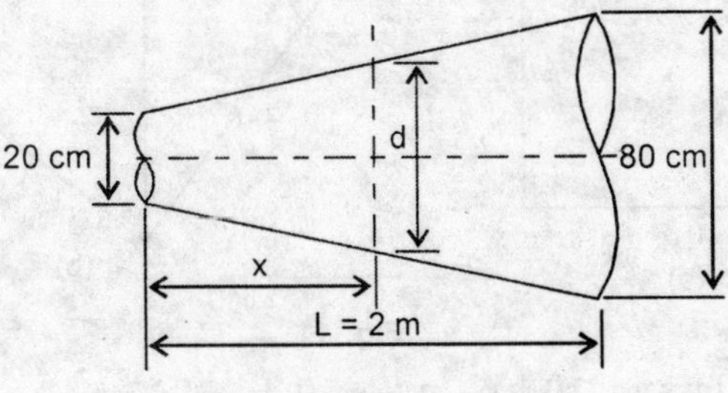

Fig. 10

(*i*) Local acceleration,

$$\frac{\partial v}{\partial t} = \frac{\partial\left(\frac{Q}{A}\right)}{\partial t} = \left(\frac{1}{A}\right)\left(\frac{\partial Q}{\partial t}\right)$$

$$= \left(\frac{1}{0.33183}\right)(50 \times 10^{-3}) = 0.1507 \text{ m/s}^2$$

(*ii*) Convective acceleration,

$$a_c = v_x \left(\frac{\partial v_x}{\partial x}\right)$$

$$v_x = \frac{Q}{A_x} = \frac{0.2 \times 4}{[\pi (0.3x + 0.2)^2]} = \frac{0.25465}{(0.3x + 0.2)^2}$$

$$\frac{\partial v_x}{\partial x} = \frac{0.152788}{(0.3x + 0.2)^3}$$

At $x = 1.5$ m

$$a_c = \left[\frac{0.25465}{(0.3 \times 1.5 + 0.2)^5}\right](-0.152788) = -0.3353 \text{ m/s}^2$$

(*iii*) Total acceleration $= 0.1507 - 0.3353 = -0.1846$ m/s^2

Q.7. (*a*) A centrifugal pump has an impeller of 80 cm in diameter and it delivers 1 m^3/sec against a head of 80 m. The impeller runs at 1000 rpm and the width at outlet is 8 cm. If the leakage loss is 3 percent of the discharge, external mechanical loss is 10 kW and the hydraulic efficiency is 80 percent, calculate the blade angle at outlet, the power required and overall efficiency of the pump.
(20)

Sol.

Given : $d_2 = 80$ cm, $b_2 = 8$ cm, $Q_a = 1$ m^3/s

$H_{mano} = 80$ m, N = 1000 rpm

Leakage loss = 3% of discharge,

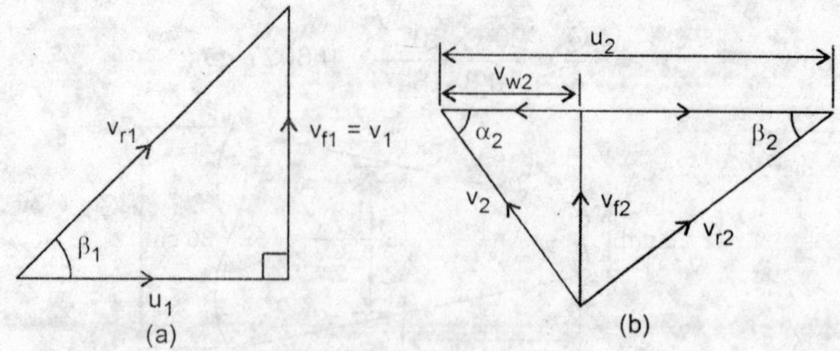

Fig. 11

$$\text{Mechanical loss} = 10 \text{ kW}, \ \eta_{hyd} = 0.8$$

$$Q_{th} = 1.3 \ Q_a = 1.3 \text{ m}^3/\text{s}$$

$$u_2 = \frac{\pi d_2 N}{60} = \frac{\pi \times 0.8 \times 1000}{60} = 41.8 \text{ m/s}$$

$$Q_{th} = \pi d_2 b_2 v_{f2},$$

$$v_{f2} = \frac{1.3}{\pi \times 0.8 \times 0.08} = 64.7 \text{ m/s}$$

$$\eta_{hyd} = \frac{gH_{mano}}{u_2 \ v_{w2}}$$

$$v_{w2} = \frac{9.81 \times 80}{41.8 \times 0.8} = 23.47 \text{ m/s}$$

$$\tan \beta_2 = \frac{v_{f2}}{u_2 - v_{w2}} = \frac{64.7}{41.8 - 23.47} = 3.535$$

$$\beta_2 = 74.2°$$

Power required

$$= \frac{(\rho g Q)\left(\dfrac{u_2 \ v_{w2}}{g}\right)}{10^3} + 10$$

$$= \frac{(1000 \times 9.81 \times 1.3)\left(\dfrac{41.8 \times 23.47}{9.81}\right)}{10^3} + 10 = 1285.4 \ kW$$

$$\eta_{mech} = \frac{1275.4}{1285.4} = 0.992 \text{ or } 99.2\%$$

$$\eta_{overall} = \eta_{mech} \times \eta_{mano} = 0.992 \times 0.8 = 0.79 \text{ or } 79\%$$

(*b*) A single acting, two stage air compressor running at 300 rpm delivers air at 20 bar, while the intake conditions are 0.98 bar and 305 K. The intermediate pressure is 5 bar and the clearance volume for low pressure compressor is 4 percent of its stroke volume. The compressor delivers 3 m³/min at 1 bar and 15°C. Estimate

(*i*) Power required to drive the compressor in kW.

(*ii*) Low pressure cylinder dimensions, if L = D.

(*iii*) Isothermal efficiency when the intercooling is perfect and the index n = 1.3 for compression and expansion. (20)

Sol.

Given : $N = 300 \text{ rpm}, p_3 = 20 \text{ bar}, p_2 = 5 \text{ bar}, p_1 = 0.98 \text{ bar},$

$$T_1 = 305 \text{ K}, r_c = 0.04, n = 1.3, \dot{V} = 3\text{m}^3/\text{min},$$

$$T_2 = T_1\left(\frac{p_2}{p_1}\right)^{(n-1)/n} = 305\left(\frac{5}{0.98}\right)^{0.3/1.3} = 444.2 \text{ K}$$

(*i*) Work input in the LP cylinder,

$$w_{LP} = \left[\frac{n}{n-1}\right] RT_2 - T_1 s$$

$$= \left(\frac{1.3}{0.3}\right) \times 0.287\,(444.2 - 305) = 173 \text{ kJ/kg}$$

For perfect intercooling,

$$T_5 = T_1 = 305 \text{ K}$$

$$T_6 = T_5\left(\frac{p_3}{p_2}\right)^{(n-1)/n}$$

$$= 305\left(\frac{20}{5}\right)^{0.3/1.3} = 419.94 \text{ K}$$

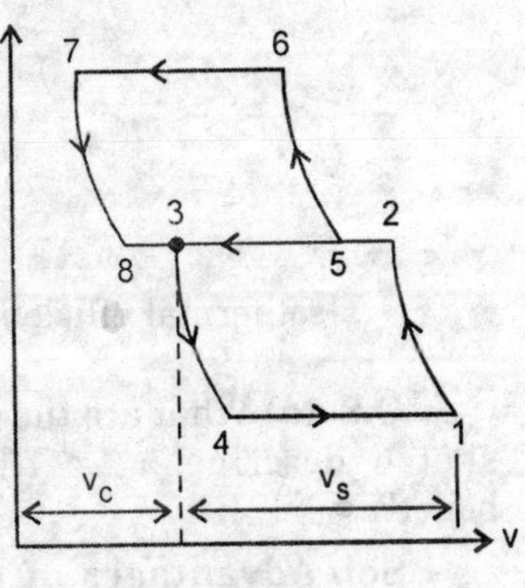

Fig. 12

Work input in HP cylinder,

$$W_{HP} = \left(\frac{1.3}{0.3}\right) \times 0.287\,(419.94 - 305)$$

$$= 142.95 \text{ kJ/kg}$$

$$\text{Total work, } w_t = 173 + 142.95 = 315.95 \text{ kJ/kg}$$

$$m = \frac{pV}{RT} = \frac{1 \times 10^5 \times 3}{287 \times 288 \times 60} = 0.0605 \text{ kg/s}$$

$$\text{Power required} = 315.95 \times 0.0605 = 19.115 \text{ kW}$$

(*ii*) Actual volume sucked,

$$V_1 = \frac{p_0 V_0 T_1}{T_0 p_1} = \frac{1 \times 3 \times 305}{288 \times 0.98} = 3.24 \text{ m}^3/\text{min}$$

Volumetric efficiency,

$$\eta_v = 1 + r_c - r_c\left(\frac{p_2}{p_1}\right)^{1/n}$$

$$= 1 + 0.04 - 0.04\left(\frac{5}{0.98}\right)^{1/1.3} = 0.899$$

Swept volume, $V_s = \frac{V_1}{\eta_v} = \frac{3.24}{0.899} = 3.604 \text{ m}^3/\text{min}$

$$V_s = \frac{\pi D^2{}_1 \times D_1 N}{4}$$

$$3.604 = \frac{\pi D^3{}_1 \times 300}{4}$$

$$D_1 = 0.2482 \text{ m or } 24.82 \text{ cm}$$

$$L = D_1 = 24.82 \text{ cm}$$

(*iii*) Isothermal workdone

$$= RT_{atm} \ln\left(\frac{p_3}{p_{atm}}\right)$$

$$= 0.287 \times 288 \times \ln\left(\frac{20}{1}\right) = 274.615 \text{ kJ/kg}$$

Isothermal efficiency $= \frac{247.615}{315.95} = 0.7837 \text{ or } 78.37\%$

Q.8. (*a*) What are the advantages of using high pressure boilers? With the help of a neat sketch describe a Loeffler boiler. What is usually the working pressure of such boiler? (20)

Sol. Advantages of using High pressure boilers.

1. The efficiency and the capacity of the plant can be increased as reduced quantity of steam is required for the same power generation if high pressure steam is used.
2. The forced circulation of water through the boiler tubes provides freedom in the arrangement of furance and water walls in addition to the reduciton in the heat exchange area.
3. The tendency of scale formation is reduced due to high velocity of water.
4. The danger of overheating is reduced as all the parts are uniformly heated.
5. The differential expansion is reduced due to uniform temperature and this reduces the possibility of gas and air leakages.

Loeffler boiler. The arrangement of the different components, and water and steam circulation are shown in the Fig. 13. The pressure feed pump draws the water through the economizer and delivers it into the evaporator drum. About 65% of the steam coming out of the superheater is passed through the evaporator drum in order to evaporate the feed water coming from the economizer.

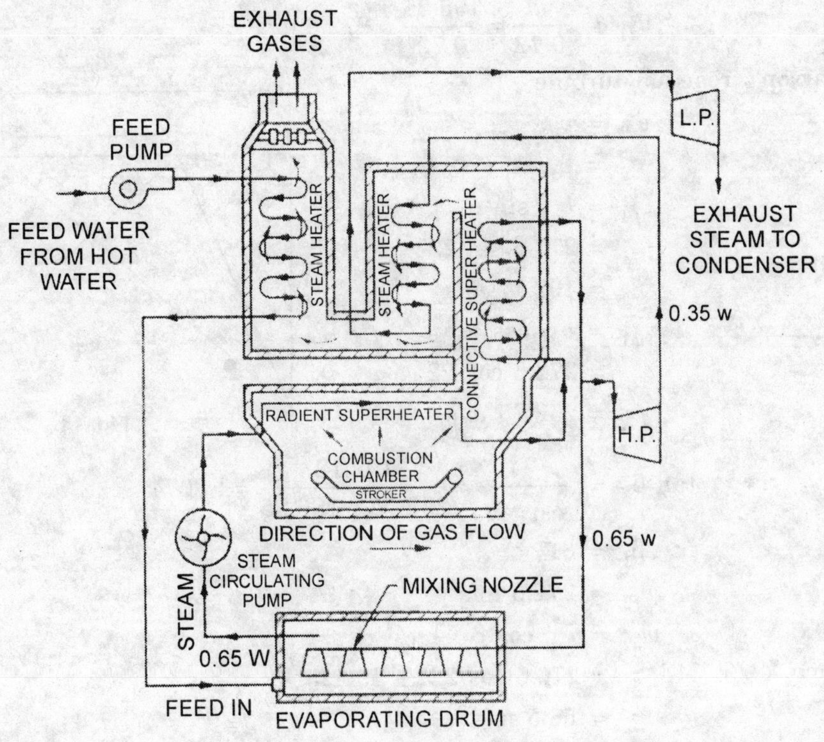

Fig. 13. Loeffler Boiler.

The steam circulating pump draws the saturated steam from the evaporator drum and and is passed through the radiant superheater and then convective superheater. About 35% of the steam coming out from the superheater is supplied to the HP steam turbine. The steam coming out coming out of the HP steam turbine is passed through the reheater before supplying to the LP steam turbine.

This boiler can carry higher salt concentration than any other boiler and is more compact. It is suitable for land or sea transport power generation. It has generating capacity of about 94.5 tonnes/hour at 140 bar.

Q.8. (*b*) In a multi-stage Parson's reaction turbine, at one of the stages the rotor diameter is 125 cm and speed ratio 0.72. The speed of the rotor is 3000 rpm. Determine.

(*i*) The blade inlet angle if the blade outlet angle is 22°.

(*ii*) Diagram efficiency.

(*iii*) Percentage increase in diagram efficiency and rotor speed if the turbine is designed to run at the best theoretical speed. **(20)**

Sol.

Given : $d = 1.25$, N = 3000 rpm, $\dfrac{u}{v_{a1}} = 0.72, \alpha_1 = \beta_2 = 22°$

(*i*) $$u = \frac{\pi d N}{60} = \frac{\pi \times 1.25 \times 3000}{60} = 196.25 \text{ m/s}$$

$$v_{a1} = \frac{u}{0.72} = \frac{196.25}{0.72} = 272.6 \text{ m/s}$$

In Parson's reaction turbine,

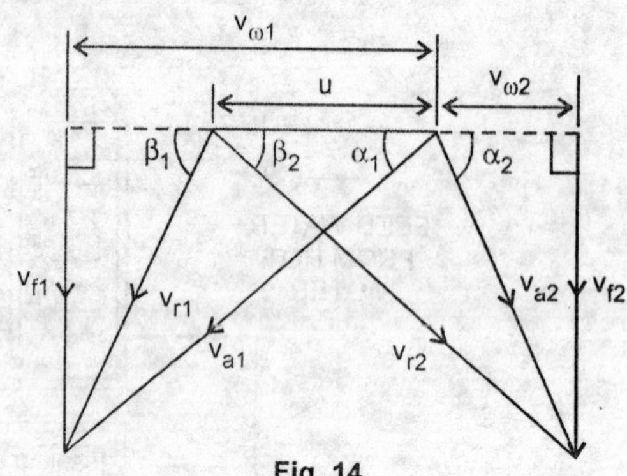

Fig. 14.

$$v_{r2} = v_{a1},\ v_{r1} = v_{a2}$$

And $$\alpha_2 = \beta_1,\ \alpha_1 = \beta_2$$

$$v_{f1} = v_{a1} \sin \alpha_1$$
$$= 272.6 \sin 22°$$
$$= 102.12 \text{ m/s}$$
$$v_{w1} = v_{a1} \cos \alpha_1$$
$$= 272.6 \cos 22°$$
$$= 252.75 \text{ m/s}$$

$$\tan \beta_1 = \frac{v_{f1}}{v_{w1} - u} = \frac{102.12}{252.75 - 196.25} = 1.8074$$

$$\beta_1 = 61°$$
$$v_{r2} = 272.6 \text{ m/s}$$

(ii) $$v_{w2} = v_{r2} \cos \beta_2 - u$$
$$= 276.6 \cos 22° - 196.25$$
$$= 56.5 \text{ m/s}$$
$$v_w = v_{w1} + v_{w2}$$
$$= 252.75 + 56.5 = 309.25 \text{ m/s}$$

$$\text{Diagram efficiency} = \frac{2uv_w}{v_{a1}^2} = \frac{2 \times 196.25 \times 309.25}{(272.6)^2} = 1.633$$

(iii) Power develped,

$$P = \frac{muv_w}{10^3} = \frac{1 \times 196.25 \times 309.25}{10^3} = 60.69 \text{ kW}$$

At best theoretical rotor sped,

$$u = v_{a1} \cos \alpha_1 = 272.6 \cos 22° = 252.75 \text{ m/s}$$

Percentage increase in rotor speed

$$= \left(\frac{252.75 - 196.25}{196.25}\right) = 2.1037$$

$$\text{Diagram efficiency} = \frac{2 \times 252.75 \times 309.25}{(272.6)^2} = 0.2882 \text{ or } 28.82\%$$

Percentage increase in diagram efficiency

$$= \left(\frac{2.1037 - 1.633}{1.633}\right) = 0.2882 \text{ or } 28.82\%$$

MECHANICAL ENGINEERING
PAPER I—2001

(Conventional)

Q.1. (*a*) Using Maxwell's relations, show that for a pure substance.

$$T.ds = c_p\,dT - Tv\beta\,dp$$

$$= c_p\,dT + T\left(\frac{\beta}{K}\right)dv$$

$$= \frac{Kc_v\,dp}{\beta} + \left[\frac{c_p}{\beta v}\right]dv$$

where b is co-efficient of cubical expansion, K is co-efficient of compressibility and c_p and c_v are specific heats at constant pressure and constant volume respectively. (15)

Sol.

$$\beta = \left(\frac{1}{v}\right)\left(\frac{\partial v}{\partial T}\right)_p$$

$$K = \left(\frac{1}{v}\right)\left(\frac{\partial v}{\partial p}\right)_T$$

Maxwell's relations are

$$\left(\frac{\partial V}{\partial s}\right)_p = \left(\frac{\partial T}{\partial p}\right)_s$$

$$\left(\frac{\partial s}{\partial p}\right)_T = -\left(\frac{\partial v}{\partial T}\right)_p$$

$$\left(\frac{\partial s}{\partial V}\right)_T = \left(\frac{\partial p}{\partial T}\right)_v$$

$$\left(\frac{\partial p}{\partial s}\right)_v = -\left(\frac{\partial T}{\partial v}\right)_s$$

$$c_p = -\left(\frac{\partial Q}{\partial T}\right)_p = \left(\frac{T\partial s}{\partial T}\right)_p = T\left(\frac{\partial s}{\partial T}\right)_p$$

Similarly, $$c_v = T\left(\frac{\partial s}{\partial T}\right)_v$$

Now taking $$s = s\,(p, T)$$

$$ds = \left(\frac{\partial s}{\partial T}\right)_p dT + \left(\frac{\partial s}{\partial v}\right)_T dp$$

$$T.ds = T\left(\frac{\partial s}{\partial T}\right)_p dT + T\left(\frac{\partial s}{\partial p}\right)_T dp$$

$$= c_p dT - T\left(\frac{\partial s}{\partial T}\right)_p dp$$

Now, taking $s = s(v, T)$

$$ds = \left(\frac{\partial s}{\partial T}\right)_v dT + \left(\frac{\partial s}{\partial T}\right)_T dv$$

$$T.ds = T\,ds = \left(\frac{\partial s}{\partial T}\right)_v dT + T\left(\frac{\partial s}{\partial T}\right)_T dv = c_v\,dT + \left(\frac{\partial s}{\partial T}\right)_v dv$$

Now
$$\frac{\beta}{K} = \frac{\left(\frac{1}{v}\right)\left(\frac{\partial v}{\partial T}\right)_p}{\left[-\left(\frac{1}{v}\right)\left(\frac{\partial v}{\partial p}\right)_T\right]} = -\left(\frac{\partial v}{\partial T}\right)_p \left(\frac{\partial p}{\partial v}\right)_T$$

Also $\left(\frac{\partial v}{\partial T}\right)_p \left(\frac{\partial p}{\partial v}\right)_T \left(\frac{\partial T}{\partial p}\right)_v = -1$

Hence
$$\frac{\beta}{K} = \left(\frac{\partial p}{\partial T}\right)_v$$

$\therefore$
$$T.ds = c_v\,dT + T(\beta/K)\,dv$$

Now $s = s(p, v)$

$$ds = \left(\frac{\partial s}{\partial p}\right)_v dp + \left(\frac{\partial s}{\partial v}\right)_p dv$$

$$T.ds = T\left(\frac{\partial s}{\partial p}\right)_v dp + T\left(\frac{\partial s}{\partial v}\right)_p dv = T\left(\frac{\partial s}{\partial p}.\frac{\partial T}{\partial p}\right)_v dp + T\left(\frac{\partial s}{\partial v}.\frac{\partial T}{\partial v}\right)_p dv$$

$$= T\left(\frac{\partial s}{\partial T}\right)_v \left(\frac{\partial T}{\partial p}\right)_v dp + T\left(\frac{\partial s}{\partial T}\right)_v \left(\frac{\partial T}{\partial v}\right)_p dv$$

$$= c_v\left(\frac{K}{\beta}\right) dp + \frac{c_p\,dv}{\beta v}$$

(*b*) State the Clausius inequality in words.

An inventor claims that he has developed a heat engine which absorbs 1200 kJ and 800 kJ from reservoirs at 800 K and 600 K respectively and rejects 600 kJ and 200 kJ as heat to reservoirs at 400 K and 300 K respectively. It develops 1200 kJ work. Determine whether the heat engine is theoretically possible. **(15)**

Sol.

Clausius inequality statement. The Clausius inequality states that whenever a system undergoes a cyclic change the algebraic sum of the ratios of heat interaction to the absolute temperature at which heat interaction occurs over the complete cycle is less than or equal to zero. Mathematicall

$$\oint \frac{dQ}{T} < 0$$

The heat engine working between four reservoirs is shown in the Fig. 1.

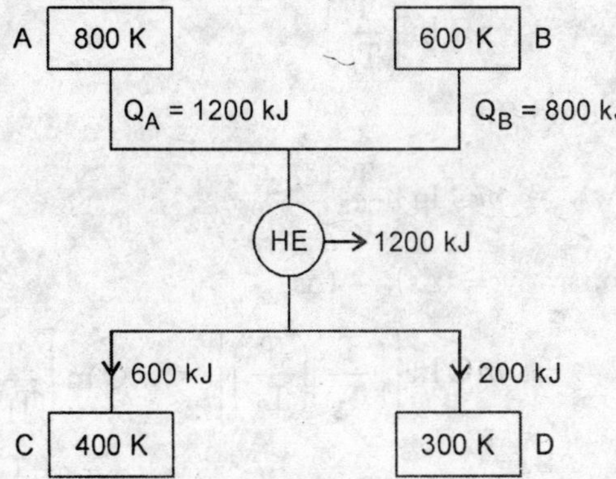

Fig. 1

Reservoir	Q, kJ	T, K	Δs = Q/T
A	– 1200	800	– 1.5
B	– 800	600	– 4/3
C	600	400	+ 1.5
D	200	300	+ 2/3

The entropy of the universe, $\sum \Delta s = -\frac{2}{3}$ kJ/kg.K

Since the net etropy change of the universe is negative, the engine is not possible as the entropy of the universe cannot decrease.

(*c*) A mass m of fluid at temperature T_1 is mixed with an equal mass of the same fluid at temperature T_2. Prove that the resultant change in entropy of the universe is

$$2mC \ln \frac{\left(\dfrac{T_1 + T_2}{2}\right)}{\sqrt{T_1 T_2}}.$$

Sol.

Let, T = temperature of the mixture

m, C, T_1	+	m, C, T_2
FLUID AT T_1		FLUID AT T_2

Fig. 2.

From heat balance, we have

$$mC\,(T_1 - T) = mC\,(T - T_2)$$

$$T = \frac{T_1 + T_2}{2}$$

Change in entropy of fluid 1,

$$(\Delta s)_1 = mC \ln\left(\frac{T}{T_1}\right)$$

Change in entropy of fluid 2,

$$(\Delta s)_2 = mC \ln\left(\frac{T}{T_2}\right)$$

Resultant change in entropy $= (\Delta s)_1 + (\Delta s)_2$

$$= mC \ln\left[\left(\frac{T}{T_1}\right)\left(\frac{T}{T_2}\right)\right] = mC \ln\left[\frac{T^2}{T_1 T_2}\right]$$

$$= 2mC \ln\left[\frac{T}{(T_1 T_2)^{1/2}}\right] = 2mC \ln\left[\frac{\left\{\dfrac{T_1 + T_2}{2}\right\}}{(T_1 T_2)^{1/2}}\right]$$

Q.2. (*a*) Explain clearly the terms pre-ignition, detonation and diesel knock. (10)

Sol.

Pre-ignition. After ignition, once a stable flame nucleus is formed, the flame will propagate from molecule to molecule. When the flame reaches the far end of the combustion chamber the peak pressure is normally attained. The end mixture experiences rapid compression first by the piston during the compression stroke and later by the advancing flame front. Thus the end charge experiences a very high pressure and temperature. If the temperature exceeds the self-ignition temperature of the mixture, pre-ignition or auto-ignition takes place.

Detonation. It is a term used to signify an unusual sharp metallic sound which is clearly audible under abnormal conditions of engine operation.

Diesel Knock. If the delay period is long a large amount of fuel will be injected and accumulated in the combustion chamber of the engine. The auto-ignition of this large amount of fuel may cause high rate of pressure rise and high maximum pressure which may cause knocking in diesel engines, usually called *diesel knock*.

(*b*) What is supercharging? Explain the thermodynamic cycle for a supercharged I.C. engine. Which engine is more suitable for supercharging-spark ignition engine or compression ignition engine? Why? (10)

Sol.

Supercharging. The power output of an engine depends upon the amount of air inducted into the cylinder per unit time, the degree of utilizaiton of this air and the thermal efficiency of the engine. The amount of air inducted per unit time can be increased by increasing the engine speed or by increasing the density of air at intake. They volumetric efficiency decreases as the speed is increased. The method of increasing the inlet air density is called

supercharging. It is generally used to increase the power output of the engine. This is done by supplying air at pressure higher than the pressure at which the engine naturally aspirates air from the atmosphere by using a pressure boosting device called a *supercharger*.

Thermodynamic cycle. The *p-v* diagram for an ideal Otto-cycle supercharged engine is shown in the Fig. 3 (*a*). The pressure p_1 represents the supercharging pressure and p_5 is the exhuast pressure. Area 5-6-7-0-1-5 represents the work done by the supercharger in supplying air at a pressure p_1 while the area 1-2-3-4-1 is the output of the engine. Area 0-1-6-7-0 represents the gain in work during the gas exchange process due to supercharging. However the area 1-6-5-1 cannot be recovered and represents a loss of work. This loss of work causes the ideal thermal efficiency of the supercharged engine to decrease with an increase in supercharging pressure.

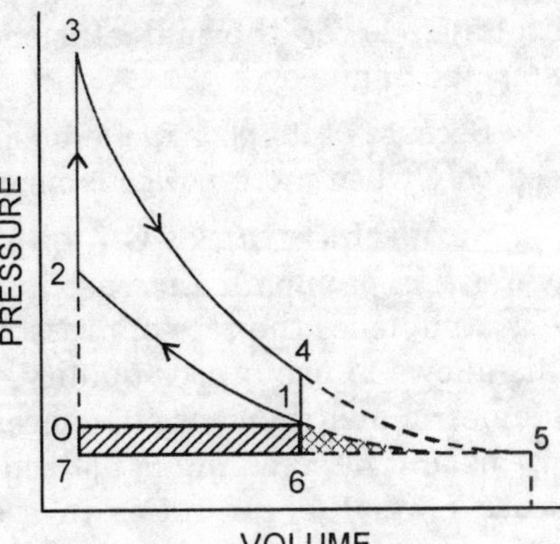

(a) p-v diagram for an ideal Otto cycle supercharged engine.

Fig. 3 (a).

Fig. 3 (*b*) shows an ideal dual combusiton cycle supercharged engine. The pressure p_1 represents the supercharging pressure and $p_6 = p_7$, is the exhaust pressure. The engine is supercharged by the compressor. Area 9-10-1-11-9 represents the work done on the supercharger in supplying air at pressure p_1. Thus for engine, 8-1 represents induction process, 1-2 as the compression process, 2-3-4 as the heat addition process, 4-5 as the expansion process, 5-6 as the blow down to atmosphere or heat rejection process and 6-7 as the process.

Work done my mass m by the supercharged engine,

$$W = \text{Area (1-2-3-4-5-1)} + \text{Area (8-1-6-7-8)} - \text{Area (9-10-1-11-9)}.$$

Thus part of supercharger work is recovered, however curve enquivalent to Area (9-10-1-11-9) – Area (8-1-6-7-8) is not recoverable and represents loss of work. This loss of work causes the ideal thermal efficiency of supercharged engine to decrease with an increase efficiency of supercharging pressure.

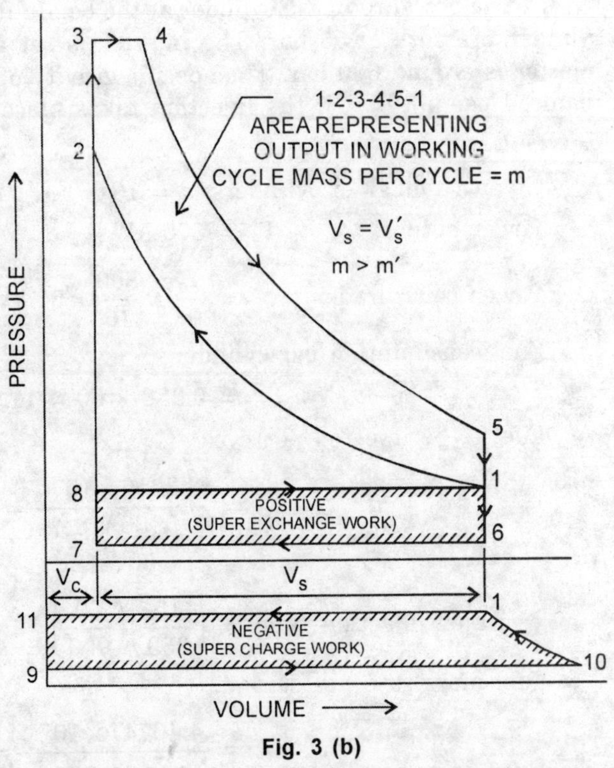

Fig. 3 (b)

Supercharging of S.I. engine. As far as S.I. engines are concerned, supercharging is employed only for aircraft and racing car egines. This is because the increase in supercharging pressure increases the tendency to detonate and pre-ignite.

Apart from increasing the volumetric efficiency of the engine supercharging results in increase in the intake temperature and pressure of the engine. Both of these reduces the ignition delay and increase flame speed. This results in greater tendency to detonate or pre-ignite. For this reason, the supercharged S.I. engines use lower compression ratio which results in lower thermal efficiency. The fuel consumption is also greater than naturally aspirated engines.

Because of its poor fuel economy, supercharging of petrol engines is not very popular and is used only when more power is needed or when more power is need to compensate altitutde loss.

Supercharging of C.I. engines. Supercharing of C.I. engines does not result into any combustion problems. Increase in pressure and tempreature of the intake air reduces ignition delay and hence the rate of pressure rise results in a better, quieter and smoother combustion. This allows to use a poor quality of fuel. The increase in intake air temperature reduces the volumetric and thermal efficiences but the increase in density due to pressure compensates for this and intercooling is not required except for highly supercharged engines. It is possible to use lower fuel-air ratios in a supercharged engine resulting in lower temperature and reduces smoke from the egine. This results in increased life of the engine.

Thus C.I. engine is more suitable for supercharging.

(*c*) A 16-cylinder diesel engine has a power output of 800 kW at 900 revolutions per minute. The engine works on the four stroke cycle and has a fuel consumption of 0.283 kg/kW.hr. The pressure in the cylinder at the beginning of injection is 32.4 bar and the maximum cylinder pressure is 55 bar. The injector is set at 214 bar and maximum pressure at the injector is around 600 bar. The specific gravity of the fuel is 0.86. Calculate the orifice area required per injector, if the injection takes place over 10 degree crank angle. (20)

Sol.

Given Number of cylinders, $n = 16$

Power output, $P_t = 800$ kW

Power per cylinder $= \dfrac{P_t}{n} = \dfrac{800}{16} = 50$ kW

Fuel consumption per cylinder

$$= 0.238 \times 50 = 11.9 \text{ kg/h or } 3.3056 \times 10^{-3} \text{ kg/s}$$

Fuel to be injected per cycle

$$= \frac{3.3056 \times 10^{-3}}{\left(\frac{900}{60}\right)\left(\frac{1}{2}\right)} = 4.40747 \times 10^{-4} \text{ kg}$$

Fuel injection time $= \left(\dfrac{10}{360}\right)\left(\dfrac{60}{900}\right) = 1.852 \times 10^{-3}$ s

Fuel mass rate per second,

$$m_f = \frac{4.40747 \times 10^{-4}}{1.852 \times 10^{-3}} = 0.238 \text{ kg/s}$$

Pressure difference in the beginning,

$$\Delta p_i = 214 - 32.4 = 181.6 \text{ bar}$$

Pressure difference at the end,

$$\Delta p_e = 600 - 55 = 545 \text{ bar}$$

Average pressure differences,

$$\Delta p_a = \frac{\Delta p_i + \Delta p_e}{2} = \frac{181.6 + 545}{2} = 363.3 \text{ bar}$$

Now

$$m_f = C_d A_i [2 \rho_f \Delta p_a]^{1/2}$$

$$0.283 = 0.6 \times A_i [2 \times 0.86 \times 10^3 \times 363.3 \times 10^5]^{1/2}$$

$$A_i = 1.587 \times 10^{-6} \text{ m or } 1.587 \text{ mm}^2$$

Q.3. (*a*) A layer of 5 cm thick insulating brick having conductivity of 1.5 W/m.K is placed between two 0.5 cm thick steel plates. The conductivity of mild steel is 50 W/m.K. The faces of brick adjacent to the plates are rough having solid-to-solid contact of 30% of the total area. The average height of the asperites is 0.1 cm. If the outer plate surface temperatures are 100°C and 500°C respectively, calculate the rate of heat transfer per unit area. The conductivity of air is 0.02 W/m.K. (25)

Sol.

Let, A = area of plate, m^2

Thermal resistance, $R = \dfrac{x}{kA}$

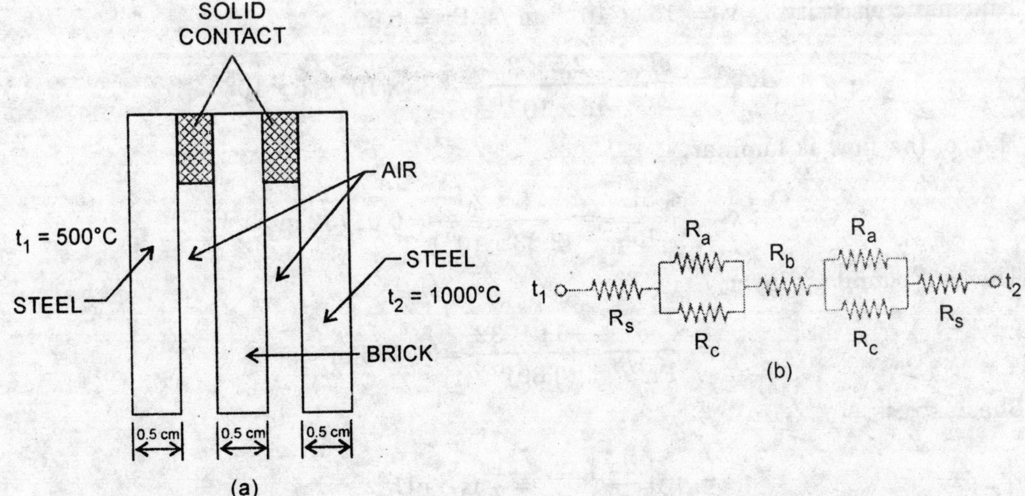

Fig. 4.

Material	Thickness, *x*, m	Conductivity, kW/m.K	R = *x*/(*k*A)
Brick	0.05	1.5	0.05/(1.5A)
Steel	0.005	50	0.005/(50A)
Air	0.001	0.02	0.001/(0.02 × 0.7A)
Solid	0.001	1.5	0.001/(1.5 × 0.3A)

$$\frac{1}{R_{eq}} = \frac{1}{R_a} + \frac{1}{R_c} = \frac{0.014A}{0.001} + \frac{0.45A}{0.001} = \frac{0.464A}{0.001}$$

$$R_{eq} = \frac{2.1552 \times 10^{-3}}{A}$$

$$R_{net} = 2\ R_s + 2\ R_{eq} + R_b$$

$$= \frac{0.01}{50A} + \frac{4.31 \times 10^{-3}}{A} + \frac{0.05}{1.5A} = \frac{0.03784}{A}$$

$$q = \frac{Q}{A} = \frac{500 - 100}{R_{net}} = \frac{400}{0.03784}$$

$$= 10569.9 \text{ W/m}^2 = 10.5699 \text{ kW/m}^2$$

(*b*) Air at 25°C flows over a thin plate with a velocity of 2.5 m/sec. The plate is 2 m long and 1 m wide. Estimate the thermal boundary layer thickness at the trailing edge of the plate and total drag force experienced by the plate. At 25°C, the density of air is 1.2 kg/m³ min and kinematic viscosity is 15 × 10⁻⁶ m²/s. Prandtl number for air is 0.69. **(15)**

Sol.

Given : Velocity of air, $v = 2.5$ m/s.

Plate length, $L = 2$, Plate width, $b = 1$ m, Air density, $\rho = 1.2$ kg/m³,

Kinematic viscosity, $\nu = 15 \times 10^{-6}$ m²/s, Pr = 0.69

$$Re_L = \frac{vL}{\nu} = \frac{2.5 \times 2}{15 \times 10^{-6}} = 3.33 \times 10^5 < 5 \times 10^5$$

Hence, the flow is laminar,

$$\delta = \frac{5L}{\sqrt{Re_L}} = \frac{5 \times 2}{(3.33 \times 10^5)} = 0.01732 \text{ m}$$

Thermal boundary layer,

$$\delta_t = \frac{\delta}{Pr^{1/3}} = \frac{0.01732}{(0.69)^{1/3}} = 0.0196 \text{ m}$$

Shear stress,

$$\tau_s = \mu . \left[\frac{du}{dy}\right]_{v=0} = \frac{1}{2} . C_{fx}\ \rho U_\infty^2$$

$$C_D = \frac{0.664}{\sqrt{Re_L}} = \frac{0.664}{(3.33 \times 10^5)^{0.5}} = 1.15 \times 10^{-3}$$

$$C_{fx} = 2C_D = 2.3 \times 10^{-3}$$

$$\tau_s = 2.3 \times 10^{-3} \times 0.5 \times 1.2 \times 2.5^2$$

$$= 8.625 \times 10^{-3} \text{ N/m}^2$$

$$\text{Drag force} = \tau_s\ A = 8.625 \times 10^{-3} \times 2 \times 1$$

$$= 0.01725 \text{ N}$$

Q.4. (*a*) Air enters a chamber at 5°C dry bulb temperature (DBT) and 2.5°C wet bulb temperature at the rate of 100 m³/min at the pressure of 1 bar. While passing through the chamber, the air absorbs 50 kW heat and picks up 35 kg/hr of saturated stream at 110°C. Show the process on a psychrometric chart and find the dry and wet bulb temperatures of the leaving air. At 110°C, enthalpy of saturated stream is 2691.3 kJ/kg. (20)

Sol.

Given : DBT, $t_1 = 5°C$, WBT $t_1' = 2.5°C$, $V_1 = 100$ m³/min

$$v_{s1} = 0.792 \text{ m}^3/\text{kg d.a.}$$

$$m_a = m_{a1} = m_{a2} = \frac{V_1}{v_{s1}} = \frac{100}{0.792} = 126.26 \text{ kg/min}$$

From energy balance :

$$m_{a1}.h_1 + m_s.h_s + Q_H = ma_2.h_2$$

$$\frac{126.269 \times 14 + 35 \times 2691.3}{60} + 50 \times 60 = 126.26 \times h_2$$

$$h_2 = 50.2 \text{ kJ/kg d.a.}$$

From sensible heat balance :

$$m_{a1}.h_1 + Q_H = m_a.h_2'$$

$$126.26 \times 14 + 50 \times 60 = 126.26\ h_2'$$

$$h_2' = 37.76 \text{ kJ/kg.d.a.}$$

From moisture balance :

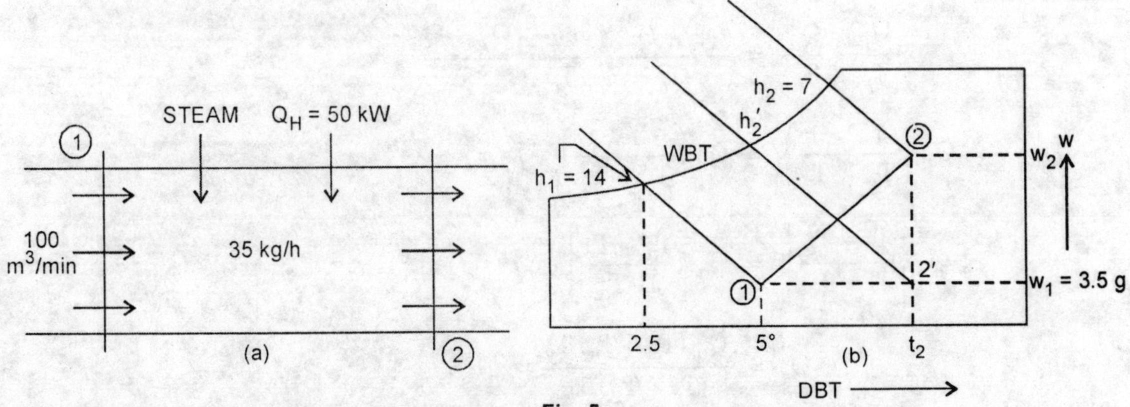

Fig. 5

$$m_{a1}.w_1 + w = m_{a2}.w_2$$

Knowing h_2' and ω_2, point 2 can be located on the psychrometric chart for leaving air.

$$\text{DBT} = 25.5°C \text{ WBT} = 17.5°C$$

DBT at point 2 can also be found out by sensible heat balance as :

$$m_{a1}.c_p.t_1 + 50 \text{ kJ/s} = m_{a2}.c_p.t_2$$

Dry air mass balance gives, $m_{a1} = m_{a2} = m_a$

$$126.26 \times 1.005 \times 5 + 50 \times 60 = 126.26 \times 1.005 \times t_2$$

$$t_2 = 28.64°C$$

(*b*) Develop an expression to calculate ideal COP of a vapour absorption refrigeration system. Describe its components.

A vapour absorption system has a source temperature of 110°C and the refrigeration temperature of – 5°C and ambient temperature of 35°C. Calculate its COP. (20)

Sol.

The components of a vapour absorption refrigeration system are shown in the Fig. 6.

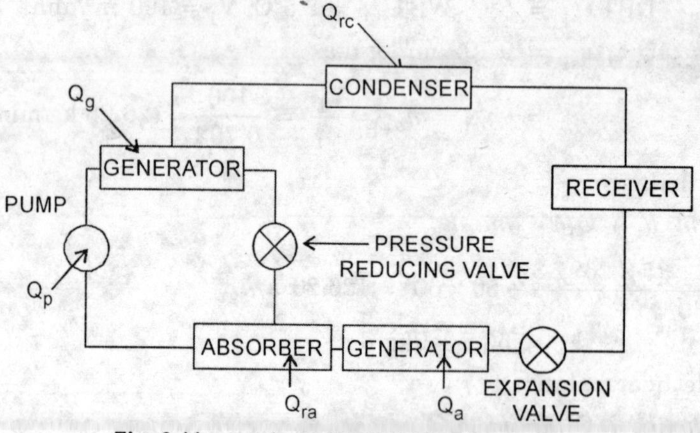

Fig. 6. Vapour absorption Refrigeration system.

COP = Heat absorbed in evaporator /[Work done by pump + Heat supplied in generator]

From heat balance, neglecting Q_p, we have

$$Q_r = (Q_{rc} + Q_{ra}) = Q_g + Q_a$$

$$\frac{Q_g}{T_g} + \frac{Q_a}{T_a} = \frac{Q_{rc} + Q_{ra}}{T_r} + \frac{Q_r}{T_r} + \frac{Q_g + Q_a}{T_r}$$

$$Q_g\left(\frac{1}{T_g} - \frac{1}{T_r}\right) = Q_a\left(\frac{1}{T_r} - \frac{1}{T_a}\right)$$

$$Q_g\left[\frac{T_r - T_g}{T_g\, T_r}\right] = Q_a\left[\frac{T_a - T_r}{T_r\, T_a}\right]$$

$$Q_g = \frac{Q_a\,(T_a - T_r)\,T_g}{(T_r - T_g)\,T_a}$$

$$\text{COP} = \frac{Q_a}{Q_g} = \frac{\left(\dfrac{T_r}{T_g} - 1\right)}{\left(1 - \dfrac{T_r}{T_a}\right)}$$

For the given data :

$$T_r = 273 + 35 = 308 \text{ K},$$
$$T_g = 273 + 110 = 383 \text{ K},$$
$$T_a = 273 - 5 = 268 \text{ K}$$

$$\text{COP} = \frac{\left(\frac{308}{383} - 1\right)}{\left(1 - \frac{308}{268}\right)} = 1.315$$

Q.5. (*a*) What is meant by boundary layer thickness, displacement thickness, momentum thickness and energy thickness? (10)

Sol.

Boundary layer thickness. It is the distance from the boundry in the laminar flow over a flat plate where the velocity differs by 1% from the free stream velocity, *i.e.*, boundary layer, $y = \delta$ for $u = 0.99\ U_\infty$.

Displacement thickness,

$$\delta^* = \int_0^\delta \left(1 - \frac{u}{U_\infty}\right) dy$$

Momentum thickness, $\theta = \int_0^\delta \left(\frac{u}{U_\infty}\right)\left(1 - \frac{u}{U_\infty}\right) dy$

Energy thickness, $\delta_a = \int_0^\delta \left(\frac{u}{U_\infty}\right)\left(1 - \frac{u^2}{U^2_\infty}\right) dy$

(*b*) What is the physical significance of Reynolds number and Mach number? Give examples of thier area of applications. (10)

Sol.

Reynolds number,

$$\text{Re} = \frac{\text{Inertia force}}{\text{Viscous force}} = \frac{\rho l^2 v^2}{\mu v l} = \frac{\rho v l}{\mu} = \frac{vl}{\nu}$$

Reynolds number helps us to determine whether the flow is laminar or turbulent. It has applications in model analysis for the criteria of dynamic similarity in the testing of ships, etc.

Match number,

$$M = \left[\frac{\text{Inertia force}}{\text{Elastic force}}\right]^{0.5} = \left[\frac{\rho l^2 v^2}{\text{EI}^2}\right] = \left[\frac{v^2}{\text{E}/\rho_0}\right]^{0.5} = \frac{v}{v_s}$$

where v_s = velocity of sound.

It helps us in comparing the velocity of a moving object with that of the sound velocity. It has applications in the model analysis of aircrafts, missiles and rotary compressors, etc.

(*c*) Oil of viscosity 0.97 Poise and 0.9 specific gravity passes through a horizontal circular pipe of 100 mm diameter and 10 m length. Calculate the pressure difference between the ends of the pipe. What is the type of flow?

Sol.

Velocity distribution of flow through a pipe is ;

$$v_x = \left[\frac{-1}{4\mu}\right]\left(\frac{dp}{dx}\right)(r_0^2 - r^2)$$

$$Q = \int_0^{r_0} 2\pi r v_x \, dr = \left[\frac{1}{8\mu}\right]\left(\frac{-dp}{dx}\right)\pi r_0^2 = \left[\frac{1}{8\mu}\right]\left(-\frac{\Delta p}{L}\right)\pi r_0^4$$

$$\Delta p = \frac{8\mu QL}{\pi r_0^4} = \frac{128\mu QL}{\pi d^4}$$

Since Q is not given, therefore, it is not possible to calculate the pressure difference and the type of flow.

$$Re = \frac{vd}{\nu}$$

If Re < 2300, then the flow is laminar.

Q.6. (*a*) Distinguish different types of flows prevailing in fluid problems. (10)

Sol.

Steady and Unsteady flow. If the fluid and flow characteristics such as density, velocity, pressure, acceleration etc., at a point do not change with time, the flow is said to be steady. Thus for steady flow, we havve

$$\frac{\partial \rho}{\partial t} = 0, \frac{\partial v}{\partial t} = 0, \frac{\partial p}{\partial t} = 0,$$

and so on.

In unsteady flow, one or more of the above variables change with time. Thus

$$\frac{\partial \rho}{\partial t} \neq 0, \frac{\partial v}{\partial t} \neq 0, \frac{\partial p}{\partial t} \neq 0,$$

and so on.

Unform and Non-uniform flow. If the velocity vector at all points in the flow is same at any instant of time, the flow is said to be uniform. Thus, for a uniform flow, we have

$$\left(\frac{\partial v}{\partial s}\right)_{t_0} = 0$$

The flow is said to be non-uniform if the velocity vector varies from point to point at any instant of time. Thus, for a non-uniform flow, we have

$$\left(\frac{\partial v}{\partial s}\right)_{t_0} \neq 0$$

Laminar and Turbulent flow. In laminar flow, the fluid particles move along smooth, regular paths which can be predicted well in advance. The streamlines are straight in laminar flow. The turbulent flow is characterized by random and erratic movement of fluid particles resulting in the formation of eddies.

(*b*) What assumptions are made in deriving Bernoulli's equation? (10)

Sol.

The following assumptions are made in deriving Bernoulli's equation :

1. The flow is steady.
2. Flow is irrotational.

3. Fluid is incompressible.

4. Fluid is non-viscous (ideal).

(c) Find the discharge of water flowing through an inclined venturimeter of 30 cm main size and 15 cm throat size. The difference between the main and the throat pressure measured by an inverted liquid manometer with the liquid of specific gravity 0.6 gives a reading of 30 cm. The loss of head between the main and the throat is 0.2 time the K.E. of the main. (20)

Sol.

Given : $d_1 = 30$ cm, $d_2 = 15$ cm, $s = 0.6$, $\Delta h = 30$ cm

$$a_1 = \frac{\pi d_1^2}{4} = \frac{\pi \times 0.3^2}{4} = 0.07065 \text{ m}^2$$

$$a_2 = \frac{\pi d_2^2}{4} = \frac{\pi \times 0.15^2}{4} = 0.01766 \text{ m}^2$$

$$h_1 - h_2 = (1 - s)\, \Delta h = (1 - 06)\, 0.3 = 0.12 \text{ m}$$

$$h_L = \frac{0.2\, v_1^2}{2g} = 0.2 \frac{\left(\frac{a_2\, v_2}{a_1}\right)^2}{2g} = 0.2\left(\frac{15}{30}\right)^4 \frac{v_2^2}{(2g)} = \left(\frac{0.2}{16}\right)\frac{v_2^2}{(2g)}$$

For the inclined venturimeter, we have

$$\frac{p_1}{\gamma} + \frac{v_1^2}{2g} + z_1 = \frac{p_2}{\gamma} + \frac{v_2^2}{2g} + z_2 + h_L$$

Also $a_1\, v_1 = a_2\, v_2$

$$\frac{v_2^2}{2g}\left(1 - \frac{a_2^2}{a_1^2}\right) = \left(\frac{p_1 - p_2}{\gamma}\right) + (z_1 - z_2) - h_L = (h_1 - h_2) - h_L$$

or

$$\frac{v_2^2}{2g}\left(1 - \frac{1}{16}\right) = 0.12 + \left(\frac{0.2}{16}\right)\frac{v_2^2}{2g}$$

$$\frac{v_2^2}{2g}\left[\frac{15}{16} - \frac{0.2}{16}\right] = 0.12$$

$$v_2^2 = 2.5453 \text{ or } v_2 = 1.5954 \text{ m/s}$$

$$Q = a_2\, v_2 = 0.01766 \times 1.5954 = 0.0282 \text{ m}^3\text{/s or } 28.2 \text{ lps}$$

Q.7. (*a*) Explain fluid coupling and torque converter and their area of application. (10)

Sol.

Fluid coupling. Fluid coupling consists of a radial pump impeller keyed to a driving shaft A, as shown in Fig. 7(*a*), and a radial type reaction turbine runner keyed to driven shaft B. There is no mechanical connection between the driving and driven shafts. The impeller and turbine runner together form a casing completely filled with ordinary mineral oil as the fluid with which the shafts A and B are supposed to be coupled. Now if the shaft A is made to revolve slowly, the oil due to forced vortex will flow out from the impeller and will strike the turbine runner blades. After sufficient head has been built up by increasing the speed of A, the fluid will drive the turbine runner and thus set the shaft B in motion. When at full speed, the two shafts are supposed to be rotating at the same speed, but owing to slip, the speed of the driven shaft is about 2% less.

The fluid coupling is employed in rail road and automobiles to transmit power from the internal combustion engine to the moving wheel. Also, they are used for power driven excavators.

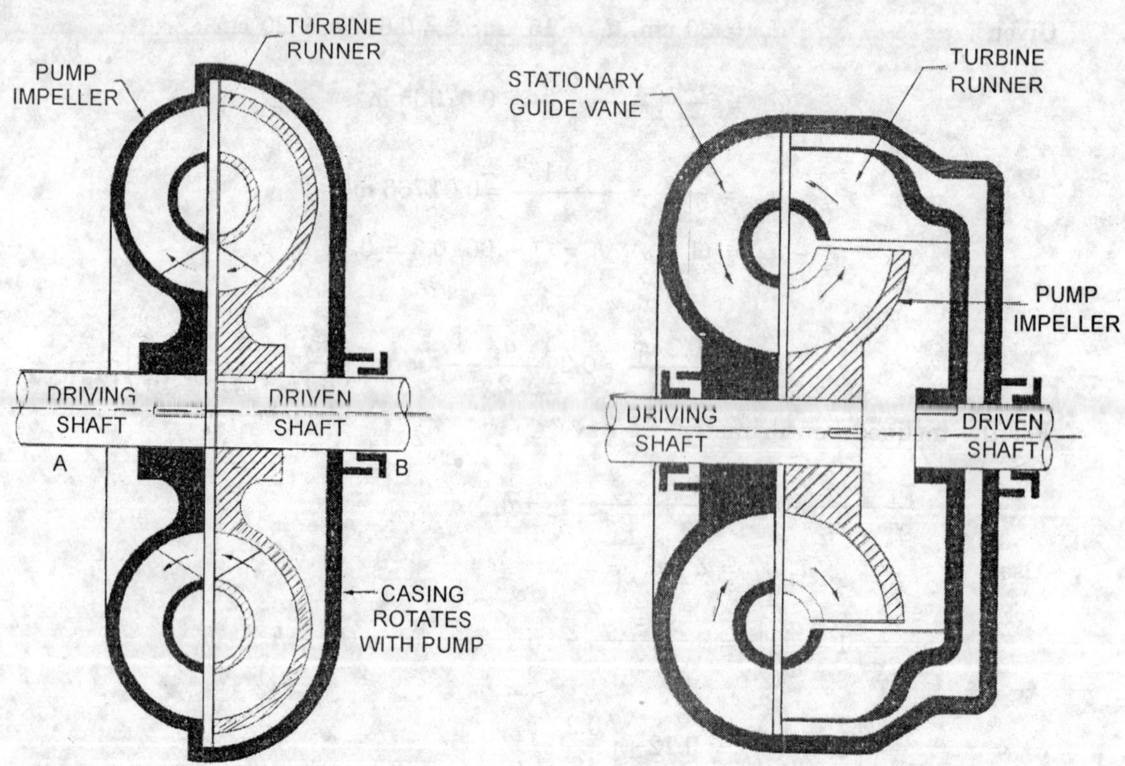

Fig. 7 (*a*). Fluid or Hydrautic Coupling.

Fig. 7 (*b*). Fluid or Hydraulic Torque Converter.

Turque converter. A torque converter is used to reduce the torque. It is similar to fluid coupling in which a third member called stator is inserted, as shown in Fig. 7 (*b*), betwen the pump impeller and the turbine runner. The stator changes the direction of the fluid. It consists of a series of fixed guide vanes. As the fluid flowing out of the impeller rim strikes the turbine runner and the blades mounted on the stationary housing, it releases its energy in the form of torque and speed. As a result of the fluid reacting upon the turbine and stationary blades, the input or engine torque is increasingly multiplied, and the speed of the output falls. Thus the stationary blades redirecting the fluid flow multiply the torque delivered by the engine and form a torque converter.

The torque converter is used as a variable gear box for the automobile and locomotive drives. It is used in bulldozers and scrapers to run at full torque output while the vehicle is stalled against an obstacle.

(*b*) What is the importance of goverting? Expalin with sketch a simple govering mechanism for a turbine. (10)

Sol.

Importance of Governing. A hydraulic turbine is coupled to an electric generator. To maintain the frequency of the generator the turbine must be run at a constant speed. Due to variation of load the speed of the turbine tends to fluctuate affecting the frequency. Thus the need for governing of the turbine.

Governing of Pelton Turbine. The combined spear-deflector regulation mechanism for a Pelton turbine is shown in the Fig. 8. The speed is regulated by spear and the pressure by deflector arrangement. As the load on the turbine is decreased, the pendulum balls occupy the top position, thus moving the bell-crank lever towards bottom with which the deflector is brought in front of the nozzle, diverting whole or part of the water jet direct to the tail race.

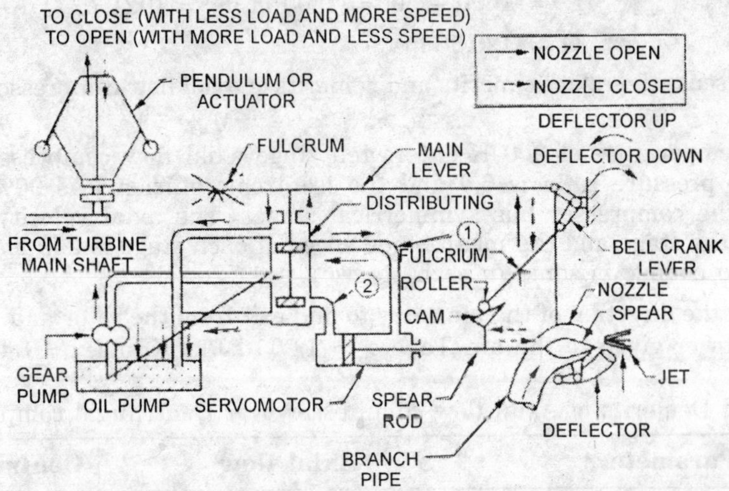

Fig. 8. Govering Impulse Turbines by Double Regulation system in Nautral Position-Simple Line Diagram.

(*c*) A pipe of 300 mm diameter and 1000 m length connects two reservoirs kept at 15 m level difference. Find the flow through the pipe. If an additional length of 500 m of 300 mm diameter pipe is connected at the later half of the existing pipe, find the increase in discharge. Neglect minor losses. Take f = 0.02. (20)

Sol.

Given : $L = 1000$ m, $d = 0.3$ m, $z_1 - z_2 = 15$ m,

$L_a = 500$ m, $d_a = 0.3$ m, $f = 0.02$

$$\frac{p_1}{\gamma} + \frac{v_1^2}{2g} + z_1 = \frac{p_2}{\gamma} + \frac{v_2^2}{2g} + z_2 + h_L$$

$$p_1 = p_2 = p_{atm}$$

$$\frac{v_2^2 - v_1^2}{2g} = (z_1 - z_2) - h_L$$

$$v_2^2 - v_1^2 = 2g\,(15 - h_L)$$

$$h_L = \frac{f\,Lv^2}{2gd} = \frac{0.02 \times 1000 \times 4Q^2}{(2 \times 9.81 \times \pi^2 \times 0.3^5)} = 681.17\,Q^2$$

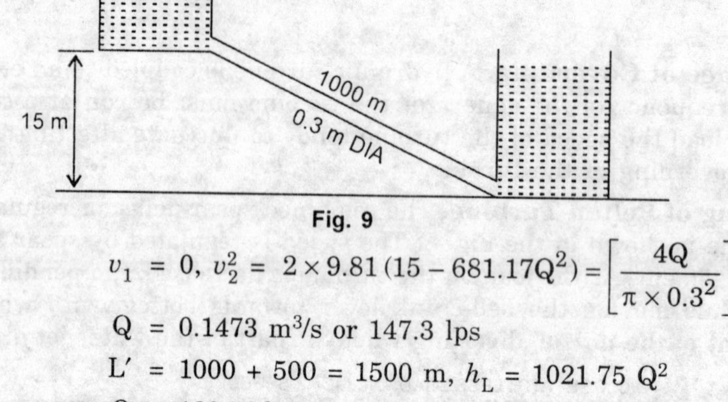

Fig. 9

For $v_1 = 0,\ v_2^2 = 2 \times 9.81\,(15 - 681.17Q^2) = \left[\dfrac{4Q}{\pi \times 0.3^2}\right]$

$Q = 0.1473\ m^3/s$ or 147.3 lps

For $L' = 1000 + 500 = 1500$ m, $h_L = 1021.75\ Q^2$

$Q = 120.56$ lps

Q.8. (*a*) Discuss the relative merits and demerits of axial flow compressor over centrifugal compressor.

Air at a temperature of 300 K enters ten stage axial flow compressor at the rate of 3.5 kg/sec. The pressure ratio is 6.0 and the isentropic efficiency is 90%. The process is adiabatic and the compressor has symmetrical stages. The axial velocity of 120 m/sec is uniform across the stage and the mean blade speed of each stage is 200 m/sec. Assume that the temperature change is same in each stage.

Determine the direction of the air entry to and exit from the rotor and the stator blades. Also find the power given to the air. Take $c_p = 1.005$ kJ/kg.K and $\gamma = 1.4$. (20)

Sol.

Merits and Demerits of Axial flow compressor over Centrifugal compressor :

SI.NO.	Parameter	Axial flow	Centrifugal
1.	Pressure ratio	1.2:1, require more number of stages for higher pressure ratio	4:1 per stage
2.	Isentropic efficiency	85 - 88%	70
3.	Power required per kg air flow rate	Lower than centrifugal	Higher than axial
4.	Flexibility of operation	Less	More
5.	Frontal area	More	Less
6.	Effect of deposit formation on the impeller	Effects High	No effect Low
7.	Starting torque	High	Low
8.	Efficiency v's speed curve	More steep	More flat
9.	Suitability for supercharging	No	Suitable
10.	Applications	Large gas turbine	Refrigerators and industrial gases.

Given :

$$n = 10,\ T_1 = 300 \text{ K},\ m = 3.5 \text{ kg/s},\ r_p = 6,\ \eta_i = 0.9$$

$$v_f = 120 \text{ m/s},\ u = 200 \text{ m/s}$$

$$(\Delta T)_{stage} = \frac{(\Delta T)_{overall}}{n} = \left[\frac{T_1}{n\eta_i}\right][r^{(\gamma-1)/\gamma} - 1]$$

$$= \left[\frac{300}{10 \times 0.9}\right][0^{0.4/1.4} - 1] = 22.284 \text{ K}$$

Increase in temperature per stage,

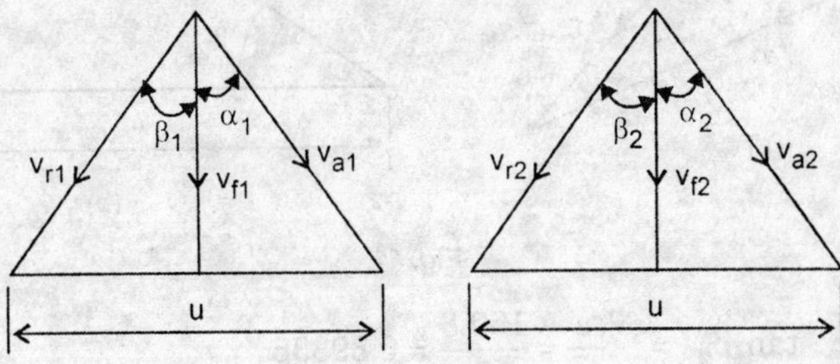

Fig. 10

$$(\Delta T)_{stage} = \left(\frac{uv_f}{c_p}\right)(\tan \beta_1 - \tan \beta_2)$$

$$\tan \beta_1 - \tan \beta_2 = \frac{22.284 \times 1.005 \times 10^3}{200 \times 120} = 0.93313$$

Also $\tan \beta_1 + \tan \beta_2 = \frac{u}{v_f} = \frac{200}{120} = 1.667$

$$\tan \beta_1 = 1.2999,$$

$$\beta_1 = 52.43°,$$

$$\beta_2 = 20.14°$$

Power supplied $= nmc_p\ (\Delta T)_{stage} = 10 \times 3.5 \times 1.005 \times 22.284 = 783.84$ kW

Q.8. (*b*) A De-Laval turbine has a mean blade velocity of 180 mps. The nozzles are inclined at 17° to the tangent. The steam flow velocity through the nozzle is 550 m/sec. For a mass flow of 3300 kg/hour and for axial exit conditions, find

(*i*) The inlet and outlet angles of the blade system

(*ii*) The power output

(*iii*) Diagram efficiency. **(20)**

Sol.

Given :

$$u = 180 \text{ m/s},\ v_{a1} = 550 \text{ m/s},$$

$$m = 3300 \text{ kg/h},\ \alpha_1 = 17°,\ \alpha_2 = 90°$$

(*i*)
$$v_{f1} = v_{f2}$$
$$v_{f1} = v_{a1} \sin \alpha_1 = 550 \sin 17°$$
$$= 160.8 \text{ m/s}$$

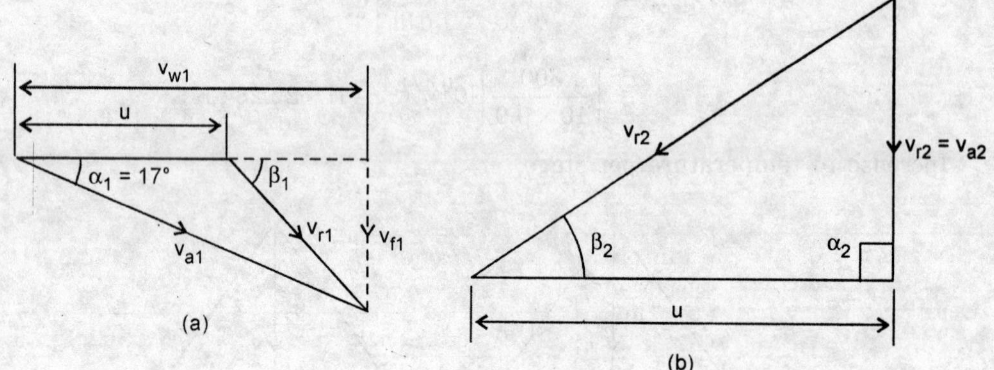

Fig. 11

$$\tan \beta_2 = \frac{v_{f2}}{u} = \frac{160.8}{180} = 0.89336$$
$$\beta_2 = 41.78°$$
$$\tan \beta_1 = \frac{v_{f1}}{v_{a1} \cos \alpha_1 - u} = \frac{160.8}{(550 \cos 17° - 180)} = 0.46478$$
$$\beta_1 = 24.93°$$

(*ii*) Power output,

$$P = muv_{w1} = \left(\frac{3300}{3600}\right) \times \left(\frac{180 \times 500 \cos 17°}{10^3}\right) = 86.785 \text{ kW}$$

(*iii*) Diagram efficiency

$$= \frac{2uv_{w1}}{v_{a1}^2} = \frac{2 \times 180 \times 550 \cos 17°}{550^2} = 0.626 \text{ or } 62.6\%$$

MECHANICAL ENGINEERING
PAPER I—2002

(Conventional)

Q.1. (***a***) An insulated tank of 1 m^3 volume contains air at 0.1 MPa and 300 K. **The tank** is connected to a high pressure line in which air at 1 MPa and 600 K flows. The tank is quickly filled with air by opening the valve between the tank and high pressure line. If the final pressure of air in the tank is 1 MPa, determine the mass of air which enters the tank and the entropy change associated with filling process.

The Universal gas constant, $\bar{R}$ = 8.314 kJ/kg.mol.K. **(20)**

Sol.

Given : $p_1 = 0.1$ MPa, $V_1 = 1\ m^3$, $T_1 = 300$ K, $p_2 = 1$ MPa

$T_2 = 600$ K, $V_2 = 1\ m^3$

$$m_1 = \frac{p_1 V_1}{RT_1} = \frac{0.1 \times 10^6 \times 1 \times 29}{8.314 \times 10^3 \times 300} = 1.1627 \text{ kg}$$

$$m_2 = \frac{p_2 V_2}{RT_2} = \frac{1 \times 10^6 \times 1 \times 29}{8.314 \times 10^3 \times 300} = 5.8134 \text{ kg}$$

Mass of air entering the tank

$$= m_2 - m_1 = 5.8134 - 1.1627 = 4.6507 \text{ kg}$$

(***b***) **By using Maxwell's relations of thermodynamics, show that Joule-Thomson co-efficient,** μ of a gas can be expresses as,

$$\mu = \left(\frac{\partial T}{\partial p}\right)_h = \left(\frac{T^2}{C_p}\right)\left[\left(\frac{\partial}{\partial T}\right)\left(\frac{v}{T}\right)\right]_p \quad \textbf{(10)}$$

Sol. $dh = T.ds + v.dp$

$$= c_p.dT - T\left(\frac{\partial v}{\partial t}\right)_p .dp + v.dp$$

$$= c_p.dT - \left[T\left(\frac{\partial v}{\partial t}\right)_p - v\right].dp$$

For an ideal gas, $dh = c_p.ds$

$$\mu_j = \left(\frac{\partial T}{\partial_p}\right)_h \left(\frac{1}{c_p}\right)\left[T\left(\frac{\partial v}{\partial T}\right)_p - v\right]$$

$$= \left(\frac{T^2}{c_p}\right)_h \left[\left(\frac{1}{T}\right)\left(\frac{\partial v}{\partial T}\right)_p - \frac{v}{T^2}\right]$$

$$= \left(\frac{T^2}{c_p}\right)_h \left[\left(\frac{1}{T}\right)\left(\frac{\partial v}{\partial T}\right)_p + \left(\frac{\partial}{\partial T}\right)\left(\frac{1}{T}\right)\right]$$

$$= \left(\frac{T^2}{c_p}\right)\left[\left(\frac{\partial}{\partial T}\right)\left(\frac{v}{T}\right)\right]_p$$

(c) A heat driven refrigeration system absorbs heat from low temperature T_g and rejects it to temperature T_c. This is run by heat supplied from a high temperature source at temperature T_H, $T_H > T_c > T_g$. Using first and second laws of thermodynamics derive the expression for maximum COP of refrigeration system in terms of tempereatures. (10)

Sol.

By First law of thermodynamics, we have

$$Q_H + Q_g = Q_c$$

By Second law, we have

$$\oint \frac{dQ}{T} = 0$$

$$\frac{Q_H}{T_H} + \frac{Q_g}{T_g} = \frac{Q_c}{T_c} = \frac{Q_H + Q_g}{T_c}$$

$$Q_H\left(\frac{1}{T_H} - \frac{1}{T_c}\right) = Q_g\left(\frac{1}{T_c} - \frac{1}{T_g}\right)$$

$$\text{COP} = \frac{T_g}{T_H} = \frac{Q_g}{Q_H} = \left(\frac{T_g}{T_H}\right)\left[\frac{T_H - T_c}{T_c - T_g}\right]$$

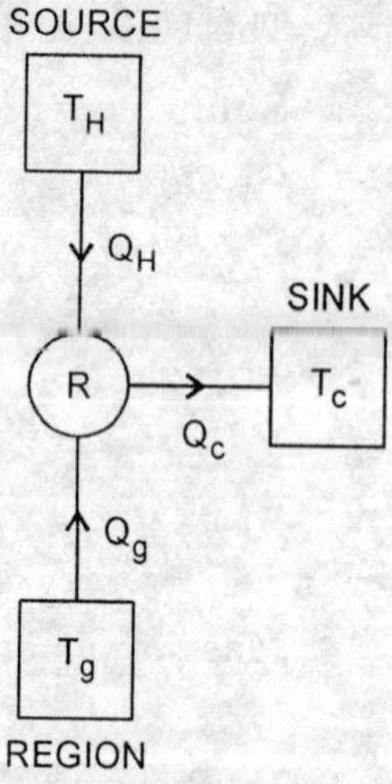

Fig. 1

Q.2. (a) An automobile carburettor having its float chamber vented to the atmosphere is tested at sea level conditions in the factory without an air cleaner. The main metering system of this carburettor is found to yield a fuel-air ratio of 0.065. The venturi/throat pressure is 0.84 bar. This carburettor is now installed in an automobile and an air cleaner is placed on the inlet to the carburettor. The air flow with and without the air filter is 230 kg/hr. The pressure drop through the filter is found to be 0.035 bar at sea level conditions. Assuming $z = 0$ and orifice co-efficient to be constant, calculate

(*i*) The venturi/throat pressure with the air cleaner

(*ii*) Fuel-air ratio with the air cleaner.

Assume incompressible flow. (15)

Sol.

(*i*) Mass of air flow, $m_a = C_{da}\, A\, [2p_a.\Delta p_a]^{0.5}$

Without air cleaner, $\Delta p_a = 1.013 - 0.840 = 0.173$ bar

With air cleaner, let p_1 = throat pressure

$$\Delta p_a' = (1.013 - 0.035 - p_1) = (0.978 - p_1) \text{ bar}$$

For the same air flow and constant co-efficients,

$$\Delta p_a = \Delta p_a'$$

$$0.173 = 0.978 - p_1$$

$$p_1 = 0.805 \text{ bar.}$$

(*ii*) Without air cleaner,

$$\Delta p_f = \Delta p_a = p_1 = 0.805 \text{ bar}$$

With air cleaner, $\Delta p_f = 1.013 - 0.805 = 0.208$ bar

As Δp_f has increased, more fuel will flow making the mixture richer.

$$\text{Now A/F ratio} = \left(\frac{1}{0.065}\right)\left[\frac{0.173}{0.208}\right]^{0.5} = 14.03$$

or $$\text{F/A ratio} = \frac{1}{14.03} = 0.07172$$

(***b***) Explain the working of a representative Nuclear power plant with a suitable diagram. List the various nuclear fuels that can be used. (15)

Sol.

Representative nuclear power plant. The main components of a nuclear power plant are :

1. Nuclear reactor.
2. Steam generator.
3. Steam turbine.
4. Condenser.
5. Electric generator.

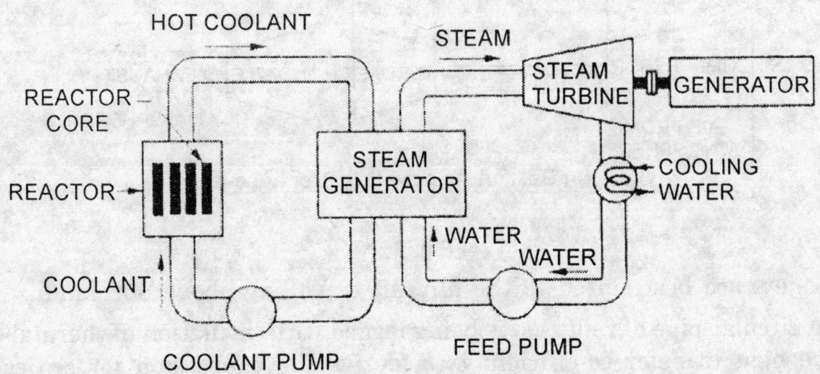

Fig. 2 Nuclear power plant.

The nuclear power plant is shown schematically in Fig. 2. The reactor performs the same function as that of the furnance of steam power plant. The heat liberated in the reactor as a result of the nuclear fission of the fuel is taken up by the coolant circulating through the reactor core. Hot coolant leaves the reactor at the top and then flows through the tubes of steam generator and passes on its heat to the feed water. The steam so produced expands in the steam turbine, producing work and thereafter is condensed in the condenser. The steam turbine in turn runs an electric generator thereby producing electrical energy. In order to maintain the flow of coolant, condensate and feed water, pumps are provided.

Nuclear fuels. The various nuclear fuels are :

Uranium-	: U^{233}, U^{235}
Uranium ceramics	
Thorium	: Th^{232}
Thorium oxide	
Plutonium	: Pu^{239}

(c) Explain the working principle of a Turbo-Jet engine with a suitable figure. (10)

Sol.

In jet propulsion, the work output of the gas turbine plant is used to produce high velocity jet of hot gases to propel the vehicles in which the systems are mounted. In the turbo-jet unit, the turbine of the gas turbine power plant produces just sufficient power to drive the compressor by partial expansion in the turbine. The gases coming out from the turbine, which are at higher pressure than atmosphere are expanded in a nozzle and produce a very high velocity jet which gives a forward motion to the aircraft by the jet reaction. The line diagram of the turbo-jet is shown in Fig. 3 (*a*) and plant used in Fig. (*b*).

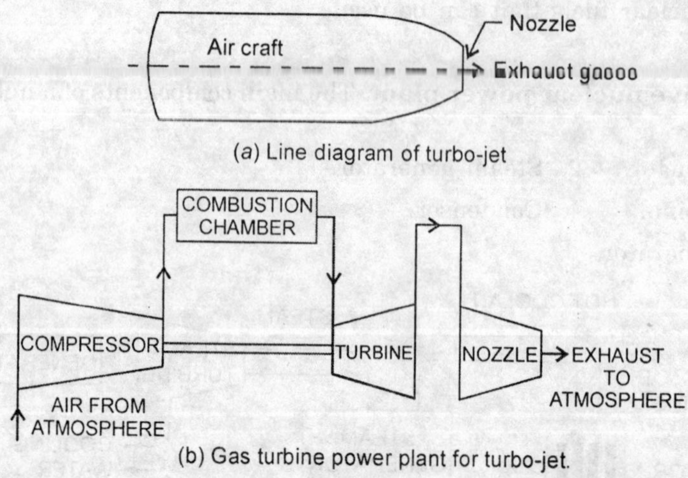

Fig. 3

The turbo-jets are best suited to the aircraft travelling above 800 km/h.

Q.3. (*a*) A circular pipe of radius r_0 is being lagged with insulation of thermal conductivity k. If convective heat transfer co-efficient is h, derive the expression for critical insulation thickness. (5)

Sol.

Consider a circular pipe of radius r_0 and length lagged with insulation of thickness t so that outer radius of the pipe becomes r_c.

Thermal resistance,

$$R_t = \left[\frac{1}{2\pi kl}\right] \ln\left(\frac{r}{r_0}\right) = \left[\frac{1}{2\pi rlh}\right] = 0$$

For critical insulation thickness,

$$\frac{dR_t}{dr} = \left[\frac{1}{2\pi kl}\right] - \left[\frac{1}{2\pi lhr^2}\right] = 0$$

$$r_c = \frac{k}{h}$$

Critical thickness, $t_c = r_c - r_0$

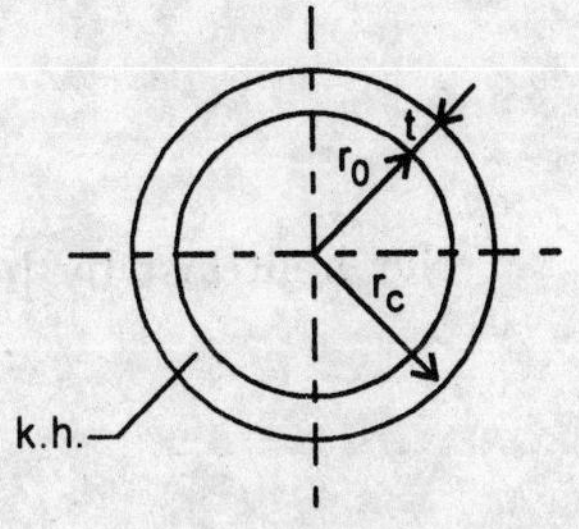

Fig. 4

(*b*) Air with an average velocity of 10 m/s at 300 K enters a copper tube of 11.2 mm diameter and 2.5 m length. The tube wall is maintained at 373 K by condensing steam at atmospheric pressure. Using LMTD method, determine the temperature of air at outlet of tube. Average properties of air are

$$k = 0.02624 \text{ W/m.K}$$

$$c_p = 1.005 \text{ kJ/kg.K}$$

$$\rho = 1.174 \text{ kg/m}^3$$

$$\nu = 1.568 \times 10^{-5} \text{ m}^2/\text{s}$$

$$\text{Pr} = 0.7$$

$$\text{Nu} = 3.66 + \frac{0.668\,(d/\text{L})\text{ Re Pr}}{1 + 0.04\,(d \text{ Re Pr})}, \text{Re} < 2300$$

$$\text{Nu} = 0.023 \text{ Re}^{0.8} \text{ Pr}^{0.4}, \text{ Re} > 2300$$

where d and L are diameter and length of tube respectively. Assume heat transfer co-efficient to be constant and neglect conduction thermal resistance of copper. **(20)**

Sol.

Given : $v = 10$ m/s, $T_1 = 300$ K, Pr = 0.7

$d = 11.2$ mm, $l = 2.5$ m, $T_w = 373$ K, $\nu = 1.568 \times 10^{-5}$ m²/s

$$\text{Re} = \frac{vd}{\nu} = \frac{10 \times 0.0112}{1.568 \times 10^{-5}} = 7142 > 2300$$

$$\text{Nu} = 0.023 \text{ Re}^{0.8} \text{ Pr}^{0.4} = 0.023 \times 7142^{0.8} \times 0.7^{0.4}$$

$$= 24.144$$

$$\text{Nu} = \frac{hl}{k}, h = \frac{24.144 \times .02624}{2.5} = 0.2534 \text{ W/m}^2\text{.K}$$

Let outside air temperature be T K. Assuming it to be a counter flow heat exchanger, we have

$$t_{h1} = t_{h2} = 373 \text{ K}, t_{c1} = 300 \text{ K}, t_{c2} = \text{T K}$$

$$\theta_1 = 373 - \text{T} \;\; \theta_2 = 373 - 300 = 73 \text{ K}$$

$$\theta_m = \frac{(373 - \text{T}) - 73}{\ln\left[\frac{373 - \text{T}}{73}\right]} = \frac{300 - \text{T}}{\ln\left[\frac{373 - \text{T}}{73}\right]}$$

$$Q = m\,c_p\,(T - 300)$$

$$m = \rho A v$$

$$= 1.174 \times \pi \times (0.0112)^2 \times \frac{10}{4} = 1.1566 \times 10^{-3} \text{ kg/s}$$

Net heat lost by heat exchanger,

$$Q = hA\theta_m = \frac{0.2534 \times \pi \times (0.0112)^2 \times (300 - T)}{\ln\left[\frac{373 - T}{73}\right]}$$

$$= mc_p\,\Delta t$$

$$= 1.1566 \times 10^{-3} \times 1.005 \times 10^3 \times (T - 300)$$

$$T = 301.38 \text{ K}$$

(*c*) The cross-section of a very long black body enclosure consists of a semi-circle with its diameter D as base. The temperature of semi-circle is 1000 K and that of diameter is 500 K. Determine the shape factors for diameter semi-circle combination and the radiation heat transfer rate per unit width (in terms of D).

Stephan-Boltzman constant = 5.64×10^{-8} W/m².K. (15)

Sol.

By summation rule,

$$F_{11} + F_{12} = 1$$

$$F_{21} + F_{22} = 1$$

$$F_{12} = 1$$

By reciprocity rule,

$$A_1F_{12} = A_2F_{21}$$

$$F_{21} = \left(\frac{A_1}{A_2}\right) F_{12}$$

Fig. 5

$$\frac{A_1}{A_2} = \frac{DL}{\left(\frac{\pi DL}{2}\right)} = \frac{2}{\pi} = 0.6366$$

$$F_{22} = 1 - F_{21} = 1 - 0.6366 = 0.3634$$

$$Q = A_1\,F_{12}\,\sigma_b\,(T_1^4 - T_2^4)$$

$$= DL \times 1 \times 5.64 \times 10^{-6} \times 10^{-8} \times (1000^4 - 5000^4)$$

$$\frac{Q}{L} = 5.2875 \times 10^4 \text{ D W/m}$$

Q.4. (*a*) If gas is used as refrigerant in reverse refrigeration cycle, draw T-*s* and *p*-*v* diagrams with heat rejection and heat asborption temperatures of T_g and T_c respectively. Show that this cycle requires two compressors and two expanders. Find their work requirement and the COP of the cycle. (10)

Sol.

The T-*s* and *p*-*v* diagrams are shown in the Fig. 6. Processes 1-2 and 2-3 are isentropic compression and processes 3-4 and 4-1 are isentropic expansion. Therefore, it sequires two compressors and two expanders.

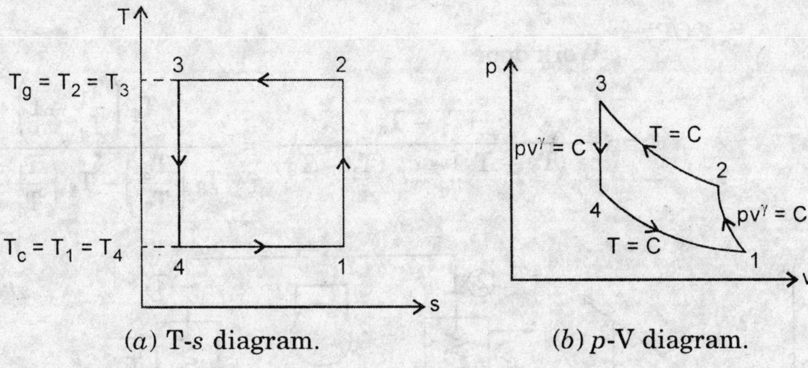

(*a*) T-*s* diagram. (*b*) *p*-V diagram.

Fig. 6

Work done per cycle per kg of air,

$$q = q_r - q_a$$

$$= q_{2-3} - q_{4-1}$$

$$= (T_g - T_c)(s_2 - s_3)$$

$$\text{Work requirement} = mc_p (T_g - T_c)$$

$$(\text{COP})_{\text{ref}} = \frac{q_a}{\text{Work done}} = \frac{q_{4-1}}{q}$$

$$= \frac{T_c (s_2 - s_3)}{(T_g - T_c)(s_2 - s_3)} = \frac{T_c}{T_g - T_c}.$$

(***b***) Dense air is used as refrigerrant in Reverse Brayton or Bell-Coleman or Joule cycle. Draw T-*s* and *p*-*v* diagrams for the cycle. Derive the expression for COP in terms of pressure ratio. If temperature at the end of heat absorption and heat rejection are 0°C and 30°C respectively, the pressure ratio is 4 and the pressure in the cooler is 4 bar, determine the tempereature at all state points and volume flow rates at inlet to compressor and at exit of turbine for 1 TR cooling coapcity. **(15)**

Sol.

The T-*s* and *p*-*v* diagrams are shown in Fig. 7. (*a*)

$$q_r = g_{2-3} = c_p (T_2 - T_3)$$

$$q_a = q_{4-1} = c_p (T_1 - T_4)$$

$$\text{Work done} = q_r - q_a = c_p (T_2 - T_3) - c_p (T_1 - T_4)$$

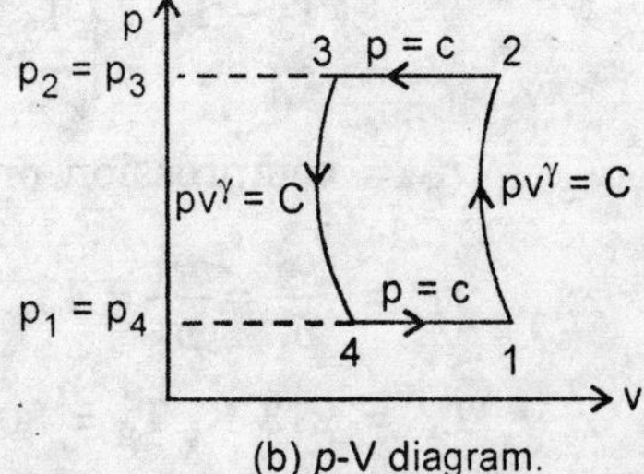

(*a*) T-*s* diagram. (b) *p*-V diagram.

Fig. 7.

$$\text{COP} = \frac{q_a}{\text{Work done}}$$

$$= \frac{c_p\,(T_1 - T_4)}{c_p\,(T_2 - T_3) - c_p\,(T_1 - T_4)} = \frac{T_4\left(\frac{T_1}{T_4} - 1\right)}{T_3\left(\frac{T_2}{T_3}\right) - T_4\left(\frac{T_1}{T_4} - 1\right)}$$

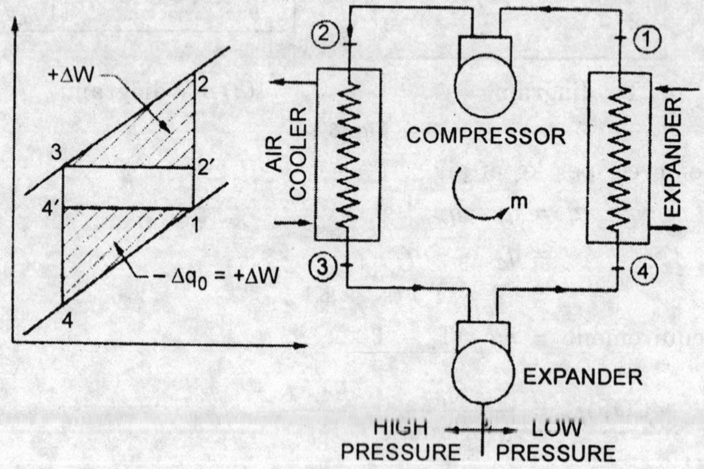

Fig. (*c*) Schematic diagram of Bell-Coleman Cycle.

For isentropic compression process 1–2,

$$\frac{T_2}{T_1} = \left(\frac{p_2}{p_1}\right)^{(\gamma - 1)/\gamma}$$

Similarly, $$\frac{T_3}{T_4} = \left(\frac{p_3}{p_4}\right)^{(\gamma - 1)/\gamma} = \left(\frac{p_2}{p_1}\right)^{(\gamma - 1)/\gamma}$$

Hence $$\frac{T_2}{T_1} = \frac{T_3}{T_4}$$

or $$\frac{T_2}{T_3} = \frac{T_1}{T_4}$$

$$\text{COP} = \frac{T_4}{T_3 - T_4} = \frac{1}{\left(\frac{T_3}{T_4} - 1\right)} = \frac{1}{(r_p)^{\frac{\gamma - 1}{\gamma}} - 1}$$

where r_p = compression or expansion ratio

$$r_p = \frac{p_2}{p_1} = \frac{p_3}{p_4} = 4$$

$$T_1 = 273 \text{ K},\ T_3 = 303 \text{ K}$$

$$T_4 = \frac{303}{(4)^{0.4/1.4}} = 203.82 \text{ K}$$

Heat extracted per minute = 1 × 211 = 211 kJ/min

Heat extracted from cold chamber per kg of air

$$= c_p\,(T_1 - T_4) = 1.005 \times (273 - 203.82) = 69.526 \text{ kJ/kg}$$

Mass flow rate of air,

$$m_a = \frac{211}{69.526} = 303 \text{ kg/min or } 182.1 \text{ kg/h}$$

Volume handled by the compressor,

$$V_c = \frac{m_a RT_1}{p_1} = \frac{182.1 \times 287 \times 273}{1 \times 10^5} = 142.68 \text{ m}^3\text{/h}$$

Voluem handled by expander,

$$V_c = \frac{m_a RT_3}{p_3} = \frac{182.1 \times 287 \times 273}{4 \times 10^5} = 39.59 \text{ m}^3\text{/h}$$

(*c*) In an air-conditioned space, 50 kga/s of fresh air at 45°C dry-bulb temperature and 30° relative humidity is introduced. The room air at 25° dry-bulb temperature and 50% relative humidity is recriculated at 450 kga/s. The mixed air flows over a cooling coil which has apparatus dew point of 12°C and bypass factor of 0.15. Determine the conditions at outlet of cooling coil, RSH, RLH, the cooling load of coil and the condensate rate. The saturation pressure of water at required temperature are

t°C	p_{ws} (bar)
12	0.014016
25	0.03166
45	0.09584

$$h = 1.005\,t + w\,(2500 + 1.88\,t) \text{ kJ/kga}$$

t is in °C and kga refers to kg of dry air. (15)

Sol.

From heat balance, we get

$$m_{a1}.c_p\,t_1 + m_{a2}.c_p.t_2 = (m_{a1} + m_{a2})\,.\,c_p.t_3$$

$$50 \times 45 + 450 \times 25 = (50 + 450) \times t_3$$

$$t_3 = \frac{13500}{500} = 27°\text{C}$$

$$\text{Bypass factor} = \frac{t_5 - t_4}{t_3 - t_4}$$

$$= \frac{t_5 - 12}{27 - 12} = 0.15$$

$$t_5 = 14.25°\text{C}$$

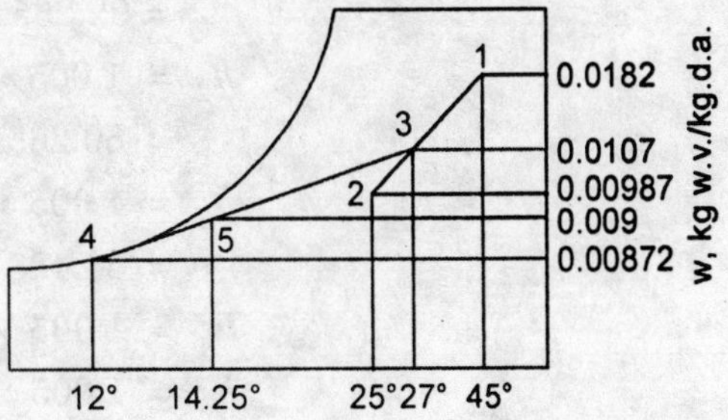

Fig. 8

Therefore, DBT at outlet of cooling coil = 14.25°C

From psychrometric chart for air, WBT = 13.25°C and RH = 88%

Now, $p_v = p_{ws} \times \text{RH}$ and $w = \dfrac{0.622}{p - p_v}$

where $p = 1.01325$ bar

$$w_1 = \frac{0.622 \times 0.09584 \times 0.30}{1.01325 \times 0.09584 \times 0.30} = 0.0182 \text{ kg w.v./kg.d.a.}$$

$$w_2 = \frac{0.622 \times 0.03166 \times 0.50}{1.01325 \times 0.03166 \times 0.50} = 0.00987 \text{ kg w.v./kg.d.a.}$$

$$w_4 = \frac{0.622 \times 0.04016 \times 1.00}{1.01325 \times 0.014016 \times 1.00} = 0.00872 \text{ kg w.v./kg.d.a.}$$

By interpolation,

$$w_3 = w_2 + \left[\frac{w_1 + w_2}{t_1 - t_2}\right](t_3 - t_2)$$

$$= 0.00987 + \left[\frac{0.0182 - 0.00987}{45 - 26}\right](27 - 25)$$

$$= 0.00987 + 0.000833 = 0.0107 \text{ kg.w.v./kg.d.a.}$$

$$w_5 = \omega_4 + \left[\frac{w_3 - w_4}{t_3 - t_4}\right](t_5 - t_4)$$

$$= 0.00872 + \left[\frac{0.0107 - 0.00872}{27 - 12}\right](14.25 - 12)$$

$$= 0.00872 + 0.000297 = 0.009 \text{ kg.w.v./kg.d.a.}$$

$$\text{RSH} = m_{a2}.c_p.(t_1 - t_2) = 450 \times 1.005 \times (45 - 25) = 9045 \text{ kJ/s}$$

$$\text{RLH} = m_{a2}.(h_{fg})_0\,(w_1 - w_2)$$

$$= 450 \times 2500 \times (0.0182 - 0.00987) = 9371.25 \text{ kJ/s}$$

$$h_1 = 1.005 \times 45 + 0.0182\,(2500 + 1.88 \times 25)$$

$$= 92.265 \text{ kJ/kga}$$

$$h_2 = 1.005 \times 25 + 0.00987\,(2500 + 1.88 \times 25)$$

$$= 50.265 \text{ kJ/kga}$$

$$h_3 = 1.005 \times 12 + 0.0107\,(2500 + 1.88 \times 27)$$

$$= 54.428 \text{ kJ/kga}$$

$$h_4 = 1.005 \times 12 + 0.00872\,(2500 + 1.88 \times 12) = 34.056 \text{ kJ/kga}$$

$$h_5 = 1.005 \times 14.25 + 0.009\,(2500 + 1.88 \times 14.25)$$

$$= 37.062 \text{ kJ/kga}$$

Cooling load of coil

$$= h_3 - h_5 = 54.428 - 37.062 = 17.366 \text{ kJ/kga}$$

Condensate rate

$$= m_a\,(w_3 - w_5) = 500\,(0.0107 - 0.009) = 0.85 \text{ kg/s}$$

Q.5. **(*a*)** Describe the working principle of a Pitot-static tube with the help of a neat sketch and explain how it can be used to measure the flow rate ? (10)

Sol.

Pitot tube. The Pitot tube is a device used for velocity measurement. It measures the stagnation pressure at any point in the flow. The static pressure is measured by a piezometer. The measurement of stagnation and static pressures are necessary and sufficient to evaluate the velocity at the point under consideration.

The pitot tube Fig. 9. consists of a tube having a 90° bend of shorter length, opened at both ends. The bent leg is directed upstream, so that a stagnation pint is created immediately ahead of its inlet, as shown in the figure. On account of the stagnation pressure so caused, the liquid rises in the vertical limb. A piezometer installed on the pipe boundary indicates the static pressure, p_o. The difference between the stangnation pressure and the static pressure is the dynamic pressure, so that

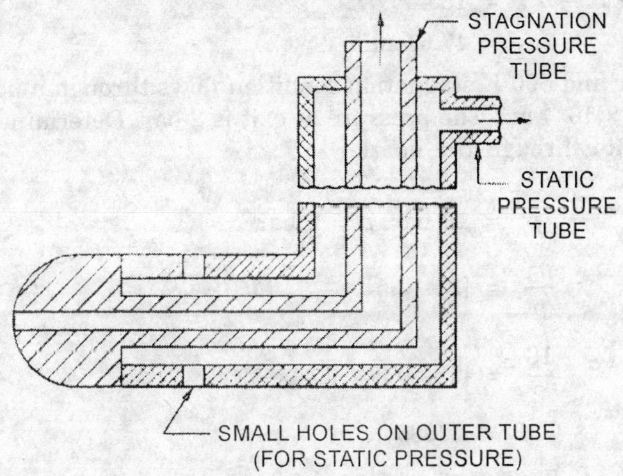

Fig. 9. Pitot static tube.

$$p_s = p_0 + \frac{1}{2}.\rho v_0^2$$

$$v_0 = \left[\frac{2(p_s - p_o)}{\rho}\right]^{0.5} = \left[\frac{2g(p_s - p_o)}{\gamma}\right]^{0.5} = [2gh]^{0.5}$$

where h = difference between the stagnation head and static head

Flow rate, $Q = Av_o$

where A = Area of pipe.

(*b*) Determine the throat diameter of a ventunrimeter in an installation of 100 mm diameter pipe carrying water. The maximum range available in water- Hg differential pressure gauge is 50 cm of Hg deflection. Find the maximum throat diameter which will induce full gauge deflection when the flow rate is 20 l/s. Assume C_d = 0.984. (20)

Sol.

Given : d_1 = 100 mm, Q = 20 lps = 0.02 m³/s. Assume C_d = 0.984

$$Q = \left[\frac{a_1 a_2}{(a^2{}_1 - a^2{}_2)^{0.5}}\right] C_d\,(2gh)^{0.5}$$

$$h = x\left(\frac{S_{mercury}}{S_{water}}\right) = 0.5\left[\frac{13.6}{1} - 1\right] = 6.3 \text{ m}$$

$$0.02 = \left[\frac{\left\{\left(\frac{\pi}{4}\right)0.1^2 \times \left(\frac{\pi}{4}\right)d_2^2\right\}}{\left\{\left(\frac{\pi}{4}\right)(0.1^4 - d_2^4)^{0.5}\right\}}\right] \times 0.984 \times [2 \times 9.81 \times 6.3]^{0.5}$$

$$0.02 = \frac{0.85922\, d^2{}_2}{(0.1^4 - d^4{}_2)^{0.5}}$$

$$0.1^4 - d^4{}_2 = 18.456\, d^4{}_2$$

$$d_2 = 47.61 \text{ mm}$$

(c) Air at 10 bar and 500 K stagnation condition flows through a nozzle. The area at the exit of nozzle is 0.25×10^{-4} m^2. The pressure at exit is 2 bar. Determine the velocity, specific volume and mean rate through the nozzle. (10)

Sol.

Given :

$$p_o = 10 \text{ bar},\ p = 2 \text{ bar}$$

$$\frac{p_o}{p} = [1 + 0.5\,(\gamma - 1)\,M^2]^{\gamma(\gamma - 1)}$$

$$\frac{10}{2} = [1 + 0.5 \times 0.4 \times M^2]^{3.5}$$

$$M = 1.7085$$

$$\frac{T_o}{T} = 1 + 0.5\,(\gamma - 1)\,M^2 = 1 + 0.5 \times 0.4 \times (1.7085)^2 = 1.5838$$

$$T = \frac{500}{1.5838} = 315.69 \text{ K}$$

$$v = M\,[\gamma RT]^{0.5} = 1.7085\,[1.4 \times 287 \times 315.69]^{0.5} = 608.5 \text{ m/s}$$

$$\rho = \frac{p}{RT} = \frac{2 \times 10^5}{287 \times 315.69} = 2.2074 \text{ kg/m}^3$$

$$\text{Specific volume} = \frac{1}{\rho} = \frac{1}{2.2074} = 0.453 \text{ m}^3/\text{kg}$$

Mean flow rate

$$\rho A v = 2.2074 \times 0.25 \times 10^{-4} \times 608.5 = 0.0335 \text{ kg/s}$$

Q.6. (a) Considering isentropic flow in a nozzle, show that

$$\frac{p^*}{p_o} = \left\{\frac{2}{\gamma + 1}\right\}^{\gamma/(\gamma - 1)}$$

$$\frac{\rho}{\rho^*} = \left\{\frac{2}{\gamma + 1}\right\}^{1/(\gamma - 1)}$$

$$\frac{T^*}{T_o} = \left\{\frac{2}{\gamma + 1}\right\}$$

where *refers to M = 1 and subscript zero refers to stagnation condition.

Sol.

Stagnation enthalpy, $h_o = h + 0.5v^2$

For an ideal gas, $h = h\,(T)$ and c_p = constant.

$$h_o - h = c_p\,(T_o - T) = 0.5v^2$$

$$\frac{T_o}{T} = 1 + \frac{0.5v^2}{c_p T} = 1 + \frac{0.5v^2\,(\gamma - 1)}{v_s^2} = 1 + 0.5\,(\gamma - 1)\,M^2$$

$$\frac{p_o}{p} = \left(\frac{T_o}{T}\right)^{\gamma/(\gamma - 1)} = [1 + 0.5\,(\gamma - 1)\,M^2]^{\gamma/(\gamma - 1)}$$

Putting M = 1, we get

$$\frac{T_o}{T^*} = 1 + 0.5\,(\gamma - 1)$$

or

$$\frac{T^*}{T_o} = \frac{2}{\gamma + 1}$$

$$\frac{p^*}{p_o} = \left[\frac{2}{\gamma + 1}\right]^{\gamma/(\gamma - 1)}$$

$$\frac{p^*}{p_o} = \left[\frac{2}{\gamma + 1}\right]^{1/(\gamma - 1)}$$

(***b***) Derive Darcy-Weisbach equation for head loss in pipe due to friction. **(17)**

Sol.

Consider a horizontal pipe of diameter d through which a fluid of viscosity μ and density ρ is flowing with a velocity v as shown in Fig. 10. Consider a control volume of the fluid bounded by the pipe surface and the two sections at a distance dx apart. The flowing fluid exerts *am* equal and opposite shear stress t_o on the fluid. The other forces acting on the control volume are the pressure forces on its end faces. For steady uniform flow, the equation of motion in the flow direction gives,

$$\left(\frac{\pi}{4}\right)d^2 p - \left(\frac{\pi}{4}\right)d^2\left[p + \left(\frac{dp}{dx}\right).dx\right] - \pi d.dx.\tau_o = 0$$

$$\left(\frac{d}{4}\right).\left(\frac{dp}{dx}\right) + \tau_o = 0$$

Now, $$\frac{dp}{dx} = -\frac{\gamma h_f}{dx}$$

Fig. 10

where γ = specific weight of the fluid

$$\tau_o = \left(\frac{d}{4}\right).\frac{\gamma h_f}{dx}$$

But $$\tau_o = \frac{1}{2}.f\,\rho v^2$$

$$h_f = \frac{1}{2}.\left(\frac{f v^2}{g}\right)\left(\frac{4}{g}\right).dx$$

For a pipe of length L, head loss

$$h_f = \frac{4 f L v^2}{2gd}$$

This is known as the Darcy-Weisbach equation.

(*c*) Give the expressions for the following dimensionless numbers and highlight their physical significance :

Reynolds number (RE)
Mach number (M)
Euler number (E)
Weber number (W)
Froude number (Fr) **(10)**

Sol.

Given : $$\text{Re} = \frac{\text{Inertia force}}{\text{Viscous force}} = \frac{v\text{L}}{\nu}$$

Reynolds number is used for ascertaining the Laminar flow. It is used for dynamic similarity in model testing.

$$\text{M} = \left[\frac{\text{Inertia force}}{\text{Compressibility force}}\right]^{0.5} = \frac{v}{v_s}$$

where v_s = Velocity of sound.

It is used in model testing of aircrafts and missiles.

$$\text{E} = \left[\frac{\text{Inertia force}}{\text{Pressure force}}\right]^{0.5} = \frac{v}{\left(\frac{\Delta p}{\rho}\right)^{0.5}}$$

It is used in cavitation studies where the pressure force is important in addition to the viscous force, and the dynamic similarly is obtained when the Euler and Reyonlds numbers are kept the same for the model and the prototype.

$$\text{W} = \left[\frac{\text{Inertia force}}{\text{Surface force}}\right]^{0.5} = \frac{v}{\left(\frac{\sigma}{\rho\text{L}}\right)^{0.5}}$$

It is used in river and harbour modeling.

$$\text{Fr} = \left[\frac{\text{Inertia force}}{\text{Gravity force}}\right]^{0.5} = \frac{v}{(gL)^{0.5}}$$

It is used in modeling of spilways, wiers, stilling basins, channel transitions, sluices, etc.

Q.7. (*a*) Derive the Euler's equation for rotating machines and explain the physical significance of each term. (10)

Sol.

Let points on the fluid's path at suction inlet, impeller inlet, impeller outlet and casing outlet be denoted by i, 1, 2 and d respectively, as shown in Fig. 11. The equation of flow between any two consecutive points can be obtained by applying Bernoulli's theorem.

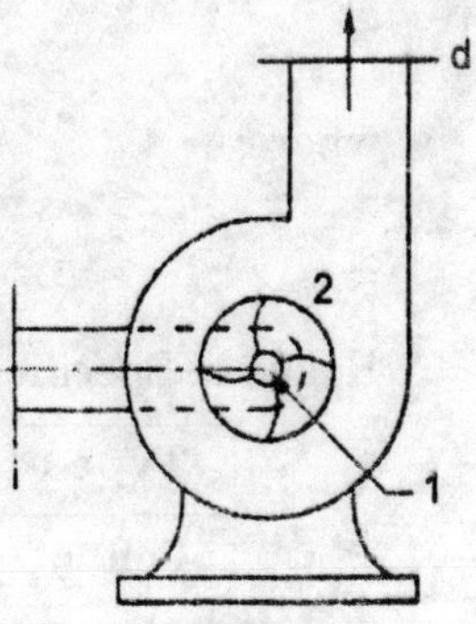

Fig. 11

Thus for flow from i to 1, we have

$$\frac{v_i^2}{2g} + \frac{p_i}{\gamma} + z_i = \frac{v_1^2}{2g} + \frac{p_1}{\gamma} + z_1 + H_{L(i-1)} \quad ...(1)$$

From 1 to 2, we have

$$\frac{v_{r1}^2}{2g} + \frac{u_1^2}{2g} + \frac{p_1}{\gamma} + z_1 = \frac{v_{r2}^2}{2g} + \frac{u_2^2}{2g} + \frac{p_2}{\gamma} + z_2 + H_{L(1-2)} \quad ...(2)$$

From 2 to d, we have

$$\frac{v_2^2}{2g} + \frac{p_2}{\gamma} + z_2 = \frac{v_d^2}{2g} + \frac{p_d}{\gamma} + z_d + H_{L(2-d)} \quad ...(3)$$

Adding (1) to (3), we get

$$\frac{v_2^2 - v_1^2}{2g} + \frac{v_{r2}^2 - v_{r1}^2}{2g} + \frac{u_2^2 - u_1^2}{2g}$$

$$= \left(\frac{v_d^2}{2g} + \frac{p_d}{\gamma} + z_d\right) - \left(\frac{v_i^2}{2g} + \frac{p_i}{\gamma} + z_i\right) + [H_{L(i-1)} + H_{L(1-2)} + H_{L(2-d)}]$$

$$= H_{mano} + \Delta H_{mano}$$

this is known as the Euler's equation for a rotating machine.

(*b*) What is the effect of blade outlet angle on the performance of centrifugal compressors? (10)

Sol.

Slightly backward curved impeller blades (*i.e.*, $\beta_2 < 90°$) gives optimum efficiency because the degree of reaction increases with decreasing blade tip angles. With increasing degree of reaction increasing part of kinetic energy is transferred into pressure energy within the impeller. By selecting very small tip blade angles β_2, the length of impeller channel increases and consequently the friction loss increase considerably, thereby decreasing the beneficial effect and reducing the efficiency. The optimum blade angle is about 45°.

(c) Define the specific speed for turbines stating the approximate range for various types of turbines. What is the physcial significance of specific speed ? (10)

Sol.

Specific speed of a turbine. The specific speed of a turbine is the speed of a geometrically similar turbine working under unit head and delivering unit brake power. Specific speed of a turbine.

$$N_s = N \frac{\sqrt{P_t}}{H^{5/4}}$$

where N = the normal working speed of turbine in rpm,

P_t = turbine brake power output in kW,

H = effective head in m.

Range of Specific Speeds

Type of turbine	Type of runner	Specific speed, N_s
Pelton	Slow	0.5 – 17
	Normal	17 – 24
	Fast	24 – 30
	Multiple jets	30 – 50
Francis	Slow	51 – 102
	Normal	102 – 153
	Fast	153 – 255
Kaplan	–	255 – 850

Physical significance of Specific speed. Specific speed is used in comparing different types of turbines working under varying heads and generating different powers while running at different speeds.

(d) What is meant by cavitation and what are the methods of a avoiding it ? (10)

Sol.

Cavitation. According to Bernoulli's equation, if the velocity of flow increases, the pressure will fall. In case of fluids, the pressure cannot fall below vapour pressure which depends upon the temperature and height above mean sea level of the site. Whenever the pressure in any turbine part drops below the evaporation pressure, the fluid boils and a large number of small bubbles of vapour are formed. It may happen that a stream of water cuts short of its path giving rise to eddies and vortices which may contain voids or bubbles. These bubbles mainly formed on account of low pressure, are carried by the stream to higher pressure zones where the vapours condense and the bubbles suddenly collapse, as the vapours are condensed to fluid form again. This results in the formation of a cavity and the surrounding fluid rushes to fill it. The steams of fluid coming from all directions collide at the centre of cavity giving rise to a very high local pressure. Formation of cavity and high pressure are repeated many thousand times a second. This causes pitting on the metallic surface of runner blades or draft tube. The material then fails by fatigue and added corrosion. This phenomenon which manifests itself in the pitting of the metallic surfaces of turbine parts is known as cavitation.

Methods to Avoid cavitation ;

1. Installation of turbine below tail race level.
2. Cavitation-free runner of reaction turbine.
3. Use of material-stainless steel is the best of all materials.
4. Use of machining-polished surfaces are resistant to cavitation.

(*e*) What is the purpose of draft tube in turbines ? (5)

Sol.

The draft tube is a conduit which connects the runner exit to the tail race. The functions of draft tube are

1. To increase the workable head on the turbine by an amount equal to the height of the runner outlet above the tail race.
2. Makes it possible to install the turbinc above the tail race level without loss of head.
3. Converts the residual kinetic energy at the exit into pressure energy.

Q.8. (*a*) find the optimum intermediate pressure of a two stage reciprocating compressor, if intercooling is done up to a temperature T_2' which is greater than the inlet temperature. (10)

Sol.

Work done in the LP cylinder, (Fig. 12)

$$W_1 = \left[\frac{n}{n-1}\right].RT_1\left[\left(\frac{p_2}{p_1}\right)^{(n-1)/n} - 1\right]$$

Work done in the HP cylinder,

$$W_2 = \left[\frac{n}{n-1}\right].RT_2'\left[\left(\frac{p_3}{p_2}\right)^{(n-1)/n} - 1\right]$$

Fig. 12.

Total work done,

$$W = W_1 + W_2 = \left[\frac{n}{n-1}\right].R$$

$$\left[T_1\left(\frac{p_2}{p_1}\right)^{(n-1)/n} + T_2'\left(\frac{p_3}{p_2}\right)^{(n-1)/n} - T_1 - T_2'\right]$$

For maximum work, $\dfrac{dW}{dp_2} = 0$

$$\left[\frac{n}{n-1}\right]\frac{T_1\, p_2^{-1/n}}{p_1^{(n-1)/n}} - T_2'\, p_3^{(n-1)/n}\left[\frac{n-1}{n}\right] p_2^{(2n-1)/n} = 0$$

or
$$p_2 = [p_1\, p_3]^{1/2}\left[\frac{T_2'}{T_1}\right]^{n/(2n-2)}$$

(*b*) Show the Reheat cycle and Regenerative feed water heating cycle on T-*s* diagram. Highlight thier significance on the performance of steam power plant. (10)

Sol.

Significance of Reheat cycle :

1. Erosion and corrosion problems in steam turbine are avoided.
2. The output of the turbine is increased.
3. Final dryness fraction is increased.
4. Nozzle and blade efficiencies are increased.

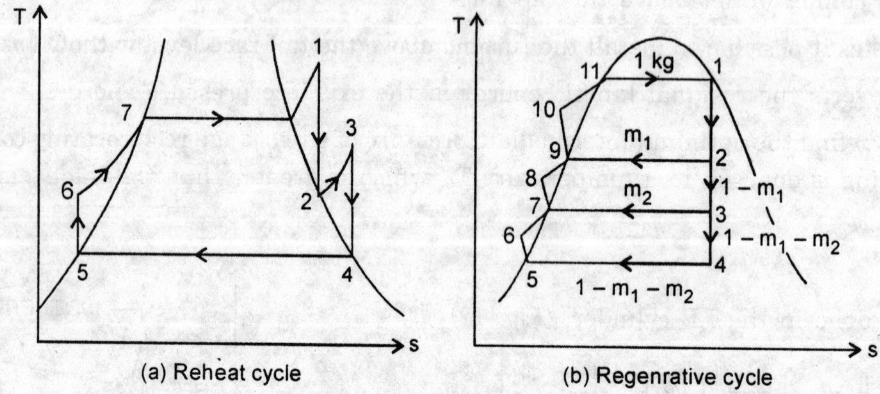

Fig. 13.

Significance of Regenerative cycle :

1. Heating process in the boiler tends to become reversible.
2. Average temperature of heat addition to the cycle is increased.
3. Temperature range in the boiler is reduced thus reducing the thermal stresses set up in the boiler.
4. It requires a reduced condenser capacity.
5. Less amount of steam is passed through the low pressure stages thereby reducing the height of the blades which results in reduced cost of the LP turbine.

(*c*) The following particulars refer to a two row velocity compounded impulse steam turbine (wheel) :

Steam velocity at nozzle exit	= 600 m/s
Nozzle angle	= 60°
Mean blade velocity	= 120 m/s
Exit angles : First row of moving blades	= 18°
Fixed guide blade	= 22°
Second row of moving blades	= 36°
Steam flow rate	= 5 kg/s
Blade friction coefficient	= 0.85

Dtermine :

(*i*) the tangential thrust,

(*ii*) the axial thrust,

(*iii*) the power developed, and

(*iv*) the diagram efficiency. (25)

Sol.

1st row :

$$v_{w1} = v_{a1} \cos \alpha_1 = 600 \cos 16° = 576.75 \text{ m/s}$$
$$v_{f1} = v_{a1} \sin \alpha_1 = 600 \sin 16° = 165.38 \text{ m/s}$$

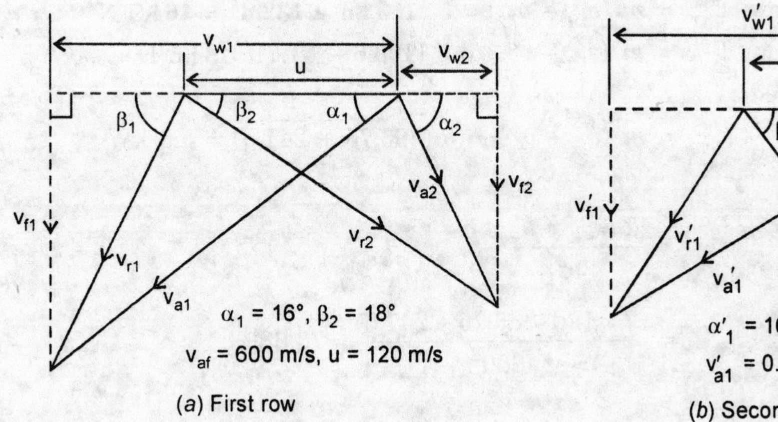

(*a*) First row

(*b*) Second row.

Fig. 14.

$$v_{r1} = \left[v_{f1}^2 + (v_{w1} - u)^2\right]^{0.5}$$
$$= [165.38^2 + (576.75 - 120)^2]^{0.5} = 504 \text{ m/s}$$
$$v_{r2} = 0.85\ v_{r1} = 0.85 \times 504 = 428.4 \text{ m/s}$$
$$v_{f2} = v_{r2} \sin \beta_2 = 428.4 \sin 18° = 132.4 \text{ m/s}$$
$$v_{w2} = v_{r2} \cos \beta_2 - u = 428.4 \cos 18° - 120 = 287 \text{ m/s}$$

$$v_{a2} = \left[v_{f2}^2 + v_{w2}^2\right]^{0.5} = [132.4^2 + 287^2]^{0.5} = 316 \text{ m/s}$$

$$v_w = v_{w1} + v_{w2} = 576.75 + 287 = 863.7 \text{ m/s}$$

2nd row :

$$v'_{a1} = 0.85\, v_{a2} = 0.85 \times 316 = 268.6 \text{ m/s}$$

$$v'_{w1} = v'_{a1} \cos 22° = 268.6 \cos 22° = 249 \text{ m/s}$$

$$v'_{f1} = v'_{a1} \sin 22° = 268.6 \sin 22° = 100.6 \text{ m/s}$$

$$v'_{r1} = \left[\left(v'_{f1}\right)^2 + \left(v'_{w1} - u\right)^2\right]^{0.5}$$

$$= \left[100.6^2 + (249 - 120)^2\right]^{0.5} = 163.6 \text{ m/s}$$

$$v'_{r2} = 0.85\, v'_{r1} = 139 \text{ m/s}$$

$$v'_{w2} = u - v'_{r2} \cos 36° = 120 - 139 \cos 36° = 7.5 \text{ m/s}$$

$$v'_{f2} = v'_{r2} \sin 36° = 139 \sin 36° = 81.7 \text{ m/s}$$

$$v'_w = v'_{w1} - v'_{w2} = 249 - 7.5 = 241.5 \text{ m/s}$$

(*i*) Tangential thrust

1st row, $F_{t1} = mv_w = 5 \times 863.75 = 4318.75 \text{ N}$

2nd row, $F_{t2} = mv'_w = 5 \times 241.5 = 1207.5 \text{ N}$

(*ii*) Axial thrust

1st row, $F_{a1} = m\,(v_{f1} - v_{f2}) = 5\,(165.38 - 132.4) = 164.9 \text{ N}$

2nd row, $F_{a2} = m\,(v'_{f2} - v'_{f2}) = 5\,(100.6 - 81.7) = 94.5 \text{ N}$

(*iii*) Power developed :

$$P = m\,(v_w + v'_w) = 5\,(863.75 + 241.5) = 5.5 \text{ kW}$$

(*iv*) Diagram efficiency

$$= \frac{2u\,(v_w + v'_w)}{v_{a1}^2}$$

$$= \frac{2 \times 120\,(863.75 + 241.5)}{600^2} = 73.68\%$$

MECHANICAL ENGINEERING
PAPER I—2003

(Conventional)

Q.1. (*a*) Explain the terms :

(*i*) Co-efficient of cubical expansion, β and

(*ii*) Co-efficient of compressibility, K. Hence, show that

$$\frac{\beta}{K} = \left(\frac{\partial p}{\partial T}\right)_v \qquad (5)$$

Sol.

Co-efficient of Cubical Expansion. It is defined as the increase of volume per unit volume for unit increase in temperature when the gas in maintained at constant pressure. If the volume increases from v to $(v + dv)$ when the increase in tempreature is from T to $(T + dT)$ at constant pressure, the coefficient of expansion is given by

$$\beta = \left(\frac{1}{v}\right)\left(\frac{\partial v}{\partial T}\right)_p .$$

The $\left(\frac{\partial v}{\partial T}\right)_p$ is the slope of the curve representing the variation of v w.r.t. T at constant pressure on v-T diagram as shown in Fig. 1. (*a*).

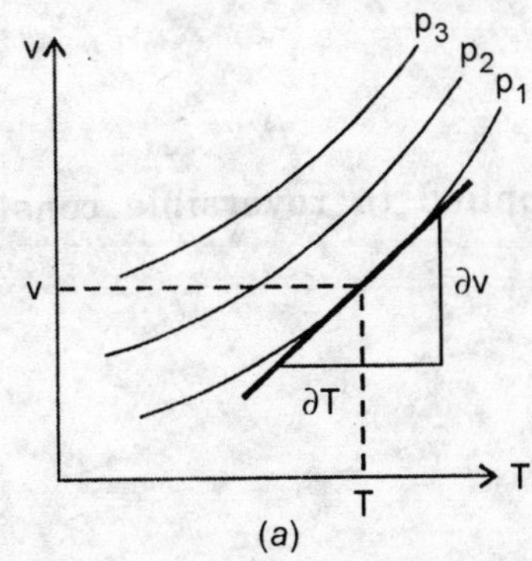

(*a*)

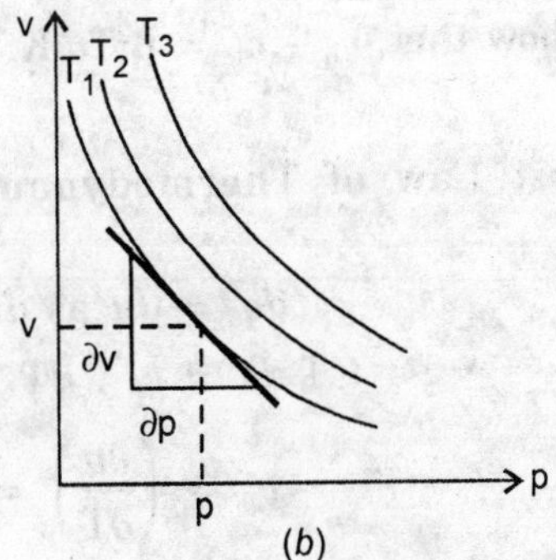

(*b*)

Fig. 1

Co-efficient of Compressibility. It is decrease of volume per unit volume for unit increase in pressure. It has two values, one when tempreature remains constant and the other when entropy remains constant.

Isothermal compressbility,

$$K_T = \left(\frac{1}{v}\right)\left(\left(\frac{\partial v}{\partial p}\right)_T\right.$$

Adiabatic compressibility,

$$K_s = -\left(\frac{1}{v}\right)\left(\frac{\partial v}{\partial p}\right)_s$$

The –ve sign is used to make RHS +ve as $\left(\frac{\partial v}{\partial p}\right)_T$ is always –ve. The slope of the curve representing the variation of v w.r.t. p is given by $\left(\frac{\partial v}{\partial p}\right)_T$ and is shown in p-v diagram for constant tempreature in Fig. 1 (b).

Now, $\left(\frac{\partial y}{\partial x}\right)_z \cdot \left(\frac{\partial z}{\partial y}\right)_x \left(\frac{\partial x}{\partial z}\right)_y = -1$

Let $x = p, y = v, z = T$, then

$$\left(\frac{\partial v}{\partial p}\right)_T \cdot \left(\frac{\partial T}{\partial v}\right)_p \cdot \left(\frac{\partial p}{\partial T}\right)_y = -1$$

$$\therefore \quad \left(\frac{\partial p}{\partial T}\right)_v = -\frac{\left(\frac{\partial v}{\partial T}\right)_p}{\left(\frac{\partial v}{\partial p}\right)_T} = -\frac{v\beta}{(-vK)} = \frac{\beta}{K}$$

Hence $\qquad \frac{\beta}{K} = \left(\frac{\partial p}{\partial T}\right)_v$

(**b**) Using Maxwell's and other equations, show that

$$c_p - c_v = \left\{p + \left(\frac{\partial u}{\partial v}\right)_T\right\}\left(\frac{\partial v}{\partial T}\right)_p$$

Hence show that, $c_p - c_v = \beta^2 Tv/K$. (10)

Sol.

The First Law of Thermodynamics as applied to reversible constant volume process is,

$$dq = du \text{ as } dw = 0$$

$$T.ds = c_v \, . \, dT$$

$$c_v = \left(\frac{\partial u}{\partial T}\right)_v = T\left(\frac{\partial s}{\partial T}\right)_v \qquad \text{...(1)}$$

$$\left(\frac{\partial c_v}{\partial v}\right)_T = T\left(\frac{\partial^2 s}{\partial T \partial v}\right) = T\left(\frac{\partial}{\partial T}\right)\left(\frac{\partial s}{\partial v}\right)$$

The third Maxwell relation is

$$\left(\frac{\partial s}{\partial v}\right)_T = \left(\frac{\partial p}{\partial T}\right)_v$$

$$\left(\frac{\partial c_v}{\partial v}\right)_T = T\left(\frac{\partial^2 p}{\partial T^2}\right)_v \qquad \text{...(2)}$$

First Law of Thermodynamics applied to reversible constant pressure process is

$$dq = du + p.dv = dh$$

$$T.ds = c_p \, .dT$$

$$c_p = \left(\frac{\partial h}{\partial T}\right)_p = T\left(\frac{\partial s}{\partial T}\right)_p \quad ...(3)$$

$$\left(\frac{\partial c_p}{\partial p}\right)_T = T\left(\frac{\partial^2 s}{\partial T \partial p}\right) = T\left(\frac{\partial}{\partial T}\right)\left(\frac{\partial s}{\partial p}\right)_p$$

The fourth Maxwell relation is

$$\left(\frac{\partial s}{\partial p}\right)_T = -\left(\frac{\partial v}{\partial T}\right)_p$$

$$\therefore \quad \left(\frac{\partial c_p}{\partial p}\right)_T = -T\left(\frac{\partial^2 v}{\partial T^2}\right)_p \quad ...(4)$$

Let $\quad u = f(T, v)$

$$du = \left(\frac{\partial u}{\partial T}\right)_v .dT + \left(\frac{\partial u}{\partial v}\right)_T .dv$$

Now, $\quad T.ds = du + p.dv$

$$= \left(\frac{\partial u}{\partial T}\right)_v .dT + \left(\frac{\partial u}{\partial v}\right)_T .dv + p.dv = \left[p + \left(\frac{\partial u}{\partial v}\right)_T\right].dv + \left(\frac{\partial u}{\partial T}\right)_v .dT$$

$$= \left[p + \left(\frac{\partial u}{\partial v}\right)_T\right].dv + c_v \, .dT \quad ...(5)$$

Again let $\quad s = f(T, p)$

$$ds = \left(\frac{\partial s}{\partial T}\right)_p .dT + \left(\frac{\partial s}{\partial p}\right)_T .dp$$

$$T.ds = T\left(\frac{\partial s}{\partial T}\right)_p .dT + T\left(\frac{\partial s}{\partial p}\right)_T .dp$$

Now, $\quad T.\left(\frac{\partial s}{\partial T}\right)_p = C_p$ and $\left(\frac{\partial s}{\partial p}\right)_T .dp = -\left(\frac{\partial v}{\partial T}\right)$

$$\therefore \quad T.ds = c_p.dT - T\left(\frac{\partial v}{\partial T}\right)_p .dp \quad ...(6)$$

Comparing Eqs. (5) and (6), we get

$$\left[p + \left(\frac{\partial u}{\partial v}\right)_T\right].dv + c_v.dT = c_p.dT - T\left(\frac{\partial v}{\partial T}\right)_p .dp$$

or $\quad (c_p - c_v)\, dT = \left[p + \left(\frac{\partial u}{\partial v}\right)_T\right].dv + T\left(\frac{\partial v}{\partial T}\right)_p .dp$

Now $$T = f(p, v)$$

$$dT = \left(\frac{\partial T}{\partial v}\right)_p .dv + \left(\frac{\partial T}{\partial v}\right)_v .dp$$

$\therefore$ $$(c_p - c_v)\left(\frac{\partial T}{\partial v}\right)_p .dv + \left(\frac{\partial T}{\partial v}\right)_v .dp \qquad ...(6)$$

$$= \left[p + \left(\frac{\partial u}{\partial v}\right)_T\right].dv + T\left(\frac{\partial v}{\partial T}\right)_p .dp$$

Compraring first term, we get

$$c_p - c_v = \left[p + \left(\frac{\partial u}{\partial v}\right)_T\right]\left(\frac{\partial v}{\partial T}\right)_p \qquad ...(7)$$

From second term, we have

$$c_p - c_v = T\left(\frac{\partial v}{\partial T}\right)_p .\left(\frac{\partial p}{\partial T}\right)_v \qquad ...(8)$$

Also $$\left(\frac{\partial v}{\partial p}\right)_T .\left(\frac{\partial T}{\partial v}\right)_p .\left(\frac{\partial p}{\partial T}\right)_v = -1$$

$\therefore$ $$c_p - c_v = -T\left(\frac{\partial v}{\partial t}\right)_p^2 .\left(\frac{\partial p}{\partial v}\right)_T$$

Now, $$\beta = \left(\frac{1}{v}\right)\left(\frac{\partial v}{\partial T}\right)_T$$

$\therefore$ $$\left(\frac{\partial v}{\partial T}\right)_p = \beta v$$

and $$K = -\left(\frac{1}{v}\right)\left(\frac{\partial v}{\partial T}\right)_T$$

$\therefore$ $$\left(\frac{\partial v}{\partial T}\right)_T = -Kv$$

$$c_p - c_v = -T(v\beta)^2\left(-\frac{1}{Kv}\right) = \frac{Tv\beta^2}{K}$$

(*c*) A reversible engine 'A' absorbs energy from a reservoir at temperature T_1 and rejects to a reservoir at temperature T_2. A second engine 'B' absorbs the same amount of energy as rejected by engine 'A', from the same reservoir at temperature T_2 and rejects energy to a reservoir at tempreature T_3. What will be the relation between T_1, T_2 and T_3 if

(*i*) The efficiencies of both the engines 'A' and 'B' are the same, and

(*ii*) The work deliveries by both the engines is the same ? (15)

Sol.

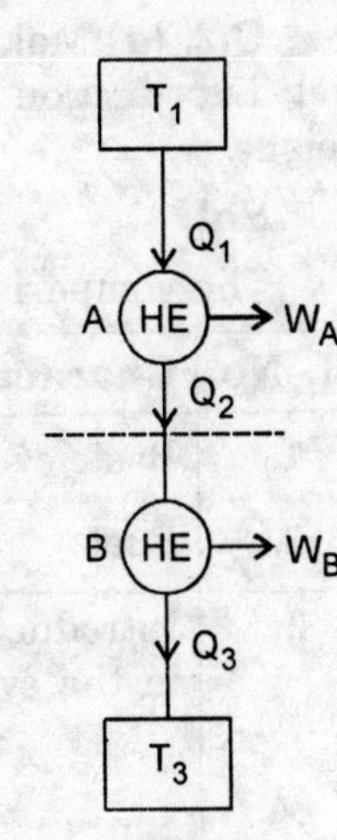

Fig. 2

$$\eta_A = \frac{W_A}{Q_1} = \frac{Q_1 - Q_2}{Q_1} = \frac{T_1 - T_2}{T_1}$$

$$W_A = Q_1 - Q_2$$

$$\eta_B = \frac{T_2 - T_3}{T_2}$$

$$W_B = Q_2 - Q_3$$

(*i*)

$$\eta_A = \eta_B$$

$$\frac{T_1 - T_2}{T_1} = \frac{T_2 - T_3}{T_2}$$

$$T_1T_2 - T^2{}_2 = T_1T_2 - T_1T_3$$

$$T_2 = [T_1T_3]^{0.5}$$

(*ii*)

$$W_A = W_B$$

$$Q_1 - Q_2 = Q_2 - Q_3$$

$$2Q_2 = Q_1 + Q_3$$

$$Q_2 = \frac{Q_1 + Q_3}{2}$$

$$T_2 = \frac{T_1 + T_3}{2}$$

(***d***) An ideal gas is heated at constant volume unitl its tempreature is 3 times the original temperature, then it is expanded isothermally till it reaches its original pressure. The gas is then cooled at constant pressure till it is restored to the original state. Determine the net work done per kg of gas if the initial temperature is 350 K. (10)

Sol.

$$T_1 = 350 \text{ K},\ T_2 = T_3 = 3T_1 = 1050 \text{ K},\ p_1 = p_3$$

Process 1-2 :

$$\frac{p_1}{p_2} = \frac{T_1}{T_2} = \frac{1}{3}$$

Process 2-3 :

$$\frac{p_3}{p_2} = \frac{V_2}{V_3} = \frac{p_1}{p_2} = \frac{1}{3}$$

Fig. 3

$$W_{1-2} = \int p.dv = 0$$

$$W_{2-3} = RT_2 \int_2^3 \frac{dV}{V} = RT_2 \ln\left(\frac{V_3}{V_2}\right)$$

$$= 0.287 \times 1050 \times \ln 3 = 331 \text{ kJ/kg}$$

$$W_{3-1} = \int_3^1 p.dV = p_1V_1 - p_3 V_3 = p_1 (V_1 - V_3) = R(T_1 - T_3)$$

$$= 0.287\ (350 - 1050) = -200.9 \text{ kJ/kg}$$

$$W_{net} = 0 + 331 - 200.9 = 130.1 \text{ kJ/kg}$$

Q.2. (*a*) Make a detailed comparison of S.I. and C.I. engines with respect to basic cycle, fuel, introduction of fuel in the cylinder, ignition, compression ratio, speed, efficiency and weight. (15)

Sol.

The comparison of S.I. and C.I. engines is given in the following table.

Sl. No.	Characteristic	S.I. Engine	C.I. Engine
1.	Basic cycle	Otto cycle	Diesel cycle
2.	Fuel	Petrol (Gasoline)	Diesel oil
3.	Introduction of fuel in the cylinder.	Air and fuel is mixed in proper proportions in the carburettor A throttle controls the quantity of the mixture introduced.	Air alone is introduced in the cylinder during the suction stroke, compressed and the fuel injected by a fuel pump at the end of compression.
4.	Ignition	Through a spark plug	High fuel control valve controls the quantity of fuel introduced.
5.	Compression ratio	6-10.5	14-20
6.	Speed	3000-5000 rpm	400-1200 rpm (1200-3000 rpm for high speed engine).
7.	Efficiency	Lower efficiency due to low compression ratio.	Higher efficiency due to high compression ratio.
8.	Weight	0.6 – 4.3 kg/kW, lighter.	3.2 – 12.2 kg/kW, heavier.

(*b*) Discuss the emission from S.I. engines and C.I. engines. On what factors do these emissions depend ?

Discuss how these emissions can be controlled. (10)

Sol.

Emission from S.I. engine. The various exhaust emissions from petrol engine are :

1. **Unburnt hydrocarbonds (HC).** The causes for the emissions of HC are :

   Incomplete combustion, crevice volumes and flow in crevices, leakage past the exhaust valve, valve overlap, deposits on walls and oil on combustion chamber walls.

   The reasons for incomplete combustion are : improper mixing of air and fuel, and flame quenching at the walls of the cylinder. Low load and idle conditions increase HC.

2. **Oxides of Carbon (CO and CO_2).** Carbon monoxide is generated with a rich fuel-air ratio mixture. This happens during starting and accelerating under load. Poor mixing, local rich regions, and incomplete combustion are the sources of CO emissions.

3. **Oxides of Nitrogen (NO_x).** NO_x are created mostly from nitrogen in the air and fuel blends. In addition to temperature, the formation of NO_x depends on pressure, A/F ratio, combustion duration and location of spark plug.

4. Oxides of Sulphur (SO_x). These are generated due to presence of sulphur in the fuel. SO_2 and SO_3 react with water to give rise to H_2SO_3 and H_2SO_4 which causes acid rain.

The variation of emission from a petrol engine with A/F ratio is shown in Fig. 4.

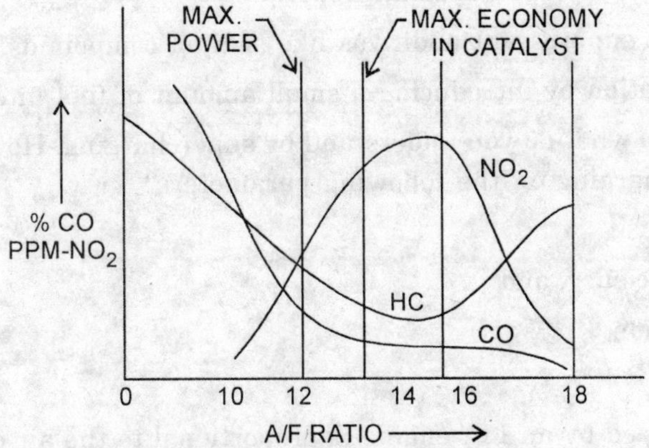

Fig. 4. Variation of emissions with A/F.

Emission from C.I. engine. Emissions from CI engine can be classified in the same categories as those from the SI engine but the level of emission in these categories vary considerably. The constitutes of CI engine exhaust are :

Hydrocarbons (HC), NO_x, RCHO, and CO. The engine type and the mode of operation are the two main factors which influence the exhaust emission from a CI engine.

Control of Emission in SI Engine. An emission control programme aims at reducing the concentration of CO, HC and NO_x. The main approaches adopted are :

(*a*) **Engine design modification.** The following steps reduce the exhaust emission by engine design modification :

(*i*) Use leaner air-fuel ratio.

(*ii*) Retard ignition timing.

(*iii*) Avoid flame quenching time by reducing the surface to volume ratio of the combustion chamber.

(*iv*) Lower compression ratio.

(*v*) Reduce valve overlap.

(*vi*) Modify the induction system by using high velocity carburettors or multi-choke carburettors.

(*b*) **Exhaust gas oxidation.** The devices used to reduce HC/CO emission are :

(*i*) Use of after-burner.

(*ii*) Use of exhaust manifold reactor.

(*iii*) Use of catalytic converter.

(*c*) **Fuel modification.** Air-fuel ratios leaner than stoichiometric results in almost insignificant amount of CO and reduce HC with reduced specific fuel consumption.

Control of Emission in CI Engine. The smoke level can be controlled by the following methods.

(*i*) Derating-*i.e.* running the engine at lower loads which however means a loss of output.

(*ii*) Maintain the injection system properly.

(*iii*) Use smoke suppressant additives like sodium compounds.

(*iv*) Use fumigation by introducing a small amount of fuel into the intake manifold.

(c) Explain clearly what do you understand by supercharging. How is it achieved? What is the effect of supercharging on the following parameters?

(*i*) Power output.

(*ii*) Mechanical efficiency, and

(*iii*) Fuel consumption. (15)

Sol.

The power developed by an I.C. engine is proportional to the air consumption. Because as the air consumption increases a greater quantity of fuel is added. The quantity of air consumption can be increased by increasing the density of the charge so that a greater mass of charge is sucked by the engine. This method of increasing the air capacity of the engine is known as supercharging. It is widely used in aircraft engines.

The apparatus used for increasing the air density is known as a supercharger. Besically a supercharger is an air compression which may be positive displacement, reciprocating (piston-cylinder), or rotary type (roots blower) or dynamic compressors (centrifugal or axial flow type). The centrifugal compressor is widely used as a supercharger for I.C. engines. The function of supercharger is either to produce more power from an engine of a given cylinder size, or to compensate the power loss at high altitudes due to rarified atmosphere.

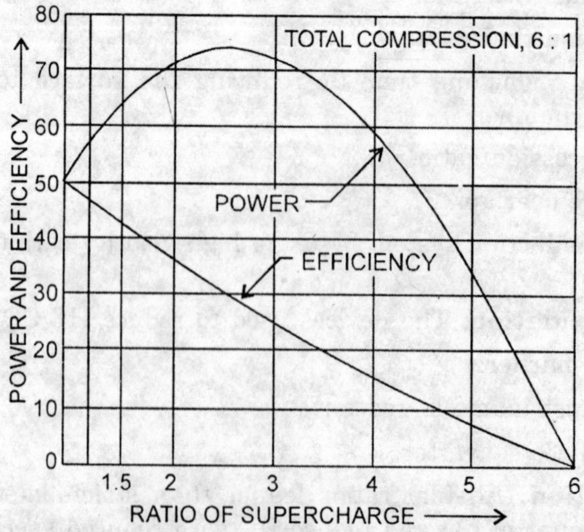

Fig. 5. Effect of supercharging ratio on power and efficiency.

Effects of Supercharging :

(i) **Power output.** Supercharging produces more power because the supercharger supplies air at a pressure and density higher than atmospheric which has the effect of increasing volumetric efficiency.

(ii) **Mechanical efficiency.** The mechanical efficiency of a supercharged engine is higher than that of one not supercharged.

(iii) **Fuel consumption.** It provides better mixing of fuel and air which results in a specific reduction of fuel consumption and the thermal efficiency increases.

Q.3. ***(a)*** An iron plate of thickness L and thermal conductivity k is subjected to a constant heat flux q_0 W/m^2 at the boundary surface at $x = 0$. From the other boundary surface at $x = $ L, the heat is dissipated by convection into a fluid at temperature T_∞ with a heat transfer co-efficient h. Develop the expression for the surface temperatures T_1 and T_2 at the surface $x = 0$ and $x = $ L respectively.

For the following data, calculate the surface temperatures T_1 and T_2 if L = 2 cm, k = 20 W/m.°C, q_0 = 10^5 W/m^2, T_∞ = 50°C and h = 500 W/m^2.°C.

Sol.

At $$x = L, \; q_0 = h(T_2 - T_\infty)$$

$$T_2 = \frac{q_0}{h} + T_\infty = \frac{10^5}{500} + 50 = 250°C$$

At $$x = 0, \; q_0 = \frac{k(T_1 - T_2)}{L}$$

$$\frac{q_0 L}{k} = T_1 - T_2$$

$$T_1 = \frac{q_0 L}{k} + T_2 = \frac{10^5 \times 0.02}{20} + 250 = 350°C$$

(b) Explain clearly what do you understand by the terms velocity boundary layer and thermal boundary layerl. Give their importance in the field of engineering.

The velocity profile $U_{x,y}$ for boundary layer flow over a flat plate is given by

$$U_{x,y} = U_\infty \left[\frac{3}{2}\left(\frac{y}{\delta_x}\right) - \frac{1}{2}\left(\frac{y}{\delta_x}\right)^3 \right]$$

If the boundary layer thickness is given as

$$\delta_x = \sqrt{\frac{280}{13} \frac{U_x}{U_\infty}}$$

then develop the expression for the local drag coefficient C_m over a distance x = L from the leading edge of the plate. (15)

Sol.

Velocity boundary layer. At very large Reynolds numbers the viscous effects are important only in the immediate vicinity of a solid boundary. This region into which the effects of viscosity are confined is known as *boundary layer*.

Consider the flow of a fluid over a flat plate at zero incidence. The viscous forces retard the fluid flow eithin a thin layer near the solid boundary. Inside the boundary layer the fluid velocity increases from zero at the boundary to free stream velicity U_∞ corresponding to the ideal fluid flow. The velocity boundary layer is defined by the region in which the fluid flow velocity changes from 0 to 0.99 U_∞, *i.e.*, $u/U_\infty = 0.99$.

Thermal boundary layer. Whenever the flow of a fluid takes place past a heated or cold surface, a temperature field is set up in the region next to the surface. The zone or the layer within which the temperature gradient exists is called the thermal *boundary layer.*

$$\frac{t_s - t}{t_s - t_\infty} = 0.99$$

$$U_{x,y} = U_\infty \left[\left(\frac{3}{2}\right)\left(\frac{y}{\delta_x}\right) - \left(\frac{1}{2}\right)\left(\frac{y}{\delta_x}\right)^3\right]$$

$$\frac{dU_{x,y}}{dy} = U_\infty \left[\frac{3}{2\delta_x} - \frac{3y^2}{2\delta_x}\right]$$

$$\tau = \mu \left|\frac{dU}{dy}\right|_{y=0} = \mu\, U_\infty \left[\frac{3}{2\delta_x} - \frac{3y^2}{2\delta_x^3}\right]_{y=0}$$

$$= \frac{3\,\mu\, U_\infty}{2\delta_x} = \left(\frac{3}{2}\right)\mu\, U_\infty \left[\frac{13 U_\infty}{280\, v\, x}\right]^{0.5}$$

$$C_f = \frac{\tau}{1/2.\rho U_\infty^2} = \left(\frac{3}{2}\right)\frac{\mu\, U_\infty \left[\dfrac{13\, U_\infty}{280\, v\, x}\right]^{0.5}}{1/2.\rho\, U_\infty^2} = 3\left(\frac{13}{280}\right)^{0.5}\left[\frac{v}{U_\infty\, x}\right]^{0.5}$$

At $$x = L,\ C_f = \frac{3\left(\dfrac{13}{280}\right)^{0.5}}{[\mathrm{Re_L}]^{0.5}} = \frac{0.646}{[\mathrm{Re_L}]^{0.5}}$$

(*c*) **Determine the net readiant interchange between two parallel oxidized iron plates, placed at a distance of 30 mm, having size of 4 m × 4 m. The surface temperature of the two plates are 120°C and 40°C respectively. The emissivity of both the plates is 0.736. (10)**

Sol.

$$Q_{1-2} = (F_g)_{12}\, \sigma_b\, A\, (T_1^4 - T_2^4)$$

$$(F_g)_{12} = \frac{1}{\left[\dfrac{(1-\varepsilon_1)}{\varepsilon_1} + \dfrac{1}{F_{12}} + \left(\dfrac{A_1}{A_2}\right)\dfrac{(1-\varepsilon_2)}{\varepsilon_2}\right]}$$

$$F_{12} = 1,\ A_1 = A_2$$

$$(F_g)_{12} = \frac{1}{\left[\dfrac{1}{\varepsilon_1} + \dfrac{1}{\varepsilon_2} - 1\right]} = \frac{1}{\left[\dfrac{2}{0.736} - 1\right]} = 0.582$$

$$Q_{12} = 0.582 \times 5.67 \times 10^{-8} \times 4 \times 4 \times (393^4 - 313^4) = 7.52 \text{ kW}$$

Q.4. (*a*) An ideal vapour compression refrigerator using Freon-12 operates between temperature limits of-10°C and 40°C. The refrigerant leaves the condenser dry saturated. The rate of flow of refrigerant through the unit is 150 kg/hr.

Calculate the refrigerating effect per kg of refrigerant, the co-efficient of performance and the capacity of the plant, if

(*i*) The refrigerant leaves the condenser dry saturated as stated,

(*ii*) The refrigerant is sub-cooled to 20°C before throttling. Assume that the enthalpy of refrigerant before throttling is approximately equal to the enthalpy of liquid refrigerant at undercooled temperature of 20°C and take it as h_f = 50.59 kJ/kg. Give your comments on the result. Solve the problem by using tables. (25)

Sol.

$$m = 150 \text{ kg/h}$$

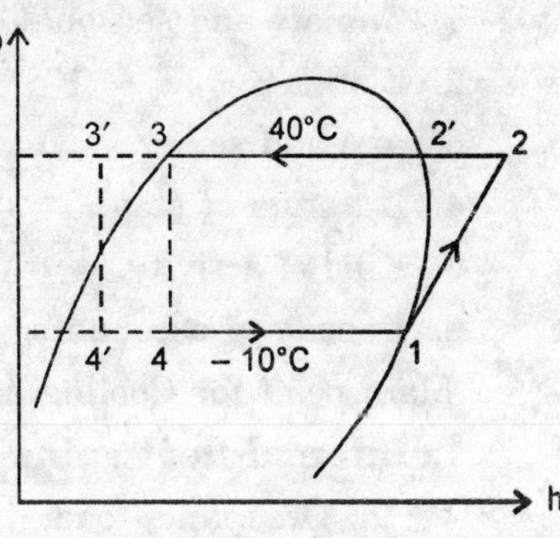

Fig. 6

From *p-h* chart for Freon-12, we have

$$h_1 = 183.05 \text{ kJ/kg},$$

$$s_1 = s_2 = 0.7014 \text{ kJ/kg.K}$$

$$h_3 = h_4 = 74.52 \text{ kJ/kg},$$

$$h_2 = 210 \text{ kJ/kg}$$

(*i*) Refrigeration effect,

$$q_0 = h_1 - h_4 = 183.05 - 74.52$$

$$= 108.53 \text{ kJ/kg}$$

$$\text{COP} = \frac{h_1 - h_4}{h_2 - h_1} = \frac{108.53}{(210 - 183.05)} = 4.02$$

Plant capacity

$$= \frac{mq_0}{3600} = \frac{150 \times 108.53}{3600} = 4.52 \text{ kW } or \text{ } 1.2853 \text{ Ton}$$

(*ii*) $$h_1 = 183.05 \text{ kJ/kg}, \; h_2 = 210 \text{ kJ/kg}$$

$$h_3 = h_{f3} = 50.59 \text{ kJ/kg} = h'_4 \text{ (Given)}$$

Refrigeration effect,

$$q'_0 = h_1 - h'_4 = 183.05 - 50.59 = 132.46 \text{ kJ/kg}$$

$$\text{COP} = \frac{h_1 - h'_4}{h_2 - h_1} = \frac{132.46}{210 - 183.05} = 4.91$$

Plant capaicty

$$= \frac{mq'_0}{3600} = \frac{150 \times 132.46}{3600} = 5.52 \text{ kW or } 1.57 \text{ Ton}$$

With subcooling, refrigerating capacity, COP and plant capacity increase.

(*b*) Define human comfort. Give the factors governing optimum effective temperature. Outline the main points considered for cooling load estimate and heating load estimate. (15)

Sol.

Human comfort. It is defined as the state of human body when the heat produced by metabolism of human body is equal to the sum of the heat dissipated to the surroundings and the heat stored in the human body by raising the temperature of body tissues.

During summer, for comfort air-conditioning :

DBT = 25 ± 1°C, RH = 50 ± 5%, Air velocity = 0.4 m/s

During winter : DBT = 21°C, RH = 50%, Air velocity = 0.15 – 0.20 m/s

Factors govering optimum effective temperature :

1. Climate and seasonal differences.
2. Clothing.
3. Age and sex.
4. Duration of stay.
5. Kind of activity.
6. Density of occupants.

Main point for Cooling and Heating load estimates :

1. Internal heat gains :

(*a*) Occupancy load.

(*b*) Lighting load.

(*c*) Appliances load.

(*d*) Piping, tanks and steam load.

(*e*) Heat of respiration.

(*f*) Process load.

2. System heat gains :

(*a*) Supply air duct gain and leakage loss.

(*b*) Air-conditioning fan load.

(*c*) Return air duct heat and leakage gain load.

(*d*) Heat gain from dehumidifier pump and piping.

3. Safety factor : 5% additional load.

Q.5. (***a***) A ship is sailing in the ocean. The captain of the ship observes an iceberg floating in the sea at a distance and he estimates the visible volume of the iceberg, above sea, level as 600 m^3. If the density of the iceberg is 915 kg/m^3, determine the total volume and weight of the iceberg.

The density of sea water may be taken as 1025 kg/m^3. (5)

Sol.

Volume of iceberg above sea level,

$$V_i = 600 \text{ m}^3$$

Density of ice, $\rho_i = 915 \text{ kg/m}^3$

Let total volume of iceberg = V

Weight of iceberg, $W_i = \rho_i V = 915 \text{ V}$

Density of sea water, $\rho_w = 1025 \text{ kg/m}^3$

Weight of sea water displaced by iceberg

$$= \rho_w \times (V - V_i) = 1025\,(V - 600)$$

$$= \text{Weight of iceberg}$$

$$\therefore \quad 1025\,(V - 600) = 915\,V$$

$$V = 5590 \text{ m}^3$$

$$W_i = \rho_i V g = 915 \times 5590 \times 9.81 = 50.1767 \times 10^6 \text{ N}$$

(*b*) Explain practical utility of venturimeter and pitot tube. Starting from first principles, derive the expression for discharge of a liquid though the venturimeter.

If the co-efficient of discharge of a venturimeter is 0.96, the venturimeter constant is 0.3 $m^{2.5}$ sec and ventrui head is 0.2 m, find discharge through the venturemeter. (15)

Sol.

Venturimeter. It is a device to measure the rate of flow of a fluid flowing through a pipe.

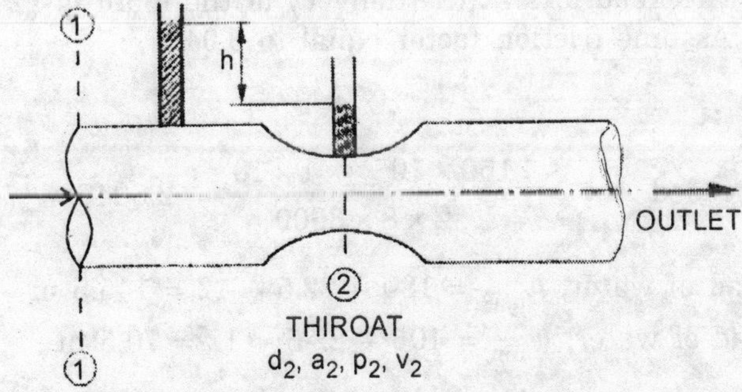

Fig. 7

Pitot tube. It is a device to measure the velocity of flow at any point in a pipe or a channel.

Discharge through a Venturimeter. Consider a horizontal pipe with a venturimeter as shown in Fig. 7. Applying Bernoulli's theorem at points 1 and 2, we have

$$\frac{p_1}{\gamma} + \frac{v_1^2}{2g} + z_1 = \frac{p_2}{\gamma} + \frac{v_2^2}{2g} + z_2$$

Now $z_1 = z_2$ for a horizontal pipe

$$\frac{p_1 - p_2}{\gamma} = \frac{v_2^2 - v_1^2}{2g} = h$$

Also $$a_1 v_1 = a_2 v_2$$

$$v_2^2 \left[1 - \left(\frac{a_2}{a_1}\right)^2\right] = 2gh$$

$$v_2^2 = \frac{2gh}{1-\left(\frac{a_2}{a_1}\right)^2}$$

$$Q = a_2 v_2 = \left[\frac{a_1 a_2}{a_1^2 - a_2^2}\right] C_d\,[2gh]^{1/2} = C_v\,C_d\,h^{1/2}.$$

where C_v = Venturi constant $= \left[\frac{a_1 a_2}{a_1^2 - a_2^2}\right](2g)^{1/2}$

Venturi constant $= 0.3\ m^{2.5}$, Q = ?, $C_d = 0.96$, $h = 0.2$ m

$$Q = \left[\frac{a_1 a_2}{\left(a_1^2 - a_2^2\right)^{0.5}}\right] C_d\,(2gh)^{0.5}$$

$$= 0.3 \times 0.96 \times [0.2]^{0.5} = 0.1287\ m^3/s$$

(*c*) A city has one million population and water supply to the city is from a reservoir 10 km away. The minimum required water supply is at the rate of 150 litres per head per day. The supply is made in two shifts, each being for 8 hours duration. The full supply level of the reservoir is R.L. 180.00 and its lowest level is R.L. 105.00. The delivery end of the main is at R.L. 22.5 m. The head loss at the delivery of the main is 12 m. Find tho diamotor of the pipe required. Assume friction factor equal to 0.04. (20)

Sol.

Total water needed $= \frac{150 \times 10^{-3} \times 1 \times 10^6}{2 \times 8 \times 3600} = 2.6042\ m^3/s$

Maximum head of water, $h_{max} = 180 - 22.5 - 12 = 145.5$ m

Minimum head of water, $h_{min} = 105 - 22.5 - 12 = 70.5$ m

$$h_{max} = \frac{4f_1\,L\,v^2}{2gd} = \frac{f'\,LQ^2_{max}}{2gd\,A^2}$$

$$145.5 = \frac{0.04 \times 10^4 \times Q^2_{max} \times 16}{2 \times 9.81 \times \pi^2 \times d^5}$$

$$Q^2_{max} = 4.4023\,d^5$$

$$Q_{max} = 2.1\,d^{2.5}$$

$$h_{min} = \frac{4f_1\,L\,v^2}{2gd} = \frac{f'\,LQ^2_{min}}{2gd\,A^2}$$

$$Q^2_{min} = 2.133\,d^5$$

$$Q_{min} = 1.46\,d^{2.5}$$

Mean discharge, $Q_m = 0.5\,(Q_{max} + Q_{min})$

$$= 0.5\,(2.1 + 1.46)\,d^{2.5} = 1.78\,d^{2.5} = 2.6042$$

$$d = 0.164\ m$$

Q.6. (*a*) A circular cylinder of height 200 mm and radius equal to 80 mm is open at the top. It is fixed on a table at its centre which can be rotated by a motor. Determine the speed of rotation of the table so that one-third of the area of the bottom of the cylinder gets exposed. At the beginning of the rotation the cylinder was filled with a liquid. (10)

Sol.

Let, r = radius at the base of the cylinder which gets exposed

$$\pi r^2 = \frac{\pi R^3}{3}$$

$$r = \frac{R}{\sqrt{3}}$$

$$z_1 = \frac{\omega^2 r^2}{2g}$$

$$z_2 = \frac{\omega^2 R^2}{2g} = \frac{3\,\omega^2 r^2}{2g}$$

$$z_2 - z_1 = \left[\frac{\omega^2 r^2}{2g}\right][3 - 1] = \frac{\omega^2 r^2}{g}$$

$$= \frac{\omega^2 R^2}{3g} = 0.2$$

$$\omega^2 = \frac{0.2 \times 3 \times 9.81}{0.08^2} = 919.6875$$

$$\omega = 30.326 \text{ rad/s or } 289.59 \text{ rpm}$$

Fig. 8

(*b*) A cylindrical tank is placed with its axis vertical. A circular orifice of 40 mm diameter fitted at its bottom. The supply of water into the tank is at uniform rate and at the same time water is discharged through the orifice. A manometer tube is fitted on the side of the tank to indicate the level of the water. An observer starts the stop watch when water level was at 400 mm. It was found that water level becomes 800 mm after one minute. Again the observer starts the stop watch when water level was at 1 m and it was found that water level becomes 1.1 m in one minute. Calculate the diameter of the tank by assuming coefficient of discharge for the orifice equal to 0.6. (25)

Sol.

Let, h = Height of water level at any instant

dh = Fall (or rise) in water level in time dt

Q = Uniform rate of supply of water into the tank

A = Area of cross-section of tank

Area of orifice, $a = \frac{\pi d^2}{4}$

C_d = Coefficient of discharge for the orifice

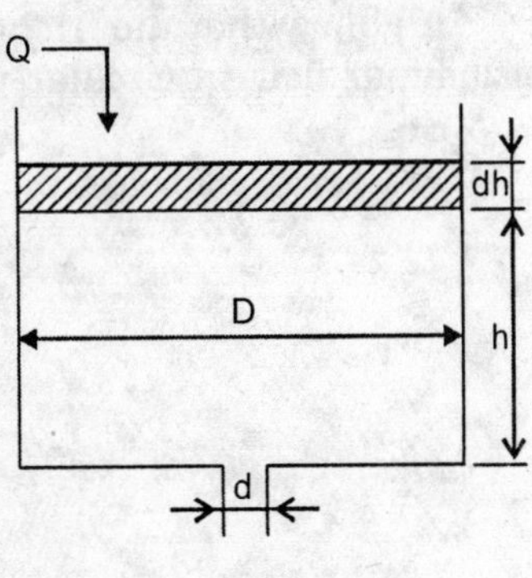

Fig. 9

Then $[Q - C_d\, a\, (2gh)^{0.5}]\,.\,dt = A.dh$

$$\int_{t_0}^{t_1} dt = \frac{\int_{h_1}^{h_2} A.dh}{[Q - C_d\, a\, (2gh)^{0.5}]}$$

Let, $$I = \int \frac{A.dh}{[Q - C_d\, a\,(2gh)^{0.5}]} = \int \left[\frac{A}{\{Q - C_d\, a\,(2g)^{0.5}.h^{0.5}\}}\right].dh$$

And $$h = z^2,\ dh = 2z.dz$$

$$I = \int \frac{2Az.dz}{[Q - C_d\, a\,(2g)^{0.5}.z]}$$

$$\left| t \right|_{t_0}^{t_1} = \left(\frac{2.A}{X}\right)\left[\left[Q \times \ln \frac{\{Q - C_d\, a\,(2gh)^{0.5}\}}{\{-C_d\, a\,(2g)^{0.5}\}}\right] - h^{0.5}\right|_{h_0}^{h_1}$$

where $$X = C_d\, a\,(2g)^{0.5}$$

$$= \frac{0.6 \times \pi \times 40^2 \times 10^{-6} \times (2 \times 9.81)^{0.5}}{4}$$

$$= 3.34 \times 10^{-3}$$

$$t_1 - t_0 = \left(\frac{2A}{X}\right)\left[-\left(\frac{Q}{X}\right)\ln\left\{\frac{Q - X\,(h_2)^{1/2}}{Q - X\,(h_1)^{1/2}}\right\} - \left\{(h_2)^{1/2} - (h_1)^{1/2}\right\}\right]$$

(*i*) $$60 = \left(\frac{2A}{X}\right)\left[-\left(\frac{Q}{X}\right)\ln\left\{\frac{Q - X\,(0.8)^{1/2}}{Q - X\,(0.6)^{1/2}}\right\} - \left\{(0.8)^{1/2} - (0.6)^{1/2}\right\}\right]$$

(*ii*) $$60 = \left(\frac{2A}{X}\right)\left[-\left(\frac{Q}{X}\right)\ln\left\{\frac{Q - X\,(1.1)^{1/2}}{Q - X\,(1.0)^{1/2}}\right\} - \left\{(1.1)^{1/2} - (1.0)^{1/2}\right\}\right]$$

Equating (*i*) and (*ii*), we get

$$Q = 10 \text{ m}^3/\text{s}$$

Then $$A = 2953 \text{ m}^2 \text{ and } D = 61.32 \text{ m}$$

(*c*) Show that the friction factor is inversely proportional to the Reynolds number in case of laminar flow in circular pipes. **(5)**

Sol.

Pressure head loss in a pipe of diameter d and length L is,

$$h_f = \frac{4f\,Lv^2}{2gd} = \frac{32\,\mu v L}{\rho g d^2}$$

$$f = \frac{16\,\mu}{\rho v d} = \frac{16}{Re}$$

or $$f \propto \frac{1}{Re}$$

Q.7. (*a*) Explain the terms : stable, unstable and neutral equilibrium in reference to the floating bodies. (6)

Sol.

Stable equilibrium. A floating body is said to be in stable equilibrium when the metacentre lies above the centre of gravity of the body.

Ustable equilibrium. A floating body is said to be in unstable equilibrium when the metacentre lies below the centre of gravity of the body.

Neutral equilibrium. A floating body is said to be in neutral equilibrium when the metacentre coincides with the centre of gravity of the body.

(*b*) Compare the velocity profiles for laminar flow and turbulent flow in pipes and comment on them. (4)

Sol.

Laminar flow in a pipe. The velocity profile for laminar flow in a pipe is parabolic and the velocity distribution surface is a paraboloid of revolution. The velocity distribution is given by

$$u = \left(\frac{1}{4\mu}\right)\left(\frac{dp}{dx}\right)(R_0^2 - r^2)$$

$$U_{max} = \left(\frac{1}{4\mu}\right)\left(\frac{dp}{dx}\right)R_0^2$$

Turbulent flow in a pipe. The turbulent flow is characterized by random, irregular and haphazard movement of the fluid particles. This type of motion gives rise to velocity fluctuations in all directions. The pressure and velocity vary with time. Turbulent flow velocity in the pipe is logarithmic in nature. The velocity distribution is more uniform than laminar flow. The Fig. 10 shows the velocity distribution in laminar and turbulent flows at same Reynolds number.

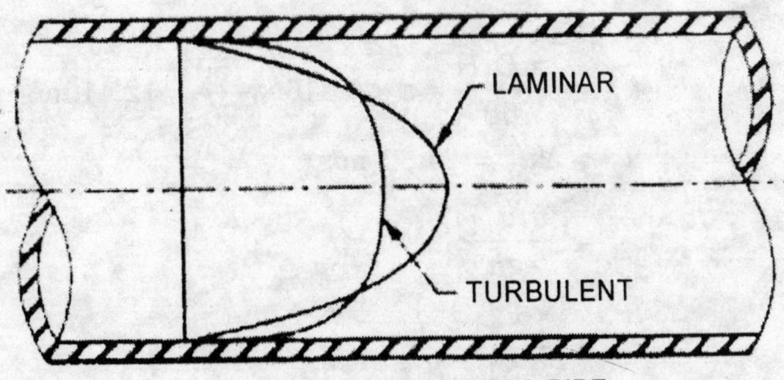

VELOCITY PROFILES IN A PIPE

(*c*) A reaction turbine has to be designed to develop 370 kW under a head of 70 m while running at speed equal to 750 rpm.

Ratio of width of runner to the outer diameter of runner = 0.1

Ratio of inner diameter to the outer diameter = 0.5

Flow ratio = 0.15

Hydraulic efficiency = 95%

Mechanical efficiency = 84%

Circumferential area occupied by thickness of vanes = 5%

Assuming velocity of flow constant, calculate guide vane angle at inlet and runner vane angles at inlet and outlet. The radial velocity is equal to 3.5 m/sec. (30)

Sol.

Overall efficiency, $\eta_0 = \eta_{mech}\ \eta_{hyd} = 0.84 \times 0.95 = 0.798$

$$Q = \frac{P}{\gamma H\ \eta_0} = \frac{370 \times 10^3}{9810 \times 70 \times 0.798} = 0.675\ m^3/s$$

$$v_{f1} = v_{f2} = C_f\ [2gH]^{0.5} = 0.15\ [2 \times 9.81 \times 70]^{0.5} = 5.56\ m/s$$

$$Q = \eta_{hyd}\ \pi\ D_2 B_2 v_{f2}$$

$$0.675 = 0.95 \times \pi \times D_2\ B_2 \times 5.56$$

$$D_2\ B_2 = 0.040$$

$$0.1\ D^2{}_2 = 0.040$$

$$D_2 = 0.631\ m$$

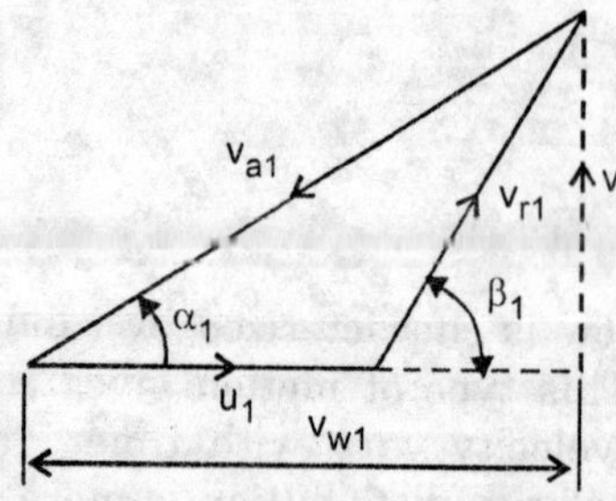

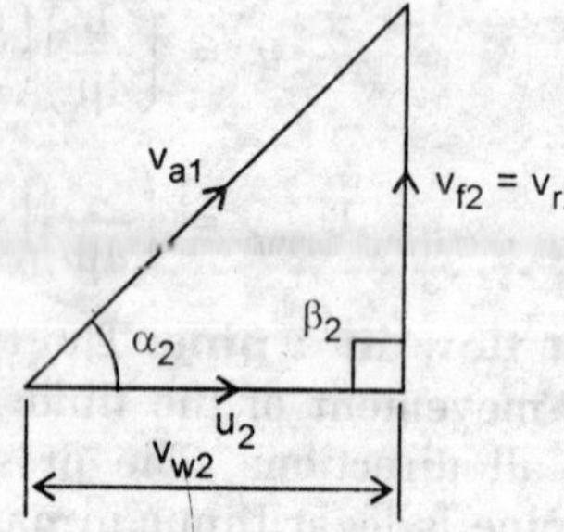

Fig. 11

$$D_1 = \frac{D_2}{2} = 0.316\ m$$

$$u_1 = \frac{\pi\ D_1 N}{60} = \pi \times 0.316 \times \frac{750}{60} = 12.41\ m/s$$

$$u_2 = 2u_1 = 24.82\ m/s$$

$$\eta_{hyd} = \frac{u_1 v_{w1}}{gh}$$

$$0.95 = \frac{12.41\ v_{w1}}{9.81 \times 70}$$

$$v_{w1} = 52.56\ m/s$$

$$\tan \alpha_1 = \frac{v_{f1}}{v_{w1}} = \frac{5.56}{52.56},\ \alpha_1 = 6.03°$$

$$\tan \beta_1 = \frac{v_{f1}}{v_{w1} - u_1} = \frac{5.56}{52.56 - 12.41},\ \beta_1 = 7.88°$$

$$\tan \alpha_2 = \frac{v_{f2}}{v_{w2}} = \frac{v_{f1}}{u_2} = \frac{5.56}{24.82},\ \alpha_2 = 12.63°$$

Q.8. (*a*) Discuss the importance of dimentional analysis. Explain clearly Buckingham's π-theorem method and Rayleigh's methods of dimentional analysis.

A circular cylinder of a given Length/Diameter ratio is kept in steady rotation at N revol./sec. in a uniform stream of fluid of velocity v. Assuming that the power required to maintain the motion depends on density ρ, kinematic viscosity ν of the fluid, and the diameter D of the cylinder. Using Rayleigh's method show that,

$$P = \left(\frac{\rho v^3}{D}\right) f\left(\frac{vD}{\nu}, \frac{ND^2}{\nu}\right) \quad (20)$$

Sol.

Dimensional analysis. It is a method which describes a natural phenomenon by a dimensionally correct equation among certain variables which affect the phenomenon. It reduces the number of variables and arrange them into dimensionless groups. It finds extensive applications in fluid mechanics and heat transfer. It is of great help in model analysis.

Buckingham-π theorem. If a phenomenon involves n variables and all the n variables are described by m fundamental dimensions, they may be grouped into $(n\text{-}m)$ dimensionless π-terms.

Rayleigh's method. If gives relationship among the dimensionless groups and does not provide any information regarding the number of dimensionless groups to be obtained as a result of dimensionless analysis.

Let,
$$P = f(\rho^a, v^b, D^c, \nu^d, N^e)$$

$$ML^2T^{-3} = f\,[ML^{-3}]^a\,[L^2T]^b\,[L]^c\,[LT^{-1}]^d\,[T^{-1}]^e$$

$$= f\,[M^a\,L^{-3a+2b+c+d}\,T^{b-d-e}]$$

$$a = 1$$

$$b = 3 - d - e$$

$$c = 2 = 3 - 6 + 2d - 2e - d = -1 + d + 2e$$

$$P = f\,[\rho v^{3-d-e}\,D^{-1+d+2e}\,\nu^d\,N^e]$$

$$= f\,[(\rho D^{-1} v^3)\,(v^{-d}\,D^d\,\nu^d)\,(v^{-e}\,D^{2e}\,N^e)]$$

$$= f\left[\left(\frac{\rho v^3}{D}\right)\left(\frac{vD}{\nu}\right)\left(\frac{ND^2}{\nu}\right)^e\right] = \left(\frac{\rho v^3}{D}\right) f\left(\frac{vD}{\nu}, \frac{ND^2}{\nu}\right)$$

(*b*) An aeroplane flies at an altitude with a velocity of 800 km/hr. The pressure and temperature at that altitude are 1.206×10^4 N/m² and 217 K respectively. Calculate

(*i*) The maximum possible temperature on the aeroplane skin.

(*ii*) The maximum possible pressure intensity on the aeroplane body,

(*iii*) The critical velocity of air relative to the aeroplane. (8)

Sol.

Given :
$$v_0 = 800 \text{ km/h} = \frac{800 \times 10^3}{3600} = 222.2 \text{ m/s}$$

$$p = 1.206 \times 10^4 \text{ N/m}^2,\ T_1 = 217 \text{ K}$$

(*i*) Stagnation temperature,

$$T_0 = T_1 + \frac{v_0^2}{2c_p} = 217 + \frac{(222.2)^2}{2 \times 1.005 \times 10^3} = 241.56 \text{ K}$$

(*ii*)
$$p_0 = p_1\left(\frac{T_0}{T_1}\right)^{(\gamma-1)/\gamma} = 1.206 \times 10^4 \left(\frac{241.56}{217}\right)^{0.4/1.4}$$
$$= 1.7552 \times 10^4 \text{ N/m}^2$$

(*iii*)
$$(v_s)_{max} = [\gamma RT_0]^{0.5} = [1.4 \times 287 \times 241.56]^{0.5}$$
$$= 311.542 \text{ m/s or } 1121.65 \text{ km/h}$$

(*c*) Define Froude number, Reynolds number, Mach number, Euler number and Weber number. (5)

Sol.

Froude number, $$Fr = \left[\frac{\text{Inertia force}}{\text{Gravity force}}\right]^{0.5} = \frac{v}{[gL]^{0.5}}$$

Reynolds number, $$Re = \frac{\text{Inertia force}}{\text{Viscous force}} = \frac{vL}{\nu}$$

Mach number, $$M = \left[\frac{\text{Inertia force}}{\text{Compressibility force}}\right]^{0.5} = \frac{v}{v_s}$$

Euler number, $$= \left[\frac{\text{Inertia force}}{\text{Pressure force}}\right]^{0.5} = \frac{v}{[\Delta p/\rho]^{0.5}}$$

Weber number $$= \left[\frac{\text{Inertia force}}{\text{Surface tension force}}\right]^{0.5} = \frac{\rho v^2 L}{\sigma}$$

(*d*) Considering Froude number as the criterion of dynamic similarity for a certain flow situation, work out the scale factors for velocity, time, discharge, acceleration, force, work and power in terms of the scale factor for length. (7)

Sol.

Froude number, $$Fr = \frac{v}{[gL]^{0.5}}$$

$$\frac{v_m}{[gL_m]^{0.5}} = \frac{v_p}{[gL_p]^{0.5}}$$

$$v_r = \frac{v_p}{v_m} = \left[\frac{L_p}{L_m}\right]^{0.5} = (L_r)^{0.5}$$

Time, $$T_r = \frac{L_r}{v_r} = \frac{L_r}{(L_r)^{0.5}} = (L_r)^{0.5}$$

Discharge, $$Q_r = v_r\ L_r^2 = L_r^{5/2}$$

Accelerationl, $$a_r = \frac{v_r}{T_r} = 1$$

Force, $$F_r = \rho_r\ v_r^2\ L_r^2 = \rho_r\ L_r^3$$

Work, $$W_r = \text{Force} \times \text{Distance} = \rho_r\ L_r^3\ L_r = \rho_r\ L_r^4$$

Power, $$P_r = \text{Force} \times \text{Velocity} = \rho_r\ (L_r)^{7/2}$$

MECHANICAL ENGINEERING
PAPER I—2004

(Conventional)

Q.1. ***(a)*** Derive the following Clapeyron and Clausius Clapeyron equations :

$$\frac{dp}{dT} = \frac{h_{fg}}{\left[T(v_g - v_f)\right]}$$

and $$\frac{dp}{p} = \left[\frac{h_{fg}}{RT^2}\right] dT$$

Explain the physical significance of these equations. (10)

Sol.

From Maxwell's reations, we have

$$\left(\frac{\partial p}{\partial T}\right)_v = \left(\frac{\partial s}{\partial v}\right)_T$$

For change of state from saturated liquid to saturated vapour of a pure substance at constant temperature, we have

$$\frac{dp}{dT} = \frac{s_g - s_f}{v_g - v_f}$$

Now, $$s_g - s_f = \frac{h_g - g_f}{T} = \frac{h_{fg}}{T}$$

$\therefore$ $$\frac{dp}{dT} = \frac{h_{fg}}{T(v_g - v_f)}$$

At very low pressure, $v_g >> v_{fg}$ and $pv = RT$

$$\frac{dp}{dT} = h_{fg} \cdot \frac{p}{RT^2}$$

or $$\frac{dp}{p} = \left[\frac{h_{fg}}{RT^2}\right] . dT$$

Clapeyron equation can be used to determine the properties in a two phase region and is also valid for the change from solid to liquid and vapour.

(b) A heat pump operates between two identical bodies. In the beginning, the bodies are at the same tempreature T_1 but operation of heat pump cools down one of the body to temperature T_2. Show that for the operation of heat pump the minimum work input needed by the heat pump for unit mass is given by

$$W_{min} = c_p \left[\frac{T_1^2}{T_2^2} + T_2 - 2T_1 \right]$$

Where c_p is the specific heat of bodies.

Discuss whether the system satisfies first and second law of thermodynamics or not. (10)

Sol.

Let T_3 = Final temperature of body A.

$$Q_B = C_p (T_1 - T_2)$$

$$Q_A = W + Q_B$$

$$= c_p (T_3 - T_1)$$

$$W = c_p (T_3 - T_1) - c_p (T_1 - T_2)$$

$$= c_p (T_2 + T_3) - 2C_p T_1$$

$$\Delta s_A = \int_{T_1}^{T_3} c_p \frac{dT}{T} = c_p \ln\left(\frac{T_3}{T_1}\right)$$

$$\Delta s_B = \int_{T_1}^{T_2} c_p \frac{dT}{T} = c_p \ln\left(\frac{T_2}{T_1}\right)$$

Fig. 1

$$(\Delta s)_{universe} = \Delta s_A + \Delta s_B = c_p \ln\left(\frac{T_3}{T_1}\right) + c_p \ln\left(\frac{T_2}{T_1}\right)$$

$$= c_p \ln\left(\frac{T_2 T_3}{T_1^2}\right)$$

Now $(\Delta s)_{universe} \geq 0$

$$c_p \ln\left(\frac{T_2 T_3}{T_1^2}\right) \geq 0$$

For W_{min}, T_3 is minimum

$$\therefore \quad c_p \ln\left(\frac{T_2 T_3}{T_1^2}\right) = 0 = \ln 1$$

or

$$T_2 T_3 = T_1^2$$

$$T_3 = \frac{T_1^2}{T_2}$$

$$W_{min} = c_p \left(\frac{T_1^2}{T_2} + T_2 - 2T_1\right)$$

(c) A large vessel contains steam at a pressure of 20 bar and a tempreature of 350°C. This large vessel is connected to a steam turbine through a valve followed by a small initially evacuated tank with a volume of 0.8 m^3. During emergency power requirement, the valve is opened and the tank fills with steam until the pressure is 20 bar. The temperature of the tank is then 400°C. Assume that the filling process takes place adiabatically and the changes in potential and kinetic energies are negligible. By drawing the control volume, calculate the amount of work developed by the turbine in kJ. (20)

Sol.

For steam at 20 bar and 350°C, from steam tables, we have

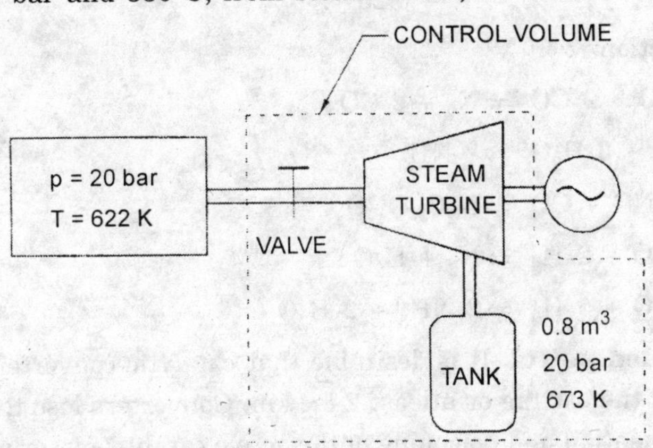

Fig. 2

$$h_1 = 3138.6 \text{ kJ/kg}, \; v_1 = 0.13856 \text{ m}^3\text{/kg}$$

At 20 bar and 400°C, we have

$$h_2 = 3246.7 \text{ kJ/kg}, \; v_2 = 0.15113 \text{ m}^3\text{/kg}$$

$$Q = W + (U_2 - U_1)$$

Here $Q = 0$. Therefore,

$$W = U_2 - U_1$$

$$U_1 = mu_1, \; U_2 = mu_2$$

$$m = \frac{0.8}{0.15113} = 5.293456 \text{ kg}$$

$$u_1 = h_1 - p_1v_1 = 31.386 - 200 \times 0.13856 = 2861.48 \text{ kJ/kg}$$

$$u_2 = h_2 - p_2v_2 = 3246.7 - 200 \times 0.15113 = 2944.44 \text{ kJ/kg}$$

$$W = m(u_2 - u_1)5.293456\,(2944.44 - 2861.48) = 439 \text{ kJ}$$

Q.2. (*a*) What does a catalyst perform in a catalytic convertor? Discuss the oxidation and reduction reactions that are taking place in a converter. Why it is preferred to use unleaded petrol in cars equipped with a catalytic converter system? (15)

Sol.

Function of a Catalyst in a Catalytic onverter. The catalyst controls the level of various exhaust pollutants from the engine by changing the chemical characteristics of the exhaust gages. Catalyst materials such as platinum or platinum and palladium, rhodium are used in the converter. CO and HC are oxidized to CO_2 and H_2O by Palladium and platinum and NO_x is reduced by rhodium. An oxidation catalyst is placed down stream of the reduction catalyst.

Oxidation reactions :

$$2\ CO + O_2 \Rightarrow 2\ CO_2$$

$$CO + H_2O \Rightarrow CO_2 + H_2$$

$$C_xH_y + z\ O_2 \Rightarrow x\ CO_2 + y\ H_2O$$

where $z = x + 0.25\ y.$

Reduction reactions :

$$2\ NO + 2\ CO \Rightarrow N_2 + 2\ CO_2$$

$$2\ NO + 5\ CO + 3\ H_2O \Rightarrow 2\ NH_3 + 5\ CO_2$$

$$2\ NO + CO \Rightarrow N_2O + CO_2$$

$$2\ NO + 2\ H_2 \Rightarrow N_2 + H_2O$$

$$2\ NO + 5\ H_2 \Rightarrow 2\ NH_3 + 2\ H_2O$$

Use of Unleaded petrol. It is desirable that catalytic converter have an effective life time equal to half of the car life or atleast 2 lac km. Converters lose their efficiency with age due to thermal degradation and poisoning of the active catalyst material. Just a small amount of lead on a catalyst site can reduce HC emission reduction by a factor of two or three. Use of leaded gasoline filled two times (full tank) would completely poison a converter and make it totally useless. Therefore, leaded gasoline cannot be used in engines equipped with catalytic converters.

(*b*) "Factors which increase detonation in S.I. engine tends to reduce knocking in C.I. engines", discuss the validity of the statement. (5)

Sol.

The knocking in SI engines and CI engines is fundamentally due to the auto – ignition lag of the fuel – air mixture. There are following differences in the knocking phenomenon of SI and CI engines :

1. In the SI engine, the detonation occurs near the end of combustion whereas in CI engine detonation occurs near the beginning of combustion.
2. The detonation in SI engine is of a homogenous charge causing very high rate of pressure rise and very high maximum pressure. In CI engine the rate of pressure rise is normally lower than that in the SI engine.
3. There is no question of pre-ignition in a CI engine as the fuel is injected in the cylinder at the end of the compression stroke.
4. In the SI engine it is relatively easy to distinguish between knocking and non-knocking operation as the human ear can easily find the distinction. But in the CI engine there is no definite distinction between normal and knocking combustion.

The comparison of knocking between SI and CI engines is given in the following table:

Table 1. Factors tending to Reduce knocking is SI and CI engines.

Sl. No.	Factors	SI engine	CI engine
1.	Self-ignition temperature of fuel	High	Low
2.	Time lag or delay period for fuel	Long	Short
3.	Compression ratio	Low	High
4.	Inlet temperature	Low	High
5.	Inlet pressure	Low	High
6.	Combustion chamber wall temperature	Low	High
7.	Speed	High	Low
8.	Cylinder Size	Small	Large

Hence the statement is valid.

(*c*) A six cylinder, four stroke spark-ignition of 10 cm × 12 cm (bore × stroke) with a compression ratio 6 is tested at 4800 rpm on a dynamometer of arm 55 cm. During a 10 minutes test, the dynamometer reads 45 kg and the engine consumed 5 kg of petrol of calorific value 45 MJ/kg. The carburettor receives the air at 29°C and 1 bar at the rate of 10 kg/min. Calculate:

(*i*) the brake power,

(*ii*) the brake mean effective pressure,

(*iii*) the brake specific fuel consumption,

(*iv*) the brake specific air consumption,

(*v*) the brake thermal efficiency,

(*vi*) the air-fuel ratio. (30)

Sol.

(*i*) Brake power, $$BP = \frac{2\pi NT}{60 \times 10^3}$$

$$T = 45 \times 9.81 \times 55 \times 10^{-2} = 242.79 \text{ N.m}$$

$$BP = \frac{2\pi \times 4800 \times 242.79}{60 \times 10^3} = 122 \text{ kW}$$

(*ii*) $$BP = p_{bm} \ [(LA) \times N/i] \times \text{Number of cylinders}$$

$$122 \times 10^3 = p_{bm} \times \left(\frac{\pi \times 0.1 \times 0.12}{4}\right) \times \frac{4800 \times 6}{2 \times 60}$$

$$p_{bm} = 5.39 \text{ bar}$$

(*iii*) $$\dot{m}_f = \frac{5}{10} \text{ kg/ min or 30 kg/h}$$

$$BSFC = \frac{30}{122} = 0.45 \text{ kg/kW.h}$$

(*iv*) $\dot{m}_a = 10$ kg/min or 600 kg/h

$$\text{BSFC} = \frac{600}{122} = 4.918 \text{ kg/kW.h}$$

(*v*)
$$\eta_{bt} = \frac{\text{BP}}{m_f \times \text{CV}}$$

$$= \frac{122 \times 10^3}{\frac{30 \times 45 \times 10^6}{3600}} = 0.325 \text{ or } 32.5\%$$

(*vi*)
$$\text{A/F ratio} = \frac{\dot{m}_a}{\dot{m}_f} = \frac{600}{30} = 20:1$$

Q.3. (*a*) List the factors which are considered in the optimuum design of fins. (3)

Sol.

Factors for Optimum design of a Fin :

1. Cost.

2. Manufacturing difficulties.

3. Geometrical configuration of component where fin is to be fitted.

4. Weight of fins.

5. Temperature drop caused by fins.

(*b*) An electric semi-conductor device generates heat equal to 480×10^{-3} W. In order to keep the surface temperature at the upper safe limit of 70°C, the generated heat has to bedissipated to the surroundings which is at 30°C. To accomplish this task, aluminium fins of 0.7 mm square and 12 mm long are attached to the surface. The thermal conductivity of aluminium fins is 170 W/m.K. If the heat transfer co-efficient is 12 W/m^2.K, calculate the number of fins required. Assume no heat loss from the tip of fins. (7)

Sol.

$$Q_g = 480 \times 10^{-3} \text{ W}$$

$$Q_{fin} = [hPAk]^{0.5}\,(t - t_\infty)$$

For square section, $P = 4 \times 0.7 = 2.8$ mm

$$Q_{fin} = [12 \times 2.8 \times 10^{-3} \times 0.7^2 \times 10^{-6} \times 170]^{0.5} \times (70 - 30)$$

$$= 0.06692 \text{ W}$$

$$\text{Number of fins required} = \frac{Q_g}{Q_{fin}} = \frac{480}{66.92} = 7.17 \approx 7$$

(*c*) A commercial aeroplane is modelled as a flat plate which is 1.5 m wide and 8 m long in size. It is maintained at 20°C. The aeroplane is flying at a speed of 800 km/h in air at 0°C and 60 m of Hg pressure. Calculate the heat loss from the wing. Take thermo-physical properties of air at the mean film temperature of 10°C as

$$k = 2.511 \times 10^{-2} \text{ W/m.K}$$

$$\nu = 14.16 \times 10^{-6} \text{ m}^2/\text{s},$$

$$\text{Pr} = 0.705$$ (15)

Sol.

Velocity of aeroplane,

$$v = \frac{800 \times 10^3}{3600} = 222.23 \text{ m/s}$$

$$Re_L = \frac{vL}{\nu} = \frac{222.23 \times 1.5}{14.16 \times 10^{-6}} = 23.54 \times 10^6 > 5 \times 10^5$$

Hence, the flow is turbulent,

Mean film temperature,

$$t_{fm} = \frac{0 + 20}{2} = 10°C$$

$$h = \left(\frac{k}{L}\right)[0.036 \, Re^{0.8} - 836] \, Pr^{0.33}$$

$$\left(\frac{2.511 \times 10^{-2}}{1.5}\right)[0.036 \times (23.54 \times 10)^{60.8} - 836] \times (0.705)^{0.33} = 411.569 \text{ W/m}^2\text{.K}$$

Heat loss

$$= hA.\Delta t = 411.569 \times 1.5 \times 8 \; (20 - 0)] = 98.86 \text{ kW}$$

(*d*) An electrically heated industrial furance cavity is modeled in the form of a cylinder having diameter 10 cm and length 20 cm. It is opened at one end of surroundings that are at a temperature of 300 K. The electrically heated sides and the bottom of the cavity which are well insulated and may be approximated as black bodies are maintained at a tempareture of 1800 K and 2000 K respectively. By showing the sketch of the furnace, find the power required to maintain the surface at this condition. Take shape factor from the bottom surface to surroundings as 0.06. (15)

Sol.

Given : $T_1 = 1800$ K, $T_2 = 2000$ K, $T_3 = 3000$ K

$D = 10$ cm, $H = 20$, $F_{23} = 0.06$

Net heat lost from the furance,

$$Q = Q_{13} + Q_{23}$$

$$= A_1F_{13} \, \sigma \, (T_1^4 - T_3^4) + A_3F_{23} \, \sigma \, (T_2^4 - T_3^4)$$

Now $F_{21} + F_{23} = 1$

$$F_{21} = 1 - 0.06 = 0.94$$

Using reciprocity rule, w have

$$A_1F_{12} = A_2F_{21}$$

$$F_{12} = \frac{A_2F_{21}}{A_1} = \left[\frac{\pi R^2}{2\pi \, RH}\right] F_{21}$$

$$= \left[\frac{R}{2H}\right] F_{21} = \left(\frac{5}{40}\right) \times 0.94 = 0.117$$

$$Q = (\pi \times 0.1 \times 0.2) \times 0.117 \times 5.67 \times 10^{-8} \times [1800^4 - 300^4] + \pi \times (0.05)^2 \times 0.06 \times 5.67 \times 10^{-8} \times [2000^4 - 300^4]$$

$$= 4372.23 + 427.29 = 4799.52 \approx 4.8 \text{ kW}$$

Q.4. (*a*) Discuss in detail the factors affecting the performance of vapour compression refrigeration cycle.

Sol.

(*i*) **Effect of suction pressure.** The *p-h* diagram is shown in Fig. 3 (*a*) for the vapour compression cycle. If the suction pressure decreases from p_1 to p_1' as shown in Fig. 3 (*b*), then

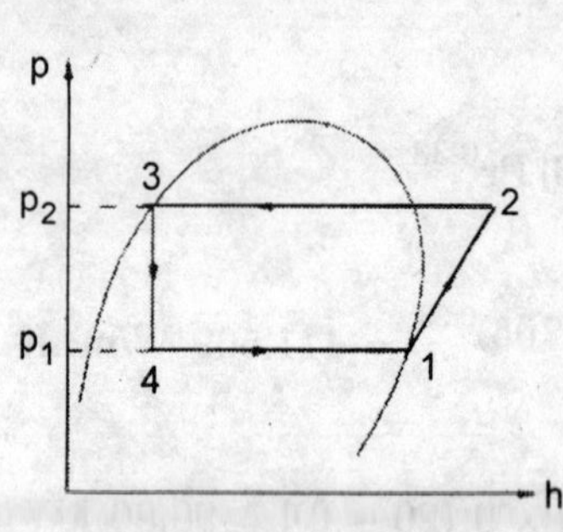

(*a*) p-h diagram.

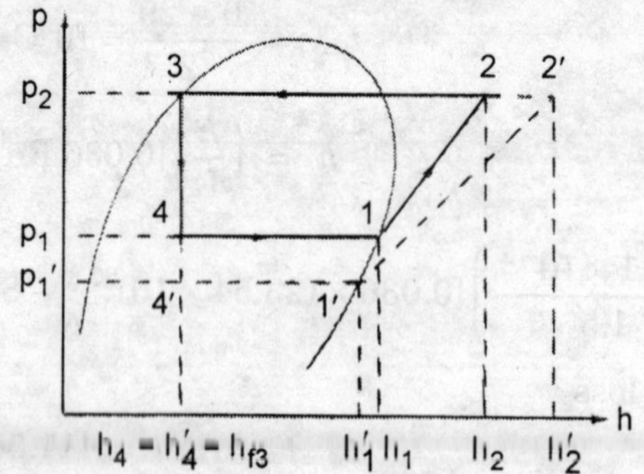

(*b*) Effect of decrease of suction pressure.

1. Refrigeration effect decreases from $(h_1 - h_4)$ to $(h_1' - h_4')$.
2. Work done for compression increases from $(h_2 - h_1)$ to $(h_2' - h_1')$.
3. COP decreases.
4. Refrigeration cost increases.

(*ii*) **Effect of discharge pressure.** If the discharge pressure increase from p_2 to p_2' as shown in Fig. 3. (*c*), then

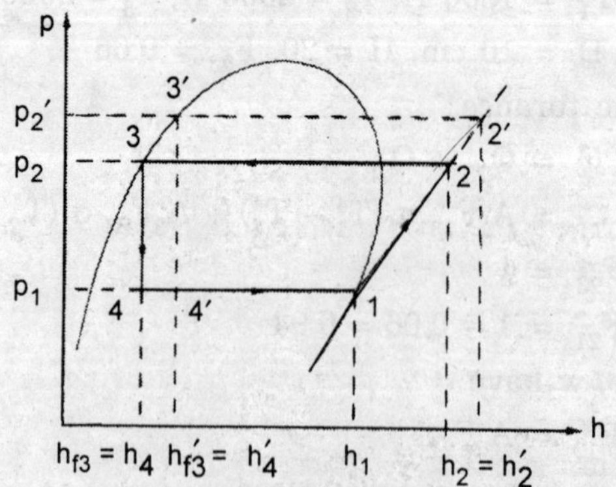

Fig. 3 (*c*) Effect of discharge pressure.

1. Refrigeration effect decreases from $(h_1 - h_4)$ to $(h_1 - h_4')$.
2. Work required for compression increases from $(h_2 - h_1)$ to $(h_2' - h_1')$.
3. COP decreases.

***(iii)* Effect of supercharging.** The supercharged cycle is shown in Fig. 3. (*d*).

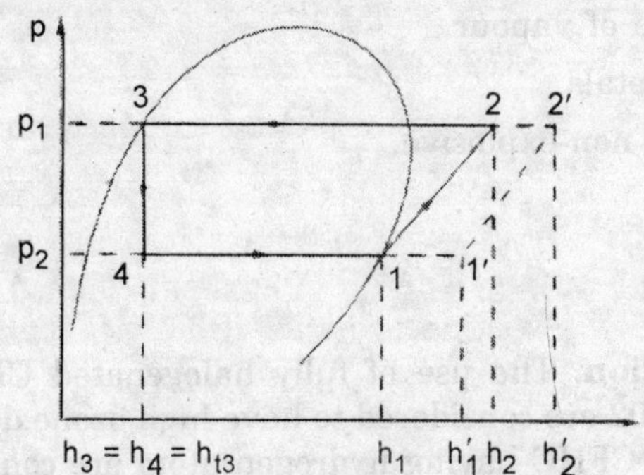

Fig. 3 (*d*) Effect of super charging.

1. Refrigeration effect increases from $(h_1 - h_4)$ to $(h_1' - h_4)$.

2. Compression work increases from $(h_2 - h_1)$ to $(h_2' - h_1')$.

3. COP increases.

***(iv)* Effect of subcooling.** The subcooled cycle is shown in Fig. 3. (*e*).

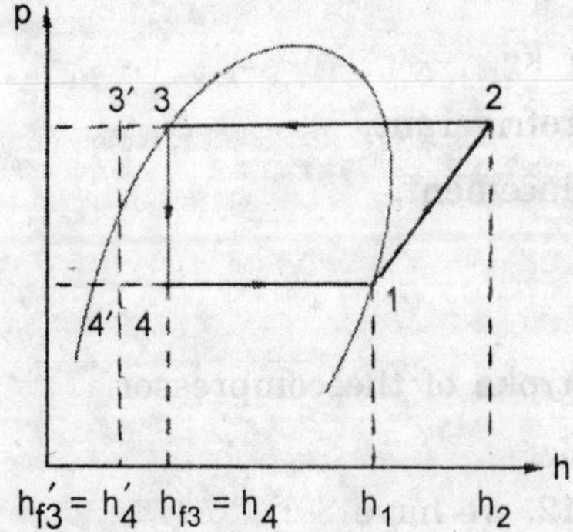

Fig. 3 (*e*) Effect of subcooling.

1. Refrigeration effect increases from $(h_1 - h_4)$ to $(h_1 - h_4')$.

2. Compression work remains same.

3. COP increases.

(*b*) Explain how a refrigerant produces cooling effect. What are the good qualities of refrigerant? Name the refrigerant which does not destroy ozone layer. (10)

Sol.

A refrigerant is a heat carrying medium which during its cycle in the refrigeration system absorbs heat from a lower temperature system and discards the heat so absorbed to a higher temperature system.

Good qualitees of a refrigerant :

1. Low boiling point.
2. High critical temperature.
3. High latent heat of vaporisation.

4. Low specific heat of liquid.
5. Low specific volume of vapour.
6. Non-corrosive to metals.
7. Non-flammable and non-explosive.
8. Non-toxic.
9. Low cost.
10. Mixes well with oil.

Ozone layer protection. The use of fully halogenated CFC's refrigerants like R 11, R–12, R–13, R–114 and R–502 are considered to have high ozone depletion potential (ODP). All refrigerants like HCF's and CFHC having hydrogen atom are considered safe for ozone layer.

(*c*) 10 tons of refrigeration is required for a food storage space to keep the food stuff cool.

The refrigerant used in the plant is Freon-12 and is sub-cooled by 5°C before entering to the expansion valve. The temperature of evaporator and conenser are – 8°C and 30°C respectively. The vapour is superheated to 6°C before leaving the evaporator coil. The compressor is of two cylinders, single acting type having stroke equal to 1.5 time the bore. The compressor runs at 800 rpm and compression is assumed to be isentropic. Estimate the following :

(*i*) Refrigerating effect per kg,
(*ii*) Mass flow rate of the refrigerant,
(*iii*) Theoretical piston displacement,
(*iv*) Theoretical power,
(*v*) C.O.P., and
(*vi*) Theoretical bore and stroke of the compressor. (15)

Sol.

From *p-h* diagram for R-12, we have

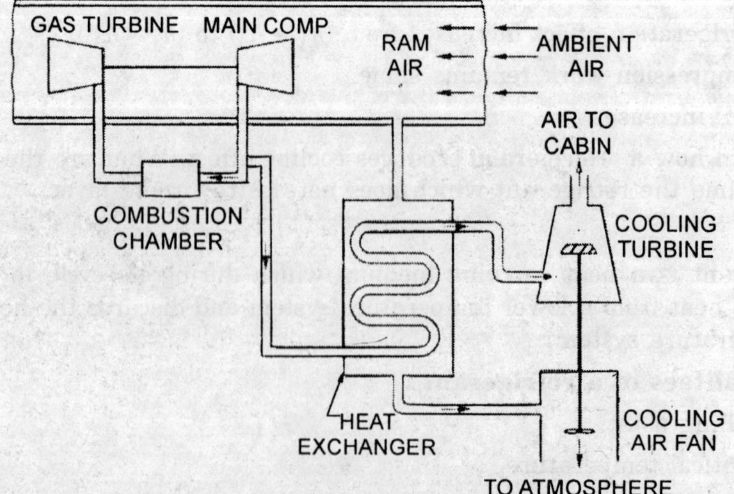

Fig. 4 (*a*) Simple air-conditioning system.

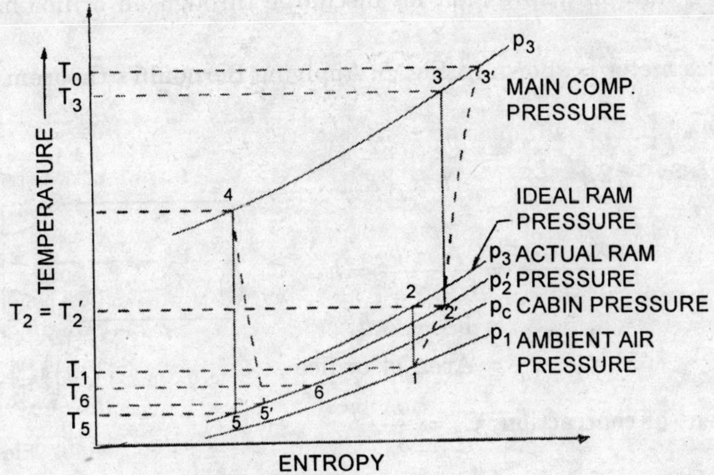

Fig. 4

$$h_1 = 355 \text{ kJ/kg}, h_3 = h_4 = 225 \text{ kJ/kg}, h_2 = 380 \text{ kJ/kg}$$

(*i*) Refrigerating effect, $q_0 = h_1 - h_4$

$$= 355 - 225 = 130 \text{ kJ/kg}$$

(*ii*) Mass flow rate of refrigerant, $\dot{m}_r = \dfrac{\text{TR}}{q_0} = \dfrac{10 \times 211}{130} = 16.23 \text{ kg/min}$

(*iii*) $v_{g1} = \dfrac{1}{13} = 0.0769 \text{ m}^3\text{/kg}$

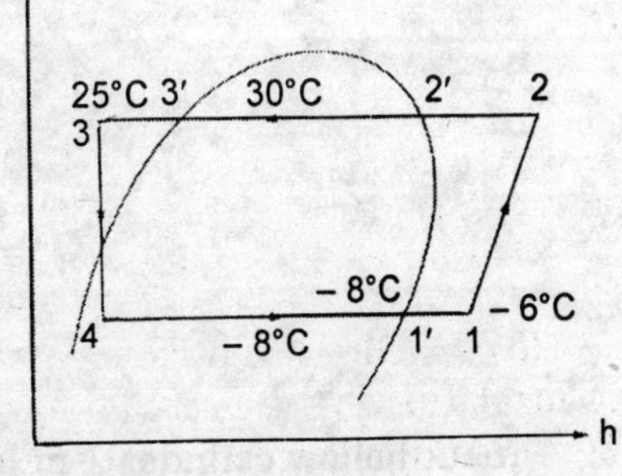

Fig. 5

Theoretical piston displacement

$$= m_r . v_{g1} = \frac{16.23}{13}$$

$$= 1.24846 \text{ m}^3\text{/min}$$

(*iv*) Theoretical power

$$= m_r (h_2 - h_1)$$

$$= 16.23 (380 - 355)$$

$$= 405.75 \text{ kJ/min or } 6.76 \text{ kW}$$

(*v*) $\text{COP} = \dfrac{q_0}{w} = \dfrac{h_1 - h_4}{h_2 - h_1} = \dfrac{130}{25} = 5.2$

(*vi*) Volume flow rate $= 1.24846 \text{ m}^3\text{/min}$

$$\text{L/D} = 1.5, \text{ N} = 800 \text{ rpm}$$

Cylinder volume $= \dfrac{2 \times \pi \text{ D}^2\text{LN}}{4}$

$$\text{D}^3 = 6.6206 \times 10^{-4}$$

$$\text{D} = 0.087 \text{ m or } 87 \text{ mm}$$

$$\text{L} = 130 \text{ mm}$$

Q.5. (*a*) Obtain an expression for discharge through an orifice meter. (10)

Sol.

The orifice meter is shown in Fig. 6. Applying Bernoulli's theorem at pionts 1 and 2, we have

$$\frac{p_1}{\gamma} + \frac{v_1^2}{2g} + z_1 = \frac{p_2}{\gamma} + \frac{v_2^2}{2g} + z_2$$

Here $z_1 = z_2$

$$\frac{v_2^2 - v_1^2}{2g} = \frac{p_1 - p_2}{\gamma} = h$$

$$v_2 = [v_1^2 + 2gh]^{0.5}$$

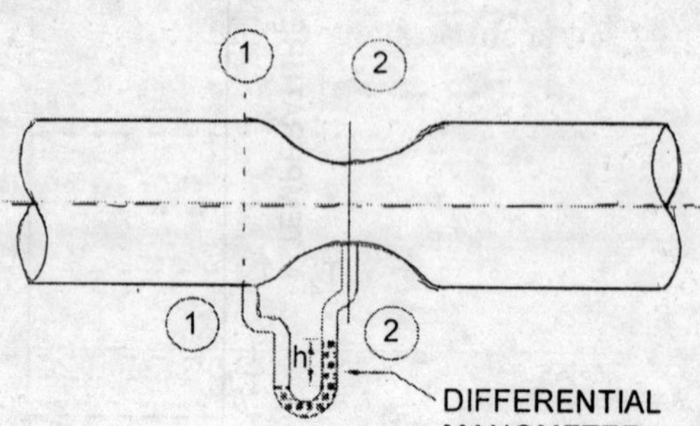

Fig. 6

Let, a_o = Area of orifice

Co-efficient of contraction, $C_c = \frac{a_2}{a_o}$

Also $a_1 v_1 = a_2 v_2,\ v_1 = \frac{a_2 v_2}{a_1} = \frac{a_o C_c v_2}{a_1}$

$$v_2 = \left[\left(\frac{a_o C_c v_2}{a_1}\right)^2 + 2gh\right]^{0.5}$$

or $$v_2^2\left[1 - \left(\frac{a_o C_c v_2}{a_1}\right)^2\right] = 2gh$$

$$v_2 = \frac{(2gh)^{0.5}}{\left[1 - \left(\frac{a_o C_c}{a_1}\right)^2\right]^{0.5}}$$

$$Q = C_d a_2 v_2 = C_d\, C_c\, a_o \left[\frac{(2gh)^{0.5}}{\left\{1 - \left(\frac{a_o C_c}{a_1}\right)^2\right\}^{0.5}}\right]$$

(*b*) A hollow cylinder 1 m long has an internal diameter and external diameter equal to 0.4 m and 0.6 m respectively, and both ends are open. Assuming the weight of cylinder as 700 N, analyse whether the cylinder would be stable while floating in water with its axis vertical. (10)

Sol.

Weight of volume of water displaced by the cylinder

= Weight of cylinder

$$\left(\frac{\pi}{4}\right) \times (0.6^2 - 0.4^2) \times 1 \times 9.81 \times 1000 \times h = 700$$

$$h = 0.454 \text{ m}$$

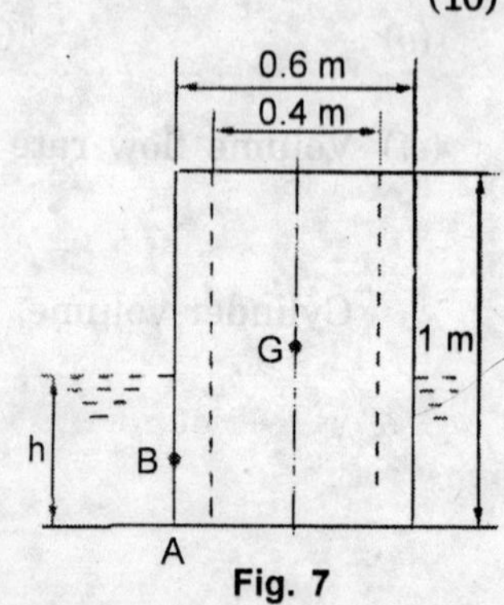

Fig. 7

Centre of buoyancy, B from the bottom A,

$$AB = \frac{h}{2} = 0.227 \text{ m}$$

C.G. from A, $AG = \frac{1}{2} = 0.5 \text{ m}$

$$BG = 0.5 - 0.227 = 0.273 \text{ m}$$

Metacentric height, $GM = \frac{I}{V} - BG$

$$I = \frac{\pi(0.6^4 - 0.4^4)}{64} = 5.1 \times 10^{-3} \text{ m}^4$$

$$V = \left(\frac{\pi}{4}\right) \times (0.6^4 - 0.4^2) \times 0.454 = 0.0713 \text{ m}^3$$

$$GM = \frac{5.1 \times 10^{-3}}{0.0713} - 0.273 = -0.201 \text{ m}$$

As GM is – ve, the point M lies below G and the cylinder will folat in unstable equilibrium.

(*c*) Show that the hydraulic efficiency of a Pelton turbine is maximum when peripheral wheel velocity is half the absolute velocity of jet at inlet.

Further show that, $(\eta_{hyd})_{max} = \frac{1 + K\cos\beta_2}{2}$

where K is the friction factor.

Mention the criteria for selecting water turbines for hydel power plant. **(20)**

Sol.

$$u_1 = u_2 = u$$

$$v_{r1} = v_{a1} - u;\ v_{w1} = v_{a1}$$

$$v_{w2} = v_{r2}\cos\beta_2 - u_2$$

$$F = \rho A v_{a1}(v_{w1} + v_{w2})$$

$$\text{Work done} = \rho A v_{a1}(v_{w1} + v_{w2}) \,.\, u$$

$$(\text{K.E.})_{\text{jet}} = \frac{1}{2} \,.\, \rho A v_{a1} \,.\, v_{a1}^2$$

$$= \frac{1}{2} \,.\, \rho A \,.\, v_{a1}^3$$

Fig. 8

$$\eta_{hyd} = \frac{\text{Work done}}{(\text{K.E.})_{\text{jet}}} = \frac{\rho A v_{a1}(v_{w1} + v_{w2})}{\frac{1}{2} . \rho A . v_{a1}^3}$$

$$= \frac{2u(v_{r1} + v_{r2}\cos\beta_2)}{v_{a1}^2} = \frac{2uv_{r1}(1 + K\cos\beta_2)}{v_{a1}^2}$$

where $K = \frac{v_{r2}}{v_{r1}}$

For hydraulic efficiency to be maximum,

$$\frac{d\eta_{hyd}}{du} = 0$$

$$\frac{2(v_{a1} - 2u)(1 + K\cos\beta_2)}{v_{a1}^2} = 0$$

$$v_{a1} = 2u$$

$$(\eta_{hyd})_{max} = \frac{1 + K\cos\beta_2}{2}$$

Criteria for selection of turbine :

1. Water head availablity.
2. Amount of discharge.
3. High head, low discharge – Pleton turbine

Medium head, high discharge – Francis turbine

Low head, high discharge – Kaplan turbine

Q.6. ***(a)*** Develop an expression for the pressure loss across a conical contraction of diameter D_1 and D_2 in a given length L for a Laminar flow. Explain the physical significance of equation. **(10)**

Sol.

Head loss for length dx,

$$dh_f = \frac{f\,dx\,v^2}{2gd}$$

$$D = D_1 - \left(\frac{D_1 - D_2}{L}\right)x$$

Fig. 9

Let $\frac{D_1 - D_2}{L} = k$ so that $D = D_1 - kx$

Also $f = \frac{64}{Re} = \frac{64\mu}{\rho v D} = \frac{16\pi\mu D}{\rho Q}$

$$v = \frac{Q}{A} = \frac{4Q}{\pi D^2}$$

$$\int dh_f = \int \frac{16\pi\mu D}{\rho Q} . dx . \left\{\frac{Q}{\pi D^2}\right\}^2 . \left\{\frac{1}{2gD}\right\}$$

$$h_f = \left[\frac{128\,\nu Q}{\pi g}\right] \int_{D_1}^{D_2} \frac{dx}{(D_1 - kx)^4} = \left[\frac{128\,\nu QL}{3\pi g\,(D_1 - D_2)}\right]\left[\frac{1}{D_2^3 - D_1^3}\right]$$

(b) Do you need priming in a gear pump? Give the reasons for your answer. **(5)**

Sol.

Priming is not required in a gear pump because the fluid entrained in the spaces between the teeth and the casing are carried round between gears from suction port to discharge port. The fluid cannot slip back into the inlet side due to meshing of gears, thus air cannot be entrapped in the pump.

(*c*) Explain the separation of flow and discuss the methods used for controlling of separation in diffuser. (10)

Sol.

In the boundary layer, the moving fluid particles are decelerated because of viscous friction resulting in decrease in their kinetic energy. If there is a pressure drop in the direction of flow, the pressure can overcome the deceleration the particles so that they combine to move along the wall. A positive pressure gradient however, will tend to further decelerate the particles in the boundary layer. Due to this the boundary layer thickness increases and the particles slow down. A little distance after that the particles stream stop and further it may reverse their direction. A reverse flow layer tends to push itself between the wall and the boundary layer. In this way the flow separates from the wall. This phenomenon which separates the boundayr layer from the wall is calles *separation of flow*.

A diffuser is an expanding passage. To control separation of flow in the diffuser, we use a large size diffuer so that energy dissipation is minimum.

(*d*) A centrifugal pump is to discharge 0.2 m³/s at a speed of 1400 rpm against a head of 25 m. The impeller diameter is 0.3 m, its width at outlet is 0.05 m and the manometric efficiency is 80%.

Determine the vane angle at the outer periphery of the impellor and lable the angle by sketching the impeller. (15)

Sol.

$$v_{f2} = \frac{Q}{\pi D_2 B_2} = \frac{0.2}{\pi \times 0.3 \times 0.05} = 4.25 \text{ m/s}$$

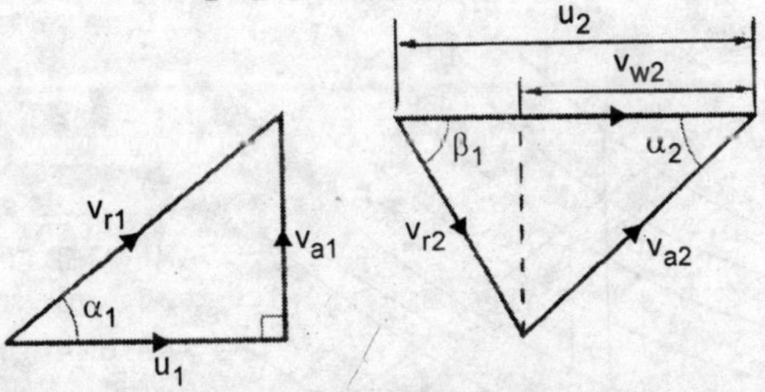

Fig. 10

$$u_2 = \frac{\pi D_2 N}{60} = \frac{\pi \times 0.3 \times 1400}{60} = 22 \text{ m/s}$$

$$v_{w2} = \frac{gH}{u_2 \eta_{mano}} = \frac{9.81 \times 25}{22 \times 0.8} = 14 \text{ m/s}$$

$$u_2 - v_{f2} \cot \alpha_2 = v_{w2}$$

$$22 - 4.25 \cot \alpha_2 = 14$$

$$\alpha_2 = 28°$$

Q.7. (*a*) What do you mean by supersaturated flow through steam nozzle? Discuss the causes of super saturation phenomena. (10)

Sol.

When steam flows through a nozzle it would normally be expected that the discharge of steam through the nozzle would be slightly less than the theoretical value. But it has been observed experimetally that the flow of wet steam is slightly greater than that calculated by the formula. The explanation for this phenomenon is that the converging part of the nozzle is so short and the steam velocity so high that the molecules of steam have insufficient time to collect and form droplets so that normal condensation does not take place. Such rapid expansion is said to be metastable and produces a supersaturated state. In this state of supersaturation the steam is undercooled to a temperature less than that corresponding to its pressure; consequently the density of steam increases and hence the weight of discharge.

The ideal expansion of superheated steam from pressure p_1 to p_b can be represented by a line AE on the Mollier diagram as shown in Fig. 11 (*a*). During the expansion, the change of phase must start to occur at pressure p_2, where the expansion line meets the saturation line at point B. But in the nozzle, under certain conditions, this phenomenon of condensation does not occur at point B as the time available is very short due to the very high velocity of steam passing through the nozzle. The equilibrium between the liquid and vapour phase is therefore delayed and the vapour continues to expand in dry state beyond point B. This is represented by BB_1. The pressure at the point B_1 can be found by extending the superheated constant pressure line p_3 upto point B_1, shown as dotted line. The steam during the expansion BB_1 remains dry and condensation is suppressed. The vapour between pressure p_2 and p_3 is said to be supersaturated or supercooled. A limit to the supersaturated state was observed by Wilson and a line drawn on the chart through the observed points is known as Wilson line. This line becomes the saturation line for all practical purposes. The condensation occurs at constant enthalpy, the pressure remaining constant. This is shown by B_1C. Line CD represents isentropic expansion to the exit pressure p_b.

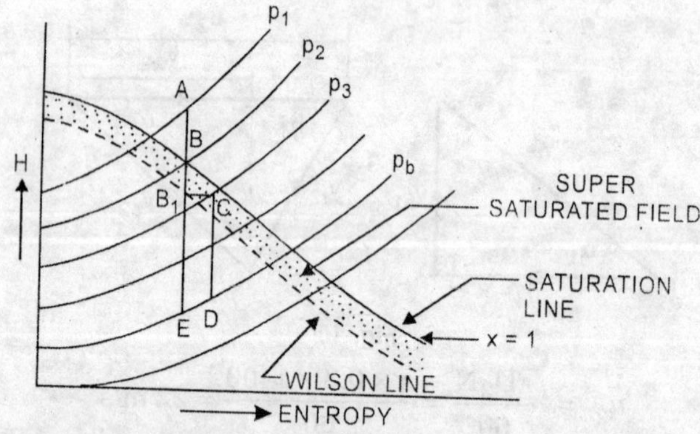

Fig. 11

(*b*) What is the basic principle of regenerative feed heating in steam power plant cycle? Sketch a practical feed heating system for 500 MW steam turbine and lable it. (10)

Sol.

Principle of regenerative feed heating. In the regenerative feed heating system, methods are adopted to heat the feed water from the hot well of the condenser irreversibly by interchange of heat within the system. It improves the thermal efficiency of the plant.

The principle of regeneration can be practically utilized by extracting steam from the turbine at several locations and supplying it to the regenerative heaters. The resulting cycle is known as the regenerative or bleeding cycle. A practical regenerative feed water heating system is shown in Fig. 12.

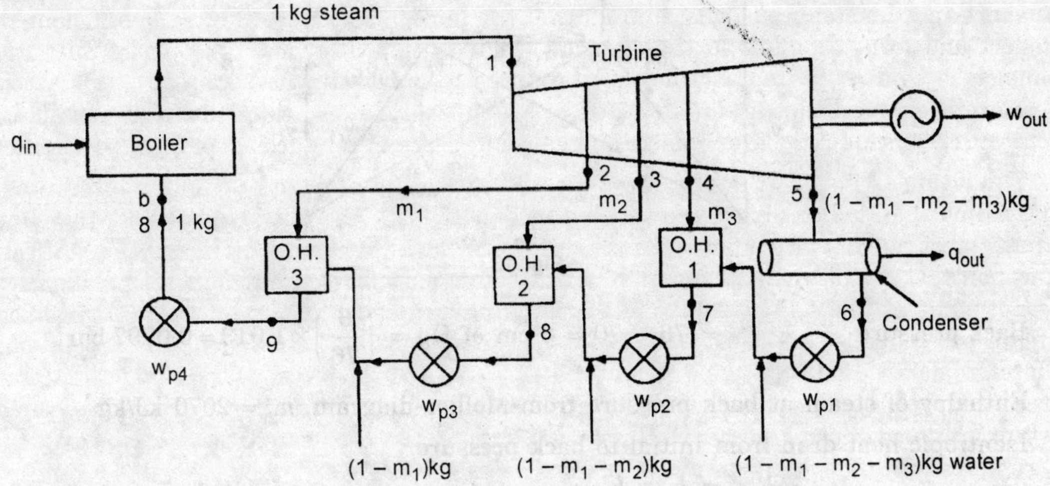

Fig. 12. Regenerative feed water heating with open heater.

Advantages :

1. The heating process in the boiler tends to become reversible.
2. The thermal stresses set up in the boiler are minimized.
3. The thermal efficiency is improved.
4. Heat rate is reduced.
5. Turbine erosion decreases.
6. Size of condenser reduces.
7. Turbine size is reduced.

(*c*) A blade group in a Parson's reaction steam turbine is of constant height. The turbine develops 60 MW running at 3000 rpm. The initial steam pressure and temperature entering the turbine are 50 bar and 400°C and back pressure is 70 cm Hg. The efficiency ratio of turbine is 70%. The mean blade velocity in the group is 120 m/s. The average specific volume of steam in the group may be taken as 0.62 m^3/kg. By drawing the velocity diagram, calculate the following :

(*i*) Mass flow rate of steam

(*ii*) Height of blade

(*iii*) Drum diameter

(*iv*) Moving outlet blade angle.

Assume the ratio of blade velocity to steam velocity as 0.55 and the nozzle (fixed blade) angle is 20°. Neglect the tip leakage and machanical losses. (20)

Sol.

Enthalpy of steam at initial pressure of 50 bar and temperature 400°C entering the turbine from steam tables, h_1 = 3198. kJ/kg

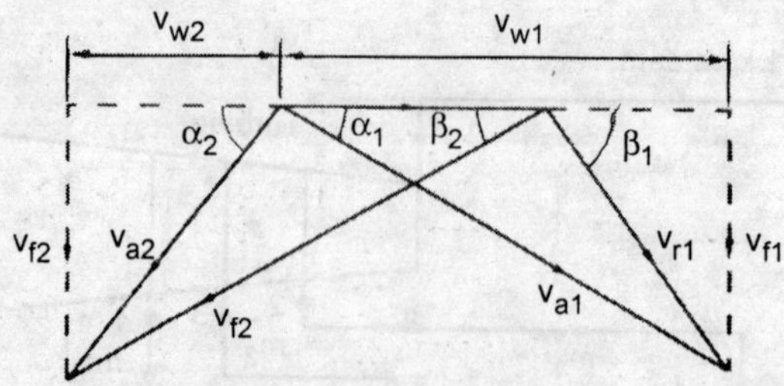

Fig. 13

Back pressure $= 76 - 70 = 6$ cm of Hg $= \left(\frac{6}{76}\right) \times 1.013 = 0.0797$ bar

Enthalpy of steam at back pressure from Mollier diagram, $h_2 = 2070$ kJ/kg

Isentropic heat drop from initial to back pressure,

$$\Delta h_{ise} = h_1 - h_2$$

$$= 3198.3 - 2070 = 1128.3 \text{ kJ/kg}$$

Efficiency ratio, η_r = Actual coupling work/Isentropic heat drop

$$= \frac{\Delta h_{cw}}{\Delta h_{ise}} - 0.7$$

$$\Delta h_{cw} \quad 1128.3 \times 0.7 = 789.8 \text{ kJ/kg}$$

In the absence of any mechanical loss being given,

Coupling work = Internal work.

(*i*) Power developed = Mass flow rate of steam, $\dot{m}_s$ × Internal work

$$\dot{m}_s = \frac{60 \times 10^3}{789.8} = 75.968 \text{ kg/s.}$$

(*ii*) Mean blade velocity, $u = 120$ m/s

$$\frac{u}{v_{a1}} = 0.55$$

Steam velocity, $v_{a1} = \frac{120}{0.55} = 218.2$ m/s

Now $u = \frac{\pi DN}{60}$

$$D = \frac{120 \times 60}{\pi \times 3000} = 0.764 \text{ m or } 764 \text{ mm}$$

Using continuity condition, we have

$$\pi D h v_{f1} = m_s V_s$$

Velocity of flow at inlet,

$$v_{f1} = v_{a1} \sin \alpha_1 = 218.2 \sin 20°$$

Height of blades,

$$h = \frac{75.968 \times 0.62}{\pi \times 0.764 \times 218.2 \sin 20°}$$

$$= 0.263 \text{ m or } 263 \text{ mm}$$

(*iii*) Drum diameter, $d_r = D - h = 764 - 263 = 501$ mm

(*iv*) In Parson's turbine, $\beta_2 = \alpha_1$

Hence $\beta_2 = 20°$

Q.8. (*a*) Why should no moisture flow along with steam from the drum to the superheater and why is no vapour bubble desired to flow along saturated water from the drum to downcomers? (7)

Sol.

No moisture should flow along with steam from the drum to the superheater so that the temperature and enthalpy of superheated steam should not get lowered. The downcomers carry superheated steam which has high specific violume. If vapour bubbles flow along with the saturated water then there is every possibility of collapsing these bubbles under high pressure and temperature. This may cause accident in the boiler.

(*b*) With the help of neat sketch discuss the working principle of Velox high pressure boiler. (15)

Sol.

Velox boiler. This boiler is based on the fact that when the gas velocity exceeds the sound velocity, the heat is transferred from the gas at a much higher rate as compared to at sub sonic flow.

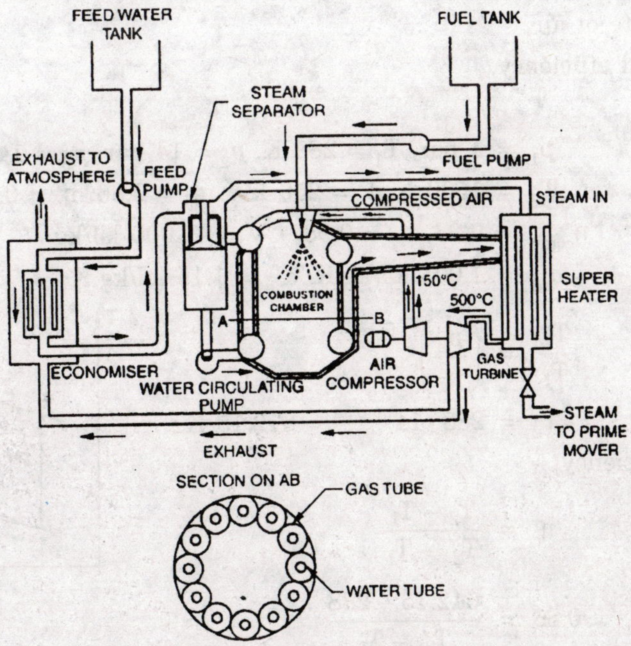

Fig. 13. Velox boiler

The arrangement of the components of this boiler is shown in Fig. 14. Air is compressed to 2.5 bar with the help of a compressor run by a gas turbine to get the supersonic velocity of the gases before passing through the combustion chamber and gas tubes. The burned gases in the combustion chamber are passed through the annulus of the tubes and transfer heat to water to generate steam. The mixture of water and steam thus formed then passes into a separator with a spiral flow. The steam is separated from water due to centrifugal action. The separated steam is further passed through a superheater and then supplied to the prime mover. The water removed from steam is again passed into the water tubes with the help of a pump. The gases coming out from the annulus at the top are further passed over the superheater where its heat is used for superheating the steam. The gases coming out of the superheater are used to run a gas turbine which in turn is used to run the air comressor. The exhaust coming out from the gas turbine are passed through the economizer to utilize the remaining heat of gases. An electric motor supplies additional power required to run the compressor.

(c) A simple open cycle gas turbine has a compressor turbine and a free power turbine. It develops electrical power output of 250 MW. The cycle takes in air at 1 bar and 288 K. Total compressor pressure ratio is 14. The turbine inlet temperature is 1500 K. The total to total isentropic efficiency of compressor and turbine are 0.86 and 0.89 respectively. The mechanical efficiency of each shaft is 0.98. Combustion efficiency is 0.98 while combustor pressure loss is 0.03 bar. Alternator efficiency is 0.98. Take calorific value of fuel equal to 42,000 kJ/kg,

$$c_{pa} = 1.005 \text{ kJ/kg.K and } c_{pg} = 1.15 \text{ kJ/kg}$$

Calculate the following :

(*i*) Air-fuel ratio,

(*ii*) Specific work output,

(*iii*) Specific fuel consumption,

(*iv*) Mass flow rate of air,

(*v*) Cycle thermal efficiency, (20)

Sol.

Given :

$$p_1 = 1 \text{ bar}, T_1 = 288 \text{ K}, p_2 = 14 \text{ bar}, p_3 = 14 - 0.03 = 13.97 \text{ bar}$$

$$T_3 = 1500 \text{ K}, P = 250 \text{ kW}. \eta_c = 0.86, \eta_t = 0.89, \eta_{mech} = 0.98$$

$$\eta_{comb} = 0.98, \eta_a = 0.98, CV = 42{,}000 \text{ kJ/kg}$$

$$c_{pa} = 1.005 \text{ kJ/kg.K}, c_{pg} = 1.15 \text{ kJ/kg.K}$$

$$\frac{T_2}{T_1} = \left(\frac{p_2}{p_1}\right)^{(\gamma-1)/\gamma}$$

$$T_2 = 288\,(14)^{0.4/1.4} = 612.15 \text{ K}$$

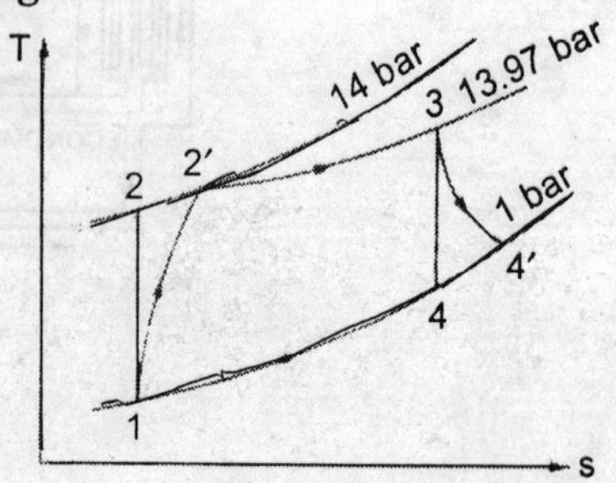

Fig. 15

Compressor efficiency,

$$\eta_c = \frac{T_2 - T_1}{T_2' - T_1}$$

$$0.86 = \frac{612.15 - 288}{T_2' - T_1}$$

$$T_2' - T_1 = 376.92 \text{ K}, \ T_2' = 664.92 \text{ K}$$

$$T_4 = \frac{T_3}{\left(\frac{p_3}{p_4}\right)^{(\gamma-1)/\gamma}} = \frac{1500}{\left(\frac{13.97}{1}\right)^{0.4/1.4}} = 706.14 \text{ K}$$

Turbine efficiency,

$$\eta_t = \frac{T_3 - T_4}{T_3 - T_4'}$$

$$T_3 - T_4' = \frac{1500 - 706.14}{0.89} = 891.97 \text{ K}$$

Actual net work done, w_n = Turbine work, w_t – Compressor work, w_c

$$= \frac{c_{pg}\,(T_3 - T_4')\,\eta_{mech} - c_{pa}\,(T_2' - T_1)}{\eta_{mech}}$$

$$= \frac{1.15 \times 891.97 \times 0.98 - 1.005 \times 376.92}{0.98} = 618.72 \text{ kJ/kg of air}$$

Effective heat supplied,

$$q_e = c_{pg}\,(T_3 - T_2') = 1.15\,(1500 - 664.92) = 960.34 \text{ kJ/kg}$$

Actual heat supplied, $q_a = \dfrac{q_e}{\eta_{mech}} = \dfrac{960.34}{0.98} = 979.94$ kJ/kg

(*i*) Air/fuel ratio,

$$A/F = \frac{\text{CV of fuel}}{q_a} = \frac{42000}{979.94} = 42.86:1.$$

(*ii*) Specific work output = 618.72 kJ/kg of air

(*iii*) Specific fuel consumption

$$= \frac{\text{Specific air consumption}}{(A/F)} = \frac{\left(\frac{3600}{\text{Specific work output}}\right)}{(A/F)}$$

$$= \frac{\left(\frac{3600}{618.72}\right)}{42.86} = 0.1357 \text{ kg/kWh}$$

(*iv*) Mass flow rate of air, $\dot{m}_a$

= Electric power developed × Alternator efficiency

$$m_a = \frac{250 \times 10^3 \times 0.98}{618.72} = 395.98 \text{ kg/s}$$

(*v*) Cycle thermal efficiency

$$= \frac{\text{Actual net work done}}{\text{Actual heat supplied}} = \frac{618.72}{979.94} = 63.14\%$$

MECHANICAL ENGINEERING
PAPER I—2005

(Conventional)

Q.1. (*a*) Derive the expression for $(\Delta h)_T$ for a substance that obeys the equation of state given by

$$p = \frac{RT}{v} = \frac{a}{v^2} \tag{10}$$

Sol.

$$p = \frac{RT}{v} = \frac{a}{v^2} \qquad ...(a)$$

Now,
$$dh = c_p\, dT + \left[v - T\left(\frac{\partial v}{\partial T}\right)_p\right] dp$$

or
$$\left(\frac{\partial h}{\partial p}\right)_T = 0 + v - T\left(\frac{\partial h}{\partial T}\right)_p \qquad ...(1)$$

Consider
$$p = f(v, T)$$

$$dp = \left(\frac{\partial p}{\partial v}\right)_T .dv + \left(\frac{\partial p}{\partial T}\right)_v .dT$$

$$(dp)_T = \left(\frac{\partial p}{\partial v}\right)_T .dv + 0 \text{ as } dT = 0 \qquad ...(2)$$

From (1), we have
$$(dp)_T = \left[v - T\left(\frac{\partial v}{\partial T}\right)_p\right](dp)_T$$

Substituting the value of $(dp)_T$ from (2), we get

$$(dh)_T = \left[v - T\left(\frac{\partial v}{\partial T}\right)_p\right]\left(\frac{\partial p}{\partial v}\right)_T .dv$$

$$= \left[v.\left(\frac{\partial p}{\partial v}\right)_T - T\left(\frac{\partial v}{\partial T}\right)_p .\left(\frac{\partial p}{\partial v}\right)_T\right].dv \qquad ...(3)$$

Now cyclic relation for p, v, T is

$$\left(\frac{\partial v}{\partial T}\right)_p .\left(\frac{\partial T}{\partial p}\right)_v .\left(\frac{\partial p}{\partial v}\right)_T = -1$$

$$\left(\frac{\partial v}{\partial T}\right)_p \cdot \left(\frac{\partial p}{\partial v}\right)_T = -\left(\frac{\partial p}{\partial T}\right)_v$$

Substituting this in (3), we get,

$$(dh)_T = \left[v\left(\frac{\partial p}{\partial v}\right)_T + \left(\frac{\partial p}{\partial T}\right)_v\right].dv \quad ...(4)$$

From (a), we have

$$\left(\frac{\partial p}{\partial v}\right)_T = \left(\frac{\partial}{\partial v}\right)\left[\frac{RT}{v} - \frac{a}{v^2}\right]_T$$

$$= \frac{RT}{v^2} + \frac{2a}{v^3} \quad ...(5)$$

and
$$\left(\frac{\partial p}{\partial t}\right)_v = \left(\frac{\partial}{\partial T}\right)\left[\frac{RT}{v} - \frac{a}{v^3}\right] = -\frac{R}{v} \quad ...(6)$$

Substituting in (4), we get,

$$(dh)_T = \left[v\left(\frac{RT}{v^2} + \frac{2a}{v^3}\right) + T\left(\frac{R}{v}\right)\right].dv$$

$$\int_1^2 (dh)_T = -RT\int_1^2 \frac{dv}{v} + 2a\int_1^2 \frac{dv}{v^2} + RT\int_1^2 \frac{dv}{v}$$

$$(h_2 - h_1) = -RT\ln\left(\frac{v_2}{v_1}\right) - 2a\left(\frac{1}{v_2} - \frac{1}{v_1}\right) + RT\ln\left(\frac{v_2}{v_1}\right)$$

$$= -2a\left(\frac{1}{v_2} - \frac{1}{v_1}\right)$$

(***b***) 4 kg of water at 40°C are mixed with 6 kg of water at 100°C in a steady flow process.

Calculate :

(i) the temperature of resulting mixture,

(ii) the change in entropy, and

(iii) the unavailable energy with respect to the energy receiving water at 40°C. (10)

Sol.

(i) Let t = temperature of mixture.

For heat balance, we have

$$m_1 c_p \cdot dt_1 = m_2 c_p \cdot dt_2$$

$$4\,(t - 40) = 6\,(100 - t)$$

$$10\,t = 760$$

$$t = 76°C$$

(*ii*) $$\Delta s = m_1 c_p \int_{313}^{349} dT + m_2 c_p \int_{313}^{349} dT$$

$$= 4.187 \left[4 \ln \left(\frac{349}{313} \right) + 6 \ln \left(\frac{349}{313} \right) \right]$$

$$= 0.1526 \text{ kJ/kg}$$

(*iii*) At 40°C, for water

$$\Delta s = m_1 c_p \int_{313}^{349} dT = 4.187 \left[4 \ln \left(\frac{349}{313} \right) \right]$$

$$= 1.823 \text{ kJ/K}$$

Fig. 1

Unavailable energy $= T_1.\Delta s = 313 \times 1.823 = 570.7$ kJ

(*c*) An inventor claims to have developed a device which requires no energy transfer by work or heat transfer, yet able to produce hot and cold streams of air from a single stream of air at an intermediate temperature. The inventor provides steady-state test data indicating temperature that when air enters at a temperature of 21°C and a pressure of 5.2 bar, separate streams of air exit at a temperature of – 19°C to 81°C respectively, and each at a pressure of 1 bar. Sixty percent of the mass entering the device exists at the lower temperature. Evaluate the inventor's claim, employing the ideal gas model for air and ignoring changes in the kinetic and potential energies of the streams from inlet to exit. **(20)**

Sol.

The schematic of the given data is shown in Fig. 2.

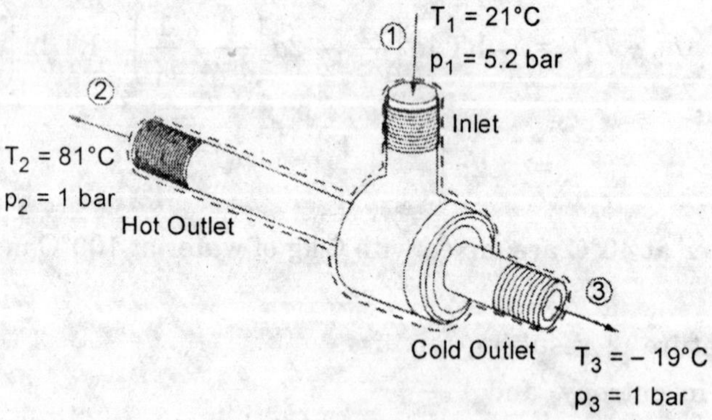

Fig. 2

The assumption made are :

1. The control volume shown is at steady state.
2. For the control volume, $W_{cv} = 0$ and $Q_{cv} = 0$.
3. Changes in KE and PE from inlet to exit are ignored.
4. The air is assumed as an ideal gas with constant.

$$c_p = 1.005 \text{ kJ/kg.K or R} = 0.287 \text{ kJ/kg.K.}$$

For the device to operate as claimed, the conservation of mass and energy principles must be satisfied. The second law of thermodynamics also must be satisfied; and in particular the rate of entropy production cannot be negative. Accordingly, the mass, energy and entropy rate balances are considered in turn.

With assumptions 1-3, the mass and energy rate balances reduce, respectively, to

$$\dot{m}_1 = \dot{m}_2 + \dot{m}_3$$

$$O = \dot{m}_1 h_1 - \dot{m}_2 h_2 - \dot{m}_3 h_3$$

Since $\dot{m}_3 = 0.6\,\dot{m}_1$ it follows from the mass rate balance that $\dot{m}_2 = 0.4\,\dot{m}_1$. By combining the mass and energy rate balances and evaluating changes in specific enthalpy using constant c_p, the energy rate balance is also satisfied. That is

$$\begin{aligned} 0 &= (\dot{m}_2 + \dot{m}_3)\,h_1 - \dot{m}_2 h_2 - \dot{m}_3 h_3 \\ &= \dot{m}_2\,(h_1 - h_2) + \dot{m}_3\,(h_1 - h_3) \\ &= 0.4\,\dot{m}_1\,[c_p(T_1 - T_2)] + 0.6\,\dot{m}_1\,[c_p\,(T_1 - T_3)] \\ &= 0.4\,(21 - 81) + 0.6\,(21 + 19) \\ &= -\,24 + 24 = 0 \end{aligned}$$

Since no significant heat transfer occurs, the entropy rate balance at steady state gives.

$$\sum_i \frac{Q_i}{T_i} + \dot{m}_1 s_1 - \dot{m}_2 s_2 - \dot{m}_3 s_3 + \dot{s}_{gen} = 0$$

Combining the mass and entropy rate balances, with $\sum_i \frac{Q_i}{T_i} - 0$, we get

$$\begin{aligned} 0 &= (\dot{m}_2 + \dot{m}_3)s_1 - \dot{m}_2 s_2 - \dot{m}_3 s_3 + \dot{s}_{gen} \\ &= \dot{m}_2\,(s_1 - s_2) + \dot{m}_3\,(s_1 - s_3) + \dot{s}_{gen} \\ &= 0.4\,\dot{m}_1\,(s_1 - s_2) + 0.6\,\dot{m}_1\,(s_1 - s_3) + \dot{s}_{gen} \end{aligned}$$

Solving for $\dot{s}_{gen} / \dot{m}_1$ and using expressions for changes in specific entropy, we get

$$\begin{aligned} \frac{\dot{s}_{gen}}{\dot{m}_1} &= 0.4\left[c_p \ln\frac{T_2}{T_1} - R\ln\frac{p_2}{p_1}\right] + 0.6\left[c_p \ln\frac{T_3}{T_1} - R\ln\frac{p_3}{p_1}\right] \\ &= 0.4\left[1.005\ln\frac{354}{294} - 0.287\ln\frac{1}{5.2}\right] \\ &\qquad + 0.6\left[1.005\ln\frac{254}{294} - 0.287\ln\frac{1}{5.2}\right] \\ &= 0.26392 + 0.19571 = 0.45963 \text{ kJ/kg.K} \end{aligned}$$

Thus, the second law of thermodynamics is also satisfied as $\dot{s}_{gen} > 0$. Therefore, the inventor's claim is valid.

Q.2. (*a*) List the objectives of supercharging. Sketch the various configurations of supercharger. Why spuercharging is not preferred with single cylinder engine? (10)

Sol.

Objectives of Supercharging :

1. To overcome the effect of high altitudes, as in the case of aircrafts and stationary engines in mountains.
2. To reduce the weght of an engine per kW of power developed, as in the case of racing cars.
3. To reduce the size of the engine to fit into a limited space, as in the case of locomotives or marine engines.
4. To increase the power output of an existing engine to meet greater power demands.

Configurations for Supercharger :

1. **Compressor coupled to Engine shaft.** The compressor is operated directly by the engine with set-up gearing to increase the speed of centrifugal compressor. Part of engine output is used to drive the supercharger and net output is calculated by deducing this power from the gross power of the engine.
2. **Turbo-charger.** The energy of the exhaust gases from the engine is used to develop power in a turbine which directly runs the supercharger. There is no mechanical connection between the engine and the supercharger.
3. **Direct coupling between Engine, Compressor and Turbine.** The advantages of this arrangement is, when the turbine output is insufficient to run the compressor, extra power required is taken from the engine. Additional power from the turbine can also be fed to the engine.
4. **Compressor geared with Engine and Free turbine.** The engine drives the compressor but the energy in the exhaust is utilized to develop the power from a separate turbine.

The various configurations are shown in Fig. 3 (*a*) to (*d*).

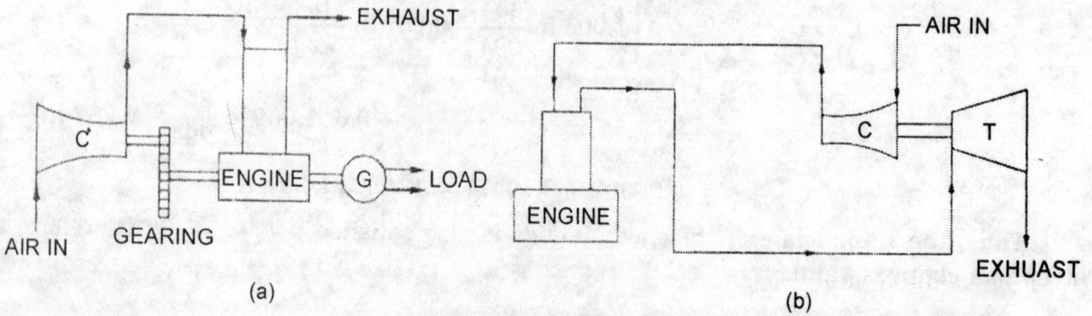

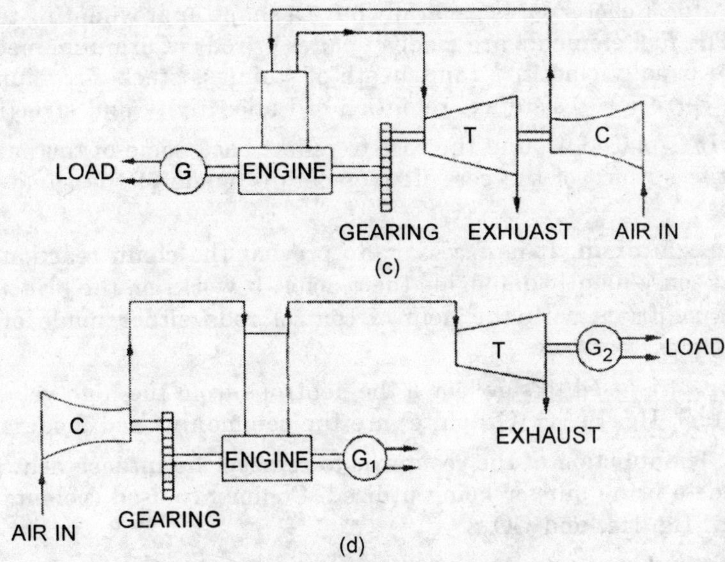

Fig. 3 Arrangement for super charging.

Supercharging of single cylinder engines. Supercharging of single cylinder engines is not carried out because the thermal efficiency drops to zero when the ratio of supercharger is about 6. Further the power is maximum when the ratio of supercharger is nearly 2 to 5 and then drops to zero at a ratio of about 6.

(b) Sketch neatly a simple nuclear reactor and label it. Explain briefly the function of each element of the nuclear reactor. (15)

Sol.

Nuclear reactor. A nuclear reactor is an apparatus in which nuclear fission is produced in the form of a controlled self-sustaining chain reaction which produces heat, neutrons and radioisotopes. A nuclear reacotr is shown in Fig. 4. The elements of a nuclear reactor are :

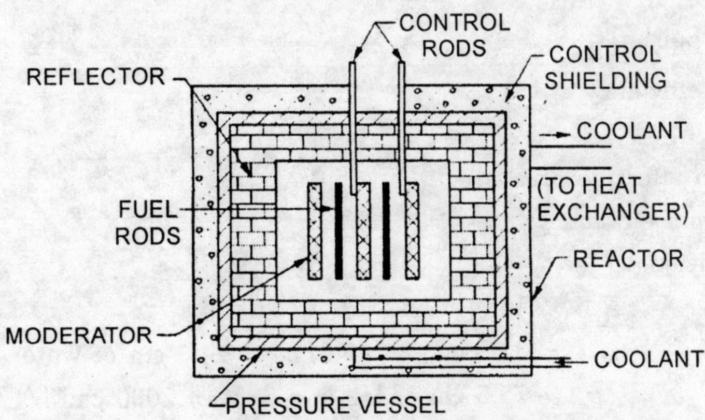

Fig. 4. Nuclear reactor.

1. **Reactor core.** In this, fission chain reaction is made to occur and where fission energy is liberated in the form of heat for operating power conversion equipment. The core of the reactor consists of an assemblage of fuel elements, control rods, coolant

and moderator. Reactor cores generally have a shape approximating to a right circular cylinder. The fuel elements are made of plates or rods of uranium metal. These plates or rods are usually clad in a thin sheath of stainless steel, zirconium or aluminium to provide corrosion resistance, retention of radioactivity and structural support.

2. **Refletor.** It is placed around the core to reflect back some of the neutrons that leak out from the surface of the core. It is generally made of the same material as the moderator.
3. **Control mechanism.** It is necessary to prevent the chain reaction from becoming violent and consequently damaging the reactor. It works on the principle of absorbing the excess neutrons with the help of control rods either made of boron steel or cadmium strips.
4. **Moderator.** It is used to slow down the neutrons from the high velocities. H_2O, D_2O (heavy water), He, Be, C (Graphite) are the commonly used moderators.
5. **Coolants.** The function of the coolant is to remove the intense heat produced in the reactor and to bring out for being utilized. Commonly used coolants are : Santiwax R (organic), Hg, He, and CO_2.
6. **Measuring instruments.** Main instrument required is for the purpose of measuring thermal neutron flux which determines the power developed by the reactor.
7. **Shielding.** It is necessary to protect the walls of the reactor vessel from radiation damage and to protect the operating personnel from exposure to radiation. Thermal shield is provided through the steel lining and the external shielding is generally made of thick concrete surrounding the reactor installation.

(c) A sharp edged orifice of diameter 3.8 cm and coefficient of discharge as 0.6 is used to measure air-consumption of a four stroke petrol engine. Pressure drop through the orifice is 145 mm of water and barometer reads 75.5 cm of Hg. The compression ratio of the engine is 6 and the piston displacement volume is 2000 cm^3.

The temperature of air is taken to be 26°C. At 2600 rpm, the engine brake power recorded is 29.5 kW. The fuel consumption is 0.14 kg/min and the calorific value of fuel used is 43960 kJ/kg.

Calculate the following :

(*i*) Volumetric efficiency

(*ii*) Air fuel ratio

(*iii*) Brake mean effective pressure

(*iv*) Brake thermal efficiency

(*v*) Relative efficiency. **(30)**

Sol.

Given :

$$d = 3.8 \text{ cm}, C_d = 0.6, h_w = 14.5 \text{ cm of water}$$

$$p_a = 75.5 \text{ cm of Hg}, r = 6. V_s = 2000 \text{ cm}^3$$

$$N = 2600 \text{ rpm}, BP = 29.5 \text{ kW}$$

$$\dot{m}_f = 0.14 \text{ kg/min}, CV = 43960 \text{ kJ/kg},$$

Density of air at atmospheric conditions,

$$\rho_a = \frac{m}{V} = \frac{p}{RT}$$

$$= \frac{1 \times 10^5}{287 \times 299} = 1.1653 \text{ kg/m}^3$$

$$\text{Head in } m \text{ of air, H} = \left(\frac{h_w}{100}\right)\left(\frac{\rho_w}{\rho_a}\right)$$

$$= \frac{14.5 \times 1000}{100 \times 1.1653} = 124.43 \text{ m of air}$$

Velocity of air passing through orifice,

$$c_a = [2gH]^{0.5} = [2 \times 9.81 \times 124.43]^{0.5} = 49.41 \text{ m/s}$$

Volume of air passing through orifice,

$$V_a = A_o c_a C_d$$

$$= \frac{\pi}{4}.(3.8)^2 \times 10^{-4} \times 49.41 \times 0.6$$

$$= 0.033622 \text{ m}^3\text{/s or } 2.0173 \text{ m}^3\text{/min.}$$

(*i*) Volumetric efficiency

$$= \frac{V_a}{V_s} = \frac{2.0173}{2000 \times 10^{-6} \times 260} = 0.3879 \text{ or } 38.79\%$$

(*ii*) Mass flow rate of air, m_a

$$= V_a \rho_a = 2.0173 \times 1.1653 = 2.35 \text{ kg/min}$$

$$\text{A/F ratio} = \frac{m_a}{m_f} = \frac{2.35}{0.14} = 16.79{:}1.$$

(*iii*) $$\text{BMEP} = \frac{BP}{V_s} = \frac{29.5 \times 10^3}{\frac{2000 \times 10^{-6} \times 2600}{60}} = 0.3404 \text{ MPa}$$

(*iv*) $$\eta_{bt} = \frac{BP}{m_f \times CV}$$

$$= \frac{29.5 \times 3600}{0.14 \times 60 \times 43960} = 0.2895 \text{ or } 28.95\%$$

(*v*) Air standard efficiency,

$$\eta_a = 1 - \frac{1}{r^{\gamma-1}} = 1 - \frac{1}{6^{0.4}} = 0.5116 \text{ or } 51.16\%$$

(*vi*) $$\text{Relative efficiency} = \frac{\eta_{bt}}{\eta_a} = \frac{28.85}{51.16} = 0.5639 \text{ or } 56.39\%$$

Q.3. (*a*) What do you mean by critical thickness of insulation for pipes? Derive an expression for optimum critical radius and show the variation of heat transfer with radius. Give some practical examples of using critical thickness of insulation. (10)

Sol.

Critical thickness of Insulation. The insulation radius at which the resistance to heat flow is minimum, is called the *critical radius*. For a pipe, critical radius of insulation,

$$r_c = k/h_o$$

where k = thermal conductivity of insulation,

h_o = outside film co-efficient.

Critical thickness of insulation,

$$t_c = r_c - r$$

where r = outside pipe radius.

Consider a thin-walled pipe of length l, radius r_i, transporting a fluid at temperature t_i, which is higher than the ambient temperature t_o, as shown in Fig. 5 (*a*). The pipe is surrounded by an insulation of thickness, $t = (r - r_i)$ and thermal conductivity, k.

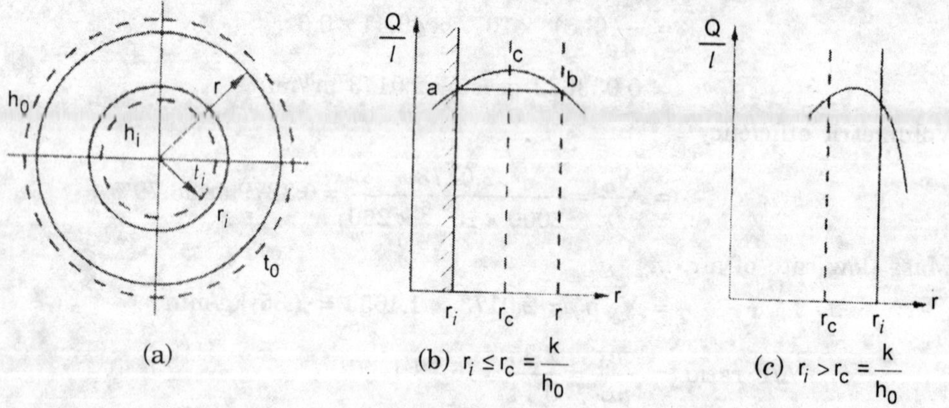

(a) (b) $r_i \le r_c = \dfrac{k}{h_0}$ (c) $r_i > r_c = \dfrac{k}{h_0}$

Fig. 5.

For steady state conditions, negligible thermal resistance due to pipe wall and one-dimensional heat flow in radial direction, heat transmission can be expressed as

$$Q = \frac{t_i - t_o}{\left[\dfrac{1}{2\pi r_i l h_i} + \dfrac{\ln\left(\dfrac{r}{r_i}\right)}{2\pi k l} + \dfrac{1}{2\pi r l h_o}\right]}$$

where h_i, h_o = film coefficients at the inner and outer surface respectively.

Now, k, r_i, l, h_i and h_o are constants. Therefore, Q will depend on r. With increase n r, the thermal resistance of isulation increases but the resistance due to convection coefficient at the outer surface drops. Therefore, addition of insulation material can either increase or decrease the rate of heat transmission depending on a change in the total resistance with r.

Total resistance, $$R_t = \left[\frac{1}{2\pi r_i l h_i} + \frac{\ln\left(\dfrac{r}{r_i}\right)}{2\pi k l} + \frac{1}{2\pi r l h_o}\right]$$

For R_t to be maximum or minimum, $\frac{dR_t}{dr} = 0$

$$\text{or} \quad \frac{1}{2\pi klr} - \frac{1}{2\pi r^2\, lh_o} = 0$$

$$r_c = \frac{k}{h_o}$$

$$\text{Also} \quad \frac{d^2R_t}{dr^2} = -\frac{1}{2\pi klr^2} + \frac{1}{\pi r^3 lh_o}$$

$$\text{At} \quad r = \frac{k}{h_o}, \frac{d^2R_t}{dr^2} = \frac{h^2}{2\pi k^3 l}$$

Which is positive. Hence $r = k/h_o$ gives minimum resistance and maximum heat flow rate.

The variation of Q/l with r is shown in Fig. 5 (*b*) and (*c*).

(*i*) When $r_i < r_c$, as shown in Fig. 5 (*b*), the heat loss increases between $r < r_c$ and decreases beyond $r > r_c$. The phenomenon of increase in heat transmission with addition of insulation is used in the insulation of electrical wires and cables.

(*ii*) When $r_i > r_c$, then overall thermal resistance increases and obstructs the flow of heat, as shown in Fig. 5 (*c*). This phenomenon is used in steam and refrigeration pipes.

(*b*) A process industry employs a counter-flow heat exchanger to cool 0.8 kg/s of oil (c_p = 2.5 kJ/kg.K) from 120°C to 40°C by the use of water entering at 20°C. The overall heat transfer coefficient is estimated to be 1600 W/m^2.K. It is assumed that the exit temperature of water will not exceed 80°C. Using NTU method and taking NTU = 4 in this case, calculate the following :

(*i*) Mass flow rate of water.

(*ii*) Surface area required.

(*iii*) Effectiveness of heat exchanger. (10)

Sol.

For heat balance, we have

$$m_c c_c\,(t_{c2} - t_{c1}) = m_h c_h\,(t_{h1} - t_{h2})$$

$$m_c \times 4.2\,(80 - 20) = 0.8 \times 2.5\,(120 - 40)$$

$$m_c = \frac{0.8 \times 2.5 \times 80}{4.2 \times 60} = 0.635 \text{ kg/s}$$

$$C_h = m_h c_h \times 0.8 \times 2.5 = 2$$

$$C_c = m_c c_c = 0.635 \times 4.2 = 2.667$$

$$C_{min} = C_h = 2 \text{ kJ/(s.K)}$$

$$\text{NTU} = \frac{UA}{C_{min}}$$

$$4 = \frac{1600A}{2 \times 10^3}$$

$$A = \frac{8 \times 10^3}{1600} = 5 \text{ m}^2$$

$$C = \frac{C_{min}}{C_{max}} = \frac{2}{2.667} = 0.75$$

$$\varepsilon = \frac{1 - \exp\{-\text{NTU}(1 - C)\}}{1 - C\exp\{-\text{NTU}(1 - C)\}}$$

$$= \frac{1 - \exp\{-4(1 - 0.75)\}}{1 - 0.75\exp\{-4(1 - 0.75)\}}$$

$$= 0.8728 \text{ or } 87.28\%$$

(c) An industrial furfance employs a hollow brick lining. The inside and outside surfaces of hollow brick lining are maintained at 900 K and 430 K by placing the radiation shields in between the hollow space. The heat loss of the furnace surroundings at 300 K is both by radiation and natural convection. By sketching the arrangement, calculate the number of radiation shields needed. The emissivity of the wall and the shields may be taken as 0.85. The convective heat transfer coefficient is governed by the expression,

$$h = 1.5\,(\Delta T)^{0.33} \text{ W/m}^2.\text{K}. \qquad \textbf{(20)}$$

Sol.

$$Q_{convection} = h_{conv}\,(T_2 - T_3)$$

$$= 1.5\,(T_2 - T_3)^{0.33}\,(T_2 - T_3)$$

$$= 1.5\,(430 - 300)^{1.33} = 971.93 \text{ W/m}^2$$

Radiation loss to surroundings,

$$Q_{rad} = (F_g)_{23}\,\sigma_b\,(T_2 - T_3)^4$$

$$(F_g)_{23} = \frac{1}{\left[\frac{1-\varepsilon_1}{\varepsilon_1} + \frac{1}{F_{23}} + \left\{\frac{1-\varepsilon_3}{\varepsilon_3}\right\}\left(\frac{A_2}{A_3}\right)\right]}$$

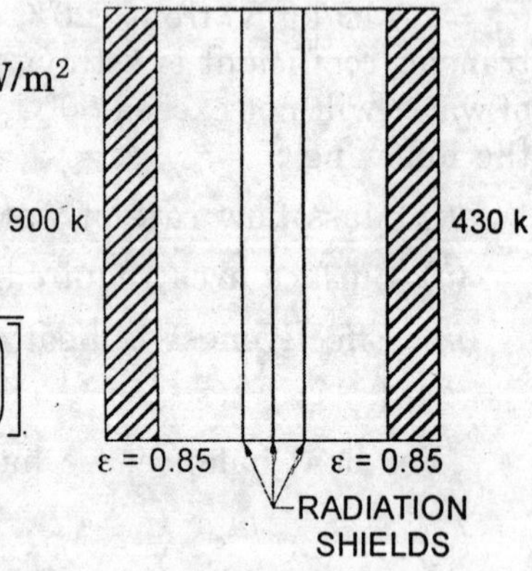

$$A_2 << A_3 \text{ and } F_{23} = 1$$

$$(F_g)_{23} = \varepsilon_1 = 0.85$$

$$Q_{rad} = 0.85 \times 5.67 \times 10^{-8}\,(430^4 - 300^4)$$

$$= 1257.3 \text{ W/m}^2$$

Fig. 6

Let n = number or radiation shields (Fig. 6).

Heat exchanger,

$$Q_{1\text{-}2} = \left[\frac{1}{n+1}\right]\sigma_b\,({T^4}_1 - {T^4}_2)/\left(\frac{2}{\varepsilon} - 1\right)$$

$$= \left[\frac{1}{n+1}\right] \times 5.67 \times 10^{-8}\,\frac{(900^4 - 430^4)}{\left(\frac{2}{0.85} - 1\right)}$$

$$= \frac{26063.5}{n+1} \text{ W/m}^2 = Q_{conv} + Q_{rad}$$

$$\frac{26063.5}{n+1} = 971.93 + 1257.3$$

$$n = 10.69 \approx 11$$

Q.4. (***a***) What is the purpose of undercooling in vapour compression refrigeration system? Differentiate between dry and wet compression. What are the advantages of one over the other ? (5)

Sol.

Undercooling. It is the cooling of refrigerant below the saturation temperature T_3' to T_3 before expansion by throttling, as shown in Fig. 7 (*a*). The purpose of undercooling in vapour compression refrigeration system is to achieve the following :

1. To reduce flashing of liquid during expansion and to increase refrigeration effect and COP.
2. Piston displacement and power per ton are reduced for all refrigerants.
3. Howevver, work of compression is increased.

Dry v's Wet compression. Thc compression process is shown in Fig. 7 (*b*). The wet compression involves the compression of wet refrigerant vapour at 1′ to dry saturated vapour at 1 or even slightly supersaturated is known as dry compression. The state of vapour at the end of compression is at 2 at saturation pressure p_k corresponding to the condensing temperature T_k. The refrigerant leaves the compressor superheated.

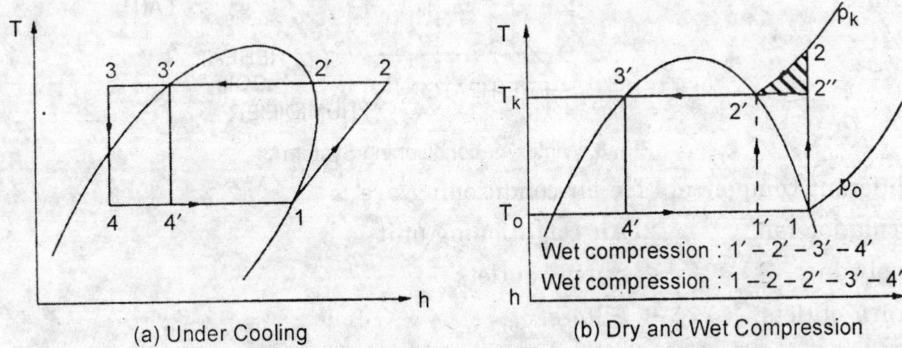

(a) Under Cooling (b) Dry and Wet Compression

Fig. 7. Vapour Compression Refrigeration System.

The advantages of one over the other are :

1. Power consumption per ton refrigeration in wet compression is less in case of ammonia as compared with dry compression.
2. In wet compression, liquid refrigerant may be trapped in the head of the cylinder and may damage the compressor valves and the cylinder, which is not so in dry compression.
3. Liquid refrigerants droplets in wet compression may wash away the lubricating oil from the walls of the compressor cylinder, thus increasing wear.
4. Wet compression is suitable for screw compressors rather than reciprocating compressors.

(*b*) Draw a neat sketch of air-conditioning system required in winters. Explain the working of different components in the circuit. **(15)**

Sol.

Air-conditioning system for Winter. The schematic arrangement for winter air-conditioning system is shown in Fig. 8. The air is heated followed by humidification. The outside air flows through the damper and mixes up with the recirculated air. The mixed air is passed through a filter to remove dirt, dust and other impurtities. The air now passes through a preheat coil in order to prevent the possible freezing of water and to control the evaporation of water in the humidifier. After that the air is made to pass through a reheat coil to bring the air to the designed DBT. Now the conditioned air is supplied to the conditioned space by a fan. From the conditioned space, a part of the used air is exhausted to the atmosphere by the exhaust fans or ventilators. The remaining part of the used air is again conditioned. The outside air is sucked and made to mix with recirculated air in order to make up for the loss of conditioned air through exhaust fans or ventilators from conditioned space.

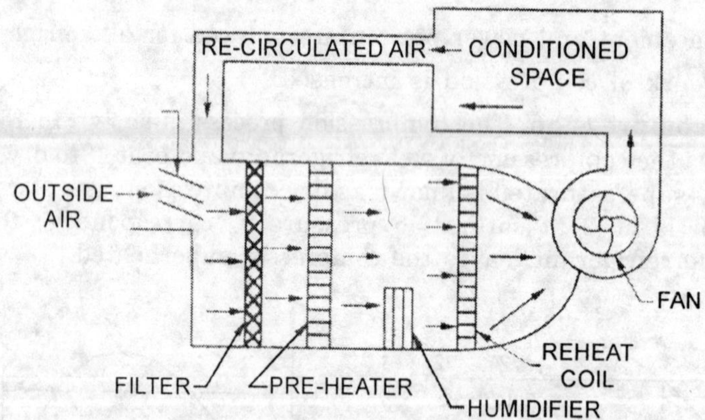

Fig. 8. Winter Air conditioning System

The different components for air-conditioning are :

1. Circulation fan
2. Air-conditioning unit
3. Supply duct
4. Supply outlets
5. Return outlets
6. Filters.

(*c*) The compressed air from the main compressor of an aircraft cooling system is bled off at 4.5 bar and 200°C. It is then passed through the heat exchanger in which the ram air is forced through for cooling purposes by the fan driven by the cooling turbine. The condition of the inlet to the cooling turbine is 4 bar and 30°C. The air expands in the cooling turbine is 80% and the flow rate in the turbine is 30 kg/min. By drawing the system and showing the process on T-s diagram, find

(*i*) The actual exit temperature from the cooling turbine.

(*ii*) The power delivered to the air blow fan.

(*iii*) The tons of refrigeration, if the cold air is tempered by mixing with by-passed warm air and delivered to the cabin cockpit area where it warms upto 25°C before exhausting out to waste. **(20)**

Sol.

Aircraft cooling system . The T-s diagram for the given data is shown in Fig. 9.

(*i*)
$$\frac{T_5}{T_4} = \left(\frac{p_5}{p_4}\right)^{(\gamma-1)/\gamma}$$

$$T_5 = 303\left(\frac{0.7}{4}\right)^{0.4/1.4}$$

$$= 184.15 \text{ K}$$

$$\text{Isentropic efficiency} = \frac{T_4 - T_5'}{T_4 - T_5}$$

$$0.8 = \frac{(303 - T_5')}{303 - 184.15}$$

$$T_5' = 207.92 \text{ K or } -65.08°\text{C}$$

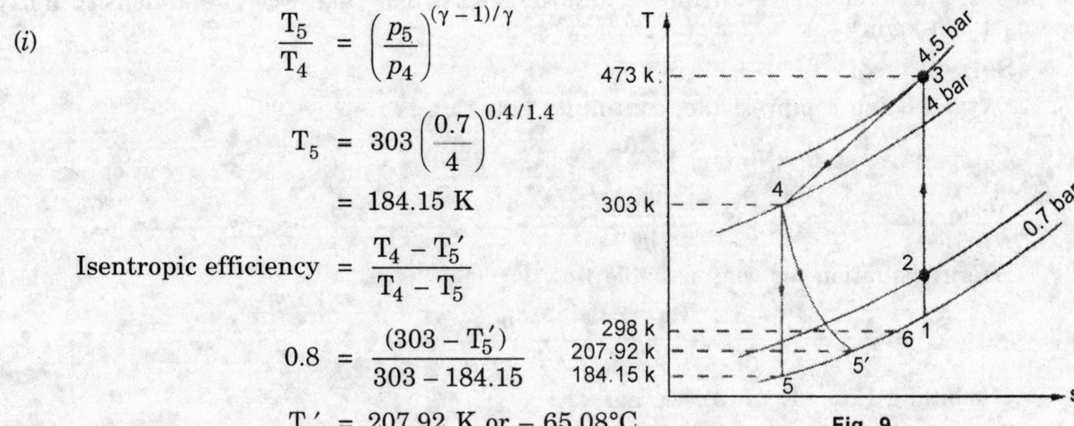

Fig. 9

(*ii*) Power delivered to the ram air blow fan

$$= m_a c_p (T_6 - T_5') = \left(\frac{30}{60}\right) \times 1.005\,(303 - 207.92) = 47.78 \text{ kW}$$

(*iii*) Refrigeration effect

$$= m_a c_p (T_6 - T_5') = 30 \times 1.005\,(298 - 207.92) = 2715.9 \text{ kJ/min}$$

or
$$= \frac{2715.9}{211} = 12.87 \text{ TR}$$

Q.5. (*a*) The quantity of water flowing through a channel is measure by rectangular notch weir. The water head over the notch is thrice the breadth of the rectangular notch. If the rectangular notch is replaced by a V-notch, calculate the angle of V-notch, other conditions remaining the same. The co-efficient of discharge for V-notch is 0.6 and for a rectangular notch is 0.64. (5)

Sol.

Rectangular notch,

$$Q_1 = \frac{2}{3}.C_{d1}\, b(2g)^{0.5}\, H^{3/2}$$

$$b = \frac{H}{3}$$

$$Q_1 = \frac{2}{9}.C_{d1}\,(2g)^{0.5}\, H^{5/2}$$

V-notch,
$$Q_2 = \frac{8}{15}.C_{d2}\tan\theta\,(2g)^{0.5}\, H^{5/2}$$

For $Q_1 = Q_2$, we get

$$\frac{2}{9} \times 0.64 = \frac{8}{15} \times 0.6 \times \tan\theta$$

$$\tan\theta = 0.444$$

$$\theta = 23.96° = 24°$$

(*b*) Calculate the mass flow rate of oxygen through a 100 mm diameter horizontal pipe if the inlet pressure is 1.06 m of mercury. The diameter at the throat is 30 mm. The ratio of pressure between the venturimeter and the inlet is 0.9. Take $\gamma = 1.4$ and density of oxygen gas is 1.429 kg/m^3. (15)

Sol.

Oxygen being compressible, continuity equation is :

$$\rho_1 A_1 v_1 = \rho_2 A_2 v_2 \quad ...(1)$$

Also,

$$\frac{p_1}{p_2} = \left(\frac{\rho_1}{\rho_2}\right)^{\gamma} \quad ...(2)$$

Energy equation for compressible flow is :

$$\frac{1}{2}v_1^2 + \frac{\gamma}{\gamma-1}.\frac{p_1}{\rho_1} = \frac{1}{2}v_2^2 + \frac{\gamma}{\gamma-1}.\frac{p_2}{\rho_2} \quad ...(3)$$

Combining Eqn. (1) to (3), we get

Rate of mass flow,

$$m = C_d A_2 \rho_2 \left[\frac{\left(\frac{2\gamma}{\gamma-1}\right).\frac{p_1}{\rho_1}\left\{1-\left(\frac{p_2}{p_1}\right)^{\frac{\gamma-1}{\gamma}}\right\}}{1-\left(\frac{A_2}{A_1}\right)^2\left(\frac{p_2}{p_1}\right)^{\frac{2}{\gamma}}}\right]^{\frac{1}{2}}$$

From Eqn. (2), we have

$$\rho_2 = \rho_1\left(\frac{p_2}{p_1}\right)^{\frac{1}{\gamma}}$$

Assuming $C_d = 1.0$

$$\dot{m} = \frac{\pi}{4}\times(30\times10^{-3})^2\times1.429\times(0.9)^{1/1.4} \times\left[\frac{\frac{2\times1.4}{0.4}\times\frac{1.394\times10^5}{1.429}\times\{1-0.9\}^{0.286}}{1-\left(\frac{30}{100}\right)^2\times(0.9)^{1.42}}\right]^{0.5}$$

$$= 0.1388 \text{ kg/s}$$

(*c*) Two inward flow reaction turbines have the same diameters viz 0.7 m and the efficiency. Both the runners work under the same head and they have the same velocity of flow, viz 5.8 m/s. One of the runners, A revolves at 500 rpm and the inlet blade angle is 65°. If the other runner, B has an inlet blade angle of 110°, at what speed should it run? (20)

Sol.

Ist runner :

$$u_1 = \frac{\pi\times0.7\times500}{60} = 18.326 \text{ m/s}$$

$$v_{u1} = u_1 + \frac{v_{f1}}{\tan \beta_1} = 18.326 + \frac{5.8}{\tan 65°} = 21.03 \text{ m/s}$$

$$\text{Work done} = v_{u1}\, u_1 = 21.03 \times 18.326$$

$$= 385.4 \text{ N.m}$$

2nd runner :

$$v'_{u1}\, u'_1 = 385.4 \text{ N-m}, \; v'_{f1} = 5.8 \text{ m/s}$$

$$\frac{v'_{f1}}{u'_1 - v'_{u1}} = \tan(180° - 110°)$$

$$u'_1 - v'_{u1} = \frac{5.8}{\tan 70°} = 2.111$$

$$(u'_1)^2 - \frac{385.4}{u'_1} = 2.111$$

$$(u'_1)^2 - 2.111\, u'_1 - 385.4 = 0$$

$$u'_1 = 0.5\,[2.111 + (2.111^2 + 4 \times 385.4)^{0.5}]$$

$$= 20.175 \text{ m/s}$$

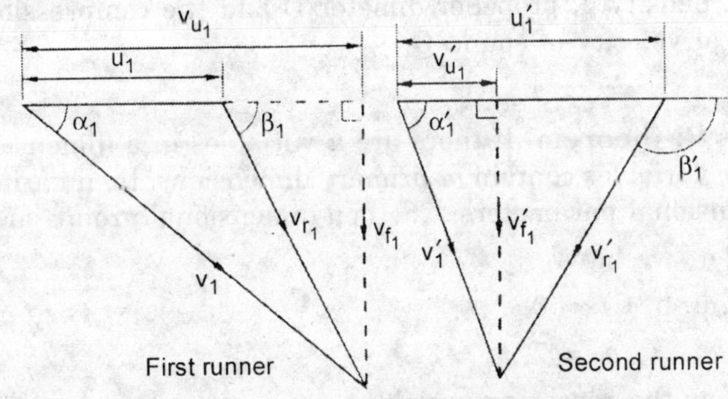

Fig. 10

$$\frac{\pi \times 0.7\, l\, N'_1}{60} = 20.715$$

$$N'_1 = 565.2 \text{ rpm}$$

Q.6. (*a*) Explain the working principle of Hot-wire Anemometer to measure the velocity of fluid with the help of a sketch. (10)

Sol.

Hot Wire Anemometer. It is used for the measurement of mean and fluctuating velocity components in a flow field. The sensor is a 5 micron diameter platinum-tungsten wire welded between the two prongs of the probe and heated electrically as a part of Wheatstone bridge circuit. When the probe is introduced into the flowing fluid, it tends to be cooled by the instantaneous velocity and consequently there is a tendency for the electrical resistance to decrease, which is a direct measure of the flow velocity. A hot wire probe is shown Fig. 11 (*a*).

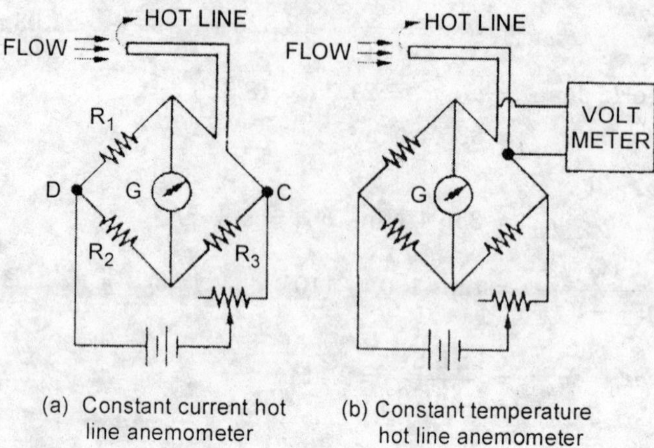

(a) Constant current hot line anemometer (b) Constant temperature hot line anemometer

Fig. 11. Hot wire Probe.

A hot wire anemometer may be operated either on the constant current mode or constant temperature mode, as shown in Fig. (*b*).

(*b*) State the Buckingham π-theorem and hence obtain an expression for the thrust developed by a propeller which depends upon the angular velocity ω, approach velocity V, dynamic viscosity μ, density ρ, propeller dimeter D and the compressibility of the medium measured by the local velocity of sound C. **(15)**

Sol.

Buckingham's Pi-theorem. If there are n variables in a dimensionally homogeneous equation and if these variables contain m primary dimensions, then variables can be grouped into $(n\text{-}m)$ non-dimensional parameters. The non-dimensional groups are called π-terms.

For the propeller, we have

$$F(T, \omega, V, D, \mu, \rho, C) = 0$$

Here $n = 7,\ m = 3$

Choose V, D, ρ as the repeating variables.

(*i*)
$$\pi_1 = V^{a1}\ D^{b1}\ \rho^{c1}\ T$$
$$M^0L^0T^0 = [LT^{-1}]^{a1}\ [L]^{b1}\ [ML^{-3}]^{c1}\ \ MLT^{-2}$$
$$= M^{c1+1}\ L^{a1+b1-3c+1}\ T^{a1-2}$$
$$0 = c_1 + 1;\ 0 = a_1 + b_1 - 3c + 1;\ 0 = a_1 - 2$$
$$a_1 = -2,\ b_1 = -2,\ c_1 = -1$$
$$\pi_1 = V^{-2}\ D^{-2}\ \rho^{-1}\ T = \frac{T}{(V^2D^2\rho)}$$

(*ii*)
$$\pi_2 = V^{a2}\ D^{b2}\ \rho^{c2}\ \mu$$
$$M^0L^0T^0 = [LT^{-1}]^{a2}\ [L]^{b2}\ [ML^{-3}]^{c2}\ \ ML^{-1}T^{-1}$$
$$= M^{c2+1}\ L^{a2+b2-3c2+1}\ T^{a2-1}$$
$$0 = c_2 + 1;\ 0 = a_2 + b_2 - 3c_2 - 1;\ 0 = -a_2 - 1$$
$$a_2 = -1,\ b_2 = -1$$

$$\pi_2 = V^{-1} D^{-1} \rho^{-1} \mu = \frac{\mu}{(VD\rho)}$$

(*iii*)

$$\pi_3 = V^{a3} D^{b3} \rho^{c3} \omega$$

$$M^0L^0T^0 = [LT^{-1}]^{a3} [L]^{b3} [ML^{-3}]^{c3} T^{-1}$$

$$= M^{c3} L^{a3 + b3 - 3c3 - 1} T^{a3 - 1}$$

$$0 = c_3;\ 0 = a_3 + b_3 - 3c_3;\ 0 = - a_3 - 1$$

$$a_3 = -1,\ b_3 = 1,\ c_3 = 0$$

$$\pi_3 = V^{-1} D^1 \rho^0 \omega = \frac{D\omega}{V}$$

(*iv*)

$$\pi_4 = V^{a4} D^{b4} \rho^{c4} C$$

$$M^0L^0T^0 = [LT^{-1}]^{a4} [L]^{b4} [ML^{-3}]^{c4} T^{-1}$$

$$= M^{c4} L^{a4 + b4 + 3c4 + 1} T^{-a4 - 1}$$

$$0 = c_4;\ 0 = a_4 + b_4 - 3c_4;\ 0 = - a_4 - 1$$

$$a_4 = -1,\ b_4 = 0$$

$$\pi_4 = V^{-1} D^1 \rho^0 C = \frac{C}{V}$$

$$\phi = \left[\frac{\mu}{(VD\rho)}, \frac{D\omega}{V}, \frac{C}{V}\right]$$

$$\frac{T}{(V^2D^2\rho)} = \phi\left[\frac{\mu}{(VD\rho)}, \frac{D\omega}{V}, \frac{C}{V}\right]$$

$$T = (V^2D^2)\,\phi\left[\frac{u}{(VD\rho)}, \frac{D\omega}{V}, \frac{C}{V}\right] = (V^2D^2)\,\phi\left[\frac{1}{Re}, \frac{D\omega}{V}, \frac{1}{M}\right]$$

(**c**) A Pelton turbine is to work at the foot of a dam whose reservoir level is 220 m. The head at the full opening at the turbine nozzle is 200 m and the co-efficient of velocity is 0.98. The turbine is to operate at 200 rpm and develop a power of 3.7 MW. Assuming the blade to jet speed ratio as 0.46, estimate the wheel diameter. If the blade outlet angle is 16°, determine the blade and hydraulic efficiencies Neglect frictional losses. (15)

Sol.

Given : $H = 200$ m, $K_v = 0.98$, $N = 200$ rpm,

$$P = 3.7 \text{ MW},\ \rho = \frac{u}{v_1} = 0.46,\ \beta_2 = 16°$$

$$v_1 = K_v (2gH)^{0.5} = 0.98\ (2 \times 9.81 \times 200)^{0.5}$$

$$= 61.389 \text{ m/s}$$

$$Q = A_1 v_1 = \frac{\pi}{4}.0.2^2 \times 61.389 = 1.9286 \text{ m}^3$$

$$P = wQH\eta_t$$

Turbine efficiency,

$$\eta_t = \frac{3.7 \times 10^6}{9810 \times 1.9286 \times 200} = 0.9778 \text{ or } 97.78\%$$

$$u = 0.46 \times 61.389 = 28.239 \text{ m/s}$$

$$= \frac{\pi DN}{60}$$

$$D = \frac{28.239 \times 60}{\pi \times 200} = 2.697 \text{ m}$$

Hydraulic efficiency,

$$\eta_{hyd} = 2(\rho - \rho^2)(1 + K \cos \beta_2) = 2(0.46 - 0.46^2)(1 + 1 \times \cos 16°)$$

$$= 0.9743 \text{ or } 97.43\%$$

Q.7. (*a*) What is the purpose of speed governing in steam turbine ? Discuss hydromechanical governing loop? **(10)**

Sol.

Speed governing of steam turbines. The purpose of speed governing is to maintain the speed of a turbine fairly constant irrespective of the load. This is essential to keep the frequency of the output electric power constant. The principal methods of steam turbine governing are : throttle governing, nozzle governing, by-gass governing, and combination of above methods.

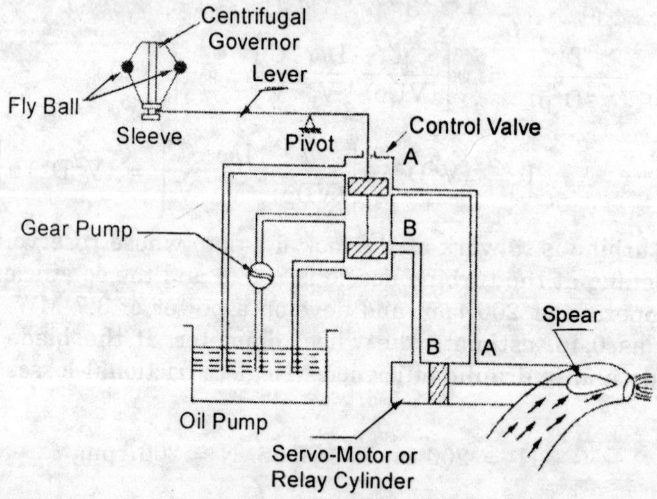

Fiig. 12. Hydromechanical loop for Throttle govering of steam turbines.

Hydromechanical Governing Loop. The hydromechanical loop for throttle governing is shown in Fig. 12. The throttle valve will open to take care of the full rated load at constant speed. If the load is reduced, the energy supplied to the turbine will be in excess and the turbine rotor will accelerate. Thus the governor sleeve will lift and the pilot piston valve spindle will get lifted opening the upper port to oil pressure and lower port to oil return. The relay piston will thus close the throttle valve partially. The lowering of throttle valve spindle will lower the pilot piston spindle and close the ports. As soon as the ports are closed, the relay piston gets stabilized in one position corresponding to the reduced load.

(*b*) Derive the expressions for maximum blade efficiency and maximum power output from a single stage impulse turbine. Give physical explanation. (10)

Sol.

Maximum blade efficiency and Power of a Single stage. The velocity diagrams of an impulse turbine as shown in Fig. 13.

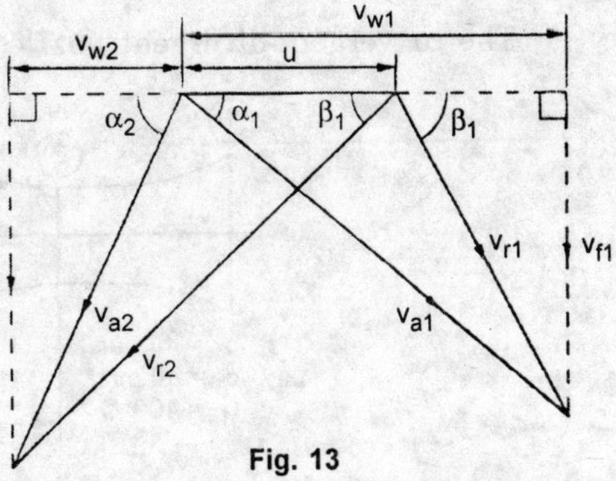

Speed ratio, $\rho = \dfrac{u}{v_{a1}}$

Blade velocity co-efficient,

$$K = \frac{v_{r2}}{v_{r1}}$$

$$C = \frac{\cos\beta_2}{\cos\beta_1}$$

Fig. 13

Work done per kg of steam,

$$w = (v_{w1} + v_{w2})\,u$$

Power developed, $P = \dfrac{(v_{w1} + v_{w2})u}{10^3}$ kW

Diagram or blade efficiency,

$$\eta_b = \frac{\text{work done on blades}}{\text{energy supplied}} = \frac{(v_{w1} + v_{w2})\,u}{\left[\frac{1}{2}.v^2_{a1}\right]} = \frac{2u\,(v_{w1} + v_{w2})}{v^2_{a1}}$$

Now,

$$\begin{aligned} v_{w1} + v_{w2} &= (v_{r1}\cos\beta_1 + u) + (v_{r2}\cos\beta_2 - u) \\ &= v_{r1}\cos\beta_1 + v_{r2}\cos\beta_2 = v_{r1}\cos\beta_1\,[1 + KC] \\ &= (v_{a1}\cos\alpha_1 - u)\,(1 + KC) \end{aligned}$$

$$\eta_b = \frac{2u\,(v_{a1}\cos\alpha_1 - u)(1 + KC)}{v^2_{a1}} = 2\rho\,(\cos\alpha_1 - \rho)\,(1 + KC)$$

For blade efficiency to be maximum,

$$\frac{d\eta_b}{d\rho} = 0$$

or $2(1 + KC)(\cos\alpha_1 - 2\rho) = 0$

$$\rho = 0.5\cos\alpha_1$$

$$(\eta_b)_{max} = 0.5\cos^2\alpha_1\,(1 + KC)$$

For symmetrical blading, $\beta_1 = \beta_2$ and C = 1. With no friction loss, K = 1.

$$(\eta_b)_{max} = \cos^2\alpha_1$$

Work done for K = 1 and C = 1 is,

$$w_{max} = 2u^2$$

Thus, maximum blade efficiency is dependent on nozzle angle only and maximum work done on u only.

(*c*) A steam nozzle receives steam at 40 bar and 400°C at an initial velocity of 40 m/s. The final pressure of steam is 10 bar. The mass flow rate of steam is 2 kg/s. The nozzle efficiency is 90%. The cross-section of the nozzle is circular. The angle of divergence is 6°. Calculate the throat and exit diameters and the length of divergent portion. Show the representation of process on h-*s* diagram and sketch the nozzle and label the dimensions calculated. (20)

Sol.

The convergent-divergent nozzle is shown in Fig. 14 and *h-s* diagram in Fig. 15.

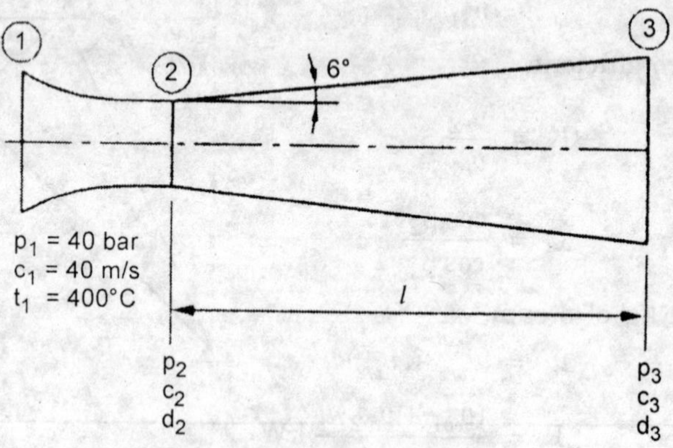

Fig. 14.

From steam tables and Mollier diagram, we have

$$h_1 = 3215.7 \text{ kJ/kg for } p_1 = 40 \text{ bar and } t = 400°\text{C}$$

$$p_2 = 0.5457\, p_1 = 0.5457 \times 40 = 21.828 \text{ bar,}$$

for superheated steam.

$$h_2 = 3025 \text{ kJ/kg, } v_{s2} = 0.125 \text{ m}^3\text{/kg}$$

$$c_2 = [2\,(h_1 - h_2) \times 10^3 + c^2_{\ 1}]^{0.5}$$

$$= [2\,(3215.7 - 3025) \times 10^3 + 40^2]^{0.5} = 618.87 \text{ m/s}$$

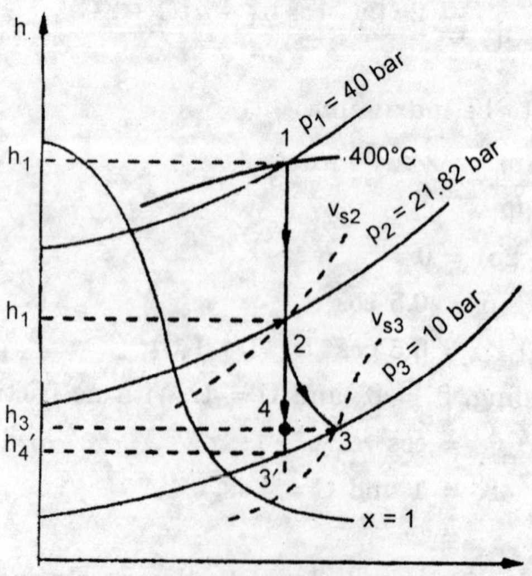

Fig. 15. *h-s* diagram

$$m_s = \frac{A_2 c_2}{v_{s2}}, A_2 = \frac{2 \times 0.125}{618.87} = 4.04 \times 10^{-4} \text{ m}^2$$

$$d_2 = \left[\frac{4 A_2}{\pi}\right]^{0.5} = \left[\frac{4 \times 4.04 \times 10^{-4}}{\pi}\right]^{0.5} = 22.68 \text{ mm}$$

$$2 - 4 = 0.9 (2 - 3')$$

$$h_3 = 2880 \text{ kJ/kg}, v_{s3} = 0.214 \text{ m}^3\text{/kg}$$

$$c_3 = [2 (3025 - 2880) \times 10^3 + (618.87)^2]^{0.5}$$

$$= 820 \text{ m/s}$$

$$A_3 = \frac{2 \times 0.214}{820} = 5.22 \times 10^{-4} \text{ m}^2$$

$$d_3 = 25.8 \text{ mm}$$

$$l = \frac{d_3 - d_2}{2 \tan 6°} = \frac{25.8 - 22.68}{2 \tan 6°} = 14.74 \text{ mm}$$

Q.8. (*a*) Why are downcomers fewer in number and bigger in diameter, while risers are more in number and smaller in diameter in a steam generator? (5)

Sol.

The hot flue gases rise upwards through the risers. For a better heat transfer from hot gases to boiler water larger surface area is provided by having number of risers of smaller size.

The downcomers are few in number and larger is size so that the heat transfer from the superheated steam to surroundings is minimum. More area is also required for superheated steam to accomodate more specific volume.

(*b*) A steam turbine is governed by throttling. The full load output of a steam turbine, measured at the coupling is 5 MW and the losses due to bearing friction, the governor and oil pump drive, etc., is 200 kW. The steam is supplied at a pressure is 20 bar and 300°C. The exhaust pressure of steam is 0.07 bar. The internal efficiency ratio at full load is 0.75. Calculate the coupling power of turbine when the steam flow through the turbine is 20% of that at full load. Assume that the external losses are the same as that at full load, exhaust pressure is the same and the internal efficiency ratio is reduced to 70% at this load. (15)

Sol.

From Mollier diagram, at 20 bar pressure and 300°C, h_1 = 3020 kJ/kg.

At 0.07 bar pressure, h_2 = 2100 kJ/kg

At full load :

Internal efficiency, $\eta_i = \frac{h_1 - h_2'}{h_1 - h_2} = \frac{\Delta h_i}{\Delta h_{isen}}$

$$0.75 = \frac{3020 - h_2'}{3020 - 2100}$$

$$h_2' = 2330 \text{ kJ/kg}$$

$$\Delta h_i = 690 \text{ kJ/kg}$$

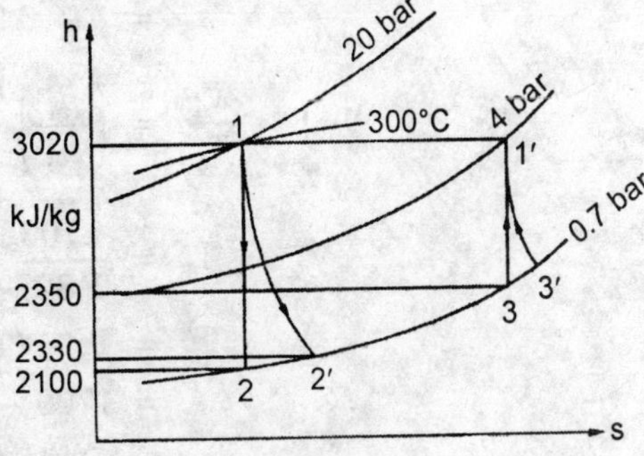

Fig. 16

Internal power developed,

$$P_i = \dot{m}\,.\Delta h_i = 5000$$

$$m = \frac{5000}{690} = 7.246 \text{ kg/s}$$

Nozzle box pressure at part load,

$$p_1' = \text{Steam mass flow ratio} \times p_1$$
$$= 0.2 \times 20 = 4 \text{ bar}$$
$$\Delta h'_{isen} = h_1 - h'_3 = 3020 - 2350 = 670 \text{ kJ/kg}$$
$$P_1' = \eta'_i \,.\, m' \,.\, \Delta h'_{isen}$$
$$= 0.7 \times 7.246 \times 0.2 \times 6701$$
$$= 679.71 \text{ kW}$$

Coupling power $= 679.71 - 200 = 479.71$ kW

(*c*) An opne cycle gas turbine employs a regenerative arrangement. The air enters the compressor at 1 bar and 288 K and is compressed to 10 bar with a compression efficiency of 85%. The air is heated in the regenerator and the combustion chamber till its temperature is raised to 1700 K and during the process pressure falls by 0.2 bar. The air is then expanded in the turbine and passes to regenerator which has 75% effectiveness and causes a pressure drop of 0.2 bar. The isentropic efficiency of turbine is 86%. By sketching the gas turbine system and showing the process on T-s diagram, calculate the thermal efficiency and power output mass flow rate alternator efficiency as 98% each, c_{pg} = 1.15 kJ/kg.K and c_{pa} = 1.005 kJ/kg.K. (20)

Sol.

$$T_2 = T_1\left(\frac{p_2}{p_1}\right)^{(\gamma-1)/\gamma} = 288\left(\frac{10}{1}\right)^{0.4/1.4} = 556 \text{ K}$$

$$T_2' = T_1 + \frac{(T_2 - T_1)}{\eta_c}$$

$$= 288 + \frac{(556 - 288)}{0.85} = 603.3 \text{ K}$$

$$p_3' = 10 - 0.2 = 9.8 \text{ bar} = p_4$$

$$p_5 = 1 + 0.2 = 1.2 \text{ bar}$$

$$\frac{T_4}{T_5} = \left(\frac{p_4}{p_5}\right)^{(\gamma-1)/\gamma} = \left(\frac{9.8}{1.2}\right)^{0.4/1.4} = 1.822$$

$$T_5 = \frac{1700}{1.822} = 932.96 \text{ K}$$

$$T_5' = T_4 - (T_4 - T_5)\,\eta_t$$
$$= 1700 - (1700 - 930.96) \times 0.86$$
$$= 1040.34 \text{ K}$$

Work consumed by the compressor,

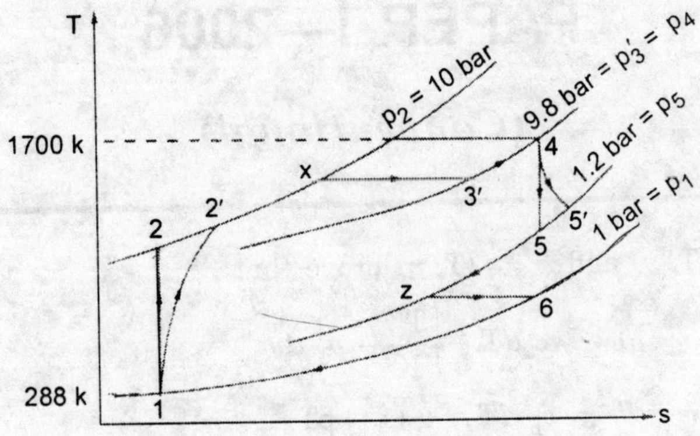

Fig. 17

$$w_c = c_{pa}(T_2' - T_1)$$
$$= 1.005\,(603.3 - 288)$$
$$= 316.876 \text{ kJ/kg}$$

Work done by turbine,

$$w_t = c_{pg}(T_4 - T_5')$$
$$= 1.15\,(1700 - 1040.34)$$
$$= 758.61 \text{ kJ/kg}$$

Net work $= w_t - w_c = 758.61 - 316.876 = 441.73$ kJ/kg

Regeneration effectiveness,

$$\varepsilon_r = \frac{T_3' - T_2'}{T_5' - T_2'} = 0.75$$

$$T_3' = 0.75\,(1040.34 - 603.3) + 603.3 = 931.08 \text{ K}$$

Heat supplied

$$= c_{pg}(T_4 - T_3') = 1.15\,(1700 - 931.08) = 884.26 \text{ kJ/kg}$$

Thermal efficiency = Net work/Heat supplied

$$= \frac{441.73}{884.26} = 0.4995 \text{ or } 49.95\,\%$$

Power developed $= 100 \times 441.73 \times 0.98 \times 0.98 \times 0.4995 \times 10^{-3}$

$= 21.19$ MW

MECHANICAL ENGINEERING
PAPER I—2006

(Conventional)

Q.1. **(*a*)** If $u = f(\mathrm{T}, v)$ and $h = f(\mathrm{T}, p)$ prove that

$$du = c_v d\mathrm{T} + \left\{\frac{\mathrm{T}\beta}{\mathrm{K}} - p\right\} dv$$

$$dh = c_p\, d\mathrm{T} + v\,(1 - \beta\mathrm{T}) \qquad \textbf{(15)}$$

Sol.]

(*a*)

$$u = f(\mathrm{T}, v)$$

$$du = \left(\frac{\partial u}{\partial \mathrm{T}}\right)_s \times d\mathrm{T} + \left(\frac{\partial u}{\partial v}\right)_\mathrm{T} \times dv = c_v \times d\mathrm{T} \times \left(\frac{\partial u}{\partial v}\right)_\mathrm{T} \times dv \qquad \text{...(1)}$$

Because $\left(\frac{\partial u}{\partial \mathrm{T}}\right)_v = c_v$

Also

$$du = \mathrm{T} \times ds - p \times dv$$

$$= \mathrm{T}\left[\left(\frac{\partial}{\partial \mathrm{T}}\right)_v \times d\mathrm{T} + \left(\frac{\partial s}{\partial v}\right)_\mathrm{T} \times dv\right] - p \times dv \qquad \text{...(2)}$$

Comparing Eqns. (1) and (2) and equating coefficient, we have

$$c_v = \mathrm{T}\left(\frac{\partial s}{\partial \mathrm{T}}\right)_v$$

$$\left(\frac{\partial u}{\partial v}\right)_\mathrm{T} = \mathrm{T}\left(\frac{\partial s}{\partial v}\right)_\mathrm{T} - p = \mathrm{T}\left(\frac{\partial p}{\partial \mathrm{T}}\right)_v - p$$

Equation (1) becomes

$$du = c_v \times d\mathrm{T} + \left[\mathrm{T}\left(\frac{\partial p}{\partial \mathrm{T}}\right)_v - p\right] dv$$

Now volume expansivity, $\beta = \left(\frac{1}{v}\right)\left(\frac{\partial v}{\partial \mathrm{T}}\right)_p$

$$\frac{\beta}{k} = \left(\frac{\partial p}{\partial \mathrm{T}}\right)_v$$

$$du = c_v \times d\mathrm{T} + \left[\frac{\mathrm{T}\beta}{k} - p\right] dv \qquad \text{...(3)}$$

Now $h = f(\mathrm{T}, p)$

$$dh = \left(\frac{\partial h}{\partial \mathrm{T}}\right)_p \times d\mathrm{T} + \left(\frac{\partial h}{\partial p}\right) \times dp$$

$$= c_p \times d\mathrm{T} + \left(\frac{\partial h}{\partial p}\right)_p \times dp \qquad ...(4)$$

Because $\left(\frac{\partial h}{\partial \mathrm{T}}\right)_p = c_p$

Also $dh = \mathrm{T} \times ds + v \times dp$

$$\mathrm{T} = \left[\left(\frac{\partial s}{\partial \mathrm{T}}\right)_p \times d\mathrm{T} + \left(\frac{\partial s}{\partial p}\right)_\mathrm{T} \times dp\right] + v \times dp \qquad ...(5)$$

Comparing Eqns. (4) and (5) and equatinbg co-efficient, we have

$$c_p = \mathrm{T}\left(\frac{\partial s}{\partial \mathrm{T}}\right)_p$$

$$\left(\frac{\partial h}{\partial p}\right)_\mathrm{T} = \mathrm{T}\left(\frac{\partial s}{\partial p}\right)_\mathrm{T} + v = \mathrm{T}\left(\frac{\partial s}{\partial p}\right)_\mathrm{T} + v$$

From Maxwell relations,

$$\left(\frac{\partial s}{\partial p}\right)_\mathrm{T} = v - \left(\frac{\partial v}{\partial \mathrm{T}}\right)_p$$

$$\left(\frac{\partial h}{\partial p}\right)_\mathrm{T} = v - \left(\frac{\partial v}{\partial \mathrm{T}}\right)_p$$

Eqn. (4) becomes

$$dh = c_p \times d\mathrm{T} + \left[v - \mathrm{T}\left(\frac{\partial v}{\partial \mathrm{T}}\right)\right] dp$$

$$= c_p \times d\mathrm{T} + v\,[1 - \beta\mathrm{T}]\,dp \qquad ...(6)$$

(*b*) A house is to be maintained at 25°C in summer as well as in winter. For this purpose, it is proposed to use a reversible device as a refrigerator in summer and as heat pump in winter. The ambient temperature is 40°C in summer and 3°C in winter. The enrgy losses as heat from the roof and the walls is estimated at 5 kW per degree Celsius temperature difference between the room and the ambient conditions. Calculate the power required to operate the device in summer and in winter. (10)

Sol.

For the cooled body, we have [Fig. 1 (*a*)]

$$(\mathrm{COP})_\mathrm{R} = \frac{\mathrm{T_L}}{\mathrm{T_H} - \mathrm{T_L}} = \frac{298}{313 - 298} = 19.867$$

$$\mathrm{Q_L} = 5 \times (40 - 25) = 75 \text{ kW}$$

Maximum power required $= \dfrac{Q_L}{(COP)_R}$

$$\frac{75}{19.867} = 3.775 \text{ kW}$$

For the heated body, we have [Fig. 1 (*b*)]

$$(COP)_P = \frac{T_L}{T_H - T_L}$$

$$= \frac{298}{298 - 276} = 13.54$$

$$Q_L = 5 \times (25 - 3) = 110 \text{ kW}$$

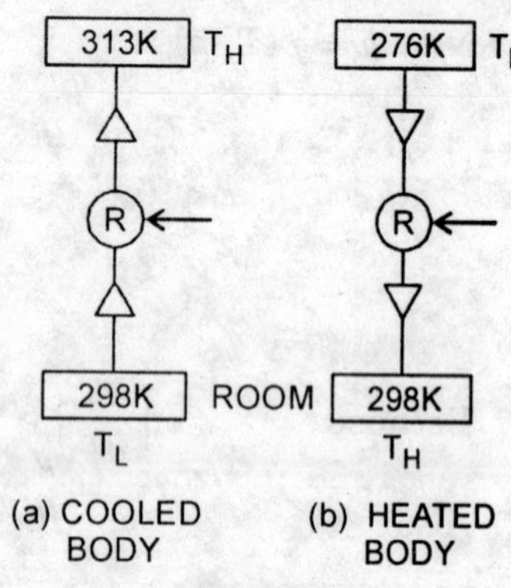

Fig. 1

Maximum power required $= \dfrac{Q_L}{(COP)_P} = \dfrac{110}{113.54} = 8.12 \text{ kW}$

(*c*) A cylindrical water tank of 50 cm diameter and 120 cm height is initially filled with water. The top of the tank is open to the atmosphere A discharge plug is fitted at the bottom of the tank. When the plug near the bottom of the tank is pulled out, a water jet streams out of the tank with a jet diameter of 1 cm. The mean velocity of the jet is given by $v = \sqrt{2gh}$ where h is the height of the water in the tank measured from the centre of the hole. Determine how long will it take for the water level in the tank to drop to 60 cm level from the bottom. Neglect the distance between the bottom of the tank and the centre of the hole compared to the total water height. **(15)**

Sol.

H_1 = initial height of liquid in the tank = 120 cm

H_2 = final height of liquid in the tank = 60 cm

t = time taken for liquid level to fall from H_1 to H_2, s

h = height of liquid at any instant above the orifice

dh = fall in height in time dt (See Fig. 2)

A = area of tank

$$= \left(\frac{\pi}{4}\right) D^2 = \left(\frac{\pi}{4}\right) \times 50^2 \times 10^{-4} \text{ m}^2$$

a = area of orifice

$$= \left(\frac{\pi}{4}\right) d^2 = \left(\frac{\pi}{4}\right) \times 1^2 \times 10^{-4} \text{ m}^2$$

Fig. 2

Volume of liquid leaving the tank in time dt = – A × dh

Discharge through orifice in time dt = $a\,(2gh)^{0.5} \times dt$

$$a\,(2gh)^{0.5} \times dt = -\text{A} \times dh$$

$$\int_0^t dt = \left(\frac{A}{a}\right)\left[\frac{1}{\sqrt{2g}}\right]\int_{60}^{120} \frac{dh}{h^{0.5}}$$

$$t = -\left(\frac{2500}{1}\right)\left[\frac{1}{(2\times 981)^{0.5}}\right][2h^{0.5}]_{60}^{120}$$

$$t = \left(\frac{2500\times 2}{44.294}\right)\times(120^{0.5} - 60^{0.5})\times = 362\ s$$

Q.2. (*a*) What are the advantages and disadvantages of using hydrogen as an I.C. Engine fuel? Explain one method by which hydrogen can be used in C.I. Engine. (10)

Sol. Hydrogen as I.C. Engine Fuel

Advantages ;

1. Low emission. Essentially no CO or HC in the exhaust as there is no carbon in the fuel. Most exhaust would be H_2O and N_2.
2. Fuel availability. There are a number of different ways of making hydrogen, including electrolysis of water.
3. Fuel leakage to environment is not a pollutant.
4. High energy content per volume when stored as a liquid. This would give a large range vehicle range for given fuel tank capacity.

Disadvantages :

1. Requirement of heavy, bulky, fuel storage both in vehicle and at the service station. Hydrogen can be sotred as a cryogenic liquid or as a compressed gas. If stored as a liquid, it would have to be kept under pressure at a very low temperature. This would require a thermally super - insulated fuel tank. Storing in a gas phase would require a high pressure vessel with limited capacity.
2. Difficult to refuel.
3. Poor engine volumetric efficiency. Any time a gaseous fuel is used in an engine, the fuel will displace some of the inlet air and poorer volumetric efficiency will result.
4. Fuel cost would be high at present-day tenchnology and availability.
5. High NO_x emissions because of high flame temperature.
6. Can detonate.

Use of Hydrogen in C.I. Engines. There are two methods by which hydrogen can be used in C.I. engines.

1. By introducting hydrogen with air and using a spray of diesel oil to ignite the mixture, that is, by the duel fuel mode. The limiting conditions are when the diesel quantity is too small to produce effective ignition, that is, failure of ignition and when the hydrogen-air mixture is so rich that the combustion becomes unacceptably violent. In between these limits a wide range of diesel to hydrogen proportions can be tolerated.
2. By introducing hydrogen directly into the cylinder at the end of compression. Since the self-ignition temperature of hydrogen is very high, the gas spray is made to impinge on a hot glow plug in the combustion chamber, that is, by surface ignition. It is also possible to feed a very lean hydrogen-air mixture during the intake into an engine and then inject the bulk of the hydrogen towards the end of the compression stroke.

(*b*) Mention three problems related to environment that are created by exhaust emissions from the I.C. Engine. What are the major exhaust emissions? (6)

Sol. Environmental Problems Created by Exhaust Emission from I.C. Engines The emissions exhausted into sorroundings pollute the atmosphere and cause the following problems :

1. Global warming.
2. Acid rain.
3. Smog.
4. Odours.
5. Respiratory and other health hazards.

Major exhaust emissions are :

(*i*) Unburnt hydrocarbons, (HC)

(*ii*) Oxides of carbon, (CO and CO_2)

(*iii*) Oxides of nitrogen, (NO and NO_2)

(*iv*) Oxides of sulphur, (SO_2 and SO_3)

(*v*) Particulates

(*vi*) Soot and smoke.

(*c*) What is delay period? Mention the various factors affecting the delay period. Discuss any four of them. (10)

Sol. Delay Period. The ignition delay period is the phase during which some fuel has already been admitted but has not yet ignited. The fuel does not ignite immediately upon injection into the combustion chamber. There is a definite period of inactivity between the time when the first droplet of fuel hits the hot air in the combustion chamber and the time it starts through the actual burning phase. This period is known as the *ignition daly period.* In Fig. 3, the delay period is shown on pressure crank angle (or time) diagram between points '*a*' and '*b*'. Point '*a*' represents the time of injection and point '*b*' represents the time at which the pressure curve first separates from the motoring curve.

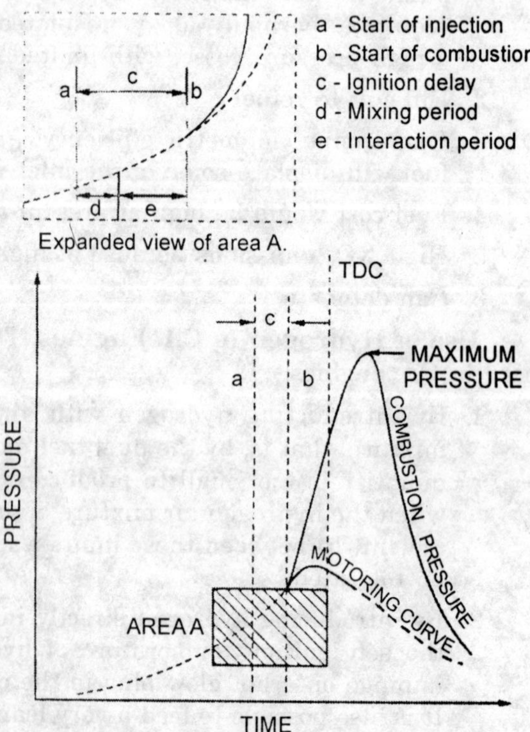

Fig. 3. Pressure-Time Diagram illustraiting Ignition Delay.

Factors Affecting Delay Period. The important factors are :

(*i*) Compression raito

(*ii*) Engine speed

(*iii*) Output

(*iv*) Atomization of fuel and duration of injection

(*v*) Injection timing

(*vi*) Quality of the fuel

(*vii*) Intake temperature

(*viii*) Intake pressure.

(*i*) Compression raito. With increase in compression ratio the minimum auto ignition temperature of a fuel decreases due to increased density of the compressed air. This results in a closer contact between the molecules of fuel and oxygen reducing the time of reaction. The increase in the compression temperature as well as the decrease in the minimum auto ignition temperature decrease the delay period. The peak pressure during the combustion process is only marginally affected by the compression ration (because delay period is shorter with higher compression ratio and hence the pressure rise is lower).

On of the practical disadvantages of using a very high compression ratio is that the mechanical efficiency tends to decrease due to increase in weight of the reciprocating parts. Therefore, in practice, the engine designers always try to use a lower compression ratio which helps in easy cold starting and light load running at high speeds.

(*ii*) Engine speed. The delay period decreases with increase in engine speed in a variable speed operation with a given fuel. With increase in engine speed, the loss of heat during compression decreases, resulting in the rise of both the temperature and pressure of the compressed air thus reducing the delay period.

(*iii*) Output. With an increase in engine output the air-fuel ratio decreases, operating temperatures increase and hence delay period decreases.

(*iv*) Quality of fuel. A lower self-ignition temperature results in a lower delay period. Also, fuels with higher cetane number give lower delay period and smoother engine operation. Other properties like volatility, latent heat, viscosity and surface tension also affect the delay period.

(***d***) Discuss the advantages and disadvantages of using LPG in SI Engine. (6)

Sol. Use of LPG in SI Engine.

Advantages :

(*i*) LPG contains less carbon than petrol. Therefore, emission is much reduced by its use.

(*ii*) LPG mixes with air at all temperature.

(*iii*) In multi-cylinder engines a uniform mixture can be supplied to all cylinders.

(*iv*) There is no crankcase dilution, because the fuel is in the form of vapour.

(*v*) Automobile engines can use propane, the fuel is in the form of vapour.

(*vi*) LPG has high antiknock characteristics.

(*vii*) Its heat energy is about 80% of gasoline, but its high octane value compensates the thermal efficiency of the engine.

(*viii*) Running on LPG translates into a cost saving of about 50%.

(*ix*) The engine may have a 50% longer life.

Disadvantages :

(*i*) Engines are normally designed to take in a fixed volume of the mixture and air. Therefore LPG will produce 10% less power for a given engine at full throttle.

(*ii*) The ignition temperature of LPG is somewhat higher than petrol. Therefore, running on LPG could lead to a 5% reduction in valve life.

(*iii*) A good cooling system is quite necessary, because LPG vaporizer uses engine coolant to provide the heat to convert the liquid LPG to gas.

(*vi*) The vehicle weight is increased due to the use of heavy pressure cylinder for storing LPG.

(*v*) A special fuel feed system is required for liquid petroleum gas.

(e) Air at 25 kPa and 230 K enters a turbojet engine with a velocity of 250 m/sec. The pressure ratio across the compressor is 12. The turbine inlet temperature is 1400 K and the pressure at the nozzle exit is 25 kPa. The diffuser, compressor, turbine and nozzle processes are isentropic and there is no pressure drop for flow through the combustor. Under steady state operating conditions, determine (*i*) the velocity at the nozzle exit, and (*ii*) the pressures and temperatures at each state. State the assumptions made. Draw the line diagram indicating all the componets and show the processes on a T–*s* diagram. (8)

Sol.

Diffuser – compressor. The line diagram of turbo-jet and the T-S diagram are shown in Fig. 4.

$$T_1 = T_0 + \frac{v_0^2}{2c_p} = 230 + \frac{250^2}{2 \times 1000 \times 1.005} = 261.1 \text{ K}$$

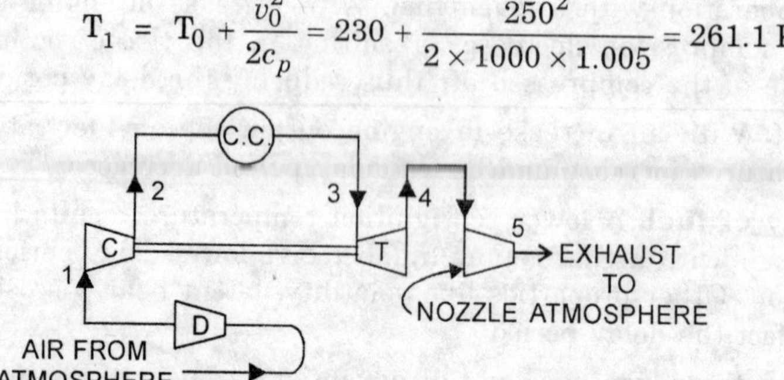

(a) LINE DIAGRAM OF TURBOJECT

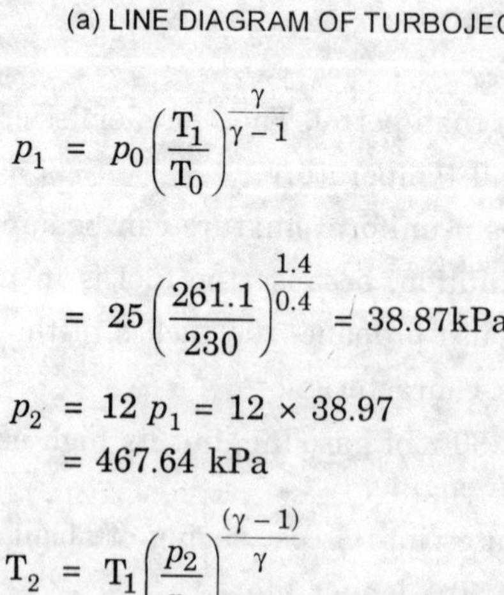

$$p_1 = p_0\left(\frac{T_1}{T_0}\right)^{\frac{\gamma}{\gamma-1}}$$

$$= 25\left(\frac{261.1}{230}\right)^{\frac{1.4}{0.4}} = 38.87 \text{kPa}$$

$$p_2 = 12\, p_1 = 12 \times 38.97$$

$$= 467.64 \text{ kPa}$$

$$T_2 = T_1\left(\frac{p_2}{p_1}\right)^{\frac{(\gamma-1)}{\gamma}}$$

$$= 261.1\ (12)0.4/1.4 = 531.05 \text{ K}$$

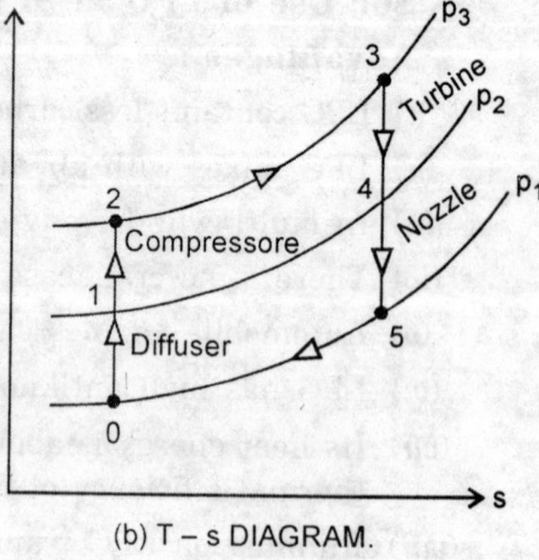

(b) T – s DIAGRAM.

Fig. 4

Combustor :

$$p_3 = p_2 = 467.64 \text{ kPa}$$

$$T_3 = 1400 \text{ K}$$

Turbine : Assuming compressor work = turbine work, we have

$$c_p\,(T_2 - T_1) = c_p\,(T_3 - T_4)$$

$$531.05 - 262.1 = 1400 - T_4$$

$$T_4 = 1130.05 \text{ K}$$

$$p_4 = p_3\left(\frac{T_4}{T_3}\right)^{\frac{\gamma}{(\gamma-1)}} = 467.64\left(\frac{1130.05}{1400}\right)^{\frac{1.4}{0.4}} = 220.95 \text{ kPa}$$

Nozzle :
$$\left(\frac{T_5}{T_4}\right) = \left(\frac{p_5}{p_4}\right)^{\frac{\gamma-1}{\gamma}}$$

$$T_5 = 1130.05\left(\frac{25}{220.95}\right)^{\frac{0.4}{1.4}} = 606.3 \text{ K}$$

$$h_4 + \frac{c_4^2}{2} = h_5 + \frac{c_5^2}{2}$$

$c_4 \approx 0$ as compared to c_5.

$$c_5 = [2\,(h_4 - h_5)]^{0.5} = [2 \times C_p\,(T_4 - T_5)]^{0.5}$$

$$= (2 \times 1000)^{0.5} \times [1.005\,(1130.5 - 606.3)]^{0.5}$$

$$= 1026 \text{ m/s}$$

Q.3. (*a*) Heat losses from the windows are to be reduced by covering them from inside with a polystyrene insulation (k_{ins} = 0.027 W/m-K). Consider application of 25 mm-thick insulation panels to 6 mm-thick windows (k_w = 1.4 W/m.K). The contact resistance between the glass and the insulation may be approximated as $R_{t,c}^{*}$ = 0.002 m²-K/W), while the converction coefficient at the outside surface of the window is nominally losing heat (h_0 = 20 W/m²-K). With the insulation, the convection co-efficient at the inner surface is h_i = 2 W/m²-K; without the insulation it is h_i = 5 W/m²-K.

(*i*) What is the percentage reduction in heat loss assocated with the use of insulation?

(*ii*) If the total surface area of the window is A_s = 12 m², what are the heat losses associated with insulated and unisulated windows for interior and exterior temperatures of $T_{\infty,\,i}$ = 20° an $T_{\infty,\,o}$ = 12°C ?

(*iii*) If the home is heated by a gas furance operating at an efficiency of η_f = 0.80 and the natural gas is priced at c_g = Re. 1 per MJ, what is the daily saving associated with convering windows for 12 hours ?

Sol. Without insulation. Overall heat transfer co-efficient,

$$\frac{1}{U} = \frac{1}{h_i} + \frac{\Sigma\Delta s}{k_w} + \frac{1}{h_0} = \frac{1}{5} + \frac{6 \times 10^{-3}}{1.4} + \frac{1}{20} = 0.2543$$

$$U = 3.9126 \text{ W/m}^2\text{-K}$$

Heat transferred,
$$Q = UA_s \times \Delta\theta$$

$$= 3.9326 \times 12 \times (20 + 12)$$

$$= 1510.08 \text{ W}$$

With insulation (See Fig. 5) :

$$\frac{1}{U} = \frac{1}{h_i} + \frac{\Sigma \Delta x}{k_w} + R^*_{t,c} + \frac{1}{h_0}$$

$$= \frac{1}{2} + \frac{6 \times 10^{-3}}{1.4} + \frac{25 \times 10^{-3}}{0..027} + 0.002 + \frac{1}{20} = 1.4822$$

$$U = 0.67467 \text{ W/m}^2\text{-K}$$

Heat transferred, $Q = UA_s \times \Delta\theta$

$$= 0.67467 \times 12 \times (20 + 12) = 259.08 \text{ W}$$

Fig. 5

(*i*) Reduction in heat loss = 1510.08 − 259.08 = 1251 W

(*ii*) Percentage reduction in heat loss

$$= \frac{1251}{1510.08} = 0.8284 \text{ or } 82.84\%$$

(*iii*) Daily saving $= \frac{1251 \times 12 \times 3600}{0.8 \times 10^6} = \text{Rs. } 67.5$

(***b***) Water at 20°C flows normal to the axis of a circular tube with a velocity of 1.5 m/sec. The diameter of the tube is 25 mm. Calculate the average heat transfer co-efficient if the tube surface is maintained at a uniform temperature of 80°C. Also estimate the heat transfer rate per unit length of the tube. Properties of water at

$$T_f = \frac{20 + 80}{2} = 50°\text{C are :}$$

Specific heat, c_p = 4.1813 kJ/kg-K

Kinematic viscosity, $\nu = 0.586 \times 10^{-6}$ m²/sec

Thermal conductivity, k = 0.6395 W/m-K

Prandtl number, Pr = 3.68

Density, ρ = 990 kg/m³

Dynamic viscosity of water (μ_w) at 80°C

$= 3.5456 \times 10^{-4}$ kg/m-sec

μ_∞ at 20°C $= 1.006 \times 10^{-3}$ kg/m-sec

$$\overline{NU} = (0.4\,Re^{0.5} + 0.06\,Re^{2/3})\,Pr^{0.4}\left\{\frac{\mu_\infty}{\mu_w}\right\}^{1/4} \quad (10)$$

Sol. Reynolds number, $Re = \frac{v d \rho}{\mu_\infty}$

$$= \frac{1.5 \times 25 \times 10^{-3} \times 990}{1.006 \times 10^{-3}} = 36903$$

Nusselt number, $Nu = (0.4\,Re^{0.5} + 0.06\,Re^{2/3})\,Pr^{0.4}\left(\frac{\mu_\infty}{\mu_\infty}\right)^{1/4}$

$$= [0.4 \times (36903)^{0.5} + 0.06 \times (36903)^{2/3}] \times 3.68^{0.4}$$

$$\times \left[\frac{1.006 \times 10^{-3}}{3.5456 \times 10^{-4}}\right]^{1/4} = 313.28$$

Now $\bar{N}u = \frac{\bar{h}d}{k} = \frac{\bar{h} \times 25 \times 10^{-3}}{0.6395} = 313.28$

$$\bar{h} = 8013.8 \text{ W/m}^2\text{.K}$$

$$\frac{Q}{L} = \pi d\bar{h} \times \Delta\theta = \pi \times 25 \times 10^{-3} \times 8013.74 \times (80 - 20)$$

$$= 27763.8 \text{ W/m length}$$

(*c*) An electrically heated sphere of 1.5 cm diameter is cooled in a quiescent medium of air 315 K. In order to maintain the surface temperature of the sphere at 385 K, estimate the amount of heat to be supplied by the electrical heater.

$$D = 1.5 \times 10^{-2} \text{ m}$$

$$T_\infty = 273 + 20 = 293 \text{ K}$$

The properties of air at $\frac{315 + 385}{2} = 350$ K are:

Kinematic viscosity, $\nu = 2.076 \times 10^{-5}$ m²/sec

Prandtl number, $Pr = 0.697$

Thermal conductivity, $k = 0.03$ W/m-K

Coefficient of thermal expansion,

$$\beta = \frac{1}{T_f} = 2.86 \times 10^{-3} \text{ K}^{-1}$$

Use the relationship

$$\overline{Nu} = 2 + 0.43\, Ra_D^{1/4}$$ (5)

Sol. Reyleight number, $Ra_D = Gr \times Pr$

Grashoff number, $Gr = \frac{d^3 g\beta \times \Delta\theta}{\nu^2}$

$$= (1.5 \times 10^{-2})^3 \times 9.81 \times 2.86 \times 10^{-3} \times \frac{(385 - 315)}{(2.076 \times 10^{-5})^2}$$

$$= 15379.85$$

Prandtl number, $Pr = 0.697$ (given)

$$Ra_D = 15379.85 \times 0.697 = 10719.17$$

$$\bar{N}u = 2 + 0.43 \times (10719.76)^{1/4} = 6.3754$$

$$\frac{\bar{h}D}{k} = 6.3754$$

$$\bar{h} = \frac{6.3754 \times 0.03}{1.5 \times 10^{-2}} = 12.75 \text{ W/m}^2.\text{K}$$

$$Q = A\bar{h} \times \Delta\theta = 4\pi\left(\frac{D}{2}\right)^2 \times \Delta\theta$$

$$= 4\pi \times (0.75 \times 10^{-2})^2 \times 12.75 \times (385 - 315) = 0.63 \text{ W}$$

(*d*) The configuration of a furance can be aproximated as an equilateral triangular duct which is sufficiently long so that end effects are negligible. The hot wall is maintained at T_1 = 1000 K and has an emissivity ε_1 = 0.8. The cold wall is at T_2 = 500 K and has an emissivity ε_2 = 0.8. The third wall is reradiating zone for which Q_3 = 0. Calculate the net radiation flux leaving the hot wall (10)

Sol. Radiation from surface 1 (See Fig. 6):

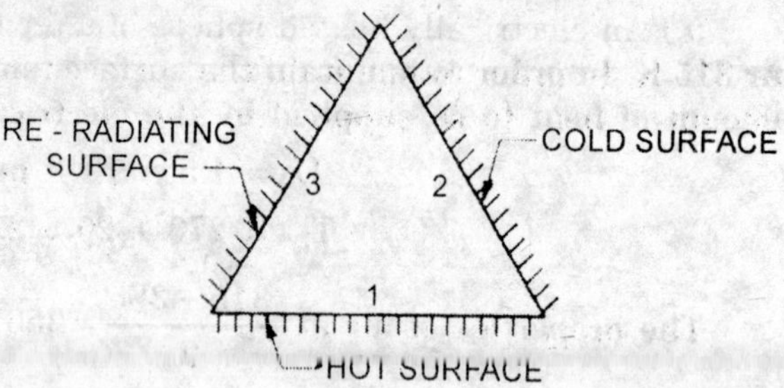

$$F_{1-1} + F_{1-2} + F_{1-3} = 1$$
$$F_{1-1} = 0$$
$$F_{1-2} + F_{1-3} = 1$$

By symmetry, $F_{1-2} = F_{1-3} = 0.5$

RE - RADIATING SURFACE
COLD SURFACE
3
2
1
HOT SURFACE

Fig. 6

Radiation from surface 2 :

$$F_{2-1} + F_{2-2} + F_{2-3} = 1$$
$$F_{2-2} = 0$$
$$F_{2-1} + F_{2-3} = 1$$
$$F_{2-3} = 1 - F_{2-1}$$
$$A_1F_{1-2} = A_2F_{2-1}$$

For $A_1 = A_2$, we get,

$$F_{1-2} = F_{2-1}$$
$$F_{2-3} = 1 - 0.5 = 0.5$$

For wall 3 to be re-radiating zone,

$$\bar{F}_{1-2} = F_{1-2} + \frac{F_{1-3} \times F_{3-2}}{1 - F_{3-3}} = 0.5 + \frac{0.5 \times 0.5}{1 - 0} = 0.75$$

$$\frac{q_{1-2}}{A} = \sigma\bar{F}_{1-2}\frac{(T_1^4 - T_2^4)}{\left[\frac{1}{\varepsilon_1} + \frac{1}{\varepsilon_2} - 1\right]}$$

$$= 5.67 \times 10^{-8} \times 0.75 \times \frac{(1000^4 - 500^4)}{\left[\frac{1}{0.8} + \frac{1}{0.8} - 1\right]} = 26578 \text{ W/m}^2$$

Q.4. (*a*) How can the solar energy be used to obtain refrigeration effect. Explain with a neat and clean sketch. A geothermal wall at 130°C supplies heat at a rate of 100, 500 kJ/h to an absorption refrigeration system. The environment is at 30°C and the refrigerated space is maintained at –22°C. Determine the maximum possible heat removal from the refrigerated space. (15)

Sol. Solar Refrigeration. Solar energy can be used to obtain refrigeration effect by either the vapour compression refrigeration system or the vapour absorption refrigeration system.

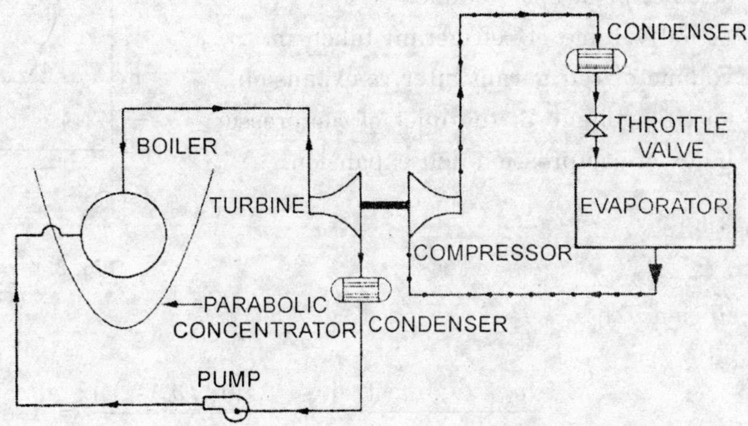

Fig. 7. Solar Vapour-Compression Refrigeratin System.

A line diagram of vapour compression system using solar energy is shown in Fig. 7. The solar energy is converted directly into mechanical work for operating conventional compressor of a conventional vapour compression refrigeration system. The shaft work to the compressor is provided by a steam turbine supplied with steam which is generated by a steam generator placed at the focus of a large parabolic mirror.

$$T_0 = 273 - 22 = 251 \text{ K}$$

$$T_H = 273 + 130 = 403 \text{ K}$$

$$T_L = 273 + 30 = 303 \text{ K}$$

$$\text{Maximum COP} = \left[\frac{T_H - T_L}{T_H}\right]\left[\frac{T_0}{T_L - T_0}\right]$$

$$= \left[\frac{403 - 303}{403}\right]\left[\frac{251}{303 - 251}\right] = 1.193$$

$$\text{Maximum possible heat removal} = \frac{100,500}{1.193}$$

$$= 84241.4 \text{ kJ/h or } 23.4 \text{ kW}$$

(***b***) Obtain an expression for the capacity of a refrigeration system (tonnage) in terms of the rate of upper and lower pressure limits, expansion index (n), v_c/v_s (v_c = clearance volume and v_s = swept volume), N (rpm), v_1 (specific volume at the inlet to compressor) and the refrigeration effect ($h_1 - h_4$).

If the pressire ratio is 6.5, $n = 1.1$, $v_c/c_s = 0.025$, N = 900 rpm, $v_s = 600$ cc, $v_1 = 0.078$ m^3/kg and refrigeration effect = 150 kJ/kg, calculate capacity of the system.

Sol. Consider the pressure – volume diagram of a single stage reciprocating compressor with clearance as shown in Fig. 8.

Let, p_1 = suction pressure of compressor.

p_2 = delivery pressure of compressor.

v_c = clearance volume.

v_s = strome (or swept) volume.

v_a = actual volume of refrigerant taken in.

v_e = volume of refrigerant after re-expansion.

v_1 = specific volume at the inlet of compressor.

n = index of compression and expansion.

Fig. 8. p-v diagram for air compressor.

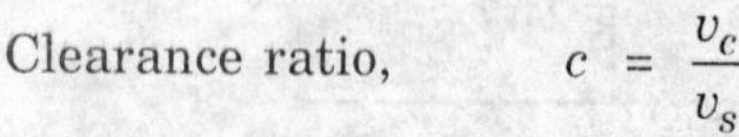

Clearance ratio, $c = \dfrac{v_c}{v_s}$

Volumetric efficiency, $\eta_v = \dfrac{v_a}{v_s}$

$$= \frac{[v_s - (v_e - v_c)]}{v_s} = \frac{[v_s - v_c\ \{(v_c / v_c) - 1\}]}{v_s} = 1 - \left(\frac{v_c}{v_s}\right)\left(\frac{v_e}{v_c} - 1\right)$$

Applying the law of expansion to points 3 and 4, we have

$$p_2\, v_c^n = p_1\, v_e^n$$

$$\frac{v_e}{v_c} = \left(\frac{p_2}{p_1}\right)^{1/n}$$

$$\eta_v = 1 - c\left[\left(\frac{p_2}{p_1}\right)^{1/n} - 1\right] = 1 + c - c\left(\frac{p_2}{p_1}\right)^{1/n}$$

Volume flow rate of piston,

$$V_s = \left(\frac{\pi d^2}{4}\right) \times L \times N/60 \text{ m}^3/\text{s}$$

where d = diameter of piston of compressor, m

L = length of stroke, m

N = speed of compressor, rpm

Mass flow rate, $m_r = \dfrac{V_s \times \eta_v}{v_1}$ kg/s

Capacity of the refrigeration system = $\dfrac{\dot{m}_r \times \text{refrigeration effect}}{3.5167}$ TR

For the given data, we have

$$\eta_v = 1 + 0.025 - 0.025\ (6.5)^{1/1.1} = 0.888$$

$$V_s = v_s \times N/60 = 600 \times 10^{-6} \times \frac{900}{60} = 0.009 \text{ m}^3/\text{s}$$

Actual volume flow rate,

$$V_a = V_s\ \eta_v = 0.009 \times 0.888 = 7.992 \times 10^{-3} \text{ m}^3/\text{s}$$

Mass flow rate, $\dot{m}_r = \frac{V_a}{v_1} = \frac{7.992 \times 10^{-3}}{0.078} = 0.10246$ kg/s

Capacity of the refrigeration system = $\dot{m}_r$ × refrigeration effect/3.5167

$$= \frac{0.10246 \times 150}{3.5167} = 4.37 \text{ TR}$$

(c) (i) List the various refrigeration and air-conditioning controls and their functions. (4)

Sol. Refrigeration and Air-conditioning Controls :

S.No.	RAC control	Functions
1.	Expapansion devices (*i*) Automatic expansion valve (*ii*) Thermostatic expansion valve	(*i*) It serves the thermodynamic function of expanding the liquid refrigerant from the condenser pressure to the evaporator pressure. (*ii*) To control the supply of the liquid to the evaporator at the rate at which it is evaporated.
2.	Cut-outs for regfrigeration devices (*i*) High pressure cut-out (*ii*) Low pressure cut-out	To check the rise and fall of pressure in a refrigeration system.
3.	Back pressure valves	To control the back pressure in the compressor.
4.	Liquid level regulating devices (*i*) High-side float valve (*ii*) Low-side float valve	To check the liquid level on the high and low sides of the compressor.
5.	Flow-regulating devices (*i*) Solenoid valves (*ii*) Check valves (*iii*) Water and brine valves	To regulate the flow of water and brine in the refrigeration system.
6.	Thermostats (*i*) Evaporator thermostat (*ii*) Room thermostat	To control the temperature.
7.	Humidity controls	To control the moisture content
8.	Motor controls	Act as a safety device for the electric motor by controlling its speed.m
9.	Modulating motors	For proportional or floating control.

(ii) Explain adiabatic saturation and thermodynamic wet-bulb temperature. When are the adiabatic saturation and wet-bulb temperatures equivalent for atmospheric air? (3)

Sol. Adiabatic Saturation Temperature, t^*. For any state of unsaturated moist air, there exists a temperature at which the air becomes adiabatically saturated by the evaporation of water into air, at exactly the same temperature, called the *adiabatic saturation temperature.*

Thermodynamic Wet Bulb Temperature, t'. The temperature measured by the wick-covered bulb of a thermometer indicates the temperature of liquid water in the wick and is called the wet bulb temperature.

$t^* = t'$ when the dimensionless quantity $\frac{f_g}{k_w c_p}$, called the Lewis number, is unity ;

where
f_g = heat transfer co-efficient,
k_w = diffusion co-efficient for mass transfer, and
c_p = specific heat at constant pressure.

(iii) What is metabolism? What is the range of metabolic rate for an average man? Why are we interested in the metabolic rate of the occupants of a building when we deal with heating and air-conditioning ? (3)

Sol. Metabolism. It is the process of generation of heat within the body of human beings on account of food consumed and respiration. The metabolism rate for an average man varies from 115 W to 265 W.

The occupants in a conditioned space give out heat at a metabolic rate that more of less depends on their rate of working. The relative proportion of the sensible and latent heats given out depends on the ambient DBT. The lower the DBT, the greater the heat given out as sensible heat. The metabolic rate adds cooling and heating loads for the building to be air-conditioned and therefore is of interest to the air-conditioning designer.

Q.5. *(a)* Show that for a flow governed only by gravity, inertia and pressure forces, the ratio of rates of flow in two dynamically similar systems equals to the 5/2 power of the length ratio. (8)

Sol. Gravity forces, $F_g = M \times g = \rho l^3 \times g$

Inertia force, $F_i = m \times a = \rho l^2 V^2$

Pressure forces, $F_p = \Delta_p \times l^2$

$$\frac{F_i}{F_g} = \frac{\rho l^2 \times V^2}{\rho l^3 \times g} = \frac{V^2}{lg}$$

Froude number, $$F = \frac{V}{\sqrt{lg}}$$

Euler number, $$E = \frac{F_i}{F_p} = \frac{\rho l^2 V^2}{\Delta p \times l^2} = \frac{\rho V^2}{\Delta p}$$

The discharge depends on velocity and area of cross-section.

From Froude number, for dynamic similarity, we have

Velocity ratio, $V_r = \frac{V_p}{V_m} = \sqrt{\frac{l_p}{l_m}} = \sqrt{l_r}$ ...(1)

where $l_r = \frac{l_p}{l_m}$ = length ratio

Time ratio, $t_r = \frac{l_p}{l_m} = \left(\frac{l_p}{l_m}\right) \times \left(\frac{V_m}{V_p}\right) = \frac{l_r}{\sqrt{l_r}} = \sqrt{l_r}$ ...(2)

$$Q = AV = \frac{l^2 \times l}{t} = \frac{l^3}{t}$$

Discharge ratio, $Q_r = \frac{l_r^3}{t_r} = \frac{l_r^3}{\sqrt{l_r}} = l_r^{5/2}$

Hence proved.

(*b*) The rate of flow Q of a liquid having mass density ρ and viscosity μ is to be measured by means of an orificemeter. The diameter of the orificemeter is d which is installed in a pipeline of diameter D. The pressure drop across the orifice plate is Δp. Obtain the expression for Q by using the method of dimensional analysis. (14)

Sol. $Q = f(d, D, \rho, \mu, \Delta p)$

Number of variables, $n = 6$

Number of fundamental dimensions, $m = 3$.

Number of dimensionless groups (π-terms)

$$= n - m = 6 - 3 = 3$$

Let us select r, Q, d as repeating variables.

$$\pi_1 = [\rho]^{a_1} [Q]^{b_1} [d]^{c_1} [D]^1$$

$$M^0 L^0 T^0 = [ML^{-3}]^{a_1} [L^3 T^{-1}]^{b_1} [L]^{c_1} [L]^1$$

$$= [M]^{a_1} [L]^{-3a_1 + 3b_1 + c_1 + 1} [T]^{-b_1}$$

For dimensional homogeneity, we have

$$a_1 = 0$$

$$-3a_1 + 3b_1 + c_1 + 1 = 0$$

$$-b_1 = 0$$

Thus $a_1 = 0$, $b_1 = 0$, $c_1 = 1$

$$\pi_1 = \left[\frac{D}{d}\right]$$

$$\pi_2 = [\rho]^{a_2} [Q]^{b_2} [d]^{c_2} [\mu]^1$$

$$M^0 L^0 T^0 = [ML^{-3}]^{a_2} [L^3 T^{-1}]^{b_2} [L]^{c_2} [LM^{-1} T^{-1}]^1$$

$$= [M]^{a_2 + 1} [L]^{-3a_2 + 3b_2 + c_2 - 1} [T]^{-b_2 - 1}$$

For dimensional homogeneity, we have

$$a_2 + 1 = 0$$

$$-3a_2 + 3b_2 + c_1 - 1 = 0$$

$$-b_2 = 0$$

Thus $a_2 = -1, b_2 = -1, c_2 = 1$

$$\pi_2 = \left[\frac{\mu d}{\rho Q}\right]$$

$$\pi_3 = [\rho]^{a_3} [Q]^{b_3} [d]^{c_3} [\Delta p]^1$$

$$M^0 L^0 T^0 = [ML^{-3}]^{a_3} [L^3 T^{-1}]^{b_3} [L]^{c_3} [LM^{-1} T^{-2}]^1$$

$$= [M]^{a_3 + 1} [L]^{-3a_3 + 3b_3 + c_3 - 1} [T]^{-b_3 - 1}$$

For dimensional homogeneity, we have

$$a_3 + 1 = 0$$

$$-3a_3 + 3b_3 + c_3 - 1 = 0$$

$$-b_3 = 0$$

Thus $a_3 = -1, b_3 = -2, c_3 = 4$

$$\pi_3 = \left[\frac{d^2 \Delta p}{\rho Q^2}\right]$$

We may write that,

$$\phi(\pi_1, \pi_2, \pi_3) = 0$$

or, $$\phi\left(\frac{D}{d}, \frac{\mu d}{\rho Q}, \frac{d^4 \Delta p}{Q^2 \rho}\right) = 0$$

or, $$\phi\left(\frac{d}{D}, \frac{\rho Q}{\mu d}, \frac{Q^2 \rho}{d^4 \Delta p}\right) = 0$$

or, $$\frac{Q^2 \rho}{d^4 \Delta p} = f_1\left(\frac{\rho Q}{\mu d}, \frac{d}{D}\right)$$

or, $$\frac{Q}{\left[d^2 \left(\frac{\Delta p}{\rho}\right)^{1/2}\right]} = f_2\left(\frac{\rho Q}{\mu d}, \frac{d}{D}\right)$$

or, $$Q = Cd^2 \left[\frac{\Delta p}{\rho}\right]^{1/2}$$

where C is dependent on $\frac{\rho Q}{\mu d}$ and $\frac{d}{D}$.

(*c*) Derive Prandtl boundary layer equations for steady two-dimensional laminar incompressible flow. Explain the significance of Pradtl boundary layer equations in comparison with Navier-Stokes equations. (14)

Sol. Prandtl Boundary Layer Equations. Fluids such as water and air which are encountered in most of the engineering problems of practical utility have very small viscosity. Their flow is characterized by a very small viscosity. Their flow is characterized by a very high Reynolds number. As we know that for flow at a very large Reynolds number the effects of velocity are confined to a very thin layer in the immediate vicinity of the solid boundary.

Prandtl introduced a concept of the boundary layer by emphasizing that the essential influence of viscosity is concentrated near the solid boundaries and showed how the Navier Stokes equation could be simplified to yield aproximate solutions for two-dimensional flow problems.

The flow at a very high Reynolds number past a flat plate of infinite extent is essentially a two-dimensional flow. The flow, as suggested by Prandtl may be divided into two regions, though the division between them is not very sharp.

1. Region of boundary layer which lies in the immediate neighbourhood of solid houndary and within which about 99% of the velocity changes takes place. The transition from zero velocity at the plate surface to the full magnitude U_∞ at some distance from the wall takes place within a very short distance-the so called boundary layer. In this boundary layer, the velocity gradient normal to the solid boundary, $\frac{\partial u}{\partial y}$, is very large with the result that even a low viscosity fluid exerts a very large shear stress, $\tau = \mu \frac{\partial u}{\partial y}$. In this region the laws of viscous flow, *i.e.*, the Navier-Stokes equations apply.
2. The region outside the boundary layer, which consitutes the main body of flowing fluid, has velocities of the order of the free-stream velocity, U_∞.

   The flow in this zone is governed by the potential flow theory, the viscous influence is negligibly small and the velocity grandients are also very small. The flow in this region is, therefore, assumed frictionless and the laws of potential flow are applicable.

The boundary conditions for analysing the flow in the boundary layer are :

(*i*) Existence of no - slip condition at the solid boundary, that is for $y = 0$, $u = v = 0$ and

(*ii*) Outside the boundary layer the velocity u tends to become equal to the free - stream velocity U_∞, that is for $y \geq \delta$, $u = U_\infty$. The assumption that the dimensionless boundary layer thickness $\frac{\delta}{x}$ is very small as compared to unity is necessary for the derivation of boundary layer equations.

In the region outside the boundary layer, $u = U_\infty$, and since there are no large velocity grandients in this region the viscous terms from the Navier-Stokes equations vanish. Thus, ignoring the body force components, the Navier - Stokes equation in the x-direction for flow outside the boundary layer reduces to :

$$\frac{\partial U_\infty}{\partial t} + \frac{U_\infty \partial U_\infty}{\partial x} = -\left(\frac{1}{\rho}\right)\frac{\partial p}{\partial x} \qquad ...(1)$$

For steady flow the pressure p is a function of x only ; the partial derivative $\frac{\partial p}{\partial x}$ may, therefore, be replaced by the total derivative $\frac{dp}{dx}$, and Eqn. (1) is expressed in the form,

$$\frac{U_\infty \, \partial U_\infty}{\partial x} = -\left(\frac{1}{\rho}\right)\frac{dp}{dx} \quad ...(2)$$

Integrating, we get,

$$p + \left(\frac{1}{2}\right)\rho U_\infty^2 = \text{constant} \quad ...(3)$$

The pressure distribution in a known potential flow may be determined by applying Eqn. (1), which expresses the Bernoulli's relationship in differential form, along a stream line at the solid boundary.

The simplified Navier-Stokes equation may be written as :

$$\frac{\partial u}{\partial t} + u\frac{\partial u}{\partial x} + v\frac{\partial u}{\partial y} = -\left(\frac{1}{\rho}\right)\frac{\partial p}{\partial x} + \nu\frac{\partial^2 u}{\partial y^2} \quad ...(4)$$

For steady flow using Eqns. (2) and (4), we get

$$u\frac{\partial u}{\partial x} + v\frac{\partial u}{\partial y} = U_\infty \frac{\partial U_\infty}{\partial x} + \nu\frac{\partial^2 u}{\partial y^2} \quad ...(5)$$

The Navier-Stokes equation in the y-direction, after ascertaining order - of - magnitude of various quantities involved, results in the following :

$$\frac{\partial p}{\partial y} = 0 \quad ..(6)$$

and the continuity equation for the two-dimensional steady flow is :

$$\frac{\partial u}{\partial x} + \frac{\partial v}{\partial y} = 0 \quad ...(7)$$

Eqs. (5) and (7) are known as the Prandtl's *boundary layer equations* and are subject to the following boundary conditions :

(*i*) at $y = 0$, $u = v = 0$, and

(*ii*) at $y = \infty$, $u = U_\infty (x, y)$ ...(8)

Although the above equations are meant for rectilinear flow they are applicable to curved flow as well provided that the radius of curvature is large enough as compared to the boundary layer thickness.

(*d*) The 1/7th power law for turbulent boundary layer is not applicable very near to the solid boundary. Do you agree? If yes, why? Justify your answer. (4)

Sol. It is true that the 1/7th power law for turbulent boundary layer is not applicable near the solid boundary. The regions of the boundary layer which lies in the immediate vicinity of the solid boundary, about 99% of the velocity changes take place in this region. The transition from zero velocity at the plate surface to the full magnitude U_∞ at some distance from the wall takes place within a very short distance-the so called boundary layer. In this boundary layer the velocity gradient normal to solid boundary, *i.e.*, $\frac{\partial u}{\partial y}$ is very large with the result that even a low viscosity fluid exerts a very large shear stress, $\tau = \mu \frac{\partial u}{\partial y}$. In this region the laws of viscous flow, *i.e.*, Navier - Stokes equation apply. This layer is called *laminar sublayer* in which $1/7^{th}$ power law is not applicable.

Q.6. (*a*) What do you mean by Fanno flow? Show that the upper and lower branches of a Fanno curve represent subsonic and supersonic flows respectively. Prove that at the maximum entropy point Mach number is unity and all processes approach this point. (15)

Sol. Fanno flow. Let us consider shock waves which occur in a plane normal to the direction of flow, called *normal shock waves*. The flow process through the shock waves is highly irreversible and cannot be approximated as being isentropic.

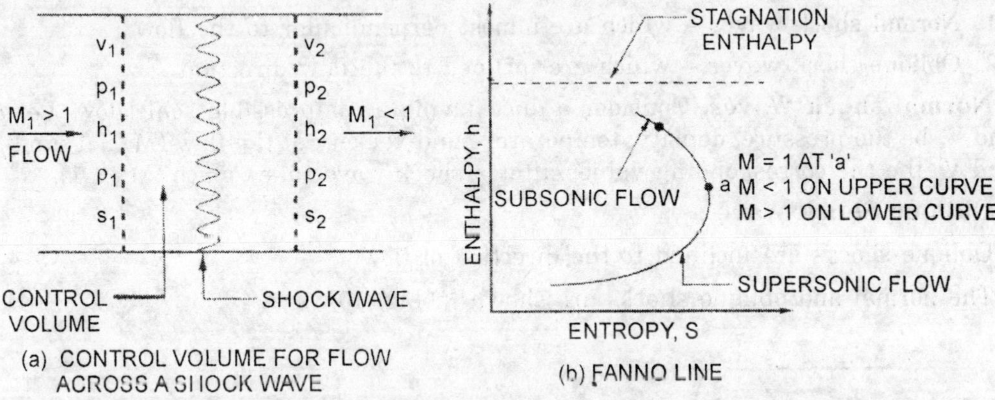

Fig. 9

Consider the control volume surrounding the shock wave as shown in Fig. 9 (*a*). Assuming steady flow with no heat and work interactions, and no potential energy changes, we have

Conservation of mass :

$$\rho_1 A_1 V_1 = \rho_2 A_2 V_2$$

With $A_1 = A_2$, we have

$$\rho_1 V_1 = \rho_2 V_2 \quad \text{...(1)}$$

Conservation of energy :

$$h_1 + \frac{V_1^2}{2} = h_2 + \frac{V_2^2}{2}$$

or

$$h_{01} = h_{02}$$

where h_0 = stagnation enthalpy

Conservation of momentum :

$$A(p_1 - p_2) = m(V_2 - V_1)$$

$$p_1 - p_2 = \rho_2 V_2^2 - \rho_1 V_1^2 \quad \text{...(3)}$$

Entropy :

$$s_2 - s_1 \geq 0 \quad \text{...(4)}$$

Conservation of mass and energy Eqns. (1) and (2) can be combined with property relations for the particular fluid to give the equation that when plotted on an $h - s$ diagram is called a *Fanno line*, as shown in Fig. 9 (*b*). The point of maximum entropy on this line (point '*a*') corresponds to M = 1. Upper part of the curve is subsonic (M < 1) and lower part supersonic (M > 1).

(*b*) Derive the Rankine-Hugoniot equations for flow through a normal shock. How does a normal shock differ from an oblique shock? Mention two useful applications of a normal shock. **(15)**

Sol. Shock Waves. Whenever a supersonic flow (compressible) abruptly changes to subsonic flow, a shock wave is produced resulting in a sudden rise in pressure, density, tempreature and entropy. A shock wave takes place in the diverging section of a nozzle, in a diffuser, throat of a supersonic wind tunnel, in front of sharp nosed bodies.

Shock waves are of two types :

1. Normal shock waves – which are almost perpendicular to the flow.
2. Oblique shock waves – which are inclined to the flow direction.

Normal shock Waves. Consider a duct having a compressible sonic flow. Let p_1, ρ_1, T_1 and V_1 be the pressure, density, temperature and velocity of the flow ($M_1 > 1$) and p_2, ρ_2, T_2 and V_2 the the corresponding values after a shock wave takes place ($M_2 < 1$).

Oblique Shock Wave :

Oblique shocks are inclined to the direction of flow.

The normal and oblique shocks are shown in Fig. 10.

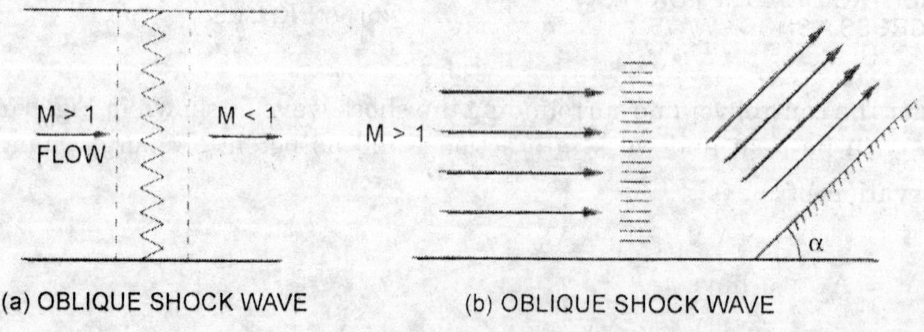

(a) OBLIQUE SHOCK WAVE (b) OBLIQUE SHOCK WAVE

Fig. 10

Rankine - Hugoniot Equations. Consider a control volume enclosing a normal shock as shown in Fig. 10 (*a*). The control volume is assumed to be at steady state with W = 0, Q = 0 and negligible effects of potential energy. The thickness of the shock is very small so that there is no significant change in flow across the shock. The upstream and downstream locations are denoted by 1 and 2 respectively.

The upstream and downstream states are related by the following equation :

Conservation of mass :

$$m = \rho_1 A_1 V_1 = \rho_2 A_2 V_2$$

With $A_1 = A_2$, we have

$$m = \rho_1 V_1 = \rho_2 V_2$$

Conservation of momentum :

$$A(p_1 - p_2) = m(V_2 - V_1) \quad ...(1)$$

$$p_1 + \rho_1 V_1^2 = p_2 + \rho_2 V_2^2 \quad ...(2)$$

Conservation of energy :

Considering the flow across the shock wave as adiabatic, we have

$$\left[\left(\frac{\gamma}{\gamma - 1}\right)\right]\left(\frac{p_1}{\rho_2}\right) + \frac{V_1^2}{2} = \left[\left(\frac{\gamma}{\gamma - 1}\right)\right]\left(\frac{p_2}{\rho_2}\right) + \frac{V_2^2}{2}$$

Because $z_1 = z_2$ for duct being in horizontal position. ...(3)

Combining continuity and momemtum equations, we get

$$p_1 + \frac{(\rho_1 V_1)^2}{\rho_1} = p_1 + \frac{(\rho_2 V_2)^2}{\rho_2} \quad ...(4)$$

This equation is known as 'Rankine Line Equation'.

Now consider continuity and energy equations.

$$\left[\frac{\gamma}{\gamma - 1}\right]\left(\frac{p_1}{p_1}\right) + \frac{(\rho_1 V_1)^2}{2\rho_1^2} = \left[\frac{\gamma}{\gamma - 1}\right]\left(\frac{p_2}{p_2}\right) + \frac{(\rho_2 V_2)^2}{2\rho_2^2}$$

This equation is called 'Fanno Line Equation'.

Further combining Eqns. (1) and (2) and (3) and solving for $\frac{p_2}{p_1}$, we get,

$$\frac{p_2}{p_1} = \frac{\left[\left\{\frac{\gamma + 1}{\gamma - 1}\right\}\left(\frac{p_2}{p_1}\right) - 1\right]}{\left[\left\{\frac{\gamma + 1}{\gamma - 1}\right\} - \left(\frac{p_2}{p_1}\right)\right]} \quad ...(6)$$

Solving for density ratio, we get,

$$\frac{\rho_2}{\rho_1} = \frac{V_1}{V_2} = \frac{\left[1 + \left\{\frac{\gamma + 1}{\gamma - 1}\right\}\left(\frac{p_2}{p_1}\right)\right]}{\left[\left\{\frac{\gamma + 1}{\gamma - 1}\right\} + \left(\frac{p_2}{p_1}\right)\right]} \quad ...(7)$$

Eqns. (6) and (7) are called 'Rankine-Hugoniot Equations'.

(c) Mention two most important characteristics of turbulent flow. Explain the concept of Prandtl's mixing length hypothesis. (10)

Sol. Characteristics of Turbulent Flow :

1. At a given point velocity and pressure are not constant with time but exhibit very irregular fluctuations of high frequency.
2. Turbulent motion, once established, has a random, haphazard and irregular nature which makes exact description difficult. The velocity distribution for the turbulent flow is not parabolic but follows the $1/7^{th}$ power law.

Prandtl Mixing Length Theory for Turbulent Flow :

Shear Stress. We know that turbulent shear stress is given by :

$$\bar{\tau} = \overline{-\rho u'v'} \quad ...(1)$$

where u', v' = fluctuating component of velocity in the direction of x and y respectively due to turbulence.

It is very difficult to calculate the value of $\bar{\tau}$ as measurement of $\overline{u'v'}$ is very difficult.

To overcome this difficulty Prandtl presented a mixing length hypothesis which can be used to express turbulent shear stress in terms of measurable quantities.

According to Prandtl, the mixing length 'l' is that distance between two layers in the transverse direction such that the lumps of fluid particles from one layer could reach the other layer and the particles are mixed in the other layer in such a way that the momentum of the particles in the direction of x is same. He also assumed that the velocity fluctuations in the x - direction u' is related to the mixing length 'l' as :

$$u' = l\left(\frac{du}{dy}\right)$$

and v', the flucuation component of velocity in the y-direction is of same order of magnitude as u' and hence, we have

$$v' = l\left(\frac{du}{dy}\right)$$

Now multiplying u' and v', we get

$$\overline{u'v'} = l^2\left(\frac{du}{dy}\right)^2$$

Substituting in Eqn. (1), we get,

$$\bar{\tau} = \rho l^2\left(\frac{du}{dy}\right)^2 \quad ...(2)$$

Thus the total shear stress at any point in turbulent flow is the sum of shear due to viscous shear and turbulent shear and can be written as :

$$\bar{\tau} = \mu\left(\frac{du}{dy}\right) + \rho l^2\left(\frac{du}{dy}\right)^2 \quad ...(3)$$

But the viscous shear stress is negligible except near the boundary. Eqn. (3) is used for most of turbulent flow problems for determining shear stress in the turbulent flow.

Q.7. **(*a*)** The clearance volume of a single cylinder single acting air compressor is 10% of the swept volume. The compressor has a bore of 120 mm and a stroke of 152 mm and runs at 400 rpm. Suction conditions are 1 bar and 24°C and the delivery pressure is 10 bar. Compression and re-expansion curves follow the law $pV^{1.3}$ = constant. Calculate the diagram volumetric efficiency, the rate of air delivery, and the power input required to drive the compressor assuming a mechanical efficiency of 74%. Take R = 0.287 kJ/kg-K. (10)

Sol.

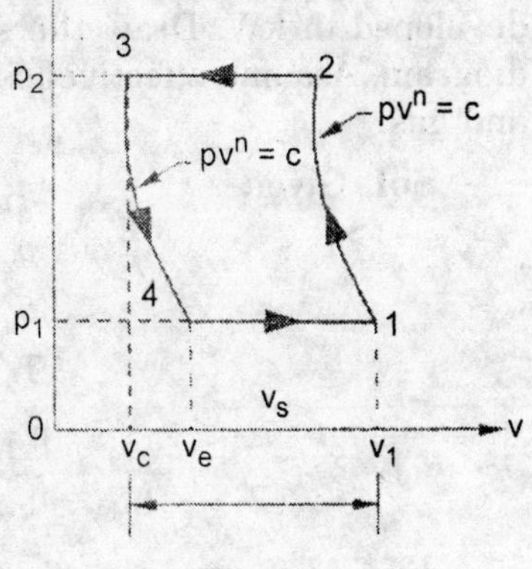

Fig. 11

Given :
$$c = \frac{v_c}{v_s} = 0.1,\ D = 0.12\ m,$$
$$L = 0.152\ m,\ N = 400\ rpm$$
$$p_1 = 1\ bar,\ p_2 = 10\ bar,$$
$$T_1 = 273 + 24 = 297\ K,$$
$$n = 1.3,\ \eta_{mech} = 0.74$$

The p–v diagram with clearance for the reciprocating compressor is shown in Fig. 11.

Volumetric efficiency of compressor,

$$\eta_v = 1 + c - c\left(\frac{p_2}{p_1}\right)^{1/n}$$
$$= 1 + 0.1 - 0.1\ (10)^{1/1.3} = 0.5122 \text{ or } 51.22\%$$

Swept volume, $$v_s = \pi D^2 \frac{L}{4} = \pi \times (0.12)^2 \times \frac{0.152}{4} = 1.719 \times 10^{-3}\ m^3$$

Air delivered per stroke $$= v_s \times \eta_v \times \frac{N}{60}$$

$$V_1 = 1.719 \times 10^{-3} \times 0.5122 \times \frac{400}{60} = 5.87 \times 10^{-3}\ m^3/s$$

Specific volume at inlet, $$v_{s1} = \frac{RT_1}{p_1} = \frac{287 \times 297}{1 \times 10^5} = 0.85239\ m^3/kg$$

Mass flow rate of air at inlet,

$$\dot{m}_1 = \frac{V_1}{v_{s1}} = \frac{5.87 \times 10^{-3}}{0.85239} = 6.886 \times 10^{-3}\ kg/s$$

Work done on the compressor,

$$W_c = \left[\frac{n}{n-1}\right] \dot{m} RT_1 \left[\left(\frac{p_2}{p_1}\right)^{(n-1)/n} - 1\right]$$
$$= \left(\frac{1.3}{0.3}\right) \times 6.886 \times 10^{-3} \times 0.287 \times 297 \times [(10)^{0.3/1.3} - 1]$$
$$= 1.7836\ kJ/s$$

Actual work done, $$W_a = \frac{W_c}{\eta_{mech}} = \frac{1.7836}{0.74} = 2.41\ kW$$

(*b*) Air enters the compressor of a gas turbine at 100 kPa, 300 K with a volumetric flow rate of 5 m^3/sec. The air is compressed in two stages to 1200 kPa with intercooling to 300 K between stages at a pressure of 350 kPa. The turbine inlet temperature is 1400 K and the expansion occurs in two stages with reheat to 1340 K between the stages at a pressure of 350 kPa. The compressor and turbine stage efficiencies are 87% and 85% respectively.

Determine (*i*) the thermal efficiency of the cycles, (*ii*) the back work ratio, (*iii*) the net power developed in kW. Draw the schematic diagram of the cycle and indicate the process on T-s diagram. Assume effectiveness of the regenerator as 80%. Assume c_p = 1.0045 kJ/kg-K for air and gas. **(20)**

Sol. Given :

$$p_1 = p_9 = 100 \text{ kPa}, T_1 = 300 \text{ K}, V_1 = 5 \text{ m}^3/\text{s}.$$
$$p_2 = p_3 = p_7 = p_8 = 350 \text{ kPa},$$
$$p_4 = p_6 = 1200 \text{ kPa}$$
$$T_3 = 300 \text{ K}, T_6 = 1400 \text{ K}, T_8 = 1340 \text{ K}$$
$$\eta_{c1} = \eta_{c2} = 0.87,$$
$$\eta_{t1} = \eta_{t2} = = 0.85, \varepsilon_r = 0.8$$
$$c_{pa} = c_{pg} = 1.0045 \text{ kJ/kg.K}$$

The line diagram of the system and the T-s diagram are shown in Fig. 12.

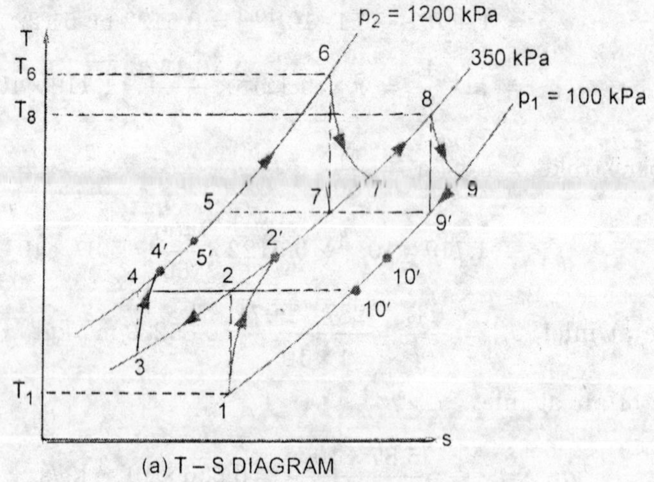

(a) T – S DIAGRAM

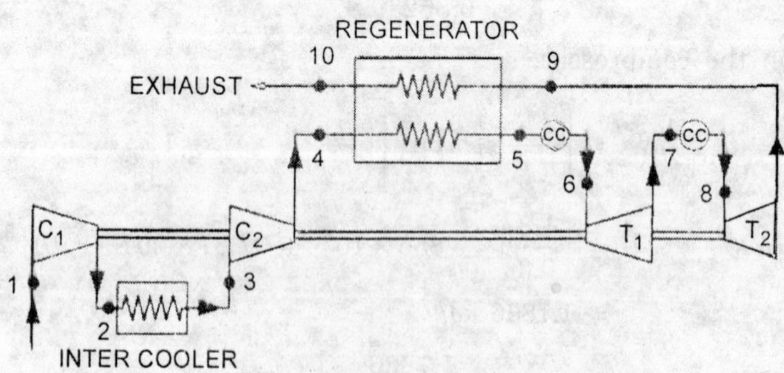

(b) SCHEMATIC DIAGRAM OF CYCLE

Fig. 12

Compressor 1st stage :

$$T_2 = T_1\left(\frac{p_2}{p_1}\right)^{(\gamma-1)/\gamma}$$

$$= 300\left(\frac{350}{100}\right)^{(1.4-1)/1.4} = 429.11 \text{ K}$$

$$\eta_{c1} = \frac{429.11 - 300}{T_2' - 300}$$

$$T_2' = 448.4 \text{ K}$$

Work done on the 1st stage of compressor.

$$w_{c1} = c_{pa}(T_2' - T_1) = 1.0045\ (448.4 - 300) = \mathbf{149.07\ kJ/kg.}$$

Compressor 2nd stage :

$$T_4 = T_3\left(\frac{p_4}{p_3}\right)^{(\gamma-1)/\gamma}$$

$$= 300\left(\frac{1250}{350}\right)^{(1.4-1)/1.4}$$

$$= 426.59 \text{ K}$$

$$\eta_{c2} = \frac{T_4 - T_3}{T_4' - T_3}$$

$$0.87 = \frac{426.59 - 300}{T_2' - 300}$$

$$T_4' = 445.50 \text{ K}$$

Work done on the 2nd stage of compresor,

$$w_{c2} = c_{pa}(T_4' - T_3) = 1.0045\ (445.50 - 300) = \mathbf{146.15\ kJ/kg}$$

Total work done on the compressor, $w_c = w_{c1} + w_{c2}$

$$= 149.07 + 146.16 = 295.23 \text{ kJ/kg}$$

Turbine 1st stage : $T_7 = T_6\left(\frac{p_7}{p_6}\right)^{(\gamma-1)/\gamma}$

$$= 1400\left(\frac{350}{1200}\right)^{(1.4-1)/1.4} = 984.55 \text{ K}$$

$$\eta_{t1} = \frac{T_6 - T_7'}{T_6 - T_7}$$

$$0.85 = \frac{1400 - T_7'}{1400 - 984.55}$$

$$T_7' = 1046.87 \text{ K}$$

Work done by the 1st stage of turbine,

$$W_{c1} = C_{pg}(T_6 - T_7') = 1.0045\ (1400 - 1046.87)$$

$$= 354.72 \text{ kJ/kg}$$

Turbine 2nd stage : $T_9 = T_8\left(\frac{p_9}{p_8}\right)^{(\gamma-1)/\gamma} = 1340\left(\frac{100}{350}\right)^{(1.4-1)/1.4} = 936.82\text{ K}$

$$\eta_{t2} = \frac{T_8 - T_9'}{T_8 - T_9}$$

$$0.85 = \frac{1340 - T_9'}{1340 - 396.82}$$

$$T_9' = 997.3\text{ K}$$

Work done by the 2nd stage of turbine,

$$w_{t1} = c_{pg}(T_8 - T_9') = 1.0045\,(1340 - 997.3)$$
$$= 344.24\text{ kJ/kg}$$

Total work done by the turbine, $w_t = w_{t_1} + w_{t_2}$

$$= 354.72 + 344.24 = 698.96\text{ kJ/kg}$$

Net work done by turbine, $w_{\text{net}} = w_t - w_c$

$$= 698.96 - 295.23 = 403.73\text{ kJ/kg}$$

Regenerator effectiveness, $\varepsilon_r = \frac{T_5' - T_4'}{T_9' - T_4'}$

$$0.8 = \frac{T_5' - 445.5}{997.3 - 445.5}$$

$$T_5' = 886.94\text{ K}$$

Heat supplied, $q_s = c_{pg}\,[(T_6 - T_5') + (T_8 - T_7')]$

$$= 1.0045\,[(1400 - 886.84) + (1340 - 1046.87)]$$
$$= 809.818\text{ kJ/kg}$$

(i) Thermal supplied, $\eta_{th} = \frac{w_{\text{net}}}{q_s} = \frac{403.73}{809.818} = 0.4985$ or 49.85%

(ii) Work ratio $= \frac{w_{\text{net}}}{w_t} = \frac{403.73}{809.818} = 0.5776$

(iii) Now $p_1V_1 = mRT_1$

$$m = \frac{100 \times 10^5 \times 5}{287 \times 300} = 580.72\text{ kg/s}$$

Net power developed $= m \times w_{\text{net}} = 580.72 \times 403.73 = 234454\text{ kW}$

(c) The turbine rotor has a mean blade ring diameter of 500 mm and the blade angles are equal. The nozzle angle is 20° and the steam leaves the nozzles with a velocity of 900 m/sec. Assuming a blade friction factor of 0.85, determine the

(i) best blade angles,

(ii) turbine speed in rpm,

(*iii*) steam consumption rate in kg/hr if the turbine generates 10 kW, and

(*iv*) maximum blade efficiency. (10)

Sol.

Given : $\alpha_1 = 20°$, $d = 0.5$ m, $v_{a1} = 900$ m/s, K = 0.85, $\beta_1 = \beta_2 = \beta$

The velocity diagrams are shown in Fig. 13.

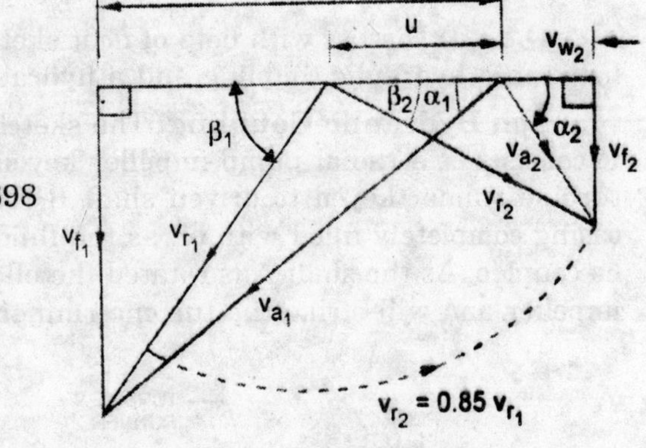

Fig. 13

For maximum blade efficiency,

Velocity ratio, $$\frac{u}{v_{a_1}} = \frac{\cos \alpha_1}{2}$$

$$= \frac{\cos 20°}{2} = 0.4698$$

$$u = 0.4698 \times 900$$

$$= 422.86 \text{ m/s}$$

Also, $$u = \frac{\pi d N}{60}$$

$$N = \frac{422.86 \times 60}{\pi \times 0.5} = 16152 \text{ rpm}$$

$$v_w = v_{w_1} + v_{w_2} = v_{r_1} \cos \beta_1 + v_{r_2} \cos \beta_2$$

$$= v_r \cos \beta\ (1 + 0.85) = 1.85\ v_{r_1} \cos \beta$$

$$u = v_{a_1} \cos \alpha_1 - v_{r_1} \cos \beta$$

$$v_{r_1} \cos \beta = 900 \cos 20° - 422.86 = 422.86$$

$$v_w = 1.85 \times 422.86 = 782.254 \text{ m/s}$$

Power developed, $$P = mu \frac{v_w}{10^3}$$

Mass flow rate of steam,

$$m = \frac{10 \times 10^3}{422.86 \times 782.2541} = 0.03023 \text{ kg/s or } 108.83 \text{ kg/h}$$

Maximum blade efficiency,

$$(\eta_b)_{max} = (1 + KB) \frac{\cos^2 \alpha_1}{2}$$

Here, $$B = \frac{\cos \beta_1}{\cos \beta_2} = 1$$

$$(\eta_b)_{max} = (1 + 0.85) \cos^2 \frac{20°}{2} = 0.8168 \text{ or } 81.68\%$$

Now $$v_{r1}^2 = v_{a1}^2 + u^2 - 2uv_{a_1} \cos \alpha_1$$

$$= 900^2 + (422.86)^2 - 2 \times 422.86 \times 922.86 \times 900 \times \cos 20°$$

$$= 273564$$

$$v_{r_1} = 523.03 \text{ m/s}$$

$$\cos\beta = \frac{422.86}{v_{r_1}} = \frac{422.86}{523\ 03} = 0.80844$$

$$\beta = 36°$$

Q.8. (*a*) Discuss with help of neat sketches the main difference in principle of operation between a hydraulic coupling and a hydraulic converter. (8)

Sol. Hydraulic Coupling. The sketch of a hydraulic coupling is shown in Fig. 14 (*a*). It consists of a radial pump impeller keyed to a driving shaft A and a radial type reaction turbine runner keyed to driven shaft B. The impeller and turbine runner together form a casing completely filled with oil as the fluid with which the shafts A and B are supposed to be coupled. As the shaft A is rotated the oil due to forced vortex will flow out from the pump impeller and will strike the turbine runner blades, and shaft B will start rotating.

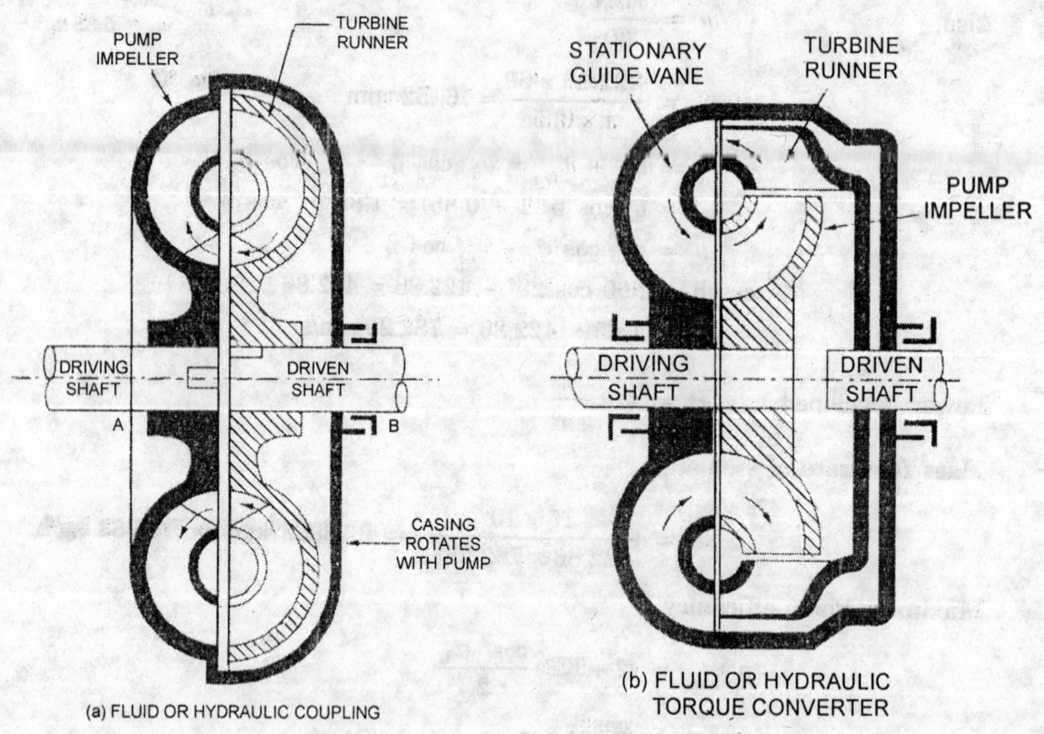

Fig. 14

Hydraulic Turque Converter. It is a device to increase or reduce the torque as shown in Fig. 14 (*b*). It is accomplished by inserting a stator between the pump impeller and the turbine runner. The sector helps in changing the direction of fluid. As the fluid flowing out of the impeller rim strikes the turbine runner and blades mounted on the stationary housing, it releases its energy in the form of torque and speed resulting in the engine torque to be increasingly multiplied.

(*b*) Mention three possible causes of trouble in case of a centrifugal pump if

(*i*) No water is delivered, and

(*ii*) Pump absorbs higher power. (6)

Sol. Troubles in Centrifugal Pump.

Sl. No.	Trouble		Causes
(*i*)	No water is delivered	1.	Excessive suction lift.
		2.	Air pockets in suction line.
		3.	Speed too low.
(*ii*)	Pump absorbs higher power	1.	Speed too high.
		2.	Manometric head of system more than manometric head of punp
		3.	Pump casing clogged.

(*c*) A centrifugal pump is required to discharge 600 litres of water per second and develop a head of 15 m when the impeller rotates at 750 rpm. The manometric efficiency is 80%. The loss of head in the pump due to fluid resistance being assumed to be 0.027 V^2 m of water, where V is the velocity with which water leaves the impeller Water enters the impeller without shock or whirl and the velocity of flow is 3.2 m/sec.

Determine :

(*i*) the impeller diameter and outlet area,

(*ii*) blade angle at the outlet edge.

Sol. Given :

$$Q = 0.6 \text{ m}^3/\text{s},\ H_{nano} = 15 \text{ m}$$

$$N = 750 \text{ rpm},\ \eta_{mano} = 0.8$$

$$v_{f2} = 3.2 \text{ m/s},\ h_f = 0.027\ V^2$$

(*i*) Impeller diameter D_2 :

$$\frac{u_2\, v_{u_2}}{g} = \frac{H_{mano}}{\eta_{mano}} = \frac{15}{0.8} = 18.75 \text{ m}$$

Also,

$$H_{mano} = \frac{u_2\, v_{u_2}}{g} - \text{ loss of head in the pump, } h_f$$

Loss of head in the pump, $h_f = 18.75 - 15 = 3.75$ m

$$0.027\, v^2 = 3.75$$

$$v^2 = \frac{3.75}{0.27} = 138.88$$

$$v = 11.78 \text{ m/s}$$

Velocity of flow at outlet,

$$v_{f2} = \frac{Q}{\text{area of flow}}$$

$$3.2 = \frac{0.6}{\text{area of flow}}$$

Area of flow, $$A_f = \frac{0.6}{3.2} = 0.1875 \text{ m}^2$$

Tangential velocity of impeller at outlet,

$$u_2 = \frac{\pi D_2 N}{60} = \pi \times D_2 \times \frac{750}{60} = 39.27\ D_2$$

Also, $$\frac{u_2\, v_{u_2}}{g} = 1875$$

$$v_{u_2} = \frac{9.81 \times 18.75}{39.27 D_2} = \frac{4.684}{D_2}$$

From Fig. 15, we have

$$v_{f2}^2 = v_{a2}^2 - v_{u2}^2$$

$$(3.2)^2 = (11.785)^2 - \left(\frac{4.684}{D_2}\right)^2$$

$$10.24 = \frac{138.89 - 21.94}{D_2^2}$$

$$D_2^2 = 0.17054$$

$$D_2 = 0.413 \text{ m or } 413 \text{ mm}$$

$$v_{u_2} = \frac{4.684}{0.413} = 11.342 \text{ m/s}$$

$$u_2 = 39.27 \times 0.413$$

$$= 16.218 \text{ m/s}$$

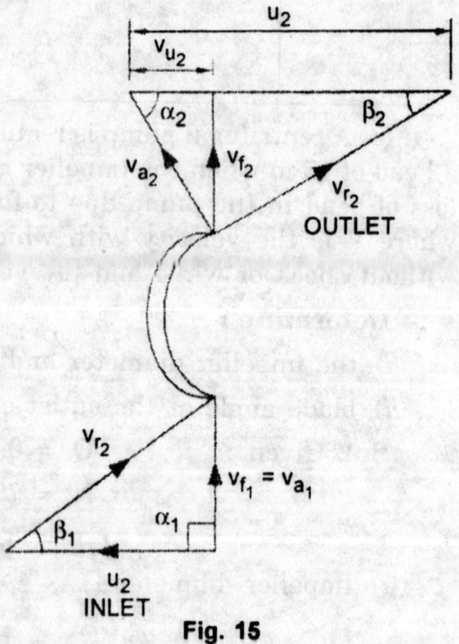

Fig. 15

(*ii*) $$\tan \beta_2 = \frac{v_{f_2}}{(u_2 - v_{u_2})} = \frac{3.2}{(16.218 - 11.342)} = 0.65627$$

$$\beta_2 = 33.276°$$

(*d*) Laboratory tests are required to be performed on a 1/10 scale modle to find the best design for a hydraulic turbine. The turbine is required to develop 30,000 kW under a net head of 50 m running at 300 rpm. The head available at the laboratory is 12 m. Determine.

(*i*) the corresponding speed

(*ii*) Power, and

(*iii*) Discharge.

The overall effiviency is 88%.

Sol. Overall efficiency, $$\eta_0 = \frac{\text{Shaft power}}{\text{Water power}} = \frac{P}{wQH}$$

$$Q = \frac{P}{\eta_0 \, wH} = \frac{30000 \times 10^3}{0.88 \times 9.81 \times 10^3 \times 50} = 69.5 \text{ m}^3/\text{s}$$

Specific speed, $$N_s = N\sqrt{\frac{P}{H^{5/4}}} = 300\sqrt{\frac{30000}{50^{5/4}}} = 390.8$$

(*i*) $$\left[\frac{H}{N^2D^2}\right]_m = \left[\frac{H}{N^2D^2}\right]_p$$

$$N_m^2 = N^2{}_p\left(\frac{D_p}{D_m}\right)^2\left(\frac{H_m}{H_p}\right)$$

$$N_m = N_p\left(\frac{D_p}{D_m}\right)\left(\frac{H_m}{H_p}\right)^{1/2}$$

$$= 300 \times 10 \times \left(\frac{12}{50}\right)^{1/2} = 1469.7 \text{ rpm}$$

(*ii*) $$\left[\frac{P}{N^3D^5}\right]_m = \left[\frac{P}{N^3d^5}\right]_p$$

$$= 30000\left(\frac{1469.5}{300}\right)^3\left(\frac{1}{10}\right)^5$$

$$= 35.273 \text{ kW}$$

(*iii*) $$\left[\frac{Q}{ND^3}\right]_m = \left[\frac{Q}{ND^3}\right]_p$$

$$Q_m = Q_p\left(\frac{N_m}{N_p}\right)\left(\frac{D_m}{D_p}\right)^3$$

$$= 69.5\left(\frac{1469.5}{300}\right)\left(\frac{1}{10}\right)^3 = 0.34 \text{ m}^3/\text{s}$$

MECHANICAL ENGINEERING
PAPER I—2007

(Conventional)

Q.1. (*a*) Prove the following relations :

$$\left(\frac{\partial u}{\partial v}\right)_T = T\left(\frac{\partial p}{\partial T}\right)_v - p$$

and $$\left(\frac{\partial u}{\partial v}\right)_T = v - T\left(\frac{\partial v}{\partial T}\right)_p$$ ...(15)

Sol. Internal energy, $u = f(T, v)$

$$\therefore \quad du = \left(\frac{\partial u}{\partial T}\right)_v .dT + \left(\frac{\partial u}{\partial v}\right)_T .dv$$

Now, $$c_v = \left(\frac{\partial u}{\partial T}\right)_v$$

$$\therefore \quad du = c_v .dT + \left(\frac{\partial u}{\partial v}\right)_T .dv$$

According to the First and Second law of Thermodynamics,

$$dq = T.ds = du + p.dv$$

$$\therefore \quad du = T.ds - p.dv$$

$$\left(\frac{\partial u}{\partial v}\right)_T = T\left(\frac{\partial s}{\partial v}\right)_T - p$$

As per Maxwell third relations, we have

$$\therefore \quad \left(\frac{\partial s}{\partial v}\right)_T = \left(\frac{\partial p}{\partial T}\right)_v$$

Substituting this in the above equation, we get,

$$\left(\frac{\partial u}{\partial v}\right)_T = T\left(\frac{\partial p}{\partial T}\right)_v - p$$

which proves the first relationship.

Now enthalpy, $h = f(T, p)$

$$\therefore \quad dh = \left(\frac{\partial h}{\partial T}\right)_p .dT + \left(\frac{\partial h}{\partial p}\right)_T .dp$$

As
$$c_p = \left(\frac{\partial h}{\partial T}\right)_p$$

$$\therefore \quad dh = c_p .dT + \left(\frac{\partial h}{\partial p}\right)_T .dp$$

According to First and Second law of Thermodynamics,

$$dq = T.ds = dh - v.dp$$

$$\therefore \quad dh = T.ds + v.dp$$

$$\therefore \quad \left(\frac{\partial h}{\partial p}\right)_T = T\left(\frac{\partial s}{\partial p}\right)_T + v$$

As per Maxwell fourth relation, we have

$$\left(\frac{\partial s}{\partial p}\right)_T = -\left(\frac{\partial v}{\partial T}\right)_p$$

Substituting this in the above equation, we get,

$$\left(\frac{\partial h}{\partial p}\right)_T = v - T\left(\frac{\partial v}{\partial T}\right)_p$$

which poves the second relationship.

(b) **An air compressor is used to fill rapidly a 3 m^3 tank at 20°C and 1 atm. The filling process is governed by the law $pv^{1.4}$ = constant. The kinetic and potential energy effects are negligible. The ratio of the final to the initial mass of air in the tank is 4. Work out the following :**

(*i*) Sketch the system and list the assumptions made;

(*ii*) Work input to the compressor. **(15)**

Sol.

(*i*) The system is shown in Fig. 1.

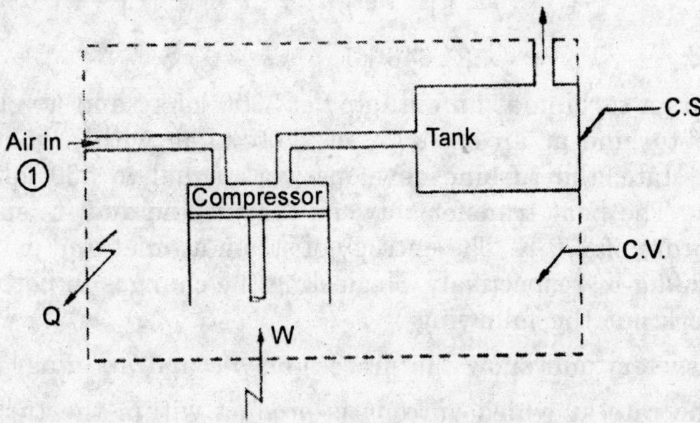

Fig. 1

Assumptions :

1. $\Delta PE = 0$ and $\Delta KE = 0$
2. $Q = 0$

$$V_f = V_i = 3 \text{ m}^3,\ T_i = 273 + 20 = 293 \text{ K},$$

$$p_i = 1 \text{ atm},\ m_f = 4\ m_i$$

Now,

$$m_i = \frac{p_i V_i}{RT_i} = \frac{1 \times 10^5 \times 3}{287 \times 293} = 3.5675 \text{ kg}$$

$$m_f = 4\ m_i = 14.2702 \text{ kg}$$

for the tank charging process,

$$(m_f - m_i)\ h_1 = m_f\ u_f - m_i\ u_i$$

$$(m_f - m_i)\ c_p \left(\frac{T_i + T_f}{2}\right) = (m_f\ T_f - m_i\ T_i)\ c_v$$

$$(m_f - m_i)\ \gamma \left(\frac{T_i + T_f}{2}\right) = m_f\ T_f - m_i\ T_i$$

$$(14.2702 - 3.5675) \times 1.4 \times \left(\frac{293 + T_i}{2}\right)$$

$$= 14.2702\ T_f - 3.5675 \times 293$$

$$\left(\frac{10.7027 \times 1.4}{2}\right)(293 + T_f) = 14.2702\ T_f - 1045.2775$$

Final temperature, $T_f = 477.84$ K or 204.84°C

Final pressure in tank,

$$p_f = \frac{m_f RT_f}{v_f} = \frac{14.2702 \times 287 \times 477.84}{3} = 6.5234 \text{ bar}$$

Work input to compressor,

$$W_c = \frac{p_i V_i}{\gamma - 1}\left[1 - \left(\frac{p_f}{p_i}\right)^{\frac{\gamma - 1}{\gamma}}\right] = \frac{1 \times 10^5 \times 3}{0.4}\left[1 - \left(\frac{6.5234}{1}\right)^{\frac{1}{3.5}}\right]$$

$$= -\ 531.646 \text{ kJ}$$

(*c*) Steam enters a turbine at an enthalpy of 3300 kJ/kg and a velocity of 180 m/s. The steam comes out of turbine at an enthalpy of 2700 kJ/kg with a velocity 120 m/s. At the condition of steady state, the turbine develops work equal to 550 kJ/kg of steam flowing through the turbine. The heat transfer between the turbine and its surroundings occurs at an average temperature of 370 K. The entropy of steam at inlet and exit of turbine are 6.932 kJ/kg-K and 7.361 kJ/kg-K, respectively. Neglecting the changes in potential energy between inlet and outlet, work out the following :

(*i*) Sketch the system and show the process on *p-v* and *h-s* diagrams.

(*ii*) Calculate the rate at which entropy is product within the turbine per kg of steam flowing. (10)

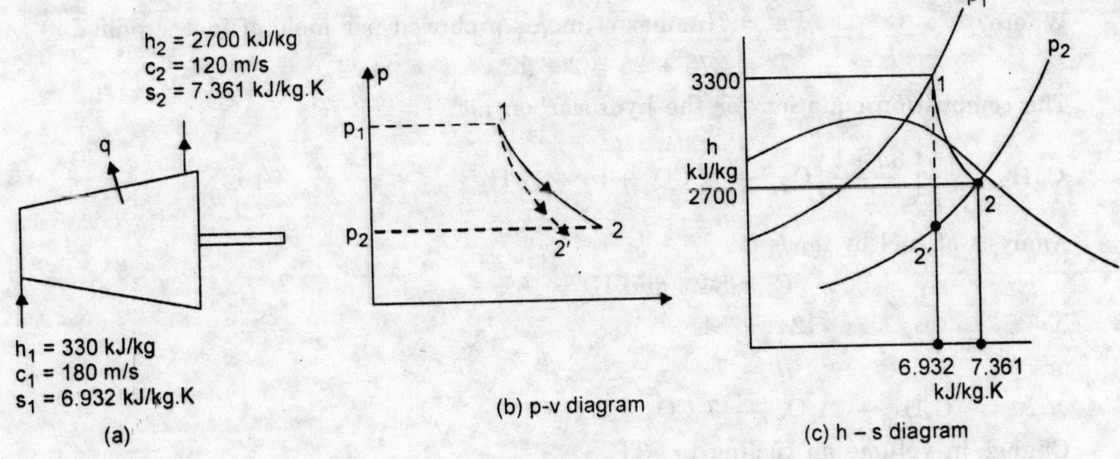

Fig. 2.

Sol.

(*i*) The system is shown in Fig. 2 (*a*), $p - v$ diagram in Fig. 2 (*b*) and h – s diagram in Fig. 2 (*c*).

(*ii*) SFEE gives,

$$h_1 + \frac{c_1^2}{2} + q_{1-2} = h_2 + \frac{c_2^2}{2} + w_{1-2}$$

$$300 + \frac{180^2}{2 \times 10^3} + q_{1-2} = 2700 + \frac{120^2}{2 \times 10^3} + 550$$

$$q_{1-2} = 59 \text{ kJ/kg}$$

Entropy produced within the turbine

$$= (s_2 - s_1) + \frac{q_{1-2}}{T_0}$$

$$= (7.361 - 6.932) - \frac{59}{370}$$

$$= 0.27 \text{ kJ/kg.K}$$

Q.2. (*a*) The lower calorific value of a liquid fuel at constant pressure is 44000 kJ/kg. The analysis of fuel by mass is 84% carbon and 16% hydrogen. Determine the higher calorific value at constant pressure, and the lower and higher calorific values at constant volume. (15)

Take h_{fg} for water at 25°C as 2442 kJ/kg.

Sol. HCV at constant pressure

$$= \text{LCV at constant pressure} + \frac{\%H_2 \times 9}{100} \times h_{fg}$$

$$= 44000 + \frac{16 \times 9}{100} \times 2442 = 47516.48 \text{ kJ/kg}$$

LCV at constant volume = LCV at constant pressure + $8.314nT$

Where n = Number of moles produced per mole of fuel supplied.

$T = 273 + 25 = 298$ K

The combustion equation for the hydrocarbon is

$$C_nH_{2n+2} + \left(\frac{3n+1}{2}\right)O_2 \rightarrow n\ CO_2 + (n+1)\ H_2O$$

Analysis of fuel by mass is

$$C = 84\% \text{ and } H_2 = 16\%$$

$$\therefore \quad 12n = 84$$

or $$n = 7$$

$$\therefore \quad C_7H_{16} + 11\ O_2 \rightarrow 7\ CO_2 + 8\ H_2O$$

Change in volume on cooling to STP

$$= 8 \text{ moles} - 11 \text{ moles} = -3 \text{ moles per mol of fuel}$$

At constant volume,

$$\text{LCV} = 44000 - \frac{8.314 \times 3 \times 298}{100} = 43927 \text{ kJ/kg}$$

$$\text{HCV} = 47516.48 - \frac{8.314 \times 3 \times 298}{100} = 47444 \text{ kJ/kg}$$

(*b*) **Derive an expression for the diameter of the injector orifice to spray fuel Q cm³/cycle /cylinder in terms of injection pressure, p_{inj} (kN/cm²), combustion chamber pressure, p_{cyl} (kN/m²), density of fuel, ρ_f (kg/cm³) and period of injection τ seconds.**

Calculate the diameter of the injector orifice of a six-cylinder, 4-stroke CI engine using the following data :

Brake power = 250 kW ; Engine speed = 1500 rpm ; BSFC = 0.3 kg/kW; Cylinder pressure = 35 bar ; Injection pressure = 200 bar ; Specific gravity of fuel = 0.88 ; Co-efficient of discharge of the fuel orifice = 0.92 ; Duration of injection = 36 ° of crank angle. (15)

Sol. The cylinder piston arrangement is shown in Fig. 3. 1.

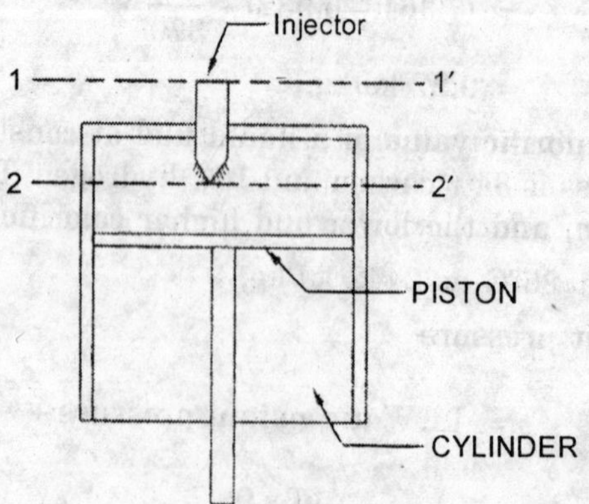

Fig. 3.

$$p_1 = p_{inj} = \text{injection pressure, kN/m}^2$$
$$p_2 = p_{cyl} = \text{combustion chamber pressure, kN/m}^2$$
$$\rho_f = \text{density of fuel, kg/cm}^3$$
$$\tau = \text{period of injection, s}$$
$$Q = \text{fuel sprayed, cm}^3\text{/cycle/cylinder}$$
$$d_f = \text{diameter of injector}$$

Let
$$v_f = \text{specific volume of fuel (assumed incompressible)}$$
$$c_1 = \text{velocity at section } 1 - 1'$$
$$c_2 = \text{velocity at section } 2 - 2'$$

Thus,
$$\frac{c_1^2}{2} + p_1 v_f = \frac{c_2^2}{2} + p_2 v_f$$

Neglecting c_1, being very small compared with c_2,

$$c_2 = \sqrt{2v_f\,(p_1 - p_2)} = \sqrt{\frac{2\,(p_1 - p_2)}{\rho_f}}$$

Actual fuel velocity, $c_f = C_{df} \,.\, C_2 = c_{df}\sqrt{\frac{2\,(p_1 - p_2)}{\rho_f}}$

where C_{df} = co-efficient of discharge for the fuel injector.

Fuel consumed per hour, m_f = power in kW × SFC in kg/kWh.

Fuel consumed per cylinder per cycle (for 4-stroke engine),

$$m_{fc} = \frac{m_f}{\frac{N}{2} \times \text{Number of cylinders} \times 60}$$

where N = rpm.

Volume of fuel injected per cylinder per cycle,

$$Q = \frac{m_{fc} \times 10^{-3}}{\text{Specific gravity of fuel}}\,\text{m}^3$$

Injection period

$$\tau = \frac{\text{Period of injection in degrees of crask angle} \times 60\text{s}}{360 \times N}$$

Nozzle hole area, $\frac{\pi}{4} d_f^2 = \frac{m_{fc}}{c_f \times \tau \times \rho_f}$

Given data :

BP = 250 kW, N = 1500 rpm , BSFC = 0.3 kg/kWh

p_1 = 200 bar, p_2 = 35 bar,

Specific gravity of fuel = 0.88, C_{df} = 0.92,

Duration of injection = 36° of crank angle,

Number of cylinders = 6, Number of strokes = 4

Fuel consumed per hour = BP x BSFC = 250 × 0.3 = 75 kg/h

Fuel consumed per cylinder per cycle

$$= \frac{75 \times 2}{1500 \times 6 \times 60} = \frac{1}{3600} \text{ kg}$$

Volume of fuel injected per cylinder per cycle

$$= \frac{10^{-3}}{3600 \times 0.88} = 0.3156 \times 10^{-6} \text{ m}^3$$

Injection period $= \frac{36}{360} \times \frac{1}{1500} \times 60 = \frac{1}{250}$ s

Density of fuel, ρ_f = 1000 x 0.88 = 880 kg/m^3

$$c_2 = \sqrt{\frac{2\,(200 - 35) \times 10^5}{880}} = 193.65 \text{ m/s}$$

Nozzle hole area, $\frac{\pi}{4} d_f^2 = \frac{\text{Mass delivered per cycle}}{C_{df} \times c_2 \times \text{Time} \times \rho_f}$

$$= \frac{1 \times 250}{3600 \times 0.92 \times 193.65 \times 1 \times 880}$$

Diameter of cylinder orifice, d_f = 0.75 mm

(c) Explain the mechanism of formation of NO_x in an SI engine. Discuss the effect of

(*i*) Compression ratio

(*ii*) Air-fuel ratio

(*iii*) Retarding spark timing

(*iv*) Exhaust gas recirculation

(*v*) Temperature and pressure of the inlet charge on the concentration of NO_x in the exhaust of an SI engine. **(10)**

Sol. Mechanism of formation of NO_x in an SI engine. Oxides of nitrogen (NO_x) occur in the engine exhaust are a combination of nitric oxide (NO) and nitrogen dioxide (NO_2). Nitrogen and oxygen react at relatively high temperatures. Therefore, high temperatures and availability of oxygen are the two main reasons for the formation of NO_x. When the proper amount of oxygen is available the higher the peak combustion temperature the more is the NO formed. The NO_x is formed in the atmosphere as NO oxidizes.

Effect of variables:

(*i*) **Compression ratio.** NO_x concentration increases with increasing compression ratio. Lower compression ratio reduces NO_x emissions due to lower maximum temperature. However, reducing the compression ratio results in some loss in power and fuel economy.

(*ii*) **Air-fuel ratio.** The maximum NO_x is formed at A/F ratios between 14:1 and 16 : 1. At lean and rich A/F mixtures the NO_x concentration is comparatively low.

(*iii*) **Retarding spark timing.** Retarding the spark increases the time for fuel burning and reduces NO_x emissions by decreasing the maximum temperatures. However, retarding the ignition timing results in greater cooling requirements and there is some loss in power and fuel economy.

(*iv*) **Exhaust gas recirculation (EGR).** A portion of the exhaust gas is recirculated to the cylinder intake charge. This reduces the peak combustion temperature, since the inert gas serves as a heat sink. This also reduces the quantity of oxygen available for combustion. It has been observed that about 15 percent recycle will reduce NO_x emission by about 80 percent.

(*v*) **Temperature and pressure of the inlet charge.** Increased temperature and pressure of the inlet charge increases NO_x emission.

Q.3. (*a*) A refrigerated truck carrying foodstuff is speeding on a highway at 90 km/h in a desert area where the ambient air temperature is 55°C. The body of the truck may be modelled as a rectangular box measuring 11 m long, 4 m wide and 3 m high. Consider the boundary layer on the four walls to be turbulent and the heat transfer only from the four surfaces. The wall surfaces of the truck are maintained at 10°C. Assume the flow to be parallel to 11 m long side. The thermo-physical properties at the mean-film temperature of 32.5°C are :

$$\rho = 1.165 \text{ kg/m}^3,\ c_p = 1.005 \text{ kJ/kg-K},$$
$$k = 2.673 \times 10^{-2} \text{ W/m-K},$$
$$\nu = 16 \times 10^{-6} \text{ m}^2\text{/s and Pr} = 0.701$$

Work out the following :

(*i*) Sketch the system and show the boundary layer development along length.

(*ii*) Heat loss from the four surfaces.

(*iii*) Tonnage of refrigeration needed.

(*iv*) Power required to overcome the resistance acting on four surfaces. **(15)**

Sol.

(*i*) A sketch of the system and boundary layer development are shown in Fig. 4.

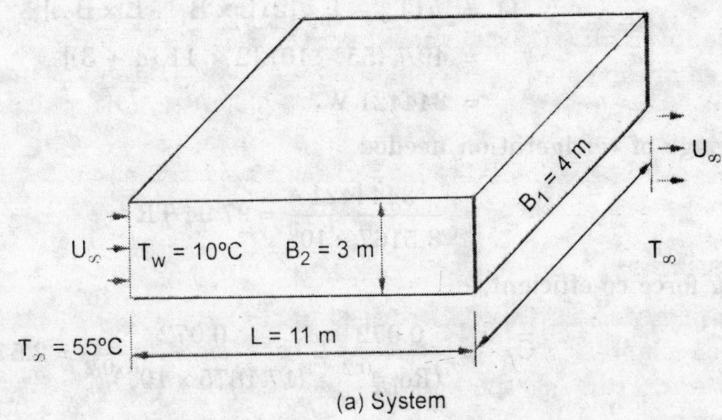

(a) System

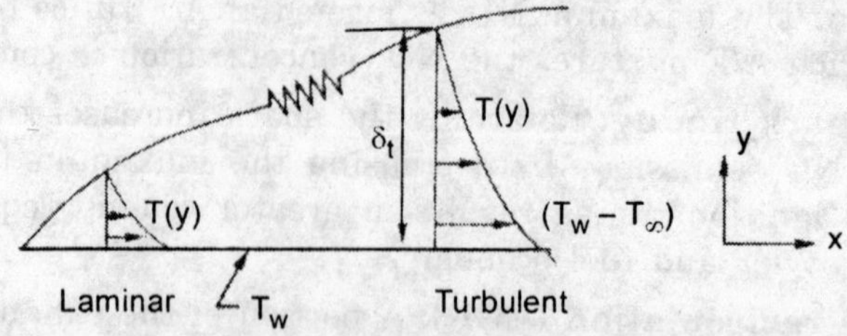

(b) Boundary layer development

Fig. 4.

(*ii*)
$$U_\infty = \frac{90 \times 10^3}{3600} = 25 \text{ m/s}$$

$$L = 1\ 1 \text{ m},\ B_1,\ 4 \text{ m},\ B_2 = 3 \text{ m}$$

$$t_f = \frac{1}{2}(10 + 55) = 32.5°\text{C}$$

$$\rho = 1.165 \text{ kg/m}^3$$

$$c_p = 1.005 \text{ kJ/kg.K},$$

$$k = 2.673 \times 10^{-2} \text{ W/m.K},$$

$$\nu = 16 \times 10^{-6} \text{ m}^2\text{/s},\ \text{Pr} = 0.701$$

$$\text{Re}_L = \frac{U_\infty L}{\nu} = \frac{25 \times 11}{15 \times 10^{-6}} = 17.1875 \times 10^6$$

For turbulent boundary layer on a flat plate,

$$\overline{N}u = \frac{\bar{h}L}{k} = 0.0375\ \text{Re}_L^{0.8}\ \text{Pr}^{1/3}$$

$$\bar{h} = \frac{2.673 \times 10^{-2}}{11} \times 0.0375 \times (17.1875 \times 10^6)^{0.8} \times (0.701)^{1/3}$$

$$= 49.7 \text{ W/m}^2\text{°C}$$

Heat loss from the four surfaces,

$$Q = \bar{h}(T_\infty - T_w)\,[2\,(L \times B_1 + L \times B_2)]$$

$$= 49.7\ (55 - 10)\ [2 \times 11\ (4 + 3)]$$

$$= 344421 \text{ W}$$

(*iii*) **Tonnage of refrigeration needed**

$$= \frac{3444421}{3.5167 \times 10^3} = 97.94 \text{ TR}$$

(*iv*) **Drag force co-efficient,**

$$\bar{C}_f = \frac{0.072}{(\text{Re}_L)^{0.2}} = \frac{0.072}{(17.1875 \times 10^6)^{0.2}} = 2.572 \times 10^{-3}$$

Drag force on four faces,

$$F_D = \bar{C}_f \times \frac{1}{2} \rho U_\infty^2 \; [2L(B_1 + B_2)]$$

$$= 2.572 \times 10^{-3} \times \frac{1}{2} \times 1.165 \times 25^2 \times 2 \times 11 \times 7 = 144.2 \text{ N}$$

Power required $\dfrac{F_D \times U_\infty}{10^3} = \dfrac{144.2 \times 25}{10^3} = 3.605 \text{ kW}$

(*b*) A flat plate solar collector with no cover plate is used to collect the solar radiation to heat water in a commercial installation. The surface emissivity of the absorber is 0.12 while its solar absorptivity is 0.95. At a given time of the day the absorber surface temperature is 130°C when the solar irradiation is 850 W/m². The effective sky temperature is – 8°C and the ambient temperature is 27°C. Assume that the heat transfer convection co-efficient for the calm dry condition is given by,

$$h = 1.23\,(T_s - T_\infty)^{1/3} \text{ W/m}^2\text{K}$$

where T_s is the surface temperature and T_∞ is free-stream ambient temperature. Assume steady-state condition, botttom surface is well insulated and the absorber surface is diffuse. Work out the following :

(*i*) Sketch the system and show the control volume;

(*ii*) The useful heat removal rate in W/m from the collector;

(*iii*) The efficiency of collector;

(*iv*) Comment on the results.

Sol.

(*i*) The control volume is shown in Fig. 5.

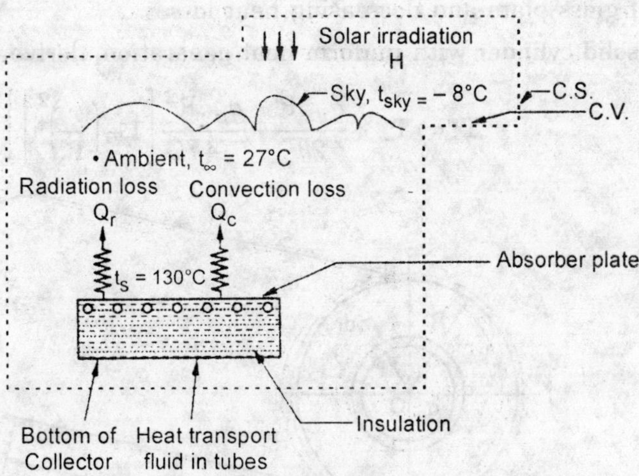

Fig. 5. Flat plate collector.

(*ii*) Solar absorptivity, $\alpha = 0.95$

Solar irradiation, $H = 580 \text{ W/m}^2$

Surface emissivety of absorber, $\varepsilon = 0.12$

Heat absorbed by collector plate,

$$Q_a = \alpha H = 0.95 \times 850 = 807.5 \text{ W/m}^2$$

Absorbed plate surface temperature,

$$T_s = 273 + 130 = 403 \text{ K}$$

Ambient temperature, $T_\infty = 273 + 27 = 300$ K

Sky temperature, $T_{sky} = 273 - 8 = 265$ K

Heat transfer convection co-efficient.,

$$h = 1.23\,(T_s - T_\infty)^{1/3} = 1.23\,(403 - 300)^{1/3}$$
$$= 5.7657 \text{ W/m}^2\text{K}$$

Heat lost by convection,

$$Q_c = h(T_s - T_\infty) = 5.7657\,(403 - 300) = 593.865 \text{ W/m}^2$$

Heat lost by radiation,

$$Q_r = \varepsilon\sigma\,(T_s^4 - T_{sky}^4) = 0.12 \times 5.67 \times 10^{-8}\,(403^4 - 265^4)$$
$$= 145.913 \text{ W/m}^2$$

Total heat lost, $Q_l = Q_c + Q_r = 593.865 + 145.913$

$$= 739.78 \text{ W/m}^2$$

Useful heat removal rate,

$$Q_u = Q_a - Q_l = 807.5 - 739.78 = 67.72 \text{ W/m}^2$$

(*iii*) Efficiency of collector,

$$\eta_c = \frac{Q_u}{Q_a} = \frac{67.72}{807.5} \times 100 = 8.387\%$$

(*iv*) The collector efficiency is very low. The efficiency can be increased by providing a cover of glass plate and decreasing heat losses.

(*c*) For a solid cylinder with uniform heat generation, derive the following expression :

$$T = T_\infty + \frac{q_g''' . R}{2h} + \frac{q_g''' . R^2}{4k}\left[1 - \left(\frac{r}{R}\right)^2\right]. \qquad (10)$$

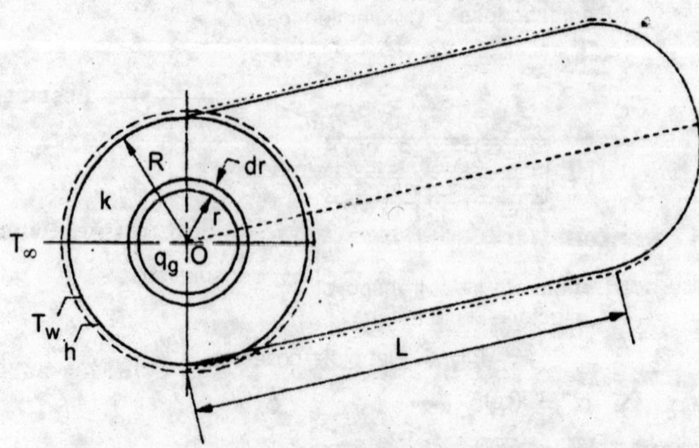

Fig. 6.

Sol. Consider a solid cylinder of radius, R and length, L in which one-dimensional radial conduction is taking place under steady state conditions.

Let k = thermal conductivity of cylinder material

h = heat transfer co-efficient

q_g = uniform heat generation per unit volume per unit time

T_∞ = ambient temperature.

Consider an element of the cylinder of radius, r and thickness, dr as shown in Fig. 6.

Heat conducted in at radius, r,

$$Q_r = -k.2\pi r L.\frac{dT}{dr}$$

Heat generated in the element,

$$Q_g = q_g . 2\pi r L.dr$$

Heat conducted out at radius, $r + dr$,

$$Q_{r+dr} = Q_r + \frac{d}{dt}(Q_r).dr$$

Under steady state conditions,

$$Q_r + Q_g = Q_{r+dr} = Q_r + \frac{d}{dr}(Q_r)\, dr$$

$$\therefore \quad Q_g = \frac{d}{dt}(Q_r)\, dr$$

$$q_g . 2\pi r L.dr = \frac{d}{dr}\left(-k.2\pi r L.\frac{dT}{dr}\right).dr$$

or
$$\frac{d}{dr}\left[r.\frac{dT}{dr}\right] = -\frac{q_g}{k}.r \qquad ...(1)$$

Integrating Eqn. (1) twice, we get,

$$r.\frac{dT}{dr} = \frac{q_g}{k}\frac{r^2}{2} + C_1$$

or
$$\frac{dT}{dr} = \frac{q_g}{k}.\frac{r}{2} + \frac{C_1}{r} \qquad ...(2)$$

$$T = \frac{q_g}{k}.\frac{r^2}{4} + C_1 \ln r + C_2 \qquad ...(3)$$

where C_1 and C_2 are constants of integration to be determined from boundary conditions.

Boundary conditions are :

(*i*) At r = R, T = T_w, the cylinder wall temperature.

(*ii*) Heat generated = Heat lost by conduction at the cylinder surface.

$$q_g \times \pi R^2 L = -k.2\pi RL.\frac{dT}{dr}\bigg]_{r=R}$$

Also at $r = 0, \quad \frac{dT}{dt} = 0$

From Eqn. (2). $$\left.\frac{dT}{dt}\right|_{r=R} = -\frac{q_g}{k}.\frac{R}{2} + \frac{C_1}{R} \qquad ...(a)$$

Also from boundary condition (*ii*), we have

$$\left.\frac{dT}{dt}\right|_{r=R} = -\frac{q_g}{k}.\frac{R}{2} \qquad ...(b)$$

Comparing Eqns. (*a*) and (*b*), we get,

$$C_1 = 0$$

Applying boundary condition (*i*) to Eqn. (3), we get,

$$T_w = \frac{q_g}{k}.\frac{R^2}{4} + C_2$$

$$\therefore \quad C_2 = T_w + \frac{q_g}{k}.\frac{R^2}{4}$$

Substituting the values of C_1 and C_2 in Eqn. (3), we get,

$$T = T_w + \frac{q_g}{4k}(R^2 - r^2) \qquad ...(4)$$

At $r = 0$, $$T_{max} = T_w + \frac{q_g}{4k}R^2 \qquad ...(5)$$

Combining Eqns. (4) and (5), we get,

$$\frac{T - T_w}{T_{max} - T_w} = 1 - \left(\frac{r}{R}\right)^2 \qquad ...(6)$$

Energy generated within the cylinder

= Energy dissipated by convection at the cylinder boundary

$$q_g \times \pi R^2 L = h \times 2\pi RL\,(T_w - T_\infty)$$

$$\therefore \quad T_w = T_\infty + \frac{q_g}{2h}R \qquad ...(7)$$

Substituting the value of Tw from Eqn. (7) in Eqn. (4), we get,

$$T = T_\infty + \frac{q_g}{2h}.R + \frac{q_g}{4k}R^2\left[1 - \left(\frac{r}{R}\right)^2\right] \qquad ...(8)$$

which is the desired expression.

Q.4. **(*a*)** In order to keep off foodstuff in better condition, a food storage locker requires 20 tons of refrigeration. The evaporator temperature is – 8°C and the condenser temperature is 30°C. The refrigerant used in the refrigeration plant is Fr-12. This is subcooled by 5°C before entering to the expansion valve. The vapour is superheated to 6°C before leaving the evaporator coil. The compressor employed is of two-cylinder, single acting having stroke equal to 1.5 times the bore. The speed of the compressor is 900 rpm. The compression process may be assumed to be isentropic. Work out the following :

(*i*) Sketch the system and show the process on T-*s* and *p*-*h* diagrams ;

(*ii*) The refrigerating effect per kg of refrigerant ;

(*iii*) The mass flow rate of the refrigerant circulated ;

(*iv*) The theoretical piston displacement ;

(*v*) The power input to compressor if mechanical efficiency is 0.9 ;

(*vi*) The co-efficient of performance ;

(*vii*) The bore and stroke if volumetric efficiency is 90%.

Use the following properties for the solution :

$$h_1 = 356 \text{ kJ/kg}$$
$$h_2 = 377 \text{ kJ/kg}$$
$$h_{fs} = h_4 = 225.5 \text{ kJ/kg}$$
$$v_1 = 0.077 \text{ m}^3\text{/kg}$$

where 1, 2, 3 and 4 refer to inlet to compressor, outlet of compressor, exit of subcoolrer and capillary tube/expansion valve exit respectively. (15)

Sol.

(*i*) A sketch of the system is shown in Fig. 7. The T-s and p-h diagrams are shown in Fig. 8.

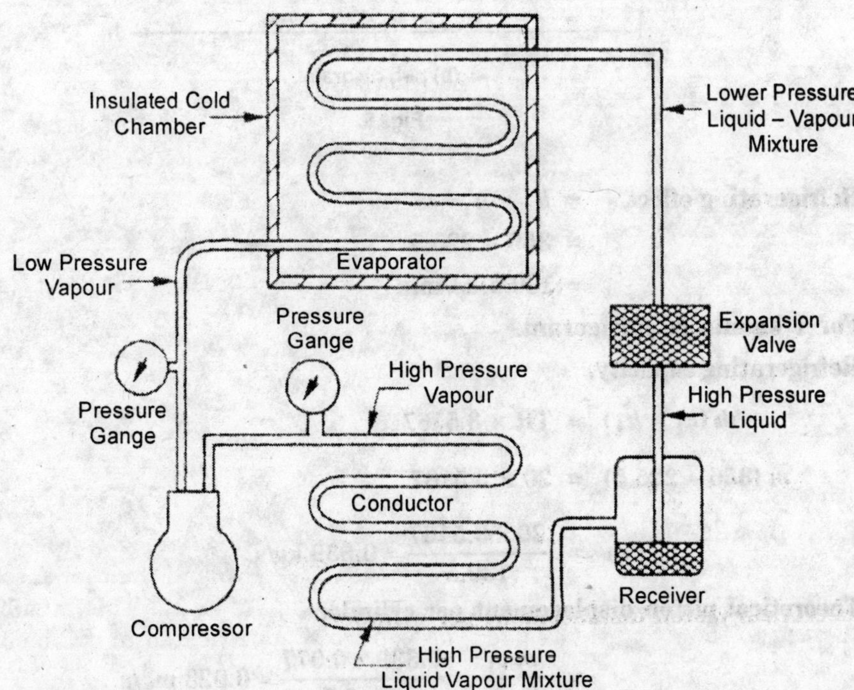

Fig. 7. Simple vapour compression refrigeration system.

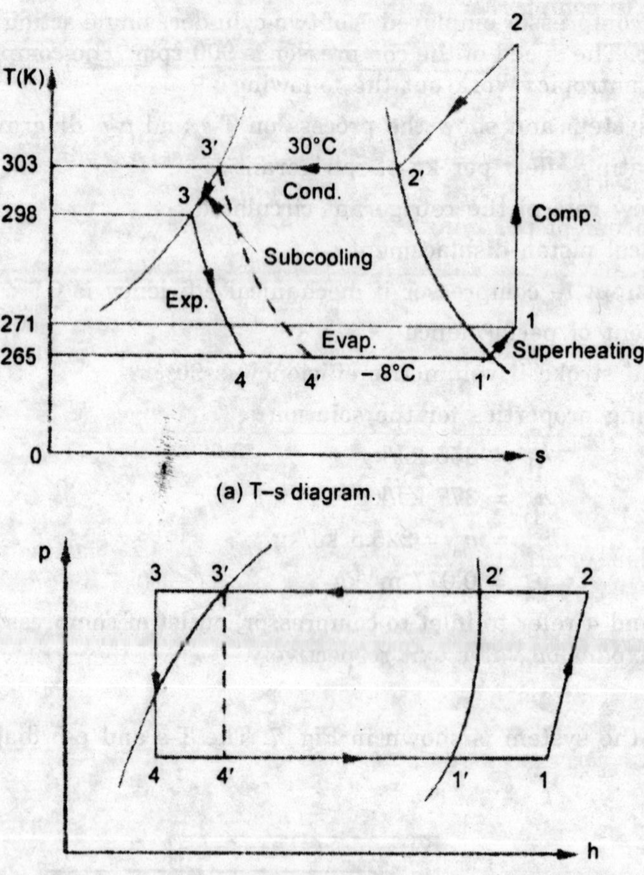

(a) T–s diagram.

(b) p–h diagram

Fig. 8.

(ii) Refrigerating effect $= h_1 - h_4$

$= 356 - 225.5$

$= 130.5$ kJ/min

For 1 kg/min of refrigerant.

(iii) Refrigerating capacity,

$$\dot{m}\,(h_1 - h_4) = \text{TR} \times 3.5167$$

$$\dot{m}\,(356 - 225.5) = 20 \times 3.5167$$

$$\dot{m} = \frac{20 \times 3.5167}{130.5} = 0.539 \text{ kg/s.}$$

(iv) Theoretical piston displacement per cylinder

$$= \frac{\dot{m}\,v_1}{2 \times \eta_v} = \frac{0.539 \times 0.077}{2 \times 0.9} = 0.023 \text{ m}^3\text{/s}$$

(*v*) Power input to compressor

$$= \frac{\dot{m}\,(h_2 - h_1)}{\eta_{mech}} = \frac{0.539\,(377 - 356)}{0.9} = 12.58 \text{ kW}$$

(*vi*) $$\text{COP} = \frac{h_1 - h_{f3}}{h_2 - h_1} = \frac{356 - 225.5}{377 - 356} = \frac{130.5}{21} = 6.21$$

(*vii*) Piston displacement per minute

$$= \text{Piston area} \times \text{Stroke} \times \text{rpm}$$

$$= \frac{\pi}{4} D^2 L \times N = \frac{\pi}{4} D^2 \times 1.5\, D \times 900$$

$$= 1060.2875\, D^3$$

$$= 0.023 \times 60$$

$$D = 109.3 \text{ mm},$$

$$L = 163.9 \text{ mm}$$

(*b*) With the help of a neat sketch, discuss the working principle of a practical vapour absorption system. Derive the expression for its co-efficient of performance. (15)

Sol. A simple vapour absorption system consists of a condenser, an expansion device, an evaporator, an absorber, a pump, a generator and a pressure reducing valve. The two commonly used refrigerant-absorbent pairs are : ammonia-water (aqua ammonia) and lithium bromide-water. The practical system is fitted with an analyser, a rectifier and a heat exchanger as shown in Fig. 9. These accessories help to improve the performance and working of the plant.

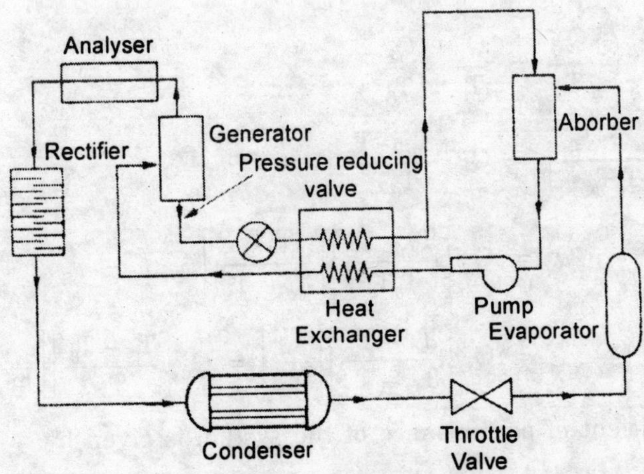

Fig. 9. Practical vapour absorption system.

The working principle of a practical ammonia absorption system is described below :

Ammonia vapours are produced in the generator at high pressure from aqua ammonia by external heating. The water vapour are removed by passing through an analyser and rectifier. The dehydrated ammonia enters the condenser where vapours are condensed. The sub-cooled high pressure liquid ammonia is passed through a throttle valve to further lower its temperature. Now they enter the evaporator and leave as saturated vapour. The saturated vapour goes to absorber to become strong after absorbing ammonia vapour and the strong solution is pumped to the generator through the pump and heat exchanger. The solution in the generator becomes weak as ammonia vapour comes out of it. The weak ammonia solution from the generator is passed to the heat exchanger through the pressure reducing valve. The cycle is repeated.

Co-efficient of performance :

Let Q_g = heat given to the refrigerant in the generator at temperature, T_g

Q_c = heat discharged to cooling water from the condenser and absorber at temperature, T_c

Q_e = heat absorbed by the refrigerant in the evaporator at temperature, T_e

Q_p = heat added to the refrigerant due to pump work.

Neglecting the heat due to pump work, Q_p, we have according to the First law of Thermodynamics,

$$Q_c = Q_g + Q_e \qquad ...(1)$$

Since the vapour absorption system can be considered as a perfectly reversible system, therefore, the initial entropy of the system must be equal to the entropy of the system after the change in its condition.

$$\frac{Q_g}{T_g} + \frac{Q_e}{T_e} = \frac{Q_c}{T_c} \qquad ...(2)$$

$$= \frac{Q_g + Q_e}{T_c} \qquad \text{[From Eqn. (1)]}$$

or

$$\frac{Q_g}{T_g} - \frac{Q_g}{T_c} = \frac{Q_c}{T_c} - \frac{Q_e}{T_e}$$

$$Q_g\left(\frac{T_c - T_g}{T_c T_g}\right) = Q_e\left(\frac{T_e - T_c}{T_e T_c}\right)$$

$$Q_g = Q_e\left[\frac{T_e - T_c}{T_e T_c}\right]\left[\frac{T_c T_g}{T_c - T_g}\right]$$

$$= Q_e\left[\frac{T_c - T_e}{T_g - T_c}\right]\left[\frac{T_g}{T_e}\right] = Q_e\left[\frac{T_c - T_e}{T_e}\right]\left[\frac{T_g}{T_g - T_c}\right] \qquad ...(3)$$

Maximum co-efficient of performance of the system is given by,

$$(\text{COP})_{\text{max}} = \frac{Q_e}{Q_g} = \left[\frac{T_e}{T_c - T_e}\right]\left[\frac{T_g - T_c}{T_g}\right] \qquad ...(4)$$

In case, the heat is discharged at different temperatures in condenser and absorber, then

$$(\text{COP})_{\text{max}} = \left[\frac{T_e}{T_c - T_e}\right]\left[\frac{T_g - T_a}{T_g}\right] \qquad ...(5)$$

where T_a = Temperature at which heat is discharged in the absorber.

(*c*) Discuss the various loads for estimating the load of an auditorium. (10)

Sol. Cooling load of an auditorium :

1. Sensible heat gain :

(*i*) The heat flowing into the building by conduction through exterior walls, floors, ceilings, doors and windows due to temperature difference on their two sides.

(*ii*) The heat received from solar radiation. It consists of,

(*a*) The heat transmitted directly through glass of windows, ventilators or doors, and

(*b*) The heat absorbed by walls and roofs exposed to solar radiation and later on transferred to the room by conduction.

(*iii*) The heat given off by lights, motors, fans, machinery, etc.

(*iv*) The heat liberated by the occupants,

(*v*) The heat carried by infiltrating air through the cracks in doors, windows and through their frequent openings,

(*vi*) The heat gain through the walls of ducts carrying conditioned air through unconditioned space in the building.

(*vii*) Heat of respiration.

2. Latent heat gain :

(*i*) The heat gain due to moisture in the outside air entering by infiltration.

(*ii*) The heat gain due to condensation of moisture from occupants.

3. Safety factor : 5% additional load.

Q.5. (*a*) What is the difference between the Bernoulli's equation and the steady flow energy equation ? Derive the steady flow energy equation and reduce it to the Bernoulli's equation. (10)

Sol. Bernoulli's equation. This equation is applicable for a steady and irrotational flow such that the fluid is incompressible and non-viscous. According to this equation,

Pressure head + Velocity head + Datum head = Constant

or $$\frac{p}{\gamma} + \frac{c^2}{2g} + z = \text{constant}$$

or $$\frac{p}{\rho} + \frac{c^2}{2} + zg = \text{constant.}$$

Steady flow energy equation (SFEE). This equation is based on the First law of Thermodynamics, which is the law of conservation of energy. This equation is applicable for steady flow, uniform composition fluid, constant fluid state with time, and only interaction between the system and surroundings being work and heat. It considers only potential, kinetic and flow energies. SFEE may be written as

$$\dot{m}\left(u_1 + p_1v_1 + \frac{c_1^2}{2} + gz_1\right) + \dot{Q}$$

$$= \dot{m}\left(u_2 + p_2v_2 + \frac{c_2^2}{2} + gz_2\right) + \dot{W}$$

Now, $h = u + pv$

For a unit mass, SFEE becomes,

$$q - w = (h_2 - h_1) + \frac{1}{2}(c_2^2 - c_1^2) + g\,(z_2 - z_1)$$

Derivation of SFEE. Consider the control volume (C.V.) of a thermodynamic open system as shown in Fig. 10, through which a working fluid is passing at a steady rate.

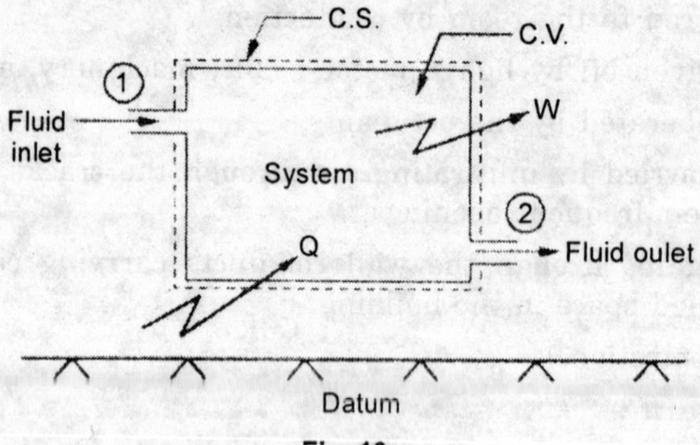

Fig. 10.

As per the law of conservation of energy,

$$E_1 \pm W \pm Q = E_2$$

where E represents the sum of all the energies associated with the fluid.

For a thermodynamics system, the above equation can be written as

$$m\left[u_1 + p_1v_1 + \frac{c_1^2}{2} + gz_1\right] \pm W \pm Q = m\left[u_2 + p_2v_2 + \frac{c_2^2}{2} + gz_2\right]$$

where gz = potential energy per unit mass

$\frac{1}{2}c^2$ = kinetic energy per unit mass

u = internal energy per unit mass

pv = flow work per unit mass.

By definition, $h = u + pv$

$$\therefore\; m\left[h_1 + \frac{c_1^2}{2} + gz_1\right] \pm W \pm Q = m\left[h_2 + \frac{c_2^2}{2} + gz_2\right]$$

For a steady flow process if heat is supplied to the system (+ Q) and work is developed by the system (– W), then

$$Q - W = m\,[(u_2 - u_1) + (p_2v_2 - p_1v_1) + \frac{1}{2}(c_2^2 - c_1^2) + g\,(z_2 - z_1)$$

$$= \Delta U + \Delta\,(pv) + \Delta KE + \Delta PE$$

For a unit mass, we get,

$$u_1 + p_1v_1 + \frac{c_1^2}{2} + gz_1 + q = u_2 + p_2v_2 + \frac{c_2^2}{2} + gz_2 + w$$

If $\Delta u = 0, q = 0$ and $w = 0$,

then $$p_1v_1 + \frac{c_1^2}{2} + gz_1 = p_2v_2 + \frac{c_2^2}{2} + gz_2$$

or $$\frac{p_1}{\rho_1} + \frac{c_1^2}{2} + gz_1 = \frac{p_2}{\rho_2} + \frac{c_2^2}{2} + gz_2$$

or $$\frac{p}{\rho} + \frac{c_1^2}{2} + gz = \text{constant}$$

which is Bernoulli's equation.

(b) Show that for sonic flow, the deviation between the compressible and incompressible flow values of the pressure co-efficient of a perfect gas ($\gamma = 1.4$) is about 27.5%. (10)

Sol. Neglecting forces due to gravity, Bernoulli's equation,

$$\frac{p}{\rho} + \frac{c^2}{2} = \text{constant}$$

or $$p + \frac{1}{2}\rho c^2 = \text{constant.}$$

The term $\frac{1}{2}\rho c^2$ is called *dynamic pressure for incompressible fluid.*

Let $$p_d = \frac{1}{2}\rho c^2$$

$$= \frac{1}{2} \cdot \frac{p}{RT} c^2 = \frac{\gamma p c^2}{2\,\gamma RT} = \frac{\gamma p c^2}{2\,c_s^2} = \frac{1}{2}\gamma p\,M^2$$

where $c_s = \sqrt{\gamma RT}$ = Sonic velocity.

Consider now a compressible fluid flowing with velocity c is decelerated isentropically to stagnation state. The pressure increases due to deceleration to dynamic pressure. For a compressible fluid,

$$\frac{p_0}{p} = \left[1 + \frac{\gamma - 1}{2} M^2\right]^{\frac{\gamma}{\gamma - 1}}$$

$$\frac{p_0 - p}{p} = \frac{p_0 - p}{\left(\frac{2p_d}{\gamma M^2}\right)} = \left[1 + \frac{\gamma - 1}{2} M^2\right]^{\frac{\gamma}{\gamma - 1}} - 1$$

Put $$\frac{\gamma - 1}{p_d} M^2 = x \text{ and } \frac{\gamma - 1}{\gamma} = n$$

For $x < 1$, expanding Binomially

$$\frac{p_0 - p}{p_d} = \frac{2}{\gamma M^2}[(1+x)^n - 1]$$

$$= \frac{2}{\gamma M^2}\left[1 + nx + \frac{n(n-1)x^2}{2!} + ... - 1\right]$$

$$= \frac{2}{\gamma M^2}\left[\frac{\gamma}{2}M^2 + \frac{\gamma M^4}{8} + ...\right]$$

Pressure co-efficient

$$= \frac{p_0 - p}{\frac{1}{2}\rho c^2} = 1 + \frac{M^2}{4} + \frac{2-\gamma}{24}M^4 + ...$$

For Sonic flow, $M = 1.$

Pressure co-efficient $= 1 + \frac{1}{4} + \left(\frac{2-1.4}{24}\right) \times 1 = 1.275$

Deviation in pressure co-efficient

$= 1.275 - 1 = 0.275$ or 27.5%

(c) A convergent-divergent nozzle has an exit area to throat area ratio of 2. Air enters the nozzle with a stagnation pressure of 950 kPa and a stagnation temperature of 350 K. The throat area is 490 mm^2. Determine the mass flow rate, exit pressure, exit temperature, exit Mach number and exit velocity for the following conditions :

(*i*) Sonic velocity at the throat, diverging section acting as a nozzle.

(*ii*) Sonic velocity at the throat, diverging section acting as a diffuser.

Table below gives the one-D, isentropic, compressible flow functions for an ideal gas represents the critical values at the throat where the Mach number is unity :

M	M*	A/A*	p/p_0	T/T_0
0.308	0.326	2.00	0.936	0.9812
1.00	1.00	1.00	0.528	0.8333
2.197	1.717	2.00	0.0939	0.5089

(20)

Fig. 11.

Sol.

(*i*) The convergent-divergent nozzle is shown in Fig. 11. For the diverging section acting as a nozzle the flow is supersonic and M* > 1.

For $\frac{A_e}{A^*} = 2, M_e = 2.197, \frac{p_e}{p_0} = 0.0939, \frac{T_e}{T_0} = 0.5089$

$\therefore$ $p_e = 0.0939 \times 950 = 89.205$ kPa

$T_e = 0.5089 \times 350 = 178.115$ K

Sonic velocity at exit, $c_{se} = \sqrt{\gamma RT_e} = \sqrt{1.4 \times 287 \times 178.115}$

$= 267.52$ m/s

Exit velocity, $c_e = M_e \times c_{se} = 2.197 \times 267.52 = 587.74$ m/s

For $M^* = 1$ at the throat, $\frac{T^*}{T_0} = 0.8333, \frac{p^*}{p_0} = 0.528$

$T^* = 350 \times 0.8333 = 291.65$ K

$p^* = 950 \times 0.528 = 501.6$ kPa

$c_{st} = c_t = \text{J}\gamma\text{RT}^* = \sqrt{\gamma RT^*} = \sqrt{1.4 \times 287 \times 297.65} = 342.3$ m/s

Density of air at throat,

$$\rho^* = \frac{p^*}{RT^*} = \frac{501.6}{0.287 \times 291.65} = 5.9926 \text{ kg/m}^3$$

Mass flow rate,

$\dot{m} = \rho^* A_t c_t = 5.9926 \times 490 \times 10^{-6} \times 342.3$

$= 1.005$ kg/s.

(*ii*) For the diverging section acting as a diffuser, M* < 1 at the throat.

$M_e = 0.308, \frac{p_e}{p_0} = 0.936, \frac{T_e}{T_0} = 0.9812$

$p_e = 0.936 \times 950 = 889.2$ kPa

$T_e = 0.9812 \times 350 = 343.42$ K

$c_{st} = \sqrt{1.4 \times 287 \times 343.42} = 371.46$ m/s q

$c_e = 0.308 \times 371.46 = 114.41$ m/s

Q.6. (*a*) In a compressor, the frictional torque is T in the impeller. If the diameter is D, speed N, fluid viscosity n and density p, prove using Buckingham's n-theorem of dimensional analysis that

$$T = \rho.N^2.D^5.f\left[\frac{u}{\rho ND^2}\right]$$

Sol. Frictional torque in a compressor,

$T = f(D, N, \mu, \rho)$

Here $n = 5, m = 3$

Number of dimensionless parameters = $n - m = 5 - 3 = 2$

The dimensions of variables are

$$\begin{aligned} D &: L \\ N &: T^{-1} \\ \mu &: ML^{-1} L^{-1} \\ \rho &: ML^{-3} \\ T &: ML^2 T^{-2} \end{aligned}$$

Let D, p, N be the repeating variables

$$\pi_1 = D^{a1} \rho^{b1}, N^{c1} \mu$$

$$M^0 L^0 T^0 = [L]^{a_1} . [ML^{-3}]^{b_1} . [T^{-1}]^{c_1} ML^{-1} T^{-1}$$

$$= L^{a_1 - 3b_1 - 1} M^{b_1 + 1} T^{c_1 - 1}$$

$$b_1 + 1 = 0, b_1 = -1$$

$$a_1 - 3b_1 - 1 = 0, a_1 = -2$$

$$-c_1 - 1 = 0, c_1 = -1$$

$$\pi_1 = D^{-2} \rho^{-1} N^{-1} \mu = \frac{\mu}{\rho ND^2}$$

$$\pi_2 = D^{a_2} \rho^{b_2} N^{c_2} T$$

$$M^0 L^0 T^0 = [L]^{a_2} . [ML^{-3}]^{b_2} . [T^{-1}]^{c_2} ML^2 T^{-2}$$

$$= M^{b_2 + 1} L^{a_2 - 3b_2 + 2} T^{-c_2 - 2}$$

$$b_2 + 1 = 0, b_2 = -1$$

$$a_2 - 3b_2 + 2 = 0, a_2 = -5$$

$$-c_2 - 2 = 0, c_2 = -2$$

$$\pi_2 = D^{-5} \rho^{-1} N^{-2} \mu = \frac{\mu}{\rho N^2 D^5}$$

Now,

$$f(\pi_1, \pi_2) = 0$$

$$\pi_2 = f(\pi_1)$$

$$\frac{T}{\rho N^2 D^5} = f\left(\frac{\mu}{\rho ND^2}\right)$$

$$T = \rho N^2 D^5 f\left(\frac{\mu}{\rho ND^2}\right)$$

(***b***) Determine the position in which a solid cylindrical block of wood of diameter 0.3 m and length 0.4 m will float in water. Take specific gravity of wood as 0.5. **(20)**

Sol. The log of wood floating in water is shown in Fig. 12.

Density of wood, ρ = specific gravity of wood × 1000

$= 0.5 \times 1000 = 500 \text{ kg/m}^3$

Weight of volume of water displaced by cylinder

= Weight of cylinder

Let d = diameter of cylinder

h = length of cylinder in water

Then, $$\frac{\pi}{4} d^2 h \times g \times \rho_w = \frac{\pi}{4} d^2 l \times \rho g$$

$$h = 0.5\ 1 = 0.5 \times 0.4 = 0.2 \text{ m}$$

Fig. 12

Centre of buoyancy B from the bottom A $= \frac{h}{2} = 0.1$ m

C.G of cylinder from A $= \frac{l}{2} = 0.2$ m

$$BG = 0.2 - 0.1 = 0.1 \text{ m}$$

Metacentric height, $GM = \frac{1}{V} - BG$

$$I = \frac{\pi d^4}{64} = \frac{\pi \times 0.4^4}{64} = 3.976 \times 10^{-4} \text{ m}^4$$

$$V = \frac{\pi}{4} \times 0.3^2 \times 0.2 = 0.01414 \text{ m}^3$$

$$GM = \frac{3.976 \times 10^{-4}}{0.01414} - 0.1 = -0.072 \text{ m}$$

Since GM is negative, point M lies below G and the cylinder will float in unstable position.

(c) Describe the working principle of a Pitot tube with the help of a neat sketch and explain how it can be used to measure the stagnation pressure. (10)

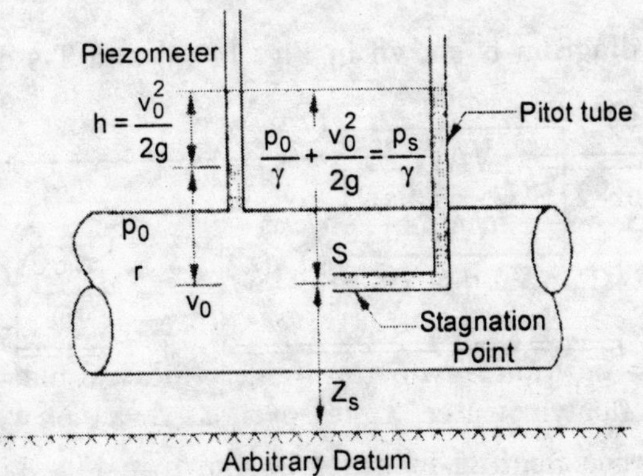

Fig. 13. Pitot tube.

Sol. Working principle of a Pitot tube. The Pilot-tube is a device used for velocity measurement. It, in fact, measures the stagnation pressure at any point in the flow. The static pressure may be measured by a piezometer.

Stagnation pressure, $p_s = p_0 + \frac{1}{2}\rho v_0^2$

where p_0 = static pressure,

v_0 = undisturbed flow velocity.

The Pitot-tube shown in Fig. 13 consists of a tube having 90° bend of shorter length, opened at its both ends. The bent leg is directed upstream, so that a stagnation point is created immediately ahead of its inlet. On account of the stagnation pressure so caused, the liquid rises in the vertical limb. A piezometer installed on the pipe boundary indicates the static pressure, p_0.

$$v_0 = \sqrt{\frac{2(p_s - p_0)}{\rho}} = \sqrt{\frac{2g(p_s - p_0)}{\gamma}} = \sqrt{2gh}$$

where h = difference between the stagnation head $\frac{p_s}{\gamma}$ and the static head $\frac{p_0}{\gamma}$.

Q.7. (*a*) In an open-cycle gas turbine plant, the air enters at 15°C and 1 bar, and is compressed in a compressor to a pressure ratio of 15. The air from the exit of compressor is first heated in a heat exchanger which is 75% efficient by turbine exhaust gas and then in a combustor to a temperature of 1600 K. The same gas expands in a two-stage turbine such that the expansion work is maximum. The exhaust gas from HP turbine is reheated to 1500 K and then expands to LP. turbine. The isentropic efficiencies of compressor and turbine may be taken as 86% and 88% respectively. The mechanical efficiencies for compressor and turbine are 97% each. The alternator efficiency is 98%. The output of turbo-alternator is 250 MW. Work out the following :

(*i*) Sketch the system and show the process on T-s diagram,

(*ii*) The cycle thermal efficiency,

(*iii*) The work ratio,

(*iv*) The specific power output,

(*v*) The mass flow rate of air. (15)

Sol.

(*i*) The schematic diagram is shown in Fig. 14 (a) and T-s diagram in Fig. 14 (*b*).

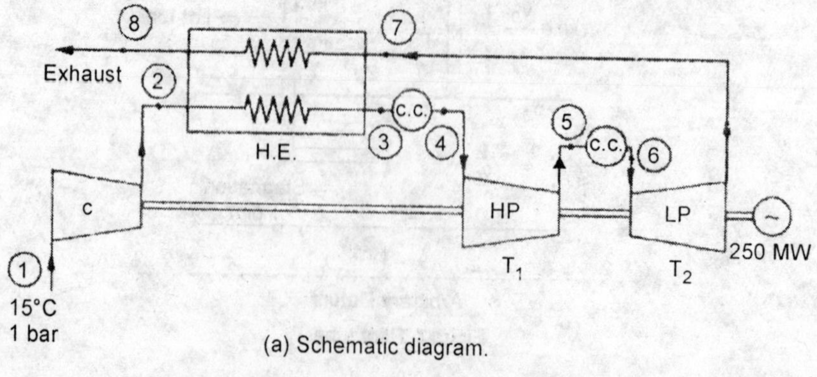

(a) Schematic diagram.

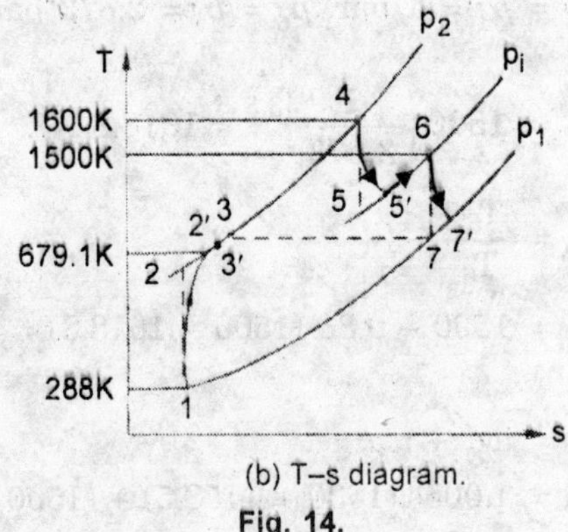

(b) T–s diagram.

Fig. 14.

(*ii*)

$$\eta_c = 0.86,\ \eta_t = 0.88,$$

$$\eta_{alt} = 0.98$$

$$(\eta_{mech})_c = (\eta_{mech})_t = 0.97$$

$$T_2 = T_1 \left(\frac{p_2}{p_1}\right)^{\frac{\gamma-1}{\gamma}}$$

$$= 288\,(15)^{\frac{1}{3.5}} = 624.3 \text{ K}$$

$$T_2' = T_1 + \frac{T_2 - T_1}{\eta_c} = 288 + \frac{624.3 - 288}{0.86} = 679.1 \text{ K}$$

Compressor work,

$$w_c = \frac{c_p\,(T_2' - T_1)}{(\eta_{mech})_c} = \frac{1.005\,(679.1 - 288)}{0.97} = 405.2 \text{ kJ/kg}$$

For maximum expansion work,

$$p_i = \sqrt{p_1 p_2} = \sqrt{1 \times 15} = 3.873 \text{ bar}$$

$$p_5 = p_i = 3.873 \text{ bar}$$

$$T_5 = T_4 \left(\frac{p_5}{p_4}\right)^{\frac{\gamma-1}{\gamma}} = 1600\left(\frac{3.873}{15}\right)^{\frac{1}{3.5}}$$

$$= 1086.7 \text{ K}$$

$$\eta_{t1} = \frac{T_4 - T_5'}{T_4 - T_5}$$

$$T_5' = T_4 - \eta_{t1}\,(T_4 - T_5)$$

$$= 1600 - 0.88\,(1600 - 1086.7) = 1158.5 \text{ K}$$

$$T_7 = T_6 \left(\frac{p_7}{p_6}\right)^{\frac{\gamma-1}{\gamma}}$$

$$p_7 = p_1 = 1 \text{ bar}, p_6 = p_i = 2.873 \text{ bar}$$

$$T_7 = 1500\left(\frac{1}{3.873}\right)^{\frac{1}{3.5}} = 1018.8 \text{ K}$$

$$\eta_{t2} = \frac{T_6 - T_7'}{T_6 - T_7}$$

$$T_7' = 1500 - 0.88\ (1500 - 1018.8) = 1076.5 \text{ K}$$

Turbine work,

$$w_t = w_{t1} + w_{t2}$$
$$= 1.005\ [(1600 - 1158.5) + (1500 - 1076.5)] \times 0.97$$
$$= 843.245 \text{ kJ/kg}$$

Net work, $w_{net} = w_t - w_c$

$$= 843.245 - 405.2 = 438.045 \text{ kJ/kg}$$

$$\text{Heat exchanger efficiency} = \frac{T_3' - T_2'}{T_7' - T_2'}$$

$$0.75 = \frac{T_3' - 679.1}{1076.5 - 679.1}$$

$$T_3' = 977.15 \text{ K}$$

Heat supplied, $= 1.005\ [(1600 - 977.15) + (1500 - 1\ 158.5)] = 969.17 \text{kJ/kg}$

Thermal efficiency,

$$\eta_{th} = \frac{w_{net}}{q_s} = \frac{438.045}{969.17} = 0.452 \text{ or } 45.2\%$$

(*iii*) Work ratio $= \dfrac{w_{net}}{w_t} = \dfrac{438.045}{843.245} = 0.5195.$

(*iv*) Specific power output

$$1 \times w_{net} = 1 \times 438.045 = 438.045 \text{ kW}$$

(*v*) Mass flow rate of air $= \dfrac{\text{MW} \times \eta_{alt}}{w_{net}}$

$$= \frac{250 \times 10^3 \times 0.98}{438.045} = 559.3 \text{ kg/s}$$

(*b*) A centrifugal pump is required to lift 9100 lit/ min. of water against a head of 6m while running at 500 rpm. The velocity of flow through the wheel is 2 m/s and the manometric efficiency is 60%. The angle of vane tip with the direction of motion is 150°. Work out the following :

(*i*) Sketch an impeller of centrifugal pump and draw the velocity triangle,

(*ii*) The diameter of impeller,

(*iii*) The width of impeller. **(15)**

Sol.

(*i*) The sketch of the impeller and velocity triangles are shown in Fig. 15.

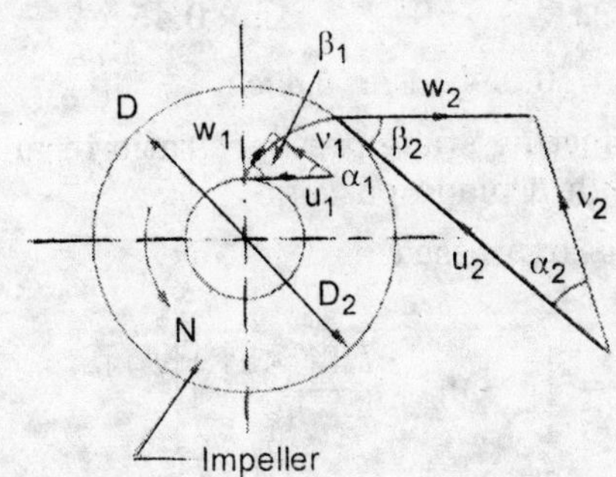

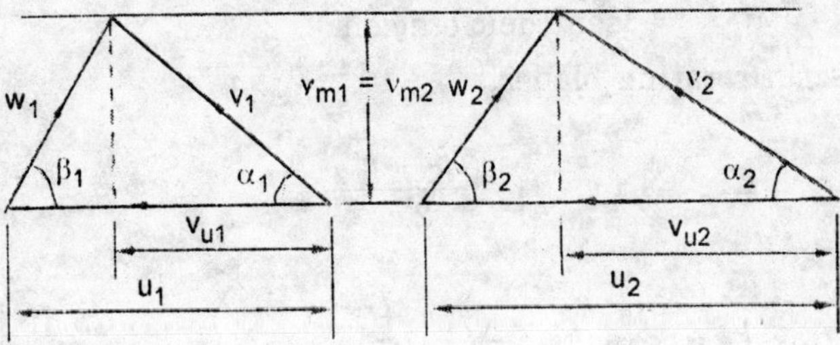

Fig. 15.

(*ii*)
$$\frac{H_{mano}}{\eta_{mano}} = \frac{u_2 v_{u2}}{g}$$

$$u_2 v_{u2} = \frac{6 \times 9.81}{0.6} = 98.1 \quad ...(1)$$

$$u_2 = \frac{\pi D_2 N}{60} - \frac{\pi D_2 \times 500}{60} = 26.18 \text{ D}_2$$

$$\beta_2 = 180° - 150° = 30°, v_{m2} = 2 \text{ m/s}$$

$$\tan \beta_2 = \frac{v_{m2}}{u_2 - v_{u2}}$$

$$\tan 30° = \frac{2}{26.18 \text{ D}_2 - v_{u2}}$$

$$v_{u2} = 26.18 \text{ D}_2 - 3.464 \quad ...(2)$$

Substituting Eqn. (2) in (1), we get,

$26.18 \text{ D}_2 (26.18 \text{ D}_2 - 3.464) = 98.1$

$D_2^2 - 0.1323 \text{ D}_2 - 0.1431 = 0$

$$D_2 = 0.1323 + \sqrt{(0.1323)^2 + 4 \times 1 \times 0.1431}$$

$$= 0.45 \text{ m or } 45 \text{ cm}$$

(*iii*)

$$Q = \pi D_2 B_2 v_{m2}$$

$$B_2 = \frac{9100 \times 10^{-3}}{60} \times \frac{1}{\pi \times 0.45 \times 2}$$

$$= 0.054 \text{ m or } 5.4 \text{ cm}$$

(*c*) Discuss the factors affecting stage pressure ratio in an axial flow compressor. Draw the velocity distribution through a blade passage. (10)

Sol. The pressure ratio is expressed as,

$$\frac{p_2}{p_1} = \left[1 + \eta_{isen}\left(\frac{T_2 - T_1}{T_1}\right)\right]^{\frac{\gamma}{\gamma - 1}}$$

where T_1, T_2 = Temperature of working fluid at inlet and outlet respectively,

η_{isen} = Isentropic efficiency.

Temperature rise in rotating blades,

$$\Delta T_r = T_2 - T_1 = \frac{v_{r1}^2 - v_{r2}^2}{2\, c_p}$$

$$\therefore \quad \frac{p_2}{p_1} = \left[1 + \eta_r \frac{\Delta T_r}{T_1}\right]^{\frac{\gamma}{\gamma - 1}}$$

Temperature rise in stationary blades,

$$\Delta T_s = T_3 - T_2 = \frac{v_{a2}^2 - v_{a3}^2}{2\, c_p}$$

$$\therefore \quad \frac{p_2}{p_1} = \left[1 + \eta_s \frac{\Delta T_s}{T_2}\right]^{\frac{\gamma}{\gamma - 1}}$$

Hence pressure rise in stationary blades,

$$\Delta p_s = p_3 - p_2 = p_2\left[\left\{1 + \eta_s \frac{\Delta T_s}{T_2}\right\}^{\frac{\gamma}{\gamma - 1}} - 1\right]$$

and in rotating blades,

$$\Delta p_r = p_2 - p_1 = p_1\left[\left\{1 + \eta_r \frac{\Delta T_r}{T_1}\right\}^{\frac{\gamma}{\gamma - 1}} - 1\right]$$

Pressure rise in a stage,

$$\Delta p_{st} = \Delta p_r + \Delta p_s$$

and

$$\Delta T_{st} = \Delta T_r + \Delta T_s$$

Stagnation pressure ratio

$$\frac{p_{02}}{p_{01}} = \left[1 + \eta_{isen}\frac{\Delta T_0}{T_0}\right]^{\frac{\gamma}{\gamma-1}}$$

If the pressure ratio per stage is same, then stage pressure ratio,

$$(r_p)_{stage} = \frac{p_{02}}{p_{01}} = \frac{p_{03}}{p_{02}}.$$

Thus, stage pressure ratio depends on the following factors:

1. Inlet and outlet temperatures of working fluid.
2. Isentropic efficiency.
3. Adiabatic constant.

Q.8. (***a***) What is the purpose of steam generator control ? With the help of sketches, discuss the various methods of steam generator control. (15)

Sol. For the safe operation of a steam generator, the boilers are equipped with boiler mountings. Boiler mountings are those machine components which are mounted over the body of the boiler itself for the safety of the boiler and for complete control of the process of steam generation. The various mountings are :

1. **Water level indicator.** Indicates the level of water in the boiler. Generally two indicators are fitted.
2. **Pressure gauge.** Measures the steam pressure of the boiler.
3. **Steam stop valve.** Shuts off the steam or regulate the flow of steam to requirements.
4. **Feed check valve.** Allows feed water under pressure to be passed to the boiler, but stop any water escaping back.
5. **Blow-down cock.** Removes sludge or sediments collected at the bottom most point in the space in a boiler, while the boiler is steaming.
6. **Fusible plug.** Protects the fire box crown plate, or fire tube from burning when the level of water in the shell falls abnormally low.
7. **Pressure safety valve.** Prevents the steam pressure in the boiler exceeding the desired rate pressure.
8. **High steam and low water safety valve.** Blows out if steam pressure is higher than the working pressure or water level in the boiler is low.

The sketches of various mountings may be referred to any book on Thermal Engineering.

(***b***) Discuss the need of governing of steam turbine. With the help of a sketch, discuss the working principle of hydromechanical speed governing loop. List the various speed governors used in practice. (15)

Sol. The purpose of speed governing is to maintain the

speed of a turbine fairly constant irrespective of the load. This is essential to keep the frequency of the output electric power constant. The principal methods of steam turbine governing are : throttle governing, nozzle governing, bypass governing and combination of above methods.

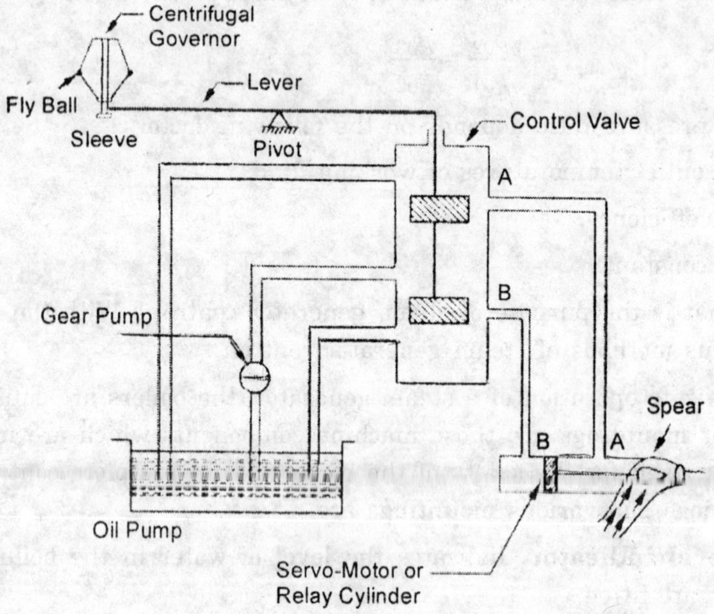

Fig. 16. Hydromechanical loop for Throttle governing of steam turbines.

Hydromechanical governing loop. The hydromechanical loop for throttle governing is shown in Fig. 16. The throttle valve will open to take care of the full rated load at constant speed. If the load is reduced, the energy supplied to the turbine will be in excess and the turbine rotor will accelerate. Thus the governor sleeve will lift and the pilot piston valve spindle will get lifted opening the upper port to oil pressure and lower port to oil return. The relay piston will thus close the throttle valve partially. The lowering of throttle valve spindle will lower the pilot piston spindle and close the ports. As soon as the ports are closed, the relay piston gets stabilized in one position corresponding to the reduced load.

(*c*) With the help of a sketch, discuss the working principle of boiling water reactor and explain in brief the function of main elements. (10)

Sol. The arrangement of the system for the boiling water reactor is shown in Fig. 17. In this reactor, ordinary water is used as coolant and enriched uranium as fuel. The steam is directly allowed to be generated in the reactor itself.

The essential elements of a nuclear reactor are :

(*i*) **Reactor core.** Fission chain reaction is made to occur here and fission energy is liberated in the form of heat for operating power conversion equipment. It consists of an assemblage of fuel elements, control rods, coolant and moderator. Its shape is approximately a right circular cylinder. The fuel elements are made of rods of uranium metal clad in a thin sheath of stainless steel, zirconium or aluminium.

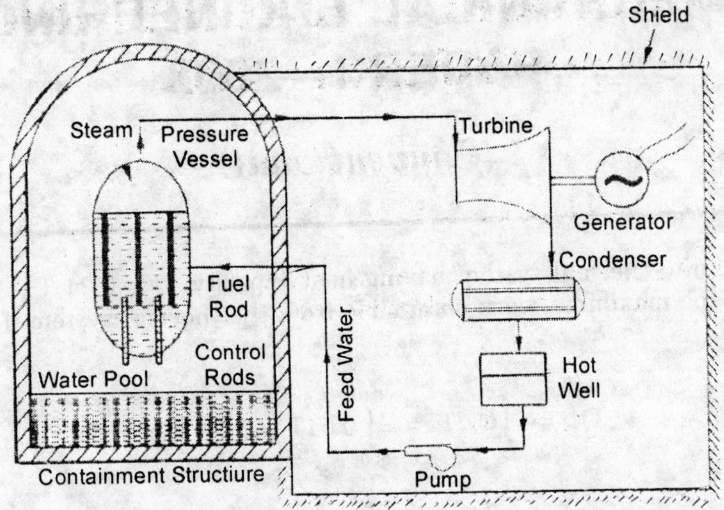

Fig. 17. Boiling water reactor.

(*ii*) **Reflector.** It is placed round the core to reflect back some of the neutrons that leak out from the surface of the core.

(*iii*) **Control mechanism.** It serves to start the reactor, maintaining at that level and for shutting down the reactor.

(*iv*) **Moderator.** Its function is to slow down the neutrons from the high velocities but not to absorb them. H_2O, D_2O (heavy water), He (gas), Be and C (graphite) are the commonly used moderators.

(*v*) **Coolants.** Its function is to remove the intense heat produced in the reactor and to for being utilised. Commonly used coolants are : Santiwax R (Organic, Hg, He, CO_2).

(*vi*) **Measuring instruments.** To measure thermal neutron flux.

(*vii*) **Shielding.** To protect the walls of the reactor vessel from radiation damage and operating personnel from exposure to radiation. It is generally made of thick concrete surrounding the reactor installation.

(*viii*) **Power generation unit.** It consists of a steam turbine, generator, condenser, hot well and a pump for feed water.

MECHANICAL ENGINEERING
PAPER I—2008

(Conventional)

Q.1. (*a*) A finite thermal system having heat capacity, C = 0.04 T^2, J/K is initially at 600 K. Estimate the maximum work obtainable from the thermal system if the surrounding is at 300 K.

Sol.

$$Q_1 = \int_{T_1}^{T_2} C.dT = \int_{600}^{300} 0.04T^2 dT$$

$$= \frac{0.04}{3}\left|T^3\right|_{600}^{300} = \frac{0.04}{3}(300^3 - 600^2)$$

$$= -2.52 \text{ MJ}$$

$$(\Delta S)_{system} = \int_{T_1}^{T_2} C.dT = \int_{600}^{300} 0.04T^2 dT$$

$$= 0.02\left|T^3\right|_{600}^{300} = 0.02(300^2 - 600^2)$$

$$= -5.4 \text{ kJ/K}$$

$$(\Delta S)_{surroundings} = \frac{Q_1 - W}{T_{surr}} = \frac{2.52 \times 10^6 \times W}{300} \text{ J/K}$$

$$(\Delta S)_{univ} = (\Delta S)_{system} + (\Delta S)_{surr}$$

$$= -5.4 \times 10^3 + \frac{2.52 \times 10^6 - W}{300} \geq 0$$

$$\frac{2.52 \times 10^6}{300} - \frac{W}{300} \geq 5.4 \times 10^3$$

$$8400 - \frac{W}{300} > 5400$$

$$\frac{W}{300} < (8400 - 5400)$$

$$W < 9 \times 10^5 \text{ J} = 900 \text{ kJ}$$

or $$W_{max} = 900 \text{ kJ}$$

(*b*) State Gouy-Stodola theorem of irreversitbility. Estimate the irreversibility associated with the expansion process of air through a very small constriction in a pipe from pressure and temperature, respectively of 8 bar and 600 K to pressure of 1.2 bar. Assume air to be an ideal gas. he temperature of surroundings is 25°C.

Sol. Gouy–Stodola theorem. The Gouy–Stodola theorem states that the irreversibility for a non-flow and steady flow process between the equilibrium states, when the system exchanges heat only with the environment is greater than or equal to zero.

The Gouy-Stodola theorem may also be stated as follows:

The rate of loss of available energy in a process is proportional to the rate of entropy generation.

$$\dot{I} = T_0 \dot{S}_{gen} = T_0(\Delta S)_{univ}$$

Given: $p_1 = 8$ bar, $T_1 = 600$ K, $p_2 = 1.2$ bar,

$$T_0 = 273 + 25 = 298 \text{ K}$$

$$I = T_0\left[(\Delta S)_{system} + (\Delta S)_{surr}\right]$$

$$= T_0 \; (\Delta S)_{univ} = T_0(\Delta S)gen$$

$$S_{gen} = \frac{Q}{T_2} - \frac{Q}{T_1} = \frac{Q(T_1 - T_2)}{T_1 T_2}$$

$$I = \frac{QT_0(T_1 - T_2)}{T_1 \, T_2}$$

The expansion process of air through a very small constriction in a pipe is a throttling process, which is an iso-enthalpic process.

Therefore, $h_1 = h_2$, $u_1 = u_2$ and $T_1 = T_2$.

Hence, $I = 0$

(*c*) A tank contains 50 kg of water initially at a temperature of 30°C. Water at the rate of 200 kg/h and temperature of 30°C enters the tank through an inlet pipe. A cooling coil immersed in the tank removes heat energy from water at the rate of 8 kW. A mechanical stirrer ensures thorough mixing of water in the tank so as to maintain a uniform temperature of water at any instant and in the process add heat energy at the rate of 0.3 kW to water.

Neglecting kinetic and potential energy changes and taking the average specific heat of water as 4.2 kJ/kg-K, derive an expression for the variation of instantaneous temperature of water in te tank with respect to time.

Sol. From the First law of thermodynamics for unsteady flow process, the rate equation is

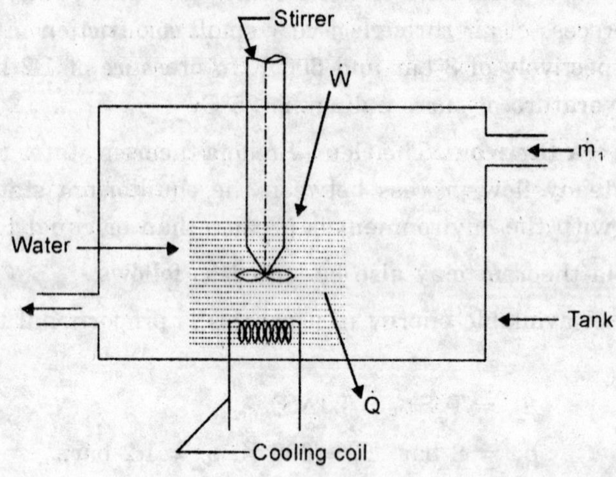

Fig. 1.

$$\dot{Q} - \dot{W} + \dot{m}_f (h_i - h_e) = \frac{dE}{dt} = \frac{dU}{dt}$$

where m_f = mass of fluid flowing, kg/s

h_i, h_e = enthalpies at inlet and exit of control volume, kJ/kg

Here $$\frac{dU}{dt} = \frac{d(mu)}{dt} = m.\frac{du}{dt} + u.\frac{dm}{dt}$$

where m = mass of water in the tank

Since $$\frac{dm}{dt} = 0$$

Hence $$\frac{dU}{dt} = m.\frac{du}{dt}$$

Let T = temperature of water in the tank at time t.

Assuming T to be uniform, we have

$$m.c_p \frac{dT}{dt} = \dot{Q} - \dot{W} + \dot{m}_f \, c_p \, (T_i - T)$$

Given: m = 50 kg, c_p = 4.2 kJ/kg.K, $\dot{Q}$ = – 8 kW

$\dot{W}$ = – 0.2 kW, m_f = 200 kg/h, T_i = 303 K

$$50 \times 4.2 \times \frac{dT}{dt} = -8 - (-0.2) + \frac{200}{3600} \times 4.2\,(303 - T)$$

$$210 \times \frac{dT}{dt} = -8 + 0.2 + \frac{303 - T}{4.2857}$$

$$= -7.8 + \frac{303 - T}{4.2857} = \frac{269.57 - T}{4.2857}$$

$$\frac{dT}{269.57 - T} = \frac{dt}{900}$$

Integrating, we get,

$$-ln(269.57 - T) = \frac{t}{900} + C$$

or $$ln(269.57 - T) = -\frac{t}{900} - C$$

where C = constant of integration

At $t = 0$, $T = 303$ K

$$ln(269.57 - 303) = 0 - C$$

$$-C = ln\,(-33.43)$$

or $$C = ln\,33.43$$

$$ln(269.57 - T) = -\frac{t}{900} - \ln 33.43$$

$$ln\,(269.57 - T) + ln\,33.43 = -\frac{t}{900}$$

$$ln[(269.57 - T) \times 33.43 = -\frac{t}{900}$$

$$(269.57 - T) \times 33.43 = \exp\left(-\frac{t}{900}\right)$$

$$T = 269.57 - \frac{1}{33.43}\exp\left(\frac{-t}{900}\right)K$$

(*d*) Show that the properties at the critical state for a gas obeying van der Walls equation of state:

$$p = \frac{RT}{V - b} - \frac{a}{V^2}$$

are given by $$p_c = \frac{a}{27b^2}, V_c = 3b \text{ and } T_c = \frac{a}{27\,Rb}$$

Hence show that the co-efficients 'a' and 'b' are expressed as :

$$a = \frac{27R^2T_c^{\,2}}{64p_c} \text{ and } b = \frac{RT_c}{8p_c}$$

and the critical coefficient for the vander waals gas is 2.66.

Sol. Van-der Walls equation:

$$p = \frac{RT}{V-b} - \frac{a}{V^2} \quad ...(1)$$

At the critical point,

$$\left(\frac{\partial p}{\partial V}\right)_c = 0 \text{ and } \left(\frac{\partial^2 p}{\partial V^2}\right)_c = 0$$

$$\left(\frac{\partial p}{\partial V}\right)_c = \frac{-RT_c}{(V_c - b)^2} + \frac{2a}{V_c^3} = 0$$

$$RT_c = \frac{2a}{V_c^3}(V_c - b)^2 \quad ...(2)$$

$$\left(\frac{\partial^2 p}{\partial V^2}\right)_c = \frac{2RT_c}{(V_c - b)^3} - \frac{6a}{V_c^4} = 0$$

$$RT_c = \frac{3a}{V_c^4}(V_c - b)^3 \quad ...(3)$$

Equating Eqs. (2) and (3), we get,

$$\frac{2a}{V_c^3}(V_c - b)^2 = \frac{3a}{V_c^4}(V_c - b)^3$$

$$2V_c = 3(V_c - b)$$

$$V_c = 3b \quad ...(4)$$

Eqn. (2) becomes,

$$RT_c = \frac{2a}{27b^3}(3b - b)^2 = \frac{8a}{27b}$$

$$T_c = \frac{8a}{27bR} \quad ...(5)$$

Substituting Eqns. (4) and (5) in (1), we get,

$$p_c = \frac{R}{(3b - b)} \times \frac{8a}{27bR} - \frac{a}{9b^2} = \frac{a}{27b^2} \quad ...(6)$$

The values of a and b becomes,

$$a = 3p_c V_c^2 = \frac{9}{8} RT_c V_c == \frac{27}{64} \frac{R^2 T_c^2}{p_c}$$

$$b = \frac{V_c}{3} = \frac{RT_c}{8p_c}$$

$$R = \frac{8}{3}\frac{p_c V_c}{T_c}$$

$$\text{Critical co-efficient} = \frac{8}{3} = 2.66$$

Q.2. ***(a)*** What is supercharging of diesel engine power plant ? What are its objectives Discuss various methods of supercharging. (15)

Sol. Supercharging of Diesel Engine Power Plant:

Refer to Q. 2 (*a*) of IESE-2005, Conventional Paper-I.

(*b*) The following data are known for a four cylinder four stroke petrol engine: cylinder dimensions : 11 cm bore, 13 cm stroke; engine speed : 2250 rpm; brake power : 50 kW; friction power : 15 kW; fuel consumption rate : 10.5 kg/h ; calorific value of fuel : 50,000 kJ/kg; air inhalation rate: 300 kg/h; ambient condition : 15°C, 1.03 bar. Estimate (*i*) brake mean effective pressure, (*ii*) volumetric efficiency, (*iii*) brake thermal efficiency, and (*iv*) mechanical efficiency.

(10)

Sol. Given:

$$D = 11 \text{ cm}, \; L = 13 \text{ cm},$$
$$N = 2250 \text{ rpm},$$
$$BP = 50 \text{ kW},$$
$$FP = 15 \text{ kW}, \; \dot{m}_f = 10.5 \text{ kg/h},$$
$$CV = 50{,}000 \text{ kJ/kg},$$
$$\dot{m}_a = 300 \text{ kg/h}, \; t_a = 15°\text{C},$$
$$p_a = 1.03 \text{ bar}, \; i = 4$$

(*i*)
$$\frac{BP}{i} = \frac{p_{bm} LA(N/2)}{60 \times 10^3}$$

$$\text{B.M.E.P., } p_{bm} = \frac{50 \times 60 \times 10^3 \times 2}{4 \times 13 \times 10^{-2} \times \frac{\pi}{4}(0.11)^2 \times 2250}$$

$$= 0.5396 \text{ MPa}$$

(*ii*) Volumetric efficiency

$$= \frac{\text{Actual volume inhaled during suction stroke at NTP}}{\text{Swept volume of piston}}$$

$$= \frac{V_a}{V_s}$$

$$V_s = \frac{\pi}{4} D^2 L = \frac{\pi}{4} \times (0.11)^2 \times 0.13 = 1.2354 \times 10^{-3} \text{ m}^3$$

Density of air, $$\rho_a = \frac{p_a}{RT_a} = \frac{1.03 \times 10^5}{287 \times 288} = 1.246 \text{ kg/m}^3$$

Swept mass of air per second

$$= V_s \times \rho_a \times \frac{N}{2 \times 60} \times i$$

$$= \frac{1.2354 \times 10^{-3} \times 1.246 \times 2250 \times 4}{2 \times 60} = 0.1154 \text{ kg/s}$$

$$\eta_v = \frac{300}{3600} \times \frac{1}{0.1154} = 0.7221 \text{ or } 72.21\%$$

(*iii*) Brake thermal efficiency,

$$\eta_{bt} = \frac{BP}{m_f \times CV} = \frac{50 \times 10^3 \times 3600}{10.5 \times 50{,}000 \times 10^3} = 0.3428 \text{ or } 34.28\%$$

(*iv*) Indicated power,

$$IP = BP + FP = 50 + 15 = 65 \text{ kW}$$

$$\eta_{mech} = \frac{BP}{IP} = \frac{50}{65} = 0.7692 \text{ or } 76.92\%$$

(*b*) Using a simple vertical sectional diagram, explain the function of principal components of the nuclear reactor of a pressurised water reactor (PWR) nuclear power plant. List the advantages and disadvantages of PWR nuclear power plant. (15)

Sol. Pressurised water reactor. Refer Q. 5 (b) of IESE-1998, Conventional Paper-I and Q. 2 (b) of IESE-2005 Conventional Paper-I

Q.3. (*a*) A copper sphere weighing 3 kg is heated in a furnace to a temperature of 300°C and is suddenly taken out and allowed to cool in ambient air at 25°C.

If it takes 60 min for the copper sphere to cool down to 35°C, what is the average surface heat transfer co-efficient? Take density of copper sphere = 8950 kg/m³ and specific heat c_p = 0.383 kJ/kg°C.

State the assumptions made and derive the relation used. (10)

Sol. Mass of sphere,

$$m = 3 \text{ kg}, \; T_i = 300°C, \; T_\infty = 25°C, \; T = 35°C,$$

$$t = 60 \text{ min}, \; \rho = 8950 \text{ hg/m}^3, \; c_p = 0.383 \text{ kJ/kg °C}$$

For cooling of an object in a fluid,

$$\frac{T - T_\infty}{T_i - T_\infty} = \exp\left[\left(\frac{-h_c A_s}{\rho V}\right)t\right]$$

Volume of sphere, $V = \frac{4}{3}\pi r^3$

Mass of sphere, $m = \rho V$

$$3 = 8950 \times \frac{4}{3} \times \pi \times r^3$$

$$r = 0.043 \text{ m}$$

Surface area of sphere, $A_s = 4\pi r^2$

$$\frac{V}{A_s} = \frac{r}{3} = \frac{0.043}{3} = 0.0143$$

$$\frac{35 - 25}{300 - 25} = \exp\left[\left\{\frac{-h_c}{8950 \times 0.0143 \times 383}\right\} \times 3600\right]$$

$$0.03636 = \exp\left[\frac{-h_c}{13.616}\right]$$

$$ln\ 0.03636 = -\frac{h_c}{13.616} - 3.3142$$

$$h_c = 45.127 \text{ W/m}^2\text{°C}$$

(*b*) From the Von Karman's integral momentum equation for the laminar boundary layer under forced convective conditions, obtain the expressions for the hydrodynamic boundary layer thickness and the drag co-efficient for the case of a linear velocity profile in the boundary layer. **(10)**

Sol. Von-Karman's Integral Momentum equation is given by,

$$\frac{\tau_0}{\rho U^2} = \frac{d\theta}{dx} \quad ...(1)$$

where q = momentum thickness

$$= \int_0^\delta \frac{u}{U}\left(1 - \frac{u}{U}\right)dy$$

U = free stream velocity

u = velocity in the fluid at a distance y from the boundary

τ_o = shear stress.

Drag force on a small distance dx of a plate is given by,

$$\Delta F_D = \text{shear stress} \times \text{area}$$
$$= \tau_0 \times (B \times dx)$$

where B = Width of plate

Total drag on a plate of length, L,

$$F_D = \int \Delta F_D = \int_0^L \tau_0 B dx$$

Drag co-efficient, $$C_D = \frac{\tau_0}{\frac{1}{2}\rho A U^2} \quad \text{...(2)}$$

where ρ = mass density of fluid,

A = area of plate surface.

Let the linear velocity profile be represented by,

$$\frac{u}{U} = f\left(\frac{y}{\delta}\right) = a + b\left(\frac{y}{\delta}\right)$$

The constants 'a' and 'b' can be determined from the boundary conditions.

At $y = 0$, $u = 0$ and at $y = \delta$, $u = U$

$\therefore$ $$0 = a$$

$$1 = b$$

$\therefore$ $$\frac{u}{U} = \frac{y}{\delta} \quad \text{...(3)}$$

Substituting Eqn. (3) in Eqn. (1), we get

$$\tau_0 = \rho U^2 \frac{d}{dx}\left[\int_0^\delta \frac{y}{\delta}\left(1 - \frac{y}{\delta}\right) dy\right]$$

$$= \rho U^2 \frac{d}{dx}\left[\int_0^\delta \left\{\frac{y}{\delta} - \left(\frac{y}{\delta}\right)^2\right\} dy\right]$$

$$= \rho U^2 \frac{d}{dx}\left[\left|\frac{y^2}{2\delta} - \frac{y^3}{3\delta^2}\right|_0^\delta\right] = \rho U^2 \frac{d}{dx}\left[\frac{\delta}{2} - \frac{\delta}{3}\right]$$

$$= \frac{\rho U^2}{6} \frac{d\delta}{dx}$$

Newton's law of viscosity, at the solid surface gives

$$\tau_0 = \mu\left(\frac{du}{dy}\right)_{y=0} = \mu\left[\frac{d}{dt}\left(\frac{Uy}{\delta}\right)\right] = \frac{\mu U}{\delta} \quad ...(4)$$

$$\therefore \quad \frac{\rho U^2}{6}\cdot\frac{d\delta}{dx} = \frac{\mu U}{\delta}$$

$$\delta.d\delta = \frac{6\mu}{\rho U}dx$$

Integrating, we get

$$\frac{\delta^2}{2} = \frac{6\mu x}{\rho U} + C$$

At $x = 0$, $\delta = 0$ gives $C = 0$

$$\frac{\delta^2}{2} = \frac{6\mu}{\rho U}dx$$

or

$$\delta^2 = \frac{12\mu x}{\rho U}$$

$$\frac{\delta}{x} = \sqrt{\frac{12\mu}{x\rho U}} = \frac{3.464}{\sqrt{Re_x}} \quad ...(5)$$

where

$$Re_x = \frac{x\rho U}{\mu} = \text{Reynolds number}$$

Now

$$\tau_0 = \frac{\mu U}{\delta} = \frac{\mu U\sqrt{Re_x}}{3.464x}$$

$$= \frac{\mu U Re_x}{3.464x.\sqrt{Re_x}} = \frac{\mu U\, x\, \rho U}{\mu} \times \frac{1}{3.464\, x\sqrt{Re_x}}$$

$$= \frac{\rho U^2}{2} \times \frac{2}{3.464} \times \frac{1}{Re_x} = \frac{1}{2}\rho U^2 \times \frac{0.5774}{\sqrt{Re_x}}$$

$$C_D = \frac{\tau_0}{\frac{1}{2}\rho U^2} = \frac{0.5774}{\sqrt{Re_x}}$$

Average value of drag co-efficient,

$$C_D = \frac{1}{L}\int_0^L C_D dx = \frac{1}{L}\int_0^L \frac{0.5774}{\sqrt{\frac{\rho U}{\mu}}} \times \frac{dx}{\sqrt{x}}$$

$$= \frac{1}{L}\left[\frac{0.5774}{\sqrt{\frac{\rho U}{\mu}}} \times \left|\frac{\sqrt{x}}{\left(\frac{1}{2}\right)}\right|_0^L\right] = \frac{1.155}{\sqrt{Re_L}}.$$

(*b*) The base of a rectangular enclosure (150 cm × 100 cm) and height 150 cm is maintained at 500°C. The top surface is held at 350°C. If the side walls are perfectly insulaed and the surfaces are diffuse grey with an emissivity of 0.7, calculate (i) the net rate of heat supply to the base, (ii) If the skin temperature of the outside of the top wall is at 70°C and heat loss from this surface to a big factory shed at 30°C, what is the convenctive heat transfer co-efficient at this surface ? **(20)**

Sol. Given.

$$\varepsilon_1 = \varepsilon_2 = \varepsilon_3 = 0.7,$$

$$T_2 = 273 + 500 = 773 \text{ K},$$

$$T_1 = 373 + 350 = 623 \text{ K}$$

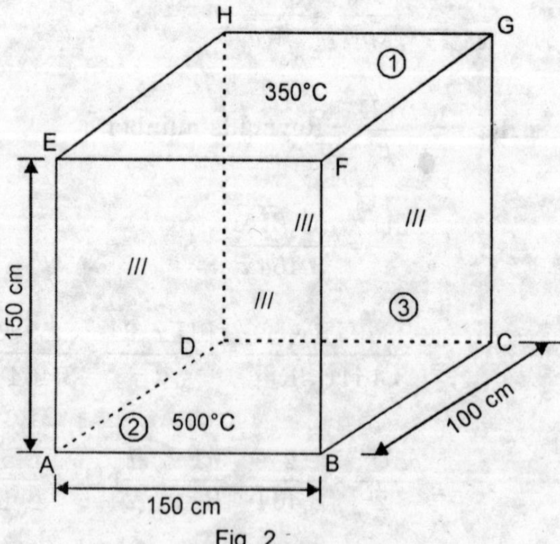

Fig. 2.

Net heat exchange between base and top surface, $Q_{1\text{-}2}$

$$F_{1-1} = F_{2-2} = 0$$

$$X = 150 \text{ cm}, Y = 100 \text{ cm}, L = 150 \text{ cm}$$

$$\frac{X}{L} = \frac{150}{150} = 1, \frac{Y}{L} = \frac{100}{150} = 0.667$$

Shape factor, F_{1-2} for $\frac{X}{L} = 1$ and $\frac{Y}{L} = 0.667$ is 0.14

$$\therefore \quad F_{1-2} = F_{2-1} = 0.14$$

$$F_{1-2} + F_{1-3} = 1$$

or $$0.14 + F_{1-3} = 1$$

$$F_{1-3} = 0.86$$

$$F_{2-1} + F_{2-3} = 1$$

or $$0.14 + F_{2-3} = 1$$

$$F_{2-3} = 0.86$$

$$(Q_{1-2})_{net} = \frac{A_1\sigma\left(T_1^4 - T_2^4\right)}{\frac{1-\varepsilon_1}{\varepsilon_1} + \left(\frac{1-\varepsilon_2}{\varepsilon_2}\right)\frac{A_1}{A_2} + \left[\frac{A_1 + A_2 - 2A_1F_{1-2}}{A_2 - A_1F_{1-2}^2}\right]}$$

$$= \frac{A_1\sigma\left(T_1^4 - T_2^4\right)}{\frac{1-\varepsilon_1}{\varepsilon_1} \times 2 + \frac{2(1-F_{1-2})}{(1-F_{1-2}^2)}} = \frac{A_1\sigma\left(T_1^4 - T_2^4\right)}{2\left[\frac{1-\varepsilon_1}{\varepsilon_1} + \frac{1}{1+F_{1-2}}\right]}$$

[$A_1 = A_2$ and $\varepsilon_1 = \varepsilon_2$]

$$(Q_{1-2})_{net} = \frac{0.15 \times 0.10 \times 5.67\left[\left(\frac{623}{100}\right)^4 - \left[\frac{773}{100}\right]^4\right]}{2\left[\frac{1-0.7}{0.7} + \frac{1}{1+0.14}\right]}$$

$$= -\ 67.22 \text{ W}$$

– ve sign indicates that heat is gained by top surface.

$$(Q_{1-2})_{net} = h_c \text{ A } \Delta T$$

$$67.22 = h_c \times 0.15 \times 0.10 \times (70 - 30)$$

$$h_c = 112 \text{ W/m}^2.\ °\text{C}$$

Q.4. (*a*) The discharge Q over a V-shaped notch is known to depend on the angle θ of the notch, the head H of the water suface, the velocity of approach V_o and the gravity g. Find the dimensionless form of the discharge equation.

Sol. $$Q = f(\theta, H, V_o, g) \quad ...(1)$$

or $$f_1(Q, \theta, H, V_o, g) = 0 \quad ...(2)$$

Total number of variables, $n = 5$

Writing dimensions of each variable, we have

$$Q = \left[L^3\ T^{-1}\right]$$

$$\theta = \left[M^0L^0T^0\right]$$

$$H = [L]$$

$$V_o = \left[LT^{-1}\right]$$

$$g = \left[LT^{-2}\right]$$

Number of fundamental dimensions, $m = 2$

Number of π-terms $= n - m = 5 - 2 = 3$

Eqn. (2) can be written as : $f_1(\pi_1, \pi_2, \pi_3) = 0$ ...(3)

Choosing H and g as repeating variables, we have

$$\pi_1 = H^{a1}\, g^{b1}\, Q$$

$$\pi_2 = H^{a2}\, g^{b2}\, V_o$$

$$\pi_3 = H^{a3}\, g^{b3}\, \theta$$

π_1-term:

$$\pi_1 = H^{a1}\, g^{b1}\, Q$$

$$M^0L^0T^0 = [L]^{a1} + [LT^{-2}]^{b1}\, [L^3\, T^{-1}]$$

$$= [L]^{a1 + b1 + 3}\, [T]^{-2b1\; b-1}$$

$$a_1 + b_1 + 3 = 0$$

$$-2b_1 - 1 = 0$$

$$b_1 = -\frac{1}{2}$$

$$a_1 = -3 + \frac{1}{2} = -\frac{5}{2}$$

$$\pi_1 = H^{-5/2}\, g^{-1/2}\, Q = \frac{Q}{g^{1/2}\, H^{5/2}}$$

π_2-term:

$$\pi_2 = H^{a2}\, g^{b2}\, V_o$$

$$[M^0L^0T^0] = [L]^{a2}\, [LT^{-2}]^{b2}\, [LT^{-1}] = [L]^{a2 + b2 + 1}\, [T]^{-2b2\, -1}$$

$$a_2 + b_2 + 1 = 0$$

$$-2b_2 - 1 = 0$$

$$b_2 = -\frac{1}{2}$$

$$a_2 = -1 + \frac{1}{2} = -\frac{1}{2}$$

$$\pi_2 = H^{-1/2}g^{-1/2}V_o = \frac{V_o}{\sqrt{gH}}$$

π_2-term:

$$\pi_3 = H^{a_3} g^{b_3} \theta$$

$$[M^0L^0T^0] = [L]^{a_3} + [LT^{-2}]^{b_3} [LT^{-1}] [M^0L^0T^0]$$

$$= [L]^{a_3 + b_3} [T]^{-2ab3}$$

$$a_3 + b_3 = 0$$

$$-2b_3 = 0$$

$$b_3 = 0$$

$$a_3 = 0$$

$$\pi_3 = H^0g^0\theta = \theta$$

Substituting the values of π_1, π_2 and π_3 in Eqn. (3), we get,

$$f_1\left(\frac{Q}{H^{5/2}g^{1/2}}, \frac{V_o}{\sqrt{gH}}, \theta\right) = 0$$

$$\frac{Q}{H^{5/2}g^{1/2}} = \phi\left(\frac{V_o}{\sqrt{gH}}, \theta\right)$$

$$Q = \sqrt{g}H^{5/2}\phi\left(\frac{V_o}{\sqrt{gH}}, \theta\right)$$

(b) If gravity, viscosity and surface tension are equally important in a model, show that for dynamic similarity, the relationship between viscosity ratio μ_r, surface tension ratio σ_r and model scale ratio L_r is given by

$$\frac{\mu_r L_r^{1/2}}{\sigma_r} = 1 \quad (15)$$

Sol. Froude number, $F = \frac{F_i}{F_g} = \frac{V^2}{Lg} = \frac{V}{\sqrt{Lg}}$

Reynolds number, $Re = \frac{F_i}{F_u} = \frac{\rho VL}{\mu}$

Weber number, $W = \frac{F_i}{F_\sigma} = \frac{\rho V^2 L}{\sigma}$

For dynamic similarity between model and prototype, we have

(*i*) $$F_m = F_p$$

$$\left(\frac{V}{\sqrt{Lg}}\right)_m = \left(\frac{V}{\sqrt{Lg}}\right)_p$$

$$V_p = V_m\sqrt{\frac{L_p}{L_m}} = V_m\sqrt{L_r} \; as \; g_m = g_p.$$

(*ii*) $$(Re)_m = (Re)_p$$

$$\left(\frac{\rho VL}{\mu}\right)_m = \left(\frac{\rho VL}{\mu}\right)_p$$

$$m_r = \frac{\mu_p}{\mu_m} = \frac{\sigma_p}{\sigma_m} \times \frac{V_p}{V_m} \times \frac{L_p}{L_m}$$

$$= \frac{\rho_p}{\rho_m} \times \sqrt{L_r} \times L_r = \frac{\rho_p}{\rho_m} \times L_r^{3/2}$$

(*iii*) $$W_m = W_p$$

$$\left(\frac{\rho V^2 L}{\sigma}\right)_m = \left(\frac{\rho V^2 L}{\sigma}\right)_p$$

$$\sigma_r = \frac{\sigma_p}{\sigma_m} = \frac{\rho_p}{\rho_m}\left(\frac{V_p}{V_m}\right) \times \frac{L_p}{L_m}$$

$$= \frac{\rho_p}{\rho_m} \times L_r \times L_r = \frac{\rho_p}{\rho_m} \times L_r^2$$

$$\frac{\mu_r L_r^{1/2}}{\sigma_r} = \frac{\rho_p}{\rho_m} \times L_r^{3/2} \times L_r^{1/2} \times \frac{\rho_m}{\rho_p} \times \frac{1}{L_r^2} = 1$$

Hence proved.

(*b*) A thin rectangular plate 10 m long and 2 m wide is to be towed in sea water at a steady speed of 5 m/s. Both the surfaces of the plate are rough, whose roughness magnitude is assumed as 5 mm. Estimate the power required to tow the plate. Assume $\rho = 1020$ kg/m^3 and $\mu = 0.0018$ Pa.s. **(15)**

Sol. Given $L = 10$ m, $B = 2$ m,

$U_\infty = 5$ m/s, $\rho = 1020$ kg/m^3,

$\mu = 0.0018$ Pa.s.

Reynolds number,

$$Re_L = \frac{U_\infty L\rho}{\mu} = \frac{5 \times 10 \times 1020}{0.0018} = 283.3 \times 10^5$$

Since $Re_L > 5 \times 10^5$, the boundary layer is turbulent.

The average drag co-efficient is given by,

$$C_f = \frac{0.074}{Re_L^{1/2}} - \frac{1700}{Re_L}$$

$$= \frac{0.04}{(283.3 \times 10^5)^{1/2}} - \frac{1700}{283.3 \times 10^5} = 2.33 \times 10^{-3}$$

Total drag force for both sides of the plate to be rough,

$$F_D = 2A\tau_0 = 2LB \times \frac{1}{2} C_f \rho U_\infty^2$$

$$= 2 \times 10 \times 2 \times \frac{1}{2} \times 2.33 \times 10^{-3} \times 1020 \times 5^2$$

$$= 1188.3 \text{ N}$$

Power required to tow the plate,

$$P = F_D U_\infty = 1188.3 \times 5 = 5941.5 \text{ W}$$

Q.5. (*a*) By using the energy equation,

$$Vdv + \frac{dp}{\rho} + d(losses) = 0$$

the continuity equation ρAV = constant, and

$c = \sqrt{\dfrac{dp}{d\rho}}$, show that for subsonic flow in a pipe, the velocity must increase in the downstream direction. (10)

Sol. Energy equation,

$$vdv + \frac{dp}{\rho} + d\,(losses) = 0 \quad ...(1)$$

Continuity equation :

$$\rho Av = \text{const.} \quad ...(2)$$

and

$$c = \sqrt{\frac{dp}{d\rho}} \quad ...(3)$$

By logarithmic differentiation of Eqn. (2), we have

$$\frac{d\rho}{\rho} + \frac{dA}{A} + \frac{dv}{v} = 0$$

$$\therefore \quad \frac{dA}{A} = -\frac{dv}{v} - \frac{d\rho}{\rho} \quad \text{...(4)}$$

From Eqn. (1), we have

$$dp = -\rho v \, dv$$

$$\frac{dp}{dv} = -\rho v < 0$$

As pressure decreases, velocity increases and *vice-versa*. Substituting Eqn. (5) in (4), we get,

$$\frac{dA}{A} = \frac{dp}{\rho v^2} - \frac{d\rho}{\rho} = \frac{dp}{\rho v^2}\left[1 - v^2 \frac{d\rho}{dp}\right]$$

$$= \frac{dp}{\rho v^2}\left[1 - \frac{v^2}{c^2}\right] \qquad \left[\because c = \sqrt{\frac{dp}{d\rho}}\right]$$

$$= \frac{dp}{\rho v^2}(1 - M^2) \quad \text{...(6)}$$

where Mach number, $M = \frac{v}{c}$

Also $$\frac{dA}{A} = (M^2 - 1)\frac{dv}{v} \quad \text{...(7)}$$

When M < 1, the inlet velocity is subsonic. As flow area A decreases, the pressure decreases and velocity increases.

(***b***) Show that oblique and normal shock waves in a gas are analogous to open-channel waves when the open channel width is constant.

Sol. Normal and Oblique shock waves in a Gas : Refer to Q. 6 (b) of IESE-2006

(***c***) Helium enters a 100 mm-1D pipe from a converging-diverging nozzle at M = 1.30, p = 14 kPa, T = 225 K. Estimate for an isothermal flow : (*i*) the maximum length of pipe for no choking, (*ii*) the downstream conditions, and (*iii*) the length from the exit to the section where M = 1.0, f = 0.016.

(15)

Sol. (*i*) For isothermal flow with friction,

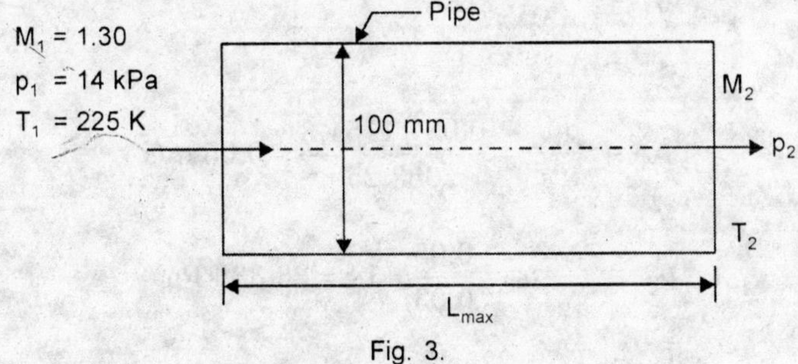

Fig. 3.

$$\frac{4fL_{max}}{D} = \frac{1 - kM_1^2}{kM_1^2} + \ln(kM_1^2)$$

where

$$D = 100 \text{ mm} = 0.1 \text{ m}$$

$$f = 0.016$$

$$k = 1.68 \text{ for Helium.}$$

$$\frac{4 \times 0.016 \times L_{max}}{0.1} = \frac{1 - 1.68 \times (1.3)^2}{1.68 \times (1.3)^2} + \ln[1.8 \times (1.3)^2]$$

$$= -0.6478 + 1.0435 = 0.3957$$

$$L_{max} = 0.6183 \text{ m}$$

(*ii*) For Isothermal flow, $\frac{p_1}{\rho_1} = \frac{p_2}{\rho_2}$

$$\rho_1 = \frac{p_1}{RT_1}$$

For Helium gas, $R = 2075$ J/kg.K

$$\rho_1 = \frac{14 \times 10^3}{2075 \times 225} = 0.03 \text{ kg/m}^3$$

$$c = \sqrt{kRT_1} = \sqrt{1.68 \times 2075 \times 225} = 885.6 \text{ m/s}$$

$$M_1 = \frac{v_1}{c}, v_1 = cM_1 = 885.6 \times 1.3 = 1151.3 \text{ m/s}$$

For limiting case of no choking,

$$M_2 = \sqrt{\frac{1}{k}} = \sqrt{\frac{1}{1.68}} = 0.7715$$

$$v_2 = 885.6 \times 0.7715 = 683.25 \text{ m/s}$$

From continuity equation, we have

$$\rho_1 A_1 V_1 = \rho_2 A_2 V_2$$

For $A_1 = A_2$

$$\rho_2 = \frac{\rho_1 v_1}{v_2} = \frac{0.03 \times 1151.3}{683.25} = 0.05 \text{ m/s}$$

$$p_2 = \frac{\rho_2}{\rho_1} p_1 = \frac{0.05}{0.03} \times 14 = 23.33 \text{ kPa}$$

$$T_2 = T_1 = 225 \text{ K}$$

$$M_2 = 0.7715$$

(iii) $$\frac{4fL}{D} = \frac{1 - kM^2}{kM^2} + \ln kM^2$$

For M = 1

$$\frac{4fL}{D} = \frac{1-k}{k} + \ln k$$

$$\frac{4 \times 0.016 \times L}{0.1} = \frac{1 - 1.68}{1.68} + \ln 1.68$$

$$= -\,0.40476 + 0.51879 = 0.11403$$

$$L = 0.1782 \text{ m}$$

Length from the exit = $L_{max} - L = 0.6183 - 0.1782 = 0.44$ m

Q.6. (*a*) A horizontal pipe of diameter D_1 has a sudden expansion to a diameter D_2. At what ratio D_1/D_2 would the differential pressure on either side of the expansion be maximum ? What is the corresponding loss of head and differential pressure head ?

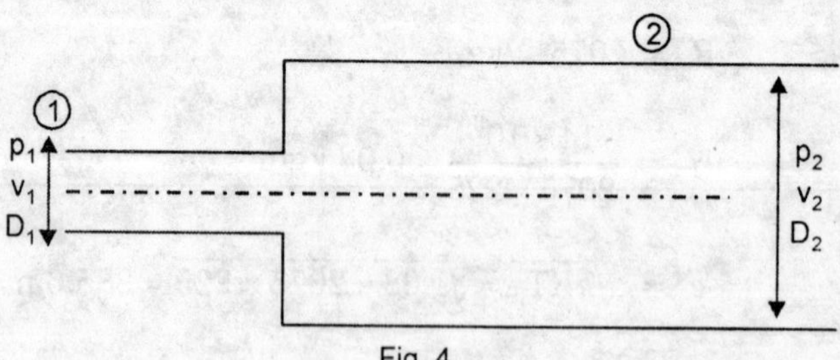

Fig. 4.

Sol. Applying Bernoulli's equation between section 1 and 2, considering head loss due to velocity change, we have

$$\frac{p_1}{\gamma} + \frac{v_1^2}{2g} = \frac{p_2}{\gamma} + \frac{v_2^2}{2g} + \frac{(v_1 - v_2)^2}{2g}$$

Pressure rise : $$\frac{\Delta p}{\gamma} = \frac{p_2 - p_1}{\gamma} = \frac{v_1^2 - v_2^2}{2g} - \frac{(v_1 - v_2)^2}{2g}$$

From the continuity equation, we have

$$Q = A_1 v_1 = A_2 v_2$$

$$v_2 = \frac{A_1}{A_2} v_1 = \left(\frac{D_1}{D_2}\right)^2 v_1$$

$$\frac{\Delta p}{\gamma} = \frac{v_1^2}{2g}\left[1 - \left(\frac{v_2}{v_1}\right)^2 - \left(1 - \frac{v_2}{v_1}\right)^2\right]$$

$$= \frac{v_1^2}{2g}\left[1 - \left(\frac{D_1}{D_2}\right)^4 - \left\{1 - \left(\frac{D_1}{D_2}\right)^2\right\}^2\right]$$

$$= \frac{v_1^2}{2g}\left[1 - \left(\frac{D_1}{D_2}\right)^4 - 1 - \left(\frac{D_1}{D_2}\right)^4 + 2\left(\frac{D_1}{D_2}\right)^2\right]$$

$$= \frac{v_1^2}{2g}\left[\left(\frac{D_1}{D_2}\right)^2 - \left(\frac{D_1}{D_2}\right)^4\right]$$

$$\Delta p = \rho v_1^2\left[\left(\frac{D_1}{D_2}\right)^2 - \left(\frac{D_1}{D_2}\right)^4\right]$$

For maximum pressure rise,

$$\frac{d(\Delta p)}{d\left(\frac{D_1}{D_2}\right)} = 0$$

$$2\left(\frac{D_1}{D_2}\right) - 4\left(\frac{D_1}{D_2}\right)^3 = 0$$

$$\left(\frac{D_1}{D_2}\right)^2 = \frac{1}{2}$$

or $$\frac{D_1}{D_2} = \frac{1}{\sqrt{2}}$$

$$(\Delta p)_{max} - \rho {v_1}^2 \left[\frac{1}{2} - \frac{1}{4}\right] = \frac{1}{4}\rho v_1^2$$

(***b***) In an isentropic flow, for flow conditions termed critical at the throat section, show that for an air flow having $k = 1.4$, the absolute temperature drops about 17% from reservoir to throat, the critical pressure is 52.8% of the reservoir pressure and the density is reduced by 37%. **(15)**

Sol.

$$\frac{T^*}{T_0} = \frac{2}{k+1} = \frac{2}{1.4+1} = \frac{2}{2.4} = 0.83$$

Drop in temperature = 1 – 0.83 = 0.17 or 17%

$$\frac{p^*}{p_0} = \left(\frac{2}{k+1}\right)^{\frac{k}{k-1}} = \left(\frac{2}{2.4}\right)^{\frac{1.4}{0.4}} = 0.528 \text{ or } 52.8\%$$

$$\frac{\rho^*}{\rho_0} = \left(\frac{2}{k+1}\right)^{\frac{k}{k-1}} = \left(\frac{2}{2.4}\right)^{\frac{1}{0.4}} = 0.634$$

Reduction in density = 1 – 0.634 = 0.366 or ≃ 37%

(***b***) The frictional torque T of a disc of diameter D rotating at a speed N in a fluid of viscosity μ and density ρ in a turbulent flow is given by

$$T = D^5 N^2 \rho Q \left[\frac{\mu}{D^2 N \rho}\right]$$

Prove this by the method of dimensions. **(15)**

Sol. $T = f(D, N, \mu, \rho)$ ...(1)

or $f_a(T, D, N, m, r) = 0$ ...(2)

Total number of variables, $n = 5$

writing dimensions of each variable, we have

$$T = [ML^2T^{-2}]$$
$$D = [L]$$
$$N = [T^{-1}]$$
$$\mu = [ML^{-1}T^{-1}]$$
$$\rho = [ML^{-3}]$$

Number of fundamental dimensions, $m = 3$

Number of π-terms = $n - m = 5 - 3 = 2$

Eqn. (2) can be written as : $f_1(\pi_1, \pi_2) = 0$

Choosing D, N, ρ as repeating variables, we have

$$\pi_1 = D^{a1}\, N^{b1}\, \rho^{c1}\, T$$
$$\pi_2 = D^{a2}\, N^{b2}\, \rho^{c2}\, \mu$$

π_1-term: $\pi_1 = D^{a1}\, N^{b1}\, \rho^{c1}\, T$

$$M^0L^0T^0 = [L]^{a1}\ [T^{-1}][ML^{-3}]^{C1}[ML^2T^{-2}]$$

$$= [M]^{c1+1} [L]^{a1+3c1+2} [T]^{-b1-2}$$

$$c_1 + 1 = 0$$

$$a_1 - 3c_1 + 2 = 0$$

$$\therefore \quad -b_1 = -2$$

$$c_1 = -1$$

$$a_1 = 3c_1 - 2 = 3 \times (-1) - 2 = -5$$

$$\pi_1 = D^{-5} N^{-2} \rho^{-1} T = \frac{T}{D^5 N^2 \rho}$$

π_1-term:

$$\pi_2 = D^{a2} N^{b2} \rho^{c2} \mu$$

$$M^0L^0T^0 = [L]^{a2} [T^{-1}]^{b2} [ML^{-3}]^{C2} [ML^{-1}T^{-1}]$$

$$= [M]^{c2+1} [L]^{a2+3c2-1} [T]^{-b2-1}$$

$$c_2 + 1 = 0$$

$$a_2 - 3c_2 + 1 = 0$$

$$-b_2 - 1 = 0$$

$$\therefore \quad b_2 = -1$$

$$c_2 = -1$$

$$a_2 = 3c_2 + 1 = 3(-1) + 1 = -2$$

$$\pi_2 = D^{-2} N^{-1} \rho^1 \mu = \frac{\mu}{D^2 N \rho}$$

Eqn. 3. becomes,

$$f_1\left(\frac{T}{D^5N^2\rho}, \frac{\mu}{D^2N\rho}\right) = 0$$

$$\frac{T}{D^5N^2\rho} = \phi\left(\frac{\mu}{D^2N\rho}\right)$$

or

$$T = D^5N^2\rho \, \phi\left(\frac{\mu}{D^2N\rho}\right)$$

Q.7. (*a*) A jet of water makes an angle with the direction of motion of a series of moving blades. If the blade angle at inlet and outlet are θ and ϕ with the direction of motion, derive an expression for the work done. If there is no shock at entry, show that the maximum efficiency that can be achieved from the system is,

$$\eta_{max} = \frac{\cos^2 \alpha}{2}\left[1 + \frac{K \cos\phi}{\cos\theta}\right]$$

Assume K = $\left(\dfrac{v_{r_2}}{v_{r_1}}\right)$. Also state what would be the maximum efficiency for semi-circular blades when K = 1.

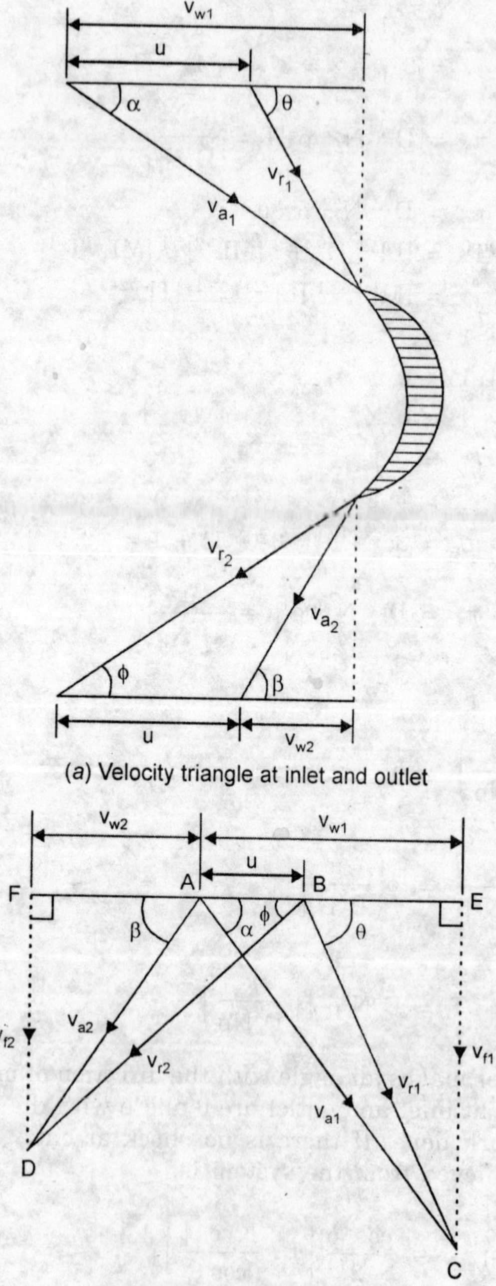

(a) Velocity triangle at inlet and outlet

(b) Combined velocity triangle

Fig. 5.

Sol. Work done per kg of water,

$$w = \text{Rate of change of momentum} \times \text{Distance travelled}$$

$$= [v_{w1} - (-v_{w2})]u$$

$$= (v_{w_1} + v_{w_2})u$$

Blade efficiency,

$$\eta_b = \frac{\text{Work done per kg of water}}{\text{Energy supplied per kg of water}}$$

$$= \frac{(v_{w_1} + v_{w_2})u}{\frac{1}{2}v_{a_1}^2} = \frac{2u(v_{w_1} + v_{w_2})}{v_{a_1}^2}$$

Form Fig. 5 (*b*), we have

$$v_{w_1} + v_{w_2} = \text{AE} + \text{AF} = (\text{AB} + \text{BE}) + (\text{BF} - \text{AB})$$

$$= \text{BE} + \text{BF}$$

$$= v_{r_1} \cos\theta + v_{r_2} \cos\phi$$

$$= v_{r_1} \cos\theta \left[1 + \frac{v_{r_2}}{v_{r_1}} \cdot \frac{\cos\phi}{\cos\theta}\right]$$

$$= v_{r1} \cos\theta \; [1 + \text{K C}]$$

where $\text{K} = \frac{v_{r_2}}{v_{r_1}}$ and $\text{C} = \frac{\cos\phi}{\cos\theta}$

Now $\text{BE} = \text{AE} - \text{AB} = v_{a1} \cos\alpha - u$

$$\eta_b = \frac{2u(v_{a_1} \cos\alpha - u)(1 + \text{KC})}{v_{a_1}^2}$$

$$= \frac{2u}{v_{a_1}}\left[\left(\cos\alpha - \frac{4}{v_{a_1}}\right)(1 + \text{KC})\right]$$

$$= 2\rho \; [\cos\alpha - \rho)\; (1 + \text{KC})$$

where $\rho = \frac{u}{v_{a_1}} = \text{speed ratio}$

for η_b to be maximum, $\frac{d}{d\rho}(\eta_b) = 0$

$$\cos\alpha - 2\rho = 0$$

$$\rho = \frac{\cos\alpha}{2}$$

$$(\eta_b)_{max} = \cos\alpha\left[\left(\cos\alpha - \frac{\cos\alpha}{2}\right)(1 + KC)\right]$$

$$= \frac{\cos^2\alpha}{2}\left[1 + K\frac{\cos\phi}{\cos\theta}\right]$$

Hence proved

For K = 1,

$$(\eta_b)_{max} = \frac{\cos^2\alpha}{2}\left[1 + K\frac{\cos\phi}{\cos\theta}\right]$$

(b) Discuss briefly the classification of a hydroelectric plant based on (i) availability of water head, (ii) nature of load capacity. (10)

Sol. Classification of Hydroelectric plants:

(i) Based on availability of water head:

1. **High head water power plants:** These plants work under heads ranging from 100 m and above. Water is usually stored up in lakes on high mountains during the rainy season or when the snow melts. From one end of the lake, tunnels are constructed which lead the water into smaller reserviors known as forebays. The forebays distribute the water to penstocks through which it is lead to the turbine. They are used to run Pelton turbines.
2. **Medium head water power plants:** These plants work under 30 to 500 m water head. A penstock joining the forebay to the turbine inlet is built up. They are used for operating Francis tubines.
3. **Low head water power plants:** They work from 25 to 80 m of water head. They are used for operating Kaplan turbines.

(ii) Based on Nature of load capacity:

1. **Base load plants:** These plants cater for the base load of the system. Such plants are required to supply a constant power when connected to the gird. Thus they run without stop and are often remote controlled with which least staff is required for such plants. Run off river plants without pondage may sometimes work as base load plants.
2. **Peak-load plants:** These plants can supply the power during peak loads. Some of such plants supply the power during average load but also supply peek load as and when it is there; whereas other peak load plants are required to work during peak load hours only. The run off river plants may be made for the peak load by providing pondage.

(c) During a trial run on a pump, cavitation occurred when the sum of static pressure and velocity head at inlet was reduced to 3.2 m. The pump total head was 35 m at a discharge of 0.045 m³/sec. The vapour pressure of water = 1.8 kPa and the barometer pressure = 750 mm of Hg. If the pump is to operate at some other location where the atmospheric pressure was reduced to 620 mm of Hg and the temperature is so reduced that the vapour pressure of water is 830 Pa, what is the value of cavitation parameter when the pump develops the same total head and discharge? Is it necessary to reduce the height of the pump and if so, by how much? (10)

Sol. $H_s + \frac{v_s^2}{2g} = 3.2 \text{ m}, H_{mano} = 35 \text{ m}, Q = 0.045 \text{ m}^3/\text{s},$

$$p_{vap} = 1.8 \text{ kPa}, p_{atm} = 750 \text{ mm of hg}$$

$$H_{vap} = \frac{1.8 \times 10^3}{9.81 \times 10^3} = 0.1835 \text{ m of } H_2O$$

$$H_{atm} = 750 \times 10^{-3} \times 13.6 = 10.2 \text{ m of } H_2O$$

$$\text{Cavitation parameter} = \frac{H_{atm} - \left(H_{vap} + H_s + \frac{v_s^2}{2g}\right)}{H_{mano}}$$

$$= \frac{10.2 - (0.1835 + 3.2)}{35} = 0.19476$$

At the other location, we have

$$p_{atm} = 620 \text{ mm of Hg}$$

$$H_{atm} = 620 \times 10^{-3} \times 13.6 = 8.432 \; m \text{ of } H_2O$$

$$0.19476 = \frac{8.432 - (0.0846 + H)}{35}$$

$$H = 1.53 \text{ m}$$

It is necessary to reduce the height of pump by,

$$\Delta H = 3.20 - 1.53 = 1.67 \text{ m}$$

Q.8. (a) A two stage compression with intercooling in between stages and a single stage turbine with regeneration is employed in an open cycle gas turbine plant. Air at 1 bar and 15°C enters the compressor and the maximum pressure ratio is 5 and the maximum temperature in the cycle is 800°C. The rate of air flow through the cycle is 250 kg/sec and the calorific value of the fuel used is 42 MJ/kg. The isentropic efficiencies of both the compressors is 0.8 and the effectivencess of the regenerator is 0.7, and the isentropic efficiency of the turbine is 0.9.The combustion efficiency is 0.95, the mechanical efficiency is 0.96 and the generator efficiency is 0.75. Take c_p of air = 1.005 kJ/kg K and γ = 1.4 and for gases c_p = 1.08 kJ/kgK and γ = 1.33.

Assuming perfect intercooling and neglecting pressure and heat losses, determine

(i) The air fuel ratio

(ii) The cycle efficiency

(iii) The power supplied by the plant

(iv) The specific fuel consumption and the fuel consumption per hour. (15)

Sol. p_1 = 1 bar, T_1 = 288 K

$$\frac{p_3}{p_1} = 5,\ T_s = 1073 \text{ K},\ m_a = 250 \text{ kg/s}$$

$$\text{CV} = 42 \text{ MJ/kg},\ \eta_{c_1} = \eta_{c_2} = 0.8,$$

$$\varepsilon_r = 0.7,\ \eta_t = 0.9,\ \eta_{comb} = 0.95,$$

$$\eta_{mech} = 0.96,\ \eta_g = 0.75$$

$$c_{pa} = 1.005 \text{ kJ/kg. K},\ \gamma_a = 1.4,$$

$$c_{pg} = 1.08 \text{ kJ/kg.K},\ \gamma_g = 1.33$$

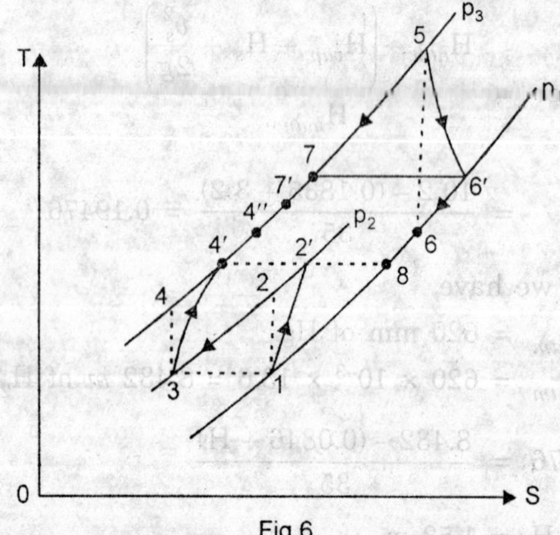

Fig.6.

Compressor 1st stage:

For perfect intercooling,

$$p_2 = \sqrt{p_1 p_3} = \sqrt{1 \times 5} = 2.236 \text{ bar}$$

$$T_2 = T_1 \left(\frac{p_2}{p_1}\right)^{\frac{\gamma_a - 1}{\gamma_a}} = 288 \left(\frac{2.236}{1}\right)^{\frac{0.4}{1.4}} = 362.4\text{K}$$

$$\eta_{c_1} = \frac{T_2 - T_1}{T_2' - T_1}$$

$$0.8 = \frac{362.4 - 288}{T_2' - 288}$$

$$T_2' = 381\text{K}$$

Work done, $w_{c_1} = c_{pa}(T_2' - T_1)$

$$= 1.005\ (381 - 288)$$

$$= 93.465\ \text{kJ/kg}$$

Compressor 2nd stage:

$$T_3 = T_1$$

$$T_4 = T_3\left(\frac{p_3}{p_2}\right)^{\frac{\gamma_a - 1}{\gamma_a}} = 288\left(\frac{5}{2.236}\right)^{\frac{0.4}{1.4}} = 362.4\ \text{K}$$

$$\eta_{c_2} = \frac{T_4 - T_3}{T_4' - T_3}$$

$$0.8 = \frac{362.4 - 288}{T_4' - 288}$$

$$T_4' = 381\ \text{K}$$

Work done, $w_{c_2} = c_{pa}\ (T_4' - T_3) = 1.005\ (381 - 288)$

$$= 93.465\ \text{kJ/kg}$$

Total work done on the compressor,

$$w_c = w_{c_1} + w_{c_2}$$

$$= 2 \times 93.465 = 186.93\ \text{kJ/kg}$$

Actual compressor work,

$$w_{ca} = \frac{w_c}{\eta_{mech}} = \frac{186.93}{0.96} = 194.72\ \text{kJ/kg}$$

Turbine:

$$T_6 = T_5\left(\frac{p_6}{p_5}\right)^{\frac{\gamma_g - 1}{\gamma_g}} = T_5\left(\frac{p_1}{p_5}\right)^{\frac{\gamma_g - 1}{\gamma_g}} = \left(\frac{1}{5}\right)^{\frac{0.35}{1.33}} \times 1073 = 719.7\ \text{K}$$

$$\eta_t = \frac{T_5 - T_6'}{T_5 - T_6}$$

$$0.9 = \frac{1073 - T_6'}{1073 - 719.7}$$

$$T_6' = 754.7\text{K}$$

Turbine work, $w_t = c_{pg}\ (T_5 - T_6')$

$$= 1.08\ (1073 - 754.7) = 343.76 \text{ kJ/kg}$$

Net work done by turbine,

$$w_{net} = w_t - w_{ca} = 343.76 - 194.72 = 149.04 \text{ kJ/kg}$$

Combustion efficiency,

$$\eta_{comb} = \frac{T_5 - T_4''}{T_5 - T_4'}$$

$$0.95 = \frac{1073 - T_4''}{1073 - 381}$$

$$T_4'' = 415.6 \text{ K}$$

Regenerator effectiveness,

$$\varepsilon_r = \frac{T_7' - T_4''}{T_6' - T_4''}$$

$$0.7 = \frac{T_7' - 415.6}{754.7 - 415.6}$$

$$T_7' = 653 \text{ K}$$

Heat supplied, $q_s = c_{pg}\ (T_5 - T_7')$

$$= 1.08\ (1073 - 653) = 453.6 \text{ kJ/kg}$$

(ii) Thermal efficiency:

$$\eta_{th} = \frac{w_{net}}{q_s} = \frac{149.04}{453.6} = 0.3286 \text{ or } 32.86\%$$

Cycle-efficiency,

$$\eta_c = \eta_{th} \times \eta_g = 0.3286 \times 0.75 = 0.2464 \text{ or } 24.64\%$$

(iii) Power supplied by plant,

$$P = 1 \times w_{net} \times \eta_g = 149.04 \times 0.75 \times 1 = 111.78 \text{ kW/kg}$$

(i) Air fuel ratio:

$$\dot{m}_f \times CV = \dot{m}_a \times c_{pg}(T_5 - T_7')$$

$$\frac{\dot{m}_a}{\dot{m}_f} = \frac{42 \times 10^3}{1.08(1073 - 653)} = 92.6$$

(iv) Specific fuel consumption,

$$\dot{m}_f = \frac{c_{pg}(T_5 - T_7')}{CV} = \frac{1.08 \times 420}{42 \times 10^3} = 0.0108 \text{ kg/kg of air}$$

$$\text{Fuel consumed per hour} = \frac{\dot{m}_f \times 3600}{w_{net}}$$

$$= \frac{0.0108 \times 3600}{149.04} = 0.26 \text{ kg/kWh}$$

(*b*) What is a supercritical boiler? What are its merits and demerits?

Sol. Supercritical boiler: A boiler working in the range of 125 atm and **510°C to 300** atm and 600°C is called a supercritical boiler. It requires a pre-heater and a **superheater.** They are used above 300 MW capacity units.

Merits:

1. Large heat transfer rates.
2. Owing to less heat capacity of the generator the pressure level is more stable and therefore gives better response.
3. Because of absence of two phase mixture the problems of erosion and corrosion are minimised.
4. More adaptable to load fluctuations because of great ease of operation, simplicity and flexibility.
5. The turbo-generator coupled to super-critical boilers can generate peak loads by changing the pressure of operation.

(*c*) In a reaction turbine with radial vanes at inlet, the velocity of flow at outlet is 'K' times that at inlet. Prove that: (10)

$$\eta_h = \frac{2}{2 + K^2 \tan^2 a}$$

and
$$u_1 = \sqrt{\frac{2gH}{2 + K^2 \tan^2 a}}$$

Sol.

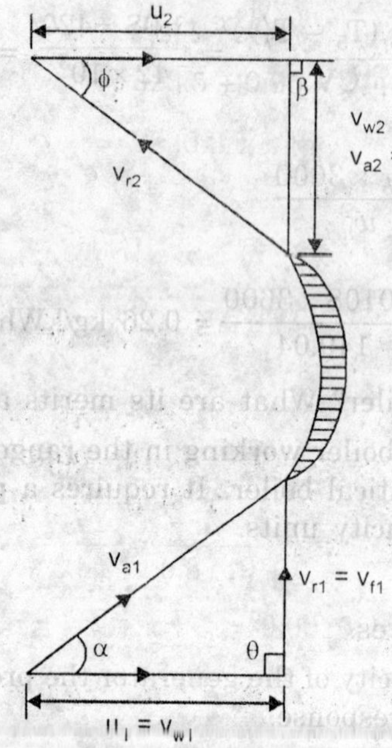

Fig. 7.

$$K = \frac{v_{f_2}}{v_{f_1}}$$

Head supplied, H = work done + Kinetic head at exit

$$H = \frac{u_1 v_{w_1} \pm u_2 v_{w_2}}{g} + \frac{v_{a_2}^2}{2g}$$

From velocity traingles, we have

$$u_1 = v_{w1}$$

$$v_{f_1} = u_1 \tan \alpha$$

For maximum efficiency, $\beta = 90°$, so that

$$v_{w2} = 0$$

Also $$v_{a_2} = v_{f_2} = K\ v_{f_1} = K\ u_1 \tan \alpha$$

$$H = \frac{u_1\ v_{w_1}}{g} + \frac{v_{a_2}^2}{2g}$$

$$= \frac{u_1^2}{g} + \frac{(K u_1 \tan \alpha)^2}{2g}$$

$$= \frac{u_1^2}{2g}(2 + K^2 \tan^2 \alpha)$$

Hydraulic efficiency,

$$\eta_h = \frac{\text{Power developed by runner}}{\text{Power supplied to turbine}}$$

$$= \frac{\frac{\gamma Q}{g} \times u_1 \, u_{w_1}}{\gamma \, QH} = \frac{u_1 \, v_{w_1}}{gH} = \frac{u_1^2}{gH}$$

$$= \frac{u_1^2}{\frac{u_1^2}{2}(2 + K^2 \tan^2 \alpha)} = \frac{2}{2 + K^2 \tan^2 \alpha}$$

Peripheral speed, $u_1 = \sqrt{\frac{2gH}{2 + K^2 \tan^2 \alpha}}$

(*d*) Why are steam turbines compounded ? What are the different methods of compounding ?

Sol. Necessity of Compounding steam turbines. If the entire pressure drop from boiler pressure to condenser pressure is carried out in a single stage nozzle, then the velocity of steam entering the turbine blades will be very high and consequently the turbine speed will be very high. Such high speed turbine rotors are not useful for practical purposes and a reduction gearing is necessary between the turbine and generator driven by it. There is also a danger of structural failure of the blades due to excessive centrifugal stresses. Further, the velocity of steam at the exit of the turbine is also very high when a single stage of blades are used. This gives rise to a considerable loss of kinetic energy reducing the efficiency of the unit.

Compounding is a method employed for reducing the rotational speed of impulse turbine to practical limits by using more than one stage. There are three methods of compounding impulse turbine :

1. Velocity compounding.
2. Pressure compounding.
3. Pressure and velocity compounding.

MECHANICAL ENGINEERING
PAPER I—2009

(Conventional)

Q.1. (*a*) The heat capacity at constant pressure of a certain system is a function of temperature only and may be expressed as

$$c_p = 2.093 + \frac{41.87}{t+100} \text{ J/°C}$$

where t is the temperature in °C. The system is heated while it is maintained at a pressure of 1 atmosphere until its volume increases from 2000 cm³ to 2400 cm³ and its temperature increases from 0°C to 100°C.

(*i*) Find the magnitude of heat interection.

(*ii*) How much does the internal energy of the system increase ?

Sol. Refer to Fig. 1

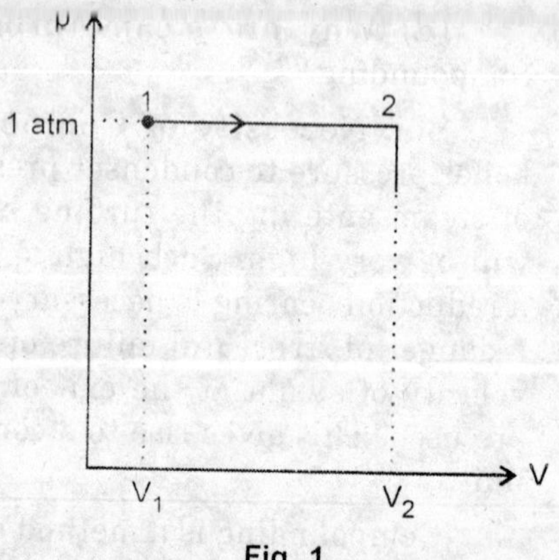

Fig. 1

(*i*) $p = 1 \text{ atm} = 1.01325 \times 10^5 \text{ N/m}^2,$

$V_1 = 2000 \text{ cm}^3, V_2 = 2400 \text{ cm}^3;$

$t_1 = 0°C, t_2 = 100°C$

$$Q_{1-2} = \int_{t_1}^{t_2} c_p \cdot dt = \int_0^{100} \left[2.093 + \frac{41.87}{t+100}\right] dt$$

$$= |2.093t + 41.87\ln(t+100)|_0^{100}$$

$$= 2.093 \times 100 + 41.87 \ln\left(\frac{200}{100}\right)$$

$$= 209.3 + 29.02 = 238.32 \text{ J}$$

(*ii*) $W_{1-2} = p(V_2 - V_1) = 1.01325 \times 10^5 (2400 - 2000) \times 10^{-6} = 40.53 \text{ J}$

According to first law of thermodynamics,

$$Q_{1-2} = (U_2 - U_1) + W_{1-2}$$

$$U_2 - U_1 = 238.32 - 40.53 = 197.79 \text{ J}$$

(*b*) Derive the equations :

(*i*) $c_p = T = \left[\frac{\partial V}{\partial T}\right]_p \left(\frac{\partial P}{\partial T}\right)_s$ (*ii*) $\left[\frac{\partial P}{\partial T}\right]_s = \frac{c_p}{V\beta T}$ (*iii*) $\frac{\left(\frac{\partial P}{\partial T}\right)_s}{\left(\frac{\partial P}{\partial T}\right)_v} = \frac{\gamma}{\gamma - 1}$

Sol.

(*i*) Let $s = s(\mathrm{T}, p)$

Then $$ds = \left(\frac{\partial s}{\partial \mathrm{T}}\right)_p \cdot d\mathrm{T} + \left(\frac{\partial s}{\partial p}\right)_\mathrm{T} \cdot dp$$

or $$\mathrm{T}\cdot ds = \mathrm{T}\left(\frac{\partial s}{\partial \mathrm{T}}\right)_p \cdot d\mathrm{T} + \mathrm{T}\left(\frac{\partial s}{\partial p}\right)_\mathrm{T} \cdot dp$$

Now $\mathrm{T}\left(\frac{\partial s}{\partial \mathrm{T}}\right)_p = c_p$ = heat capacity at constant pressure

and $$\left(\frac{\partial s}{\partial p}\right)_\mathrm{T} = -\left(\frac{\partial v}{\partial \mathrm{T}}\right)_p$$

$$\therefore \quad \mathrm{T}\cdot ds = c_p \cdot d\mathrm{T} - \mathrm{T}\left(\frac{\partial v}{\partial \mathrm{T}}\right)_p \cdot dp$$

For constant s, $ds = 0$

$$\therefore \quad 0 = \mathrm{C}_p \cdot d\mathrm{T} - \mathrm{T}\left(\frac{\partial v}{\partial \mathrm{T}}\right)_p \cdot dp$$

or $$c_p = \mathrm{T}\left(\frac{\partial v}{\partial \mathrm{T}}\right)_p \cdot \left(\frac{dp}{d\mathrm{T}}\right)_s \quad (1)$$

Hence proved.

(*ii*) From Eq. (1), we have

$$\left(\frac{dp}{d\mathrm{T}}\right)_s = \frac{c_p}{\mathrm{T}\left(\frac{\partial v}{\partial \mathrm{T}}\right)_p}$$

How volume expansivity,

$$\beta = \frac{1}{v}\left(\frac{\partial v}{\partial \mathrm{T}}\right)_p$$

$$\therefore \quad \left(\frac{\partial v}{\partial \mathrm{T}}\right)_p = \beta v$$

$$\therefore \quad \left(\frac{dp}{d\mathrm{T}}\right)_s = \frac{c_p}{v\beta\mathrm{T}}$$

Hence proved.

(*iii*) For an ideal gas,

$$pv = \mathrm{RT}$$

Differentiating w.r.t. T, we get

$$v\left(\frac{\partial p}{\partial \mathrm{T}}\right)_v = \mathrm{R}$$

$$\left(\frac{\partial p}{\partial \mathrm{T}}\right)_v = \frac{\mathrm{R}}{v}$$

Also $$\left(\frac{\partial p}{\partial \mathrm{T}}\right)_s = \frac{c_p}{v\beta \mathrm{T}}$$

$$\frac{\left(\frac{\partial p}{\partial \mathrm{T}}\right)_s}{\left(\frac{\partial p}{\partial \mathrm{T}}\right)_v} = \frac{\mathrm{C}_p}{v\beta \mathrm{T}} \times \frac{v}{\mathrm{R}} = \frac{\mathrm{C}_p}{\beta \mathrm{TR}}$$

For an ideal gas,

$$\beta = \frac{1}{\mathrm{T}}$$

$$\therefore \quad \frac{\left(\frac{\partial p}{\partial \mathrm{T}}\right)_s}{\left(\frac{\partial p}{\partial \mathrm{T}}\right)_v} = \frac{c_p}{\mathrm{R}} = \frac{c_p}{c_p - c_v} = \frac{(c_p/c_v)}{(c_p/c_v)^{-1}} = \frac{\gamma}{\gamma - 1}$$

Hence proved.

(c) Show that for an ideal gas, the slope of the constant volume line on the T–s diagram is more than that of the constant pressure line.

Sol. Consider the heating of 1 kg of a gas (Fig. 2) at constant volume from state 1 to 2′ and at constant pressure from state 1 to 2. Let the temperature rises from T_1 to T_2. Then at constant volume, heat added,

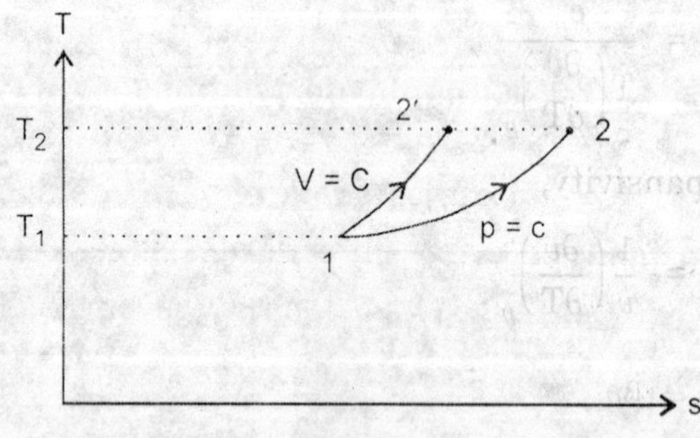

Fig. 2

$$\mathrm{Q} = c_v(\mathrm{T}_2 - \mathrm{T}_1)$$

or $$d\mathrm{Q} = c_v \cdot d\mathrm{T}$$

Dividing both sides by T, we have

$$\frac{d\mathrm{Q}}{\mathrm{T}} = c_v \cdot \frac{d\mathrm{T}}{\mathrm{T}}$$

or $$ds = c_v \cdot \frac{d\mathrm{T}}{\mathrm{T}}$$

$$\left(\frac{dT}{ds}\right)_v = \frac{T}{c_v} \qquad (1)$$

Similarly
$$\left(\frac{dT}{ds}\right)_p = \frac{T}{c_p} \qquad (2)$$

Dividing Eq. (1) by (2), we get

$$\left(\frac{dT}{ds}\right)_v / \left(\frac{dT}{ds}\right)_p = \frac{c_p}{c_v} > 1 \quad \text{as } c_p > c_v \text{ for an ideal gas}$$

Hence proved

(*d*) Explain the phenomenon of bounday layer separation, Describe four methods of controlling bounday layer separation.

Sol. When a solid body is immersed in a flowing fluid, a thin layer of fluid called boundary *layer* is formed, adjacent to the solid body. In this thin layer of fluid, the velocity varies from zero to *free stream velocity* in the direction normal to the solid body. Along the length of the solid body, the thickness of the boundary layer increases. *The fluid layer adjacent to the solid surface has to do work against surface friction at the expense of kinetic energy. This loss of kinetic energy is recovered from the immediate fluid layer in contact with the layer adjacent to the solid surface through momentum exchange process. Thus the velocity of the layer goes on decreasing.* Along the length of solid body, *at a certain point a stage may come when the boundary layer may not be able to keep sticking to the solid body, if it cannot provide kinetic energy to overcome the resistance offered by the solid body.* In other words, the boundary layer will be *separated from the surface*. This phenomenon is called the **boundary layer separation**. *The point on the body at which the boundary layer is on the verge of separation from the surface is called* **point of separation.**

Methods to control separation

1. Motion of solid boundary : By rotating a circular cylinder lying in a stream of fluid, so that the upper side of cylinder where the fluid as well as the cylinder move in the same *direction, the boundary layer does not form*. However on the lower side of cylinder where the fluid *motion is opposite* to that of cylinder *separation would occur* (Fig. 3).

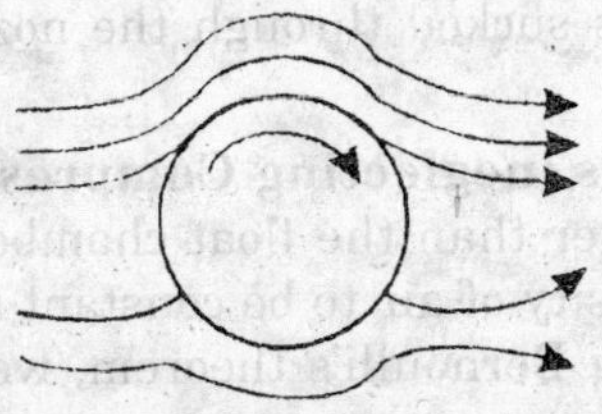

Fig. 3

2. Acceleration of fluid in the boundary layer : This method of controlling separation consists of *supplying additional energy to particles of fluid which are being retarded in the boundary layer. This may be achieved either by injecting the fluid into the region of boundary layer from the interior of the body with the help of some available device as shown in Fig. 4 or by diverting a portion of fluid of the stream from the region of high pressure to the retarded region of boundary layer through a slot provided in the body* (Fig. 5).

Fig. 4. Injecting fluid into boundary layer. **Fig. 5. Slotting wing.**

3. Suction of fluid from the boundary layer : In this method, the slow moving fluid in the boundary layer is removed by suction through slots or through a porous surface as shown in the Fig. 6.

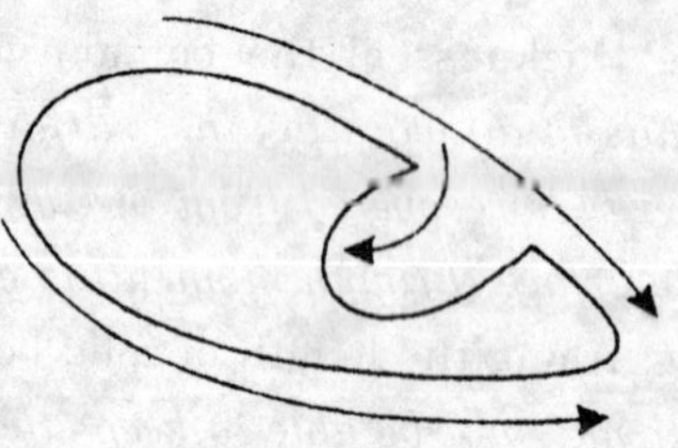

Fig. 6.

4. Streamlining of body shapes : By the use of suitably shaped bodies, the point of transition of the boundary layer from laminar to turbulent can be *moved down stream which results in the reduction of the skin friction drag.* Furthermore by *streamlining of body shapes, the separation may be eliminated.*

Q.2. (*b*) Derive an expression for air/fuel ratio of a carburettor by

(*i*) neglecting compressibility of air

(*ii*) taking compressibility effects into account.

Sol. The air from the atmosphere is sucked through the carburettor by the pressure difference across it created when the piston moves on its suction stroke. The velocity of air increases as it passes through the venturi and reaches maximum at venturi throat (Fig. 7). The pressure also changes and is maximum at section 2–2, because this is minimum area in the induction track. The fuel is sucked through the nozzle because of suction of created in the venturi.

(*i*) Approximate Analysis (neglecting Compressibility of air) : Let z be the height in metre of fuel nozzle tip higher than the float chamber level. Assuming initial velocity of air to be negligible ($c_1 = 0$), density of air to be constant and considering sections at entrance and venturi throat, by applying Bernoulli's theorem, we get

$$\frac{p_1}{\rho_a} = \frac{p_2}{\rho_a} + \frac{c_2^2}{2}$$

where p_1, p_2 = pressure at sections 1 and 2 respectively, N/m^2

c_2 = velocity of air at section 2, m/s

ρ_a = density of air, kg/m^3

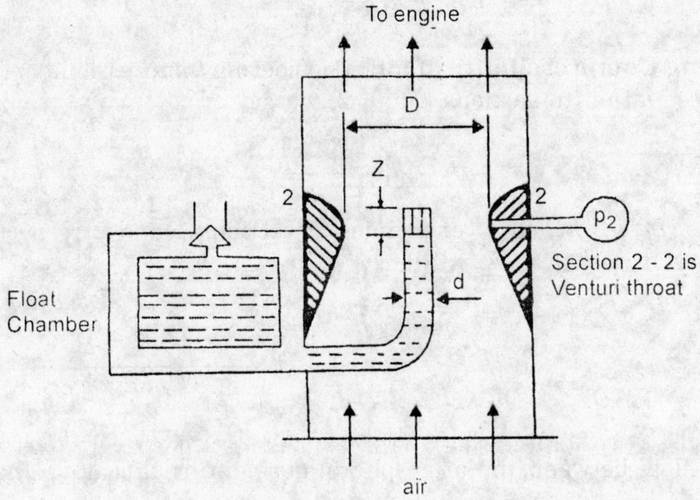

Fig. 7. Principles of a simple carburettor.

$$\therefore \quad c_2 = \sqrt{\frac{2(p_1 - p_2)}{\rho_a}}$$

Mass of air per second, $\dot{m}_3 = A_2 c_2 \rho_a = A_2\sqrt{2\rho_a(p_1 - p_2)}$ (1)

where A_2 = area of venturi throat, m^2

Similarly, for the flow of fuel, we have

$$\frac{p_1}{\rho_f} = \frac{p_2}{\rho_f} + \frac{c_f^2}{2} + zg$$

where ρ_f = constant density of fuel, kg/m^3

c_f = velocity of flow of fuel, m/s

Velocity of fuel, $c_f = \sqrt{\frac{2(p_1 - p_2 - zg\rho_f)}{\rho_f}}$

Mass of fuel per second, $\dot{m}_f = A_f C_f \rho_f = A_f\sqrt{2\rho_f(p_1 - p_2 - zg\rho_f)}$ (2)

where A_f = cross-sectional area of fuel nozzle, m^2

Fuel-air ratio, $\frac{\dot{m}_f}{\dot{m}_a} = \frac{A_f}{A_2} = \sqrt{\frac{\rho_f}{\rho_a}} \cdot \sqrt{1 - \frac{zg\rho_f}{p_1 - p_2}}$ (3)

Taking the coefficient of discharge of fuel nozzle and venturi into account, we have

$$\frac{\dot{m}_f}{\dot{m}_a} = \frac{C_{df}}{C_{da}} \times \frac{A_f}{A_2} \times \sqrt{\frac{\rho_f}{\rho_a}} \times \sqrt{1 - \frac{zg\rho_f}{p_1 - p_2}} \quad (4)$$

where C_d = coefficient of discharge.

If $z = 0$, then Eq. (4) reduces to,

$$\frac{\dot{m}_f}{\dot{m}_a} = \frac{C_{df}}{C_{da}} \times \frac{A_f}{C_{da}} \times \sqrt{\frac{\rho_f}{\rho_a}} \quad (5)$$

***(ii)* Considering Compressibility of air :** Considering compressibility of air and applying steady flow energy equation to sections 1 and 2, we get

$$Q - W = h_2 - h_1 + \frac{c_2^2 - c_1^2}{2}$$

where h_1, h_2 = enthalpies at sections 1 and 2 respectively

Since $Q = 0$, $W = 0$ and $c_1 = 0$

$$\therefore \quad c_2 = \sqrt{2(h_1 - h_2)}$$

$$= \sqrt{2c_p(T_1 - T_2)} = \sqrt{2c_p T_1\left(1 - \frac{T_2}{T_1}\right)} \quad (6)$$

For isentropic flow between the atmosphere and venturi throat, we have

$$\frac{T_2}{T_1} = \left(\frac{p_2}{p_1}\right)^{\frac{\gamma-1}{\gamma}} \quad (7)$$

$$\therefore \quad c_2 = \sqrt{2c_p T_1\left[1 - \left(\frac{p_2}{p_1}\right)^{\frac{\gamma-1}{\gamma}}\right]} \quad (8)$$

Now
$$\dot{m}_a = \frac{A_1 c_1}{v_2} = \frac{A_2 c_2}{v_2} \quad (9)$$

where v = specific volume

Since
$$p_1 v_1^\gamma = p_2 v_2^\gamma$$

$$v_2 = v_1\left(\frac{p_1}{p_2}\right)^{\frac{1}{\gamma}} = \frac{RT}{p_1}\left(\frac{p_1}{p_2}\right)^{\frac{1}{\gamma}} \quad (10)$$

Eq. (9) becomes

$$\dot{m}_a = \frac{A_2 p_1}{RT_1}\left(\frac{p_2}{p_1}\right)^{1/\gamma}\sqrt{2c_p T_1\left[1 - \left(\frac{p_2}{p_1}\right)^{\frac{\gamma-1}{\gamma}}\right]} = \frac{A_2 p_1}{R\sqrt{T_1}}\sqrt{2c_p\left[\left(\frac{p_2}{p_1}\right)^{\frac{2}{\gamma}} - \left(\frac{p_2}{p_1}\right)^{\frac{\gamma+1}{\gamma}}\right]} \quad (11)$$

$$\therefore \quad \frac{\dot{m}_a}{\dot{m}_f} = \frac{C_{da}}{C_{df}} \times \frac{A_2}{A_f} \times \frac{p_1}{R\sqrt{T_1}} \times \frac{\sqrt{2c_p\left[\left(\frac{p_2}{p_1}\right)^{\frac{2}{\gamma}} - \left(\frac{p_2}{p_1}\right)^{\frac{\gamma+1}{\gamma}}\right]}}{\sqrt{2\rho_f(p_1 - p_2 - zg\rho_f)}} \quad (12)$$

If $z = 0$, then

$$\frac{\dot{m}_a}{\dot{m}_f} = \frac{C_{da}}{C_{df}} \times \frac{A_2}{A_f} \times \frac{p_1}{R\sqrt{T_1}} \times \frac{\sqrt{2c_p\left[\left(\frac{p_2}{p_1}\right)^{2/\gamma} - \left(\frac{p_2}{p_1}\right)^{\frac{\gamma+1}{\gamma}}\right]}}{\sqrt{2\rho_f(p_1 - p_2)}} \quad (13)$$

(*b*) A four stroke diesel engine of 30000cc capacity develops 14 kW per m^3 of free air induced per minute. When running at 3500 rev/min it has a volumetric efficiency of 85 per cent referred to free air-conditions of 1.013 bar and 27°C. It is proposed to boost the power of the engine by supercharging by a blower (driven mechanically from the engine) of pressure ratio 1.7 and isentropic efficiency of 80 per cent. Assuming that at the end of induction the cylinders contain a volume of charge equal to the swept volume, at the pressure and temperature of the delivery from the blower, estimate the increase in bp to be expected from the engine. Take overall mechanical efficiency as 80 per cent, ρ for air = 1.4, R = 0.287 kJ/kg K.

Sol.

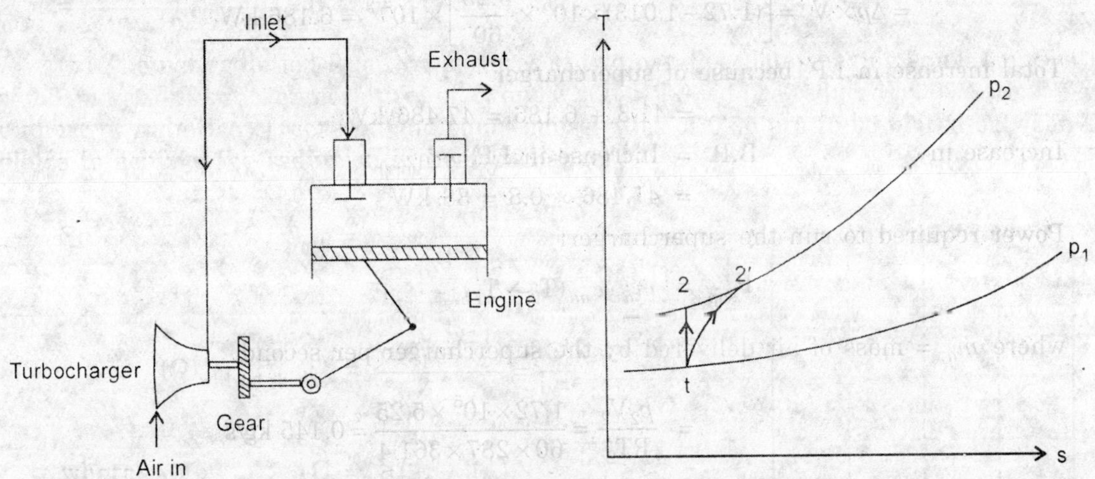

Fig. 8

Swept volume of engine per minute,

$$V_s = 3 \times 10^{-3} \times \frac{3500}{2} = 5.25 \text{ m}^3/\text{min}$$

Unsupercharged inducted volume

$$= V_s \times \eta_{vol} = 5.25 \times 0.85 = 4.4625 \text{ m}^3/\text{min}$$

The actual and isentropic compression processes of supercharger are shown in Fig. 8(*b*).

$$T_2 = T_1\left(\frac{p_2}{p_1}\right)^{\frac{\gamma-1}{\gamma}} = (273 + 27)\,(1.7)^{\frac{0.4}{1.4}} = 349.1 \text{ K}$$

Isentropic efficiency of supercharger,

$$\eta_c = \frac{T_2 - T_1}{T_{2'} - T_1}$$

$$0.8 = \frac{349.1 - 300}{T_{2'} - 300}$$

$$T_{2'} = 361.4 \text{ K}$$

The supercharger delivers 5.25 m^3/min at p_2 = 1.013 × 1.7 = 1.72 bar and 361.4 K to the engine. This volume referred to 1.013 bar and 300 K is,

$$V = \frac{1.72}{1.013} \times \frac{300}{361.4} \times 5.25 = 7.41 \text{ m}^3/\text{min}$$

Increase in inducted volume = V – V_s = 7.41 – 4.4625 = 2.95 m^3/min

As indicated power (I.P.) is directly proportional to induced volume, therefore, increase in I.P. due to increase in inducted volume of air

= 14 × 2.95 = 41.3 kW

Increase in I.P. due to increase in inducted air pressure because of supercharger

$$= \Delta p \times V_s = \left[(1.72 - 1.013) \times 10^5 \times \frac{5.25}{60}\right] \times 10^{-3} = 6.186 \text{ kW}$$

Total increase in I.P. because of supercharger

= 41.3 + 6.186 = 47.486 kW

Increase in B.P. = Increase in I.P. × η_{mech}

= 47.486 × 0.8 = 38 kW

Power required to run the supercharger.

$$P_{sup} = \dot{m}_a \cdot c_{pa}(T_2' - T_1)$$

where $\dot{m}_a$ = mass of air delivered by the supercharger per second.

$$= \frac{p_2 V_2}{RT_2'} = \frac{1.72 \times 10^5 \times 5.25}{60 \times 287 \times 361.4} = 0.145 \text{ kg/s}$$

$$P_{sup} = 0.145 \times 1.005 \times (361.4 - 300) = 8.95 \text{ kW}$$

Net increase in B.P. = 38 – 8.95 = 29.05 kW

$$\text{Percentage increase in B.P.} = \frac{29.05 \times 100}{8.95} = 324.6\%$$

(c) A liquid fuel C_7H_{16} is burned with 10% more air than the stoichiometric air. Assuming complete combustion, calculate

(*i*) the mass of air supplied per kg of fuel and

(*ii*) the volumetric analysis of the dry products of combustion. Assume air contains 21 per cent O_2 by volume.

Sol.

(*i*) The chemical equation of combustion for the given fuel is:

$$C_7H_{16} + 11O_2 = 7CO_2 + 8H_2O$$

$$100 \text{ kg} + 11 \times 32 \text{ kg} = 308 \text{ kg} + 144 \text{ kg}$$

$$1 \text{ kg of fuel requires } = \frac{11 \times 32}{100} \text{ kg of } O_2$$

$$= \frac{11 \times 32}{100} \times \frac{100}{23}$$

$$= 15.3 \text{ kg of air/kg of fuel}$$

When 10% more air than the stoichiometric air is burned, then actual

Mass of air supplied = 1.1 × 15.3 = 16.83 kg of air/kg of fuel

$$\text{Mass of } CO_2 \text{ formed } = \frac{308}{100} = 3.08 \text{ kg/kg of fuel}$$

O_2 comes in exhaust from complete air supplied,

$$N_2 = 16.83 \times \frac{77}{100} = 12.96 \text{ kg/kg of fuel}$$

Q_2 comes in exhaust from only excess air supplied,

$$O_2 = 15.3 \times 0.1 \times \frac{23}{100} = 0.352 \text{ kg}$$

(*ii*) The mass is converted into volume as listed below in the table :

Products of combustion	**Mass kg (a)**	**Molecular weight M (b)**	**Proportional volume** $c = \frac{a}{b}$	**Percentage by volume** $= \frac{c}{\Sigma c} \times 100$
N_2	12.96	28	0.463	85.11
CO_2	3.08	44	0.070	12.87
O_2	0.352	32	0.011	2.02

$$\Sigma C = 0.544$$

Q.3. (*a*) In a refrigeration system brine solution having a viscosity 165 N-s/m^2 and thermal conductivity 0.85 W/mK is flowing through a long pipe 2.5 cm inner diameter at a velocity of 6.1 m/s. Under these conditions the heat transfer coefficient was found to be 1135 W/m^{2}K for a brine temperature of –1°C and pipe temperature of 18.3°C. Find the temperature rise of brine per metre length of pipe if the velocity is doubled and same heat transfer takes place. Assume sp. heat of brine is 3768 J/kg K and density is 1000 kg/m^3. Assume fully developed flow.

Sol. Given : μ = 16.5 N·s/m^2, k = 0.85 W/m·K, d_i = 2.5 cm, v = 6.1 m/s,
h = 1135 W/m^2·K, t_i = –1°C, t_0 = 18.3°C, c_{pb} = 3768 J/kg·K, ρ_b = 100 kg/m^3

Reynolds number of brine solution,

$$Re = \frac{\rho v d_i}{\mu} = \frac{100 \times 6.1 \times 2.5 \times 10^{-2}}{16.5} = 9.242 < 2000$$

Hence flow is laminar.

Rate of heat transfer, $\dot{Q} = A_s h(t_o - t_i) = \dot{m}_b c_{pb} \cdot \Delta t$

$$\dot{m}_b = \rho_b A \times 2v = 100 \times \frac{\pi}{4}(2.5 \times 10^{-2})\, 2 \times 6.1 = 6 \text{ kg/s}$$

$$A_s = \pi d_i l = \pi \times 2.5 \times 10^{-2} \times 1 = 0.0785 \text{ m}^2$$

$$0.0785 \times 1135 \times (18.3 + 1) = 6 \times 3768 \times \Delta t$$

$$\Delta t = 0.075°C$$

(*b*) A solid sphere of diameter 10 cm is heated to 1000°C and suspended in a room whose walls are at 30°C. Compute the following :

(*i*) Rate of heat transfer due to radiation only neglecting other losses.

(*ii*) Time taken by the sphere to cool to 500°C assuming emissivity for the sphere = 0.1 and density 8.68 gm/cc. Sp. heat 0.098 J/kg K.

Sol. Given :

$$d = 10 \text{ cm},$$

$$T_1 = 273 + 100 = 1273 \text{ K},$$

$$T_2 = 273 + 30 = 303 \text{ K}$$

(*i*)

$$Q = \sigma A(T_1^4 - T_2^4)$$

$$= 5.67 \times 10^{-8} \times [4\pi \times (5 \times 10^{-2})^2] \times [(1273)^4 - (303)^4]$$

$$= 4662.84 \text{ W}$$

(*ii*)

$$\varepsilon = 0.1,\ \rho = 8.68 \text{ gm/c}\cdot\text{c or } 8680 \text{ kg/m}^3,$$

$$c_p = 0.098 \text{ J/kg}\cdot\text{K}$$

Heat transfer coefficient for radiation,

$$h_{rad} = \varepsilon\sigma(T_1^2 + T_2^2)(T_1 + T_2)$$

$$= 0.1 \times 5.67 \times 10^{-8}(1273^2 + 303^2)(1273 + 303) = 15.3 \text{ W/m}^2\cdot°\text{C}$$

$$\frac{t - t_a}{t_i - t_a} = \exp\left[-\frac{h_{rad} A_s}{\rho V c_p}\tau\right]$$

$$\frac{V}{A_s} = \frac{r}{3}$$

$$\frac{h_{rad} A_s}{\rho V c_p} = \frac{15.3 \times 3}{8680 \times 5 \times 10^{-2} \times 0.098} = 1.0792$$

$$\frac{t - t_a}{t_i - t_a} = \frac{500 - 30}{1000 - 30} = \frac{470}{970} = 0.48454$$

$$0.48454 = \exp(-1.0792\,\tau)$$

$$\ln 0.4854 = -1.0792\,\tau$$

$$-0.72456 = -1.0792\,\tau$$

$$\tau = 0.67 \text{ s}$$

(*c*) Find the surface area required for a surface condenser dealing with 25000K kg of saturated steam per hour at a pressure of 0.5 bar. Temperature of condensing water is 25°C. Cooling water is heated from 15°C to 25°C while passing through the condenser. Assume a heat transfer coefficient of 10 kW/m^{2}K. The condenser has 2 water passes with tubes of 19 mm OD and 1.2 mm thickness. Find the length and number of tubes per pass. Assume velocity of water is 1 m/s. Assume correction factor for 2 tube pass exchanger as 0.86. At 0.5 bar saturation temperature is 32.55° and latent heat is 2560 kJ/kg. Sp. heat of water is 4.18 kJ/kg K and density is 1000 kg/m^3.

Sol. Given : $\dot{m}_s$ = 25000 kg/h, x = 1.0, p = 0.5 bar, t_c = 25°C, t_i = 15°C, t_o = 25°C, U = 10 kW/m^2·K, d_0 = 19 mm, d_i = 19 – 2 × 1.2 = 16.6 mm, v_w = 1 m/s, F = 0.86, t_s at 0.5 bar = 32.55°C, h_{fg} = 2560 kJ/kg, c_{pw} = 4.18 kJ/kg·K, ρ_w = 1000 kg/m^3

Heat lost by steam = Heat gained by cooling water

$$\dot{m}_s(h_1 - h_c) = \dot{m}_w c_{pw}(t_o - t_i)$$

At p = 0.5 bar, $h_1 = h_f + xh_{fg} = c_{pw} \times t_s + xh_{fg}$ = 4.18 × 32.55 + 1.0 × 2560 = 2696 kJ/kg

$$h_c = c_{pw} \times t_c = 4.18 \times 25 = 104.5 \text{ kJ/kg}$$

$$\left(\frac{25,000}{3600}\right)(2696 - 104.5) = \dot{m}_w \times 4.18\,(25 - 15)$$

$$\dot{m}_w = 401.7 \text{ kg/s}$$

Let N = number of tubes per pass.

$$N \times \frac{\pi}{4} d_i^2 \times v_w \times \rho_w = \dot{m}w$$

$$N \times \frac{\pi}{4}(16.6 \times 10^{-3})^2 \times 1 \times 1000 = 401.7$$

$$N = 1856$$

Total number of tubes required = 2 × 1856 = 3712

Refer to Fig. 9.

$$\theta_1 = 32.55 - 15 = 17.55°C$$

$$\theta_2 = 32.55 - 25 = 7.55°C$$

$$\theta_m = \frac{\theta_1 - \theta_2}{\ln\left(\frac{\theta_1}{\theta_2}\right)} = \frac{17.55 - 7.55}{\ln\left(\frac{17.55}{7.55}\right)} = \frac{10}{0.843} = 11.85°C$$

$$Q = \left(\frac{25000}{3600}\right)(2696 - 104.5) = 17996.5 \text{ kJ/s}$$

$$Q = UA_s\theta_m F$$

$$A_s = \frac{17996.5}{10 \times 11.85 \times 0.86} = 176.59 \text{ m}^2$$

$$A_s = \pi d_0 L \times (2N)$$

$$L = \frac{176.59}{\pi \times 19 \times 10^{-3} \times 37/2} = 0.797 \text{ m}$$

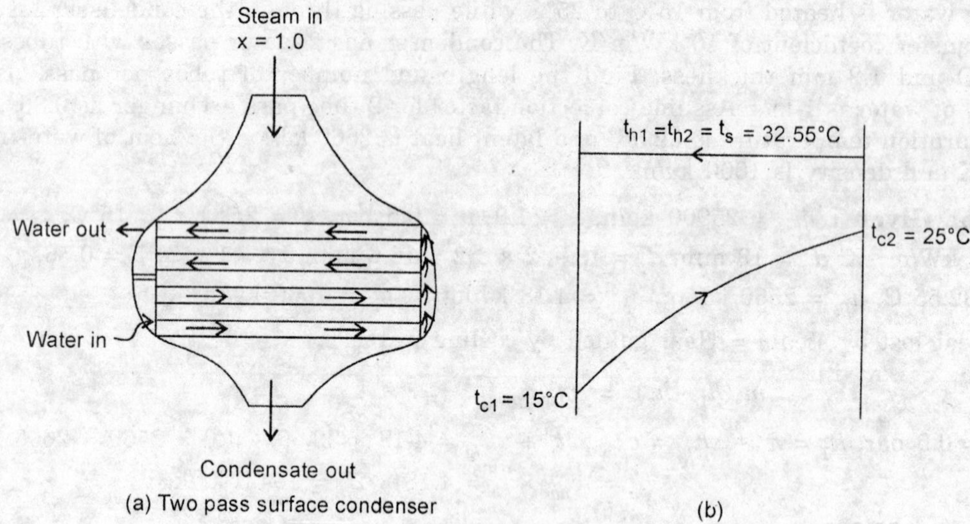

(a) Two pass surface condenser (b)

Fig. 9

(*d*) Estimate the coefficient of heat transfer from a vertical plate 2m × 2m to the surrounding air at 25°C. The plate surface temperature is 150°C. Also calculate the rate of heat transfer from the plate. For air assume the kinematic viscosity as 1.6×10^{-5} m²/s. The properties of air at film temperature are density 0.972 kg/m³, sp. heat 1.009 kJ/kg K, thermal conductivity 3.13×10^{-2} W/mK, Prandtl No. 0.69. The constants 'C' and '*n*' in Nusselt no. equation are 0.15 and 1/3 respectively.

Sol. Given : t_s = 150°C, t_∞ = 25°C, $\nu = 1.6 \times 10^{-5}$ m²/s, ρ = 0.972 kg/m³, c_p = 1.009 kJ/kg·K, $k = 3.13 \times 10^{-2}$ W/m·K, Pr = 0.69, L = 2m, C = 0.15, $n = \frac{1}{3}$

Mean film temperature,

$$t_f = \frac{1}{2}(t_s + t_\infty) = \frac{1}{2}(150 + 25) = 87.5°\text{C}$$

$$\beta = \frac{1}{T} = \frac{1}{(273 + t_f)} = \frac{1}{(273 + 87.5)} = \frac{1}{360.5} = 0.002774$$

Grashof number,
$$Gr_L = \frac{L^3 g\beta(t_s - t_\infty)}{\nu^2}$$

$$= \frac{2^3 \times 9.81 \times 0.002774\,(150 - 25)}{(1.6 \times 10^{-5})^2} = 1063 \times 10^8$$

Rayleigh number, $Ra_L = Gr_L \times Pa = 1063 \times 10^8 \times 0.69 = 73.35 \times 10^9 > 10^9$

The flow is turbulent

$$\bar{N}u_L = 0.15(Gr_L \times Pr)^{1/3} \text{ for } 10^9 < Gr \times Pr < 10^{12}$$

$$\frac{\bar{h}L}{k} = 0.15\ (1063 \times 10^8 \times 0.69)^{1/3} = 627.894$$

$$\bar{h} = \frac{627.894 \times 3.13 \times 10^{-2}}{2} = 9.826\ \text{W/m}^2 \cdot {}^\circ\text{C}$$

Rate of heat transfer from both side of the plate,

$$Q = 2\bar{h}\ A_s(t_s - t_\infty)$$

$$= 2 \times 9.826 \times 2 \times 2\ (150 - 25)$$

$$= 9826\ \text{W}$$

Q.4. (*a*) What factors are considered in selecting a refrigerant for the following application, mentioning the name of the refrigerant in each case

(*i*) Household refrigerator

(*ii*) An ice plant

(*iii*) Air conditioning plant for a cinema hall.

Sol. The factors to be conidered in selecting a refrigerant are:

(*i*) Household refigerator

1. It should be non-toxic, non-flammable, and non-explosive.
2. It should be fully oil miscible.
3. It should have high COP.
4. It should condense at moderate pressure and under atmospheric conditions.
5. It should have no effect on the quality of stored food.
6. It should be odourless.
7. Ease of leak detection.
8. Availability and low cost.
9. Ease of handling.
10. High electrical insulation.

Refrigerant : R–12 : Dichloro-difluoro methane (CCl_2F_2)

R–22 : Monochloro-difluoro ethane ——— ($CHClF_2$)

R–114 : Dichloro-tetafluoro ethane ($C_2Cl_2F_4$)

(*ii*) Ice Plant

1. Excellent thermal properties.
2. High refrigerating effect per kg of refrigerant.
3. Low volumetric displacement.
4. Low cost.
5. High efficiency.
6. Ease of leak defection.
7. Low weight of liquid circulated per ton of refrigeration.
8. Low boiling point and high latent heat of vaporisation.

Refrigerants : R–717 : Ammonia (NH_3)

***(iii)* Cinema hall air-conditioning**

1. Colourless and odourless.
2. Non-toxic and non-flammable.
3. Non-explosive and non-corrosive.
4. High operating pressure.
5. Moderate refrigerating effect.
6. Low cost.
7. Easy handling.

Refrigerants : R–744 : Carbon dioxide (CO_2)

R–113 : Trichloro-trifluomethane ($C_2Cl_3F_3$)

R–22 : Monochloro-difluoromethane ($CHClF_2$)

(b) A dense air refrigeration machine operating on Bell-Coleman cycle operates between 3.4 bar and 17 bar. The temperature of air after the cooler is 15°C and after the refrigerator is 6°C. If the refrigeration capacity is 6 tons, calculate :

(*i*) temperature after compression and expansion

(*ii*) air circulation per min

(*iii*) work of compressor and expander

(*iv*) theoretical COP.

(*v*) Rate of water circulation required in the cooler in kg/min if the rise in temperature is limited to 30°C.

Sol.

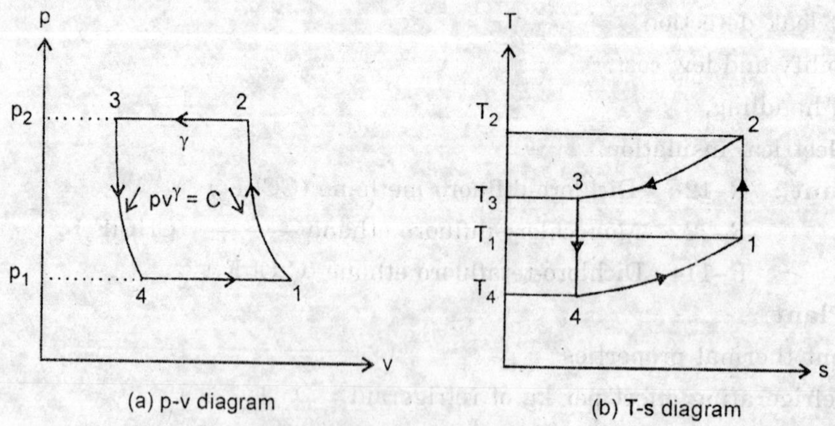

Fig. 10. Bell-Coleman cycle.

Refer to Fig. 10

Given :

$$p_1 = 3.4 \text{ bar}, \; p_2 = 17 \text{ bar},$$

$$T_3 = 273 + 15 = 288 \text{ K},$$

$$T_1 = 273 + 6 = 279 \text{ K},$$

$$Q = 6 \text{ TR}.$$

(*i*) $$\frac{T_2}{T_1} = \left(\frac{p_2}{p_1}\right)^{\frac{\gamma-1}{\gamma}} = \left(\frac{17}{3.4}\right)^{\frac{0.4}{1.4}} = 1.58382$$

$$T_2 = 279 \times 1.58382 = 441.9 \text{ K or } 68.9°\text{C}$$

$$\frac{T_3}{T_4} = \left(\frac{p_3}{p_4}\right)^{\frac{\gamma-1}{\gamma}} = \left(\frac{p_2}{p_1}\right)^{\frac{\gamma-1}{\gamma}} = \left(\frac{17}{3.4}\right)^{\frac{0.4}{1.4}} = 1.58382$$

$$T_4 = \frac{288}{1.58382} = 181.8 \text{ K or } -91.16°\text{C}$$

(*ii*) Heat extracted per min = 6 × 210 = 1260 kJ/min

$$\dot{m}_a c_p(T_1 - T_4) = 1260$$

$$\dot{m}_a = \frac{1260}{1.005\,(279 - 181.8)} = 12.9 \text{ kg/min}$$

(*iii*) Compresser work,

$$w_{1-2} = \left(\frac{\gamma}{\gamma-1}\right) R\,(T_2 - T_1)$$

$$= \frac{1.4}{0.4} \times 0.287\,(441.9 - 279) = 163.63 \text{ kJ/kg}$$

Work of expansion,

$$w_{3-4} = \left(\frac{\gamma}{\gamma-1}\right) R(T_3 - T_4)$$

$$= \left(\frac{1.4}{0.4}\right) \times 0.287\,(288 - 181.8) = 106.68 \text{ kJ/kg}$$

(*iv*) Theoretical $$\text{COP} = \frac{\text{Heat extracted}}{\text{Net workdone}} = \frac{q}{w_{net}}$$

$$q = c_p(T_1 - T_4) = 1.005(279 - 181.8) = 97.686 \text{ kJ/kg}$$

$$w_{net} = w_{1-2} - w_{3-4} = 16.363 - 106.68 = 56.95 \text{ kJ/kg}$$

$$(\text{COP})_{th} = \frac{97.686}{56.95} = 1.715$$

(*v*) $$\dot{m}_w \times c_{pw} \times \Delta t = \dot{m}_a c_p (T_2 - T_3)$$

$$\dot{m}_w = \frac{12.9 \times 1.005\,(441.9 - 288)}{4.18 \times 30} = 15.91 \text{ kg/min}$$

(*c*) A simple saturation cycle using R_{12} is designed for taking a load of 10 tons. The refrigerant and ambient temperatures are 0°C and 30°C respectively. A minimum temperature difference of 5°C is required in evaporator and condenser for heat transfer. Find the

(*i*) mass flow rate through the system

(*ii*) power required in kW

(*iii*) cylinder dimensions assuming L/D ratio of 1.2 for a single cylinder single acting compressor if it runs at 300 rpm with a volumetric efficiency of 0.9. The following properties are taken for R_{12}. Sp. heat of R_{12} vap = 0.95 kJ/kg K.

Sat temp °C	Sat. pr bar	Sp. vol.		Enthalpy		Entropy	
		Sat. liq $\times 10^{-3}$	Sat. vap m³/kg	Sat. liq kJ/kg	Sat. vap	Sat. liq kJ/kg·K	Sat. vap
–5°C	2.61	0.71	0.0650	31.4	185.4	0.1251	0.6991
0°C	3.08	0.72	0.0554	36.1	187.5	0.1420	0.6966
30°C	7.45	0.77	0.0235	64.6	199.6	0.2399	0.6854
35°C	8.47	0.79	0.0206	69.5	201.5	0.2559	0.6839

Sol. Given :

$$Q = 10 \text{ TR}, t_1 = -5°\text{C}, t_2 = 30°\text{C}, t_3 = 0°\text{C}$$

$$L/D = 1.2, N = 300 \text{ rpm}, \eta_{vol} = 0.9, c_p = 0.95 \text{ kJ/kg·K}$$

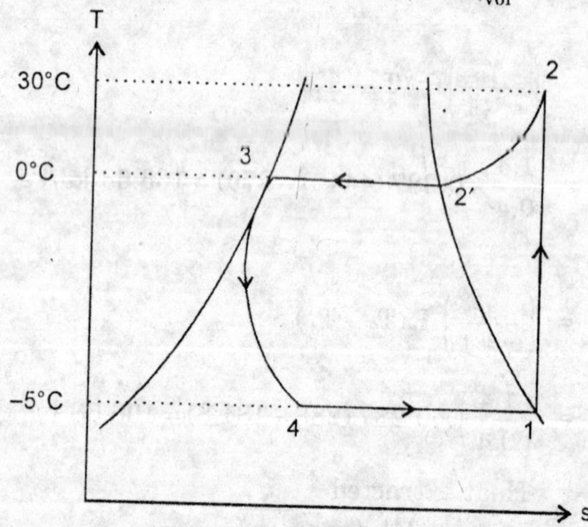

Fig. 11. T-S diagram.

(*i*) The T-*s* diagram is shown in Fig. 11.

$$h_1 = 185.4 \text{ kJ/kg}, \; h_{f_3} = h_4 = 36.1 \text{ kJ/kg}$$

Refrigerating effect,

$$\text{RE} = h_1 - h_{f_3} = 185.4 - 36.1 = 149.3 \text{ kJ/kg}$$

Refrigeration capacity of plant,

$$Q = 10 \times 14000 = \frac{14 \times 10^5}{149.3} = 38.89 \text{ kJ/s}$$

Mass flow rate of refrigerant,

$$\dot{m}_a = \frac{Q}{\text{RE}} = \frac{38.89}{149.3} = 0.2605 \text{ kg/s}$$

(*ii*) $$h_2 = h_1 + c_p(t_2 - t_1) = 185.4 + 0.95\ (30 + 5) = 218.65 \text{ kJ/kg}$$

Power required,

$$P = \dot{m}_a(h_2 - h_1)$$
$$= 0.2605(218.65 - 185.4)$$
$$= 8.66 \text{ kW}$$

(*iii*) Specific volume of saturated vapour at point 1,

$$v_1 = 0.0650 \text{ m}^3\text{/kg}$$

Assuming that there is no clearance, piston displacement per second,

$$V = \frac{\dot{m}_R \times v_1}{\eta_v} = \frac{0.2605 \times 0.0650}{0.9} = 0.0188 \text{ m}^3\text{/s}$$

Also $$V = \frac{\pi}{4} D^2 L \frac{N}{60} = \frac{\pi}{4} \times 1.2\ D^3 \times \frac{300}{60} = 4.7124\ D^3 \text{ m}^3\text{/s}$$

$$\therefore \quad 4.7124\ D^3 = 0.0188$$
$$D = 0.1586 \text{ m or } 158.6 \text{ mm}$$
$$L = 1.2\ D = 1.2 \times 158.6\ 190.3 \text{ mm}$$

(*d*) (*i*) What are the factors affecting comfort air condtioning?

(*ii*) Explain the air conditioning system used in winter season.

(*iii*) Show how air washer can be used for year round air conditioning?

(*iv*) A spray cooling coil is used to operate under the following conditions–Air inlet 28° DBT/21° WBT, Air outlet 10° DBT/6° WBT. Total air flow rate 2000 m³/mt. Chilled water inlet and outlet temperatures are 7°C and 12°C. Find

(*a*) Cooling load on the coil

(*b*) Water flow rate through the coil.

Properties of moist air from psychometric chart are as follows :

DBT	**WBT**	**Sp. humidity gm/kg**	**Sp. Vol. m³/kg**	**Enthalpy kJ/kg**
28	21.0	12.95	0.87	61.0
10	6.0	4.2	0.81	21.0

Sol. (*i*) Factors affecting comfort air-conditioning : The factors affecting comfort air-conditioning are :

1. **Temperature of air :** It is the temperature maintained within an enclosed space. This is accomplished either by the addition or removal of heat from the enclosed space. Human beings feel comfortable when the air is at 21°C with 56% relative humidity.
2. **Humidity of air :** The humidity of air is controlled by either decreasing or increasing the moisture content of air. In general, for summer air-conditioning, the relative humidity should not be less than 60%, whereas for winter air – conditioning, it should not be more than 40%.

3. **Purity of air :** To feel comfortable while breathing, air should not be contaminated. Purity in air can be achieved by proper filtration and cleaning to make it free from dust and other impurities.

4. **Motion of air :** The motion or circulation of air is essential to keep constant temperature throughout the conditioned space. Recommended air velocity for comfort air – conditioning is 0.4 m/s during summer and 0.15 to 0.20 m/s during winter.

***(ii)* Winter Air Conditioning System :** In winter air conditioning, the air is heated, which is generally accompained by humidification. The schematic arrangement of the system is shown in Fig. 12.

The outside air flows through a damper and mixes up with the *recirculated air* (which is obtained from the conditioned space). The mixed air passes through a filter to remove dirt, dust and other impurities. The air now passes through a preheat coil in order to prevent the possible freezing of water and to control the evaporation of water in the humidifier. After that, the air is made to pass through a reheat coil to bring the air to the designed dry bulb temperature. Now, the conditioned air is supplied to the conditioned space by a fan. From the conditioned space, a part of the used air is exhausted to the atmosphere by the exhaust fans or ventilators. The remaining part of the used air (known as recirculated air) is again conditioned as shown in Fig. 12.

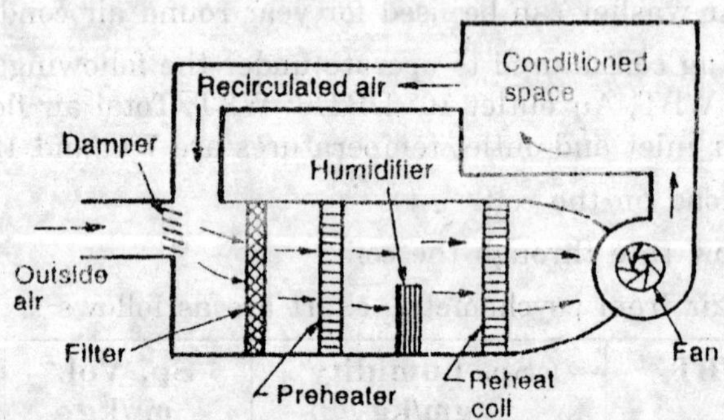

Fig. 12. Winter air conditioning system.

The outside air is sucked and made to mix with recirculated air, in order to make up for the loss of conditioned (or used) air through exhaust fans or ventilation from the conditioned space.

***(iii)* Year-Round Air Conditioning System Using air washer :** An air washer involves the flow of air through a spray of water. During the of flow the air may be cooled or heated, humidified or dehumidified, or simply adiabatically saturated, depending upon the mean surface temperature of water.

The year-round air conditioning system should have equipment for both the summer and winter air conditioning. The schematic arrangement of a modern summer year round air conditioning system is shown in Fig. 13.

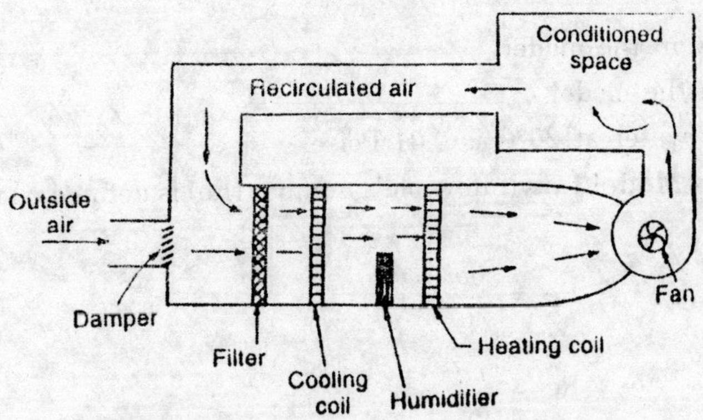

Fig. 13. Year-round air conditioning system.

The outside air flows through the damper and mixes up with the recirculated air (which is obtained from the conditioned space). The mixed air passes through a filter to remove dirt, dust and other impurities. In summer air conditioning, the cooling coil operates to cool the air to the desired value. The dehumidification is obtained by operating the cooling coil at a temperature lower than the dew point temperature (apparatus dew point). In winter, the cooling coil is made inoperative and the heating coil operates to heat the air. The spray type humidifier is also made use in the dry season to humidify the air.

(*iv*) (*a*) Mass of air flowing through the cooling coil,

$$\dot{m}_a = \frac{V_1}{v_{s_1}} = \frac{2000}{0.87} = 2298.85 \text{ kg/min}$$

Capacity of cooling coil

$$= \dot{m}_a(h_1 - h_2)$$

$$= 2298.85\ (61 - 21) = 91954 \text{ kJ/min}$$

$$= \frac{91954}{60} = 1532.56 \text{ kW}$$

Amount of water vapour removed per minute

$$= m_a(w_1 - w_2)$$

$$= 2298.85\ (12.95 - 4.2) \times 10^{-3} = 20.1 \text{ kg/min}$$

(*b*) Let $\dot{m}_w$ = water flow rate through the coil, kg/min

$$91954 = \dot{m}_w \times c_{pw} \times \Delta t = \dot{m}_w \times 4.18 \times (12 - 7)$$

$$\dot{m}_w = 4399.7 \text{ kg/min or } 73.33 \text{ kg/s}$$

Q.5. (*a*) (*i*) Explain what do you mean by Reynold's model law. Give two examples where Reynold's model can be applied.

(*ii*) An oil of specific gravity 0.92 and viscosity 0.03 Poise is to be transported at the rate of 2500 L/s through a 1.2 m diameter pipe. Tests were conducted on a 12 cm diameter pipe using water at 20°C.

Determine,

1. Velocity of flow in the model
2. Rate of flow in the model

Given viscosity of water at 20°C is 0.01 Poise.

Sol. (*i*) Reynold's Model Law : Reynold's number (Re) is defined as the ratio of inertia force to viscous force.

$$\text{Re} = \frac{\rho v l}{\mu} = \frac{vl}{\nu}$$

For pipe flow,

$$\text{Re} = \frac{vd}{\nu}$$

Where v = flow velocity, d = pipe diameter, and ν = kinematic viscosity. According to Reynold's model law :

$$(\text{Re})_m = (\text{Re})_p$$

The Reynolds number is taken as the criterion of dynamic similarity in the flow situations where the viscous forces predominate. Reynolds model can be applied in the following cases:

1. Motion of submarine completely under water.
2. Low velocity motion around automobiles and aeroplanes.
3. Incompressible flow through pipes of smaller sizes.
4. Flow through low speed turbomachies.

(*ii*)

$$S_0 = 0.92,\ \mu_0 = 0.03 \text{ poise},$$

$$Q = 2500 \text{ L/s},\ d_0,\ P = 1.2 \text{ m},$$

1.

$$d_m = 12 \text{ cm} \quad \text{or} \quad 0.12 \text{ m},$$

$$\mu_w = 0.01 \text{ Poise}$$

Density oil,

$$\rho_0 = S_0 \rho_w = 0.92 \times 1000 = 920 \text{ kg/m}^3$$

Velocity of oil,

$$v_0 = \frac{Q}{A} = \frac{2500 \times 10^{-3} \times 4}{\pi \times (1.2)^2} = 2.21 \text{ m/s}$$

Now

$$(\text{Re})_0 = (\text{Re})_w$$

$$\frac{\rho_0 v_0 d_0}{\mu_0} = \frac{\rho_w v_w d_w}{\mu_w}$$

$$\frac{920 \times 2.21 \times 1.2}{0.03} = \frac{1000 \times v_w \times 0.12}{0.01}$$

Velocity of flow in the model,

$$v_w = 6.78 \text{ m/s}$$

2.

$$\frac{\rho_0 d_0}{\mu_0} \times \left(\frac{4Q_0}{\pi d_0^2}\right) = \frac{\rho_w d_w}{\mu_w} \times \left(\frac{4Q_w}{\pi d_w^2}\right)$$

$$\frac{\rho_0 Q_0}{\mu_0 d_0} = \frac{\rho_w Q_w}{\mu_w d_w}$$

Rate of flow in the model,

$$Q_w = \frac{\rho_0}{\rho_w} = \frac{d_w}{d_0} \times \frac{\mu_w}{\mu_0} \times Q_0 = \frac{920}{1000} \times \frac{0.12}{1.2} \times \frac{0.01}{0.03} \times 2500$$

$$= 76.67 \text{ L/s}$$

Q.5. (*b*) A rough pipe of diameter 0.1 m carries water at 20°C at the rate of 50 L/s. If the average height of the roughness projections on the pipe surface is 0.15 mm, determine the

(*i*) friction factor

(*ii*) shear stress at the pipe surface

(*iii*) shear velocity

(*iv*) maximum velocity

Take for water at 20°C,

γ = Kinematic viscosity = 1×10^{-6} m²/s.

ρ = Mass density = 1000 kg/m³.

Sol. Given : d = 0.1 m, t = 20°C, Q = 50 L/s, k_s = 0.15 mm, $\nu = 1 \times 10^{-6}$ m²/s, ρ = 1000 kg/m³.

(*i*) Velocity of water in pipe,

$$v = \frac{4Q}{\pi d^2} = \frac{4 \times 50 \times 10^{-3}}{\pi \times (0.1)^2} = 6.37 \text{ m/s}$$

$$\text{Reynolds number, Re} = \frac{vd}{\nu} = \frac{6.37 \times 0.1}{1 \times 10^{-6}} = 637000 > 2300$$

The flow is turbulent.

The Karman-Prandtl resistance equation for turbulent flow in rough pipes is given by

$$\frac{1}{\sqrt{f}} = 2\log_{10}\frac{r_0}{k_s} + 1.74 = 2\log_{10}\left(\frac{0.05}{0.15 \times 10^{-3}}\right) + 1.74 = 6.78575$$

Friction factor, $f = 0.0217$

(*ii*) Shear stress at the pipe surface,

$$\tau_0 = \frac{f}{8}\rho v^2 = \frac{0.0217}{8} \times 1000 \times (6.37)^2 = 110.065 \text{ N/m}^2$$

(*iii*) Shear velocity, $u* = v\sqrt{\frac{f}{8}} = 6.37\sqrt{\frac{0.0217}{8}} = 0.332$ m/s

(*iv*) Maximum velocity,

$$\frac{u_{\max}}{v} = 1 + 1.33\sqrt{f}$$

$$= 1 + 1.33\sqrt{0.0217} = 1.196$$

$$u_{\max} = 6.37 \times 1.196 = 7.618 \text{ m/s}$$

Q.5. (*c*) Explain with the help of diagrams the phenomenon of Karman Vortex Street. How is Strouhal number, a dimensionless number, related to the above phenomenon?

Sol. In the immediate vicinity of the solid boundary the curvature of the velocity profile depends only on the pressure gradient. In accelerated flow $dp/dx < 0$, yields $(\partial^2 u/\partial y^2)_{y=0} < 0$ and therefore for the entire thickness of the boundary layer $\partial^2 u/\partial y^2 < 0$. In the decelerated region $dp/dx > 0$, which results in $(\partial^2 u/\partial y^2)_{y=0} > 0$. Since, within the boundary layer at some distance from the boundary $\partial^2 u/\partial y^2 < 0$, and in the retarded flow $\partial^2 u/\partial y^2 > 0$ at and near the boundary, it follows that there must exist point for which $\partial^2 u/\partial y^2 = 0$. This point is known as the **point of inflexion.**

At some distance downstream along the boundary, a point is reached at which the velocity gradient is zero, *i.e.*, $(\partial u/\partial y)_{y=0} = 0$. Downstream of this point reverse-flow exists and this point can be considered a "separation point of the boundary-layer". Beyond this point velocity gradient is negative and back-flow occurs. The location of the separation point can be found from the condition.

$$(\partial u/\partial y)_{y=0} = 0$$

However this requires integration of the boundary-layer equations to obtain $u = u(x, y)$. The line of zero velocity dividing the forward flow and the back flow leaves the boundary surface at the separation point, and is known as the *'separation streamline'* (which is a slipline). The region between the separation streamline and the boundary surface is known as the *'wake'*. The back flow is the main cause of formation of wake which is characterized by large size irregular eddies (or vortices) resulting in a considerable energy loss. The vortices formed in the wake zone behind a blunt body are not shed in pairs but alternately from top and bottom and move downstream in the fluid. The formation of vortices and their subsequent movement downstream changes completely the flow-field in the wake. The pressure distribution in the wake suffers a radical change as compared to that in an ideal fluid flow.

At some distance downstream of the body it is possible to visualize a regular pattern of two rows of vortices which move alternately clockwise and anticlockwise. This system of vortices is known as a Karman vortex street and is shown in Fig. 14.

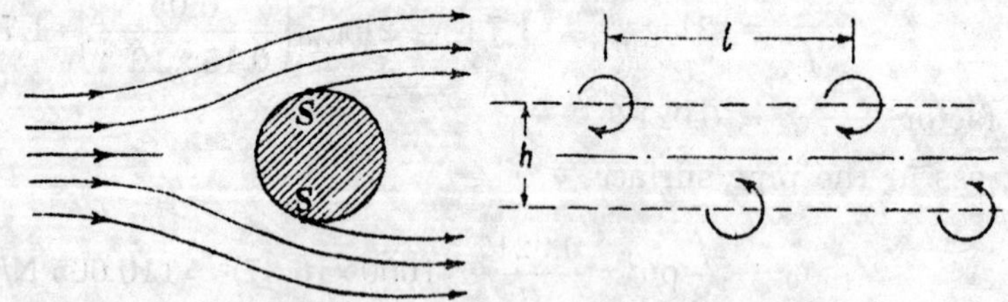

Fig. 14. Karman vortex street behind a circular cylinder.

The frequency of vortex shedding is expressed in terms of a dimensionless parameter known as the Strouhal number,

$$S = \frac{nd}{U_\infty}$$

where n = frequency of vortex shedding.

As the Reynolds number increases, the frequency of shedding increases until at $Re = 10^3$, S = 0.21. For $Re < 10^3$, the Strouhal number remains practically constant at 0.21. It varies from 0.12 to 0.21 for Re range of 50 to 10^3 respectively.

Q.5. (*d*) (*i*) Explain what do you mean by Temporal on Local acceleration, and convective acceleration.

(*ii*) The velocity field in a fluid medium is given by $\bar{V} = 3xy^2\hat{i} + 2xy\hat{j} + (2zy + 3t)\hat{k}$ determine, the rotational velocity vector at (1, 2, 1) and t = 3.

Sol.

(*i*) For unsteady and non-uniform flow,

Tangential acceleration,

$$a_s = \frac{\partial v}{\partial t} + v\frac{\partial v}{\partial s}$$

and normal acceleration, $$a_n = \frac{\partial v_n}{\partial t} + \frac{v^2}{r}$$

The partial derivatives $\frac{\partial v}{\partial t}$ and $\frac{\partial v_n}{\partial t}$ express variations of tangential and normal velocities with time, and are known as local accelerations in the respective directions.

The other partial derivative, $\frac{v\partial v}{\partial s}$ takes into account the change in velocity with space, and is termed as convective acceleration. Thus each component of total acceleration consists of two parts; first one is due to unsteadiness of flow and is known as local acceleration, and the second part is due to non-uniformly of flow and is called convective acceleration.

(*ii*) $$u = 3xy^2,\ v = 2xy,\ w = 2zy + 3t$$

$$\omega_x = \frac{1}{2}\left(\frac{\partial w}{\partial y} - \frac{\partial v}{\partial z}\right) = \frac{1}{2}(2z - 0) = z = 1$$

$$\omega_y = \frac{1}{2}\left(\frac{\partial u}{\partial z} - \frac{\partial w}{\partial x}\right) = \frac{1}{2}(0 - 0) = 0 = 0$$

$$\omega_z = \frac{1}{2}\left(\frac{\partial v}{\partial x} - \frac{\partial u}{\partial y}\right) = \frac{1}{2}(2y - 6xy) = \frac{1}{2}\ (2 \times 2 - 6 \times 1 \times 2) = -4$$

Rotational velocity vector,

$$\vec{\omega} = 1\hat{i} - 4\hat{k}$$

Q.6. (*a*) The velocity profile for a laminar boundary layer is given by

$$\frac{u}{U} = \sin\left(\frac{\pi}{2}\frac{y}{\delta}\right)$$

where u is the velocity at a distance y from the plate, U is the free stream velocity. δ is the boundary layer thickness,

Determine expressions in terms of Reynolds number for the following :

(*i*) boundary layer thickness

(*ii*) drag force on both sides of the plate

(*iii*) co-efficient of drag

Sol.

(*i*) Boundary layer thickness :

$$\frac{\tau_0}{\rho U^2} = \frac{d}{dx}\left[\int_0^\delta \frac{u}{U}\left(1-\frac{u}{U}\right)dy\right]$$

$$= \frac{d}{dx}\left[\int_0^\delta \sin\left(\frac{\pi}{2}\cdot\frac{y}{\delta}\right)\left\{1-\sin\left(\frac{\pi}{2}\cdot\frac{y}{\delta}\right)\right\}dy\right]$$

$$= \frac{d}{dx}\left[\int_0^\delta \left\{\sin\left(\frac{\pi}{2}\cdot\frac{y}{\delta}\right)-\sin^2\left(\frac{\pi}{2}\cdot\frac{y}{\delta}\right)\right\}dy\right]$$

$$= \frac{d}{dx}\left[\int_0^\delta \left\{\sin\left(\frac{\pi}{2},\frac{y}{\delta}\right)-\frac{1-\cos\left(\frac{\pi y}{\delta}\right)}{2}\right\}dy\right]$$

$$= \frac{d}{dx}\left|\frac{-\cos\left(\frac{\pi y}{2\delta}\right)}{\left(\frac{\pi}{2\delta}\right)}-\frac{y}{2}+\frac{\sin\left(\frac{\pi y}{\delta}\right)}{\left(\frac{\pi}{2\delta}\right)}\right|_0^\delta$$

$$= \frac{d}{dx}\left[-\left(\frac{2\delta}{\pi}\right)\left\{\cos\frac{\pi}{2}-\cos\theta\right\}-\frac{1}{2}(\delta-0)+\frac{2\delta}{\pi}\left\{\sin\pi-\sin\theta\right\}\right]$$

$$= \frac{d}{dx}\left[-\left(\frac{2\delta}{\pi}\right)(0-1)-\frac{1}{2}\times\delta+\left(\frac{2\delta}{\pi}\right)\times 0\right]$$

$$= \frac{d}{dx}\left[\frac{2\delta}{\pi}-\frac{\delta}{2}\right]=\left(\frac{4-\pi}{2\pi}\right)\frac{d\delta}{dx}$$

$$\tau_0 = \left(\frac{4-\pi}{2\pi}\right)\rho U^2\cdot\frac{d\delta}{dx} \qquad (1)$$

Now $\tau_0 = \mu\left(\frac{du}{dy}\right)_{y=0} = \mu\left[\frac{d}{dy}\left\{U\sin\left(\frac{\pi y}{2\delta}\right)\right\}\right]_{y=0}$

$$= \mu U\left[\cos\left(\frac{\pi y}{2\delta}\right)\times\left(\frac{\pi}{2\delta}\right)\right]_{y=0}$$

$$= \mu U\times\left(\frac{\pi}{2\delta}\right)=\frac{\mu U\pi}{2\delta} \qquad (2)$$

Equating Eqs. (1) and (2), we have

$$\left(\frac{4-\pi}{2\pi}\right)\rho U^2\cdot\frac{d\delta}{dx} = \frac{\mu U\pi}{2\delta}$$

$$\delta \cdot d\delta = \left(\frac{\mu U x}{2}\right)\left(\frac{2\pi}{4-\pi}\right) \times \frac{1}{\rho U^2} \times dx$$

$$= \left(\frac{\pi^2}{4-\pi}\right) \times \frac{\mu}{\rho U} \times dx$$

Integrating both sides, we get

$$\frac{\delta^2}{2} = \left(\frac{\pi^2}{4-\pi}\right) \times \frac{\mu}{\rho U} \times x + C$$

where C = constant of integration

At $x = 0, \delta = 0$. Therefore, $C = 0$

$$\delta^2 = \left(\frac{2\pi^2}{4-\pi}\right)\frac{\mu x}{\rho U} = 22.995 \times \frac{\mu x}{\rho U}$$

$$\delta = 4.795 \cdot \sqrt{\frac{\mu}{\rho U x}} \cdot x = \frac{4.795x}{\sqrt{Re_x}}$$

(*iii*) Shear stress, $\tau_0 = \dfrac{\mu U \pi}{2\delta}$

$$= \frac{\mu U \pi \sqrt{Re_x}}{2 \times 4.795x} = 0.327 \frac{\mu U}{x}\sqrt{Re_x}$$

Coefficient of drag,

$$C_D = \frac{F_D}{\frac{1}{2}\rho A U^2}$$

where $F_D = \int_0^L \tau_0 \times B \times dx = \int_0^L 0.327 \frac{\mu U}{x_L}\sqrt{Re_x} \times B \cdot dx$

$$= 0.327 \ \mu U B \int_0^L \sqrt{\frac{\rho U x}{\mu}} \times \frac{1}{x} \cdot dx$$

$$= 0.327 \ \mu U B \cdot \sqrt{\frac{\rho U}{\mu}} \int_0^L x^{-1/2} dx = 0.327 \ \mu U B \sqrt{\frac{\rho U}{\mu}} \left|2x^{1/2}\right|_0^L$$

$$= 0.654 \ \mu U B \sqrt{\frac{\rho U L}{\mu}}$$

$$C_D = \frac{0.654 \ \mu U B \sqrt{\frac{\rho U L}{\mu}}}{\frac{1}{2}\rho L B U^2} = 1.31 \frac{\mu}{\rho L U}\sqrt{\frac{\rho U L}{\mu}} = \frac{1.31}{\sqrt{\frac{\rho L U}{\mu}}} = \frac{1.31}{\sqrt{Re_L}}$$

(*iii*) Drag force on both sides of the plate

$$F_D = 2 \times \frac{1}{2} \rho A U^2 \times C_D = \rho A U^2 \times \frac{1.31}{\sqrt{Re_L}}$$

Q.6. (*b*) A Pitot-Static tube is used to monitor the velocity of an air stream. At the location of insertion of the probe, the static pressure is 1.5 bar and temperature is 35°C. Calculate the reading of a mercury manometer connected differentially across the static and total pressure openings of the probe, if the air stream velocity is

(*i*) 60 m/s

(*ii*) 200 m/s

(*iii*) 500 m/s

Take into consideration the compressibility characteristics of the flow wherever applicable. The following values may be used.

Isentropic table for perfect gas, k = 1.4

M	p/p_0
0.54	0.820
0.56	0.808
0.58	0.796
0.70	0.720
0.72	0.708
0.74	0.695

Normal shock table for perfect gas, k = 1.4

M_x	M_y	p_y/p_x	ρ_y/ρ_x
1.38	0.748	2.055	1.655
1.40	0.739	2.120	1.689
1.42	0.731	2.185	1.724
1.44	0.723	2.252	1.759

Sol. Given : p_s = 1.5 bar,

t = 35°C, k = 1.4

(*i*) v_0 = 60 m/s

Velocity of sound, $c = \sqrt{kRT} = \sqrt{1.4 \times 237 \times (273 + 35)} = 351.8$ m/s

Max number, $M = \frac{v_0}{c} = \frac{60}{351.8} = 0.1706 < 1.0$

$$\frac{p_0}{p_s} = \left[1 + \frac{k-1}{2} M^2\right]^{k/(k-1)} = \left[1 + \left(\frac{1.4-1}{2}\right)(0.1706)^2\right]^{\frac{1.4}{0.4}} = 1.0205$$

p_0 = 1.5 × 1.0205 = 1.53 bar

(*ii*)
$$v_0 = 200 \text{ m/s}$$

$$M = \frac{200}{351.8} = 0.5685 < 1.0$$

The flow is subsonic.

$$\frac{p_0}{p_s} = \left[1 + \left(\frac{1.4-1}{2}\right) \times (0.5685)^2\right]^{3.5} = 1.2451$$

$$p_0 = 1.5 \times 1.2451 = 1.8677 \text{ bar}$$

or from isentropic table, by linear interplation for

$$m = 0.5685,$$

$$\frac{p_s}{p_0} = 0.808 - \left(\frac{0.808 - 0.796}{0.58 - 0.56}\right) \times 0.0085 = 0.8029$$

$$p_0 = \frac{1.5}{0.8029} = 1.868 \text{ bar}$$

(*iii*)
$$v_0 = 500 \text{ m/s}$$

$$M = \frac{500}{351.8} = 1.4213 > 1$$

The flow is supersonic.

For normal shocks, we have

$$\frac{p_2}{p_1} = \frac{2kM_1^2 - (k-1)}{(k+1)} = \frac{2 \times 1.4(1.4213)^2 - (1.4-1)}{(1.4+1)} = 2.19$$

$$p_2 = 1.5 \times 2.19 = 3.285 \text{ bar}$$

or from normal shock table, for $M_x = 1.42$, $p_y/p_x = 2.185$

$$\therefore \quad p_y = 1.42 \times 2.185 = 3.10 \text{ bar}$$

Q.6. (*c*) Derive an expression for the area velocity relationship for a compressible fluid flow in the form of

$$\frac{dA}{A} = \frac{dV}{V}[1 - M^2].$$

Explain properly, with the help of diagrams, what are the important concousions derived from the above relationship.

Sol. For a cross-section of area over which velocity V and density ρ are constant, the continuity equation for steady flow is :

$$\rho AV = \text{const.}$$

Taking ln, we have

$$\ln \rho + \ln A + \ln V = \ln \text{const.}$$

Taking differential, we have

$$\frac{d\rho}{\rho} + \frac{dA}{A} + \frac{dV}{V} = 0 \qquad (1)$$

Euler's equation of motion for steady, frictionless flow neglecting the gravity and body forces, may be written as,

$$VdV + \frac{dp}{\rho} = 0 \tag{2}$$

Multiplying the second term by $\frac{dp}{d\rho}$, and noting that velocity of sound, $c = \sqrt{\frac{dp}{d\rho}}$, we get

$$\frac{d\rho}{\rho} + \frac{VdV}{c^2} = 0$$

Substituting for $\frac{d\rho}{\rho}$ in Eq. (1), we get

$$-\frac{VdV}{C^2} + \frac{dA}{A} + \frac{dV}{V} = 0$$

or
$$\frac{dA}{A} = \frac{dV}{V}\left(\frac{V^2}{c^2} - 1\right) = -\frac{dV}{V}(1 - M^2) \tag{3}$$

where
$$M = \frac{V}{c} = \text{Mach number}$$

Hence proved.

From Eq. (2), we have

$$\frac{dV}{V} = -\frac{dp}{\rho V^2}$$

Eq. (1) then becomes,

$$\frac{d\rho}{\rho} - \frac{dp}{\rho V^2} + \frac{dA}{A} = 0$$

$$\frac{dA}{A} = \frac{dp}{\rho V^2} - \frac{d\rho}{\rho} = \frac{dp}{\rho V^2}\left(1 - V^2\frac{d\rho}{dp}\right) = \frac{dp}{dV^2}\left(1 - \frac{V^2}{c^2}\right) = \frac{dp}{\rho V^2}(1 - M^2)$$

$\therefore$
$$dp = \frac{\rho V^2}{A} \cdot \frac{1}{1 - M^2} \cdot dA \tag{4}$$

Also
$$c^2 = dp/d\rho \quad \text{or} \quad d\rho = \frac{1}{c^2}dp \tag{5}$$

Important Conclusions

1. Subsonic flow (M < 1) :

   (*a*) Eq. (3) becomes, $\frac{dA}{A} = -\frac{dV}{V}$. Therefore, on increase of cross-sectional area causes a reduction of velocity and vice versa.

   (*b*) Eq. (4) becomes, $dp \propto dA$. Therefore, increase in cross sectional area causes an increase in pressure and vice-versa.

2. Supersonic flow (M > 1) :

(*a*) Eq. (3) shows that $dA \propto dV$. Therefore, an increase in cross-sectional area is accompanied by an increase in velocity and vice-versa.

(*b*) Eq. (4) shows that $dp \propto -dA$. Therefore, an increase in cross-sectional area is accompaned by a decrease in pressure and vice-versa.

3. Sonic flow (M = 1) :

(*a*) Therefore from Eq. (3), dA must be zero. Since second derivative is positive, A must be minimum. Therefore, M will be unity where the cross-sectional area is minimum (*i.e.* throat of a nozzle).

(*b*) From Eq. (4), either dp must be infinite or dA must also be zero simultaneously. In that case dp shall become indeterminate. This condition could occur when a duct changes from a converging duct to a diverging duct or vice versa.

	Conclusions		Conclusions
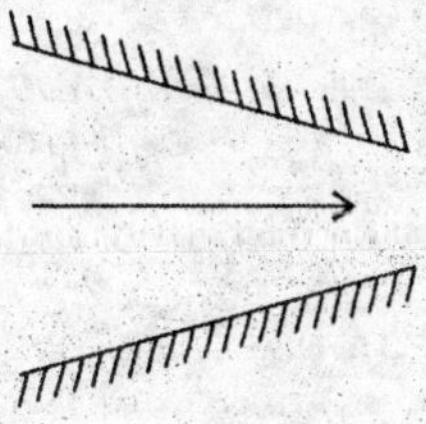 (*a*) Converging duct	V increases p decreases ρ decreases T decreases	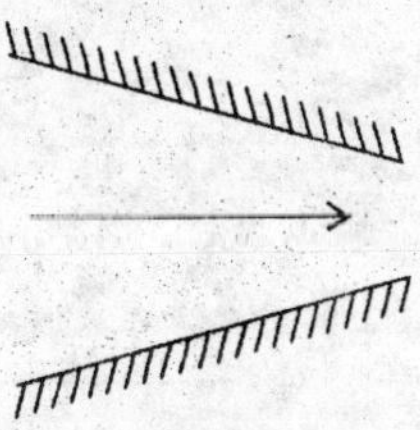 (*a*) Converging duct	V decreases p increases ρ increases T increases
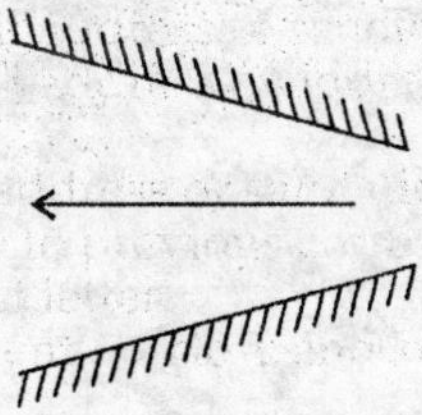 (*b*) Diverging duct	V decreases p increases ρ increases T increases	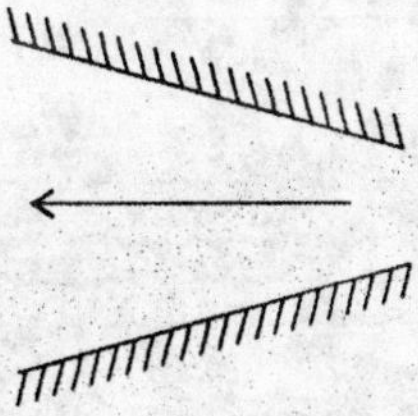 (*b*) Diverging duct	V increases p decreases ρ decreases T decreases
1. Subsonic flow (M < 1)		2. Supersonic flow (M > 1)	

Variation of flow variables with area.

Q.7. (*a*) A three jet Pelton turbine is required to generate 10000 kW under a net head of 400 m. The balde angle at the outlet is 15° and the decrease in the relative velocity while passing over the blade is 5%. Determine,

(*i*) The diameter of the jet

(*ii*) The force exerted by a jet on the buckets.

Given,

overall efficiency of turbine = 80%

co-efficient of velocity = 0.98

speed ratio = 0.46

Further, if the jet ratio is not to be less than 10, calculate

(*iii*) Speed of the wheel for a frequency of 50 Hertz.

(*iv*) Corresponding wheel diameter.

Sol. Given : $n = 3, \rho = 10{,}000$ kW, $H = 400$ m, $\beta_2 = 15°$, $v_{r2} = 0.95\, v_{r1}$,

$$\eta_0 = 0.80,\ C_v = 0.98,\ \rho = \frac{u}{v_1} = 0.46,\ m = \frac{D}{d} = 10,\ f = 50 \text{ Hz}$$

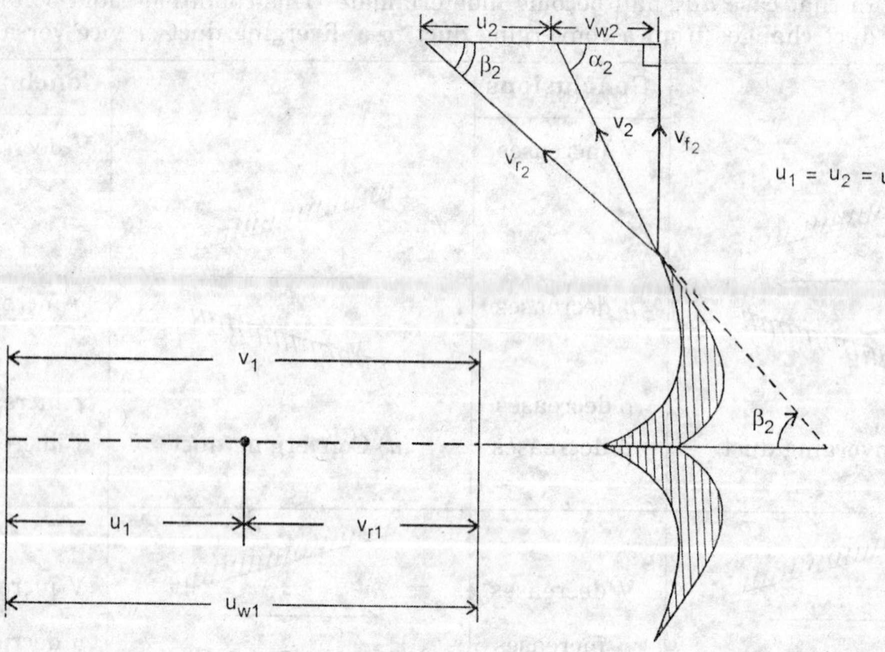

Fig. 15

(*i*) Refer to Fig. 15

Jet velocity, $v_1 = C_v \sqrt{2gH} = 0.98\sqrt{2 \times 9.80 \times 400} = 86.82$ m/s

Wheel velocity, $u = \rho v_1 = 0.46 \times 86.82 = 39.93$ m/s

Overall efficiency, $$\eta_0 = \frac{P}{\gamma QH}$$

$$0.8 = \frac{10{,}000 \times 10^3}{9.81 \times 10^3 \times Q \times 400}$$

$$Q = 3.186 \text{ m}^3/\text{s}$$

Discharge per jet $$= \frac{Q}{n} = \frac{3.186}{3} = 1.062 \text{ m}^3/\text{s}$$

Now
$$\frac{Q}{n} = \frac{\pi}{4} d^2 v_1$$
$$1.062 = \frac{\pi}{4} d^2 \times 86.82$$
$$d = 0.125 \text{ m} \quad \text{or} \quad 125 \text{ mm}$$

(*ii*)
$$v_{r1} = v_1 - u = 86.82 - 39.93 = 46.89 \text{ m/s}$$
$$v_{r2} = 0.95 \times 46.89 = 44.54 \text{ m/s}$$
$$v_{w1} = v_1 = 86.82 \text{ m/s}$$
$$v_{w2} = v_{r2} \cos \beta_2 - u = 44.54 \cos 15° - 39.93 = 3.097 \text{ m/s}$$

Force exerted by 3 jets on the buckets,
$$F_u = \rho Q (v_{w1} + v_{w2})$$
$$= 10^3 \times 3.186\,(86.82 + 3.097)$$
$$= 286.47 \text{ kN}$$

Force exerted by one jet $= \dfrac{286.47}{3} = 95.492$ kN

(*iii*)
$$u = \frac{\pi DN}{60} = \frac{\pi \times 1.25 \times N}{60} = 39.93$$
$$N = 610 \text{ rpm}$$

Synchronous speed, $N_{syn} = \dfrac{50 \times 60}{p} = \dfrac{3000}{p}$

where p = pair of poles

For $p = 5$, $N_{syn} = \dfrac{3000}{5} = 600$ rpm

Revised diameter of wheel,
$$D_1 = \frac{610}{600} \times 1.25 = 1.27 \text{ m}$$

Jet ratio $= 1.27/0.125 = 10.17$

(*iv*) Wheel diameter, $D = 10\,d = 10 \times 125 = 1250$ mm or 1.25 m

Q.7. (*b*) A centrifugal pump is to deliver 4.5 m^3/s when running at 750 rpm. The diameter of the impeller at inlet is 53 cm and at outlet is 76 cm. It may be assumed that the air enters radially with a speed of 15 m/s. The vanes are set backwards at outlet at 70° to the tangent, and width at outlet is 10 cm. The volute casing gives at 30% recovery of the outlet velocity head. The losses in the impeller may be taken as equivalent to 25% of the outlet velocity head. Blade thickness effects may be neglected. Determine.

(*i*) The manometric efficiency, and

(*ii*) The pressure at the discharge.

Sol. Given : $Q = 4.5$ m^3/s, $N = 750$ rpm, $D_1 = 53$ cm, $R_1 = 0.265$ m, $D_2 = 76$ cm, $R_2 = 0.38$ m, $\alpha_1 = 90°$, $\beta_2 = 70°$, $v_1 = 15$ m/s, $B_2 = 10$ cm, $h_{volute} = +0.3 \dfrac{v_2^2}{2g}$, $h_{loss} = -0.25 \dfrac{v_2^2}{2g}$

(*i*) Refer to Fig. 16.

$$v_1 = \frac{\pi D_1 N}{60} = \frac{\pi \times 0.53 \times 750}{60} = 20.81 \text{ m/s}$$

$$v_2 = \frac{\pi D_2 N}{60} = \frac{\pi \times 0.76 \times 750}{60} = 29.84 \text{ m/s}$$

$$Q = \pi D_1 \beta_1 v_{f1} = \pi D_2 \beta_2 v_{f2}$$

$$4.5 = \pi \times 0.76 \times 0.1 \times v_{f2}$$

$$v_{f2} = 18.85 \text{ m/s}$$

$$\tan \beta_2 = \frac{v_{f2}}{v_2 - v_{w2}}$$

$$\tan 70° = \frac{18.85}{29.84 - v_{w2}}$$

$$2.74748 = \frac{18.85}{29.84 - v_{w2}}$$

Fig. 16

$$v_{w2} = 22.98 \text{ m/s}$$

$$\tan \alpha_2 = \frac{v_{f2}}{v_{w2}} = \frac{18.85}{22.98} = 0.8203$$

$$\alpha_2 = 39.36°$$

cos $\alpha_2 = \frac{v_{w2}}{v_2} = 0.77316$

$$v_2 = 29.72 \text{ m/s}$$

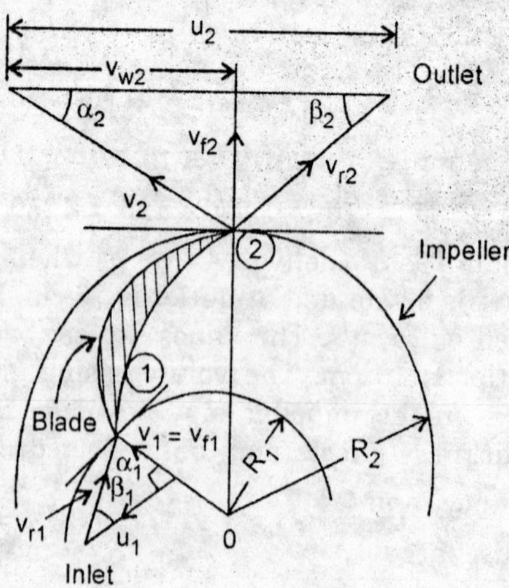

Fig. 16

Applying Bernoulli's theorem at inlet and outlet of impeller, we have

$$\frac{p_1}{\gamma}+\frac{v_1^2}{2g} = \frac{p_2}{\gamma}+\frac{v_2^2}{2g}-\frac{u_2 v_{w2}}{g}$$

$$\frac{p_2-p_1}{\gamma} = \frac{v_1^2-v_2^2}{2g}+\frac{u_2 v_{w2}}{g} = \frac{15^2-(29.72)^2}{2\times 9.81}+\frac{29.84\times 22.98}{9.81}$$

$$= -\ 33.55 + 69.90 = 36.35 \text{ m of air}$$

Velocity head at outlet $= \dfrac{v_2^2}{2g} = \dfrac{(29.72)^2}{2\times 9.81} = 45.02$ m of air

Head recovered in volute casing

$= 0.3 \times 45.02 = 13.51$ m of air

Head loss in impeller $= 0.25 \times 45.02 = 11.26$ m of air

Net pressure rise $= 36.35 + 13.51 - 11.26 = 38.6$ m of air

$$\eta_{mano} = \frac{\text{Net pressure rise}}{\dfrac{u_2 v_{w2}}{g}} = \frac{38.6}{69.90} = 0.5522 \text{ or } 55.22\%$$

Q.7. (*c*) Discuss how the volumetric efficiency varies with the clearance and the pressure ratio in an air compressor.

Sol. Volumetric efficiency of reciprocating compressor is given by,

$$\eta_{vol} = 1 + c - c(r_p)^{1/n} = 1+c-c\left(\frac{p_2}{p_1}\right)^{\frac{1}{n}}$$

Let n = 1.2

c, % ↓	η_{vol}				
r_p →	4	5	6	8	10
3	0.93	0.91	0.89	0.84	0.82
4	0.91	0.88	0.86	0.81	0.77
5	0.89	0.86	0.83	0.77	0.71
6	0.87	0.83	0.79	0.72	0.65
7	0.85	0.80	0.76	0.67	0.59
8	0.83	0.77	0.72	0.63	0.53
10	0.78	0.72	0.65	0.53	0.42

The variation of volumetric efficiency is shown in Fig. 17.

The volumetric efficiency decreases as percentage clearance and pressure ratio increase.

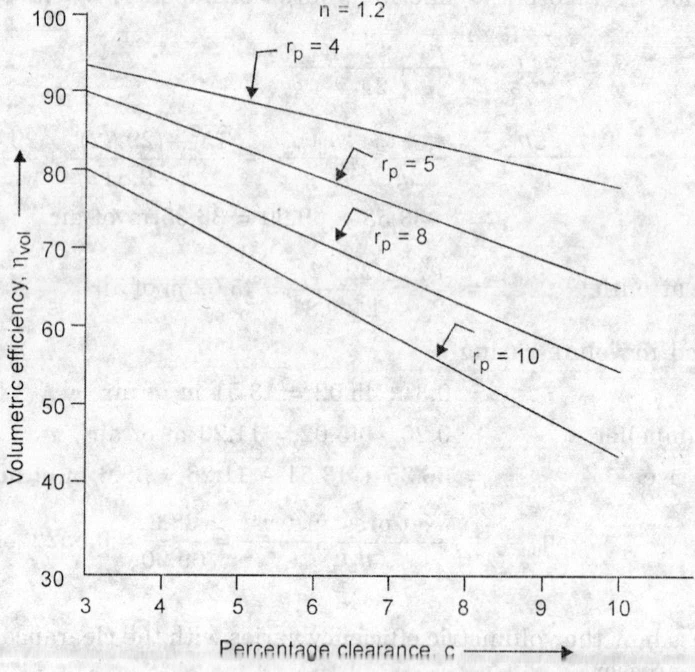

(b) v,s c

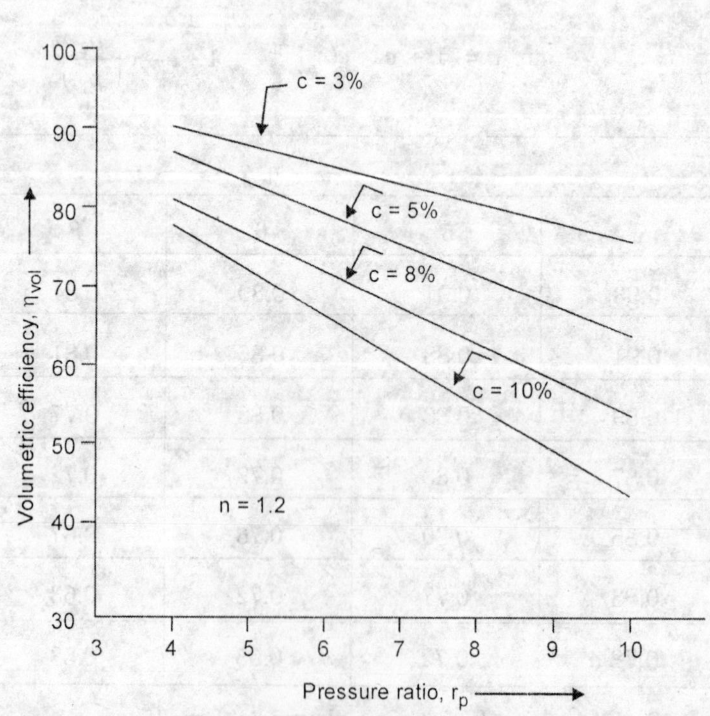

Fig. 17. Variation of volumetric efficiency of reciprocating compressor with clearance ratio and pressure ratio.

Q.8. (*a*) With the help of a neat Schematic explain the working of a Bubbling type Fluidized Bed Boiler. What are its advantages?

Sol. A fluidized bed may be defined as the bed of solid particles behaving as a fluid. The working of a bubbling type fluidized bed boiler is given below.

When a gas is passed through a pached bed of finely divided solid particles, it experiences a pressure drop across the bed. At low gas velocities, this pressure drop is small and does not disturb the particles. But if the gas velocity is increased further, a stage is reached, when *particles are suspended in the gas stream and the packed bed becomes a 'fluidised bed'*. With further increase in gas velocity, the bed becomes turbulent and rapid mixing of particles occurs. In general, the behaviour of this mixture of solid particles and gas is like a fluid. Burning of a fuel in such a state is known as a *fluidised bed combustion*.

Fig. 18 shows the arrangement of the FBC system.

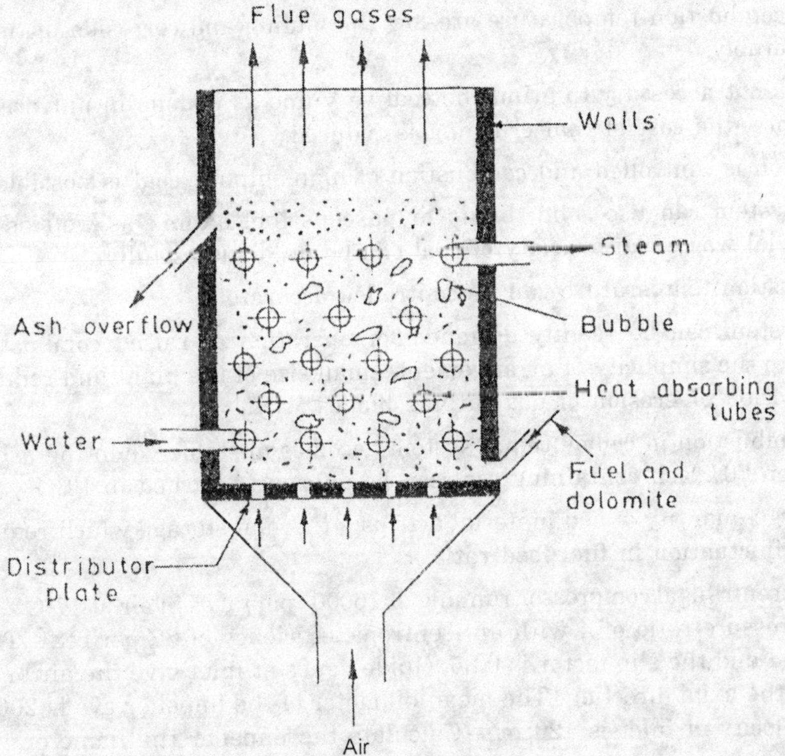

Fig. 18. Basic FBC system.

On the distributor plate are fed the fuel and inert material *dolomite* and from its bottom air is supplied. The high velocity of air keeps the solid feed material in suspending condition during burning. The generated heat is rapidly transferred to the water passing through the tubes immersed in the bed and generated steam is taken out. During the burning sulphur dioxide formed is absorbed by the dolomite and prevents its escape with the exhaust gases. The molten slag is tapped from the top surface of the bed.

The primary object of using the *inert material* is to control the bed temperature, it accounts for 90% of the bed volume. It is very necessary that the selection of an inert material should be done judiciously as it remains with the fuel in continuous motion and at high temperature to the tune of 800°C. Moreover, the inert material should not distintegrate coal, the parent material of the bed.

The cost economics shows that a saving of about 10% in operating cost and 15% capital cost could be achieved for a unit rating of 120 MW and it may be still higher for bigger units.

Advantages

1. As a result of better heat transfer, the unit size and hence the *capital costs are reduced.*
2. In can respond rapidly to changes in load demand (since thermal equilibrium between air and coal particles in the bed is quickly established).
3. Low combustion temperatures (800 to 950°C) inhibits the formation of nitrogen oxides like nitric oxide and nitrogen dioxide.
4. Since combustion temperature are low the fouling and corrosion of tubes is reduced considerably.
5. As it is not necessary to grind the coal very fine as is done in pulverised fuel firing, therefore, the cost of coal crushing is reduced.
6. Pollution is controlled and combustion of high-sulphur coal is possible.
7. FBC system can use solid, liquid or gaseous fuel or mix as well as domestic and industrial waste. Any variety of coal can be used successfully.
8. Combustion temperature can be controlled accurately.
9. The system can be readily designed for operation at raised combustion, pressure, owing to the simplicity of arrangement, small size of the plant and reduced likelihood of corrosion or erosion of gas turbine blades.
10. The combustion in conventional system becomes unstable when the ash exceeds 48% but even 70% ash containing coal can be efficiently burned in FBC.
11. The large quantity of bed material acts as a thermal storage which reduces the effect of any fluctuation in fuel feed ratio.

Q.8. (*b*) A centrifugal compressor running at 16000 rpm takes in air at 17°C and compresses it through a pressure ratio of 4 with an isentropic efficiency of 82 per cent. The blades are radially inclined and the slip factor is 0.85. Guide vanes at inlet give the air an angle of pre-whirl of 20° to the axial direction. The mean diameter of the impeller eye is 200 mm and the absolute air velocity at inlet is 120 m/s. Calculate the impeller tip diameter.

Sol. Given :

$$N = 16000 \text{ rpm}$$

$$T_1 = 273 + 17 = 290 \text{ K}$$

$$r_p = 4$$

$$\eta_{isen} = 0.82$$

$$\phi_s = 0.85$$

$$\alpha_1 = 70°$$

$$D_1 = 0.2 \text{ m}$$

$$v_1 = 120 \text{ m/s}$$

Refer to Fig. 19.

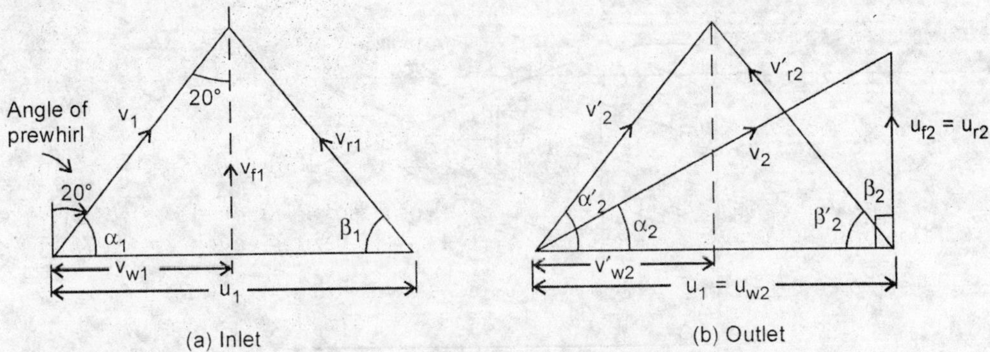

Fig. 19. Velocity triangles with prewhirl.

Temperature after isentropic compression, $T_2 = T_1\left(\frac{p_2}{p_1}\right)^{\frac{\gamma-1}{\gamma}} = 290 \times 4^{\frac{0.4}{1.4}}$ = 430.99 K.

Isentropic temperature rise, $(\Delta T)_{isen}$ = 430.94 − 290 = 140.94 K.

Actual temperature rise, $(\Delta T)_{act} = \frac{(\Delta T)_{isen}}{\eta_{isen}} = \frac{140.94}{0.82} = 171.88$ K.

Power input per unit mass flow rate = $C_p \times (\Delta T)_{act}$ = 1.005 × 171.88 = 172.74 kJ/kg

$$u_1 = \frac{\pi D_1 N}{60} = \frac{\pi \times 0.2 \times 16000}{60} = 167.58 \text{ m/s}$$

Let D_2 = impeller tip diameter, mm

$$u_2 = \frac{\pi D_2 N}{60} = \frac{\pi D_2 \times 16000}{60 \times 10^3} = 0.83776\ D_2 \text{ m/s}$$

For radial discharge, $v_{w2} = u_2$

Slip factor $\phi_s = \frac{v'_{w2}}{v_{w2}} = 0.85$

Power input per unit mass flow rate

$$= u_2 v'_{w2} - u_1 v_{w1} = 172.74$$

$$= u_2 \times 0.85 u_2 - 167.58 \times 41.04 = 172.74 \times 10^3$$

$$0.85\, u_2^2 = 179617$$

$$u_2 = 459.7 \text{ m/s} = 0.83776\ D_2$$

$$D_2 = 548.7 \text{ mm}$$

Q.8. (*c*) Explain briefly stage efficiency, internal efficiency and Reheat factor in a multistage steam turbine. Find out the relation between these three factors.

Sol. The expansion of steam through a number of stages of turbine is shown in Fig. 20. A_1B_1 represents isentropic expansion in the first stage. The actual state of steam with frictional reheating is shown by point A_2. Therefore the actual heat drop in first stage is A_1C_1. Similarly the isentropic and actual heat drops in the succeeding stages are shown by A_2B_2, A_3B_3..., and A_2C_2, A_3C_3, ..., and so on. A_1D represents the isentropic heat drop between supply and condenser state of steam.

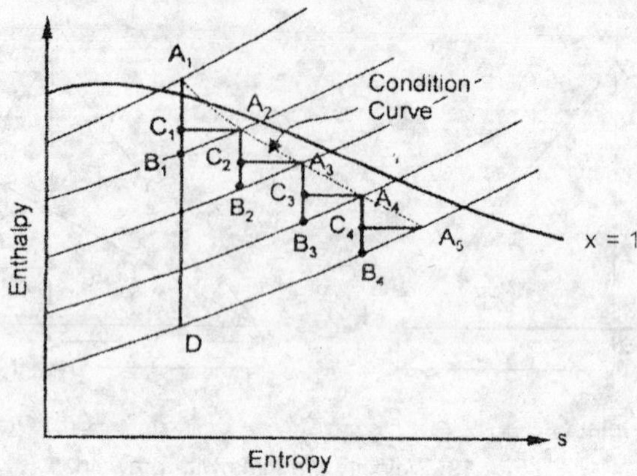

Fig. 20. Expansion of steam through a number of turbine stages.

$$\text{Stage efficiency, } \eta_s = \frac{\text{Actual enthalpy drop}}{\text{Isentropic enthalpy drop}}$$

$$\text{For first stage, } \eta_{s1} = \frac{A_1C_1}{A_1B_1}$$

$$\text{For second stage, } \eta_{s2} = \frac{A_2C_2}{A_2B_2} \text{ and so on.}$$

$$\text{Reheat factor} = \frac{\text{Cumulative enthalpy drop}}{\text{Isentropic heat drop}}$$

$$R_f = \frac{A_1B_1 + A_2B_2 + ...}{A_1D} = \frac{\sum_{i=1}^{n} A_iB_i}{A_1D} \quad (1)$$

$$\text{Turbine internal efficiency, } \eta_i = \frac{\text{Total actual turbine work}}{\text{Isentropic work}}$$

$$\eta_i = \frac{A_1C_1 + A_2C_2 + ...}{A_1D} = \frac{\sum_{i=1}^{n} A_iC_i}{A_1D} \quad (2)$$

If $\eta_{s1} = \eta_{s2} = ... = \eta_s$, then

$$\eta_s = \frac{A_1C_1}{A_1B_1} = \frac{A_2C_2}{A_2B_2} = ... = \frac{A_1C_1 + A_2C_2 + ...}{A_1B_1 + A_2B_2 + ...}$$

or

$$\eta_i = \frac{\Sigma A_iC_i}{\Sigma A_iB_i} \quad (3)$$

From Eqs. (2) and (3), we get

$$\eta_i = \frac{\Sigma A_iB_i}{A_1D} \times \eta_s \quad (4)$$

Combining Eqs. (1) and (4), we get

Internal efficiency = Stage efficiency × Reheat factor.

MECHANICAL ENGINEERING
PAPER I—2010

(Conventional)

Q.1. (*a*) 10 kg of pure ice at –10°C is separated from 6 kg of pure water at +10°C in an adiabatic chamber using a thin adiabatic membrane. Upon rupture of the membrane, ice and water mix uniformly at constant pressure. At this pressure, the melting temperature of ice is 0°C and the latent heat of melting is 335 kJ/kg. The mean specific heat at constant pressure for ice and water are respectively 2.1 kJ/kg-K and 4.2 kJ/kg-K.

(*i*) Sketch the systems before and after mixing.

(*ii*) What is the final equilibrium temperature of the system after the completion of the mixing process?

(*iii*) Estimate the change of entropy of the universe due to the mixing.

(*iv*) Is the final phase of the system solid, ice, liquid water or ice-water mixture?

Sol. Given : m_1 = 10 kg, T_1 = 273 – 10 = 263 K,

m_2 = 6 kg, T_2 = 273 + 10 = 283 K,

L_1 = 335 kJ/kg, c_{p1} = 2.1 kJ/kg·K, c_{p2} = 4.2 kJ/kg

(*i*) The mixing process is shown in Fig. 1.

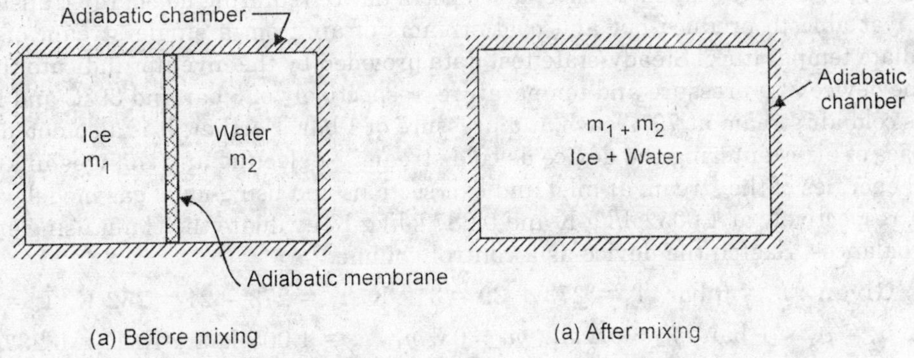

Fig. 1

(*ii*) Let t_f = find temperature of mixture after mixing

For heat energy balance, we have

Total energy before mixing = Total energy after mixing

$$m_1[c_{p1}(0 - t_1) + L_1 + c_{p2}\,[t_2 - 0] + m_2 c_{p2}(t_2 - 0) = (m_1 + m_2)\,c_{p2}(t_f - 0)$$

$$10[2.1\{0 - (-10)\} + 3.35 + 4.2(10 - 0)] + 6 \times 4.2\ (10 - 0) = (10 + 6) \times 4.2\ (t_f - 0)$$

$$10[21 + 335 + 42] + 252 = 67.2\ t_f$$

$$t_f = 62.97°\text{C}$$

Final equilibrium temperature of system after mixing = 62.97°C

(*iii*) Entropy change of ice as it gets heated from –10°C to 0°C,

$$(\Delta S)_1 = \int_{263}^{273} m_1 c_{p1} \frac{dT}{T} = 10 \times 2.1 \left| \ln T \right|_{263}^{273} = 0.78367 \text{ kJ/K}$$

Entropy change of ice at 0°C to water at 0°C,

$$(\Delta S)_2 = \frac{L}{273} = \frac{335}{273} = 1.22711 \text{ kJ/K}$$

Entropy change of ice and water at 0°C to 10°C,

$$(\Delta S)_3 = \int_{273}^{283} (m_1 + m_2) c_{p2} \frac{dT}{T}$$

$$= 16 \times 4.2 \left| \ln T \right|_{273}^{283} = 2.41753 \text{ kJ/K}$$

Entropy change of water,

$$(\Delta S)_4 = \int_{273}^{283} m_2 c_{p2} \frac{dT}{T} = 6 \times 4.2 \left| \ln T \right|_{273}^{283} = 0.90659 \text{ kJ/K}$$

Change in entropy $= (\Delta S)_1 + (\Delta S)_2 + (\Delta S)_4 - (\Delta S)_3$

$= 0.78367 + 1.22711 + 0.90657 - 2.41753 = 0.49982$ kJ/K

(*iv*) The final phase of system will be ice-water mixture.

Q.1. (*b*) An inventor claims to have developed a device requiring no energy transfer by heat or work, yet able to produce hot and cold streams of air from a single stream of air at an intermediate temperature. Steady-state test data provided by the inventor indicate that the air enters the device at a pressure and temperature, respectively of 5 bar and 39°C and leaves the device as cold air stream at 79°C each at a pressure of 1 bar. Further, it is also noted that 40% of the mass of air is entering the device as cold stream. Neglecting any changes in kinetic and potential energies of the stream at inlet and exit sections and using ideal gas model with c_p and R for air, respectively at 1.005 kJ/kg-K and 0.287 kJ/kg-K, evaluate the claim using energy and entropy balances. Sketch the device as a control volume.

Sol. Given : $p_1 = 5$ bar, $T_1 = 273 + 39 = 312$ K, $T_2 = 273 - 21 = 252$ K, $T_3 = 273 + 79 = 352$ K, $p_2 = p_3 = 1$ bar, $\dot{m}_2 = 0.4\,\dot{m}_1$, $\dot{m}_3 = 0.6\,\dot{m}_1$, $c_p = 1.005$ kJ/kg-K, R = 0.287 kJ/kg·K.

The assumptions made are the following:

1. The control volume shown is at steady state.
2. For the control volume, W = 0 and Q = 0.
3. The changes in KE and PE from inlet to exit are ignored.
4. The air is assumed as an ideal gas with constant specific heat and characteristic gas constant.

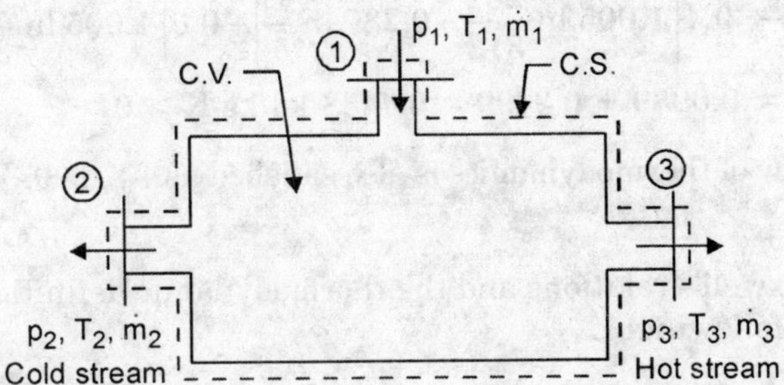

Fig. 2

For the device to operate as claimed, the conservation of mass and energy principles must be satisfied. The second law of thermodynamics also must be satisfied, and in particular the rate of entropy production cannot be negative. Accordingly, the mass, energy and entropy rate balances are considered in turn.

With assumptions 1 and 3, the mass and energy rate balances reduce, respectively, to

$$\dot{m}_1 = \dot{m}_2 + \dot{m}_3$$

$$\dot{Q} = \dot{m}_1 h_1 - \dot{m}_2 h_2 - \dot{m}_3 h_3$$

Now $\quad \dot{m}_2 = 0.4\,\dot{m}_1$, therefore, $\dot{m}_3 = 0.6\,\dot{m}_1$

$$\begin{aligned}\dot{Q} &= (\dot{m}_2 + \dot{m}_3)\,h_1 - \dot{m}_2 h_2 - \dot{m}_3 h_3 \\ &= \dot{m}_2(h_1 - h_2) + \dot{m}_3(h_1 - h_3) \\ &= 0.4\,\dot{m}_1[c_p(T_1 - T_2)] + 0.6\,\dot{m}_1[c_p(T_1 - T_3)] \\ &= 0.4(312 - 252) + 0.6(312 - 352) \\ &= 24 - 24 = 0\end{aligned}$$

Since no significant heat transfer occurs, the entropy rate balance at steady state gives,

$$\sum_i \frac{Q_i}{T_i} + \dot{m}_1 s_1 - \dot{m}_2 s_2 - \dot{m}_3 s_3 + \dot{s}_{gen} = 0$$

Combining the mass and entropy rate balances with $\sum_i \frac{Q_i}{T_i} = 0$, we get

$$\begin{aligned}0 &= (\dot{m}_2 + \dot{m}_3)s_1 - \dot{m}_2 s_2 - \dot{m}_3 s_3 + \dot{s}_{gen} \\ &= \dot{m}_2(s_1 - s_2) + \dot{m}_3(s_1 - s_3) + \dot{s}_{gen} \\ &= 0.4\dot{m}_1(s_1 - s_2) + 0.6\dot{m}_1(s_1 - s_3) + \dot{s}_{gen}\end{aligned}$$

Solving for $\dot{s}_{gen}/\dot{m}_1$ and using expressions for changes in specific entropy, we get

$$\frac{\dot{s}_{gen}}{\dot{m}_1} = 0.4\left[c_p \ln\frac{T_2}{T_1} - R\ln\frac{p_2}{p_1}\right] + 0.6\left[c_p \ln\frac{T_3}{T_1} - R\ln\frac{p_3}{p_1}\right]$$

$$= 0.4\left[1.005 \ln \frac{252}{312} - 0.287 \ln \frac{1}{5}\right] + 0.6\left[1.005 \ln \frac{352}{312} - 0.287 \ln \frac{1}{5}\right]$$

$$= 0.0989 + 0.3499 = 0.4488 \text{ kJ/kg·K} > 0$$

Thus the second law of thermodynamics is also satisfied as $\dot{s}_{gen} > 0$. Therefore, the inventor's claim is valid.

Q.1. (*c*) Using Maxwell's relations and the thermodynamic definition for c_p and c_v in terms of gradients, show the following:

(*i*)
$$\text{TDS} = c_v dT + T\left(\frac{\partial p}{\partial T}\right)_v dv$$
$$= c_p dT - T\left(\frac{\partial v}{\partial T}\right)_p dp$$

(*ii*) Joule-Thomson coefficient

$$\mu_j = \frac{1}{c_p}\left[T\left(\frac{\partial v}{\partial T}\right)_p - v\right]$$

(*iii*) There is no change in temperature when an ideal gas is made to undergo Joule-Thomoon oxpancion.

Sol.

(*i*) Entropy,
$$s = f(T, v)$$

$$\therefore \quad ds = \left(\frac{\partial s}{\partial T}\right)_v \cdot dT + \left(\frac{\partial s}{\partial v}\right)_T \cdot dv$$

$$T.ds = T\left(\frac{\partial s}{\partial T}\right)_v \cdot dT + T\left(\frac{\partial s}{\partial v}\right)_T . dv \quad (1)$$

For a reversible constant volume change,

$$dq = c_v(dT)_v = T(ds)_v$$

$$\text{or} \quad c_v = T\left(\frac{ds}{dT}\right)_v \quad (2)$$

$$\text{But} \quad \left(\frac{\partial s}{\partial v}\right)_T = \left(\frac{\partial p}{\partial T}\right)_v \quad \text{(Maxwell's equation)}$$

Putting in Eq. (1), we get

$$T\cdot ds = c_v \cdot dT + T\left(\frac{\partial p}{\partial T}\right)_v \cdot dv$$

$$\text{Now} \quad s = f(T, p)$$

$$\therefore \quad T\cdot ds = T\left(\frac{\partial s}{\partial T}\right)_p \cdot dT + T\left(\frac{\partial s}{\partial p}\right)_T \cdot dp \quad (3)$$

and $$c_p = T\left(\frac{\partial s}{\partial T}\right)_p \quad (4)$$

Also $$\left(\frac{\partial s}{\partial p}\right)_T = -\left(\frac{\partial v}{\partial T}\right)_p \quad \text{(Maxwell's equation)}$$

Putting in Eq. (3), we get

$$T\cdot ds = c_p \cdot dT - T\left(\frac{\partial v}{\partial T}\right)_p \cdot dp$$

(*ii*) We have,

$$T\cdot ds = c_p \cdot dT - T\left(\frac{\partial v}{\partial T}\right)_p \cdot dp$$

$$\therefore \quad dh = c_p \cdot dT - \left[T\left(\frac{\partial v}{\partial T}\right)_p - v\right] dp \quad (5)$$

For constant enthalpy, $dh = 0$

$$\therefore \quad 0 = (c_p \cdot dT)_h + \left[\left\{v - T\left(\frac{\partial v}{\partial T}\right)_p\right\} dp\right]_h$$

$$\therefore \quad (C_p \cdot dT)_h = \left[\left\{T\left(\frac{\partial v}{\partial T}\right)_p - v\right\} dp\right]_h$$

$$\mu_j = \left(\frac{\partial T}{\partial p}\right)_h = \frac{1}{c_p}\left[T\left(\frac{\partial v}{\partial T}\right)_p - v\right]$$

(*iii*) We have, $$pv = RT$$

$$v = \frac{RT}{p}$$

$$\left(\frac{\partial v}{\partial T}\right)_p = \frac{R}{p} = \frac{v}{T}$$

$$\mu_j = \frac{1}{c_p}\left(T \times \frac{v}{T} - v\right) = 0$$

Therefore, if an ideal gas is throttled, there will not be any change in temperature.

Q.1. (*d*) Describe the step-by-step prodcedure to experimentally determine the calorific value of a solid fuel using bomb calorimeter. Draw a section view of the calorimeter.

Sol. Refer to Q.2(*a*) of IESE—1999.

Q.2. (*a*) A four-cylinder engine of an automobile is converted to run on propane (C_3H_8) fuel. A dry analysis of engine exhaust gives volumetric percentage of CO, CO_2 and O_2, respectively at 9.79%, 4.90% and 2.45%. Write the resulting chemical reaction and find the equivalence ratio.

Sol. Given : Volumetric analysis : CO = 9.7%, CO_2 = 4.90%, O_2 = 2.45%

The volumetric analysis of exhaust analysis can be converted to mass analysis as follow:

Constituent	Percentage by volume, (*a*)	Molecular weight, (*b*)	Proportional mass $c = (a) \times (b)$	Percentage by mass $= \dfrac{C \times 100}{\Sigma C}$
CO	9.79	28	274.12	9.49
CO_2	4.90	44	215.6	7.46
O_2	2.45	32	78.4	2.71
N_2 (by difference)	82.86	28	2320.08	80.34
			$\Sigma C = 2888.2$	

Percentage of C in $C_3H_8 = \dfrac{3 \times 12}{3 \times 12 + 8 \times 1} = \dfrac{36}{44} = 0.8182$ or 81.82%

Percentage of H in C_3H_8 = 18.18%

Mass of air supplied per kg of fuel,

$$m_a = \frac{NC}{33\,(CO + CO_2)} = \frac{82.86 \times 81.82}{33(9.79 + 4.90)} = 13.98 \text{ kg}$$

$$A : F = 13.98 : 1$$

$$C_3H_8 + 5O_2 = 3CO_2 + 4H_2O$$

$$(3 \times 12 + 8 \times 1) + 5 \times 2 \times 16$$

$$44 + 160$$

Oxygen required for complete combustion of 1 kg of fuel

$$= \frac{160}{44} = 3.636 \text{ kg/ kg of fuel}$$

or $$\frac{3.636 \times 100}{23} = 15.81 \text{ kg of air}$$

Minimum air required for complete combustion of fuel

$$= \frac{1}{23}\left[\frac{8}{3}C + 8H\right]$$

$$= \frac{1}{23}\left[\frac{8}{3} \times 81.82 + 8 \times 18.18\right] = 15.81 \text{ kg/kg of fuel}$$

Equivalence ratio, $$\phi = \frac{\text{Stoichiometric A/F ratio}}{\text{Actual A/F ratio}} = \frac{15.81}{13.98} = 1.131$$

Q.2. (*b*) The spark plug is fixed at 18° before top dead centre (TDC) in an SI engine running at 1800 r.p.m. It takes 8° of rotation to start combustion and get into flame propagation mode. Flame termination occurs at 12° after TDS. Flame front can be approximated as a sphere moving out from the spark plug which is offset 8 mm from the centre line of the cylinder whose bore diameter is 8.4 cm. Calculate the effective flame front speed during flame propagation. The engine speed is increased to 34000 r.p.m. and subsequently as a result of which the effective flame front speed increases at a rate such that it is directly proportional to 0.85 times of engine speed. Flame development after spark plug firing still takes 8° of engine rotation. Calculate how much engine rotation must be advanced such that the flame termination again occurs at 12° after TDC.

Sol. Given : $\theta_1 = 8°$, $\theta_2 = 10° + 12° = 22°$, $N_1 = 1800$ rpm, e = 8 mm, D = 8.4 cm, $N_2 = 34000$ rpm,

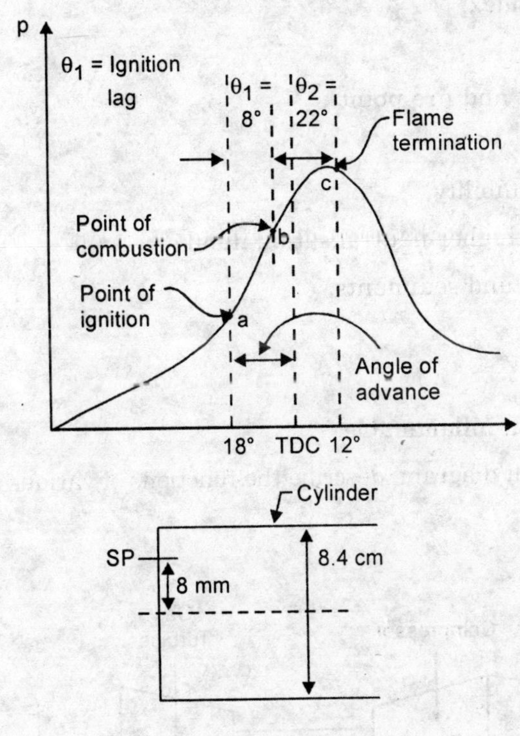

Fig. 3.

$$\omega_1 = \frac{2\pi N_1}{60} = \frac{2\pi \times 1800}{60} = 60\pi \text{ rad/s}$$

$$\theta_2 = \frac{\pi}{180} \times 22 = 0.122 \times \pi \text{ rad}$$

Time taken, $$t = \frac{0.122\,\pi}{60\,\pi} = 0.002037 \text{ s}$$

Flame front speed, $$v_1 = \frac{0.008}{0.002037} = 3.92 \text{ m/s}$$

When engine speed is increased to 34000 rpm, then

$$\omega_2 = \frac{2\pi N_2}{60} = \frac{2\pi \times 34000}{60} = 1133.3 \times \pi \text{ rad/s}$$

$$\upsilon_2 = 3.92 \times 0.85 \times \frac{34000}{1800} = 62.94 \text{ m/s}$$

$$\theta_2 = \frac{0.008}{\upsilon_2} = \frac{0.008}{62.94} = 1.27 \times 10^{-4} \text{ rad or } 2.6°$$

Q.2. (*c*) Discuss the basic properties that a lubricant should possess to meet the lubrication requirement of internal combustion engines.

Sol. Basic properties of a lubricant required for I.C. engines are :

1. High viscosity index.
2. Low pour point.
3. High flash point and fire point.
4. Higher oiliness.
5. High oxidation stability.
6. Low acidity and higher neutralisation number.
7. Free from water and sediments.
8. Low cost.
9. Cleaning ability.
10. Non-toxic and non-inflammable.

Q.2. (*d*) Using a layout diagram, describe the functions of various components of a turboject engine.

Sol.

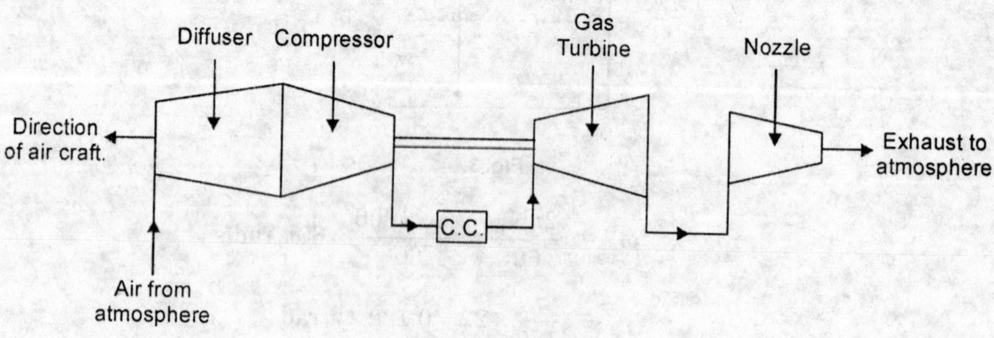

Fig. 4. Turbo-jet engine.

The schematic diagram of turbo-jet engine is shown in Fig. 4. The functions of various components are:

1. **Diffuser :** It converts kinetic energy of entering air into static pressure rise.

2. **Compressor :** Further increases the pressure of air.
3. **Combustion chamber :** Fuel is added to high pressure air and ignited.
4. **Gas Turbine :** Develops required power to run the moving parts.
5. **Nozzle :** The exhaust gases from turbine are expanded in the nozzle to produce a very high velocity jet which gives forward motion to the aircraft by the jet reaction.

Q.3. (*a*) Explain Reynolds analogy and derive the expression to evaluate the heat transfer coefficient using it. Give physical meaning of the expression.

Sol. Reynolds Analogy : The Reynolds analogy is the relationship between fluid friction and Newton's law of viscosity.

The wall shear stress according to Newton's law of viscosity is given by,

$$\tau_0 = \mu \frac{du}{dy} \tag{1}$$

Fourier's equation for heat flow along y-direction is,

$$Q = -kA\frac{dt}{dy} \tag{2}$$

When Prandtl number, $Pr = 1$, then temperature and velocity profiles are identical.

$$\therefore \quad Pr = \frac{\mu c_p}{k} = 1$$

$$\text{or} \quad c_p = \frac{k}{\mu} \tag{3}$$

Combining Eqs. (1) to (3), we get

$$Q = -C_p A\tau_0 \cdot \frac{dt}{du} \tag{4}$$

At the plate surface : $u = 0$ and $t = t_s$

At the outer edge of boundary layer : $u = U_\infty$ and $t = t_\infty$

$$\frac{Q}{c_p A\tau_0}\int_0^{U_\infty} du = -\int_{t_s}^{t_\infty} dt$$

$$\frac{Q}{c_p A\tau_0}\cdot U_\infty = t_s - t_\infty$$

$$\frac{Q}{A(t_s - t_\infty)} = \tau_0 \frac{c_p}{U_\infty} \tag{5}$$

$$\text{Now} \quad \frac{Q}{A(t_s - t_\infty)} = h_x \text{ and } \tau_0 = C_{fx} \times \frac{1}{2}\rho U_\infty^2$$

where C_f = skin friction coefficient

Substituting in Eq. (5), we get

$$h_x = C_{fx} \times \frac{1}{2} \rho U_\infty^2 \times \frac{c_p}{U_\infty} = \frac{1}{2} C_{fx} (\rho\, c_p U_\infty)$$

or
$$\frac{h_x}{\rho c_p U_\infty} = \frac{C_{fx}}{2} \tag{6}$$

$$\frac{h_x}{\rho c_p U_\infty} = \text{Stanton number, } St_x$$

$$= \frac{Nu_x}{Re_x \cdot Pr} = \frac{C_{fx}}{2} \quad \text{...(7)}$$

Eq. (7) is called Reynolds analogy.

Q.3. (*b*) Differentiate between fin efficiency and fin effectiveness.

Sol. Fin efficiency,
$$\eta_{fin} = \frac{\text{Heat lost from the fin}}{\text{Heat lost from the fin if whole surface of fin is maintained at base temperature}}$$

$$= \frac{Q_{fin}}{Q_{max}}$$

For an infinitely long fin,

$$\eta_{fin} = \sqrt{\frac{Ph\, k\, A_c(t_0 - t_a)}{h\, Pl(t_0 - t_a)}} = \sqrt{\frac{hA_c}{hPl^2}} = \frac{1}{ml}$$

where
$$m = \sqrt{\frac{hP}{k\, A_c}}$$

For a fin insulated at the tip,

$$\eta_{fin} = \frac{\text{Fin heat transfer rate}}{\text{Heat transfer rate without a fin}}$$

$$= \frac{Q_{with\ fin}}{Q_{without\ fin}} = \sqrt{\frac{Pk}{h\, A_c}}$$

Q.3. (*c*) Show that the differential equation governing conduction heat transfer in a solid sphere with heat generation is given by

$$\frac{d^2T}{dr^2} + \frac{2}{r} \times \frac{dT}{dr} + \frac{q'''}{k} = 0,$$

where T is the temperature at any radius r, q''' is the heat generated per unit volume and k is the thermal conductivity of the solid sphere. Show the general nature of the temperature distribution in this case.

Sol. Refer to Q. 6(*a*) of IESE—1992.

Q.3. (*d*) A counterflow heat exchanger is to be designed to cool 900 kg/hr of oil from 60°C to 32°C using a fluid with sp. heat 1.0 kJ/kg-K at 15°C. The sp. heat of the oil is 0.5 kJ/kg-K and the maximum allowable exit temperature of the cooling fluid is 27°C. Work out the following:

(*i*) Sketch the system and show the temperature distribution.

(*ii*) Find NTU.

If the length of the tube is 2 m through which the cooling fluid passes and the overall heat-transfer coefficient is 200 W/m^2-K, find the number of tubes required. Assume density of the cooling fluid as 250 kg/m^3. If the maximum velocity through the tube cannot exceed 2 m/s and the maximum length of the exchanger is limited to 12 m due to space restriction, find the configuration of the exchanger and sketch the final design.

Sol. Given : $\dot{m}_h$ = 900 kg/hr, t_{h_i} = 60°C, t_{h_0} = 32°C, c_{pc} = 1.0 kJ/kg-K, t_{ci} = 15°C, c_{ph} = 0.5 kJ/kg.K, t_{co} = 27°C, L = 2m, U = 200 W/m^2. K, ρ = 250 kg/m^3, v = 2 m/s, L = 12 m.

(*i*) The heat exchanger system is shown in Fig. 5.

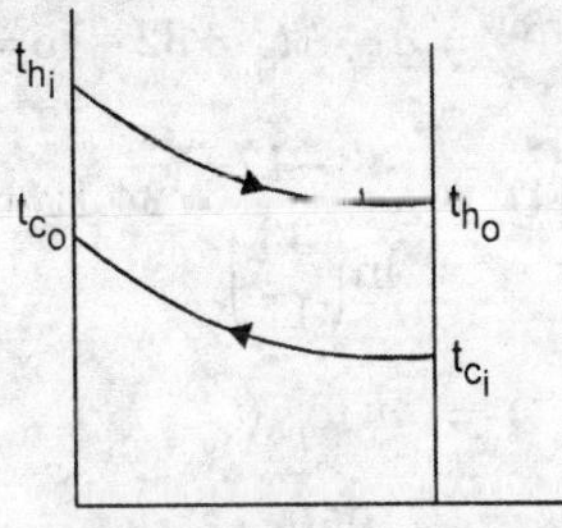

Fig. 5.

(*ii*) For heat balance,

$$\dot{m}_h\, c_{ph}(t_{h_i} - t_{h_0}) = \dot{m}_c\, c_{pc}(t_{co} - t_{ci})$$

$$900 \times 0.5(60 - 32) = \dot{m}_c \times 1.0(27-15)$$

$$\dot{m}_c = 1050 \text{ kg/hr}$$

Thermal capacities,

$$C_h = \dot{m}_h c_{ph} = 900 \times 0.5 = 45 \text{ kJ/hr}\cdot\text{K}$$

$$C_c = \dot{m}_c c_{pc} = 1050 \times 1.6 = 1050 \text{ kJ/hr}\cdot\text{K}$$

$$C_h < C_c$$

$\therefore$

$$C_{\text{min}} = C_h = 450 \text{ kJ/hr}\cdot\text{K}$$

$$C_{\text{max}} = C_c = 1050 \text{ kJ/hr}\cdot\text{K}$$

Capacity ratio,

$$CR = \frac{C_{\text{min}}}{C_{\text{max}}} = \frac{450}{1050} = 0.4286$$

Effectiveness of heat exchanger,

$$\varepsilon = \frac{t_{h_i} - t_{h_0}}{t_{h_i} - t_{c_i}} = \frac{60-32}{60-15} = \frac{28}{45} = 0.6222$$

Also

$$\varepsilon = \frac{1-\exp[-\text{NTU}(1-\text{CR})]}{1-\text{CR}\exp[-\text{NTU}(1-\text{CR})]}$$

$$0.6222 = \frac{1-\exp[-\text{NTU}(1-0.4286)]}{1-0.4286\ \exp[-\text{NTU}(1-0.4286)]}$$

$$= \frac{1-\exp(-0.5714\ \text{NTU})}{1-0.4286\exp(-0.5714\ \text{NTU})}$$

$$0.6222 - 0.2666\exp(-0.517\ \text{NTU}) = 1 - \exp(-0.5714\ \text{NTU})$$

$$0.7334\exp(-0.5714\ \text{NTU}) = 0.3778$$

$$\text{NTU} = 1.16$$

$$\text{LMTD} = \frac{\theta_1-\theta_2}{\ln\frac{\theta_1}{\theta_2}}$$

$$\theta_1 = t_{h_i} - t_{c_0} = 60-27 = 33°\text{C}$$

$$\theta_2 = t_{h_0} - t_{c_i} = 32-15 = 17°\text{C}$$

$$\text{LMTD} = \frac{33-17}{\ln\left(\frac{33}{17}\right)} = 24.12°\text{C}$$

$$Q = \dot{m}_c c_{pc}(t_{c_0} - t_{c_i})$$

$$= \frac{1050}{3600}\times 1\times 10^3(27-15) = 3500\ \text{W}$$

$$Q = \text{UA} \times \text{LMTD}$$

$$A = N \times \pi d L$$

$$A = \frac{3500}{200\times 24.12} = 0.7255\ \text{m}^2$$

Volume flow rate of cooling fluid

$$= \frac{\dot{m}_c}{\rho} = \frac{1050}{3600\times 250} = 1.167\times 10^{-13}\ \text{m}^3/\text{s}$$

$$= \frac{\pi}{4}d^2\times v = \frac{\pi}{4}d^2\times 2$$

$$d = 0.027\ \text{m}$$

$$0.7255 = N\times\pi\times 0.027\times 2$$

$$N = 4.27 \simeq 5$$

Q.3. (*e*) Find the average film coefficient heat transfer on the water side of a single pass steam condenser. The inner diameter of the tube is 23 mm and cooling water enters at 15°C and leaves at 25°C. The average water velocity is 2.1 m/s. Sketch the system and show the temperature distribution. Properties of water are given below. Find the heat transfer per metre length of the tube for the above case :

Temp. (°C)	10	20	30
Density (kg/m³)	1000	1000	1000
Sp. heat (kJ/kg-K)	4.192	4.183	4.174
Thermal Conductivity × 10^2 (W/m-K)	57.498	59.780	61.345
Viscosity × 10^2 (kg/hr)	469	361.892	288.650
Pr	9.52	7.02	5.42
ν × 10^6 (m²/s)	1.306	1.006	0.805

Sol. Given : d_i = 23 mm, t_{c_i} = 15°C, t_{c_0} = 25°C, ν = 2.1 m/s, $\rho_{15°C}$ = 1000 kg/m³, μ = 2.57×10^{-5}

$$Re = \frac{\rho v d}{\mu} = \frac{1000 \times 2.1 \times 23 \times 10^{-3}}{2.57 \times 10^{-5}} = 18.79 \times 10^5$$

$$Pr = \frac{\mu\ c_p}{h} = \frac{2.57 \times 10^{-5} \times 4.19}{58.45} = 1.84 \times 10^{-6}$$

$$Nu = \frac{hd}{k} - 0.023 \times (Re)^{0.8} \times (Pr)^{0.4}$$

$$h = \frac{58.45}{0.023} \times 0.023 \times (18.79 \times 10^5)^{0.8} \times (1.84 \times 10^{-6})^{0.4}$$

$$= 31035 \text{ W/m}^2\text{-K}$$

Heat transfer per unit length, Q = $hA{\cdot}\Delta t$ = 31035 × π × 0.023 × 1 × (25 – 15)

= 22417.6 W/m length

Q.4. (*a*) Saturated ammonia vapour enters a 15 cm dia × 14 cm stroke twin-cylinder single-acting compressor at 0.2365 MPa whose volumetric efficiency is 79% and speed 420 r.p.m. The delivery pressure is 1.1672 MPa. Liquid NH_3 at 21°C enters the expansion valve. For ideal cycle, find

(*i*) the ammonia circulated in kg/min,

(*ii*) the refrigeration in tons and

(*iii*) COP of the cycle.

Assume sp. heat of NH_3 as 219 kJ/kg-K.

Properties of NH_3 are given below :

Pressure (MPa)	Sat. temp. (°C)	Sp. volume of vap. at sat. (m³/kg)	Enthalpy (kJ/kg)		Entropy (kJ/kg-K)		Specific heat kJ/kg-K	
			Sat. liq.	Sat. vap.	Sat. liq.	Sat. vap.	Sat. liq	Sat. vapour
0.2365	–15	0.5106	112.39	1426.58	0.4572	5.5497	4.396	2.303
1.1672	+30	0.1107	323.08	1468.87	1.2037	4.9842	4.60	2.19

Sol. Given : D = 15 cm, L = 14 cm, n = 2, η_v = 0.79, N = 420 rpm, $T_2' = T_3'$ = 303 K, T_3 = 294 K, h_1 = 1426.58 kJ/kg, h_2' = 1468.87 kJ/kg, h_3' = 323.08 kJ/kg, c_{pv} = 2.19 kJ/kg·K, c_{pl} = 4.60 kJ/kg·K.

The T-s diagram is shown in Fig. 6.

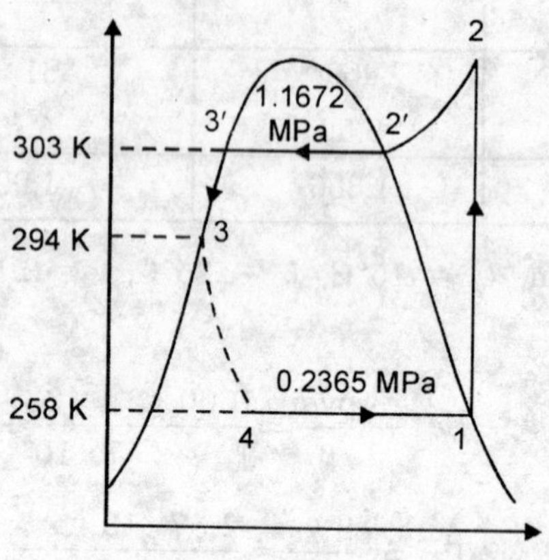

Fig. 6

$$s_1 = s_2 - s_2' + c_{pv} \ln \frac{T_2}{T_2'}$$

$$5.5497 = 4.9842 + 2.19 \ln \frac{T_2}{303}$$

$$T_2 = 392.27 \text{ K}$$

$$h_2 = h_2' + c_{pu}(T_2 - T_2') = 1468.87 + 2.19(392.27 - 303)$$

$$= 1664.37 \text{ kJ/kg}$$

$$h_{f3} = h_{f3}' - c_{pl}(T_3' - T_3) = 323.08 - 4.60\ (303 - 294) = 281.68 \text{ kJ/kg}$$

(*i*) Volume of ammonia circulated,

$$V = \frac{\pi}{4} D^2 LN \times n \times \eta_v = \frac{\pi}{4}(0.15)^2 \times 0.14 \times 420 \times 2 \times 0.79$$

$$= 1.64 \text{ m}^3/\text{min}$$

Specific volume of ammonia at entry,

$$v_1 = 0.5106 \text{ m}^3/\text{kg}$$

Mass of ammonia circulated,

$$\dot{m} = \frac{V}{v_1} = \frac{1.64}{0.5106} = 3.212 \text{ kg/min}$$

(*ii*) Refrigeration effect

$$RE = \dot{m}(h_1 - h_4) = \dot{m}(h_1 - h_{f3})$$

$$= 3.212\ (1426.58 - 281.68) = 3677.4 \text{ kJ/min}$$

$$= \frac{3677.4 \times 60}{14000} = 15.76 \text{ ton}$$

(*iii*) Work done = $h_2 - h_1$ = 1664.37 – 1626.58 = 237.79 kJ/kg

$$\text{COP} = \frac{\text{RE}}{\dot{m} \times (h_2 - h_1)} = \frac{3677.4}{3.212 \times 237.79} = 4.81$$

Q.4. (*b*) A Bell-Coleman refrigeration system is used to produce 10 tons of refrigeration. The cooler and refrigerator pressure are 4.2 bar and 1.4 bar. Air is cooled in the cooler to 45°C and temperature of air at inlet of the compressor is –20°C. For an ideal cycle, calculate COP, mass of air circulated/min, theoretical piston displacement of compressor and power required per ton of refrigeration. Assume c_p for air as 1.005 kJ/kg-K. Find the cylinder dimensions if the compressor is single-acting single-cylinder with L/D ratio of 1.2 and runs at 600 r.p.m.

Sol. Given : p_1 = 1.4 bar, p_2 = 4.2 bar, T_1 = 273 – 20 = 253 K, T_3 = 273 + 45 = 318 K, c_p = 1.005 kJ/kg·K,

$$\frac{\text{L}}{\text{D}} = 1.2, \text{ N} = 600 \text{ rpm}$$

The T–*s* diagram for Bell-Coleman system is shown in Fig. 7.

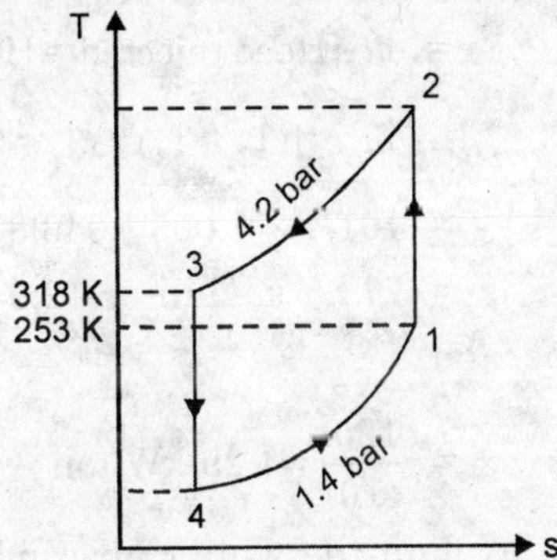

Fig. 7

$$\frac{\text{T}_2}{\text{T}_1} = \left(\frac{p_2}{p_1}\right)^{\frac{\gamma-1}{\gamma}}, \text{ T}_2 = 253\left(\frac{4.2}{1.4}\right)^{\frac{1.4-1}{1.4}} = 346.3 \text{ K}$$

$$\frac{\text{T}_4}{\text{T}_3} = \left(\frac{p_1}{p_2}\right)^{\frac{\gamma-1}{\gamma}}, \text{ T}_4 = 318 \times 0.7306 = 232.3 \text{ K}$$

$$\text{COP} = \frac{\text{T}_1 - \text{T}_4}{(\text{T}_2 - \text{T}_3) - (\text{T}_1 - \text{T}_4)} = \frac{253 - 232.3}{(346.3 - 318) - (253 - 232.3)} = 2.723$$

Refrigeration capacity = 10 TR

Heat extracted per min = 10 × 211 = 2110 kJ/min

Heat extracted from refrigerator per kg of air

$$= c_p(\text{T}_1 - \text{T}_4) = 1.005(253 - 232.3) = 20.80 \text{ kJ/kg}$$

Rate of air circulation, $\dot{m}_a = \dfrac{2110}{20.80} = 101.4$ kg/min

$$p_1V_1 = \dot{m}_a RT_1$$

$$V_1 = \frac{101.4 \times 287 \times 253}{1.4 \times 10^5} = 52.59 \text{ m}^3/\text{min}$$

$$V_4 = \frac{V_1 V_4}{T_1} = \frac{52.59 \times 232.3}{253} = 48.29 \text{ m}^3/\text{min}$$

$$\frac{\pi}{4} D^2 L \times N = V_1$$

$$\frac{\pi}{4} \times 1.2 \times D^3 \times 600 = 52.59$$

$$D = 0.453 \text{ m}$$

$$L = 1.2 \times 0.453 = 0.5437 \text{ m}$$

Piston displacement of compressor = 0.5437 m

Net work done on the refrigeration machine per minute

$$= \dot{m}_a[\text{Heat rejected} - \text{Heat extracted}]$$

$$= \dot{m}_a c_p[(T_2 - T_3) - (T_1 - T_4)]$$

$$= 101.4 \times 1.005\ [(346.3 - 318) - (253 - 232.3)]$$

$$= 774.5 \text{ kJ/min or } 12.9 \text{ kW}$$

Theoretical power required per ton

$$= \frac{12.9}{10} = 1.29 \text{ kW/ton}$$

Q.4. (*c*) With a neat sketch, explain the winter air-conditioning system. Why a single psychrometric process cannot be applied in winter air-conditioning?

Sol. Refer to Q. 4(*d*)—(*ii*) of IESE—2009.

Q.4. (*d*) Calculate all the psychrometric properties of air at 1 bar and 25°C dbt and 15°C wbt. The following properties of water may be assumed:

Temp (°C)	Sat. pressure (bar)	Sp. volume of vapour (m^3/kg)	Enthalpy (kJ/kg)	
			Sat. liq.	Sat. vap.
25	0.03166	43.40	—	2547.3
15	0.01703	77.98	62.94	—
10	0.01078	—	—	—
0	0.01002	—	—	—

The following expression may be used, if necessary:

$$p_v = (p_{vs})_{wb} - \frac{[p_t - (p_{vs})_{wb}](t_{db} - t_{wb})}{1527.4 - 1.3\, t_{wb}}$$

Sol. Given : t_{db} = 25°C, t_{wb} = 15°C, p_t = 1 bar

$$p_v = (p_{vs})_{wb} - \frac{[p_t - (p_{vs})_{wb}](t_{db} - t_{wb})}{1527.4 - 1.3\, t_{wb}}$$

Corresponding to 15°C, $(p_{vs})_{wb}$ = 0.01703 bar

$$p_v = 0.01703 - \frac{[1 - 0.01703](25° - 15°)}{1527.4 - 1.3 \times 15} = 0.0105 \text{ bar}$$

(*i*) Specific humidity, $w = \frac{0.622\, p_v}{p_t - p_v} = \frac{0.622 \times 0.0105}{(1 - 0.0105)} = 6.6 \times 10^{-3}$ kg/kg d.a.

(*ii*) Relative humidity, $\phi = \frac{p_v}{p_w} = \frac{0.0105}{0.01703} = 0.6165$ or 61.65%

(*iii*) Vapour density, $\phi_v = \frac{p_v}{R_v T_v} = \frac{0.0105 \times 10^5}{\frac{8314.3}{18} \times (273 + 25)} = 7.628 \times 10^{-3}$ kg/m³

(*iv*) Dew point temperature, t_{dp} = 10.5°, corresponding to p_v = 0.0105 bar from steam tables.

(*v*) Enthalpy of mixture, $h = 1.005\, t_{db} + w\,(2500 + 1.88\, t_{db})$

$= 1.005 \times 25 - 6.6 \times 10^{-3}\,(2500 + 1.88 \times 25)$

$= 41.93$ kJ/kg of d.a.

Q.5. (*a*) Considering the T-*s* diagram of Rayleigh flow and using the differential forms of the conservation equations and property relations, show that the

(*i*) Mach number is unity at the point of maximum entropy and

(*ii*) Mach number is $\frac{1}{\sqrt{\gamma}}$ at the point of maximum temperature.

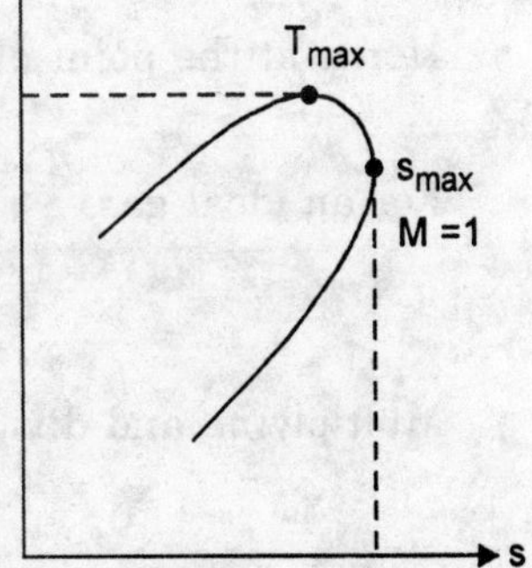

Fig. 8. T-s diagram for Rayleigh flow

Sol.

1. Continuity equation : $\dot{m} = \rho A v$

For a constant area duct,

$$\frac{\dot{m}}{A} = \rho v = G = \text{const.}$$

where G = mass velocity (1)

2. Momentum equation :

$$p + \rho v^2 = \text{const.}$$

$$p + \frac{G^2}{\rho} = \text{const.}$$

and $$p + VG^2 = \text{const.} \qquad \left[\because V = \frac{1}{\rho}\right] \qquad (2)$$

This is the equation of Rayleigh line. Applying the equation of Rayleigh line between any two points in a constant area duct, we have

$$p_1 + V_1G^2 = p_2 + V_2G^2$$

$$\frac{p_2 - p_1}{V_1 - V_2} = G^2$$

or $$\frac{dp}{dV} = \frac{p_2 - p_1}{V_2 - V_1} = -G^2 = -\rho^2 v^2 \qquad (3)$$

For constant entropy, $$\left(\frac{dp}{dV}\right)_s = \text{const}$$

$$V = \frac{1}{\rho}$$

$$dV = -\frac{d\rho}{\rho^2}$$

$$\left(\frac{dp}{dV}\right)_s = -\rho^2\left(\frac{dp}{d\rho}\right)_s = -\rho^2 c^2 \qquad (4)$$

where c = local sound velocity

By comparings Eqs. (3) and (4), we get

$$c^2 = v^2$$

or $$M = 1$$

Hence at the point of maximum entropy,

$$M = 1$$

For an ideal gas, $$pV = RT$$

$$\frac{dp}{dV} = -\frac{RT}{V^2} = -\frac{p}{V} = -p\rho = -\rho^2 RT$$

Multiplying and dividing by γ on the R.H.S., we get

$$\frac{dp}{dV} = -\frac{\rho^2(\gamma RT)}{\gamma}$$

Now $$c = \sqrt{\gamma RT}$$

$\therefore$ $$\frac{dp}{dV} = -\frac{\rho^2 c^2}{\gamma} \qquad (5)$$

Comparing Eqs. (3) and (5), we get

$$-\rho^2 v^2 = -\frac{\rho^2 c^2}{\gamma}$$

$$\frac{v^2}{c^2} = \frac{1}{\gamma}$$

or

$$M^2 = \frac{1}{\gamma}$$

$$M = \frac{1}{\sqrt{\gamma}}$$

Hence at the point of maximum temperature,

$$M = \frac{1}{\sqrt{\gamma}}$$

Q.5. (*b*) Air at 1 MPa and 600°C enters a converging nozzle with a velocity of 150 m/s. Determine the mass flow rate through the nozzle for a nozzle throat area of 50 cm^2 when the back pressure is

(*i*) 0.7 MPa, and

(*ii*) 0.4 MPa.

Assume that the flow through the nozzle is steady, one-dimensional and isentropic.

You may use the following table for one-dimensional isentropic flow (for an ideal gas with $\gamma - 1.4$) :

M	$\frac{p}{p_0}$	$\frac{\rho}{\rho_0}$	$\frac{T}{T_0}$	$\frac{A}{A^*}$	M*
0.74	0.695	0.771	0.901	1.068	0.770
0.76	0.682	0.761	0.896	1.057	0.788
0.78	0.669	0.750	0.892	1.047	0.807
0.80	0.656	0.740	0.887	1.038	0.825
0.82	0.643	0.729	0.881	1.030	0.843

Sol. Given : p = 1 MPa, T_E = 273 + 600 = 873 K, c_E = 150 m/s, A_t = 50 cm^2

Velocity of sound, $\quad c_s = \sqrt{k R T_C} = \sqrt{1.4 \times 287 \times 873} = 592.26$ m/s

Match number, $\quad M = \frac{c_v}{c_s} = \frac{150}{592.26} = 0.253$

$$\frac{T_0}{T} = 1 + \left(\frac{k-1}{2}\right) M^2 = 1 + \left(\frac{1.4-1}{2}\right) \times (0.253)^2 = 1.0128$$

$$T_0 = 873 \times 1.0128 = 884.2 \text{ K}$$

$$\frac{p_0}{p} = \left[1 + \left(\frac{k-1}{2}\right) M^2\right]^{\frac{k}{k-1}} = \left[1 + \left(\frac{1.4-1}{2}\right)(0.253)^2\right]^{\frac{1.4}{0.4}} = 1.0455$$

$$p_0 = 1 \times 1.0455 = 1.0455 \text{ MPa}$$

Critical pressure ratio,

$$\frac{p^*}{p_0} = \left(\frac{2}{k+1}\right)^{\frac{k}{k-1}} = \left(\frac{2}{2.4}\right)^{3.5} = 0.528$$

Critical pressure, $p^* = 1.0455 \times 0.528 = 0.552$ MPa

(i) $\frac{p_b}{p_0} = \frac{0.7}{1.0455} = 0.6695, \frac{T_t}{T_0} = 0.892$ from given table, $M^* = 0.807$

$$T_t = 884.2 \times 0.892 = 788.7 \text{ K}$$

Sonic velocity at throat,

$$c_{ts} = \sqrt{kRT_t} = \sqrt{1.4 \times 287 \times 788.7} = 562.9 \text{ m/s}$$

Velocity at throat, $c_t = M^* \times c_{ts} = 0.807 \times 562.9 = 454.2$ m/s

Density of air at throat,

$$\rho_t = \frac{p_t}{RT_t} = \frac{0.74 \times 10^6}{287 \times 788.7} = 3.27 \text{ kg/m}^3$$

Mass flow rest, $\dot{m} = \rho_t A_t c_t$

$$= 3.27 \times 50 \times 10^{-4} \times 454.2 = 7.426 \text{ kg/s}$$

(ii) $p_b = 0.4$ MPa

$$\frac{p_b}{p_o} = \frac{0.4}{1.0455} = 0.38259$$

$\frac{p}{p_0}$	$\frac{T}{T_0}$	M^*
0.41238	0.77640	1.1583
0.38259	0.75960	1.2004
0.36091	0.74738	1.2311

For $\frac{p_b}{p_0} = 0.38259$, $\frac{T}{T_0} = 0.75960$ by interpolation and $M^* = 1.2004$

$$T_t = 0.75960 \times 884.2 = 671.6 \text{ K}$$

$$c_{ts} = \sqrt{1.4 \times 287 \times 671.6} = 519.4 \text{ m/s}$$

$$c_t = 1.2004 \times 519.4 = 623.5 \text{ m/s}$$

$$\rho_t = \frac{0.4 \times 10^6}{287 \times 671.6} = 2.07 \text{ kg/m}^3$$

$$\dot{m} = 2.07 \times 50 \times 10^{-4} \times 623.5 = 6.45 \text{ kg/s}$$

Q.5. (c) Explain the phenomenon of boundary layer separation over a curved surface. Discuss various methods of controlling boundary layer separation.

Sol. Refer to Q. 7(*d*) of IESE—1997.

Q.6. (*a*) Explain what you mean by the specific speed of a turbine. Using Buckingham-π-theorem and variables such as power (P), speed (N), head (H), diameter of turbine (D), density of fluid (ρ) and acceleration due to gravity (g), obtain the expression for the specific speed for a turbine.

Sol. The specific speed of a turbine is defined as the speed of a geometrically similar turbine which would develop unit power when working under a unit head.

Power, $$P = f(N, H, D, \rho, g) \quad (1)$$

or $$f_1(P, N, H, D, \rho, g) = 0 \quad (2)$$

Number of variables, $n = 6$

Number of fundamental dimensions,

$$m = 3$$

Number of π-terms $= n - m = 6 - 3 = 3$

Eq. (2) can be written as :

$$f_1(\pi_1, \pi_2, \pi_3) = 0 \quad (3)$$

Each π-term contains $(m + 1)$ variables, where $m = 3$ and is also equal to repeating variables. Choosing D, N, ρ as repeating variables, we get three π-terms as :

$$\pi_1 = D^{a_1} \cdot N^{b_1} \cdot \rho^{c_1} \cdot \rho$$

$$\pi_2 = D^{a_2} \cdot N^{b_2} \cdot \rho^{c_2} \cdot H$$

$$\pi_3 = D^{a_3} \cdot N^{b_3} \cdot \rho^{c_3} \cdot g$$

π_1-term :

$$M^0L^0T^0 = L^{a_1} \cdot T^{-b_1} \cdot [ML^{-3}]^{c_1} \cdot [ML^2T^{-3}]$$

$$= [L]^{a_1-3c_1+2} \cdot [M]^{c_1+1} \cdot [T]^{-b_1-3}$$

$$c_1 + 1 = 0, \; c_1 = -1$$

$$-b_{1-3} = 0, \; b_1 = -3$$

$$a_1 - 3c_1 + 2 = 0, \; a_1 + 3 + 2 = 0, \; a_1 = -5$$

$$\pi_1 = D^{-5} \cdot N^{-3} \cdot \rho^{-1} \cdot \rho = \frac{\rho}{D^5N^3\rho}$$

π_2-term :

$$M^0L^0T^0 = L^{a_2} \cdot T^{-b_2} \cdot [ML^{-3}]^{c_2} \cdot L^1$$

$$= [M]^{c_2} \cdot [L]^{a_2-3c_2+1} \cdot [T]^{-b_2}$$

$$c_2 = 0$$

$$b_2 = 0$$

$$a_2 - 3c_2 + 1 = 0, \quad a_2 = 3c_2 - 1 = -1$$

$$\pi_2 = D^{-1}H = \frac{H}{D}$$

π_3**-term :**

$$M^0L^0T^0 = L^{a_3} \cdot T^{-b_3} \cdot [ML^{-3}]^{c_3} \cdot [LT^{-2}] = M^{c_3} \cdot [L]^{a_3-3c_3+1} \cdot [T]^{-b_3-2}$$

$$c_3 = 0$$

$$a_3 - 3c_3 + 1 = 0, \quad a_3 = 3c_3 - 1 = -1$$

$$-b_3 - 2 = 0, \quad b_3 = -2$$

$$\pi_3 = D^{-1}N^{-2}g = \frac{g}{DN^2}$$

$$f_1\left[\frac{P}{D^5N^3\rho}, \frac{H}{D}, \frac{g}{DN^2}\right] = 0$$

$$P = \rho N^3D^5\ \phi\left(\frac{H^{DN^2}}{D}, \frac{g}{DN^2}\right) = \rho N^3D^5\phi\left(\frac{D^2N^2}{gH}\right)$$

$$N_s = \frac{H\sqrt{P}}{H^{5/4}}.$$

Q.6. (*b*) Explain what you mean by momentum correction factor. The velocity distribution in a pipe is given by

$$\frac{u}{U} = \left(1-\left(\frac{r}{R}\right)^n\right)$$

where, U = maximum velocity at the centre of the pipe

u = local velocity along r

R = radius of the pipe

Find the momentum correction factor, if $n = 0.20$.

Sol. Momentum Correction Factor : Momentum correction factor is defined as the ratio of momentum of the flow per second based on actual velocity to the momentum of flow per second based on average velocity across a section.

Consider an elementary ring of fluid between radii r and $r + dr$ of the pipe of radius R, as shown in Fig. 8

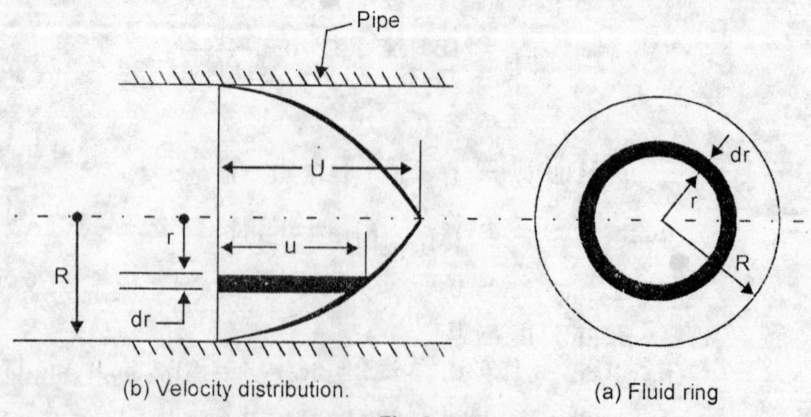

(b) Velocity distribution. (a) Fluid ring

Fig. 8

Let u = velocity of fluid at radius r.

U = Maximum velocity at the centre of pipe.

Rate of flow through the ring,

$$dQ = \text{area of ring} \times u = 2\pi r \cdot dr \times u$$

Total flow,

$$Q = \int_0^R 2\pi r \times U\left[1-\left(\frac{r}{R}\right)^n\right] dr = 2\pi U \int_0^R \left[r - \frac{r^{n+1}}{R^n}\right] dr$$

$$= 2\pi U\left[\frac{r^2}{2} - \frac{r^{n+2}}{(n+2)R^n}\right]_0^R = 2\pi U\left[\frac{R^2}{2} - \frac{R^2}{n+2}\right]$$

$$= 2\pi UR^2\left[\frac{n}{2(n+2)}\right] = \left(\frac{n}{n+2}\right)\pi UR^2$$

Let $\bar{u}$ = average velocity of flow.

Then

$$Q = \pi R^2 \bar{u} = \left(\frac{n}{n+2}\right)\pi U R^2$$

$$\bar{u} = \left(\frac{n}{n+2}\right) U$$

Momentum correction factor,

$$\beta = \frac{1}{A\bar{u}^2}\int u^2 dA$$

$$= \frac{1}{A\bar{u}^2}\int_0^R U^2\left[1-\left(\frac{r}{R}\right)^n\right]^2 2\pi r \cdot dr$$

$$= \frac{2\pi U^2}{A\bar{u}^2}\int_0^R \left[1+\left(\frac{r}{R}\right)^{2n} - 2\left(\frac{r}{R}\right)^n\right] r\, dr$$

$$= \frac{2\pi U^2}{A\bar{u}^2}\int_0^R \left[r + \frac{r^{2n+1}}{R^{2n}} - \frac{2}{R^n}\cdot r^{n+1}\right] dr$$

$$= \frac{2\pi U^2}{A\bar{u}^2}\left|\frac{r^2}{2} + \frac{r^{2n+2}}{(2n+2)\,R^{2n}} - \frac{2}{R^n}\cdot\frac{r^{n+2}}{(n+2)}\right|_0^R$$

$$= \frac{2\pi U^2}{A\bar{u}^2}\left[\frac{R^2}{2} + \frac{R^2}{(2n+2)} - \frac{2}{(n+2)}\cdot R^2\right]$$

$$= \frac{2\pi R^2 U^2}{A\bar{u}^2}\left[\frac{1}{2} + \frac{1}{2n+2} - \frac{2}{n+2}\right]$$

$$= \frac{2\pi R^2 U^2}{Au^2}\left[\frac{(2n+2)(n+2) + 2(n+2) - 8(n+1)}{2(2n+2)(n+2)}\right]$$

$$= \frac{2\pi R^2 U^2}{A\bar{u}^2}\left[\frac{2n^2 + 6n + 4 + 2n + 4 - 8n - 3}{4(n+1)(n+2)}\right]$$

$$= \frac{2\pi R^2 U^2}{A\bar{u}^2}\left[\frac{2n^2}{4(n+1)(n+2)}\right] = \frac{\pi R^2 U^2 n^2}{A\bar{u}^2 (n+1)(n+2)}$$

$$A\bar{u}^2 = \pi R^2 \times \left(\frac{n}{n+2}\right)^2 U^2$$

$$\beta = \frac{n^2}{(n+1)(n+2)} \times \frac{(n+2)^2}{n^2} = \frac{n+2}{n+1}$$

For $n = 0.20$

$$\beta = \frac{2.2}{1.2} = 1.833$$

Q.6. (c) (*i*) Define degree of turbulence.

(*ii*) Explain, for boundary layer flow, whether the curve representing δ(boundary layer thickness) as a function of X over a flat plate is a streamline of flow or not.

Sol.

(*i*) Degree of turbulence :

It is defined by, $\frac{1}{U_\infty}\sqrt{(\bar{u}'^2 + \bar{v}'^2 + w'^2)/3}$

where U_∞ = free stream velocity

$\bar{u}', \bar{v}', \bar{w}'$ = temporal mean velocity fluctuations.

(*ii*) The growth of boundary layer over a flat plate is shown in Fig. 9. The curve representing boundary layer thickness δ as a function of x from the leading edge is a stream line flow.

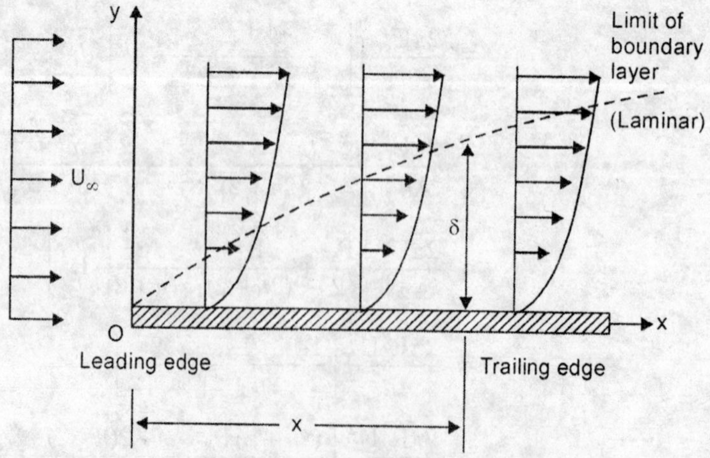

Fig. 9. Growth of boundary layer over the flat plate.

Q.6. (*d*) Distinugish, with the help of neat sketches, between a hydro-dynamically rough surface and a hydrodynamically smooth surface.

Sol. The smooth and rough boundaries are shown in Fig. 10. If k is the average height of the irregularities of the surface of a boundary, then in general the boundary is said to be rough if value of k is high and smooth if k is low. However, for proper classification of smooth and rough boundaries, besides the boundary characteristics, the flow and fluid characteristics need to be considered.

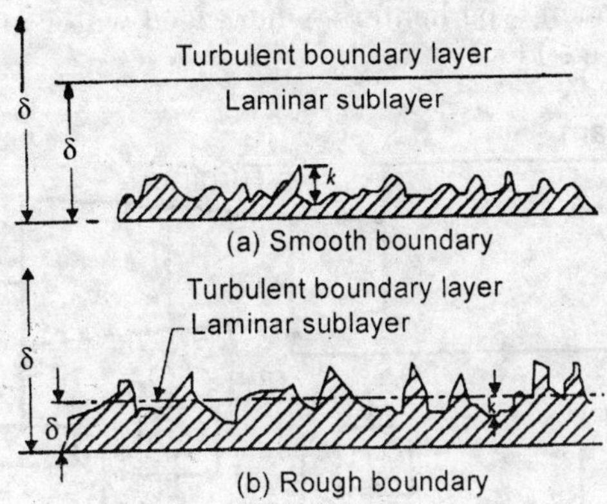

Fig. 10. Smooth and rough boundaries.

As shown in Fig. 10 when the average height k of the irregularities (projecting from its surface) is *much less than* the thickness of the laminars sublayer δ' the flow outside the laminar sublayer is turbulent; the eddies of various sizes present, try to penetrate the laminar sublayer. These eddies cannot reach the surface irregularities/projections due to greater thickness of the laminar sublayer and so the boundary acts as **smooth boundary**. As the Reynolds number (Re) increases the thickness of the sublayer decreases (can even become much less than k), the irregularities will then project through the laminar sublayer and eventually the laminar sublayer is destroyed completely. Subsequently, the eddies will come in contact with the surface irregularities and there will be a large amount of energy loss. This type of boundary is known as hydrodynamically **rough boundary**.

Through experiments Nikuradse found that boundary behaves as;

(*i*) Hydrodynamically smooth boundary ...when $\left(\frac{k}{\delta'}\right) < 0.25$,

(*ii*) Hydrodynamically rough boundary ...when $\left(\frac{k}{\delta'}\right) > 6.0$, and

(*iii*) Boundary in transition ...when $0.25 < \left(\frac{k}{\delta'}\right) < 6.0$.

In terms of roughness Reynolds number $\frac{u_f k}{\nu}$:

(*i*) For smooth boundary ... $\frac{u_f k}{\nu} < 4$,

(*ii*) For rough boundary ... $\frac{u_f k}{\nu} > 100$, and (*iii*) For boundary in transition stage ... $\frac{u_f k}{\nu}$ lies between 4 and 100.

Q.7. (*a*) Give a neat sketch of practical feed heating arrangement of a 600 MW unit of steam turbine. Write the salient features of this system with reasonings for selecting its main parts.

Sol. Feed heating system : The regeneration effect can be achieved by bleeding or extracting small quantities of stream at different points during the expansion and exploiting the energy of bled stream rather than whole steam. Thus, that part of the steam which continues to expand and to do work does not condense excessively. A regenerative cycle with three bleeding points and, therefore, three open type heaters, where feed water and steam mix, is shown in a schemematic diagram in Fig. 11.

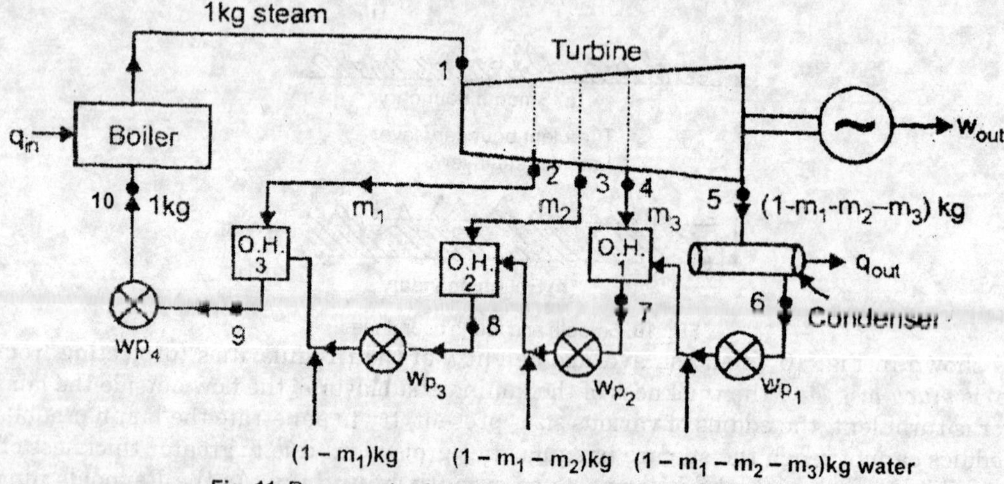

Fig. 11. Regenerative feed water heating with open heaters.

Q.7. (*b*) Discuss the need of governing of steam turbine. With the help of a neat sketch, discuss the working principle of hydromechanical speed-governing loop of a steam turbine by showing the characteristics on torque and frequency versus time and torque or load versus frequency.

Sol. The basic need for governing of steam turbine is to maintain the speed of the turbine fairly constant irrespective of the load.

Working principle of hydromechanical speed-governing loop of steam turbine

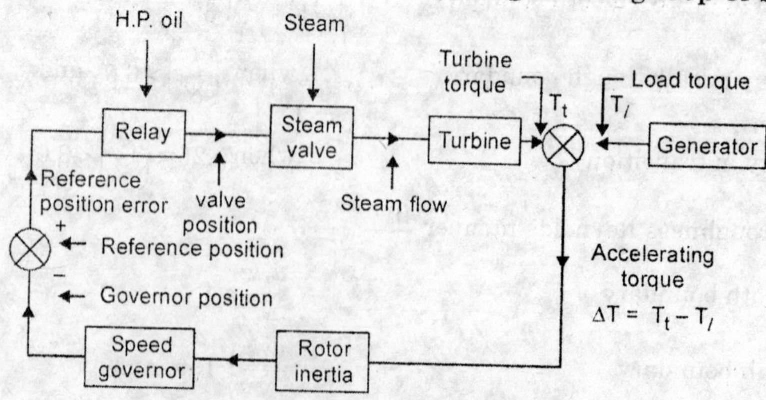

Fig. 12 Hydromechanical speed governing loop.

Fig. shows the block diagram of a hydromechanical speed governing loop. Let T_e and T_t be the load torque and turbine torque respectively. Then accelerating torque, $\Delta T = T_t - T_e$.

Rate of acceleration, $\alpha = \frac{\Delta T}{I}$

where I = machine inertia.

As the speed of turbine increases, the speed governor will respond and through the control system close the steam value partially/fully until the turbine torque T_t is equal to the new load demand.

The response of the turbine to a sudden change in load and the resultant change in speed is shown in Fig. 13 below.

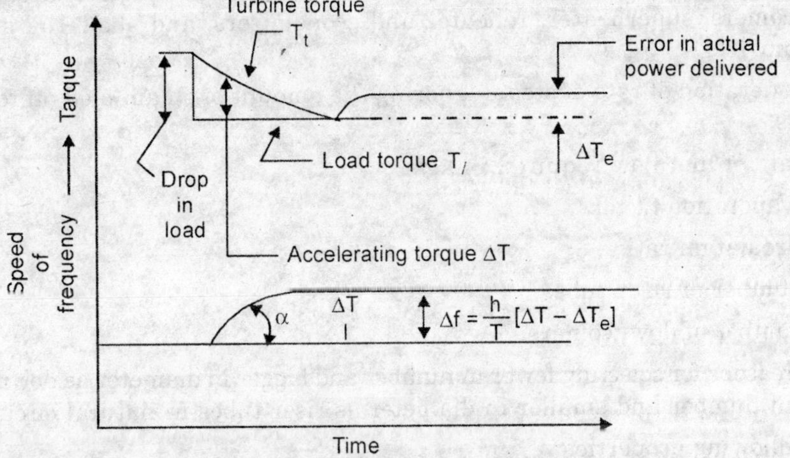

Fig. 13 Torque and speed v's Time.

The change in load and frequency is given by,

$$\Delta f = \frac{h}{T}(\Delta T - \Delta T_e)$$

where ΔT_e = error in actual power delivered to the consumer. The variation of T_l and frequency is shown in Fig. 14.

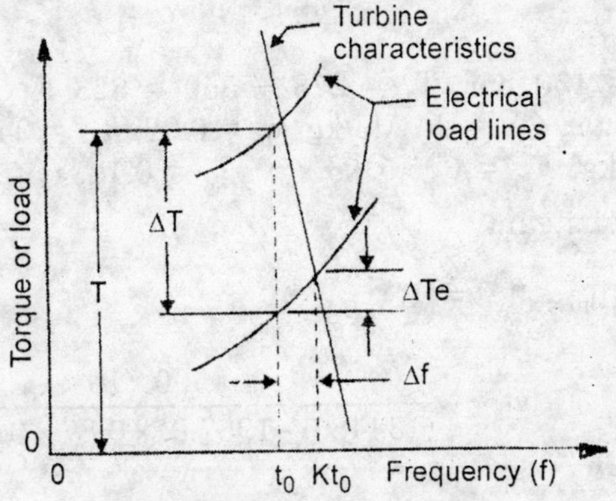

Fig. 14. Torque vs. Frequency

Q.7. (*c*) A steam power plant generating 500 MW of electrical power employs a natural circulation boiler which supplies steam at a pressure of 150 bar temperature of 550°. The condenser pressure is 0.05 bar. The turbine, mechanical and generator efficiencies are 87%, 98% and 99% respectively. The boiler uses pulverized coal having a calorific value of 26 MJ/kg and yields 92% efficiency. The feedwater passing through the feed heaters enters the boiler at 160°C. The risers of the furnace are 55 m high while the downcomers are placed outside the furnace for producing natural circulation. The quality of steam at the top of the riser is 12% and a minimum exit velocity of mixture leaving the risers and entering the drum is 1.4 m/s. The dimensions of the riser tubes are 65 mm OD and 3 mm wall thickness while the dimensions of downcomers are 185 mm OD and 8 mm thick. Assume no pressure drop and heat loss to the risers. Work out the following:

(*i*) Sketch the layout of natural circulation boiler unit showing furnace, drum, risers, downcomers, superheater, reheater and economizers, and show the process on T-*s* diagram.

(*ii*) The generation of rate of steam ignoring the amount of steam bled off to feed heaters in kg/s.

(*iii*) The rate of fuel flow required in kg/s.

(*iv*) The evaporation factor.

(*v*) The circulation ratio.

(*vi*) The number of riser tubes.

(*vii*) The number of downcomers.

(*viii*) State reasons for selecting fewer in number and bigger in diameter as downcomers while more in number and smaller in diameter as riser tubes in natural circulation boiler.

Take the following properties :

$$h_1 = 3448.6 \text{ kJ/kg}, \; h_{2s} = 1987 \text{ kJ/kg}$$
$$h_{f3} = 137.8 \text{ kJ/kg}, \; h_{f5} = 675.5 \text{ kJ/kg}$$
$$\rho_{\text{riser top}} = 396 \text{ kg/m}^3$$
$$\rho_{\text{downcomer inlet}} = 603 \text{ kg/m}^3$$

where suffixes 1, 2, 3 and 5 denote boiler exit (or turbine inlet), condenser exit, pump inlet and boiler inlet respectively.

Sol. Given :

(*i*) P = 500 MW, p_1 = 150 bar, T_1 = 273 + 550 = 823 K, p_2 = 0.05 bar, η_t = 0.87, η_{mech} = 0.98, η_g = 0.99, CV = 26 MJ/kg, η_b = 0.92, h_1 = 344.86 kJ/kg, h_2 = 1987 kJ/kg, h_{f3} = 137.8 kJ/kg, h_{f5} = 675.5 kJ/kg, x_{top} = 0.12, v_i = 1.4 m/s.

(*ii*) Generation of rate of steam

$$\dot{m}_s(h_1 - h_2)\eta_t\eta_{mech}\eta_g = P$$

$$\dot{m}_s = \frac{500 \times 10^3}{(3448.6 - 1987.0) \times 0.87 \times 0.98 \times 0.99} = 405.28 \text{ kg/s}$$

(*iii*) Rate of fuel flow.

$$\dot{m}_f \times CV \times \eta_b = \dot{m}_s(h_1 - h_{f5})$$

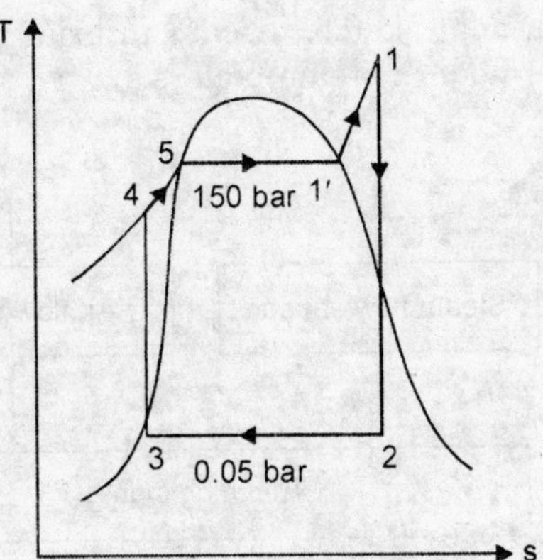

Fig. 15. T-s diagram.

$$\dot{m}_f = \frac{405.28(3448.6 - 675.5)}{26 \times 10^3 \times 0.92} = 46.98 \text{ kg/s}$$

(*iv*) Evaporation factor $= \dfrac{\dot{m}_s}{\dot{m}_f} = \dfrac{405.28}{46.98} = 8.62$

(*v*) Circulation ratio, $\text{CR} = \dfrac{1}{\text{Quality of steam at the top of the riser} = x_{\text{top}}}$

$$= \frac{1}{0.12} = 8.33$$

(*vi*) Number of riser tubes.

Rate of steam formation in one riser :

$$\dot{m}_{rs} = A_i \rho_{\text{top}} v_i x_{\text{top}}$$

$$A_i = \frac{\pi}{4}(65 - 6)^2 \times 10^{-6} = 2.734 \times 10^{-3} \text{ m}^2$$

$$\dot{m}_{rs} = 2.734 \times 10^{-3} \times 396 \times 1.4 \times 0.12 = 0.1819 \text{ kg/s}$$

Number of riser tubes, $n_{rt} = \dfrac{\dot{m}_s}{\dot{m}_{rs}} = \dfrac{405.28}{0.1819} = 2228$ tubes

(*vii*) Number of down comers.

Mass flow rate of feed water through one downcomer,

$$\dot{m}_{dw} = A_i \rho_d v_i$$

$$A_i = \frac{\pi}{4}(185 - 16)^2 \times 10^{-6} = 0.02243 \text{ m}^2$$

$$\dot{m}_{dw} = 0.02243 \times 603 \times 1.4 = 18.937 \text{ kg/s}$$

Number of downcomers, $n_{dw} = \dfrac{405.28}{18.937} = 21.4 \simeq 22$

Q.7. (*d*) With the help of a neat sketch, discuss in brief the working principle of steam pressure control system used in modern steam boilers.

Sol.

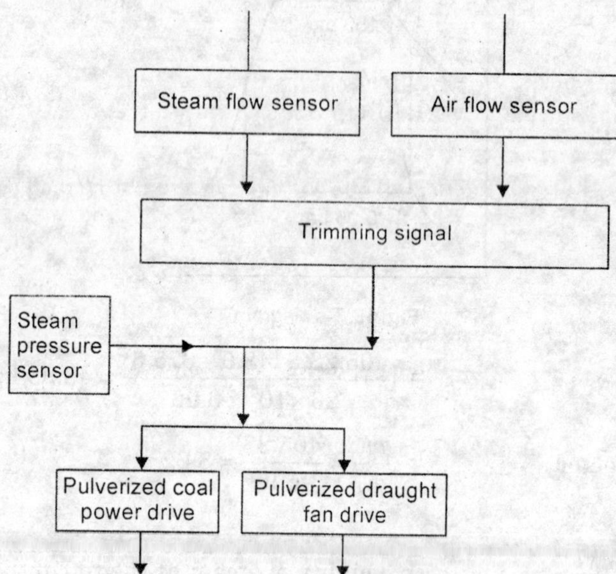

Fig. 16. Steam pressure control

Fig. shows the steam pressure control system used in modern steam boilers. The steam pressure is controlled by adjusting fuel and combustion air flows. There are three sensors incorporated in the system to control steam pressure. There are : steam flow sensor, air flow sensor and steam pressure sensor. Accordingly as per demand the pulverized coal power dried and forced draught fan drive affect the desired changes.

Q.8. (*a*) What do you mean by cavitation phenomena associated with hydraulic turbomachines? Discuss the causes of cavitation and its prevention with reference to hydraulic turbines. Mention the location of cavitation in hydraulic turbupumps and turbines and give reasons for this.

Sol. Refer to Q. 7(*d*) of IESE—2002.

Q.8. (*b*) An axial-flow compressor employed in gas turbine plant delivers air at the rate of 300 kg/s and develops a total pressure ratio of 20. The inlet stagnation conditions are 300 K and 1 bar. The isentropic efficiency of the compressor is 87%. The compressor is having 18 stages and the blade speed is kept at 200 m/s to minimize noise generation. The stage degree of reaction at the mean blade height is 50%. The axial velocity of flow is 160 m/s. The work done factor is 0.88. The hub-tip diameter ratio is 0.8. Assume actual temperature rise in each stage. Take R = 0.287 kJ/kg-K and c_p = 1.005 kJ/kg-K. Work out the following :

(*i*) Sketch the system, show the process on T-*s* diagram and draw velocity diagram.

(*ii*) All the fluid angles of the first stage.

(*iii*) The hub and tip diameters including balde height.

(*iv*) State the reasons why the pressure rise per stage in axial-flow compressor is less than that of centrifugal compressor.

Sol. Given :

$$\dot{m}_a = 300 \text{ kg/s}, r_p = 20, T_{01} = 300 \text{ k},$$
$$p_{01} = 1 \text{ bar}, \eta_{isen} = 0.87,$$
$$z = 18, U = 200 \text{ m/s},$$
$$R_d = 0.5, \phi_w = 0.88, \text{hub-tip ratio} = 0.8,$$
$$R = 0.287 \text{ kJ/kg-K},$$
$$C_p = 1.005 \text{ kJ/kg-K},$$
$$v_f = 160 \text{ m/s}.$$

(*i*) The T-*s* and velocity diagrams are shown in Fig.

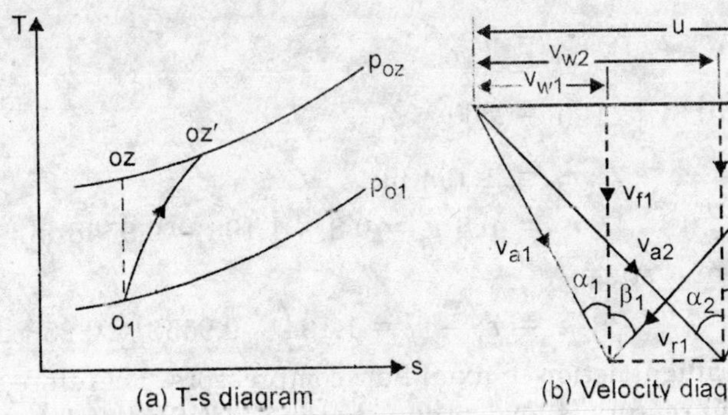

(a) T-s diagram (b) Velocity diagrams.

Fig. 17.

$$T_{oz} = T_{01}(r_p)^{\frac{\gamma-1}{\gamma}} = 300(20)^{\frac{1.4-1}{1.4}} = 706.06 \text{ K}$$

$$\eta_{isen} = \frac{T_{0z} - T_{01}}{T_{0z'} - T_{01}}$$

$$0.87 = \frac{706.06 - 300}{T'_{02} - 300}$$

$$T_{0z'} = 766.7 \text{ K}$$

$$v_{w2} = v_f \tan \alpha_2, \; v_{w1} = v_f \tan \alpha_1$$

For $\qquad R_d = 0.5, \alpha_2 = \beta_1 \text{ and } \alpha_1 = \beta_2$

Work consumed by compressor per kg of air

$$= u(v_{w2} - v_{w1})\, \phi_w \times z$$

$$= c_p(T_{0z'} - T_{01})$$

$$200(\tan \alpha_2 - \tan \alpha_1)\; 160 \times 0.88 \times 18 = 1.005\;(766.7 - 300)$$

$$\tan \alpha_2 - \tan \alpha_1 = 9.25 \times 10^{-4}$$

or $\qquad \tan \beta_1 - \tan \alpha_1 = 9.25 \times 10^{-4} \qquad (1)$

Also $\qquad \tan \alpha_1 + \tan \beta_1 = \frac{u}{v_f} = \frac{200}{160} = 1.25 \qquad (2)$

Adding Eqs. (1) and (2), we get

$$2 \tan \beta_1 = 1.250925$$

$$\beta_1 = 32.02° = \alpha_2$$

$$\tan \alpha_1 = 1.25 - 0.62546 = 0.62453$$

$$\alpha_1 = 31.98° = \beta_2$$

(*iii*) Density of approaching air to first stage,

$$\rho_1 = \frac{p_{01}}{RT_{01}} = \frac{1 \times 10^5}{287 \times 300} = 1.1614 \text{ kg/m}^3$$

Applying continuity equation,

$$\rho_1 A_f v_f = \dot{m}_a$$

$$1.1614 \times \pi r_1^2 [1 - (0.8)^2] \times 160 = 300$$

Tip radius, $r_1 = 1.194$ m

Hub radius, $r_h = 0.8\ r_1 = 0.8 \times 1.194 = 0.9558$ m

Blade height of first stage,

$$h = r_1 - r_h = 1.194 - 0.9558 = 0.238 \text{ m}$$

(*iv*) There is no centrifugal action in axial flow compressors. Therefore, the pressure ratio in axial flow compressors is lower than that of centrifugal compressors.

Q.8. (*c*) A gas turbine power plant developing 250 MW of electrical power employs a single-shaft gas turbine reheat cycle having the following data :

Total compressor pressure ratio = 30

Total ambient condtions = 1 bar and 300 K

Polytropic efficiencies for both compressor and turbine = 0.9 each

Total turbine inlet temperature of both turbines = 1600 K each

Pressure loss in both combustors = 2% of entry pressure each

Total turbine exhaust pressure = 1.05 bars

Mechanical efficiency of assembly = 0.98

Combustion efficiency of both combustors = 0.97

Alternator power output = 250 MW

Actual air-fuel ratio = 25

Lower calorific value of fuel = 42 MJ/kg

Take, c_{pa} = 1.005 kJ/kg-K and c_{pg} = 1.16 kJ/kg-K

Work out the following :

(*i*) Sketch the system and show the process on T-s diagram.

(*ii*) The plant specific work in kJ/kg.

(*iii*) The mass flow rate of air required in kg/s.

(*iv*) The specific fuel consumption in kg/kWh.

(*v*) The actual thermal efficiency.

Sol.

$$P_a = 250 \text{ MW}, \eta_a = 0.97, r_{pc} = 30,$$
$$p_1 = 1 \text{ bar}, T_1 = 300 \text{ K},$$
$$\eta_{pc} = \eta_{pt} = 0.9, T_3 = T_5 = 1600 \text{ K},$$
$$(\Delta p)_{comb} = 20\%, p_6 = 1.05 \text{ bar}, \eta_{mech} = 0.98,$$
$$\eta_{comb} = 0.97, A : F = 25, LCV = 42 \text{ MJ/kg},$$
$$c_{pa} = 1.005 \text{ kJ/kg·K}, c_{pg} = 1.16 \text{ kJ/kg·K}.$$

The T-*s* diagram of the system is shown in Fig. 18.

For $\eta_{t1} = \eta_{t2}$ and $T_3 = T_5$, intermediate pressure of turbines,

$$p_4 = \sqrt{p_3 p_6}$$
$$p_2 = r_{pc} p_1 = 30 \times 1 = 30 \text{ bar}$$
$$p_3 = 0.98\, p_2 = 0.98 \times 30 = 29.4 \text{ bar}$$

Compressor :

$$\frac{T_2}{T_1} = (r_{p_c})^{\frac{\gamma-1}{\gamma}}, T_2 = 300(30)^{\frac{1.4-1}{1.4}} = 792.78 \text{ K}$$

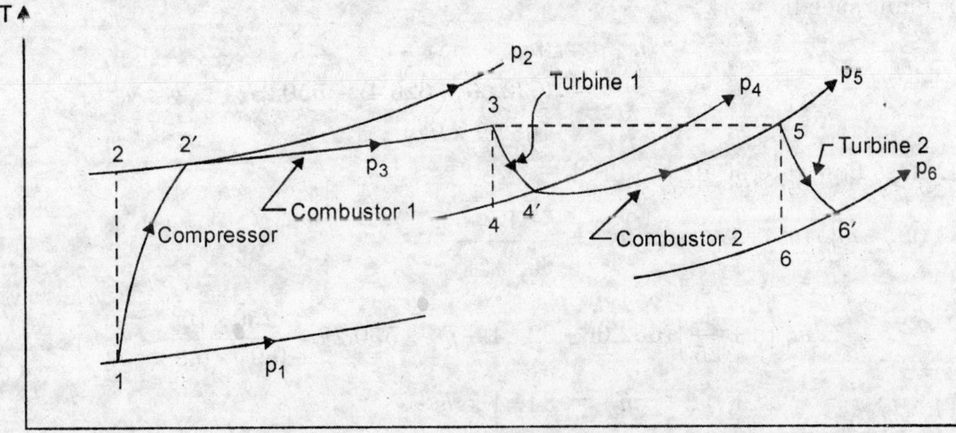

Fig. 18.

$$\eta_1 = \frac{T_2 - T_1}{T_2' - T_1}, \; 0.9 = \frac{792.78 - 300}{T_2' - 300}, \; T_2' = 847.53 \text{ K}$$

$$w_c = c_{pa}(T_2' - T_1) = 1.005\,(847.53 - 300) = 550.27 \text{ kJ/kg}$$

Turbine 1 :

$$p_4 = \sqrt{29.4 \times 1.05} = 5.556 \text{ bar}$$

$$\frac{T_3}{T_4} = \left(\frac{p_3}{p_4}\right)^{\frac{\gamma-1}{\gamma}} = \left(\frac{29.4}{5.556}\right)^{1/3.5} = 1.6097$$

$$T_4 = \frac{1600}{1.6097} = 994 \text{ K}$$

$$\eta_{t1} = \frac{T_3 - T_4'}{T_3 - T_4}, 0.9 = \frac{1600 - T_4'}{1600 - 994}, \; T_4' = 1054.6 \text{ K}$$

$$w_{t1} = c_{pq}(T_3 - T_4') = 1.16(1600 - 1054.6) = 632.66 \text{ kJ/kg}$$

Turbine 2 :

$$p_5 = 0.98 \times 5.556$$
$$= 5.445 \text{ bar}$$

$$\frac{T_5}{T_6} = \left(\frac{p_5}{p_6}\right)^{1/3.5} = \left(\frac{5.445}{1.05}\right)^{1/3.5} = 1.60$$

$$T_6 = \frac{1600}{1.60} = 1000 \text{ K}$$

$$\eta_{t2} = \frac{T_5 - T_6'}{T_5 - T_6}, \; 0.9 = \frac{1600 - T_6'}{1600 - 1000}, \; T_6' = 1060 \text{ K}$$

$$w_{t2} = c_{pg}(T_5 - T_6') = 1.16(1600 - 1060) = 626.4 \text{ kJ/kg}$$

(*ii*) Plant specific work.

$$w_{\text{net}} = (w_{t1} + w_{t2}) - w_c$$
$$= (632.66 + 626.4) - 550.27$$
$$= 708.79 \text{ kJ/kg}$$

(*iii*) Mass flow rate of air required.

$$(\dot{m}_a + \dot{m}_f)(w_{t1} + w_{t2}) - \dot{m}_a w_c = \frac{P_a}{\eta_a \eta_{\text{mech}}}$$

$$\dot{m}_a\left(1 + \frac{1}{25}\right)(632.66 + 624.4) - \dot{m}_a \times 550.27 = \frac{250 \times 10^3}{0.97 \times 0.98}$$

$$\dot{m}_a = 347.4 \text{ kg/s}$$

(*iv*) Specific fuel consumption.

$$\dot{m}_f = \frac{\dot{m}_a}{25} = \frac{347.4}{25} = 13.89 \text{ kg/s}$$

$$\text{SFC} = \frac{13.89 \times 3600}{250 \times 10^3} = 0.2 \text{ kg/kWh}$$

(*v*) Actual thermal efficiency,

$$\eta_{\text{th}} = \frac{(\dot{m}_a + \dot{m}_f)(w_{t1} + w_{t2}) - \dot{m}_a w_c}{\dot{m}_f \times \text{LCV}}$$

$$= \frac{(1 + 25)(632.66 + 626.4) - 25 \times 550.27}{42{,}000}$$

$$= 0.4518 \text{ or } 45.18\%$$

Q.8. (*a*) With the help of a simple sketch, discuss in brief the working principle of hydromechanical speed-governing system (*prime control*) of *a water* reaction turbine.

Sol. Governing of Reaction Turbine : The discharge in reaction turbines is regulated by changing the gate opening of guide vanes which controls the area of flow passage. The governing mechanism is designed to change the position of guide vanes uniformly, and thus controls the area of flow passage which controls the area of flow passage. The mechanism changes the position of guide vanes to suit the load on the turbine. The system consists of a centrifugal governing mechanism, linkages, servomotor with its oil pressure governor and the guide wheel. the guide vanes are pivoted and connected by levers and links to the regulating ring. Two regulating rods are attached to the regulating ring at one end and to the regulating lever at the other, as shown in Fig. 19. The regulating lever is connected to a regulating shaft which is operated by the servomotor piston of the oil pressure governor. The motion of the servomotor piston is transmitted to the regulating ring which causes all the guide vanes to turn simultaneously in one direction through the same angle. Thus the area of flow passage is increased or reduced according to the load on the turbine.

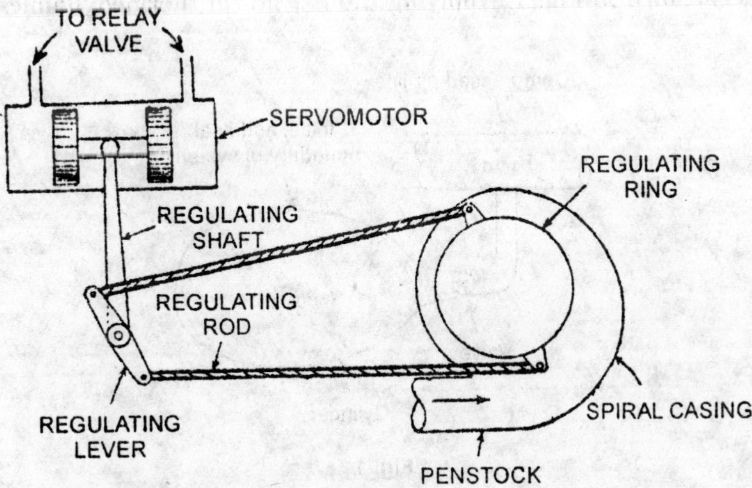

Fig. 19. Governing mechanism for Reaction Turbine.

MECHANICAL ENGINEERING
PAPER I—2011

(Conventional)

Q.1. **(*a*)** A pressure cylinder of volume V contains air at pressure p_0 and temperature T_0. It is to be filled from a compressed air line maintained at constant pressure p_1 and temperature T_1. Show that the temperature of the air in the cylinder after it has been charged to the pressure of the line is given by

$$T = \frac{\gamma T_1}{1 + \frac{p_0}{p_1}\left(\gamma \frac{T_1}{T0} - 1\right)}.$$

Sol. Let m_1 = initial mass in the cylinder

$m_2 - m_1$ = mass which shall flow into the cylinder from the compressed air line

The system is shown in Fig. 1. Applying the fist law of thermodynamics to the system, we have

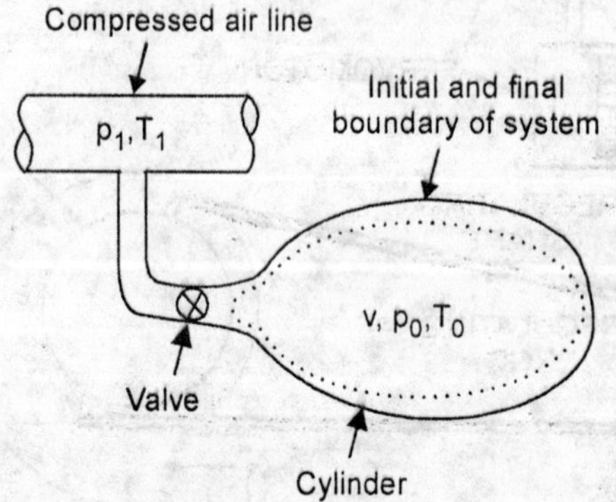

Fig. 1

Transitional energy + Net energy entering the cylinder
= change of energy in the control volume

$$Q \text{ or } W \text{ or both} + (m_2 - m_1)\left(h_{\text{air line}} + \frac{c_1^2}{2}\right) = m_2u_2 - m_1u_1$$

where u = internal energy, c = velocity, h = enthalpy.

Here Q and W = 0, c_1 = 0.

$$\therefore \quad (m_2 - m_1)h = m_2u_2 - m_1u_1$$

or

$$(m_2 - m_1)\, c_pT_1 = m_2c_vT_1 - m_1c_vT_0$$

$$(m_2 - m_1)\,\gamma T_1 = m_2 T_1 - m_1 T_0$$

Now
$$pV = mRT$$

or
$$mT = \frac{pV}{R}$$

$$\left(\frac{p_1 V}{RT} - \frac{p_0 V}{RT_0}\right)\gamma T_1 = \frac{p_1 V}{R} - \frac{p_0 V}{R}$$

or
$$p_1 - p_0 = \left(\frac{p_1}{T} - \frac{p_0}{T_0}\right)\gamma T_1 = \frac{p_1 \gamma T_1}{T} - \frac{p_0 \gamma T_1}{T_0}$$

$$\frac{p_1 \gamma T_1}{T} = p_1 - p_0 + \frac{p_0 \gamma T_1}{T_0} = p_1 + p_0\left(\frac{\gamma T_1}{T_0} - 1\right)$$

$$T = \frac{\gamma T_1}{1 + \frac{p_0}{p_1}\left(\frac{\gamma T_1}{T_0} - 1\right)}$$

Hence proved.

Q.1. (*b*) Consider an engine in outer space which operates on the Carnot cycle. The only way in which heat can be transferred from the engine is by radiation. The rate at which heat is radiated is proportional to the fourth power of the absolute temperature and to the area of the radiating surface. Show that for a given power output and a given T_1, the area of the radiator will be a minimum when $\frac{T_2}{T_1} = \frac{3}{4}$.

Sol. Refer to Fig. 2

Amount of heat radiated, $Q_2 = \sigma A T_2^4$

For a reversible engine,
$$\frac{Q_1}{T_1} = \frac{Q_2}{T_2} = \frac{Q_1 - Q_2}{T_1 - T_2} = \frac{W}{T_1 - T_2}$$

$$Q_2 = \frac{WT_2}{T_1 - T_2} = \sigma A T_2^4$$

$$A = \frac{WT_2}{\sigma T_2^4 (T_1 - T_2)} = \frac{W}{\sigma}\left[\frac{1}{T_1 T_2^3 - T_2^4}\right]$$

Fig. 2

For area to be minimum,

$$\frac{dA}{dT_2} = 0$$

$$\frac{d}{dT_2}\left[\frac{1}{T_1 T_2^3 - T_2^4}\right] = 0$$

$$3T_1 T_2^2 - 4T_2^3 = 0$$

$$T_2 = \frac{3}{4} T_1$$

or

$$\frac{T_2}{T_1} = \frac{3}{4}$$

Hence proved.

Q.1. (*c*) A lead storage battery used in an automobile is able to deliver 5.2 MJ of electrical energy. This energy is available for starting the car. Let compressed air be considered for doing an equivalent amount of work in starting the car. The compressed air is to be stored at 7 MPa, 25°C. What is the volume of the tank that would be required to let the compressed air have an availability of 5.2 MJ? For air, $pv = 0.287$ T, where T is in K, p in kPa, and v in m³/kg.

Sol. Since electrical energy is high grade energy, total available electrical energy,

$$AE = 5.2 \times 10^6 \text{ J}$$

Available energy of compressed air,

$$(AE)_{air} = (U_1 - U_0) - T_0(S_1 - S_0)$$

$$= mc_v(T_1 - T_0) - T_0(S_1 - S_0)$$

$$s_1 - s_0 = c_p \ln\frac{T_1}{T_0} - R \ln\frac{p_1}{p_0}$$

$$\Delta w = RT_0 \ln\frac{p_1}{p_0}$$

Here $T_1 = T_0 = 25°C = 298$ K

$p_0 = p_{atm} = 1$ bar, $p_1 = 7$ MPa

$$\Delta w = 0.287 \times 298 \ln\left(\frac{7 \times 10}{1}\right) = 363.36 \text{ kJ/kg}$$

Mass of air required, $m_a = \dfrac{5.2 \times 10^3}{363.36} = 14.311$ kg

Also $pV = mRT$

Specific volume, $v = \dfrac{V}{m} = \dfrac{RT}{p_1} = \dfrac{0.287 \times 298}{7000} = 0.012218 \text{ m}^3/\text{kg}$

Total volume of tank required

$$= m_a v$$

$$= 14.311 \times 0.012218$$

$$= 0.1749 \text{ m}^3$$

Q.1. (*d*) Obtain an expression for the specific work output of a gas turbine unit in terms of pressure ratio, isentropic efficiencies of the compressor and turbine, and the maximum and minimum temperatures, T_3 and T_1. Hence show that the pressure ratio r_p for maximum power is given by

$$r_p = \left[\eta_T \cdot \eta_C \frac{T_3}{T_1}\right]^{\frac{\gamma}{2(\gamma-1)}}$$

Sol. Maximum Pressure ratio for maximum specific work : It is not possible to achieve the isentropic compression and expansion in the actual Brayton cycle due to inevitable losses due to friction and turbulence in the compressor and turbine. Therefore, the temperature at the end of compression and expansion are higher than that in an ideal cycle for the same pressure ratio. The actual Brayton cycle is shown in Fig. 3.

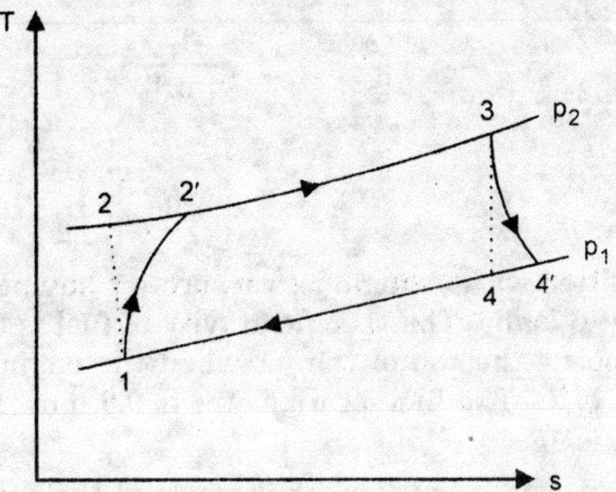

Fig. 3. Actual Brayton cycle with machine efficiencies.

Isentropic efficiency of compressor,

$$\eta_c = \frac{T_2 - T_1}{T_2' - T_1}$$

Isentropic efficiency of turbine,

$$\eta_t = \frac{T_3 - T_4'}{T_3 - T_4}$$

Net specific work available,

$$w_{net} = w_t - w_c = c_p(T_3 - T_4') - c_p(T_2' - T_1)$$

$$= c_p\eta_t(T_3 - T_4) - c_p\frac{(T_2 - T_1)}{\eta_c}$$

$$= c_p\eta_t T_3\left(1 - \frac{T_4}{T_3}\right) - \frac{c_p T_1}{\eta_c}\left(\frac{T_2}{T_1} - 1\right)$$

$$= c_p\eta_t T_3\left[1 - \frac{1}{\gamma_p^{\frac{\gamma-1}{\gamma}}}\right] - \frac{c_p T_1}{\eta_c}\left[\gamma_p^{\frac{\gamma-1}{\gamma}} - 1\right]$$

For specific work output to be maximum for given temperature limits,

$$\frac{d}{dr_p}(w_t) = 0$$

$$\eta_t T_3\left[\left(\frac{\gamma-1}{\gamma}\right) r_p^{\frac{\gamma-1}{\gamma}-1}\right] - \frac{T_1}{\eta_c}\left(\frac{\gamma-1}{\gamma}\right) r_p^{\frac{\gamma-1}{\gamma}-1} = 0$$

$$\eta_t T_3 r_p^{\frac{\gamma-1}{\gamma}-1} - \frac{T_1}{\eta_c} r_p^{\frac{-1}{\gamma}} = 0$$

$$r_p^{2\left(\frac{1-\gamma}{\gamma}\right)} = \frac{T_1}{\eta_c \eta_t T_3}$$

$$r_p = \left[\eta_c \eta_t \left(\frac{T_3}{T_1}\right)\right]^{\frac{\gamma}{2(\gamma-1)}}$$

Q.2. (*a*) An engine fitted with a single jet carburettor having a jet diameter of 1.25 mm has a fuel consumption of 6 kg/hr. The specific gravity of fuel is 0.7. The level of fuel in the float chamber is 5 mm below the top of the jet when the engine is not running. Ambient conditions are 1 bar and 17°C. The fuel jet diameter is 0.6 mm. The discharge coefficient of air is 0.85. Air-fuel ratio is 15.

Determine the critical velocity of flow at throat and the throat diameter. Express the pressure at throat in mm of water column. Neglect compressibility effect. Assume discharge coefficient of fuel flow is 0.60.

Sol. Given : d_j = 1.25 mm, $\dot{m}_j$ = 6 kg/hr, S_f = 0.7, z = 5 mm, p_1 = 1 bar, T_1 = 273 + 17 = 290 K, d_f = 0.6 mm, C_{da} = 0.85, A : F = 15, C_{df} = 0.60.

$$pV = mRT$$

$$\rho_a = \frac{m}{V} = \frac{p_1}{RT_1} = \frac{1\times 10^5}{287\times 290} = 1.20 \text{ kg/m}^3$$

$$S_f = \frac{\rho_f}{\rho_a} = 0.7, \quad \rho_f = 0.7\times 1.20 = 0.84 \text{ kg/m}^3$$

Applying Bernoulli's equation for air flow at entrance and venturi throat, we have

$$\frac{p_1}{\rho_a} = \frac{p_2}{\rho_a} + \frac{c_{a2}^2}{2}$$

$$c_{a2} = \sqrt{\frac{2(p_1 - p_2)}{\rho_a}} = \sqrt{\frac{2\Delta p}{\rho_a}} \text{ or } \Delta p = \frac{c_{a2}^2 \rho_a}{2}$$

Mass of air flow per second,

$$\dot{m}_a = C_{da} A_2 c_{a2} \rho_2 = c_{da} A_2 \sqrt{2\rho_a (p_1 - p_2)}$$

$$= C_{da} A_2 \sqrt{2\rho_a \cdot \Delta p}$$

For the flow of fuel, we have

$$\frac{p_1}{\rho_f} = \frac{p_2}{\rho_f} + \frac{c_{f2}^2}{2} + zg$$

$$c_{f2} = \sqrt{\frac{2(p_1 - p_2 - zg\rho_f)}{\rho_f}} = \sqrt{2\left[\frac{\Delta p - zg\rho_f}{\rho_f}\right]}$$

Mass of fuel flow per second,

$$\dot{m}_f = C_{df}A_f c_{f2} p_f = C_{df}A_f\sqrt{2\rho_f(p_1 - p_2 - zg\rho_f)} = C_{df}A_f\sqrt{2\rho_f(\Delta p - zg\rho_f)}$$

F:A ratio, $$\frac{\dot{m}_f}{\dot{m}_a} = \frac{C_{df}A_f}{C_{da}A_2}\sqrt{\frac{\rho_f}{\rho_a}}\cdot\sqrt{1 - \frac{zg\rho_f}{p_1 - p_2}}$$

Critical velocity of flow at throat,

$$c_{a2} = \sqrt{\frac{2(p_1 - p_2)}{\rho_a}} = \sqrt{2gz\frac{p_f}{\rho_a}} \qquad [\because p_1 - p_2 = gz\rho_f]$$

$$= \sqrt{2 \times 9.81 \times 5 \times 10^{-3} \times 0.7} = 0.262 \text{ m/s}$$

Now $$p_1 - p_2 = gz\rho_f = 9.81 \times 5 \times 10^{-3} \times 0.84 = 0.0412$$

$$p_2 = 1 \times 10^5 - 0.0412$$

$$= 0.99999 \times 10^5 \text{ N/m}^2$$

Pressure at throat, $$h = \frac{p_2}{9.81 \times 10^3} = \frac{0.99999 \times 10^5}{9810} = 10.194 \text{ m of } H_2O$$

$$\dot{m}_f = \frac{6}{360} = 1.66 \times 10^{-3} \text{ kg/s}$$

$$\dot{m}_a = 15\,\dot{m}_f = 0.025 \text{ kg/s}$$

$$\dot{m}_a = C_{da} \times \frac{\pi}{4}d_2^2 \times c_{a2} \times \rho_a$$

$$0.025 = 0.85 \times \frac{\pi}{4}d_2^2 \times 0.262 \times 1.2$$

$$d_2 = 0.345 \text{ m}$$

Q.2. (*b*) Find the percentage increase in the efficiency of a Diesel Cycle having a compression ratio 'r' of 16 and cut-off ratio 'r_c' is 10% of the swept volume, if c_v decreases by 2%. Take c_v = 0.717 kJ/kg.K and γ = 1.4.

Sol. Given : $r = 16$, $r_c = 10\%$ of V_s, $\Delta c_v = -2\%$,

$c_v = 0.717$ kJ/kg.K, $\gamma = 1.4$.

Refer to Fig. 4.

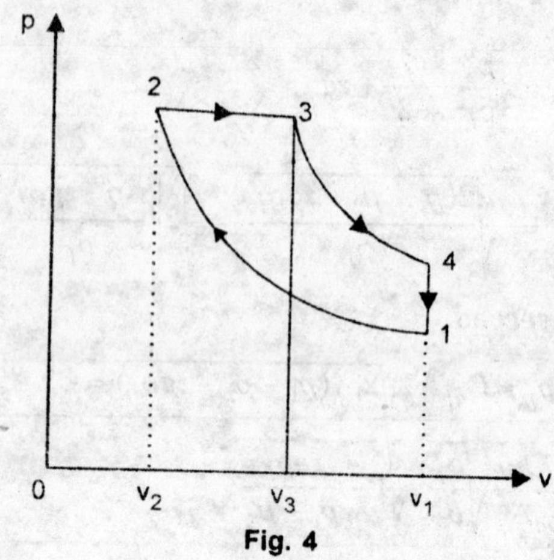

Fig. 4

Efficiency of diesel cycle,

$$\eta = 1-1-\frac{1}{r^{\gamma-1}}\left[\frac{r_c^{\gamma}-1}{\gamma(r_c-1)}\right]$$

$$V_3 - V_2 = 0.1\ (V_1 - V_2)$$

$$\frac{V_3}{V_2}-1 = 0.1\left(\frac{V_1}{V_2}-1\right)$$

$$r_c - 1 = 0.1\ (r - 1)$$

$$r_c = 0.1(16 - 1) + 1 = 2.5$$

$$\eta = 1-\frac{1}{(16)^{1.4-1}}\left[\frac{(2.5)^{1.4}-1}{1.4\ (2.5-1)}\right]$$

$$= 0.5905 \quad \text{or } 59.05\%$$

Decreased value of $c_v = 0.98 \times 0.717 = 0.70266$ kJ/kg·K

$$c_p - c_v = R,\ c_p = 0.287 + 0.70266$$

$$= 0.98966 \text{ kJ/kg·K}$$

$$\gamma = \frac{c_p}{c_v} = \frac{0.98966}{0.70266} = 1.4085$$

$$\eta = 1-\frac{1}{(16)^{1.4085-1}}\left[\frac{(2.5)^{1.4085}}{1.4085\ (2.5-1)}\right]$$

$$= 0.5981 \quad \text{or } 59.81\%$$

% change in efficiency $= 59.81 - 59.05 = 0.76\%$

Q.2. (*c*) Explain the knocking phenomenon in a CI engine. Compare it with that of SI engines. Discuss the effect of operating variables on delay period and diesel knowck.

Sol. Refer to Q. 2(*b*) of IESE—2000 and Q. 2(*c*) of IESE—2006.

Q.3 (*a*) A steel pipe having internal diameter of 2 cm, outer diameter of 2.4 cm and thermal conductivity of steel of 54 W/mK carries hot water at 95°C. Heat transfer coefficient between the inner surface of steel pipe and the hot water is 600 W/m²K. An asbestos insulation with thermal conductivity of 0.2 W/mK and thickness 2 cm is put on the steel pipe. Heat is lost from the outer surface of the asbestos insulated pipe to the surrounding air at 30°C, heat transfer coefficient for the outer surface of the insulation being 8 W/m²K.

Determine :

(*i*) The rate of heat transfer per meter length of the pipe.

(*ii*) Determine the temperatures at the inner, outer surfaces of the steel pipe and the outer surface of the insulation.

(*iii*) What do you understand by the term "critical radius of insulation"? What is the value of critical radius in the above question? What is the rate of heat loss, if thickness of insulation were to correspond to critical radius? Comment on the results.

Sol. Given : $r_1 = 1$ cm, $r_2 = 1.2$ cm, $r_3 = 3.2$ cm, $k_1 = 54$ W/m·K, $t_i = 95$°C,
$h_i = 600$ W/m²·K, $k_2 = 0.2$ W/m·K, $t_a = 30$°C, $h_0 = 8$ W/m²·K.

The insulated pipe is shown in Fig. 5(*a*) and thermal circuit in Fig. 5(*b*).

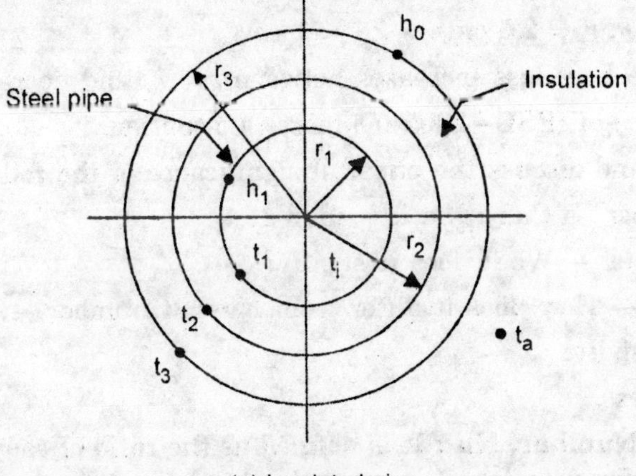

(a) Insulated pipe

Fig.5

Thermal resistance,

$$R_{th} = \frac{1}{600\,(2\pi \times 0.01 \times 1)} + \frac{\ln(1.2/1)}{2\pi \times 54 \times 1} + \frac{\ln(3.2/1.2)}{2\pi \times 0.2 \times 1} + \frac{1}{8(2\pi \times 0.032 \times 1)}$$

$$= 0.0265 + 5.3736 \times 10^{-4} + 0.7805 + 0.6217$$

$$= 1.429 \text{ K/W}$$

(*i*) Rate of heat transfer per metre length of pipe.

$$q = \frac{t_i - t_a}{R_{th}} = \frac{95 - 30}{1.429} = 45.486 \text{ W/m}$$

(*ii*)

$$q = \frac{t_i - t_1}{\dfrac{1}{h_i(2\pi r_1 L)}}$$

$$45.486 = \frac{(95 - t_1)}{\dfrac{1}{600 \times 2\pi \times 0.01 \times 1}} = \frac{95 - t_1}{0.0265}$$

$$t_1 = 93.79°C$$

$$q = \frac{t_1 - t_2}{\dfrac{\ln (r_2/r_1)}{2\pi\, k_1 L}} = \frac{93.79 - t_2}{5.3736 \times 10^{-4}} = 45.486$$

$$t_2 = 93.76°C$$

$$q = \frac{t_2 - t_3}{\dfrac{\ln (r_3/r_2)}{2\pi k_2 L}} = \frac{93.76 - t_3}{0.7805} = 45.486$$

$$t_3 = 58.26°C$$

(*iii*) Critical radius of insulation,

$$r_c = \frac{k_2}{h_0} = \frac{0.2}{8} = 0.025 \text{ m or } 2.5 \text{ cm}$$

$$r_c(= 2.5 \text{ cm}) > r_1(= 1 \text{ cm})$$

or $r_1 < r_c$. The heat loss increases between $r < r_c$ and decreases beyond $r > r_c$. Refer to Q. 3(*a*) of IESE—2005 for more information.

Q.3. (*b*) Define and discuss the physical significance of the following :

(*i*) Nusselt Number — Can it be less than 1?

(*ii*) Prandtl Number — What does it signify?

(*iii*) Biot Number — How does it differ from Nusselt number?

(*iv*) Thermal diffusivity.

(*v*) Fin effectiveness.

Sol. (*i*) **Nusselt Number, Nu :** It is defined as the ratio of convective heat transfer to conduction heat transfer in a layer. It may also be defined as the ratio of characteristics length L to the thickness Δx of a stationary fluid layer conducting heat at the same rate under the same temperature difference as in the case of convection process.

$$\text{Nu} = \frac{Q_{conv}}{Q_{cond}} = \frac{h}{\left(\dfrac{k}{L}\right)} = \frac{hL}{k}$$

or $$Q = k\frac{\Delta T}{\Delta x} = h \cdot \Delta T$$

or $$\Delta x = \frac{h}{k}$$

$$\text{Nu} = \frac{L}{\Delta x} = \frac{Lh}{k}$$

Significance : Nusselt number is a measure of the convective heat transfer coefficient. For a given value of Nusselt number, the convective heat tranfer coefficient is directly proportional to the thermal conductivity of the fluid and inversely proportional to the significant length parameter.

(*ii*) **Prandtl Number, Pr :** It is defined as the ratio of kinematic viscosity (ν) to thermal diffusivity (α) of the fluid.

$$Pr = \frac{\nu}{\alpha}$$

$$= \frac{\mu c_p}{k} = \frac{\rho \nu c_p}{k} = \frac{\nu}{\left(\frac{k}{\rho c_p}\right)} = \frac{\nu}{\alpha}$$

Kinematic viscosity indicates the impulse transport through molecular friction whereas thermal diffusivity indicates the heat energy transport by conduction process.

Significance : Prandtl number provides a measure of the relative effectiveness of the momentum and energy transport by diffusion. A fluid having higher Prandtl number has better heat transfer characteristic. Prandtl number is a connecting link between hydrodynamic boundary layer thickness (δ) and thermal boundary layer thickness (δ_t).

if
$$Pr > 1, \quad \text{then} \quad \delta_t < \delta$$
$$Pr = 1, \quad \text{then} \quad \delta_t = \delta$$
$$Pr < 1, \quad \text{then} \quad \delta_t > \delta$$

(*iii*) **Biot Number, Bi :** It is defined as the ratio of internal (conduction) resistance to surface (convection) resistance.

$$Bi = \frac{hL_c}{k}, \text{ where characteristic length, } L_c = \frac{\text{Volume}}{\text{Surface area}} = \frac{V}{A_s}$$

Significance : If gives an indication of the ratio of internal (conduction) resistance to surface (convection) resistance. When the value of Bi is small, it indicates that the system has a small internal (conduction) resistance, *i.e.*, relatively small temperature gradient on the existence of practically uniform temperature within the system. The convective resistance then predominates and the transient phenomenon is controlled by the convective heat exchange.

In Biot number, thermal conductivity of solid is considered while in Nusselt number thermal conductivity of fluid is considered.

(*iv*) **Thermal Diffusivity, α :** It is defined as the ratio of thermal conductivity of the solid its thermal capacity.

$$\alpha = \frac{k}{\rho_c}$$

It is the ability of the material to allow thermal energy to get diffused or passed through the medium. Larger the value of α, faster will be the heat diffusion through the materialj and its temperature will change faster with time.

(*v*) **Fin Effectiveness, ε_{fin}** : It is defined as the ratio of fin heat transfer rate to the heat transfer rate without a fin.

$$\varepsilon_{fin} = \frac{Q_{with\ fin}}{Q_{wihout\ fin}}$$

Q.3. (*c*) (*i*) State and explain Lambert law. How is it used in radiative heat transfer claculations? For a diffused flat surface having emissivity of 0.7 at a temperature of 800°C, calculate normal intensity of radiation.

What would be the intensity of radiation at an angle of 30° with respect to normal direction?

(*ii*) For a very long semicircular duct having surface areas A_2 and A_1 as shown below in Fig. 5.

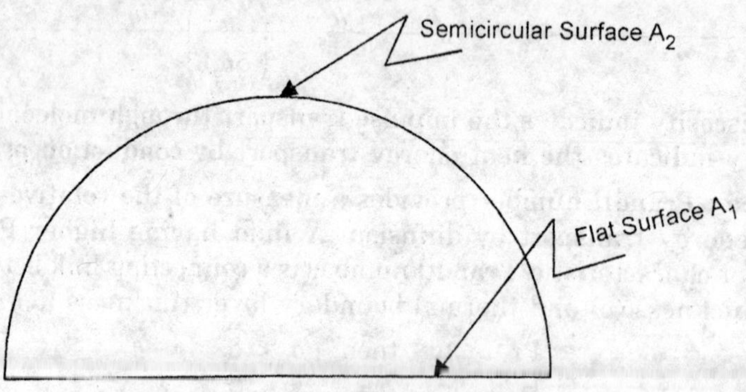

Fig. 6 Determine shape factors F_{12}, F_{21}, F_{22}.

Sol. (*i*) **Lambert's Cosine Law :** The law states that the total emissive power E_θ from a radiating plane surface in any direction is directly proportional to the cosine of the angle of emission. The angle of emission θ is the angle subtended by the normal to the radiating surface and the direction vector of emission of the receiving surface. If E_n be the total emissive power of the radiating surface in the direction of its normal, then (see Fig. 7).

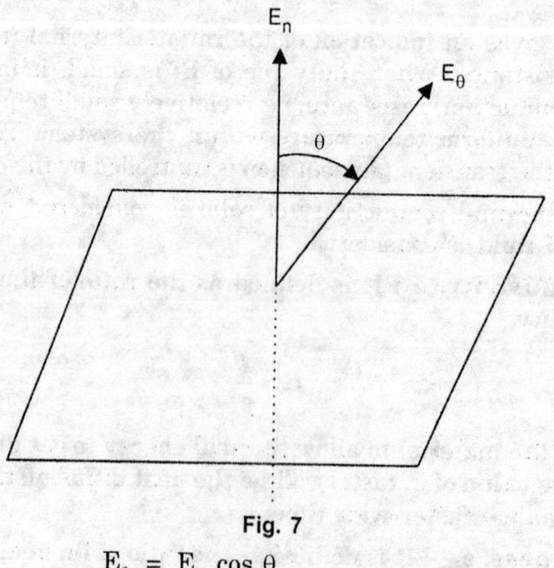

Fig. 7

$$E_\theta = E_n \cos\theta$$

This equation is valid only for diffuse radiation surface. The radiation emanating from a point on a surface is termed diffused if the intensity I is constant. This law is also called as Lambert's law of diffuse radiation.

Now total emissive power of diffuse surface,

$$E = \pi I$$

or

$$I = \frac{E}{\pi}$$

Given : $\varepsilon = 0.7$, $T = 273 + 800 = 1073$ K

$$E = \varepsilon\sigma T^4 = 0.7 \times 5.67 \times 10^{-8} \times 1073^4 = 52{,}611.4 \text{ W/m}^2$$

Normal intensity,

$$I_n = \frac{E}{\pi} = \frac{52{,}611.4}{\pi} = 16{,}746.7 \text{ W/m}^2 \text{ staradian}$$

$$I_{\theta=30°} = I_n \cos 30° = 16{,}746.7 \times \cos 30° = 14{,}503 \text{ W/m}^2 \text{ staradian}$$

(*ii*) By summation rule, (see Fig. 8)

$$F_{11} + F_{12} = 1$$

Now $F_{11} = 0$

$\therefore$ $F_{12} = 1$

By reciprocality rule, we have

$$A_1F_{12} = A_2F_{21}$$

$$F_{21} = \frac{A_1}{A_2} F_{12} = \frac{\pi r^2}{2\pi r^2} \times 1 = 0.5$$

Fig. 8

By summation, rule,

$$F_{21} + F_{22} = 1$$

$$F_{22} = 1 - 0.5 = 0.5$$

Q.3. (*d*) (*i*) What is fouling in heat exchangers? How is it specified? How does the fouling affect the rate of heat transfer in heat exchangers?

(*ii*) Distinguish and differntiate between direct tranfer heat exchangers and storage type heat exchangers with the help of simple sketches giving their advantages and disadvantages.

Sol. (*i*) **Fouling in heat exchangers :** The formation of scale or deposits (ash, dirt, soot, etc.) in heat exchangers is called fouling. The reciprocal of the scale heat transfer coefficient is called fouling factor $\left(\frac{k}{A}\right)$. The thermal resistance increases due to scale formation on the tubes of heat exchanger which results in lower performance. It is difficult to ascertain thickness and thermal conductivity of scale deposits, therefore, the effect of scale on heat flow is considered by specifying an equivalent scale heat transfer coefficient.

Let h_i, h_0 = heat transfer coefficient for scale deposits on the inside and outside surfaces respectively

The thermal resistances to scale formation are :

$$R_i = \frac{1}{A_i h_i}, \text{ on inside surface}$$

and

$$R_0 = \frac{1}{A_0 h_0}, \text{ on outside surface}$$

(*ii*) **Direct transfer heat exchanger :** In this type of heat exchanger, the cold and hot fluids do not mix, and flow in opposite directions. Heat is transfered through the separating walls of tubes. Such a heat exchanger is shown in Fig. 9(*a*).

Advantages

1. Generally suitable for stationary plants.
2. Provide more surface area for heat transfer.
3. They are economical and easy to construct.
4. Most favoured for heating and cooling of fluids.

Disadvantages

1. Lower heat transfer coefficient.
2. Less generating capacity.
3. Sooting problems in pipes.

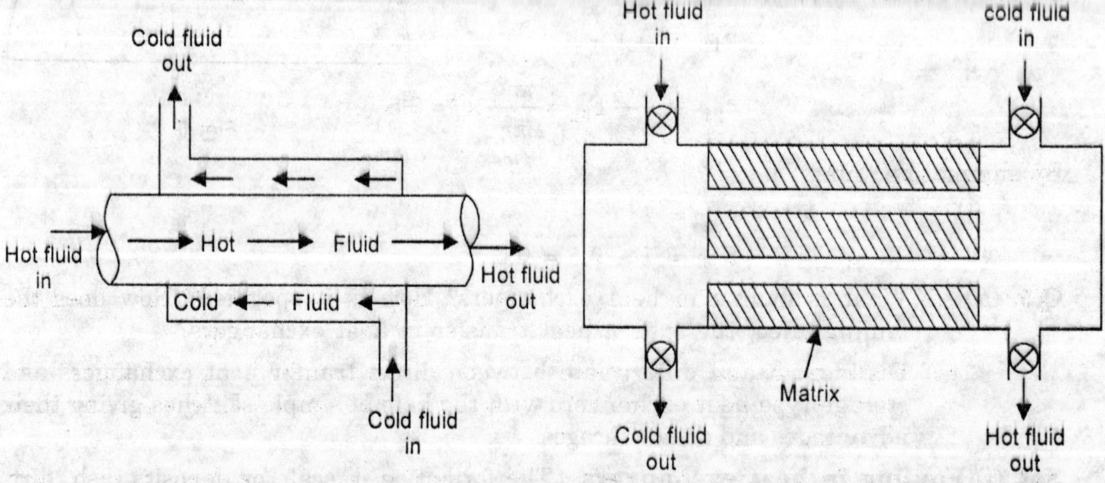

(a) Direct transfer heat exchanger (b) Single matrix storage type heat exchanger

Fig. 9

Storage type heat exchanger

In this type of heat exchanger, shown in Fig. 9(*b*), the hot and cold fluids flow alternately through a solid material of high heat capacity. When hot fluid flows through the matrix then heat is transferred from the fluid to the matrix which stores it in the form of an increase in internal energy. This stored energy is then transferred to the cold fluid as it flows through the matrix. Thus the matrix is subjected to periodic heating and cooling.

Advantages

1. High efficiency.
2. High heat transfer coefficient.

3. Quick response to load.
4. More compact arrangement.
5. Provide more surface area for heat transfer per unit volume.

Disadvantages

1. Prone to leakage.
2. Perfect sealing is required.
3. Costlier as compared to other type of heat exchangers.

Q.4. (*a*) A refrigeration system is to be designed for a cooling capacity of 7.5 tons of refrigeration at saturation pressure corresponding to –20°C. It uses refrigerant R-22. Condenser pressure is to be corresponding to saturation temperature of 40°C. The refrigeration system uses a liquid-vapour regenerative heat exchanger. The refrigerant vapours coming out of the evaporator are superheated by 5°C and then pass through the regenerative heat exchanger to cool the liquid refrigerant coming from the condenser. The temperature of the refrigerant vapours at the exit of heat exchanger is 20°C. The liquid refrigerant is subcooled to 36°C at the exit of the condenser.

Make a sketch of the system and represent it on *p-h* and T-S diagrams.

The system uses two single-acting cylinders with bore to stroke ratio of 0.8, speed of compressor is 1420 rpm, clearance factor for compressors is 0.04 and polytropic index of compression is 1.1. Mechanical Efficiency of the compressor is 0.8.

Determine :

(*i*) temperature of liquid refrigerant at the exit of regenerative heat exchanger, assuming specific heat of the liquid to be 1.37 kJ/kg K.

(*ii*) mass flow rate of the refrigernat.

(*iii*) the temperature of the refrigerant vapours at the exit of compressor, assuming ideal gas behviour in the superheat region, specific heat of refrigerant vapours in the superheat region may be taken as 0.85 kJ/kg K.

(*iv*) Power input to the compressor.

(*v*) COP of the system.

(*vi*) Volumetric efficiency of the compressor, and

(*vii*) Bore and stroke of the compressor.

The following properties of R-22 are given :

Sat. temp. T_s(°C)	Sat. pressure p_s(bar)	Sp. Vol. of Sat. vapour (m³/kg)	Enthalpy (kJ/kg)	
			Sat. Liquid h_f	Sat. vapour h_g
–20	2.448	0.0928	177.4	397.5
40	15.335	0.0151	249.08	415.95

For superheated refrigerant vapour at 2.448 bar and saturation temperature of –20°C.

Temperature °C	Specific volume (m^3/kg)	Enthalpy kJ/kg
–15	0.0951	400.7
20	0.1107	423.9

Sol. Given : RE = 7.5 TR, i = 2, $\frac{D}{L}$ = 0.8, c = 0.04, n = 1.1, η_{mech} = 0.8, N = 1420 rpm

Refer to Fig. 10.

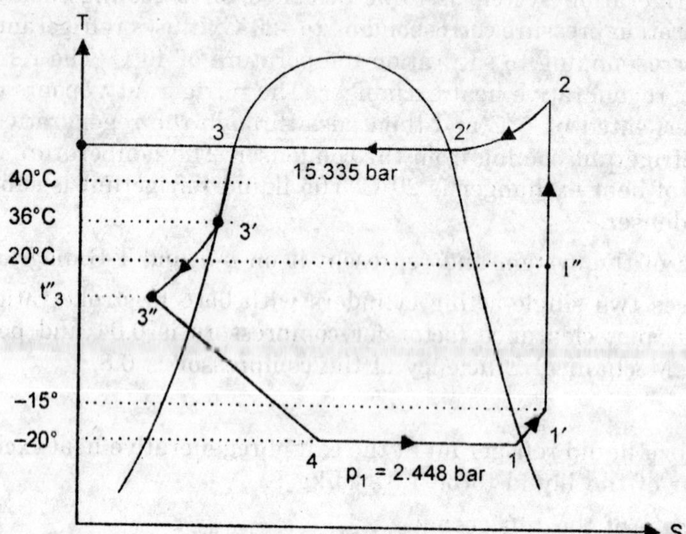

Fig. 10. T-s diagram.

1 – 1′ : Superheating
1′ – 1″ : Heat exchanger
1″ – 2 : Compressor
2′ – 3 : Condenser
3 – 3′ : Subcooling
3′ – 3″ : Heat exchanger
3″ – 4 : Throttling
4 – 1 : Evaporator

$$\eta_v = 1 + c - c\left(\frac{p_2}{p_1}\right)^{\frac{1}{n}} = 1 + 0.04 - 0.04\left(\frac{15.335}{2.448}\right)^{1/1.1}$$

$$= 0.8279$$

From given table, $v_1 = 0.0928 \text{ m}^3/\text{kg}$

$$\eta_v = \frac{\dot{m} v_1}{\frac{\pi}{4} D^2 L \frac{N}{60} i}$$

$$\dot{m} = 0.8279 \times \frac{\pi}{4} \times 1.25^3 D \times \frac{1420}{60} \times 2 \times \frac{1}{0.0928}$$

$$\dot{m}/D^3 = 414.57$$

(*i*) $$c_{pl} = 1.37 \text{ kJ/kg·K}, \; c_{pv} = 0.85 \text{ kJ/kg·K}$$

From heat balance for the heat exchanger, we have

$$c_{pl}(T_3' - T_3'') = c_{pv}(T_1'' - T_1'')$$

$$1.37\,(309 - T_3'') = 0.85(293 - 258)$$

$$T_8'' = 287.28 \text{ K}$$

$$RE = \dot{m}(h_1 - h_4)$$

$$h_1 = 397.5 \text{ kJ/kg}$$

$$h_4 = h_3'' = h_{f3} - C_{pl}(T_3 - T_3'') = 249.08 - 1.37(313 - 287.28)$$

$$= 213.84 \text{ kJ/kg}$$

$$7.5 \times 3.5 = \dot{m}(397.5 - 213.84)$$

$$\dot{m} = 0.1429 \text{ kg/s}$$

(*ii*) Power input, $P = \dot{m}(h_2 - h_1'')$

$$s_1' = s_2$$

$$s_1 + c_{pv} \ln \frac{T_1'}{T_1} = s_2' + c_{pv} \ln \frac{T_2}{T_2'}$$

$$0.9638 + 0.85 \ln \frac{258}{253} = 0.8822 + 0.85 \ln \frac{T_2}{313}$$

$$T_2 = 351.3 \text{ K}$$

$$P = \dot{m}\,c_{pv}(T_2 - T_1'')$$

$$= 0.1429 \times 0.85\,(351.3 - 293)$$

$$= 7.08 \text{ kW}$$

(*iii*) $$COP = \frac{RE}{P} = \frac{7.5 \times 3.5}{7.08} = 3.70$$

(*iv*) $$\frac{\dot{m}}{D^3} = 414.57$$

$$D^3 = \frac{0.1429}{414.57}, \; D = 0.070 \text{ m or } 70 \text{ mm}$$

$$L = 1.25 \times 0.07$$

$$= 0.087 \text{ m or } 87 \text{ mm}$$

Q.4. (*b*) (*i*) Discuss the measurement of wet bulb temperature by means of a wick-covered bulb of a thermometer. How is it related to dry-bulb temperature and specific humidity of moist air and on what other factors does it depend?

(*ii*) Define thermodynamic wet bulb temperature. How is it determined? How does it differ from measured wet-bulb temperature?

Sol.

(*i*) WBT is measured by a thermometer whose bulb is covered by a cotton wick saturated with water. As air stream flows past the wick, some water evaporates taking away the latent heat from the wick, thus decreasing its temperature. In turn, energy is transferred to the wick from the surrounding air. When equilibrium condition as reached, there is a balance between energy removed from water film of wick by evaporation and energy supplied to the wick by heat transfer from surrounding air. The temperature indicated by the thermometer at this instant is called the wet bulb temperature.

The WBT is always less than the DBT except when air is saturated. For saturated air, WBT = DBT.

The other factors affecting WBT are :

1. Dry bulb temperature.
2. Dew point temperature.
3. Barometric pressure.

(*ii*) **Thermodynamic WBT :** The thermodynamic WBT (or adiabatic saturation temperature) is the temperature at which air can be brought to saturation state adiabatically by the evaporation of water into the flowing air. It can be measured by adiabatic saturation process. The equipment (see Fig. 11) used for adiabatic saturation of air consists of an insulated chamber containing adequate quantity of water. The make-up water is sprayed from the top of chamber. The unsaturated air enters from one side of the chamber, passes over a long sheet of water. The water evaporates and the specific humidity of air increases. The water level is kept constant by make-up water spray. Both the air and water cool as evaporation takes place. This process continues until the energy transferred from the air to water is equal to the energy required to vaporise the water. When steady state conditions are reached the leaving air from the chamber is saturated with water vapour. The temperature of saturated air at exit from the chamber is known as thermodynamic WBT. The thermodynamic WBT is taken equal to the WBT for all practical purposes.

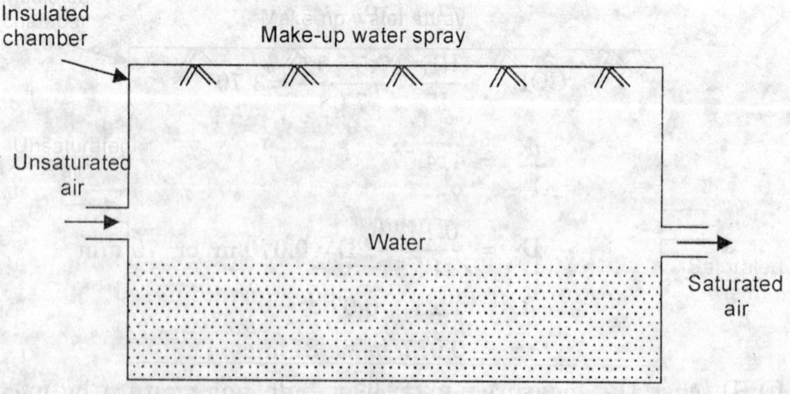

Fig. 11

Q.4. (*c*) How do you define effective temperature as an index of comfort? On what factors does it depend? What optimum inside design conditions are recommended for comfort for summer airconditioning including ventilation requirements?

Sol. Effective Temperature : The degree of warmth and cold felt by a human body depends mainly on the following three factors :

1. DBT 2. RH, and 3. Air velocity

In order to evaluate the combined effect of these factors, the term effective temperature is employed. It is defined as that index which corelates the combined effects of air temperature, relative humidity and air velocity on human body. The effective temperature depends on the following factors :

1. DBT, 2. RH, 3. Air velocity, and 4. Mean radiation temperature.

Optimum inside design conditions for comfort summer air-conditioning are :

Effective temperature = 21.7°C

DBT = 25 ± 1°C

RH = 50 ± 5%

Q.5. (*a*) A hydraulic lift of the type commonly used for greasing automobiles consists of a 280 mm diameter ram that slides in a 280.18 mm cylinder. The annular space between the ram and cylinder is filled with oil having a kinematic viscosity of 0.00042 m²/s and specific gravity of 0.86. If the rate of travel of the ram is 0.22 m/s, find the frictional resistance when 2 m of the ram is engaged in the cylinder.

Sol. Given : d_r = 280 mm, d_c = 280.18 mm, ν = 0.00042 m²/s, s = 0.86, u = 0.22 m/s, l_r = 2m

Specific gravity, $$s = \frac{\rho_{oil}}{\rho_{water}}$$

Density of oil, $$\rho_{oil} = 0.86 \times 100 = 860 \text{ kg/m}^3$$

$$\mu = \rho\nu = 860 \times 0.00042 = 0.3612 \text{ kg/(m·s)}$$

$$\tau = \mu\frac{dv}{dy}$$

$$dy = \frac{1}{2}(280.18 - 280) = 0.09 \text{ mm}, \; du = 0.22 \text{ m/s}$$

$$\tau = 0.3612 \times \frac{0.22}{0.09 \times 10^{-3}} = 882.93 \text{ N/m}^2$$

Frictional force, $$F = \tau A_s = 882.93 \times \pi \times 280 \times 10^{-3} \times 2$$
$$= 1553.3 \text{ N}$$

Q.5. (*b*) A solid, half-cylinder-shaped log of 0.48 m radius and 2.5 m long, floats in water with the flat face up.

(*i*) If the immersion depth of the lowest point is 0.3 m, what is the uniform specific weight of the log?

(*ii*) The log tillts about its axis (zero net applied force), by less than 22°. Is it in stable equilibrium? Justify your answer with a sketch and logic.

(*iii*) If the log tills by 18° (left side down; zero net applied force), what is the magnitude and sense of any moment that results

Sol. (*i*) **Given :** R = 0.48 m, L = 2.5 m, h = 0.3 m

Refer to Fig. 12.

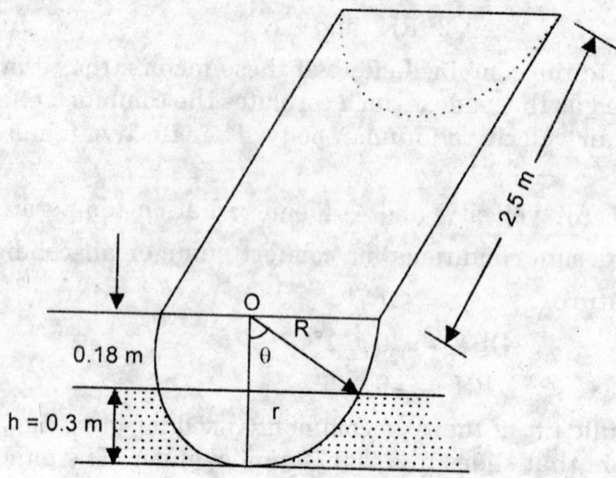

Fig. 12

$$r = \sqrt{R^2 - (R-h)^2} = \sqrt{(0.48)^2 - (0.48-0.3)^2} = 0.445 \text{ m}$$

$$\sin\theta = \frac{r}{R} = \frac{0.445}{0.48} = 0.927, \ \theta = 67.98^\circ \text{ or } 1.186 \text{ rad}$$

$$\cos\theta = \frac{0.18}{0.48} = 0.375$$

Area of circular segment, $A = R^2(\theta - \sin\theta\cos\theta)$

$= (0.48)^2 (1.186 - 0.927 \times 0.375) = 0.193 \text{ m}^2$

Weight of water displaced, $W = AL\gamma$

$= 0.193 \times 2.5 \times 9810 = 4733.3 \text{ N}$

$$\text{Specific weight of log} = \frac{W}{\text{Volume of log}} = \frac{W}{\frac{\pi R^2}{2} \times L}$$

$$= \frac{4733.3}{\frac{\pi(0.48)^2}{2} \times 2.5} = 5231.4 \text{ N/m}^3$$

(*ii*) Height of centre of buoyancy B above bottom of log (see Fig. 13)

$$= \frac{h}{2} = \frac{0.3}{2} = 0.15 \text{ m}$$

Distance of centre of gravity G of log from the top

$$= \frac{4R}{3\pi} = \frac{4 \times 0.48}{3\pi} = 0.2037 \text{ m}$$

$$\overline{BG} = (0.48 - 0.15) - 0.2037 = 0.1263 \text{ m}$$

$$\overline{BM} = \frac{I}{V}$$

$$I = 0.11\ R^4 + \frac{\pi R^2}{2}\left[(0.48 - 0.15) - 0.2037\right]^2$$

$$= 0.11 \times (0.48)^4 + \frac{\pi}{2}(0.48)^2\left[0.33 - 0.2037\right]^2$$

$$= 0.3111\ m^4$$

$$V = AL = 0.193 \times 2.5 = 0.4825\ m^3$$

$$\overline{BM} = \frac{0.3111}{0.4825} = 0.644\ m$$

$$\overline{GM} = \overline{BM} - \overline{BG} = 0.644 - 0.1263 = 0.5177\ m$$

Since M lies above G, the log is in stable equilibrium

$$\overline{GM} = 0.7625\ m$$

Restoring couple $= W \times \overline{GM} \tan\theta$

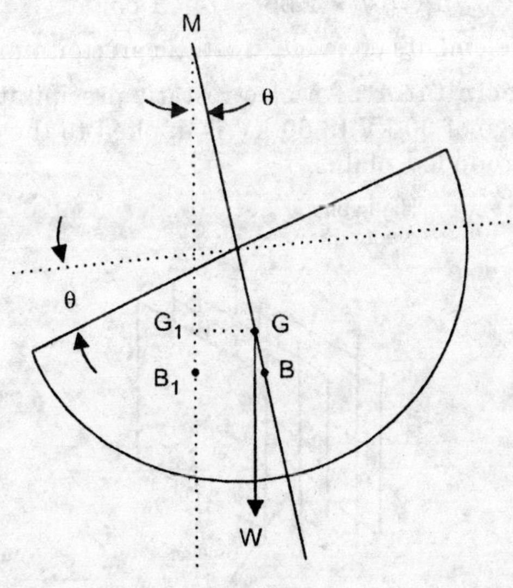

Fig. 13

Q.5. **(*c*)** Air enters into a constant area frictionless duct with M = 3, p = 7 bar and T = 288 K. It is desired to reduce the flow Mach number to 2 at the exit of the duct. Determine the amount of heat added and the corresponding change in pressure. For air, C_p = 1.003 kJ/kgK.

Take :

M	T_0/T_0^*
3	0.6434
2	0.7934

Sol. Given : $M_1 = 3$, $p_1 = 7$ bar, $T_1 = 288$ K, $M_2 = 2$, $C_p = 1.003$ kJ/kg·K

$$\frac{T_2}{T_1} = \left[\frac{1+kM_1^2}{1+km_2^2}\right] \cdot \left(\frac{M_2}{M_1}\right)^2$$

$$T_2 = 288\left[\frac{1+1.4\times 3^2}{1+1.4\times 2^2}\right]^2 \times \left(\frac{2}{3}\right)^2 = 543.5 \text{ K}$$

From the given table, we have

For $\quad M = 3, \ \frac{T_{01}}{T_1} = 0.6534, \ T_{01} = 288 \times 0.6534 = 188.18$ K

For $\quad M = 2, \ \frac{T_{02}}{T_2} = 0.7934, \ T_{02} = 543.5 \times 0.7934 = 431.21$ K

$$Q = C_p(T_{02} - T_{01}) = 1.003(431.21 - 188.18) = 243.76 \text{ kJ/kg}$$

$$\frac{p_2}{p_1} = \frac{1+kM_1^2}{1+kM_2^2} = \frac{1+1.4\times 3^2}{1+1.4\times 2^2} = 2.06$$

$$p_2 = 7 \times 2.06 = 14.42 \text{ bar}$$

Q.5. (*d*) Show the basic elements of an electrostatic precipitator and explain its operation.

Sol. Electrostatic Precipitator : An electrostatic precipitator is shown in Fig. 14. In this device a very high voltage of 30 kV to 60 kV is applied to the wires suspended in a gas-flow passage between two grounded plates.

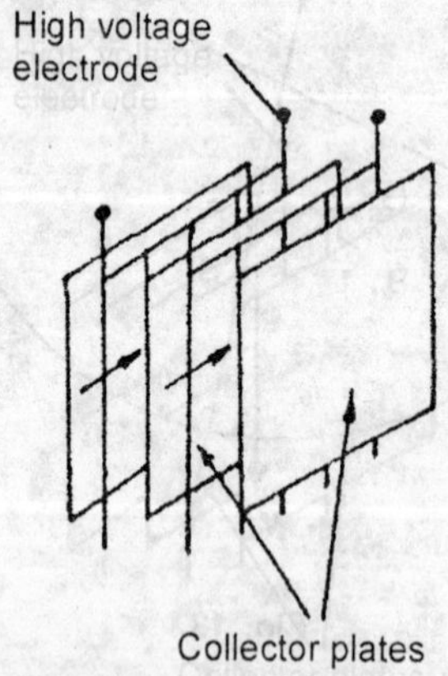

Fig. 14. Electrostatic precipitator.

The particles in the gas stream acquire a charge from the negatively charged wires and are then attracted to the ground plates. The grounded plates are periodically rapped by a steel plug which is raised and dropped by an electromagnet and dust is collected in the hoppers below.

– In this type of collector, *core* must be taken to see that *large quantity of unburnt gases do not enter the precipitator*. If such a mixture enters, power should be turned off, otherwise there could be *explosion because of constant sparking between wires and plates.*

- The collection efficiency is about 99 per cent.
- Electrostatic precipitators are suitable for power plants where *fly-ash content is high.* Fly-ash having high electrical resistivity *does not* separate in the electrostatic precipitator. This problem can be solved by injecting *sulphur trioxide* into the exhaust gas which *improves the conductivity of fly-ash*. This again poses a problem of discharging objectionable sulphur trioxide into the atmosphere; this needs a wet scrubber after the electrostatic precipitator.

Q.6. (*a*) A flow field is defined by $u = 2y$ and $v = xy$. Derive expressions for the acceleration components. Find the magnitude of the velocity and acceleration at the point (2, 3). Specify units in terms of L and T.

Sol. Given : $u = 2y,\ v = xy$

In a 2-dimensional flow, components of acceleration are :

$$a_x = u\frac{\partial u}{\partial x} + v\frac{\partial u}{\partial y}$$

$$= 2y \times 0 + xy \times 2 = 2xy$$

$$a_y = u\frac{\partial v}{\partial x} + v\frac{\partial v}{\partial y}$$

$$= 2y \times y + xy \times x = 2y^2 + x^2y$$

Acceleration, $\vec{a} = a_x\hat{i} + a_y\hat{j}$

$$= 2xy\hat{i} + (2y^2 + x^2y)\,\hat{j}$$

At point (2, 3) : $u = 6,\ v = 2 \times 3 = 6$

$$\vec{V} = u\hat{i} + v\hat{j} = 6\hat{i} + 6\hat{j}$$

$$V = \sqrt{6^2 + 6^2} = 6\sqrt{2} = 8.484 \text{ m/s}$$

$$\vec{a} = 2\times2\times3\hat{i} + (2\times9 + 4\times3)\hat{j} = 12\hat{i} + 30\hat{j}$$

$$a = \sqrt{12^2 + 30^2} = 32.31 \text{ m/s}^2$$

Units of velocity $= \text{m/s} = \text{LT}^{-1}$

Units of acceleration $= \text{m/s}^2 = \text{LT}^{-2}$

Q.6. (*b*) Derive an expression for small flow rates over a spillway, in the form of a function including dimensionless quantities. Use dimensional analysis with the following parameters :

Height of spillway = P

Head of spillway = H

Viscosity of liquid = μ

Density of liquid = ρ

Surface tension = σ

Acceleration due to gravity = g.

Sol. Given : $Q = f(P, H, \mu, \rho, \sigma, g)$

or $f(Q, P, H, \mu, \rho, \sigma, g) = 0$

Number of variables, $n = 7$

Number of fundamental quantities,

$$m = 3$$

Number of π-terms $= n - m = 7 - 3 = 4$

Repeating variables must contain geometric property, flow property and fluid property. We select H, g, μ as repeating variables.

π_1-term :

$$\pi_1 = H^{a_1} \cdot g^{b_1} \cdot \mu^{c_1} \cdot Q$$

$$M^\circ L^\circ T^\circ = [L]^{b_1} \cdot [LT^{-2}]^{b_1} \cdot [ML^{-1}T^{-1}]^{c_1} \cdot [L^3T^{-1}]$$

$$= [M]^{c_1} \cdot [L]^{a_1+b_1-c_1\times 3} \cdot [T]^{-2b_1-c_1-1}$$

$c_1 = 0$

$$-2b_1 - c_1 - 1 = 0, \; b_1 = -\frac{1}{2}$$

$$a_1 + b_1 - c_1 + 3 = 0, \; a_1 = \frac{1}{2} - 3 = -\frac{5}{2}$$

$$\pi_1 = H^{-5/2} \cdot g^{-1/2} \cdot \mu^0 \cdot Q = \frac{Q}{\sqrt{g\mu^5}}$$

π_2 – term :

$$\pi_2 = H^{a_2} \cdot g^{b_2} \cdot \mu^{c_2} \cdot p$$

$$M^\circ L^\circ T^\circ = [L]^{a_2} \cdot [LT^{-2}]^{b_2} \cdot [ML^{-1}T^{-1}]^{c_2} \cdot [L]^1$$

$$= [M]^{c_2} \cdot [L]^{a_2+b_2-c_2+1} \cdot [T]^{-2b_2-c_2}$$

$$c_2 = 0$$

$$-2b_2 - c_2 = 0, \quad b_2 = 0$$

$$a_2 + b_2 - c_2 + 1 = 0, \quad a_2 = -1$$

$$\pi_2 = \frac{P}{H}$$

π_3 = term

$$\pi_3 = H^{a_3} g^{b_3} \mu^{c_3} \rho$$

$$M^\circ L^\circ T^\circ = [L]^{a_3} \cdot [LT^{-2}]^{b_3} \cdot [ML^{-1}T^{-1}]^{c_3} \cdot [ML^{-3}]$$

$$= [M]^{c_3+1} \cdot [L]^{a_3+b_3-c_3-3} \cdot [T]^{-2b_3-c_3}$$

$$c_{3+1} = 0, \quad c_3 = -1$$

$$-2b_3 - c_3 = 0, \quad b_3 = \frac{1}{2}$$

$$a_3 + b_3 - c_3 - 3 = 0, \quad a_3 = -\frac{1}{2} - 1 + 3 = \frac{3}{2}$$

$$\pi_3 = H^{3/2} g^{1/2} \mu^{-1} \rho = \frac{\rho\sqrt{gH^3}}{\mu}$$

π_3 – term

$$\pi_4 = H^{a_4} \cdot g^{b_4} \mu^{c_4} \cdot \sigma$$

$$M°L°T° = [L]^{a_4} \cdot [LT^{-2}]^{b_4} \cdot [ML^{-1}T^{-1}]^{c_4} \cdot [ML^{-2}]$$

$$= [M]^{c_4+1} \cdot [L]^{a_4+b_4-c_4} \cdot [T]^{-2b_4-c_4-2}$$

$$c_{4+1} = 0, \quad c_4 = -1$$

$$-2b_4 - c_4 - 2 = 0, \quad -2b_4 = -1 + 2 = 1, \; b_4 = -\frac{1}{2}$$

$$a_4 + b_4 - c_4 = 0, \quad a_4 = \frac{1}{2} - 1 = -\frac{1}{2}$$

$$\pi_4 = H^{-1/2} \cdot g^{-1/2} \cdot \mu^{-1} \cdot \sigma = \frac{\sigma}{\mu\sqrt{gH}}$$

$$\pi_1 = \phi(\pi_2, \pi_3, \pi_4)$$

$$\frac{Q}{\sqrt{gH^5}} = \phi\left[\frac{P}{H}, \frac{\rho}{\mu}\sqrt{gH^3}, \frac{\sigma}{\mu\sqrt{gH}}\right]$$

Q.6. (*c*) Assume the velocity profile for turbulent flow in a circular pipe to be approximated by a parabola from the axis to a point very close to the wall where the local velocity is $u = 0.6u_m$, where u_m is the maximum velocity at the axis. The equation for this parabola is $u = u_m[1 - 0.4(r/r_0)^2]$. Find the value of kinetic energy correction factor.

Sol. Kinetic energy correction factor,

$$\alpha = \frac{1}{A}\int\left(\frac{u}{\bar{u}}\right)^3 dA$$

$$= \frac{1}{A\bar{u}^3}\int_0^{r_0} u_0^3\left[1 - 0.4\left(\frac{r}{r_0}\right)^2\right]^3 \times 2\pi r \cdot dr$$

$$= \frac{2\pi u_m^3}{A\bar{u}^3}\int_0^{r_0}\left[1 - \frac{0.064 r^6}{r_0^6} - 1.2\frac{r^2}{r_0^2} + \frac{0.48 r^4}{r_0^4}\right] r\, dr$$

$$= \frac{2\pi u_m^3}{A\bar{u}^3}\int_0^{r_0}\left[r - \frac{0.064}{r_0^6}\cdot r^7 - \frac{1.2}{r_0^2}\cdot r^3 + \frac{0.48}{r_0^4} r^5\right] dr$$

$$= \frac{2\pi u_3^3}{A\bar{u}^3}\left|\frac{r^2}{2} - \frac{0.008r^8}{r_0^6} - \frac{0.3r^4}{r_0^2} + \frac{0.08r^6}{r_0^4}\right|_0^{r_0}$$

$$= \frac{2\pi u_m^3 r_0^2}{A\bar{u}^3}\left[\frac{1}{2} - 0.008 - 0.3 + 0.08\right] = \frac{2\pi u_m^3}{A\bar{u}^3} \times 0.272$$

$$Q = \int_0^{r_0} 2\pi r \cdot u_m\left[1 - 0.4\left(\frac{r}{r_0}\right)^2\right] dr = 2\pi u_m \int_0^{r_0}\left[r - 0.4\frac{r^3}{r^2}\right] dr$$

$$= 2\pi u_m \left|\frac{r^2}{2} - \frac{0.1r^4}{r_0^2}\right|_0^{r_0} = 2\pi u_m r_0^2\left[\frac{1}{2} - 0.1\right] = 2\pi u_m r_0^2 \times 0.4$$

$$\bar{u} = \frac{Q}{\pi r_0^2} = \frac{2\pi u_m r_0^2 \times 0.4}{\pi r_0^2} = 0.8\, u_m$$

$$\alpha = \frac{2\pi u_m^3 r_0^2 \times 0.272}{\pi r_0^2 \times (0.3u_m^0)^3} = \frac{2 \times 0.272}{(0.8)^3} = 1.0625$$

Q.6. (*d*) With the help of simple schematic flow diagrams, explain the difference between once-through steam generators and the drum type steam generators.

Sol. Once-through Steam Generator : A line diagram of once-through steam generator is shown in Fig. 15(*a*). The feed water goes through the economiser, furnace walls and superheater. They require high purity feed water and are suitable for super-critical pressure operation.

Drum-type steam generator : A line diagram of drum-type boiler is shown in Fig. 15(*b*). The feed water is heated in superheater, sent to the steam drum and recirculated to the boiler. The superheated steam is passed through the economiser for further use.

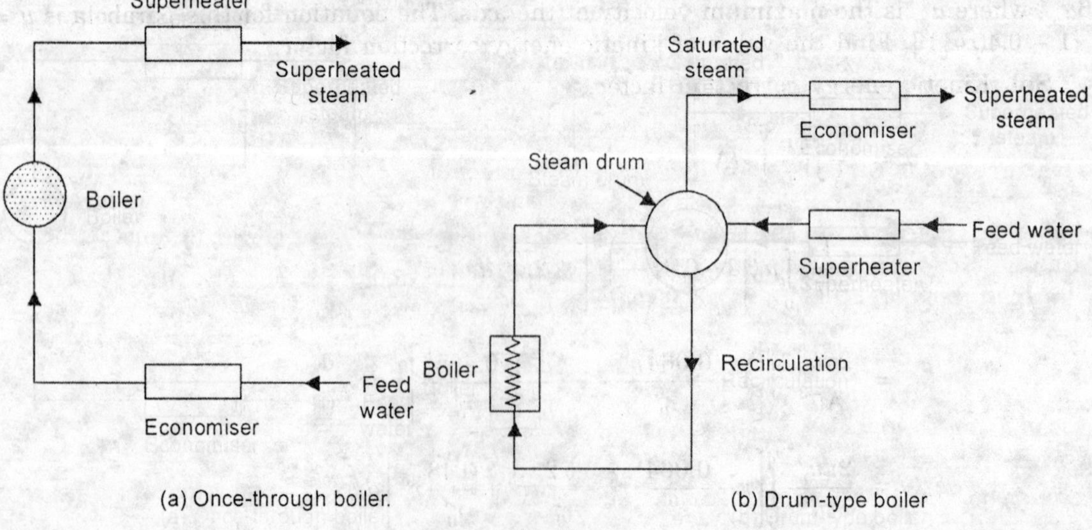

Fig. 15

Q.7. (*a*) For the velocity distribution prescribed by $u/u_0 = (y/\delta)^{1/7}$, show that ratio of displacement thickness (δ^*) to momentum thickness (θ) is 1.285.

Sol. Given :
$$\frac{u}{u_0} = \left(\frac{y}{\delta}\right)^{1/7}$$

Displacement thickness,
$$\delta^* = \int_0^\delta \left(1 - \frac{u}{u_0}\right) dy = \int_0^\delta \left[1 - \left(\frac{y}{\delta}\right)^{1/7}\right] dy$$
$$= \left| y - \frac{1}{\delta^{1/7}} \frac{y^{8/7}}{(8/7)} \right|_0^\delta = \delta - \frac{7}{8}\cdot\delta = \frac{\delta}{8}$$

Momentum thickness,
$$\theta = \int_0^\delta \frac{u}{u_0}\left(1 - \frac{u}{u_0}\right) dy = \int_0^\delta \left(\frac{y}{\delta}\right)^{1/7} \left[1 - \left(\frac{y}{\delta}\right)^{1/7}\right] dy$$
$$= \int_0^\delta \left[\left(\frac{y}{\delta}\right)^{1/7} - \left(\frac{y}{\delta}\right)^{2/7}\right] dy$$
$$= \left| \frac{7}{8} \times \frac{1}{\delta^{1/7}} \cdot y^{8/7} - \frac{7}{9} \cdot \frac{1}{\delta^{2/7}} \cdot y^{9/7} \right|_0^\delta$$
$$= \frac{7}{8}\cdot\delta - \frac{7}{9}\delta = \delta\left(\frac{7}{8} - \frac{7}{9}\right) = \frac{7}{72}\delta$$
$$\frac{\delta^*}{\theta} = \frac{\delta}{8} \times \frac{72}{7} \times \frac{1}{\delta} = \frac{9}{7} = 1.285$$

Q.7. (*b*) A runner of a centrifugal pump has an outer diameter of 25 cm and runs at 1500 rpm. it has 10 blades, each of 5 mm thickness. They are backward facing at 30° to the tangent and the breadth of the flow passages at outlet is 12.5 mm. Pressure gauges are fitted close to the pump casing on the suction and discharge pipes, both are 2.5 m above water level in supply sump. When discharge is 26 litres per second, the gauge readings are 4 m vacuum and 16.5 m of water respectively. Assume that there is no whirl at inlet and no whirl slip. If 50% of velocity head at outlet from the runner is recovered as static head in the volute, estimate:

(*i*) theoretical head

(*ii*) manometric efficiency

(*iii*) losses in the impeller, and

(*iv*) capacity of the motor to derive the pump, if mechanical efficiency is 0.9

Sol. Given : $D_2 = 0.25$ m, $N = 1500$ rpm, $n = 10$, $t = 5$ mm, $\beta_2 = 30°$, $B_2 = 12.5$ mm,
$$\frac{p_s}{\gamma} = -4\text{m},\ \frac{p_d}{\gamma} = 16.5 \text{ m},\ H_s = 2.5 \text{ m},\ \eta_{mech} = 0.9$$

Thickness of vane measured along the circle,
$$s = \frac{t}{\sin\beta_2} = \frac{5}{\sin 30°} = 10 \text{ mm}$$

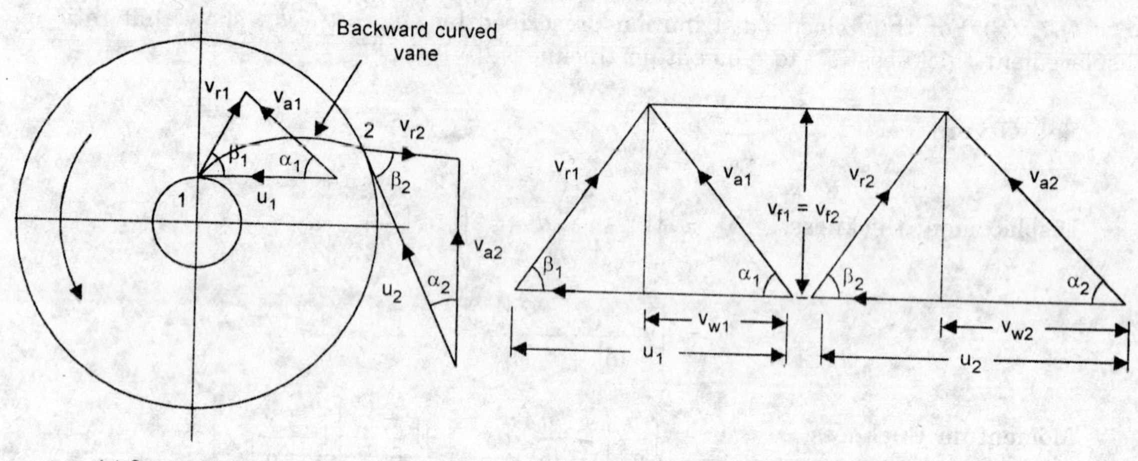

(a) Centrifugal impeller

(b) Velocity diagrams

Fig. 16

$$Q = (\pi D_2 - ns) B_2 v_{f2}$$

$$26 \times 10^{-3} = (\pi \times 0.25 - 10 \times 10 \times 10^{-3}) \times 12.5 \times 10^{-3} \times v_{f2}$$

$$v_{f2} = 3.03 \text{ m/s}$$

$$u_2 = \frac{\pi D_2 N}{60} = \frac{\pi \times 0.25 \times 1500}{60} = 19.635 \text{ m/s}$$

$$v_{w2} = u_2 - \frac{v_{f2}}{\tan \beta_2} = 19.635 - \frac{3.03}{\tan 30°} = 14.38 \text{ m/s}$$

Applying Bernoulli's equation between the water surface in the sump and inlet of pump where pressure gauge is fitted, we have

$$0 = H_s + \frac{p_s}{\gamma} + \frac{v_s^2}{2g}$$

$$0 = 2.5 - 4 + \frac{v_s^2}{2g}$$

$$\frac{v_s^2}{2g} = 1.5 \text{ m}$$

Now applying Bernoulli's equation between the impeller outlet where delivery pressure gauge is fitted and the outlet of delivery pipe.

$$\frac{p_d}{\gamma} + \frac{v_d^2}{2g} + 0.5 \times \frac{v_{a_2}^2}{2g} = H_d + \frac{v_d^2}{2g}$$

$$H_d = \frac{p_d}{\gamma} + 0.5\frac{v_{a_2}^2}{2g}$$

Now
$$v_{a_2}^2 = v_{f2}^2 + v_{w2}^2 = (3.03)^2 + (14.38)^2 = 215.96$$

$$v_{a2} = 14.65 \text{ m/s}$$

$$H_d = 16.5 + \frac{0.5 \times 215.96}{2 \times 9.81} = 22 \text{ m}$$

$$H = H_s + H_d + \frac{v_s^2}{2g} = 2.5 + 22 + 1.5 = 26 \text{ m}$$

(*i*) Theoretical head = 26 m

(*ii*) Manometric efficiency, $\eta_{mano} = \frac{gH}{u_2 v_{w2}}$

$$= \frac{9.81 \times 26}{19.635 \times 14.38} = 0.9033 \text{ or } 90.33\%$$

(*iii*) Losses in the impeller $= \frac{u_2 v_{w2}}{g} - H$

$$= \frac{19.635 \times 14.38}{9.81} - 26 = 2.78 \text{ m}$$

(*iv*) Capacity of motor to drive the pump

$$P = \frac{\gamma Q_H}{\eta_{mech}} = \frac{9810 \times 26 \times 10^{-3} \times 26}{0.9 \times 10^3} = 7.37 \text{ kW}$$

Q.7. (*c*) Show the velocity diagrams of a 50% reaction turbine operating with maximum blade efficiency and highlight its salient features.

Sol.

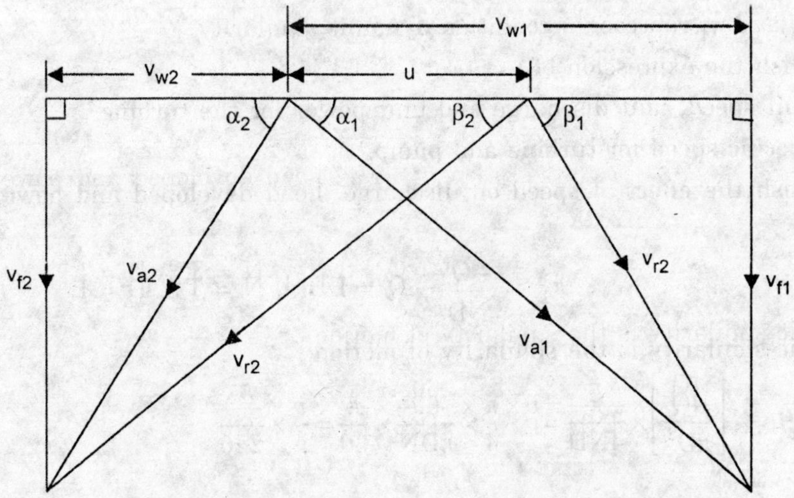

Fig. 16. Velocity diagrams for reaction turbine.

Degree of reaction, $R_d = \frac{(\Delta h)_{\text{moving blade}}}{(\Delta h)_{\text{moving blade}} + (\Delta h)_{\text{fixed blade}}}$

For R_d = 50% : $\alpha_1 = \beta_1$ and $\alpha_2 = \beta_1$ (see Fig. 17)

Blade efficiency, $\eta_b = \dfrac{\text{Work done}}{\text{Energy supplied to a stage}} = \dfrac{W}{\Delta h} = \dfrac{2\rho(\cos\alpha_1 - \rho)}{1 + 2\rho\cos\alpha_1 - \rho^2}$

where $\rho = \dfrac{u}{v_{a1}}$ = speed ratio

For maximum blade efficiency, $\rho = \cos\alpha_1$

$$(\eta_b)_{max} = \frac{2\cos^2\alpha_1}{1+\cos^2\alpha_1}$$

Q.8. (***a***) Application of dimensional analysis technique to incompressible flow rotodynamic machines gives following dimensionless ratios :

(*i*) $\pi_1 = Q/ND^3$

(*ii*) $\pi_2 = \dfrac{gH}{N^2D^2}$

(*iii*) $\pi_3 = \dfrac{\rho ND^2}{\mu}$

(*iv*) $\pi_4 = \dfrac{P}{\rho N^3D^5}$.

Using them only :

(*a*) Show that

(*i*) π_1 and π_2 represents condition of kinematic similarity.

(*ii*) π_3 and π_4 represents condition dynamic similarity.

(*b*) Establish the expression of :

(*i*) unit speed, unit discharge and unit power for the turbine.

(*ii*) specific speed for turbine and pump.

(*c*) Establish the effect of speed on discharge, head developed and power required by pump.

Sol. (*a*) (*i*) $\pi_1 = \dfrac{Q}{ND^3}$, $Q = L^3T^{-1}$, $N = T^{-1}$, $D = L^1$

Kinematic similarity is the similarity of motion.

$$\frac{Q}{ND^3} = \left[\frac{4Q}{\pi D^2}\right] \times \frac{\pi}{4ND} = v \times \frac{\pi}{4} \times \frac{60}{\pi DN} \times \frac{\pi}{60} = \frac{v}{u} \times \frac{\pi^2}{240}$$

$$\therefore \quad \frac{Q}{ND^3} = \frac{v}{u} \times \text{const.}$$

or $\left(\dfrac{v}{u}\right)_m = \left(\dfrac{v}{u}\right)_p$

∴ It represents condition of kinematic similarity.

$$\pi_2 = \frac{gH}{N^2D^2}$$

or $$\pi_2 = \frac{ND}{\sqrt{gH}} = \left(\frac{\pi DN}{60} \times \frac{60}{\pi}\right) \times \frac{\sqrt{2}}{\sqrt{2gH}}$$

$$= \frac{u}{v} \times \frac{60\sqrt{2}}{\pi} = \frac{u}{v} \times \text{const}$$

Hence π_2 represents π kinematic similarity condition of

(*ii*) $$\pi_3 = \frac{PND^2}{\mu} = \frac{\rho}{\mu} \cdot \left(\frac{\pi DN}{60}\right) \times \left(\frac{60}{\pi}\right) \times D = \frac{\rho v D}{\mu} \times \text{const}$$

$$= Re \times \text{cosnt.}$$

$\therefore$ $\pi_3 = \frac{\text{Inertia force}}{\text{Viscous force}}$. It represents condition of dynamic similarity.

$\pi_4 = \frac{\rho}{\rho N^3 D^5}$ represents a condition of dynamic similarity.

(*b*) (*i*) **Unit speed for a turbine**

Let D = diameter of runner

N = rpm of runner

Peripheral speed of runner,

$$u = \frac{\pi DN}{60}$$

$$u \ \alpha \ \sqrt{2gH}$$

$\therefore$ $$N \ \alpha \ u \alpha \sqrt{H}$$

or $$N = k\sqrt{H}$$

By definition, unit speed,

$$N_1 = k_1\sqrt{1} = k_1 = \frac{N}{\sqrt{H}}$$

Unit discharge

$$Q \ \alpha \ Av_f \quad \alpha \ A\sqrt{2gH}$$

$\therefore$ $$Q \ \alpha \ \sqrt{H} = k_2\sqrt{H}$$

Unit discharge, $$Q_1 = k_2\sqrt{1} = k_2 = \frac{Q}{\sqrt{H}}$$

Unit power

$$P = \frac{\gamma \, QH}{10^3}$$

$$P \ \alpha \ Q \cdot H$$

$$Q \alpha \sqrt{H}$$

α $$P \alpha H^{3/2}$$

or $$P = k_3 H^{3/2}$$

Unit power, $$P_1 = k_3(1)^{3/2} = k_3 = \frac{P}{H^{3/2}}$$

(*ii*) **Specific speed for turbine**

$$u = \pi DN$$

and $$u \propto \sqrt{H}$$

$\therefore$ $$D \propto \frac{\sqrt{H}}{N}$$

$$P \propto QH$$

$$Q \propto D^2\sqrt{H}$$

$\therefore$ $$P \propto D^2H^{3/2} \quad \alpha \quad \frac{H}{N^2} \cdot H^{3/2} \; \alpha \; \frac{H^{5/2}}{N^2}$$

or $$N \propto \frac{H^{5/4}}{\sqrt{P}}$$

Specific speed is the speed of a geometrically similar turbine working under unit head and developing unit power.

or $$N = N_s \frac{H^{5/4}}{\sqrt{P}},$$

where $$N_s = \frac{N\sqrt{P}}{H^{5/4}}$$

when $$P = 1, \; H = 1, N_s = N$$

Specific speed of a pump:

$$u = \pi DN$$

$$u \propto \sqrt{H}$$

$$D \propto \frac{\sqrt{H}}{N}$$

$$Q \propto D^2\sqrt{H} \propto \frac{H}{N^2} \cdot \sqrt{H} \propto \frac{H^{3/2}}{N^2}$$

$$N = \frac{H^{3/4}}{\sqrt{Q}} = N_s \frac{H^{3/4}}{\sqrt{Q}}$$

$$N_s = \frac{N\sqrt{Q}}{H^{3/4}}$$

When $$Q = 1 \text{ and } H = 1, \; N_s = N$$

Q.7. (*c*)

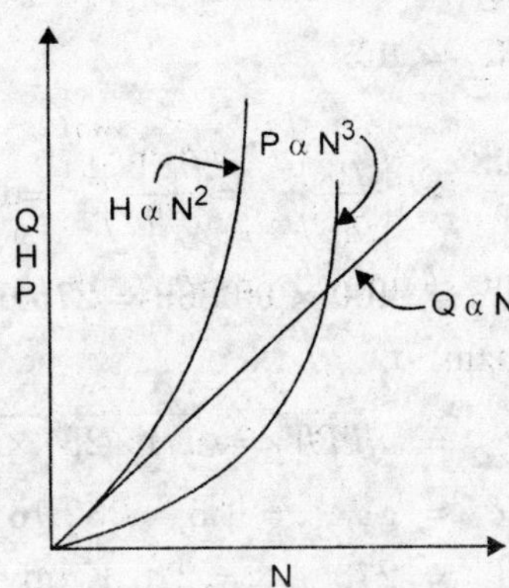

Fig. 16. Q, H, P *v's* N curves for contrifugal pump.

Q.8. (*b*) A constant area circular duct of 0.001 m^2 cross-sectional area is connected to the convergent divergent nozzle exit. The air enters the nozzle from a tank at a pressure of 7 bar (*abs*) and temp. of 127°C. The pressure at the nozzle exit is 1.9 bar. If the temperature of the air is 3°C at the end of the duct, and the duct length is 17.5 diameter of the duct, find the friction co-efficient of the duct. Consider flow is adiabatic through a duct and isentropic in the nozzle.

Use :

M	$4\bar{f}$ L m/D
3	.522
1.5	.136

Sol.

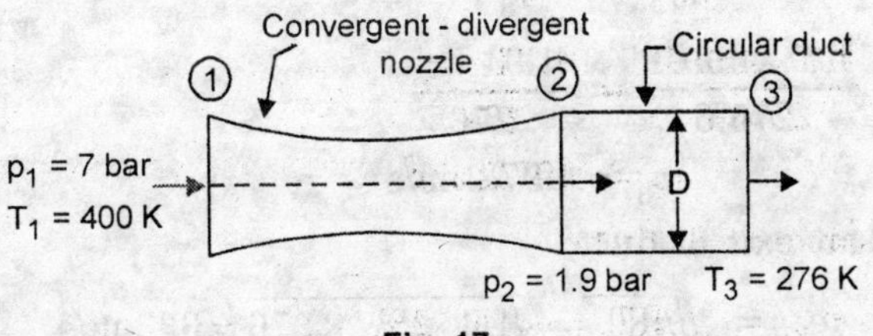

Fig. 17

Sonic velocity at inlet of nozzle

$$c_{s1} = \sqrt{kRT_1} = \sqrt{1.4 \times 287 \times 400} = 400.9 \text{ m/s}$$

$$\frac{p_1}{p_2} = \left[1 + \left(\frac{k-1}{2}\right)M_2^2\right]^{\frac{k}{k-1}}$$

$$\frac{7}{1.9} = \left[1 + \left(\frac{1.4-1}{2}\right)M_2^2\right]^{\frac{1.4}{1.4-1}}$$

$$\frac{7}{1.9} = \left[1 + 0.2M_2^2\right]^{3.5}$$

$$1 + 0.2M_2^2 = 1.45148$$

$$M_2 = 1.5$$

$$\frac{T_2}{T_1} = \left(\frac{p_2}{p_1}\right)^{\frac{k-1}{k}} = \left(\frac{1.9}{7}\right)^{\frac{0.4}{1.4}} = 0.6889$$

$$T_2 = 400 \times 0.6889 = 275.6 \text{ K}$$

Sonic velocity of exit of nozzle,

$$c_{s2} = \sqrt{kRT_2} = \sqrt{1.4 \times 287 \times 275.6} = 332.76 \text{ m/s}$$

$$c_2 = M_2 c_{s2} = 1.5 \times 332.76 = 499.1 \text{ m/s}$$

$$T_3 = 273 + 3 = 276 \text{ K (given)}$$

For adiabatic flow in duct,

$$\frac{p_3}{p_2} = \left(\frac{T_3}{T_2}\right)^{\frac{\gamma}{\gamma-1}} = \left(\frac{276}{275.6}\right)^{\frac{1.4}{0.4}} = 1.005$$

$$p_3 = 1.9 \times 1.005 = 1.91 \text{ bar}$$

For continuity of flow in duct of constant area of cross-section, mass velocity remains constant (Fanno flow is adiabatic flow in a duct with friction).

$$\rho_2 c_2 = \rho_3 c_3$$

or
$$\frac{p_2}{RT_2} \times c_2 = \frac{p_3}{RT_3} \times c_3$$

or
$$\frac{p_2 c_2}{T_2} = \frac{p_3 c_3}{T_3}$$

$$\frac{1.9 \times 499.1}{275.6} = \frac{1.91 \times c_3}{276}$$

$$c_3 = 497.2 \text{ m/s}$$

Velocity of sound at exit of duct,

$$c_{s3} = \sqrt{kRT_3} = \sqrt{1.4 \times 287 \times 276} = 333 \text{ m/s}$$

$$M_3 = \frac{c_3}{c_{s3}} = \frac{497.2}{333} = 1.493 \simeq 1.5$$

$$m = \rho_3 A_3 c_3 = \frac{p_3}{RT_3} \times A_3 c_3 = \frac{1.91 \times 10^5}{287 \times 276} \times 0.001 \times 497.2 = 1.199 \text{ kg/s}$$

For $M_3 = 1.5,$

$$\frac{4\bar{f}\, Lm}{D} = 0.136$$

$$\therefore \quad 4\bar{f} \times 17.5 \times 1.199 = 0.136$$

$$\bar{f} = 0.00162$$

Q.8. (*c*) Explain supersaturated expansion in case of flow through nozzle and discuss briefly the factors causing it. Represent the phenomenon on h-s diagram indicating superheated zone. State the effects of supersaturation.

Sol. Supersaturated flow through a nozzle : The isentropic expansion of superheated steam from supply pressure p_1 to back pressure p_b can be represented on the h-s diagram by the line AE, as shown in Fig. 18. During expansion, change of phase starts at point B where the pressure line p_2 meets the saturationj line. But in nozzles, under certain conditions, this phenomenon of condensation does not occur at point B as the time available is very short due to the very high velocity of steam through the nozzle. The equilibrium between the vapour phase and liquid is therefore delayed and the vapours continue to expand in dry state even beyond point B upto point B_1. The pressure at B_1 can be found by extending the superheated constant pressure line p_3 upto B_1. BB_1 remains dry and condensation is suppressed.

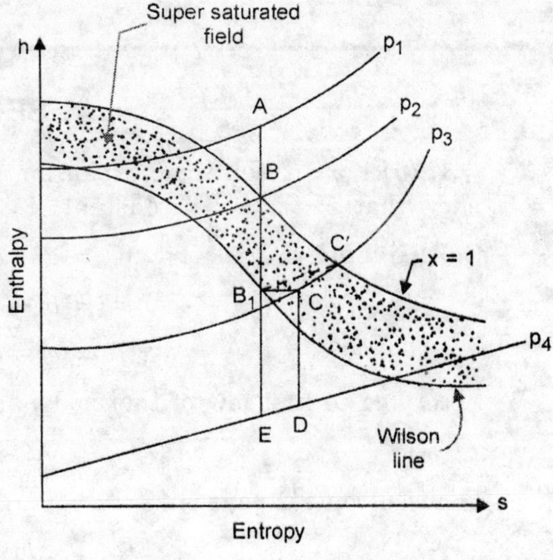

Fig. 18

The vapours between pressures p_2 and p_3 are said to be supersaturated or supercooled. Such a flow in nozzles is called supersaturated or metastable flow of steam. A limit to the supersaturated state was observed by Wilson and the line drawn on the Mollier chart (h-s diagram) through the observed points is known as Wilson line. For all practical purposes this line has become the saturation line for nozzle flow.

The flow is also called supercooled flow because at any pressure between p_2 and p_3 the temperature of the vapour is always lower than the saturation temperature corresponding to that pressure. The difference in this temperature is called the degree of undercooling.

When the expansion reaches at point B_1 on the Wilson line the condensation occurs at constant enthalpy, as shown by horizontal line B_1C. Further isentropic expansion to the exit pressure is represented by line CD. The ratio of saturation pressures corresponding to the temperatures at B and B_1 is called the degree of supersaturation.

Effects of Supersaturation

1. Entropy increases at constant enthalpy and specific volume decreases.
2. Heat drop decreases from AE to AD. This decreases exit velocity of steam.
3. Due to supercooling the density of steam increases which increases the mass of steam discharged.

MECHANICAL ENGINEERING
PAPER I—2012

(Conventional)

SECTION-A

Q.1. **(*a*)** For an isentropic expansion of a gas with $c_p = a + k\text{T}$, $c_v = b + k\text{T}$ and $c_p - c_v = \text{R}$, show that $\text{T}^b\, v^{a-b}\, e^{k\text{T}} = \text{constant}$. (5)

Sol. Given :

$$c_p = a + k\text{T}$$

$$c_v = b + k\text{T}$$

$$\text{R} = c_p - c_v = (a + k\text{T}) - (b + k\text{T}) = a - b$$

According to first law of thermodynamics, for a unit mass of gas,

$$dq = dw + du$$

For isentropic expansion,

$$dq = 0$$

$$\therefore \quad q = p \cdot dv + c_v \cdot d\text{T}$$

or

$$\frac{q}{\text{T}} = p\frac{dv}{\text{T}} + c_v \cdot \frac{d\text{T}}{\text{T}}$$

$$= \text{R} \cdot \frac{dv}{v} + c_v \cdot \frac{d\text{T}}{\text{T}}$$

$$= (a-b)\,\frac{dv}{v} + (b + k\text{T}) \cdot \frac{d\text{T}}{\text{T}}$$

Integrating, we get

$$\int \frac{q}{\text{T}} = \int (a-b)\frac{dv}{v} + \int (b + k\text{T})\frac{d\text{T}}{\text{T}}$$

or $\quad (a - b) \ln v + b \ln \text{T} + k\text{T} = \text{const.}$

$$\ln v^{(a-b)} + \ln \text{T}^b + \ln e^{k\text{T}} = \text{const.}$$

$$\therefore \quad \text{T}^b \cdot v^{a-b} \cdot e^{k\text{T}} = \text{const.}$$

Hence proved

(*b*) A small flexible bag contains 0.1 kg ammonia at –10°C and 3 bar. The bag material is such that the pressure inside varies linearly with volume. The bag is left in open space where the incident solar radiation is 75 W. The heat energy lost to the ground and surrounding air from bag is at the rate of 25 W. After a while, it is found that the bag is heated to 30°C at which time the pressure of ammonia is 10 bar. Estimate (*i*) the amount of heat energy infiltrated into the bag and (*ii*) the elapsed time.

Properties of ammonia :

(*a*) Compressed liquid ammonia at –10°C, 3 bar : specific volume of saturated liquid, V_f = 0.001002 m³/kg, specific internal energy of saturated liquid, U_f = 134 kJ/kg.

(*b*) Superheated ammonia vapour at 30°C, 10 bar : specific volume = 0.1321 m³/kg, specific internal energy = 1347 kJ/kg. (10)

Sol. (*i*) Amount of heat energy infiltrated into the bag

$$= 75 - 25 = 50 \text{ W}$$

(*ii*) According to first law of thermodynamics,

$$\delta Q = dU + \delta W$$

Here $\delta w = 0$

$\therefore$ $\delta Q = dU$

$\therefore$ $dQ = U_2 - U_1$

$$= m(u_2 - u_1)$$

$$= 0.1(1347 - 134)$$

$$= 121.3 \text{ kJ}$$

Elapsed time, $\tau = \dfrac{121.3 \times 10^3}{50 \times 60} = 40.4$ min

(*c*) Steam enters a 15 cm diameter horizontal pipe as saturated vapour at 5 bar with a velocity of 10 m/s and exit at 4.5 bar and a quality of 0.95. Heat is transferred to surroundings at 300 K from the pipe surface which is at an average temperature of 400 K. Under the steady state operating conditions, determine

(*i*) the exit velocity,

(*ii*) the rate of heat transfer from pipe surface in kW.

(*iii*) the rate of entropy production in kW/K, for the control volume comprising of only pipe and its contents and

(*iv*) the rate of entropy production for the enlarged control volume that includes pipe, its contents and the immediate surroundings. (10)

Properties of steam at saturation condition :

p Bar	t_{sat} **°C**	**Sp. volume, m³/kg**		**Sp. enthalpy, kJ/kg**		**Sp. entropy, kJ/kg.K**	
		liquid (v_f)	**vapour (v_g)**	**liquid (h_f)**	**vapour (h_g)**	**liquid (s_f)**	**vapour (s_g)**
4.5	147.93	0.001088	0.4140	623.5	2744	1.8207	6.8565
5.0	151.86	0.001093	0.3749	640.0	2749	1.8607	6.8213

Sol. SFEE gives,

$$h_1 + \frac{c_1^2}{2} + q = h_2 + \frac{c_2^2}{2} + w$$

$$h_1 = 2749 \text{ kJ/kg}$$

$$h_2 = h_{f2} + x_2 h_{g2}$$
$$= 623.5 + 0.95 \times 2744$$
$$= 3230.3 \text{ kJ/kg}$$

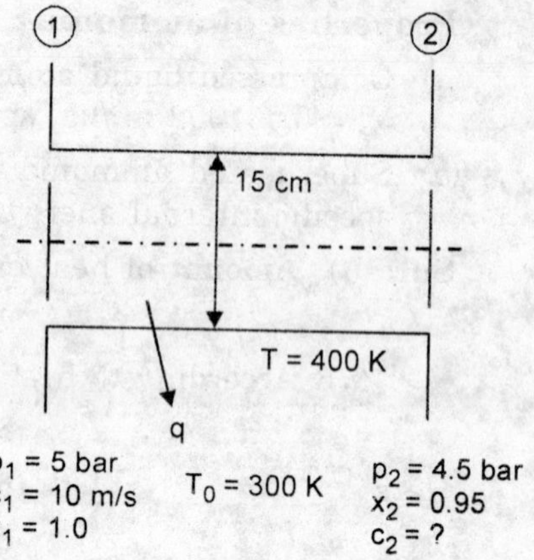

Fig. 1: 15 cm; T = 400 K; q; $p_1 = 5$ bar, $c_1 = 10$ m/s, $x_1 = 1.0$; $T_0 = 300$ K; $p_2 = 4.5$ bar, $x_2 = 0.95$, $c_2 = ?$

Fig. 1

(*i*) Mass flow rate :

Refer to Fig. 1

$$\dot{m}_1 = \dot{m}_2$$

$$A_1 c_1 \rho_1 = A_2 c_2 \rho_2$$

Here $\quad A_1 = A_2$

$$\rho_1 = \frac{1}{v_{g1}} = \frac{1}{0.3749} \text{ kg/m}^3$$

$$v_2 = v_{f2} + x_2 v_{g2} = 0.001088 + 0.95 \times 0.4140 = 0.39439 \text{ m}^3\text{/kg}$$

$$\rho_2 = \frac{1}{v_2}$$

$$\frac{10}{0.3749} = \frac{c_2}{0.39439}$$

$$c_2 = 10.52 \text{ m/s}$$

(*ii*) SFEE gives,

$$h_1 + \frac{c_1^2}{2} + q = h_2 + \frac{c_2^2}{2} + w$$

$$2749 + \frac{10^2}{2} \times 10^{-3} + q = 3230.3 + \frac{(10.52)^2}{2} \times 10^{-3} + 0$$

$$q = 481.3 \text{ kJ/kg}$$

$$\dot{m} = \frac{\pi}{4}(0.15)^2 \times 10 \times \frac{1}{0.3749} = 0.4714 \text{ kg/s}$$

$$\dot{Q} = \dot{m}q = 0.4714 \times 481.3 = 226.87 \text{ kw}$$

(*iii*)
$$\dot{S}_{gen} = \frac{\dot{Q}}{T} = \frac{226.87}{400} = 0.57 \text{ kW/K}$$

(*iv*)
$$\dot{S}_{gen} = \frac{\dot{Q}}{T_0} = \frac{226.87}{300} = 0.756 \text{ kW/K}$$

(*d*) (*i*) A certain gas obeys the equation of state, $p(v - a) = RT$. Show that the ratio of its volume expansivity, β with that of an ideal gas, β_{ideal} and the ratio of its isothermal compressibility, K_T with that of an ideal gas, $K_{T,\ ideal}$ are given by

$$\frac{\beta}{\beta_{ideal}} = \frac{RT}{RT + ap} \quad \text{and}$$

$$\frac{K_T}{K_{T,\,ideal}} = 1-\left(\frac{a}{v}\right)$$

(*ii*) For a substance with volume expansivity, β > 0, show that at every point of a single phase region (vapour region) on a Mollier diagram, the slope of constant pressure line is greater than the slope of constant temperature line but less than that of constant volume line. **(15)**

Sol. (*i*) Volume expansivity,

$$\beta = \frac{1}{v}\left(\frac{\partial v}{\partial T}\right)_p$$

For an ideal gas,

$$pv = RT$$

or

$$v = \frac{RT}{p}$$

$$\left(\frac{\partial v}{\partial T}\right)_p = \frac{R}{p}$$

$$\beta_{ideal} = \frac{1}{v}\times\frac{R}{p} = \frac{R}{pv} = \frac{R}{RT} = \frac{1}{T} \quad (1)$$

For the real gas,

$$p(v - a) = RT$$

$$v = \frac{RT}{p} + a = \frac{RT + ab}{p}$$

$$\left(\frac{\partial v}{\partial T}\right)_p = \frac{R}{p}$$

$$\beta = \left(\frac{p}{RT + ap}\right)\times\frac{R}{p} = \frac{R}{RT + ap} \quad (2)$$

$$\frac{\beta}{\beta_{ideal}} = \frac{R}{RT + ap}\times T = \frac{RT}{RT + ap}$$

Hence proved.

Isothermal compressibility,

$$k_T = -\frac{1}{v}\left(\frac{\partial v}{\partial p}\right)_T$$

For an ideal gas,

$$pv = RT$$

$$v = \frac{RT}{p}$$

$$\left(\frac{\partial v}{\partial p}\right)_T = RT\times\frac{-1}{p^2}$$

$$(k_T)_{ideal} = -\frac{p}{RT}\times\left(\frac{-RT}{p^2}\right) = \frac{1}{p}$$

For the real gas, $p(v - a) = \text{RT}$

$$v = \frac{\text{RT}}{p} + a = \frac{\text{RT} + ap}{p}$$

$$\left(\frac{\partial v}{\partial p}\right)_{\text{T}} = \frac{-\text{RT}}{p^2}$$

$$k_{\text{T}} = -\frac{p}{\text{RT} + ap} \times \left(\frac{-\text{RT}}{p^2}\right) = \frac{\text{RT}}{p(\text{RT} + ap)}$$

$$\frac{k_{\text{T}}}{(k_{\text{T}})_{\text{ideal}}} = \frac{\text{RT}}{p(\text{RT} + ap)} \times p = \frac{\text{RT}}{\text{RT} + ap} = \frac{p(v - a)}{p(v - a) + ap}$$

$$= \frac{p(v - a)}{pv} = \frac{v - a}{v} = 1 - \left(\frac{a}{v}\right)$$

Hence proved.

(*ii*)
$$\beta = \frac{1}{v}\left(\frac{\partial v}{\partial \text{T}}\right)_p > 0 \text{ (given)}$$

Now
$$\text{T}\cdot ds = du + p\cdot dv$$
$$= c_v \cdot d\text{T} + p\cdot dv$$

$\therefore$
$$\left(\frac{\partial \text{T}}{\partial s}\right)_v = \frac{\text{T}}{c_v}$$

Also
$$\text{T}\cdot ds = dh - v\cdot dp$$
$$= c_p \cdot d\text{T} - v\cdot dp$$

$$\left(\frac{\partial \text{T}}{\partial s}\right)_p = \frac{\text{T}}{c_p}$$

$$c_p > c_v$$

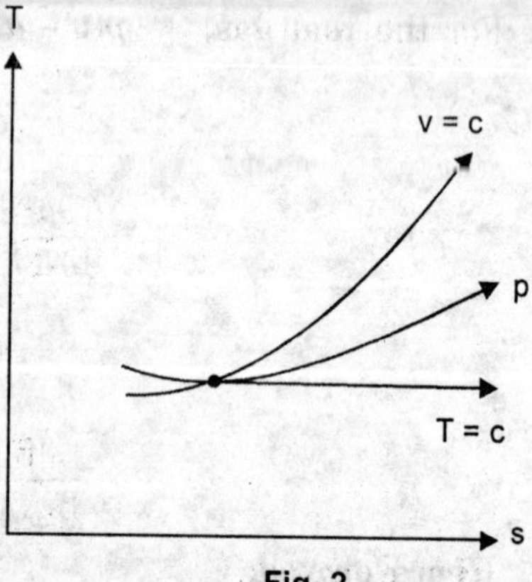

$\therefore$
$$\frac{\text{T}}{c_v} > \frac{\text{T}}{c_p}$$

$\therefore$
$$\left(\frac{\partial \text{T}}{\partial s}\right)_v > \left(\frac{\partial \text{T}}{\partial s}\right)_p$$

or
$$\left(\frac{\partial \text{T}}{\partial s}\right)_p < \left(\frac{\partial \text{T}}{\partial s}\right)_v$$

Fig. 2

Slope of constant pressure line is less than that of constant volume line.

From the Fig. it is obvious that slope of constant temperature line is zero.

$\therefore$ Slope of constant pressure line is more than that of constant temperature line.

Hence proved.

SECTION-B

Q.2. (*a*) Explain with neat sketch, the working principle of a hybrid rocket engine. What are advantages of this engine. (10)

Sol. In the hybrid rocket engine, a combination of solid and liquid propellants are used. The oxidizer is carried in liquid form and the fuel in solid form. A schematic diagram of hybrid rocket engine is shown in Fig. 3. The oxidizer is stored in a propellent tank and the solid fuel is cast inside the combustion chamber. During engine operation, the oxidizer is sprayed on the solid fuel. When ignited, the combustion proceeds. Due to the heat generated by the combustion process in the solid fuel core the solid fuel produces further fuel vapour. The gases generated by combustion are expanded in the nozzle producing the required thrust.

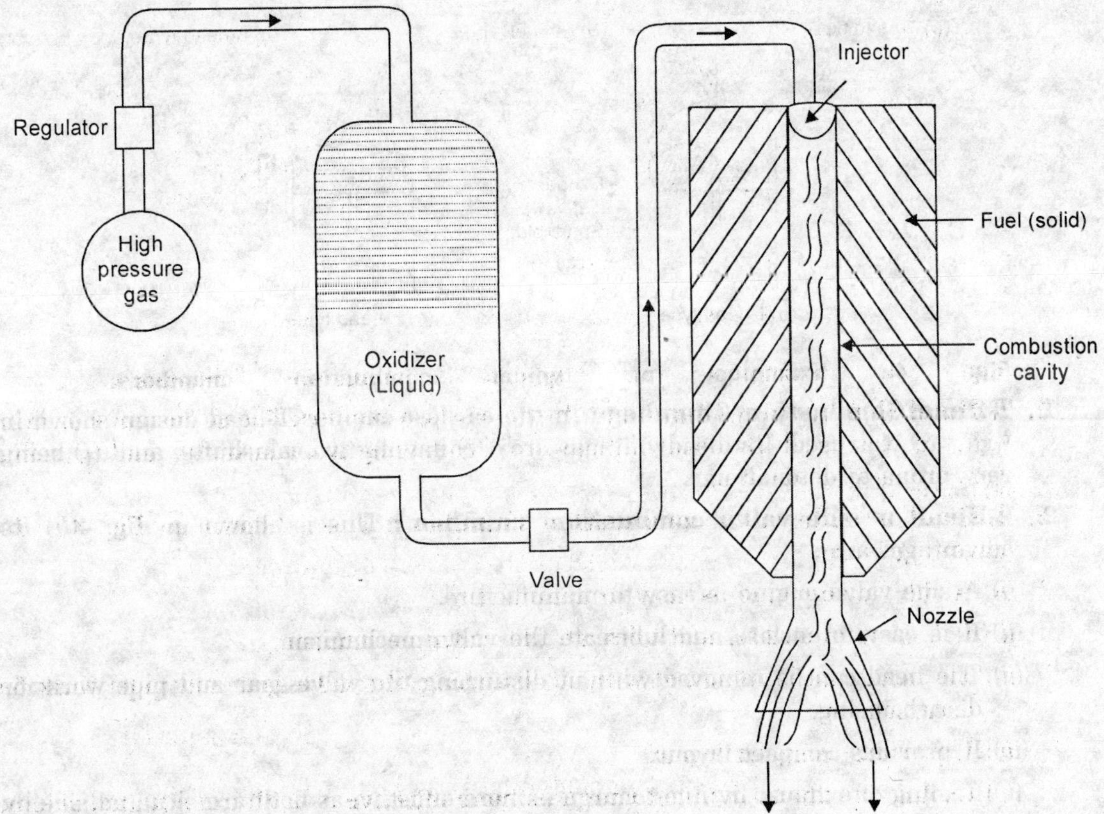

Fig. 3. Schematic diagram of hybrid rocket engine.

Advantages

1. The specific impulse is more than that of a solid but less than a liquid rocket engine.
2. The system reliability lies between the solid and liquid system.
3. Easier operation than solid rocket engine.
4. Storage of hybrids are safe as compared to solid engine.

(b) What are the different types of combustion chambers in C.I. engine? Explain with a neat sketch, an open combustion chamber. What are the merits and demerits of the open combustion chamber? **(10)**

Sol. The various types of combustion chambers are shown in Fig. 4.

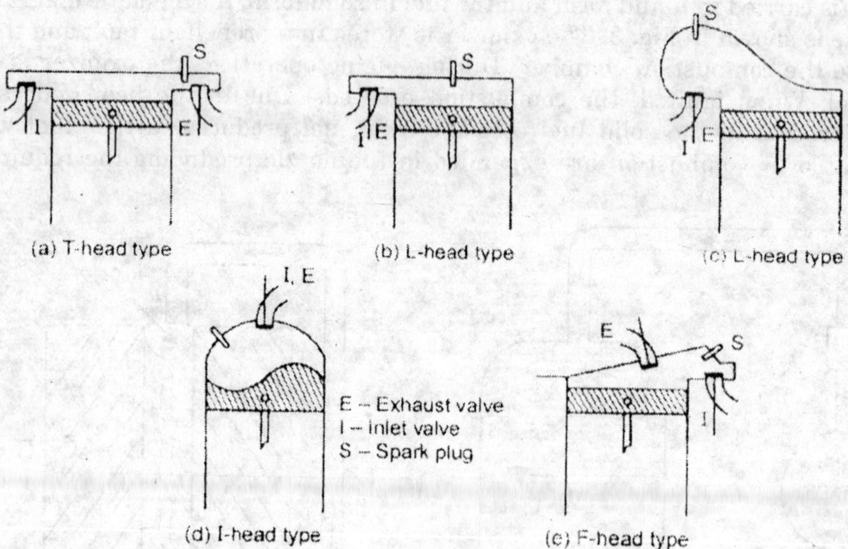

Fig. 4. Examples of typical combustion chambers.

1. **T-Head Combustion Chamber :** In the earliest engines T-head design shown in Fig. 4 (*a*) was used. Its disadvantages are : (*i*) having two camshafts, and (*ii*) being very prone to detonation.

2. **L-Head or side valve combustion chamber :** This is shown in Fig. 4(*b*) Its advantages are :

   (*i*) A side valve engine is easy to manufacture.

   (*ii*) It is easy to enclose and lubricate the valve mechanism.

   (*iii*) The head can be removed without disturbing the valve-gear and pipe work for decarbonising.

   (*iv*) It provides compact layout.

   (*v*) Cooling of exhaust by inlet charge is more effective as both are situated side by side.

   Its shortcoming are :

   (*i*) Lack of turbulence in the incoming charge.

   (*ii*) Extremely prone to knocking as the flame has to travel long distance.

   (*iii*) Extremely sensitive to ignition timing due to slow combustion.

3. **Ricardo Turbulent L-Head side valve Design :** This arrangement is shown in Fig. 4(*c*). This gives higher flame speed and to reduce the knocking tendency. The entire combustion space is concentrated over the inlet and exhaust valves. This

provides sufficient turbulence during the compression stroke and increased flame velocity. The ratio of area of volume of the space above the piston is large which provides sufficient quench area essential to avoid knocking.

4. **Overhead valve or J-head combustion chamber :** In this arrangement, both the valves are located in the cylindrical head as shown in Fig. 4(*d*). This arrangement is superior to side valve or L-head arrangement when high compression ratios are used, for the following reasons :

(*i*) Lower pumping losses and higher volumetric efficiency from better breathing of the engine from larger valves or valve lifts and more direct passage ways.

(*ii*) Less distance for the flame to travel and therefore greater freedom from knock.

(*iii*) Less force on the head bolts and therefore less possibility of leakage.

(*iv*) Absence of exhaust valve from block results in more uniform cooking of cylinder and piston.

(*v*) Lower surface-volume ratio and, therefore, less heat loss and less air pollution.

(*vi*) Easier to cast and hence lower casting cost.

5. **F-Head combustion chambers :** A combustion chamber in which one valve is located in the head and other in the block is known as F-head combustion chamber, as shown in Fig. 4(*e*). This arrangement is a compromise between L-head and I-head combustion chambers. Fig. 4(*e*) shows exhaust valve in head and inlet valve in block. This gave a rather poor performance. The modern F-head engines have inlet valve in the head and exhaust valve in the block. The valves are inclined with plug located in the flat roof. This provides the shortest flame travel and ensures minimum knocking tendency.

Combustion Chambers for C.I. Engines

The most important function of the CI engine combustion chamber is to provide proper mixing of fuel and air in a short time. For this purpose, an organised air movement, called air swirl, is provided to produce high relative velocity between the fuel droplets and air.

There are three basic methods of generating swirl in a CI engine combustion chamber.

1. By directing the flow of the air during its entry to the cylinder, known as induction swirl. This method is used in open combustion chambers shown in Fig. 5.

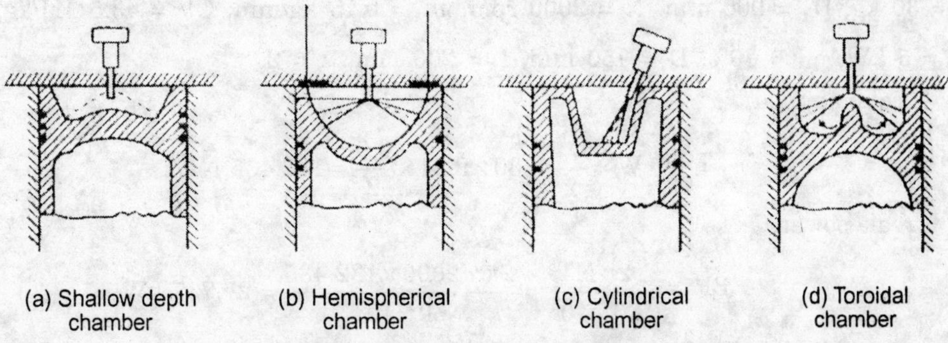

(a) Shallow depth chamber (b) Hemispherical chamber (c) Cylindrical chamber (d) Toroidal chamber

Fig. 5. Open combustion chambers

The advantages and disadvantages of open combustion chambers are :

Advantages

1. Simple in construction.
2. Higher thermal efficiency due to less heat losses.
3. Easy starting of engine.
4. Gives higher mechanical efficiency.
5. Less costly fuel can be used in low speed engines.

Disadvantages

1. Requires large excess air.
2. Requires higher injection pressure with multiple nozzles.
3. There is possibility of clogging the holes of nozzle, being too small.

(c) A four stroke single cylinder petrol engine mounted on a motor cycle was put to load test. The load measured on dynamometer was 30 kg with drum diameter and speed respectively at 900 mm and 2000 rpm. The engine consumed 0.15 kg of fuel in one minute, the calorific value of fuel being 43.5 MJ/kg. The fuel supply to the engine was stopped and was driven by a motor which needed 5 kW of power to keep it running at the same speed, the efficiency of the motor being 80%. The engine cylinder bore and stroke are respectively at 150 mm and 200 mm.

Calculate

(*i*) brake power,

(*ii*) Indicated power,

(*iii*) mechanical efficiency,

(*iv*) brake thermal efficiency,

(*v*) indicated thermal efficiency,

(*vi*) brake mean effective pressure and

(*vii*) indicated mean effective pressure.

Sol. Given : 4-stroke single cylinder petrol engine,

$W = 30$ kg, $D_d = 900$ mm, $N = 2000$ rpm, $\dot{m}_f = 0.15$ kg/min, $CV = 43.5$ MJ/kg,

$P_m = 5$ kW, $\eta_m = 80\%$, $D = 150$ mm, $L = 200$ mm, $n = 2$.

(*i*) Torque applied,

$$T = W \times \frac{D_d}{2} = 30 \times 9.81 \times \frac{0.9}{2} = 132.435 \text{ N} \cdot \text{m}$$

Break power,

$$BP = \frac{2\pi NT}{60 \times 10^3} = \frac{2\pi \times 2000 \times 132.435}{60 \times 10^3} = 27.737 \text{ kW}$$

(*ii*) Indicated power,

$$IP = BP + FP$$

Frictional power,

$$FP = \frac{P_m}{\eta_m} = \frac{5}{0.8} = 6.25 \text{ kW}$$

Indicated power, $IP = 27.737 + 6.25 = 33.987$ kW

(*iii*) Mechanical efficiency,

$$\eta_{mech} = \frac{BP}{IP} = \frac{27.737}{33.987} = 0.8161 \text{ or } 81.61\%$$

(*iv*) Brake thermal efficiency,

$$(\eta_{th})_b = \frac{BP}{\dot{m}_f \times CV} = \frac{27.737 \times 60}{0.15 \times 43.5 \times 10^3} = 0.2550 \text{ or } 25.5\%$$

(*v*) Indicated thermal efficiency,

$$(\eta_{th})_i = \frac{IP}{\dot{m}_f \times CV} = \frac{33.987 \times 60}{0.15 \times 43.5 \times 10^3} = 0.3125 \text{ or } 31.25\%$$

(*vi*)

$$BP = \frac{p_{bm} \times LAN}{60 \times 10^3 \times n}$$

$$27.737 = \frac{p_{bm} \times 0.2 \times \frac{\pi}{4}(0.15)^2 \times 2000}{60 \times 10^3 \times 2}$$

$$\text{BMEP, } p_{bm} = 4.7 \text{ bar}$$

(*vii*)

$$IMEP = \frac{4.7 \times 33.987}{27.737} = 5.77 \text{ bar}$$

(*d*) The overall thermal efficiency of a 5 MW nuclear power plant for a submarine is 30%. Calculate the amount of natural uranium, U^{235}, needed to generate the power if the average energy release per fission for U^{235} is 190 MeV. Take 1 W = 6.241×10^{12} MeV/s. Avogadro's Number is 6.02×10^{23}. (10)

Sol. Given $(\eta_{th})_{overall} = 30\%$, P = 5 MW, E = 190 MeV/fission,
1 W = 6.241×10^{12} Mev/s, AN = 6.02×10^{23}

Amount of nuclear power required to be generated

$$= \frac{P}{(\eta_{th})_{overall}} = \frac{5}{0.3} = 16.667 \text{ MW}$$

$$= 16.667 \times 10^6 \times 6.241 \times 10^{12}$$

$$= 104.017 \times 10^{18} \text{ MeV/s}$$

Number of atoms of U^{235} to be fissioned

$$= \frac{104.017 \times 10^{18}}{190} = 0.54745 \times 10^{18}$$

Mass of U^{235} required,

$$m = \frac{0.54745 \times 10^{18}}{AN} = \frac{0.54745 \times 10^{18}}{6.02 \times 10^{23}}$$

$$= 0.0909 \times 10^{-5} \text{ kg}$$

Q.3. (*a*) Cooling water at a steady rate of 0.5 kg/s flows through an inner tube having inner diameter of 25 mm and length of 10 m of a tube-in-tube condenser. The mean inlet temperature of cooling water is 10°C. Saturated steam condenses in the annulus at a uniform rate such that the inner surface temperature of the tube is constant throughout the length of the tube at 40°C. The average condensing side heat transfer coefficient is 1000 W/m²-K. Neglect the thickness of the heat exchanger tube. Calculate the effectiveness of the heat exchanger and the exit water temperature.

Properties of water are given below :

Specific heat = 4180 J/kg-K,

Denisty = 990 kg/m³,

Dynamic viscosity = 0.8×10^{-3} Pascal. sec,

Thermal conductivity = 0.57 W/m-K,

You may use the relation

$$\text{Nu} = 0.023\ \text{Re}_d^{0.8}\ \text{Pr}^{0.4},$$

Sol. Given : $\dot{m}_c = 0.5$ kg/s, $d_i = 25$ mm, $l_i = 10$ m, $t_{ic} = 10°\text{C}$, $t_{ih} = 40°\text{C}$, $h_0 = 10{,}000$ W/m²·K, $t_{oc} = ?$, $\varepsilon = ?$

$c_{pc} = 4180$ J/kg·K, $\rho_c = 990$ kg/m³, $\mu_c = 0.8 \times 10^{-3}$ Pa·s, $k_c = 0.57$ W/m·K, $\text{Nu} = 0.023\ \text{Re}_d^{0.8}\ \text{Pr}^{0.4}$

$$\dot{m}_c = \frac{\pi}{4} d_i^2\ \upsilon\rho$$

$$0.5 = \frac{\pi}{4}(0.025)^2 \times \upsilon \times 990$$

$$\upsilon = 1.029 \text{ m/s}$$

$$\text{Nu} = \frac{h_i d_i}{k_c} = \frac{h_i \times 25 \times 10^{-3}}{0.57} = h_i \times 43.86 \times 10^{-3}$$

$$\text{Pr} = \frac{\mu_c c_{pc}}{k_c} = \frac{0.8 \times 10^{-3} \times 4180}{0.57} = 5.867$$

$$\text{Re} = \frac{\rho_c \upsilon d_i}{\mu_c} = \frac{990 \times 1.029 \times 25 \times 10^{-3}}{0.8 \times 10^{-3}} = 31{,}834$$

$$43.86 \times 10^{-3} \times h_i = 0.023 \times (31{,}834)^{0.8} \times (5.867)^{0.4} = 186.8$$

$$h_i = 4259.3 \text{ W/m}^2\text{.K}$$

$$\frac{1}{\text{U}} = \frac{1}{h_i} + \frac{1}{h_0} = \frac{1}{4259.3} + \frac{1}{10{,}000} = 3.3478 \times 10^{-4}$$

$$\text{U} = 2987 \text{ W/m}^2\text{.K}$$

$$\theta_1 = t_{ih} - t_{ic} = 40 - 10 = 30°\text{C}$$

$$\theta_2 = t_{oh} - t_{oc} = 40 - t_{oc}$$

$$\text{AMTD} = \frac{\theta_1 + \theta_2}{2} = \frac{30 + (40 - t_{co})}{2} = 35 - 0.5\, t_{oc}$$

Heat transfer rate,

$$\dot{Q} = \dot{m}_c c_{pc}(t_{oc} - t_{ic}) = U \times A_s \times \text{AMTD}$$

$$0.5 \times 4180 \times (t_{oc} - 10) = U \times \pi d_i l_i \times \text{AMTD}$$

$$= 2987 \times \pi \times 25 \times 10^{-3} \times 10 \times (35 - 0.5\, t_{oc})$$

$$t_{oc} - 10 = 1.1225\,(35 - t_{oc})$$

$$t_{oc} = 23.2°\text{C}$$

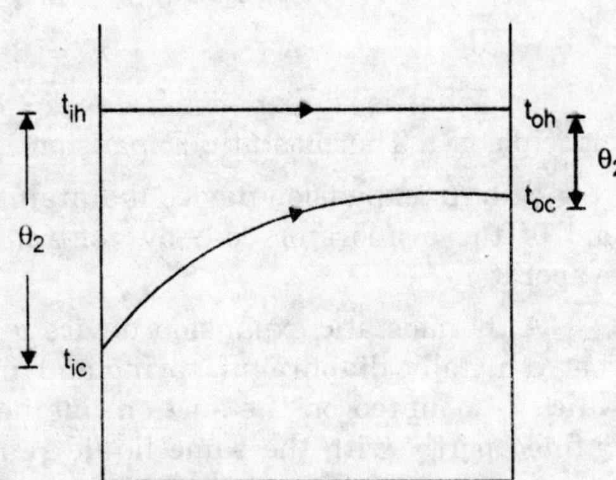

Fig. 6

Heat exchanger effectiveness,

$$\varepsilon = \frac{t_{oc} - t_{ic}}{t_{ih} - t_{ic}}$$

$$= \frac{23.2 - 10}{40 - 10} = \frac{13.2}{30} = 0.44 \text{ or } 44\%$$

(*b*) The net radiation from the surfaces of two parallel plates having equal emissivities of 0.8 and at different temperatures of T_1 and T_2 ($T_1 > T_2$) is to be reduced by 99%. How many number of radiation screens having equal emissivities of 0.05 are to be placed between the plates to achieve the reduction in heat exchanger?

Sol.

$$(Q_{1-2})_{net} = \frac{\sigma(T_1^4 - T_2^4)}{\frac{1}{\varepsilon_1} + \frac{1}{\varepsilon_2} + 2\sum_{i-1}^{n} \frac{1}{\varepsilon_{si}} - (n+1)}$$

$$(Q)_{\text{without shields}} = \frac{\sigma(T_1^4 - T_2^4)}{\frac{2}{0.8} - 1} = \frac{C}{1.5}$$

where $C = \sigma\left(T_1^4 - T_2^4\right)$

$$(Q)_{\text{with shields}} = \frac{\sigma(T_1^4 - T_2^4)}{\frac{2}{0.8} + 2\left(\frac{n}{0.05}\right) - (n+1)}$$

$$= \frac{C}{2.5 + 40n - n - 1} = \frac{C}{1.5 + 39n}$$

T₁ εs1 = 0.05 εsn = 0.05 T₂

ε₁ = 0.8 ε₂ = 0.8

Fig. 7

Percentage reduction in heat flow

$$= \left[1 - \frac{(Q)_{\text{with shields}}}{(Q)_{\text{without shields}}}\right] \times 100$$

$$= \left(1 - \frac{1.5}{1.5 + 39n}\right) \times 100 = 99$$

$$1 - \frac{1}{1 + 26n} = 0.99$$

$$\frac{1}{1+26n} = 0.01$$

$$0.01 + 0.26n = 1$$

$$n = 3.8 \simeq 4$$

(*c*) What is an expansion device? Explain with the help of a neat sketch the working principle of a Thermostatic Expansion Device (TEV). (10)

Sol. An expansion device maintains constant superheat of the vapour refrigerant at the end of the evaporator coil, by controlling the flow of the liquid refrigerant through the evaporator.

A thermostatic expansion device is shown in Fig. 8. It consists of a needle valve and a seat, a metallic diaphragm, spring and an adjusting screw. It has also a feeler or thermal bulb which is mounted on the suction line near the outlet of the evaporator coil. The feeler bulb is filled partly with the same liquid refrigerant as used in the refrigeration system.

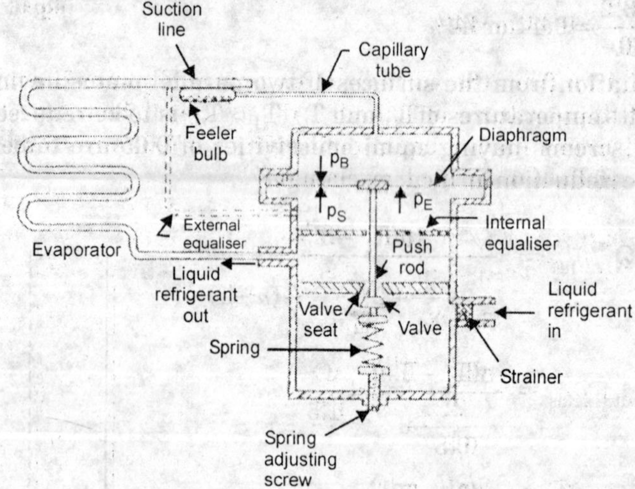

Fig. 8. Thermostatic expansion device.

The feeler bulb will be at the same temperature as the refrigerant at that point. Any change in the temperature of refrigerant will cause a change in the feeler bulb which will be transmitted to the top of the diaphragm.

Under normal operating conditions, the feeler bulb pressure acting at the top of the diaphragm is balanced by the spring pressure and the evaporator pressure acting at the bottom of the diaphragm. The force tending to close the valve is dependent upon the spring pressure and the evaporator pressure which, in turn, depends upon the saturation temperature of the refrigerant in the evaporator coil. The force tending to open the valve depends upon the feeler bulb pressure which, in turn, depends upon the temperature of refrigerant in the bulb. Thus the operation of valve is controlled by the difference between the two temperatures (*i.e.*, saturation temperature and feeler bulb temperature) which is the superheat. The degree of superheat of the vapour refrigerant leaving the evaporator depends upon the initial setting of the spring tension, which can be changed with the help of spring adjusting screw. When the valve is set for a certain superheat, then it maintains that setting under all load conditions on the evaporator.

If the load on the evaporator increases, it causes the liquid refrigerant to boil faster in the evaporator coil. The temperature of the feeler bulb increases due to early vaporisation of the liquid refrigerant. Thus the feeler bulb pressure increases and this pressure is transmitted through the capillary tube to the diphragm. The diaphragm moves downwards and open the valve to admit more quantity of liquid refrigerant to the evaporator. This continues till the pressure equilirbium on the diaphragm is reached. On the other hand, when the load on the evaporator decreases, less liquid refrigerant evaporates in the evaporator coil. The excess liquid refrigerant flows towards the evaporator outlet which cools the feeler bulb with the result the feeler bulb pressure decreases due to decrease in its temperature. The low feeler bulb pressure is transmitted through the capillary tube to the diaphragm and moves it upward. This reduces the opening of the valve and thus the flow of liquid refrigerant to the evaporator. The evaporator pressure decreases due to reduced quantity of liquid refrigerant flowing to the evaporator. This continues till the evaporator pressure and the spring pressure maintains equilibrium with the feeler bulb pressure.

Q.4. (*a*) An R-12 vapour compression plant producing 10 tonnes of refrigeration operates with condensing and evaporating temperatures of 35°C and –10°C respectively. A suction line heat exchanger is used to subcool the saturated liquid leaving the condenser. Saturated vapour leaving the evaporator is superheated in the suction line heat exchanger to the extent that a discharge temperature of 60°C is obtained after isentropic compression, Calculate

(*i*) The subcooling achieved in the heat exchanger,

(*ii*) The refrigerant flow rate in kg/s,

(*iii*) The cylinder dimensions of the two-cylinder compressor, if the speed is 900 rpm, stroke-to-bore ratio is 1 : 1 and the volumetric efficiency is 80%,

(*iv*) The COP of the plant and

(*v*) The power required to drive the compressor in kW.

Draw the cycles on p-h and T-s diagrams.

You may use the following table :

Temperature (°C)	Specific Volume (m^3/kg)	Specific enthalpy KJ/kg		Specific entropy KJ/kg-K	
	Vapour (v_g)	Liquid (h_f)	Vapour (h_g)	Liquid (s_f)	Vapour (s_g)
–10	0.0766	190.88	347.13	0.9660	1.5600
+35	0.0206	233.50	365.92	1.1139	1.5419

Average specific heat of desuperheating vapour at condenser = 0.796 KJ/kg-K.

Average specific heat of vapour between evaporator outlet and compression suction = 0.658 KJ/kg-K.

Sol. Given : Q = 10 TR, T_1' = –10°C = – 10 + 273 = 263 K, $T_2' = T_3'$ = 35°C = 35 + 273 = 308 K,

$$T_2 = 60°C = 60 + 273 = 333 \text{ K}, N = 900 \text{ rpm}, \frac{L}{D} = 1, \eta_{vol} = 80\%,$$

$v_1' = 0.0766$ m^3/kg, $v_2' = 0.0206$ m^3/kg, $h_{f1}' = 190.88$ kJ/kg,
$h_{f3}' = 233.50$ kJ/kg, $h_1' = 347.13$ kJ/kg, $h_2' = 365.92$ kJ/kg,
$s_{f1}' = 0.9660$ kJ/kg·K, $s_{f3}' = 1.1139$ kJ/kg·K, $s_1' = 1.5600$ kJ/kg·K, $s_2' = 1.5419$ kJ/kg·K,
$c_{pl} = 0.796$ kJ/kg·K, $c_{pv} = 0.658$ kJ/kg·K

The *p-h* and T-*s* diagrams are shown in Fig.

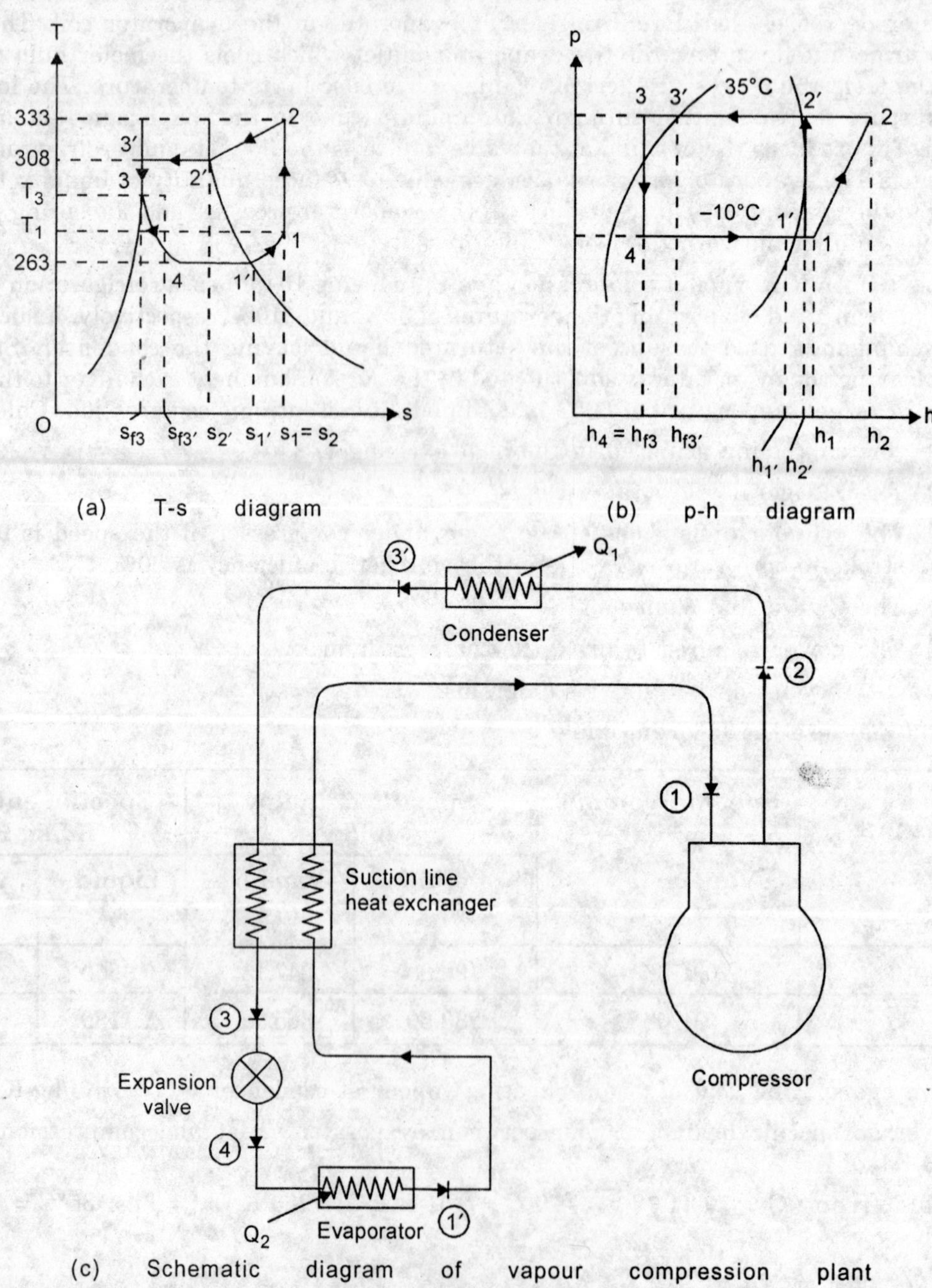

(a) T-s diagram (b) p-h diagram

(c) Schematic diagram of vapour compression plant

Fig. 9

Suction line heat exchanger : For heat balance, we have

$$h_{f3}' - h_{f3} = h_1 - h_1'$$

$$233.5\ h_{f3} = h_1 - 347.13 \quad (1)$$

$$s_1 = s_2$$

$$s_1 = s_1' + c_{pv} \ln \frac{T_1}{T_1'} = 1.5600 + 0.658 \ln \frac{T_1}{263}$$

$$s_2 = s_2' + c_{pv} \ln \frac{T_2}{T_2'} = 1.5419 + 0.658 \ln \frac{333}{308} = 1.5932 \text{ kJ/kg} \cdot \text{K}$$

$$\therefore \quad 1.5600 + 0.658 \ln \frac{T_1}{263} = 1.5932$$

$$T_1 = 276.6 \text{ K or } 3.6°\text{C}$$

$$h_1 = h_1' + c_{pv}(T_1 - T_1') = 347.13 + 0.658\ (276.6 - 263)$$

$$= 356.08 \text{ kJ/kg}$$

Eq. (1) gives,

$$233.5\ h_{f3} = 356.08 - 347.13$$

$$h_{f3} = 224.55 \text{ kJ/kg}$$

Now

$$h_{f3} = h_{f3}' - c_{pl}\ (T_3' - T_3)$$

$$224.55 = 233.50 - 0.796\ (308 - T_3)$$

$$T_3 = 296.76 \text{ K} \quad \text{or} \quad 23.76°\text{C}$$

(*i*) Subcooling achieved in the heat changer

$$= T_3' - T_3$$

$$= 308 - 296.76 = 11.24°\text{C}$$

(*ii*) Refrigerating effect,

$$\text{RE} = h_1 - h_{f3} = 356.08 - 224.55 = 131.53 \text{ kJ/kg}$$

Refrigerating capacity of plant

$$Q = 10 \text{ TR} = 10 \times 211 = 2110 \text{ kJ/kg}$$

Mass flow rate of refrigerant,

$$\dot{m}_r = \frac{Q}{\text{RE}} = \frac{2110}{131.53} = 16.04 \text{ kg/min}$$

(*iii*) Theoretical suction volume or piston displacement per minute

$$V_1 = \dot{m}_r \times v_1 \times \frac{1}{\eta_{vol}}$$

Now

$$\frac{v_1}{T_1} = \frac{v_1'}{T_1'}$$

or

$$v_1 = v_1' \times \frac{T_1}{T_1'} = 0.0766 \times \frac{276.6}{263} = 0.0806 \text{ m}^3\text{/kg}$$

$$V_1 = 16.04 \times 0.0806 \times \frac{1}{0.8} = 1.616 \text{ m}^3/\text{min}$$

Number of cylinders = 2

∴ Piston displacement per cylinder

$$= \frac{1.616}{2} = 0.808 \text{ m}^3/\text{min}$$

$$= \frac{\pi}{4} \times D^2 \times D \times N = \frac{\pi}{4} D^3 \times 900$$

$$= 706.86 \text{ D}^3 \text{ m}^3/\text{min}$$

$$\therefore \quad L = D = 0.1046 \text{ m or } 104.6 \text{ mm}$$

(*iv*)

$$\text{COP} = \frac{h_1 - h_{f3}}{h_2 - h_1}$$

$$h_2 = h_2' + c_{pv} (T_2 - T_2') = 365.92 + 0.658 (333 - 308)$$
$$= 382.37 \text{ kJ/kg}$$

$$\text{COP} = \frac{356.08 - 224.55}{382.37 - 356.08} = \frac{131.53}{26.29} = 5.003$$

(*v*) Work done during compression of refrigerant

$$= \dot{m}_r (h_2 - h_1) = 16.04 (382.37 - 356.08) = 421.69 \text{ kJ/min}$$

Theoretical power required

$$= \frac{421.69}{60} = 7.028 \text{ kW}$$

(***b***) A rectangular fin of length 30 cm, width 30 cm and thickness 2 mm is attached to a surface at 300°C. The fin is made of aluminium (K = 204 W/m-K) and is exposed to air at 30°C. The fin end is uninsulated and can lose heat through its end also. The convection heat transfer coefficient between the fin surface and air is 15 W/m²-K.

Determine :

(*i*) the temperature of the fin at 30 cm from the base,

(*ii*) the rate of heat transfer from the fin, and

(*iii*) fin efficiency

Sol. Given : l = 30 cm, b = 30 cm, y = 2 mm, t_0 = 300°C, k = 204 W/m·K, t_a = 30°C, k = 15 W/m²·K, x = 30 cm.

Area of cross-section of fin,

$$A_{cs} = by = 30 \times 10^{-12} \times 2 \times 10^{-3} = 6 \times 10^{-4} \text{ m}^2$$

Perimeter of fin, $P = 2(b + y) = 2(300 + 2) \times 10^{-3} = 0.604$ m

$$m = \sqrt{\frac{hP}{kA_{cs}}} = \sqrt{\frac{15 \times 0.604}{204 \times 6 \times 10^{-4}}} = 8.603$$

(*i*) For a rectangular fin losing heat from the end also,

$$t - t_a = (t_0 - t_a)\left[\frac{\cos h\,\{m(l-x)\} + \dfrac{h}{km}\sin h\{m(l-x)\}}{\cos h\,(ml) + \dfrac{h}{km}\sin h\,(ml)}\right]$$

$$\frac{h}{km} = \frac{15}{204 \times 8.603} = 8.547 \times 10^{-3}$$

$$m(l - x) = 8.60^3\ (30 - 30) = 0$$

$$ml = 8.603 \times 30 \times 10^{-2} = 2.5809$$

$$t - 30 = (300 - 30)\left[\frac{\cos h\,0 + 8.547 \times 10^{-3} \times \sin h\,0}{\cos h\,2.5809 + 8.547 \times 10^{-3} \times \sin h\,2.5809}\right]$$

$$= 270\left[\frac{1+0}{6.6985}\right] = 40.3$$

$$t = 70.3°\text{C}$$

(*ii*) $$Q_{fin} = \sqrt{hPkA_{cs}}\ (t_0 - t_a)\left[\frac{\tan h\,(ml) + \dfrac{h}{km}}{1 + \dfrac{h}{km}\tan h\,(ml)}\right]$$

$$= \sqrt{15 \times 0.604 \times 204 \times 6 \times 10^{-4}} \times (300 - 30)\left[\frac{\tan h\,2.5809 + 8.547 \times 10^{-3}}{1 + 8.547 \times 10^{-3}\ \tan h\,2.5809}\right]$$

$$= 1.053 \times 270\left[\frac{0.997}{1.008}\right]$$

$$= 281.2\ \text{W}$$

(*iii*) $$\eta_{fin} = \frac{\tan h\left[\sqrt{\dfrac{2h}{ky}}\left(l + \dfrac{y}{2}\right)\right]}{\sqrt{\dfrac{2h}{ky}}\left(l + \dfrac{y}{2}\right)}$$

$$\sqrt{\frac{2h}{ky}} = \frac{2 \times 15}{204 \times 2 \times 10^{-3}} = \sqrt{73.529} = 8.575,\ l + \frac{y}{2} = 300 + \frac{2}{2} = 301\ \text{mm or } 0.301\ \text{m}$$

$$\eta_{fin} = \frac{\tan h\,[8.575 \times 0.301]}{8.575 \times 0.301} = \frac{0.9886}{2.581} = 0.383 \quad \text{or} \quad 38.3\%$$

(*c*) Brief describe the various methods of air-conditioning duct design.

Sol. The following methods are used for air-conditioning duct design :

1. **Velocity reduction method :** In this method, the velocities in the ducts are assumed such that they progressively decrease as the flow proceeds. The pressure drops are calculated for these velocities for respective branches and the main duct. The duct sizes are determined for assumed velocities and known quantities of air to be supplied through the respective ducts. The pressure at the outlet is adjusted by

dampers in the respective ducts. The fan is designed to overcome the pressure losses along any single run including losses of the main duct, branch duct, elbows, enlargements and contractions of areas etc. In case the fan is already selected, the velocities in reducing order are adjusted to consume pressure available in the longest run or the run in which the maximum pressure loss is expected. The pressure in the remaining branches is adjusted by dampers.

This method is the easiest in sizing ducts and the velocities can be adjusted to avoid noise. The major disadvantage of this system is that considerable experience and judgement is required in selecting velocities so as to make the system optimum in economy and power.

The velocity reduction method of designing ducts is usually adopted for very simple systems.

2. **Equal pressure drop (or friction loss) method :** In this method, the size of duct is decided to give equal pressure drop (or friction loss) per metre length in all ducts. If the layout of the ducts is symmetrical giving the same length of the various runs, this method gives equal pressure loss in various branches and no dampering is required to balance them. In case the runs are of different lengths, then the shortest run will have minimum loss and consequently high pressure at the outlet. It is, therefore, necessary to reduce this high pressure by heavy dampering or modifying this method to provide higher velocities in shorter runs. But the high velocities in short run to reduce high pressure may create objectionable noise. Thus noise absorbing outlets and fittings must be provided. The dampers if provided near the main duct will help in reducing the noise as the branch duct will dissipate some noise.

   The velocities, in this method are automatically reduced in the branch ducts as the flow is decreased. This method does not however balance the pressures at the outlets if the branches are of different lengths and hence dampers are required for balancing the pressure drops in various branches.

   A modification of this method is to design the main duct for equal friction and branch ducts for consuming the pressure available at the take off from the main duct. In such a design, the pressures at the outlet will be same and no dampering is required for balancing the pressure drops in various runs. However, dampers are provided for small adjustment.

3. **Static region method :** In this method, the size of the duct is decided to give equal pressure at all outlets, for perfect balancing of the air duct layout system. This may be done by equalising the pressure losses in various branches. This is possible if the friction loss in each branch is made equal to the gain in pressure due to reduction in velocity. The gain in pressure (or static pressure regain) due to change in velocity is given by

$$\text{SPR} = \text{R}\,\rho_a \left[\frac{(\text{V}_1)^2 - (\text{V}_2)^2}{2}\right] = \text{R}(p_{v1} - p_{v2}) \qquad \ldots\left(\because\ p_v = \frac{\rho_a \text{V}^2}{2}\right)$$

where R is the static regain factor.

It may not be possible to design economically very long branches and the branches very near to the fan for complete regain. In such cases, it is sufficient to design the main duct for complete regain and provide same pressure at all outlets from the main duct for branches.

The partial regain may be considered a good practice for a few outlets from the main duct. so that same pressure loss is allowed in the beginning.

This method allows for balancing but reducing velocity increases duct size and it should not be taken beyond the economic limit.

SECTION-C

Q.5. (*a*) A tank with the vertical sides measuring 3 m × 3 m contains water to a depth of 1.2 m. An oil of density 900 kg/m³ was poured in the tank up to a depth of 0.8 m. The vertical wall can withstand the thrust of 58 kN. Calculate the actual thrust on the wall and centre of pressure. If the oil level is increased up to 0.9 m, what will be stability of the wall?

Sol. Given : $h_w = 1.2$ m, $h_0 = 0.8$ m, $\rho_0 = 900$ kg/m³

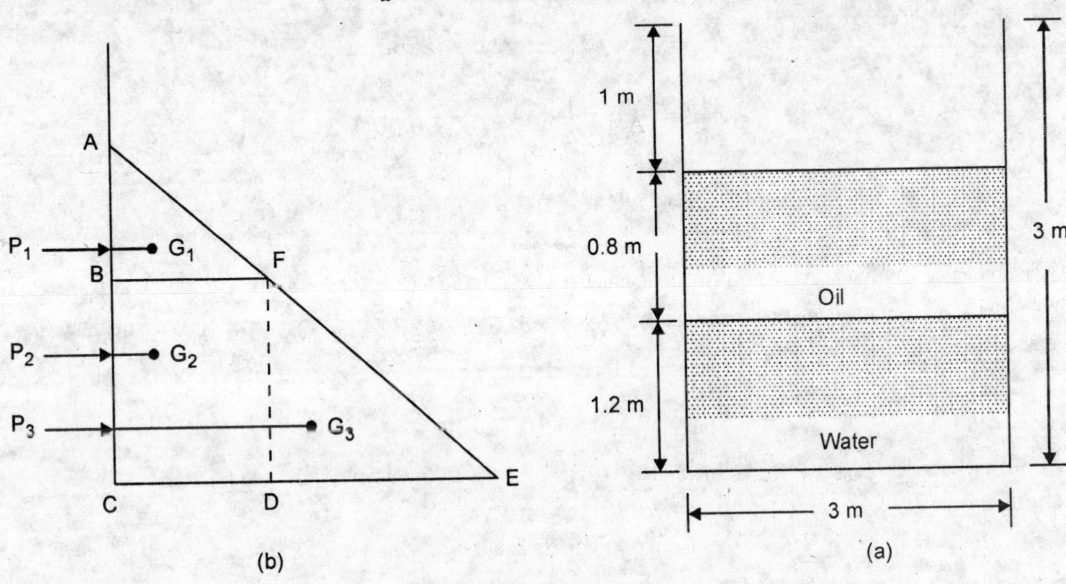

Fig. 10

Pressure on BF, the interface of water and oil,

$$p_B = \rho_0 g h_0$$
$$= 900 \times 9.81 \times 0.8$$
$$= 7063.2 \text{ N/m}^2$$

Pressure on base CE,

$$p_c = \rho_0 g h_0 + \rho_w g h_w$$
$$= 7063.2 + 1000 \times 9.81 \times 1.2 = 18835.2 \text{ N/m}^2$$

Force due to oil, $P_1 = \Delta ABF \times \text{width of tank}$

$$= \frac{1}{2} AB \times BF \times 3$$

$$= \frac{1}{2} \times 0.8 \times 7063.2 \times 3 = 8{,}475.84 \text{ N}$$

Force, P_2 = area of rectangle BCDF × width of tank

$= 1.2 \times 7063.2 \times 3 = 25{,}427.52$ N

Force, $P_3 = \Delta DEF = \frac{1}{2} \times FD \times DE \times \text{width of tank}$

$= \frac{1}{2} \times 1.2 \times (18835.2 - 7063.2) \times 3$

$= 21{,}189.6$ N

Total force, $P = P_1 + P_2 + P_3 = 8{,}475.84 + 25{,}427.52 + 21{,}189.6$

$= 55{,}092.96 \text{ N} < 58 \times 10^3$ N

Centre of pressure : Hence the wall is stable

$$\bar{h}_1 = \frac{2}{3} AB = \frac{2}{3} \times 0.8 = 0.533 \text{ m}$$

$$\bar{h}_2 = AB + \frac{1}{2} BC = 0.8 + 0.5 \times 1.2 = 1.4 \text{ m}$$

$$\bar{h}_3 = AB + \frac{2}{3} BC = 0.8 + \frac{2}{3} \times 1.2 = 1.6 \text{ m}$$

$$P \times h_c = P_1\bar{h}_1 + P_2\bar{h}_2 + P_3\bar{h}_3$$

$$55{,}092.96 \times \bar{h} = 8475.84 \times 0.533 + 25{,}427.52 \times 1.4 + 21{,}189.6 \times 1.6$$

$$= 74{,}019.51$$

$$h_c = 1.343 \text{ m from top}$$

When $h_0 = 0.9$ m,

then $p_B = 7946.1$ N/m²

$p_C = 19718.1$ N/m²

$P_1 = 10727.2$ N

$P_2 = 28606$ N

$P_3 = 21{,}189.6$ N

$P = 60522.8 \text{ N} > 5800$ N

$\bar{h}_1 = 0.6$, $\bar{h}_2 = 1.5$ m, $\bar{h}_3 = 1.7$ m

$$60522.8 \times h_c = 10727.2 \times 0.6 + 28606 \times 1.5 + 21{,}189.6 \times 1.7$$

$$h_c = 1.41 \text{ m}$$

Hence the wall shall be unstable.

(b) For a rate of flow exceeding certain value, the co-efficient of discharge for a venturimeter used for measuring the discharge of an incompressible fluid is found to be constant. Prove that the loss of head in the convergent portion of the venturi can be expressed as KQ^2 under these conditions, where K is a constant (function of C_d, areas of the venturimeter) and Q is flow rate in m³/s. What will be the value of K, assuming the C_d constant.

Sol. Applying Bernoulli's equation to the entrance of convergent portion of the venturi and throat considering loss of head, we have

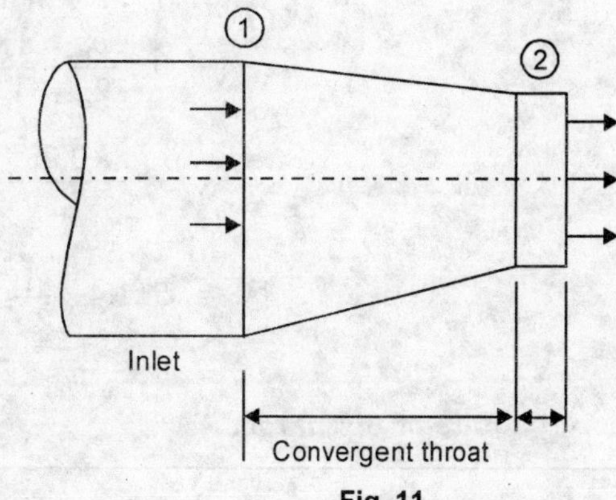

Fig. 11

$$\frac{p_1}{\gamma}+\frac{v_1^2}{2g} = \frac{p_2}{\gamma}+\frac{v_2^2}{2g}+h_f$$

$$\frac{p_1-p_2}{\gamma} = \frac{v_2^2-v_1^2}{2g}+h_f$$

$$h_f = \frac{p_1-p_2}{\gamma}-\frac{v_2^2-v_1^2}{2g} = h-\frac{v_2^2-v_1^2}{2g}$$

$$v_1 = \frac{Q}{A_1},\ v_2 = \frac{Q}{A_2},\quad A_1v_1 = A_2v_2$$

$$Q = c_d\frac{A_1A_2}{\sqrt{A_1^2-A_2^2}}\cdot\sqrt{2gh},\quad h = \frac{p_1-p_2}{\gamma}$$

$$Q \propto K$$

$$h_f \propto v^2 \propto Q^2 = KQ^2$$

(*c*) A straight inclined pipeline 300 m long discharges freely at a point 50 m lower than the water surface at intake. The pipe intake projects into the reservoir (k_e = 0.8). The first 200 m of the pipe is of 350 mm diameter and the remaining 100 m is of 250 mm diameter (k_e = 0.21).

(*i*) Find the rate of discharge assuming f = 0.06.

If the junction point C of the two sizes of the pipe is 40 m belo the intake water surface level, find the pressure head.

(*ii*) just upstream of C and

(*iii*) just downstream of C.

Assume sudden contraction at C.

Sol. Given : l_1 = 200 m, d_1 = 350 mm, l_2 = 100 mm, d_2 = 250 mm, H = 50 m,

$$k_{e1} = 0.8,\ k_{e2} = 0.21,\ f_1 = 0.006 = f_2$$

(*i*) Minor losses :

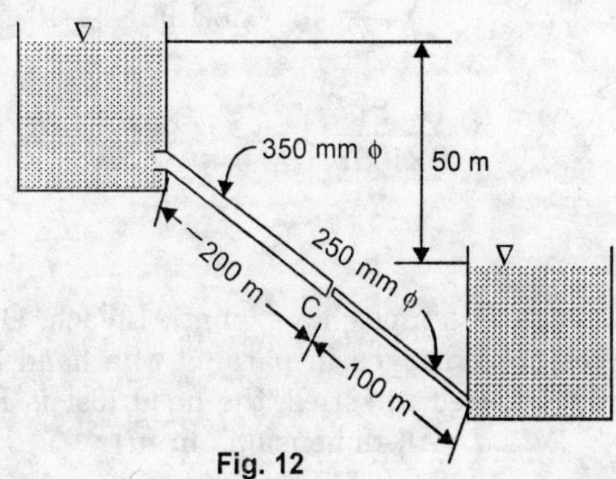

Fig. 12

$$A_1v_1 = A_2v_2$$

$$v_2 = \left(\frac{d_1}{d_2}\right)^2 v_1 = \left(\frac{350}{250}\right)^2 v_1 = 1.96\ v_1$$

$$H = k_{e1}\frac{v_1^2}{2g}+\frac{4f_1l_1v_1^2}{2gd_1}+k_{e2}\frac{v_2^2}{2g}+\frac{4f_2l_2v_2^2}{2gd_2}+\frac{v_2^2}{2g}$$

$$= 0.8\frac{v_1^2}{2g} + \frac{4\times0.06\times200\times v_1^2}{2g\times0.35} + \frac{0.27\times(1.96)^2\times v_1^2}{2g}$$

$$+ \frac{4\times0.06\times100\times(1.96)^2\times v_1^2}{2g\times0.25} + \frac{(1.96)^2 v_1^2}{2g}$$

$$50 = 511.4\times\frac{v_1^2}{2g}$$

$$v_1 = 1.38 \text{ m/s}$$

Rate of discharge,

$$\dot{Q} = v_1 A_1 = 1.38\times\frac{\pi}{4}(0.35)^2 = 0.1328 \text{ m}^3/\text{s}$$

(*ii*) Applying Bernoulli's equation at section 1 and upstream of 2 (point C), we get

$$\frac{p_1}{\gamma} + \frac{v_1^2}{2g} + z_1 = \frac{p_2}{\gamma} + \frac{v_2^2}{2g} + z_2$$

$$\frac{v_1^2}{2g}[1-(1.96)^2] + (z_1 - z_2) = \frac{p_2}{\gamma}$$

$$\frac{(1.38)^2}{2\times9.81}\left[1-(1.96)^2\right] + 40 = \frac{p^2}{\gamma}$$

$$\frac{p_2}{\gamma} = 39.72 \text{ m}$$

(*ii*) Applying Bernoulli's equation at section 1 and just downstream of point C, we have

Loss of head due to sudden contraction,

$$h_c = 0.5\frac{v_2^2}{2g}$$

$$\frac{p_1}{\gamma} + \frac{v_1^2}{2g} + z_1 = \frac{p_2}{\gamma} + \frac{v_2^2}{2g} + z_2 + \frac{0.5\,v_2^2}{2g}$$

$$\frac{v_1^2}{2g}\left[1-1.5(1.96)^2\right] + 40 = \frac{p_2}{\gamma}$$

$$\frac{(1.38)^2}{2\times9.81}\left[1-1.5\,(1.96)^2\right] + 40 = \frac{p_2}{\gamma}$$

$$\frac{p_2}{\gamma} = 39.54 \text{ m}$$

(***d***) Two pipes have length L each. One of them has a diameter of D and other *d*. If the pipes are arranged in parallel, the head loss is *h* for a discharge of Q m³/s. When the pipe are connected in series, the head lost is H for same discharge. Find the ratio of H to *h*, for D = 2.25 *d*. (All dimensions in *m*) **(10)**

Sol. 1. Pipes connected in parallel

$$Q = Q_1 + Q_2$$

Loss of head in each pipe,

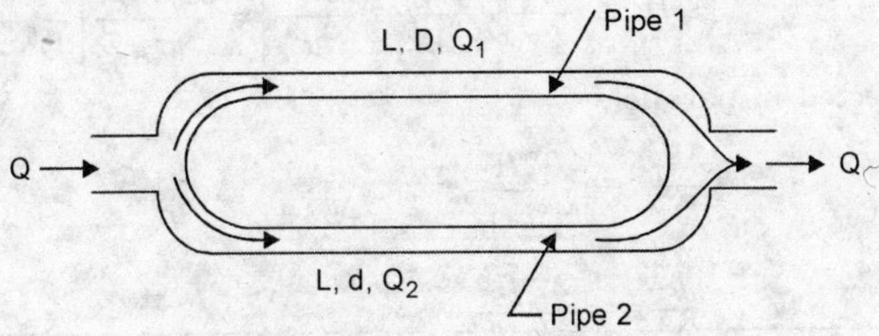

Fig. 13(a). Pipes connected in parallel

$$h_{l1} = \frac{4fLv_1^2}{2gD}$$

$$v_1 = \frac{4Q_1}{\pi D^2}$$

$$h_{l1} = \frac{4fL}{2gD}\left(\frac{4Q_1}{\pi D^2}\right)^2 = \frac{32\, fLQ_1^2}{\pi^2 gD^5}$$

$$h_{l2} = \frac{32fLQ_2^2}{\pi^2 gd^5}$$

But $\quad h_{l1} = h_{l2} = h$

$$\therefore \quad \frac{Q_1^2}{D^5} = \frac{Q_2^2}{d^5}$$

$$\frac{Q_1}{Q_2} = \left(\frac{D}{d}\right)^{2.5} = (2.25)^{2.5} = 7.594$$

$$Q = (7.594 + 1)\, Q_2$$

$$Q_2 = \frac{Q}{8.594} = 0.116\ Q,\ Q_1 = Q - Q_2 = Q - 0.116\ Q = 0.884\ Q$$

2. Pipes connected in series

Fig. 13(b). Pipes connected is series.

$$H = \frac{4fLv_1^2}{2gD} + \frac{4fLv_2^2}{2gd}$$

$$v_1 = \frac{4Q}{\pi D^2}, \quad v_2 = \frac{4Q}{\pi d^2}$$

$$H = \frac{4fL}{2g}\left[\frac{16Q^2}{\pi^2 D^5} + \frac{16Q^2}{\pi^2 d^5}\right] = \frac{32fLQ^2}{\pi^2 g}\left[\frac{1}{D^5} + \frac{1}{d^5}\right]$$

For pipes connected in parallel,

$$\frac{h}{Q_1^2} = \frac{32fL}{\pi^2 g D^5}$$

and

$$\frac{h}{Q_2^2} = \frac{32\,fL}{\pi^2 g d^5}$$

$$\therefore \quad H = h\left(\frac{Q^2}{Q_1^2} + \frac{Q^2}{Q_2^2}\right)$$

$$\frac{H}{h} = Q^2\left[\frac{1}{(0.884Q)^2} + \frac{1}{(0.116Q)^2}\right] = 75.60$$

Q.6. (*a*) Develop (in general terms) an expression for the per cent of error in Q over a triangular weir if there is a small error in the measurement of the vertex angle. Assume there is no error in the weir coefficient. Compute the per cent error in Q, if there is 1° error in the measurement of total vertex angle of a triangular weir which is having a total vertex angle of 60°. (10)

Sol. Discharge over a triangular weir is given by,

$$Q = \frac{8}{15} C_d \sqrt{2g}\,\tan\frac{\theta}{2} \times H^{5/2}$$

$$= k\tan\frac{\theta}{2} \qquad (1)$$

where

$$k = \frac{8}{15} C_d \sqrt{2g}\, H^{5/2}$$

Differentiating Eq. (1) w.r.t. θ, we get

$$dQ = k\sec^2\frac{\theta}{2}\cdot\frac{1}{2}d\theta \qquad (2)$$

Dividing Eq. (2) by (1), we get

$$\frac{dQ}{Q} = \frac{k\sec^2\frac{\theta}{2}\times\frac{1}{2}d\theta}{k\tan\frac{\theta}{2}} = \frac{d\theta}{2\cos^2\frac{\theta}{2}\times\frac{\sin\frac{\theta}{2}}{\cos\frac{\theta}{2}}} = \frac{d\theta}{2\sin\frac{\theta}{2}\cos\frac{\theta}{2}}$$

$$= \frac{d\theta}{\sin\theta} = \text{cosec}\,\theta\cdot d\theta$$

$$\theta = 60° \text{ and } d\theta = 1° \text{ (given)}$$

$$\frac{dQ}{Q} = \frac{\pi}{180} \times 1° \times \text{cosec } 60° = \frac{\pi}{180} \times \frac{2}{\sqrt{3}} = 0.02 \quad \text{or} \quad 2\%$$

$\therefore$ Error in $\quad Q = 2\%$

(*b*) What is the hydraulic jump in the flow in open channel? What are its types and characteristics with respect to Froude's number of flow (10)

Sol. Hydraulic Jump : In an open channel, when flow changes from super-critical state to subcritical state, the depth of flow is increased. This increase in depth occurs in a short length of the channel, and is known as the hydraulic jump. A hydraulic jump is, therefore, an abrupt rise in the depth of flow signifying a change of flow from super-critical state to sub-critical state. It involves violent mixing due to high turbulence and air entrainment.

Hydraulic jump generally occurs at such places as the foot of spillway and downstream of the sluice gate where the velocities are very high. The Hydraulic jump converts kinetic energy of rapidly flowing stream into potential energy. It is responsible for high loss of energy and is frequently employed as an effective energy dissipating device. The hydraulic jump is also known as a *standing wave.*

Analysis of Hydraulic Jump : The following assumptions are made in the analysis of hydraulic jump :

1. The flow is uniform and the pressure distribution is hydrostatic before and after the jump.
2. The channel is horizontal or of very small slope.
3. Loss of head due to friction in the channel is negligible.
4. The momentum correction factor is unity.

The hydraulic jump may be analysed by applying the continuity equation, momentum principle and the energy equation. For simplicity in mathematical treatment, we consider a horizontal channel of rectangular section.

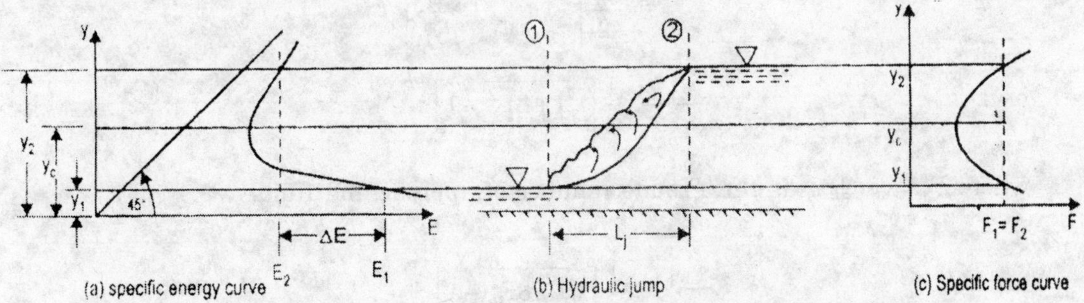

Fig. 14. Hydraulic jump in a horizontal rectangular channel.

Applying momentum equation to the body of water enclosed between sections (1) and (2) of the hydraulic shown in Fig. 14(*b*), we have

$$\frac{\gamma Q}{g}(v_2 - v_1) = P_1 - P_2 \tag{1}$$

where

Hydrostatic force at section 1,

$$P_1 = \gamma z_1 A_1 = \frac{\gamma y_1}{2} \times b y_1 = \frac{1}{2} \gamma b y_1^2$$

Hydrostatic force at section 2,

$$P_2 = \frac{1}{2} \gamma b y_2^2$$

b = width of rectangular channel

From continuity equation, we have

$$Q = b_1 y_1 v_1 = b_2 y_2 v_2 \quad (2)$$

If q = discharge per unit width

Then $Q = bq$

Eq. (1) becomes,

$$\frac{\gamma(bq)}{g}\left(\frac{q}{y_2} - \frac{q}{y_1}\right) = \frac{1}{2}\gamma b y_1^2 - \frac{1}{2}\gamma b y_2^2$$

which simplifies to,

$$\frac{2q^2}{g} = y_1 y_2 (y_1 + y_2) \quad (3)$$

or

$$y_2^2 + y_1 y_2 - \frac{2q^2}{g y_1} = 0$$

$$y_2 = \frac{y_1}{2}\left[-1 \pm \sqrt{1 + \frac{8q^2}{g y_1^3}}\right]$$

$$= \frac{y_1}{2}\left[-1 \pm \sqrt{1 + \frac{8v_1^2}{g y_1}}\right]$$

or

$$\frac{y_2}{y_1} = \frac{1}{2}\left[\sqrt{1 + 8F_1^2} - 1\right]$$

where $F_1 = \frac{v_1}{\sqrt{g y_1}}$, is the Froude number of approaching fluid.

and

$$\frac{y_1}{y_2} = \frac{1}{2}\left[\sqrt{1 + 8F_2^2} - 1\right]$$

where $F_2 = \frac{v_2}{\sqrt{g y_2}}$, is the Froude number downstream of the jump.

(*i*) For $F_1 < 1$, $\frac{y_2}{y_1} < 1$, This is the condition for a falling liquid surface and the flow is sub-critical.

(*ii*) For $F_1 = 1$, $\frac{y_2}{y_1} = 1$, and the flow is critical.

(*iii*) For $F_1 > 1$, $\frac{y_2}{y_1} > 1$, It is essential for the formation of hydraulic jump.

(*c*) Air flows isothermally in a long pipe. At one section the pressure is 600 kPa abs, the temperature is 25°C and the velocity is 30 m/s. At a second section (at some distance from the first section) the pressure is 100 kPa abs. Find the energy head loss due to friction and determine the thermal energy that must have been added to or taken from the fluid between the two sections. Assume the diameter of pipe to be constant. (10)

Sol. $p_1 = \rho_1 RT_1$

$$\rho_1 = \frac{600 \times 10^3}{287 \times 298} = 7.015 \text{ kg/m}^3$$

For isothermal process,

$$T_2 = T_1$$

$$\rho_2 = \frac{p_2}{RT_2} = \frac{100 \times 10^3}{287 \times 298} = 1.169 \text{ kg/m}^3$$

Also $\rho_1 v_1 = \rho_2 v_2$, $\quad v_2 = \frac{7.015 \times 30}{1.169} = 5$ m/s

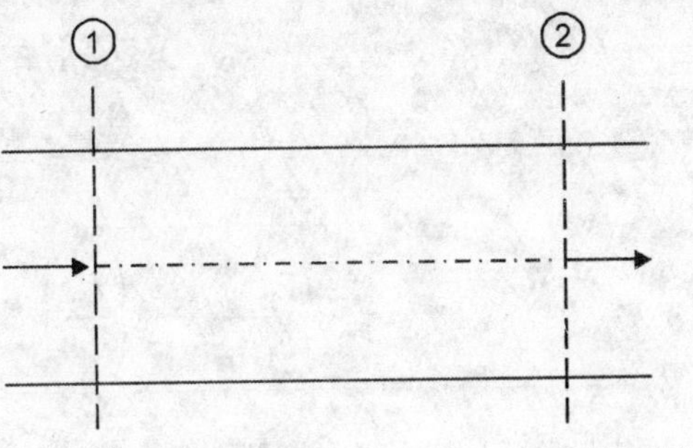

Fig. 15

Applying energy equation for isothermal flow, we have

$$\frac{p_1}{\rho_1 g} \ln p_1 + \frac{v_1^2}{2g} = \frac{p_2}{\rho_2 g} \ln p_2 + \frac{v_2^2}{2g} + h_E$$

$$\frac{600 \times 10^3}{7.015 \times 9.81} \ln(600 \times 10^3) + \frac{30^2}{2 \times 9.81} = \frac{100 \times 10^3}{1.169 \times 9.81} \ln(100 \times 10^3) + \frac{5^2}{2 \times 9.81} + h_E$$

$$116 \times 10^3 + 45.87 = 100.392 \times 10^3 + 1.274 + h_E$$

$$h_E = 15{,}652.6 \text{ m}$$

Energy equation gives,

$$h_1 + \frac{v_1^2}{2} = h_2 + \frac{v_2^2}{2}$$

$$h_1 - h_2 = \frac{v_2^2 - v_1^2}{2} = \frac{1}{2}(5^2 - 30^2) = -437.5 \text{ J/kg}$$

Hence 437.5 J/kg of thermal energy has been taken from the fluid between two sections.

(*d*) Researchers plan to test a 1 : 13 model of a ballistic missile in a high speed wind tunnel. The prototype missile will travel at 380 m/s through air at 23°C and 95.0 kPa (abs).

(*i*) If the air in the wind tunnel test section has a temperature of –20°C at a pressure of 89 kPa (abs), what must its velocity be? and

(*ii*) Estimate the drag force on the prototype if the drag force on the model is 400 N.

Sol. Given : v_p = 380 m/s, T_p = 23 + 273 = 296 K, p_p = 95 kPa (abs)

T_m = – 20 + 273 = 253 K, p_m = 89 kPa (abs), v_m = ?

$(F_D)_m$ = 400 N, $(F_D)_p$ = ?

(*i*)
$$(v_s)_m = \sqrt{\gamma R T_m},\ (v_s)_p = \sqrt{\gamma R T_p}$$

$$(v_s)_m/(v_s)_p = \sqrt{T_m / T_p} = \sqrt{\frac{253}{296}} = 0.9245$$

$$\frac{v_m}{v_p} = \sqrt{\frac{(v_s)_m}{(v_s)_p}} \text{ for } \rho_m = \rho_p$$

$$v_m = 380\sqrt{0.9245} = 365 \text{ m/s}$$

(*ii*)
$$l_r = \frac{l_m}{l_p} = \frac{1}{13}$$

$$\frac{(F_D)_m}{(F_D)_p} = \left(\frac{l_m}{l_p}\right)^2 \times \left(\frac{v_m}{v_p}\right)^2$$

$$(F_D)_p = 400 \times (13)^2 \times \left(\frac{380}{365}\right)^2 = 73,270 \text{ N}$$

Q.7. (*a*) Explain the effect of vane angle on manometric efficiency in a centrifugal pump. Why is vane angle in a centrifugal pump not kept at values below 20°? (10)

Sol. Manometric efficiency of centrifugal pump,

$$\eta_{mano} = \frac{gH_{mano}}{u_2 v_{w2}}$$

$$v_{w2} = u_2 - v_{r2} \cos \beta_2 = u_2 - v_{f2} \cot \beta_2$$

where β_2 = vane angle at outlet

As β_2 increases, v_{w2} increases and η_{mano} decreases. As β_2 decreases, v_{w2} decreases and η_{mano} increases. However β_2 cannot be decreased beyond a certain value as this will decrease the velocity of flow drastically. Hence β_2 is not kept below 20° from practical considerations.

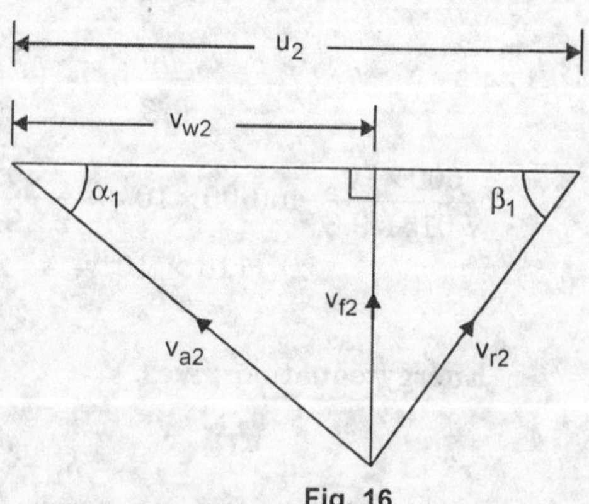

Fig. 16

(*b*) Fluid flows with a velocity of V_∞ over a flat plate located at $y = 0$. The leading edge of the plate is located at $x = 0$. The possible velocity profiles in the boundary layer having thickness $\delta(x)$ are as follows :

(*i*) $$\frac{u}{V_\infty} = 2\frac{y}{\delta} - \left(\frac{y}{\delta}\right)^2$$

(*ii*) $$\frac{u}{V_\infty} = \frac{3}{2}\frac{y}{\delta} - \frac{1}{2}\left(\frac{y}{\delta}\right)^3$$

(*iii*) $$\frac{u}{V_\infty} = \frac{y}{\delta} - \frac{1}{2}\left(\frac{y}{\delta}\right)^2$$

Find out which of these velocity profiles are feasible for laminar flow. Also, identify the smoothest velocity profile from these options. (10)

Sol. (*i*) For $$\frac{u}{V_\infty} = 2\left(\frac{y}{\delta}\right) - \left(\frac{y}{\delta}\right)^2$$

$$\delta = \frac{5.48\,x}{\sqrt{\text{Re}_x}} \text{ for laminar boundary layer}$$

(*ii*) For $$\frac{u}{V_\infty} = \frac{3}{2}\left(\frac{y}{\delta}\right) - \frac{1}{2}\left(\frac{y}{\delta}\right)^2$$

$$\delta = \frac{4.64\,x}{\sqrt{\text{Re}_x}} \text{ for laminar boundary layer}$$

(*iii*) For $$\frac{u}{V_\infty} = \left(\frac{y}{\delta}\right) - \frac{1}{2}\left(\frac{y}{\delta}\right)^2$$

$$\frac{\tau}{\rho V_\infty^2} = \frac{d}{dx}\left[\int_0^\delta \frac{u}{V_\infty}\left(1 - \frac{u}{V_\infty}\right)dy\right]$$

$$= \frac{d}{dx}\left[\int_0^\delta \left\{\frac{y}{\delta} - \frac{1}{2}\left(\frac{y}{\delta}\right)^2\right\}\left\{1 - \frac{y}{\delta} + \frac{1}{2}\left(\frac{y}{\delta}\right)^2\right\}dy\right]$$

$$= \frac{d}{dx}\left[\int_0^\delta \left\{\frac{y}{\delta} - \frac{1}{2}\left(\frac{y}{\delta}\right)^2 - \frac{y^2}{\delta^2} + \frac{1}{2}\left(\frac{y}{\delta}\right)^3 + \frac{1}{2}\left(\frac{y}{\delta}\right)^3 - \frac{1}{4}\left(\frac{y}{\delta}\right)^4\right\}dy\right]$$

$$= \frac{d}{dx}\left[\int_0^\delta \left\{\frac{y}{\delta} - \frac{3}{2}\frac{y^2}{\delta^2} + \frac{y^3}{\delta^3} - \frac{1}{4}\frac{y^4}{\delta^4}\right\}dy\right]$$

$$= \frac{d}{dx}\left|\frac{y^2}{2\delta} - \frac{1}{2}\frac{y^3}{\delta^2} + \frac{y^4}{4\delta^3} - \frac{1}{20}\frac{y^5}{\delta^4}\right|_0^\delta$$

$$= \frac{d}{dx}\left[\frac{\delta}{2} - \frac{\delta}{2} + \frac{\delta}{4} + \frac{\delta}{20}\right]$$

$$= \frac{d}{dx}\left[\frac{3}{10}\delta\right] = \frac{3}{10}\frac{d\delta}{dx}$$

$$\tau_0 = \frac{3}{10}\rho V_\infty^2 \frac{d\delta}{dx}$$

Also $$\tau_0 = \mu\left(\frac{du}{dy}\right)_{y=0}$$

But $$u = V_\infty\left[\frac{1}{\delta} - \frac{1}{2}\left(\frac{y}{\delta}\right)^2\right]$$

$$\frac{du}{dy} = V_\infty\left[\frac{1}{\delta} - \frac{y}{\delta^2}\right]$$

$$\left(\frac{du}{dy}\right)_{y=0} = V_\infty \times \frac{1}{\delta}$$

$$\tau_0 = \frac{\mu V_\infty}{\delta} = \frac{3}{10}\rho V_\infty^2 \frac{d\delta}{dx}$$

$$\delta \cdot d\delta = \frac{10}{3}\frac{\mu}{\rho V_\infty}dx$$

Integrating both sides, we get

$$\frac{\delta^2}{2} = \frac{10}{3}\frac{\mu}{\rho V_\infty}x + c$$

At $x = 0$, $\delta = 0$. $\therefore\ c = 0$

$$\delta^2 = \frac{20}{3}\frac{\mu x}{\rho V_\infty}$$

$$\delta = \sqrt{\frac{20}{3}} \cdot \sqrt{\frac{\mu x}{\rho V_\infty}}$$

$$= 2.582\sqrt{\frac{\mu}{\rho V_\infty x} \cdot x^2} = 2.582\frac{x}{\sqrt{\text{Re}_x}}$$

All the three velocity profiles are feasible for laminar flow. Third profile is most smoothest.

(*c*) Describe with neat sketch, the construction and working principle of Rotameter. What are its advantages? (10)

Sol. Rotameter : It is a commonly used device to measure discharge. In this device, shown in Fig. 17 the flow enters the bottom of a vertically placed tapered tube and causes the bob or float to move upwards. The float will rise to a point in the tube where the drag force and buoyant force is balanced by the weight of the float. The position of the float in the tube is taken as an indication of the flow rate.

The force balance equation of the float is :

$$F_{drag} + F_{buoyancy} = F_{weight}$$

$$(p_d - p_u) + \rho_f g V_b = \rho_b g V_b$$

where V_b = volume of bob

A_b = area of bob

ρ_f, ρ_b = density of flowing fluid and bob respectively

p_d, p_u = pressure at the downward and upward faces of bob respectively

$$p_d - p_u = \frac{V_b}{A_b} g(\rho_b - \rho_f)$$

$$Q_{\text{actual}} = C_d \frac{A_t(A_t - A_b)}{\sqrt{A_1^2 - (A_t - A_b)^2}} \sqrt{2g} \sqrt{\frac{p_d - p_u}{\rho_f g}}$$

where A_t = area of tube at the float level

$A_t - A_b$ = minimum annular area between tube and bob

C_d = coefficient of discharge

$$Q_{\text{actual}} = \frac{C_d(A_t - A_b)}{\sqrt{1 - \left[\frac{A_E - A_b}{A_t}\right]^2}} \sqrt{2g} \sqrt{\frac{V_b}{A_b} \frac{(\rho_b - \rho_f)}{\rho_f}}$$

If $\frac{A_t - A_b}{A_t} << 1$ and c_d is constant, then

$$Q_{\text{actual}} = k(A_t - A_b) = k\sqrt{\frac{\rho_b - \rho_f}{\rho_f}}$$

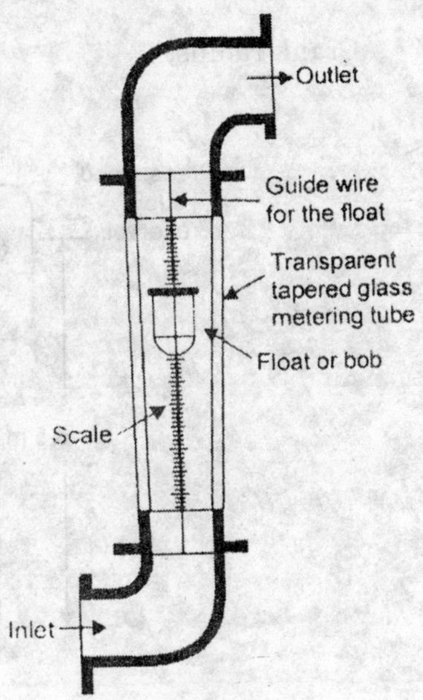

Fig. 17. Rotameter

(*d*) A single acting reciprocating pump lifts a liquid of specific weight 9.0 kN/m^3 from a pressurised storage reservoir to an overhead container. The free surface in the supply reservoir is at an elevation of 3.5 m above the centre of the pump. The ambient pressure over the liquid surface in the supply reservoir is 27 kN/m^2 (vacuum). The other relevant data relating to the pump are as follows :

Length of the suction pipe = 6.0 m

Diameter of suction pipe = 10 cm

Minimum pressure anywhere in the system = 27 kPa (abs)

Length of stroke = 50 cm

Diameter of cylinder = 20 cm

Atmospheric pressure = 99 kN/m^2 (abs)

Determine the maximum speed admissible in rpm. (10)

Sol. Given : γ = 9 kN/m^3, h = –3.5 m, p_a = 27 kN/m^2 (vacuum), L_s = 6m, d_s = 10 cm, $p_{\min}$ = 27 kPa (abs), L = 50 cm, D = 20 cm, p_{atm} = 99 kN/m^2 (abs)

$$H_{\text{atm}} = \frac{p_{\text{atm}}}{\gamma} - \frac{99}{9} = 11 \text{ m}$$

$$H_s = \frac{p_a}{\gamma} = \frac{27}{9} = 3 \text{ m(vacuum)}$$

Separation pressure head,

$$H_{sep} = \frac{p_{min}}{\gamma} = \frac{27}{9} = 3 \text{ m}$$

Crank radius, $$r = \frac{L}{2} = \frac{50}{2} = 25 \text{ cm} \quad \text{or} \quad 0.25 \text{ m}$$

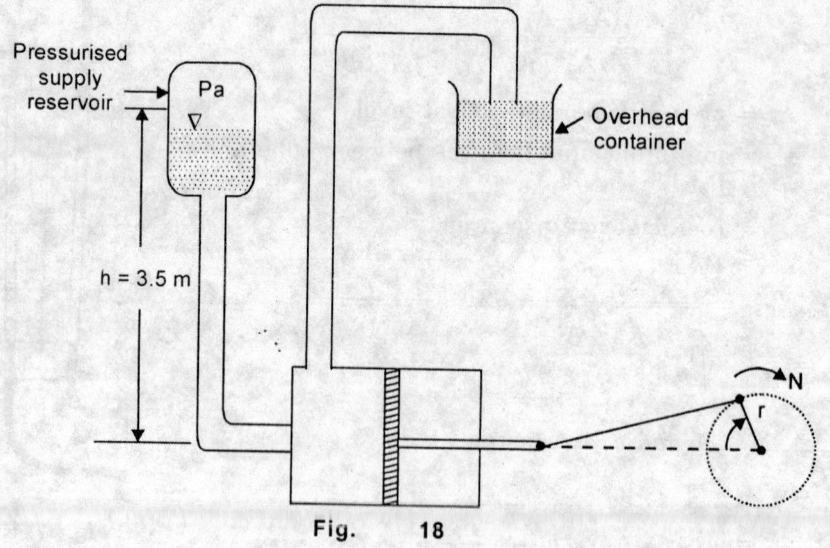

Fig. 18

The possibility of separation is only at the beginning of suction stroke,

Acceleration head in suction pipe,

$$H_{as} = \frac{L_s}{g} \times \frac{A}{A_s} \times w^2 r = \frac{6}{9.81} \times \left(\frac{20}{10}\right)^2 \times \left(\frac{2\pi \times 1}{60}\right)^2 \times 0.25$$

$$= 6.707 \times 10^{-3} \times N^2 \text{ m}$$

$$H_{atm} - (H_s + H_{as}) = H_{sep}$$

$$11 - (3 + 6.707 \times 10^{-3} \times N^2) = 3$$

$$5 = 6.707 \times 10^{-3} \times N^2$$

$$N = 27.3 \text{ rpm}$$

Q.8. (*a*) Define air rate, specific power and the cycle work ratio in a gas turbine. What is the significance of these parameters?

Sol. Air rate : It is the amount of air supplied to the turbine per unit time expressed in kg/s.

Specific power : It is the net power developed by the turbine per kg of air supplied.

Specific output = $\frac{w_{net}}{c_p T_1}$

Cycle work ratio : It is the ratio of the net turbine work to the turbine work.

Work ratio, $$R_w = \frac{w_{net}}{w_t} = \frac{w_t - w_c}{w_t} = 1 - \frac{w_c}{w_t}$$

(*b*) Justify the selection of a turbine runner with highest specific speed possible. Derive an expression for the specific speed of a pelton wheel, for speed ratio 0.86, overall efficiency 85% and co-efficient of velocity of nozzle 0.98. (10)

Sol. Specific speed of a turbine,

$$N_s = \frac{N\sqrt{P}}{H^{5/4}}$$

If N_s is more then it means with the same head available and runner speed the power developed will be more.

Given : Speed ratio, $k_v = 0.46,\ \eta_{overall} = 0.85,\ C_v = 0.98$

Wheel velocity, $u = K_u\sqrt{2gH} = 0.46\sqrt{2\times9.81\times H} = 2.037\sqrt{H}$ m/s

Jet velocity, $v_1 = C_v\sqrt{2gH} = 0.98\sqrt{2\times9.81\times H} = 4.341\sqrt{H}$ m/s

Overall efficiency, $\eta_{overall} = \dfrac{10^3\times P}{\gamma QH} = 0.85$ (P in kW)

$$u = \frac{\pi DN}{60} = 2.037\sqrt{H}$$

$$N = \frac{38.904\sqrt{H}}{D}$$

$$Q = \text{area of nozzle} \times \text{jet velocity}$$

$$= \frac{\pi}{4}d^2\times v_1 - \frac{\pi}{4}d^2\times 4.311\sqrt{H} = 3.409\sqrt{H}d^2$$

$$P = 0.85\times10^3\times9.81\times\ 3.409\sqrt{H}\ d^2\times H/10^3 = 28.426\ d^2\ H^{3/2}$$

$$N_s = \frac{38.904\sqrt{H}}{D}\times\frac{[28.426\ d^2\ H^{3/2}]^{1/2}}{H^{5/4}} = \frac{207.42}{m}$$

where $m = D/d$ = jet ratio

Otherwise

$$N = \frac{60\ K_u\sqrt{2gH}}{\pi D}\ \text{rpm}$$

$$P = \frac{\gamma QH}{10^3}\times\eta_0\ \text{kW}$$

$$Q = \frac{\pi}{4}d^2K_v\sqrt{2gH}\ \text{m}^3/\text{s}$$

$$P = \frac{9.81\times10^3\times\frac{\pi}{4}d^2\ K_v\sqrt{2g\ H}\times H\times\eta_0}{10^3}$$

$$= 7.7047\ d^2\ K_v\sqrt{2gH}\times H\times\eta_0$$

$$N_s = \frac{60\,K_u\sqrt{2g\,H}}{\pi D} \times \frac{\left[7.7047\,d^2 K_v \sqrt{2gH} \times H \times \eta_0\right]^{1/2}}{H^{5/4}}$$

$$= \frac{60 \times \sqrt{2g}}{\pi} \times 2.7757 \times (2g)^{1/4} \times \frac{K_u \sqrt{H}}{D} \times \frac{d\sqrt{K_v} \times H^{3/4} \times \eta_0^{1/2}}{H^{5/4}}$$

$$= 494.2 \times K_u \sqrt{K_v}\,\frac{d}{D} \times \eta_0^{1/2}$$

$$= 494.2 \left[\frac{K_u \sqrt{K_v}\ \eta_0^{1/2}}{m}\right]$$

(*c*) What are the shortcomings of a fire tube boiler restricting it for limited use? How are the impurities in feed water removed from the boiler? (10)

Sol. Shortcomings of a fire-tube boiler

1. Takes more time to raise stream.
2. Unsuitable for large power plants because of lower rate of stream raised.
3. Low operating pressure.
4. Prone to serious risk of damage to property in case of bursting because of large water capacity.
5. Difficult to transport because of large size.
6. Difficult to maintain.

Methods for removal of impurities from boiler feed water

Impurities from boiler feed water are removed by the following methods :

1. Blow down cock
2. Water treatment plant

(*d*) What are the constructional features of an axial flow compressor? How air is compressed and what is the method of getting higher compression ratio in such compressor?

Constructional Features of Axial Flow Compressor

The cross-section of a typical axial flow compressor is shown in Fig. 19.

It consists of adjacent rows of moving and fixed blades. The moving blades are mounted on the rotating drum and fixed blades are fixed to the casing or stator. One stage of the compressor comprises of a row of moving blades followed by a row of fixed blades.

The blades are of aero-foil section designed on the basis of aerodynamic theory. Such a design prevents losses due to shock and turbulence and are free from stalling problems. The blades are said to be stalled when the air stream fails to follow the blade profile.

The blades of a reaction gas turbine have profile formed by a number of circular arcs. This is so because the acceleration process carried out in the converging blade passage of a reaction turbine is much more efficient and stable process as compared with the diffusing or decelerating process carried out in the diverging passage between the blades of an axial flow compressor.

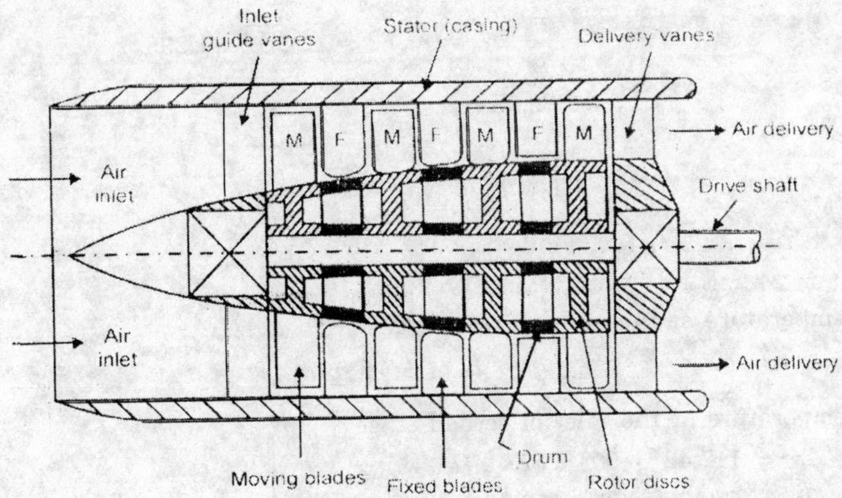

Fig. 19. Typical axial flow compressor.

To keep the flow velocity constant throughout the compressor length the annular area is reduced from inlet to outlet of the compressor. In the diverging passage of the moving blades, there is rise in temperature due to diffusion and the absolute velocity is also increased due to work input.

The fixed blades serve to convert a part of the kinetic energy of air into pressure energy by diffusion process and also guide and redirect the air stream to enter the next stage without shock.

Working Principle

The axial flow compressor is generally driven by an internal combustion engine or a turbine. The work input to the rotor shaft is transferred by the rotor moving blades to the air, thus accelerating it. The space between the moving blades form diffuser passages so that the velocity of air relative to the blades is decreased as the air passes through them and consequently the pressure rises. The air is then further diffused in the fixed blades which also form diffuser passages. In the fixed blades the air is turned throught an angle so that its direction is such that it can pass to the next ring of moving blades without shock. Generally 5 to 14 stages are used.

(e) A gas turbine utilizes two-stage centrifugal compressor. The pressure ratios for the first and second stages are 2.5 and 2.1 respectively. The flow of air is 10 kg/s, this air being drawn at 1.013 bar and 20°C. If the temperature drop in the intercooler is 60°C and the isentropic efficiency is 90% for each stage, calculate :

(*i*) the actual temperature at the end of each stage and

(*ii*) the total compressor power.

Assume $\gamma = 1.4$ and $c_p = 1.005$ kJ/kg K for air.

Sol. Given : $r_{p1} = 2.5$, $r_{p2} = 2.1$, $\dot{m}_a = 10$ kg/s, $p_1 = 1.013$ bar, $T_1 = 20 + 273 = 293$ K, $\Delta T_2 = 60°C$, $\eta_{i1} = \eta_{i2} = 0.9$, $\gamma = 1.4$, $c_p = 1.005$ kJ/kg·K.

(i) $$\frac{T_2}{T_1} = \left(\frac{p_2}{p_1}\right)^{\frac{\gamma-1}{\gamma}} = (r_{p1})^{\frac{\gamma-1}{\gamma}} = (2.5)^{\frac{0.4}{1.4}} = 1.299$$

$$T_2 = 380.68 \text{ K}$$

$$\eta_{i1} = \frac{T_2 - T_1}{T_2' - T_1}$$

$$0.9 = \frac{380.68 - 293}{T_2' - 293}$$

Temperature at the end of first stage,

$$T_2' = 390.42 \text{ K or } 117.42°\text{C}$$

Temperature at the inlet of second stage

$$= 117.42° - 60° = 57.42°\text{C}$$

$$T_3 = 57.42 + 273 = 330.42 \text{ K}$$

$$\frac{T_4}{T_3} = \left(\frac{p_3}{p_2}\right)^{\frac{\gamma-1}{\gamma}} = (r_{p2})^{\frac{\gamma-1}{\gamma}} = (2.1)^{\frac{0.4}{1.4}} = 1.236$$

$$T_4 = 408.44 \text{ K}$$

$$\eta_{i2} = \frac{T_4 - T_3}{T_4' - T_3}$$

$$0.9 = \frac{408.44 - 330.42}{T_4' - 330.42}$$

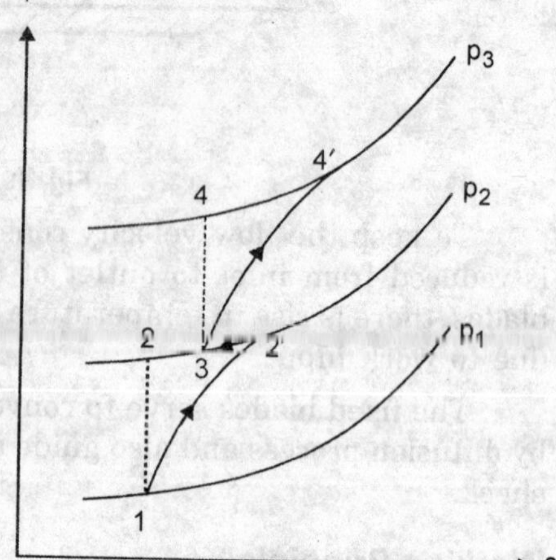

Fig. 20

Temperature at the end of second stage,

$$T_4' = 417.11 \text{ K} \quad \text{or} \quad 144.1°\text{C}$$

(ii) $$\dot{W}_{c1} = \dot{m}_a c_p (T_2' - T_1) = 10 \times 1.005\ (390.42 - 293) = 979.07 \text{ kJ/s}$$

$$\dot{W}_{c2} = \dot{m}_a c_p (T_4' - T_3) = 10 \times 1.005\ (417.11 - 330.42) = 877.23 \text{ kJ/s}$$

Total compressor power,

$$\dot{W}_c = \dot{W}_{c1} + \dot{W}_{c2} = 979.07 + 871.23 = 1850.3 \text{ kW}$$

MECHANICAL ENGINEERING
PAPER II—1992

(Conve ntional)

Time allowed : Three hours *Maximum marks : 200*

Candidates should attempt Question No. 1 in Section A which is compulsory, TWO questions from Section B and TWO questions from Section C.

Question No. 1 is of short answers type.

The number of marks carried by each subdivision of a question is indicated at the end of the subdivision / questions.

Answer must be written in ENGLISH.

Asusme suitable data, if necessary and indicate the same clearly.

Unless otherwise indicated, symbols and notations used have their usual meanings.

Neat sketches to be drawn, wherever required.

Section A

Q.1. **(*a*)** State the condition for a four bar linkage to be called a Grashof's linkage and also when this linkage would become a double–crank mechanism.

Sol. Condition for a Grashof's linkage : $(l + s) < (a + b)$

Where l = length of the longest link

s = length of the shortest link

a, b = length of other links

It would become a double crank mechanism when s is fixed.

(*b*) Define the coefficient of fluctuation of energy in case of a flywheel. How is it related to the coefficient of fluctuation of speed?

Sol. Coefficient of fluctuation of energy, K_e = Energy change/Average energy

Coefficient of fluctuation of speed, K_s = Speed change/Average speed

$$K_e = 2K_s$$

(*c*) What is hunting in a centrifugal governor? When does it occur?

Sol. Hunting is a state of high amplitude oscillations of the governor. It occurs when frequency of speed fluctuation coincides with the natural frequency of oscillation of the governor.

(*d*) When can you say that a vibratory system is critically damped? How is critical damping coefficient related to mass (m) and spring constant (k) of the system?

Sol. The minimum damping for which a vibrating system ensues aperiodic motion after being disturbed from its equilibrium position, is defined as critical damping. $c_c = 2(km)^{0.5}$, where c_c = critical damping, k = spring constant, and m = mass.

(*e*) What is a Gyro? How is gyroscopic torque (T) related to spin velocity (ω_s) and precession velocity (ω_p)?

Sol. Gyro is a spinning rotor having possibility of precessing about another axis.

$T = I\omega_s\omega_p$, where I = mass moment of inertia of the rotor.

(*f*) State the fundamental law of gearing. State when the meshing surfaces can be *called conjugate.*

Sol. The law of gearing states that the common normal at the point of contact of two meshing gears must always pass through a fixed point (pitch point) on the line of centres. When meshing surfaces provide a constant angular velocity ratio, they are called conjugate.

(*g*) Sketch international metric V-thread with flat root and express various dimensions in terms of pitch *p*.

Sol. The thread is shown in Fig. 1.

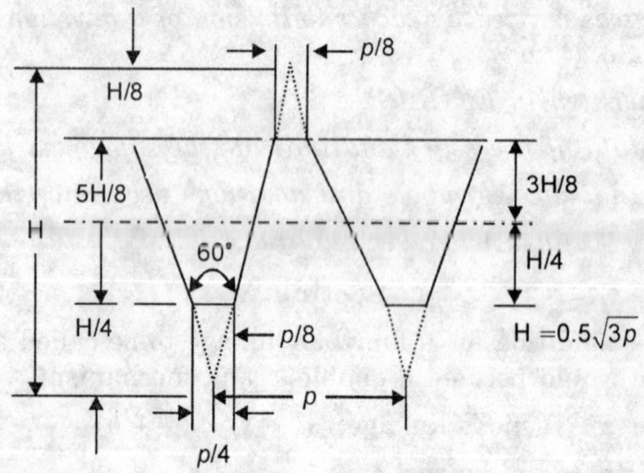

Fig. 1. V-thread with flat foot.

(*h*) Distinguish between unilateral and bilateral tolerances.

Sol. Unilateral tolerances allow variations only in one direction from design size.

Bilateral tolerances allow variation in both directions from design size.

(*i*) Define true stress and true strain in a simple tensile test. Express their relations with the corresponding nominal quantities.

Sol. True stress, $\bar{\sigma}$ = Load/instantaneous area of cross-section

True strain, $\bar{\varepsilon}$ = Change in length/instantaneous length

$= \ln(1 + \varepsilon)$

where ε = nominal (or engineering) strain

$\bar{\sigma} = \sigma(1 + \varepsilon)$, where σ = nominal (or engineering) stress.

(*j*) How is toughness different from resilience of a material? Give an example, where the former is more important.

Sol. Toughness represents the capability of a material to absorb energy before fracture while resilience represents that before yield. Thoughness is important in a situation where there is a risk of collision (heavy shock).

(*k*) State Distorition Energy Theory for failure.

Sol. It states that yielding begins at stress values, when distortion energy produced by them equals the distortion energy for simple tension at yield.

(*l*) From the aspect of corrosion explain how a tin plated iron sheet is different from a galvanised iron sheet?

Sol. Tin is cathodic to iron whereas zinc is anodic to iron. So in case of a scratch of coat, iron gets corroded in the former while it remains protected in the latter.

(*m*) Distinguish between the microstructures of pearlite and plate martensite formation.

Sol. Pearlite is a lamellar arrangement of alternate parallel plates of ferrite and cementite whereas plate martensite is arranged in a zig-zag fashion partitioning the austenite grains into small pockets.

(*n*) What effects are brought about by the following in plastics?

(a) Reinforcements (b) Plasticizers.

Sol. (a) Mechanical properties are improved.

(b) Flexibility is improved and brittleness reduced.

(*o*) What is the difference between austempering and ordinary tempering?

Sol. Austempering is the quenching of steel into bainitic bay above the martensite temperature and keeping isothermally till all the austenite is converted to lower bainite.

Ordinary tempering is the reheating of martensite to higher temperature to eliminate quenching stresses and to restore ductility.

(*p*) What do you understand from machinability of a material? List any two properties, which have a bearing on machinability.

Sol. It means ease to get machined and is quantified by the amount of material removed or energy consumed for a given amount of tool wear. Hardness and ductility have a bearing on machinability.

(*q*) Distinguish between electrochemical machining and electrochemical grinding.

Sol. ECM consists in anodic dissolution of metal whereas in ECG, metal is removed by anodic dissolution as well as by abrasion of grinding wheel.

(*r*) List any four factors which determine a plant layout.

Sol. 1. Move material by minimum distances.

2. Avoid or minimize back tracking of materials.

3. Plan for a minimum material handling.

4. Use manufacturing space economically.

5. Allow for plant expansion.

(*s*) What is ABC in inventory control?

Sol. ABC inventory system procedure consists in plotting percent of total inventory items against percent of total money value for one year. The items contributing 80% of the total money constitute class A, those corresponding to inventory value 80 – 95% form class B, and the remaining class C.

(*t*) Define the assignment problem and illustrate it by an example.

Sol. It consists in assigning m jobs (one job per machine) to n machines (m, n) at the least total cost. Let two categories of machines (m_1, m_2 in number) be assigned three types of tasks (T_1, T_2, T_3 in number) and unit costs be given, then objective is the optimal assignment of tasks to machines.

Section B

Q.2. (*a*) A steel shaft of 7.5 cm diameter has an aluminium disc of 30 cm outside diameter shrunk on it. The shrink allowance is 1 part per 1000. Find the tangential and radial stresses at the interface due to shrink-fit. Calculate also the r.p.m. of rotation, at which the shrink-fit loosens up. Neglect the expansion of the shaft caused by rotation. Take

Poisson's ratio for both metals = 0.3

E for steel = 210×10^3 MPa

E for aluminium = 71×10^3 MPa

Density of aluminium = 2.76 g/cm³

Sol. Given : $r_1 = 37.5$ mm, $r_2 = 150$ mm, $\delta r_1/r_1 = 10^{-3}$,

$\nu_s = \nu_a = 0.3$, $E_a = 71 \times 10^3$ MPa, $E_s = 210 \times 10^3$ MPa,

$\rho_a = 2.76$ g/cm³

Shrinkage Streses

Refer to Fig. 2.

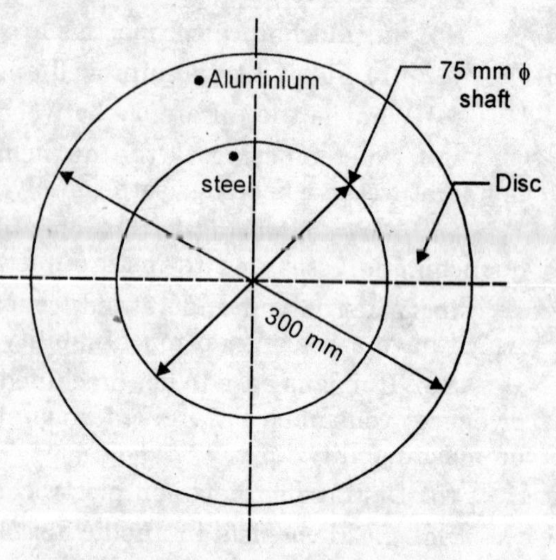

Fig. 2.

Let p_s = shrinkage pressure at the interface.

Shaft : $\sigma_r = \sigma_\theta = -p_s$

$$(\varepsilon_\theta)_s = -\frac{p_s}{E_s}(1-\nu_s)$$

Disc : At $r = 37.5$ mm, $\sigma_r = -p_s$

$$\sigma_\theta = p_s\left(\frac{k^2+1}{k^2-1}\right), \; k = \frac{r_2}{r_1} = \frac{150}{37.5} = 4$$

$$= p_s\left(\frac{16+1}{16-1}\right) = \left(\frac{17}{15}\right)p_s$$

$$(\varepsilon_\theta)_a = \frac{1}{E_a}[\sigma_\theta - \nu_a\sigma_r] = \frac{1}{E_a}\left[\frac{17}{15}+\nu_a\right]p_s$$

Resultant hoop strain,

$$\varepsilon_\theta = -(\varepsilon_\theta)_s + (\varepsilon_\theta)_a$$

$$= +\frac{p_s}{E_s}(1-\nu_s) + \frac{p_s}{E_a}\left[\frac{17}{15}+\nu_a\right]$$

$$= p_s\left[\frac{1}{210\times10^9}(1-0.3) + \frac{1}{71\times10^9}\left(\frac{17}{15}+0.3\right)\right]$$

$$= 23.52 \times 10^{-6} \times p_s$$

Now $$\varepsilon_\theta = \frac{\delta r_1}{r_1} = 10^{-3}$$

$$\therefore \quad p_s = \frac{10^{-3}}{23.52\times10^{-6}} = 42.52 \text{ MPa}$$

At the interface, $\sigma_r = \sigma_\theta = -42.52$ MPa due to shrink fitting in the shaft and $\sigma_r = -42.52$ MPa, $\sigma_\theta = \frac{17}{15} \times 42.52 = 48.18$ MPa in the disc.

Centrifugal Stresses

Let N = rpm at which shrink fit loosens up.

Hoop stress in the disc at inner radius,

$$r_1 = 37.5 \text{ mm},$$

$$\sigma_\theta = \left(\frac{3+\nu}{4}\right)\rho\omega^2\left[r_2^2 + \left(\frac{1-\nu}{3+\nu}r_1^2\right)\right]$$

$$= \left(\frac{3.3}{4}\right)\times 2760\times\omega^2\left[150^2 + \frac{0.7}{3.3}\times(37.5)^2\right]\times 10^{-6}$$

$$= 51.912\ \omega^2$$

When the shrink fit loosen up,

$$\sigma_r = 0$$

Hoop strain in the disc,

$$\varepsilon_\theta = \frac{\sigma_\theta}{\varepsilon_a} = \frac{51.912\omega^2}{71\times 10^9} = 10^{-3}$$

$$\omega = 1169.49 \text{ rad/s}$$

$$N = \frac{60\times 1169.49}{2\pi} = 11167.8 \text{ rpm.}$$

Q.2. (*b*) A uniform beam of span L and flexural rigidity EI is simply supported at its ends. It is carrying a load W at a distance b from the left hand support and subjected to an axial force P (compressive). Derive the expression for the deflection of the beam. What would be the deflection under the lateral load, when b = L/2? What is the critical value of P for which the deflection becomes very large?

Sol. $\Sigma M_B = 0$ gives, $R_A = \frac{W(L-b)}{L}$

and $R_B = \frac{Wb}{L}$

At a distance x from A (Fig. 3),

$$M_x = Py + R_A \cdot x$$

$$EI\frac{d^2y}{dx^2} = -M_x = -Py - R_Ax$$

or

$$\frac{d^2y}{dx^2} + \frac{P}{EI}y = -\frac{R_A}{EI}\cdot x \qquad ...(1)$$

The auxiliary solution of Eq. (1) is,

$$y_a = A_1 \cos kx + B_1 \sin kx$$

where $k = \sqrt{\dfrac{P}{EI}}$

The particular solution is

$$y_p = \frac{\frac{R_A}{EI}x}{D^2 + k^2} = \frac{R_A}{EI} \times \frac{x}{k^2} = -\frac{R_A}{P} \cdot x$$

The complete solution is,

$$y = y_a + y_p$$
$$= A_1 \cos kx + B_1 \sin kx - \frac{R_A}{P} \times x$$

At $x = 0, y = 0$

$\therefore$ $A_1 = 0$

$$Y = B_1 \sin kx - \frac{R_A}{P}x$$

At $x = b, \; y_1 = B_1 \sin kb - \frac{W(L-b)b}{PL}$ ...(2)

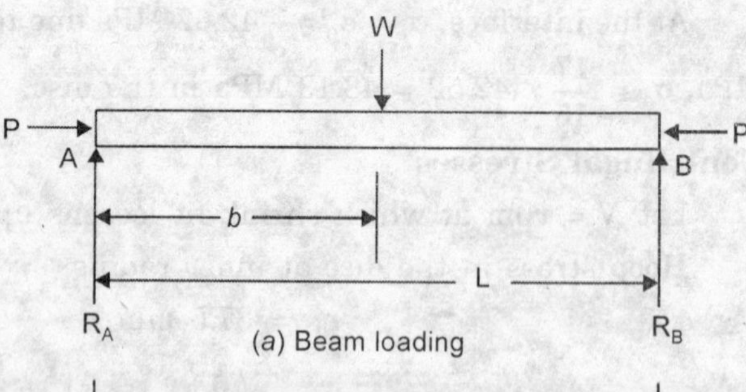

(a) Beam loading

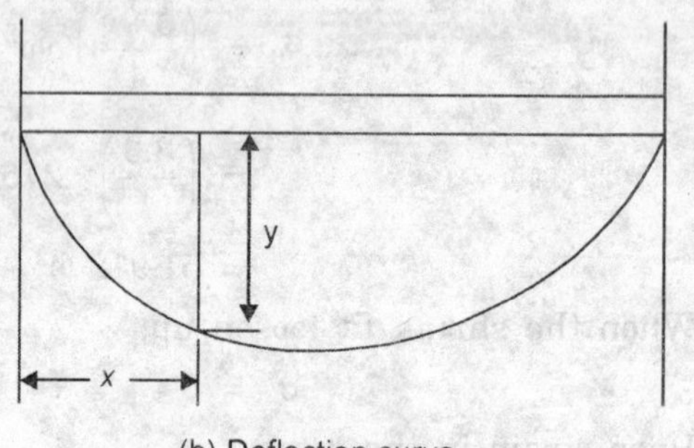

(b) Deflection curve

Fig. 3.

By taking origin at B and proceeding as above, we can get at $x = -b$,

$$y_2 = B_2 \sin k(L-b) - \frac{Wb(L-b)}{PL} \quad ...(3)$$

But $y_1 = y_2$

$$\therefore \quad B_1 \sin kb - \frac{W(L-b)b}{PL} = B_2 \sin k(L-b) - \frac{W(L-b)b}{PL}$$

$$\therefore \quad B_1 \sin kb = B_2 \sin k(L-b) \quad ...(4)$$

Now $\left(\dfrac{dy_1}{dx}\right)_{x=b} = B_1 k \cos kb - \dfrac{W(L-b)}{PL}$

$$\left(\frac{dy_2}{dx}\right)_{x=b} = B_2 k \cos k(L-b) - \frac{Wb}{PL}$$

But $\left(\dfrac{dy_1}{dx}\right)_{x=b} = -\left(\dfrac{dy_2}{dx}\right)_{x=b}$

$$\therefore \quad B_1 k \cos kb - \frac{W(L-b)}{PL} = -B_2 k \cos k(L-b) + \frac{Wb}{PL}$$

or $B_1 k \cos kb + B_2 k \cos k(L-b) = \dfrac{Wb}{PL} + \dfrac{W(L-b)}{PL}$

$$B_1\left[k\cos kb + \frac{\sin kb}{\sin k(L-b)} \times k\cos k(L-b)\right] = \frac{W}{PL}[b + L - b]$$

$$B_1k[\cos kb \times \sin k(\mathrm{L} - b) + \sin kb \times \cos k(\mathrm{L} - b] = \frac{\mathrm{W}}{\mathrm{P}} \sin k(\mathrm{L} - b)$$

$$B_1k \sin k\mathrm{L} = \frac{\mathrm{W}}{\mathrm{P}}\sin k(\mathrm{L}-b)$$

$$B_1 = \frac{\mathrm{W}}{k\mathrm{P}}\times\frac{\sin k(\mathrm{L}-b)}{\sin k\mathrm{L}}$$

$$\therefore \quad y = \frac{\mathrm{W}}{k\mathrm{P}}\times\frac{\sin k(\mathrm{L}-b)}{\sin k\mathrm{L}}\times\sin kx - \frac{\mathrm{W(L}-b)}{\mathrm{LP}}x$$

At
$$b = \frac{\mathrm{L}}{2}$$

$$y = \frac{\mathrm{W}}{k\mathrm{P}}\times\frac{\sin(k\mathrm{L}/2)}{\sin k\mathrm{L}}\times\sin kx - \frac{\mathrm{W}}{2\mathrm{P}}x$$

$$= \frac{\mathrm{W}}{k\mathrm{P}}\times\frac{\sin(k\mathrm{L}/2)\times\sin kx}{2\sin(k\mathrm{L}/2)\cos(k\mathrm{L}/2)} - \frac{\mathrm{W}}{2\mathrm{P}}x$$

$$= \frac{\mathrm{W}}{2k\mathrm{P}}\times\frac{\sin kx}{\cos(k\mathrm{L}/2)} - \frac{\mathrm{W}x}{2\mathrm{P}}$$

$$y\Big|_{x=b=\frac{\mathrm{L}}{2}} = \frac{\mathrm{W}}{2k\mathrm{P}}\times\frac{\sin(k\mathrm{L}/2)}{\cos(k\mathrm{L}/2)} - \frac{\mathrm{WL}}{4\mathrm{P}}$$

$$= \frac{\mathrm{W}}{2k\mathrm{P}}\times\tan(k\mathrm{L}/2) - \frac{\mathrm{WL}}{4\mathrm{P}}$$

$$= \frac{\mathrm{W}}{\mathrm{P}}\left[\frac{\tan(k\mathrm{L}/2}{2k} - \frac{\mathrm{L}}{4}\right]$$

Q.3. (*a*) The crank of a four-bar mechanism shown in the figure is balanced and rotating in anti-clockwise direction at a constant angular speed of 200 rad/sec. The particulars

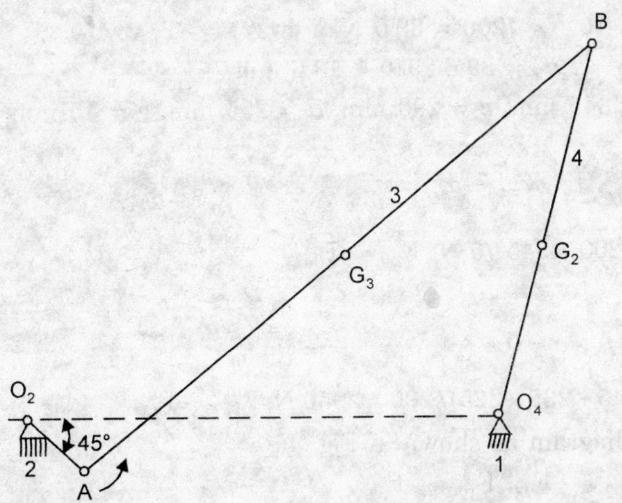

of the mechanism are O_2A = 5 cm; AB = 45 cm; AG_3 = 22.5 cm; O_4B = 20 cm; O_4G = 10 cm; O_2O_4 = 35 cm; W_3 = 1.2 kg; W_4 = 3 kg; I_3 = 68.6 kg cm^2; I_4 = 550 kg cm^2. C_3 and C_4 are mass-centres of links 3 and 4, W_3, W_4 their respective masses and I_3, I_4 their respective mass moment of inertia about their mass centres. For the given angular position of the crank 2, draw velocity and acceleration diagrams and find the angular accelerations of links 3 and 4. Determine also the forces acting at the pin-joints A, B and the external torque, which must be applied to link 2. Ignore the gravitation effects.

Sol. The configuration diagram is shown in Fig. 4(a) and the forces are shown in Fig. 4(b).

$\omega = 200$ rad/s

$v_a = O_2A \times \omega = 0.05 \times 200 = 10$ m/s

Draw the velocity diagram as shown in Fig. 4(c)

$v_{ba} = ab = 7.3$ cm $= 14.6$ m/s, $v_b = o_4b = 7.6$ cm $= 15.2$ m/s

$f^n{}_a = v^2{}_a/O_2A = (10)^2/0.05 = 2000$ m/s^2, $f^n{}_{ba} = v^2{}_{ba}/AB$ $(14.6)^2/0.45 = 473.7$ m/s^2

$f^n{}_b = v^2{}_b/O_4B = (15.2)^2/0.2 = 1155.2$ m/s^2

Draw the acceleration diagram as shown in Fig. 4(d).

$f^t{}_{ba} = 0$, Hence, $f^n{}_{ba} = f_{ba}$

$f_b = o_2b = 9.7$ cm $= 1940$ m/s^2, $f_b^t = 7.2$ cm $= 1440$ m/s^2

$\alpha_3 = f_{ba}^t / AB = 0$, $\alpha_4 = f_b^t/O_4B = 1440/0.2 = 7200$ rad/s^2 cw

$o_4G_4^l = O_4G_4 \times o_4b/O_4B = 100 \times 9.7/200 = 4.85$ cm

$aG_3^l = AG_3 \times ab/AB = 225 \times 2.368/450 = 1.184$ cm

$f_{G3} = o_2G_3^l = 9.8$ cm $= 1960$ m/s^2, $f_{G4} = o_4G_4^l = 4.85$ cm $= 970$ m/s^2

$F_3 = m_3f_{G4} = 3 \times 970 = 2910$ N, $f_4 = -F_4$

$T_3 = l_3\alpha_3 = 0$, $t_3 = -T_3 = 0$

$T_4 = l_4\alpha_4 = 55 \times 10^{-4} \times 7200 = 39.6$ N.m cw, $t_4 = -T_4$ ccw

$h_3 = t_3/f_3 = 0$, $h_4 = t_4/f_4 = 39.6/2910 = 13.6$ mm

$a = 65$ mm, $b = 345$ mm, $c = 290$ mm, $d = 255$ mm, $e = 40$ mm

$\Sigma M_{o4} = 0$ gives

$$F_{34}^t \times BO_4 = f_4 \times a$$

$$F_{34}^t = 2910 \times 65/200 = 945.75 \text{ N}, \ F_{43}^t = -F_{34}^t$$

$\Sigma M_A = 0$ gives

$$F_{43}^t \times b - F_{43}^n \times c + f_3 \times d = 0$$

$$F_{43}^n = (945.75 \times 345 + 235 \times 225)/290 = 2950 \text{ N}, \ F_{34}^n = -F_{43}^n$$

Draw the force diagram as shown in Fig. 4(e).

$F_{23} = 3500$ N, $F_{32} = -F_{23}$, $F_{12} = -F_{32}$

$T_2 = F_{32} \times e = 3500 \times 0.04 = 140$ N.m cw

Torque exterted on O_2A = 140 N.m cw

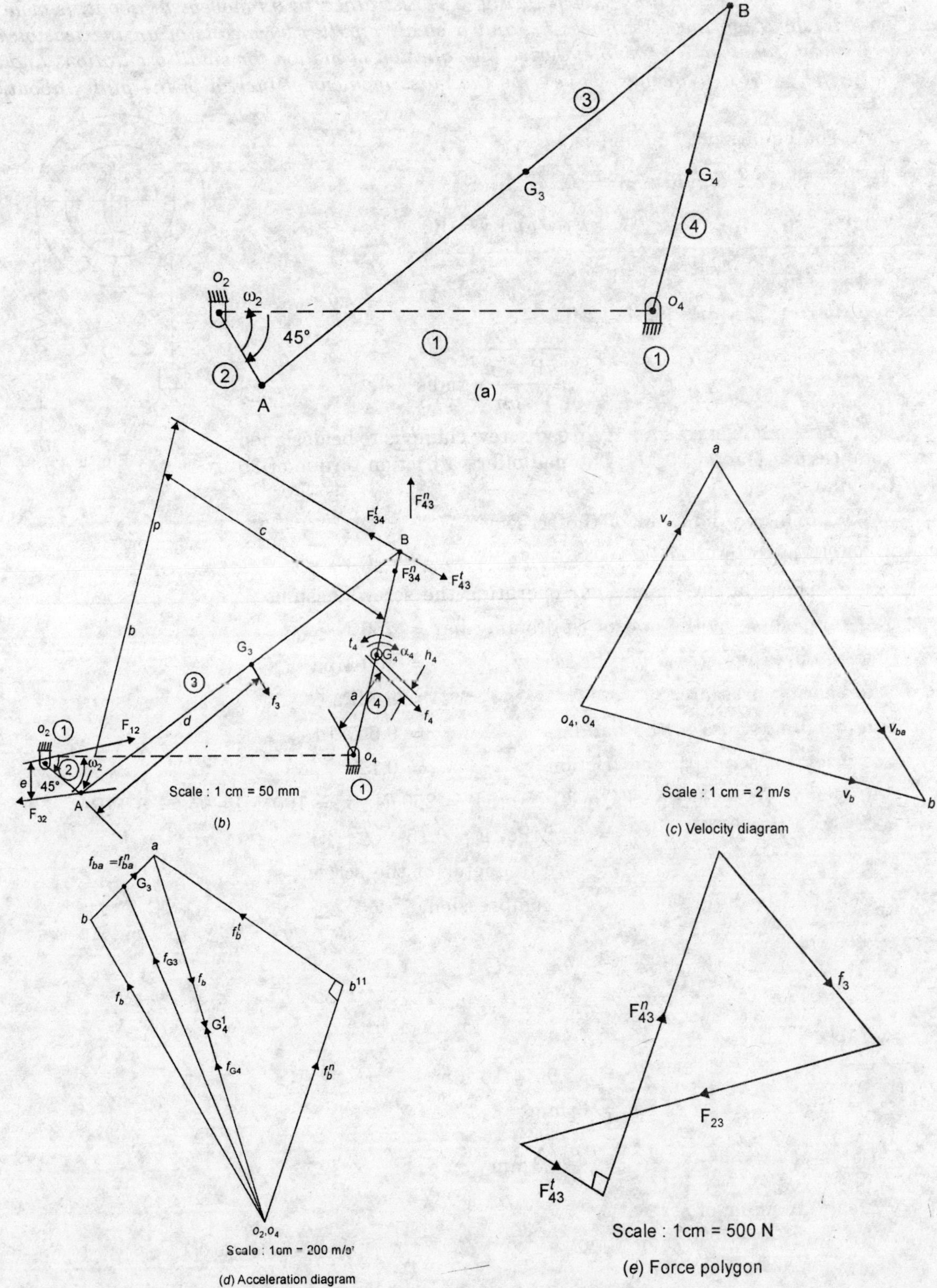
B
3
G3
G4
4
O2
ω2
45°
O4
1
2
A
1
(a)
F43n
F34t
p
c
B
F34n
F43t
b
G3
t4
G4
α4
h4
3
f3
a
4
f4
d
o2
1
F12
ω2
2
o4
e
45°
A
F32
Scale : 1 cm = 50 mm
1
(b)
a
va
o4, o4
vba
Scale : 1 cm = 2 m/s
vb
b
(c) Velocity diagram
fba = fban
a
G3
b
fbt
fG3
fb
b11
fb
G41
fG4
fbn
o2,o4
Scale : 1cm = 200 m/s²
(d) Acceleration diagram
f3
F43n
F23
F43t
Scale : 1cm = 500 N
(e) Force polygon

Fig. 4

Q.3. (*b*) *An integral pulley shown in Fig. 5 is restrained in its movement about its own axis by a torsional spring of stiffness k_1, and a smaller pulley by means of an inextensible string. Using D'Alembert's principle, derive the equation of motion for small oscillations and determine the natural frequency. Take J as the mass moment of inertia of the pulley about its axis.*

Sol. The equation of motion is,

$$J\ddot{\theta} + m\ddot{x}r + kyR + k_t\theta = 0$$

where $\ddot{x} = r\ddot{\theta}$ and $y = R\theta$

$$J\ddot{\theta} + mr^2\ddot{\theta} + (kR^2 + k_t)\theta = 0$$

or $$(J + mr^2)\ddot{\theta} + (kR^2 + k_t)\theta = 0$$

$$\omega_n = \sqrt{\frac{kR^2 + k_t}{J + mr^2}} \text{ rad/s}$$

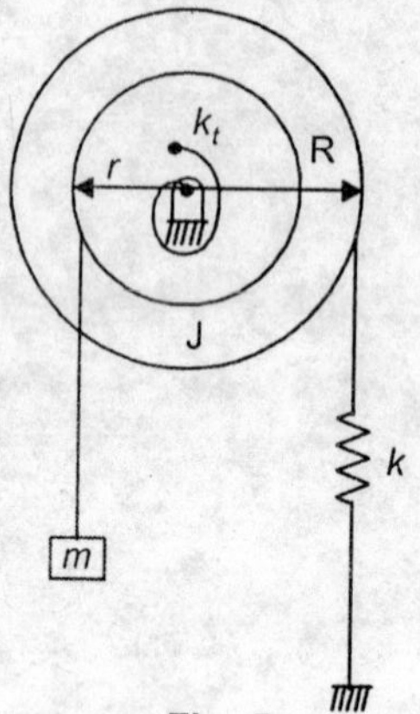

Fig. 5

Q.4. (*a*) A single start square thread screw clamp is to be designed for a compressive load of 20 kN. The pad offers a friction torque of 30 Nm. Calculate

(*i*) the pitch and diameter of the screw;

(*ii*) height of the nut and

(*iii*) dimensions of the tommy bar operating the screw. Assume

Maximum force at the end of the tommy bar	= 250 N
Safe compressive stress for the screw	= 100 Mpa
Safe bearing pressure	= 8 MPa
Safe bending stress in the handle	= 9.5 MPa
Coefficient of screw thread friction	= 0.15

Sol. Given : W = 20 kN, T_{fp} = 30 N·*m*, P = 250 N, σ_c = 100 MPa, p_b = 8 MPA, σ_b = 95 MPa, μ = 0.15

Let d_c = core diameter of the screw.

1. Considering failure of screw in compression,

$$\frac{\pi}{4}d_c^2\sigma_c = W$$

$$\frac{\pi}{4}d_c^2 \times 100 = 20 \times 10^3$$

$$d_c = 15.95 \simeq 16 \text{ mm}$$

Let pitch p = 6 mm

Depth of threads $= \frac{p}{2} = 3\,\text{mm}$

Mean diameter of screw,

$$d_m = d_c + \frac{p}{2} = 16 + 3 = 19 \text{ mm}$$

Outside diameter of screw

$$d_o = d_c + p = 16 + 6 = 22 \text{ mm}$$

2. Nut size

Let n = number of threads

Bearing area of each thread,

$$A_b = \frac{\pi}{4}(d_o^2 - d_c^2)$$

$$= \frac{\pi}{4}(22^2 - 16^2) = 179.1 \text{ mm}^2$$

Considering bearing of threads,

$$n \times A_b \times p_b = W$$

$$n \times 179.1 \times 8 = 20 \times 10^3$$

$$n = 14$$

Height of nut, $h = np = 14 \times 6 = 84$ mm

Outside diameter of nut $= 2 \times d_o = 2 \times 22 = 44$ mm

Size of tommy bar :

Helix angle of screw threads,

$$\alpha = \tan^{-1}\left(\frac{p}{\pi d_m}\right) = \tan^{-1}\left(\frac{6}{\pi \times 19}\right) = 5.74°$$

Angle of friction, $\phi = \tan^{-1} \mu = \tan^{-1} 0.15 = 8.53°$

Torque required to overcome friction of threads,

$$T_f = \left(\frac{W d_m}{2}\right)\tan(\alpha + \phi)$$

$$= \left(\frac{20 \times 10^3 \times 19}{2}\right) \times \tan(5.74° + 8.53°)$$

$$= 48324.5 \text{ N·mm}$$

Torque to be applied to the tommy bar, $T = T_{fp} + T_f = 30{,}000 + 48{,}324.5 = 78{,}3245$ N·mm

Now $P \times L = T$

$$250 \times L = 78{,}324.5$$

$$L = 313.3 \text{ mm} \simeq 315 \text{ mm}$$

Let d_b = diameter of tommy bar

Considering failure of the tommy bar in bending,

$$\frac{\pi}{32} d_b^3 \sigma_b = T$$

$$\frac{\pi}{32}d_b^3 \times 95 = 78{,}324.5$$

$$d_b = 20 \text{ mm}$$

Q.4. (*b*) A shaft running at 500 r.p.m. carries a pulley 100 cm diameter, which drives another pulley in the same direction with a speed reduction of 2 : 1 by means of ropes. The drive transmits 250 H.P. Angle of grooves is 40°. The distance between pulley centres is 200 cm. The coefficient of friction between the rope and pulley is 0.2. The rope weighs 1.2 kg/m and has a safe allowable stress of 175 N/cm². It is recommended that the initial tension in the rope should not exceed 800 N. Find the number of ropes required and rope diameter. Calculate also the length of each rope.

Sol. Given : $N_1 = 500$ rpm, $D_1 = 1$ m, $i = 2 : 1$, P = 250 HP = 186.5 kW, $2\beta = 40°$, C = 2m, $\mu = 0.2$, $m_r = 1.2$ kg/m, $\sigma = 1.75$ N/mm², $T_o = 800$ N

Peripherel velocity of rope,

$$v = \frac{\pi D_1 N_1}{60} = \frac{\pi \times 1 \times 500}{60} = 26.18 \text{ m/s}$$

Let n = number of ropes

Power transmitted,
$$P = \frac{n(T_1 - T_2)v}{10^3}$$

$$186.5 = \frac{n(T_1 - T_2) \times 26.18}{10^3}$$

$$\therefore \quad n(T_1 - T_2) = 7123.76$$

$$D_2 = \frac{D_1 N_1}{N_2} = \frac{1 \times 500}{250} = 2 \text{ m}$$

Angle of lap of rope on pulley,

$$\theta = \pi - 2\sin^{-1}\left(\frac{D_2 - D_1}{2C}\right)$$

$$= \pi - 2\sin^{-1}\left(\frac{2-1}{2\times 2}\right)$$

$$= \pi - 29° = 151°$$

Effective coefficient of friction between rope and pulley,

$$\mu_e = \frac{\mu}{\sin\beta} = \frac{0.2}{\sin 20°} = 0.58476$$

$$\frac{T_1}{T_2} = e^{\mu_e \theta} = \exp\left[0.58476 \times \frac{\pi}{180} \times 151\right] = \exp(1.541) = 4.67$$

Initial tension,
$$T_0 = \frac{1}{2}(T_1 + T_2)$$

$$800 = \frac{1}{2} \times T_2(4.67 + 1)$$

$$T_2 = \frac{1600}{5.67} = 282.2 \text{ N}$$

$$T_1 = 1317.8 \text{ N}$$

$$\therefore \quad n(1317.8 - 282.2) = 7123.76$$

Number of ropes, $n = 6.88 \simeq 7$

Centrifugal tension, $T_c = m_r v^2 = 1.2 \times (26.18)^2 = 822.5$ N

Maximum tension, $T_{max} = T_1 + T_c = 1317.8 + 822.5 = 2140.3$ N

Cross-sectional area of rope

$$A_r = \frac{T_{max}}{\sigma} = \frac{2140.3}{1.75} = 1223 \text{ mm}^2$$

Rope diameter, $d_r = \sqrt{\frac{1223 \times 4}{\pi}} = 39.46 \simeq 40$ mm

Length of rope, $L_r = \frac{\pi}{2}(D_1 + D_2) + 2C + \frac{(D_2 - D_1)^2}{4C}$

$$= \frac{\pi}{2}(1+2) + 2 \times 2 + \frac{(2-1)^2}{4 \times 2}$$

$$= 8.837 \text{ m}$$

Section C

Q.5. (*a*) Draw continuous-cooling-Transformation (C-C-T) diagram for an eutectoid steel and explain how the cooling rate affects the microstructure of the transformation products.

Sol. The information on the change of phase with the cooling rate can be conveniently displayed with the help of a continuous-cooling transformation diagram (commonly known as the CCT diagram). In such a diagram, the temperature is plotted along the vertical axis (using a linear scale), whereas the abscissa represents the time on a logarithmic scale. The TTT diagram for carbon steel is given in Fig. 6. When austenite is brought to a temperature θ_1 from θ_0 (in essentially zero time) and thereafter held at θ_1, the transformation to pearlite begins after a lapse of time t_1, as shown by the point *A* in the figure. Such a transformation, taking place at a constant temperature is shown as an isothermal transformation. The point *B* indicates a time t_2, after which the transformation is complete.

In Fig. 6, the transformations corresponding to other temperatures, viz., θ_2 and θ_3, are also shown. At about 600°C, the transformation starts after a minimum lapse of time, and this part of the diagram is called the nose. Below this temperature, austenite transforms into bainite which is an intimate mixture of ferrite and cementite (cementite exists in the form of tiny spheroids). Bainite cannot be produced by continuous cooling.

When the temperature of isothermal transformation is decreased (above 600°C), the time required for the transformation reduces. This results in a finer grain structure as less time is available for the growth of new nuclei.

It may be noted that the curve, indicating the beginning of the transformation, does not exist below about 220°C. Below this temperature, austenite instantaneously starts transforming into martensite.

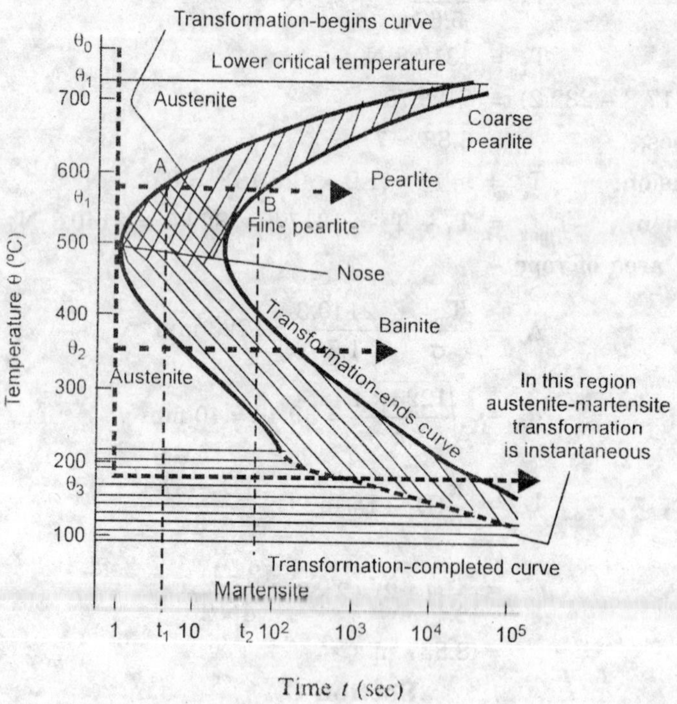

Fig. 6. C-C-T diagram for carbon steel.

Q.5. ***(b)*** **What properties are expected of the alloys used in bearings? Describe the role of hard particles and soft matrix in the functioning of a bearing alloy.**

Sol. Desirable properties of bearing alloys are :

1. High compressive strength to withstand high pressure and prevent permanent deformation.
2. High fatigue strength to withstand repeated loads without developing surface fatigue cracks.
3. Adequate conformability to accommodate shaft deflections and bearing inaccuracies.
4. Adequate embeddability to accommdate small particles of dust, dirt, etc., without scoring the material.
5. Quality of bondability with other materials.
6. High corrosion resistance against lurbicants.
7. High thermal conductivity for rapid dissipation of generated heat.
8. Low coefficient of thermal expansion to check change in bearing clearance.

Role of Hard Particles and Soft Matrix in Bearing Alloys

Soft matrix in bearing alloys provides favourable frictional properties. During bearing operation, the soft matrix is easily extruded out to form a thin lubricating film. The hard particles disperse in the soft matrix to impart hardness to the alloy. They do not affect its frictional properties.

Q.5. (*c*) Machine components are received from two vendors. Vendor A supplies 75% of the components known to include 5% defective and Vendor B's supply of 25% components includes 10% defective. When a sample of size 5 is inspected, only one defective component is found. By this information, determine the posterior probability that the components are delivered from Vender A, from Vender B.

Sol.

$$p(A) = 0.75$$

$$p(B) = 0.25$$

$$p(D/A) = 0.05$$

$$p(D/B) = 0.10$$

$$p(A) \times p(D/A) = 0.75 \times 0.05 = 0.0375$$

$$p(B) \times p(D/B) = 0.25 \times 0.10 = 0.0250$$

By Baye's theorem,

$$p(A/D) = \frac{0.0375}{0.0375 + 0.0250} = \frac{0.0375}{0.0625} = \frac{3}{5}$$

$$p(B/D) = \frac{0.025}{0.0375 + 0.0250} = \frac{0.025}{0.0625} = \frac{2}{5}$$

Q.6. (*a*) Explain the difference between orthogonal and oblique cutting. In an orthogonal cutting process, the following observations were made :

Depth of cut	= 0.25 mm
Chip thickness ratio	= 0.45
Width of cut	= 4 mm
Cutting velocity	= 40 m/min
Cutting force component parallel to cutting velocity vector	= 1150 N
Cutting force component normal to cutting velocity vector	= 140 N
Rake angle of the tool	= 18°

Determine resultant cutting force, power of cutting, shear plane angle, friction angle and force component parallel to shear plane.

Sol. Orthogonal cutting : The cutting edge of the tool is arranged perpendicular to the cutting velocity vector.

Oblique cutting : The cutting edge of the tool is set at some angle other then 90° to the cutting velocity vector.

Given : $t = 0.25$ mm, $r = \dfrac{t}{t_c} = 0.45$, $b = 4$ mm, $v_c = 40$ m/min, $F_c = 1150$ N, $F_t = 140$N,

$\alpha = 18°$

$$\text{Resultant cutting force, R} = \sqrt{F_c^2 + F_t^2} = \sqrt{1150^2 + 140^2} = 1158.49\,\text{N}$$

$$\text{Power of cutting,}\quad P = \frac{F_c \times v_c}{10^3} = \frac{1150 \times 40}{60 \times 10^3} = 0.767\ \text{kW}$$

Shear plane angle, $\phi = \tan^{-1}\left[\frac{r\cos\alpha}{1-r\sin\alpha}\right] = \tan^{-1}\left[\frac{0.45\cos 18°}{1-0.45\sin 18°}\right] = 26.43°$

Force of friction, $F = F_c \sin\alpha + F_t \cos\alpha = 1150 \sin 18° + 140 \cos 18° = 488.52$ N

Normal force, $N = F_c \tan\alpha - F_t \sin\alpha = 1150 \cos 18° - 140 \sin 18° = 1050.45$ N

Friction angle, $\beta = \tan^{-1}\left(\frac{F}{N}\right) = \tan^{-1}\left(\frac{488.52}{1050.45}\right) = 24.94°$

Force parallel to shear plane,

$$F_s = F_c \cos\phi - F_t \sin\phi$$
$$= 1150° \cos 26.43° - 140 \sin 26.43°$$
$$= 967.5 \text{ N}$$

Q.6. (*b*) List the various gear cutting processes. How are they classified? Describe with simple sketches the principle and working of a gear shaper.

Sol. Gear Cutting Processes

The main gear cutting methods are :

1. Profiling or Form cutter method.
2. Generation.

1. Profiling or Form Cuttinger Method : In this method, the spaces between the teeth are cut by means of a cutter conforming to the gear tooth form. This method necessitates the use of special cutter for each application and so is not very suitable for large volume production. The various methods under this category are :

1. Gear cutting on milling machine with formed disc cutter or end-milling cutter.
2. Gear cutting on planer or shaper with single point formed tool.
3. Cutting external or internal teeth with formed cutter in broaching machine.
4. Gear cutting with form tool blades by shear speed process.

2. Gear generating process : This process is based on the principle that any two involute gears, or any gear and a rack, of the same module will mesh together properly. If one of the gears (or the rack) is made into a cutter by proper sharpening and it meshes with the gear blank, then the teeth on blank are developed or generated by the relative rolling motion of the cutter and the blank. This process in which the cutter is similar to a mating pinion or a rack is called "Gear Shaping". If the generating tool (cutter) is threaded and gashed, it is called a "hob" and the process is called as "Gear Hobbing".

The major advantages of gear generation is that the same cutter of a particular module can cut gears of different number of teeth without profile deviation. Due to this, this is the most popular method of producing gears in mass.

Gear Shaper

In gear shaper the cutter reciprocates rapidly. The teeth are cut by the reciprocating motion of the cutter and because of this, these machines are called "gear shapers". The cutter can either be a 'rack-type cutter', or a 'rotary pinion cutter', Fig. 7. The main drawback of 'rack-type cutter' is that once the length of rack is covered by the gear blank, the cutting process is interrupted to index the blank back to strating point. In the case of 'rotating pinion type cutter', such an indexing is not required, therefore, this type is more productive and so common. The principle of a gear shaper is shown in Fig. 8. The cutter is pinion shaped with a clearance on the tooth face. the gear blank mounted on a vertical spindle and the cutter on the end of a second, vertical, reciprocating spindle. Cutting takes place either on the downward stroke or on the upward stroke of the cutter (depending upon the design of the machine). During each return stroke, either the table with the blank is widthrawn from the cutter, or the cutter is withdrawn from the blank. Its purpose is to prevent rubbing and resulting wear of the cutting edges of the cutter and damage to the tooth profiles of the gear being cut.

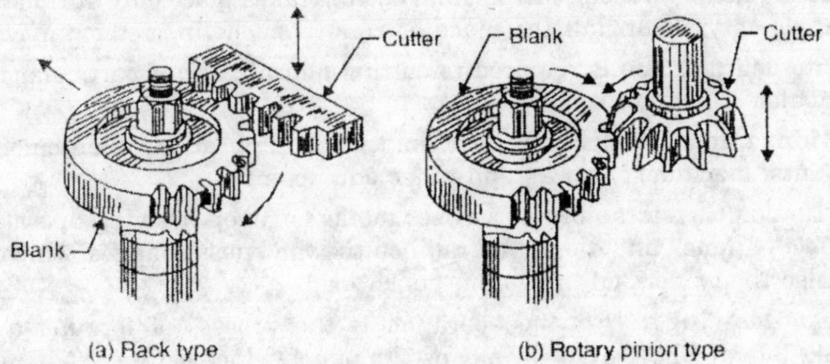

Fig. 7. Gear Shaper Cutters.

To start cutting a gear, the tool is fed into the gear blank before each cutting stroke. When a required depth is reached, the inward feed stops and the cutter and blank slowly start rotating as if they were in mesh. The gear is cut in one complete rotation of the blank. The rotary motions of the cutter and the blank are co-ordinated through change gears so that the surface speeds of the cutter and the blank are the same. The cutter reciprocates at about 100 strokes per minute for an average job and the strokes can be upto 2000 per minute for fine tooth gears. Both straight and helical-tooth gears can be cut on gear shapers. For cutting helical-tooth gears, both the tool and the gear blank are given an oscillating rotational motion, during each stroke of the cutter, turning in one direction during the cutting stroke and in the opposite direction during the return stroke.

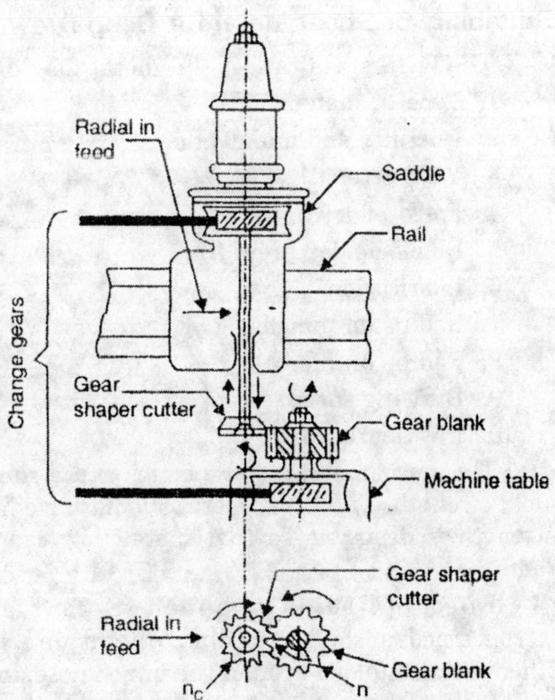

Fig. 8. Principle of a Gear Shaper.

Advantages

1. One cutter can be used for cutting all spur gears of the same module, *irrespective* of number of teeth on the gear.
2. It can cut teeth, up close to a shoulder with only a slight clearance recess.
3. Because the cutting stroke can be adjusted, gear shapers are particularly suitable for cutting cluster gears.

4. The method is versatile and can cut spur, helical, herring-bone, internal and cluster gears, racks, spines, pawls and many others.
5. The cutter has a generated profile which is more accurate in shape than some cutters, so, the gears produced are accurate.
6. The method is suitable for medium and large batch production.

Limitations

1. The cutting action is reciprocating, and cutting takes place only during one half of the stroke, so, only about half the machine time is spent in metal removal.
2. Separate helical guide is required to cutting helical gear of particular helix angle in a particular direction.

Applications : Gears produced by gear shaper find applications in automobiles, machine tools, instruments, machinery, clocks and other equipment.

Q.6. (*c*) How is the suitability of a sheet metal for deep drawing operation assessed? Discuss the effect of punch and die profile radii on the thickness changes, which occur during drawing and also their effect on maximum punch load.

Sol. Deep Drawing : Drawing operation is the process of forming a flat piece of material (blank) into a hollow shape by means of a punch which causes the blank to flow into the die cavity. If the depth of the formed up exceeds the diameter, it is called deep drawring.

Suitability of Sheet Metal for Deep Drawing

The factors that affect the metal flow during drawing are as follows :

1. Type of material.
2. Ductility and malleability.
3. Yield strength.
4. Force of friction.
5. Blank holder pressure.
6. Lubrication.
7. Radius on punch.
8. Radius on die.
9. Drawing speed.
10. Die clearance.

The most important properties of the sheet metal for deep drawing are its ductility and yield strength. The sheet metal should have high ductility and low yield strength. Low yield strength is desirable so that metal flow can begin easily without tearing near the punch radius.

Effect of Radius on punch : So far radius on punch is concerned a sharper radius will require higher forces when metal is folded around the punch nose and may result in excessive thinning or tearing at the bottom of the cup. The commonly used value of radius on punch (R_1) is as follows :

$$R_1 = 4t \text{ to } 10t$$

where t = Metal thickness

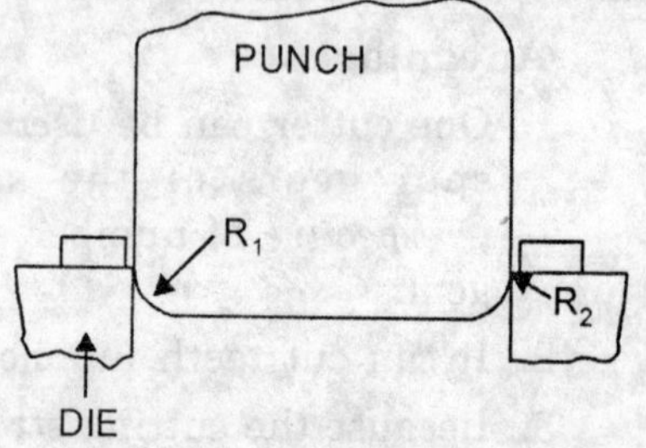

Fig. 9

Effect of Radius on die : The radius on draw die should be as large as possible to permit full freedom of metal flow as it passes over the radius. Generally

$$R_2 = \text{Radius on draw die (Fig. 9)}$$
$$= 4t$$

For drawing shallow cups of heavy gauge metals without blank holder.

$$R_2 = 6t \text{ to } 8t$$

However if the draw radius is too large the metal will be released by the blank holder too soon and wrinkling will result. On the other hand a too sharp radius will hinder normal flow of metal and will cause uneven thinning of the cup wall with resultant tearing.

Following table indicates drawing radius for some values of stock thickness.

Table : Drawing radius

Stock thickness (mm)	Drawing Radius (mm)
0.4	1.6
0.8	3.2
1.6	6.3
2	10
2.5	11.2

Analysis of Load

During drawing operation the total force to be extерted by the press P = Drawing force + Blank holding force + ironing force.

The curve representing the force required during cupping reveals many characteristics of drawing. Fig. 10 shows the force curve during cupping operation. The maximum force occurs at the first instant of cupping which consists of a high instantaneous force which immediately reduces after metal flow has been started, a small force will draw the remaining metal into the cup contour.

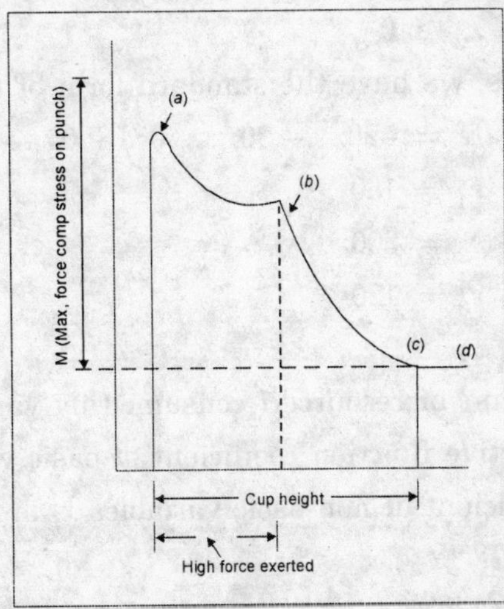

Fig. 10

With reference to figure above, point 'a' inducates the drawing just in the beginning, point '*b*' indicates drawing nearly completed and point '*c*' indicates drawing completed. It is clear from the curve that initial contact of punch with sheet metal blank is the point of the severest stress or strain. If wrinkless occurs, they start while high forces are applied. If the cup breaks, failure occurs while these high forces are applied. Once metal flow has started and the forces reduced the chance of wrinkles, of cup breakages are limited, l-cup failure occurs after this point, the failure can usually be attributed to inclusions or defects in the sheet metal.

Q.7. (*a*) Two products are manufactured by passing sequentially through three machines. Time per machine allocated to the two products are limited to 10 hours per day. The production time and profit per unit of each product are

Production Time (Minutes)

Product	Machine 1	Machine 2	Machine 3	Profit (Rs.)
1	12	10	5	20
2	4	8	10	30

(*i*) Find the optimal mix of the two products using Simplex Method.

(*ii*) Identify the machine(s) with abundant capacity at the optimum solution.

(*iii*) For each machine with full utilization, determine the worth per unit increase in its capacity.

(*iv*) Which of the three machines should be given highest priority for capacity increase.

Sol. Let x_1, x_2 = Number of units produced of products 1 and 2 respectively.

Maximize : $z = 20x_1 + 30x_2$

Subject to :

$$12x_1 + 4x_2 \le 600 \quad \text{or} \quad 3x_1 + x_2 \le 150$$
$$10x_1 + 8x_2 \le 600 \quad \text{or} \quad 5x_1 + 4x_2 \le 300$$
$$5x_1 + 10x_2 \le 600 \quad \text{or} \quad x_1 + 2x_2 \le 120$$
$$x_1, x_2 \ge 0$$

1. Adding slack variables, we have the standard form of the problem as :

Minimize : $z = -20x_1 - 30x_2 + 0s_1 + 0s_2 + 0s_3$

Subject to :

$$3x_1 + x_2 + s_1 = 150$$
$$5x_1 + 4x_2 + s_2 = 300$$
$$x_1 + 2x_2 + s_3 = 120$$
$$x_1, x_2, s_1, s_2, s_3 \ge 0$$

Notations : a_{ij} = amount of resource i consumed by variable j.

c_i = objective function coefficient of basic variables.

c_j = coefficient of non-basic variables x_j in the objective function.

$$c_j' = c_j - \sum_{i=1}^{m} a_{ij}c_i; \; j = 1 \text{ to } n$$

2. Construct tableau

c_i	Basic	Non-Basic		Basic			b_i	Replacement Ratio (RR)
	s_i	x_1	x_2	s_1	s_2	s_3		(b_i/a_{ir}) for $a_{ir} > 0$
0	s_1	3	1	1	0	0	150	150/1 = 150
0	s_2	5	4	0	1	0	300	$\frac{300}{4} = 75$
0	s_3	1	2	0	0	1	120	$\frac{120}{2} = \boxed{60}$ ←
	c_j	–20	–30	0	0	0	0	
	Solution	0	0	150	300	120		
	c'_j	–20	$\boxed{-30}$ ↑	0	0	0		

Key column; Pivot element; Key row; Departing variable (Least + ve ratio)

Entry variable (Maximum–ve value)

$$c'_1 = c_1 - (a_{11}c_1 + a_{21}c_2 + a_{31}c_3)$$
$$= -20 - [3 \times 0 + 5 \times 0 + 1 \times 0]$$
$$= -20$$
$$c'_2 = c_2 - (a_{12}c_1 + a_{22}c_2 + a_{32}c_3)$$
$$= -30 - (1 \times 0 + 4 \times 0 + 2 \times 0)$$
$$= -30$$
$$c'_3 = c'_4 = c'_5 = 0$$

Solution elements are written from the columns where there is 1 in one row only and zeros in all other rows.

3. Apply Gauss-Jordan procedure to obtain a column vector with 2 as the pivotal element and zeros elsewhere.

(*i*) Divide key row by 2 and subtract from 1st row.

(*ii*) Muliply key row by 2 and subtract from 2nd row.

c_i	Basic	x_1	x_2	s_1	s_2	s_3	b_i	(b_i/a_{ir}) for $a_{ir} > 0$
0	s_1	5/2	0	1	0	–1/2	90	$90 \times \frac{2}{5} = 36$
0	s_2	3	0	0	1	–2	60	60/3 = (20) ←
–30	x_2	$\frac{1}{2}$	1	0	0	$\frac{1}{2}$	60	60 × 2 = 120
	c_j	–20	–30	0	0	0	1800	
	Solution	0	60	90	60	0		
	c'_j	$\boxed{-5}$ ↑	0	0	0	15		

4. Continue the solution as all c'_j are not non-negative.

(i) Multiply 2nd row by 5/6 and subtract from 1st row.

(ii) Multiply 2nd row by 1/6 and subtract from 3rd row.

c_i	Basic	x_1	x_2	s_1	s_2	s_3	b_i	(b_i/a_{ir}) for $a_{ir} > 0$
0	s_1	0	0	1	–5/6	7/6	40	
–20	x_1	1	0	0	1/3	–2/3	20	
–30	x_2	0	1	0	–1/6	5/6	50	
	c_j	–20	–30	0	0	0	1900	
	Solution	20	50	40	0	0		
	c'_j	0	0	0	1.67	13.3		

(i) Since c'_j are all non-negative or zeros, therefore, optimum solution is :

$$x_1 = 20 \text{ and } x_2 = 50$$

$$z_{max} = 20 \times 20 + 30 \times 50 = \text{Rs. } 1900$$

(ii) Capacity of machines

M/c-1 : $12 \times 20 + 4 \times 50 = 440$ minutes < 600 minutes

M/c-2 : $50 \times 20 + 4 \times 50 = 300 = 300$

M/c-3 : $20 + 2 \times 50 = 120$

m/c-1 has abundant capacity at the optimum solution. Spare capacity of M/c-1 = 160 minutes

(iii) $12x_1 + 4x = 600$

Spare capacity on M/c-1 = 160 minutes

$12x_1 + 4x = 160$

Let $x_1 = 10$, then $x_2 = 10$

Increase in profit $= 20 \times 10 + 30 \times 10 = \text{Rs. } 500$

(iv) M/c-1 should be given highest priority for capacity increase.

Q.7. (b) What are essential components of Production Planning? Explain them with simple examples.

Sol.

1. Predetermination of future achievements and requirements.
2. Planning the product design if the product is new or modifications in design of existing products.
3. Technology selection for manufacturing or improvements in existing technology.
4. Planning procurement of new materials.
5. Planning the standard of quality to be achieved and means to it.

6. Planning capacity of equipment and break-even volumes.
7. Planning for sales and marketing strategy and logistics etc.
8. To plan standards of safety for accident free production.
9. To plan cleanliness of work place.
10. To plan labour turnover.

Q.7. (*c*) Write briefly on

(*i*) Surface integrity of machined parts,

(*ii*) Value engineering.

Sol. (*i*) Surface Integrity of Machined Parts : The surface characteristics, namely, the surface roughness, metallurgical properties and hardness play an important role in the reliability and durability of the component parts. They influence the friction and wear, lubrication and physical properties, corrosion resistance, thermal conductivity and the behaviour of the assembled components.

The study of surface phenomena, recently termed as surface integrity, is the sum of all the elements that describe all the conditions existing on or at the surface of a piece of finished hardware. This has two aspects, firstly, the surface topography which describes the roughness, lay or texture of the outer most layer of the workpiece and secondly the *surface* metallurgy which describes the nature of the altered layer below the surface with respect to the base or matrix material. It is the assessment of the impact of manufacturing process on the properties of the workpiece material. We shall discuss only the surface texture or topography briefly.

To define the surface topography, five distinct parameters are necessary. These are :

1. Root mean square
2. Centre-line average
3. Peak to trough roughness
4. Spacing
5. Shape factor

Surface texture may be classifined under two headings :

(*a*) Macroscopic roughness measurements.

(*b*) Microscopic roughness measurements.

Macroscopic roughness measurements is used mostly for metallic surfaces and is accomplished by using a mechanical stylus.

Microscopic surface roughness measurements are usually confined to non-metallic surfaces. More recent ways of rougness measurements are the statistical evaluation of profile, probability density and distribution function, auto-correlation functions and power spectral density function.

(*ii*) Value Engineering : Value engineering may be defined as the systematic application of recognised techniques which identify the function of a product or service, establish a monetary value for that function, and provides the necessary function reliably at the lowest overall cost.

The value engineering can be defined simply as : It is a systematic step-by-step approach, to achieve the desired functions of a system, service, process or product at an overall minimum cost, without deteriorating the quality, performance, reliability, safety or environment.

Value engineering : It assures cost effectiveness :

– It is cost avoidance instead of cost reduction

– It locates all frills and gold platings

- It segregates the necessary from unnecessary
- It is an organised study of functions and cost
- It is not a crash cost reduction method
- It is a combined effort of several departments
- It provides necessary functions and avoids unnecessary functions
- It is not a sacrifice of quality
- It is not a mere criticism of existing system, method, design or process but an appraisal of practical alternatives.

Value Engineering is Engineering for Value.

In Value Engineering Challenge Everything.

In Value Engineering Nothing is Taken for Granted.

Value Engineering Applications

The value engineering technique can be applied to these following cases :

(*i*) Project and construction cost reduction

(*ii*) Energy conservation

(*iii*) Optimal use of energy

(*iv*) Production and productivity improvement

(*v*) Economic recycling of waste products

(*vi*) Import substitution

(*vii*) Economic use of residual life

(*viii*) Substitution of scarce and costly items

(*ix*) Streamlining systems and procedures etc.

MECHANICAL ENGINEERING
PAPER II—1993

(Conventional)

Time allowed : Three hours *Maximum marks : 200*

Section A

Q.1. **(*a*)** What is coriolis acceleration? Give its formula.

Sol. If we consider two coincident *points* on a moving link such that one point has a sliding motion with respect to the other then the total acceleration will contain an additional component, which is called the coriolis acceleration.

Coriolis acceleration = $2v\omega$

where v = velocity of sliding, and ω = angular speed.

(*b*) Name mechanisms for generating accurate and approximate straight line motion.

Sol. For generating accurate straight line : Peaucellier mechanism, Hart mechanism, and Scott - Russel mechanism.

For generating approximate straight line : Grasshopper mechanism, Watt mechanism, and Tchebicheff mechanism.

(*c*) How can the quality of a governor be ascertained?

Sol.

(*i*) Sensitiveness = Range of speed/Mean speed

(*ii*) Stability - to have one equilibrium speed for a given configuration.

(*iii*) Effort - is the average force that acts on the sleeve for a given percentage change of speed.

(*iv*) Power - is the work done on the sleeve for a given percentage change of speed.

(*d*) State the conditions for complete balancing of reciprocating parts of an engine.

Sol.

1. Primary force polygon must close.
2. Primary couple polygon must close.
3. Secondary force polygon must close
4. Secondary couple polygon must close.

(*e*) Explain the meaning of the following :

(*a*) Hexagonal screw M 12 × 1.5 × 70

(b) T - bolt 22 × 100.

Sol.

(a) Hexagonal screw of major diameter 12 mm with a fine pitch 1.5 mm and length 70 mm.

(b) T-bolt of width 22 mm and length 100 mm.

(*f*) Why more than one full length leaves are provided in a leaf spring? State the ratio of stress between full length and graduated leaves.

Sol. To take the transverse shear at the ends. The ratio of stresses between full and graduated leaves is 1.5.

(*g*) State the units of dynamic and kinematic viscosities in the SI units.

Sol. Dynamic (absolute) viscosity : Nsm^{-2} (or mPa.s).

Kinematic viscosity : m^2s^{-1}.

(*h*) What do you understand by basic static load rating of a bearing?

Sol. It is that radial load which corresponds to a total permanent deformation of rolling element and race at the most heavily stressed contact of 10^{-3} of the ball or roller diameter.

(*i*) State the standard value of pulley groove angle and the minimum recommended angle of arc of contact for a V-belt.

Sol. 40° and 120°.

(*j*) Define diameteral quotient and thermal capacity of a worm gear set.

Sol. Diametral quotient is the ratio of the reference diameter to the axial module, *i.e.* $q = d/m$.

Thermal capacity, $H_d = h_{cr} A_c \Delta t/1000$ kW.

where h_{cr} = combined heat transfer coefficient $(W/m^2 \cdot K)$

A_c = area of housing exposed to ambient air = $14.4\, a^{1.7}$

a = centre distance, m

Δt = temperature difference between oil and ambient air, K.

(*k*) A timber beam is to be reinforced with two steel flitches of same size. They can be attached either at the top and bottom or symmetrically on the sides. In which case will the moment of resistance be larger?

Sol. In the first case when attached at the top and bottom.

(*l*) What type of rivetted joints are used for the longitudinal joints of a cylindrical pressure vessel and why?

Sol. Circumferential joint - Lap joint

Longitudinal joint - Butt joint.

To maintain the shape of the cylinder circular and obtain higher efficiency and strength of joints.

(*m*) What is standardisation? Give its four advantages.

Sol. Standardisation aims at establishing obligatory norms to which the different types, grades, parameters, etc. should conform. Its advantages are :

1. Helps in obtaining design for manufacture.
2. Ensures better quality and reliability.
3. Helps in easy maintenance.
4. Helps in reducing time and effort needed to manufacture a new machine.

(*n*) Name four types of steel castings.

Sol.

1. Unalloyed steel
2. Alloy steel
3. Heat resistant steel
4. Corrosion resistant steel.

(*o*) What are preferred numbers? Give their main advantages.

Sol. Preferred numbers form a general basis for standardising and grading a series of similar dimensions, characteristics or articles. They are arranged in geometrical series. They contain the powers of 10, such as 1, 10, 100, ... The intermediate numbers are obtained by dividing each decimal range into 5, 10, 20, 40 geometrically equal steps.

The greatest advantage of preferred numbers is that they greatly reduce unnecessary variations in the sizes of an article, thus reducing the variety of production and of the material in the store.

(*p*) Name four locating devices for jigs and fixtures.

Sol.

1. Jack and supporting pins
2. Location pins
3. Diamond pin locators
4. Conical locators

(*q*) List four advantages of numerical control.

Sol.

1. Machine utilisation is very high due to flexibility.
2. Provides consistency in quality due to automation.
3. Reduction in handling time due to reduction in number of set ups.
4. Design changes in the products can easily be incorporated.

(*r*) Write four objectives of plant layout.

Sol.

1. Reduce material handling
2. Provide increased production
3. Effective utilisation of space
4. Worker convenience and job satisfaction.

(*s*) What is production planning? State its levels.

Sol. Production planning is the function concerned with planning directing and controlling the methods to be used to make products, and the way in which the production facilities like building, machines, equipment etc. should be laid out in the space available for production. The levels of planning are :

Factory planning, Process planning, and Operation planning.

(*t*) Explain standard rating and standard performance.

Sol. Rating is the assessment of the workers rate of working relative to the observer's concept of the rate corresponding to standard pace. Standard rating is the average rate at which a qualified worker will naturally work at a job when using the correct method and motivated to apply himself to his work.

Standard performance is the rate of output which qualified workers will naturally achieve without over exertion as an average over the working day.

Section B

Q.2. (*a*) A disc of mass 5 kg is mounted in the middle of a simply supported horizontal shaft of span 0.5 m and of diameter 10 mm. The centre of gravity of the disc is displaced 2.5 mm from the geometric centre. The equivalent viscous damping for the system may be taken as 50 $N.s.m^{-1}$. The shaft rotates at 740 rpm. Calculate (*i*) dynamic stress in the shaft (*ii*) static stress in the shaft, and (*iii*) power required to drive the shaft. Take E = 2 × 10^5 N.$(mm)^{-2}$.

Sol. Given : m = 5 kg, L = 0.5 m, d = 10 mm, e = 2.5 mm, c = 50 $N.s.m^{-1}$,

N = 740 rpm, E = 2 × 10^5 N.$(mm)^{-2}$.

$$I = \frac{\pi d^4}{64} = \frac{\pi \times 10^4}{64} = 490.87 \text{ mm}^4$$

For a simply supported shaft carrying central load, stiffness,

$$k = \frac{48 \text{ EI}}{L^3} = \frac{48 \times 2 \times 10^5 \times 490.87}{(500)^3} = 37.699 \text{ N/mm}$$

$$w_n = \sqrt{\frac{k}{m}} = \sqrt{\frac{37.699 \times 10^3}{5}} = 86.8 \text{ rad/s}$$

$$w = \frac{2\pi N}{60} = \frac{2\pi \times 740}{60} = 77.49 \text{ rad/s}$$

$$\beta = \frac{w}{w_n} = \frac{77.49}{86.8} = 0.893$$

$$c_c = 2mw_n = 2 \times 5 \times 86.8 = 868 \text{ N.s/m}$$

$$\tau = \frac{c}{c_c} = \frac{50}{868} = 0.0576$$

$$\frac{r}{e} = \frac{\beta^2}{\sqrt{(1-\beta^2)^2 + (2\zeta\beta)^2}} = \frac{(0.893)^2}{\sqrt{\{1-(0.893)^2\}^2 \times (2 \times 0.0576 \times 0.893)^2}} = 3.51$$

$$r = 2.5 \times 3.51 = 8.77 \text{ mm}$$

Dynamic force,

$$F_d = \sqrt{(kr)^2 + (cwr)^2} = 8.77 \times 10^{-3} \sqrt{(37.699 \times 10^3)^2 + (50 \times 77.49)^2} = 332.4 \text{ N}$$

Total load,

$$F = F_d + mg = 332.4 + 5 \times 9.81 = 381.4 \text{ N}$$

Maximum bending stress in the shaft,

$$\sigma_{max} = \frac{\frac{FL}{4} \times \frac{d}{2}}{I} = \frac{\frac{381.4 \times 0.5}{4} \times \frac{10 \times 10^{-3}}{2}}{490.87 \times 10^{-12}} = 485.6 \text{ MPa}$$

(*i*) Dynamic stress in the shaft,

$$\sigma_{dyn} = \frac{\frac{F_d L}{4} \times \frac{d}{2}}{I} = \frac{485.6}{381.4} \times 332.4 = 423.2 \text{ MPa}$$

(*ii*) Static stress in the shaft,

$$\sigma_{st} = \frac{\frac{mg\,L}{4} \times \frac{d}{2}}{I} = \frac{485.6}{381.4} \times 49.05 = 62.45 \text{ MPa}$$

(*iii*) Damping force $= cwr = 50 \times 77.49 \times 8.77 \times 10^{-3} = 34$ N

Damping torque, $T_d = 34 \times 8.77 \times 10^{-3} = 0.298$ N.m

Power required, $P = T_d w = 0.298 \times 77.49 = 23.1$ Watts

Q.2. (*b*) The roller follower of a cam moves with SHM during ascent and with uniformly accelerated and decelerated motion during descent. The cam rotates at 370 rpm. Draw the cam profile for the following data :

Least radius of cam	= 60 mm
Angle of ascent	= 54°
Angle of dwell between ascent and descent	= 48°
Angle of descent	= 66°
Lift of follower	= 50 mm
Roller dia	= 30 mm
Offset	= 20 mm

Determine the maximum velocity and acceleration of follower during descent.

Sol. Given : $N = 370$ rpm, $\theta_1 = 54°$, $\theta_2 = 48°$, $\theta_3 = 66°$, $h = 50$ mm, $r_c = 60$ mm, $r_r = 15$ mm, offset = 20 mm

$$w = \frac{2\pi N}{60} = \frac{2\pi \times 370}{60} = 38.75 \text{ rad/s}$$

Maximum velocity during descent with uniform acceleration and deceleration,

$$v_{max} = \frac{2hw}{\theta_3} = \frac{2 \times 50 \times 38.75 \times 10^{-3}}{\frac{\pi}{180} \times 66} = 3.364 \text{ m/s}$$

Maximum acceleration during descent,

$$f_{max} = \frac{4w^2 h}{\theta_3^2} = \frac{4(38.75)^2 \times 50 \times 10^{-3}}{\left(\frac{\pi}{180} \times 66\right)^2} = 226.32 \text{ m/s}^2$$

Cam Profile

1. Draw the displacement diagram as shown in Fig. 1(*a*).
2. Draw the base circle with 60 mm radius. Draw another circle with 75 mm radius to represent the path traced out by the roller centre.
3. Draw a circle of 20 mm radius, equal to the offset.

4. Divide the angle of ascent and angle of descent into six equal parts.
5. Draw tangents at the circumference of offset circle at the above points.
6. From the circle of the path of the roller centre, measure off distances from the displacement diagram along the tangential lines.
7. Draw circles at these points equal to the roller radius.
8. Draw a smooth curve tangential to these circles to get the cam profile as shown in Fig. 1(*b*).

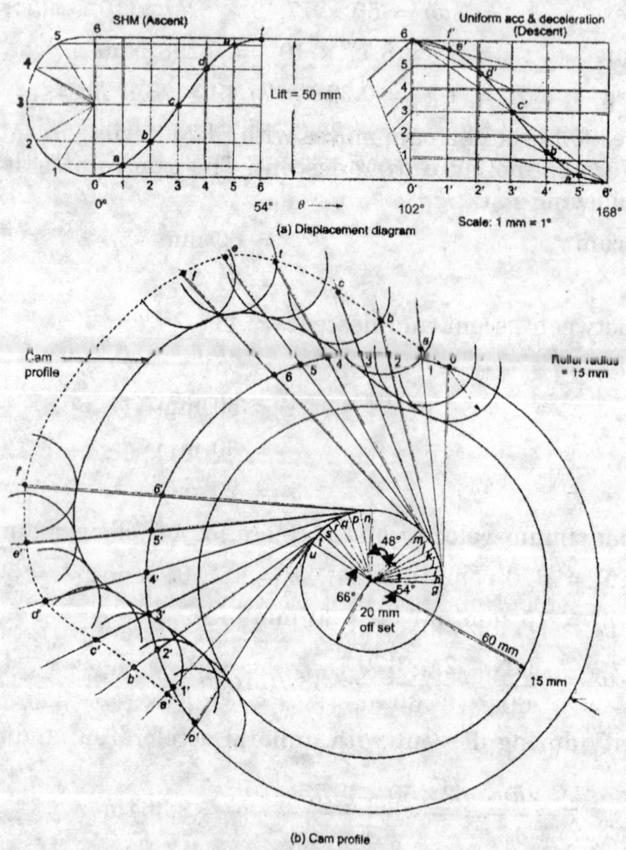

Fig. 1. Cam with offset roller follower having SHM during ascent uniform acceleration and retardation during descent

Q.3. (*a*) A spur pinion is driven by an electric motor of power 15 kW running at 740 rpm. Pitch diameter of pinion is 108 mm and speed reduction is 3 : 1. Module is 6 mm and pressure angle 20°. the pinion is mounted on a shaft located in symmetrical bearings over a span of 300 mm. The gear is overhanging by 50 mm on the left of two bearings 200 mm apart. Calculate the bearing reactions for the gear when the pinion rotates clockwise.

Sol. $P = 15$ kW, $N_p = 740$ rpm, $d_p = 108$ mm, $i = 3$, $m = 6$ mm, $\alpha = 20°$,

$l_{sp} = 300$ mm, $l_{sg} = 200$ mm, $l_0 = 50$ mm

Tangential force,

$$F_t = \frac{10^3 \times P \times 60}{\pi d_p N_p} = \frac{10^3 \times 15 \times 60}{\pi \times 0.108 \times 740} = 3584.6 \text{ N}$$

Radial force, $F_r = F_t \tan \alpha = 3584.6 \tan 20° = 1304.7$ N

Refer to Fig. 2

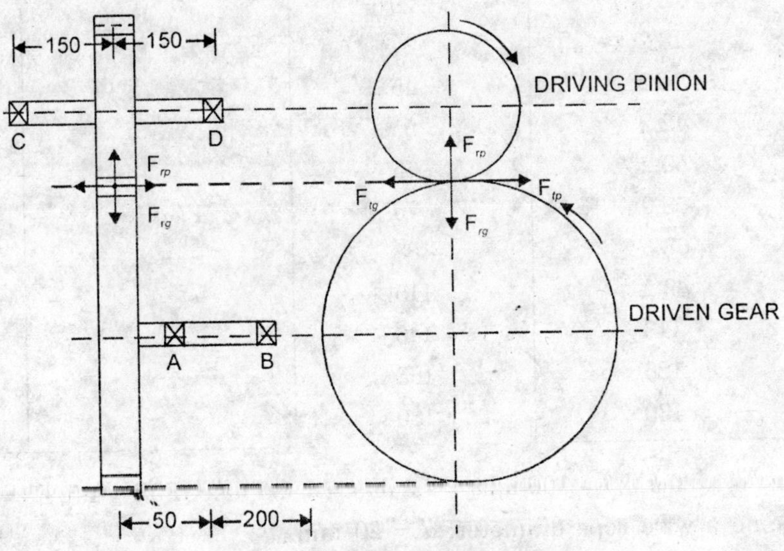

Fig. 2. Forces on pinion and gear

Calculation of bearing reactions for the gear.

Force	Magnitude	Bearing Reaction	
		Bearing A	Bearing B
Tangential	$F_{tg} = 3584.6$ N ⊙	$R_{ta} = \frac{3584.6 \times 250}{200}$ = 4480.75 N ⊕	$R_{tb} = \frac{3584.6 \times 50}{200}$ = 896.15 N ⊙
Radial	$F_{rg} = 1304.7$ N ↓	$R_{ra} = \frac{1304.7 \times 250}{200}$ = 1630.87 N ↑	$R_{rb} = \frac{1304.7 \times 50}{200}$ = 326.17 N ↓
Resultant bearing load	$\sqrt{F_{tg}^2 + F_{rg}^2}$	$R_a = \sqrt{(4480.75)^2 + (1630.87)^2}$ = 4768.32 N	$R_b = \sqrt{(896.15)^2 + (326.17)^2}$ = 953.66 N

⊕ indicates force acting indwards, normal to the plane of paper.

⊙ indicates force acting outwards, normal to the plane of paper.

Q.3. (***b***) An 8 × 19 (9/9/1) steel wire rope is used to lift a load of 20 kN from a depth of 1 km. The maximum speed of rope is 2.5 ms^{-1} and the acceleration is 1.5 ms^{-2} when starting under no slack conditions. Determine the size of the rope required for a desired factor of safety of 4.5. Modified modulus of elasticity for rope is 68600 MPa.

Table 1. Data for 8 × 19 (9/9/1) steel wire rope construction.

Nominal dia, mm d	Approximate mass kg/100 m	Breaking load (min) kN	Recommended sheave dia	Area of rope mm^2
8	22.3	29		
9	28.2	37		
10	34.2	45		
11	42.2	55		
12	50.2	65		
13	58.9	76	17 d	0.4252 d^2
14	68.3	88		
16	89.2	115		
18	113	146		
19	126	163		
20	139	180		

Sol. Given : W = 20 kN, l = 100 m, v = 2.5 m/s, a = 1.5 m/s², FOS = 4.5, E' = 68600 MPa

Let us assume a wire rope diameter, d = 20 mm

Recommended sheave diameter from Table,

$$D = 17d = 17 \times 20 = 340 \text{ mm}$$

Area of rope $A = 0.4252d^2 = 0.4252 \times (20)^2 = 170 \text{ mm}^2$

Now $d = 1.5d_w \sqrt{i}$, where d_w = wire diameter, i = number of wires in rope.

$$20 = 1.5d_w\sqrt{8 \times 19}$$

$$d_w = 1.08 \text{ mm} \approx 1 \text{ mm}$$

Mass of rope, $M_r = 139 \text{ kg/100 m} = \frac{139}{100} \times 1000 = 1390 \text{ kg}$

Weight of rope, $W_r = m_r g = 1390 \times 9.81 = 13{,}636 \text{ N}$

Acceleration load,

$$W_a = (W + W_r)\frac{a}{g} = (20{,}000 + 13636) \times \frac{1.5}{9.81} = 5143 \text{ N}$$

Bending load, $W_b = \left(\frac{E' d_w}{D}\right) \times A = \left(\frac{68600 \times 1}{340}\right) \times 170 = 34{,}300 \text{ N}$

Starting Load

Since there is no slack in the rope while starting, hence, starting load,

$$W_{st} = 2(W + W_r) = 2(20{,}000 + 13{,}636) = 67{,}272 \text{ N}$$

Effective load,

$$W_{eff} = W_{st} + W_b = 67{,}272 + 34{,}300 = 1{,}01{,}572 \text{ N}$$

Working factor of safety

$$= W_{br}/W_{ef} = \frac{1{,}80{,}000}{1{,}01{,}572} = 1.77$$

Effective load during acceleration

$$W_{eff} = W + W_r + W_a + W_b$$
$$= 20{,}000 + 13{,}636 + 5143 + 34{,}300 = 73{,}079 \text{ N}$$

Working factor of safety

$$= \frac{W_{br}}{W_{eff}} = \frac{180 \times 10^3}{73{,}079} = 2.46$$

Effective load during uniform velocity

$$W_{ef} = W + W_r + W_b$$
$$= 20{,}000 + 13{,}636 + 34{,}300 = 67{,}936 \text{ N}$$

Working factor of safety

$$= \frac{180 \times 10^3}{67{,}936} = 2.65$$

Sudden load

When the rope is being lowered down and is stopped suddenly, then

$$W_{sud} = \sigma\left[1 + \sqrt{\frac{v^2}{gx_s}}\right] \times A$$

where $\sigma = \dfrac{W + W_r}{A} = \dfrac{29000 + 13{,}636}{170} = 197.86 \text{ N/mm}^2$

$$x_s = \frac{\sigma l}{E'} = \frac{197.86 \times 1000 \times 10^3}{68{,}600} = 2884.2 \text{ mm}$$

$$W_{sud} = 197.86\left[1 + \sqrt{\frac{(2.5)^2}{9.81 \times 2884.2 \times 10^{-3}}}\right] \times 170$$

$$= 49{,}445 \text{ N}$$

Effective load during sudden slopping,

$$W_{eff} = W + W_r + W_b + W_{sud}$$
$$= 20{,}000 + 13{,}636 + 34{,}300 + 49{,}445 = 1{,}17{,}381 \text{ N}$$

Working factor of safety

$$= \frac{180 \times 10^3}{1{,}17{,}381} = 1.53$$

Minimum factor of safety = 1.53

To obtain a factor of safety of 4.5, we shall have to use 3 ropes of 20 mm diameter.

Q.4. (*a*) A thick spherical pressure vessel of inner radius 150 mm is subjected to an internal pressure of 80 MPa. Calculate its wall thickness based upon the

(*i*) maximum principal stress theory, and

(*ii*) total strain energy theory

Poisson's ratio = 0.30, yield strength = 300 MPa.

Sol. Given : r_1 = 150 mm, p = 80 MPa, σ_y = 300 MPa, ν = 0.3

(*a*) $$t = r_1\left[\left(\frac{\sigma_y + p}{\sigma_y - 0.5p}\right)^{1/3} - 1\right] = 150\left[\left(\frac{300+80}{300-40}\right)^{1/3} - 1\right] = 20.227 \text{ mm}$$

(*b*) $$m = \frac{\sigma_y}{p} = \frac{300}{80} = 3.75$$

$$k = \frac{r_2}{r_1}$$

$$\frac{k^3+2}{k^3-1} = \frac{-2\nu + \sqrt{4\nu^2 - 2(1-\nu)(1-m^2)}}{1-\nu} = \frac{-2\times 0.3 + \sqrt{4\times 0.09 - 2\times 0.7[1-(0.75)^2]}}{0.7}$$

$$= 5.31182$$

$$k = 1.1925$$

$$r_2 = 150 \times 1.1925 = 178.87 \text{ mm}$$

$$t = r_2 - r_1 = 28.87 \text{ mm}$$

Q.4. (*b*) Construct the bending moment and shearing force diagrams for the beam shown in Fig. 3.

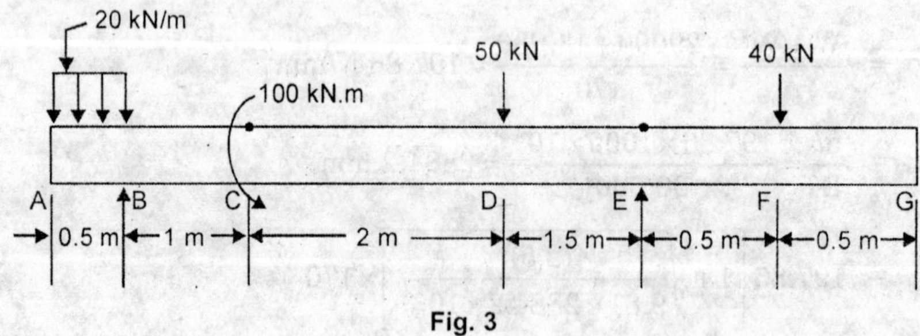

Fig. 3

Sol. Refer to Fig. 4.

$$R_B \times 4.5 = 20 \times 0.5 \times 4.75 + 100 + 50 \times 1.5 - 40 \times 0.5$$

$$= 202.5$$

$$R_B = 45 \text{ kN}$$

$$R_E = 100 - 45 = 55 \text{ kN}$$

SFD : AB $$F_x = -20x$$

$$F_B = -20 \times 0.5 = -10 \text{ kN}$$

BCD : $$F_x = -10 + 45 = 35 \text{ kN}$$

$$F_B = F_D = 35 \text{ kN}$$

$$DE : F_x = -10 + 45 - 50 = -15 \text{ kN}$$
$$EF : F_x = -10 + 45 - 50 + 55 - 40 \text{ kN}$$

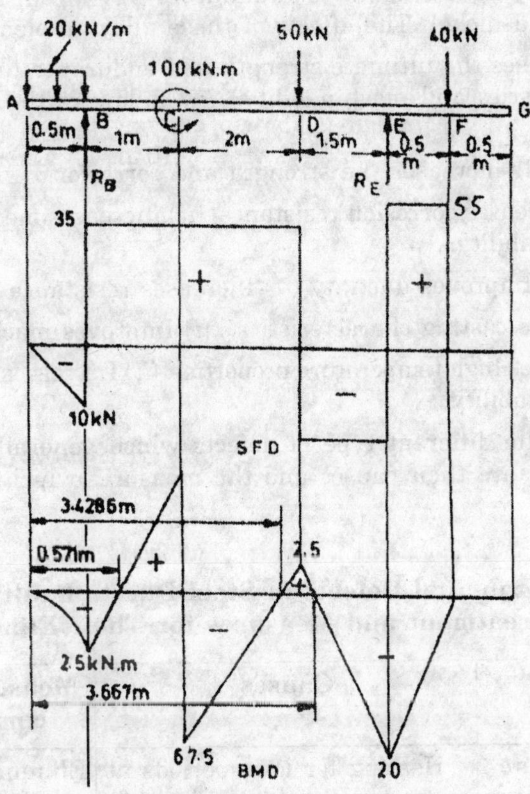

Fig. 4

$$\text{BMD : AB :} \quad M_x = -\frac{20x^2}{2} = 10x^2$$
$$M_B = -2.5 \text{ kN.m}$$
$$\text{BC :} \quad M_x = -20 \times 0.5 \times (x - 0.25) + 45(x - 0.5)$$
$$= -10x + 2.5 + 45x - 22.5$$
$$= 35x - 20$$

For $M_x = 0,\ x = 0.571$ m from A

$$M_c = 35 \times 1.5 - 20 = 32.5 \text{ kN.m}$$

CD : $M_x = -10(x - 0.25) + 45(x - 0.5) - 100 = 35x - 120$

For $M_x = 0,\ x = 3.4286$ m

DE : $M_x = -10(x - 0.25) + 45(x - 0.5) - 100 = -50(x - 3.5) = -15x + 55$

$$M_E = -15 \times 5 + 55 = -20 \text{ kN.m}$$

For $M_x = 0,\ x = 3.667$ m, At point F, $M_F = 0$

Section C

Q.5. (*a*) What are the major alloying elements to aluminium? Discuss their effects.

Sol. The major alloying elements to aluminium are : copper, silicon, manganese, magnesium, iron, zinc and nickel. The effects of these alloying elements are :

1. **Copper :** (*i*) It raises the ultimate strength and endurance limit. (*ii*) It improves the casting characteristics and machinability. (*iii*) It lowers the resistance to corrosion somewhat.
2. **Manganese :** (*i*) It increases the strength and corrosion resistance.
3. **Silicon :** (*i*) It increases corrosion resistance, (*ii*) Increases mechanical properties. (*iii*) Decreases machinability.
4. **Magnesium :** (*i*) Improves ductility (*ii*) Increases resistance to impact.
5. **Zinc :** (*i*) Improves casting characteristics. (*ii*) Improves machining property.
6. **Nickel :** (*i*) Improves high temperature properties. (*ii*) Decreases coefficient of expansion. (*iii*) Improves weldability.

Q.5. (*b*) What are the different type of defects which generally develop due to heat treatment of steels? What are their causes and the measures which can be taken for their prevention?

Sol.

Table 2. Principal Defects in Steel Due to Faulty Heat Treatment and Measures for Their Elimination

Defect and its characteristics	Causes	Measures for avoiding or correcting the defect
1. Overheating—coarse-grained microstructure and fracture, Widmanstatten structure in the annealed steel, coarse crystallic martensite in the hardened steel, Reduced ductility and, especially, impact strength (Fig. 167, *a*)	Heating for long periods at temperatures considerably exceeding the normal values	Elimination : 1. For slight overheating—apply normal annealing and normalising, 2. for considerable overheating—apply double annealing or normalising (first annealing or normalising at a temperature of Ac_3 + 100° to 150°, second at normal temperature)
2. Burning—grain boundaries have: (*i*) Regions enriched in carbon—1st stage of burning, (*ii*) nonoxidised cavities and blow-holes—2nd stage,	Heating for a long period at a high temperature in an oxidising atmosphere or at temperatures near the melting point	1. Ist stage—use homogenising followed by double annealing (see overheating), 2. 2nd stage—use forging followed by annealing, 3. 3rd stage—defect cannot be remedied

(*iii*) iron oxide inclusions—3rd stage (Fig. 167, *b*). Stone-like fracture. Low ductility		
3. Oxidation, Thick layer of scale on the surface of the steel article	Oxidising atmosphere in the heating furnace	Prevention of the defect : 1. Heat in furnaces with reducing, neutral, or protective atmospheres, 2. heat in boxes with used carburising agent or cast iron chips, 3. heat in molten salt baths
4. Decarburisation—loss of carbon in the surface layers of the article. Lower hardness after quenching, lower fatigue limit	Ditto	Ditto Correction of the defect—machine off the decarburised layer if the machining allowance is sufficiently large
5. Excessive hardness of hot-worked annealed steel	Excessive cooling rate for ordinary annealing or insufficient holding time for isothermal annealing	Repeat annealing with cooling at specified rate
6. Black fracture—free graphite (annealing carbon) inclusions in the steel	Excessive heating time and slow cooling after annealing	Heat the steel to a high temperature and forge it thoroughly
7. Quenching cracks. Usually appear at the grain boundaries and are zigzag in form. They may be external and intermal. Cracks may be longitudinal, arc-like, etc.	1. Internal stresses due to volume changes at low temperatures, caused by the martensitic transformation. 2. The martensitic transformation is not simultaneous throughout the article being hardened	This defect cannot be corrected. It may be prevented by : 1. avoiding sharp projections, sharp corners, and sudden transitions from thick to thin sections in designing the article, 2. the article should be free of stresses before being hardened, stresses should be relieved by previous annealing, 3. heat to the minimum suitable temperature for hardening, 4. cool slowly in the martensitic range (apply interrupted quenching, oil quenching, martempering; etc.), 5. apply austempering, 6. temper immediately after quenching

8. Deformation and volume (dimension) changes after hardening	The volume of the steel increases due to the martensitic transformation. the higher the hardenability of the steel, the more severe the deformation in hardening will be	1. Use alloy steels that are only slightly deformed by quenching, 2. cool slowly in the martensitic range, 3. apply surface hardening (when feasible)
9. Warping—asymmetrical deformation of the article in quenching	1. Volume changes in cooling (or heating), 2. nonuniform heating (or cooling) of the article, 3. internal stresses in the article before heating. 4. The article is lowered into the quenching bath in an inclined position	Ditto, and also 1. anneal, normalise, or temper at a high temperature before hardening, 2. heat uniformly for hardening, 3. quench as uniformly as possible in hardening, 4. hold the article in the proper position when lowering it into the quenching bath, 5. use special quenching jigs. The defect may be corrected by straightening or grinding (if the grinding allowance is sufficiently large)
10. Insufficient hardness after quenching	1. Hardening temperature is too low, 2. holding time is insufficient at the hardening temperature, 3. cooling rate is too low	The defect may be corrected by normalising or annealing followed by hardening with the proper specified procedure
11. Soft spots—zones on the surface of hardened articles with lower hardness	1. Vapour accumulates (forming bubbles) on the surface of the quenched articles reducing the cooling rate at the given places, 2. localised decarburisation, 3. inhomogeneity of initial structure in solidification of the steel	The defect may be prevented by : 1. using a more effective quenching medium, 2. obtaining a more homogeneous structure, employing annealing or normalisation before hardening, 3. protecting against decarburisation in heating. The defect may be corrected by normalising followed by quenching at a higher cooling rate.

12. Excessive hardness after tempering	Low temperature or insufficient holding time in tempering	Corrected by a second tempering with proper temperature and holding time
13. Insufficient hardness after tempering	Tempering temperature was too high	Anneal, reharden, and temper at normal temperature
14. Erosion—reduction in size of the article or distortion of its form due to loss of metal from its surface	Chemical reaction of chlorous salts and oxidation of metals heated in molten salt baths	Preventive measures : 1. deoxidise salt bath with borax or ferro-silicon. 2. properly position the article in reference to the electrodes of electrical salt baths
15. Corrosion—pitting or groove-type destruction of the surface of an article	In heating in molten salt baths : (a) high content [of sulphuric salts (over 0.7-0.8 per cent). (b) bath has become rich in oxygen or iron oxides. In heating in flame furnaces—nonuniform scale formation	Preventive measures : 1. Carefully control salt composition; 2. deoxidise the bath; 3. eliminate the oxidising atmosphere when heating in flame furnaces

Q.5. (*c*) Discuss the underlying principles for the design of jigs and fixtures.

Sol. Underlying Principles for the Design of Jigs and Fitures.

1. Since the total machining time for a workpiece includes work-handling time, the methods of location and clamping should be such that the idle time is minimum.
2. The design of jig and fixture should allow easy and quick loading and unloading of the workpiece. This will also help in reducing the idle time to minimise.
3. The jig and fixture should be as open as possible to minimize chip or burr accumulation and to enable the operator to remove the chips easily with a brush or an air jet.
4. **Fool proofing :** It can be defined as the incorporation of design features in the jig or fixture, that will make it impossible to load the workinto the jig or fixture in an improper position but will not interfere with proper loading and locating the workpiece. There are many foolproofing devices such as fouling pegs, blocks or pins which clear correctly positioned parts but prevent incorrectly loaded parts from entering the jig or fixture body.
5. **Clearance :** Clearance is provided in the jig or fixture body for two main reasons :
   (*i*) to allow for any variation in component sizes, especially castings and forgings.
   (*ii*) to allow for hand movements so that the owrkpiece can easily be placed in the jig or fixture and removed after machining.
6. **Rigidity :** Jigs and fixtures should be sufficiently sitff to secure the preset accuracy of machining.

7. **Trunnions :** To simplify the handling of heavy jigs or fixtures, the following means can be adopted :
   (*i*) Eye-bolts, rings or lifting lugs can be provided for the lifting of the jig or fixture.
   (*ii*) If the workpiece is also heavy, then the jig design should allow for side loading and unloading by sliding the workpiece on the machine table.
8. **Burr grooves :** A burr raised on the work at the start of a cut is termed a 'minor burr' and that at the end of a cut a 'major burr'. Jigs should be designed so that the removal of the workpiece is not obstructed by these burrs. For this, suitable clearance grooves or slots should be provided as shown in Fig. 1.35.
9. **Ejectors :** the use of ejection devices to force the workpiece out from the jig or fixture is important in two situations :
   (*i*) the workpiece is heavy.
   (*ii*) machining pressure forces the workpiece to the sides or base of the jig or fixture and the pressure and oil or coolant film will cause the work to stick and be difficult to remove.
10. **Inserts :** To avoid any damage to fragile and soft workpieces and also to the finished surfaces of a workpiece while clamping, inserts of some soft material such as copper, lead, fibre, leather, hard rubber, plastic or felt should be fitted to the faces of the clamps.
11. **Design for safety :** Jigs/fixtures must be safe and convenient in use. Following are some of the factors for the safety of the worker working with a jig/fixture :
    (*i*) Sharp corners on the body of the jig/fixture should be avoided.
    (*ii*) Sighting surfaces should be clear.
    (*iii*) Bolts and nuts should be inside the body of the jig/fixture and not protrude on the surface.
12. **Sighting Surface :** Machining on a workpiece must be clearly visible to the worker. He should not be required to bend his neck for seeing the work surface.
13. **Simplicity in Design :** Design of the jig/fixtures should be a simple one. A complicated design require a large maintenance. They should be cheap in manufacture and should lend themselves readily to maintenance and replacement of worn-out parts.
14. **Economical :** Jig/fixture should be simple in construction, give high accuracy, be sufficiently rigid and light in weight. To satisfy all these conditions, an economical balance has to be made.
15. They should be easy to *set in the* machine tool, which is so important in quality production where jigs/fixtures are replaced at intervals.

Q.6. (*a*) Show with a neat sketch the forces acting on a chip in orthogonal machining. Derive an expression to calculate the coefficient of friction between tool chip interface.

Sol. Orthogonal cutting is the case where the cutting tool generates a plane surface parallel to an original plane of the surface being cut and set with its cutting edge perpendicular to the direction of relative motion of the tool and the workpiece. The forces acting on a chip in orthogonal cutting are :

F_s = Resisting force in shear of the metal in forming the chip.

F_n = Force normal to the shear plane.

N = Force acting on the chip, normal to the cutting face of the tool and is provided by the tool.

F = Frictional resistance of the tool acting on the chip.

R = Resultant force of F_s and F_n.

R′ = Resultant force of F and N.

= R, for equilibrium conditions.

ϕ = Shear angle.

λ = Angle of friction.

F_c = Horizontal cutting force of the tool on the workpiece.

F_t = Vertical cutting force required to hold the tool against the work.

Fig. 5 shows the forces acting on a chip. Since the angle between F_s and F_n is a right angle, the intersection of these forces lie on a circle with diameter R (Fig. 6*a*).

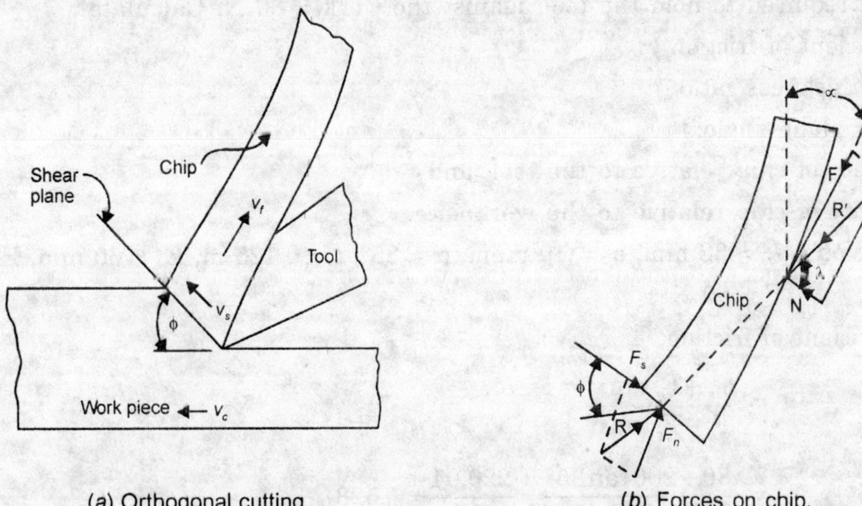

(*a*) Orthogonal cutting (*b*) Forces on chip.

Fig. 5

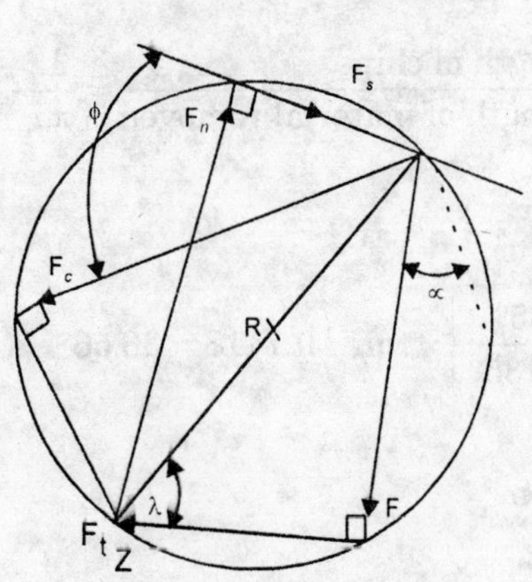

(*a*) (*b*)

Fig. 6. Forces in machining.

From Fig. 6 (*b*) we find that

$$F = ED = CA = CB + BA = HG = BA$$
$$= F_c \sin\alpha + F_t \cos\alpha$$
$$N = DA = EC = EH - CH - EH - BG$$
$$= F_c \cos\alpha - F_t \sin\alpha$$

Coefficient of friction

$$\mu = \frac{F}{N} = \tan\lambda$$

$$= \frac{F_t \cos\alpha + F_c \sin\alpha}{F_c \cos\alpha - F_t \sin\alpha} = \frac{F_t + F_c \tan\alpha}{F_c - F_t \tan\alpha}$$

Q.6. (*b*) A seamless tube 32 mm outside diameter is turned on a lathe. Cutting velocity of the tool relative to the workpiece is 10 m/min. Rake angle = 35°, depth of cut = 0.125 mm, length of chip = 60 mm, horizontal cutting force of the tool on the workpiece = 200 N, vertical cutting force required to hold the tool against the work = 80 N. Calculate :

(*i*) coefficient of friction.

(*ii*) chip thickness ratio.

(*iii*) shear plane angle.

(*iv*) velocity of chip relative to the tool, and

(*v*) velocity of chip relative to the workpiece.

Sol. Given : d_0 = 32 mm, v_c = 10 mpm, α = 35°, t = 0.125 m, L_c = 60 mm, F_c = 200N, F_t = 80 N

(*i*) Coefficient of friction,

$$\mu = \frac{F_t + F_c \tan\alpha}{F_c - F_t \tan\alpha}$$

$$= \frac{80 + 200\tan 35°}{200 - 80\tan 35°} = \frac{220.04}{143.98} = 1.528$$

(*ii*) Chip thickness ratio,

$$r = \frac{t}{t_c} = \frac{L_c}{L} = \frac{\text{Length of chip}}{\text{Corresponding length of material removed}} = \frac{L_c}{\pi d_0}$$

$$= \frac{60}{\pi \times 32} = 0.596$$

(*iii*) Shear plane angle, $\phi = \tan^{-1}\left[\frac{0.596\cos 35°}{1 - 0.596\sin 35°}\right] = \tan^{-1} 0.7418 = 36.56°$

(*iv*) Velocity of chip relative to the tool,

$$v_f = rv_c = 0.596 \times 10 = 5.96 \text{ mpm}$$

(*v*) Velocity of chip relative to the workpiece,

$$v_s = \frac{v_c \cos\alpha}{\cos(\phi - \alpha)} = \frac{10 \cos 35°}{\cos(36.56 - 35°)} = 8.2 \text{ mpm}$$

Q.6. **(*c*)** Differentiate between the decimal and binary code systems. Find the decimal and binary codes for the number 140.

Sol. 1. Decimal Number System : The decimal number system is a number system which has a base (or radix) of 0. All positional number systems have the highest numerical symbol having a value one less than the base (or radix), that is, in decimal system, 10 symbols from 0 to 9. In the decimal system, the successive positions to the left of the decimal point represent units, tens, hundreds, thousands, etc. Thus an 'n' digit number N is

$$N_{10} = d_n\, 10^{n-1} + d_{n-1}\, 10^{n-2} + \ldots + d_2\, 10^1 + d_1\, 10^0$$

$$= \sum_n d_n 10^{n-1} \quad \ldots(a)$$

where N_{10} = Any decimal number

d_n = digit in the nth position to the left of the decimal point.

Fractional part of a decimal number are the digits to the right to the decimal point. The fraction part F_{10} is represented as

$$F_{10} = f_1\, 10^{-1} + f_2\, 10^{-2} + \ldots + f_m\, 10^{-m}$$

$$= \sum_m f_m 10^{\,m} \quad \ldots(b)$$

f_m = the m digit to the right of the decimal point.

Combining (*a*) and (*b*) we can write generalized decimal number of

$$N_{10}F_{10} = (d_n 10^{n-1} + d_{n-1} 10^{n-2} + \ldots + d_2 10^1 + d_1 10^0)\,(f_1 10^{-1} + f_2 10^{-2} + \ldots f_{m-1}\, 10^{m-1} + f_m\, 10^{-m}$$

$$= \left(\sum_n d_n 10^{n-1}\right)\left(\sum_m f_m 10^{-m}\right) \quad \ldots(c)$$

2. Binary Number System : The electrical devices generally have two stable states. Either they are passing current (*i.e.* are ON or are in state 1 or are not passing current (that is are OFF or are in state 0). Hence, if this fact is kept in mind, using a binary system of number, will simplify the design, reduce the cost and improve the reliability of computer systems. Binary number system uses a base of 2. The possible symbols therefore are just 0 & 1. Each position in a number represents a power of the base 2.

Decimal Code for the number 140

$$N_{10} = \sum_n d_n 10^{n-1}$$

Here $N_{10} = 140,\ n = 3,\ d_3 = 1,\ d_2 = 4,\ d_2 = 0$

$\therefore$ $140 = 1 \times 10^2 + 4 \times 10^1 + 0 \times 10^0$

Binary code for the number 140

10001100

2	140	Remainder
2	70	0
2	35	0
2	17	1
2	8	1
2	4	0
2	2	0
	1	0

Q.7. (*a*) The annual fixed costs of a product are known to be Rs. 2×10^5 and the annual net profit is Rs. 4×10^4, the average monthly sales being 820 units. A new design is contemplated involving an expenditure of preparations amounting to Rs. 80×10^4 to be returned in two years. It is expected that with new production methods the product to volume ratio may be increased by 5 percent. What would be the new annual sales figure for the new product so that :

(*i*) same net profit is realized.

(*ii*) in addition to this profit a yield of 10 percent on the capital involved will be obtained.

Sol. Given : Fixed cost, F = Rs. 2×10^5/year

Net profit, P = Rs. 4×10^4/year

Average sales, Q_1 = 820 units/month

Preparations expenditure to be returned in 2 years, S = Rs. 8×10^4

Increase in P/V ratio, r = 5%

Yield on capital involved = 10%

Contribution = Sales – Variable cost = Fixed cost + Profit

$$\frac{P}{V} \text{ ratio, } r = \frac{\text{Contribution}}{\text{Sales}} = \frac{\text{Fixed cost} + \text{Profit}}{\text{Sales}}$$

(*i*) Initial P/V ratio, $r_1 = \dfrac{F + P}{12\, Q_1 \times x}$

where x = sales price per unit

$$r_1 = \frac{2 \times 10^5 + 4 \times 10^4}{12 \times 820 \times x} = \frac{24 \times 10^4}{9840 \times x} \qquad ...(1)$$

When P/V ratio increases by 5%, then r_2 = 1.05

$$r_2 = \frac{F + P + \dfrac{S}{2}}{12 Q_2 \times x}$$

$$1.05r_1 = \frac{2\times10^5 + 4\times10^4 + \dfrac{8\times10^4}{2}}{12Q_2 \times x} = \frac{28\times10^4}{12Q_2\times x} \quad ...(2)$$

where Q_2 = new average sales

Dividing (2) by (1), we have

$$1.05 = \frac{28\times820}{Q_2\times24}$$

$$Q_2 = 911.11 \text{ units/month or } 10933 \text{ units/year.}$$

(*ii*) 10% return on capital involved

$$= 0.1\left(2\times10^5 + \frac{8\times10^4}{2}\right) = \text{Rs. } 24{,}000$$

For 5% P/V ratio rise,

$$\frac{1.05r_1}{r_1} = \frac{28\times10^4 + 24{,}000}{12Q_3\times x} \times \frac{9840\times x}{24\times10^4}$$

$$Q_3 = 989.2 \text{ units/month or } 11870 \text{ units/year}$$

Q.7. (*b*) A building project consists of 10 activities. Their estimated duration is given below :

Activity	Duration
1 – 2	5
2 – 3	2
2 – 4	6
3 – 5	4
3 – 6	4
4 – 5	2
4 – 7	3
5 – 8	7
6 – 8	8
7 – 8	2

Draw the network and compute :

(*i*) event times.

(*ii*) activity times.

(*iii*) total float, and determine the

(*iv*) critical path.

Sol.

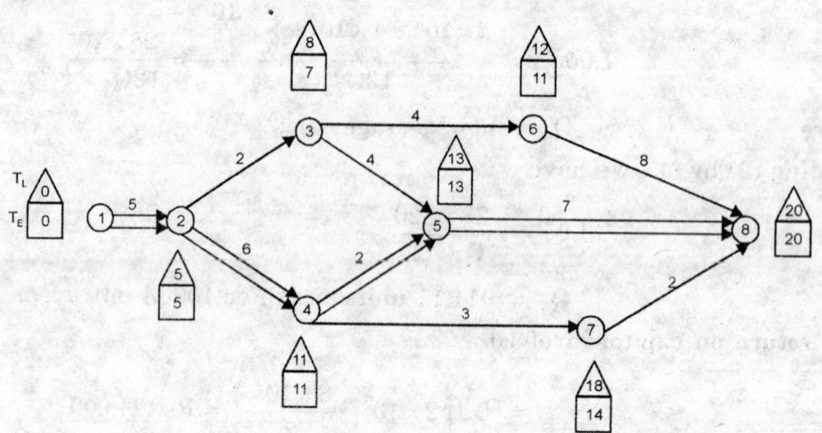

Fig. 7

□ : Forward pass

△ : Backward pass

Even	T_E	T_L
1	0	0
2	5	5
3	7	8
4	11	11
5	13	13
6	11	12
7	14	18
8	20	20

Activity	Start time		Finish time		Total Float
	Earliest	Latest	Earliest	Latest	
1 – 2	0	0	5	5	0
2 – 3	5	6	7	8	1
2 – 4	5	5	11	11	0
3 – 5	7	7	11	13	2
3 – 6	7	8	11	12	1
4 – 5	11	11	13	13	0
4 – 7	11	15	14	18	4
5 – 8	13	13	20	20	0
6 – 8	11	12	19	20	1
7 – 8	14	18	16	20	4

Critical path : 1 – 2 – 4 – 5 – 8

MECHANICAL ENGINEERING
PAPER II—1994

(Conventional)

Time allowed : Three hours *Maximum marks : 200*

Section A

Q.1. Give brief answers :

(*a*) Give Klein's construction and its use.

Sol. It is used to draw the acceleration diagram for slider crank mechanism. It helps in finding the velocity and acceleration of the piston, and acceleration of the connecting rod.

(*b*) What is the phenomenon of creep in belt drive ?

Sol. It is due to unequal tension in the tight and slack sides of the belt. The belt on the tight side gets stretched more as compared to the belt on the slack side. As a result of it the driving pulley gets more length of the belt and the belt must jump forward. The reverse happens on the driven pulley.

(*c*) Why is the pressure angle for involute gears always constant ?

Sol. A tangent to the base circle is a normal to the involute. Hence a common tangent to the two base circles, is a common normal to the involute on the tooth profile of the gears in mesh. Thus the path of contact is a straight line. In other words, the pressure angle for involute gears is constant.

(*d*) What is a friction circle? What is the magnitude of its radius ?

Sol. It is the circle at which the coefficient of friction acts when two bodies are rolling together. The radius of the friction circle, a = Fr/W, where r = radius of rolling body, W = weight of rolling body, and F = force of friction.

(*e*) Differentiate between eutectic and eutectoid.

Sol. Eutectic is a mechanical mixture of two or more phases, which solidify simultaneously from the liquid alloy.

Eutectoid is similar to a eutectic except that the initial phase is a solid solution.

(*f*) Distinguish between chaplets and internal chills. What is common between both?

Sol. Chaplet is a metal distance piece inserted in a mould either to prevent a core from shifting its position or to give extra support to the core.

Chills are metallic objects which are placed in the mould to increase the cooling rate of castings and to provide directional solidification.

(*g*) Distinguish between permeability and porosity.

Sol. Permeability is the easeness with which the sand mould allows the gases to escape through it. Porosity is the ratio of air voids to sand grains.

(*h*) Distinguish between hot and cold chamber die casting machine.

Sol. In hot chamber die casting, the liquid metal is transferred to the mould directly from the holding pot by a submerged pump at pressures up to 40 MPa.

In cold chamber process, separate melting facilities are used. Melt of a quantity sufficient for one shot is individually transferred often automatically, to the hot chamber from which the plugner squirts it into the die cavity at pressures up to 150 MPa.

(*i*) Distinguish between carbonitriding and cyaniding.

Sol. It is a process in which both carbon and nitrogen are added to the surface layer of steel by heating the steel in a gas to increase its hardness and wear resistance. Cyaniding is a process in which both carbon and nitrogen are added to the surface layer of steel by heating the steel in a liquid or solid media.

(*j*) What is the method of mass production of toothpaste tubes (metallic)?

Sol. Metallic toothpaste tubes are made by deep drawing process.

(*k*) Distinguish between blank holder and ejector pins.

Ans. A blank holder is a device to hold work in the correct relative position in the fixture and to ensure that the job is not displaced under cutting forces.

(*l*) Give two functions of a coolant.

Sol.

1. Cooling down of the chip - tool - work zone by carrying away some of the generated heat.
2. Reducing the coefficient of friction at the chip - tool interface due to the formation of a weaker compound at the interface.
3. Washing away the chips and clearing the machining zone.
4. Protecting the finished surface from corrosion.

(*m*) Distinguish between flank wear and crater wear of a single point cutting tool.

Sol. Flank wear is primarily due to abrasion and adhesion, whereas crater wear develops due to diffusion at a high speed since the temperature in the rake face is much higher than that in the flank surface.

(*n*) Define Poisson's ratio. What is its value for isotropic materials?

Sol. Poisson's ratio is the ratio of lateral strain to longitudinal strain. For isotropic materials its value varies from 0.25 to 0.34.

(*o*) What is creep ?

Sol. Creep is the phenomena of continuous extension of a material at constant temperature under constant load.

(*p*) What is event slack time ?

Sol. It is the difference between the earliest and latest times at which an event can take place without affecting the duration time of the project.

(*q*) What is break even analysis ?

Sol. It makes use of the fixed and variable nature of costs to indicate the range of volume necessary for profitable operation. At the break even point the volume of sales is such that the total net revenue after variable costs just equals the total pool of fixed costs.

(*r*) Explain the abbreviations (*a*) EOQ, and (*b*) EDM.

Sol. (*a*) Economic Order Quantity (*b*) Electro - Discharge Machining.

(*s*) What are therbligs ?

Sol. Therbligs : These are a set of 17 elementary motions commonly found in manual operations and developed by Frank Gilbreth. These are : Search, Find, Select, Grasp, Hold, Transport Load, Position, Assemble, Use, Disassemble, Inspect, Pre-position, Release Load, Transport Empty, Rest for overcoming fatigue, avoidable delay, Plan.

(*t*) Distinguish between thermoplastics and thermosetting plastics.

Sol. Thermoplastics require cooling to harden them after softening by heat; can be resoftened and rehardened.

Thermosetting plastics get hardened by chemical changes produced by heating, remain hardened without cooling and do not soften when reheated.

Section B

Q.2. (*a*) In Fig.1, a uniform load '*q*' covers only a part of the span. Calculate the reactions R_1 and R_2 and draw the shear force and bending moment diagrams.

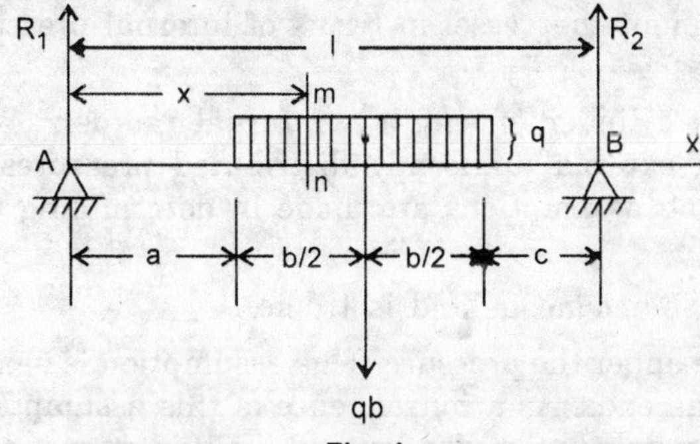

Fig. 1

Sol. Refer to Fig. 2.

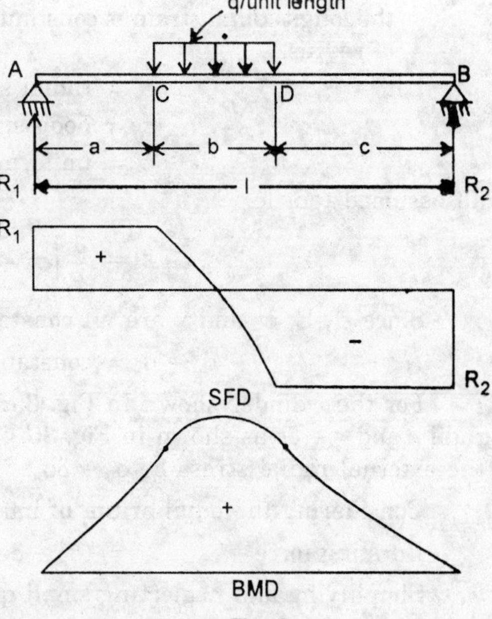

Fig. 2

$$R_1 \times l = qb\left(c + \frac{b}{2}\right)$$

$$R_1 = \frac{qb}{l}(c + 0.5b)$$

$$R_2 = qb - \frac{qb}{l}(c + 0.5b) = \frac{qb}{l}[l - (c - 0.5b)]$$

$$= \frac{qb}{l}\left[a + b + c - c - \frac{b}{2}\right] = \frac{qb}{l}\left(a + \frac{b}{2}\right)$$

SFD :

AC : $F_x = R_1$

CD : $F_x = R_1 - q(x - a)$

For $F_x = 0$,

or $$x = \frac{1}{q}\left[\frac{qb}{l}(c+0.5b)+qa\right] = \frac{b}{l}(c+0.5b)+a$$

$$F_D = R_1 - qb$$

DB : $F_x = R_1 - qb$

BMD : AC : $M_x = R_1 x$

CD : $M_x = R_1 x - \frac{q(x-a)^2}{2}$

DB : $M_x = R_1 x - qb(x - a - 0.5b)$

$$M_{max} = \frac{qb}{l}(c+0.5b)\left\{\frac{b}{l}(c+0.5b)+a\right\} - qb\left\{\frac{b}{l}(c+0.5b)+a-a-0.5b\right\}$$

$$= \frac{qb}{l}(c+0.5b)\left\{\frac{b}{l}(c+0.5b)+a\right\} - \frac{qb^2}{l}\{(c+0.5b)-0.5l\}$$

Q.2. (*b*) What are Lame equations?

For a long hollow thick walled cylinder subjected to a high internal pressure, determine the radial and circumferential stresses throughout the vessel in terms of internal pressure '*p*' and internal, external radii.

Sol. Lame's Theory : Consider a thick cylinder of length *l*, internal radius r_1 and external radius r_2, subjected to *internal* and external uniformly distributed pressures of intensities p_1 and p_2 respectively. The following assumptions are made in determining the distribution of stresses in the cylinder :

(*i*) The material of the cylinder is linear, homogneous and isotropic.

(*ii*) Plane transverse sections remain plane under the pressure. This assumption is nearly true at a considerable distance from the ends. As a consequence of this assumption, the longitudinal strain is constant at all points in the cylinder wall, *i.e.* it is independent of radius.

Let σ_r = radial stress at any radius r

σ_θ = hoop stress at the same radius r

σ_z = uniform longitudinal stress.

all assumed tensile.

$$\therefore \quad \varepsilon_z = \frac{1}{E}[\sigma_z - \nu(\sigma_\theta - \sigma_r)]$$

Since ε_z, E, σ_z and ν are all constants,

$$\therefore \quad \sigma_\theta + \sigma_r = \text{constant} = 2A \text{ (say)} \quad ...(1)$$

For the cylinder shown in Fig. 3(*a*), consider an annular ring of the cylinder between radii r and $r + \delta r$ as shown in Fig. 3(*b*). Let the internal radial stress in this ring be σ_r and the external radial stress be $\sigma_r + \delta\sigma_r$.

Considering the equilibrium of half the ring, we find that,

Bursting force $= (\sigma_r + \delta\sigma_r) \times 2(r + \delta r)l - \sigma_r \times 2rl.$

Simplifying and neglecting small quantities, we get

$$= 2\sigma_r \times \delta r l + 2\delta\sigma_r \times rl$$

Resisting force

For equilibrium of the ring, we get

$$2\sigma_\theta \times l\delta r = 2\sigma_r \times \delta r l + 2\delta\sigma_r \times rl$$

$$\sigma_\theta = \sigma_r + r \times \frac{\delta\sigma_r}{\delta_r}$$

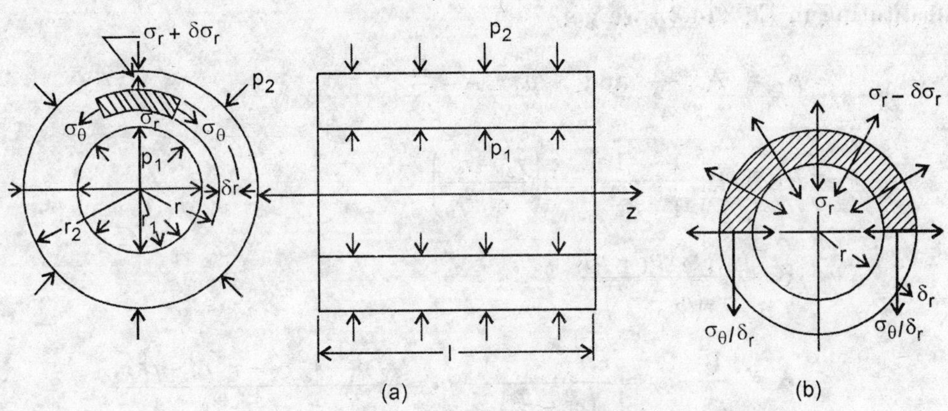

Fig. 3. Thick cylinder under preassure.

In the limiting case, we get $\sigma_\theta = \sigma_r + r\dfrac{d\sigma_r}{dr} = \dfrac{d}{dr}(r\sigma_r)$ ...(2)

Substituting in Eq. (1) we get,

$$\frac{d}{dr}(r\sigma_r) + \sigma_r = 2\text{A} \quad \text{or} \quad r\frac{d\sigma_r}{dr} = 2(\text{A} - \sigma_r)$$

$$\therefore \quad \frac{d\sigma_r}{dr} = \frac{2(\text{A} - \sigma_r)}{r} \text{ or } \frac{\sigma_r}{(\text{A} - \sigma_r)} = \frac{2dr}{r}$$

Interating, we get $\displaystyle\int \frac{d\sigma_r}{\text{A} - \sigma_r} = \int \frac{2dr}{r}$

$$-\ln(\text{A} - \sigma_r) = 2\ln r - \ln \text{B}$$

$$\ln(\text{A} - \sigma_r) = -2\ln r + \ln \text{B}$$

$$\ln(\text{A} - \sigma_r) = \ln\frac{\text{B}}{r^2}$$

$$\therefore \quad \text{A} - \sigma_r = \frac{\text{B}}{r^2} \quad \text{...(3)}$$

where B is another constant.

Substituting Eq. (3) in (1), we get

$$\sigma_\theta + \text{A} - \frac{\text{B}}{r^2} = 2\text{A}$$

$$\sigma_\theta = \text{A} + \frac{\text{B}}{r^2} \quad \text{...(4)}$$

Eqs. (3) and (4) are known as the Lame's equations.

The constant A and B can be determined from the boundary conditions at $r = r_1$ and $r = r_2$.

(*i*) At $r = r_1,\ \sigma_r = -p_1$

(*ii*) At $r = r_2,\ \sigma_r = -p_2$

Substituting in Eq. (14.3), we get

$$-p_1 = A - \frac{B}{r_1^2} \text{ and } -p_2 = A - \frac{B}{r_2^2}$$

$$p_1 - p_2 = B\left(\frac{1}{r_1^2} - \frac{1}{r_2^2}\right) = B\left(\frac{r_2^2 - r_1^2}{r_1^2 r_2^2}\right)$$

$$B = \frac{r_1^2 r_2^2 (p_1 - p_2}{\left(r_2^2 - r_1^2\right)}$$

$$A = \frac{B}{r_1^2} - p_1 = \frac{r_2^2 (p_1 - p_2)}{\left(r_2^2 - r_1^2\right)} - p_1 = \frac{r_2^2 p_1 - r_2^2 p_2 - r_2^2 p_1 + r_1^2 p_1}{\left(r_2^2 - r_1^2\right)}$$

$$= \frac{p_1 r_1^2 - p_2 r_2^2}{\left(r_2^2 - r_1^2\right)}$$

$$\sigma_r = \frac{p_1 r_1^2 - p_2 r_2^2}{\left(r_2^2 - r_1^2\right)} - \frac{r_1^2 r_2^2 (p_1 - p_2)}{r^2\left(r_2^2 - r_1^2\right)}$$

$$\sigma_\theta = \frac{p_1 r_1^2 - p_2 r_2^2}{\left(r_2^2 - r_1^2\right)} + \frac{r_1^2 r_2^2 (p_1 - p_2)}{r^2\left(r_2^2 - r_1^2\right)}$$

Now we will consider some particular cases.

(*a*) Internal pressure only, *i.e.* $p_2 = 0$, then

$$\therefore \quad \sigma_r = \frac{p_1 r_1^2}{\left(r_2^2 - r_1^2\right)} - \frac{r_1^2 r_2^2 p_1}{r^2\left(r_2^2 - r_2^2\right)} = \frac{p_1 r_1^2}{\left(r_2^2 - r_1^2\right)}\left[1 - \frac{r_2^2}{r^2}\right] \quad \text{...(5)}$$

$$\sigma_\theta = \frac{p_1 r_1^2}{\left(r_2^2 - r_1^2\right)} + \frac{r_1^2 r_2^2 p_1}{r^2\left(r_2^2 - r_1^2\right)} = \frac{p_1 r_1^2}{\left(r_2^2 - r_1^2\right)}\left[1 + \frac{r_2^2}{r^2}\right] \quad \text{...(6)}$$

Eqs. (5) and (6) give the radial and hoop stresses in the cylinder wall under internal pressure p_1.

Q.3. (*a*) In the Proel governor shown in Fig. 4 (below) each ball weighs 3 kg and the central sleeve weight is 25 kg. The arms are of 20 cm length and pivoted about axes displaced from the central axis of rotation by 37.5 mm, y = 238 mm, x = 303.5 mm, CE = 85 mm, MD = 142.5 mm.

Determine the equilibrium speed.

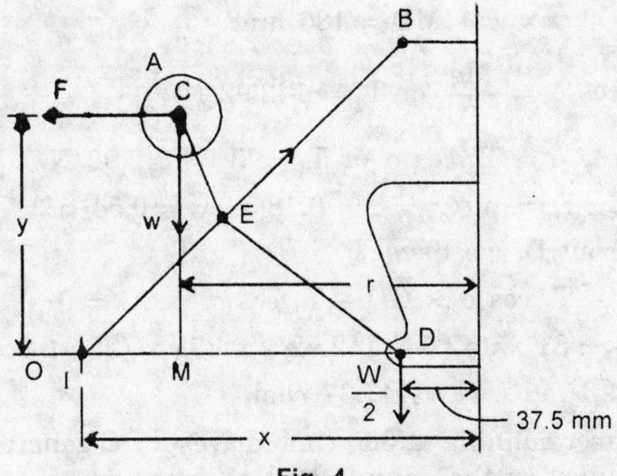

Fig. 4

Sol. Given : m = 3 kg, M = 25 kg, l = 200 mm, c = 37.5 mm, x = 303.5 mm, y = 238 mm, CE = 85 mm, MD = 142.5 mm

Refer to Fig. 5.

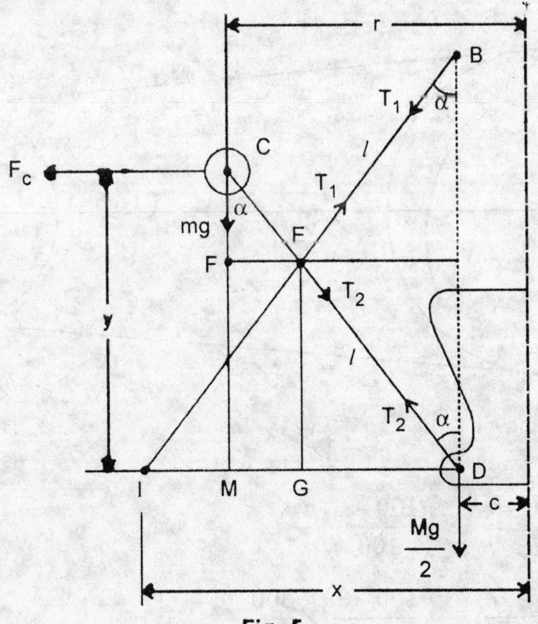

Fig. 5

$$\tan \alpha = \frac{MD}{CM} = \frac{MD}{y} = \frac{142.5}{238} = 0.59874$$

$\alpha = 30.91°$

$\cos \alpha = 0.858$

FE = CE $\sin \alpha$ = 43.66 mm

GD = MD − FE = 98.84 mm

EG = FM = CM − CF = y − CE $\cos \alpha$ = 165.07 mm

$$r = c + \text{MD} = 180 \text{ mm}$$

At point D, $\quad T_2 \cos \alpha = \dfrac{Mg}{2}$ or $T_2 = 142.92$ N

At point E, $\quad T_1 \cos \alpha = T_2 \cos \alpha$ or $T_1 = T_2 = 142.92$ N

$$F_c = mr\omega^2 = 3 \times 0.18 \times \omega^2 = 0.54\ \omega^2$$

Taking moments about D, we have

$F_c \times y + mg \times \text{MD} - T_1 \cos \alpha \times \text{GD} - T_1 \cos \alpha \times \text{EG} = 0$

$0.54\ \omega^2 \times 238 + 3 \times 9.81 \times 142.5 - 142.92 \times 0.858 \times (98.84 + 165.07) = 0$

$\omega^2 = 219.17$, $\omega = 14.8$ rad/s, N = 141.37 rpm

Q.3. (*b*) A prime mover running at 300 rpm, drives a DC generator at 500 rpm by a belt drive. Diameter of the pulley on the output shaft of the prime mover is 600 mm. Assuming a slip of 3 percent, determine the diameter of the generator pulley if the belt running over it is 6 mm thick.

Sol. Given : $n_1 = 300$ rpm, $n_2 = 500$ rpm, $d_1 = 600$ mm, $s = 3\%$, $t = 6$ mm, $d_2 = ?$

Refer to Fig. 6

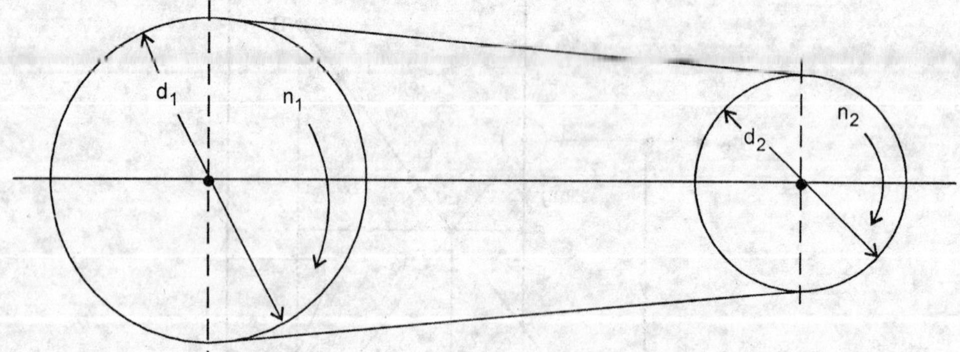

Fig. 6

$$\frac{d_2 + t}{d_1 + t} = \left(\frac{100 - s}{100}\right)\frac{n_1}{n_2}$$

$$\frac{d_2 + 6}{600 + 6} = \left(\frac{100 - 3}{100}\right) \times \frac{300}{500}$$

$$d_2 + 6 = 352.7$$

$$d_2 = 346.7 \text{ mm}$$

Q.4. (*a*) Explain the following terms, with the help of a neat sketch, pertaining to the Helical gears.

(*i*) Axial pitch (*ii*) Normal circular pich

(*iii*) Lead (*iv*) Helix angle

Sol. The cross-sections of a helical gear in the axial plane *x*-*x* and normal plane *y*-*y* are shown in Fig. 7

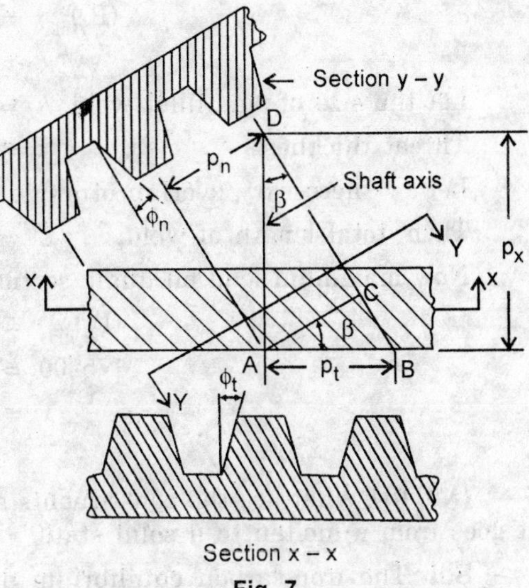

Fig. 7

(*i*) Axial pitch p_x : It is the distance measured parallel to the shaft axis. AD represents the axial pitch.

$$p_x = \left(\frac{\pi d}{z}\right)\cot\beta = p_t \cot\beta$$

(*ii*) Normal circular pitch, p_n : It is the distance between corresponding points of adjacent teeth as measured in a plane perpendicular to the tooth elements. This is shown by plane *y-y* in Fig. 7, which is perpendicular to teeth elements. AC represents normal incular pitch.

$$p_n = p_t \cos\beta = \left(\frac{\pi d}{z}\right)\cos\beta$$

where p_t = transverse circular pitch

β = helix angle of teeth

d = pitch circle diameter

z = number of teeth

(*iii*) Lead, L : It is the distance measured between two consecutive intersections of a helix and a straight generator of the cylinder on which it lies. It is defined as the distance of a point on one tooth to the corresponding point on the same helix, measured on the pitch surface in an axial direction.

Lead, L – number of starts × p_x

(*iv*) Helix angle, β **:** It is the angle between a line drawn through one of the teeth and the centre line of the shaft on which the gear is mounted.

Q.4. (*b*) An equal angle section 100 mm × 100 mm × 8 mm is welded to a gusset plate to serve as a tension member. Calculate the minimum amount of overlapping by the angle section on the gusset plate with a view to limit the induced stress to 100 mega Pascal in the angle section when only the side welds are used. The effective area of angle may be taken as the area of the connected leg plus half the area of the unconnected leg. The safe stress in the weld is 90 mega Pascal and the shock factor is 1.5.

Sol. Given : σ_a = 100 MPa, σ_w = 90 MPa, k_s = 1.5

Refer to Fig. 8.

Effective area of angle = Area of connected leg

$+ \frac{1}{2} \times$ Area of the unconnected leg

$$A_a = 100 \times 8 + \frac{1}{2}\,(100 - 8) \times 8 = 1168 \text{ mm}^2$$

Strength of angle section,

$$P_a = \sigma_a \times A_a = 100 \times 1168 = 116800 \text{ N}$$

Maximum load on angle section,

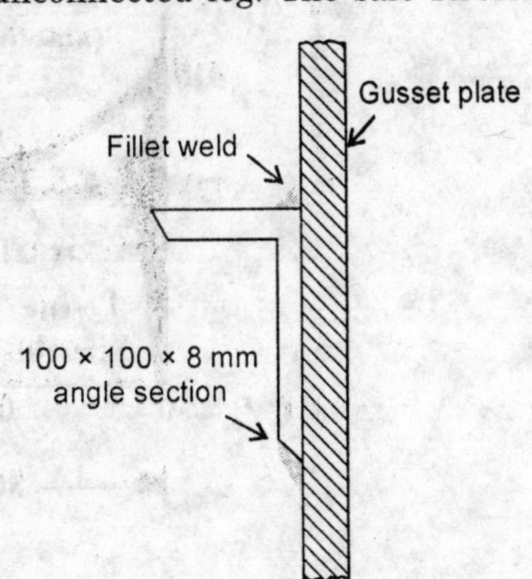

Fig. 8

$$(P_a)_{max} = P_a \times \text{shock factor}, k_s$$
$$= 116800 \times 1.5 = 175200 \text{ N}$$

Let the size of side fillet weld, $h = 6$ mm

Throat thickness, $t = 0.707\ h = 0.707 \times 6 = 4.242$ mm

Let x = necessary overlap of angle section

Then, total length of weld, $L = (2x + 100)$ mm

Now maximum load on angle section = Maximum strength of weld

$$(P_a)_{max} = P_w = \sigma_w\ tL$$
$$175200 = 90 \times 4.242 \times (2x + 100)$$
$$x = 179.45 \simeq 180 \text{ mm}$$

Section C

Q.5. **(*a*)** What are the constituents and lattice structure of a steel with 0.2% carbon as it goes from a molten to a solid state.

Sol. The iron-carbon equilibrium diagram is shown in Fig. 9 and microstructure of various phases of steel with varying percentage of carbon is shown in Fig. 10.

0.2% carbon steel while cooling from molten state upto around 1000°C mainly consists of austenite. Austenite is the solid solution of carbon with iron in which carbon completely dissolves in iron. As the steel is further cooled down, its structure gets transformed from austenite to ferrite + pearlite. With increased cooling rates, austenite gets converted into martensite.

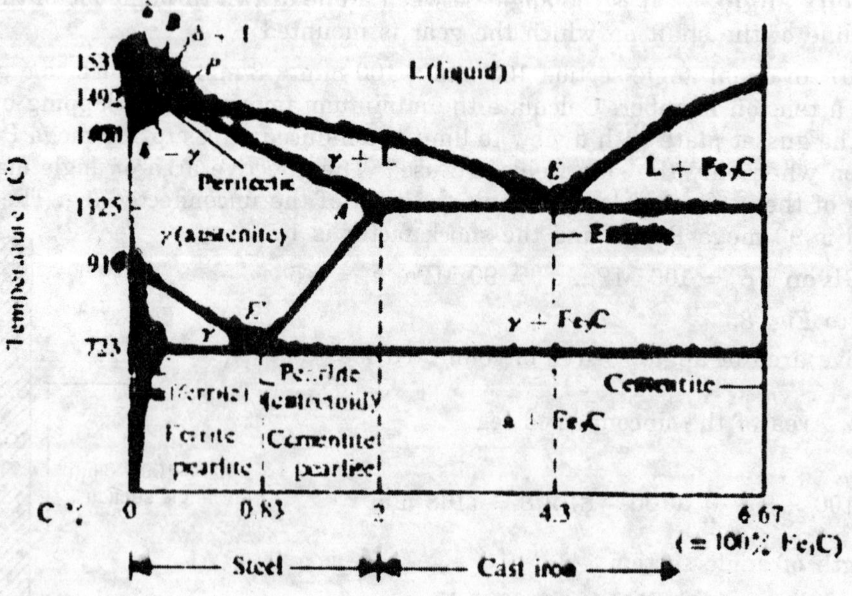

Fig. 9. Iron-carbon equilibrium diagram.

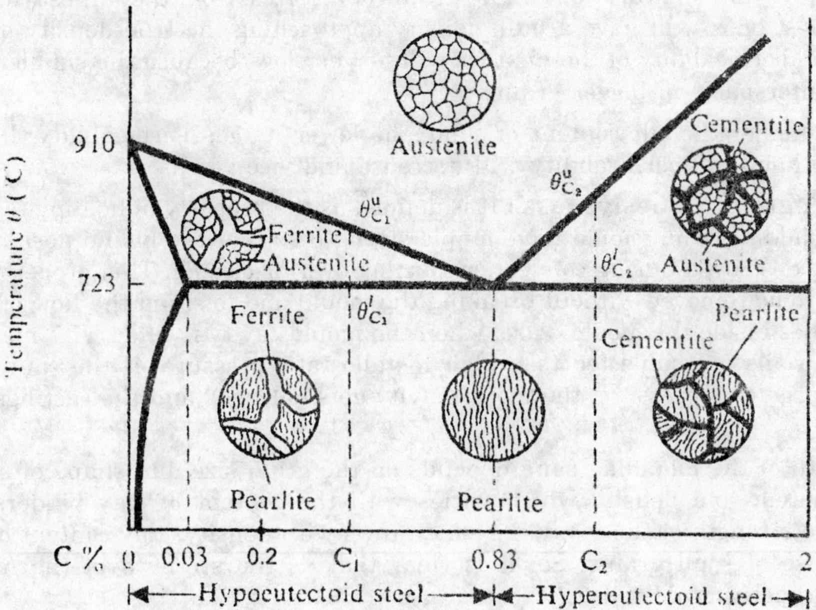

Fig. 10. Microstructure of various phases of steel.

Q.5. (*b*) "Moulding sand is different from the river sand." Give your comments in support of the above statement and also indicate the properties of moulding sand which are desirable for marking sound castings. How do you determine four of these properties in the laboratory?

Sol. The material for moulding sand is a mixture of sand, clay, water, and some organic additives, *e.g.*, wood floor, sea coal, etc. Sand is an inexpensive refractory material, but river bed sand may not have all the desirable qualities of a moulding material. The river bed sand has higher clay content than desired. The specific grain size distribution is also not there.

Desirable Properties of Moulding Sand for Making Sound Castings

1. **Permeability :** Permeability or porosity of the moulding sand is the measure of its ability to permit air to flow through it.

Molten metal always contains a certain amount of dissolved gases which try to leave it when the metal solidifies. Also, when the mould is poured, water vapours and other gases are generated when the molten metal comes in contact with the moulding sand and the cores. If all these gases and vapours are not able to escape completely through the walls of the mould, they may penetrate the liquid metal where, after solidification, they form gas holes and pores. To avoid these defects, the moulding sand should have good gas permeability.

Permeability of the moulding sand is influenced by : quality and quantity of clays and quality in the mixture, moisture content, degree of compaction etc. The higher the percentage of silica content of a moulding sand mixture (that is lower clay content), the better its permeability and vice versa. The larger the grain size, the better its permeability and vice versa. The distribution of *grain size* in a particular sand mix, greatly influenced the permeability. If all the sand particles in a mix are of the same size, then the bulk density is low and the permeability is higher because of the voids between the sand particles. Angular grains that are randomly oriented will have a lower bulk density than spherical grains and thus higher permeability. A sand mixture containing grains of all sizes, from the largest that can be used to the smallest ones, will have a bulk density approaching the true density of the sand. In this case, the permeability of the mixture will be very low, because the smaller particles will fill up the interspaces of bigger grains.

Again, higher the silt content of sand, the lower its gas permeability. If the mould is rammed too hand, its permeability will decrease and vice versa.

2. Strength or Cohesiveness : It is defined as the property of holding together of sand grains. A moulding sand should have ample strength so that the mould does not collapse or get partially destroyed during conveying, turning over or closing. This property also enables the pattern to be removed without breaking the mould and to stand the flow of molten metal when it rushes inside the mould. Also, when the mould cavity is filled with molten metal, its bottom and walls are subjected to a high metallostatic pressure. If the sand is not strong enough to resist this pressure, the mould cavity gets enlarged and the castings obtained are oversize.

Strength of the moulding sand depends on the grain size and shape of sand particles, moisture content and density which varies with the content of clay binders in the sand mixture. The strength of the moulding sand grows with density, clay content of the mix and decreased size of sand grains. So, it is clear that as the strength of the moulding sand increases, its porosity decreases.

3. Refractoriness : It is the ability of the moulding sand mixture to withstand the heat of melt without showing any signs of softening or fusion. This property is greatly influenced by the purity of the sand particles and their size. It increases with the grain size of sand and its content and with the diminished amount of impurities and silt.

4. Plasticity or Flowability : It is the measure of the moulding sand to flow around and over a pattern during ramming and to uniformly fill the flask. This property may be enhanced by adding clay and water to the silica sand.

5. Collapsibility : This is the ability of the moulding sand to decrease in volume to some extent under the compressive forces developed by the shrinkage of metal during freezing and subsequent cooling. This property is especially important for cores. This property permits the moulding sand to collapse easily during shake out and permits of the core to collapse easily during its knockout from the cooled casting. Lack of collapsibility in the moulding sand and core may result in the formation of cracks in the casting. This property depends on the amount of quartz sand and binders and their type.

6. Adhesiveness : This is the property of sand mixture to adhere to another body (here, the moulding flasks). The moulding sand should cling to the sides of the moulding boxes so that it does not fall out when the flasks are lifted and turned over. This property depends on the type and amount of binder used in the sand mix.

7. Co-efficient of Expansion : The sand should have low co-efficient of expansion.

Sand Testing

Moulding sands should be of proper quality to meet the required level of accuracy and surface finish of castings. Proper quality of te moulding sand will result in sound castings that will decrease the cost/unit and increase the production.

The following tests have been recommended by B.I.S.

1. Moisture content test

2. Clay content test
3. Permeability test
4. Fineness test or sand grain size test (Sieve analysis)
5. Strength test
6. Mould hardness test.

1. Moisture content test : Moisture is an important constituent of moulding sand. Its content in the moulding sand can be estimated as given below :

(*a*) **By the loss of weight after evaporation :** Test sample is carefully weighed 50 gm of moulding sand. It is heated for 2 hours at 105°C. It is then allowed to cool to room temperature on a desicator, after which it is reveighed. The difference in weight divided by the weight of the sample (50 gm) and multiplied by 100 gives the percentage moisture content.

(*b*) **By Moisture teller :** In this method, the sample is placed on the teller pan which has 500 mesh screen in the bottom. Hot air is blown through the sand. Moisture gets removed in only about 5 minutes and so the method is quite fast. The sand sample is weighed directly in the teller pan and the loss in weight is accordingly determined.

(*c*) **By moisture teller based on chemical reaction :** In this method, a measured weight of sand (10 gm) in a cap is placed in the moisture teller alongwith a container containing a measured amount of CaC_2 (Sand and CaC_2 should not come into contact with each other). The reactions of CaC_2 with moisture in the sand will produce C_2H_2,

$$CaC_2 + 2H_2O \longrightarrow C_2H_2 + Ca(OH)_2$$

The *amount* of C_2H_2 produced is proportional to the content of moisture in the sand. The moisture content is directly measured from a calibrated scale on the instrument.

2. Clay content test : Under this test, the clay content is measured by washing it off the sand sample. The standard dried sand sample (50 gm) is put in a mixing device along with distilled water and 1% NaOH solution. The mixture is stirred for 5 minutes and is then allowed to settle for 10 minutes. The sand settles down at the bottom. The dirty water is siphoned off from the top. The operation is repeated until the water above the sand in the mixer is clean. This water is drained off. The sand is dried and weighed. The loss of weight gives the clay content.

3. Permeability test : As already defined, permeability or porosity of the moulding sand is the measure of its ability to permit air to flow through it. It is measured in terms of "Permeability number". In this test, a specimen of moulding sand (50.8 mm dia. × 50.8 mm long) is placed in a tube. Time taken for 2000 cm^3 of air at a pressure of 10 gm/cm^2 to pass through the specimen is noted. Then, the permability number is given as :

$$P = \frac{V \times h}{p \times A \times t}$$

Where, P = Permeability number

V = Volume of air passing through test specimen

h = Height of specimen

p = pressure of air

A = Cross-sectional area of specimen

t = time taken by air to flo through the specimen

Understandard conditions, the formula becomes

$$P = \frac{2000 \times 5.08}{10 \times \frac{\pi}{4} \times (5.08)^2 \times t}$$

$$= \frac{50.128}{t} \quad (t \text{ is in minutes})$$

or $$P = \frac{501.28}{p \times t}$$

4. Fineness Test : Fineness test of the sand specimen determines the size of grains and the distribution of gains of different sizes in the moulding sand. Their effect on the properties of moulding sand has been discussed in Art. 3.4.5. The test is performed on completely dry and clay free sand. The test apparatus consists of a set of graded sieves placed one above the other, with the coarsest sieve at the top and the finest sieve at the bottom, next to the pan. The sand is put in the top sieve and the set of sieve is shaken continuously for 15 minutes with the help of an electric motor. At the end of the shaking operation, the sieves are taken out and the sand retained on each is carefully weighed. Each weight is converted to a percentage basis. Each percentage is multiplied by a weighting factor and these are added to get the sum of products. Then the grain fineness number (GFN) is given as,

$$\text{GFN} = \frac{\text{Sum of products}}{\text{Sum of percentages of sand retained on each sieve and pan}}$$

As per AFS (American Foundrymen's Society), the various sievemesh numbers and the corresponding weighting factors (Multiplying factors) are given below :

Sieve number	6	12	20	30	40	50	70	100	140	200	270	pan
Weightage factor	3	5	10	20	30	40	50	70	100	140	200	300

or $$\text{GFN} = \frac{7113}{100} = 71.13$$

5. Strength Test : These tests are carried out to check the holding powers of various binders used in Green moulding sand, dry moulding sand and core sand. These tests are performed on a U.T.M. The various tests carried out include :

(*i*) Compressive strength test

(*ii*) Shear strength test

(*iii*) Tensile strength test

(*iv*) Bending strength test

Out of these, compressive strength test is most important.

The test specimen is 5.08 cm ϕ × 5.08 long. The test is performed by applying a uniformly increasing load on sand specimen until it breaks. The compressive strength is read directly on the scale of the machine. The other tests are also similarly carried out by using suitable jaws.

The green strength of moulding sands ranges from 30 to 160 k Pa

The dry strength of moulding sands ranges from 140 to 1800 k Pa

The green shear of moulding sands ranges from 10 to 50 k Pa.

6. Hardness Test : The hardness of the prepared mould surface is tested to check the ramming density of the sand. The instrument resembles a dial indicator, and carries a spring loaded steel bell of 5 mm. The ball is pressed into the rammed surface. The depth of penetration of the ball is indicated on the dial of the indicator (which is divided into 100 parts) in terms of hardness units. The value of hardness units for sands rammed to different degrees are approximately :

40 for lightly rammed sand

50 for medium rammed sand

and 70 for hard rammed sand

Q.5. (*c*) For manual arc welding of mild steel, what characteristics of current and voltage will be desired? Will the requirements change if we switch over to automatic welding?

Sol. (*a*) Manual Electric Arc Welding of Mild Steel : The desirable characteristics of current and voltage are :

Current = 10 – 2000 Amperes

Voltage – 10 – 50 Volts

(*i*) Selection of Static Volt-Ampere Characteristics

The four types of characteristics are :

1. Steeply drooping type,
2. Gradually drooping type,
3. Flat or constant voltage type,
4. Rising voltage type.

All these types of power source characteristics with the volt-ampere characteristics of the welding arc superimposed on them are shown in Fig. 11.

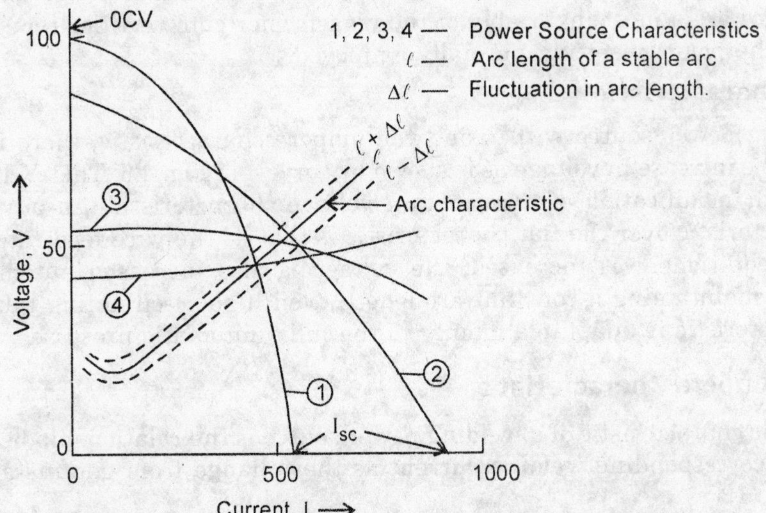

Fig. 11. Volt-ampere characteristics of different welding power sources and the welding arc.

1. Steeply Drooping V-I characteristic

The welding power source with steeply drooping volt-ampere characteristic has a high open circuit voltage and low short-circuit current as shown by curve 1 in Fig. 11. It is evident that as the arc length changes between $l - \delta l$ and $l + \delta l$, the change in current is very little. This type of volt-ampere characteristic is best suited for SMAW that is manual metal arc welding with coated electrodes because a slight change in arc length due to the intrinsic movement of the human hand during welding operation will not affect the melting rate of the electrode. Also, high open circuit voltage ensures easy initiation and maintenance of the welding arc.

2. Gradually Drooping V-I Characteristic

The power source with gradually drooping static volt-ampere characteristic, as shown by curve 2 in Fig. 4.6, can supply high short-circuit current as is needed for submerged arc welding with thick electrodes particularly for electrode diameter more than 3-5 mm. A power source with this type of volt-ameter characteristic reqires some technique of arc initiation similar to that of touch and draw used for SMAW or alternatively steel wool may be used to provide a momentary short-circuit between the electrode and the workpiece. The open circuit voltage may be slightly lower than that in the case of steeply drooping V-I characteristic. This feature helps in providing some sort of self-regulation of the arc length during welding as for the same change in arc length the change in arc current is considerably more than in the case of steeply drooping volt-ampere characteristic.

3. Flat V-I Characteristic

In a constant voltage welding power source for a small change in arc length there is a large change in welding current which makes it quite sensitive and hence helps in maintaining a steady arc length with consequential consistent quality welds. This is generally referred to as the self-regulation of the arc length and is an essential requirement for the success of gas metal arc welding. The change in arc length is inevitable particularly in semi-automatic GMAW hence constant voltage static volt-amere characteristic is very useful for fine wire welding processes. However, the flat V-I characteristic as shown by curve 3 in Fig. 4-6, is not truly flat but normally droops at 1-3 volt per 100 ampere. All welding power sources with flat V-I characteristics are almost invariably of the trnasformer cum rectifier type and electrode positive (ep) is the polarity setting normally employed.

4. Rising V-I characteristic

In a welding power source with rising volt-ampere characteristic, there is an increase in current with the increase in voltage as is shown by curve 4 in Fig. 4.6. This V-I characteristic is based on slight modification of the constant voltage characteristic. An advantage of the rising V-I characteristic over the flat characteristic is that as the wire feed rate is increased, the amperage requirement is increased, the voltage is also increased automatically. This feature helps in maintaining a constant arc length even if short-circuiting takes place. The rising V-I characteristic is adaptable mainly to the fully automatic processes.

Dynamic Volt-Ampere Characteristic

The dynamic characteristic of a welding power source is the relationship between the arc voltage and the corresponding welding current as they change from one instant to another as shown in Fig. 12.

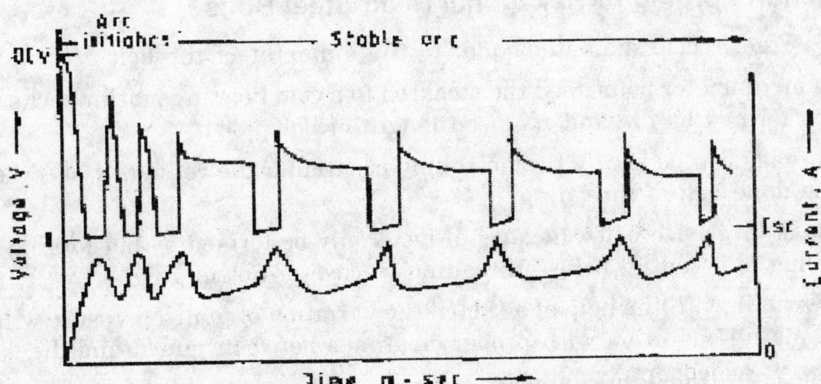

Fig. 12. Dynamic volt-ampere characteristics of a d.c. power souce with very short arc length.

It is imperative to know the nature of dynamic characteristic to determine the rate of rise of current after a short-circuit which affects the rate of electrode melting and the weld spatter.

A typical arc length-voltage characteristics of a d.c. arc are :

$$V = 24 + 4l \text{ for welding arc}$$

$$V = 80 - \left(\frac{80}{600}\right)l \text{ for power source}$$

(*b*) Automatic arc Welding

In automatic arc welding process, the arc is maintained between the workpiece and a wire which is driven forward at a constant speed as it melts away from the tip. An increase in arc length increases the voltage and, as a result, the current falls. An increase in the arc length causes a decrease in the melting rate. If the reduction in the current due to an increase in the arc length is significant, the melting rate decreases considerably so that the arc length returns to its original value. Hence, in this case, for a stable operation, a flat characteristics of the power source is desirable.

Q.6. (*a*) What are the advantages of forging over casting?

How will you mass produce a steel bolt with a hexagonal head?

Sol.

Advantages of Forging over Casting

1. Structural homogeneity is obtained as the impurities in the metal in the form of inclusions are broken up and distributed throughout the metal.
2. Parosity, voids and blow holes in the metal are largely eliminated giving sound structure.
3. Coarse grains are refined resulting increased yield strength and ultimate strength.
4. Directional flow of forged components imports better mechanical properties.
5. Superior machining qualities are obtained.
6. Relatively smoother surface of forging.
7. Rapid duplication of components is possible by forging.

Mass Production Forging of Hexagonal Head Steel Bolts

1. Take a steel rod of diameter equal to the diameter of the bolt.
2. For small diameter bolts, feed the steel rod to a cold heading machine where hexagonal head is formed by die and required length of bolt is cut.
3. The threads can be cut by thread rolling or circular die rolling. If need be, annealing may be done before threading.
4. In case of large size bolts, hexagonal head may be formed by hot forging using a die. Annealing is essential before threading in case of hot forging.

Q.6. (*b*) Explain with the help of a sketch the variation of punch pressure with the punch travel for a blanking operation with proper clearance between punch and die.

Why is shear provided on a punch?

Sol. The operations of blanking and piercing are carried out on a press and consists of cutting the outside contour of a stamping. It requires a single action press equipped with tools comprising of a punch, a corresponding die and stripper to keep the sheet from following the punch on its up stroke. In blanking the cut out piece is the desired part. In this case, punch is made smaller in size and the die is made of exact size. Thus clearance is provided on the punch.

The cutting force in blanking is given by,

$$F_{max} = \pi d t \cdot \tau_u \text{ for round pieces}$$

$$= Lt \cdot \tau \text{ for other contours}$$

where d = hole diameter, mm

t = material thickness, mm

τ_u = ultimate shear strength of material, MPa or N/mm^2

L = shear length, mm

= $2(b + l)$ for rectangular blank of length l and width b.

The energy consumed in blanking operation is given by,

$$E = F_{max} \times K \times t \times C$$

where K = percentage of penetration required to cause rupture

c = a factor to allow for energy lost in machine friction, etc.

$\approx$ 1.16

The variation of cutting force with punch travel, as calculated by above formulae, is shown in Fig. 13(*a*). It has been assumed that the bottom of the punch and top of the die block lie in parallel planes. This process is characterized by very high punch forces exerted over a very short time, resulting in shock or impulse conditions.

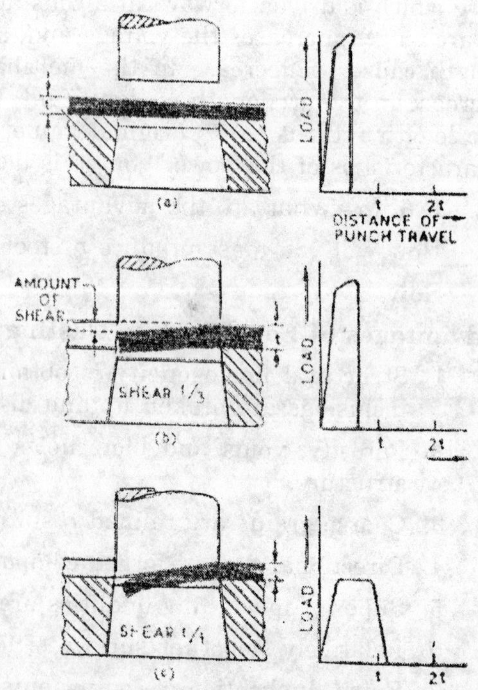

Fig. 13. Effect of shear on cutting force

Necessity of Shear on a Punch

To reduce cutting forces and to smooth out the shock impact of heavy loads, gradual cut instead of sudden cut of the stock is arranged. This is achieved by grounding off the working faces of the punch or die so that these do not remain parallel to the horizontal plane but are inclined to it. The anlge of inclination is called the 'shear'. This has the effect of reducing the area in shear at any one time and the maximum blanking force is greatly reduced. The effect of providing shear on maximum cutting load is shown in Fig. 13(*b*) and (*c*).

Q.6. (*c*) Distinguish between orthogonal and oblique cutting. Derive an expression for the shear plane angle in a turning operation on a lathe. Explain all the assumptions made to arrive at the solution.

Sol. Types of Metal Cutting Process : There are two types of metal cutting processes.

1. Orthogonal cutting process : It is a two-dimensional cutting process in which the major cutting edge of the tool is presented to the work piece perpendicular to the direction of cutting velocity vector. the orthogonal cutting process is shown in Fig. 14.

2. Oblique cutting process : It is a three-dimensional cutting process in which the major edge of cutting tool is presented to the workpiece at an angle other than 90° to the cutting velocity vector. The oblique cutting process is shown in Fig. 15.

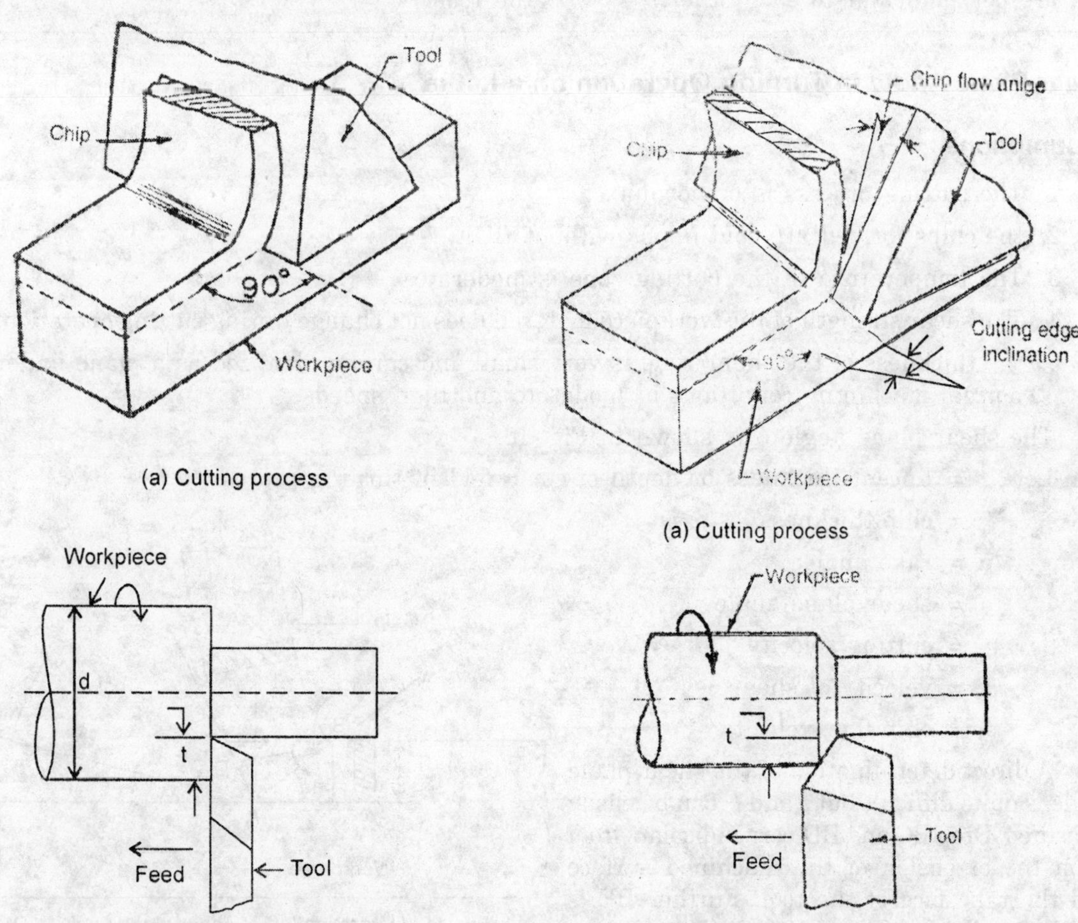

(a) Cutting process

(b) Turning operation

Fig. 14. Orthogonal cutting.

(a) Cutting process

(b) Turning operation

Fig. 15. Oblique cutting.

The comparison between orthogonal and oblique cutting is given in Table 1.

t = depth of cut

d = workpiece diameter

Table 1. Comparison of orthogonal and oblique cutting

Orthogonal cutting	Oblique cutting
(*i*) The cutting edge of the tool remains normal to the direction of tool feed.	The cutting edge of the tool is inclined at an acute angle to the direction of tool feed.
(*ii*) The direction of chip flow velocity is normal to the cutting edge of the tool.	The direction of ship flow velocity is at an angle with the normal to the cutting edge of the tool.
(*iii*) The cutting edge clears the width of the workpiece on either ends.	The cutting edge may or may not clear the width of the workpiece.
(*iv*) Only two components of cutting forces act on the tool. These two components are perpendicular to each other.	Three mutually perpendicular components of cutting forces act at the cutting edge of the tool.

Shear Plane Angle in Turning Operation on a Lathe

Assumptions :

1. The cutting process is orthogonal.
2. the chips formed are continuous without BUE.
3. The temperature in the cutting zone is moderate.
4. The shear strength of the workpiece material does not change during cutting operation.
5. the thickness of the shear zone is very small and can be idealized as a plane under normal machining conditions at moderate and high speeds.

The shear plane angle ϕ is shown in Fig. 16.

Let t = uncent thickness or depth of cut = feld in turning process

t_c = chip thickness

α = rake angle

ϕ = shear plane angle

v_c = cutting velocity

v_s = velocity of shear

v_f = chip flow velocity

A direct determination of the shear plane angle is quite difficult but t and t_c can be easily measured. Drop BC and BD perpendiculars from B on the extension of the machined surface and the rake face of the tool. Further BE is drawn parallel to AC and F is the intersection of BE with the normal drawn at A on AC.

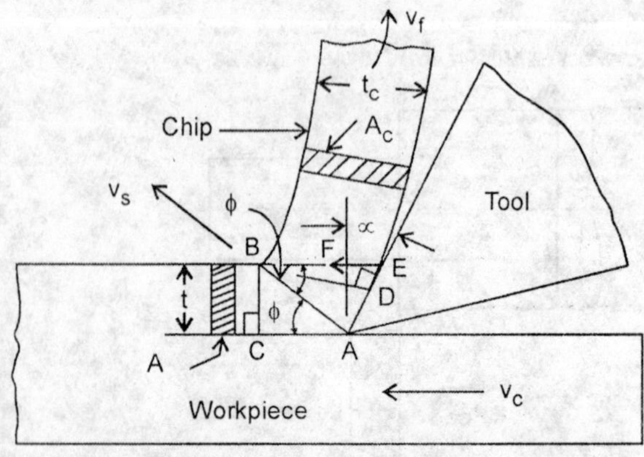

Fig. 16. Determination of shear plane angle

Now $\angle EBD = \angle EAF = \alpha$

or $\angle DBA = \angle EBA - \angle EBD < \phi - \alpha$

Again $$AB = \frac{BD}{\cos(\phi - \alpha)} = \frac{t_c}{\cos(\phi - \alpha)} = \frac{BC}{\sin\phi} = \frac{t}{\sin\phi}$$

$$\therefore \quad \frac{t}{t_c} = \frac{\sin\phi}{\cos(\phi - \alpha)}$$

Now $\frac{t}{t_c} = r =$ chip thickness ratio

$$\therefore \quad r = \frac{\sin\phi}{\cos(\phi - \alpha)} = \frac{\sin\phi}{\cos\phi\cos\alpha + \sin\phi\sin\alpha}$$

or $r \cos\phi \cos\alpha + r \sin\phi \sin\alpha = \sin\phi$

tan $\phi = r \cos\alpha + r \tan\phi \sin\alpha$

$$\tan\phi = \frac{r\cos\alpha}{1 - r\sin\alpha}$$

Q.7. (*a*) It is a problem for a seed packager to minimize the cost of satisfying the nutritional requirements. Unlimited quantities of seeds can be purchased at the costs indicated below.

Nutritional Item	**Proportional Content**		**Total Requirement**
	Buck Wheat	**Sunflower Wheat**	
Fat	0.04	0.06	≥ 480 kg.
Protein	0.12	0.10	≥ 1200 kg.
Roughage	0.10	0.15	≤ 1500 kg.
Cost per kg.	Rs. 18	Rs. 10	

Find the optimal solution.

Sol. Let x_1, x_2 = quantities of seeds of buck wheat and sunflower wheat respectively punchased.

Then Min : $z = 18x_1 + 10x_2$

Subject to : $0.04x_1 + 0.06x_2 \geq 480$ or $4x_1 + 6x_2 \geq 48{,}000$

$0.12x_1 + 0.10x_2 \geq 1200$ or $12x_1 + 10x \geq 1{,}20{,}000$

$0.10x_1 + 0.15x_2 \geq 1500$ or $10x_1 + 15x_2 \leq 1{,}50{,}000$

The various constraints are plotted on $x_1 - x_2$ condinesg in Fig. 17. The bound points are:

A (7500,300) Point B : $12x_1 + 10x_2 = 1{,}20{,}000$

B (3750,7500) $10x_1 + 15x_2 = 1{,}50{,}000$

C (15,000,0) or $120x_1 + 100x_2 = 12{,}0{,}000$

D (0,12000) $120x_1 \times 180x_2 = 18{,}00{,}000$

$x_2 = 7500$

$x_1 = 3750$

Point A :

$$12x_1 + 10x_2 = 1,20,000$$
$$4x_1 + 6x_2 = 48,000$$
or $$12x_1 + 18x_2 = 1,44,000$$
$$8x_2 = 24,000$$
$$x_2 = 3,000$$
$$x_1 = 7500$$

Z_A = 18 × 7500 + 10 × 3000 = Rs. 165000

Z_B = 18 × 3750 + 10 × 7500 = Rs. 142500

Z_C = 18 × 15000 = Rs. 2,70,000

Z_D = 10 × 12000 = Rs. 120000

Z_{min} = Rs. 2,70,000 for x_1 = 15,000, x_2 = 0

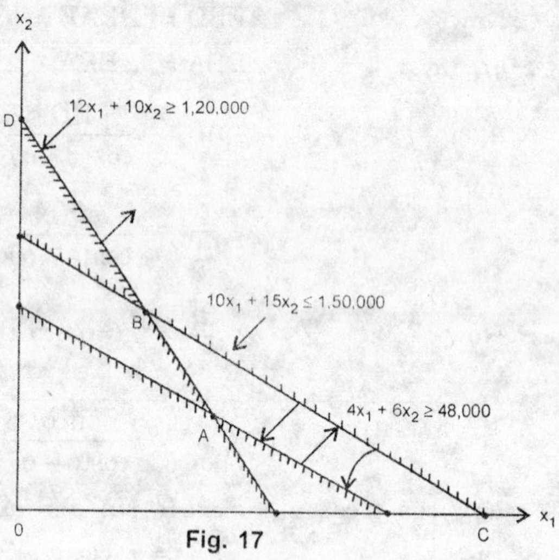

Fig. 17

Q.7. (b) If Kelvinator produces refrigerators in batches, how many units in a batch should they produce? In each batch once the production starts they can make 80 unhits per day. The demand during the production period is 60 units per day. Estimated demand for the year is 10,000 units. Set up cost of the manufacturing process is Rs. 3000/unit. Carrying cost is Rs. 15 per unit per year.

Sol. Given : k = 80 × 365 = 29200 units/year, D = 60 × 365 = 21900 units/year

R = 10,000 units/year, C_o = Rs. 300/unit, C_i = Rs. 15 per unit per year

Optimum lot size, $$Q^* = \sqrt{\frac{2kC_oR}{C_i(k-R)}}$$

$$= \sqrt{\frac{2 \times 29,200 \times 3000 \times 10,000}{15(29,200 - 10,000)}}$$

$$= 2466$$

MECHANICAL ENGINEERING PAPER II—1995

(Conventional)

Time allowed : Three hours *Maximum marks : 200*

Section A

Q.1. **(*a*)** Distinguish between austempering and martempering of steel.

Sol. Austempering is carried out by heating steel above the critical temperature and then cooling at slow rates in an oil bath to 430°C so as to ensure complete austenite decomposition into bainite or acicular troosite.

Martempering is carried out at a fast cooling rate in air or oil to 240° – 300°C so that austenite is converted into martensite.

(*b*) What do you understand by the term proof stress? What is the value of strain for 0.2 percent proof stress?

Sol. Proof stress is the stress at which the non-proportional elongation amounts to a specified percentage of original gauge length. The proof stress is generally specified by 0.2% of original gauge length.

(*c*) Explain the following :

(*a*) CPM, and (*b*) PERT.

Sol.

(*a*) Critical Path Method (CPM) is a basic technique for planning, scheduling, and controlling projects of all types. It includes a method for expediting projects at minimum cost.

(*b*) Programme Evaluation and Review Technique (PERT) is a technique for scheduling special projects by making three time estimates, *i.e.* pessimistic, optimistic, and the most likely times. Its objectives are to find the earliest date of completion of the project, activities likely to have bottlenecks, and the effect of transferring resources from one activity to the other.

(*d*) Give two methods of joining two pieces of metal by heating alone.

Sol. Forge welding and friction welding.

(*e*) What is Scleroscope?

Sol. It is a device to measure the hardness of materials by dropping a tup from a height on to the material. The height of rebound of the tup is a measure of the hardness of the material.

(*f*) What is 'killed' steet?

Sol. Killed steels are those steels in which gas reaction is prevented (killed or deoxidized) by the addition of aluminium, silicon, etc. In these steels, segregation is virtually absent, properties are uniform throughout, and grain size can be controlled in the finished product.

(*g*) What is lost wax method?

Sol. Lost wax method is also called the investment casting process. In this method, pattern is made of wax complete with feeding system, and the refractory slurry is poured around this. In solid investment, fine refractory slurry followed by coarser refractory slurry is poured around the wax pattern. When the slurry gets gelled, the mould is dried in an oven in an upside - down position to allow the wax to run out.

Before casting the mould is fired at 700° 1000°C to impart strength. In the ceramic shell - moulding process, the wax pattern is covered with refractory in a fluidized bed.

(*h*) What are mitre wheels?

Sol. Mitre gears are two bevel gears of the same size and mounted on two shafts whose axes are at 90°.

(*i*) What is hunting of a governor?

Sol. It is a condition in which the speed of the engine controlled by the governor fluctuates continuously above and below the mean speed.

(*j*) In which machine tool the Whitworth quick return mechanism is used?

Sol. Shaper, slotter, and gear planer.

(*k*) What is coefficient of restitution?

Sol. It is the negative ratio of the relative velocities of two colliding bodies after and before collision. Coefficient of restitution = $-(v_1 - v_2)/(u_1 - u_2)$.

(*l*) Kinetically equivalent systems must satisfy three requirements, give any two of them.

Sol.

1. $m = m_1 + m_2$ 2. $m_1 a_1 = m_2 a_2$

3. $m\,K_G^2 = m_1 a_1^2 + m_2 a_2^2$

(*m*) The method which gives the acceleration of the elements of a connecting rod of a slider crank mechanism by a simple graphical form is known as ....

Sol. Klein's construction.

(*n*) Indicate the arc welding process which is more suitable for overhead welding. Justify your answer.

Sol. Shielded metal arc welding is quite suitable for overhead welding. It uses coated electrodes of 2.5 - 6.35 mm diameter and 300 - 450 mm long held in an electrode holder. It can use A.C. or D.C. constant current type power source. D.C. reverse polarity gives more heat at electrode (+ve).

(*o*) What is a dummy activity in network models?

Sol. An activity which has a duration time of zero and does not consume any resources. It is inserted for the sole purpose of preserving the precedence relationship.

(*p*) Expand the abbreviations (*a*) DNC, and (*b*) CNC.

Sol.

(*a*) Direct (or Distributed) Numerical Control

(*b*) Computer Numerical Control.

(*q*) Expand the abbreviations (*a*) ECM, and (*b*) HSS.

Sol.

(*a*) Electro - Chemical Machining (*b*) High Speed Steel.

(*r*) Name four parts of an automobile where cast iron is used.

Sol.

1. Engine block
2. Brake drum
3. Piston rings
4. Magneto - bush.

(*s*) How sand cores are supported?

Sol. Sand cores are supported on core prints and chaplets.

(*t*) What is Kennedy's theorem?

Sol. This theorem states that if three bodies have relative motion with respect to each other, their relative instentaneous centres lie on a straight line.

Section B

Q.2. (*a*) A motor drives a main shaft by means of a flat belt. The diameters of the pulleys on the motor shaft and the main shaft are 450 mm and 750 mm respectively. Another pulley of diameter 500 mm drives a counter shaft having a pulley of diameter 700 mm. The pulley of diameter 500 mm is mounted on the main shaft. If the slip on each drive is 3 per cent, calculate the speed of counter shaft if motor runs at 1400 rpm.

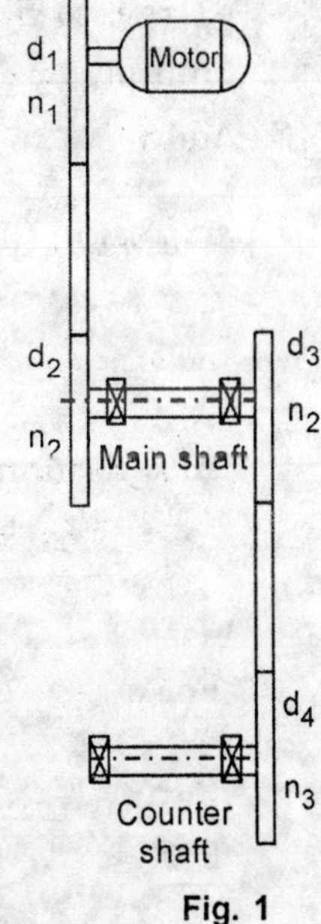

Fig. 1

Sol. Given : $d_1 = 450$ mm, $d_2 = 750$ mm, $d_3 = 500$ mm, $d_4 = 700$ mm, $s_1 = s_2 = 3\%$, $n_1 = 1400$ rpm

Refer to Fig. 1.

$$n_2 = n_1\left(\frac{d_1}{d_2}\right)\left(\frac{100 - s_1}{100}\right)$$

$$= 1400\left(\frac{450}{750}\right)\left(\frac{100-3}{100}\right) = 814.8 \text{ rpm}$$

$$n_3 = n_2\left(\frac{d_3}{d_4}\right)\left(\frac{100 - s_2}{100}\right)$$

$$= 814.8 \times \left(\frac{500}{700}\right)\left(\frac{100-3}{100}\right)$$

$$= 564.5 \text{ rpm}$$

Q.2. (*b*) In an epicyclic gear of the sun and planet type shown in Fig. 2 the pitch circle diameter of the annular wheel 'A' is to be nearly equal to 216 mm, and the module is 4 mm. When the annular wheel is stationary, the spider which carries three planet gears 'P' of equal size, has to make one revolution for every five revolutions of the driving spindle carrying 'S' gear. Determine the number of teeth on all the wheels and also the exact pitch circle diameter of 'A'.

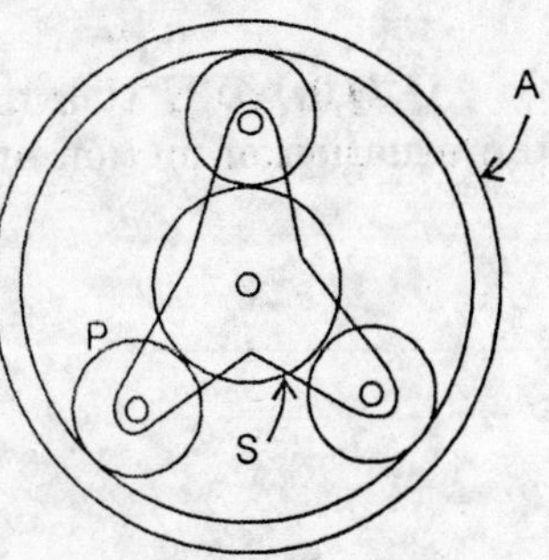

Fig. 2

Sol. Given : $d_A = 216$ mm, $m = 4$ mm,

$$z_A = \frac{d_A}{m} = \frac{216}{4} = 54$$

Now
$$d_A = d_S + 2d_P$$
$$m.z_A = m.z_s + 2mz_P$$
or
$$z_A = z_S + 2z_P$$
or
$$z_S + 2z_P = 54$$

Operation	Revolutions of			
	Spider	Sun gear S	Planets P	Annular wheel A
1. Spider fixed, +1 rev. to S, ccw.	0	+ 1	$-z_S/z_P$	$-z_S/z_A$
2. Multiply by x.	0	$+x$	$-x\,z_S/z_P$	$-xz_S/z_A$
3. Add y	$+y$	$x+y$	$-xz_S/z_P + y$	$-xz_S/z_A + y$

For +1 revolutions of spider, the sum gear has to make + 5 revolutions.

$$x + y = 5$$
and
$$y = 1$$
$\therefore$
$$x = 4$$

Also for annular wheel fixed,

$$-x\,z_S/z_A + y = 0 \quad \text{or} \quad 1 - 4\,z_S/z_A = 0 \quad \text{or} \quad z_A = 4z_S$$

$$z_S = \frac{54}{4} = 13.5 \quad \textit{i.e.} \quad 13 \text{ or } 14$$

For $z_S = 13$, $z_A = 52$

$\therefore$ $$z_P = \frac{z_A - z_S}{2} = \frac{52-13}{2} = 19.5, \text{ which is not practiable}$$

For $z_S = 14$, $z_A = 56$

$$z_P = \frac{56-14}{2} = 21$$

$$d_A = 4 \times 56 = 224 \text{ mm}$$

Q.3. (*a*) A 'U' tube as shown in Fig. 3 contains a liquid of density one kg/m^3. Determine the equation of motion and find the natural frequency of the liquid column.

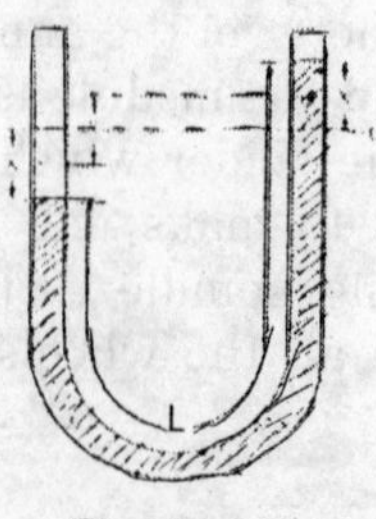

Fig. 3

Sol. Let A = uniform cross-sectional area of tube, m²

ρ = density of liquid, kg/m³

x = displacement of liquid column in the tube.

Total mass of liquid, $m = \rho$ AL kg

Kinetic energy, $T = \frac{1}{2}(\rho AL)\dot{x}^2$

Potential energy, $U = (\rho Ax) \times g \times x = \rho Agx^2$

Now $\frac{d}{dt}(T + U) = 0$

$$\rho AL \cdot \dot{x} \cdot \ddot{x} + 2\rho Ag \cdot x \cdot \dot{x} = 0$$

or $$\ddot{x} + \left(\frac{2g}{L}\right)x = 0$$

$$w_n = \sqrt{\frac{2g}{L}} \text{ rad/s}$$

or $$f_n = \frac{1}{2\pi}\sqrt{\frac{2g}{L}} \text{ Hz}$$

Q.3. (*b*) The arms of a Porter governor are pivoted on the governor axis and are each 250 mm log. Mass of each ball is 0.5 kg and mass of sleeve is 2 kg. The arms are inclined at an angle of 30° to the governor axis in the lowermost position, of the sleeve. Lift is equal to 50 mm. Determine the force of friction if the speed at the moment the sleeve starts lifting from the lowermost position is the same as the speed at the moment it falls from the uppermost position.

Sol. l = 250 mm, m = 0.5 kg, M = 2 kg, $\alpha = \beta$ = 30°, lift, h = 50 mm

Refer to Fig. 4

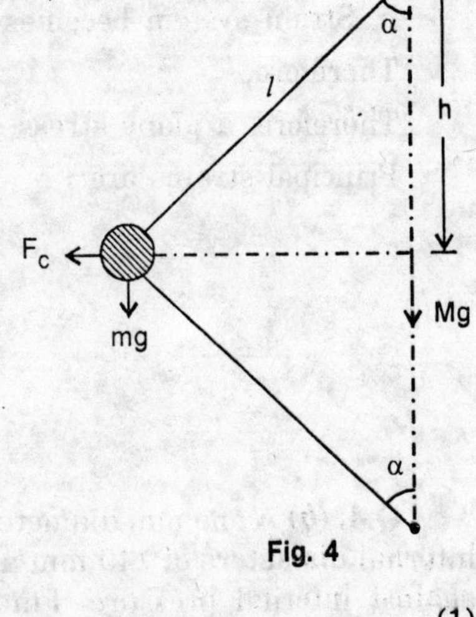

Fig. 4

$$h_1 = l \cos \alpha = 250 \cos 30° = 216.5 \text{ mm}$$

$$h = 2(h_1 - h_2)$$

$$50 = 2(216.5 - h_2)$$

$$h_2 = 191.5 \text{ mm}$$

When the sleeve starts lifting from lowest position

$$h_1 = \left(\frac{g}{w^2}\right)\left[1 + \frac{Mg + F}{mg}\right]$$

$$0.2165 = \left(\frac{9.81}{w^2}\right)\left[1 + \frac{2 \times 9.81 + F}{0.5 \times 9.81}\right]$$

$$w^2 = 45.31\ [1 + 4 + 0.203875\ F]$$

$$= 226.55 + 9.238\ F \qquad ...(1)$$

When the sleeve starts lowering down,

$$h_2 = \left(\frac{g}{w^2}\right)\left[1+\frac{mg-F}{mg}\right]$$

$$0.1915 = \left(\frac{9.81}{w^2}\right)\left[1+\frac{2}{0.5}-\frac{F}{0.5\times 9.81}\right]$$

$$w^2 = 51.227\ [1 + 4 - 0.203875\ F]$$

$$= 256.135 - 10.444\ F \qquad ...(2)$$

Comparing Eqs. (1) and (2), we get

$$226.55 + 9.238\ F = 256.135 - 10.444\ F$$

$$F = 1.5\ N$$

Q.4. (***a***) Do you think that a plane stress system gives rise to a plane strain system always? Give reasons for your answer. Three principal stresses σ_{11}, σ_{22}, σ_{33} are acting on the three faces of a cube respectively. Express the resulting three principal strains in terms of the principal stresses.

Sol. A plane stress system is defined by :

$$\sigma_x,\ \sigma_y,\ \tau_{xy} \text{ and } \sigma_z = 0,\ \ \tau_{yz} = 0,\ \tau_{xz} = 0$$

$$\varepsilon_x = \frac{1}{E}[\sigma_x - \nu\sigma_y]$$

$$\varepsilon_y = \frac{1}{E}[\sigma_y - \nu\sigma_x]$$

$$\varepsilon_z = \frac{1}{E}[\sigma_z - \nu(\sigma_x + \sigma_y)] = -\frac{\nu}{E}(\sigma_x + \sigma_y)$$

$$\gamma_{xy} = \frac{\tau_{xy}}{G} = \frac{2(1+\nu)\,\tau_{xy}}{E}$$

∴ Strain system becomes : ε_x, ε_y, ε_z, γ_{xy}

Therefore, $\varepsilon_z \neq 0$

Therefore, a plane stress system does not give rise to plane strain system alwasy.

Principal strains are :

$$\varepsilon_{11} = \frac{1}{E}\left[\sigma_{11} - \nu\left(\sigma_{22} + \sigma_{33}\right)\right]$$

$$\varepsilon_{22} = \frac{1}{E}\left[\sigma_{22} - \nu\left(\sigma_{11} + \sigma_{33}\right)\right]$$

$$\varepsilon_{33} = \frac{1}{E}\left[\sigma_{33} - \nu\left(\sigma_{11} + \sigma_{22}\right)\right]$$

Q.4. (***b***) A one mm diameter steel wire is wound around a copper tube with external and internal diameters of 140 mm and 120 mm respectively, to increase the strength of the tube against internal pressure. Find, what initial tension must be given to the wire, so that maximum allowable stresses for the tube and wire materials namely 90 and 200 mega Pascals respectively are reached simultaneously. Assume 'μ' for copper and steel to be 0.3 and wire winding as a thin cylinder. Modulus of elasticity for copper and steel may be assumed as 10^5 and 2×10^5 mega Pascals respectively. The tube is open at the ends.

Sol. Given : d_w = 1 mm, d_o = 140 mm, d_i = 120 mm, $(\sigma_c)_p$ = 90 MPa, $(\sigma_w)_p$ = 200 MPa,
$\mu_c = \mu_s$ = 0.3, $E_c = 10^5$ MPa, $E_s = 2 \times 10^5$ MPa

Mean diameter of tube, $$d_m = \frac{1}{2}(d_i + d_o) = \frac{1}{2}(120 + 140) = 130 \text{ mm}$$

Thickness of tube, $$t = \frac{1}{2}(d_o - d_i) = \frac{1}{2}(140 - 120) = 10 \text{ mm}$$

Longitudinal stress in tube,

$$\sigma_l = \frac{pd_m}{4t} = \frac{p \times 130}{4 \times 10} = 3.25\,\text{p MPa}$$

Let T = initial tension (stress) in the wire, MPa

Then compressive stress in tube due to winding,

$$(\sigma_c)_w = \frac{\pi}{4} \times \frac{d_w}{t} \times T = \frac{\pi}{4} \times \frac{1}{10} \times T = \frac{\pi T}{40} \text{ MPa}$$

Now $$d_m \times p = (\sigma_c)_p \times 2t + (\sigma_w)_p \times \frac{\pi}{2} d_w$$

$$130p = 90 \times 2 \times 10 + 200 \times \frac{\pi}{2} \times 1$$

$$p = 16.263 \text{ MPa}$$

$$\frac{(\sigma_w)_p - T}{E_w} = \frac{1}{E_c}\left[(\sigma_c)_p - (\sigma_c)_w - \nu_c \sigma_l\right]$$

$$\frac{200 - T}{2 \times 10^5} = \frac{1}{10^5}\left[90 - \frac{\pi T}{40} - 0.3 \times 3.25 \times 16.263\right]$$

$$100 - 0.5\text{ T} = 74.14 - 0.0785\text{ T}$$

$$\text{T} = 61.358 \text{ mPa}$$

Section C

Q.5. (*a*) Give the percentages of carbon in the following :

(*i*) Low carbon steels

(*ii*) Medium carbon steels

(*iii*) High carbon steels

Sol.

Steel	Percentage of carbon
(*i*) Low carbon steel	0.05 – 0.30
(*ii*) Medium carbon steel	0.30 – 0.60
(*iii*) High carbon steel	0.60 – 1.50

Q.5. (*b*) Give the percentages of the alloying elements in

(*i*) Low alloy steels (*ii*) High alloy steels

Sol.

(*i*) Low alloy steels

Mn = upto 0.8%

Si = upto 0.3%

S = upto 0.05%

P = upto 0.05%

(*ii*) High alloy steels

Ni = 0.5%

Cr = upto 4%

Mo = 0.2 – 5%

W = 2 – 3%

Mn = 1.2 – 1.4%

Si = upto 5%

V = upto 2%

Q.5. (*c*) All cast irons contain at least six elements. Name these elements.

Sol. Elements in C.I. are :

C, Si, Mn, S, P, and Fe

Q.5. (*d*) When will you recommend carburizing of steel? Give the salient features of the process.

Sol. Carburizing of steel is recommended for parts used under heavy impact loads, like gears, worms, cams, etc. Such parts must have a soft core and hard surface.

Carburizing is the process of saturating the surface layer of low carbon steels at 850° – 950°C with carbon from a carbonaceous source capable of giving up its carbon to the metal. The carbon content in the surface layer is increased to 0.75 – 1.2%. The depth of carbon penetration ranges from 0.5 to 2 mm.

Q.5. (*e*) Give the composition and uses of the following alloys :

(*i*) Bronze (*ii*) Duralumin

Sol.

Alloy	Composition	Uses
(*i*) Bronze	Cu = 75 – 95% Sn = 5 – 25%	Bushes, plungers, feed pump rods, worm gears, gears, bearings, springs, pump parts.
(*ii*) Duralumin	Cu = 2.2 – 5.2% mg = upto 1.75% Mn = upto 1% Si = upto 1% Fe = 1% Al = upto 95%	Pistons for I.C. engines, vanes, screws, housings, pump impellers, rings, covers, disks etc.

Q.6. (*a*) Name five casting defects. How they are caused and what are the remedies?

Sol.

Casting Defects	**Causes**	**Remedies**
1. Blow holes	(*i*) Excessive evolved gases are not able to flow through the mould (*ii*) Excessive moisture content in sand, chills or metal inserts (*iii*) Inadequate gas permeability of moulding sand (*iv*) Poor venting of mould (*v*) Insufficient drying of mould and cores (*vi*) Low pouring temperature	(*i*) Adjust the moisture content in sand (*ii*) Use proper grain size of sand (*iii*) Ramming of sand should not be too hard (*iv*) Provide adequate venting
2. Misrun or short run	(*i*) Inadequate metal supply (*ii*) Too low mould or melt temperature (*iii*) Improperly designed gases	(*i*) Provide adequate metal supply (*ii*) Melt temperature should be adequate (*iii*) Check design of gates
3. Mismatch	(*i*) Mould or core shift (*ii*) Improper assembly of cope and drag (*iii*) Worn out pins and pin bushes (*iv*) Dimensional discrepancy between core prints of pattern and core	(*i*) Ensure proper alignment of pattern (*ii*) Align cope and drag properly (*iii*) Check locating pins before use
4. Scab	(*i*) Too fine sand (*ii*) Low permeability of sand (*iii*) High moisture content of sand (*iv*) Uneven mould ramming.	(*i*) Check sand fineness (*ii*) Provide adequate sand permeability (*iii*) Reduce moisture content of sand (*iv*) Ramming of sand should be even
5. Small	(*i*) Low strength of mould (*ii*) High moisture content (*iii*) Insufficient mould ramming	(*i*) Reduce moisture content (*ii*) Ramming of mould should be adequate

Q.6. (***b***) Explain the following forging operations :

(*i*) Drawing out (*ii*) Upsetting

Sol. (*i*) Drawing out forging : Drawing out is an operation in which the stock size is reduced only at one end. The heated stock is slowly progressed by each blow to obtain the required length.

(*ii*) Upsetting forging : Upsetting forging operation is used to increase the cross-sectional area usually by pressing or hammering. Upset forging forces the end of a heated bar into a desired shape. The bar is heated, clamped and upset into the die opening.

(***c***) How is acetylene stored?

Sol. Storage of Acetylene : Acetylene is stored in metallic containers from which it can be supplied at desired pressure. The nipples have left hand threads for accommodating pressure regulators and are pointed maroon to distinguish from oxygen cylinders. To compress free acetylene to a pressure more than one atmosphere is not safe. Therefore, acetylene cylinders contain porous spongy material such as charcoal, asbestos, balsa wood, silicon, etc. The main purpose of porous material is to fill the space within the cylinder. The cylinder is partly filled with acetone to dissolve acetylene.

(***d***) What is the principle of resistance welding? What are the various components of the joint resistance?

Sol. Resistance Welding

Principle : In resistance welding, the heat is generated at the interface of contacting workpieces due to resistance offered to the flow of electric current. No flux or filler metal is used in welding. The two workpieces are fused tegether by the application of pressure.

Components of Joint Resistance

Heat generated, $H_g = I^2Rt$, J

where I = current, Amperes

R = resistance, ohms,

t = time of current flow, s

Q.7. (***a***) Ahuja furniture company manufactures tables and chairs using wood and labour only. Wood for one table is 30 units and for chair is 20 units and the labour spent on table is 5 units and for chair is 10 units. Total units of wood available are 300, and of labour are 110. The unit profit for chair is Rs. 8 and for table is Rs. 6. How many chairs and tables should be made to get maximum profit?

Sol. Let x_1 = number of tables

x_2 = number of chairs

Max : $z = 6z_1 + 8z_2$

Subject to : $30x_1 + 20x_2 \leq 300$

$5x_1 + 10x_2 \leq 110$

Bound points are : (Fig. 5)

Point A : (10, 0)

Point B : (4, 9)

$$x_1 + 2x_2 = 22$$
$$3x_1 + 2x_2 = 30$$
$$2x_1 = 8$$
$$x_1 = 4$$
$$x_2 = \frac{1}{2}(22-4) = 9$$

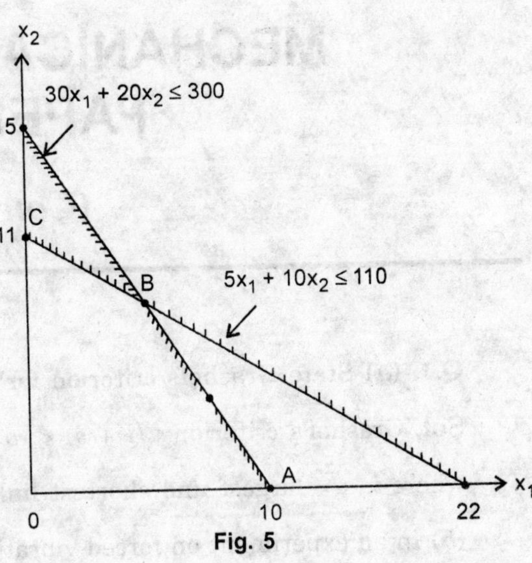

Fig. 5

Point C : (0, 11)

$$z_A = 6 \times 10 + 0 = \text{Rs. } 60$$
$$z_B = 6 \times 4 + 8 \times 9 = \text{Rs. } 96$$
$$z_C = 0 + 8 \times 11 = \text{Rs. } 88$$

Optimum solution is :

Tables = 4, chairs = 9,

Maximum profit = Rs. 96

Q.7. (*b*) A store sells 5200 cases of a cold drink per year. The supplier charges Rs. 100 for each delivery, regardless of how many cases have been ordered; delivery always occurs the day after ordering and the average carrying cost is Rs. 10.40 per unit per year. Find the number of cases per order.

Sol. Given : D = 5200/year, C_o = Rs. 100/order, C_c = Rs. 10.40/unit

Economic order quantity, $Q^* = \sqrt{\frac{2DC_o}{C_c}}$

$$= \sqrt{\frac{2 \times 5200 \times 100}{10.40}}$$

$$= 316.22 \simeq 316$$

Number of cases per order $= \frac{5200}{316} = 16.45 \simeq 16$

MECHANICAL ENGINEERING PAPER II—1996

(Conventional)

Section A

Q.1. ***(a)*** State Grashofs criterion for a movability of a four bar linkage.

Sol. Grashof's criterion : $(l + s) < (p + q)$

where l, s = largest and shortest link lengths, and p, q = other link lengths.

(b) In an experiment on forced vibration response of a single degree of freedom system, it is found that half-power points lie at frequencies 40 and 44 Hz. Find the damping factor of the system.

Sol. Damping factor = $(\omega_2 - \omega_1)/2\omega_m$ = (44 – 40)/(2 × 42) = 0.048

(c) Name any two centrifugal governors, which are gravity controlled. State which one is more sensitive.

Sol. Porter and Proell. Proell is more sensitive.

(d) In reference to a cam follower mechanism, define (*a*) prime circle, and (*b*) pressure angle.

Sol.

(*a*) Prime circle is the smallest circle drawn from the centre of the cam and tangent to the pitch curve.

(*b*) Pressure angle is the anlge between the direction of follower motion and normal to the pitch curve.

(e) The elastic and shear modulli of an elastic material are respectively 200 GPa and 80 GPa. Determine Poisson's ratio of the material.

Sol. $G = E/[2(1 + \nu)]$ or $\nu = 200/(2 \times 80) - 1 = 0.25$

(f) A uniform beam of length L is carrying a uniformly distributed load w per unit length and is simply supported at its ends. What would be the maximum bending moment and where does it occur?

Sol. Maximum B.M. = $wL^2/8$ and occurs at midspan.

(*g*) Euler's critical load for a column with both ends hinged is found as 40 kN. What would be the change in critical load, if its both ends are fixed?

Sol. Critical load for both ends fixed = $4 \times 40 = 160$ kN.

(*h*) A thin cylindrical vessel of internal diameter d and thickness t is closed at both ends and is subjected to an internal pressure p. How much would be the hoop stress and longitudinal stress in the material?

Sol. Hoop stress = $pd/2t$; Longitudinal stress = $pd/4t$.

(*i*) A power screw has a helix angle α and coefficient of friction μ. Express the condition to be satisfied by the screw to be self locking.

Sol. $\mu > \tan \alpha$

(*j*) State the modes of loading on a steel key connecting a belt driven pulley to a shaft.

Sol. 1. Shear 2. Crushing or compression

(*k*) What are the main constituting elements of brass and phosphor bronze?

Sol. Brass : Cu, Zn

Phosphor bronze : Cu, Sn, P

(*l*) Group the following materials into thermoplastics and thermosets :

PVC, Bakelite, Teflon, Formica.

Sol. Thermoplastics : PVC, Teflon

Thermosets : Bakelite, Formica

Fig. 1

(*m*) How much surface roughness is denoted by the following (Fig. 1).

Sol. 1.6 to 8 Ra μm.

(*n*) Name the form of welds, which are represented symbolically as (Fig. 2) :

(a) (b)

Fig. 2

Sol. (*a*) Square or rectangular butt (*b*) Single bevel butt.

(*o*) In a turning operation, determine the material removal rate, if depth of cut = 2 mm, feed per revolution = 0.5 mm and the surface velocity of the job piece = 50 m/minute.

Sol. Material removal rate = $2 \times 0.5 \times 50 = 50$ cm^3/min.

(*p*) Name any two advantages, which a hot working process has over the cold working process.

Sol.

1. No change in hardness or ductility. 2. Grain reformation.
3. Metal is made tougher. 4. Less force needed.

(*q*) List any two advantages of Rowan plan of wage incentive for direct workers.

Sol.

1. Provides guarantee of minimum wage.
2. Provides bonus on relative time saved.
3. Fresh workers are not penalised.
4. Management shares a part of bonus.

(*r*) In a PERT analysis, the critical path of a project is of 120 days with a variance of 16(days). Determine the 95% confidence limits for the project completion time.

Sol. Standard deviation, $\sigma = (16)^{0.5} = 4$

95% confidence limits : $120 \pm 2\sigma = 120 \pm 2 \times 4 = 128$ and 112

(*s*) There are two car manufacturing factories A and B, producing same model of cars. They are employing 1200 and 3000 workers and producing 15 and 50 cars per month respectively. Find labour productivity of each firm.

Sol. A : 15/1200 = 1/80

B = 50/3000 = 1/60

(*t*) What is a compiler in a computer?

Sol. A compiler is a software link which translates a symbolic language into the form suitable to computer. It performs the function of an assembler.

In other words, it converts high level language to low level language.

Section B

Q.2. (*a*) A solid shaft of diameter 3 cm is fixed at one end. It is subjected to a tensile force of 10 kN and a torque of 60 Nm. At a point on the surface of the shaft, determine the principal stresses and the maximum shearing stress.

Sol. Given : d = 3 cm, P = 10 kN, T = 60 Nm

Tensile stress, $$\sigma = \frac{4\text{ P}}{\pi d^2} = \frac{4 \times 10 \times 10^3}{\pi(3 \times 10)^2} = 14.147 \text{ N/mm}^2$$

Torsional shear stress, $$\tau = \frac{16\text{ T}}{\pi d^3} = \frac{16 \times 60 \times 10^3}{\pi(30)^3} = 11.317 \text{ N/mm}^2$$

Principal stresses, $$\sigma_{1,2} = \frac{1}{2}\sigma \pm \frac{1}{2}\sqrt{\sigma^2 + 4\tau^2}$$

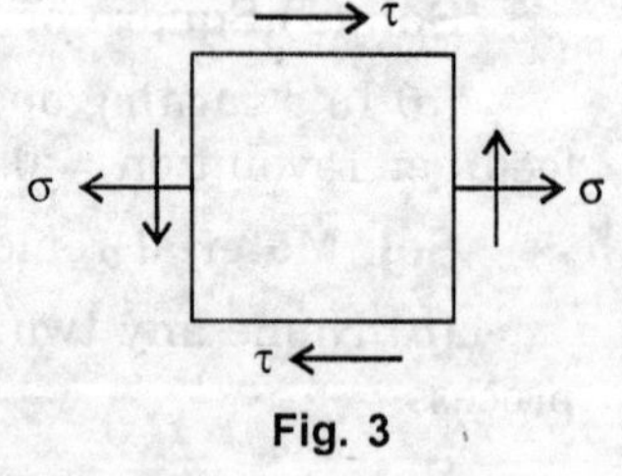

Fig. 3

$$= \frac{14.147}{2} \pm \frac{1}{2}\sqrt{(14.147)^2 + 4(11.317)^2}$$

$$= 7.073 \pm 13.346$$

$$\sigma_1 = 20.419 \text{ N/mm}^2$$

$$\sigma_2 = -6.273 \text{ N/mm}^2$$

Maximum shear stress, $$\tau_{max} = \frac{1}{2}(\sigma_1 - \sigma_2) = 13.346 \text{ N/mm}^2$$

Q.2. (*b*) The equation of the turning moment for a three-crank engine is

$$\text{T}_e = 25.0 - 7.5 \sin 3\theta \text{ kN m}$$

where θ is the crank angle measured from inner dead centre. The resisting torque exerted by the driven machine is given by

$$\text{T}_r = 25.0 + 3.6 \sin \theta \text{ kN m}$$

The moment of inertia of the flywheel is 360 kg m^2 and the mean engine speed is 450 rpm. Calculate

(*i*) power of the engine

(*ii*) maximum fluctuation of flywheel energy per cycle,

and (*iii*) the coefficient of fluctuation of speed.

Sol. Change in torque, $T = T_r - T_e = (25.0 + 3.6 \sin \theta) - (25.0 - 7.5 \sin 3\theta)$

$= 3.6 \sin \theta + 7.5 \sin 3\theta$

Work done per crank per cycle

$$= \int_{0^\circ}^{120^\circ} (3.6 \sin\theta + 7.5 \sin 3\theta)\, d\theta$$

$$= \left|-3.6 \cos\theta - 2.5 \cos 3\theta\right|_{0^\circ}^{120^\circ}$$

$$= [-3.6 (\cos 120^\circ - \cos 0^\circ) - 2.5 (\cos 360^\circ - \cos 0^\circ]$$

$$= -3.6 (-0.5 - 1) - 2.5 (1 - 1) = 5.4 \text{ kN.m}$$

(*i*) Power of engine $= \dfrac{3 \times 5.4 \times 450}{60} = 121.5$ kW

(*ii*) Mean torque, $T_m = \dfrac{5.4 \times 180}{120 \times \pi} = 3.578$ kN·m

For $T_m = T$, $3.6 \sin \theta + 7.5 \sin 3\theta = 0$

or $3.6 \sin \theta + 7.5(3 \sin \theta - 4 \sin^3\theta) = 0$

$26.1 \sin \theta - 30 \sin^3 \theta = 0$

$\sin \theta (26.1 - 30 \sin^2\theta) = 0$

Either $\sin \theta = 0^\circ, 180^\circ, 360^\circ$

or $\sin \theta = \sqrt{\dfrac{26.1}{30}} = 0.93273$

$\theta = 68.866^\circ, 111.134^\circ$

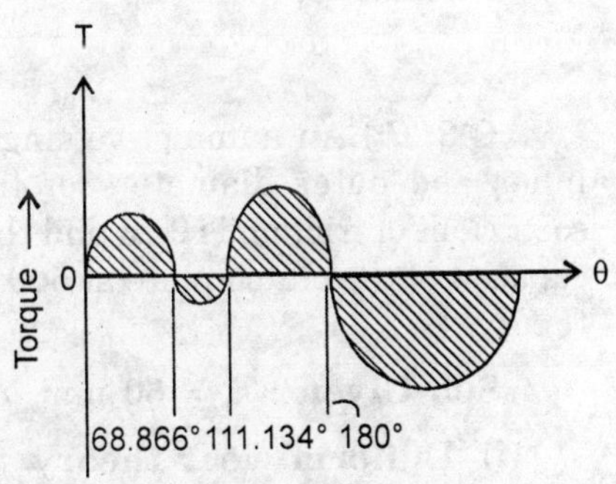

Fig. 4

T-θ diagram is shown in Fig. 4.

Maximm fluctuation of energy,

$$E_f = \int_{111.134^\circ}^{180^\circ} (3.6 \sin\theta + 7.5 \sin 3\theta) d\theta$$

$$= \left|-3.6 \cos \theta - 2.5 \cos 3\theta\right|_{111.134^\circ}^{180^\circ}$$

$$= -3.6(\cos 180^\circ - \cos 111.134^\circ) - 2.5(\cos 540^\circ - \cos 323.402^\circ)$$

$$= 2.302 + 4735 = 7.037 \text{ kN·m}$$

(*iii*) Mass of flywheel,

$$m = \frac{900\, E_f}{\pi^2 K^2 N_m^2 C_s}$$

$$C_s = \frac{900 \times 7.037}{\pi^2 \times 360 \times 450^2} = 0.0088 \text{ or } 0.88\%$$

Q.3. (*a*) The link of a mechanism is subjected to axial compressive force. It has solid circular cross-section with diameter 9 mm and length 200 mm. The two ends of the link are hinged. It is made of steel having yield strength = 400 N/mm^2 and elastic modulus = 200 kN/mm^2. Calculate the critical load that the link can carry. Use Johnson's equation.

Sol. Given : d = 9 mm, l = 200 mm, σ_{cy} = 400 N/mm^2, E = 200 kN/mm^2, P_{cr} = ?

Johnson's equation
$$\frac{P_{cr}}{A} = \sigma_{cy} - a\left(\frac{l}{k}\right)^2$$

where
$$A = \frac{\pi d^2}{4} = \frac{\pi \times 9^2}{4} = 63.62 \text{ mm}^2$$

$$a = \text{constant}$$

$$= \frac{\sigma_{cy}^2}{4\pi^2 E} = \frac{400^2}{4\pi^2 \times 2 \times 10^5} = 0.0203 \text{ N/mm}^2$$

$$K = \sqrt{\frac{I}{A}} = \sqrt{\frac{\pi d^4}{64} \times \frac{4}{\pi d^2}} = \frac{d}{4} = \frac{9}{4} = 2.25 \text{ mm}$$

$$P_{cr} = 63.62\left[400 - 0.0203\left(\frac{200}{2.25}\right)^2\right]$$

$$= 15244 \text{ N}$$

Q.3. (*b*) An automotive single-plate clutch consists of a pair of contacting surfaces. The inner and outer diameters of friction plate are 120 mm and 250 mm respectively. The coefficient of friction is 0.25 and the total axial force is 15 kN. Calculate the power transmitting capacity of the clutch plate at 500 rpm using (*i*) uniform wear theory, and (*ii*) uniform pressure theory.

Sol. Given : r_1 = 60 mm, r_2 = 125 mm, μ = 0.25, W_a = 15 kN, N = 500 rpm, n = 1

(*i*) Uniform wear theory

Torque transmitted,
$$T = n\mu W_a\left[\frac{r_1 + r_2}{2}\right]$$

$$= 1 \times 0.25 \times 15 \times \frac{1}{2}(60 + 125)$$

$$= 346.875 \text{ N·m}$$

Power transmitted,
$$P = \frac{2\pi NT}{60 \times 10^3}$$

$$= \frac{2\pi \times 500 \times 346.875}{60 \times 10^3} = 18.162 \text{ kW}$$

***(ii)* Uniform pressure theory**

$$T = n \times \frac{2}{3} \times \mu W_a \left[\frac{r_2^3 - r_1^3}{r_2^2 - r_1^2}\right]$$

$$= 1 \times \frac{2}{3} \times 0.25 \times 15 \left[\frac{125^3 - 60^3}{125^2 - 60^2}\right]$$

$$= 361.15 \text{ N·m}$$

$$P = \frac{2\pi \times 500 \times 361.15}{60 \times 10^3} = 18.91 \text{ kW}$$

Q.4. *(a)* An epicyclic train as shown in Fig. 5 has a sun wheel S of 30 teeth and two planet wheels P-P of 45 teeth. The planet wheels mesh with the internal teeth of a fixed annulus A. The driving shaft carrying the sun wheel transmits 4 kW at 360 rpm. The driven shaft is connected to an arm, which carries the planet wheels. Determine the speed of the driven shaft and the torque transmitted, if the overall efficiency is 95%.

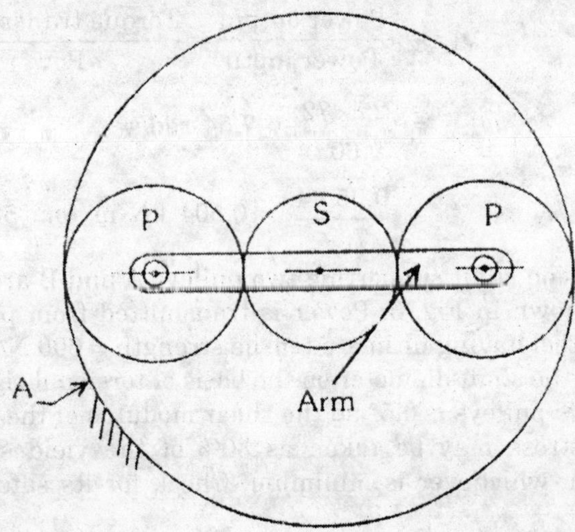

Fig. 5

Sol. Given : $z_s = 30$, $z_p = 45$, $P = 4$ kW, $N_s = 360$ rpm, $\eta_0 = 0.95$, $N_A = 0$

Operation	Revolutions of			
	Arm	**Sun wheel**	**Planets**	**Annulus**
1. Arm fixed, +1 revolutions to S, CCW.	0	+1	$-\frac{z_s}{z_p}$	$-z_s/z_a$
2. Multiply by x	0	$+x$	$-xz_s/z_p$	$-xz_s/z_a$
3. Add y	$+y$	$x + y$	$-xz_s/z_p + y$	$-\frac{xz_s}{z_a} + y$

Now $z_a = z_s + 2z_p = 30 + 2 \times 45 = 120$

Now $N_a = 0$

$$\therefore \quad -\frac{x \times 30}{120} + y = 0$$

$$-\frac{x}{4} + y = 0$$

or $x = 4y$

Also $N_s = 360$

$$x + y = 360$$

$$4y + y = 360$$

Speed of arm, $y = \frac{360}{5} = 72$ rpm

$$x = 4 \times 72 = 288 \text{ rpm}$$

Overall efficiency, $\eta_0 = \frac{\text{Power output}}{\text{Power inptu}} = \frac{\text{Torque transmitted} \times w_{arm}}{\text{Power input}}$

$$w_{arm} = \frac{2\pi \times 72}{60} = 7.54 \text{ rad/s}$$

Torque transmitted $= \frac{0.95 \times 4}{7.54} = 0.504$ kN·m or 504 N·m

Q.4. (*b*) A transmission shaft supporting two pulleys A and B are mounted between two bearings C_1 and C_2 as shown in Fig. 6. Power is transmitted from pulley A to B. The shaft is made of plain carbon steel having ultimate tensile strength = 600 N/mm^2 and yield strength = 380 N/mm^2. Determine the shaft diameter on the basis of torsional rigidity, if the permissible angle of twist between two pulleys is 0.5° ad the shear modulus of the material is 80 kN/mm^2. The permissible shear stress may be taken as 30% of the yield strength or 18% of the ultimate tensile strength, whichever is minimum. Check for its safety.

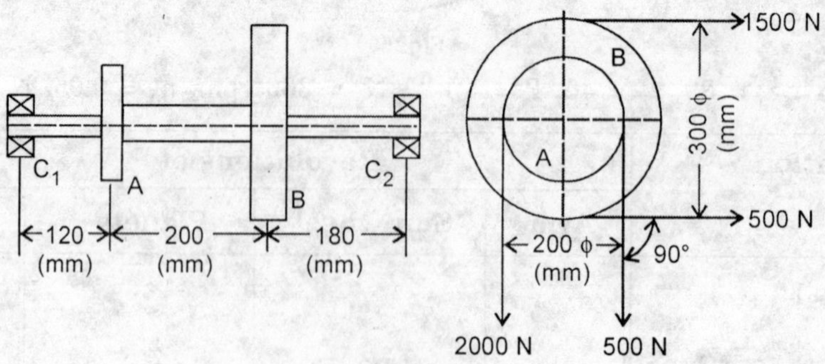

Fig. 6

Sol. Given : $\sigma_{ut} = 600$ N/mm^2, $\sigma_{yt} = 380$ N/mm^2, $\theta = 0.5°$, G = 80 kN/mm^2,

$\tau_w = 0.3, \sigma_{yt}$ or $0.18\ \sigma_{ut}$, whichever is minimum.

$F_1 = 2000$ N, $F_2 = 500$ N, $F_3 = 1500$ N, $F_4 = 500$ N, $r_A = 0.1$ m, $r_B = 0.15$ m

Vertical Forces : Pulley A

Vertical force, $F_A = F_1 + F_2 = 2000 + 500 = 2500$ N

Torque, $T_A = (F_1 - F_2)\, r_A = (2000 - 500) \times 100 \times 10^{-3} = 150$ N·m

$\Sigma M_{c2} = 0 :$ $(R_{c1})_v \times 500 = 2500 \times 380$

$$(R_{c1})_v = 1900 \text{ N}, (R_{c2})_v = 2500 - 1900 = 600 \text{ N}$$

$$(M_A)_v = (R_{c1})_v \times 120 \times 10^{-3} = 1900 \times 0.12 = 228 \text{ N·m}$$

$$(M_B)_v = 600 \times 0.180 = 108 \text{ N·m}$$

Horizontal Forces : Pulley B

$$F_B = F_3 + F_4 = 1500 + 500 = 2000 \text{ N}$$

$$T_B = (F_3 - F_4)\, r_B = (1500 - 500) \times 0.15 = 150 \text{ N·m}$$

$$(R_{c1})_H \times 500 = 2000 \times 180$$

$$(R_{c1})_H = 720 \text{ N}, (R_{c2})_H = 2000 - 720 = 1280 \text{ N}$$

$$(M_A)_H = 720 \times 0.12 = 86.4 \text{ N·m}$$

$$(M_B)_H = 1280 \times 0.18 = 230.4 \text{ N·m}$$

Resultant Bending Moments

$$M_A = \sqrt{(M_A)_V^2 + (M_A)_H^2} = \sqrt{(228)^2 + (86.4)^2} = 243.82 \text{ N·m}$$

$$M_B = \sqrt{(M_B)_V^2 + (M_B)_H^2} = \sqrt{108^2 + (230.4)^2} = 254.45 \text{ N.m}$$

Equivalent Torque

$$(T_e)_A = \sqrt{(243.82)^2 + 150^2} = 286.26 \text{ N.m}$$

$$(T_e)_B = \sqrt{(254.45)^2 + 150^2} = 295.37 \text{ N.m} \quad \text{(Maximum)}$$

Permissible shear stres, $\tau_w = 0.3 \times 380 = 114$ N/mm²

or $= 0.18 \times 60 = 108$ N/mm² (Minimum)

$\therefore$ $\tau_w = 108$ N/mm²

$\therefore$ $$(T_e)_B = \frac{\pi}{16} d^3 \tau_w$$

$$295.37 \times 10^3 = \frac{\pi}{16} d^3 \times 108$$

$$d = 24.06 \simeq 24 \text{ mm}$$

Also $$\frac{T}{J} = \frac{G\theta}{l}$$

$$\theta = \frac{Tl}{GJ} \quad (l = 200 \text{ mm})$$

$$0.5 \times \frac{\pi}{180} = \frac{295.37 \times 10^3 \times 200 \times 32}{80 \times 10^3 \times \pi d^4}$$

$$d = 30.47 \text{ mm} \simeq 30 \text{ mm}$$

Adopt shaft diameter, $d = 30$ mm

Section C

Q.5. (*a*) Discuss the factors, which govern the selection of a material for a machine component.

Sol. Factors governing the selection of materials for machine components

1. Type of load : Static or fluctuating or impact.
2. Operating conditions : normal, corrosive, abrasive, etc.
3. Temperature : Cryogenic, normal or high.
4. Vibrations : Shock, random or transient.
5. Wear
6. Stiffness and rigidity
7. Availability, and
8. Economics

(b) Describe briefly different stages of tempering process in steels.

Sol. Different stages of tempering process in steels : Tempering consists of reheating below the critical temperature the quenched metal to restore some of the ductility and reduce the brittleness. Increased toughness is obtained at the expense of high strength.

The different stages of tempering process in steels are :

1. Reheating after hardening below the critical temperature of steel (723°C).
2. Holding it there for 1 to 2 hours.
3. Slow cooling to 350° – 450°C for components to be subjected to impact load. For components to be subjected to high stress and impact, tempering is carried out at 500° 650°C.

(c) What are the principal uses of (*i*) low carbon, (*ii*) medium carbon and (*iii*) high carbon steels? Mention also the amount of carbon present in these steels.

Sol.

Type of steel	% Carbon	Principal Uses
(*i*) Low carbon steel	0.10 – 0.20	Structures, rolled steel sections, drop forgings, screws, case hardening purposes.
	0.20 – 0.30	Machine structures, shafting and forging, gears
(*ii*) Medium carbon steel	0.3 – 0.4	Shaftings, axles, crane hooks, connecting rods.
	0.4 – 0.5	Axles, gears, shafts, rotors, tyres, crank shafts
	0.5 – 0.6	Rails, loco tyres, wire ropes
(*iii*) High carbon steel	0.6 – 0.7	Drop hammers, dies, jaws, screw drivers
	0.7 – 0.8	Hammers, anvils, wrenches, leaf springs, cable wires, band saws, dies
	0.8 – 0.9	Shear blades, punches, rock drills, cold chisels

Q.6. **(*a*)** What is upset forging and how is it done? What are its limitations?

Sol. Upset Forging : Upset forging is the process in which volume of metal at one end only of the workpiece is increased. this method is used for the mass production of nuts, bolts, rivets, wood screws, roller and ball bearings.

The steps involved in upset forging are illustrated in Fig. 7.

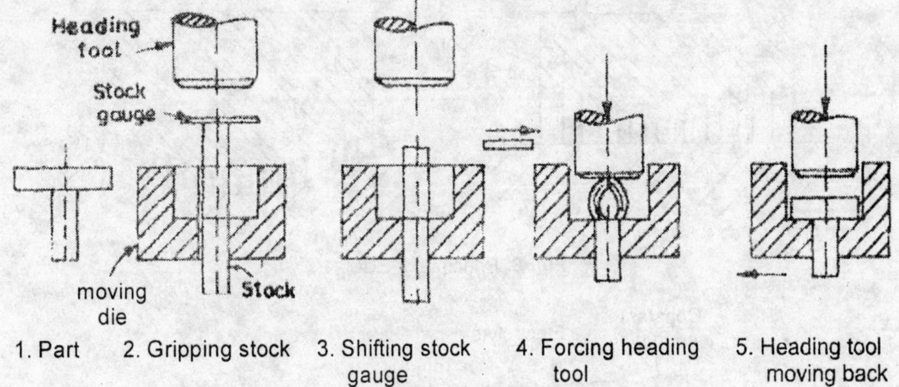

Fig. 7. Steps in upset forging

Step 1. Part to be forged.

Step 2. Gripper die closes against the stationary die to tightly grip the stock.

Step 3. Automatic stock gauge moves aside.

Step 4. Header slide moves forward and forces the heading tool against the projecting stock, moving the stock into the die cavity.

Step 5. Metal completely fills the die impression, both the moving die and heading tool start moving back to their original positions.

Limitations of Upset Forging

1. It is not convenient to forge heavier jobs due to the material handling difficulties.
2. The maximum diameter of the stock which can be upset is limited.
3. Intricate, non-symmetric and heavy jobs are difficult to be forged on a forging machine.
4. The tooling cost may be high.

(*b*) Explain the working principle of ECM. Discuss how the machining rate in this process is affected by the following :

(*i*) rate of flow of electrolyte

(*ii*) temperature of electrolyte

(*iii*) applied voltage between tool and workpiece

Sol. Working principle of ECM : The ECM process is the reverse of electroplating. In ECM the objective is to remove metal. Therefore, the workpiece is connected to the positive terminal, and the tool to the negative terminal. Fig. 8 shows a workpiece and a suitably shaped tool, the gap between the tool and the job being full of a suitable electrolyte. When the current is passed, the dissolution of the anode occurs. Now, if the tool is given a downward motion, the work surface tends to take the same shape as that of the tool.

In the ECM process, the tool is provided with a constant feed motion. The electrolyte is pumped at a high pressure through the tool and the small gap between the tool and workpiece, as shown in Fig. 9. The electrolyte is so chosen that the anode is dissolved but no deposition takes place on the cathode (tool).

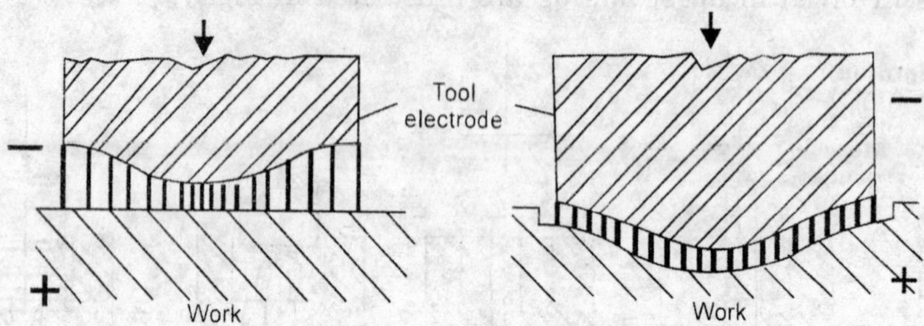

Fig. 8. Principle of ECM.

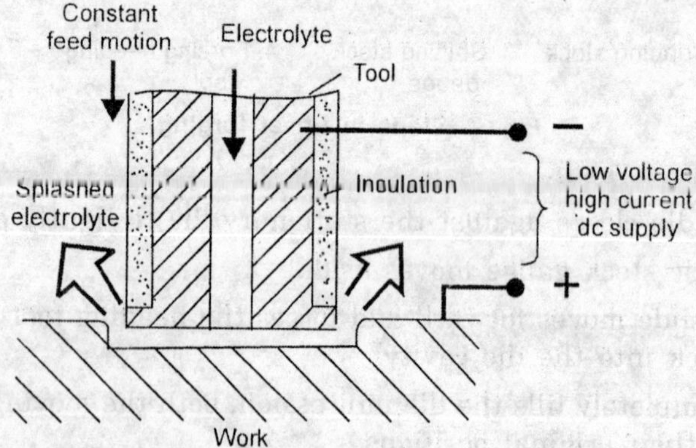

Fig. 9. Electrochemical maching.

Parameters affecting machining rate in ECM

(*i*) Rate of flow of electrolyte :

The machining rate increases with the increase in rate of flow of electrolyte.

(*ii*) Temperature of electrolyte.

The machining rate decreases with the increase in temperature of electrolyte.

(*iii*) Applied voltage between tool and workpiece.

The machining rate increases as the applied voltage between tool and workpiece is increased.

(*c*) Describe briefly the operation of a machining centre.

Sol. Operation of Machining Centre : A machining centre is having a large number of tools doing different jobs. The tools are mounted on a tool magazine. The tools are fed to the workpiece stationed at a particular location in sequence, related to the manufacturing programme or operational layout. A machining centre may have one or two machine tools with the specific feature of automatic tool changer. The machine tools are numerically controlled.

Q.7. (*a*) If in a lot 10 per cent components are found defective, find the probability, that in a random sample of 5 components not more than one is defective.

Sol. Assuming Poisson's distribution, we have

$$e^{-z} = 1 + z + \frac{z^2}{2!} + \frac{z^3}{3!} + \ldots \text{ for 1, 2 and 3 defectives}$$

where $z = np$

p = probability of defective = 0.10

n = sample size = 5

$$p(0) = e^{-0.5} = 0.6065$$

$$p(1) = e^{-0.5} \times \frac{1}{2} = 0.3033$$

$$p = p(0) + p(1) = 0.9098$$

(*b*) Describe clearly the function of routing, scheduling and dispatching.

Sol. Routing : It is defined as the procedure for preparing a complete operation sheet or how the work is to be done most economically. It specifies the sequence of operations to be performed on a product.

Scheduling : It is preparing a time schedule for all production operations according to producing a given factory order. It determines when the actual production should start and end.

Despatching : It is defined as a routine of setting production activities in motion through the release of orders, instructions in accordance with the previously planned time and sequences as embodied in route sheets. It determines by whom work shall be done.

(*c*) What is ABC analysis? Describe briefly the procedural steps taken in this analysis.

Sol. ABC Analysis (Always Better Control) : In ABC analysis, more attention is paid to the item value.

Pareto's Law : It states that a fewer items of high usage having high investment value should be paid more attention than a bulk of items having low usage value and having a low investment in capital.

A-items-High Value : These are relatively few items whose value accounts for 60-70% of the total value of the inventory. These will usually be 10-15% of the total items.

B-item-Medium value : A large number of items in the middle of the list, usually about 20-25% of the items whose total value accounts for about 20-25% of the total value.

C-item-Low Value : The bulk of items usually about 60-70% whose total inventory values are almost negligible. They account for 10-15% of the total value.

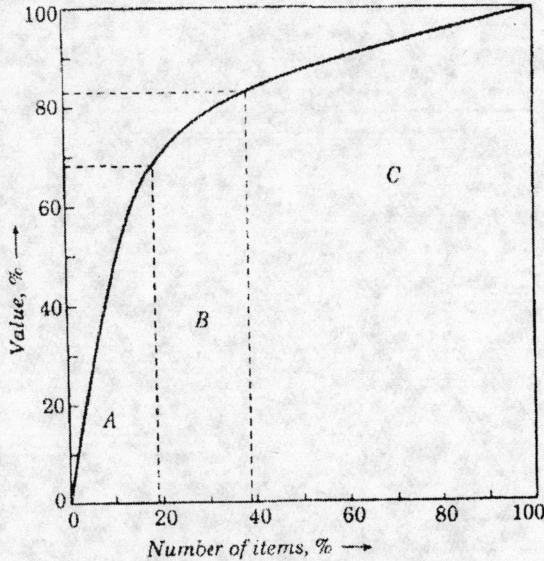

Fig. 10. ABC analysis.

For the inventory control, close attention is paid to items falling in category A and least to items in category C. The term ABC means always Better Control. The variation of value v's number of items is shown in Fig. 10.

Steps in ABC Analysis

1. Estimate the quantity of each material expected to be used in a given period.
2. Calculate the money value of the items of materials by multiplying the quantity of each item with the price.
3. Rearrange the items in the descending order of their values irrespective of their quantities.
4. Take the running total of all the values and items. Convert the figure so obtained into percentages of the gross total.
5. Distribute the items into A, B and C categories.
6. Plot percentage of total items on x-axis and the corresponding usage value on y-axis.

MECHANICAL ENGINEERING
PAPER II—1997

(Conventional)

Section A

Q.1. **(*a*)** Write the number of degree of freedom constrained in each case by the kinematic paris :

(*a*) revolute pair, (*b*) cylindrical pair, (*c*) screw pair, and (*d*) spherical pair.

Sol. Degrees of freedom (Constrained degrees of freedom) are :

(*a*) 5 (1) (*b*) 4 (2)

(*c*) 5 (1) (*d*) 3 (3).

(*b*) A rigid link placed on a smooth horizontal plane is struck by an impulsive force at a point 20 cm distant from its c.g., such that it starts rotating about a point 5 cm away from the c.g. Determine the radius of gyration of the link and comment on the relative location of three points, viz. point of force application, c.g. and point of rotation.

Sol. *Equivalent length,* $l_e = l + K^2/l$

$$25 = 5 + K^2/5$$

or $K = 10$ cm

The point of force application, C.G., and point of rotation are collinear.

(*c*) State the conditions for a rigid rotor to be dynamically balanced.

Sol. $\Sigma F = 0$ and $\Sigma M = 0$

(*d*) Show graphically the variation of the ratio of damped frequency to undamped frequency against damping ratio in case of a damped oscillator.

Sol. $\omega_d/\omega_n = [1 - \zeta^2]^{0.5}$. The variation is shown in Fig. 1.

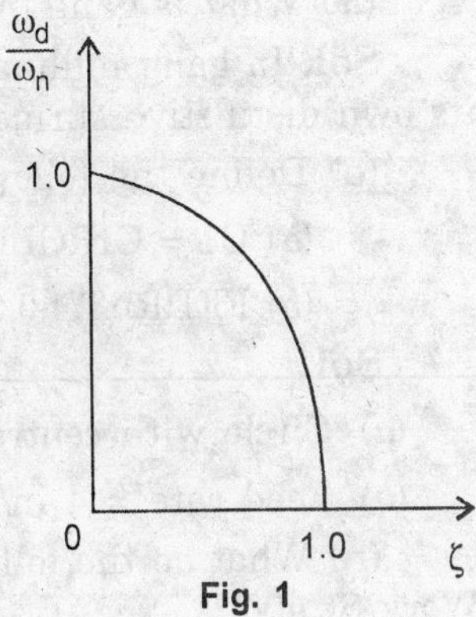

Fig. 1

(*e*) The tight side and slack side tensions in a running belt drive are respectively T_1 and T_2. What would be the belt tension, when the drive stops?

Sol. Belt tension when the drive stops = $(T_1 + T_2)/2$.

(*f*) Determine the torsional stiffness of a hollow shaft of length L and having outside diameter equal to 1.5 times inside diameter *d*. The shear modulus of the material is G.

Sol. Torsional stiffness = $T/\theta = J\,G/L = \pi[(1.5\,d)^4 - d^4]$ G/[32 L]

$$= 0.4\ Gd^4/L$$

(*g*) A uniform cantilever beam (EI constant) of free length L is carrying a concentrated load P at its free end. What would be its slope at the (*a*) free end, and (*b*) built-in end.

Sol. (*a*) $PL^2/(2EI)$ (*b*) Zero.

(h) If the Poisson's ratio of an elastic material is 0.3, determine the ratio of its elastic modulus to its shear modulus. Write the formula you use.

Sol. E/G = 2(1 + ν) = 2(1 + 0.3) = 2.6

(i) How are unilateral and bilateral tolerances indicated on dimensions of a component? Illustrate each with an example.

Sol. Unilateral : $50^{+0.02}_{-0.00}$ or $50^{+0.02}_{-0.01}$, *i.e.* one tolerance limit is zero.

Bilateral : $50^{-0.01}_{+0.02}$ or $50^{+0.02}_{-0.01}$, *i.e.* there is variation in both directions from normal size.

(j) Two materials as per BIS system of designation are

(*a*) FG 35 Si 15, and (*b*) Cu Zn 42 Pb 2.

Name them with complete specifications.

Sol.

(*a*) Special grey iron with minimum carbon of 0.35% and average silicon of 1.5%.

(*b*) Brass with average Zn = 42%, Pb = 2%, and residual Cu = 56%.

(k) Specify the ball size and the magnitude of load applied in case of Brinell hardness tester.

Sol. Ball size, *d* = 10 mm dia. P/d^2 = 30, where P = load.

(l) What are Vents and Chills in mould?

Sol. Vents are small holes made by perforating the sand just short of pattern. Chills are metal shapes inserted in moulds to speed up the solidification.

(m) In a surface grinding operation of a workpiece of size 20 cm (length) × 10 cm (width), the speed of table is 25 m/min and feed is 0.2 mm/stroke.

Determine the time for grinding single surface in a single cut.

Sol. Time of grinding = (0.2 × 100)/(25 × 0.2) = 4 min.

(n) What is gang milling?

Sol. In gang milling, two or more milling cutters are mounted on *an arbor* to produce its own distinctive surface.

(o) Define the following statements as given in APT language :

(*a*) C1 = CIRCLE/CENTRE, P1, TANTO, L1

(*b*) FEDRAT/4.0

Sol.

(*a*) Circle with centre at P_1 and tangent to line L_1.

(*b*) Feed rate = 4 in/min

(p) What do the following symbols (Fig. 2) Imply in Work Study?

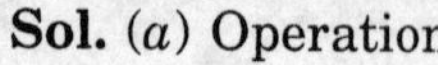

Sol. (*a*) Operation (*b*) Transport

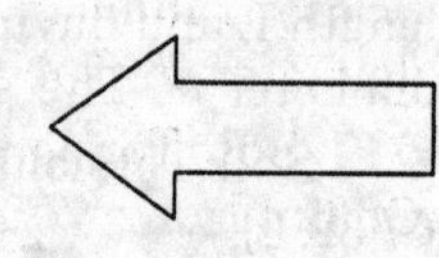

Fig. 2

(q) An item can be produced at a fixed cost of Rs. 80,000 per annum and a unit variable cost of Rs. 40. If the selling price per item is Rs. 60, determine the break even production volume.

Sol. Break even production volume = 80000/20 = 4000 items/year

(*r*) Give two situations, where you would prefer CPM to PERT.

Sol.

1. Minimum overall cost is of primary importance.
2. Better utilisation of resources.

(*s*) Distinguish between Quality Assurance and Quality Control.

Sol. Quality assurance is concerned with the activities to maximise the probability that the product will be manufactured within the design specifications. It starts at product design area and continue in manufacturing planning. Quality control is concerned with the activities related to inspection and checking quality.

(*t*) Name the two important constituents of CUP in a computer.

Sol.

1. Control unit
2. Arithmetic-logic (ALU)

Section B

Q.2. (*a*) A square threaded bolt of mean diameter 24 mm and pitch 5 mm is tightened by screwing a nut, whose mean diameter of bearing surface is 50 mm. If the coefficient of friction for nut and bolt is 0.1 and that for nut and bearing surface is 0.15, find the force required at the end of a spanner 0.5 m long, when the load on the bolt is 12 kN.

Sol. Given : d_m = 24 mm, p = 5 mm, D_b = 50 mm,

μ = 0.1, μ_b = 0.15, l = 0.5 m,

W = 12 kN

Helix angle of threads,
$$\alpha = \tan^{-1}\left[\frac{p}{\pi d_m}\right] = \tan^{-1}\left[\frac{5}{\pi \times 24}\right] = 3.794°$$

Angle of friction between nut and bolt,
$$\phi = \tan^{-1}\mu = \tan^{-1} 0.1 = 5.71°$$

Friction torque between nut and bolt,
$$T_f = \frac{Wd_m}{2} \times \tan(\alpha + \phi)$$
$$= 12 \times 12 \times \tan(3.794° + 5.71°) = 24.11 \text{ N.m}$$

Assuming uniform wear theory, torque between nut and bearing surface,
$$T_b = \mu_b W \times \frac{1}{4}(D_b + d_m)$$
$$= 0.15 \times 12 \times \frac{1}{4}(50 + 24) = 33.3 \text{ N} \cdot m$$

Total torque,
$$T = T_f + T_b$$
$$= 24.11 + 33.3 = 57.41 \text{ N·m}$$

Let F = force applied at the end of spanner.

Then
$$F \times l = T$$
$$F = \frac{57.41}{0.5} = 114.82 \text{ N}$$

Q.2. (*b*) In a spring controlled governor of the Hartung type, the lengths of the horizontal and vertical arms of the bell crank lever are 100 mm and 80 mm respectively. The fulcrum of the bell crank lever is at a distance of 120 mm from the axis of the governor. Each revolving mass is 8 kg. The stiffness of the spring is 20 kN/m. If the length of each spring is 120 mm, when the radius of rotation is 70 mm and the equilibrium speed is 380 rpm, find the free length of the spring if the radius of rotation increases to 120 mm, What will be the corresponding percentage increase in speed? Ignore sleeve mass.

Sol. Given: a = 80 mm, b = 100 mm, m = 8 kg, k = 20 kN/m, r_1 = 70 mm, N_1 = 380, l_1 = 120 mm, r_2 = 120 mm, l_2 = ?, % ΔN = ?

Refer to Fig. 3

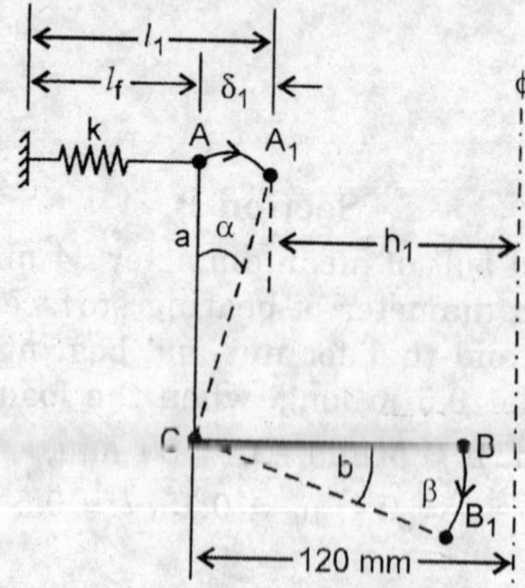

Fig. 3

$$\omega_1 = \frac{2\pi N_1}{60} = \frac{2\pi \times 380}{60} = 39.79 \text{ rad/s}$$

Let δ_1 = extension of spring, mm

Spring force, $S_1 = k\delta_1 = 20\delta_1 N$

Contifugal force, $F_{c1} = mr_1\omega_1^2 = 8 \times 70 \times 10^{-3} \times (39.79)^2 = 886.8 \text{ N}$

For equilibrium of governor,

$$S_1 = F_{c1}$$
$$20\delta_1 = 886.8$$
$$\delta_1 = 44.34 \text{ mm}$$

Free length of spring, $l_f = l_1 - \delta_1$

$$= 120 - 4434 = 75.66 \text{ mm}$$

$$F_{c2} = 8 \times 120 \times 10^{-3} \times \omega_2^2 = 0.96\omega_2^2 \text{ N}$$
$$S_2 = k[(r_1 - r_2) + \delta_1]$$
$$= 20\ [(120 - 70) + 44.34] = 188.68 \text{ N}$$

For $F_{c2} = S_2$, we have

$$0.96\ \omega_2^2 = 188.68$$
$$\omega_2 = 44.333 \text{ rad/s}$$

$$N_2 = \frac{60 \times 44.333}{2\pi} = 423.35 \text{ rpm}$$

$$\text{Percentage increase in speed} = \left(\frac{N_2 - N_1}{N_1}\right) \times 100$$

$$= \left(\frac{423.35 - 380}{380}\right) \times 100 = 11.4\%$$

Q.3. (*a*) A V-belt of 6.0 cm^2 cross-section has a groove angle of 40° and an angle of lap of 150°, μ = 0.1. The mass of belt per meter run is 1.2 kg. The maximum allowable stress in the belt is 850 N/cm^2. Calculate the horse power that can be transmitted at a belt speed of 30 m/s.

Sol. Given : A = 6 cm^2, 2β = 40°, θ = 150°, μ = 0.1, m = 1.2 kg/m, σ_w = 850 N/cm^2, v = 30 m/s, P = ?

$$\mu_e = \frac{\mu}{\sin\beta} = \frac{0.1}{\sin 20°} = 0.2924$$

$$\exp(\mu_e) = \exp\left(0.2924 \times \frac{\pi}{180} \times 150\right) = 2.15$$

$$F_c = mv^2 = 1.2 \times 30^2 = 1080 \text{ N}$$

$$F_{max} = \sigma_w A = 850 \times 6 = 5100 \text{ N}$$

$$F_1 = F_{max} - F_c = 5100 - 1080 = 4020 \text{ N}$$

$$F_2 = \frac{F_1}{\exp(\mu_e \theta)} = \frac{4820}{2.15} = 1869.77 \text{ N}$$

$$P = \frac{(F_1 - F_2)v}{10^3} = \frac{(4020 - 1869.77)30}{10^3} = 65.5 \text{ kW or } 87.8 \text{ HP}$$

Q.3. (*b*) A solid right cone of axial length h is made of a material having density ρ and elasticity modulus E. It is suspended from its circular base. Determine its elongation under its own weight.

Sol. Consider a solid right circular cone of base radius b and axial length h. The radius r of an elementary strip at a distance x from base is (Fig. 4),

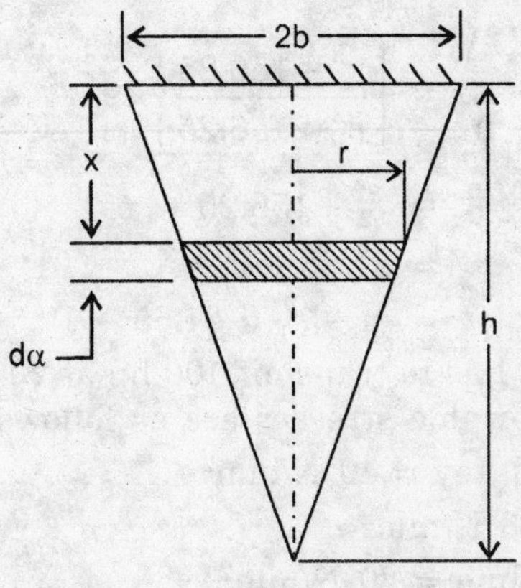

Fig. 4

$$r = (h-x)\frac{b}{h}$$

Load acting on the elementary strip,

$$\Delta W = \frac{1}{3} \times \pi r^2 \times \rho g(h-x)$$

$$= \frac{\pi}{3} \times \rho g(h-x) \times \frac{b^2}{h^2}(h-x)^2$$

$$= \frac{\pi}{3} \times \frac{\rho g b^2}{h^2}(h-x)^3$$

Area of elementary strip, $\Delta A = \pi r^2 = \frac{\pi b^2}{h^2}(h-x)^2$

Stress in the elementary strip,

$$\sigma = \frac{\Delta W}{\Delta A} = \frac{1}{3} \times \rho g(h-x)$$

Extension of elementary strip of thickness

$$dx = \frac{\sigma \cdot dx}{E} = \frac{1}{3E} \times \rho g(h-x) \cdot dx$$

Extension of cone, $\delta = \frac{\rho q}{3E}\int_0^h (h-x)\, dx$

$$= \frac{\rho g}{6E}\left| -(h-x)^2 \right|_0^h = \frac{\rho g h^2}{6E}$$

Q.4. (*a*) The pressure within the cylinder of a hydraulic press is 9 MPa. The inside diameter of the cylinder is 25 mm. Determine the thickness of the cylinder wall, if the permissible tensile stress is 18 N/mm².

Sol. Given : p = 9 MPa, r_1 = 12.5 mm, σ_θ = 18 N/mm², t = ?

$$\sigma_\theta = \left[\frac{r_2^2 + r_1^2}{r_2^2 - r_1^2}\right] p$$

$$18 = \left[\frac{r_2^2 + 156.25}{r_2^2 - 156.25}\right] \times 9$$

$$2r_2^2 - 312.5 = r_2^2 + 156.25$$

$$r_2 = 21.65 \text{ mm}$$

Wall thickness of cylinder, $t = r_2 - r_1 = 21.65 - 12.5 = 9.15$ mm

Q.4. (*b*) A mild steel shaft has to transmit 100 hp at 200 rpm. Design a cast iron flange coupling for the shaft. The allowable stresses are as follows :

Shear stress for shaft and key = 40 N/mm²

Shear stress for bolts = 28 N/mm²

Shear stress for C.I. coupling = 20 N/mm²

Take bearing stress as twice the shear stress value and number of bolts for the coupling as 6.

Sketch the coupling showing various dimensions.

Sol. Given : P = 100 HP or 74.6 kW, N = 200 rpm, $\tau_s = \tau_k = 40$ N/mm², $\tau_b = 28$ N/mm², $\tau_{CI} = 20$ N/mm², $\sigma_b = 2\tau$, $n = 6$.

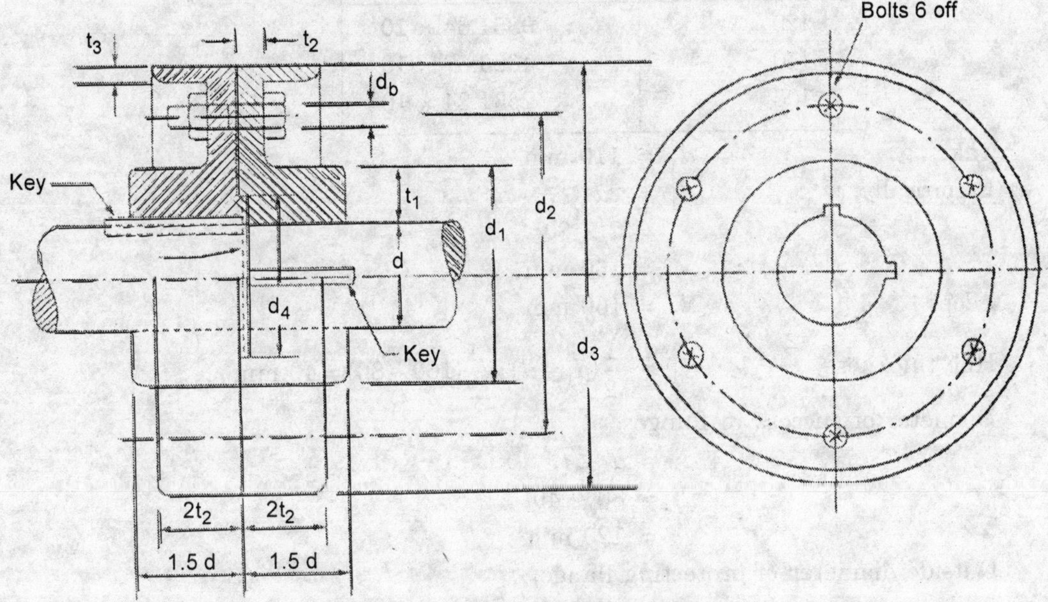

Fig. 5. Protected type rigid flange coupling

A protected type rigid flange coupling is shown in Fig. 5.

1. Shaft dimeter, *d* :

$$P = \frac{2\pi NT}{60 \times 10^3}$$

$$T = \frac{74.6 \times 60 \times 10^3}{2\pi \times 200} = 3561.9 \text{ N}\cdot\text{m}$$

$$\frac{\pi}{16} d^3 \tau_s = T$$

$$d = \left[\frac{16 \times 3561.9 \times 10^3}{\pi \times 40}\right]^{1/3} = 76.83 \simeq 80 \text{ mm}$$

2. Hub diameter, d_1 :

$$\frac{\pi}{16}\left[\frac{d_1^4 - d^4}{d_1}\right]\tau_h = T$$

Taking $\tau_h = \tau_{CI} = 20$ N/mm²

$$\frac{\pi}{16}\left[\frac{d_1^4 - 80^4}{d_1}\right] \times 20 = 3561.9 \times 10^3$$

$$d_1^4 - 4096 \times 10^4 = 907030\, d_1$$

or $$d_1^4 - 409070\, d_1 = 4096 \times 10^4$$

d_1, mm	$d_1^4 - 907030\, d_1$
120	9851.64×10^4
110	4663.67×10^4
105	2631.24×10^4

Take $d_1 = 110$ mm

Empirically, $d_1 = 2d$

$= 2 \times 80$

$= 160$ mm

Adopt $d_1 = 160$ mm

Hub thickness, $t_1 = \frac{1}{2}(d_1 - d) = \frac{1}{2}(160 - 80) = 40$ mm

Diameter of success in flanges,

$d_4 = d + t_1$

$= 80 + 40$

$= 120$ mm

Outside diameter of protecting flange,

$d_3 = 4d = 4 \times 80$

$= 320$ mm

3. Hub length l_h :

Width of key, $b = \frac{d}{4} = \frac{80}{4} = 20$ mm

Length of key, $l_h = 1.5\, d$

$= 1.5 \times 80$

$= 120$ mm

Thickness of key $t = \frac{d}{6} = \frac{80}{6} \simeq 14$ mm

Length of hub, $l_h = l_k = 120$ mm

4. Bolt diameter, d_b :

Radius of bolt circle, $r_2 = \frac{d_1}{2} + 5\text{ mm} + 1.85\, d_b + 8\text{ mm}$

$= 1.5\, d$

$= 120$ mm (empiricably)

Shearing face on each bolt,

$$F = \frac{T}{r_2 n} = \frac{3561.9 \times 10^3}{120 \times 6} = 4947 \text{ N}$$

The bolts are in single shear.

$$\frac{\pi}{4} d_b^2 \tau_b = F$$

$$\frac{\pi}{4} d_b^2 \times 28 = 4947$$

$$d_b = 15 \text{ mm}$$

5. Flange thickness, t_2

Considering bearing failure of flange,

$$F = n \times d_b t_2 \times \sigma_b$$

$$4947 = 6 \times 15 \times t_2 \, (2 \times 20)$$

$$t_2 = 1.37 \simeq 2 \text{ mm}$$

Empirically, $$t = \frac{1}{2} t_1 + 6\text{mm} = \frac{40}{2} + 6 = 26 \text{ mm}$$

6. Thickness of pratcing flange,

$$t_3 = 0.5\, t_2 = 13 \text{ mm}$$

7. Hub overlap = 3 mm

SECTION-C

Q.5. (*a*) Discuss the effect of Manganese and Nickel as alloying elements in steels and cast irons.

Sol.

Alloying element	Effect on steels	Effect on cast irons
1. Manganese	1. Increases hardness and hardenability 2. Increases toughness 3. Increases wear resistance 4. Increases tensile strength 5. Reduces ductility and weldability	1. Increases hardness 2. Tends to whiten cast irons
2. Nickel	1. Increases hardenability and hardness 2. Increases toughness 3. Increases tensile strength and shock resistance 4. Increases wear resistance	1. Improves machinability 2. Increases strength and hardness 3. Increases corrosion resistance

(*b*) Distinguish between

(*i*) Pearlite and Martensite transformation

(*ii*) White and Nodular cast iron

Sol. (*i*) Pearlite and Martensite Transformation : The T-T-T diagramshown in Fig. 6. It represents the transformation of austenite at varying cooling rates.

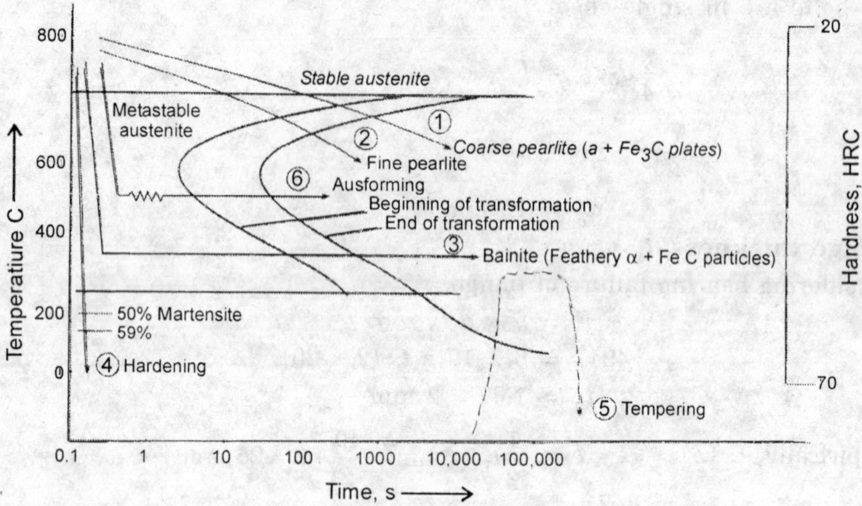

Fig. 6. T-T-T diagram.

Pearlite is the eutectoid mixture of ferrite and cementite. If a hypoeutectoid steel is heated to a temperature above its critical point and then cooled slowly to normal temperature than pearlite is obtained. Also when steel containing 0.85% carbon (Eutectoid steel) is heated to the austenite state and then slowly cooled to normal temperature, only pearlite remains. Slow cooling rates (1) and (2) corresponds to pearlite transformation in Fig. 6.

Martensite is a supersaturated solid solution of carbon in α-iron. It is a needed like structure heaving high hardness and low ductility. Rapid cooling or quenching of austenite to about 200°C converts it into martensite (curve 4). Martensite transformation begins at 240°C and ends at –50°C.

***(ii)* White and Nodular Cast Iron :** In white cast iron, carbon is present in combined form (Fe_3C) as cementile, which looks white. It is brittle and hard. Due to this, it is unmachinable except by grinding.

In nodular cast iron, carbon or graphite nodules are present in spherical shapes. It is less brittle. It has high fluidity with excellent combination of strength and ductility.

(*c*) Write a FORTRAN program to read in the elements of the $n \times m$ matrix A and also that of the $m \times n$ matrix B, and then perform the matrix multiplication operation

$$C = AB$$

Sol.

```
READ (7, *) ((A(I, J), J = 1, M), I = 1, N)
READ (7, *) ((B(I, J), J = 1, N), I = 1, M)
          DO 10 I = 1, N
          DO 10 J = 1, N
                S = 0
          DO 20 K = 1, M
             20 S = S + A(I, K) * B(K, J)
         10 C(I, J) = S
```

Q.6. (*a*) Describe Thermit Welding and mention iits applications.

Sol. Thermit Welding (TW) : Thermit welding is "A group of welding processes wherein coalescence is produced by heating with superheated liquid metal and slag resulting from a chemical reaction between a metal oxide and aluminium with or without the application of pressure. Filler metal, when used, is obtained from liquid metal".

Thermit is a mixture of powdered aluminium (1 part) and *iron oxide* in the form of roll scale (3 parts). When the thermit mixture is heated or ignited with a torch or ignition powder, to the temperature of 1100°C, it burns according to the following reaction :

$$8Al + 3Fe_3O_4 \rightarrow 9Fe + 4Al_2O_3 + \text{heat}$$

The reaction is nonexplosive and requires about a minute to go to completion. A temperature of about 2760°C and higher is reached in the combustion of thermit mixture. The reaction results in the formation of liquid reduced iron and slag (Al_2O_3) which are used for welding the pieces.

There are two types of thermit welding processes :

(*i*) plastic or pressure welding process.

(*ii*) fusion or non-pressure welding process.

Pressure Welding Process : The parts to be welded are butted tightly together and enclosed in a removable mould which provides a space between the inner surfaces of the mould and the parts. The products of the reaction are then poured into the mould in such a manner that slag enters the mould first. On contact with the metal parts, the slag cools rapidly and provides a layer of brittle, glasslike material so that the thermit iron which flows does not fuse with or adhere to the surface of the parts. The heavier thermit iron sinks to the lower half of the mould. This and the aluminium slag on top will both give up their heat to the pieces being welded. When the ends of the pieces to be welded reach the welding heat, they are forced together by means of clamps to make a pressure butt weld. The mould is then removed and the thermit iron and slag are knocked off from around the weld.

Non-pressure Welding Process : Here, the sections to be joined are lined up and a parallel gap or vee joint is cut at the fractured interface. A wax pattern is then formed between and around the ends which are to be joined. Then a sand mould complete with sprue, gate and riser is built around the joint area. The mould is then heated to melt the wax, causing it to flow into the sand mould and leaving a void space at the joint. The thermit powder is ignited in a crucible placed on top of the mould and the resulting iron is tapped-through a bottom hole of the crucible, directly into the mould. This highly superheated iron results in a deep penetration weld between the sections to be joined, (Fig. 7). After solidification, the mould is destroyed and the still hot excess iron is chiselled off.

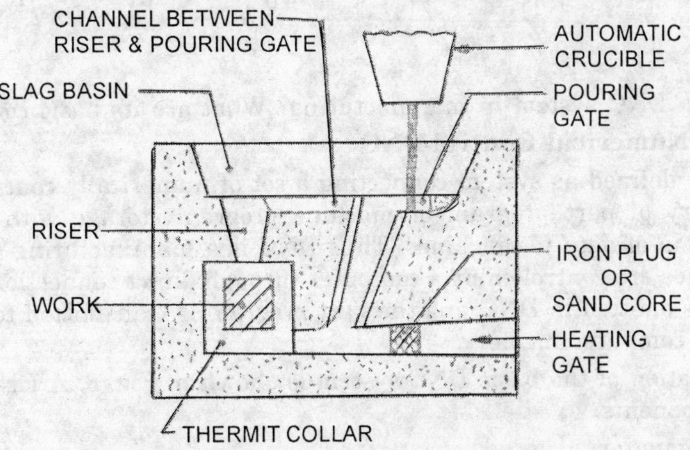

Fig. 7. Thermit Welding.

Applications of Thermit Welding

1. Welding of roils
2. Casting repairs
3. Welding reinforcing bars
4. Welding electrical conductors
5. Ship building
6. Making large crankshafts and propeller shafts

(*b*) A cylindrical block of diameter d_o and height h_o is forged to a disc of diameter d and height h in an open die forging operation. Determine.

(*i*) the final disc diameter, d

(*ii*) the true strain in the longitudinal direction, ε_h

(*iii*) show also that the true longitudinal strain, ε_h is twice the true radial strain, ε_r.

Sol.

(*i*) For volume constancy,

$$\frac{\pi}{4}d_0^2 h_0 = \frac{\pi}{4}d^2 h$$

$$d = \sqrt{\frac{h_0}{h}}\cdot d_0$$

(*ii*) True strain in the longitudinal direction,

$$\varepsilon_h = \int_{h_0}^{h}\frac{dh}{h} = \ln\left(\frac{h}{h_0}\right) = \ln(1+\varepsilon)$$

where $$\varepsilon = \frac{h-h_0}{h_0}$$

(*iii*) True radial strain,

$$\varepsilon = \int_{d_0}^{d}\frac{-dD}{D} = -\ln\left(\frac{d}{d_0}\right) = -\ln\sqrt{\frac{h_0}{h}} = \frac{-1}{2}\ln\left(\frac{h_0}{h}\right) = \frac{1}{2}\ln\left(\frac{h}{h_0}\right) = \frac{1}{2}\varepsilon_h$$

$$\therefore \quad \varepsilon_h = 2\varepsilon_r$$

(*c*) What is a DNC system in manufacturing? What are its basic components?

Sol. Direct Numerical Control-DNC

DNC may be defined as system connecting a set of numerically controlled machines to a common memory of part program or machine program storage with provision for on-demand distribution of data to machines. Thus DNC is a manufacturing system in which a number of machines are controlled by a computer through direct connection and in real time. The tape reader is omitted in DNC and the part program is transmitted to the machine tool directly from the computer memory.

The configuration of the basic DNC system is shown in Fig. 8. A DNC system consists of four basic components.

1. Central computer.
2. Bulk memory to store the NC part programs.

3. Telecommunication lines.
4. Machine tools.

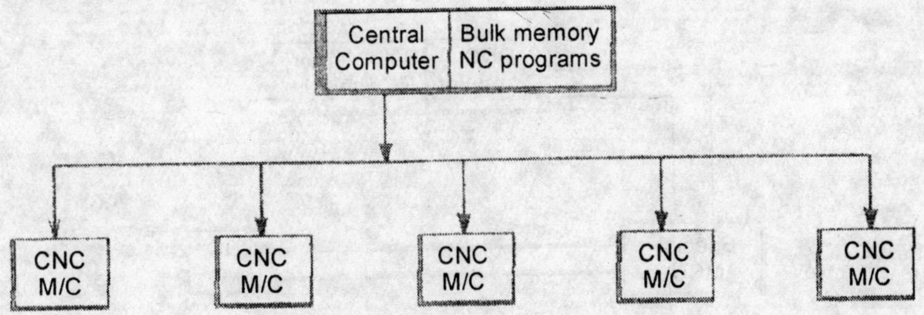

Fig. 8. General configuration of DNC system.

Depending on the number of machines and the computational requirements, it is sometimes necessary to make use of satellite computers, as shown in Fig. 9.

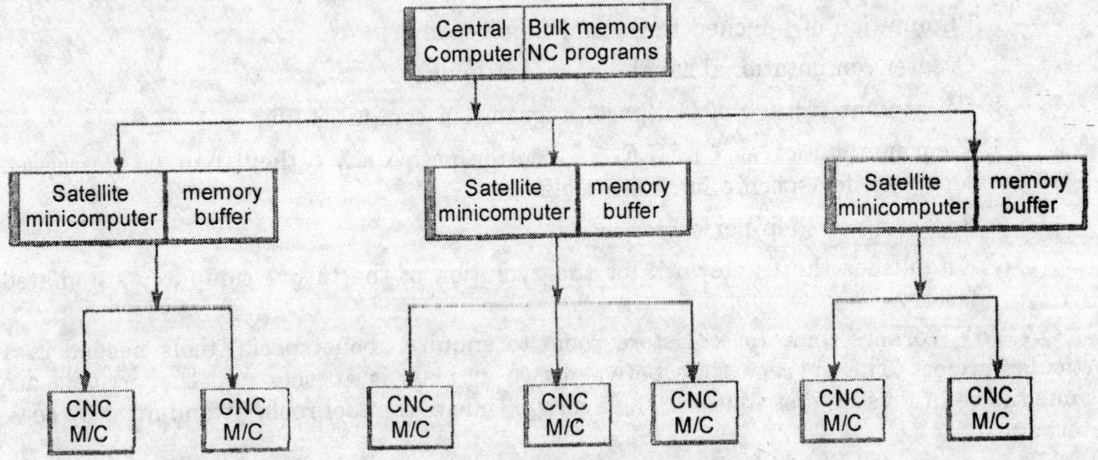

Fig. 9. DNC with satellite mini computer.

There are two alternative system configurations by which communication link is established between the control computer and the machine tool.

1. Behind-The-Tape-Reader (BTR) System. In this system (Fig. 10), the computer is linked directly to the regular NC controller unit. The connection with the computer is made between the tape reader and the controller unit-behind the tape reader. The tape reader is replaced by telecommunication lines.

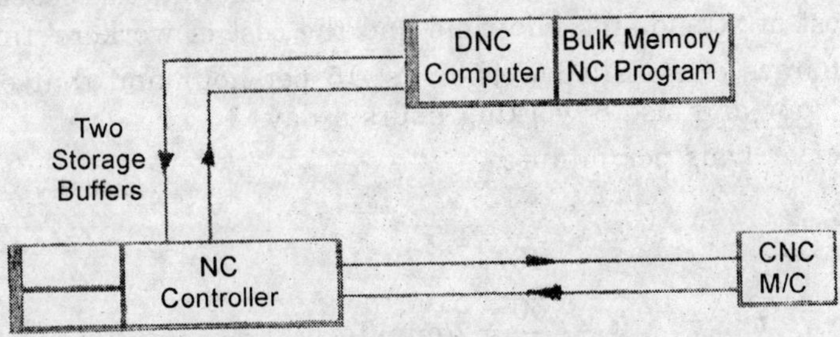

Fig. 10. DNC with BTR configuration.

2. Special machine control unit. The regular NC controller is eliminated altogether and replaced by a special MCU (Fig. 11). The special MCU communicates between the machine tool and the computer.

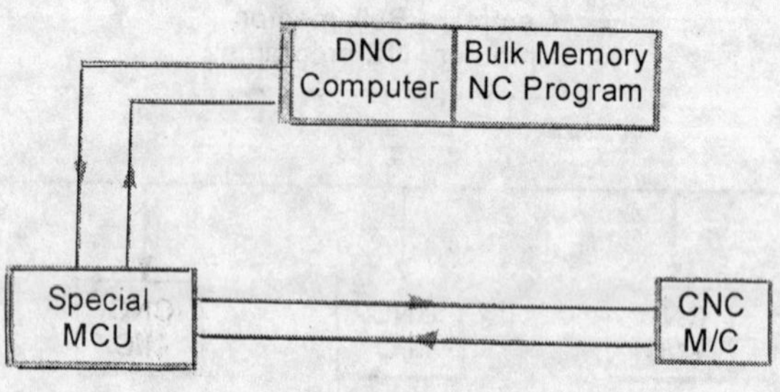

Fig. 11. DNC with special MCU.

Advantages of DNC

The major advantages of DNC system are :

1. Elimination of punched tapes and tape readers.
2. Greater computational capability and flexibility.
3. Convenient storage of NC part programs in computer files.
4. Programs stored as CLFILE, *i.e.*, cutter path data rather then postprocessed programs for specific machine tools.
5. Reporting of shop performance.
6. Establishes the framework for the evolution of the future computer automated factory.

Q.7. (*a*) Workers come to tool store room to enquire about special tools needed in a particular project. The average time between two arrivals is 40 secs and the arrivals are assumed to be in Poisson distribution. The average time of the tool room attendant is 30 secs. Determine

(*i*) average queue length

(*ii*) average length of non-empty queues

(*iii*) average number of workers in system including the worker being attended.

(*iv*) mean waiting time of an arrival

(*v*) average waiting time of an arrival, who waits

(*vi*) whether additional number of tool store room attendants is needed to minimize the combined cost of attendant's idle time and the cost of workers' time.

Assume the charges of a skilled worker Rs. 15 per hour and that of tool store room attndant Rs. 4 per hour and also 8 working hours a day.

Sol. Number of arrivals per minute,

$$\lambda = \frac{60}{40} = 1.5$$

Service rate, $$\mu = \frac{60}{30} = 2\text{/minute}$$

(*i*) Average queue length,

$$L_q = \frac{\lambda^2}{\mu(\mu-\lambda)} = \frac{(1.5)^2}{2(2-1.5)} = 2.25$$

(*ii*) Average length of non-empty queue

$$= \frac{\mu}{\mu-\lambda} = \frac{2}{2-1.5} = 4$$

(*iii*) Average number of workers in the system including the worker being attended

$$= \frac{\lambda}{\mu-\lambda} = \frac{1.5}{2-1.5} = 3$$

(*iv*) Mean waiting time of an arrival

$$= \frac{\lambda}{\mu(\mu-\lambda)} = \frac{1.5}{2(2-1.5)} = 1.5$$

(*v*) Average waiting time of an arrival, who waits

$$= \frac{1}{\mu-\lambda} = \frac{1}{2-1.5} = 2$$

(*b*) Write short notes on

(*i*) Dynamic Programming

(*ii*) Product Planning.

Sol. (*i*) Dynamic Programming : In some of the practical problems, the decisions are to be made sequentially. These are called sequential decision problems or multistage decision problems. Dynamic programming is a mathematical technique well suited for the optimization of multistage decision problems. In the dynamic programming technique, the multistage decision problem is decomposed as a sequence of single stage decision problems. The decomposition is done in such a way that the optimal solution of the original multivariable problem can be obtained from the optimal solutions of the one-dimensional problems. The technique of dynamic programming has been applied to a variety of problems such as the problem of inventory control, network analysis, production control etc.

(*ii*) Product Planning : It may be defined as – *Evaluation of the range, mix, specification and pricing of existing and new products in relation to present and future market requirements and competition; planning of product range, mix, specification and pricing to satisfy company objectives; and specifying the research design and development support required.*

From the above definition it is clear that both existing and potential products must be included in the product planning activity, and it follows that the activity itself should deal with proper balance between old and new product-mix so far as the future course of the business is concerned. The evaluation and planning activities referred to, and particularly the emphasis on satisfying the company's objectives, implies the existence of an overall corporate plan which provides a framework within which these activities may take place.

Within the overall corporate plan as set down by senior management, the product plan usually represents the activity which links the company with its market and so is most directly concerned with the forward development of the company as a whole.

Product planning is the prerogative of top management. It is a function which is most usually discharged by line management. Successful product planning can only occur within a clear frame work of goals and objectives laid down by the board of directors. Further, senior management must ensure direct lines of communication between itself and the product planning function. Product planning serves as an input to process design.

Need for Product Planning

(*a*) Needs determined from current deficiencies

This area primarily relates to those needs which are stimulated as a result of some problem or deficiency occuring in a product today. The resolution is often achieved through the development, production and use of a new product or process.

(*b*) Needs determined from anticipated deficiencies.

Through the prediction of societal, political, ecological, technological, and/or economic trends, anticipated needs can often be determined *e.g.* we suspect that economic constraints in the future will likely prohibit the use of certain system curently in use. As a result, a replacement system that exhibits a lower overall life-cycle cost is needed to accomplish the designated function(s).

Constituents of Product Planning

1. Marketing and Marketing analysis
2. Performance of feasibility studies, and
3. Advanced planning.

MECHANICAL ENGINEERING PAPER II—1998

(Conventional)

SECTION-A

Q.1. (*a*) The hexagonal bolt proportions for pitch, head thickness, nut thickness, head and nut diameters, etc. are given in terms of nominal diameter. Why ?

Sol. A hexagonal bolt is subjected to tension, bending or torsional loads. The strength of the bolt under any one of these loads or a combination of these is based on nominal diameter of the bolt. Other dimensions are then specified as proportions of nominal diameter.

(*b*) What are the common profiles used for splines in machine elements ?

Sol.

1. Straight-sided splines.
2. Triangular splines.
3. Involute splines.

(*c*) What is self-locking taper and what is self-releasing taper ? Give an example of each where they are used.

Sol. If the taper provided is in excess (upto 1:8), there is a possibility of the attached component slipping back. Such a taper is called self releasing taper. If the taper is less than 1 : 8, it is called a self locking taper. The gib and cotter joint has a self locking type of taper.

(*d*) What are extreme pressure lubricants ? What are the main additives and how do they act ?

Sol. The extreme pressure lubricants can resist very large pressure in the bearing. Solid lubricants are predominantly used for this purpose. These are lamellar solids like Molybdenum sulphide, Tungsten sulphide, Graphite ; polymers like PTFCE, Nylons, Acetals etc.; Oxides like Molybdenum oxide, Lead oxide, Silicon oxide, etc. The solid lubricants act by resisting the penetration of peaks on the journal under extreme loads.

(*e*) When using strain gauge system for stress/force/ displacement measurements, how are in-built magnification and temperature compensation achieved ?

Sol. The output of strain gauges in the adjacent arms of a Wheatstone bridge subtract and in the opposite arms add. Thus by knowing the nature of strain being measured, the in-built magnification can be achieved. Self temperature compensation can be achieved by putting strain gauges in such a way that temperature strains cancel out or by using dual element gauges having opposite coefficient of expansion.

(*f*) Specific strength of materials is very high when they are in fibre size but lower when they are in bar form. Why?

Sol. The fibres are very strong and stiff in tension because the molecules in the fibres are oriented in the longitudinal direction, and their cross-sections are so small that the probability is low that any defects exist in the fibre. Glass fibres, for example, can have tensile strength as high as 4600 MPa, whereas the strength of glass in bulk form is much lower.

(*g*) How are porous bearings made and where are they used ?

Sol. Porous bearings are made of powder metals like lead, bronzes, alloys of iron, copper etc. They require little or no attention for lubrication. These bearings are used for home appliances, small motors, machine tools and aircraft automotive accessories etc. These bearings are made as follows :

1. Powdered metals are blended into uniform mixture which is pressed under high pressure in tungsten carbide dies and then sintered below its melting point into a coherent mass.
2. Bore, outside diameter and faces are sized in a burnishing operation.
3. Bearing is placed in a pool of lubricating oil to fill voids.

(*h*) What are the ingredients of stainless steel and what are their approximate percentages?

Sol. Austenitic stainless steels :

Carbon = 0.12 – 0.14%

Chromium = 17 – 20%

Nickel = 8 – 11%

Ferritic stainless steels :

Carbon = 0.12%

Chromium = 14 – 18%

Martensitic stainless steels :

Carbon = 0.12%

Chromium = 12 – 14%.

(*i*) For winches and cranes, wire ropes are preferred to wires. Why ?

Sol. Wire ropes are flexible, more stronger than the strength of same number of wires, can withstand shocks, more durable, gives silent operation and are reliable.

(*j*) In general terms, what is pressure angle ? How is it defined for a cam ?

Sol. The pressure angle represents the angle at whidh the resultant force between two mating surfaces, moving over the other, acts. For a cam, it is the angle between the direction of the follower motion and a normal to the pitch curve.

(*k*) The best shape of a runner in sand casting is an inverted frustrum of a cone. Why?

Sol. For a mould made of a permeable material like sand, the pressure anywhere in the liquid metal stream should not fall below atmospheric pressure to avoid entering of gases originating from baking of organic compounds in the mould, into the molten metal stream, producing porous castings. To avoid negative pressure in the pouring basin, the runner should be tapered.

(*l*) Multiple coated disposable carbide tips have more-or- less replaced brazed carbide tipped tools in CNC applications. why?

Sol. For machining of different materials in CNC machines, we require different type of tool materials. Therefore, disposable tips are preferable as compared to brazed carbide tipped tools so that we need not replace the whole tool.

(***m***) State standard point angle and helix angle of twist drill.

Sol. For steels and cast irons :

Point angle = 118°

Helix angle = 22° to 33° (lower value for smaller size).

(*n*) State two products each produced by forward extrusion and reverse extrusion.

Sol.

Type of exttrusion	**Products**
Forwar	Tubes, rods of various shapes
Reverse	Collapsible tubes for tooth paste, Liquid containers

(***o***) What are the best geometrical shapes suggested for sheet metal drawing and spinning?

Sol.

Process	**Geometrical shapes**

Drawing : Cylindrical or box shaped parts like cups

Spinning : Axi-symmetrical parts.

(***p***) Expand MRP I and MRP II indicating the differences and similarity.

Sol.

MRP I. Material requirement planning.

MRP II. Manufacturing-resource planning.

MRP II is an attempt to tie MRP I to the financial system of a company.

MRP I is a method for managing inventories and issuing orders for parts and materials.

MRP II is concerned with all facets of business including sales, production, inventories and cash flow.

(***q***) Indicate the similarity and differences between EOQ and EBQ in manufacturing.

Sol. EOQ (Economic order quantity). It represents how much to order for purchase for no loss of money.

EBQ (Economic batch quantity). It represents how much to produce for maximum profit in a batch size.

(***r***) State the similarities and differences between a fixture and a jig.

Sol. Fixture. It is a structure for holding and locating a component or workpiece in a definite position for a specific operation. It does not guide the tool.

Jig. It addition to locating, holding and supporting a component, a jig also guides the tool.

(*s*) State, sequentially the elements of a canned CNC drilling cycle.

Sol. NG81 XYZR

where G81 : canned cycle

X, *Y* : centre co-ordinates of the points of drilling

Z : depth of drilled holes

R : position of the clearance plane.

(*t*) What is FORTRAN and for what types of engineering problems is it suited ?

(2 × 20 = 40)

Sol. FORTRAN. Formula Translation. It is suitable for solving problems in areas such as mathematics, physics, basic sciences, operational research, statistics and engineering.

SECTION-B

Q.2. (*a*) A thin cylinder with closed ends has an internal diameter of 50 mm and a wall thickness of 2.5 mm. It is subjected to axial pull of 10 kN and a torque of 500 N.m. while under an internal pressure of 6 MN / m^2.

(*i*) Determine the principal stresses in the tube and maximum shear stress.

(*ii*) Represent the stress configuration on a square element taken in the load direction with direction and magnitude indicated (schematic).

(*iii*) Sketch the Mohr's stress circle. 20

Sol. (*i*) Longitudinal stress, $\sigma_l = \frac{pd}{4t} = \frac{6 \times 50}{4 \times 2.5} = 30$ MPa

Hoop stress, $\sigma_\theta = \frac{pd}{2t} = \frac{6 \times 50}{4 \times 2.5} = 60$ MPa

Direct axial stress,

$$\sigma_d = \frac{F}{\pi dt} = \frac{10 \times 10^3}{\pi \times 50 \times 2.5 \times 10^{-6}} = 25.465 \text{ MPa}$$

$$\sigma_x = \sigma_l + \sigma_d$$

$$= 30 + 25.465 \times 10^{-6} = 55.465 \text{ MPa}$$

$$\sigma_y = \sigma_\theta = 60 \text{ MPa}$$

$$\tau_{xy} = \frac{2T}{\pi d^2 t} = \frac{2 \times 500}{\pi \times 500^2 \times 2.5 \times 10^{-9}} = 50.93 \text{ MPa}$$

Principal stresses,

$$\sigma_{1,2} = 0.5\,(\sigma_x + \sigma_y) \pm 0.5\,[(\sigma_x - \sigma_y)^2 + 4\tau^2_{xy}]^{0.5}$$

$$= 0.5\,(55.465 + 60) \pm 0.5\,[(55.465 - 60)^2 + 4 \times (50.93)^2]^{0.5}$$

$$= 57.73 \pm 50.98 = 108.61 \text{ MPa}, 6.75 \text{ MPa}$$

Maximum shear stress = 50.98 MPa

(*ii*) The state of stress on a square element is shown in Fig- 1 (*a*).

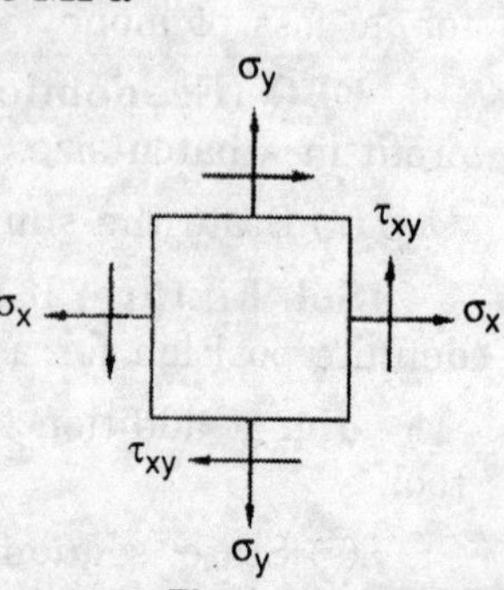

Fig 1 (a)

(*iii*) The Mohr's stress circle is shown in Fig. 1 (*b*).

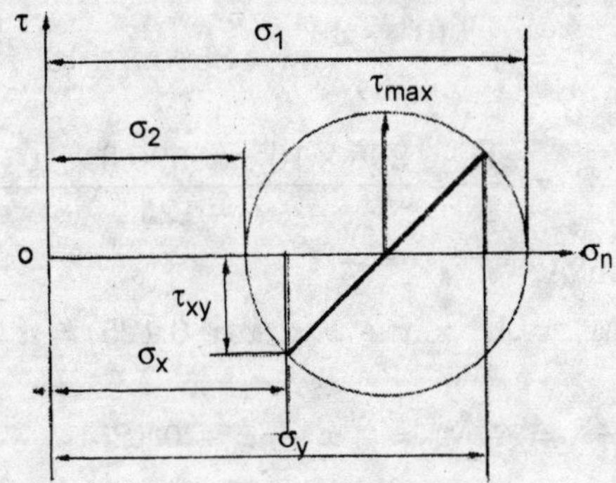

Fig. 1. (b)

(*b*) A tube 40 mm outside diameter, 5 mm thick and 1.5 m long is simply supported at 125 mm from each end and carries a concentrated load of 1 kN at each extreme end.

(*i*) Neglecting the weight of the tube, sketch the shearing force and bending moment diagrams, and

(*ii*) Calculate the radius of curvature and the deflection at mid-span. Take the modulus of elasticity of the material as 208 GN/m^2.

(*iii*) Give an example where maximum deflection and maximum stress do not occur at the same point,

(*iv*) Give an example where they occur at the same point. (20)

Sol. (*i*) The loading diagram is shown in Fig. 2 (*a*).

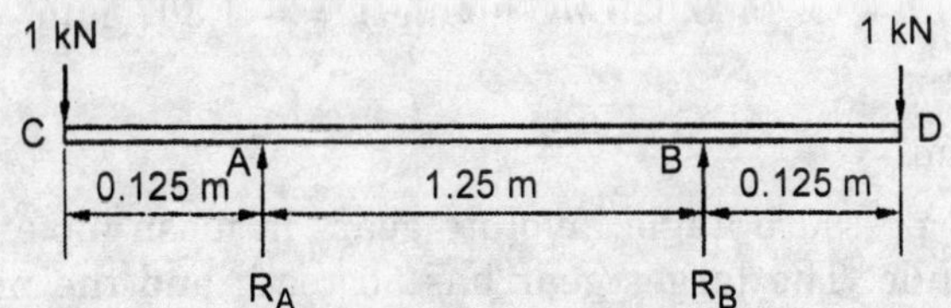

Fig. 2 (a) Loading diagram

$$\sum M_B = 0 \text{ gives}$$

$$1.25\, R_A = 1 \times 1.375 - 1 \times 0.125 = 1.25$$

$$R_A = 1 \text{ kN.}$$

BMD: $M_A = -1 \times 0.125 = -0.125$ kN-m

$M_C = M_D = 0$

The BMD is shown in Fig. 2 (*b*).

SFD:

Span CA : F = – 1 kN

Span AB : F = 0

Span ED : F = 1 kN

The SFD is shown in Fig. 2 (*c*).

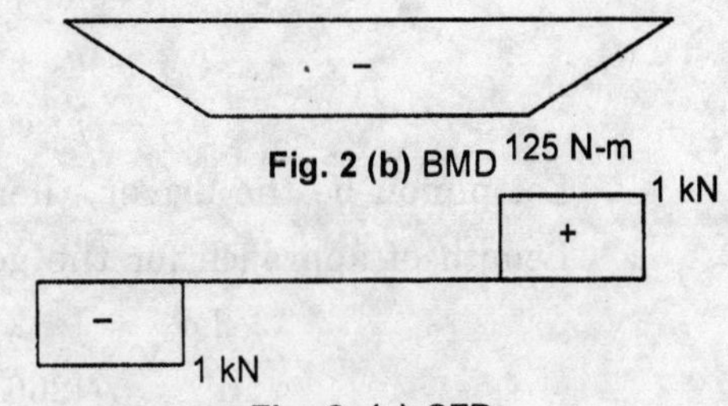

Fig. 2 (b) BMD

Fig. 2. (c) SFD

(*ii*)
$$I = \pi \frac{(40^4 - 30^4)}{64} = 8590.3 \times 10^{-12}\ m^4$$

$$R = \frac{EI}{M} = \frac{208 \times 10^9 \times 8590.3 \times 10^{-12}}{-125} = -142.94\ m$$

At a distance x from A,

$$M_x = R_A \times x - 1 \times (x + 0.125) = x - (x + 0.125)$$

$$\frac{EId^2y}{dx^2} = -M_x = -x + (x + 0.125)$$

$$\frac{EIdy}{dx} = \frac{x^2}{2} + \frac{(x+0.125)^2}{2} + C_1$$

$$EIy = -\frac{x^3}{6} + \frac{(x+0.125)^3}{6} + C_1x + C_2$$

At $x = 0$, $y = 0$, giving $C_2 = -\frac{(0.125)^3}{6} = -3.255 \times 10^{-4}$

At $x = 1.25$ m, $y = 0$, giving $C_1 = -0.08595$

$$EIy = -\frac{x^3}{6} + \frac{(x+0.125)^3}{6} - 0.08595x - 3.255 \times 10^{-4}$$

At mid-span, $x = 0.625\ m$, giving $y = -1.367$ mm

(*iii*) At the supports.

(*iv*) At the mid-span.

Q.3. (*a*) Two 20° pressure angle involute gears in mesh have a module of 10 mm. The addendum is one module. The larger gear has 50 teeth and the pinion 13 teeth.

(*i*) Does interference occur ?

(*ii*) If it occurs to what value should the pressure angle be changed to eliminate interference?

(*iii*) Represent schematically the meshing of two gears showing the base, pitch and addendum circles and the path of contact. (20)

Sol.
$$d_1 = mz_1 = 10 \times 13 = 130 \text{ mm}$$
$$d_2 = mz_2 = 10 \times 50 = 500 \text{ mm}$$

(*i*)
$$r_{a2} = r_2 + h_{a2} = 250 + 10 = 260\text{mm},$$
$$r_{b2} = r_2 \cos \alpha = 250 \cos 20° = 234.92 \text{ mm}$$

Let pinion be the driver, then

Length of approach for the gear,

$$La_2 = [r^2_{a2} - r^2_{b2}]^{0.5} - r_2 \sin \alpha$$
$$= [260^2 - (234.92)^2]^{0.5} - 250 \sin 20°$$
$$= 25.907 \text{ mm}$$

Maximum length of approach

$$= r_1 \sin \alpha$$
$$= 65 \sin 20°$$
$$= 22.23 \text{ mm}$$

Since length of approach is more than the maximum length of approach, therefore, interference will occur.

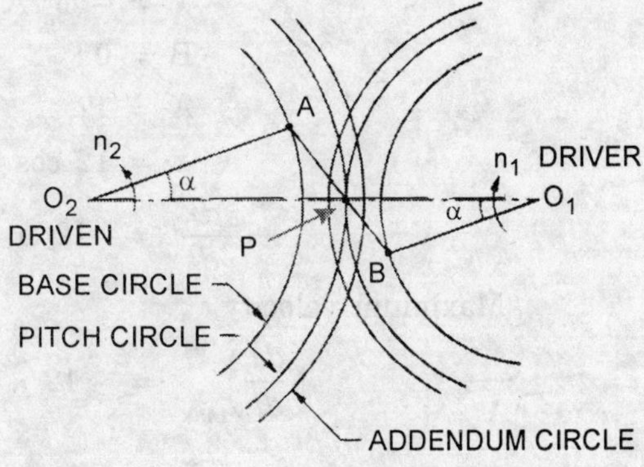

Fig. 3.

(*ii*) $[260^2 - (250 \cos \alpha)^2]^{0.5} - 250 \sin \alpha$

$$= 22.23$$
$$\sin \alpha = 0.41438$$
$$\alpha = 24.48°.$$

(*iii*) The meshing of two gears is shown in Fig. 3.

(***b***) The stiffness of a closed-coil spring is such that it deflects 36 mm when an axial force of 10 N is applied on the end of the hook.

(*i*) Calculate the mass to be hung on the hook so that when set vibrating, it will make one complete oscillation per second. The mass of the spring is 0.6 kg and the effect of this when vibrating is to be taken as equivalent to a mass on the end of the hook equal to one third of the mass of the spring.

(*ii*) Calculate also the maximum velocity, and

(*iii*) Maximum acceleration of the vibrating mass when initially displaced 12 mm from the position of equilibrium.

(*iv*) What is the natural frequency of a vibrating system ?

(*v*) Can you suggest an application for the approach indicated above in practice ? (20)

Sol. (*i*) Stiffness of spring,

$$k = \frac{10}{36 \times 10^{-3}} = 277.778 \text{ N/m}$$

Natural frequency, $f_n = \left(\frac{1}{2\pi}\right)\left[\frac{k}{m + m_s/3}\right]^{0.5}$

$$1 = (1/2\pi)\,[277.778/(m + 0.2)]^{0.5}$$
$$m = 6.836 \text{ kg}$$

(*ii*) The equation of motion is

$$\left(m + \frac{m_s}{3}\right)\ddot{x} + k\,x = 0$$

The solution of this equation is

$$x(t) = \text{A} \cos \omega_n x + \text{B} \sin \omega_n x$$

When $t = 0,\ x = 12 \text{ mm, and } \frac{dx}{dt} = 0$

$$12 = \text{A} \cos 24\pi + \text{B} \sin 24\pi$$

$$x = -\omega_n A \sin \omega_n x + B \omega_n \cos \omega_n x = 0$$
$$B = 0$$
$$A = 12$$
$$x = 12 \cos \omega_n x$$
$$\frac{dx}{dt} = -12 \omega_n \sin \omega_n x$$

Maximum velocity,

$$\left(\frac{dx}{dt}\right)_{max} = -12 \times \pi = -75.4 \text{ mm/s}$$

(*iii*)
$$\frac{d^2x}{dt^2} = -12 \omega^2{}_n \cos \omega_n x$$

Maximum acceleration,

$$\left(\frac{d^2x}{dt^2}\right)_{max} = -12 \times (2\pi)^2 = 473.74 \text{ mm/s}^2$$

(*iv*) The natural frequency of a vibrating system is the frequency at which it vibrates of its own without the help of any external agency.

(*v*) This approach can be used to determine the stiffness of a spring.

Q.4. (*a*) Two shafts A and D are in the same line (axes in one line). They are geared together ' through an intermediate parallel shaft carrying wheels B and C which mesh with the wheels A and D respectively. Wheels A and B have module of 4 mm and the wheels C and D have a module of 9 mm. The number of teeth on any wheel is to be not less than 15 and the speed of D is to be about, but not greater than 1/12 the speed of A and the ratio of each reduction is the same. Find

(*i*) Suitable number of teeth for the wheels.

(*ii*) The actual reduction.

(*iii*) The distance of the intermediate shaft from the axes of the shafts A and D (centre distance). Indicate the configuration with a sketch.

(*iv*) What is addendum modification related to correction of gears and when are they used in practice ?

(Gear A is on shaft A and gear D on shaft D). (20)

Sol.
$$m_a = m_b = 4 \text{ mm}, \ m_c = m_d = 9 \text{ mm}$$
$$z_{min} = 15, \frac{n_a}{r_d} = 12$$

Now
$$\frac{n_a}{n_b} = \frac{n_c}{n_d}$$

But
$$n_b = n_c$$

Therefore
$$n^2b = n^2{}_c = n_a \, n_d \text{ or } n_b = n_c = \sqrt{n_a \, n_d} = \sqrt{12} \, n_d$$

$$\frac{n_c}{n_d} = \frac{n_b}{n_d} = \sqrt{12} = 3.464$$

For same centre distance,

$$4\ (z_a + z_b) = 9\ (z_c + z_d)$$

$$4\ (z_a + z_b) = 2.25\ (z_c + z_d)$$

$$z_a\ (1 + 3.5) = 2.25\ z_c\ (1 + 3.5)$$

$$z_a + 2.25\ z_c$$

Let $z_c = 15$, then $z_a = 33.75 \cong 34$

Now $\dfrac{z_d}{z_c} = 3.5,\ z_d = 51.96 \cong 52$

$$34 + z_b = 2.25(15 + 52)$$

$$z_b = 116.75 = 117$$

$$\frac{n_A}{n_B} = (3.464)^2 = 12,\ C = 4\ (34 + 117) = 604 \text{ mm}$$

Fig. 4.

(*b*) The semi-cone angle of a cone clutch is 12.5° and the contact surfaces have a mean diameter of 80 mm. The co-efficient of friction is 0.32.

(*i*) What is the minimum torque required to produce slipping of the clutch for an axial force of 200 N ?

If the clutch is used to connect an electric motor with a stationary flywheel

(*ii*) What is the time needed to attain the full speed, and

(*iii*) The energy lost during slipping ?

Motor speed is 900 rpm and the moment of inertia of the flywheel is 0.4 kg-m².

(*iv*) What are the considerations in the selection of plate clutches and cone clutches ? (20)

Sol. $\alpha = 12.5°,\ d_m = 80\text{mm},$

$$\mu = 0.32,\ W = 200N,\ I = 0.4 \text{ kg-m}^2.$$

(*a*) $T_f = \mu W r_m \text{ cosec } \alpha$

$$= 0.32 \times 200 \times 40 \times 10^{-3} \times \text{cosec } 12.5° = 11.83 \text{ N-m}$$

$$T_f = I.\alpha$$

$$\text{Angular acceleration, } \alpha = \frac{11.83}{0.4} = 29.575 \text{ rad/s}^2$$

$$\omega = \frac{2\pi \times 900}{60} = 94.248 \text{ rad/s}$$

(*b*) Time taken to obtain full speed,

$$t = \frac{\omega}{\alpha} = \frac{94.248}{29.575} = 3.18 \text{ s}$$

Angle turned through by the driving shaft during slipping period,

$$\theta_1 = \omega t = 94.248 \times 3.18 = 299.7 \text{ rad}$$

Angle turned by the driven shaft,

$$\theta_2 = 0.5\ \alpha\ t^2 = 0.5 \times 29.575 \times (3.18)^2 = 149.54 \text{ rad}$$

Energy lost in friction due to clutch slip

$= T_f (\theta_1 - \theta_2)$

$= 11.83 \times (299.7 - 149.54) = 1776.4$ N-m.

(c) Kinetic energy of driven shaft $= 0.5\ I\omega^2$

$= 0.5 \times 0.4 \times (94.248)^2 = 1776.53$ N-m

Total energy supplied during slipping

$= 1776.53 + 1\ 776.4 = 3552.93$ N-m

SECTION-C

Q.5. (*a*) (*i*) What is High-efficiency range or High -economic range (Hi-E) in machining?

(*ii*) Why is it called so ?

(*iii*) The results of machining steel with two grades of tool materials are given below :

Tool	Taylor's exponent	Cutting speed for one-minute tool life, m/min
A	0.20	100
B	0.25	120

What are the standard tool life equations for these materials ?

(*iv*) For a 180 minute tool life, which tool is recommended and why ?

(*v*) If the tool grinding and changing time is 15 minutes, which cutting speed for the tool 'A' has to be chosen from the two available speeds, 45 m/min, 50 m/min ? (20)

Sol. (*i*) Optimum cutting speed for minimum cost,

$$v_o = \frac{c\,(L_m + B_m)^n}{[(1/n - 1)\{T_c(L_m + B_m) + T_g\,(L_g + B_g) + D_c\}]^n} \quad ...(1)$$

where

L_m = labour rate per hour

B_m = overhead charge of the machine

T_c = total change time

T = tool grind time

L = labour rate per hour of tool grinder operator

B_g = overhead charge of tool grinder

D_c = depreciation of tool per grind, Rs.

n = index of $vT^n = C$

Optimum cutting speed for maximum production rate,

$$v_o = \frac{c}{[(1/n - 1)\,T_c]^n} \quad ...(2)$$

Plotting Eqns. (1) and (2), we get the curves as shown below. Join the lowest cost of per piece curve and highest point of production rate curve. The rectangular area (shown shaded) is the Hi-E range in machining, as shown in Fig. 5.

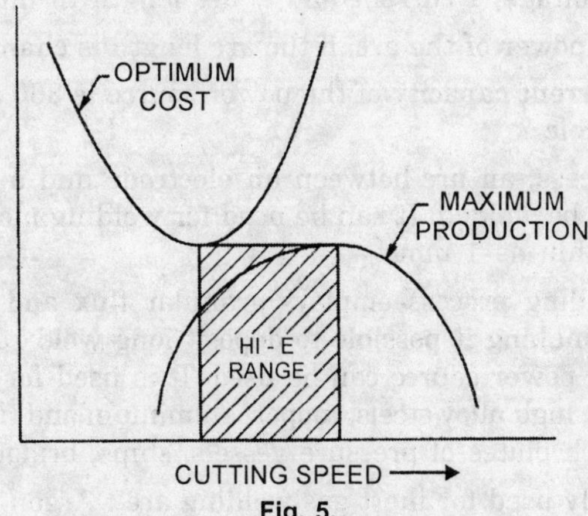

Fig. 5.

(ii) Hi – E range gives a compromise between the minimum cost and maximum production. Hence, it gives the highest efficiency for best economy.

(iii) Taylor's tool life equation is $vT^n = C$

Tool A : $100 \times 1^{0.2} = C$

$$C = 100$$
$$vT^{0.2} = 100$$

Tool B : $120 \times 1^{0.25} = C$

$$C = 120$$
$$vT^{0.25} = 120$$

(iv)
$$v_A \times 180^{0.20} = 100$$
$$v_A = 35.4 \text{m/min}$$
$$v_B \times 180^{0.25} = 120$$
$$v_B = 32.76 \text{ m/min}$$

Tool *A is* recommended because for the same tool life it can work at a higher cutting speed to give more output.

(v)
$$45 \times T^{0.2} = 100$$
$$T = 54.2 \text{min}$$
$$50 \times T^{0.2} = 100$$
$$T = 32 \text{ min}$$

Choose 50 m/min cutting speed.

(b) (i) What are the basic differences between arc welding and submerged arc welding

(ii) What are the inert gases generally used in inert gas welding?

(iii) What is weld penetration and what is weld reinforcement ? .

(iv) In MIG welding, the power source characteristic is

$$V_p = 36 - \frac{I}{60}$$

and the arc characteristic is

$$V_a = 2\,l_a + 27$$

where V_p and V_a voltage, I current and l_a arc length in mm.

Find the change in power of the arc, if the arc length is changed from 2 mm to 4 mm.

If the maximum current capacity of the power source is 360 amps, find the minimum arc length sustainable ? (20)

Sol.(*i*) In arc welding process, an arc between an electrode and a work piece is utilized to heat the joint to be welded. It can be used for welding mainly steel plates as thick as 75 mm and as thin as 1 mm.

Submerged arc welding process employs granular flux and a copper-coated wire in spooled form, thus making it possible to deposit long weld runs without interruption. Both A.C. and D.C. power source can be used. It is used for welding low carbon, low alloy stainless steels, high alloy steels, copper, aluminium and titanium. It finds extensive use for welding thick plates of pressure vessels, ships, bridges and structural work.

(*ii*) Inert gases generally used for inert gas welding are : Argon, Helium, Carbon dioxide and Nitrogen.

(*iii*) **Weld penetration.** It is the ratio of the width of the weld to its depth. It is of the order of 1.25 for gas metal arc welding, 2.5 for shielded metal arc, 5 for plasma arc and 1.5 for electron beam welding.

Weld reinforcement. To increase the strength of the weld, the weld is reinforced by its throat thickness so that its thickness is more than the plate thickness being welded. For example, for a fillet weld, the throat thickness is 0.85 t instead of 0.707 t. Both these terms are shown in the following Fig. 6.

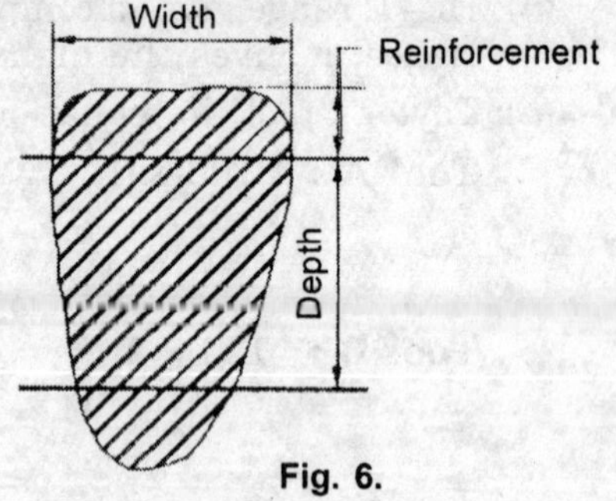

Fig. 6.

(*iv*) For

$$V_p = V_a$$

$$36 - \frac{I}{60} = 2l_a + 27$$

$$I = [36 - (2l_a + 27)] \times 60$$

$$= (9 - 2l_a) \times 60$$

Power, $P = VI = (2l_a + 27) \times (9 - 2l_a) \times 60$

For $l_a = 2$ mm, $P = (2 \times 2 + 27) \times (9 - 2 \times 2) \times 60 = 9300$ W

For $l_a = 4$ mm, $P = (2 \times 4 + 27) \times (9 - 2 \times 4) \times 60 = 2100$ W

Change in power $= 9300 - 2100 = 7200$ W

For $I = 360$ A, $(9 - 2l_a) \times 60 = 360$

$$l_a = 1.5 \text{ mm}$$

Q.6. (***a***) (*i*) What is the difference between counter-boring and counter-sinking ?

(*ii*) What are the basic differences between gear milling and gear hobbing ?

(*iii*) What are the basic differences between facing at constant speed and facing at constant feed ?

(*iv*) A surface 80 mm x 160 mm is rough machined using a face milling cutter 150 mm diameter having 10 teeth. The cutter centre is off-set by 15 mm from the line of symmetry of the surface. Estimate the time to rough machine the surface, if the feed per tooth is 0.25 mm and a cutting speed of 20 m/min is employed.

(*v*) With 5 mm approach and 5 mm over-run, what is the single pass feed time ?

(*vi*) If tool axis is symmetrical with the job, what happens to the machining time ? (20)

Sol. (*i*) Counter-boring. It is the process of enlargement of diameter of a hole for certain depth only. Ordinarily flat cutters can be effectively used to counterbore the previously bored large holes. Tools used for counterboring small holes need a pilot at the front end. The tool may be in single piece or may have inserted cutters.

Counter-sinking. This is the process of providing a conical recess at the end of a hole to make a seat for the countersunk head of a mating component, usually fasteners, like rivets and screws, etc. A countersinking tool of suitable included angle can be used for this purpose.

(*ii*) Milling is an operation by which a gear work piece is machined by means of a milling cutter having a large number of cutting edges. The feed motion is given to the work piece. The cutter is mounted on the arbor.

Hobbing is a process of generating a gear by means of a rotating cutter called a *hob*. It is a continuous indexing process in which both the cutter and work piece rotate in a constant relationship while the hob is being fed into the work. The hob is fed into the gear blank to the proper depth and the two are rotated together as if in mesh. Each hob tooth cuts its own profile resulting into gear teeth.

(*iii*) Difference between facing at constant feed and constant speed. Facing is an operation for generating flat surfaces on a lathe. The feed is given in a direction perpendicular to the axis of revolution. While facing at constant feed, the cross slide is given automatic feed. The cutting action can be started either from the circumference or from the centre of the job,The cutting speed increases linearly from the centre to the circumference, as v = *r*c>. Since the surface finish depends on the cutting speed, therefore the surface finish becomes very poor as the tool approaches the centre of the job.

The surface finish shall remain constant if facing is done at constant cutting speed. For this purpose the rotational speed has to be varied to keep the cutting speed constant.

(*iv*)

$$D = 150 \text{ mm}, f = 0.25 \text{ mm/tooth}, v = 20 \text{ m/min}, z = 10$$

$$v = \pi DN, N = \frac{20}{\pi \times 0.150} = 42.44 \text{ rpm}$$

Length of work piece, l = 160 mm

Width of work piece, b = 80 mm

Extent of cutter's first contact with work piece,

$$l_c = \frac{D}{2} = \frac{150}{2} = 75 \text{ mm}$$

With the cutter centre off-set by 15 mm, number of passes of cutter

$$= (b + 2 \times \text{overhang of 5 mm}) = \frac{80 + 10}{150} = 0.6 \approx 1$$

Machining time,

$$t = \frac{l + l_c}{f\,Nz} = \frac{160 + 150}{0.25 \times 42.94 \times 10} = 2.92 \text{ min.}$$

(*v*) Length of cut $= l + l_c + 2 \times \text{Approach} + 2 \times \text{over run}$

$$= 160 + 150 + 2 \times 5 + 2 \times 5 = 330 \text{ mm}$$

$$\text{Machining time} = \frac{330}{0.25 \times 42.94 \times 10} = 3.11 \text{ min}$$

(*vi*) With the symmetrical tool axis, number of passes shall remain the same and machining time will not change.

(***b***) In the State of Bihar, in a particular region there are 5 coal mines which produce the following output at the indicated production costs:

Mine	Output metric tons/day	Production cost units of 100 Rs. per metric ton
1	120	25
2	150	29
3	80	34
4	160	26
5	140	28

Before the coal can be sold to the steel making units, it must be cleaned and graded at one of the 3 coal preparation plants. The capacities and operating cost of these 3 plants are as follows :

Plants	Capacity metric tons/day	Operating cost units of 100 Rs. per metric ton
A	300	2
B	200	3
C	200	3

All coal is transportable by rail at a cost of Rs. 50 per metric ton kilometre and the distance in kilometre from each mine to the preparation plants are indicated below :

Preparation	Distance in kilometres to Mines				
Plant	1	2	3	4	5
A	22	44	26	52	24
B	15	16	24	42	48
C	44	32	16	16	22

(*i*) Using transportation model, determine how the output of each mine should be allocated to the three preparation plants to optimize cost.

(*ii*) Are alternative approaches possible ? If so, what is the logic of 1st allocation in these alternatives ?

(*iii*) What is degeneration and when can it happen ? (20)

Sol. (*i*) Output = 650 metric tons and capacity = 700 metric tons. Therefore, a fictitious mine 6 for plant A having 50 metric ton output is added to get the solution.

Here $m = 3, n = 6, m + n - 1 = 8$

Number of allocations = 8

Therefore, the solution is optimal.

The allocations are :

1 to A = 70 m ton, 3 to A = 40 m ton, 5 to A = 140 m ton

1 to B = 50 m ton, 2 to B = 150 m ton

3 to C = 40 m ton, 4 to C = 160 m ton

From → To ↓	1	2	3	4	5	6	Capacity	Row difference					
A	22	44	26	52	24	0	300	22	22	2	2	2	7
	70		40		140	50							
B	18	16	24	42	48	0	200	16	16	2	6	6	6
	50	150											
C	44	32	16	16	22	0	200	16	16	6	6		
			40	160									
Ouput	120	150	80	160	140	140	50	700					
Column	4	16	8	26	2	0							
Diffe-	4	16	8	–	2	–							
rence	4	16	8	–	2	–							
	4	–	8	–	2	–							
	4	–	2	–	24	–							
	4	–	2	–	–	–							

(*ii*) Yes, Sixth mine can be allocated to plant *B or C* and the solution can be obtained.

(*iii*) If number of allocations are less than $(m + n - 1)$, then the solution will degenerate.

Q.7. (*a*) (*i*) What do you understand by quality control by attributes and Q.C. by variables?

(*ii*) Where do you need a larger sample size in the above and why ?

4 batches of products are inspected at 90 minute intervals and their percentage difference in weights from the nominal are given below :

Percentage difference on Weights in Samples

Batch or Sample 1	Sample 2	Sample 3	Sample 4
1.2	0.6	0.6	2.1
1.8	0.3	1.5	0.6
1.5	0.3	1.0	0.6
0.9	0.0	0.0	2.7
0.3	0.6	1.9	2.7

Assume that normal process average variation is 0.9% and process standard deviation is 0.5%.

Assume further that range U.C.L. is 5 times S.D. Plot the $\overline{X}$, R chart indicating the various control limits.

(*iii*) Is the process under control ? Any comments. (20)

Sol. (*i*) Quality control by attributes. This involves observing the presence or absence of qualitative characteristics such as external or internal defects in machined, formed or welded parts or dents in sheet metal products, in each of the units in the group under consideration.

Quality control by variables. It is the quantitative measurement of the characteristics of a part, *e.g.,* dimensions, tolerances, surface finish, or physical or mechanical properties. Such measurements are made for each of the units in the group under consideration, and the results are compared against specifications.

(*ii*) **Sample size data for attribute.** Type data is generally larger than the variable-type data on account of the qualitative characteristic of the data.

Percentage difference on Weights in Samples

Sample 1	Sample 2	Sample 3	Sample 4
1.2	0.6	0.9	2.1
1.8	0.3	1.5	0.6
1.5	0.0	1.0	0.6
0.9	0.0	0.0	2.7
0.3	0.6	1.9	2.7

Sample No.	$x = \Sigma x_i/n$	$R = x_{max} - x_{min}$
1	5.7/5 = 1.14	1.8 – 0.3 = 1.5
2	1.8/5 = 0.36	0.6 – 0.0 = 0.6
3	5.0/5 = 1.0	1.9 – 0.0 = 1.9
4	8.7/5 = 1.74	2.7 – 0.6 = 2.1

$$\overline{\overline{X}} = \frac{\sum \overline{X}}{4} = \frac{4.24}{4} = 1.06$$

$$\overline{R} = \frac{\sum R}{4} = \frac{6.1}{4} = 1.525$$

$\overline{X}$ – Chart : $\text{UCL} = \overline{\overline{X}} + A_2\overline{R}$

$$= 1 .06 + 0.5 \times 0.9 \times 1.525 = 1.746$$

$$\text{LCL} = \overline{\overline{X}} - A_2\overline{R}$$

$$= 1.06 - 0.5 \times 0.9 \times 1.525 = 0.374$$

R-chart : UCL $= D_4 \overline{R}$

$$= 5 \times 0.5 \times 1.525 = 3.8125$$

LCL $= D_3 \overline{R} = 0$

The $\overline{X}$ and R charts are shown below in Fig. 7.

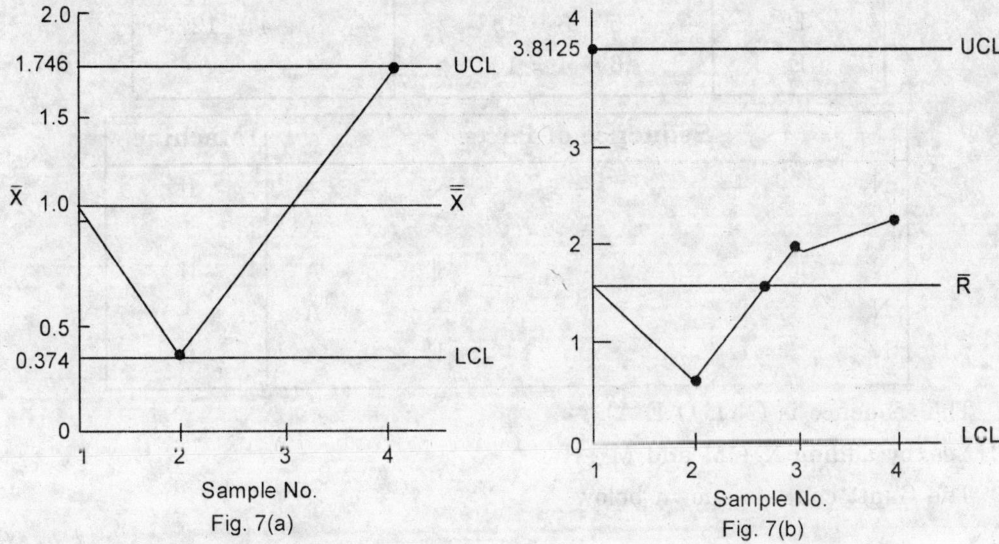

Fig. 7(a)

Fig. 7(b)

The process is under control.

(*b*) A company is producing a batch of five parts A, O, E, N and L using three machines X, W, M. All these parts are to be produced in the technological order X, M, W machines. The sum of the processing and set-up times in minutes for these parts in each of these machines is given below:

Part	Time on Machines, minutes		
	X	**W**	**M**
N	8	5	3
A	4	6	4
O	7	7	3
L	5	8	4
E	6	4	4

(i) Find the sequence in which parts are to be produced to minimize total make span.

(*ii*) Are there alternative sequences ?

(*iii*) Represent your schedule in the form of a Gantt chart with starting and ending timings. Neglect machine to machine shifting time.

(iv) What are the machine utilization percentages for each machine if no other job is taken till the completion of the batch ? (20)

Sol. (*i*) Convert the problem into a two machines problem :

Part	Machine I X + W	Machine II W + M
N	8 + 5 = 13	5 + 3 = 8
A	4 + 6 = 10	6 + 4 = 10
O	7 + 7 = 14	7 + 3 = 10
L	5 + 8 = 13	8 + 4 = 12
E	6 + 4 = 10	4 + 4 = 8

Sequence of Parts					Machine
N	–	–	–	–	II
N	–	–	–	A	I
N	–	–	E	A	II
N	L	–	E	A	I
N	L	O	E	A	II

The sequence is (N L O E A).

(*ii*) Yes, by adding X + M and M + W.

(*iii*) The Gantt chart is shown below.

(*iv*)

	X		W		M	
	In	Out	In	Out	b	Out
N	0	8	8	13	13	16
L	8	13	13	21	16	20
0	13	20	20	27	20	23
E	20	26	26	30	23	27
A	36	30	30	36	27	31

Idle time for machines

$$X = 36 - 30 = 6 \text{ min}$$

$$W = 8 + (21 - 20) + (27 - 26) = 10 \text{ min}$$

$$M = 13 + (36 - 31) = 18 \text{ min}$$

% utilization for machines :

$$X = \frac{30}{36} = 83.3\% \ 36$$

$$W = \frac{26}{36} = 72.25\% \ 36$$

$$M = \frac{18}{36} = 50\% \ 36$$

MECHANICAL ENGINEERING
PAPER II—1999

(Conventional)

SECTION-A

Q.1. (*a*) Distinguish between fatigue strength and fatigue limit.

Sol. Fatigue strength. It is the stress which a material can withstand under cyclic or repeated loading for a definite number of cycles.

Fatigue limit. It is the limiting stress below which a material will withstand repeatedly an infinite number of cycles. It is generally taken at 10^6 cycles.

(*b*) Give two examples of parts where the design criteria is rigidity and not the strength.

Sol. Those structures for which damping of vibrations is the main critorion, rigidity is the main consideration. Examples are : lathe bed, column of a radial drilling machine, or machine tool structures in general.

(*c*) A right hand helical gear is being cut on a milling machine. What changes in machine settings have to be made to cut a left hand helical gear of same pitch and number of teeth ?

Sol.

1. Change the direction of rotation of the work piece by addition of an idler gear in the milling machine gear box.
2. Moving the table of the milling machine in the opposite direction at the same helix angle.

(*d*) What is resonance ? Why is it generally avoided ?

Sol. Resonance is a phenomena in which the natural frequency of a vibrating member matches with the forcing frequency. At resonance the amplitude of vibration tends to be very large causing high stresses in the member. Generally, the members are operated much above their natural frequency to avoid resonance.

(*e*) Explain the effect of alloying Chromium and Nickel in stainless steel.

Sol.

1. Chromium added to stainless steel prevents corrosion.
2. Nickel imparts ductility to stainless steel by stabilizing austenitic structure.

(*f*) List atleast two factors that promote transition from ductile to brittle fracture.

Sol.

(*i*) High strain rate.

(*ii*) Very low temperature.

(*iii*) Presence of stress raisers causing stress-concentrations.

(*g*) Distinguish between creep and fatigue.

Sol. Creep. It may be defined as the slow and progressive deformation of a material with time under a constant stress.

Fatigue: It is the reduced tendency of a material to offer resistance to the applied stresses. This phenomenon leads to fracture under repeated or fluctuating cyclic stresses.

(*h*) Describe two distinguishing features of Robots, which make them different from conventional material handling equipments.

Sol.

(*i*) A robot is programmable.

(*ii*) It can be integrated with other manufacturing facilities.

(*i*) Explain the difference between Roughness and Waviness.

Sol. Surface roughness. It concerns all those irregularities which form surface relief and which are conventionally within the area where deviations of form and waviness are eliminated. Basically these are micro-geometrical errors of small wavelength.

Waviness. These are micro-geometrical errors of considerable wavelength of a periodic nature.

(*j*) Mention two types of dislocation.

Sol. (*i*) Screw dislocation. (*ii*) Edge dislocation.

(*k*) Give the reasons for plotting average of samples instead of individual values in $\overline{X}$ control chart.

Sol.

(*i*) Individual values are bulky and more prone to errors.

(*ii*) $\overline{X}$ follows normal distribution even though X is non-normal distribution.

(*l*) What is the direction of movement of the carriage while cutting left hand thread on a lathe?

Sol. From headstock side towards tailstock side, *i.e.,* from left to right.

(*m*) What is the principal constituents of brass ?

Sol. Copper = upto 90%

Zinc = upto 50%.

(*n*) In R.C.C., why is steel reinforcement provided at the top of the neutral axis for a cantilever beam ?

Sol. Tensile stresses are developed above the neutral axis in a cantilever beam. RCC is weak in tension and strong in compression. Therefore, steel reinforcement is provided above the neutral axis to take up tensile stresses.

(*o*) When do you prefer a belt drive to a gear drive ?

Sol. (*i*) Distance between the two shafts is comparatively large.

(*ii*) Where slip and positive transmission of motion is not the main consideration.

(*iii*) Where damping out the effects of vibrating forces is important.

(*p*) State the Kennedy's theorem with reference to instantaneous centres.

Sol. The Kennedy's theorem states that if three kinematic links have relative plane motion among themselves, their three instantaneous centres or rotation must lie on a straight line.

(*q*) State the law of gearing.

Sol. For constant angular velocity ratio of the two gears in contact, the common normal at the point of contact must always intersect the line of centres at a fixed point (pitch point) and divide this line in the inverse ratio of the angular velocities of the two gears.

(*r*) Distinguish between a jig and a fixture.

Sol. Jig: Locates and clamps the work piece, and guides the tool.

Fixture: Locates and clamps the work piece only.

(*s*) Why are hydrodynamic bearing used for grinding machine in preference to ball or roller bearings ?

Sol. Hydrodynamic bearings have better damping characteristics as compared to ball or roller bearings which helps in getting better surface finish during grinding.

(*t*) Mention the name of process by which the following are manufactured :

(i) Tungsten carbide tools (*ii*) Engine cylinder block

(*iii*) Connecting rod (*iv*) Plastic buckets. (20 × 2 = 40)

Sol.

Component	**Manufacturing method**
(*i*) Tungsten carbide tools	Powder metallurgy
(*ii*) Engine cylinder block	Casting
(*iii*) Connecting rod	Forging
(*iv*) Plastic buckets	Injection moulding

SECTION-B

Q.2. (*a*) Construct the profile of a disc cam with translating roller follower using the following data :

Base circle dia = 150 mm

Roller dia = 15 mm

Lift = 50 mm in 180° and return in 135° and dwell.

Both lift and return motion are simple harmonic. Cam rotates clockwise. (20)

Sol. Draw the displacement diagram to a scale of 1 cm = 12.5 mm, as shown in Fig. 1 (*a*). Then draw the cam profile as shown in Fig. 1 (*b*).

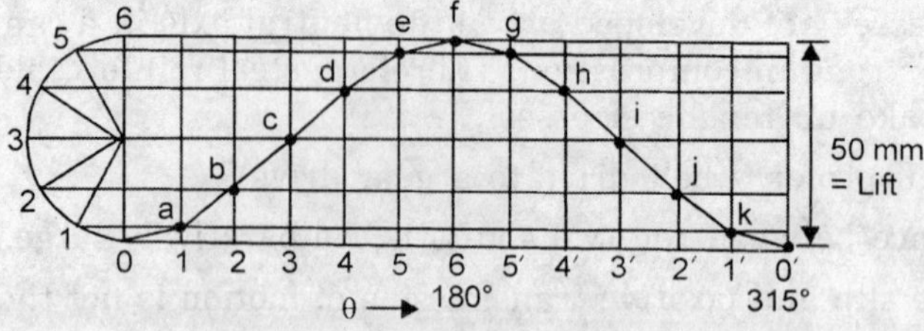

Fig. 1. (a) Displacement Diagram

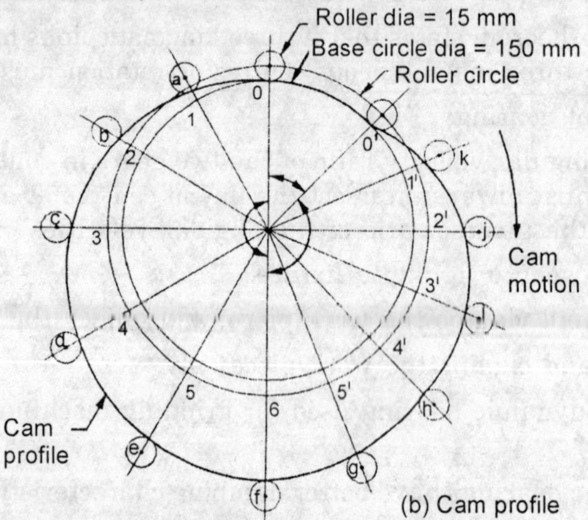

Fig. 1 (b)

(*b*) In a brick making machine, a wide roller 250 mm in dia mounted on shaft E is driven by a motor carrying a pulley A, which is 150 mm in dia. The pulley in turn drives another pulley of 1200 mm dia mounted on a shaft B by an open belt. On the same shaft B, a spur gear of 20 T is mounted , which meshes a spur gear of 160 T mounted on shaft C. The shaft C also carries a spur wheels of 20 T which drives a 30 T gear on shaft E through an idler gear of 112 T mounted on shaft D. Sketch the arrangement and determine the speed reduction. If motor rotates clockwise (viewing from right) at 1440 rpm, determine the sense of rotation of the roller. What is the surface speed of roller ? (20)

Sol. The arrangement has been sketched in Fig. 2

$$\frac{N_B}{N_A} = \frac{150}{1200} = \frac{1}{8}$$

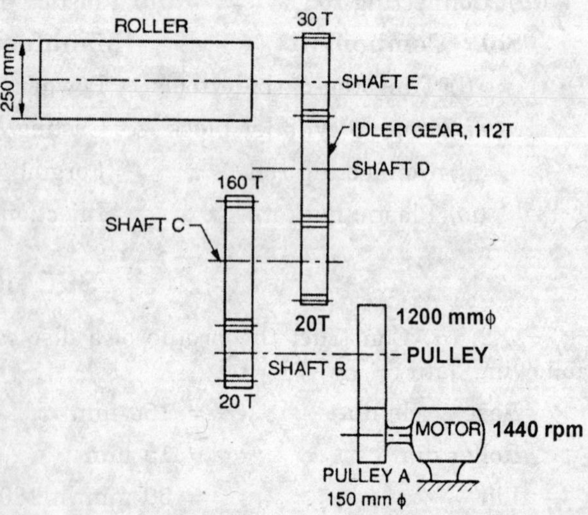

Fig. 2

$$\frac{N_C}{N_B} = \frac{z_B}{z_C} = \frac{20}{160} = \frac{1}{8}$$

$$\frac{N_E}{N_C} = \frac{20}{30} = \frac{2}{3}$$

$$\left(\frac{N_B}{N_A}\right) \times \left(\frac{N_C}{N_B}\right) \times \left(\frac{N_E}{N_C}\right) = \left(\frac{1}{8}\right) \times \left(\frac{1}{8}\right) \times \left(\frac{2}{3}\right)$$

$$\frac{N_E}{N_A} = \frac{1}{96}$$

$$\text{Speed of roller} = \frac{1440}{96} = 15 \text{ rpm}$$

$$\text{Surface speed of roller} = \pi d N_E = \pi \times 250 \times 10^{-3} \times 15 = 11.78 \text{ m/min (ccw)}$$

Q.3. (*a*) A spur steel pinion (σ_0 = 200 MN/m^2) is to drive a spur steel gear (σ_0 = 140 MN/m^2). The diameter of the pinion is to be 100 mm and the centre distance is 200 mm. The pinion is to transmit 5 kW at 900 rpm. The teeth are to be 20° full depth. Determine the necessary module and width of face to give greatest number of teeth. Design for strength only using Lewis equation.

σ_0 is the endurance strength corrected for average stress concentration ?

The allowable stress σ is given by

$$\sigma = \sigma_0 \left[\frac{3}{3 + v}\right]$$

where v is the pitch line velocity in m/s. The following table gives the form factor y for use in Lewis strength equation.

Table 1-Form Factory *y*-For use in Lewis strength equation.

Number of teeth	14.5° Full-Depth involute	20° Full-depth or composite involute	20° sub-involute
12	0.067	0.078	0.099
13	0.071	0.083	0.103
14	0.075	0.088	0.108
15	0.078	0.092	0.111
16	0.081	0.094	0.115
17	0.084	0.096	0.117
18	0.086	0.098	0.120
19	0.088	0.100	0.123
20	0.090	0.102	0.125
21	0.092	0.104	0.127
23	0.094	0.106	0.130

25	0.097	0.108	0.133
27	0.099	0.111	0.136
30	0.101	0.114	0.139
34	0.104	0.118	0.142
38	0.106	0.122	0.145
43	0.108	0.126	0.147
50	0.110	0.130	0.151
60	0.113	0.134	0.154
75	0.115	0.138	0.158
100	0.117	0.142	0.161
150	0.119	0.146	0.165
300	0.122	0.150	0.170
Rack	0.124	0.154	0.175

Sol. Let,

$$m = 5 \text{ mm}$$

$$z_p = \frac{d_p}{m} = \frac{100}{5} = 20$$

$$C = 0.5m\,(z_p + z_g)$$

$$200 = 0.5 \times 5\,(20 + z_g)$$

$$z_g = 60$$

$$v = \frac{\pi \times 100 \times 900}{60 \times 10^3} = 4.7124 \text{ m/s}$$

	σ	*y*	σ*y*
Gear	140	0.134	18.76
Pinion	200	0.102	20.40

Therefore gear is the weaker, and need to be designed.

$$P = \frac{F_t v}{10^3}$$

$$F_t = \frac{5 \times 10^3}{4.7124} = 1061 \text{ N} = \pi\,\sigma\,b\,y\,m$$

$$1061 = \pi \times 140 \times \left[\frac{3}{3 + 4.7124}\right] \times b \times 0.134 \times 5$$

$$b = 9.256 \cong 10 \text{ mm}$$

(*b*) A100 mm dia shaft operating at 200 rev /min is supported by means of a 150 mm long full journal bearing which is subjected to a radial load of 43 kN. Assume

$$\left[\frac{\mu N}{p}\right] = 30 \times 10^{-6}$$

Determine :

(*i*) the co-efficient of friction using McKee equation which is given by

$$f = 0.326\left[\frac{\mu N}{p}\right]\frac{D}{C} + K$$

Take D/C = 100, K = 0.002.

(*ii*) Bearing pressure in MN/m^2, and

(*iii*) Heat generated. (10)

Sol. (*i*) Co-efficient of friction,

$$\mu = 0.326 \times 30 \times 10^{-6} \times 100 + 0.002 = 0.002978$$

(*ii*)
$$p = \frac{W}{ld} = \frac{43 \times 10^3}{0.150 \times 0.100} = 2.867 \text{ MPa}$$

(*iii*)
$$H_g = \mu pldU = 0.002978 \times 2.867 \times 150 \times 100$$
$$\times\left(\pi \times \frac{0.100 \times 200}{60}\right) = 134 \text{ Watts}$$

Q.4. (***a***) What are principal planes ? (5)

Sol. Principal planes are those planes in a stressed body on which there are normal stresses only and the shear stress is zero. It is always possible to find three mutually perpendicular planes on which the normal stress is either maximum or minimum.

(***b***) Direct tensile stresses of 120 MN/m^2 and 70 MN/m^2 act on a body on mutually perpendicular planes. What is the magnitude of shearing stress that can be applied, so that the major principal stress at the point does not exceed 135 MN/m^2 ? Determine the value of minor principal stress and the maximum sheer stress. (15)

Sol.
$$\sigma_1 = 0.5\,(\sigma_x + \sigma_y) + 0.5\,[(\sigma_x - \sigma_y)^2 + 4\,\tau^2_{xy}]^{0.5}$$
$$135 = 0.5\,(120 + 70) + 0.5\,[(120 - 70)^2 + 4 \times \tau^2_{xy}]^{0.5}$$
$$\tau_{xy} = 31.225 \text{ MPa}$$
$$\sigma_2 = 95 - 0.5\,[2500 + 3900] = 95 - 40 = 55 \text{ MPa}$$

$$\text{Maximum shear stress} = \frac{(\sigma_1 - \sigma_2)}{2} = \frac{(135 - 55)}{2} = 40 \text{ MPa}$$

(***c***) State the limitation of Euler's formula for calculating critical load on columns. (5)

Sol. (*i*) The column in initially exactly straight.

(*ii*) The load is truly axial.

(*iii*) The slenderness ratio (ratio of length to least radius of gyration is more than 120, *i.e.,* the column is long.

(*iv*) The direct compressive stress is not accounted for.

(***d***) A hollow cast iron column of 300 mm external dia and 220 mm internal dia is used as a column 4 m long with both ends hinged. Determine safe compressive load the column can carry without buckling using Euler's formula and Rankine's formula.

E = 0.7 × 10^5 N/mm^2, Factor of safety = 4, Rankine constant,

$$a = \frac{1}{1600},$$

Crushing stress, $\sigma_c = 567 \text{ N/mm}^2.$ (15)

Sol.

$$I = \frac{\pi\left[(0.3)^4 - (0.22)^4\right]}{4} = 2.8262 \times 10^{-4} \text{ m}^4$$

$$A = \frac{\pi\left[(0.3)^2 - (0.22)^2\right]}{64} = 326.7 \times 10^{-4} \text{ m}^2$$

$$K^2 = \frac{I}{A} = 8.65 \times 10^{-3} \text{ m}^2$$

Euler's buckling load,

$$P_e = \frac{\pi^2 EI}{\ell^3 \times \text{FOS}} = \frac{\pi^2 \times 70 \times 10^9 \times 2.8262 \times 10^{-4}}{16 \times 4} = 3050.85 \text{ kN}$$

Rankine's load,

$$P = \frac{\sigma_c \dfrac{A}{\text{FOS}}}{\left[1 + a\left(\dfrac{\ell}{K}\right)^2\right]} = \left(\frac{567}{4}\right) \times \frac{326.7 \times 10^{-4}}{\left[1 + \left(\dfrac{1}{1600}\right)\left(\dfrac{16}{8.65 \times 10^{-3}}\right)\right]}$$

$$= 2178 \text{ kN}$$

SECTION-C

Q.5. (*a*) A medium force fit H7p6 is specified for an assembly of a hand wheel on a shaft of 50 mm diameter. Calculate

(i) the dimensions of the hole,

(*ii*) the dimensions of the shaft,

(*iii*) maximum interference, and

(*iv*) minimum interference.

Given the following :

Nominal	Upto including	Limits of tolerance for p6 shaft in size and microns	
		e_s	e_i
		+	+
40	50	42	51
50	65	26	32
		Limits of tolerance for H7 hole in microns	
		E_s	E_i
30	50	25	0
50	80	30	0

(10)

Sol. *(i)* $\quad$ + 0.25

Hole size – 50 H 7 : $\quad$ 50

(ii) $\quad$ + 0.000

+ 0.042

Shaft size – 50p6 : $\quad$ 50

+ 0.026

(iii) Maximum interference = Maximum size of shaft-Minimum size of hole

= 50.042 – 50.000 = 0.042 mm

Minimum interference = Minimum size of shaft -Maximum size of hole

= 50.026 – 50.025 = 0.001 mm.

Q.5. (*b*) Explain TIG (Tungsten inert gas welding) with reference to *(i)* Principle of the process *(ii)* Equipment used *(iii)* Applications. (10)

Sol. *(i)* It is an arc welding process in which the arc is maintained between a non-consumable tungsten electrode and the work piece in a protective inert gas atmosphere. Argon is most commonly used to provide the inert atmosphere. Any filler material needed is supplied externally,

(ii) The equipment used for TIG welding is shown schematically in the following Fig. 3.

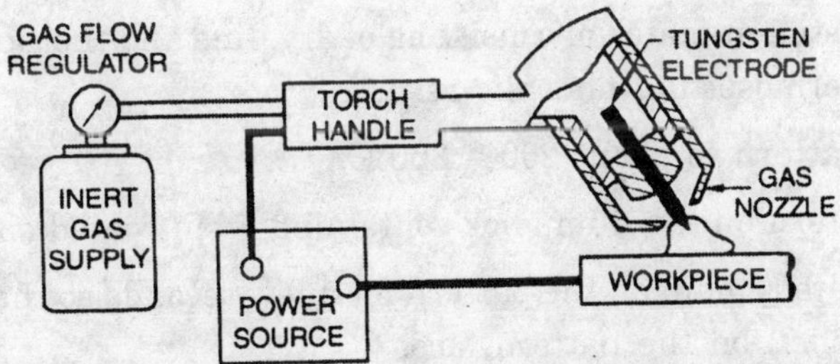

Fig. 3 TIG welding equipment.

The various components of the equipment are :

Welding torch, tungsten electrode, filler metal, power source, high frequency D.C. unit, cables, inert gas cylinder, pressure regulator, flow meter.

(iii) Welding of aluminium, magnesium and other reactive metals.

Q.5. (*c*) Explain briefly how parts are manufactured by shell moulding giving an example. (10)

Sol. The shell moulding process involves the use of match plate patterns similar to green sand casting. The process is based on the capability of thermosetting resin and sand mixture to assume the shape of a preheated metal pattern and to form a dense, quickly hardened shell mould. The two matching shell halves are clamped or glued together to form the mould cavity. Castings are obtained by pouring molten metal into this mould cavity. The mould must be broken away to facilitate the removal of the solidified casting as shown in figure. The process involves the following steps : (Fig. 4).

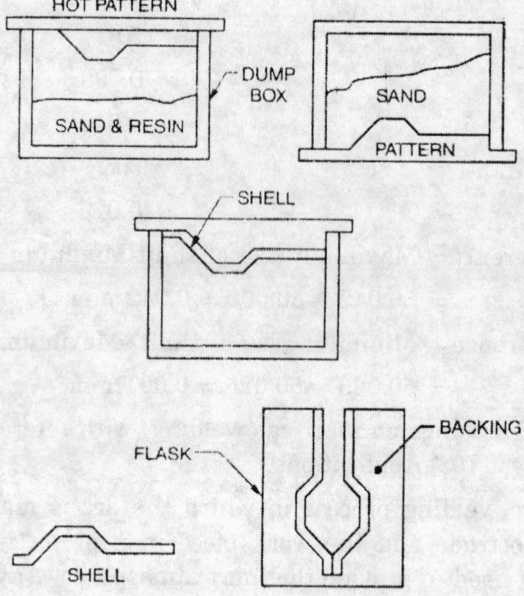

Fig 4 Shell moulding process.

1. Preparation of metal pattern.
2. Mixing of investment material consisting of dry, fine and thoroughly washed silica and 5 – 10% of thermosetting phenolic resin.
3. Heating the pattern to about 200 – 300°C.
4. Fixing the pattern on the dump box containing resin coated silica and then inverted.
5. Allowing silica to remain on the pattern for a few seconds so that resin sets in forming a semi-cured shell on the pattern face.
6. Removing the pattern plate and placing in a curing oven for about 1 – 3 minutes at a temperature of about 250 – 450°C.
7. Removing the shell from the pattern plate by the use of ejector pins.
8. Repeating steps 2 – 7 for the other half of the shell.
9. Assembling the shells by clamping or bonding together at the parting line.
10. Pouring the mould.
11. Removing the casting by breaking the shell.

Q.5. **(*d*)** With reference to cubic crystal, sketch the following:

(111), [111]. (10)

Sol. (111) is the plane ABC.

[111] is the direction equally inclined to the three planes.

The sketch is shown below in Fig. 5.

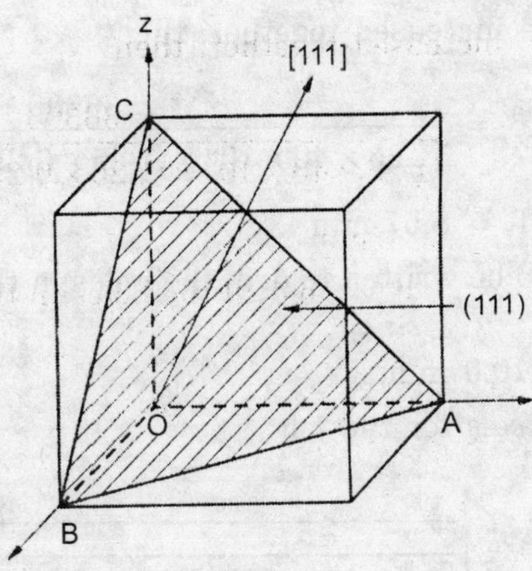

Fig. 5

Q.6. (***a***) The following equation for tool life was obtained for H.S.S. tool :

$$v\text{T}^{0.13}\, f^{0.6}\, d^{0.3} = \text{C}.$$

A 60 min tool life was obtained using the following cutting conditions :

$$v = 40 \text{ m/min}, f = 0.25 \text{ mm}, d = 2.0 \text{ mm}$$

Calculate the effect on tool life if speed, feed and depth of cut are together increased by 25% and also if they are increased individually by 25%; where f = feed, d = depth of cut, v = speed. **(10)**

Sol. (*i*) $(40 \times 10^3) \times 60^{0.13} \times 0.25^{0.6} \times 2^{0.3} = \text{C}$

$$\text{C} = 36500$$

(*ii*) When f, d and v are increased individually :

1. v increased by 25%.

$$\text{T}^{0.13} = \frac{36500}{[1.25 \times 40 \times 10^3 \times (0.25)^{0.6} \times (2)^{0.3}]} = 1.36223$$

$$\text{T} = 10.48 \text{ min}$$

2. f is increased by 25%

$$\text{T}^{0.13} = \frac{36500}{[40 \times 10^3 \times (1.25 \times 0.25)^{0.6} \times (2)^{0.3}]} = 1.4894$$

$$\text{T} = 11.46 \text{ min}$$

3. d is increased by 25%

$$\text{T}^{0.13} = \frac{36500}{[40 \times 10^3 \times (0.25)^{0.6} \times (1.25 \times 2)^{0.3}]} = 1.5925$$

$$\text{T} = 12.25 \text{ min}$$

(*iii*) When f, d and v are increased together, then

$$T^{0.13} = \frac{36500}{[1.25 \times 40 \times 10^3 \times (1.25 \times 0.25)^{0.6} \times (1.25 \times 2)^{0.3}]} = 1.11437$$

$$T = 8.57 \text{ min}$$

(*b*) A part program is to be written to drill the holes in the work part shown below. The part is 10 mm thick.

The dia of drill is ϕ = 10.0 mm.

(*i*) Define the x and y axes for the job.

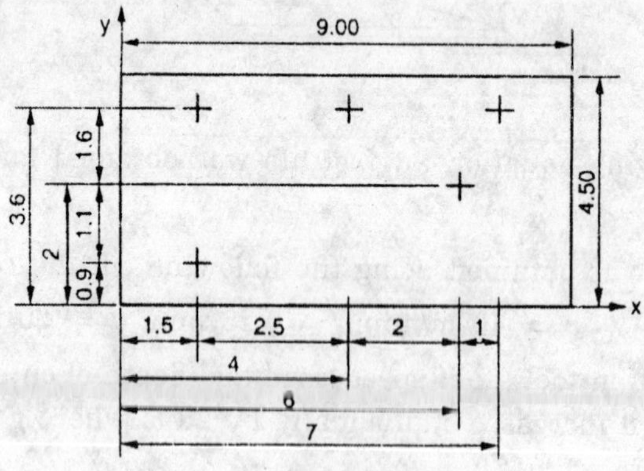

Fig. 6

(*ii*) Write manual part programming using word address format and an absolute position system. The words that must be specified for the particular drill press are $n - x - y$ and $m-$ words. The speed and feed are manually set by the operator. (10)

Sol. (*i*) The x and y axes are shown in Fig. 6.

(*ii*) n 001 x 0000 y 0000

n 002 x 1500 y 9000 m 03

n 003 x 1500.y 3600 m 03

n 004 x 4000 y 3600 m 03

n 005 x 6000 y 2000 m 03

n 006 x 7000 y 3600 m 03

n 007 x 6500 y 2000 m 05

n 008 x 0000 y 5000 m 30

(*c*) Define an Industrial robot. Sketch four important basic configurations of robots used. (10)

Sol. A robot is a programmable multi-functional manipulator designed to move materials, parts tools or specialized manufacturing implements through variable programmed motions for the performance of specific manufacturing tasks. An industrial robot has grippers attached which can grip or pick up various objects. They are used to pick and place work pieces from one place to another. The four basic configurations of a robot are shown below in Fig. 7.

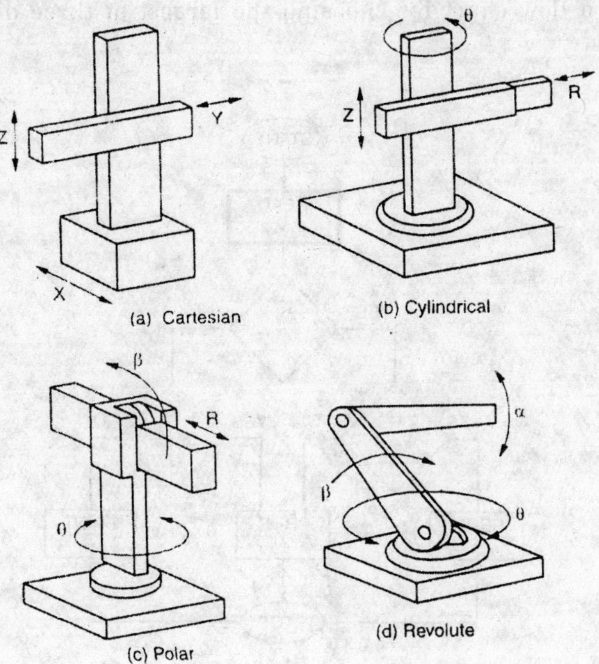

Fig. 7 Robot classification.

(*d*) Differentiate between Gear hobbing and Gear shaping with reference to various relative motions and applications. (10)

Sol. Gear hobbing. It is a generating process for gears. The hob is fed into the gear blank to the proper depth and the two are rotated together as if in mesh. The teeth of the hob cut into the work piece in a successive order and each in a slightly different position. Each hob cuts its own profile depending on the shape of the cutter, but the accumulation of these straight cuts produces a curved form of the gear tooth.

The gear cutting with a hob involves three basic motions, all of them occurring at a time. The hob and the blank have a rotating motion and the third one is the radial advancement for the hob, thus causing the cutting and indexing simultaneously.

Hobbing machines can produce spur, helical and herringbone gears, splines and gear sprockets. It cannot cut bevel and internal gears.

Gear shapping. It is also a generating method for gears. The cutter is reciprocated with its required cutting speed for stock removal along the face of the work piece, and is gradually fed radially to plunge it for correct teeth depth by means of a cam. A relieving mechanism enables the cutter to clear the work on non-cutting or return stroke. The continuous generation motion is obtained by feeding cutter to full depth and rotating the cutter and work piece slowly in the exact ratio of their respective number of teeth by means of a chain of gears.

Both internal and external gears can be cut by gear shaping. Non-conventional type of gears can also be cut. However, worm and worm gears cannot be generated by this process.

Q.7. (*a*) Write a flow chart for choosing the largest of three distinct numbers x, y, z. (10)

Sol.

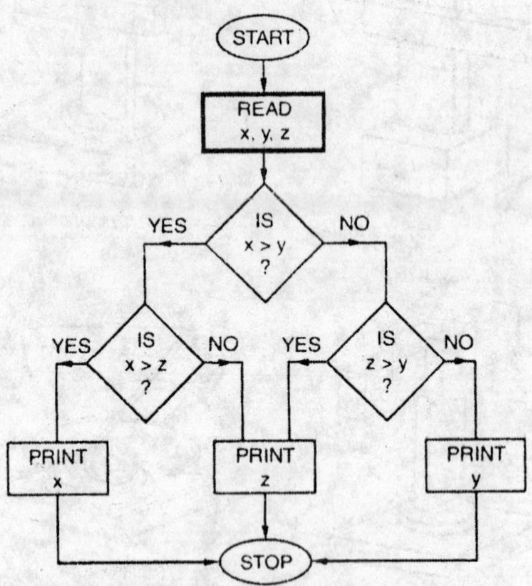

Fig. 8. Flow chart for choosing the largest of three numbers.

(*b*) What is inventory ? A company uses a certain component X at the rate of 500/year. The cost/item is Rs. 20 and it costs Rs. 200 to place an order. The annual carrying cost of inventory is 10% of the price of the item.

Storage cost is negligible. Assume zero safety stock. Calculate the Economic order quantity (EOQ). (10)

Sol. Inventory refers to any kind of resource having economic value and is maintained to meet the present and future needs of an organization.

$$D = 5000\text{/year}, \; C = \text{Rs. } 20\text{/item}$$

$$C_o = \text{Rs. } 200\text{/order}, \; C_c = 10\%$$

$$EOQ = \left[\frac{2DC_o}{CI}\right]^{0.5}$$

$$= \left[\frac{2 \times 5000 \times 200}{20 \times 0.1}\right]^{0.5} = 1000 \text{ items}$$

(*c*) Sub-groups of 5 items each are taken from a manufacturing process at regular intervals.

A certain quality characteristic is measured, and $\overline{X}$ and $\overline{R}$ values are computed for each sub-group. After 25 sub-groups $\sum \overline{X} = 357.50, \sum R = 8.80$. Compute control chart limits. All points on both charts fall within these limits. If the specification limits are 14.40 ± 0.40, what conclusions can you draw about the ability of the existing process to produce items within these specifications ? Suggest possible ways in which the situation could be improved.

$$A_2 = 0.58, \; D_3 = 0, \; D_4 = 2.11, \; d_2 = 2.326. \qquad (10)$$

Sol.
$$\bar{\bar{X}} = \frac{\Sigma \bar{X}}{25} = \frac{357.50}{25} = 14.3$$
$$\bar{\bar{R}} = \frac{\Sigma \bar{R}}{25} = \frac{8.80}{25} = 0.352$$

$\bar{\bar{X}}$-control chart :
$$UCL = \bar{\bar{X}} + A_2 \bar{R} = 14.3 + 0.58 \times 0.352 = 14.504$$
$$LCL = \bar{\bar{X}} - A_2 \bar{R} = 14.3 - 0.58 \times 0.352 = 14.096$$

R-control chart :
$$UCL = 2.11 \times 0.352 = 0.743$$
$$LCL = D_4 \bar{R} = 0$$
$$\sigma = \frac{\bar{R}}{d_2} = \frac{0.352}{2.326} = 0.151$$

Upper specification limit, USL : 14.40 + 0.40 = 14.80

Lower specification limit, LSI = 14.40 – 0.40 = 14.00
$$6\sigma = 6 \times 0.151 = 0.908$$
$$USL - LSL = 14.80 - 14.00 = 0.80$$

Process capability, $6\sigma > (USL - LSL)$

Hence, production of defectives is inevitable.

(*d*) Solve the following linear programming problem graphically :
$$2x_1 + 3x_2 \leq 6$$
$$x_1 + 4x_2 \leq 4$$
$$x_1, x_2 \geq 4$$
$$\text{Max Z} = x_1 + 1.5\, x_2$$
(10)

Sol. The constraints are shown graphically in Fig. 9.

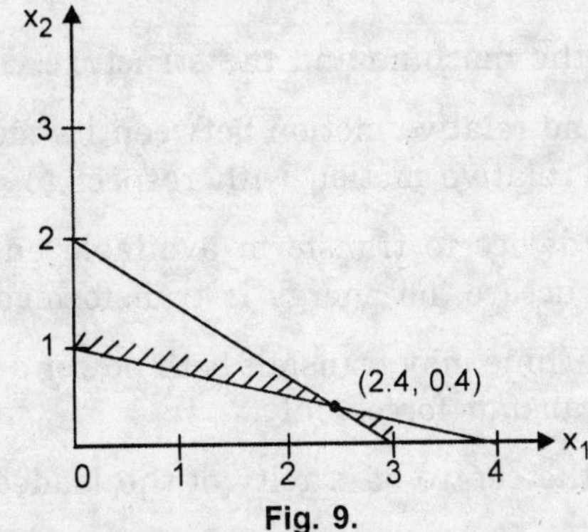

Fig. 9.

The bound points, are : (3,0), (2.4,0.4), (0,1)
$$Z_1 = 3,\ Z_2 = 3,\ Z_3 = 1.5$$
$$Z_{max} = 3$$

MECHANICAL ENGINEERING
PAPER II—2000

(Conventional)

SECTION-A

Q.1. ***(a)*** Name two situations where worm-worm gear drive is preferred.

Sol. 1. Where high reduction in speed ratio is required.

2. To transmit power between two non-intersecting shafts, generally at right angle.

(b) Write the expression for secondary unbalanced force in case of a single cylinder internal combustion engine. How many times this secondary unbalanced force reaches the maximum in one revolution of the crank ?

Sol. Secondary unbalanced force,

$$F_s = \frac{R\, r\, \omega^2 (\cos 2\theta)}{n}$$

where $n = l/r$, l = length of connecting rod, r = crank radius,

θ = crank angle,

R = mass of reciprocating parts

F_s is maximum when $\cos 2\theta = 1$

In one revolution of crank, angle turned through, $\theta = 360°$. Hence $\theta = 0°, 90°, 180°, 270°, 360°$, *i.e.*, F_s is maximum 5 times.

(c) Distinguish between the machine and the structure.

Sol. 1. A structure has no relative motion between its members, whereas members of a machine have relative motion with respect to each other.

2. A machine is a device to transform available energy into some useful work, whereas in a structure, no energy is transformed.

3. The links of a machine may transmit both power and motion, whereas members of a structure transmit forces only.

(d) List two reasons for the centre of gravity of the loaded shaft is displaced from the axis of rotation.

Sol. 1. Due to own weight and external loads acting on it.

2. Due to centrifugal forces while in rotation.

(*e*) What is the amount of clearance and the need for clearance between nominal diameter of the rivet and rivet hole?

Sol. The amount of clearance between the rivet and the rivet hole depends upon the material of the rivet and the purpose of the riveted joint. Generally the clearance lies between 1 mm to 2 mm. The clearance is provided to accommodate the lateral expansion of the rivet when head is formed on the tail end side of the rivet.

(*f*) Why is taper provided in a cotter ? What is the normal value of taper in cotter joint?

Sol. Taper on the cotter is provided to achieve self locking of the joint by the cotter. It also facilitates in making the disassembly of the joint easily. A taper of 1 : 24 is generally provided.

(*g*) At a point in a two-dimensional stress system the normal stresses on two mutually perpendicular planes are σ_x and σ_y and the shear stress is τ_{xy}. At what value of shear stress, one of the principal stress will become zero?

Sol. $$\sigma_{1,2} = 0.5\,(\alpha_x + \sigma_y) \pm 0.5\,[(\sigma_x - \sigma_y)^2 + 4\,\tau^2_{xy}]^{0.5}$$

For $$\sigma_2 = 0$$

$$(\sigma_x + \sigma_y)^2 = (\sigma_x - \sigma_y)^2 + 4\,\tau^2_{xy}$$

or $$\sigma_x\,\sigma_y = \tau^2_{xy}$$

or $$\tau_{xy} = \sqrt{\sigma_x\,\sigma_y}$$

(*h*) The ratio of inside to outside diameter of a hollow shaft is 0.6. If there is a solid shaft with same torsional strength, what is the ratio of the outside diameter of hollow shaft to the diameter of the equivalent solid shaft ?

Sol. For the hollow shaft,

$$\tau_h = \frac{16\,\mathrm{T}}{\pi d_0^3\left[1 - \left(\frac{d_i}{d_o}\right)^4\right]}$$

$$\tau_h = \frac{16\mathrm{T}}{\pi d^3{}_0\,\{1 - (0.6)^4\}} = \frac{16\mathrm{T}}{\pi d^3{}_0 \times 0.8704}$$

For the solid shaft,

$$\tau_s = \frac{16\mathrm{T}}{\pi d_s^3}$$

For equal torsional strength,

$$\tau_h = \tau_s$$

$$\frac{1}{d^3{}_0 \times 0.8704} = \frac{1}{d_s^3}$$

or $$\frac{d_o}{d_s} = \left(\frac{1}{0.8704}\right)^{1/3} = 1.047$$

(*i*) Distinguish between a beam and a short column. List two differences.

Sol. 1. A beam is a structural member whose length is very large as compared to the other two dimensions (width and thickness), whereas a short column is one whose slenderness ratio (L/K_{least}) is less than 32.

2. A beam supports transverse loads, whereas a column supports axial loads.

(*f*) Name the strongest and weakest type of atomic bonds.

Sol. Strongest atomic bond : Ionic bond

Weakest atomic bond : Dipole bond.

(*k*) Calculate the number of atoms, from first principles, per unit cell of F.C.C. and B.C.C. structure.

Sol. F.C.C. : $N = 6 \times \frac{1}{2} + 8 \times \frac{1}{8} = 4$

B.C.C : $N = 1 + 8 \times \frac{1}{8} = 2$

(**i**) List two effects of Manganese in plain carbon steels.

Sol. 1. Increases tensile strength.

2. Increases hardenability.

(*m*) Briefly explain the characteristic feature of stretch forming.

Sol. 1. Simultaneous stretching and bending.

2. No spring back.

(*n*) List four parameters to be considered in the gating design for moulds.

Sol. 1. Distribution of molten metal in the mould cavity at the proper rate.

2. Temperature loss.

3. Turbulence.

4. Entrapping gases and slags.

(*o*) Name two criterias for cutting tool life.

Sol. 1. Wear on the flank and rake surfaces.

2. Cutting speed.

(*p*) Classify the following cost elements under two types for inventory control:

(*i*) Cost of material handling.

(*ii*) Cost of inspection upon receiving.

(*iii*) Cost of advertising to select vendor.

(*iv*) Pilferage.

Sol. Carrying cost : (*i*), (*ii*), and (*iv*)

Ordering cost : (*iii*).

(*q*) Certain spare parts can be purchased at two different rates R_1 and R_2, with a boundary volume V_3. If the parts are to be manufactured, the total cost data linearly varying with volume is available.

With a break-even chart, show how the decision to buy / make can be made.

Assume $V_1 < V_3 < V_2$ and $R_1 > R_2$ where V_1 and V_2 are the break-even volumes for R_1 and R_2 respectively.

Sol. The break-even chart is shown in the Fig. 1.

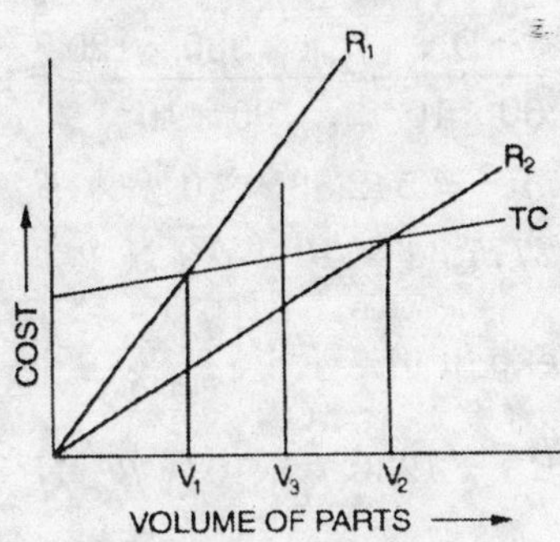

Fig. 1.

(*r*) What is the approach while allocating quantities in cells, using Vogel approximation method for solving Transportation problems ?

Sol. Allocate as much as we can in the cells where the cost is minimum in a row or a column of having maximum penalty.

(*s*) What is the standard deviation of the project completion time along the critical path, if the standard deviations of the corresponding activities are S_1, S_2 and S_3 ?

Sol. Standard deviation of the project completion time,

$$\sigma = [S^2{}_1 + S^2{}_2 + S^2{}_3]^{0.5}$$

(*t*) Name four parts of feeding devices in automation. (2 × 20 = 40)

(*i*) Spindle drive, (*ii*) Cam shaft, (*iii*) Cam, and (*iv*) Feed hopper.

SECTION-B

Q.2. (*a*) A rope pulley is designed to transmit 30 kW. Diameter of pulley = 360 mm, Speed = 120 rpm. Angle of groove = 45°, Angle of lap on smaller pulley = 170°, Co-efficient of friction = 0.27, Number of ropes = 10, Mass of the rope = 55 C^2 kg/m length and working tension of rope is limited to 125 C kN, where C is the circumference of rope in meter.

Find (*i*) the initial tension and (*ii*) diameter of each rope. (20)

Sol.

$$P = 30 \text{ kW}, D = 360 \text{ mm}, N = 120 \text{ rpm}, 2\beta = 45°$$

$$\theta = 170°, \mu = 0.27$$

$$n = 10, m = 55 \text{ C}^2 \text{ kg/m}, F_w = 125 \text{ C}^2\text{kN}$$

$$\mu_e = \frac{\mu}{\sin\beta} = \frac{0.27}{\sin 22.5°} = 0.70554$$

$$\exp(\mu_e\theta) = \exp\left(\frac{0.70554 \times \pi \times 170}{180}\right) = 8.1123$$

Let d_r = diameter of rope, mm

$$C = \pi d_r \times 10^{-3} \text{ m}$$

$$m = 55 \times \pi^2 d^2_r \times 10^{-6} = 542.83\, d^2_r \times 10^{-6} \text{ kg/m}$$

$$F_w = 125\, \pi^2 d^2_r = 1233.7\, d^2_r \times 10^{-6} \text{ kN}$$

$$= 1233.7 \times 10^{-3}\, d^2_r \text{ N}$$

$$v = \frac{\pi DN}{60 \times 10^3} = \frac{\pi \times 360 \times 120}{60 \times 10^3} = 2.262 \text{ m/s}$$

$$F_c = mv^2 = 542.83 \times 10^{-6} \times d^2_r \times (2.262)^2$$

$$= 2777.34 \times 10^{-6}\, d^2_r \text{ N}$$

$$\frac{F_1 - F_c}{F_2 - F_c} = \exp(\mu_e \theta)$$

$$F_1 = F_w = 1233.7 \times 10^{-3}\, d^2_r \text{ N}$$

$$P = (F_1 - F_2)\frac{vn}{10^3}$$

Substituting above, we get

$$F_2 = 1\,54.52 \times 10^{-3}\, d^2_r \text{ N}$$

Substituting the values, we get

$$d_r = 35 \text{ mm}$$

$$m = 0.665 \text{ kg/m}$$

$$W_r = mg = 6.523 \text{ N/m}$$

$$F_1 = 1511.3 \text{ N},\ F_2 = 189.3 \text{ N}$$

Initial tension, $F_i = \dfrac{F_1 + F_2}{2} = 850.3 \text{ N}$

(*b*) In an epicyclic gear train of the sun and planet type shown in Fig. 2, the pitch circle diameter of the internally toothed rings is 252 mm and the module is 3.5 mm. The ring D is stationary. The spider A, which carries three planet wheels P of equal size is to make one revolution in the same sense as the sun wheel S for every five revolutions of the driving spindle carrying the sun wheel S. Determine appropriate number of teeth for all the wheels. (20)

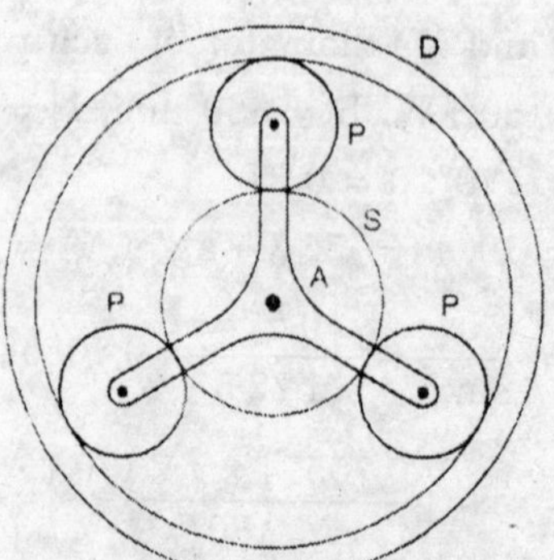

Fig. 2

Sol.

Operation	Spider A	Sun gear S	Planet gear P	Ring D
1. Spider A fixed, + 1 rev to gear S, ccw.	0	+ 1	$-z_s/z_p$	$-z_s/z_d$
2. Multiply by x	0	$+x$	$-x\,z_s/z_p$	$-x\,z_s/z_d$
3. Add y	y	$x+y$	$-xz_s/z_p+y$	$-xz_s/z_p+y$

$$y = x + 1$$

$$x + y = 5$$

$$x = 4$$

Ring gear D is stationary, $-\dfrac{xz_s}{z_d} + y = 0$

$$-\frac{4\,z_s}{z_d} = 1 = 0$$

$$\frac{z_s}{z_d} = \frac{1}{4}$$

$$d_d = 252 \text{ mm}, \; m = 3.5 \text{ mm}$$

$$z_d = \frac{252}{3.5} = 72$$

$$z_s = \frac{72}{4} = 18$$

$$z_s = 2\,z_p = z_d$$

$$z_p = \frac{(72-18)}{2} = 27$$

Q.3. (*a*) A hollow shaft with diameter ratio 0.7 is required to transmit 500 kW at 300 rpm with a uniform twisting moment. Allowable shear stress in the material is 65 MPa and the twist in a length of 2.4 m is not to exceed one degree. Calculate the minimum external diameter of the shaft satisfying these conditions. Modulus of rigidity = 8.2×10^4 N/mm^2. (20)

Sol.

$$P = \frac{2\pi\,NT}{60 \times 10^3}$$

$$T = \frac{500 \times 60 \times 10^3}{2\pi \times 300} = 15915 \text{ N.m}$$

$$J = \pi \frac{\left(d^4{}_o - d^4{}_i\right)}{32} = \pi \frac{\pi d^4{}_0 \,[1-(0.7)^4]}{32} = 0.0746\ d^4{}_o\ \text{m}^4$$

$$\frac{T}{J} = \frac{2\,\tau_{max}}{d_o}$$

$$15916 = 2 \times 65 \times 106 \times 0.0746\ d^3{}_0$$

$$d_o = 137.5 \text{ mm}$$

Minimum outside diameter of shaft = 137.5 mm

(*b*) For a symmetrical tangent cam with a roller follower, the least radius of the cam is 25 mm and the roller radius is 18 mm. The angle for outward movement is 90° and the total lift is 20 mm. The cam shaft runs at 900 rpm. Determine (*i*) principal dimensions of the cam, (*ii*) the acceleration of the follower at the beginning of the lift, where the straight flank merges into the circular nose, (*iii*) acceleration of the follower at the apex of the circular nose, i.e., when the angle turned by cam measured from the position where the roller is at the top of the nose, is zero. Assume there is no dwell between outward and inward travel of the follower. (20)

Sol. $r_1 = 25$ mm, $r_3 = 1\ 8$ mm, $\alpha = 90°$,

Lift = 20 mm, N = 900 rpm.

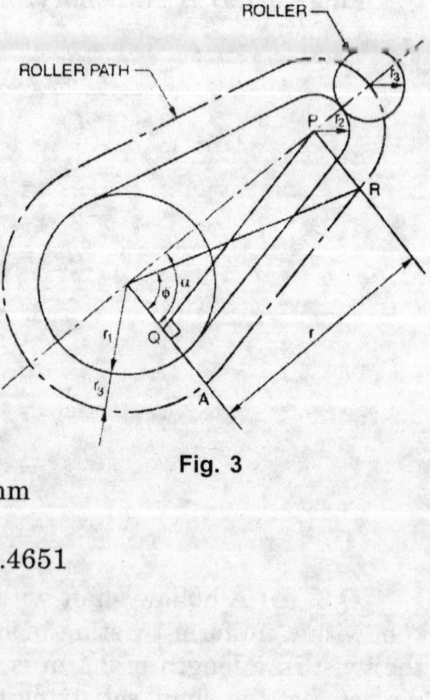

(*i*) $$\omega = \frac{2\pi \times 900}{60} = 94.26 \text{ rad/s}$$

$$OP + r_2 = r_1 = \text{Lift}$$

$$OP = 25 + 20 - r_2 = 45 - r_2$$

$$OQ + r_2 = r_1$$

$$\cos\alpha = \frac{OQ}{OP} = \frac{25 - r_2}{45 - r_2}$$

$$= \cos 90° = 0$$

$$r_2 = 25 \text{ mm}$$

$$r = OP = 45 - 25 = 20 \text{ mm}$$

$$OA = r_1 + r_3 = 25 + 18 = 43 \text{ mm}$$

$$\tan\phi = \frac{OP \sin\alpha}{OA} = \frac{20 \sin 90°}{43} = 0.4651$$

$$\phi = 24.944°$$

Fig. 3

(*ii*) Acceleration, $$a = \omega^2 (r_1 + r_3)\left[\frac{2-\cos^2\phi}{\cos^3\phi}\right]$$

$$(94.248)^2\,(43 \times 10^{-3})\left[\frac{2-\cos^2 24.944°}{\cos^3 24.944°}\right]$$

(*iii*) $$L = r_2 + r_3 = 43 \text{ mm},$$

$$r = r_2 = 25 \text{ mm}$$

Q.4. **(*a*)** Explain the functions of the four components of Flexible Manufacturing System. Indicate the situation where FMS is preferred as compared to Transfer lines. (10)

Sol. The four major components of a FMS are :

1. Machine tools.
2. Control system.
3. Material handling system.
4. Human operators.

1. Machine tools. The machine tools used include lathes, milling machines, drilling machines, and so on. The type of machine tools actually included in an FMS depends on the setting in which the system will be used. The various machine tools are :

FMS workstations : CNC machine tools :

- Machining centres : Automatic tool changing and tool storage, palatalized work parts.
- CNC control and capacity for DNC control.
- Head changers and Head indexers.
- Milling modules.
- Turning modules.
- Assembly stations.

2. Control systems. The functions performed by the FMS computer control system are:

- Control of each workstation.
- Distribution of control instructions to workstations.
- Production control.
- Traffic control.
- Shuttle control.
- Work handling system monitoring.
- Tool control.
- System performance monitoring and reporting.

3. Material handling system. It helps to hold a group of independent CNC machines into a comprehensive FMS. It performs the following functions :

(*i*) Random independent movement of work parts between workstations.

(*ii*) Handle a variety of work parts configurations?

(*iii*) Temporary storage.

(*iv*) Convenient access for loading and unloading work parts.

(*v*) Compatible with computer control.

4. Human operators. FMS require intellectuals with internal motivation, alertness and problem solving capabilities rather than physical workers. FMS require team work, communication and involvement from the human resources.

FMS and Transfer lines. Manual machine tools are very flexible but productivity is poor. A lot of time is wasted in setting up tools, fixtures, components, loading/unloading of tools and workpiece. On the other hand the transfer line is very productive but is suitable for one or two component types. The FMS compromise for above two cases.

(*b*) ABC company has to supply 30,000 switches per year to its customer. The demand is fixed and known. The customer uses its items in assembly operations and has no storage space. A shortage cost of Rs. 10/unit is incurred if the company fails to deliver the required units. The set up cost per run is Rs. 3500. Determine

(*i*) The optimum run size, q

(*ii*) The optimum level of inventory at the beginning of any period

(*iii*) The optimum scheduling period

(*iv*) The minimum total expected annual cost. (20)

Sol. $D = 30{,}000/\text{year}, C_s = \text{Rs. } 1\ 0/\text{unit},$

$C_o = \text{Rs. } 3500\ /\text{run},$

(*i*) $$\text{EOQ} = q = \left[\frac{2\,D\,C_o}{C_s}\right]^{0.5} = \left[\frac{2 \times 30{,}000 \times 3500}{10}\right]^{0.5} = 4583 \text{ switches}$$

(*ii*) Reorder cycle time, $t^* = \frac{q}{D} = \frac{4583}{30{,}000} = 55 \text{ days}$

(*iii*) Optimum level of inventory at the beginning of any period

$$= \frac{4583}{2} = 2292 \text{ switches}$$

(*iv*) Number of runs $= \frac{30{,}000}{4583} = 6.545 \cong 7$

Let cost of each switch be = Rs. 10

Annual cost 30,000 × 10 + 7 × 3500 = Rs. 3, 24,500

(*c*) Calculate the fundamental deviation and tolerance and hence the limits of size for the shaft and hole for the following fit : 64 mm H8-F7. The diameter steps are 50 mm and 80 mm. For shaft designation f, upper deviation is assumed as $-\ 5.5\ D^{0.41}$:

Data for tolerance

H8	25i
F7	16i

(10)

Sol. $D = [50 \times 80]^{0.5} = 63.25 \text{ mm}$

Tolerance unit, $i = 0.45\ (D)^{1/3} \div 0.001\ D$

$= 0.45\ (63.25)^{1/3} + 0.001 \times 63.25 = 1.8562$

Hole :

Tolerance $= 25\ i = 25 \times 1.8562 = 46.4 \text{ urn} = 0.046 \text{ mm}$

Shaft :

Tolerance $= 16\ i = 16 \times 1.8562 = 29.69 \text{ urn} = 0.030 \text{ mm}$

Fundamental upper deviation for shaft

$= -5.5 \times (63.25)^{0.41} = -30.115 \cong 30\ \mu m = 0.030$ mm

Shaft size : $64^{-0.030}_{-0.030-0.030} = 64^{-0.030}_{-0.060}$

Maximum shaft size = 63.97mm

Minimum shaft size = 63.94 mm

Hole size : $64^{+0.046}_{-0.000}$

Maximum hole size = 64.046 mm

Minimum hole size = 64.000 mm

SECTION-C

Q.5. (*a*) The following data from the orthogonal cutting test is available :

Rake angle = 10°, Chip thickness ratio = 0.35, Uncut chip thickness = 0.51 mm,

Width of cut = 3 mm, Yield shear stress of work material = 285 MPa,

Mean friction co-efficient on tool face = 0.65.

Determine the (*i*) cutting force, (*ii*) radial force, (*iii*) normal force on tool, and (*iv*) shear force on the tool. (20)

Sol. $\alpha = 10°, r = 0.35, t_1 = 0.51$ mm,

$b_1 = 3$ mm, $\tau_s = 285$ MPa, $\mu = 0.65$

$$\tan\phi = \frac{r\cos\alpha}{1 - r\sin\alpha} = \frac{0.35\cos 10°}{1 - 0.35\sin 10°} = 0.367$$

Shear plane angle, $\phi = 20.15°$

(*iv*) Shear force, $$F_s = \frac{\tau_s b_1 t_1}{\sin\phi} = \frac{285 \times 3 \times 0.51}{\sin 20.15°} = 1265.68 \text{ N}$$

Angle of friction, $\beta = \tan^{-1}\mu = \tan^{-1} 0.65 = 33°$

(*i*) Cutting force,

$$F_c = \tau_s b_1 t_1 \left[\frac{\cos(\beta - \alpha)}{\sin\phi\cos(\phi + \beta - \alpha)}\right]$$

$$= 285 \times 3 \times 0.51 \left[\frac{\cos(33 - 10)°}{\sin 20.15° \cos(20.15 + 33 - 10)°}\right]$$

$$= 1597.1 \text{ N}$$

$$F_c = F_s\cos\phi + F_n\sin\phi$$

$$1597.1 = 1265.68\cos 20.15° + F_n\sin 20.15°$$

$$F_n = 1187 \text{ N}$$

(*ii*) $F_t = F_n \cos \phi - F_s \sin \phi$

$= 1187 \cos 20.15° - 1265.68 \sin 20.15° = 678.35$ N

(*iii*) Normal force, $N = F_c \cos \alpha - F_t \sin \alpha$

$= 15976.1 \cos 10° - 678.35 \sin 10° = 1455$ N

(*b*) A Governor of Proell type has each arm 250 mm. The upper and lower ends of the arms are pivoted on the axis of the governor sleeve. Each ball has a mass of 15 kg and attached to the extension of the lower arms which are 100 mm long. The minimum and maximum radii of the governor are 125 and 175 mm. The central sleeve is of mass 75 kg. Determine the range of equilibrium speeds, assuming that the extensions of the lower arms are parallel to the governor axis, at the minimum radius. (20)

Sol. $m = 15$ kg, M = 75 kg, $r_1 = 175$ mm, $r_2 = 125$ mm

At the minimum radius, the governor position is shown in Fig. 4 (*a*).

$$\sin \alpha = \frac{125}{250} = 0.5, \alpha = 30°$$

$$BF = [2502 - 125]^{0.5} = 216.5 \text{ mm}$$

$$CF = BC + BF = 100 + 216.5 = 316.5 \text{ mm}$$

$$h_1 = 216.5 \text{ mm}$$

$$\omega^2{}_1 = \left(\frac{DF}{CF}\right)\left[1 + \frac{M}{m}\right]\left(\frac{g}{h_1}\right)$$

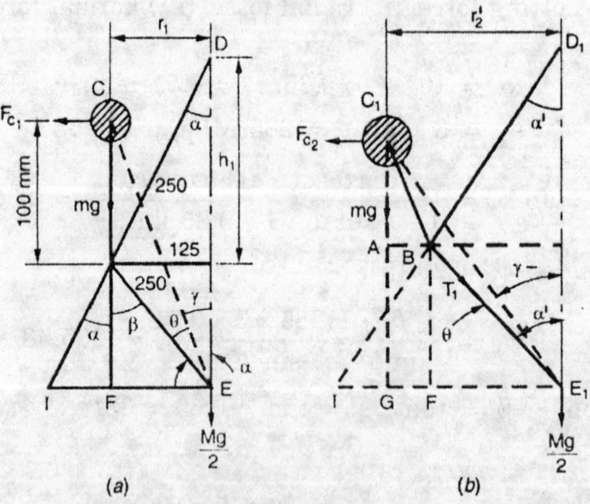

Fig. 4

$$= \left(\frac{216.5}{316.5}\right)\left[1 + \frac{75}{15}\right]\left(\frac{9.81}{0.2165}\right) = 185.97$$

$$\omega_1 = 13.617 \text{ rad/s or } N_1 = 130.2 \text{ rpm}$$

$$CE = [CF^2 + FE^2]^{0.5} = [(316.5)^2 + (125)^2]^{0.5} = 340.3 \text{ mm}$$

$$\tan \gamma = \frac{FE}{CF} = \frac{125}{316.5} = 0.395$$

$$\gamma = 21.55°$$

$$\theta = \alpha - \gamma = 30° - 21.55° = 8.45°$$

At the maximum radius, the position of the governor is shown in Fig. 4 (*b*).

$$h_2 = [l^2 - r^2_{\ 2}]^{0.5} = [250^2 - 175^2]^{0.5} = 178.53 \text{ mm}$$

$$\sin \alpha' = \frac{r_2}{l} = \frac{175}{250} = 0.7$$

$$\alpha' = 44.42°$$

$$\gamma' = \alpha' - 6 = 44.42° - 8.45° \approx 36°$$

$$C_1E_1 = \text{CiE} \cos \gamma' = 340.3 \cos 36° = 275.3 \text{ mm}$$

$$r_2' = E_1G = QE_1 \sin \gamma' = 340.3 \sin 36° = 200 \text{ mm}$$

At point E_1, we have

$$\text{T1} \cos \alpha' = \frac{Mg}{2}, T_1 = \frac{75 \times 9.81}{2 \cos 44.42°} = 515 \text{ N}$$

At point B,

$$T_2 = T_1 = 515 \text{ N}$$

$$BF = BE_1 \cos \alpha' = 250 \cos 44.42° = 178.6\text{mm}$$

$$E_1F = BE_1 \sin \alpha' = 250 \sin 44.42° = 175 \text{ mm}$$

$$IF = BF \tan \alpha' = 178.6 \tan 44.42° = 175 \text{ mm}$$

$$IE_1 = IF + FE_1 = 175 + 175 = 350 \text{ mm}$$

$$IG = IF + FE_1\ r_2' = 175 + 175 - 200 = 150 \text{ mm}$$

Taking moments about I, we have

$$mr_2' \omega^2_{\ 2} \times C_1G = mg \times IG + \frac{Mg}{2} \times IE_1$$

$$15 \times 0.2 \times \omega^2_{\ 2} \times 0.2573 = 15 \times 9.81 \times 0.15 + \frac{75 \times 9.81 \times 0.35}{2}$$

$$\omega_2 = 13.98 \text{ rad/s or } 133.5 \text{ rpm}$$

Range of speed $= 133.5 - 130.2 = 3.3$ rpm

Q.6. (*a*) List four types of resistance welding and explain any two of them with sketches. **(10)**

Sol. The four types of resistance welding processes are :

1. Percussion welding. 2. Spot welding.

3. Seam welding. 4. Projection welding.

Resistance spot welding. Resistance welding is based on the principle that heat is generated by the resistance offered to the flow of current. In the resistance spot welding process, the overlapping sheets are joined by local fusion, at the interface, by the concentration of current flowing between two electrodes, as shown in Fig. 5.(*a*) below. The workpieces are clamped between two water cooled Copper electrodes. This process is characterized by low cost, high speed and dependability.

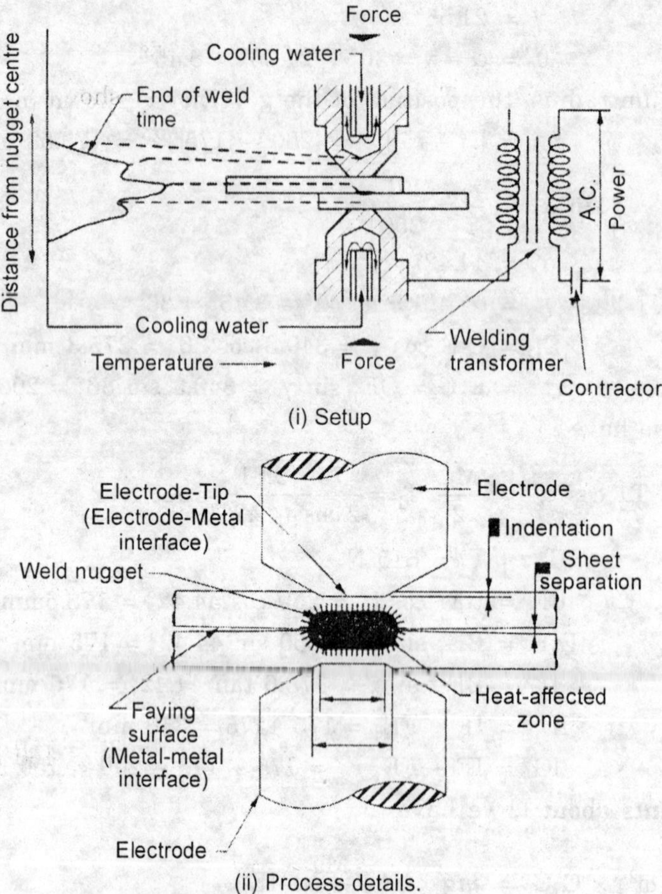

Fig. 5 (a) Resistance spot welding.

Resistance projection welding. This process is used for joining two sheets or a sheet and a thick component, or a small component like nut to a big body like automotive chasis, by making raised portions or projections on one of the components, where weld nugget is required to be made. The projections act to localize the heat of the welding circuit. The electrodes used are flat platens of hard material to cover the entire area of the work over which projection welds are to be made in one go, as shown in Fig. 5 (*b*) below. Usually two or three projections are welded at a time. The projections are made by forging, embossing or by intersection and may be button or dome type, ring type, shoulder projection, cross wire welding, and radius projection.

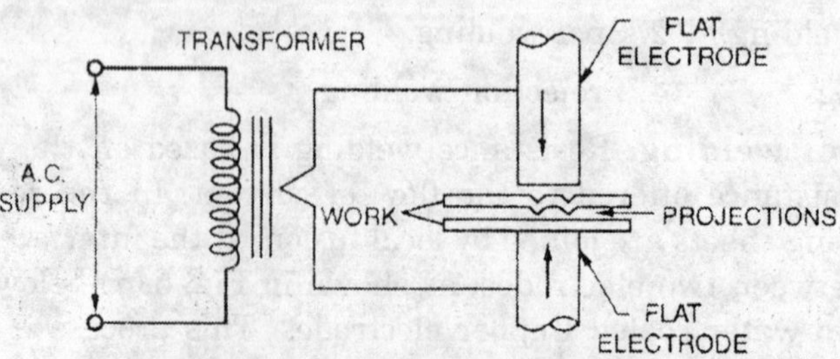

Fig. 5 (b) Resistance projection welding.

(*b*) Name four types of defects found in forging and mention the causes and remedies for two of the defects. (10)

Sol.

Type of forging defect	Causes	Remedies
1. Cold shuts and laps	Metal surface folding against itself during forging sharp corners in dies.	Avoid sharp corners in dies.
2. Pitting	Scale cause depressions in forgings.	Proper control of furnace and frequent]cleaning
3. Dies shift or mismatch	Misalignment between the top and bottom forgin dies due to loose wedges.	Wedges should be tightened properly.
4. Incomplete filling of dies.	Wrong amount of metal, insuffcient number of blows during forging, these causes. too low temperature, incorrect die design.	Ensure the correctness of all

(*c*) Two planes AB and BC which are at right angles are acted upon by tensile stress of 140 MPa and a compressive stress of 70 MPa respectively and also by shear stress 35 MPa. Determine the principal stresses and the principal planes. Find also the maximum shear stress and planes on which they act. Sketch the Mohr circle and mark the relevant data. (20)

Sol.

$$\sigma_{1,\,2} = 0.5\,(\sigma_x + \sigma_y) \pm 0.5\,[(\sigma_x - \sigma_y)^2 + 4\,\tau^2_{xy}]^{0.5}$$

$$= 0.5\,(140 - 70) \pm 0.5\,[(140 + 70)^2 + 4 \times 35^2]^{0.5}$$

$$= 35 + 110.68 = 145.68 \text{ MPa}, -75.68 \text{ MPa}$$

$$\tau_{max} = 110.687 \text{ MPa}$$

Principal planes are :

$$\tan 2\theta = \frac{2\tau_{xy}}{\sigma_x - \sigma_y} = \frac{2 \times 35}{140 + 70} = 0.3333$$

$$2\theta = 18.435°,\ \theta_1 = 9.216°\ \theta_2 = 99.216°$$

Planes of maximum shear stress are : 45° + 9.2161° = 54.216° and 144.216°

The Mohr circle has been sketched in Fig. 6 to a scale of 1 cm = 35 MPa.

$$OA = 140 \text{ MPa},\ OB = 70 \text{ MPa},\ AP = BQ = 35 \text{ MPa}$$

$$\sigma_1 = OD = 4.15 \text{ cm} = 145.25 \text{ MPa},$$

$$\sigma_2 = OE = 2.15 \text{ cm} = 75.25 \text{ MPa}$$

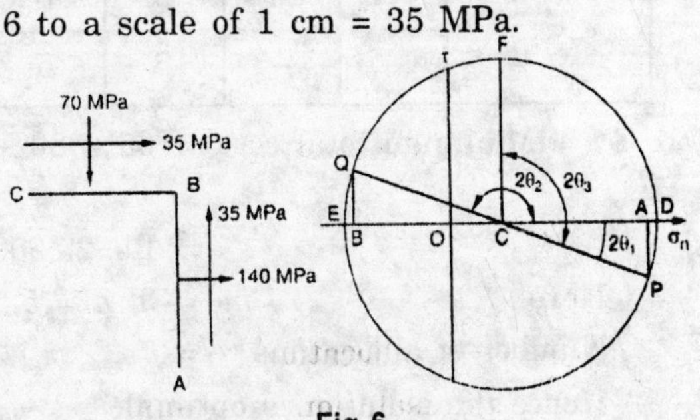

Fig 6.

$$\tau_{max} = CF = 3.15 \text{ cm} = 110.25 \text{ MPa}$$

Planes of maximum shear stress,

$$\theta_3 = 54°, \theta_4 = 144°$$

Q.7. (***a***) A company has three plants at A, B and C which of supply to warehouses located at D, E, F, G and H. Weekly plant capacities are 200, 125 and 225 tons respectively.Weekly warehouse requirments are 75, 105, 130, 155 and 85 tons respectively. Unit transportation cost matrix is given below :

To → **From ↓**	**D**	**E**	**F**	**G**	**H**
A	50	82	65	60	35
B	45	70	70	65	50
C	80	45	75	60	40

Determine the optimum cost distribution pattern and also the minimum total cost. (20)

Sol.

To → **From ↓**	**D**	**E**	**F**	**G**	**H**	**Supply**	**Row difference**				
A	50	82	65 80	60 120	35	200	15	15	10	5	5
B	45 75	70	70 50	65	50	125	5	5	20	5	5
C	80	45 105	75	60 35	40 85	225	5	20	20	15	–
Demand	75	105	130	155	85						
Column difference	5	25	5	5	5						
	5	–	5	5	–						
	5	–	5	5	–						
	–	–	5	5	–						
	–	–	5	5	–						

Minimum total cost $= 65 \times 80 + 60 \times 120 + 45 \times 75 + 70$ ” $50 + 45 \times 105 + 60 \times 35 + 40 \times 85$

$= \text{Rs. } 29{,}500$

Here $m = 3, n = 5, m + n - 1 = 3 + 5 - 1 = 7$

Number of allocations $= 7$

Hence the solution is optimal.

(*b*) A set of data obtained from inspection of castings, with a sample size of 200 castings are shown in the table:

Group No.	No. of Defects
1	76
2	68
3	76
4	89
5	47
6	69
7	83
8	93
9	70
10	79

Compute :

(*i*) The process average and control limits,

(*ii*) The process average for future production, and

(*iii*) Revised control limits. (10)

Sol. $n = 200$

(*i*) Total number of defects = 76 + 68 + 76 + 89 + 47 + 69 + 83 + 93 + 70 + 79 = 750

Total parts inspected = 200 × 10 = 2000

***p*-chart :**

$$p = \text{Total defectives/total parts inspected} = \frac{750}{2000} = 0.375$$

Standard deviation,

$$\sigma_p = \left[\frac{p(1-p)}{n}\right]^{0.5} = \left[\frac{0.375(1-0.375)}{200}\right]^{0.5} = 0.0342$$

$$\text{UCL} = p + 3\sigma_p$$
$$= 0.375 + 3 \times 0.0342$$
$$= 0.47777$$

$$\text{LCL} = p - 3\sigma_p$$
$$= 0.375 - 3 \times 0.0342$$
$$= 0.2724$$

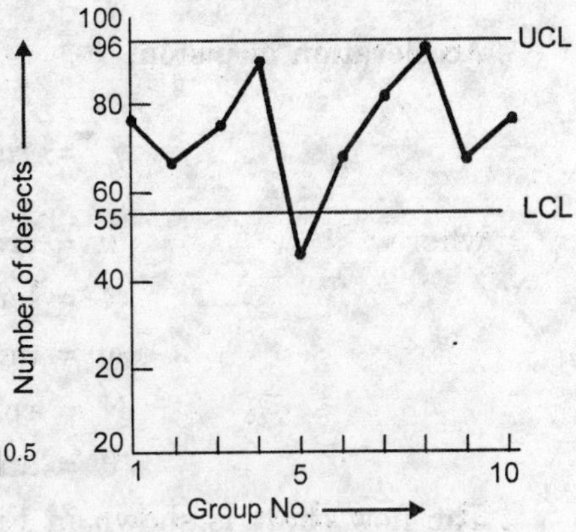

***p*-chart :**

$$\sigma_{np} = [np(1-p)]^{0.5}$$
$$= [200 \times 0.375(1 - 0.375)]^{0.5}$$
$$= 6.846$$

Fig 7.

$$\text{UCL} = np + 3\sigma_{np} = 200 \times 0.375 + 3 \times 6.846 = 95.54 \cong 96$$

$$\text{LCL} = np - 3\sigma_{np} = 200 \times 0.375 - 3 \times 6.846 = 54.46 \cong 55.$$

It is observed that group No. 5 with 47 defects is outside the LCL. We ignore this sample and recalculate the control limits.

(*ii*) Total number of defects = 703

Total parts inspected = 200 × 9 = 1800

p-chart :

$$p = \frac{\text{Total defects}}{\text{Total parts inspected}} = \frac{703}{1800} = 0.39$$

Standard deviation,

$$\sigma_p = \left[\frac{p(1-p)}{n}\right]^{0.5} = \left[\frac{0.39(1-0.39)}{200}\right]^{0.5} = 0.0345$$

$$\text{UCL} = p + 3\,\sigma_p = 0.39 + 3 \times 0.0345 = 0.4935$$

$$\text{LCL} = p - 3\sigma_p = 0.39 - 3 \times 0.0345 = 0.2865.$$

(*iii*) ***np*-chart :**

$$\sigma_{np} = [np(1-p)]^{0.5}$$

$$= [200 \times 0.39 (1 - 0.39)]^{0.5} = 6.898$$

$$\text{UCL} = np + 3\,\sigma_{np}$$

$$= 200 \times 0.39 + 3 \times 6.898 = 98.7 = 99$$

$$\text{LCL} = np - 3\,\sigma_{np}$$

$$= 200 \times 0.39 - 3 \times 6.898 = 57.3 = 58$$

Now all the sample groups are within control limits.

(*c*) Construct a flow chart to find the velocity and acceleration of the piston in a single slider mechanism.

Sol. Velocity of piston,

$$v_p = r\omega\left[\sin\theta + \frac{\sin 2\theta}{2n}\right]$$

Acceleration of piston,

$$f_p = r\omega^2\left[\cos\theta + \frac{\cos 2\theta}{n}\right]$$

where r = crank radius,

l = length of connecting rod,

ω = angular speed of crank,

N = rpm, $n = l/r$

θ = angle turned through by the crank.

The flow chart is shown in Fig. 8.

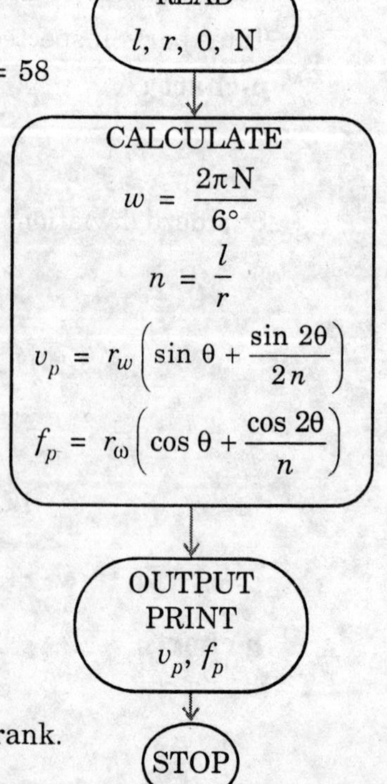

Fig. 8. Flow chart.

MECHANICAL ENGINEERING
PAPER II—2001

(Conventional)

Time allowed : Three hours *Maximum marks : 200*

1. *Candidates should attempt question 1 in Section A, which is compulsory, TWO questions from Section B and TWO questions from Section C. Question 1 is of short answers type, limiting answer of each part to 30 words.*
2. *The number of marks carried by each question is indicated at the end of the question.*
3. *Answers must be written in English.*
4. *Assume suitable data, if necessary and indicate them clearly.*

SECTION-A

Q.1. **(*a*)** Distinguish between computer education and computerized educatin.

Sol. Computer education is concerned with the study of computer fundamentals, both software and hardware. On the other hand, computerized education means imparting education with the help of computers as a tool.

(***b***) Study of a machining operation yields the cycle time of 9.0,8.0,9.0 and 10.0 minutes. The observed worker was rated at 90%. Providing an allowance factor of 15%, compute the standard time for the operation.

Sol. Average cycle time $= \dfrac{9+8+9+10}{4} = 9$ min

Normal time = Observed time × Rating = 9 × 0.9 = 8.1 min

$$\text{Standard time} = \frac{\text{Normal time}}{1 - \text{Allowance factor}} = \frac{8.1}{1-0.15} = \frac{8.1}{0.85} = 9.53 \text{ min}$$

(***c***) Differentiate between forecasting and prediction.

Sol. Forecasting is a statistical method to estimate the future trends of an activity based upon previous data, whereas prediction is made by intuition based on experience.

(***d***) What is JIT ? What is the primary benefit derived from its adoption ?

Sol. JIT: Just in time denotes providing exactly the required unit of the right quantities at the right time so as to meet the promised delivery date.

The benefits of JIT are : Low scrap cost, Low inventory cost, Zero defect products, Improved team work, Improved productivity, Reduced batch size and adherence to delivery time.

(*e*) What is concurrent engineering ?

Sol. CE deals with carrying out the design and manufacturing activities at the same time while designing the product. This allows the design and manufacturing engineers to interactively trade off parameters to give an optimum design of product and process.

(*j*) Why a large sphere solidifies and cools to ambient temperature at a much slower rate than does a smaller sphere?

Sol. Solidification time,

$$t_s \propto V/A$$

where V = volume, and A = surface area

For a large sphere of radius R,

$$t_{s1} \propto \frac{\left(\frac{4\pi R^3}{3}\right)}{(4\pi r^2)} \propto R$$

For a small sphere of radius r,

$$t_{s2} \propto r$$

As R > *r*, therefore, $t_{s1} > t_{S2}$.

Hence, a large sphere shall cool and solidify at a slower rate as compared to a small sphere.

(*g*) Comment on the machines surface characteristics obtained after the EDM process.

Sol. 1. High temperature generated by the sparks causes melting and vaporization of the metal which affects the properties of the shallow layers (2.5 to 150 μ) of the surface machines.

2. The outermost layer is rapidly chilled, and is therefore very hard. The layer below this is somewhat in tempered condition.

3. Due to hardening of the surface layer, EDM imparts better wear resistance characteristics. However, the fatigue strength reduces due to the micro cracks that develop in the surface layer during chilling.

(*h*) What is diffusion welding ?

Sol. Diffusion welding is a process in which a weld is achieved by the application of pressure, of the order of 5 to 75 MPa, while the pieces are held at a high temperature, normally about 70% of the melting point in degrees absolute. The process is based on solid-phase diffusion, which is accelerated by temperature. Diffusion in metals take place due to vacant lattice sites or along grain boundaries.

(*i*) What is temper embrittlement in alloy steels and what are its effects ?

Sol. Embrittlement is the loss of ductility of a metal either due to physical or chemical changes. Tempering consists of reheating below the critical temperature the quenched metal to restore some of the ductility and reduce brittlement. Temper embrittlement in alloy steels is carried out nearly at 200°C. This process is used in order to retain hard micro structure of martensite, but to give up brittleness.

(*j*) Differentiate between cold, warm and hot working of metals.

Sol. Cold working. It is the plastic deformation of metals at temperature below their recrystallisation temperature, which is 0.5 times the absolute melting point temperature.

Warm working. In warm working, the deformation temperature is kept between 0.3 to 0.5 times the absolute melting point of the metal.

Hot working. In hot working, the plastic deformation of metals is carried out above their recrystallisation temperature.

(*k*) Group the following into thermosets and thermo-plastics :

Acrylic, Alkyds, Polyimide, Nylon, Epoxy, Phenolic, Silicone and PVC.

Sol. Thermosets. Alkyd, Epoxy, Silicone, Phenolic.

Thermoplastics. Acrylic, Polyimide, Nylon, PVC.

(*l*) What is meant by point of contraflexure or point of inflexion in a beam ? Show the same for the beam given below.

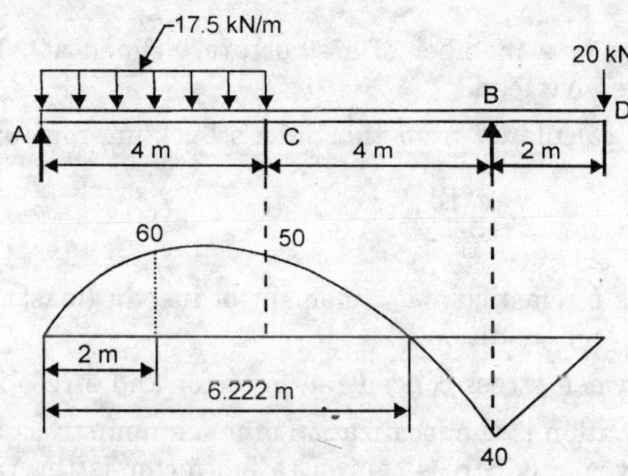

Fig. 1

Sol. The point of contraflexure is a point in a beam where the bending moment changes sign from positive to negative or *vice-versa.*

For the given beam,

$$R_A + R_B = 17.5 \times 4 + 20 = 90 \text{ kN}$$

$$R_A \times 8 = 17.5 \times 4 + 6\text{-}20 \times 2 = 380$$

$$R_A = 47.5 \text{ kN}, R_B = 42.5 \text{ kN}.$$

Span AC :

At a distance x from A, $M_x = 47.5\ x - 8.75\ x^2$

$$M_A = 0, M_C = 50 \text{ kN-m}.$$

Span CB :

$$M_x = 47.5\ x - 17.5 \times 4 \times (x - 2) = 47.5 \times - 70\ (x - 2)$$

For $M_x = 0$

$$47.5\ x - 70\ x + 140 = 0$$

$$x = \frac{140}{22.5} = 6.22 \text{ m}$$

The point of contraflexure occurs at 6.22 m from A.

The B.M. diagram is shown in the Fig. 1.

(***m***) A tension member with a cross-sectional area of 30 mm^2 resists a load of 80 kN. Calculate the normal and shear stresses on the plane of maximum shear.

Sol. Axial stress, $\sigma_x = P/A = \frac{80}{30} = 2.667 \text{kN/mm}^2$

Plane of maximum shear stress, $\theta = 45°$

Normal stress, $\sigma_n = \sigma_x \cos^2 \theta = 2.667 \times \cos^2 45° = 1.334 \text{kN/mm}^2$

Shear stress,

$$\tau = \sigma_x \sin \theta \cos \theta$$
$$= 2.667 \times \cos 45° \times \sin 45° = 1.334 \text{ kN/mm}^2$$

(***n***) Differentiate between a strut and column. What is the general expression used for determination of their critical load ?

Sol. Strut. It is a bar or a member of a structure in any position and carrying an axial compressive load.

Column. It is a bar or a member of a structure inclined at 90° to the horizontal and carrying an axial compressive load.

The critical load is calculated from the Euler's buckling formula, given by

$$P_E = \frac{\pi^2 EI}{L_e^2}$$

where E = modulus of elasticity. I = moment of inertia (least), L_e = equivalent length of column for the given end conditions.

(***o***) Distinguish between stress-concentration factor and stress-intensity factor.

Sol. Stress-concentration is the localization of maximum stress at the stress raisers like notches, holes and grooves, etc. Stress concentration factor is the ratio of maximum stress at the stress raiser to the nominal stress as calculated by the simple formula of strength of materials.

Stress-intensity factor refers to the stress-concentration at cracks and governs their growth and propagation.

(***p***) Why two universal joints with an intermediate shaft is generally used in practice for transmitting motion from one shaft to other, having angular misalignment ? Sketch the arrangement.

Sol. An intermediate shaft is used between two universal joints to compensate for the angular fluctuations of the two shafts during power transmission.

The arrangement is shown in Fig. 2.

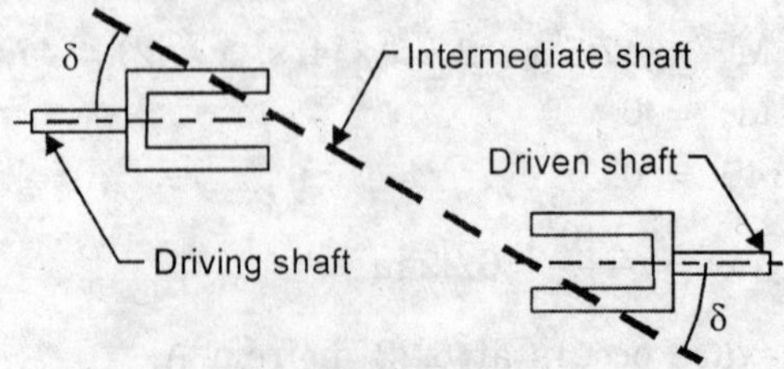

Fig. 2

(*q*) What is meant by 6 × 7 Lang lay-Right lay wire rope ?

Sol. This is a rope with 6 strands, each strand containing 7 wires.

In a Lang lay rope, the strands and wires in the strand are twisted in the same direction. In a Right lay rope, the strands coil in the same direction as a right hand screw thread.

(*r*) Two gears having an angular velocity ratio of 3 : 1 are mounted on shafts whose centres are 136 mm apart. If the module of the gears is 4 mm, how many teeth are there on each gear ?

Sol. $$\frac{n_2}{n_1} = 3,\ C = 136 \text{ mm},\ m = 4 \text{ mm}$$

$$\frac{z_2}{z_1} = \frac{n_2}{n_1} = \frac{1}{3}$$

$$C = 0.5\,(d_1 + d_2) = 0.5\text{ m }(z_1 + z_2)$$

$$136 = 0.5 \times 4\,(3 + 1)\,z_2,\ z_2 = 17,\ z_1 = 51$$

(*s*) Whether the size of the flywheel for an inline multicylinder engine of particular h.p. can be reduced by increasing the number of cylinders ? If so, why ?

Sol. The size of the flywheel of a multi-cylinder engine can be reduced by increasing the number of cylinders because by increasing the number of cylinders the turning moment diagram will be more smoother and less fluctuation of energy.

(*t*) What is the difference between whirling and critical speed of a shaft ? Mention two physical parameters of the shaft-bearing system, which can increase the critical speed of the shaft. (2 × 20 = 40)

Sol. Whirling means excessive vibrations of the shaft whereas critical speed is the speed of the shaft at which resonance occurs, *i.e.*, the forcing frequency matches with the natural frequency of the shaft-bearing system.

Natural frequency of a shaft carrying a rotor at mid-span is given by :

$$\omega_n = \sqrt{k/m}$$

where $$k = \text{stiffness of shaft} = \frac{48\text{ EI}}{L^3},\ I = \frac{\pi d^4}{64}$$

$\therefore$ $$k \propto \frac{d^4}{L^3}$$

Therefore, increasing d and decreasing L shall increase the critical speed.

SECTION-B

Q.2. (*a*) A single cylinder two stroke vertical engine has a bore of 30 cm and a stroke of 40 cm with a connecting rod of 80 cm long. The mass of the reciprocating parts is 120 kg. When the piston is at quarter stroke and moving down, the pressure on it is 70 N/cm^2. If the speed of the engine crankshaft is 250 rpm clockwise, find the turning moment on the crankshaft. Neglect the mass and inertia effects of connecting rod and crank.

Sol. $D = 30$ cm, $L = 40$ cm, $l = 80$ cm, $R = 120$ kg

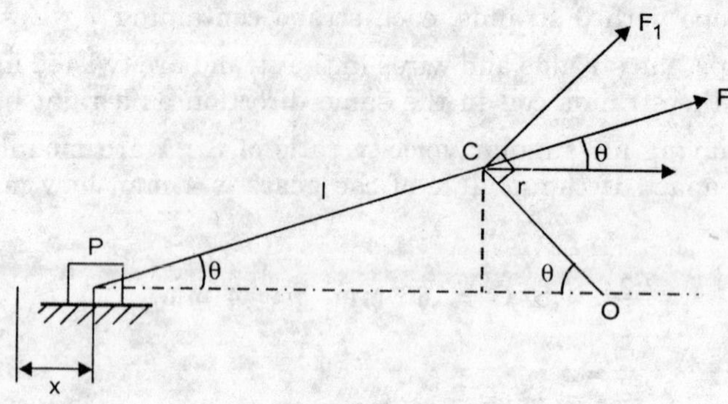

Fig. 3.

$$r = 20 \text{ cm}, n = \frac{l}{r} = \frac{80}{20} = 4$$

$$x = r(1 - \cos\theta)$$

$$\frac{L}{4} = \left(\frac{L}{2}\right)(1 - \cos\theta)$$

$$\cos\theta = \frac{1}{2} \text{ or } \theta = 60°$$

$$p = 70 \text{ N/cm}^2, \text{ N} = 270 \text{ rpm}$$

$$\omega = \frac{2\pi \times 250}{60} = 26.18 \text{ rad/s}$$

$$P_s = \frac{\pi d^2 p}{4} = \frac{\pi \times 30^2 \times 70}{4} = 49480 \text{ N}$$

Acceleration of piston,

$$a_p = \omega^2 r\left[\cos\theta + \frac{\cos 2\theta}{n}\right]$$

$$= (26.18)^2 \times 0.20\left[\cos 60° + \frac{\cos 120°}{4}\right] = 51.4 \text{ m/s}^2$$

Inertia force, $F_t = R\, a_p = 120 \times 51.4 = 6168.5$ N

Piston effort,

$$PE = P_s + Rg - F_i$$

$$= 49480 + 120 \times 9.81 - 6168.5 = 44489 \text{ N}$$

Turning moment $= PE \times r\left[\sin\theta + \dfrac{\sin 2\theta}{n}\right]$

$$= 44489 \times 0.20\left[\sin 60 + \frac{\sin 120°}{4}\right]$$

$$= 8668.9 \text{ N-m}$$

(***b***) The turning moment exerted on the crankshaft of an engine is given by :

$$T = 10{,}000 + 2000 \sin 2\theta - 1800 \cos 2\theta$$

where θ is the crank angle measured from T.D.C. Assuming the load-torque to be constant, determine

(*i*) The power of the engine when turning at 250 rpm,

(*ii*) The moment of inertia of the flywheel if the speed variation from the mean speed is not to exceed ± 0.25%,

(*iii*) The angular acceleration of the flywheel for 0 = 45°.

Sol. Mean torque,

$$T_m = \left(\frac{1}{\pi}\right)\int_0^{\pi} T.d\theta$$

$$= \left(\frac{1}{\pi}\right)\int_0^{\pi}(10{,}000 + 2000 \sin 2\theta - 1800 \cos 2\theta)\, d\theta$$

$$= \left(\frac{1}{\pi}\right)[10{,}000\,\theta - 1000 \cos 2\theta - 9000 \sin 2\theta]_0^{\pi} = 10{,}000 \text{ N-m}$$

$$\omega = \frac{2\pi \times 250}{60} = 26.18 \text{ rad/s.}$$

(*i*) Power of engine,

$$P = \frac{T_m\,\omega}{10^3} = \frac{10{,}000 \times 26.18}{10^3} = 261.8 \text{ kW}$$

(*ii*) Fluctuation of energy,

$$E_f = 0.5 \times I\,(\omega^2{}_{max} - \omega^2{}_{min})$$

$$\omega_{max} = \omega_m + \frac{0.25}{100}\,\omega_m = 1.0025\,\omega_m$$

$$\omega_{min} = \omega_m - \frac{0.25}{100}\,\omega_m = 0.9975\,\omega_m$$

$$E_f = 0.5\, I\,(26.18)^2\,[(1.0025)^2 - (0.9975)^2] = 3.427\, I$$

Change in torque $= T - T_m = 2000 \sin 2\theta - 1800 \cos 2\theta = 0$

$$\tan 2\theta = 0.9$$

$$2\theta = 42° \text{ and } 222°$$

or $\theta = 21° \text{ and } 111°$

$$E_f = \int_{21°}^{111°} (2000 \sin 2\theta - 1800 \cos 2\theta)\, d\theta$$

$$= \Big|1000 \cos 2\theta - 9000 \sin 2\theta\Big|_{21°}^{111°} = 2690.72 \text{ N-m} = 3.427$$

or $I = 785.15 \text{ kg-m}^2$

(*iii*) Angular acceleration of the flywhee

$$= \frac{T_m}{I} = \frac{10000}{785.15} = 12.73 \text{ rad/s}^2$$

(*c*) Can the above engine be completely balanced ? If yes, explain the method of balancing. If not, what is the technique adopted for minimization of unbalance. Find out the mass required at the radius of 30 cm from crank shaft for such balancing. Also find the resultant unbalance force at the crank position, as given in part (*a*).

Sol. The above engine cannot be completely balanced. Partial balancing can be resorted to for minimizing of unbalance.

Let B = Balance mass placed at radius of 30 cm.

Then, $30 \times B = R \times r$

$$B = \frac{120 \times 20}{30} = 80 \text{ kg}$$

Resultant unbalance force

$$= R\,\omega^2\, r\, [(1 - c)^2 \cos^2\theta + c^2 \sin^2\theta]^{0.5}$$

For $c = 0.5 = 120 \times (26.18)^2 \times 0.2\ [0.25 \cos^2 60° + 0.25 \sin^2 60°]$

$$= 4112.35 \text{ N}$$

Q.3. (*a*) A steel tube of inner diameter 100 mm and wall thickness 5 mm is subjected to a torsional moment of 1000 N.m. Calculate the principal stresses and orientations of the principal planes on the outer surface of the tube.

Sol. $r = 50$ mm, $l = 5$ mm, T = 1000 N.m

$$\tau_{max} = \frac{T}{2\pi r^2 t} = \frac{1000}{2\pi \times 50^2 \times 5 \times 10^{-9}} = 12.73 \text{ MPa}$$

Principal stresses, $\sigma_{1,\,2} = \pm\, \tau_{max} = \pm\, 12.73$ MPa

Principal planes are : $\theta_1 = 45°$, $\theta_2 = 135°$

(*b*) A helical spring B is placed inside the coils of a second helical spring A, having the same number of coils and free axial length and of the same material. The two springs are compressed by an axial load of 210 N which is shared between them. The mean coil diameters of A and B are 90 mm and 60 mm and the wire diameters are 12 mm and 7 mm respectively. Calculate the load shared by individual springs and the maximum stress in each spring.

Sol. $n_A = n_B$, $I_A = I_B$,

$W = 210$ N, $D_A = 90$ mm

$D_B = 60$ mm, $d_A = 12$ mm, $d_B = 7$ mm

$\delta_A = \delta_B$

$$\frac{8\, W_A\, D_A}{\pi d_A^3} = \frac{8 \times 151 \times 90}{\pi \times 12^3} = 20 \text{ MPa}$$

$$\frac{W_A}{W_B} = \left(\frac{D_B}{D_A}\right)^3 \left(\frac{d_A}{d_B}\right)^4 = \left(\frac{60}{90}\right)^3 \left(\frac{12}{7}\right)^4 = 2.55$$

Also

$$W_A + W_B = 210$$

$$W_A = 151 \text{ N}, W_B = 59 \text{ N}$$

$$\tau_A = \frac{8 W_A D_A}{\pi d_A^3} = \frac{8 \times 151 \times 90}{\pi \times 12^3} = 20 \text{ MPa}$$

$$\tau_B = \frac{8 W_B D_B}{\pi d_B^3} = \frac{8 \times 59 \times 60}{\pi \times 7^3} = 26.28 \text{ MPa}$$

(*c*) A cantilever tube of length 120 mm is subjected to an axial tension, P = 9.0 kN, a torsional moment, T = 72.0 N-m, and a bending load, F = 1.75 kN at the free end. The material is aluminium alloy with an yield strength of 276 MPa. Find the thickness of the tube limiting the outside diameter to 50 mm so as to ensure a factor of safety of 4.

Sol. Direct tensile stress,

$$\sigma_d = \frac{P}{A} = \frac{P}{\pi d_o t} = \frac{9 \times 10^3}{\pi \times 50 \times t} = \frac{57.3}{t} \text{ N/mm}^2$$

Moment of inertia, $\quad I = \frac{\pi (d_0{}^4 - d_i{}^4)}{64} \doteq \frac{\pi\, d_0{}^3 i}{8}$

Section modulus, $\quad Z = \frac{2I}{d_0} = \frac{\pi\, d_0{}^2\, t}{8}$

Bending stress,

$$\sigma_b = \frac{M}{Z} = \frac{1.75 \times 10^3 \times 120}{\left(\pi \times 50^2 \times \frac{t}{4}\right)} = \frac{106.95}{t} \text{ N/mm}^2$$

Polar moment of inertia,

$$J = 2I = \frac{\pi \times d^3 \times t}{4}$$

Polar section modulus,

$$Z_p = \frac{\pi\, d_0^2 \times t}{4}$$

Torsional shear stress,

$$\tau = \frac{T}{Z_p} = \frac{72 \times 10^3 \times 2}{\pi \times 50^2 \times t} = \frac{18.33}{t} \text{ N/mm}^2$$

Maximum principal stress,

$$\sigma_1 = 0.5\,(\sigma_d + \sigma_b) + 0.5\,[(\sigma_d + \sigma_b)^2 + 4\,\tau^2]^{0.5}$$

$$\frac{276}{4} = \frac{0.5\,(57.3 + 106.95)}{t} + \left(\frac{0.5}{t}\right)[(57.3 + 106.95)^2 + 4 \times 18.33^2]^{0.5}$$

$$69 = \frac{279}{t}$$

$$t = 4 \text{ mm}$$

Q.4. (*a*) Design a set of stepped pulleys to drive a machine from a counter-shaft that runs at 220 rpm. The centre distance between the two sets of pulleys is 2 m. The diameter of the smallest step on the counter-shaft is 160 mm. The machine is to run at 80, 100 and 120 rpm and should be able to rotate in either directions. Find the length of the belt required for both cases.

Sol. $N = 220$ rpm, $C = 2$m, $n = 80, 100, 130$ rpm

$$D_1 N = d_1 n_1$$

Largest diameter of machine shaft pulley,

$$D_1 = \frac{160 \times 220}{80} = 440 \text{ mm}$$

$$R_1 = 220 \text{ mm}$$

$$D_i + d_i = 440 + 160 = 600 \text{ mm} = \text{const.}$$

or

$$R_i + r_i = 300 \text{ mm}$$

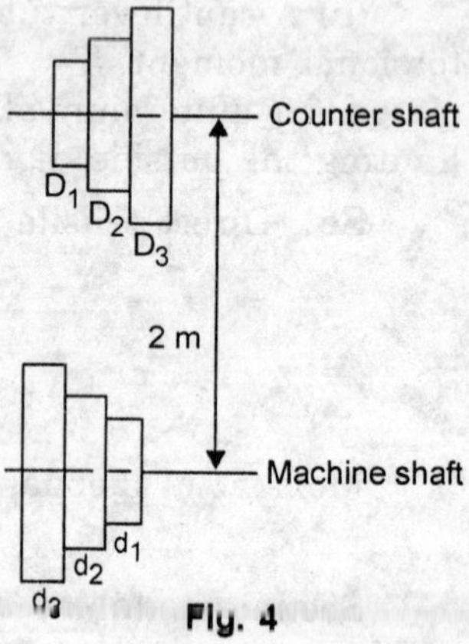

Diameters of machine pulley are

$$D_1 = 440 \text{ mm}$$

$$D_2 = \frac{80 \times 440}{100} = 352 \text{ mm}$$

$$D_3 = \frac{80 \times 440}{130} = 271 \text{ mm}$$

Diameters of countershaft pulley are

$$d_1 = 160 \text{ mm}$$

$$d_2 = 248 \text{ mm}$$

$$d_3 = 329 \text{ mm}$$

Counter shaft
D_1
D_2
D_3
2 m
Machine shaft
d_1
d_2
d_3

Fig. 4

Length of open belt,

$$L_0 = 2C + \frac{(d_2 - d_1)^2}{4C} + \frac{\pi (d_1 + d_2)}{2}$$

$$= 2 \times 2 + \frac{(440 - 160)^2 \times 10^{-6}}{4 \times 2} + \frac{\pi (160 + 440) \times 10^{-3}}{2}$$

$$= 4.9523 \text{ m}$$

Length of cross belt,

$$L_C = 2C + \frac{(d_2 - d_1)^2}{4C} + \frac{\pi (d_1 + d_2)}{2}$$

$$= 2 \times 2 + \frac{(440 + 160)^2 \times 10^{-6}}{4 \times 2} + \frac{\pi (160 + 440) \times 10^{-3}}{2}$$

$$= 4.9875 \text{ m}$$

(b) When two cylindrical parts are assembled by shrink-fitting or by press-fitting, a contact pressure is created between the two parts. If the radii of the inner cylinder are a and c and that of the outer cylinder are (c-δ) and b, δ being the radial interference, the contact pressure is given by

$$p = \left(\frac{E\delta}{c}\right)\frac{(b^2 - c^2)(c^2 - a^2)}{2c^2(b^2 - a^2)}$$

where E is the Young's modulus of the material. Can you outline the steps involved in developing this important design equation ?

Sol. For the inner cylinder :

$$\sigma_r = -p$$

$$\sigma_\theta = -p\left[\frac{c^2 + a^2}{c^2 - a^2}\right]$$

$$\varepsilon_{\theta 1} = \left(\frac{1}{E}\right)[\sigma_\theta - v\,\sigma_r] = \left(-\frac{p}{E}\right)\left[\frac{c^2 + a^2}{c^2 - a^2} - v\right]$$

$$\delta c_1 = c\ \delta\theta_1$$

For the outer cylinder :

$$\sigma_r = -p$$

$$\sigma_\theta = p\left[\frac{b^2 + a^2}{b^2 - c^2}\right]$$

$$\delta_{\theta 2} = \left(\frac{1}{E}\right)[\sigma_\theta - v\,\sigma_r] = \left(\frac{p}{E}\right)\left[\frac{b^2 + c^2}{b^2 - c^2} + v\right]$$

$$\delta c_2 = c\ \delta\theta_2$$

$$\delta c_1 + \delta c_2 = \delta = \left(\frac{cp}{E}\right)\left[\frac{b^2 + c^2}{b^2 - c^2} + \frac{c^2 + a^2}{c^2 - a^2}\right]$$

$$p = \left(\frac{E\delta}{c}\right)\frac{(b^2 - c^2)(c^2 - a^2)}{2c^2(b^2 - a^2)}$$

(c) A hollow shaft of diameter ratio 3/8 is required to transmit 600 kW at 110 rpm, the maximum torque being 20% greater than the mean. The shear stress is not to exceed 63 MPa and the twist in a length of 3 m not to exceed 1.4 degrees. Determine the diameter of the shaft. Assume modulus of rigidity for the shaft material as 84 GPa.

Sol. Power transmitted,

$$P = \frac{2\pi N T_m}{60 \times 10^3}\ \text{kW}$$

$$T_m = \frac{300 \times 60 \times 10^3}{2\pi \times 110} = 26043.5\ \text{N-m}$$

$$T_{max} = 1.2\ T_m = 31252.2 \text{ N-m}$$

$$J = \frac{\pi (d_0^4 - d_i^4)}{32} = \frac{\pi d_o^4 [1-(3/8)^4]}{32} = 0.09623\, d_0^4$$

$$\frac{T_{max}}{J} = \frac{2\tau_{max}}{d_0}$$

$$\frac{31252.2\times 10^3}{0.09623\, d_0^4} = \frac{2\times 63}{d_0}$$

$$d_0 = 137 \text{ mm}$$

Also
$$\frac{T_{max}}{J} = \frac{G\theta}{l}$$

$$\frac{31252.2\times 10^3}{0.09623\, d_0^4} = 84\times 10^3 \times \left(\frac{\pi\times 1.4}{180}\right)\times \frac{0.09623\, d_0^4}{(3\times 10^3)}$$

$$d_0 = 147.6 \text{ mm}$$

Adopt
$$d_0 = 148 \text{ mm},$$

$$d_i = \frac{148\times 3}{8} = 55.5 \text{ mm}$$

Q.5. **(*a*)** Describe the major imperfections that may be present in the crystal structure of metals. How do they influence the mechanical properties of materials ? (10)

Sol. Any geometric deviation from ideal crystalline structure is regarded as an imperfection. The imperfections may be classified as follows :

1. Point defects,
2. Line defects, and 3. Surface defects.

1. **Point Defects.** A point defect is confined to a point in a lattice. Point defects arise either due to imperfect packing during original crystallization or due to thermal vibrations of atoms at elevated temperatures. Point defects are of the following types:
   - **(*a*) Vacancies.** These involve missing atoms from a regular lattice site. Vacancies may be single or multiple.
   - (*b*) **Interstitial defects.** These are due to extra atoms in the crystal lattice at interstitial (void) position.
   - **(*c*) Substitutional defect.** These arise due to a foreign atom occupying a position meant for the parent atom.
   - **(*d*) Frenkel defect.** Whenever a missing atom occupies an interstitial position. It is a common type of defect in ionic crystals.
   - (*e*) **Schottky defect.** Whenever a pair of positive and negative ions is missing from a crystal so that the crystal becomes electrically neutral.
   - **(*f*) Phonon.** Whenever a group of atoms is displaced from its ideal location.
2. **Line defects.** A line defect is caused due to distortion of atoms along a line, in some direction. A line defect is commonly called a *dislocation.* The dislocations are of two types :

(*a*) **Edge dislocation.** Whenever a half plane of atortis is inserted between the planes of atoms in a perfect crystal, the defect produced is called an *edge dislocation.* Edge dislocation arises when there is a slight mismatch in the orientation of adjacent parts of the growing crystal either due to introduction of extra row of atoms or elimination of a row of atoms.

(*b*) **Screw dislocation.** In a screw dislocation, the atoms are displaced in two separate planes perpendicular to each other. The region of the crystal is twisted in the form of a screw around the line of dislocation if seen in its direction.

3. **Surface defects.** These defects take place on the surface of the crystal either due to imperfect packing of atoms during crystallization or defective orientation of the surface. These defects are of the following types :

(*a*) **Grain boundary.** Here the grains of different orientation separate the general pattern of atoms and exhibits a boundary. This type of defect generally arise during the solidification of the liquid metal.

(*b*) **Twin boundary.** Here the arrangement of atoms on one side of the boundary is the mirror image of the arrangement of atoms of the other side.

(*c*) **Stacking fault.** This fault arises due to the improper sequence stacking of atoms throughout the crystal.

(*b*) What are super alloys ? Discuss in brief the compositions, properties and applications of these alloys. (10)

Sol. Super alloys have iron base, cobalt base, or nickel base. They contain nickel, chromium, cobalt and molybdenum as major alloying elements.

Composition :

Super alloy	**Composition**	**Examples**
1. Iron base	Iron : 32-67%	
	Chromium: 15-22%	Inconel
	Nickel: 9-38%	
2. Cobalt base	Cobalt: 35-65%	
	Chromium: 19-30%	
	Nickel : upto 35%	
3. Nickel base	Nickel: 38-76%	
	Chromium: 27%	Inconel, Nimonic
	Cobalt: 20%	

Properties :

1. Good resistance to corrosion.
2. High resistance to mechanical and thermal fatigue and shock.
3. High resistance to creep and erosion at elevated temperatures.

Applications :

1. Manufacture of jet engines.
2. Construction of gas turbines.

(*c*) Give a step-by-step procedure for the cold chamber die-casting process. Also discuss the advantages and limitations of hot and cold chamber processes. (20)

Sol. The steps involved in the production of die-casting are :

1. Closing and locking the two halves of the dies.
2. Forcing the molten metal into the die cavity under pressure.
3. Maintaining the pressure for a short time and permitting the metal to solidify.
4. Opening the die halves.
5. Ejecting the casting with its assembly of sprue, runner and gates by pins.
6. Removing the parting lines and gates by trimming the excess metal from them.
7. Carrying out the finishing process.

Advantages:

1. Good grain structure.
2. Good surface finish and better appearance.
3. High dimensional accuracy.
4. Repeated use of moulds.
5. Rapid production rate.
6. Low scrap loss.
7. Low porosity.
8. Lesser floor space required.
9. Smaller cored-holes can be obtained with metal cores.
10. Inserts can be readily cast in place.

Limitations :

1. High initial cost of moulds and equipment and the cost of mould maintenance.
2. Shape, size and intricacy limitations.
3. Mainly used for low melting point non-ferrous alloys.

Q.6. (*a*) Discuss in brief the flank and crater wear mechanisms.

In a certain machining operation with a cutting speed of 50 ml mm, tool life of 45 minutes was observed. When the cutting speed was increased to 100 m/min, the tool life decreased to 10 minutes. Estimate the cutting speed for maximum productivity, if tool change time is 2 minutes.

Sol. There are various mechanisms of wear, *e.g.*, abrasion wear, adhesion wear, and diffusion wear. Abrasion and adhesion wear are primarily responsible for the flank wear, whereas diffusion may play an important role in the development of crater wear at a high speed since then the temperature in the rake face is much higher than that in the flank surface.

Taylor's tool life equation is $v\,T^n = C$

$$50 \times 45^n = C$$

$$100 \times 10^n = C$$

$$n \ln 45 = \ln 2 + n \ln 10$$
$$3.8067\, n = 0.69315 + 2.3026\, n$$
$$n = 0.46$$
$$C = 50 \times 45^{0.46} = 288$$
$$v \times 2^{0.46} = 288$$
$$v = 209.4 \text{ m/min}$$

(*b*) (*i*) Explain the difference between hard and soft automation. Why are they so called ?

(*ii*) What are the features of an intelligent robot ?

Sol.

(*i*) Hard automation. In hard or fixed-position automation, the sequence of assembly operations are performed as fixed by the job configuration. Sometimes the operations to be done on the job are fixed by equipment configuration also. In fixed automation, many operations on raw material, to shape it to a finished object, are integrated and performed by one piece of equipment thus making the automated system, complex and difficult to maintain. The features of hard automation arc :

(*a*) Rigid, as regards to product design changes.

(*b*) Generally meant for high volume of production.

(*c*) Equipment is made to order and has high cost.

Examples of fixed automation are : mechanized assembly lines and transfer lines. It is used for producing engine blocks, valve gears and spindles.

Soft automation. The soft (or flexible) automation manufacturing system is one which is capable of changing over to adopt to changed physical set up of tooling, machine settings and fixtures arising out of new product configuration, intended to be manufactured. Its important features are

(*a*) Flexibility to deal with variety of job configuration.

(*b*) Continuity of production is maintained even with varied product-mix jobs.

(*c*) Medium production sales.

It can produce parts of complex shapes. The soft automation machines are expensive.

(*ii*) Intelligent robot. The intelligent robot is capable of performing some of the functions and tasks carried out by human beings. It is equipped with visual (computer vision) and tactile (touching) capabilities. Much like human beings, the robot observes and evaluates the immediate environment and proximity to other objects, machinery, etc. by perception and pattern recognition. It then makes appropriate decisions for the next movement and proceeds with them Powerful computers arc required to control it.

(*c*) Describe any four common types of discontinuities that may develop in a welded joint. What remedies you suggest to avoid them ?

Sol.

Discontinuity in welded joint	**Causes**	**Remedies**
1. Porosity or voids	Trapped gases during welding, chemical reaction or contamination	Select proper electrodes and filler materials, preheat the weld zone, prevent contaminators, decreases welding speed.
2. Slaginclusions	Trapping of oxides, fluxes, and electrode coating materials in the weld zone.	Clean the weld-bead before next deposition, provide shielding gas, permit sufficient space for proper manipulation of the molten weld.
3. Incomplete fusion and penetration	Insufficient heat input or Poor design.	Increase heat input, lower speed of welding, change joint design, proper fitting of joining surfaces.
4. Undercut	Non-uniform feed of filler rod. improper position of electrode or torch or excessive heating.	Add additional metal to fill the groove.

(*d*) Discuss the technical and economic reasons leading to the development of Non-traditional machining (NTM) processes.

Sol.

1. The hardness and strength of the material is very high or the material is too brittle.
2. The work piece is too flexible, slender or delicate to withstand the cutting or grinding forces.
3. Surface finish and tolerances better than those obtainable by other processes are required. To meet better dimensional and accuracy requirements.
4. Complexity of the die-cavity shape is difficult to meet the high productivity demands.
5. Need for attaining accuracy combined with automation. The search for better ways of machining complex shapes (dies, moulds, etc.) in hand, high strength temperature resistant alloys, such as nimonics, carbides, titanium alloys, hastalloy, nitralloy, waspalloy, stainless steels and heat resistant steels etc.

Q.7. (*a*) Demand for a certain item has been as shown below:

Time	Actual demand
April	200
May	50
June	150

The forecast for April was 100 units with a smoothing constant of 0.20 and using first order exponential smoothing. What is the July forecast ? What do you think about a 0.20 smoothing constant ?

Sol. $F_{t+1} = F_t + \alpha (X_t - F_t) = F_t + \alpha e_t$

where X_t = actual demand at time t

Here $\alpha = 0.20$

Month	Actual demand X_t	Forecast demand F_t	$e_t = X_t - F_t$
April	200	100	+ 100
May	50	100 + 0.2 × 100 = 120	– 70
June	150	120 – 0.2 × 70 = 106	+ 46
July		106 + 0.2 × 46 = 115.2 ≅ 115	

Forecast for July = 115 units

Q.7. (*b*) Consider the following jobs and their processing times at corresponding machnies -

Duration Hours

Job	Machine 1	Machine 2	Machine 3
A	13	5	9
B	5	3	7
C	6	4	5
D	7	2	6

Using Johnson's rule, find the optimal sequence.

Sol. Coverting the problem into a 2 machine problem, we have

Job	Machine 1	Machine 2
A	13 + 5 = 18	5 + 9 = 14
B	5 + 3 = 8	3 + 7 = 10
C	6 + 4 = 10	4 + 5 = 9
D	7 + 2 = 9	2 + 6 = 8

Job Schedule	Sequence	Machine	Processing time Hours
B	B - - -	1	8
D	B - - D	2	8
C	B C - D	1	10
A	B C A D	2	14

The sequence is : (B C A D).

Q.7. (*c*) Assume that the following specify a generalized linear programming problem -

Maximize : $f(x) = 2x_1 + x_2$

Suject to :

$$x_1 + x_2 \leq 6$$
$$x_1 \leq 3$$
$$2x_1 + x_2 \geq 4$$
$$x_1, x_2 \geq 0$$

Graph the problem, identifying the three constraint equation lines and the fesibble zone common to all of them. Plot dotted lines for value of 3, 6, 9 and 12 for the objective function $f(x)$. What appears to the highest feasible value $f(x)$, and for what values of x_1 and x_2 does it occur ?

Sol.

$\mathbf{F(x) = 2x_1 + x_2}$	x_1	x_2	x_2	x_1
30301.5				
6	0	6	0	3
9	0	9	0	45
12	0	12	0	6

The constraint lines and the objective function lines are plotted in Fig. 5.

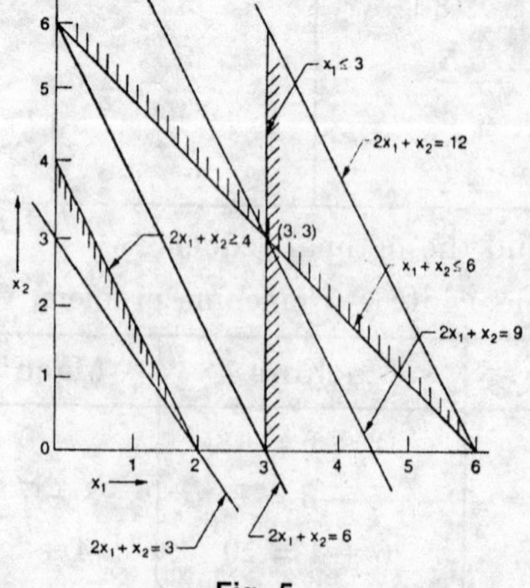

Fig. 5

The feasible optimal solution is at (3, 3). Max $f(x) = 9$.

MECHANICAL ENGINEERING
PAPER II—2002

(Conventional)

SECTION-A

Q.1. **(*a*)** Why are the bolts, subjected to impact, made longer ?

Sol. The capacity to store strain energy by a body is directly proportional to volume of the body. By making longer bolts, their volume increases and hence the capacity to store strain energy increases. This increases their capacity to absorb more shocks under impact loads.

(*b*) Why are anti-friction bearings not used at very high speed ?

Sol. At high speed, an anti-friction bearing has a higher co-efficient of friction than a journal bearing. Therefore, the heat generation is more resulting into higher power loss combined with excessive wear.

(*c*) If two cantilever beams of identical dimensions but made of mild steel and gray cast iron are subjected to same point load at the free end, within elastic limits, which one will deflect more and why ?

Sol. Deflection of free end of a cantilever due to end load,

$$\delta = \frac{WL^3}{2EI}$$

Therefore, $$\delta \propto \frac{1}{E}$$

Since $E_s > E_{CI}$, therefore, $\delta_S < \delta_{CI}$.

Thus, cast iron will deflect more than steel.

(*d*) Differentiate clearly between 'shaft' and 'axle'.

Sol. A shaft is a rotating member used for the purpose of transmitting power. It may also support transverse load to induce bending in it. On the other hand, an axle supports revolving parts and may either be fixed relative to its supports or rotate together with the parts mounted on it. It may also support transverse loads but does not transmit any power.

(e) Which principle reduces the problem of kinetics to equivalent problem of statics ?

Sol. D'Alembert's principle.

(*f*) What is the magnitude of coriolis acceleration for a slider sliding at 10 cm/sec on a link which is rotating at 60 r.p.m. ?

Sol. $\omega = \dfrac{2\pi \times 60}{60} = 2\pi$ rad/s

Coriolis acceleration $= 2v\omega = \dfrac{2 \times 10 \times 2\pi}{100} = 1.257$ m/s^2

(*g*) Why sometimes the axis of translating roller follower in a cam-follower mechanism is offset from the axis of cam rotation?

Sol. To avoid jamming of the follower.

(*h*) Name the different types of cast irons and mention, which type is the hardest.

Sol. 1. Gray cast iron,

2. White cast iron,
3. Malleable cast iron,
4. Ductile cast iron,
5. Chilled cast iron,
6. Alloy cast iron.

Chilled cast iron is the hardest.

(*i*) Compare die casting with investment casting w.r.t. production rate, melting point of work materials and ability of producing complex shaped products.

Sol.

Characteristics	Die casting	Investment casting
1. Production rate	Rapid	Less than die casting
2. Melting point of work material	More than investment casting	Less than die casting
3. Ability to produce complex shaped products	Less than investment casting	More than die casting

(*j*) Name three processes of manufacturing long low carbon steel tubes.

Sol. 1. Piercing 2. Extrusion 3. Rolling.

(*k*) What are the relative advantages of submerged arc welding over conventional open air manual arc welding and why ?

Sol. Prevents the molten steel from forming oxides and nitrides in steel due to chemical combination with oxygen and nitrogen.

(*l*) Why are the lathe-spindles made hollow ?

Sol. To pass a long feeding bar stock through the bore.

(*m*) How does application of cutting fluid help in reducing cutting forces and improving tool life and surface quality ?

Ans :

1. Carry away heat generated.
2. Keep the tool-chip-work piece cool.
3. Reduces the co-efficient of friction at tool-chip interface.
4. Increases tool life.
5. Improves surface finish.

(*n*) In which machine tools, teeth of internal spur gears can be cut and which one of those machine tools work fastest ?

Sol. 1. Broaching.

2. Pinion cutter generating technique.

Broaching method is the fastest.

(*o*) What are meant by 'A – 60 – K – 8 – V when used for a grinding wheel specifications ?

Sol. A : Abrasive type – Al_2O_3

60 : Grit size – Medium

K : Grade – Medium

8 : Structure – Dense

V : Bond – Vitrified.

(*p*) What do you understand by the term anisotropy ? State any three materials where the anisotropy is an important consideration ?

Sol. Anisotropy means variation in elastic constants with direction.

Examples : wood, asbestos, crystalline substances.

(*q*) What is the difference in the analysis of thick tubes compared to that for thin tubes ? State the basic equations describing stress distribution in a thick tube.

Sol. The thickness to diameter ratio of a thin tube is less than 1/20. The hoop and longitudinal stresses are constant throughout the thin tube. There is no radial stress in a thin tube.

The radial and hoop stresses vary parabolically in a thick tube wall. Basic equations governing the stresses in a thick tube are:

Radial stress, $$\sigma_r = \frac{B}{r^2} - A$$

Hoop stress, $$\sigma_\theta = \frac{B}{r^2} + A$$

Where *A* and *B* are constants.

(*r*) State generalized quantitate forecast model. What are the possible nature of independent variables when using extrapolation of a time series concept ?

Sol. Forecasting is a prediction for future so as to minimize the uncertainties of unknown future. The independent variables when using extrapolation of a time series concept are :

Trend (T), Cyclic variation (C), Seasonal variation (S), and Random variation (R). Forecast, Y = TCSR.

(*s*) State Johnson's rule for scheduling of n jobs requiring two operations on two separate facilities where the sequence cannot be altered.

Sol.

1. Select the smallest processing time from two machines. If there is a tie, select either of the smallest processing time.
2. (*a*) If the minimum processing time occurs on machine 1, place the associated job in the first available position in the sequence.

   (*b*) If the minimum processing time occurs on machine 2, place the associated job in the last available position in the sequence.
3. The job sequenced by either step 2 (*a*) or 2 (*b*), is removed from consideration. The process is repeated, starting from first step, until all jobs are sequenced. Ties can be broken arbitrarily.

(*t*) State Pareto's principle State some of its applications. (2 × 20 = 40)

Sol. Pareto's principle states that there are a few people with a lot of money and many people with little money. Thus, there are a vital few and useful many.

Applications:

1. A few customers account for the majority of sales.
2. A few processes account for the bulk of the scarp or rework cost.
3. A few nonconformities account for the majority of customer complaints.
4. A few suppliers account for the majority of rejected parts.
5. A few products account for the majority of the profit.
6. A few items account for the bulk of the inventory cost.

SECTION-B

Q.2. (*a*) Two bearings of a shaft at A and B are 5 m apart. The shaft carries three eccentric masses C, D and E which are 160 kg, 170 kg and 85 kg respectively. The respective eccentricity of each mass, measured from the axis of rotation is 0.5 cm, 0.3 cm and 0.6 cm, and the distance from A is 1.3 m, 3 m and 4 m respectively. Determine the angular position of each mass with respect to C, so that no dynamic force is exerted at B, and also find the dynamic force at A for this arrangement when the shaft runs at 100 r.p.m. (15)

Sol.

Rotor	Mass kg	Eccentricity, *r, m*	F_c/ω^2 = mr, kg-m	Distance from A (ℓ, m)	Mrℓ kg.m^2
C	160	0.005	0.80	1.3	1.04
D	170	0.003	0.51	3.0	1.53
E	85	0.006	0.51	4.0	2.04

Draw the couple polygon with the data from the last column, as shown in Fig. 1 (*b*). The angular positions of the three masses with respect to mass C are : $\theta_D = 75°$, $\theta_E = 225°$.

Now draw the force polygon with the data from column 4, as shown in Fig. 1 (*d*). The closing side of the force polygon is oe = 0.56.

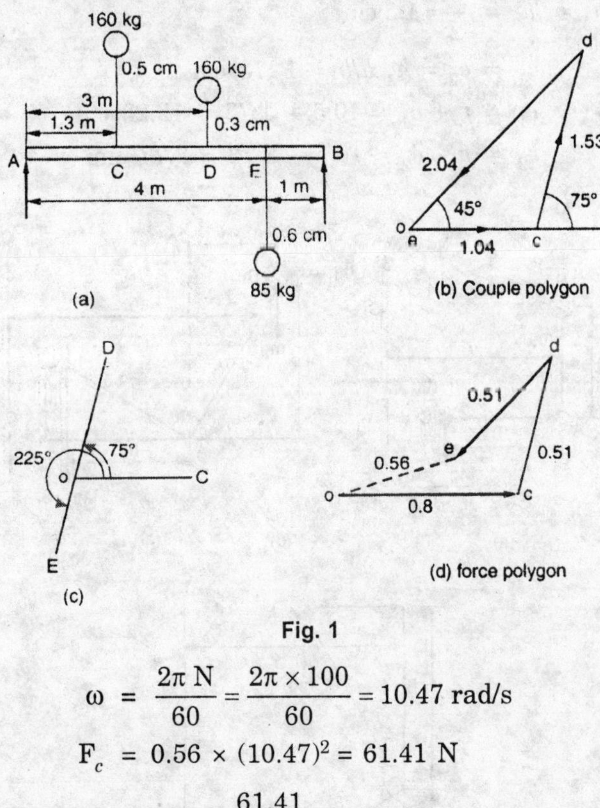

Fig. 1

$$\omega = \frac{2\pi N}{60} = \frac{2\pi \times 100}{60} = 10.47 \text{ rad/s}$$

$$F_c = 0.56 \times (10.47)^2 = 61.41 \text{ N}$$

$$\text{Unbalanced force on bearing A} = \frac{61.41}{2} = 30.70 \text{ N}$$

(***b***) The total sleeve movement in a Hartnell governor is 3 cm. The mass of the rotating balls is 1.5 kg each. At the mid-position of the sleeve, the sleeve arm, which is 6.5 cm long, is horizontal. The ball arm has a length of 7.5 cm. At the mid-position of the sleeve, the balls rotate at a radius of 10.5 cm. Due to maladjustment of the spring, the equilibrium governor speed at the topmost position of the sleeve is 415 rpm and that corresponding to the lowest position 430 rpm. Determine

(*i*) The stiffness and initial compression of the spring, and (*ii*) The required initial compression of the spring to give an equilibrium speed at the topmost position which is 10 rpm more than that at the lowest position : Neglect the moment due to the weight of the balls. (15)

Sol. $h = 3$ cm, $m = 1.5$ kg, $b = 6.5$ cm, $a = 7.5$ cm

$r = 0.5$ cm,

$N_1 = 415$ rpm, $N_2 = 430$ rpm.

(*i*)
$$k = 2\left(\frac{a}{b}\right)^2\left[\frac{F_{c1} - F_{c2}}{r_1 - r_2}\right]$$

$$r_1 - r_2 = \frac{h_a}{b} = \frac{3 \times 7.5}{6.5} = 3.46 \text{ cm}$$

$$h_1 = h_2 = \frac{3}{2} = 1.5 \text{ cm}$$

$$c_1 = c_2 = h_1\ a/b = 1\ .73 \text{ cm}$$

$$r_1 = r + c_1 = 10.5 + 1.73 = 12.23 \text{ cm}$$

$$r_2 = r - c_2 = 10.5 - 1.73 = 8.77 \text{ cm}$$

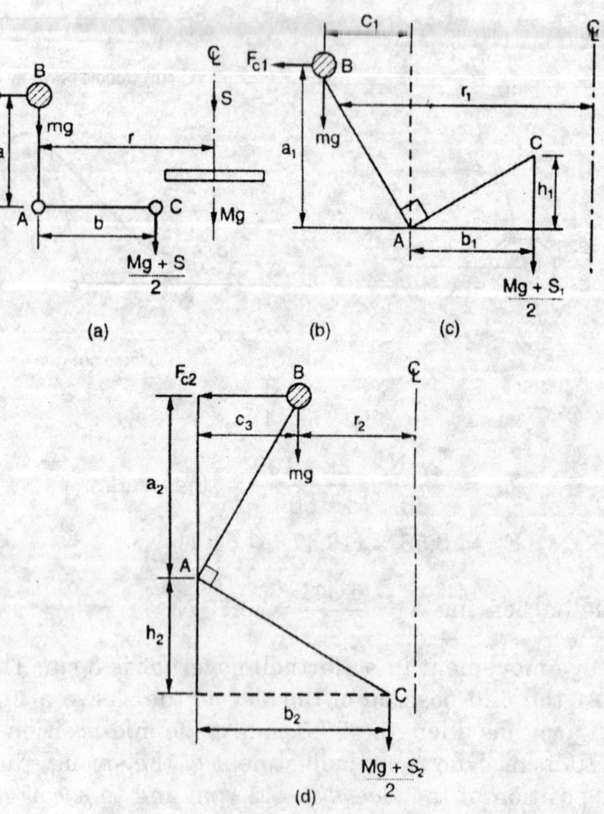

Fig. 2

$$F_{c1} = mr_1\,\omega^2{}_1 = 1.5 \times 12.23 \times 10^{-2}\left(\frac{2\pi \times 415}{60}\right)^2 = 346.474 \text{ N}$$

$$F_{c2} = mr_1\,\omega^2{}_1 = 1.5 \times 8.77 \times 10^{-2}\left(\frac{2\pi \times 430}{60}\right)^2 = 266.736 \text{ N}$$

$$k = 2\left(\frac{7.5}{6.5}\right)^2\left[\frac{346.474 - 266.738}{3.46 \times 10^{-2}}\right] = 6133.58 \text{ N/m}$$

Mean speed, $N_m = 0.5\,(415 + 430) = 422.5$ rpm

$$F_c \times a = 0.5\text{ S} \times b$$

$$mr\omega^2 \times a = 0.5\text{ S} \times b$$

$$1.5 \times 10.5 \times 10^{-2}\left(\frac{2\pi \times 422.5}{60}\right)^2 \times 7.5 = 0.5\text{ S} \times 6.5$$

$$\text{S} = 711.49 \text{ N}$$

$$\text{Initial compression} = \frac{\text{S}}{k} = \frac{711.49}{6133.58} = 11.6 \text{ cm.}$$

(*ii*) $N_{max} = N_m + 10 = 430 + 10 = 440$ rpm

$$mr_1\,(2\pi N_{max})^2 \times a = 0.5\text{ S} \times b$$

$$1.5 \times 10.5 \times 10^{-2}\left(\frac{2\pi \times 440}{60}\right)^2 \times 7.5 = 0.5\,\text{S} \times 6.5$$

$$\text{S} = 898.79 \text{ N}$$

$$\text{Initial compression} = \frac{898.79}{6133.58} = 14.65 \text{ cm}$$

(*c*) Two standard full depth gears of pressure angle 14.5° have a module of 5 mm. The pinion has 15 teeth while the gear has 60 teeth. If the addendum of the gear is equal to the module.

(*i*) Show that the gear will interfere with the pinion.

(*ii*) To what value should the pressure angle be increased in order to eliminate the interference ? (10)

Sol. $\alpha = 14.5°,\ m = 5 \text{ mm},\ z_1 = 15,\ z_2 = 60,\ h_{a2} = 5 \text{ mm}$

(*i*)

$$r_1 = \frac{mz_1}{2} = \frac{5 \times 15}{2} = 37.5 \text{ mm}$$

$$r_2 = \frac{mz_2}{2} = \frac{5 \times 60}{2} = 150 \text{ mm}$$

$$r_{a2} = r_2 + h_{a2} = 150 + 5 = 155 \text{ mm}$$

$$r_{b2} = r_2 \cos\alpha = 150 \cos 14.5° = 145.22 \text{ mm}$$

Length of approach, $L_{a2} = \left[r^2{}_{a2} - r^2{}_{b2}\right]^{0.5} - r_2 \sin\alpha$

$$= [155^2 - (145.22)^2]^{0.5} - 150 \sin 14.5° = 16.629 \text{ mm}$$

Maximum $L_{a2} = r_1 \sin \alpha = 37.5 \sin 14.5° = 9.389$ mm

Since the length of approach 16.629 mm is more than the permissible length of approach 9.389 mm, therefore interference will take place.

(*ii*) $[155^2 - (145.22)^2]^{0.5} - 150 \sin \alpha = 9.389$

$$155^2 - (150 \cos \alpha)^2 = (150 \sin \alpha + 9.3S9)^2$$

$$\sin \alpha = 0.51071$$

$$\alpha = 30.67°$$

Q.3. (*a*) A pair of straight teeth spur gears having 20° full depth involute teeth is to transmit 20 kW. The pinion runs at 300 r.p.m. and the speed ratio is 3 : 1. The following data are given :

Number of teeth on pinion = 15

Service factor, $C_s = 1$

Velocity factor, $C_v = \dfrac{3}{3+v}$

where v is the pitch line velocity in m/sec.

Tooth form factor, $y = 0.154 - \dfrac{0.912}{T}$, where T = Number of teeth.

Face width = 14 m, where m = module in mm

Allowable static stress for pinion and gear materials are 120 MPa and 100 MPa respectively.

Check the gear for wear if the surface endurance limit is 600 MPa and modulus of elasticity for pinion and gear materials are 200 GPa and 100 GPa respectively. (20)

Sol. Let m = module, mm

$z_p = 15$, $z_g = 45$

Pitch diameter of pinion, $d_p = 15\ m$ mm

Pitch diameter of gear, $d_g = 45\ m$ mm

Pitch line velocity of pinion,

$$v_p = \frac{\pi d_p n_p}{60} = \frac{\pi \times 15m \times 300 \times 10^{-3}}{60} = 0.23562\ m \text{ m/s}$$

Velocity factor, $C_v = \dfrac{3}{0.23562\ m \times 1}$

Tangential force, $F_t = \dfrac{P \times 10^3}{v \times C_s} = \dfrac{20 \times 10^3}{0.23562\ m \times 1} = \dfrac{84882}{m}$ N

Tooth form factors :

Pinion, $y_p = 0.154 - \dfrac{0.912}{15} = 0.0932$

Gear, $y_g = 0.154 - \dfrac{0.912}{45} = 0.1337$

$$\sigma_p y_p = 120 \times 0.0932 = 11.184$$

$$\sigma_g y_g = 100 \times 0.1337 = 13.37$$

$$\sigma_p y_p < \sigma_g y_g.$$

Hence, pinion is the weaker and need to be designed.

By Lewis equation, we have

$$F_t = \frac{\sigma_s \, bmy \, C_v}{K_f}$$

Assuming stress concentration factor for fatigue, $K_f = 1$, we have

$$\frac{84882}{m} = 120 \times 14\, m \times 0.0932 \times \left[\frac{3}{3 + 0.23562\, m}\right]$$

$$= \frac{469.728\, m^2}{3 + 023562\, m}$$

$$469.728\, m^3 = 84882\, (3 + 023562 \text{ m}) = 254646 + 20000\, m$$

$$\text{m}^3 - 42.5776 \text{ m} - 542.1 = 0$$

Solving, we find that, $m \approx 10$ mm

$$F_t = 8488.2 \text{ N}$$

Wear load, $F_w = d_p b K_w Q$

where $Q = \dfrac{2z_g}{z_g + z_p} = \dfrac{2 \times 45}{45 + 15} = 1.5$

$$K_w = \left(\frac{\sigma^2_{es} \sin \phi}{1.4}\right)\left[\frac{1}{E_p} + \frac{1}{E_g}\right]$$

$$= \left(\frac{600^2 \sin 20°}{1.4}\right)\left[\frac{1}{2} + \frac{1}{1}\right] \times 10^5 = 1.3192$$

$$F_w = 15 \times 10 \times 14 \times 10 \times 1.3192 \times 1.5 = 41555 \text{ N}$$

(***b***) A transmission shaft, rotating at 500 rpm, drives a milling machine which requires 3.75 kW at 750 rpm. A 300 mm diameter cast iron pulley is mounted on the transmission shaft. A preliminary design proposes using a belt 4.75 mm thick, which has a density of 970 kg/m . The allowable stress is 2 MPa. Two pulleys rotate in opposite directions and the centre distance of the shafts is 750 mm. The coefficient of friction is 0.3 for both the pulleys. Determine the width of the belt. (10)

Sol. $d_2 = \dfrac{n_1 d_1}{n_2} = \dfrac{300 \times 500}{750} = 200$ mm

Mass of the belt per metre length,

$$m = \rho bt = 970 \times b \times 4.75 \times 10^{-6} = 0.004607\ 6\text{kg/m}$$

Velocity of the belt, $v = \dfrac{\pi \times 300 \times 500}{60 \times 10^3} = 7.854$ m/s

Centrifugal tension,

$$F_c = mv^2 = 0.004607\ b \times (7.S54)^2 = 0.286 \text{ N}$$

Angle of arc of contact,

$$\theta = \pi + 2\sin^{-1}\left[\frac{300+200}{2\times 750}\right] = 218.94° = 3.8212 \text{ rad}$$

$$\exp(\mu\,\theta) = \exp(0.3 \times 3.8212) = 3.1467$$

Power transmitted,

$$F_1 = bt\sigma_a = 1\ 3.75 \times 10^3 \times b\ (9.5 - 0.284) \times 7.854$$

$$= 3.75 \times 10^3 \times b\,(9.5 - 0.284) \times 7.854\left[\frac{3.1467-1}{3.1467}\right]$$

$$b = 75.94 = 76 \text{ mm}$$

(*c*) A 5 cm diameter solid shaft is welded to a flat plate by 1 cm fillet weld. What will be the maximum torque that the welded joint can sustain if the permissible shear stress in the weld material is not to exceed 8 kN/cm^2 ? Derive the expression for the shear stress in the throat from the basic theory. (10)

Sol. Let d = diameter of the shaft, T = torque in the shaft, F = force in the weld.

Then $$T = \frac{Fd}{2}$$

or $$F = \frac{2T}{d} = \pi dt\,\tau$$

where t = throat thickness of weld = 0.7071,

l = length of weld,

t = permissible shear stress in the weld,

$$\tau = \frac{2T}{\pi d^2 t}$$

Maximum torque sustained by the weld,

$$T = \frac{\pi \times 5^2 \times 0.707 \times 1 \times 8 \times 10^3}{2 \times 10^2} = 2221 \text{ N-m}$$

Throat area of weld in shear around the shaft,

$$A_s = \pi dt$$

Shear force in the weld, $F = \tau \,.\, A_s = \tau \,.\, \pi dt$

Resisting torque due to the weld $= \dfrac{Fd}{2}$

Hence $$\frac{Fd}{2} = T$$

Shear stress, $$\tau = \frac{2\tau}{\pi d^2 t}$$

Q.4. (*a*) Derive an expression for the distortion energy per unit volume for a body subjected to a uniform stress state, given by the principal stresses o, and o_2 with the third principal stress a_3 being zero. (10)

Sol. Work done per unit volume by $\sigma_1 = \dfrac{\sigma_1^2}{2E}$

Work done per unit volume reduced due to lateral strain produced by σ_2 and σ_3 along σ_1

$$= \left(\frac{\sigma_1}{2}\right)\left(\frac{\nu}{E}\right)[\sigma_2 + \sigma_3]$$

Net work done per unit volume by σ_1,

$$w_1 = \left[\frac{1}{2E}\right][\sigma^2{}_1 - \nu(\sigma_1\sigma_2 + \sigma_1\sigma_3)]$$

Similarly, net work done per unit volume by σ_2 and σ_3 respectively are

$$w_2 = \left[\frac{1}{2E}\right][\sigma^2{}_2 - \nu(\sigma_2\sigma_3 + \sigma_1\sigma_2)]$$

$$w_3 = \left[\frac{1}{2E}\right][\sigma^2{}_3 - \nu(\sigma_1\sigma_3 + \sigma_2\sigma_3)]$$

Total work done per unit volume under the action of all the principal stresses,

$$w = w_1 + w_2 + w_3$$

= Strain energy stored per unit volume

$$u = \left[\frac{1}{2E}\right][\sigma^2{}_1 + \sigma^2{}_2 + \sigma^2{}_3 - 2\nu(\sigma_1\sigma_2 + \sigma_2\sigma_3 + \sigma_3\sigma_1)]$$

When $\sigma_3 = 0$, then $u = \left[\dfrac{1}{2E}\right][\sigma^2{}_1 + \sigma^2{}_2 - 2\nu\sigma_1\sigma_2]$

Distortion energy = Total strain energy - Strain energy due to volumetric strain

Mean stress, $\sigma_m = \dfrac{\sigma_1 + \sigma_2 + \sigma_3}{3}$

Under hydrostatic compression, $\sigma_1 = \sigma_2 = \sigma_3 = \sigma$

$$u = \left[\frac{1}{2E}\right][3\sigma^2 - 6\nu\sigma^2] = \left[\frac{3\sigma^2}{2E}\right](1 - 2\nu)$$

Volumetric strain energy per unit volume due to

$$\sigma_m = \left[\frac{3\sigma^2}{2E}\right](1 - 2\nu)$$

$$u_v = \frac{\left[(\sigma_1 + \sigma_2 + \sigma_3)^2\right](1 - 2\nu)}{6E}$$

For $\sigma_3 = 0$, $u_v = \dfrac{\left[(\sigma_1 + \sigma_2)^2\right](1 - 2\nu)}{6E}$

Shear strain energy per unit volume,

$$u_s = u - u_v$$

$$= \left[\frac{1}{2E}\right]\left[\sigma^2_1 - \sigma^2_2 - 2\nu\,\sigma_1\sigma_2\right] - \frac{(1-2\nu)}{(6E)}\left[(\sigma_1+\sigma_2)^2\right]$$

$$= \left[\frac{1}{6E}\right]\left[\sigma^2_1 - \sigma^2_2 - \sigma_1\sigma_2\right]$$

where $G = \dfrac{E}{2(1+\nu)}$

(*b*) A beam of rectangular cross-section 50 mm wide and 100 mm deep is simply supported over a span of 1500 mm. It carries a concentrated load of 50 kN, 500 mm from the left support. Calculate

(*i*) The maximum tensile stress in the beam and indicate where it occurs;

(*ii*) The vertical deflection of the beam at a point 500 mm from the right support.

E for the material of the beam = 2 × 10^5 MPa. (20)

Sol.
$$I = \frac{50 \times 1000}{12} = 4166667\ \text{mm}^3$$

Refer to Fig. 3.

(a)

Fig. 3. (b) BMD.

(*i*)
$$R_A = \frac{50 \times 1000}{1500} = \frac{100}{3}\ \text{kN-m}$$

$$R_B = \frac{50}{3}\ \text{kN}$$

$$M_c = \frac{100 \times 500 \times 10^{-3}}{3} = \frac{50}{3}\ \text{kN-m}$$

$$\sigma_{max} = \left(\frac{50 \times 10^3}{3}\right) \times \frac{0.050}{(416667 \times 10^{-12})} = 200\ \text{MPa}$$

Maximum stress occurs under the load.

(*ii*)
$$M_x = \left(\frac{100}{3}\right)x - 50\,(x - 500)$$

$$\frac{EId^2y}{dx^2} = -M_x = -\left(\frac{100}{3}\right)x + 50\,(x-500)$$

$$\frac{EIdy}{dx} = -\frac{50\,x^2}{3} + 25\,(x-500)^2 + C_1$$

$$EIy = -\frac{50\,x^2}{9} + \left(\frac{25}{3}\right)(x-500)^3 + C_1x + C_2$$

At $x = 0$ and $x = 1500$ mm, y

$$C_2 = 0$$

$$0 = -\left(\frac{50}{9}\right)(1500)^3 + \left(\frac{25}{3}\right)(1000)^3 + 1500\ C_1$$

$$C_1 = 69.45 \times 10^5$$

At $x = 1000$ mm, EIy

$$= -\left(\frac{50}{9}\right)(1000)^3 + \left(\frac{25}{3}\right)(500)^3 + 69.45 \times 10^5 \times 1500$$

$$y = 7.1 \text{ mm}$$

(*c*) The magnitudes of normal stresses on two mutually perpendicular planes, at a point in an elastic body are 60 MPa compressive and 80 MPa tensile respectively. Find the magnitudes of shearing stresses on these planes if the magnitude of one of the principal stresses is 100 MPa tensile. Find also the magnitude of the other principal stress at this point. (10)

Sol.

$$\sigma_1 = 0.5\,(\sigma_x + \sigma_y) + 0.5\left[(\sigma_x - \sigma_y)^2 + 4\,\tau_{xy}^2\right]^{0.5}$$

$$100 = 0.5\,(-60 + 80) + 0.5\left[(-60 - 80)^2 + 4\,\tau_{xy}^2\right]^{0.5}$$

$$= 10 + 0.5\left[140^2 + 4\,\tau_{xy}^2\right]^{0.5}$$

$$180^2 = 140^2 + 4\,\tau_{xy}^2$$

$$\tau_{xy} = \pm\ 56.57 \text{ MPa}$$

$$\sigma_2 = 10 - 90 = -\ 80 \text{ MPa}$$

SECTION-C

Q.5. (*a*) How does a cutting tool fail ? (10)

Sol. When a cutting tool is unable to cut, consuming a reasonable energy, and cannot produce an acceptable finish, it is considered to have failed. The failure of a cutting tool may be due to one or a combination of the following modes

1. Plastic deformation of the tool due to high temperature and large stress.
2. Mechanical breakage of the tool due to large force and insufficient strength and toughness.
3. Blunting of the cutting edge of the tool through a process of gradual wear.

The three modes of failure are shown in Fig. 4.

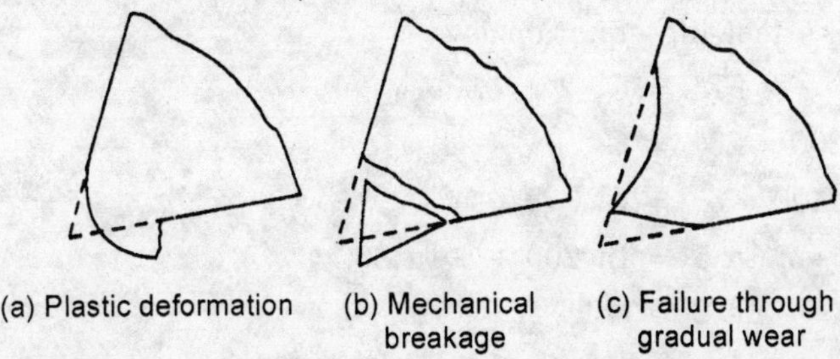

(a) Plastic deformation (b) Mechanical breakage (c) Failure through gradual wear

Fig. 4. Modes of cutting tool failure.

(*b*) List the different methods suitable for making through holes of diameter around 4 mm in a 2 mm thick glass plate. State briefly the working principle of ultrasonic machining (USM). (15)

Sol. Methods for making through holes are :

(*i*) Electro-discharge machining (EDM)

(*ii*) Electron beam machining (EBM)

(*iii*) Plasma arc machining (PAM)

(*iv*) Electro chemical machining (ECM)

(*v*) Ultra-sonic machining (USM)

(*vi*) Abrasive jet machining (AJM).

Ultra-sonic machining. The USM process involves a tool made of a ductile and tough material vibrating with a very high frequency and a continuous flow of an abrasive slurry in the small gap between the tool and the work surface. The tool is gradually fed with a uniform force. The impact of the hard abrasive grain fractures the hard and brittle work surface, resulting in the removal of the work material in the form of small wear particles which are carried away by the abrasive slurry. The tool material being tough and ductile, wears at a much slower rate.

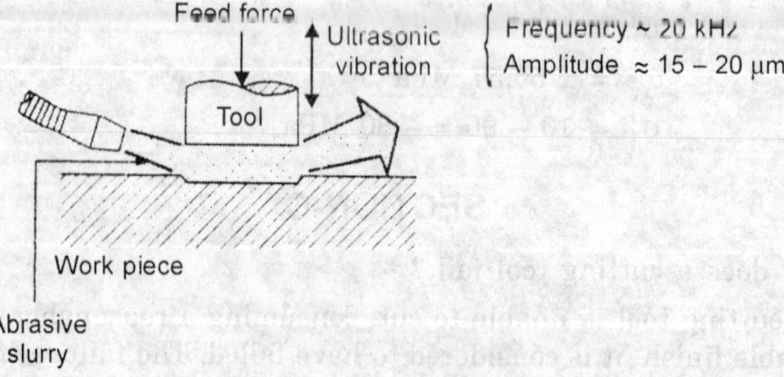

Fig. 5. Ultrasonic machining.

The abrasives used are Boron carbide, Silicon carbide, Aluminium oxide and Diamond.

It can be used for machining hard and brittle materials and their alloys, semi-conductors, non-metals like glass and ceramics.

(*c*) In a drilling operation under a given condition, the tool life was found to decrease from 20 min to 5 min due to increase in drill speed from 200 rpm to 400 rpm. What will be the tool life of the drill under the same condition if the drill speed is 300 rpm ? (15)

Sol. The Taylor's tool life equation is

$$vT^n = C$$

$$200 \times 20^n = C$$

$$400 \times 5^n = C$$

$$\text{In } 400 + n \text{ In } 5 = \text{In } 200 + n \text{ In } 20$$

$$\text{In } 2 = n \ln 4$$

$$n = \frac{\ln 2}{\ln 4} = 0.5$$

$$200 \times 20^{0.5} = C$$

or $$C = 894.43$$

$$300 \times T^{0.5} = 894.43$$

$$T = 8.89 \text{ min}$$

Q.6. (*a*) What are the major constituents of high speed steel (HSS) and uncoated single carbides which are used as cutting tool materials ? How are those carbide tool inserts manufactured? (10)

Sol. HSS : Tungsten = 18%, Chromium = 4%,

Vanadium = 1%, Carbon = 0.7%,

Carbide : Tungsten carbide = 82%,

Titanium carbide = 10%, Cobalt = 8%.

Manufacture of Carbide tool inserts. Pure tungsten powder is mixed under high heat at about 1500°C with pure carbon (lamp black) in the ratio of 94% and 6% by weight. This is then mixed with cobalt until the mass is homogeneous. The homogeneous mass is pressed at pressures from 100-400 MPa, into suitable blocks and then heated in hydrogen, boron, titanium and tantalum to form carbide. Carbide tools are made by brazing or silver soldering the formed inserts on the end of commercial tool holders.

(*b*) How much force will be required to pierce a circular hole of diameter 20 mm in a 2 mm thick mild steel plate with the help of flat ended die and punch in a press tool ? The shear strength of the work material is 350 MPa. (10)

Sol. Force, F = Sheared area x Shear strength of material

$$= n\, dt\, \tau = \pi \times 20 \times 2 \times 350 = 43982 \text{ N}$$

(*c*) Name the instruments by which diameter of machined circular holes can be measured and mention their application capabilities in terms of precision and diameter range. (10)

Sol.

Instrument	Precision	Diameter range
1. Simple bore gauge		300 mm
2. Portable indicating bore gauge		400 mm
3. Bench type indicating gauges		
4. Ball type plug gauges		More than 75 mm
5. Cylinder gauges		
6. Automobile cylinder bore gauge		
7. Pin gauge		75 mm
8. Internal micrometer	0.01mm	5 - 300 mm

(*d*) The thickness of a rectangular brass plate of length L_w and width B_w has to be reduced by *t* mm in one pass by a helical fluted plain or slab milling cutter of length l_c (>B_w), diameter D_c and number of teeth Z_c at cutting velocity, V_c m/min and feed S_c mm/tooth. How will you determine the time that will be required to accomplish the aforesaid work ? All other dimensions are in mm. (10)

Sol. Number of feeds required $= \dfrac{t}{S_c}$

Time to cut one circumferential length in single feed

$$= \frac{\pi D}{V_c \times 10^3} \text{ min}$$

Total time taken to machine whole length

$$L_w = \left[\frac{\pi D}{V_c \times 10^3}\right] \times \left(\frac{t}{S_c}\right) \times \left[\frac{L_w}{\pi D}\right] = \left[\frac{L_w}{V_c \times 10^3}\right] \times \left(\frac{t}{S_c}\right) \text{ min}$$

Q.7. (*a*) A train reservation facility has 5 counters each capable of handling 20 requests per hour. The persons coming for reservation arrive at a mean rate of 90/hour. Assume that each person comes with one request only. Calculate (*i*) the mean number of persons at any time at this facility, (*ii*) the mean time a person spends at the facility, and (*iii*) the average length of queue at each counter when (*a*) the time to serve a request is constant, and (*b*) the time to serve a request is exponentially distributed with the same mean rate *i.e.*., 20/hour. What would happen to the queue if the arrival rate was to reach 105/hour? (25)

Sol. Number of counters, $\sigma = 5$

Arrival rate, $\lambda = 90$ per hour

Service rate, $\mu = 20$ per hour

$$\rho = \frac{\lambda}{\mu} = \frac{90}{20} = 4.5$$

$$P_0 = \left[\sum_{n=0}^{s-1}\left(\frac{1}{n!}\right)\rho^n + \left(\frac{1}{s!}\right).\rho^s.\frac{s\mu}{(s\mu - \lambda)}\right]^{-1}$$

$$= \left[\sum_{n=0}^{4}\left(\frac{1}{n!}\right)4.5^n + \left(\frac{1}{5!}\right).4.5^5.\left\{\frac{5 \times 20}{5 \times 20 - 90}\right\}\right]^{-1}$$

$$= \left[\frac{1 + 4.5 + 4.5^2}{2} + \frac{4.5^2}{6} + \frac{4.5^4}{24} + \frac{120 \times 4.5^5 \times 100}{10}\right]^{-1}$$

$$= 0.00496$$

(*i*) Mean number of persons at any time at this facility,

$$L_s = L_q + \rho = \left[\frac{1}{(s-1)!}.\rho^5.\left\{\frac{\mu\,\lambda}{(s\mu - \lambda)^2}\right\}\right]\rho_0 + \rho_2$$

$$= \left(\frac{1}{4!}\right).4.5^5.\left\{\frac{20 \times 90}{(5 \times 20 - 90)^2}\right\} 0.00496 + 4.5$$

$$= 6.86 + 4.5 = 11.36$$

(*ii*) Mean time a person spends at the facility,

$$W_s = \frac{L_q}{\lambda'} + \frac{1}{\mu} = \frac{6.86}{90} + \frac{1}{20} = 0.126 \text{ hours or } 7.58 \text{ min}$$

(*iii*) (*a*) Average length of queue,

$$L_q = 6.86$$

(***b***) Why do we need inventory ? Explain why we need to optimize the order quantity.

The demand for a component is 10,000 pieces per year. The cost per item is Rs. 50 and the interest cost is 1% per month. The cost associated with placing an order is Rs. 240. What is the EOQ ? (15)

Sol. Need of inventory :

(*i*) To gain economics in purchasing beyond current requirements.

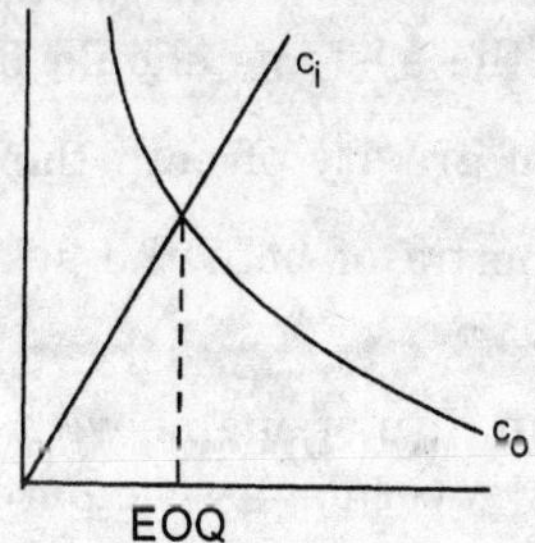

Fig. 6

(*ii*) To level out production cycles by producing to inventory.

(*iii*) To carry a reserve in order to prevent stock out or lost sales.

(*iv*) To maintain service stocks while replacement stocks are in transit.

(*v*) To protect against variation in demand.

We need to optimize the order quantity so that procurement cost is equal to the inventory carrying cost.

$$D = 10{,}000$$

$$C_0 = \text{Rs. } 240$$

$$C_e = 50 \times 0.01 \times 12 = \text{Rs. } 6$$

$$EOQ = \left[\frac{2\,DC_0}{C_c}\right]^{0.5} = \left[\frac{2 \times 10{,}000 \times 240}{6}\right]^{0.5}$$

$$= 894.4 \cong 895 \text{ pieces}$$

MECHANICAL ENGINEERING
PAPER II—2003

(Conventional)

SECTION-A

Q.1.(*a*) State the conditions, which should be satisfied to obtain a kinematically equivalent system with two lumped masses for a connecting rod executing plane motion.

Sol. (*i*) The total mass of both the systems should be same.

(*ii*) The location of centre of gravity of both the systems should be same.

(*iii*) The mass moment of inertia of both the systems about their centre of gravity should be same.

(*b*) What is meant by hunting in case of engine with governor ? Is the degree of hunting more or less in case of isochronous governor as compared to unstable governor?

Sol. This is the phenomena in which the speed of the engine continuously fluctuates above and below the mean speed. If the governor is too sensitive for a small change in speed of rotation, the sleeve moves up and down continuously causing fluctuation of speed due to continuous change in fuel supply.

Degree of hunting is more in an isochronous governor as compared to unstable governor.

(*c*) What is mean by (*i*) pressure angle, and (*ii*) pitch point of a cam-follower mechanism ?

Sol. (*i*) Pressure angle at any point of the cam profile is the angle between the radial direction of the point from the cam centre and the normal drawn to the pitch curve at the point of interest.

(*ii*) Pitch point is the point on the cam profile at which pressure angle is maximum.

(*d*) Define contact ratio of two mating gears and mention its importance in design of gear drive.

Sol. Contact ratio is the ratio of arc of contact to circular pitch of two mating gears. For design of gears drive, contact ratio should be always more than one, so as to ensure that the next pair of teeth engages even before the first pair fully disengages. Moreover the load is shared by the two pairs of teeth.

(*e*) How does damping in a single-degree freedom system affect the natural frequency of vibration ? Explain its expression in terms of system parameters.

Sol. Damped natural frequency,

$$\omega_{nd} = \omega_n \sqrt{1-\zeta^2}$$

where ω_n = natural frequency of a single-degree freedom system,

$$\zeta = \text{damping factors} = \frac{c}{c_c}.$$

As damping factor increases, damped natural frequency decreases.

(f) Mention the relationship between three elastic constants *i.e.*, elastic modulus (E), rigidity modulus (C), and bulk modulus (K), for any elastic material. How is the Poisson's ratio (μ) related to these moduli ?

Sol.
$$E = \frac{9\,KC}{3K + C}$$

(g) State what is meant by kinematic inversion of a mechanism. Name any one of the inversions of double slider-crank mechanism, as used in practice.

Sol. It is the process of transformation of one mechanism to another mechanism by choosing a different link to be a fixed link. Inversions of double slider-crank mechanism are : Scotch yoke, Oldham's coupling, Elliptical crammel.

(h) If a rod of brittle material is subjected to pure torsion, show with help of a sketch, the plane along which it will fail and state the reason for its failure.

Sol. Under pure torsion, tensile and compressive stresses are set up on planes inclined at 45° to the longitudinal axis of the shaft. Brittle material is weak in tension as compared to in compression. Therefore, it will fail along the 45° plane on which the principal stress is tensile.

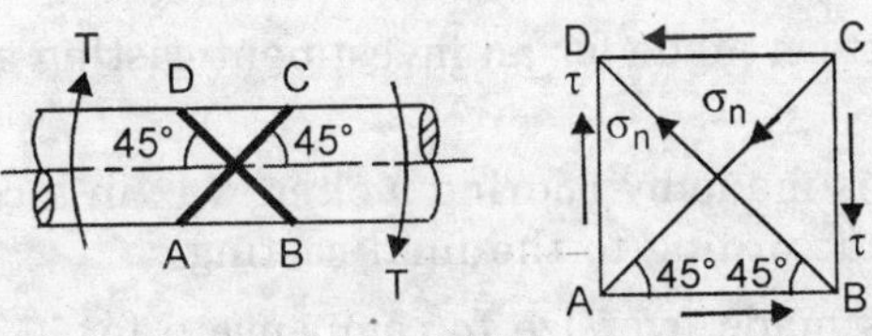

Fig. 1

As shown in the Fig. 1 tensile normal stress develop on plane AC and compressive normal stress on plane BD, such that $\sigma_n = \tau$ in magnitude. Therefore, the shaft will fail along diagonal AC.

(i) The diameter of a mild steel rod is to be reduced from 70 mm to 65 mm by plain turning in a single pass at speed of 300 rpm and feed of 0.4 mm/rev. Determine the material removal rate (MRR) in cubic centimetre (cc)/min that will occur during the aforesaid machining operation.

Sol.
$$\text{MRR} = A\,f\,N$$

$$= \frac{\pi}{4}\left(d_2^2 - d_1^2\right) \times f \times N$$

$$= \frac{\pi}{4}\left(7.0^2 - 6.5^2\right) \times 0.04 \times 300$$

$$= 63.6 \text{ cm}^3/\text{mm}$$

(j) Keeping the other parameters same, what will be the effect of increasing the diameter of the straight fluted plain milling cutter on the average uncut chip thickness and why ?

Sol. In plain milling, mean uncut chip thickness,

$$t_m = f\sqrt{\frac{d}{D}}$$

where

f = Feed per tooth

d = Depth of cut

D = Cutter diameter

If D increases, t_m will decrease.

(k) How is 'Bend Allowance' (BA) calculated in sheet metal working ?

Sol. Bend allowance,

$$BA = \frac{2\pi\theta}{360}(R_1 + Kt)$$

where

θ = angle of bend as measured on the product,

R_1 = inside radius of bend,

t = thickness of sheet metal,

K = a factor depending on severity of bend,

≃ 0.33 fordifficult bends,

≃ 0.50 for easy bends.

(l) How is the wax pattern made for an investment casting and why is the master made oversize ?

Sol. The wax pattern is made by pouring molten wax in a composite die which is made by using a master pattern according to the final casting.

The master pattern is made oversize to compensate for

(i) Shrinkage of wax and casting metal, and

(ii) Machining allowance, if required.

(m) What is meant by Basic length unit (BLU) of CNC machine tools ? Determine the BLU of a CNC machine *ool whose work-table is moved by a 4 mm pitch lead screw driven by a stepper motor having 200 steps in one rotation/revolution.

Sol. BLU is the minimum possible incremental distance by which work table can be moved linearly.

$$\text{BLU} = \frac{\text{Pitch of lead screw}}{\text{Number of steps of stepper motor per revolution}}$$

$$= \frac{4 \times 10^{-3}}{200} = 20 \text{ µm or microns.}$$

(*n*) Name any four welding processes which are employed for permanently joining two dissimilar metals.

Sol. (*i*) Friction welding,

(*ii*) Explosive welding,

(*iii*) Ultrasonic welding,

(*iv*) Resistance welding,

(*v*) Electron beam welding,

(*vi*) Percussion welding.

(*o*) How fine through holes of diameter less than 50 microns can be made in 1.0 mm thick aluminium sheets ?

Sol. Laser beam machining is an appropriate method.

(*p*) Mention the relative advantageous characteristics of polymer products and their manufacturing over those of similar metal products.

Sol. (*i*) Higher strength-to-weight ratio,

(*ii*) Wider range of design and colour possibility,

(*iii*) Lower working temperature,

(*iv*) Lower cost,

(*v*) More chemical and corrosion resistance,

(*vi*) Better surface finish,

(*vii*) Faster production.

(*q*) Enumerate four purposes of keeping inventory.

Sol.

(*i*) It helps to offset forecasting error

(*ii*) It helps in economic utilization of production facilities during lean period,

(*iii*) It permits procurement of items at economic lot sizes,

(*iv*) It eliminates adverse effects of being 'behind' or 'ahead' of schedule.

(*r*) What are the properties to be satisfied by a competitive situation to be called a "GAME" with reference to operation research ?

Sol.

(*i*) There are a finite number of participants called "players",

(*ii*) Each player has a finite number of strategies,

(*iii*) Each player employs the strategy. This constitutes "play",

(*iv*) There is an outcome of the game.

(*s*) Define "Dynamic Programming" and Bellman's principle of optimality.

Sol. Dynamic programming is a computational technique for solution of a multi-stage decision problem where decisions have to be made at successive stages.

Bellman's principle. Whatever the initial state and initial decisions are, the remaining decisions must constitute an optimal policy with regard to the state resulting from the first decision.

(*t*) Illustrate the simplest form of ALLOCATION PROBLEM by Three Mathematical equations. (2 × 20 = 40)

Sol. Maximize, $\sum_{i=1}^{n} g_i (x_i)$

Subject to, $\sum_{i=1}^{n} a_i x_i = c$

and $x_i > o$, for $i = 1$ to n.

SECTION-B

Q.2. (*a*) Two parallel shafts indicated in Fig. 2 are connected by an intermediate shaft with a Hooke's joint at each end. Show how the joints should be oriented to obtain a constant angular velocity ratio between the driving and driven shafts.

The intermediate shaft of the above arrangement has a mass moment of inertia 3 gm and is inclined at 30° to the axes of the driving and driven shafts. If the driving shaft rotates uniformly at 2400 rpm with a steady input torque of 300 N-m, determine the maximum fluctuation of output torque.

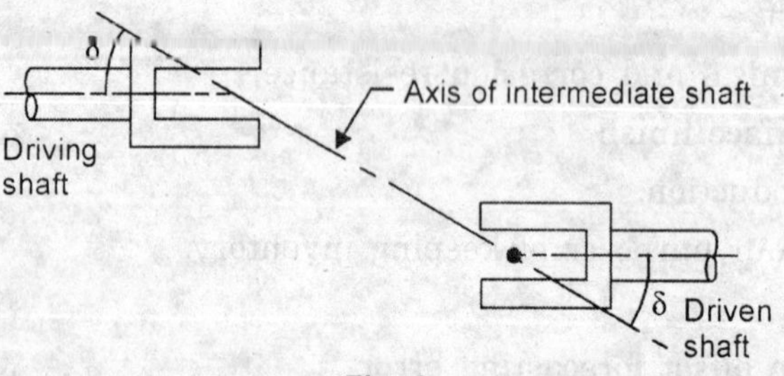

Fig. 2

Sol. To obtain a constant angular velocity ratio between the driving and driven shafts the two forks at the end of intermediate shaft should be *(i)* in the same plane, and *fti)* inclination angle at both the ends should also be same.

Let, α = angle turned by the driving shaft

γ = angle turned by the intermediate shaft

β = angle turned by the driven shaft

δ = angle between intermediate shaft and the axes of driving and driven shafts .

Then
$$\tan \alpha = \tan \gamma \tan \delta$$
$$\tan \beta = \tan \gamma \tan \delta$$
$$\tan \alpha = \tan \beta$$

or
$$\alpha = \beta$$

For maximum acceleration of intermediate shaft,

$$\cos 2\alpha = \frac{2 \sin^2 \delta}{2 - \sin^2 \delta}$$

$$= \frac{2 \sin^2 30°}{2 - \sin^2 30°} = 0.28571$$

$$\alpha = 36.7° \text{ and } 143.3°$$

Maximum acceleration of intermediate shaft,

$$\alpha_{max} = -\omega^2 \left[\frac{\cos \delta \sin^2 \delta \sin 2\alpha}{(1 - \sin^2 \delta \cos^2 \alpha)^2} \right]$$

$$= -\left(\frac{2\pi \times 2400}{60} \right) \left[\frac{\cos 30° \times \sin^2 30° \times \sin 73.4°}{(1 - \sin^2 30° \times \cos^2 36.7°)^2} \right]$$

$$= -18605 \text{ rad/s}^2$$

Maximum torque on intermediate shaft

$$= I\,\alpha_{max} = 0.003 \times 18605 = 55.815 \text{ N-m}$$

Minimum torque on intermediate shaft = – 55.815 N-m

Torque on output shaft = 300 ± 55.815 N-m

Maximum fluctuation of torque = 2 × 55.815 = 111.65 N-m

(***b***) A single cylinder horizontal reciprocating engine mechanism has a crank of 8 cm length and connecting rod 36 cm length. The engine speed is 2000 rpm clockwise. Determine the velocity and acceleration of piston when the crank is 315° from inner dead centre. Also determine the angular acceleration of connecting rod and total acceleration of its mid-point. Use relative velocity and acceleration method only. (10)

Sol. Data given :

$$OC = 8 \text{ cm}, \; CP = 36 \text{ cm}, \; \theta = 360° - 315° = 45°$$

$$\omega = \frac{2\pi N}{60} = \frac{2\pi \times 2000}{60} = 209.4 \text{ rad/s}$$

$$v_c = \omega \times OC = 209.4 \times 8 \times 10^{-2} = 16.76 \text{ m/s}$$

Velocity diagram. Draw the configuration diagram, as shown in Fig. 3 (*a*) to a scale of 1 cm = 4 cm.

Draw *oc* ⊥ OC in clockwise direction.

op || OP and *cp* ⊥ CP

Draw the velocity diagram, as shown in Fig. 3(*b*) to a scale of 1 cm = 4 m/s

Velocity of piston, v_p = 3.4 cm = 13.6 m/s

Velocity of rod, v_{pc} = 3 cm = 12 m/s

$$\omega_{pc} = \frac{v_{pc}}{cp} = \frac{12}{0.36} = 33.3 \text{ rad/s}$$

$$f^n_c = \frac{v_c^2}{oc} = \frac{(16.76)^2}{0.08} = 3511.22 \text{ m/s}^2$$

$$f^n_{cp} = \frac{v_{pc}^2}{cp} = \frac{(12)^2}{0.36} = 400 \text{ m/s}^2$$

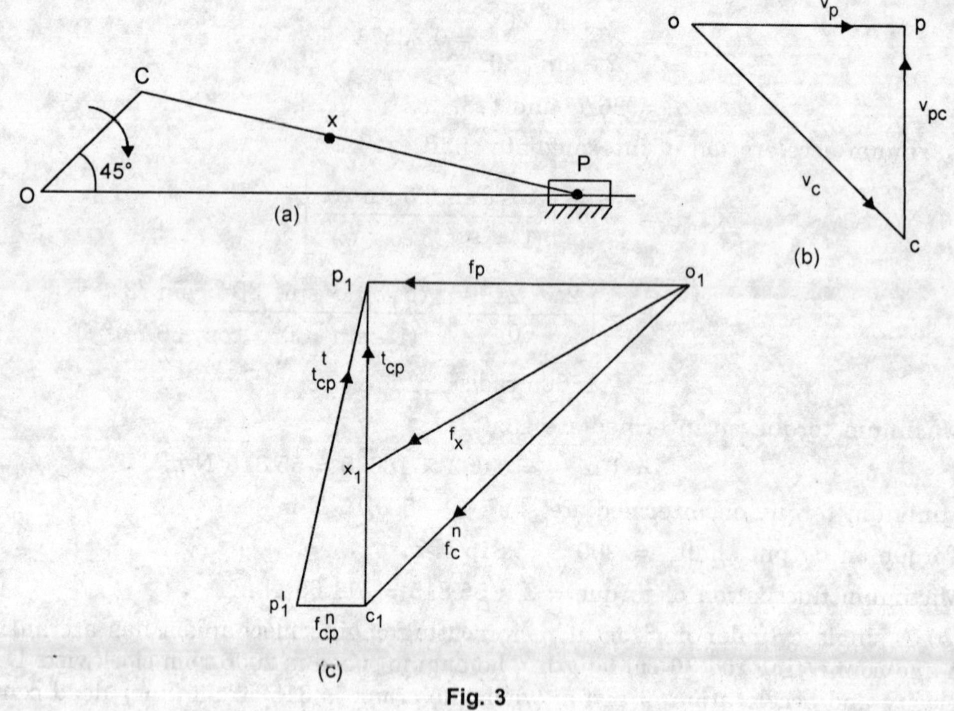

Fig. 3

Acceleration diagram:

Draw $f_c^n = o_1 c_1 \parallel$ OC and $f^n_{cp} = c_1 p_1' \parallel$ PC to a scale of

$$1 \text{ cm} = 400 \text{ m/s}^2.$$

Draw $p_1' p_1 \perp$ PC and $o_1 p_1 \parallel$ OP.

Join $c_1 p_1$. Let x_1 be the mid-point of $c_1 p_1$. Join $o_1 x_1$.

Then $o_1 x_1 = f_x$, the acceleration of mid-point X of PC.

The acceleration diagram has been draw in Fig. 3 (*c*).

Acceleration of piston = $o_1 p_1$ = 6.3 cm = 2520 m/s^2

Tangential acceleration of connecting rod,

$$f^t_{cp} = p_1' p_1 = 6.1 \text{ cm} = 2440 \text{ m/s}^2$$

Angular acceleration of connecting rod

$$\alpha_{cp} = \frac{f^t_{cp}}{pc} = \frac{2440}{0.36} = 6778 \text{ rad/s}^2$$

Total acceleration of mid-point of connecting rod

$$f_x = o_1 x_1 = 7 \text{ cm} = 2800 \text{ m/s}^2$$

(*f*) Determine the addendum of the teeth of a gear pair consisting of two spur wheels each having 30 teeth to have a minimum contact ratio equal to 2. The circular pitch is 2.5 cm and pressure angle is 20°. (10)

Sol.

$$\text{Contact ratio} = \frac{\text{Length of path of constant, } L_p}{p \cos a}$$

$$L_p = 2\left[\left(r^2_{a1} - r^2_{b1}\right)^{0.5} - r_1 \sin \alpha\right]$$

$$r_1 = \frac{pz}{2\pi} = \frac{25 \times 30}{2\pi} = 119.366 \text{ mm}$$

$$r_{b1} = r \cos \alpha$$

$$= 119.366 \cos 20° = 112.16 \text{ mm}$$

$$2 = \frac{2\left[\left\{r^2_{a1} - (112.675)\right\}^{0.5} - 119.366 \sin 20°\right]}{25 \cos 20°}$$

$$r_{a1} = 129.299 \text{ mm}$$

$$= r_1 + h_{a1}$$

$$h_{a1} = 129.299 - 119.366$$

$$= 9.933\text{mm}$$

Q.3. (*a*) A steel bolt of diameter 10 mm passes through a brass tube of internal diameter 15 mm and external diameter 25 mm. The bolt is tightened by a nut so that the length of tube is reduced by 1.5 mm. If the temperature of the assembly is raised by 40°C, estimate the axial stresses in the bolt and the tube before and after heating. Material properties for steel and brass are :

$$E_s = 2 \times 10^5 \text{ N/mm}^2,\ \alpha_s = 1.2 \times 10^{-5}/°\text{C};$$
$$E_b = 1 \times 10^5 \text{ N/mm}^2,\ \alpha_b = 1.9 \times 10^{-5}\ °\text{C}.$$

Sol.

$$A_s = \frac{\pi \times 100}{4} = 78.54 \text{ mm}^2$$

$$A_b = \frac{\pi(25^2 - 15^2)}{4} = 314.16 \text{ mm}^2$$

Before heating, $\sigma_s A_s = \sigma_b A_b$

$$\sigma_s = \left(\frac{314.16}{78.54}\right)\sigma_b = 4\sigma_b$$

$$\delta\ell = \frac{\sigma_b \ell}{E_b}$$

Length of tube or bolt is not given. Let l = 2 m

$$1.5 = \frac{\sigma_b \times 2 \times 10^3}{1 \times 10^5}$$

$$\sigma_b = 75 \text{ N/mm}^2$$

$$\sigma_s = 300 \text{ N/mm}^2$$

After heating, $\frac{\sigma_b}{E_b}+\frac{\sigma_s}{E_s} = (\alpha_b - \alpha_s)$

$$\frac{\sigma_b}{1\times 10^5}+\frac{4\,\sigma_b}{2\times 10^5} = (1.9 - 1.2)\times 10^{-5}$$

$$\sigma_b = 9.34\ \text{N/mm}^2$$

$$\sigma_s = 37.36\ \text{N/mm}^2$$

(*b*) A solid phosphor bronze shaft 60 mm in diameter is rotating at 800 rpm and transmitting power. It is subjected to torsion only. An electrical resistance strain gauge mounted on the surface of the shaft with its axis at 45° to the shaft axis, gives the strain as 3.98×10^{-4}. If the modulus of elasticity for bronze is 105 GN/mm² and Poisson's ratio is 0.3, find the power being transmitted by the shaft. Bending effect may be neglected.

Sol. For a shaft under torsion, at 45° planes with the axis of the shaft, there are two mutually perpendicular normal stresses equal in magnitude but opposite in nature. The magnitude of these stresses is equal to the shear stress. Therefore,

$$\sigma_1 = \tau \text{ and } \sigma_2 = -\tau$$

Hoop strain, $\varepsilon_\theta = \frac{\sigma_1 - \nu\sigma_2}{E}$

$$\varepsilon_{45} = \frac{\tau(1+\nu)}{E}$$

$$3.98\times 10^{-4} = \frac{\tau(1+0.3)}{105\times 10^3}$$

$$\tau = 32.146\ \text{MPa}$$

Torque, $T = \frac{\pi d^3\tau}{16} = \frac{\pi\times 60^3\times 32.146}{16} = 1363.37\ \text{N-m}$

Power transmitted, $P = \frac{2\pi\, NT}{60\times 10^3}\ \text{kW} = \frac{2\,\pi\times 800\times 1363.7}{60\times 10^3} = 114.217\ \text{kW}$

(*c*) Find the slope and deflection at the free end of a cantilever beam of length 6 m loaded as shown in Fig. 4, using method of superposition. Evaluate their numerical values using E = 200 GPa, I = 1×10^{-4} m⁴ and W = 1 kN.

Sol. Slope due to load at free end,

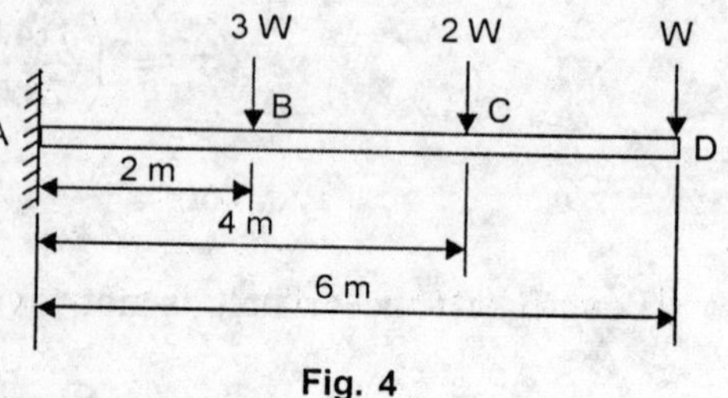

Fig. 4

$$\theta = \frac{WL^2}{2EI}$$

Deflection at free end due to load at free end,

$$\delta = \frac{WL^3}{3EI}$$

Deflection at any other point

$$= \frac{WL^3}{3EI} + \theta\times L$$

Load point	Load	Length L from A, m	Slope, 6	Length, L from D, m	Deflection
B	3 W	2	6 W/EI	4	32 W/EI
C	2 W	4	16 W/EI	2	224 W/3EI
D	W	6	18 W/EI	0	72 W/EI
		Σ	40 W/EI		536 W/EI

$$\theta_D = \frac{40 \times 10^3}{200 \times 10^9 \times 10^{-4}} = 0.002 \text{ rad}$$

$$\delta_D = \frac{536 \times 10^3}{3 \times 200 \times 10^9 \times 10^{-4}} = 9 \text{ mm}$$

Q.4. ***(a)*** A steel rod of diameter 50 mm is forced into a bronze casing of outside diameter 90 mm, producing a tensile hoop stress of 30 MPa at the outside diameter of the casing. Find (*i*) the radial pressure between the rod and the casing, (*ii*) the shrinkage allowance and (*iii*) the rise in temperature which would just eliminate the force fit

Assume the following material properties :

$$E_s = 2 \times 10^5 \text{ MPa}, \nu_s = 0.25, \alpha_s = 1.2 \times 10^{-5} \text{ /°C};$$

$$E_b = 1 \times 10^5 \text{ MPa}, \nu_b = 0.30, \alpha_b = 1.9 \times 10^{-5} \text{ /°C}. \quad (20)$$

Sol. Bronze casing,

$$\sigma_r = \frac{B}{r^2} - A$$

$$\sigma_\theta = \frac{B}{r^2} + A$$

At $r = 45$ mm, $\sigma_r = 0$ and $\sigma_\theta = 30$ MPa

$$0 = B - \frac{A}{45^2}$$

$$30 = B + \frac{A}{45^2}$$

Solving, we get, $A = 15, B = 30375$

At $r = 25\text{mm}, \sigma_r = \frac{30375}{625} - 15 = 36.6$ MPa

(*i*) $p_s = -\sigma_r = -33.6$ MPa

At $r = 25\text{mm}, \sigma_\theta = \frac{30375}{625} + 15 = 63.6$ MPa

Hoop strain at $r = 25$ mm

$$\varepsilon_\theta = \frac{1}{E_b}[\sigma_\theta - \nu_b \, \sigma_r]$$

$$= \frac{1}{1 \times 10^5}[63.6 + 0.3 \times 33.3] = 7.368 \times 10^{-4}$$

Increase in diameter of casing

$$= d_1 \varepsilon_r = 50 \times 7.368 \times 10^{-4}$$

$$= 0.03684 \text{ mm.}$$

Steel rod :

$$\sigma_\theta = \sigma_r = -33.6 \text{ MPa}$$

$$\varepsilon_\theta = \frac{1}{E_s}[\sigma_\theta - \nu_s \sigma_r]$$

$$= \frac{1}{2 \times 10^5}[-33.6 + 0.25 \times 33.6] = -1.26 \times 10^{-4}$$

Decrease in diameter of rod

$$= 50 \times 1.26 \times 10^{-4} = 0.0063 \text{ mm.}$$

(*ii*) Shrinkage allowance on radius

$$= \frac{1}{2}(0.03684 + 0.0063) = 0.02157 \text{ mm}$$

Let $\Delta\theta$ be the rise in temperature

$$r_1 (\alpha_b - \alpha_s) . \Delta\theta = 0.02157$$

$$\Delta\theta = \frac{0.002157}{25 \times 0.7 \times 10^{-5}} = 123.26 \text{ °C}$$

(***b***) A cylindrical billet of 40 mm diameter and 100 mm length is reduced by indirect (backward) extrusion to a 15 mm diameter using flat dies. If the Johnson equation has $a = 0.8$ and $b = 1.5$ and the flow curve for the work metal has A = 750 MPa and $n = 0.15$, determine :

(*i*) Extrusion ratio

(*ii*) True strain (homogeneous deformation)

(*iii*) Extrusion strain

(*iv*) Ram force. (10)

Sol.

(*i*) Extrusion ratio $= \dfrac{A_o}{A_f} = \left(\dfrac{D_o}{D_f}\right)^2 = \left(\dfrac{40}{15}\right)^2 = 7.11$

(*ii*) True strain, $\bar{\varepsilon} = \ln\left(\dfrac{A_o}{A_f}\right) = \ln 7.11 = 1.962$

(*iii*) Extrusion strain, $\varepsilon_x = a + b \ln\left(\dfrac{A_o}{A_f}\right)$

$$= 0.8 + 1.5 \text{ In } 7.11 = 3.742$$

(*iv*) Mean extrusion pressure,

$$p = K\,\varepsilon_x = 750 \times 3.742 = 2806.5 \text{ MPa}$$

Ram force $= p \times$ Billet area

$$= 2806.5 \times \frac{\pi}{4} \times 40^2 = 3526648 \text{ N}$$

(*c*) Mention the major essential constituents of any Flexible Manufacturing System (FMS) used for production by machining. Also state the basic functions of those constituents (sub systems). (10)

Sol. The major essential constituents of any FMS and their basic functions are given below :

Constituents	Basic functions
1. Host computer	(*i*) Computer-aided design (CAD)
	(*ii*) Computer-aided manufacturing (CAM)
	(*iii*) Material handling
2. CNC machines	For actual machining operation.
3. Robots	(*i*) Machine loading and unloading.
	(*ii*) Handling tools, jigs and fixtures (if needed)
	(*iii*) Automatic inspection (if required)
	(*iv*) Painting (if required)
	(*v*) Assembly.
4. Peripherals	(*i*) Conveyors
	(*ii*) Feeders
	(*iii*) Warehouse keeping tools, blanks, fixtures etc.

SECTION-C

Q.5. (*a*) Compare the solidification time of two optimum side-risers of the same volume when one has a cylindrical shape and the other is of square parallelopiped.

Sol. Cylindrical riser :

Let, d = diameter and h = height

Volume, $$V = \frac{\pi d^2 h}{4}$$

Surface area, $$A = \pi\, dh + 2 \times \frac{\pi d^2}{4} = \pi\left(dh + \frac{d^2}{2}\right)$$

Now, $$h = \frac{4V}{\pi\, d^2}$$

$$A = \frac{4V}{d} + \frac{\pi\, d^2}{2}$$

For A to be minimum, $\frac{\partial A}{\partial d} = 0$

$$-\frac{4V}{d^2} + \pi d = 0$$

$$d^3 = \frac{4V}{\pi} + d^2 h$$

or $$d = h$$

$$\frac{A}{V} = \frac{6}{d}$$

Square riser :

Let a = Side of the square and h = Height

$$V = a^2 h,\ A = 2a^2 + 4ab = 2a^2 + \frac{4a\,V}{a^2} = 2a^2 + \frac{4V}{a}$$

$$\frac{\partial A}{\partial a} = 0 \text{ gives}$$

$$4a - \frac{4V}{a^2} = 0$$

or $$a = h$$

Risers are of the same volume. Hence

$$\frac{\pi d^3}{4} = a^3$$

$$a = \left(\frac{\pi}{4}\right)^{1/3} d = 0.922\,d$$

Solidification time, $$t \propto \left(\frac{V}{A}\right)^2$$

$$t_{cyl} \propto \left(\frac{d}{6}\right)^2$$

$$t_{sq} \propto \left(\frac{a}{6}\right)^2$$

$$\frac{t_{sq}}{t_{cyl}} = \left(\frac{a}{d}\right)^2 = (0.922)^2 = 0.851$$

(b) Name the processes by which external screw threads can be produced (both manually using tools and in machines) (10)

Sol. 1. Manually operated dies :

(*i*) solid die

(*ii*) split die

(*iii*) spring die.

2. Machining :
   - (*i*) single point chasing in centre lathes
   - (*ii*) self opening die in capstan lathes
   - (*iii*) solid button die in automatic screw cutting lathes
   - (*iv*) long thread milling in lathe
   - (*v*) short thread milling.
3. Grinding (for finishing or straight in case of small threads)
   - (*i*) single thread or multi-thread grinding.
4. Thread rolling.
   - (*i*) circular die
   - (*ii*) flat die, horizontal, vertical, inclined/slant.

(*c*) For resistance spot welding of two aluminium sheets, each 2 mm thick, a current of 5000 A was passed for 0.15 sec. The total resistance was estimated to be 75 μΩ and the nugget diameter and thickness were measured to be 5 mm and 2.5 mm respectively. What would be the proportion of heat energy utilized for welding if the melting energy per unit volume for aluminium is taken as 2.9 J/mm ? (10)

Sol. Heat generated,

$$H_g = I^2Rt = (5000)^2 \times 75 \times 10^{-6} \times 0.15$$
$$= 281.25 \text{ J}$$

Volume of weld, $$V = \frac{\pi D^2 h}{4} = \frac{\pi \times 5^2 \times 2.5}{4} = 49.1 \text{ mm}^3$$

Heat required for melting,

$$H_r = 2.9 \times 49.1 = 142.4 \text{ J}$$

Proportion of heat utilized for welding

$$= \frac{H_r}{H_g} = \frac{142.4}{281.25} = 0.506 \text{ or } 50.6\%$$

Q.6. (*a*) During turning a carbon steel rod of 160 mm diameter by a carbide tool of geometry ; 0°, 0°, 10°, 8°, 15°, 75°, 0 (mm) at speed of 400 rpm, feed of 0.32 mm/rev. and 4.0 mm depth of cut, the following observations were made :

Tangential component of the cutting force,

$$P_z = 1200 \text{ N}$$

Axial component of the cutting force,

$$P_x = 800 \text{ N}$$

Chip thickness (after cut), $t_2 = 0.8$ mm.

For the above machining condition determine the values of

(*i*) Friction force, F and normal force, N acting at the chip-tool interface.

(*ii*) Yield shear strength of the work material under this machining condition.

(*iii*) Cutting power consumption in kW. (20)

Sol.

(*i*) $P_x = 800N, P_z = 1200$ N

$t_1 = 4$ mm, $t_2 = 0.8$ mm

Side cutting edge angle,

$\alpha_s = 75°$

ζ = chip reduction co-efficient

$$= \frac{t_2}{f \sin \alpha_s} = \frac{0.8}{0.32 \sin 75°} = 2.588$$

$$P_{xy} = \frac{P_x}{\sin \alpha_s} = \frac{800}{\sin 75°} = 828 \text{ N}$$

Orthogonal rake, $\gamma_0 = 0°$

$F = P_z \sin \gamma_0 + P_{xy} \cos \gamma_0$

$= P_{xy} = 828$ N

$N = P_z \cos \gamma_0 - P_{xy} \cos \gamma_0 = P_z = 1200$ N.

(*ii*) $P_z = t_1 f \tau_s (\zeta - \tan \gamma_0 + 1)$

$1200 = 4 \times 0.32 \times \tau_s (2.588 - 0 + 1)$

$\tau_s = 261$ MPa

(*iii*) Cutting power consumption,

$$P_c = P_z v_c = \frac{1200 \times \pi \times 160 \times 400}{10^3 \times 60 \times 10^3} = 4 \text{ kW}$$

(*b*) When a rectangular section, 100 mm wide and 60 mm thick, is rolled to some thickness in one pass with rolls of 400 mm diameter in hot rolling, where co-efficient of friction is 0.3,

(*i*) What is the highest possible reduction in that pass ?

(*ii*) What will be the "length of contact" with the rolls ?

(*iii*) If the mean rolling pressure is 500 MPa, what is the total rolling load ? (10)

Sol. For highest reduction, $\tan \alpha = \mu$

$\alpha_{max} = \tan^{-1} 0.3 = 16.7°$

(*i*) $\Delta h_{max} = D (1 - \cos \alpha_{max})$ (See Fig. 5)

$= 400 (1 - \cos 16.7°)$

$= 16.88$ mm

Fig. 5

(*ii*) Length of contact, $l_c = \sqrt{R.\Delta h}$

$= \sqrt{200 \times 16.88} = 58.1$ mm

(*iii*) Area of contact, $A_c = l_c \times b$

$= 58.1 \times 100 = 5810 \text{ mm}^2$

Total rolling load $= 500 \times 5810 = 2905$ kN

(c) (*i*) What are the detrimental effects of the high cutting temperature on the machined product and the cutting tool ?

(*ii*) How can such cutting temperature be reduced without sacrificing productivity ?

(6 + 4=10)

Sol. (*i*)

(*a*) On Machined product:

(*i*) Dimensional deviation (in accuracy) due to thermal distortion,

(*ii*) Deterioration of surface integrity (quality) by oxidation, burning and induction of tensile residual stresses,

(*iii*) Development of surface/sub-surface microcracks.

(*b*) On tool:

1. Reduction of tool life by

(*i*) Rapid tool wear,

(*ii*) Chances of early plastic deformation and/or fracturing at the tool tip.

2. Cutting temperature can be reduced by.

(*i*) Proper selection of speed-feed combination,

(*ii*) Proper selection of cutting tool material and geometry,

(*iii*) Proper selection and application of cutting fluicT

Q.7. (*a*) A certain company produces tea trays and ash trays out of sheet metal. Following data is given on capacity availablity and economics of each product:

Department	**Time taken for**		**Available**
	Ash tray	**Tea tray**	**time**
Stamping	10 sec	20 sec	30000 sec
Forming	15 sec	5 sec	30000 sec.

Determine the optimum production schedule. Each ash tray contributes Rs. 20 to the gross profit and each tea-tray contributes Rs. 30 to the gross profit. Total daily fixed costs amount to Rs. 45000.

Solve this problem graphically. What is the maximum net profit per day at the optimum production level, including the effect of fixed cost ? (20)

Sol. Let, x_1 = optimum number of ash trays

x_2 = optimum number of tea trays

Objective function, $z = 20\,x_1 + 30\,x_2$

Subject to : $10\,x_1 + 20\,x_2 \leq 30000$ ...(1)

$15\,x_1 + 5\,x_2 \leq 30000$ ...(2)

$x, x_2 \geq 0$

For Eq. (1), wnen $x_2 = 0, x_1 = 3000$

$x_1 = 0, x_2 = 1500$

For Eqn. (2), when $x_2 = 0,$

$$x_1 = 2000$$

$$x_1 = 0,$$

$$x_2 = 6000$$

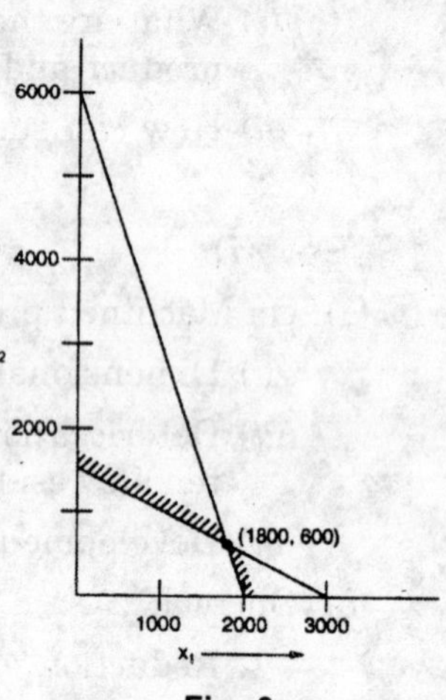

Fig. 6

Lines given by Eqn. (1) and (2) intersect at $x_1 = 1800$, $x_2 = 600$, which is the bound point. The Constraint lines have been drawn in Fig. 6.

$$Z = 20 \times 1800 + 30 \times 600$$

$$= \text{Rs. } 54000/\text{day}$$

Net profit $= 54{,}000 - 45{,}000$

$$= \text{Rs. } 9000/\text{day}$$

(***b***) What is a "Trend Line" in sales forecasting by Time series analysis ?

In a company the sales figures for a particular product for 16 consecutive quarters were recorded as follows :

Quarter	Units sold
1	1000
2	3000
3	4000
4	2000
5	1000
6	3000
7	5000
8	3000
9	2000
10	4000
11	6000
12	3000
13	2000
14	5000
15	7000
16	4000

Determine the trend line for this product. (10)

Sol. In time series analysis method of sales forecasting, a company analysis its past sales to determine if there is a "trend". This trend is then projected into the future. The trend for the given data is shown in Fig. 7.

A "Trend Line" is represented by an equation of the type :

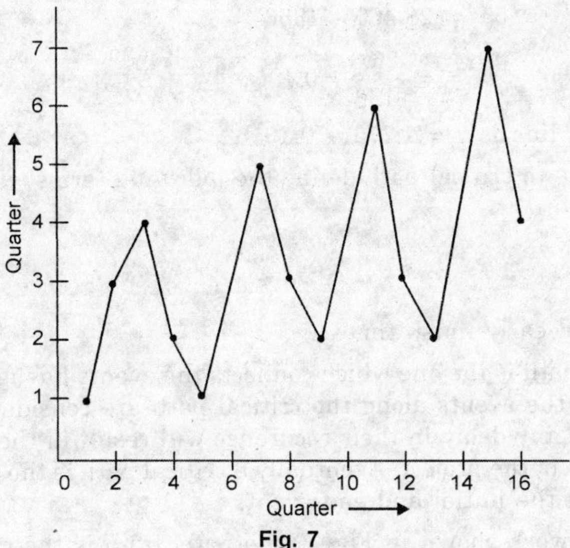

Fig. 7

$$y = a + bx$$

where y = number of units sold

x = number of time units (quarter or year)

a, b = constants.

The trend line can be obtained by analysing past data. The constants 'a' and 'b' are determined by using the least square method from the following equations :

$$\sum y = na + b\sum x$$

$$\sum xy = a\sum x + b\sum x^2$$

where n = number of data points.

Quarter, x	1	2	3	4	5	6	7	8	9	10	11	12	13	14	15	16	$\sum x = 136$
Units sold, y	1	3	4	2	1	3	5	3	2	4	6	3	2	5	7	4	$\sum y = 55$
xy	1	6	12	8	5	18	35	24	18	40	66	36	26	70	105	64	$\sum xy = 534$
x^2	1	4	9	16	25	36	49	64	81	100	121	144	169	196	225	256	$\sum x^2 = 1496$

$$55 = 16a + 136b \times 11$$
$$534 = 136a + 1496b$$
$$605 = 176a + 1496b$$
$$- \quad - \quad -$$
$$-71 = -40a$$

$$a = 1.775$$

$$55 = 28.400 + 1366$$

$$b = = \frac{26.6}{136} = 0.1956$$

Hence, the trend line is $y = 1.775 + 0.1956\ x$

(c) With reference to critical path define the following terms with a brief description of each of them :

(i) Critical path,

(ii) Project graph,

(iii) Early start and early finish times. (10)

Sol. (i) A critical path is the one which connects the events havng zero or minimum slack times. All the events along the critical path are considered to be critical in the sense that any delay in their occurence will result in the delay in the scheduled completion of the project. Eventually a critical path is the longest path (time wise) connecting the initial and end event.

In the network shown in Fig. 8, the path 1-2-5 is the critical path. In CPM, a critical path is the one which passes through critical activities. This critical path is determined on the basis of minimum float for each activity.

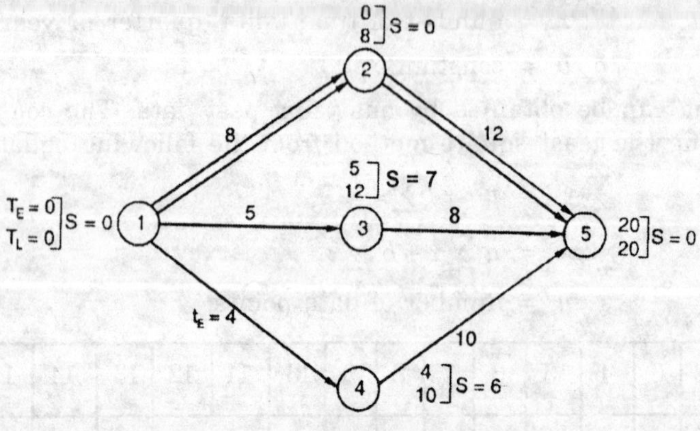

Fig. 8

(ii) Project graph is a graphical representation of the various activities of the project, showing their precedence for performance and inter-connection. Arrows are used to connect the activities. The events are numbered such that they reflect the logical sequence of the activities. The figure shown above is the project graph.

(iii) The Earliest start time (EST) for an activity is the earliest time by which it can commence. This is equal to the earliest event time associated with the tail of the activity arrow.

$$\text{EST} = T_E^2$$

The Earliest finish time (EFT) for an activity is defined as the earliest time by which it can be finished. This is equal to the earliest start time plus estimated duration of the activity.

$$\text{EFT} = T_E^i + t^{ij}$$

MECHANICAL ENGINEERING
PAPER II—2004

(Conventional)

SECTION-A

Q.1. **(*a*)** Which one of the three shafts listed here has the highest ultimate tensile strength?

What is the approximate carbon content in each steel? (*i*) Mild steel (*ii*) Cast iron (*iii*) Spring steel.

Sol. Spring steel has the highest ultimate tensile strength.

Material	Carbon content
Mild steel	0.15 – 0.45%
Cast iron	1.7 – 4.5%
Spring steel	0.45 – 1.1%.

(*b*) State the Von-Mises's theory. Also give the mathematical expression.

Sol. This theory states that a material fails when the shear strain (distortion) energy under three dimensional state of stress at a point in a material exceeds the shear strain energy at yield as calculated based upon the standard tensile test performed under uniaxial tension.

$$\left[\frac{1}{12G}\right]\left[(\sigma_1-\sigma_2)^2+(\sigma_2-\sigma_3)^2+(\sigma_3-\sigma_1)^2\right]$$

$$=\frac{\sigma_{yt}^2}{6G}$$

$$(\sigma_1-\sigma_2)^2+(\sigma_2-\sigma_3)^2+(\sigma_3-\sigma_1)^2 = 2\,\sigma^2_{yt}$$

where σ_{yt} = yield strength under simple tension.

(*c*) Why are shafts not made out of cast iron?

Sol. Shafts are subjected to transverse and horizontal shear stresses, which give rise to tensile stress of equal magnitude on 45° planes. Cast iron is very weak in tension and brittle. Therefore, it is likely to fail earlier. So it is not recommended for shafts.

(*d*) When are I-sections preferred in engineering applications? Elaborate your answer.

Sol. The section modulus in bending of I-section is very large as compared to other sections having the same area of cross-section. Thus it offers more resistance to bending moment, and is preferred.

(*e*) For the three velocity ratios specified below which one of the following gear drive would be preferable for transmitting power with reductions of 4, 20 and 70?

The drives are single stage spur gear reduction, single start worm drive and triple start worm drive.

Sol.

Reduction	**Drive**
4	Single stage spur gear
20	Single start worm drive
70	Triple start worm drive.

(*f*) In a single involute gear already manufactured of the three diameters given below two are determined. Which are those two ? Substantiate your answer. The three diameters given are : addendum circle diameter, pitch circle diameter and base circle diameter.

Sol. Pitch circle diameter and base circle diameter

$$D_b = D_p \cos \phi.$$

The pressure angle is dependent on these two diameters, which is the governing factor for the design of gears.

(*g*) In an epicyclic gear train the number of planets in any row has to be more than one. Do you agree or disagree, why?

Sol. Yes, from the point of view of balancing.

(*h*) In a typical power transmission from an induction motor running at 1440 rpm to an output speed of 1 rpm. Which order of reduction of the three drives listed below is preferred? Why?

Worm gear reduction, belt reduction and spur gear reduction.

Sol. Worm gear reduction is the only suitable drive to achieve such a heavy reduction of 1440. Other drives are not suitable as belts can be used for reductions upto 10 and single stage spur gears upto 4.

(*i*) Is the cam in mesh with the follower a case of complete constraint? Substantiate.

Sol. Yes, because the motion of the follower is constrained by the cam profile and the guides provided for the follower.

(*j*) Do you agree that in a four stroke four cylinder I.C. engine the primary forces and the primary couples are in balance ? Substantiate your answer.

Sol. In a four stroke four cylinder I.C. engine,

Primary force,

$$F_p = Rr\omega^2 [\cos\theta + \cos(180° + \theta) + \cos(180° + \theta) + \cos\theta] = 0$$

Primary couple, »

$$C_p = Rr\omega^2 [3a \,.\, \cos\theta + a.\cos(180° + \theta) - a.\cos(180° + \theta) - a.\cos\theta] = 0$$

Hence the primary forces and primary couples are in balance.

(*k*) Distinguish between unilateral and bilateral tolerances for the dimensions. Indicate the tolerance zones for each by taking an example.

Sol. In a unilateral system, the tolerance is specified on one side of the nominal size only. For example :

+ 0.001

50

+ 0.000

Whereas, in the bilateral system, the tolerances are specified on both sides of the nominal size. For example :

+ 0.005

50

0.002

(*l*) What is the hole basis and shaft basis of fits? Which one is preferred in practice and why?

Ans. Hole basis system. The design size of the hole is the basic size and the allowance is applied to the shaft.

Shaft basis system. The design size of the shaft is the basic size and the allowance is applied to the hole.

Hole basis system is preferred because hole is the most difficult to produce than a shaft due to the fixed character of the hole producing tools like drills and the reamers. The shaft basis system, however is used mostly in industries using semi-finished shafting as raw material.

(*m*) Write the chemical equation for the reaction going on in the oxy-acetylene flame. Where do you get the most part of the oxygen, is it from the cylinder or the atmosphere ?

Sol. $2\ C_2H_2 + O_2 \Rightarrow 2\ CO + H_2$

$2\ CO + O_2 \Rightarrow 2\ CO_2$

$2H_2 + O_2 \Rightarrow 2H_2O$

Most part of the oxygen is obtained from the atmosphere during oxy-acetylene welding.

(*n*) Clearances have to be provided on the press tools. What is the order of clearances for the shearing operation and deep drawing operation ? On what tool will you provide the clearance in

(*i*) Punching operation

(*ii*) Blanking operation.

Sol.

Operation	Punching clearance	Blanking clearance
Shearing	Clearance is provided on die 2-6% for round jobs and 3-8% for others, of sheet metal thickness. Angular clearance on die = 0.25° – 15° per side	Clearance is provided on the punch. About 5% of sheet metal thickness.
Deep drawing	5 – 20% of sheet metal thickness	5-20% of sheet metal thickness.

(*o*) What is a three high rolling mill ? Indicate the movement of rolls and the workpiece.

Sol. A three high rolling mill consists of three rollers mounted one over another for rolling the ingots. Top and bottom rolls move in the same directions and the middle roll moves in the opposite direction, as shown in Fig. 1 below.

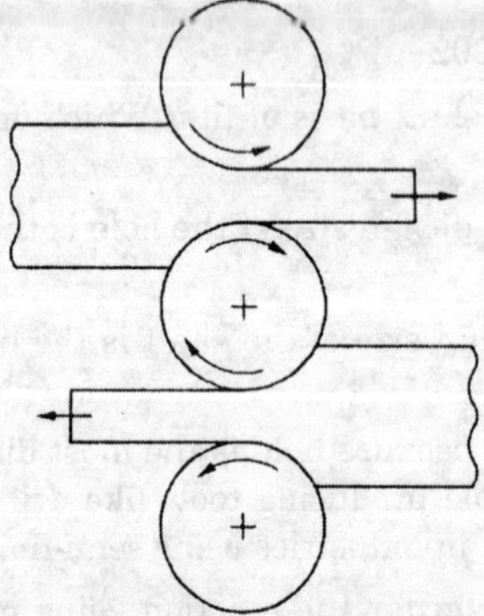

Fig. 1. Three high rolling mill

(*p*) What is the basis for Selective Inventory Control ? Mention four applications of the same in the Inventory management

Sol. Selective inventory control is based on the principle that it is impossible to manage any control every item in inventory holdings in the same way and skill so as to reduce investment in inventories and also to avoid stock-outs and shortages. Selective control concentrates on those items where it is justified due to essentiality or amount of money involved. The various selective inventory controls are :

1. A-B-C : always Better control.
2. VED analysis : Vital-essential-desirable.
3. SDE analysis : Scarce-difficult-easily.
4. MNG analysis : Moving-Nonmoving-Ghost.

(*q*) What is the distribution followed by the activity durations of a PERT network? Mention its mean and standard deviation.

Sol. PERT network employs Beta-distribution for the time estimates.

Standard deviation of each activity,

$$\sigma = \frac{t_p - t_0}{6}$$

where t_p = permissible time, and

t_0 = optimistic time.

(*r*) What do you understand by tracking signal in forecasting ? How is it computed?

Sol. Tracking signal (TS). It is the method of measuring forecast error.

It is the ratio of forecast error to mean absolute deviation.

$$TS = \frac{\sum_{t=1}^{N} (D_t - F_t)}{\left[\frac{\left\{\sum_{t=1}^{N} |D_t - F_t|\right\}}{N}\right]}$$

where D_t = actual current demand,

F_t = forecast for time t,

N = number of values included in average.

(*s*) Distinguish between P-system and Q-system in inventory control.

Sol. P-system. The overview period is constant. The order quantity would vary during each order and is determined by the state of stock level during the review. The maximum inventory level is fixed. This is used for *B* and C types of items and items having short lead times.

Q-system. The order quantity is fixed and orders are triggered whenever stock level reaches the record point. The level of safety stock depends upon the variations during the lead time and service levels fixed by the policy. Safety stock depends on accuracy, reliability, lead time variable and rate of demand. This is used for *A* type of items.

(*t*) What are the characteristics of cellular manufacturing system? (2×20=40)

Sol.

1. CMS consists of sub-systems of similar parts and machines.
2. It is a combination of both process and product layout.
3. It requires less material handling process and follows fixed path for material travel.
4. High utilization of facilities.
5. Special purpose operating facilities are incorporated.
6. Employ multi-skilled operators.

SECTION-B

Q.2 (*a*) An antifriction bearing used in a gear box has a specification 6108. What do the numbers 6, 1 and 08 signify ? (12)

Sol. 6108 is a SKF bearing

6 : Width series-Single row deep groove ball bearing

1 : Diameter series-Extremely light

08 : bore = 5 × 08 = 40 mm.

(*b*) Mention three different methods used to prevent an anti-friction bearing from sliding axially in a shaft at the inner race and in a bore at the outer race. (18)

Sol. Methods to prevent sliding of bearing axially:

(*a*) At the inner race -

(*i*) Adequate press fitting allowance (interference fit) on the shaft.

(*ii*) Lock nut and washer.

(*iii*) Snap ring.

(*iv*) Shaft shoulders.

(*v*) Use duplexing type bearing.

(*vi*) Use notches at inner race.

(*b*) At the outer race -

(*i*) Provide close interference fit in the housing.

(*ii*) Have notches on the outer race.

(*iii*) Use needle type roller bearing.

(c) Which one the three bearings listed below will you choose to take a heavy axial load on the bearing - (*i*) Self aligning ball bearing ; (*ii*) deep groove ball bearing ; (*iii*) thrust bearing. Give reasons for your selection. (10)

Sol.

Bearing type	Applications
1. Self aligning ball bearing	Supports radial load only. Accommodates slight shaft misalignment.
2. Deep groove ball bearing	Supports relatively high thrust loads in either direction.
3. Thrust bearing	Takes heavy and pure thrust loads

Therefore, the thrust bearing is the suitable bearing to take heavy thrust loads.

Q.3. A single cylinder engine and a four cylinder in -line four stroke engine (both horizontal) are running at speed of 1500 rpm. Crank radius and connecting rod for both are 100 mm and 500 mm respectively. Reciprocating parts for the single cylinder engine weigh 160 N and for each of the four cylinders of the four cylinder engine weigh 40 N. Revolving masses for both single and four cylinder engines work out to 160 N with a net eccentricity of 0.1 mm. Compute the following for both the engines :

(*a*) Magnitude, frequency and direuion of the predominant unbalanced force due to the reciprocating parts. (18)

(*b*) Magnitude, frequency and direction of the unbalanced force due to the revolving parts. (18)

(*c*) If a vibration pick up is used to measure the vibration signals on the frame of the engine, direction in which the signals will be predominant. (4)

Expression for the piston acceleration is given by

$$\omega^2 r\ (a_0 + a_1 \cos\theta + a_2 \cos 2\theta + a_4 \cos 4\omega)$$

Where ω is the crank speed in rad/s, *r* crank radius,

$$a_0 = 0,\ a_1 = 1,\ a_2 = 0.2 \text{ and } a_4 = 0.002\ \theta \text{ the crank angle.}$$

Sol. (*i*) Angular speed,

$$\omega = \frac{2\pi \times 1500}{60} = 157 \text{ rad/s}$$

$$m_i = 160/9.81 \text{ kg}$$

Acceleration, $f = \omega^2 r\ (a_0 + a_1 \cos\theta + a_2 \cos 2\theta + a_4 \cos 4\theta)$

Unbalanced force for the single cylinder engine,

$$F_1 = m_1 f$$

$$= \left(\frac{160}{9.81}\right) \times (157)^2 \times 0.1$$

$$\times [\cos\theta + 0.2 \cos 2\theta + 0.002 \cos 4\theta]$$

Predominant component,

$$F_1 = 40202.24 \cos\theta = 40202.24 \cos \omega t$$

Frequency, $f = \dfrac{\omega}{2\pi} = \dfrac{157}{2\pi} = 25$ Hz

F_1 is maximum when, θ = 0°, i.e., along the line of stroke.

For the four cylinder in-line four stroke engine, all the cranks are symmetrically placed at 90° interval.

Unbalanced force,

$$F_4 = \left(\frac{40}{9.81}\right) \times (157)^2 \times 0.1 \times \left[(\cos\theta + 0.2\cos 2\theta + 0.002 \cos 4\theta)\right]$$

$$+\left\{\cos(90° + \theta) + 0.2\cos(180° + 2\theta) + 0.002\cos(360° + 4\theta)\right\}$$

$$+\left\{\cos(180° + \theta) + 0.2\cos(360° + 2\theta) + 0.002\cos(720° + 4\theta)\right\}$$

$$+\left\{\cos(270° + \theta) + 0.2\cos(540° + 2\theta) + 0.002\cos(1080° + 4\theta)\right\}$$

$$= 80.40 \cos 4\theta \text{ N}$$

F_4 is maximum when, θ = 0°, 90°, 180°, 360°, *i.e.*, along the line of stroke.

Frequency = 4*f* = 100 Hz.

(*ii*) Revolving mass, $m_2 = \dfrac{160}{9.81} = 16.31$ kg

Revolving unbalanced force

$$= m_2\, e\, \omega^2 = 16.31 \times 0.1 \times 10^{-3} \times (157)^2 = 40.2\text{N}$$

The direction of the revolving unbalanced force is radially outwards from the centre to the plane at which unbalanced mass is present.

Frequency = 25 Hz.

(*iii*) The signal from the vibration pick up will be predominant along the line of stroke.

Q.4. A hot rolled steel shaft is subjected to a torsional load that varies from 300 kN-mm clockwise to 100 kN.mm anti- clockwise and an applied bending moment at a critical section varies from + 400 kN-mm to - 200 kN-mm. The shaft is of uniform cross-section and no key way is present at the critical section. Determine the required shaft diameter by taking factor of safety 1.5. For the material, take ultimate tensile stress as 500 MPa, stress at yield point 420 MPa, design stress is 280 MPa. Also take the modification factor as 0.62, size correction factor as 0.85, the load factor for bending as 1 and the load factor for torsion as 0.58. (40)

Sol. $T_{max} = 300$ kN-mm, $T_{min} = -100$ kN–mm

$M_{max} = +\,400$ kN-mm, $M_{min} = -200$ kN-mm

FOS $= 1.5$, $\sigma_u = 560$ MPa, $\sigma_y = 420$ MPa, $\sigma_d = 280$ MPa

$C_m = 0.62$, $C_s = 0.85$, $C_L = 1$ in bending,

$C_L = 0.58$ in torsion,

Take $K_f = 1$ in bending and torsion, $k_{fs} = 1$

Endurance limit, $\sigma'_e = 0.5\,\sigma_u = 0.5 \times 560 = 280$ MPa

Design endurance limit in bending,

$$\sigma_e = \frac{C_m C_s L_L \sigma'_e}{K_f}$$

$$= 0.62 \times 0.85 \times 1 \times 280 = 147.56 \text{ MPa}$$

Design endurance limit in torsion:

$$\tau_u = 0.8\,\sigma_u \text{ and } \tau'_e = 0.5 \times \tau_u = 0.5 \times 0.8 \times 560 = 224 \text{ MPa}$$

$$\tau_y = 0.5\,\sigma_y = 210 \text{ MPa}$$

$$\tau_e = \frac{C_m C_s L_L \sigma'_e}{K_{fs}} = 0.62 \times .85 \times 0.58 \times 224 = 68.47 \text{ MPa}$$

Alternating bending moment,

$$M_a = \frac{M_{max} - M_{min}}{2} = \frac{400 + 200}{2} = 300 \text{ kN-mm}$$

Mean bending moment,

$$M_m = \frac{M_{max} + M_{min}}{2} = \frac{400 - 200}{2} = 100 \text{ kN-mm}$$

Mean bending stress in shaft,

$$\sigma_m = \frac{32 M_m}{\pi d^3}$$

Alternating bending stress in shaft,

$$\sigma_a = \frac{32 M_a}{\pi d^3}$$

Equivalent bending stress,

$$\sigma_{me} = \sigma_m + K_f \left(\frac{\sigma_y}{\sigma_e}\right) \sigma_a$$

$$= \left(\frac{32}{\pi d^3}\right) . \left[100 + \frac{420 \times 300}{147.56}\right]$$

$$= \left(\frac{32}{\pi d^3}\right) \times 953.9 \text{ kN/mm}^2$$

Alternating torque,

$$T_a = \frac{T_{max} - T_{min}}{2} = \frac{300 + 100}{2} = 200 \text{ kN.mm}$$

Mean torque $$T_m = \frac{T_{max} + T_{min}}{2} = \frac{300 - 100}{2} = 100 \text{ kN-mm}$$

Mean torsional shear stress in shaft,

$$\tau_m = \frac{16 T_m}{\pi d^3}$$

Alternating torsional shear stress in shaft,

$$\tau_a = \frac{16 T_a}{\pi d^3}$$

Equivalent torsional shear stress,

$$\tau_{me} = \tau_m + K_{fs} . \left(\frac{\tau_y}{\tau_e}\right) \tau_a$$

$$= \left(\frac{16}{\pi d^3}\right) . \left[100 + \frac{210 \times 200}{68.47}\right]$$

$$= \left(\frac{16}{\pi d^3}\right) \times 713.4 \text{ kN/mm}^2$$

Maximum shear stress $$= 0.5\left[\sigma_{me}^2 + 4\tau_{me}^2\right]^{0.5}$$

$$= 0.5\left(\frac{32}{\pi d^3}\right) \left[(953.9)^2 + (713.4)^2\right]^{0.5}$$

$$= \left(\frac{32}{\pi d^3}\right) \times 595.58 \text{ kN-mm}^2$$

Design shear stress,

$$\tau_d = \frac{280}{2} = 140 \text{N/mm}^2$$

$$\left(\frac{32}{\pi d^3}\right) = 140$$

$$d = 35 \text{ mm}$$

SECTION-C

Q.5. **(*a*)** What are the velocities which come into existence when a metal is cut orthogonally? Show these velocities graphically on a velocity diagram and determine the mathematical relationship in terms of shear and rake angles. (20)

Sol. The three velocities are :

Cutting velocity. v_c : Shearing velocity, v_s ; and Chip flow velocity, v_f

These velocities have been shown in the Fig. 2 below.

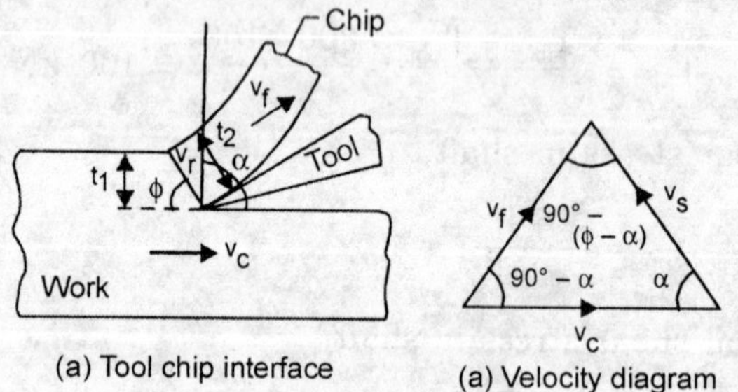

(a) Tool chip interface (a) Velocity diagram

Fig. 2

where α = rake angle, and ϕ = shear plane angle.

From the velocity triangle, we have

$$\frac{v_c}{\sin\left[90° - (\phi - \alpha)\right]} = \frac{v_s}{\sin(90° - \alpha)} = \frac{v_f}{\sin\phi}$$

$$v_f = \frac{v_c.\sin\phi}{\cos(\phi - \alpha)} \text{ or } v_s = \frac{v_c.\cos\alpha}{\cos(\phi - \alpha)}$$

(*b*) Distinguish between jig and fixture, with the help of diagrams for at least five each. (20)

Sol. A jig locates and holds the work piece and guides the tool, whereas the fixture holds and locates the work piece only.

[See Fig. 3 *(i)* for a jig and a fixture].

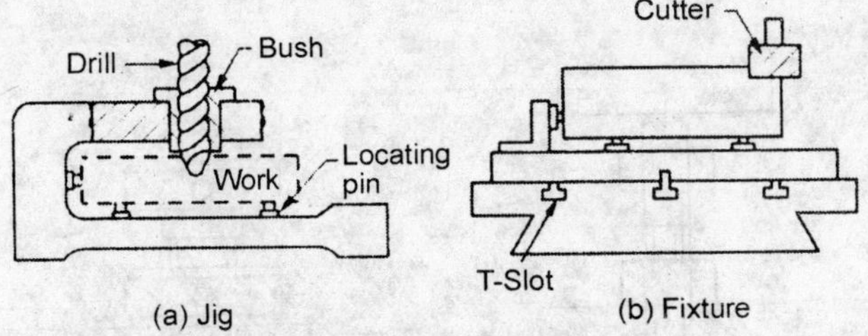

Fig. 3 (i) A Simple Jig and a Fixture.

Locating devices :

1. Locating pins-conical and cylindrical.

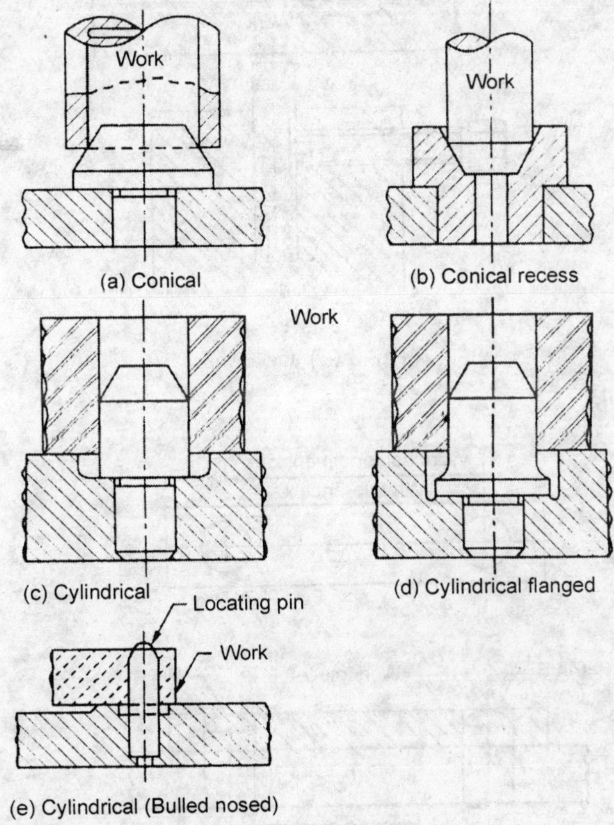

Fig. 3 (ii) Locating Pins.

2. Support pins-fixed, adjustable and jack pin.

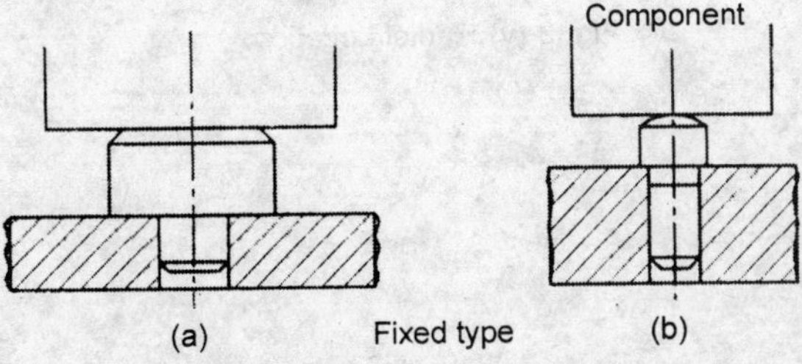

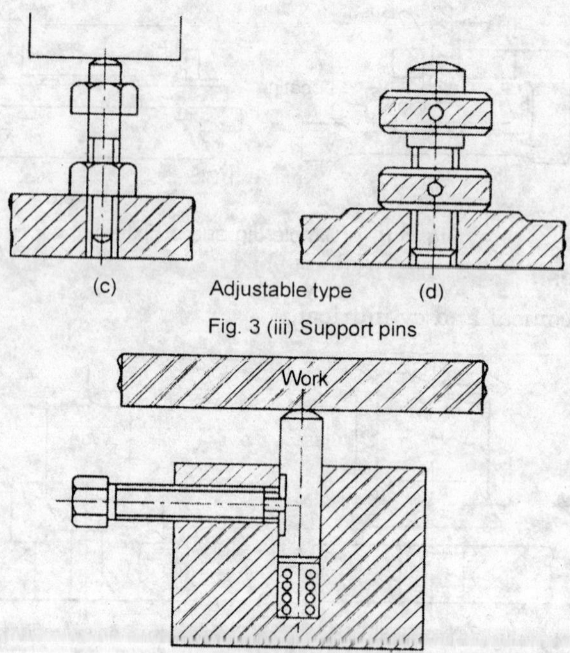

Fig. 3 (iii) Support pins

Fig. 3 (iv) Jack Pin.

3. Radial location

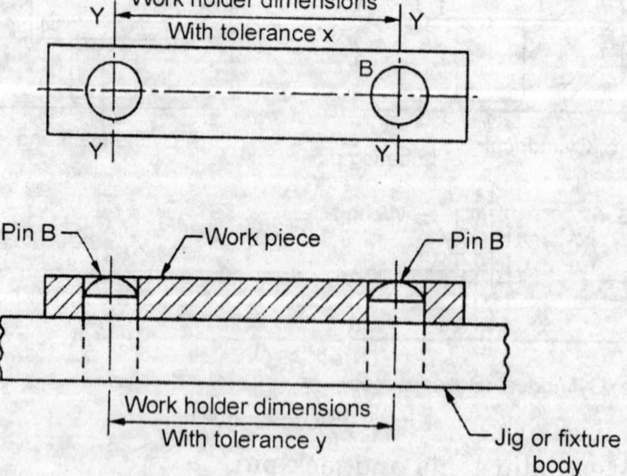

Fig. 3 (v) Radial Location.

Clamping deves:

1. Clamping screws.
2. Hook bolt clamp.
3. Lever type clamps

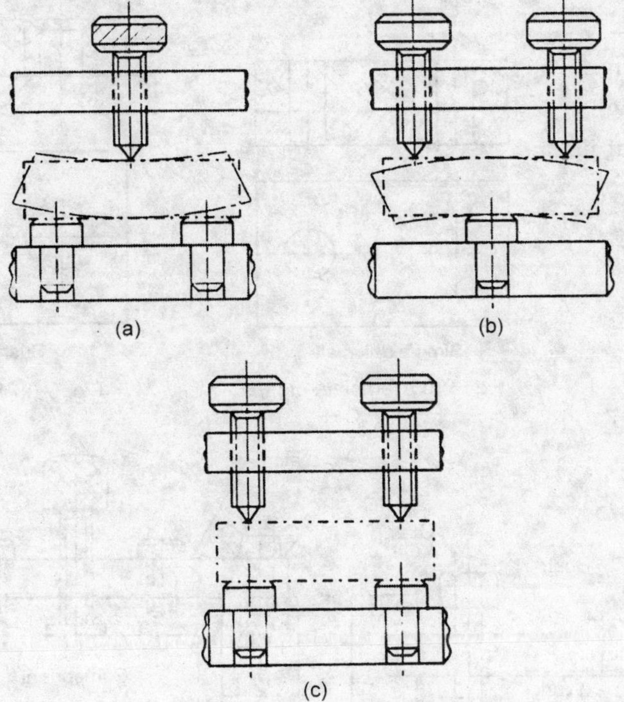

Fig. 3 (vi) Position of Clamp.

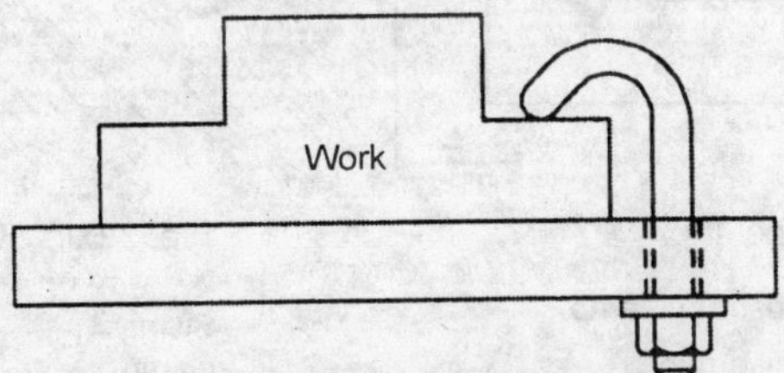

Fig. 3 (vii) Hook Bolt Clamp.

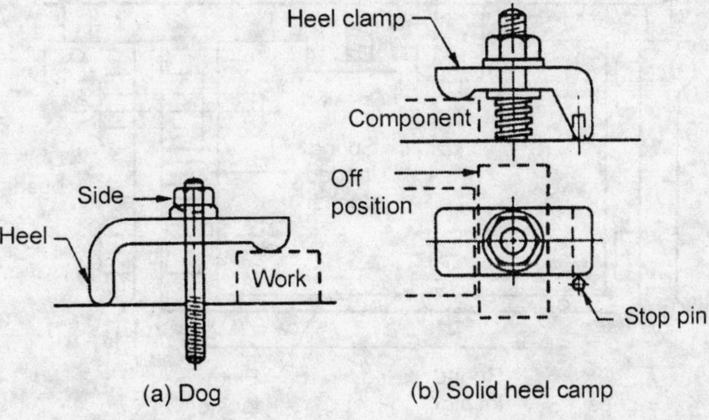

Fig. 3 (viii)

Jigs :

1. Template jig.
2. Plane type jig.
3. Open type jig.
4. Swinging leaf jig.
5. Box type jig.

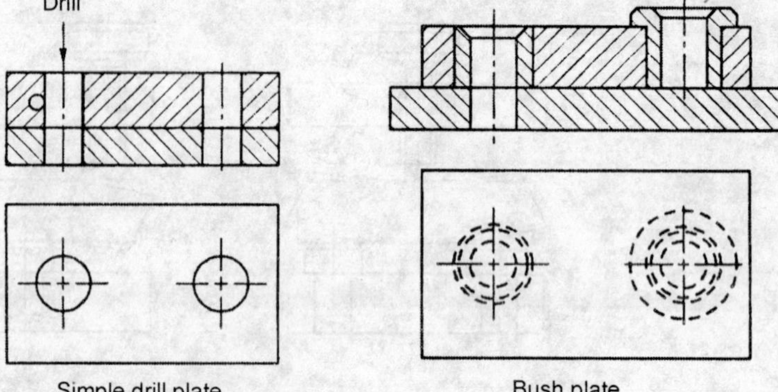

Fig. 3 (ix) Template Jig.

Fig. 3 (x) Plate Type Jig.

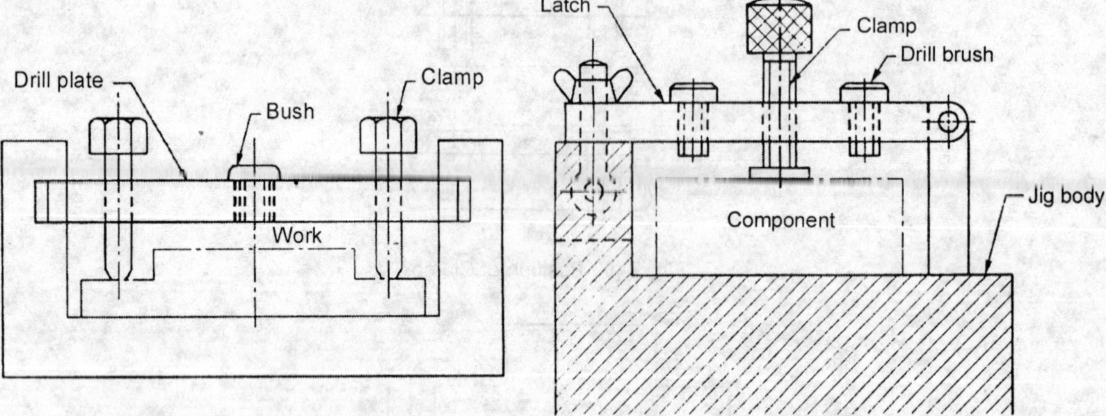

Fig. 3 (xi) Opent Type Jig.

Fig. 3 (xii) Swinging Leaf Jig.

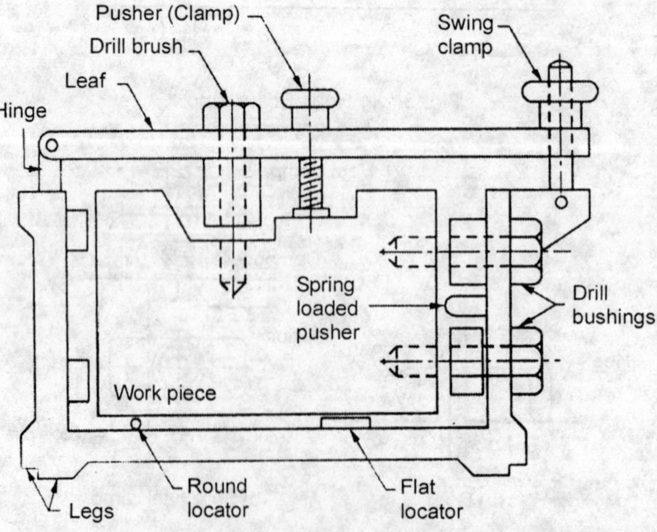

Fig. 3 (xiii) Box Type Jig.

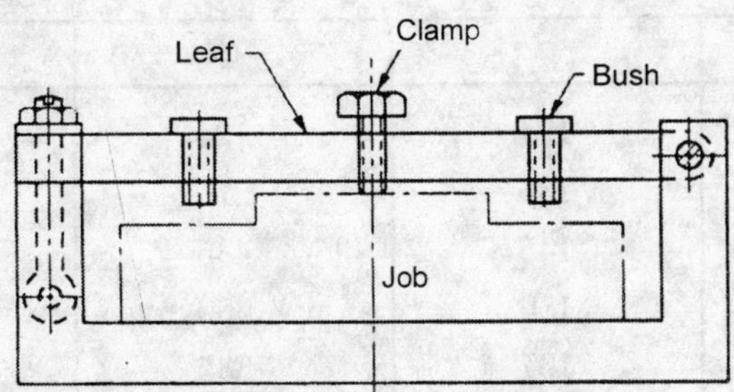

Fig. 3 (xiv) Channel Type Jig.

Q.6. (*a*) What are the steps involved in making of carbide tool bits? What do you understand by the green density of a metal powder compacted part ? What are the chief merits of a bush made by powder compaction? (20)

Sol. Making of Carbide tool bits:

1. Reduce tungsten oxide and cobalt oxide powders by separately intimately mixing them with lamp black and heating in a current of hydrogen. The size of the metallic powders is controlled by the heating time, temperature and rate of hydrogen flow.
2. Tungsten carbide is then ground, mixed with required lamp black and then the mixture is heated for several hours in reducing atmosphere at a temperature of 1600°C. By this process tungsten is converted into tungsten carbide.
3. Mix tungsten carbide and cobalt powder in suitable proportions in a mixing machine.
4. The mixture is then compacted in alloy steel dies at a pressure of about 480 MPa. The sintering is done in two stages. The preliminary sintering is done at about 900°C in hydrogen atmosphere. Cobalt fuses at this temperature whereas tungsten carbide remains in tact. These products are then shaped to the exact size by machining. The final sintering is done at about 1300°C for two hours.
5. Cooling gradually in the furnace itelf.

Green density. It is the density of the unsintered powder. It is defined as the ratio of mass of loosely heaped powder to its volume to fill a die cavity of given capacity.

Chief merit of a bush made by metal powder compaction is that lubricants can be impregnated into the bush under pressure and then it will act as a self lubricating during its operation.

(*b*) The demand for a product during the last 10 years is given below. Estimate the demand for the next two years by the method of regression. (15)

Year	1	2	3	4	5	6	7	8	9	10
Units	124	135	145	150	167	157	161	170	187	168

Sol. Using a linear forecast method, let

$$d_c = a + bt$$

where d_c = demand, t = period, and a, b are constants.

Using the method of least squares, we have

$$\sum_{i=1}^{n} d_i = na + b\sum_{i=1}^{n} t_i, \quad \sum_{i=1}^{n} d_i\, t_i = a\sum_{i=1}^{n} t_i + b\sum_{i=1}^{n} t_i^2$$

t	d	dt	t^2
1	124	124	1
2	135	270	4
3	145	435	9
4	150	600	16
5	167	835	25
6	157	942	36
7	161	1127	49
8	170	1360	64
9	187	1683	81
10	168	1680	100
$\Sigma f = 55$	$\Sigma d = 1564$	$\Sigma dt = 9056$	$\Sigma t^2 = 385$

$$100 + 556 = 1564 \quad \text{...(1)}$$

$$55a + 385b = 9056 \quad \text{...(2)}$$

or $$a + 5.5b = 156.4$$

$$a + 76 = 164.65$$

Solving, we get, $a = 126.15,\ b = 5.5$

$$d_c = 126.15, + 5.5\ t$$

For the eleventh year,

$$d_c = 126.15 + 5.5 \times 11 = 186.65 \simeq 187$$

For the twelth year,

$$d_c = 126.15 + 5.5 \times 12 = 192.15 \simeq 192$$

(c) Define Value engineering. What are the conditions conducive to take up VA/VE study? (5)

Sol. Value engineering is a tool of management which attempts to save cost from the point of view of value by studying the relationship between the design function and cost of a part, keeping in view to reduce part and cost through change in design, modification in specification of the material used by changing the source of supply and so on.

VA is normally applied to existing rather than to new products. It is a more comprehensive technique than conventional cost reduction technique since it begins with an examination of the purpose or functions of the product to establish the means by which such a purpose or function can be fulfilled best.

The conditions conducive to take up VA/VE study are:

1. To provide required use value at least cost.
2. To increase the value of the product either by increasing its utility with the same cost or decreasing cost for the same utility.
3. It can be applied for import substitution.

4. When there is a need for standardisation of materials and manufacturing processes.
5. When organisational efficiency is to be increased.
6. When existing product has to be simplified.
7. When the utility of the product has to be increased. VA/VE has been applied to military equipments, import substitution, machine tool industry, material handling equipments, and auto industries etc.

Q.7. (*a*) In a machine shop, certain type of machines breakdown at an average rate of 6 per hour. The breaking downs are in accordance with Poisson process. The estimated costs of idle machine is Rs. 15 per hour. Two repairmen A and B with different skills are being considered to be hired as repairmen. Repairman A takes six minutes on an average to repair a machine and his wages are Rs. 8 per hour, whereas the repairman B takes five minutes to repair and his wages are Rs. 10 per hour. Which repairman's services should be used and why ? Assume the work shift of 8 hours. (20)

Sol. Average rate of breakdown, λ = 6 per hour

Idle cost = Rs. 15 per hour per machine

Repairman per machine	**Repair time hour, Rs.**	**Wages per μ machines per hour**	**Service rate,**
A	6	8	60 / 6 = 10
B	5	10	60 / 5 = 12

Work shift = 8 hours

Average waiting time of a machine in breakdown :

Repairman A :

$$t_A = \frac{1}{\mu_A - \lambda} = \frac{1}{10-6} = \frac{1}{4} \text{ hours per machine}$$

Repairman B:

$$t_B = \frac{1}{\mu_B - \lambda} = \frac{1}{12-6} = \frac{1}{6} \text{ hours per machine}$$

Total number of breakdowns in a shift = 6 × 8 = 48

Idle cost of A $= \frac{1}{4} \times 15 \times 48 =$ Rs. 180

Idle cost of B $= \frac{1}{6} \times 15 \times 48 =$ Rs. 120

Total cost of A = Idle cost + Wages

= 180 + 8 × 8 = Rs. 244

Total cost of B = 120 + 8 × 10 = Rs. 200

Therefore, B should be used as its total cost is lower than that of A.

(**b**) For xyz company, the annual requirement of an item is 2400 units. Each item costs the company Rs. 6. The supplier offers a discount of 5% if 500 or more quantities are purchased. The ordering cost is Rs. 32 per order and the average inventory cost is 16%. Is it advisable to accept tne discount? Comment on the result. (20)

Sol. Annual requirement, D = 2400 units per year

Item cost, C = Rs. 6 per item, I = 16%

Ordering cost, C_0 = Rs. 32 per order

Inventory carrying cost, C_i = CI

Quantity	**Price per item, Rs.**
$0 \le Q_1 < 500$	$C_1 = 6$
$500 \le Q_2$	$C_2 = 6 \times 0.95 = 5.70$

$$Q_2^* = \left[\frac{2DC_0}{C_2 I}\right]^{0.5} = \left[\frac{2\times 2400\times 32}{5.7\times 0.16}\right]^{0.5} = 410 \text{ units}$$

$$Q^*_2 < b = 500$$

$$Q^*_2 = \left[\frac{2DC_0}{C_1 I}\right]^{0.5} = \left[\frac{2\times 2400\times 32}{6\times 0.16}\right]^{0.5} = 400 \text{ units}$$

$$Q^*_1 < b = 500$$

Total cost, TC

$$Q^*_1 = DC_1 + \frac{DC_0}{Q_1^*} + \frac{Q_1^* C_1 I}{2}$$

$$= 2400\times 6 + \frac{2400\times 32}{400} + \frac{400\times 6\times 0.16}{2} = \text{Rs.}14784$$

$$TC(b) = DC_2 + \frac{DC_0}{b} + \frac{bC_2 I}{2}$$

$$= 2400\times 5.7 + \frac{2400\times 32}{500} + \frac{500\times 5.7\times 0.16}{2}$$

$$= \text{Rs. } 14061.6$$

$$TC(Q^*_1) > TC(b)$$

$$\therefore \quad Q^* = Q^*_2 = 410 \text{ units}$$

It is advisable to accept the discount.

MECHANICAL ENGINEERING
PAPER II—2005

(Conventional)

SECTION-A

Q.1. ***(a)*** What do you mean by alloy steel? Write effects of following alloying elements on steel:

(*i*) Nickel

(*ii*) Chromium.

Sol. Alloy steel consists of iron, carbon and varying percentage of alloying elements like Ni, S, Cr, Cu, Mn, Mo, V, W, Pb, Br, Co and Ti.

Ni (0.3 - 0.5%) - increases hardenability, toughness, tensile strength, wear resistance, hardness and shock resistance. Cr (0.3 - 0.4%) - increases resistance to corrosion, abrasion, wear and oxidation; increases hardenability, high temperature strength, toughness and shock resistance.

(b) Illustrate the graphical comparison of following theories of failures for two-dimensional stress system :

(*i*) Maximum normal stress theory

(*ii*) Maximum shear stress theory

(*iii*) Distortion energy theory.

Sol. Maximum normal stress theory-a square for equal tensile and compressive yield strength.

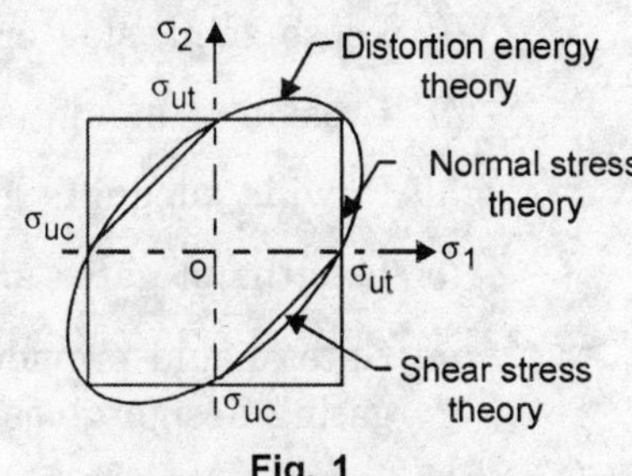

Fig. 1

Maximum shear stress theory-hexagon.

Maximum distortion theory-ellipse. ,

The three theories are shown in the Fig. 1.

(c) What do you understand by backlash in gears ? Along which circle is it measured ? How can we change backlash during mounting of gears?

Sol. Backlash in gears is the minimum distance between the non-driving side of a tooth and adjacent side of the mating tooth at the pitch circle. Backlash is measured along the pitch circle. Backlash can be changed during mounting of gears by changing the center distance slightly.

(*d*) Why is gear finishing done ? Explain 'gear shaving' and 'gear burnishing' finishing operations.

Sol. Gear finishing is done to decrease wear and for smooth and noise free operation of gears.

Gear shaving is the process of finishing gears by a rotary shaving cutter. The cutter and the gears are run in mesh. As they rotate, the gear is traversed longitudinally across the shaving cutter or *vice versa.*

Gear burnishing is the process of revolving gears in a barrel using medium balls, shots or round pins to finish the gear.

(e) Explain the pressure angle of cam. Why is high value of pressure angle not preferred?

Sol. Pressure angle of a cam is the angle between the line of motion of the follower and a line normal to cam profile at the point of contact. High value of pressure angle is not preferred in order to minimize the follower operating force.

(*f*) What do you understand by 'Selective Assembly' process? How is the range of clearance reduced ?

Sol. Selective assembly is a procedure in which parts to be assembled of any type are classified into same number of several groups according to size. Components with larger tolerances are assembled first to reduce the range of clearances.

(*g*) Explain good 'gating design' in moulding. What are its advantages?

Sol. Good gating dengn practice:

(*i*) Metal should flow with a minimum of turbulence and spiration of mould gases.

(*ii*) Loose sand, slag or oxides should be prevented from entering mould cavity.

(*iii*) Erosion of mould surfaces hould be avoided.

(*iv*) Provide adequate gates and runners.

(*v*) Lengths of gates and runners should be kept minimum.

(*vi*) Gate should provide equal distribution of metal toseveral mould cavities. A good gating design gives sound castings.

(*h*) The torque developed by an engine is given by the equation given below:

$$T = 14250 + 2200 \sin 2\theta - 1800 \cos 2\theta$$

where 'P is the torque in N-m and θ is the crank angle displacement from inner dead centre position. Determine the mean torque.

Sol. Mean torque,

$$T_m = \left[\frac{1}{2\pi}\right]\int_0^{2\pi} [14250 + 2200 \sin 2\theta - 1800 \cos 2\theta]\, d\theta$$

$$= \left[\frac{1}{2\pi}\right][14250\,\theta + 2200 \sin 2\theta - 900 \sin 2\theta]_0^{2\pi} = 14250 \text{ N.m.}$$

(*i*) What do you mean by 'autogeneous' and 'homogeneous' welding processes ? Metallurgically name the three distinct zones in a welded part. Enumerate the most important factors governing the fusion welding process.

Sol. Autogeneous welding process. No filler material is added during joining, *e.g.*, all types of solid phase welding and resistance welding.

Homogeneous welding process. The filler material used to provide the joint is the same as the parent material, *e.g.*, arc, gas and thermit welding.

Zones in a welded part:

1. Original cold worked metal
2. Recrystallised grains
3. Molten metal.

Factors governing fusion welding :

1. Characteristics of heat source.
2. Nature of deposition of filler.
3. Heat flow characteristics in the joint.
4. Gas metal or slag metal reactions in fusion zone.
5. Cooling of fusion zone with the associated contraction, residual stresses and metallurgical changes.

(*j*) Explain 'batch model' and 'mixed model' assembly lines. Enumerate the advantages of 'mixed model' assembly line over 'batch model' assembly line.

Ans. Batch model assembly lines. Only one product or one type of product is assembled in an operating area. The operations are performed in a sequence. All parts, sub-assemblies etc. are started at right time so as to be ready at the required time.

Mixed model assembly lines. It is a combination of batch, process and fixed position assembly.

Batch model	**Mixed model**
1. With the change of model, machinery also requires a change. 2. High assembly machinery cost with fall in volume of production. 3. Machinery idleness with one or more lines running light. 4. With one machine breakdown, whole assembly line has to remain idle.	Most suitable in factories where products are first manufactured and then assembled.

(*k*) Define 'value'. Differentiate between 'value engineering' and 'value-analysis'. State various phases of value-analysis job plan.

Sol. Value. It can be defined as the cost proportionate to the utility, *i.e.*,

$$\text{value} = \frac{\text{utility}}{\text{cost}}$$

Value engineering. It is a tool of management which attempts the question of saving cost from the point of view of value. Its main aim is to study the relationship between the design function and cost of a part, keeping in view to reduce part and cost through change in design, modification in specification of the material used by changing the source of supply, etc.

Value analysis. It is technique of cost reduction based on systematic and organized examination of every item of cost which goes into the manufacture of the product in terms of the value or customer satisfaction it adds to the product.

Phases of Value analysis job plan:

1. Customer wants.
2. Find best method of performing the work to be done.
3. Find relationship between cost and performance.
4. Search for better ways of performing work.

(*l*) Illustrate the following probability distributions schematically and specify their applications (*i*) Normal (*ii*) Weibull (*iii*) Exponential.

Sol. Normal Distribution. It is represented by

$$y = \left[\frac{1}{(2\pi)^{1/2}}\right] \exp\left(\frac{-z^2}{2}\right)$$

where $z = \dfrac{x - \mu}{\sigma}$

m = mean = 0

where σ = standard deviation =1

Its applications are in the following :

1. Calculation of errors made by chance in experimental measurements.
2. Computation of hit probability of a shot.
3. Statistical inference in almost every branch of science.

Weibull distribution. It is represented by

$$f(x) = \left(\frac{\alpha}{c}\right) x^{(\alpha-1)} . \; e^{-x/c}, \; x > 0, \; c > 0$$

where c = a scale parameter, $\propto$ = shape parameter.

It has applications:

1. To describe experimentally observed variation in fatigue resistance of steel and its elastic limit.
2. To study the variation of length of service of radio service equipment.

Exponential distribution. It is represented by :

$$f(x) = \lambda\, e^{-\lambda x} \text{ for } x > 0, \; \lambda = \text{a parameter.}$$

Its applications are in : 1. Reliability theory, and 2. Queuing theory.

(***m***) What do you understand by degree of freedom of planer mechanism? State the Grubler's Equation to find the d.o.f. of a mechanism.

Sol. Degree of freedom of a mechanism is the number of independent relative motions, a pair can have $F = 6 -$ Number of constraints.

Grubler's criteria:

Degrees of freedom, $F = 3(N - 1) - 2P_1 - P_2$

where, N = number of links,

P_1 = number of lower pairs

P_2 = number of higher pairs

(*n*) (*i*) Explain lower pairs and higher pairs.

(*ii*) What will be the number of inversions in a kinematic chain having '*n*' links ?

(*iii*) State whether the helical gears which are in contact are a 'higher pair' or 'lower pair'.

Sol. (*i*) **Lower pair** – is formed by two links having surface contact so that relative motion between two links is constrained.

Higher pair – is formed by two links having point or line contact while in motion. The relative motion being the combination of sliding and turning.

(*ii*) Number of inversions = Numbxer of links.

(*iii*) Higher pairs : helical gears.

(***a***) What is 3-2-1 principle of location with respect to fixture design? Explain.

Sol. 3-2-1 principle of location. There are three locating pins in the base of the fixed body, two in a vertical plane and one in another vertical plane, the three planes being perpendicular to one another, restrict 9 degrees of freedom. This is a method of locating a workpiece in a jig or a fixture and is also called the *six point location principle.*

(***p***) A hollow cylinder C.I. column, 3 m long has its internal and external diameters as 80 mm and 100 mm respectively. Calculate the safe load using Rankine formula ; if (*i*) both ends are hinged, and (*ii*) both ends are fixed. Take crushing strength of material as 600 N/mm^2,

Rankine constant = $\frac{1}{1600}$, and factor of safety = 3.

Sol.

$$P = \frac{\sigma_c A}{1 + a\left(\frac{L_e}{K}\right)^2}$$

$$A = \left(\frac{\pi}{4}\right)(100^2 - 80^2) = 2827.4 \text{ mm}^2$$

$$K^2 = \frac{I}{A} = \left(\frac{\pi}{64}\right)\frac{(100^4 - 80^4)}{\left[\left(\frac{\pi}{4}\right)(100^2 - 80^2)\right]} = 10.25 \text{ mm}^2$$

(*i*) $$P = \frac{200 \times 2827.4}{1 + \left(\frac{1}{1600}\right)\left(\frac{3000^2}{1025}\right)} = 87160\,\text{N}$$

(*ii*) $$P = \frac{200 \times 2827.4}{1 + \left(\frac{1}{1600}\right)\left(\frac{1500^2}{1025}\right)} = 238403\,\text{N}$$

(*q*) How do the slacks in PERT and floats in CPM decide on selection of critical path?

Sol. Slack is the difference between the earliest (TE) and latest (TL) times at which an event can take place without affecting the duration time of the project.

Slack, $S_j = TL_j - TE_j$

+ ve slack means ahead of schedule, -ve slack means behind schedule, and zero slack means on schedule condition.

Float is the length of time by which the duration of time of an activity can be increased without increasing the total time of the project.

(*r*) Draw a sketch showing various angles of single point cutting tool for turning aluminium alloys.

Sol. Single point turning tool for aluminium alloys :

Back rake (α_b) = 20°; Side rake (α_s) = 15°

End relief (θ_e) = 6°; Side relief (θ_s) = 6°

End cutting edge angle, (c_e) = 6°

End cutting side angle, (c_s) = 15°

Nose radius (R) = $\frac{1}{8''}$

The tool geometry is shown in Fig. 2.

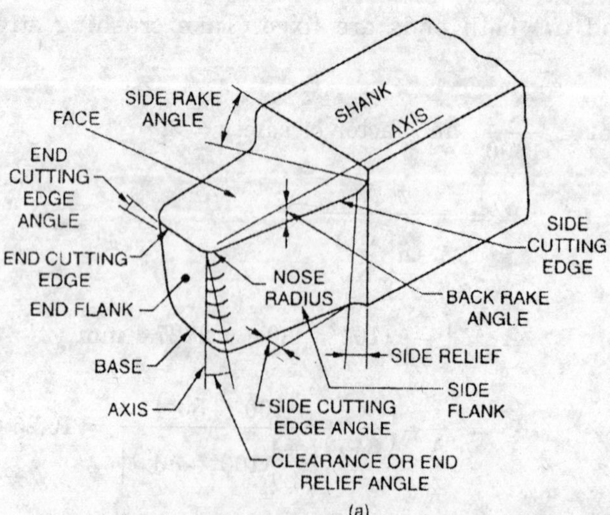

(a)

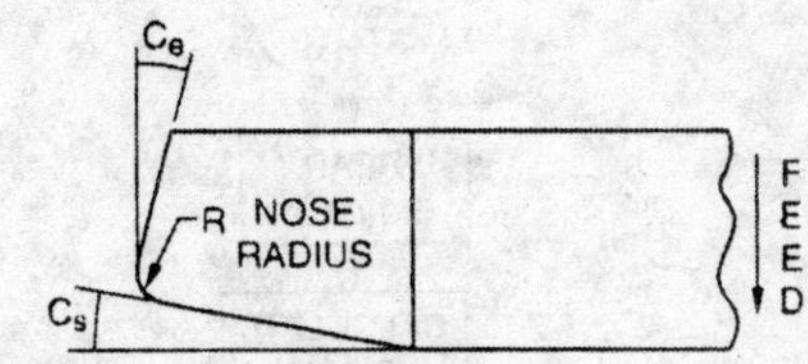

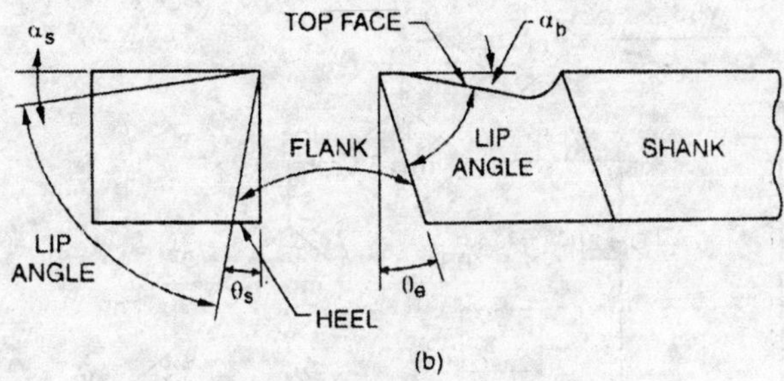

(b)

Fig. 2 Tool terminology

(s) What is meant by 'interpolation in CNC programming' ? What for are 'G' and 'M' codes used in NC part programming ?

Sol. Interpolation in CNC programming. Interpolation means the location of coordinates of start point and end point of the arc, the radius of the circle, the coordinate location of the circle center, and the direction in which the cutter is to travel.

G code : Preparatory functions.

M code : Miscellaneous functions.

(*t*) Draw a flow chart to compute crippling load for column with different end conditions, Euler's equation for crippling load is

$$P_{cr} = \frac{\pi^2 EI}{l^2}$$

l = L for both ends hinged

l = 2L for one end fixed and the other free

$l = \frac{L}{\sqrt{2}}$ for one end fixed and other end hinged

$l = \frac{L}{2}$ for both ends fixed.

Sol. Flow chart for crippling load of a column:

$$P_{cr} = \frac{\pi^2 EI}{L_e^2} = \frac{\pi^2 EI}{(cL^2)}$$

where c = a constant for end conditions of column

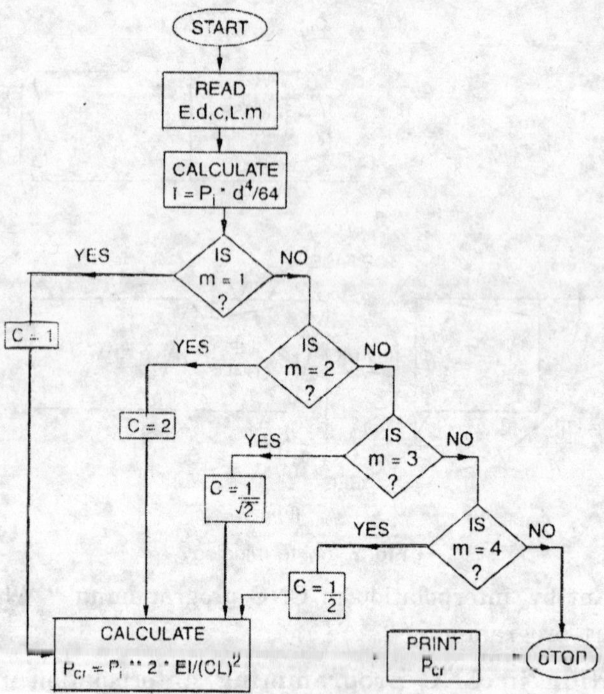

Fig. 3 Flow chart for Crippling load of Column.

SECTION-B

Q.2. (*a*) A simply supported beam of length 10 m carries a uniformly varying load whose intensity varies from a maximum value of 5 kN/m at both ends to zero at the centre of the beam. It is desired to replace the beam with another simply supported beam which will be subjected to the same maximum 'Bending Moment' and 'Shear Force' as in the case of the previous one. Determine the length and rate of loading for the second beam, if it is subjected to a uniformly distributed load over its whole length. Draw the variations of 'SF' and 'BM' in both the cases. (25)

Sol. The loading is symmetrical. At a distance x from A,

$$w_x = 5 - x$$

$$10\,R_A = 12.5\left(5 + \frac{10}{3}\right) + \frac{12.5 \times 5}{3} = 125$$

$$R_A = 12.5 \text{ KN}$$

$$M_x = 12.5\,x - (5 - x)\,\frac{x^2}{2} - \frac{x^3}{3}$$

(a)

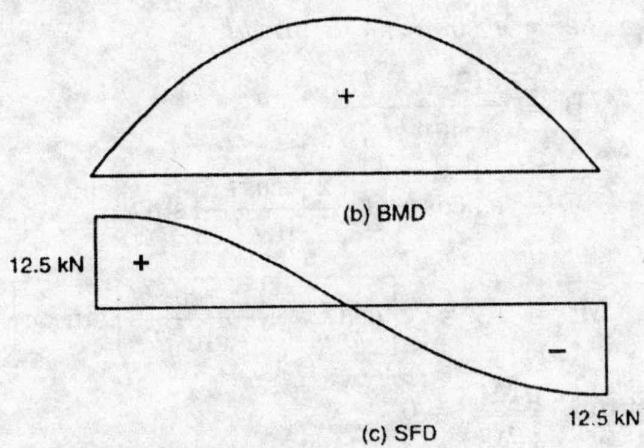

Fig. 4

At $x = 5$ m, M = 20.83 kN-m

$$F_x = 12.5 - \frac{x^2}{2} - x\,(5 - x) = 125 + 0.5\,x^2 - 5x$$

At $x = 5$ m, F = 0

For u.d.l., $\dfrac{wl^2}{8} = 20.83$ and $w\ell = 12.5$

$\ell = 6.667$ m and $w = 3.75$ kN/m

The BM and SF diagrams are shown in Fig. 4.

(b) A long strut AB of length l is of uniform section throughout. A thrust P is applied at the ends eccentrically on the same side of the centre line with eccentricity at the end B twice than that at the end A. Show that the maximum bending moment occurs at a distance x from the end A, where

$$\tan kx = \frac{2 - \cos kl}{\sin kl} \text{ and } k = \sqrt{\frac{P}{EI}} \qquad (15)$$

Sol. Consider the column having unequal eccentricity, as shown in Fig. 5.

$$M_x = Py$$

$$\frac{EI\,d^2y}{dx^2} = -M_x = -Py$$

$$\frac{d^2y}{dx^2} + \left(\frac{P}{EI}\right)y = 0$$

Let $k^2 = \dfrac{P}{EI}$

$$\frac{d^2y}{dx^2} + k^2y = 0$$

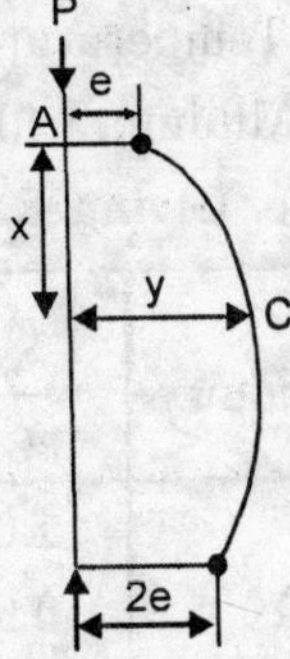

Fig. 5

Its solution is $y = A \cos kx + B \sin kx$

At $x = 0,\ y = e$ and at $x = \ell,\ y = 2e.$

$A = e$

and $$2e = e\cos kl + B\sin kl$$

$$B = \frac{e(2-\cos kl)}{\sin kl}$$

$$y = e\left[\cos kx + \left\{\frac{2-\cos kl}{\sin kl}\right\}\sin kx\right]$$

$$M_x = Py = Pe\left[\cos kx + \left\{\frac{2-\cos kl}{\sin kl}\right\}\sin kx\right]$$

For M_x to be maximum, $\frac{dM_x}{dx} = 0$

$$\frac{dM_x}{dx} = -k\sin kx + \left\{\frac{2-\cos kl}{\sin kl}\right\}(k\cos kx) = 0$$

$$\tan kx = \frac{2-\cos kl}{\sin kl}$$

Q.3. (***a***) A full journal bearing operating under a steady load has the following specifications:

1. Journal diameter : 60 m
2. Bearing length : 00 mm
3. Radial load on bearing : 2.8 kN
4. Journal speed : 1020 rpm
5. Radial clearance : 0.05 mm
6. Viscosity of oil : 80×10^{-9} N.s/mm²
7. Density of oil : 860 kg/m³
8. Specific heat of oil : 1.76 kJ/kg.°C

Using Raymonds and Boyd table given at the end of the problem, determine:

(*i*) Sommerfeld Number

(*ii*) Power loss in friction

(*iii*) Temperature rise if heat generated is entirely carried by oil

(*iv*) Minimum film thickness, and its location. (20)

Table. Giving Raymonds and Boyd data for journal bearing for L/D = 1

Attitude ε	$\frac{h_0}{c}$	S	ϕ	$\frac{r}{c}f$	$\frac{Q}{rcn_s l}$	$\frac{Q_s}{Q}$	$\frac{P}{P_{max}}$
0	1.0	00	85	00	n	0	—
0.1	0.9	1.33	79.5	26.4	3.37	0.150	0.540
0.2	0.8	0.630	74.02	12.8	3.59	0.280	0.529
0.4	0.6	0.264	63.10	5.79	3.99	0.497	0.484
0.6	0.4	0.121	50.58	3.22	4.33	0.680	0.415

Sol. Pressure, $p = \frac{W}{ld} = \frac{2.8 \times 10^3}{60 \times 60} = 0.778 \text{ N/mm}^2$

Sommerfeld number,

$$S = \left(\frac{r}{c}\right)^2 \left(\frac{\mu n_s}{p}\right)$$

$$= \left(\frac{30}{0.05}\right)^2 \left(\frac{80 \times 10^{-9} \times 1020}{60}\right) \times \left(\frac{1}{0.778}\right) = 0.63$$

$$\frac{\ell}{d} \text{ ratio} = \frac{60}{60} = 1$$

From the given Table, for S = 0.63, (r/c) f = 12.8

Co-efficient of friction,

$$f = \frac{12.8 \times 0.05}{30} = 0.0213$$

$$\frac{h_0}{c} = 0.8 \text{ (See. Fig. 6.)}$$

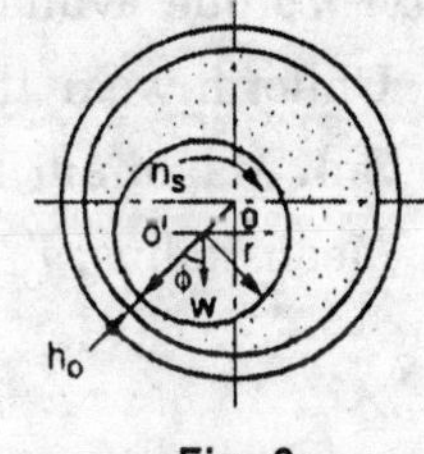

Fig. 6

Minimum oil film thickness,

$$h_0 = 0.05 \times 0.8 = 0.04 \text{ mm}$$

$$\frac{Q}{rcn_s l} = 3.59$$

Flow of lubricant,

$$Q = \frac{3.59 \times 30 \times 0.05 \times 1020 \times 60}{60} = 5492.7 \text{ mm}^3/\text{s}$$

Frictional power,

$$(kW)f = (2\pi n_s) \cdot (fW_r) \times 10^{-6}$$

$$= \frac{2\pi \times 1020 \times 0.0213 \times 2.8 \times 10^3 \times 30 \times 10^{-6}}{60} = 0.1911\text{kW}$$

Heat generated, $H_g = 0.1911$ kJ/s

Heat carried away by oil flow,

$$H_c = mc_p \cdot \Delta t$$

$$m = \rho Q \times 10^{-9} \text{ kg/s}$$

$$= 860 \times 5492.7 \times 10^{-9}$$

$$= 4.72 \times 10^{-3} \text{ kg/s}$$

For heat balance, $H_g = H_c$

$$\Delta t = \frac{H_g}{m c_p} = \frac{0.1911 \times 10^3}{4.724 \times 1.76} = 22.9°\text{C}$$

Attitude angle, $\phi = 74.02°$

(***b***) A car has a three cylinder engine which produces maximum torque of 20 N-m at 2000 rpm. The power is transmitted by a single plate friction clutch of 230 mm outside diameter. Design dimensions of friction surfaces. Further when the car is engaged at a speed of 60 km/hr, calculate the clutch slip period (time) during engagement.

Additional data are as under:

(*i*) Engine speed at beginning of clutch engagement : 2000 rpm

(*ii*) Engine torque at engagement : 100 N-m

(*iii*) Mass of car (loaded) : 1500 kg

(*iv*) Wheel diameter : 710 mm

(*v*) Coefficient of friction : 0.3

(*vi*) Permissible pressure for friction material : 0.25 N/mm^2

(*vii*) Mass moment of inertia of engine rotating parts, flywheel and driving side of clutch: 1.5 kg-m^2

(*viii*) Gear reduction at differential : 4.12

(*ix*) Torque available at the rear wheels : 105 N.m. (20)

Uniform wear theory may be used for the clutch plate design.

Sol. Maximum torque, $T_{max} = n\mu W r_m$

Mean radius, $r_m = 0.5\ (r_1 + r_2)$ for uniform wear

$$= 0.5\ (r_1 + 115)\ \text{m}$$

$$W = \pi\left(r_2^2 - r_1^2\right)p_m = \pi\left(115^2 - r_1^2\right) \times 0.25$$

$200 \times 10^3 = 2 \times 0.3 \times \pi \times (115^2 - r^2\, i\ ) \times 0.25 \times 0.5 \times (r_1 + 115)$

which gives, $r_1^3 + 115\, r_1^2 - 115^2 r_1 - 672049 = 0$

This equation is similar to $y^3 + py^2 + qy + r = 0$

Here $p = 115,\ q = -115^2,\ r = -672049$

Putting $y = \dfrac{r-p}{3}$, we get,

where $a = \dfrac{3q - p^2}{3} = \dfrac{-3 \times 115^2 - 115^2}{3} = -17633.3$

$$b = \frac{2p^3 - 9pq + 27r}{27}$$

$$= \frac{2 \times 115^3 - 9 \times 115 \times (-115^2) + 27 \times (-672049)}{27}$$

$$= -52433.3$$

$$\cos\theta = \frac{b}{2\left(\dfrac{-a^3}{27}\right)^{0.5}} = 0.05818$$

$$\theta = 86.65°$$

$$g = 2\left[\frac{-a}{3}\right]^{0.5} = 2\left[\frac{17633.3}{3}\right]^{0.5} = 153.33$$

$$y = g\cos\left(\frac{\theta}{3}\right) - \frac{p}{3} = 153.33\cos\left(\frac{86.65}{3}\right) - \frac{15}{3} = 95.9$$

$$r_1 = y = 95.9 \text{ mm}$$

$$W = \pi\,(115^2 - 95.9^2) \times 0.25$$

$$= 3163.73 \text{ N}$$

Automobile speed, $v = \dfrac{60 \times 10^3}{3600} = 16.67$ m/s

Angular speed of car wheel tyre,

$$\omega_t = \frac{v}{\text{Car wheel radius}}$$

$$= \frac{16.67}{0.355} = 46.96 \text{ rad/s}$$

Angular speed of clutch follower shaft,

$$\omega_f = \omega_t \times \text{reduction at differential}$$

$$= 46.96 \times 4.12 = 193.466 \text{ rad/s}$$

Angular speed of engine

= Angular speed of clutch input shaft

$$\omega_e = \frac{2\pi \times 2000}{60} = 209.4 \text{ rad/s}$$

Angular acceleration of engine during clutch slip period,

$$\alpha_e = \frac{T_e - T_{max}}{I_e} = \frac{100 - 200}{1.5} = -66.67 \text{ rad/s}^2$$

Angular acceleration of clutch output, $\alpha_f = \dfrac{a_f}{r_t}$

Linear acceleration of car wheel tyre, $a_t = \dfrac{F_a}{m_a}$

where mass of car, $m_a = 1500$ kg

Acceleration force on automobile,

$$F_a = \frac{T_a}{r_i} = \frac{105}{0.355} = 295.77 \text{ N}$$

$$a_t = \frac{295.77}{1500} = 0.197 \text{ m/s}^2$$

Angular acceleration of car tyre,

$$\alpha_t = \frac{a_t}{r_t} = \frac{0.197}{0.355} = 0.555 \text{ rad/s}^2$$

Angular acceleration of clutch follower shaft,

$$\alpha_f = a_t \times \text{Reduction at differential}$$
$$= 0.555 \times 4.12 = 2.288 \text{ rad/s}^2$$

At the end of engagement period,

$$\omega_e + \alpha_e \cdot \Delta t = \omega_f + \propto_f \cdot \Delta t$$

Clutch slip period, $$\Delta t = \frac{\omega_e - \omega_f}{\alpha_f - \alpha_e} = \frac{209.44 - 193.466}{2.288 - (-6.67)} = \frac{15.974}{68.958} = 0.232\text{s}$$

Q.4. (*a*) Define 'Body Centrode' and 'Space Centrode' for a link in mechanism. Explain how you will find these. (10)

Sol. Body Centrode. It is the locus of the instantaneous centre of the fixed body relative to the movable body, *i.e.*, the locus of the instantaneous centre assuming the movable body to be fixed and the fixed body to be movable.

Space centrode. It is the locus of the instantaneous centre of the moving body relative to the fixed body.

Consider a four-bar chain ABCD in which link 1 is fixed. The locus of instantaneous centre of links 1 and 3 over a range of motion of link 3 is the space centrode [Fig. 7(a)].

If the link 3 is assumed fixed and the link 1 m ovable, then the locus of instantaneous centre of link 1 and 3 is the body centrode (Fig. 7 (b)).

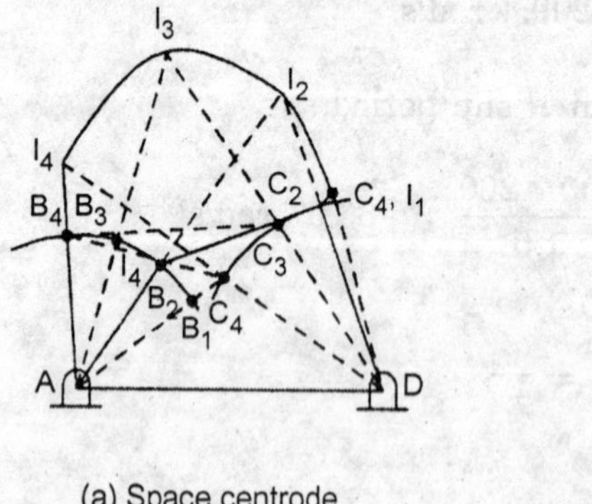

(a) Space centrode

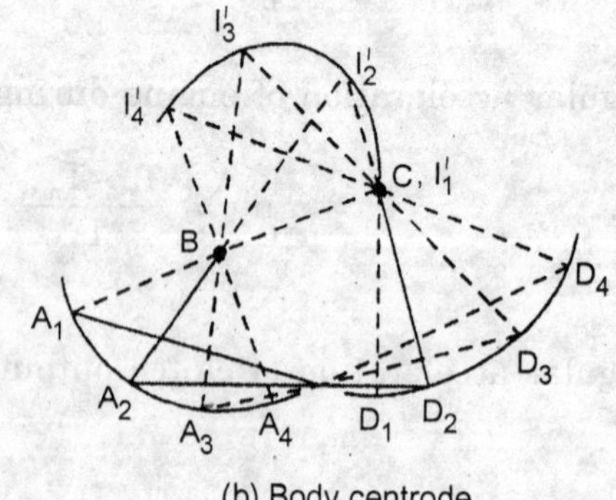

(b) Body centrode

Fig. 7.

(*b*) The crank length of a petrol engine is 50 mm and the connecting rod is 125 mm long and the crank rotates at a uniform speed of 400 rpm. Calculate the velocity and acceleration of the piston at different positions of the piston along its stroke and plot the two curves. (10)

Also find the crank position at which the piston's acceleration becomes zero. (10)

Sol. $r = 50$ mm, $\ell = 175$ mm,

$$n = \frac{l}{r} = \frac{175}{50} = 3.5, \text{N} = 40 \; rpm$$

$$\omega = \frac{2\pi N}{60} = \frac{2\pi \times 400}{60} = 41.888 \text{ rad/s}$$

Velocity of piston, $\quad v_p = \omega r\left[\sin\theta + \frac{\sin 2\theta}{2n}\right]$

$$= 41.888 \times 50 \times 10^{-3}\left[\sin\theta + \frac{\sin 2\theta}{7}\right]$$

Acceleration of piston,

$$f_p = \omega^2 r\left[\cos\theta + \frac{\cos 2\theta}{n}\right]$$

$$= (41.888)^2 \times 50 \times 10^{-3}\left[\sin\theta + \frac{\sin 2\theta}{3.5}\right]$$

$$= 87.73\left[\cos\theta + \frac{\cos 2\theta}{3.5}\right]$$

The velocity and acceleration of the piston are plotted in the Fig. 8.

θ, deg	0	30	60	90	120	180	210	240	270	300	330	360
v_p m/s	0	1.30	2.07	2.09	1.55	0.78	0	–0.78	–1.5	–2.07	–1.30	0
f_p m/s^2	11.28	88.5	31.3	–25.1	–56.4	–63.4	–62.7	–63.4	–56.4	–25.1	1.4	88.5

The velocity and aceleration of the piston are ploted in the Foig. 8.

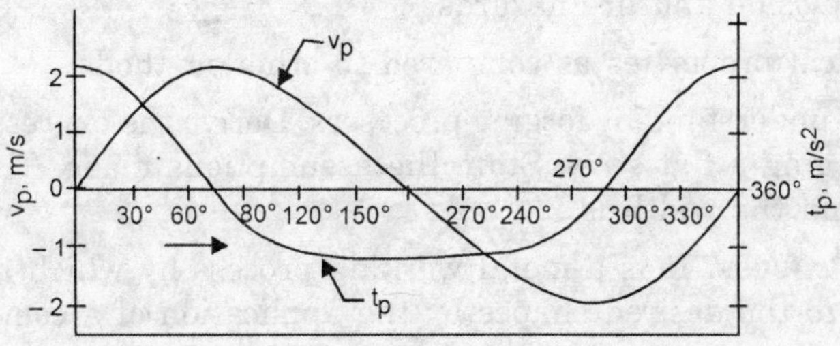

Fig. 8

Q.5. (*a*) Enumerate the steps involved in 'Powder Metallurgy' process. Discuss these steps. Name the materials used in 'Powder Metallurgy'. What are the limitations of Powder Metallurgy ? (10)

Sol. Steps involved in Powder metallurgy:

1. Preparation of powder of desired composition.

2. Mixing and blending. More than one powders are mixed thoroughly and blended to ensure their even distribution, or a powder is mixed with lubricants, additives or binders that might be used.

3. Compacting. The blended mixture is compacted to bring the finely divided particles of powder into Jose proximity while imparting the desired configuration and providing green strength to the part.

4. Sintering. The compacted mixture is heated at elevated temperature to establish permanent strong bonds between adjacent particles and thereby to impart strength to the compacts.

Materials used in Powder Metallurgy:

1. Combining metallic and non-metallic materials, *e.g.*, Cu or Ag+ graphite; Cu-Sn or Cu-Zn alloys with embedded SiO_2, $A1_2O_3$.
2. High melting point temperature castings like tungsten carbide and cobalt (cemented carbide).
3. Cermets : $A1_2O_3$ + W, Mo, Boron and Ti etc.
4. Diamond + Cu and Sn.
5. Superalloys.
6. Nuclear fuel elements + UO2.

Limitations of Power Metallurgy:

1. Dies and equipment cost is high. Therefore, the process is economically feasible only for mass production.
2. The material cost in powder form is high.
3. Metal powders are dificult to store.
4. Intricate designs in the products are difficult to obtain.
5. A uniformly high-density product is dificult to produce.
6. Part size and weight are restricted.
7. Danger of explosion and fire hazards.
8. Low mechanical properties as compared to other methods.

(***b***) What do you understand by forging processes. Derive the expression for the maximum force required for forging a flat strip. State the assumptions made. The initial thickness of the workpiece is h and the width is $2l$. (20)

Sol. Forging Process. It is a metal working process by which metals and alloys are plastically deformed to the desired shapes by the application of a compressive force.

Forging operations : 1. Upsetting 2. Heading 3. Fullering 4. Edging 5. Drawing down 6. Bending 7. Flattening 8. Blocking 9. Cut-off 10. Piercing 11. Punching 12. Swaging 13. coining.

Forging a Flat strip :

Assumptions

1. Material of strip is rigid - perfectly plastic.
2. Material is isotropic, incompressible and homogeneous.
3. Elastic strains are neglected.
4. Stresses on a plane normal to the flow direction are principal stresses.

***(i)* Sliding friction.** Consider the plain strain open-die forging of a flat strip of length 21, width b and thickness h, as shown in Fig. 9. At a distance x from the centre of strip the stresses are σ_x and $\sigma_x + d\sigma_x$ • The normal pressure is p and shear stress τ_x. The equilibrium of forces in the x-direction gives,

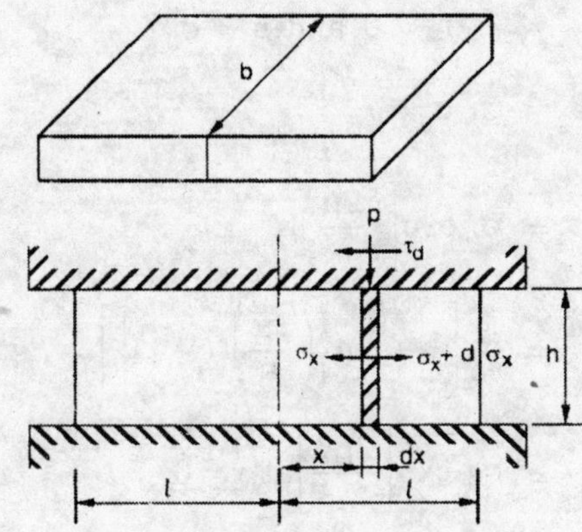

Fig. 9 Plain strain open die forging.

$$(\sigma_x + d\sigma_x)bh - \sigma_x.bh - 2\tau_x.b\,dx = 0$$

or $$\frac{d\sigma_x}{dx} - \frac{2\tau_x}{h} = 0 \quad ...(1)$$

Now $$\tau_x = \mu p$$

where μ = co-efficient of friction between strip and dio.

Therefore, Eqn. (1) becomes,

$$\frac{d\sigma_x}{dx} - \frac{2\mu p}{h} = 0 \quad ...(2)$$

According to Tresca's yield criterion, we have

$$\sigma_1 - \sigma_3 = \sigma_0 = 2k$$

where σ_0 = yield strength of strip material in simple tension

k = yield shear strength.

Here $\sigma_1 = \sigma_x$ and $\sigma_3 = -p$

Hence $$\sigma_x + p = \sigma_0$$

or $$\frac{d\sigma_x}{dx} - \frac{dp}{dx} = 0$$

or $$\frac{d\sigma_x}{dx} = -\frac{dp}{p} \quad ...(3)$$

Eqn. (2) then becomes

$$\frac{dp}{p} = -\left(\frac{2\mu}{h}\right)dx$$

Integrating, we get

$$\ln p = -\left(\frac{2\mu}{h}\right)x + C$$

where C = a constant of integration.

At $\quad x = \ell,\ \sigma_x = 0 \text{ and } p = \sigma_0$

Hence $\quad C = \ell n\ \sigma_0 + \frac{2\mu l}{h}$

which gives,

$$p = \sigma_0 \exp\left[\left(\frac{2\mu}{h}\right)(\ell - x)\right] \quad ...(4)$$

$$\sigma_x = \sigma_0\left[1 - \exp\left\{\left(\frac{2\mu}{h}\right)(\ell - x)\right\}\right] \quad ...(5)$$

$$p_{max} = \sigma_0 \exp\left[\frac{2\mu l}{h}\right] \text{at } x = 0$$

$$\approx \sigma_0\left(1 + 2\frac{\mu l}{h}\right) \quad ...(6)$$

or

$$\left(\frac{p}{2k}\right)_{max} = 1 + 2\frac{\mu l}{h}$$

$$(\sigma_x)_{max} = \sigma_0\left[1 - \exp\left(\frac{2\mu l}{h}\right)\right] \quad ...(7)$$

Total forging load,

$$P_t = 2b\int_0^l p.dx = 2bl\ \sigma_0\left(1 + \frac{\mu l}{h}\right) \quad ...(8)$$

Average pressure,

$$p_{avg} = \frac{P_t}{2bt} = \sigma_0\left[1 + \frac{\mu l}{h}\right] \quad ...(9)$$

***(ii)* Sticking friction :**

$$\tau_s = k = \frac{\sigma_0}{2}$$

$$\frac{d\sigma_x}{dx} - \frac{\sigma_0}{h} = 0$$

$$\frac{dp}{dx} - \frac{\sigma_0}{h} = 0$$

Integrating, we get,

$$p = -\left(\frac{\sigma_0}{h}\right)x + C_1$$

where $\quad C_1$ = another constant of integration.

Let sticking occurs at $x = x_s$, where $\tau_s = \mu p = \dfrac{\sigma_0}{2}$

$$\frac{\sigma_0}{2} = \mu\sigma_0 \exp\left[\left(\frac{2\mu}{h}\right)(1-x)\right]$$

$$x_s = 1-\left[\frac{h}{2\mu}\right]\text{In}\left[\frac{1}{2\mu}\right]$$

At $\quad x = x_s,\ p = p_s$

$$p_s = -\left(\frac{\sigma_0}{h}\right)x_s + C_1$$

$$C_1 = p_s + \left(\frac{\sigma_0}{h}\right)x_s$$

In sticking region, $\quad p = p_s + \left(\dfrac{\sigma_0}{h}\right)(x_s - x)$

For sticking over the whole interface, proceeding as above, we get,

$$\left(\frac{p}{2k}\right)_{\max} = 1+\frac{l}{h} \text{ at } x = 0 \qquad \text{...(11)}$$

$$p_{avg} = 2k\left(1+\frac{2l}{h}\right) \qquad \text{...(12)}$$

(*c*) A strip of lead with initial dimensions 24 mm × 24 mm × 150 mm is forged between two flat dies to a final size of 6 mm × 96 mm × 150 mm. If the coefficient of friction is 0.25, determine the maximum forging force. The average yield stress of lead in tension is 7 N/mm^2. (10)

Sol. Final dimensions :

$$h = 6 \text{ mm},\ 2\ell = 96 \text{ mm}$$

Sticking length, $\quad x_s = \ell - \left[\dfrac{h}{2\mu}\right]\text{In}\left[\dfrac{1}{2\mu}\right]$

$$= 48-\left[\frac{6}{2\times0.25}\right]\text{In}\left[\frac{1}{2\times0.25}\right]$$

$$= 48\text{-}8.318$$

$$= 39.682 \text{ mm}$$

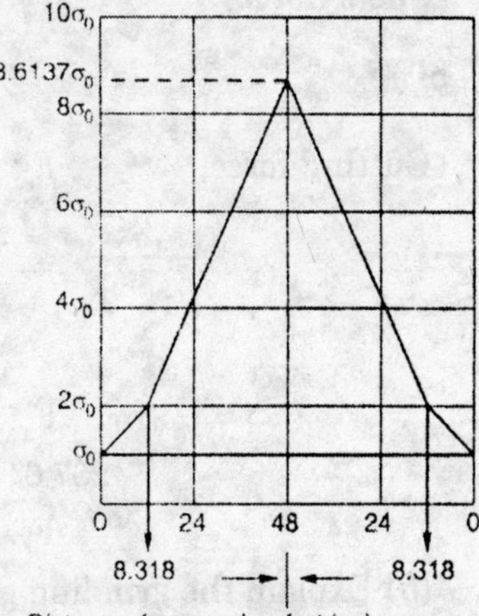

Fig. 10

In the sliding portion : 39.682 mm < x < 48 mm

$$p = \sigma_0 \exp\left[\left(\frac{2\mu}{h}\right)(l-x)\right]$$

At $\quad x = x_s,\ p_s = \sigma_0 \exp\left[\left(\dfrac{2\times0.25}{6}\right)(48-39.682)\right] = 2\sigma_0$

In the sticking region,

$$p = p_s + \left(\frac{\sigma_0}{h}\right)(x_s - x) = 2\sigma_0 + \left(\frac{\sigma_0}{6}\right)(39.682 - x)$$

At the centre, $x = 0$

$$p_0 = 2\sigma_0 + \left(\frac{\sigma_0}{6}\right)(39.682) = 8.6137\ \sigma_0$$

Forging load = Area of curve × b (see Fig. 10)

$$= 2\ [0.5 \times 8.318 \times 1 + 0.5 \times 39.682 \times 6.6137] \times \sigma_0 \times b$$

$$= 270.763 \times 7 \times 150 = 284.301 \text{ kN}$$

Q.6. (*a*) Mild steel is being machined at a cutting speed of 200 m/min with a tool rake angle of 10°. The width of cut and uncut thickness are 2 mm and 0.2 mm respectively. If the average value of co-efficient of friction between the tool and the chip is 0.5 and the shear stress of the work material is 400 N/mm², determine (*i*) shear angle, and (*ii*) cutting and thrust components of the machining force. (10)

Sol.

$$v_c = 200 \text{ m/min}, \alpha = 10°, b = 2 \text{ mm}, t = 0.2 \text{ mm}$$

$$\mu = 0.5, \tau = 400 \text{ N/mm}^2, \phi, F_c, F_t = ?$$

Angle of friction, $\beta = \tan^{-1}\mu = \tan^{-1} 0.5 = 26.56°$

According to Merchant's theory of metal cutting, shear plane angle,

$$\phi = 45° + 0.5\ (\alpha - \beta)$$

$$= 45° + 0.5\ (10° - 26.56°) = 36.72°$$

Area of shear plane,

$$A_s = \frac{bt}{\sin\phi} = \frac{2 \times 0.2}{\sin 36.72°} = 0.669 \text{ mm}^2$$

Shear force,

$$F_s = \tau \times A_s$$

$$= 400 \times 0.669 = 267.6 \text{ N}$$

Cutting force,

$$F_c = \frac{F_s}{\sin(\beta - \alpha)\cos(\phi + \beta - \alpha)}$$

$$= \frac{267.6}{\sec 16.56° \times \cos 53.28°} = 429 \text{ N}$$

$$F_s = F_c \cos\phi - F_t \sin\phi$$

$$267.6 = 429 \cos 36.72° - F_t \sin 36.72°$$

$$F_t = 127.6 \text{ N}$$

(*b*) Explain the grinding process for the following operations : (*i*) Roll grinding (*ii*) Thread grinding, and explain speed, feed and depth of cut as applicable to grinding works. (10)

Sol. Roll Grinding. Roll grinding is similar to the cylindrical grinding in which primarily cylindrical surfaces are ground. The cylindrical surfaces may be parallel or tapered as shown in Fig. 11(a) and 11(b). The depth of cut is controlled by feeding wheel into the work.

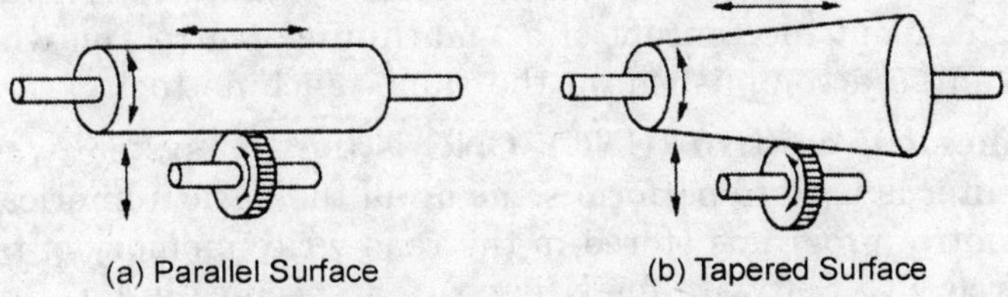

(a) Parallel Surface (b) Tapered Surface

Fig. 11 Roll Grinding.

Rolls are generally used in process industries such as r)aper, paint, steel, textile, printing metal foil, flour, soap, rubber, etc. Roll grinders usually have provision for rotating the rolls on their known bearings. The grinding wheel traverses the work on these machines and generates convex, concave or slight taper as necessary. As the metal of rolls is different depending upon the purpose for which it is intended to be used, the provision has to be made for variation in speeds, type of wheel, coolant and general treatment.

Thread grinding. It is used to produce very accurate threads on hardened materials and too soft materials. It is used to cut threads on taps, micrometer screws, thread gauges and thread milling cutters.

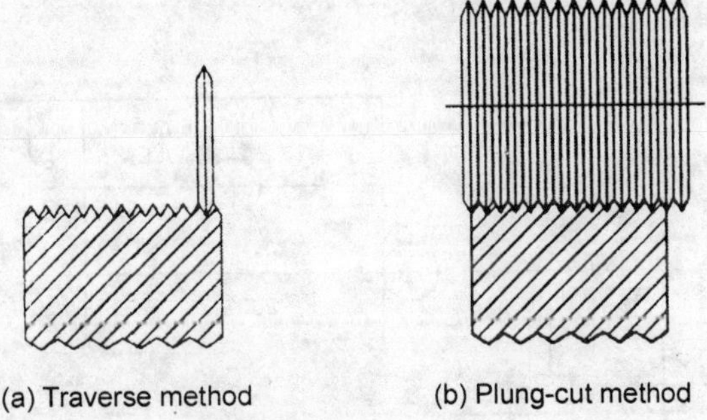

(a) Traverse method (b) Plung-cut method

Fig. 12 Thread grinding.

The grinding wheels can be single ribbed or multi-ribbed. In the case of single-ribbed wheel, the wheel turns against the rotation of the job. In addition to this rotary motion, a relative axial motion between the wheel and the job is provided with the help of a precision lead screw. The wheel is tilted by an angle equal to the helix angle of the thread, to the axis of the job. This method is known as "traverse Thread Grinding", and is used to produce long and coarse pitch threads.

A multi-ribbed wheel is slightly longer than the work and is used to cut the entire threads in one revolution of work. The wheel is fed into the work to the required depth and moves axially a distance equal to the pitch of the thread while the work revolves through one revolution. The cutter is set parallel to the axis of the job. This method is known as "Plunge Cut Grinding", and is used when production is more important than accuracy. The principle of these two methods is shown in Fig. 12. A thread grinding machine is similar to centre type cylindrical grinding machine with an arrangement for precision movement of the machine table and provision for tilting the grinding wheel at the helix angle of the thread.

(*c*) Define NC, CNC and DNC. Also explain their developments, and improvement in their applications. (10)

Sol. Numerical control (NC). NC of machine tools is a method of automatic control that uses symbolically coded instructions to cause the machine to perform a specific series of operations. Basically, it is a control of machine tools by numbers. The development of NC technology has brought about the concept of a " machining centre", on which a wide variety of machining tasks can be accomplished on the same machine tool.

Computer numerical control (CNC). CNC is the NC system wherein a dedicated, stored program computer is used to perform some or all the basic numerical control functions in accordance with control programs stored in the read-write memory of the computer. CNC is also called soft-wired NC. Software for N is written specifically for a particular machine device.

Direct numerical control (DNC). DNC is a system connecting a set of numerically controlled machines to a common computer memory for part program storage, distribution of machining data, provision for collection, display or editing of part programs, operator instructions, or data related to the NC process. DNC is employed as the heart of the control system for FMS, in which a number of NC machine tools are linked by means of electronic data communication and mechanical automation.

(*d*) Prepare a part program for machining a rectangular contour of 200 mm × 100 mm size from a suitable MS blank of 5 mm thickness. Also, a reamed hole of 16 mm has to be made centrally in the rectangular contour. Take speeds and feeds arbitrarily. (10)

Sol. Refer to Fig. 13.

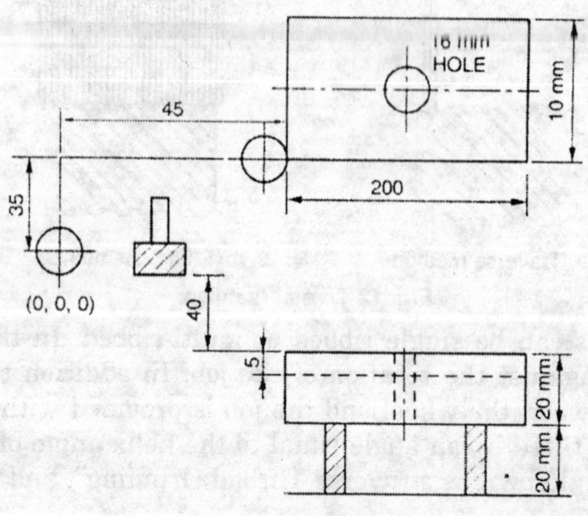

Fig. 13

N01	G92	X0	Y0	Z0>EOB>
N02	G90			
		(Absolute system)		
N03	G01	G78	X45000	Y35000
Z45000	S2000	F1000	T01	M03
N04	G79	X265000		
N05	G78	Y50000		
G06	G79	X45000		
G07	G78	Y65000		

```
N08    G79    X265000
N09    G78    Y80000
N10    G79    X45000
Nil    G78    Y95000
N12    G79    X265000
N13    G78    Yl 10000
N14    G79    X45000
N15    G01    X45    Y35    Z20
F0     M05    <EOB>
N16    M05
                     (Spindle stop )
N17    G01    X00    Y00    Z00
       <EOB>
N18    M06    T02
N19    M03    F200   S1000
```

Work piece size : 200 × 100 × 50 mm

Milling end cutter dia. = 20 mm

Depth of cut = 5 mm

T01 : Milling end cutter of 20 mm dia.

T02 : Driling tol. Drill size 16 mm

T03 : Reaming tool. Remaer size 16 mmm

```
N20    GOO    X145000    Y85000    Z-35000
N21    M07
N22    G01    Z-60000
N23    GOO    Z-35000
N24    M05
N25    M09
N26    GOO    X000       Y000      Z000
<EOB>
N27    M06    T03
N28    M03    F200       S500
N29    GOO    X145000    Y85000    Z-35000
N30    G01    Z-6000
N31    GOO    Z-35000
```

N32	M05			
N33	G00	X000	Y000	Z000
N34	M02			
		<EOB>	(End of program)	

Q.7. (*a*) A company has factories A, B and C with supply warehouses D, E, F and G Monthly factory capacities are 180, 170 and 200 units respectively. Monthly warehouse requirements are 90, 100,120 and 180 units respectively. Unit shipping cost in Rupes are given in the table below. Determine the initial feasible solution using Vogel's Approximation and find the optimum distribution for this company. (20)

		D	E	F	G	Supplies
	A	44	50	40	39	180
Factories	R	42	51	54	51	170
	C	41	40	42	45	200
Requiremen		90	100	120	180	

Sol. Requirement = 90 + 100 + 120 + 180 = 490

Supply = 180 + 170 + 200 = 550

Therefore, we introduce a dummy warehouse H with 60 units requirement and zero cost of shipping to meet the rim condition.

	D	E	F	G	H	Supply	Column difference			
A	44	50	40	39 [180]	0	180	39	1	1	1
B	42 [90]	51	54 [20]	53	0	170	42	9	11	1
C	41	40 [100]	42 [100]	45	0 [60]	200	40	1	1	3
Requirement	90	100	120	180	60	550				
Row Difference	1	10	2	6	0					
	1	10	2	6	–					
	1		2	6	–					
	–	–	2	–	–					
	–	–	2	–	–					

After carrying out Vogel's approximation for the initial feasible solution, we get the distribution as follows :

A to G = 180 units

B to D = 90

B to F = 20

B to H = 60

C to E = 100

C to F = 100

Optimum cost = 180 × 39 + 90 × 42 + 20 × 54 + 60 × 0 +100 × 40 +100 × 42 = Rs.20,080

Here m = 3, n = 5 so that $m + n - 1 = 7$

Number of occupied cells = 6

Therefore, the initial solution is degenerate. In order to remove degeneracy, we assign infinitely small units ε to unoccupied cell (AH) which has minimum cost among all unoccupied cells, as shown below :

	D	E	F	G	H	u_i
A	44 +2	50 −2	40 −14	39 [180]	0 [+ε]	u_1 = 39
B	42 [90]	51 −1	54 [20]	53 +14	0 [60]	u_2 = 39
C	41 +11	40 [100]	42 [100]	45 +18	0 +12	u_3 = 39
v_j	v_1 = 3	v_2 = 13	v_3 = 15	v_4 = 0	v_5 = 39	

Now Cij — $u_i + v_j$. Determine u_i and v_j for occupied cells, as shown above. Now maximum number of allocations (180) are in cell *AG*. Therefore, let V4 = 0.

Now calculate opportunity cost in unoccupied cells by using the relation, $d_{ij} = c_{ij} - (u_i + v_j)$. These values are also given in above table.

+ → −
40 0
+ ε ε + ε
20 − ε 60 + ε
54 0
− ← +

Fig. 14

The opportunity cost d_{13} is largest negative, therefore, it must enter the basis and the cell (A, H) must leave the basis.

	D	E	F	G	H	u_i
A	44 +16	50 +12	40 ε	39 [180]	0 [+14]	u_1 = 39
B	42 [90]	51 −1	54 [20−ε]	53 +14	0 [60+ε]	u_2 = 53
C	41 +11	40 [100]	42 [100]	45 +4	0 +12	u_3 = 41
v_j	v_1 = −1	v_2 = −1	v_3 = 1	v_4 = 0	v_5 = −53	

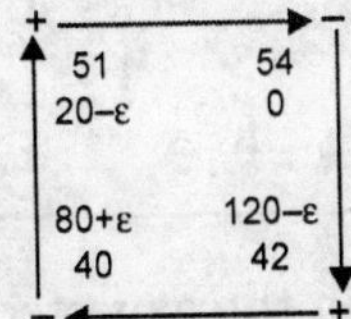

Fig. 15

Now d_{22}. i.e. (B, E) is largest negative. Hence (B, E) must enter the basis and (B, F) must leave the basis.

	D	E	F	G	H	u_i
A	44 +15	50 +12	40 [ε]	39 [180]	0 [+13]	$u_1 = 39$
B	42 [90]	51 [20–ε]	54 +1	53 +1	0 [60+ε]	$u_2 = 52$
C	41 +10	40 [80+ε]	42 [120–ε]	45 +4	0 +11	$u_3 = 41$
v_j	$v_1 = -10$	$v_2 = -1$	$v_3 = 1$	$v_4 = 0$	$v_5 = -52$	

Now all d_{ij} are positive. Therefore, it is the optimal solution.

The allocations are:

A to G = 180

B to D = 90

B to E = 20

C to E = 80

C to F = 120

$$\text{Optimum cost} = 180 \times 39 + 90 \times 42$$
$$= 20 \times 51 + 80 \times 40 + 120 \times 42$$
$$= \text{Rs. } 20060$$

(b) Draw the PERT network for the activities whose three time estimates are given in the table.

Activity	Predecessor activity	Optimistic time	Pessimistic time	Most likely time
A	–	1	5	3
B	–	2	4	3
C	–	3	5	4
D	A	2	10	9
E	C	4	6	5
F	B, D, E	5	13	6
G	A	2	6	4
H	G, F	0	6	3

From the three time estimates obtain the expected times of all the activities and slacks of all the events. Also find the critical path. (20)

Sol.

Activity	Predecessor activity	Optimistic time, t_0	Pessimistic time, t_p	Most likely time, t_m	Expected time $t_e = (t_0 + 4t_m + t_p)/6$
A	–	1	5	3	3
B	–	2	4	3	3
C	–	3	5	4	4
D	A	2	10	9	8
E	C	4	6	5	5
F	B, D, E	5	13	6	7
G	A	2	6	4	4
H	G F	0	6	3	3

The network diagram is shown in Fig. 16. Slack = $T_L - T_E$

Critical path: A – D – F – H

Critical time = 21 days

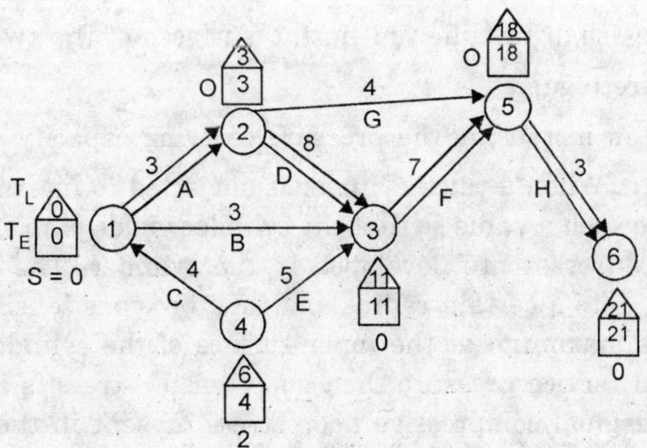

Fig. 16.

MECHANICAL ENGINEERING
PAPER II—2006

(Conventional)

SECTION-A

Q.1 (*a*) Two hollow shafts of same diameter are used to transmit same power.

One shaft is rotating at 1000 rpm while the other at 1200 rpm.

What will be the nature and magnitude of the stress on the surface of these shafts? Will it be same in two cases or different ? Justify your answer.

Sol. For $P_1 = P_2,\ T \propto \frac{1}{N}, \frac{T_1}{T_2} = \frac{N_2}{N_1}$

For $J_1 = J_2,\ \tau \propto T$

$$\frac{\tau_1}{\tau_2} = \frac{T_1}{T_2} = \frac{N_2}{N_1} = \frac{1200}{1000} = \frac{6}{5}$$

The shear stresses shall be different on the surfaces of the two shafts.

(*b*) What is autofrettage?

How does it help in increasing the pressure carrying capacity of a thick cylinder?

Sol. Autofrettage. When a thick cylinder is subjected to internal pressure then if this pressure is increased to such a value so that the cylinder yields partially or fully, on releasing this pressure residual stresses are developed in the cylinder. The hoop residual stress is compressive in nature upto a certain radius and then becomes tensile in a partially yielded cylinder. This stress is maximum at the inner surface of the cylinder. Now if this cylinder is subjected to internal service pressure then hoop tensile stress is induced in the cylinder. Due to the previous residual compressive hoop stress present in the cylinder the resultant hoop stress is reduced. Thus the cylinder can withstand higher service pressure. This phenomenon of strengthening the cylinder is called autofrettage of self-hooping.

(*c*) What are statically determinate and indeterminate beams ?

Illustrate each case through examples.

Sol. Statically Determinate Beam. It is a beam whose supporting forces can be determined by methods of regid- body mechanics. For such beams : $\Sigma F_x = 0, \Sigma F_y = 0, \Sigma F_z = 0$ and

$$\Sigma M_x = 0, \Sigma M_y = 0, \Sigma M_z = 0.$$

Examples are: 1. Simply supported beams on two supports and carrying concentrated and uniformly distributed loads or couples.

2. Cantilever beams carrying concentrated and uniformly distributed loads or couples.

Statically Indeterminate Beams. Whose supporting forces cannot be determined solely by methods of rigid-body mechanics. For such beams, we must take into account the manner of deformation in order to ascertain the supporting forces. 'i

Examples are: A beam resting on three supports (continuous beam).

(d) Define :

(*i*) Kinematic chain

(*ii*) Mechanism.

Sol. Kinematic Chain. It is an assembly of links in which the relative motion of the links is possible and the motion of each link relative to the others is definite. The last link of the kinematic chain is attached to the first link.

Mechanism. It is a combination of restraining bodies which are so shaped and connected that they move upon each other with definite relative motion. A kinematic chain becomes a mechanism if its one link is fixed.

(e) Distinguish between the functions of Flywheel and Governor.

Ans. A flywheel keeps the speed of a prime mover fairly constant in each cycle by absorbing surplus energy during power stroke and releasing it during idle stroke. A governor controls the speed of the prime mover over a number of cycles by controlling the input energy.

(f) For a free viscous damped vibration, state the condition for critical damping. The system has mass 'm', spring stiffness 'k' and damping coefficient 'c'.

Sol. The equatuion of motion for a free viscous damped vibration system is:

$$m \times \frac{d^2x}{dt^2} + c \times \frac{dx}{dt} + kx = 0$$

For critical damping, we have ;

$$\left[\frac{c_c}{2m}\right]^2 - \frac{k}{m} = 0$$

or $$c_c = 2m\omega_n = 2\ (mk)^{0.5}$$

where c_c = critical damping factor

$c\omega_n$ = natural frequency of vibration,

k = spring stiffness,

m = mass.

(*g*) Sketch a castle nut used for locking device. State two specific advantages of this locking device.

Sol. Castle Nut. The sketch of the castle nut is shown in Fig. 1. The slots are cut in a cylindrical collar provided on the top of the nut. A split pin is inserted in the slot which comes in line with the hole drilled in the bolt and then opened out at its end.

Advantages:

1. It serves as a locking device.
2. It is widely used in automobile and locomotive engines where they are subjected to sudden shocks and vibrations.

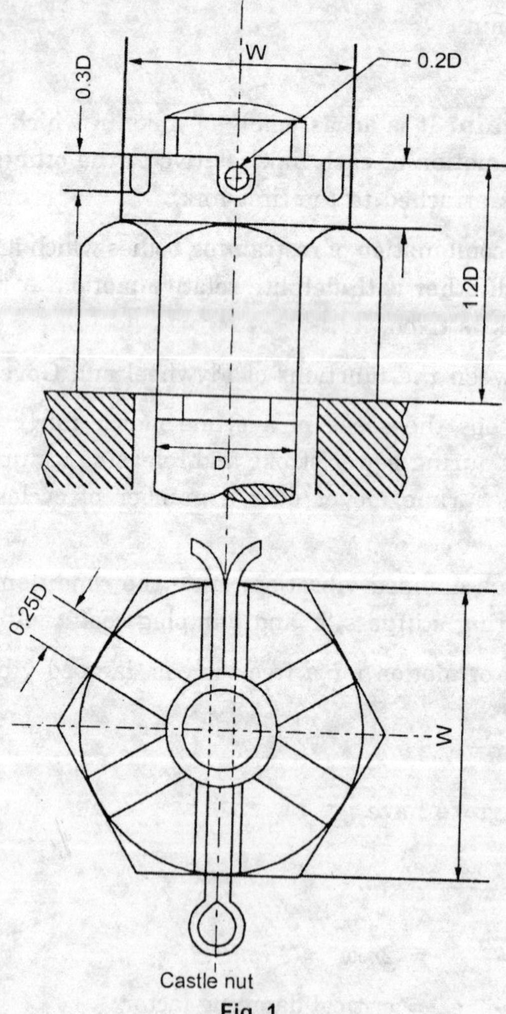

Castle nut

Fig. 1

(*h*) Between a flat belt drive and V-belt drive for the same coefficient of friction, arc of contact and allowable tension in the belts, which one will have more power transmission capability ? Give reason to your answer.

Sol. V-belt drive will have more power transmission capacity on account of the following reasons:

1. The effective coefficient of friction for the V-belt drives is $\mu/\sin\beta$, where β = semi-groove angle of pulley. It is more than a flat belt drive.
2. V-belt touches the pulley groove only on the sides and in case of any slip the belt shall go down in the groove without any slipping.
3. There are generally two or more than two V-belts giving a more positive drive.

(*i*) In journal bearings distinguish between "Bearing characteristic number"and "Bearing Modulus".

Sol. Bearing Characteristic Number. It is the non-dimensional group $\left(\frac{\mu n}{p}\right)$ for the hydrodynamic bearing.

Bearing Modulus. It is the value of $\left(\frac{\mu n}{p}\right)$ where the value of co-efficient of friction is minimum.

Where μ = dynamic viscosity of lubricant, mPa.s

n = speed of journal, rev/s

p = oil pressure, Pa.

(*j*) Why is annealing done?

What does the following represent in case of tool steels:

(*i*) T103

(*ii*) T35 Mn2 Mo 45.

Sol. Annealing is done to obtain softness, improve machinability, increase ductility and toughness, relieve internal stresses and refine grain size.

T103 : Low alloy and medium alloy tool steel having 1.03%C.

T35 Mn2 Mo 45 : Low alloy and medium alloy tool steel having C = 0.35%. ; Mn = $\frac{2}{4}$ = 0.5%, Mo = $\frac{45}{10}$ = 4.5%.

(*k*) Name any four materials used for making wire drawing dies, and explain the application of each of them.

Sol. Materials for making Wire Drawing Dies.

For cold work :

XT215 Crl2: High-carbon high chromium tool steels.

T110W2Crl: Non-deforming tool steels.

For hot work:

XT33W9Cr3V: Hot work tool steels.

T35Cr20Mol4Vll: - do -.

(*l*) Classify the advantages of powder metallurgy process in terms of economic and capabilities of the process, and properties of the parts made.

Sol. Advantages of Powder Metallurgy.

Economics:

1. Cost effectiveness for mass production.
2. Low labour cost.
3. No scrap.

Capabilities of process:

Certain type of parts can be made by powder metallurgy only, e.g.,

1. electrical contact points.
2. electrical brushes for motor.
3. cemented carbide tools. ,
4. cermets, and 5. Superalloys.

Properties of parts made:

1. Close tolerances.
2. High surface finish.
3. Residual porosity.
4. Low mechanical properties.

(*m*) What is a gating system? Name and show the various elements of a gating system by means of a diagram.

Sol. Gating system. For a casting is a series of channels which lead molten metal from ladle into the mould cavity. It may include any or all of the following:

Pouring basin, sprue, sprue base or well, runner, choke, skim bob, gates and/or ingates, and riser.

The gating sytem is shown in Fig. 2.

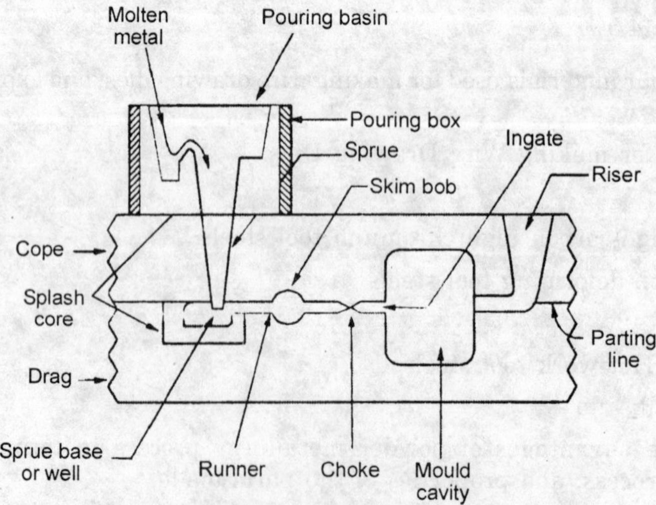

Fig. 2

(*n*) What is standard marking of grinding wheel? What each parameter indicates?

Sol. Standard marking of Grinding Wheel. The standard marking is shown in Fig. 3.

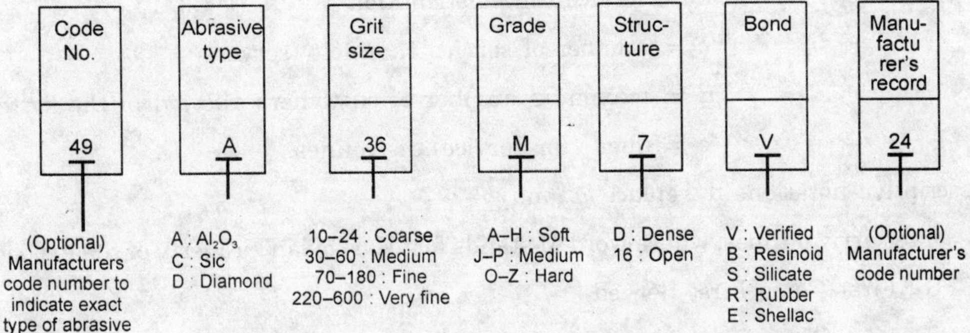

Fig. 3 Alphanumeric system for grinding wheel specification.

(*o*) Name any four flushing methods used in EDM process. What type of flushing method is recommendable in case of very small size hole drilling ?

Sol. Flushing methods in EDM.

1. Petroleum based hydrocarbon fluids.
2. Liquid paraffin.
3. White spirit.
4. Transformer oil.
5. Kerosene.
6. Mineral oil, silicon oil.

Kerosene oil is recommended as the flushing medium for drilling holes of very small size.

(*p*) What is the principle of angle dekkor auto collimator? What is its application?

Sol. Principle of Angle Auto-Collimator. It consists of a light source, telescope with eyepiece having either micrometer controlled setting lines or a scale. The principle of working of this instrument is the basic principle of reflection of light, *i.e.,* if beam of light strikes a flat reflecting surface which is perpendicular to direction of ray, it is reflected back along its original path. However, if the mirror is tilted through a small angle, the reflected light will get deflected by twice the angle of tilt of mirror.

Applications :

1. To measure small angular inclinations.
2. To check straightness, flatness and alignments or parallelism.

(*q*) What are the six categories of queuing models as per Kendall notation ?

Sol. Kendall's Queuing Models.

Kendall notation : [(a/b/c) : (d/e)}

where a = arrival distribution

b = service time distribution

c = number of service channels (servers)

d = maximum number of customers allowed in the system

e = queue (or service) disciplines.

Descriptive notations to replace 'a' and 'b'

M = inter arrival times, where word M stands for 'Markovian' property of the exponential.

FCFS : First Come First Served

LCFS : Last Come First Served

SIRO : Service In Random Order

GD : General service Discipline.

Model I [(MM/I): (∞ /FCFS)] Single server, unlimited queue model.

Model II [(M/M/I). (∞/SIRO)] Single server, service in random order.

Model III [(M/M/I): (N/FCFS)] Single server, finite (or unlimited) queue model.

Model IV (A) [(M/M/s): (∞/FCFS)] Multiple servers, unlimited queué model.

Model V [(M/M/s) : (M/GD)] Finite population (source) model.

Model VI [(M/Ek/I) : (∞/FCFS)] Eriangian service time distribution with k-phases.

(*r*) With the help of quantity-cost curve, explain the significance of Economic Order Quantity (EOQ). What are the limitations of using EOQ formula?

Sol. $\text{EOQ}^* = \left[2 \times \dfrac{\text{Annual demand} \times \text{Ordering cost}}{\text{Carrying cost}}\right]^{1/2}$

The cost *v's* order quantity is shown in Fig. 4. Limitation of EOQ is that the demand rate has to be constant and known.

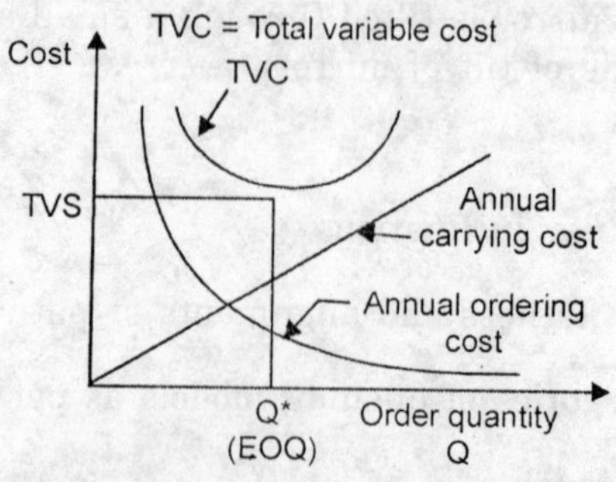

Fig. 4 Economic Order Quantity Chart

(*s*) What is a critical path? Why is the critical path of such importance in large project scheduling and control? Can a critical path change during the course of a project?

Ans. Critical Path. It is the continuous chain of critical activities in a network diagram. It is the longest path starting from first to the last event. The length of the critical path is the sum of the individual times of all the critical activities lying on it and defines the minimum time required to complete the project.

Critical path is important for directing maker's attention and effort to critical activities where improvement will pay the largest dividend.

The critical path may change during the course of a project.

(*t*) To print the values of array a_{ij}, in the order given below, write the source code in FORTRAN and in 'C"

a_{11}	a_{21}	a_{31}	a_{41}	a_{51}
a_{12}	a_{22}	a_{32}	a_{42}	a_{52}
a_{13}	a_{23}	a_{33}	a_{43}	a_{53}
a_{14}	a_{24}	a_{34}	a_{44}	a_{54} (20 × 2 = 40)

Sol. Source code in FORTRAN.

```
  READ (7, *) ((a (i,j,), j = 1, 5), i = 1, 4)
  WRITE (8, 5)
5 FORMAT (3*, 'The values of array a (i, J) are :')
  WRITE (8, 6) ((a (i,j),j = 1, 5), i = 1,4)
6 FORMAT (5*, 5f15.3)
```

Source code in C

```
printf ("Enter the elements of matrix a:");
  for (i = 1 ; i <=4; i++)
  {
        for j=l;j<=S;j++)
        {
              scanf ("%d", &A [i] [j]) ;
        }
}
```

```
printf ("You have entered the following matrix");
for(j=1; j<=4; i++)
{
        printf("\n");
for (/=l; j< = 5; j++)
{
        printf("%d/t", A [i] [j']) ;
{
                printf("n");
}
```

SECTION-B

Q.2 (*a*) A mild steel shaft of 50 mm diameter is subjected to a bending moment of 1.5 kN-m and torque *T*. If the yield point of steel in tension is 210 MPa, find the maximum value of the torque without causing yielding of the shaft material according to

(*i*) Maximum principal stress theory, and

(*ii*) Maximum shear stress theory. (20)

Sol. (*i*). Equivalent bending moment,

$$M_e = 0.5\,[M + (M^2 + T^2)^{0.5}]$$

$$= 0.5\,[1.5 + (2.25 + T^2)^{0.5}] \times 10^3$$

$$= \left(\frac{\pi}{32}\right) d^3 \sigma_y = \left(\frac{\pi}{32}\right) \times (50 \times 10^{-3})^3 \times 210 \times 10^6$$

$$(2.25 + T^2)^{0.5} = \left(\frac{\pi}{32}\right) \times (50 \times 10^{-3}) \times 210 \times 10^3 \times 2 - 1.5$$

$$= 3.654$$

$$T^2 = 11.103$$

$$T = 3.332 \text{ kN.m.}$$

(*ii*) Equivalent torque, $T_e = [M^2 + T^2]^{0.5} = [2.25 + T^2]^{0.5} \times 10^3$

$$= \left(\frac{\pi}{16}\right) d^3 \frac{\sigma_y}{2} = \left(\frac{\pi}{16}\right) \times \frac{(50 \times 10^{-3}) \times 210 \times 10^3}{2}$$

$$[2.25 + T^2]^{0.5} = 2.577$$

$$T^2 = 4.391$$

$$T = 2.095 \text{ kN.m}$$

(*b*) A timber beam 15 cm wide and 20 cm deep carries a uniformly distributed load over a span of 4 m and is simply supported.

If the permissible stresses are 30 N/mm² longitudinally and 3 N/mm² transverse shear, calculate the maximum load which can be carried by the timber beam. (20)

Sol. Bending stresses:

Let w = udl, N/m

$$M_{max} = \frac{wl^2}{8} = \frac{w \times 4^2}{8} = 2w \text{ N.m}$$

$$I = \left(\frac{15 \times 20^3}{12 \times 10^{-8}}\right) = 10^{-4} \text{ m}^4$$

Maximum bending stress,

$$\sigma_{max} = \frac{M_{max}\ y}{I} = \frac{2w \times 10 \times 10^{-2}}{10^{-4}} = 30 \times 10^6$$

Maximum transverse shear stress :

$$\tau_{max} = \frac{FA\bar{y}}{Ib}$$

$$F = 2w \text{ N}$$

$$A = \frac{15 \times 20}{2} = 150 \text{ cm}^2$$

$$\bar{y} = 5 \text{ cm}$$

$$b = 15 \text{ cm}$$

$$\frac{2w \times 150 \times 10^4 \times 5 \times 10^{-2}}{10^{-4} \times 15 \times 10^{-12}} = 3 \times 10^6$$

$$w = 30 \text{ kN/m}$$

Q.3. **(*a*)** The dimensions of different links of the crank and slotted lever quick-return mechanism shown in Fig. 5 are given below:

$$O_1O_2 = 700 \text{ mm},$$
$$O_1B = 250 \text{ mm}$$
$$O_2D = 1250 \text{ mm},$$
$$DR = 350 \text{ mm}.$$

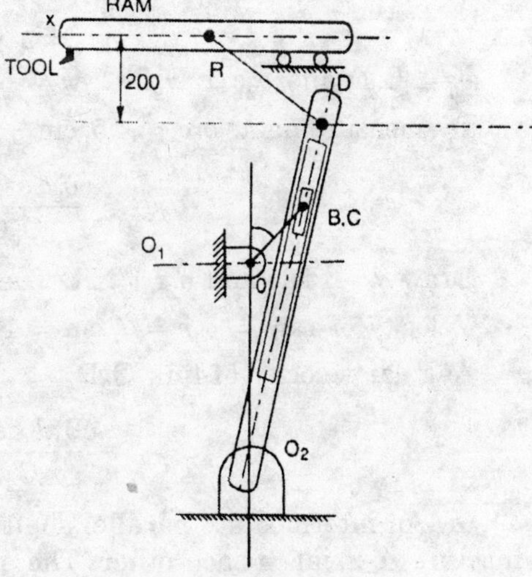

Fig. 5

The crank O_1B rotates at 40 r.p.m. in the counterclockwise direction and at the present instant of consideration makes an angle of 45° with the vertical.

Determine (*i*) Velocity of the ram R which moves in a horizonal direction (*ii*) Angular velocity of link O_2D.

Sol. Angular speed of O_1B,

$$\omega = \frac{2\pi N}{60} = \frac{2\pi \times 40}{60}$$

$$= 4.19 \text{ rad/s}$$

Linear velocity of B, $v_b = \omega \times O_1B = 4.19 \times 25 = 104.72$ cm/s

Draw the configuration diagram as shown in Fig. 6 *(a)* to a scale of 1 cm = 10 cm.

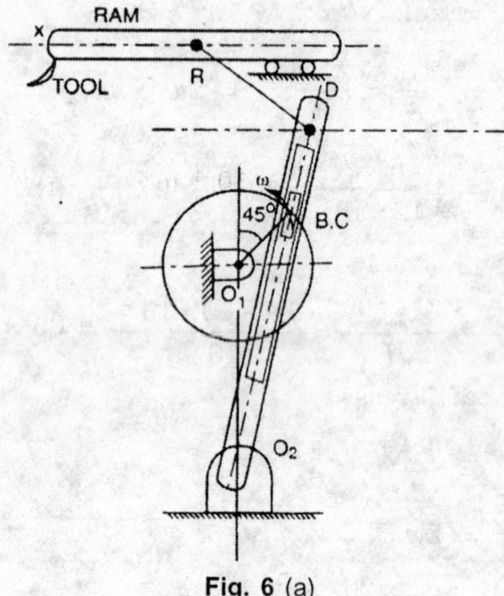

Fig. 6 (a)

Draw the velocity diagram to a scale of 1 cm = 20 cm/s, as shown in Fig. 6 *(b)*.

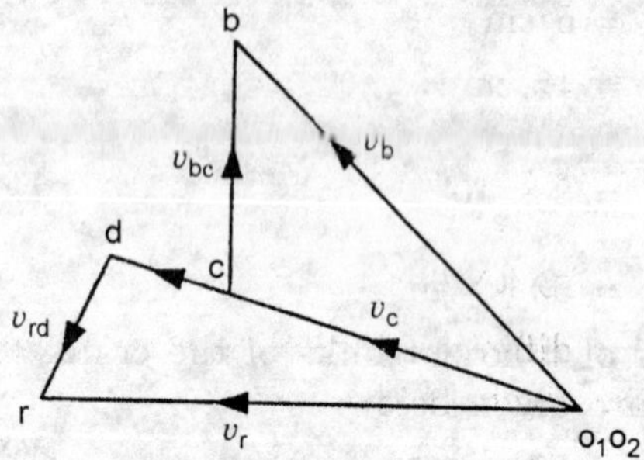

Fig. 6 (b) Velocity Diagram.

$v_b \perp O_2D$ and $o_2c \perp O_2D$

By measurement, o_2c = 4.5 cm

$$o_2d = \frac{o_2c \times O_2D}{O_2C} = \frac{4.5 \times 12.5}{9} = 6.25 \text{ cm}$$

Draw $dr \perp DR$ and $o_2r \parallel XR$. Then

Velocity of ram = o_1r = 7 cm = 140 cm/s

Angular velocity of link O_2D

$$= \frac{o_2d \times \text{scale}}{O_2D} = \frac{6.25 \times 20}{125} = 1 \text{ rad/s}$$

(b) An intermediate parallel shaft 'Z' is used for gearing together two shafts 'X' and 'Y' which are co-axial to each other. The gear wheels connecting 'X' and 'Z' have a module of 2 and those connecting 'Z' and 'Y' a moudle of 3. The speed of 'F is to be about, but less than, $\frac{1}{12}$ that of 'X'. If the two pinions have 24 teeth each, find suitable numbers of teeth for the gears and pinions, the actual velocity ratio and the corresponding distance of shaft 'Z' from shaft 'X'.

Sol. The configuration diagram of the gear drive is shown in Fig. 7.

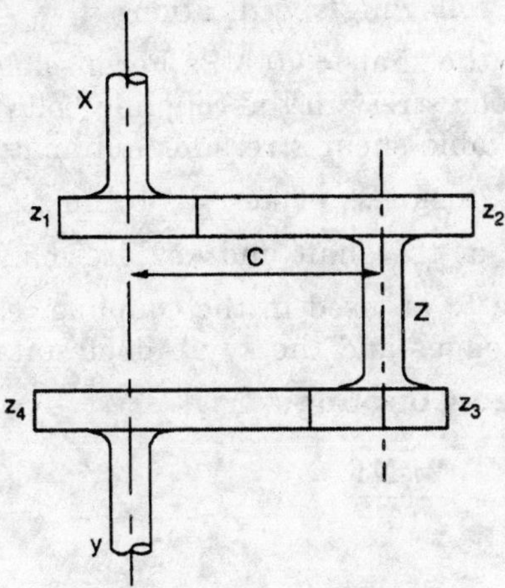

Fig. 7 Reverted gear train.

Centre distance, $\quad C = r_1 + r_2 = r_3 + r_4$

or, $$\frac{m_1(z_1+z_2)}{2} = \frac{m_2(z_3+z_4)}{2}$$

$$2z_2 - 3z_4 = 24 \qquad \text{...(1)}$$

Also, $$\frac{n_4}{n_1} = \frac{z_1 z_3}{z_2 z_4}$$

or, $$\frac{1}{12} = \frac{24 \times 24}{z_2 z_4}$$

$$z_4 = \frac{24 \times 24 \times 12}{z_2} = \frac{6912}{z_4} \qquad \text{...(2)}$$

Eqn. (1) then becomes, $2z_2 - \dfrac{3 \times 6912}{z_2}\ 24$

or, $$2z_2^2 - 24z_2 - 20736 = 0$$

$$2z_2^2 - 12z_2 - 10368 = 0$$

$$z_2 = 6 \pm 0.5\ [144 + 4 \text{ x } 10368]^{05}$$

$$= 6 \pm 102 = 108$$

$$z_4 = \frac{6912}{108} = 64$$

$$C = \frac{2(24+108)}{2} = 132 \text{ mm}$$

Actual velocity ratio, $\dfrac{n_4}{n_1} - \dfrac{24 \times 24}{108 \times 64} = \dfrac{1}{12}$

Q.4. (*a*) (*i*) Design and draw a cast iron flange coupling for a mild steel shaft transmitting 80 kW at 220 r.p.m. with the following specifications :

Allowable shear stress in the shaft = 40 MPa Permissible angle of twist < 1° in a length of 20 diameters. Allowable shear stress in the coupling bolts = 30 MPa. Rigidity modulus of shaft material = 85 GPa Allowable shear stress for hub material = 15 MPa You may adopt,

Width of key = 25 mm, Thickness of key = 14 mm.

You may also consider that the shaft and key are made of the same material.

Four number of bolts are to be used in the coupling. Find out the shaft required and the main dimensions of the flenges and the key. Adopt suitable dimensions.

Sol. (*i*) **Design of Flange Coupling:**

Power transmitted, $$P = \frac{2\pi NT}{10^3} \text{ kW}$$

Torque transmitted, $$T = \frac{80\times10^3}{2\pi\times220} = 57.845 \text{ N.m}$$

1. Shaft size:

(*i*) Based on maximum shear stress:

Let keyway effect, $k = 0.75$

d_s = diameter of shaft

$$T = k\times\left(\frac{\pi}{16}\right)d_s^3\,\tau_s$$

$$57.845 = 0.75 \times \left(\frac{\pi}{16}\right)d_s^3\times40\times10^6$$

$$d_s^3 = 9.82 \times 10^{-6}$$

$$d_s = 2.14 \times 10^{-2} \text{ m or } 21.4 \text{ mm}$$

(*ii*) Based on angle of twist :

$$\theta = \frac{TL}{GJ} = T\times20d_5\times\frac{32}{G\times\pi d_s^4}$$

$$\frac{\pi}{180} = \frac{57.845\times20\times32}{85\times10^9\times\pi\times d_s^3}$$

$$d_s^3 = = 7943.27 \times 10^{-9}$$

$$d_s = 19.95 \times 10^{-3} \text{ m or } 19.95 \text{ mm}$$

We adopt d_s = 21.5 mm

2. Hub size. Let d_h = outside diameter of hub

$$T = \left(\frac{\pi}{16}\right)\left[\frac{d_h^4 - d_s^4}{d_h}\right]\tau_h$$

Empirically $d_h = 1.75\ d_s + 6$ mm

$= 1.75 \times 21.5 + 6 = 43.625 \approx 44$ mm

$$\tau_h = \frac{57.845 \times 10^3 \times 16 \times 44}{\pi(44^4 - 21.5^4)}$$

$= 3.6675$ N/mm^2 < 15 N/mm^2

Hence safe. Adopt $d_h = 44$ mm

Thickness of hub,

$t_h = 0.5\ (d_h - d_s) = 0.5\ (4 - 21.5) = 11.25$ mm

3. Length of key:

$$\left(\frac{\pi}{16}\right) d_s^3\ \tau_5 = \frac{lbd_5\tau_s}{2}$$

$$l = \frac{\pi d_s^2}{8b} = \frac{\pi \times (21.5)^2}{8.25} = 6.26 \text{ mm}$$

Empirically, $l = 1.5\ d_s = 1.5 \times 21.5 = 32.25$ mm

Adopt length of key, $l = 32$ mm

Length of hub = length of key = 32 mm

4. Diameter of Bolts :

Radius of bolt circle, $r \approx 1.5\ d_s = 32$ mm

Shearing force in each bolt, $F = \frac{T}{rn}$

Where n = number of bolts = 4

$$F = \frac{57.845 \times 10^3}{32 \times 4} = 451.914 \text{ N}$$

Bolts are in single shear.

$$F = \left(\frac{\pi}{4}\right) d_b^2\ \tau_b$$

$$d_b^2 = \frac{451.914 \times 4}{\pi \times 30} = 19.18$$

$d_b = 4.38$ mm

Adopt $d_b = 4.5$ mm

5. Flange thickness :

Flange thickness, $t_f = 0.5\ t_h + 6$ mm

$= 0.5 \times 11.25 + 6 = 11.625 \approx 12$ mm

(ii) Draw a neat sketch for the Flange coupling with the designed dimensions.

Sol. The sketch of the flange coupling with the designed dimensions is shown in Fig. 8.

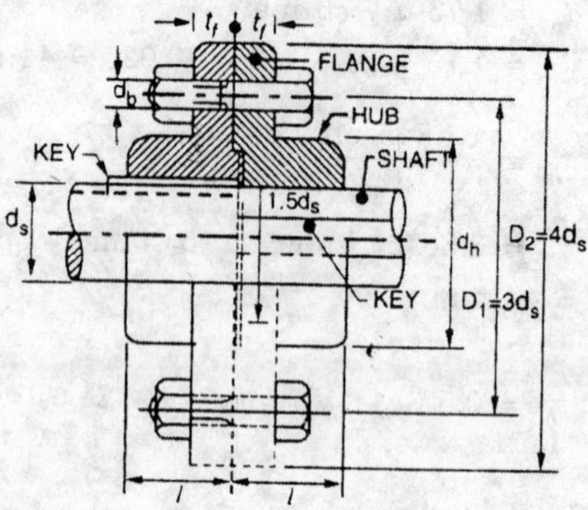

Fig. 8 Unprotected type flange coupling

(*b*) A centrifugal clutch with four shoes is used to transmit 15 kW at 720 r.p.m. The speed at which engagement begins is 80% of full speed. The inside diameter of the drum is 325 mm and outside diameter of the drum is 350 mm. The radial distance of C.G of each shoe from the axis of shaft is 120 m. The radial clearance between the shoe and the rim is 10 mm. The pressure exerted on the shoe during engagement is 0.1 MPa. The coefficient of friction between shoe lining and the drum is taken as 0.25.

Determine:

(*i*) Mass of each shoe.

(*ii*) Size of the shoe if the angle subtended by shoe at centre is 60°.

Also design the helical spring used in the shoes taking the safe shear stress of spring material as 600 MPa and modulus of rigidity of the material as 80 kN/mm_2. Use the spring index of 6.

Sol. Design of Centrigugal Clutch:

(*i*) **Mass of sloe.** Let m = mass of each shoe

Angular speed, $\omega = \dfrac{2\pi N}{60} = \dfrac{2\pi \times 720}{60} = 75.4$ rad/s

Speed at engagement of clutch,

$$\omega_e = 0.8 \times 75.4 = 60.32 \text{ rad/s}$$

Radial distance of shoe C.G. from shaft axis, $r = 0.12$ m

Centrigugal force acting on each shoe, $F_c = mr\omega_e^2$

$$= m \times 0.12 \times (75.4)^2 = 682.22 \text{ mN}$$

Centrifugal force at engagement of clutch, $F_e = mr\omega_e^2$.

$$= m \times 0.12 \times (60.32)^2 = 436.62\ m \text{ N}$$

Torque transmitted at running speed, $T = 10^3 \times P \times \dfrac{60}{2\pi N}$

$$= 10^3 \times 15 \times \frac{60}{2\pi \times 720} = 198.94 \text{ N.m}$$

where R = inside drum radius = $\frac{325}{2}$ = 162.5 mm

n = number of shoes = 4

μ = co-efficient of friction between shoe lining and drum = 0.25

$$198.94 = 0.25\ (682.22 - 436.62)\ m \times 0.1625 \times 4$$

$$m = 4.985 \text{ kg}$$

(*ii*) **Width of shoes:**

Let l = contact length of shoes, mm

b = width of shoes, mm

Now $$1 = \frac{162.5 \times \pi}{3} = 170.17 \text{ mm}$$

Area of contact of shoe, A = lb = 170. 17 b mm^2

Force exerted by shoe on the drum = Ap = 170.17 b × 0.1

$= 17.017\ b$ N

$= F_c - F_e$

$= (682.22 - 436.62)\ 4.985 = 1224.316$ N

$b = 71.95 \approx 72$ mm

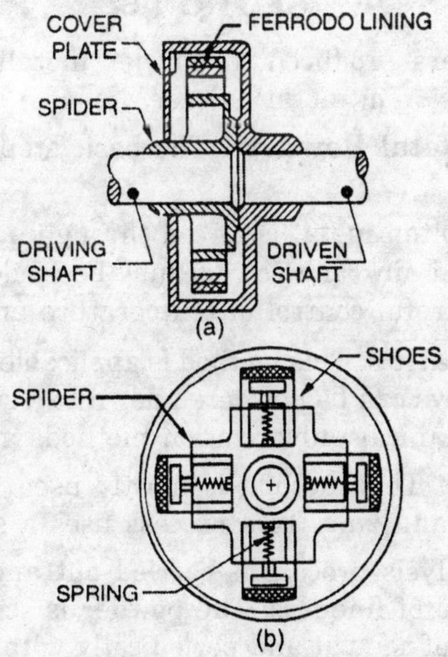

Fig. 9 Centrifugal clutch.

(*iii*) **Design of Spring.** Load on the spring

$= F_e = 436.62 \times 4.985 = 2176.55$ N

Wahl correction factor,

$$K_w = \frac{4C-1}{4C-4} + \frac{0.615}{C}$$

$$= \frac{4\times6-1}{4\times6-4}+\frac{0.615}{6} = 1.2525$$

Shear stress in spring wire, $\tau = \frac{K_w \times 8F_eC}{(\pi d^2)}$

$$600 = \frac{1.2525\times8\times2176.55\times6}{\pi d^2}$$

$$d^2 = 69.42$$

$$d = 8.33 \text{ mm}$$

$$D = 6d$$

$$= 50 \text{ mm}$$

Deflection of spring, $\delta = \frac{8F_eC^3n}{(Gd)}$

$$10 = \frac{8\times21766.55\times6^3\times4}{80\times10^3\times8.33}$$

Number of turns, $n = 1.77 \approx 2$

A centrifugal clutch is shown in Fig 9

SECTION-C

Q.5 (*a*) How are powders produced in powder metallurgy ? Name and explain the various steps in making parts from metal powder. (15)

Sol. Manufacture of Metal Powders. The basic steps for the manufacture of metal powders are :

1. **Atomization.** The molten metal is forced through a small orifice and broken up by a stream of compressed air. Fine powder may be made by this process but it requires special nozzles and careful control of temperature and pressure etc.
2. **Mechanical pulverization.** This method is applicable for brittle metals like antimony. Brittle metals and alloys can be powdered down to a size of 0.001 mm. Machining and shotting are the mechanical pulverization methods generally used.
3. **Chemical reduction.** This method is mostly used for the manufacture of copper, iron, tungsten, silver and zinc. This process uses a salt of the metal.

The conventional electrolysis process is carried out and copper is deposited as a fine powder to the required degree of fineness. The powder is scrapped from the electrodes and made to collect in the bottom of the vat and periodically withdrawn when sufficient quantity i&(accumulated. The powder is washed and .dried.

Various steps of the Process. The various steps of the process are :

1. Preparation of powder/powders of the desired composition.
2. Mixing and blending.
3. Compacting.
4. Sintering.

(*b*) A certain disc of lead of radius 150 mm and thickness 50 mm is reduced to a thickness of 25 mm by open die forging.

If the coefficient of friction between the job and die is 0.25, determine the maximum forging force. The average shear yield stress of lead can be taken as 4 N/mm². (10)

Sol. Sticking radius, $r_0 = R - \left\{\frac{t}{2\mu}\right\} \text{In}\left\{\frac{1}{\sqrt{3}\mu}\right\}$

$$= 150 - \left\{\frac{50}{2 \times 0.25}\right\} \text{In}\left\{\frac{1}{\sqrt{3} \times 0.25}\right\} = 66.3 \text{ m}$$

$$p_1 = \sqrt{K} \exp\left[\left(\frac{2\mu}{t}\right)(R - t)\right] \text{ for } r_0 \le r \le R$$

$$p_2 = \left(\frac{2K}{t}\right)(R - r) + \left(\frac{K}{\mu}\right)\left[1 + \text{In}\left(\sqrt{3}\,\mu\right)\right] \text{ for } 0 \le r \le r_0$$

Total forging force, $F = 2\pi\left[\int_0^{r_0} p_2 r \times dr + \int_{r_0}^{R} p_1 r \times dr\right]$

$$= 2\pi\left[\left[\left(\frac{2K}{t}\right) r^2 \left(\frac{R}{2}\right) - \left(\frac{r}{3}\right) + \left\{\frac{Kr^2}{2\mu}\right\}\left\{1 + \text{In}\left(\sqrt{3}\mu\right)\right\}\right]\right]_0^{r_0}$$

$$-\left(\frac{\sqrt{3}K}{4\mu^2}\right)\exp\left\{\frac{2\mu}{t}(R - r)\right\} t^2 \left(1 + \frac{2\mu r}{t}\right)\Bigg|_{r_0}^{R}$$

Here R = 150 mm, t = 50 mm, μ = 0.25, K = 4 N/mm²

Substituting these values, we get,

$$F = 853.4 \text{ kN}$$

(*c*) What are the common defects found in welding ? Explain their causes and remedies.

A projection welding of 3 spots is obtained with a current of 35,000 A for a period od 0.01 sec. If the effective resistance of the joint be taken as 120 microohms, and if the joint can be considered as a cylinder of 6 mm diameter and 2.5 mm in height, calculate the heat distributed to the surroundings.

Take heat required for melting steel as 10 J/mm³.

Sol.

S.N.	Defect in welding	Causes	Remedies
1.	Porous welds	Short arc, insufficient puddling time impaired base metal or poor electrodes.	Check impurities in base metal, allow sufficient puddling time for gases to escape, use of proper current, weaving of weld to eliminate pinholes, use of proper electrodes for the job and adjusting arc length.

2.	Poor penetration	Fast welding speed, too large electrodes, too low current, or faulty penetration.	Use enough current to get penetration, weld slowly, select proper electrodes according to welding groove size calculate penetration properly, leave proper space at bottom ofize, weld.
3.	Undercutting	Faulty electrode and too high current.	Use uniform weave in butt welding, avoid excessive weaving, use moderate current, weld slowly, hold electrode at a safe distance from vertical plane in making horizontal fillet welds, avoid too large electrodes.
4.	Distortion	Uneven heat, improper sequence, or deposited metal shrinks.	Clamp parts properly, dispose off rolling or forming strains before welding, avoid uneven heating, examine structure and use proper sequence.
5.	Cracked weld	Wrong selection of electrode, unbalanced weld and part sizes, faulty welds, faulty penetration, and rigid joints.	Heat parts before welding, eliminate rigid joints, avoid weld in string b eads, keep ends free make sound welds of good fusion, adjust weld size, allow a proper and uniform free space for joints, use low current.
6.	Poor appearance	Faulty selection of electrodes, overheating, improper use of electrodes, or wrong arc and current voltage.	Use uniform weave, avoid high current, avoid overheating and use proper welding technique.
7.	Poor fusion	Wrong speed, faulty preparation, high low current, improper electrode size.	Adjust electrode and V-size, sufficient weaving to melt sides of joints, use proper current for deposition and penetration of metal.
8.	Brittle welds	Selection of wrong electrode, faulty preheating, hardening of metal by air.	Preheat at 150° to 270° C in case of medium-carbon speed/alloy steels, make multiple layer welds, anneal the joint after welding, use stainless or low-hydrogen electrodes for increasing weld ductility.
9.	Spatter	Arc blow, too high current, length arc, or use of faulty electrodes.	Adjust current as per requirement, adjust arc length, lightening arc blow, and picking suitable electrode.

10.	Magnetic blow	Magnetic fields.	Use steel blocks to alter magnetic path around arc, use AC welding, weld in the same direction in which the arc blows.

Heat generated.

$$Q = i^2 Rt = (35000)^2 \times 10^{-6} \times 1(T^6 \times 0.01 = 1470 \text{ J}$$

$$\text{Volume of fused metal} = \left(\frac{\pi}{4}\right) d^2 h = \left(\frac{\pi}{4}\right) \times 6^2 \times 2.5 = 70.686 \text{ mm}^3$$

Heat required for melting steel = 70.686 × 10 = 706.86 J

Heat distributed to surroundings

$$= 1470 - 706.86 = 763.14\text{J}$$

Q.6 (*a*) Find the limit sizes, tolerances and allowances for a 100 mm diameter shaft and hole pair designated by $f_8\ h_{10}$. Also specify the type of fit that the above pair belongs to. Given: 100 mm diameter lies in the diameter step range of 80–120 mm. The fundamental deviation for shaft designation '*f*' is – 5.5 $D^{0.41}$.

The values of standard tolerances for grades of ITS and IT10 are 25 *i* and 64 *i* respectively. Also, indicate the limits and tolerance on a diagram. (15)

Sol. Tolerance unit, $i = 0.45\ D^{1/3} + 0.001\ D$ micron 100 mm diameter lies between 80 mm and 120 mm.

$$D = [80 \times 120]^{0.5} = 97.98 \text{ mm}$$

$$i = 0.45\ (97.98)^{1/3} + 0.001 \times 97.98$$

$$= 2.17258 \text{ microns or } 0.00217 \text{ mm}$$

For shaft *f*8, tolerance = 25 *i* = 25 × 0.00217 = 0.054 mm

For hole *h*10, tolerance = 64 *i* = 64 × 0.00217 = 0.139 mm

Fundamental deviation for shaft $f = -5.5\ D^{0.41}$

$$= -5.5 \times 97.98^{0'41} = -0.036 \text{ mm}$$

Fundamental deviation for hole *h* = 0.000 mm.

Limit of size:

Hole:

Low limit = 100 – 0.000 = 100.000 mm

High limit = Low limit + tolerance

= 100.000 + 0.139 = 100.139 mm

Shaft:

High limit = 100 – 0.036 = 99.964 mm

Low limit = High limit – tolerance

= 99.964 – 0.054 = 99.910 mm

We find that high limit of shaft is less than the low limit of hole. Therefore, it is a clearance fit.

The limits and tolerance are shown in Fig. 10.

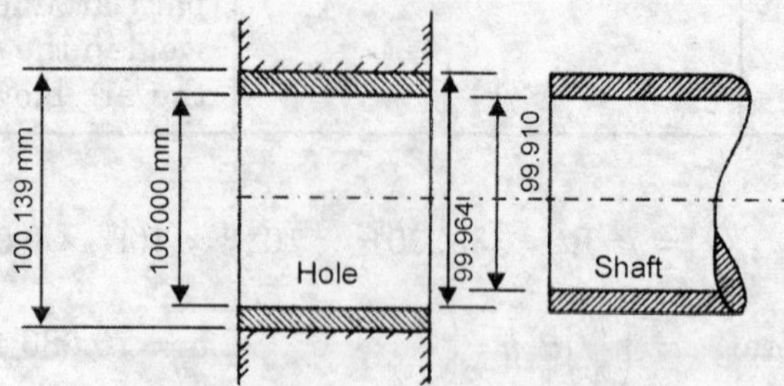

Fig. 10. Limits and tolerances.

(*b*) An HSS tool is used for turning operation. The tool life is 1 hr. When turning is carried at 30m/min. The tool life will be reduced to 2.0 min if the cutting speed is doubled. Find the suitable speed in RPM for turning 300 mm diameter so that tool life is 30 min. (10)

Sol. Taylor's equation for tool life is:

$$VT^n = C$$

For the given data, we have $30 \times 60^n = C$

$$60 \times 2^n = C$$

Equating these equations, we get

$$30 \times 60^n = 60 \times 2^n$$

$$30^n = 2$$

$$n \ln 30 = \ln 2$$

$$n = 0.2038$$

$$C = 30 \times 60^{0.2038} = 69.105$$

The Taylor's equation becomes:

$$VT^{0.2038} = 69.105$$

Now $\quad V \times 300^{0.2038} = 69.105$

$$V = 34.552 \text{ m/min}$$

Also $\quad V = \pi \times 300 \times 10^{-3} \times W = 34.552$

$$N = 36.66 \text{ rpm}$$

(*c*) Prepare part program to machine the contour shown in the Fig. 11 using APT on CNC milling machine.

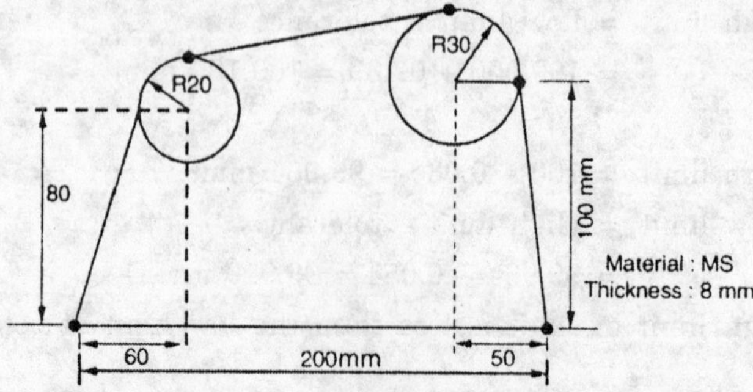

Fig. 11

Sol. APT Part Program on CNC Milling Machine. The contour to be machined by milling is shown in Fig. 12.

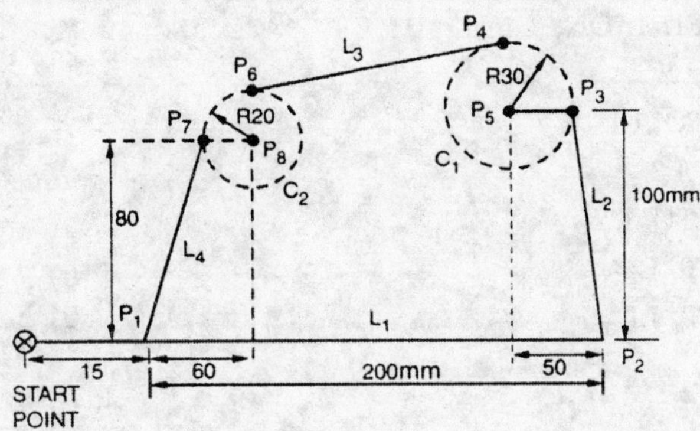

Fig. 12

```
PART NO SAMPLE PART
MILLING OPERATION
MACHINE/MILLING 02
        CLPRT
        UNITS/MM
        CUTTER/20
REMARK Part geometry points and lines are defined 10 mm below the part top surface.
PTARG = POINT / – 15, 0, – 10
  PI = POINT/0, 0, –10
  P2 = POINT/200, 0, –10
  P3 = POINT/180, 100,–10
  P4 = POINT/150, 130,–10,
  P5 = POINT/150, 100,–10
  P6 = POINT/60, 100,–10
  P7 = POINT/40, 80, –10
  P8 = POINT/60, 80,–10
  LI = LINE/PI, P2
  L2 = LINE/P2, P3
  Cl = CIRCLE/CENTER, P5, RADIUS, 30
  L3 = LINE/P4, LFT, TANTO, Cl
  C2 = CIRCLE/CENTER, P8, RADIUS, 20
  L4 = LINE/P7. LFT, TANTO, C2
  PL1 =PLANE/P1,P2, P7
REMARK Milling cutter motion statements
        FROM/PTARG
```

```
SPINDL/2000, cw
FEDRAT/15, IPM
GO/TO, LI, TO, PL 1, ON, L4
GORGT/L1,PAST,L2
GOLFT/L2, TANTO, Cl
GOFWD/C1,PAST,L3
GOFWD/L3, TANTO, C2
GOFWD/C2, PAST, L4
GOFWD/L4, PAST, LI
RAPID
GOTO/PTARG
SPINDL/OFF
FINI
```

Q.7 (*a*) Explain the need for sales forecasting . How are forcesting methods classified.? The past data about the load on a machine centre is as given below:

Month	Load. Machine-Hours
1	585
2	611
3	656
4	748
5	863
6	914
7	964

(*i*) If a five month moving average is used to forecast the next month's demand, compute the forecast of the load on the centre in the 8th month.

(*ii*). Compute a weighted three month moving average for the 8th month, where the weights are 0.5 for the latest month, 0.3 and 0.2 for the other months, respectively. (10)

Sol. Need for Sales Forecasting:

1. Majority of the activities of the industries depend upon the future sales.
2. Projected sales for the future assists in decision -making with respect to investment in plant and machinery, market planning and programmes.
3. To schedule the production activity to ensure optimum utilization of plant's capacity.
4. To prepare materials available at right quantity and right time.
5. To provide an information about the relationship between demand for different products in order to obtain a balanced production in terms of quantity required of different products as a function of time.

6. To provide a future trend which is very much essential for product design and development.

Classification of Forecasting Methods. Forcasting methods may be classified as follows:

1. Judgement techniques:
   (*i*) Opinion survey method
   (*ii*) Executive opinion method
   (*iii*) Customer and distributor surveys
   (*iv*) Marketing trials
   (*v*) Market research
   (*vi*) Delphi technique.
2. Time Series Analysis:
   (*i*) Time series forecast

$$Y = TCSR$$

where Y = forcast value,
T = secular trend,
C = cyclic variations,
S = seasonal variations,
R = irregular fluctuations.

   (*ii*) Moving Average Method
   (*iii*) Weighted Moving Average Method
   (*iv*) Exponential smoothing
3. Causal Method :
   (*i*) Least square method
   (*ii*) Regression method.

Month	1	2	3	4	5	6	7
Load, Machine–Hours	585	611	656	748	863	914	964

(*i*) Five month's moving average

= (Sum of load for month's 3 to 7)15

= (656 + 748 + 863 + 914 + 964)/5 = 829

Load on the centre in the 8th month = 829 machine-hours

(*ii*) Three month's weighted moving average

= 964 × 0.5 + 914 × 0.3 + 863 × 0.2

= 928.8 machine-hours.

(***b***) What is break-even analysis ? How is it useful to the manager ?

For a particulars product, the following information is given:

Selling price per unit : Rs. 100

Variable cost per unit : Rs. 60

Fixed costs : Rs. 10,00,000.

Due to inflation the variable costs have increased by 10% while fixed costs have increased by 5%. If the break-even quantity is to remain constaant by what percentage should the Sales price be raised ? (15)

Sol. Break-Even Analysis. Break even analysis establishes the relationship among the factors affecting profit. It indicates at what level cost and revenue are in equilibrium. It is a simple method of presenting to management the effect of changes in volume to profit.

The detailed analysis of break-even data will help the management to understand the effect of alternative decisions that convert costs from variable to fixed, the costs which increase sales volume and revenue. It is a powerful tool in evaluating alternative course of action. The break-even chart is shown in Fig. 13.

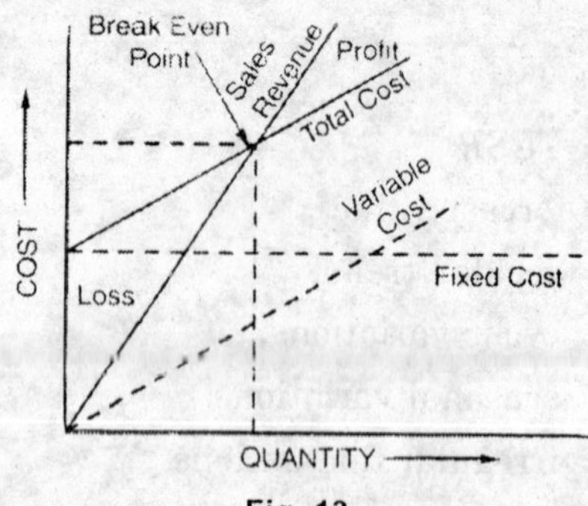

Fig. 13

Where
$$F = \text{fixed cost}$$
$$a = \text{variable cost per-unit}$$
$$b = \text{sales price per unit}$$
$$Q = \text{quantity produced and sold}$$
$$Q = \frac{F}{b-a}$$

For the given data, we have
$$F = \text{Rs. } 1000000$$
$$a = \text{Rs. } 60/\text{unit}$$
$$b = \text{Rs. } 100/\text{unit}$$
$$Q = \frac{1000000}{100-60} = 25000$$

New variable cost, $a_1 = 1.1 \times 60 = \text{Rs. } 66/\text{unit}$

New fixed cost, $F_1 = 1.05 \times 1000000 = \text{Rs. } 1050000$

$$Q = \frac{F_1}{b_1 - a_1}$$
$$25000 = \frac{1050000}{b_1 - 66}$$

$$b_1 = 66 + \frac{1050000}{25000} = \text{Rs. } 108/\text{unit}$$

$$\text{Increase in sales price} = b_1 - b = 108 - 100 = \text{Rs. } 8/\text{unit}$$

$$\% \text{ increase} = \left(\frac{8}{100}\right) \times 100 = 8\%.$$

(*c*) Prepare a flow chart and write a program in FORTRAN to compute the stresses required in the following problem :

For a two dimensional stress system, a point in a member is subjected to normal stresses σ_x and σ_y on two matually perpendicular planes accompanied with shear stress σ_{ry}. Compute the normal and shear stresses on all the planes making angles from 0 to 180° with intervals of 0.10° with plane of σ_x. Also find the principal stresses based on the maximum and minimum of the normal stresses obtained on all the planes. (15)

Sol. The normal and shear stresses on a plane making an angle θ with the σ_x. plane are (See Fig. 14):

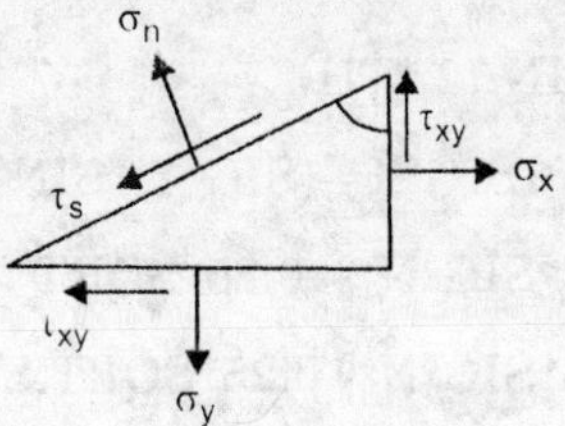

Fig. 12

$$\sigma_n = 0.5\ (\sigma_x + \sigma_y) + 0.5\ (\sigma_x - \sigma_y) \cos 2\theta - \tau_{xy} \sin 2\theta$$

$$\tau_s = 0.5\ (\sigma_x - \sigma_y) \sin 2\theta + \tau_{xy} \cos 2\theta$$

On any plane in a two-dimensional state of stress, there is only one normal stress and one shear stress, ihe other normal stress being zero. The principal stresses on this plane are:

$$\sigma_{1,2} = 0.5\ \sigma_n \pm 0.25\ \left[\sigma_n^2 + 16\tau_s^2\right]^{0.5}$$

The flow chart is shown in Fig. 15.

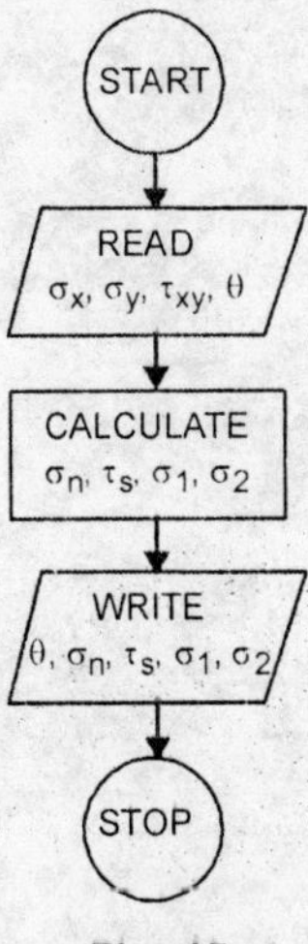

Fig. 15

Flow chart for calculating normal, shear and principal stresses.

Program in FORTRAN

```
DIMENSION SIG(200)
OPEN (UNIT = 7, FILE = 'SA.IN', STATUS = 'OLD')
OPEN (UNIT = 8, FILE = 'SA.OUT', STATUS = 'NEW')
Read (7,*) SIG X, SIG Y, SIG XY
DO 100 K = 11, 1811, 1
                  AK = (AT-11)/10.
                  PI = 3.14159
                  TH = AK*PI/180
               SIG N = 0.5*(SIG X + SIG Y) + 0.5*(SIG X - SIG Y)
1*COS (2.*TH) 1 – SIG XY *SIN (2,*TH)
               SIG S = 0.5*(SIG X – SIG Y)*SIN (2.TH) 2 + SIG XY*COS (2.*TH)
               SIG 1 = 0,5*SIG N + 0.25*SQRT (SIG N**2 + 16.*SIG S**2]
               SIG 2 = 0.5*SIG N - 0.25*SQRT(SIG N**2 + 16.*SIG S**2]
                 THD = TH*180./PI
WRITE (8, 50) THD, SIG N, SIG S, SIG 1, SIG 2
50 FORMAT (5F10.2,/)
100 CONTINUE
CLOSE (8)
STOP
END.
```

MECHANICAL ENGINEERING
PAPER II—2007

(Conventional)

SECTION-A

Q.1. **(*a*)** At a point in a loaded structure, a pure shear stress state of τ = ± 400 MPa prevails on two given planes at right angles :

(*i*) What would be the state of stress across the planes of an element taken at ± 45° to the given planes ?

(*ii*) What are the magnitudes of these stresses ?

Sol.

(*i*) There shall be normal stress on the diagonals at + 45° to the given planes, tensile stress on one diagonal and compressive stress on the other.

(*ii*) The magniture of these normal stresses shall be ± 400 MPa

(*b*) A coil spring of stiffness 'k' is cut to two halves and these two springs are assembled in parallel to support a heavy machine. What is the combined stiffness provided by these two springs in the modified arrangement ?

Sol. Stiffness of spring, $k \propto \frac{1}{n}$.

When the spring is cut to two halves the stiffness becomes 2 k. When these two halves are assembled in parallel the stiffness becomes 4 k.

(*c*) What is the value of Euler's buckling load for an axially loaded pin-ended (hinged at both ends) strut of length 'l' and flexural rigidity 'EI' ? What would be the order of Euler's buckling load carrying capacity of a similar strut but fixed at both ends in terms of the load carrying capacity of the earlier one ?

Sol. Euler's buckling load for an axially loaded both ends hinged column, ...

$$P_e = \frac{\pi^2 EI}{l^2}$$

When both ends are fixed then buckling load = $4P_e$.

(*d*) Suggest most suitable and cheapest material for the following:

(*i*) Body of screw jack

(*ii*) Piston rings

(*iii*) Belts

(*iv*) Rivets.

Sol.

(*i*) Cast steel

(*ii*) Solid woven impregnated cotton belting

(*iii*) Alloyed cast iron heat treated

(*iv*) Steel grade Fe-360 BR.

(*e*) An oil ring bearing supports a shaft of diameter 95 mm which runs at 230 rpm. The maximum load on the bearing is 18 kN. Determine the bearing characteristic number for l/d = 1.5 and absolute viscosity 0.0225 kg/ms.

Sol. $$p = \frac{W}{ld} = \frac{18 \times 10^3}{1.5 \times 95^2} = 1.33 \text{ N/mm}^2$$

$$\text{Bearing characteristic number} = \frac{\mu n}{p}$$

$$= 0.0225 \times \frac{230}{60} \times \frac{1}{1.33 \times 10^6}$$

$$= 0.06485 \times 10^{-6}$$

(*f*) Mention two best reasons why accelerometer is preferred over seismometer to measure vibration in mechanical systems ?

Sol.

(*i*) An accelerometer can measure vibration when amplitudes are even small whereas a seismometer cannot.

(*ii*) A seismometer requires a heavey mass whereas an accelerometer is a miniature instrument giving more accuracy.

(*g*) The type of cam used in cam shaft of I.C. engines is of specified contour or specified motion. State it and give reasons for your answer.

Sol. The cam used in cam shaft of *1C.* engines is of specified motion because it is required to operate the inlet and exhaust valves for a specified period of the crank shaft.

(h) A six cylinder 4-stroke multicylinder in line engine is completely balanced in respect of reciprocating masses. State 4 reasons for it.

Sol.

(*i*) Primary forces are completely balanced.

(*ii*) Primary couples are completely balanced?

(*iii*) Secondary forces are completely balanced.

(*iv*) Secondary couples are completely balanced.

(i) Mention the names of 4 plastics which are used for bearings. Also mention four advantages which plastics offer over other materials.

Sol. Names of plastics for bearings:

(*i*) Fluorocarbons (TFE)

(*ii*) Nylons

(*iii*) Acetals

(*iv*) Polyethylenes.

Advantages:

(i) Self lubricated

(ii) Low co-efficient of friction

(iii) Vibration absorption and impact resistance

(iv) Free from corrosion and abrasion action.

(j) Explain the following in brief:

(*i*) Effect of size on the tensile strength,

(*ii*) Effect of surface finish on endurance limit.

Sol.

(*i*) Tensile strength decreases with increase in size of the component. As the size of the component increases, the surface area also increases, resulting in a greater number of surface defects.

(*ii*) The endurance limit is maximum for a polished surface. As surface finish decreases, the endurance limit also decreases. Surface roughness introduces stress-concentration and alters the physical properties of surface layer of material.

(k) What is the principle of resistance welding ? Indicate where the resistance is maximum in spot welding operation.

Sol. A low voltage high amperage current flows from one adjoining plate to the other until the metal at the interface is heated to a high enough temperature to cause localized fusion which under the applied pressure squeezes the molten metal from the two parts to a homogeneous mass called the *weld nugget.* No flux or filler metal is used.

In spot welding operation, the resistance is maximum in the middle of the electrode (nugget).

(l) Sometimes the parting plane between two forging dies is not a horizontal plane, give the main reason for this design aspect, why is parting plane provided, in closed die forging?

Sol. In case of non-symmetrical and complicated parts, the parting plane between two forging dies is not a horizontal plane.

In closed-die forging, cavities or impressions are cut in the die block, in which the metal is forced to take its final shape and dimensions. Half the metal volume is in the lower die and the other half in the upper die. Therefore, a parting line is provided.

(m) What is permeability ? Permeability is more important in the basic process of sand casting than porosity? Give one important reason for this feature.

Sol. Permeability of the moulding sand is the measure of its ability to permit air to flow through it. Molten metal always contains a certain amount of dissolved gases which try to leave it when the metal solidifies. If all these gases are not able to escape completely through the walls of the mould, they may penetrate the liquid metal, where after solidification, they form gas holes and pores. To avoid these defects, the moulding sand should have good gas per-meability.

(n) Metal powders are compacted by many methods, but sintering is required to achieve which property ? What is hot isostatic pressing ?

Sol. Sintering is the bonding of adjacent surfaces of particles in a mass of metal powder, by heating below the melting point of powder. Hot isostatic pressing - In this process, the powder is encased in a deformable metal or ceramic can or container that has the shape of the desired part. The can is evacuated and then placed inside a furnace, which in turn is enclosed in a high pressure chamber. The chamber is pressurized with an inert gas (argon) upto 300 MPa or more at 480°C to 2000°C. The pressure applied is hydrostatic. Therefore, the compact will be uniformly pressed throughout its mass. The powder is compacted, densified and sintered in one step. When the container is removed, a finished product close to the final shape is obtained.

(o) Cast iron with impurities of carbide requires a particular rake angle for efficient cutting with single point tools, what is the value of this rake angle, give reasons for your answer.

Sol. Cast iron with impurities of carbide requires negative rake angle for efficient cutting with single point tools. The rake angle is generally kept 5°. Carbide impurities being hard and brittle gives rise to shock loading on the tool. The tool will fail if positive rake angles are used. Using negative rake angles, directs the force back into the body of the tool away from the cutting edge which gives protection to the cutting edge.

(p) From the data of a pilot study, the percentage of occurrence of an activity is 60%. Find the number of observations for 95% confidence level and an accuracy of ± 2%.

Sol. $p = 0.6$, S = 0.02, K = 2 for 95% confidence level,

$$\text{Sp} = \text{K}\sqrt{\frac{p(1-p)}{\text{N}}}$$

$$0.02 \times 0.6 = 2\sqrt{\frac{0.6 \times 0.4}{\text{N}}}$$

$$\text{N} = 6667$$

(q) A typical activity i-j in a CPM network has activity duration (T_{ij} of 2.5 time units. The earlier expected time (T_E^i) and latest allowable occurrence time (T_L^i) of event i are computed as 8 and 11 units respectively. The corresponding times of event j, *i.e.*, T_E^j and T_L^j are respectively 13.5 and 13.5 units. Find the three floats of the activity $i - j$.

Sol.

(*i*) Total float, $\text{FT} = \left(T_L^j - T_L^i\right) - t_{ij} = (13.5 - 8) - 2.5 = 3$

(*ii*) Free float, $\text{FF} = \left(T_E^j - T_E^i\right) - t_{ij} = (13.5 - 8) - 2.5 = 3$

(*iii*) Independent float, $\text{FID} = (T_E^j - T_L^i) - t_{ij} = (13.5 - 11) - 2.5 = 0$

$t_{ij} = 2.5$

(i) —— (j)

$T_E^i = 8$ $T_E^j = 13.5$

$T_L^i = 11$ $T_L^j = 13.5$

(*r*) State any four basic elements of a Robot. Give one line statement for each.

Sol.

(*i*) **Controller.** A special-purpose device similar to a computer.

(*ii*) **Manipulator.** It performs the principal movements in doing manufacturing work.

(*iii*) **End effectors.** It is any special device or tool attached to the tool plate to allow the robot to perform some specialised task.

(*iv*) **Power source.** It supplies power to the controller and the manipulator.

(*s*) With the following printf () statements in 'C', express how the value x = 38.4625 appears: (*i*) printf ("% 7.2f, *x*), (*ii*) printf ("% 7.4 f, x), (*iii*) printf ("% *f*, *x*) and (*iv*) printf ("% -7.2 f, x).

Sol. (*i*) –38.46 (*ii*) –38.4625 (*iii*) 38.4625 (*iv*) 38.46.

(*t*) What for FORTRAN commands REAL and DIMENSION are used and what are their equivalents in 'C' language? (20 × 2 = 40)

Sol. Real. This is an explicit specification which overrides the implicit classification of a variable and precedes all other specification statements. For example :

REAL MAT, KAT, JET

Dimension. It provides information to the compiler for reserving a specific number of memory locations for each subscripted variable appearing in the program. Its form is :

DIMENSION A1 (*n*1), A2 (*n2, n*3)

where *n*1, *n*2, *n*3 are non-zero integer constants. **SECTION-B**

Q.2. (*a*) A simply supported beam made of rolled steel joist (I-section : 450 mm x 200 mm) has a span of 5 m and it carries a central concentrated load W. The flanges are strengthened by two 300 mm x 20 mm plates, one rivetted to each flange over the entire length of the flanges. The second moment of area of the joist about the principle bending axis is 35,060 cm . Calculate :

(*i*) The greatest central load the beam will carry if the bending stress in the 300 mm × 20 mm plates is not to exceed 125 MPa.

(*ii*) The minimum length of the 300 mm plates required to restrict the maximum bending stress in the flanges of the joist to 125 MPa. (20)

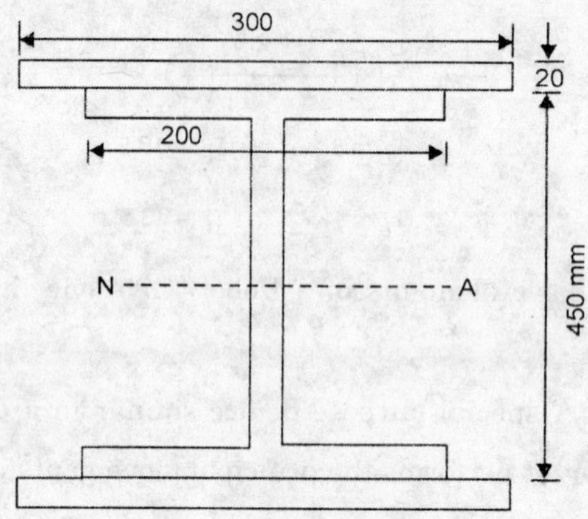

Fig. 1.

Sol. Moment of inertia of plates about N – A,

$$I_p = 2\left[\frac{30\times 2^3}{12} | 30\times 2\times(24.5-1.0)^2\right]$$

$$= 66310 \text{ cm}^4$$

For the joist, $I_j = 35060 \text{ cm}^4$

$$I = I_j + I_p$$

$$= 35060 + 66310 = 101370 \text{ cm}^4$$

Maximum bending moment,

$$M = \frac{WL}{4} = \frac{W\times 5}{4} = 1.25 \text{ W N.m}$$

Section modulus, $Z = \dfrac{I}{y} = \dfrac{101370}{24.5} = 4137.551 \text{ cm}^3.$

(*i*) $M = \sigma Z$

$$1.25\, W = 125 \times 10^6 \times 4137.551 \times 10^{-6}$$

$$W = 413755 \text{ Nor } 413.755 \text{ kN}$$

(*ii*) Let l = Length of 300 mm plates

$$I_p = 2\left[\frac{l\times 2^3}{12} + l\times 2\times(23.5)^2\right]$$

$$= 2210.3\, l\text{cm}^4$$

$$I = 35060 + 2210.3\, l \text{ cm}^4$$

$$Z = \frac{I}{y_f} = \frac{35060 + 2210.3l}{22.5} = 1558.2 + 98.2\, l \text{ cm}^3$$

$$1.25 \times 413755 = 125 \times 10^6 \times (1558.2 + 98.2l) \times 10^{-6}$$

$$l = 26.26 \text{ cm or } 262.6 \text{ mm.}$$

(b) A solid shaft is to transmit 300 kW at 120 rpm. If the shear stress is not to excess 100 MPa, find the diameter of the shaft. What percent saving in weight would be obtained if this shaft were replaced by a hollow one whose internal diameter equals 0.6 of the external diameter, the length, material and maximum allowable shear stress being the same? (20)

Sol. $P = 300$ kW, $N = 120$ rpm, $\tau = 100$ MPa

$$d_i = 0.6\, d_o$$

Power transmitted, $$P = \frac{2\pi NT}{60 \times 10^3} \text{ kW}$$

Torque, Now $$T = \frac{300 \times 50 \times 10^3}{2\pi \times 120} = 23873 \text{ N.m}$$

Now $$T = \frac{\pi}{16} d_s^3 \tau$$

Diameter of solid shaft,

$$d_s = \sqrt[3]{\frac{16T}{\pi\tau}} = \sqrt[3]{\frac{16 \times 23873}{\pi \times 100 \times 10^6}}$$

$$= 10.673 \times 10^{-2} \text{ m or } 106.73 \text{ mm}$$

For the hollow shaft,

$$J_h = \frac{\pi}{32} d_0^4 \left[1 - \left(\frac{d_i}{d_0}\right)^4\right]$$

$$= \frac{\pi}{32} d_0^4 \left[1 - (0.6)^4\right] = 0.08545\, d_0^4$$

Polar section modulus of hollow shaft,

$$Z_p = \frac{2J_h}{d_0} = 0.1709\, d_0^3$$

$$T = \tau Z_p$$

$$23873 = 100 \times 10^{-2} \times 0.1709\, d_0^3$$

$$d_0 = 11.178 \times 10^{-2} \text{ m or } 1121.8 \text{ mm}$$

$$d_i = 67.07 \text{ mm}$$

For same length, material and maximum allowable shear stress, the weight of shaft is proportional to area of cross-section.

Percentage saving in weight

$$= \left[\frac{W_s - W_h}{W_s}\right] \times 100$$

$$= \left[1 - \frac{(111.8)^2 - (67.07)^2}{(106.73)^2}\right] \times 100$$

$$= 29.76\%$$

Q.3 ***(a)*** Two gears mesh externally to provide velocity ratio 3 : 1. Both the gears have standard addendum and pressure angle is equal to 20°. Determine minimum number of teeth on the pinion to avoid interference. Instead of using a pinion of these number of teeth another pinion having three teeth less than the above number is to be used. Determine the minimum amount of stubbing of gear teeth in terms of module which is to be done now for this pinion. The velocity ratio is to be maintained same. (20)

Sol. Velocity ratio, i = 3, Pressure angle, ϕ = 20°

Minimum number of teeth on the pinion to avoid interference,

$$Z_1 = \frac{2}{i\left[\left\{1+\left(\frac{1}{i}\right)^2 \sin^2\phi + \frac{2}{i}\sin^2\phi\right\}^{0.5} - 1\right]}$$

$$= \frac{2}{3\left[\left\{1+\left(\frac{1}{3}\right)^2 \sin^2 20° + \frac{2}{3}\sin^2 20°\right\}^{0.5} - 1\right]} = 15$$

Number of teeth on stub pinion = 15 – 3 = 12

Number of teeth on gear,

$$Z_2 = 12 \times 3 = 36$$

$$Z_2 = \frac{2a_w}{\left[1+\left(\frac{1}{i}\right)^2 \sin^2\phi + \frac{2}{i}\sin^2\phi\right]^{0.5} - 1}$$

$$36 = \frac{2a_w}{\left[1+\frac{1}{9}\sin^2 20° + \frac{2}{3}\sin^2 20°\right]^{0.5} - 1}$$

$$a_w = 0.8$$

$$h_{a2} \quad a_w \text{ m} = 0.8 \text{ m}$$

(b) A machine of mass 500 kg. It is supported on helical springs which deflect by 5 mm due to the weight of the machine. The machine has rotating unbalance equal to 250 kg mm. The speed of the machine is 1200 rpm. Determine the dynamic amplitude. The damping factor of the viscous damper is 0.4.

Now this machine is mounted on a larger concrete block of mass 1200 kg. The stiffness of the springs is changed such that the static deflection is still the same with the same viscous damper as in earlier case. Determine the change in the dynamic amplitude. (20)

Ans.

$$M = 500 \text{ kg}, \ \delta = 5 \text{ mm}, \ me = 250 \text{ kg mm}$$

$$N = 1200 \text{ rpm}, \ \zeta = 0.4$$

$$\omega = \frac{2\pi N}{60} = \frac{2\pi \times 1200}{60} \quad 125.6 \text{ rad/s}$$

$$k = \frac{Mg}{\delta} = \frac{500 \times 9.81}{5} = 981 \text{ N/mm}$$

$$\omega_n = \sqrt{\frac{k}{m}} = \sqrt{\frac{981 \times 10^3}{500}} = 44.294 \text{ rad/s}$$

$$\beta = \frac{\omega}{\omega_n} = \frac{125.6}{44.294} = 2.843$$

$$\frac{X}{\frac{me}{M}} = \frac{\beta^2}{\sqrt{(1-\beta^2)^2 + (2\zeta\beta)^2}}$$

$$X = \frac{250}{500} = \frac{(2.843)^2}{\sqrt{\{1-(2.843)^2\}^2 + (2 \times 0.4 \times 2.843)^2}} = 0.543 \text{ mm}$$

Q.4. (*a*) Design a flat belt drive system for the following specifications :

Power to be transmitted = 100 kW

Pulley diameter = 0.8 m and 1 m, respectively

Centre distance between the pulleys = 3m

Belt speed = 20 m/s

Co-efficient of friction between the belt and pulleys = 0.3

Slip = 12% at each pulley

Friction loss at each shaft = 5%

Overload capacity = 20%

Density of leather belt material - 1000 kg/m^3

Safe stress for the belt = 2.5 MPa

Thickness of the belt = 12 to 16 mm.

Take into account the effect of centrifugal tension.

Find the width and the length of the belt.

Draw a schematic diagram of the belt drive. (20)

Ans. Linear speed on driver pulley,

$$v_1 = \left(1 - \frac{s_1}{100}\right) = 20\left(1 - \frac{12}{100}\right) = 17.6 \text{ m/s}$$

Linear speed on driven pulley,

$$v_2 = v_1\left(1 - \frac{s_2}{100}\right) = 17.6\left(1 - \frac{12}{100}\right) = 15.49 \text{ m/s}$$

$$P = 100 \text{ kW}$$

$$P_1 = P \times \text{Overload capacity} \times (1 - \text{Friction loss})$$

$$= 100 \times 1.2 \times (1 - 0.05) = 114 \text{ kW}$$

$$P_2 = 114 \times 0.95 = 108.3 \text{ kW}$$

Length of open belt, $L_0 = 2C + \frac{(d_2 - d_1)^2}{4C} + \frac{\pi}{2}(d_2 + d_1)$

$$= 2\times 3+\frac{(1-0.8)^2}{4\times 3}+\frac{\pi}{2}\ (1 + 0.8) = 8.83 \text{ m}$$

$$P_{max} = P_1 = 114 \text{ kW}$$

$$\sin \alpha = \frac{d_2 - d_1}{2C} = \frac{1-0.8}{2\times 3} = \frac{0.2}{6} = 0.0333$$

$$\alpha = 1.91°$$

$$\theta = 180° - 2\alpha = 176.18°$$

$$\exp(\mu\theta) = \exp\left(0.3\times\frac{\pi}{180}\times 176.18\right) = 2.516$$

Let $t = 14$ mm

Centrifugal tension,

$$T_c = mv_2^2 = \rho btv_1^2 = 1000 \times b \times 14 \times 10^{-3} \times (17.6)^2 = 4336.64\, b \text{ N}$$

$$P_1 = (T_1 - T_2)\ \frac{v_1}{10^3}$$

$$114 \times 10^3 = (T_1 - T_2) \times 17.6$$

$$T_1 - T_2 = 6477.3$$

Also, $$\frac{T_1}{T_2} = \exp(\mu\ \theta) = 2.516$$

$\therefore$ $$T_2 = 4272.6 \text{ N},\ T_1 = 10750 \text{ N}$$

$$T_1 + T_C = T_{max}$$

$$10750 + 4336.64\, b = 2.5 \times 10^6 \times b \times 14 \times 10^{-3}$$

$$30663.36\, b = 10750$$

$$b = 0.35 \text{ m or } 35 \text{ cm}$$

Q.4. (*b*) A shaft ABCD is supported at one end A and at point C The other end D consists of a pulley which has belt tensions in the vertical direction. The distance AD is 1200 mm. AC distance is 800 mm. A spur gear having pressure angle 20° is mounted at point B which is at a distance 400 mm from A. The diameter of pulley is 450 mm and that of the gear is 300 mm. The gear is suppplied power equal to 20 kW at 500 rpm by another gear at the top of it. The tension on the tight side is 3 times the tension on slack side in the belt. The combined shock and fatigue factors for bending moment and torsional moment may be taken as 1.5. The tensile strength of the material of the shaft and key is 700 MPa. Design the shaft and key for the pulley assuming factor of safety equal to 5. Assume allowance for the key way for stress as ft 75. (20)

Sol. Gear. Mean velocity,

$$v_m = \frac{\pi d_2 n_2}{60\times 10^3} = \frac{\pi\times 300\times 500}{60\times 10^3} = 7.854 \text{ m/s}$$

Power transmitted, $$P = \frac{F_t\ v_m}{10^3} \text{kW}$$

Tangential force, $F_t = \dfrac{20 \times 10^3}{7.854} = 2546.48 \text{ N} \leftarrow$

Radial force, $F_r = F_t \tan \phi = 2546.48 \tan 20° = 926.84 \text{ N}\downarrow$

Pulley.

$$v = \frac{\pi dn}{60 \times 10^3} = \frac{\pi \times 450 \times 500}{60 \times 10^3} = 11.78 \text{ m/s}$$

$$P = \frac{(T_1 - T_2)v}{10^3}$$

$$T_1 - T_2 = \frac{20 \times 10^3}{11.78} = 1697.65$$

$$\frac{T_1}{T_2} = 3$$

$$T_1 = 2546.48 \text{ N} \downarrow, \; T_2 = 848.83 \text{ N} \downarrow$$

$$T_1 + T_2 = 3395.31 \downarrow$$

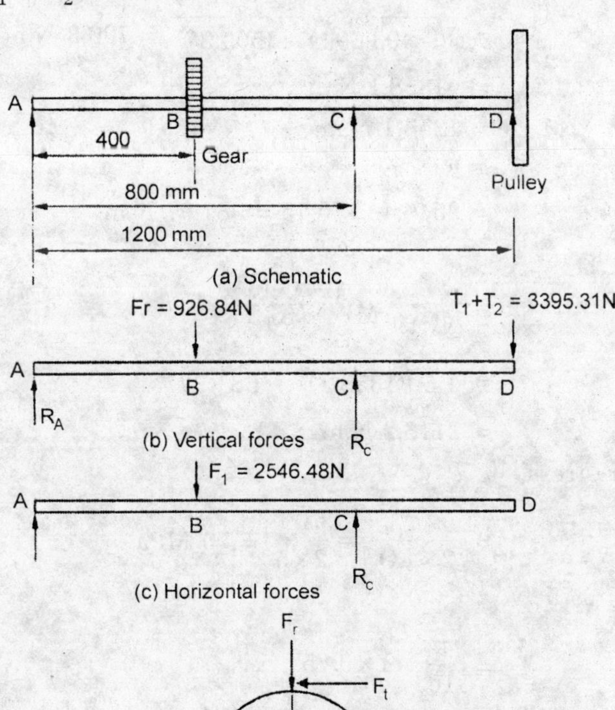

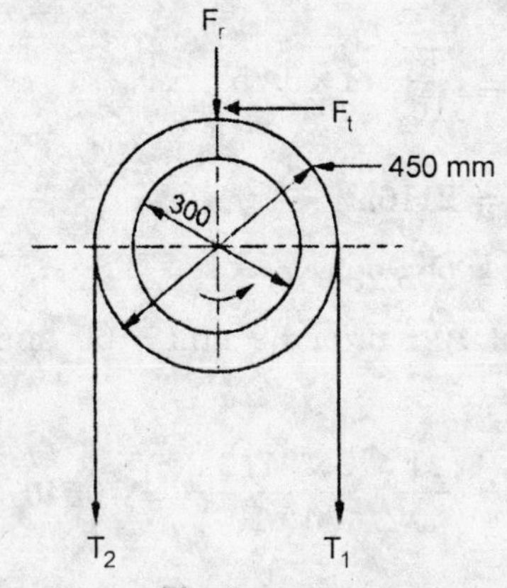

Fig. 2.

Vectical forces. The vertical forces on the shaft are shown in Fig. (*b*)

$$R_A = 800 \times 926.84 \times 400 - 3395.31 \times 40$$

$$R_A = -1234.235 \text{ N}$$

$$= 5556.238 \text{ N}$$

$$M_B = -1234.235 \times 0.4 = -493.694 \text{ N.m}$$

$$M_c = 3395.3\ 1 \times 0.4 = -1358.1 \text{ N.m}$$

Horizontal force:

$$R_A \times 800 = 2546.48 \times 400$$

$$R_A = 1273.24 \text{ N}$$

$$R_c = 1273.24 \text{ N}$$

$$M_B = 1273.24 \times 0.4 = +509.3 \text{ N.m}$$

$$M_C = 0.$$

Resultant bending moment:

$$M_B = \sqrt{(-493.694)^2 + (509.3)^2} = 709.3 \text{ N.m}$$

$$M_C = -1358.1 \text{ N.m}$$

$$\therefore \quad M_{max} = 1358.1 \text{ N.m}$$

Torque on gear, $T = F_t \times r$

$$= 2546.48 \times 0.15 = 381.97 \text{ N.m}$$

Equivalent torque,

$$T_e = \sqrt{(K_M\ M)^2 + (K_T T)^2}$$

$$= 1.5\sqrt{(1358.1)^2 + (381.97)^2}$$

$$= 2116.2 \text{ N.m}$$

Allowable shear stress,

$$\tau_a = \frac{\sigma_t}{2 \times \text{FOS}} = \frac{700}{2 \times 5} = 70 \text{ MPa}$$

Keyway effect = 0.75

$$\therefore \quad T_e = \frac{\pi}{16} d^3\, \tau \times 0.75$$

$$\frac{\pi}{16} d^3 \times 70 \times 10^6 \times 0.75 = 2116.2$$

$$d^3 = 205.29 \times 10^6$$

Shaft diameter, $d = 5.9 \times 10^{-2}$ m or 59 mm ≃ 60 mm

Key size:

$$F = \frac{2T}{d} = \frac{2 \times 2116.2}{60 \times 10^{-3}} = 70540 \text{ N}$$

Let width of key, $b = \frac{d}{4} = \frac{60}{4} = 15$ mm

thickness of square key, $t = b = 15$ m

then, $\frac{F}{lb} = \tau$

Length of key, $l = \frac{70540}{15 \times 10^{-3} \times 70 \times 10^{6}} = 67$ m $\simeq 70$ mm

SECTION-C

Q .5. (a) A cylinder of height 60 mm and diameter 100 mm is forged at room temperature between two flat dies. Find the die load at the end of compression to a height of 30 mm, using slab method of analysis. The yield strength of the work material is given as 120 N/mm and the co-efficient of friction is 0,05. Assume that volume is constant after deformation. There is no sticking. Also find the mean die pressure. **(20)**

Sol. $h_1 = 60$ mm, $h_2 = 30$ mm,

$r_1 = 50$ mm, $\sigma_y = 120$ N/mm^2

$\mu = 0.05$

For volume constancy of cylinder,

$$\pi r_1^2 h_1 = \pi r_2^2 h_2$$

$$r_2 = r_1\sqrt{\frac{h_1}{h_2}} = 50\sqrt{\frac{60}{30}} = 70.7 \text{ mm}$$

Yield shear strength, $K = \frac{\sigma_y}{2} = \frac{120}{2} = 60$ N/mm^2

In the non-sticking zone,

$$p_1 = \sqrt{3}\,K \exp\left[\frac{2\mu}{h_2}(r_2 - r)\right]$$

For no sticking, $r = 0$

$$p_1 = \sqrt{3} \times 60 \times \exp\left[\frac{2 \times 0.05 \times 70.7}{30}\right]$$

$$= 131.54 \text{ N/mm}^2$$

Forging force,

$$F = -\sqrt{3}\,K \left| \frac{\exp\left[\frac{2\mu}{h_2}(r_2 - r)\right]}{4\mu^2} h_2^2 \left(1 + \frac{2\mu r}{h_2}\right) \right|_0^{r_2}$$

$$= -\sqrt{3}\,K \left[\frac{h_2^2}{4\mu^2}\left(1 + \frac{2\mu r_2}{h_2}\right) - \frac{\exp\left(\frac{2\mu r}{h_2}\right)}{4\mu^2} \times h_2^2\right]$$

$$= -\sqrt{3}\times 60\left[\frac{30^2}{4(0.05)^2}\left(1+\frac{2\times 0.05\times 70.7}{30}\right)-\frac{\exp\left(\frac{2\times 0.05\times 70.7}{30}\right)}{4\times(0.05)^2}\times 30^2\right]$$

$$= \sqrt{3}\times 60\ [111210 - 113917.7]$$

$$= 281.4 \text{ kN}$$

(*b*) Chvorinov and Caine gave rules for solidification time and freezing ratio for a riser. Using these rules or otherwise find the size of a cylindrical riser of height to diameter ratio as one for a steel casting of size 250 × 250 × 50 mm , when the casting is fed horizontally and riser is a side one. Thickness of the casting is 50 mm. For steel a = 0.10, *b* = 0.03 and c = 1.00. Consider it as a long bar of cross-section 250 × 50. (20)

Sol. Chvorinov's rule:

Solidification time, $t = C\left(\frac{V}{SA}\right)^2$

where V = volume of casting,

SA = surface area of casting,

C = constant

Casting:

$$V = 250 \times 50 \times 250 = 3125 \times 10^3 \text{ mm}^3$$

$$SA = 2[250 \times 50 \times 2 + 250 \times 250] = 175 \times 10^3 \text{ mm}^2.$$

Riser :

Let d = diameter of riser

h = height of riser

Then, $V = \frac{\pi}{4}d^2 h = \frac{\pi}{4}d^3$

$$SA = 2\times\frac{\pi}{4}d^2 + \pi d h = \frac{\pi}{2}d^2 + \pi d^2 \text{ for } d = h$$

$$= \frac{3\pi}{2}d^2$$

$$\frac{V}{SA} = \frac{d}{6}$$

Generally $\left(\frac{V}{SA}\right)_{riser} = (1.1 \text{ to } 1.15)\left(\frac{V}{SA}\right)_{casting}$

$$\frac{d}{6} = 1.15\left(\frac{3125}{175}\right)$$

$$d = 123.2 \text{ mm} \simeq 124 \text{ mm}.$$

Caine's formula:

Relative freezing time or freezing ratio,

$$X = \frac{\left(\frac{SA}{V}\right)_{casting}}{\left(\frac{SA}{V}\right)_{riser}} = \frac{\left(\frac{175}{3125}\right)}{\frac{6}{7}} = 9.33 \times 10^{-3} d$$

Volume ratio, $$Y = \frac{(V)_{riser}}{(V)_{casting}} = \frac{\frac{\pi}{4} d^3}{3125 \times 10^3} = 2.51 \times 10^{-7} d^3$$

$$X = \frac{a}{Y - b} + c$$

where a = freezing characteristics constant = 0.10,

b = liquid solidification contraction = 0.03,

c = relative freezing rate of riser and casting = 1.00

$$9.33 \times 10^{-3} d = \frac{0.10}{2.51 \times 10^{-7} d^3 \times 0.03} + 1.00$$

$$9.33 \times 10^{3} d\,(2.51 \times 10^{-7} d^3 - 0.03)$$

$$= 0.10 + 1.00\,(2.51 \times 10^{-7} d^3 - 0.03)$$

$$23.418 \times 10^{-10} d^4 - 2.51 \times 10^{-7} d^3 - 0.28 \times 10^{-3} d - 0.07 = 0$$

or $d^4 - 107.18 d^3 - 11956\, d - 29891536 = 0$

Let $f(d) = d^4 - 107.18 d^3 - 119566\, d - 29891536$

d	125	128	129
$f(d)$	−10032598	−1533279	+1525203

We may take d = 129 mm

h = 129 mm

Q.6. (*a*) Two steel sheets of thickness one mm are welded by resistance projection welding technique. A current of 30,000 A for 0.005 second is made to flow. The effective resistance of joint can be taken as 100 micro ohms. The joint can be considered as a cylinder of diameter 5 mm ariti height 1.5 mm. The density of steel is 0.00786 gm/ mm^3. The heat needed for welding steel is 10 J/mm^3. Calculate the efficiency of welding. (20)

Sol. I = 30,000 A, t = 0.005s,

R = 100 micro ohms,

ρ = 0.00786 gm/mm^3, Q = 10 J/mm^3

d_n = 5 mm, h_n =1.5 mm

Volumc cffused metal $= \frac{\pi}{4} d_n^2\, h_n$

$$= \frac{\pi}{4} \times 5^2 \times 1.5 = 29.45 \text{ mm}^3$$

Mass of fused metal $= \text{Volume} \times \text{Density}$

$= 29.45 \times 0.00786 = 0.2315$ gm

Heat required for melting $= 10 \times 29.45 = 294.5$ J

Heat generated $= I^2 Rt = (30{,}000)^2 \times 100 \times 10^{-6} \times 0.005$

$= 450$ J

Efficiency of welding $= \dfrac{294}{250} \times 100 = 65.44\%$

(***b***) Prepare part program using APT language for milling the contour shown in Fig. 3 in a single pass. (20)

Sol.

Part program using APT language (Fig. 4)

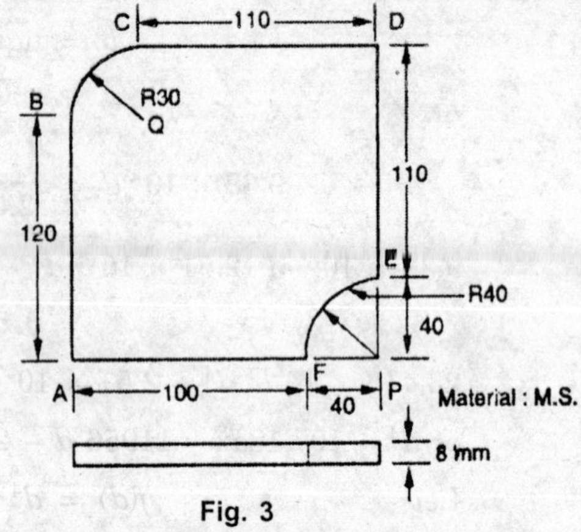

Fig. 3

```
$ $ GEOMETRY
SET PT = POINT/–50, 0, 0
PI = POINT/ 0,0, –25
P2= POINT 730, 120,  25
P3= POINT /O, 120, –25
P4= POINT/ 30, 150, –25
P5=POINT/140, 150, –25
P6= POINT / 140, 0, –25
P7= POINT /1 40,40,–25
P8= POINT /1 00,0, –25
Cl= CIRCLE / CENTRE, P2, RADIUS, 30
C2= CIRCLE / CENTRE, P6, RADIUS, 40
L1=LINE/P1,P3
L2= LINE / P4, P5
L3= LINE / P5, P7
L4=LINE/P8, PI
PL1= PLANE/ PI, P3,P8
$ $ MOTION
FROM / SETPT
LOAD TL /I, ADJUST, 1
CUTTER/ 20
SPINDL / 2000, CW
FEDRAT / 300, MMPM
RAPID, GO LI , PAST L4
```

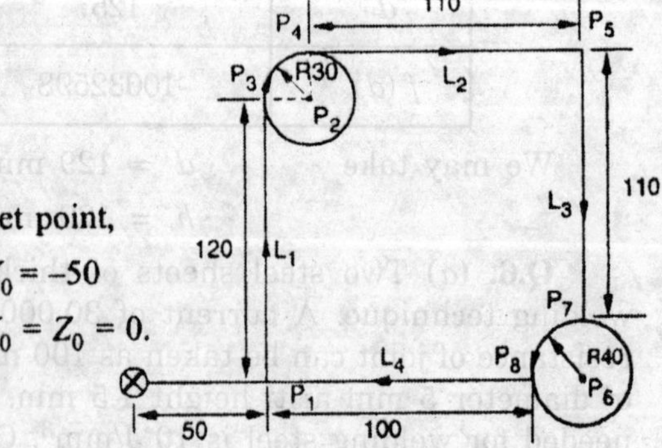

Fig. 4.

TLLFT, GO LFT /LI, PAST L2

GORGT / L 2, TANT6, C 1

GOFWD / L 2, PAST, L 3

GOFWD / C 2, PAST, L 3

RAPID, GO TO / SP

FINI.

Q.7. **(a)** Two products A and B are to be machined on three machine tools P, Q and R. Product A takes 10 hrs on machine P, 6 hrs on machine Q and 5 hrs on machine R. The product B takes 7.5 hrs on machine P, 9 hrs on machine Q and 13 hrs on machine R. The machining time available on these machine tools P, Q, R are respectively 75 hrs, 54 hrs and 65 hrs per week. The producer contemplaces profit of Rs. 60 per product A, and Rs. 70 per product B. Formulate LP model for the above problem and show the feasible region graphically/ geometrically. What are the basic feasible solutions to the above problem ? Estimate graphically/ geometrically the optimum product mix for maximizing the profit. Explain why one of the vertices of the feasible region becomes the optimum solutions point.

(**Note:** Graph sheet need not be used). (15)

Sol. Let x_1 and x_2 be the quantity of products A and B respectively,

Max $Z = 60x_1 \times 70x_2$

Subject to :

$$10x_1 + 7.5x_2 \le 75 \quad ...(1)$$

$$6x_1 + 9x_2 \le 54 \quad ...(2)$$

$$5x_1 + 13x_2 \le 65 \quad ...(3)$$

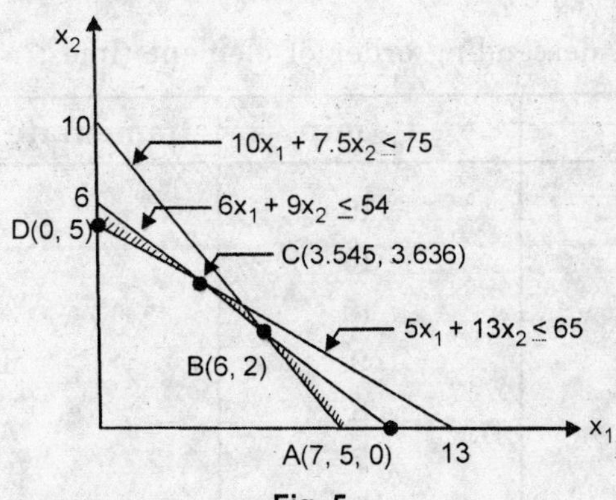

Fig. 5.

Solving Eqs. (1) and (2), we get point B (6, 2)

Solving Eqs. (2) and (3), we get point C (3.545, 3.636)

The feasible points are B (6, 2) and D (0, 5).

$$Z_s = 60 \times 6 + 70 \times 2 = \text{Rs. } 500$$

$\therefore$ Maximum profit, Z_{max} = Rs. 500

Quantity of product *A*, $x_1 = 6$

Quantity of product B, $x_2 = 2$

(***b***) An assembly of an equipment involves the tasks A to N, the precedence tasks and processing times of these tasks in minutes are given in the following table. Considering cycle time of 20 minutes at any work station, balance the assembly line and find the optimum number of work stations. Also find the idle time at each work station : (10)

Task	Predecessor	Duration
A	—	6
B	—	5
C	—	4
D	A	8
E	B	5
F	C	4
G	D	7
H	E	6
I	F	10
J	G	10
K	I	7
L	J,H	10
M	K	6
N	L,M	9

Sol. Largest Candidate rule :

1. List all elements in descending order of element time :

Task	I_e, min	Immediate predecessors
I	10	F
J	10	G
L	10	J,H
N	9	L,M
D	8	A
G	7	D
K	7	I
A	6	-
H	6	E
M	6	K
B	5	-
E	5	B
C	4	-
F	4	C

2. Assign task to stations according to Largest-candidate rule :

Station	Task	T_e, min	ΣT_e	Idle time, min
1	A	6		
	B	5		
	C	4		
	F	4	19	1
2	I	10		
	D	8	18	2
3	G	7		
	K	7		
	M	6	20	Zero
4	J	10		
	E	5	15	5
5	H	6	6	14
6	L	10		
	N	9	19	1

There are six stations.

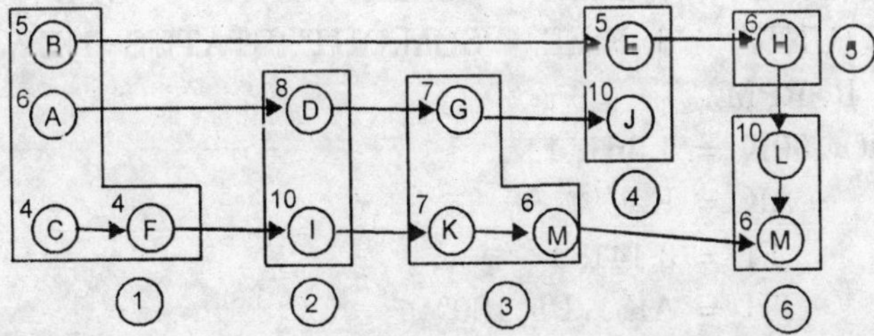

Fig. 6. Line balancing by Largest Candidate rule.

(*c*) Draw flow chart and write source code in FORTRAN to solve the following problem:

A reciprocating engine has crank radius r and crank length '*l*'. The crank is rotating with uniform angular velocity 'to'. It is required to find the velocity and acceleration of piston at different crank positions. Also find the position of crank and piston at maximum acceleration and at acceleration nearer to zero.

Hint:

Velocity of piston is $v_p = r\omega\left(\sin\theta + \dfrac{\sin 2\theta}{2n}\right)$

Acceleration of piston is $a_p = r\omega^2\left(\cos\theta + \dfrac{\cos 2\theta}{n}\right)$. (15)

The flow chart is shown in Fig. 7.

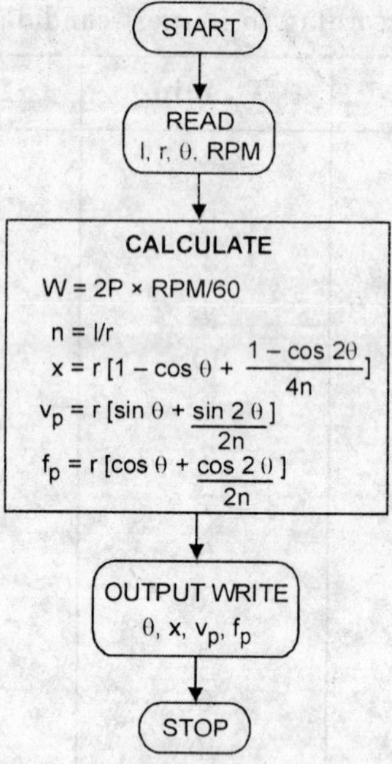

Fig. 7 Flow chart.

Sol. Program in FORTRAN :

```
DIMENSION THD (500), AX(500), VP(500), FP(500)
        OPEN (UNIT  = 7, FILE = 'SCM.IN', STATUS='OLD')
        OPEN (UNIT  = 8, FILE = 'SCM.OUT', STATUS='NEW)
READ (7, *) AL, R, RPM
             DO 100K  = 1,361, 1
                  AK  = K–l
                  PI  = 3.14159
                  TH  = AK * PI/ 180°
                  OM  = 2.* PI * RPM / 60°
                  AN  = AL / R
                  AX  = R*((1.-COS (TH)) + (l.-COS (2.*TH)) / (4*AN))
                  VP  = R*OM* (SIN (TH) + (SIN (2.*TH)) / AN)
                  FP  = R*OM**2 (cos (TA)) + (cos (2.*TH)) / AN)
                 THD  = TH*180°/PI
WRITE (8, 50) THD, AX, VP, FP
50 FORMAT (4F 10.2,/)
100 CONTINUE
CLOSE (8)
STOP
END.
```

The position of crank and piston (values of THD and AX) for maximum acceleration and zero acceleration can be known from the output.

MECHANICAL ENGINEERING
PAPER II—2008

(Conventional)

SECTION-A

Q.1. ***(a)*** Under what conditions of relative motion between links of a mechanism, Coriolis component of acceleration is generated ? Express its magnitude, direction and sense with the help of a sketch.

Sol. The total acceleration of a point with respect to another point in a rigid link is the vector sum of its normal and tangential components. If the second point, which was considered stationary, now slides, the total acceleration will contain an additional component called the coriolis component of acceleration.

Coriolis component, $f^{cr} = 2\ vw$

where v = linear velocity of slider,

ω = angular velocity of rotating link

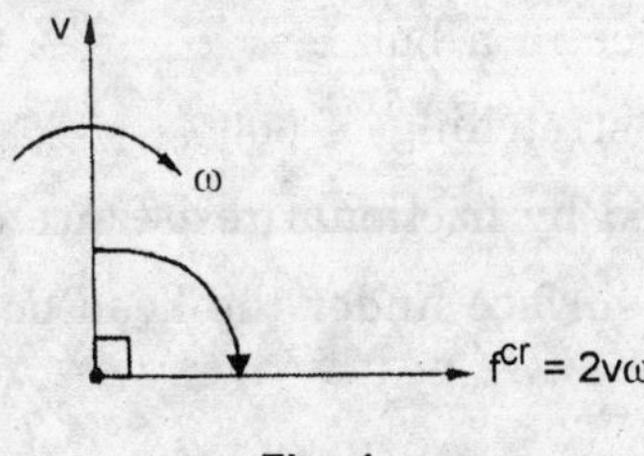

Fig. 1.

The direction of the Coriolis component of acceleration is such as to rotate the sliding velocity vector in the same sense as the angular velocity of rotating link though 90° (Fig. 1).

(b) What is law of gearing ? Express the magnitude of sliding velocity at the point of contact between two meshing gears and its value at pitch point.

Sol. The law of gearing states that for constant angular velocity ratio of two gears in mesh the common normal at the point of contact must always intersect the line of centres at a fixed point (pitch point) and divide ihis line in the inverse ratio of the angular velocities of the two gears.

$$\frac{\omega_1}{\omega_2} = \frac{O_2P}{O_1P}$$

Sliding velocity, $v_s = (\omega_1 + \omega_2) \times$ Distance of point of contact from pitch point

At the pitch point, $v_s = 0$.

(c) An accelerometer indicates that a structure is vibrating at 80 c.p.s. with a maximum acceleration of 50 g. Determine amplitude of vibration of the structure.

Sol. $f = 80$ cps, $a_{max} = 50g$

Maximum acceleration $= y_0 w^2$

Amplitude, $y_0 = \frac{a_{max}}{\omega^2} = \frac{50 \times 9.81 \times 10^3}{(2\pi \times 80)^2} = 1.94$ mm

(d) In a two-cylinder V-engine having 90° included angle between the cylinder axes has crank length equal to 10 cm and reciprocating mass for each cylinder equal to 10 kg. If the crank rotates at 500 rpm, determine primary unbalanced force in the engine.

Sol. $r = 10$ cm, M = 10 kg, N = 500 rpm, $2\propto = 90°$

Primary unbalanced force,

$$F_p = M\,\omega^2\, r$$

$$= 10 \times \left(\frac{2\pi \times 500}{60}\right)^2 \times 0.1 = 2741.6 \text{ N.}$$

(e) What different stresses set up in a bolt due to initial tightening, while used as a fastener? Name all the stresses in detail.

Sol. Initial tightening stresses in a bolt :

(*i*) Tensile stress due to stretching of bolt

(*ii*) Torsional stress caused by frictional resistance of threads

(*iii*) Bending stress if the surface under the head or nut are not perfectly normal to bolt axis.

(f) Name the four principal types of roller bearings.

Sol. Roller bearings :

(i) Plain roller

(ii) Helical roller

(iii) Self-aligning spherical roller

(iv) Taper roller.

(g) A prismatic bar in compression has a cross-sectional area A = 900 mm^2 and carries an axial load P = 90 kN, What are the stresses acting on

(*i*) A plane transverse to the loading axis ;

(*ii*) A plane at $\theta = 60°$ to the loading axis?

Sol. A = 900 mm², P = 90 kN, $\theta = 60°$

(*i*) $$\sigma = \frac{\sigma}{2}(1 + \cos 2\theta)$$

(*ii*) Normal stress, $= \frac{100}{2}(1 + \cos 120°) = 25 \text{ N/mm}^2$

Shear stress, $\tau = \frac{\sigma}{2} \sin 2\theta = \frac{100}{2} \sin 120° = 43.3 \text{ N/mm}^2$.

(*h*) A thin cylindrical pressure vessel of inside radius r and thickness of metal, *t* is subjected to an internal fluid pressure, *p*. What are the values of

(*i*) Maximum normal stress;

(*ii*) Maximum shear stress?

Sol.

(*i*) Maximum normal stress, $\sigma_\theta = \frac{pr}{t}$

(*ii*) Maximum shear stress, $\tau_{max} = \frac{1}{2}\left(\frac{pr}{t} - \frac{pr}{2t}\right) = \frac{pr}{4t}$

(*i*) A hollow shaft and a solid shaft constructed of the same material have the same length and the same outside radius. The inside radius of the hollow shaft is 0.6 times of the outside radius. Both the shafts are subjected to the same torque.

(*i*) What is the ratio of maximum shear stress in the hollow shaft to that of solid shaft ?

(*ii*) What is the ratio of the angle of twist in the hollow shaft to that of solid shaft?

Sol. (*i*) $$l_s = l_h,\ \rho_s = \rho_h,\ r_s = (r_0)_h,\ (r_i)_h = 0.6\,(r_0)_h$$

$$T_s = T_h,\ G_s = G_h$$

$$T_s = \frac{\pi}{2} r_s^3 \tau_s$$

$$T_h = \frac{\pi}{2}(r_0)_h^3\left[1 - \left(\frac{r_i}{r_0}\right)^4\right]\tau_h = \frac{\pi}{2}(r_0)_h^3\left[1 - (0.6)^4\right]\tau_h$$

$$\frac{\tau_h}{\tau_s} = \frac{1}{1-(0.6)^4} = 1.1489$$

(*ii*) *T, l* and *G* being same for solid and hollow shafts,

$$\theta \propto \frac{1}{J}$$

$$\frac{\theta_h}{\theta_s} = \frac{J_s}{J_h} = \frac{\frac{\pi}{2} r_s^4}{\frac{\pi}{2}(r_0)_h^4\left[1 - \left(\frac{r_i}{r_0}\right)_h^4\right]} = \frac{1}{1-(0.6)^4} = 1.1489$$

(*j*)

(*i*) Define 'ductility' of metals.

(*ii*) Arrange the metals mild steel, aluminium, zinc and copper in order of their diminishing ductility (highest first-lowest last).

Sol.

(*i*) Ductility of metals is defined as the ability to be drawn into wires.

(*ii*) Cu → Al → Mild steel → Zn.

(*k*) What is extrusion ? Name the process for manufacturing collapsible tubes and draw figures to show the process.

Sol. Extrusion may be defined as the manufacturing process in which a block of metal enclosed in a container is forced to flow through the opening of a die (see Fig. 2).

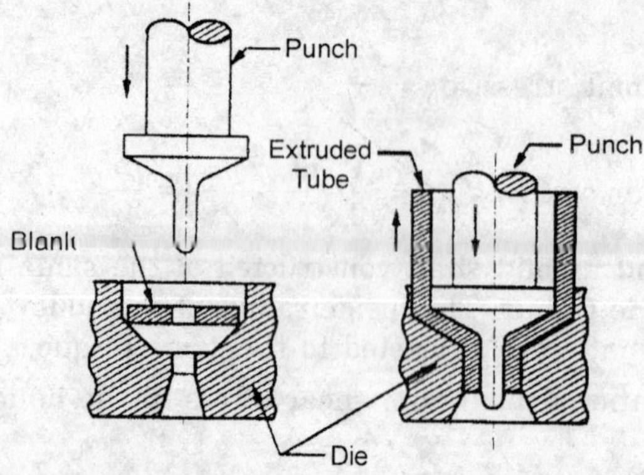

Fig. 2. Impact Extrusion

Collapsible tubes are manufactured by cold impact extrusion.

(*l*) Name the casting methods by which the following products are made:

(*i*) Gear blanks

(*ii*) Blades for gas turbines

(*iii*) Carburettors

(*iv*) Hollow pipes.

	Product	**Casting method**
(*i*)	Gear blanks	Permanent mould casting
(*ii*)	Blades for gas turbines	Lost wax method
(*iii*)	Carburettors	Die casting
(*iv*)	Hollow pipes	Centrifugal casting.

(*m*) Draw labelled figures to show the following, in case of oxy-acetylene welding :

(*i*) Carburizing flame

(ii) Oxidizing flame.

Sol.

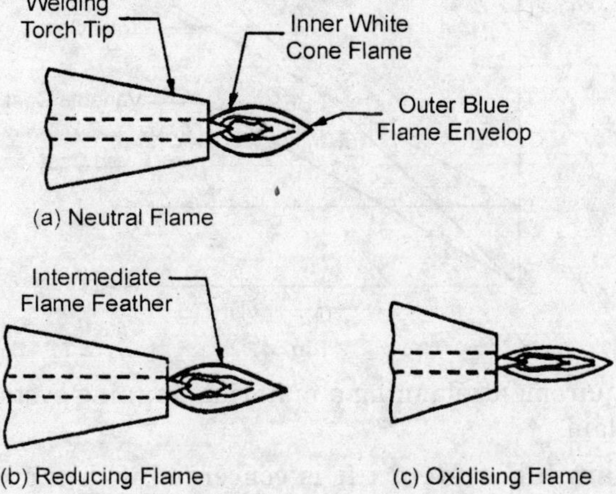

Fig. 3. OAW Flame Settings.

(n) What is the function of manipulator in a robot ? Classify robots as per their path control.

Sol. The manipulator actually performs the principal movements in doing manufacturing work. Robots may be classified as per their path control as follows :

(*i*) Non-servo controlled point-to-point.

(*ii*) Servo controlled point-to-point.

(*iii*) Servo controlled continuous path.

(o) What is meant by hot hardness, in case of tool materials ? Name any two tool materials which are not suitable for low speed applications.

Sol. Hot hardness: It is the hardness of a tool material at high temperature. Tool materials not suitable for low speed applications :

(*i*) H S S: Tungsten type –18:4: 1 and Molybdenum type.

(*ii*) Sintered carbide.

(p) Name three types of fits and show a labelled diagram of a clearance fit.

Sol. Types of fit:

(*i*) Clearance fit:

(*ii*) Transition fit, and

(*iii*) Interference fit.

Clearance fit

(q) What is meant by break-even point ? Draw a figure to illustrate your answer.

Sol. Break even point : It is the point at which the total cost of production equals the sales income.

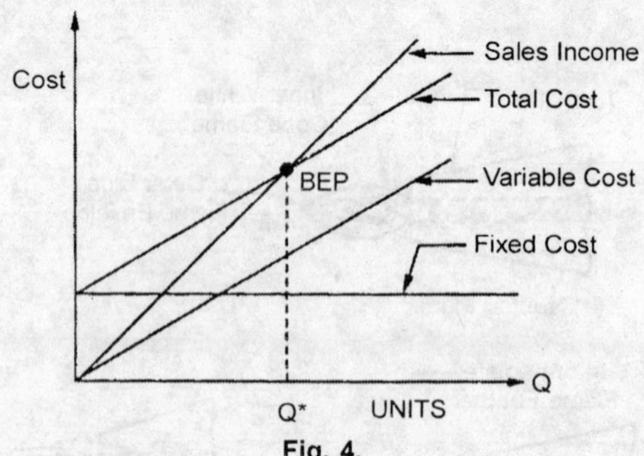

Fig. 4.

(*r*) Is Material requirement planning a material planning system, a production planning system, or both ? Explain.

Sol. Material planning system : It is concerned with the procurement of inventory for the various products to be manufactured. Accurate inventory data is vital, with accuracies of atleast 95% being desirable.

Production planning system : It represent everything the firm plans to ship : finished items, spares, etc.

(*s*) What are the differences between transportation and assignment problems, in relation to

(*i*) Structure of the problem ;

(*ii*) Procedure for solving ?

Sol. In both the transportation and assignment problems, the objective is to fulfil the targets by means of available resources which are available in specified amounts. In the transportation problem, each target can be attained in one way only, whereas in assignment problem the individual targets can be attained in different ways.

Assignment problem may be regarded as a special case of transportation problem where the facilities represent the 'sources' and the jobs represent the 'destinations'. The supply available at each source is 1, *i.e.*, $a_i = 1$ for all i. Similarly demand at each destination is 1, *i.e.*, $b_j = 1$ for all j.

		Jobs				
		1	2	.....	n	Supply a_1
Facilities	1	C_{11}	C_{12}	.....	C_{1n}	1
	2	C_{21}	C_{22}	.....	C_{2n}	1
	⋮	⋮	⋮	⋮	⋮	⋮
	m	C_{m1}	C_{m2}	.....	C_{mn}	1
Demand	b_j	1	1	.....	1	

Fig. 5.

Transportation problem is solved by Vogel approximation method and assignment problem by Hungarian method.

(*t*) What are pointers? Explain their use with a suitable program?

Sol. Pointer operator (*) : It is used to get the content of the address operator pointing to a particular memory element. Pointers are variables (or constants) that hold memory address as their values.

```
# include < iostream. h >
void main ()
{
double PI = 3.14;
double * ptr PI = and PI;
/ / Declares a variable ptr PI and
/ / initializes it to the address of the
/ / previosly declared constant PI.
cout << "value of PI = "<< PI
<< "Address of PI = "<< ptr PI << end l;
}
```

The output may be something like

Value of PI = 3.14 Address of PI = 0 × 194 a offc.

SECTION-B

Q.2.(*a*) A copper rod 6 cm in diameter is placed within a steel tube, 8 cm external diameter and 6 cm internal diameter, of exactly the same length. The two pieces are rigidly fixed together by two transverse pins 20 mm in diameter, one at each end passing through both rod and the tube. Calculate the stresses induced in the copper rod, steel tube and the pins if the temperature of the combination is raised by 50°C.

Take :

$$E = 210 \text{ GP}a,\ \alpha_s = 0.0000115/°\text{C}$$
$$E_c = 105 \textbf{ GP}\boldsymbol{a},\ \alpha_c = 0.000017/°\text{C}.$$

Sol. The co-efficient of linear expansion for copper rod is more than that of steel tube, therefore, it will elongate more than steel tube. As this free expansion is prevented, the copper bar will be under compression and the steel tube in tension (see Fig. 6).

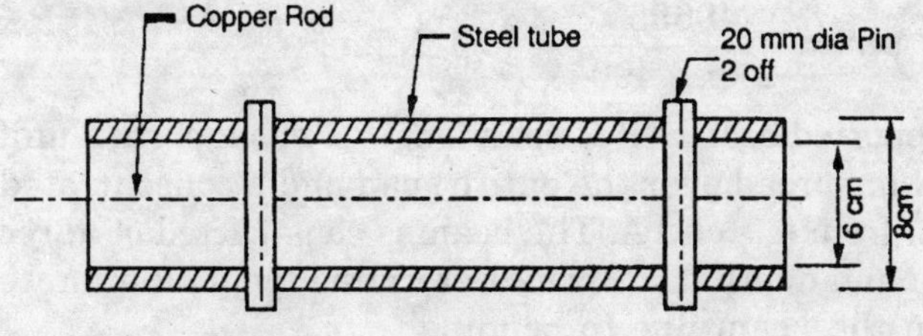

Fig. 6.

Let l = length of copper rod and steel tube, mm

Free elongation of copper rod,

$$\delta_c = \alpha_c \times \Delta t \times l$$
$$= 17 \times 10^{-6} \times 50 \times l = 85 \times 10^{-5} \times l \text{ mm}$$

Free elongation of steel tube,

$$\delta_s = \alpha_s \times \Delta t \times l$$
$$= 11.5 \times 10^{-6} \times 50 \times l = 57.5 \times 10^{-5} \times 1 \text{ mm}$$

Difference in elongation,

$$\delta = \delta_c - \delta_s = (85 - 57.5)\ 5\ 10^{-5} \times l$$
$$= 27.5 \times 10^{-5} \times l$$

If this difference in length is prevented by applying a compressive force P to copper rod and pulling the steel tube by the same force P, then contraction of copper rod,

$$\delta l_c = \frac{Pl}{A_c E_c} = \frac{4\,Pl}{\pi \times 3600 \times 1.05 \times 10^5} \text{mm}$$

Extension of steel tube,

$$\delta l_s = \frac{Pl}{A_s E_s} = \frac{4\,Pl}{\pi\,(6400 - 3600) \times 2.1 \times 10^5} \text{ mm}$$

The difference in length is eliminated, when

$$\delta l_c + \delta l_s = \delta$$

$$\frac{4\,Pl}{\pi \times 10^5}\left[\frac{1}{3600 \times 1.05} + \frac{1}{2800 \times 2.1}\right] = 27.5 \times 10^{-5} \times l$$

$$P = 49695 \text{ N}$$

Stress induced in copper rod

$$= \frac{P}{A_c} = \frac{4 \times 49695}{\pi \times 3600} = 17.5 \text{ N/mm}^2 \text{ (–ve)}$$

Stress induced in steel tube

$$= \frac{P}{A_c} = \frac{4 \times 49695}{\pi \times 2800} = 22.6 \text{ N/mm}^2 \text{ (+ ve)}$$

The pins are in double shear.

$$\text{Shear stress in pin} = \frac{49695}{2 \times \pi \times 400} = 79.1 \text{ N.mm}^2$$

(b) A simply supported beam AB of span length 4 m supports a uniformly distributed load of intensity, q = 4 kN/m spread over the entire span and a concentrated load, P = 2 kN placed at a distance of 1.5 m from left end A. The beam is constructed of a; rectangular cross-section with width b = 10 cm and depth d = 20 cm. Determine the maximum tensile and compressive stresses developed in the beam due to bending. (15)

Sol. Taking moments about B,

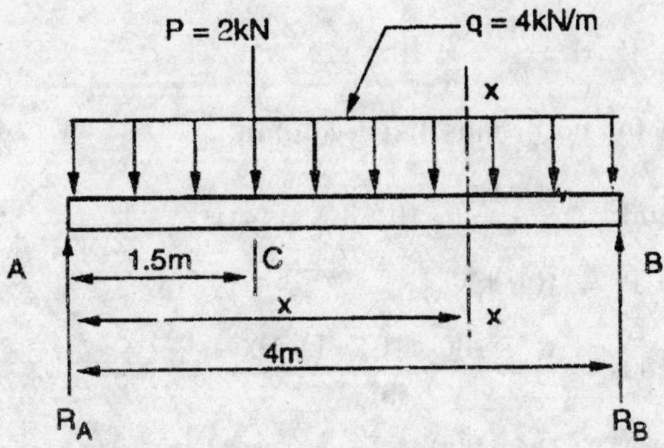

Fig. 7.

$$4\ R_A = 2 \times 2.5 + 4 \times 4 \times 2 = 37$$

$$R_A = 9.25 \text{ kN}$$

At section *x–x,* B. M. is,

$$M_x = 9.25\ x - 2\ (x - 1.5) - 2x^2$$

For M_x to be maximum,

$$\frac{dM_x}{dx} = 0$$

$$9.25 - 2 - 4x = 0$$

$$x = \frac{7.25}{4} = 1.8125 \text{ m}$$

$$M_{max} = 9{,}25 \times 1.8125 - 2\ (1.8125 - 1.5) - 2 \times (1.8125^2)$$

$$= 9.5703 \text{ kN.m}$$

Section modulus, $$Z = \frac{bd^2}{6} = \frac{10 \times 20^2}{6} = 666.67 \text{ cm}^3$$

Maximum tensile and compressive stress

$$= \frac{M_{max}}{Z}$$

$$= \frac{9.5703 \times 10^6}{666.67 \times 10^3} = 14.355 \text{ N/mm}^2$$

(c) A both ends hinged cast iron hollow cylindrical column 3 m in length has a critical buckling load of P kN. When the column is fixed at both the ends, its critical buckling load rises by 300 kN more. If ratio of external diameter to internal diameter is 1.25 and E = 100 GPa, determine the external diameter of the column. (10)

Sol.

$$l = 3 \text{ m}, \ \frac{d_0}{d_i} = 1.25, \text{ E} = 100 \text{ GPa}$$

Critical buckling load for both ends hinged column,

$$P = \frac{\pi^2 EI}{l^2} \times 10^{-3} \text{ kN}$$

Critical buckling load for both ends fixed column,

$$P + 300 = \frac{4\pi^2 EI}{l^2} \times 10^{-3} \text{ kN} = 4\text{ P}$$

$$P = 100 \text{ kN}$$

$$100 = \frac{\pi^2 \times 100 \times 10^9 \times I \times 10^{-3}}{9}$$

$$I = \frac{900}{\pi^2 \times 100 \times 10^6} = 9.1189 \times 10^{-7} \text{ m}^4$$

$$d_i = \frac{\pi}{64} \times d_i^4 \left[(1.25)^4 - 1\right] = 0.07075\, d_i^4$$

$$d_i = 0.06 \text{ m},\ d_0 = 0.075 \text{ m or } 75 \text{ mm}.$$

Q.3. (*a*) (*i*) A close-coiled helical spring has coil diameter D, wire diameter d and number of turns n. The spring material has a shearing modulus G. Derive an expression for the stiffness k of the spring. (5)

Sol. (*i*) Torque acting on spring, $T = \dfrac{WD}{2}$

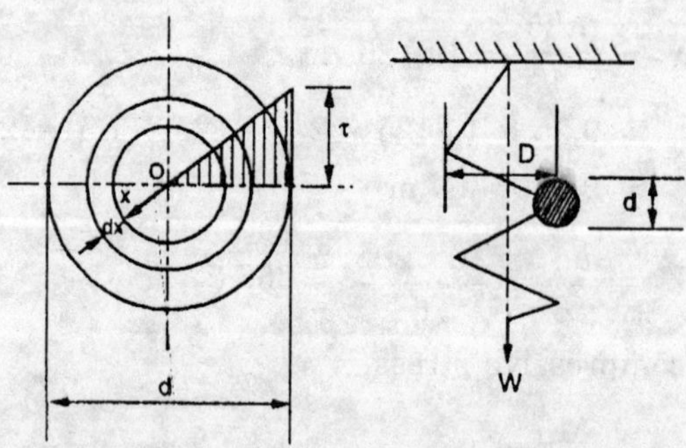

Fig. 8.

Shear stress in the wire at radius x,

$$\tau_x = \frac{2x\tau}{d}$$

$$dT = (2\pi\, x\, dx) \times \left(\frac{2x}{d}\tau\right) \times x = \frac{4\pi x^3 \tau}{d} \times dx \;\backslash$$

Total torque,

$$T = \int_0^{d/2} \frac{4\pi x^3 \tau}{d}\, dx$$

$$\frac{WD}{2} = \frac{4\pi\tau}{d} \left| \frac{x^4}{4} \right|^{d/2} = \frac{4\pi x}{4d} \times \frac{d^4}{16} = \frac{\pi d^3}{16}\tau$$

$$\tau = \frac{8WD}{\pi d^3}$$

Shear strain, $$\gamma = \frac{\tau}{G} = \frac{8WD}{\pi d^3 G}$$

Length of spring wire,

$$l = \pi Dn$$

$$d\theta = \frac{2r\,dx}{d}$$

$$\theta = \int_0^l \frac{2\tau\,dx}{d} = \int_0^{\pi Dn} \frac{16\,WD}{\pi d^4 G}\,dx = \frac{16\,WD}{\pi d^4 G}\,|x|_0^{\pi Dn} = \frac{16\,WD^2 n}{Gd^4}$$

Diflection, $$\delta = \frac{\theta DW}{2} = \frac{8WD^3 n}{Gd^4}$$

Stiffness. $$k = \frac{W}{\delta} = \frac{G\,d^4}{8\,D^3 n}$$

(*ii*) A close-coiled helical spring has coil diameter to wire diameter ratio of 6. The spring deflects 3 cm under an axial load of 500 N and the maximum shear stress is not to exceed 300 MPa. Find the diameter and the length of the spring wire required. Shearing modulus of wire material = 80 GPa. (15)

Sol.

$$\frac{D}{d} = C = 6, k = \frac{500}{30} = \frac{50}{3} \text{ N/mm}, \tau_{max} = 300 \text{ MPa},$$

$$G = 80 \text{ GPa}, W = 500 \text{ N}, d = 30 \text{ mm}$$

$$\tau = \frac{8WD}{\pi d^3} = \frac{8\,WC}{\pi d^2}$$

$$300 = \frac{8 \times 500 \times 6}{\pi d^2}$$

$$d = 5 \text{ mm}$$

$$D = 30 \text{ mm}$$

$$\delta = \frac{8W\,D^3 n}{G\,d^4} = \frac{8W\,C^3 n}{G\,d}$$

$$30 = \frac{8 \times 500 \times 6^3 \times n}{80 \times 10^3 \times 5}$$

$$n = 13.89 \simeq 14$$

Length of spring wire, $l = \pi Dn$

$$= \pi \times 30 \times 14 = 1319.5 \text{ mm.}$$

(*b*) A rotor is balanced by attaching two 2 kg trial masses in each of the planes A and B as shown in the figure below at a radius of 15 cm. Complete, dynamic balance is obtained with the angular positions shown in the same figure. Determine the position and mass of the material to be removed from each of the planes C and D at a radius of 10 cm in order to balance the rotor when the trial masses are removed. (20)

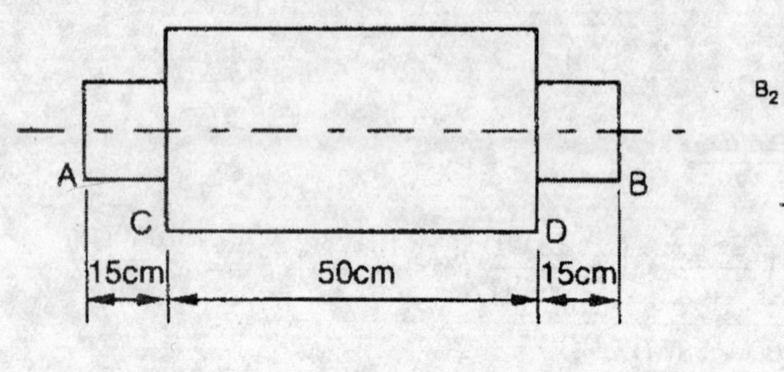

Fig. 9.

Sol.

Plane	Mass, *m*	θ	*r*	*mr*	*l*	mr*l*
	kg	deg	*m*	*kg. m*	m	*kg. m*^{2}
(1)	(2)	(3)	(4)	(5)	(6)	(7)
A_1	2	0	0.15	0.30	–0.15	–0.045
A_2	2	240	0.15	0.30	–0.15	–0.045
B_1	2	210	0.15	0.30	+0.65	+0.195
B_2	2	60	0.15	0.30	+0.65	+0.195
RP→ C	M_c	θ_c	0.10	0.1 M_c	0	0
D	M_D	θ_D	0.10	0.1 M_D	+0.50	0.05 M_D

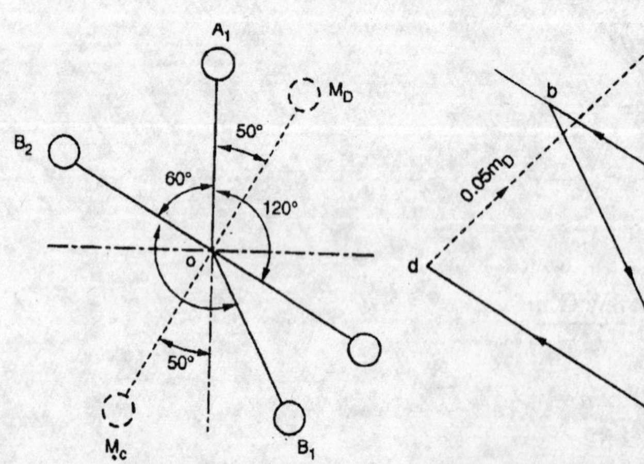

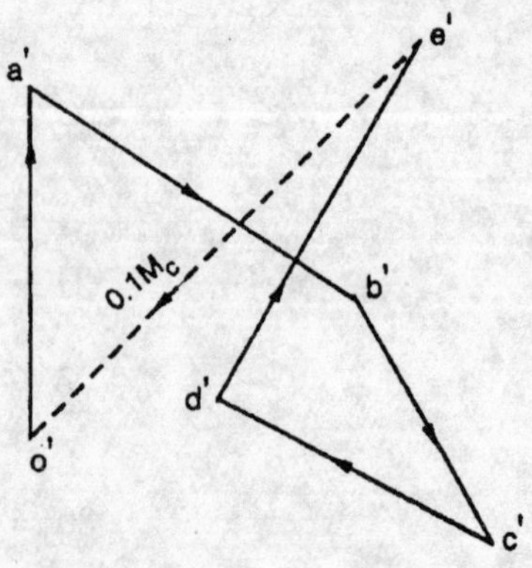

(a) Location of masses

Scale: 1 cm = 0.3 kg.m^2

(b) Couple Polygon.

Scale 1 cm = 0.1 kg.m

(c) Force Polygon

Fig. 10.

1. The location of masses is shown in Fig. 10 (a). With C as the reference plane (RP), complete the table.
2. Draw the couple polygon as shown in Fig. 10 (b), with the data from column (7).
   (i) Draw oa parallel to OA_1 downwards.
   (ii) Draw ab parallel to OA_2, be paralle to OB_t, Cd parallel to OB_2.
   (iii) Joint od. Then

$$\text{od} = 4.8\text{cm} = 0.144 \text{ kg/m}^2$$
$$= 0.05\ M_D$$
$$\therefore \quad M_D = 2.88 \text{ kg}$$

   (iv) Draw a line parallel to od in Fig. 10 (a). Angle, A_tOM_D = 50° CW from A.
3. Draw the force polygon as shown in Fig. 10 (c) with the data from column (5).
   (i) Draw o′ a′ parallel to OA_1 upwards.
   (ii) Draw o′ a′ parallel to OA_2, b′c′ parallel to OB_1, c′d′ parallel to OB_2 and d′e′ parallel to OM_D.
   (iii) Join o′e′. Then

$$o'e' = 4.4 \text{ cm} = 0.44 \text{ kg.m}$$
$$= 0.1\ M_1$$
$$M_C = 4.4 \text{ kg}$$

   (iv) Draw a line parallel to o′e′ in Fig. 10 (a). Angle M_1OM_C = 230° CW from A.

Q.4. (a) A vertical steel shaft of 15 mm diameter is held in long bearings 1 m apart and carries at its middle a disc of mass 15 kg. The eccentricity of the centre of gravity of the disc from the centre of the shaft is 0.30 mm. The modulus of elasticity for the shaft material is 200 GPa and the permissible stress is 70 MPa. Determine the critical speed of the shaft neglecting the mass of the shaft and the range of speed over which it is unsafe to run the shaft. (20)

Sol. $d = 15 \text{ mm}, l = 1 \text{ m}, m = 15 \text{ kg}, e = 0.30 \text{ mm},$

$E = 200 \text{ GP}a, \sigma = 70\ MPa$

For the shaft held in long bearings, stiffness,

$$k = \frac{192EI}{l^3}$$

$$I = \frac{\pi d^4}{64} = \frac{\pi \times 15^4}{64} = 2485 \text{ mm}^4$$

$$k = \frac{192 \times 2 \times 10^5 \times 2485}{(1000)^3} = 95.426 \text{ N/mm}$$

Natural speed, $$\omega_n = \sqrt{\frac{k}{m}} = \sqrt{\frac{95.426 \times 10^3}{15}} = 79.76 \text{ rad/s}$$

Critical speed of shaft, $\omega = \omega_n = 79.76$ rad/s

$$N = \frac{60 \times 79.76}{2\pi} = 761.6 \text{ rpm}$$

Now $$\frac{M}{I} = \frac{\sigma}{y}$$

Let F = Dynamic force causing bending of the shaft.

For a shaft having long bearings,

$$M = \frac{Fl}{8}$$

$$s = \frac{My}{I} = \frac{Fl}{8} \times \frac{d}{2} \times \frac{64}{\pi d^4}$$

$$70 = \frac{F \times 1000}{8} \times \frac{15}{2} \times \frac{64}{\pi \times 15^4}$$

$$F = 185.55 \text{ N}$$

Deflection of shaft due to dynamic force F,

$$\delta = \frac{Fl^3}{192EI} = \frac{185.55 \times 1000^3}{192 \times 2 \times 10^5 \times 2485} = 1.944 \text{ mm}$$

$$\pm\frac{\delta}{e} = \frac{\beta^2}{1-\beta^2}, \beta = \frac{\omega}{\omega_n}$$

$$\pm\frac{1.944}{0.30} = \frac{\beta^2}{1-\beta^2}$$

$$\pm 6.48 = \frac{\beta^2}{1-\beta^2}$$

$$6.48 - 6.48\,\beta^2 = \beta^2$$

$$\beta = 0.93 \text{ and } 1.0874$$

$$\omega_1 = 0.93 \times 761.6 = 708.3 \text{ rpm}$$

$$\omega_2 = 1.0874 \times 761.6 = 828.2 \text{ rpm}$$

***(b)* Determine the size of the weld required for static loading in the eccentrically loaded weld as shown in the Fig. 11 below. Allowable stress for the weld may be taken as 75 N/mm². *O* is the *CG* of the weld area. (20)**

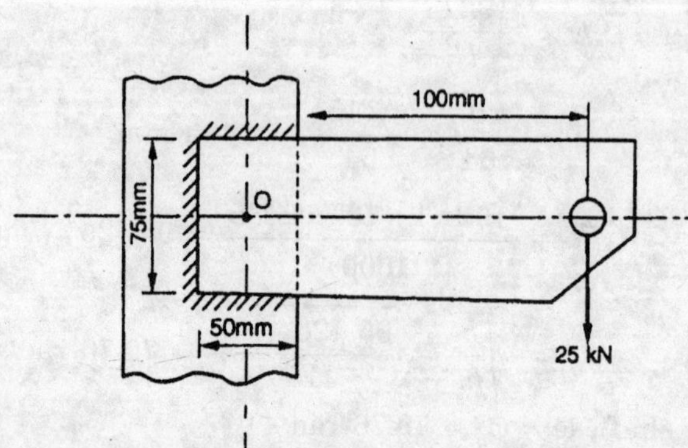

Fig. 11.

Sol. Let t = throat area of weld in *mm* (ser Fig. 12)

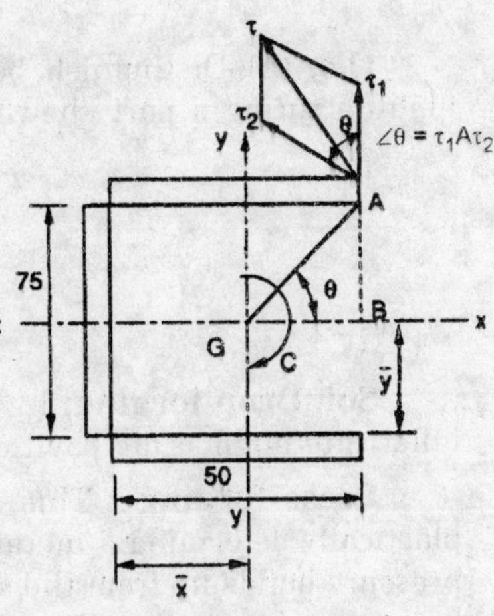

Fig. 12.

$$t\,(2 \times 50 + 75)\ \bar{x} = 2 \times 50t \times 25$$

$$175\ \bar{x} = 2500$$

$$\bar{x} = 14.286 \text{ mm}$$

Eccentricity, $e = 100 + (50 - 14.286) = 135.714$ mm

Couple, $C = Pe$

$$= 25 \times 10^3 \times 135.714 \times 10^{-3}$$

$$= 3392.85 \text{ N.m}$$

$$\bar{y} = \frac{75}{2} = 37.5 \text{ mm}$$

$$I_x = 2 \times 50t \times (37.5)^2 + \frac{t \times 75^3}{12}$$

$$= 17781 \text{ mm}$$

$$I_y = 2\left[\frac{t \times 50^3}{12} + 50t\,(25 - 14.286)^2\right] = 47619\ t \text{ mm}^4$$

$$J = I_x + I_y$$

$$= 175781\ t + 47619\ t$$

$$= 223400 \text{ t mm}^4$$

$$A = 2 \times 50 \text{ t} + 75\ t = 175\ t \text{ mm}^2$$

$$GA = r_{max} = \sqrt{(35.714)^2 + 37.5^2} = 51.78 \text{ mm}$$

Primary shear stress,

$$\tau_1 = \frac{P}{A} = \frac{25 \times 10^3}{175t} = \frac{142.86}{t} \text{N/mm}^2$$

Secondary shear stress,

$$\tau_2 = \frac{C r_{max}}{J} = \frac{3392.85 \times 10^3 \times 51.78}{223400t} = \frac{786.4}{t} \text{N/mm}^2$$

$$\cos\theta = \frac{GB}{GA} = \frac{35.714}{51.78} = 0.6897$$

Resultant shear stress,

$$\tau = \sqrt{\tau_1^2 + \tau_2^2 + 2\tau_1\tau_2\cos\theta}$$

$$= \frac{1}{t}\left[(142.86)^2 + (786.4)^2 + 2 \times 142.86 \times 786.4 \times 0.6897\right]^{1/2}$$

$$= \frac{890.96}{t} = 75$$

$$t = 11.88 \simeq 12 \text{ mm}$$

SECTION-C

Q.5. (a) Distinguish between drop forging and press forging. Explain the process of manufacturing a part shown in the Fig. 13 by drop forging. Draw suitable sketches also. (15)

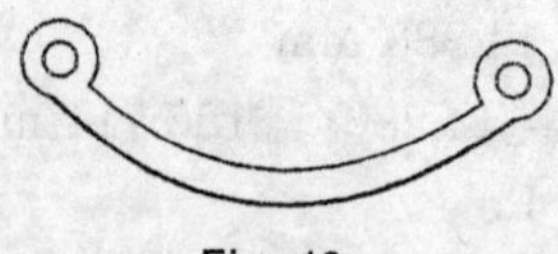

Fig. 13.

Sol. Drop forging: In this process, the forging is made by hammering a heated bar or billet into aligned die cavities. The forging equipment used for this is at drop hammer'.

Press forging : This process also uses impression dies but the parts are made by plastically deforming a metal blank into die-cavities by a slow squeezing action. The forging pressure builds up from the start to the end of stroke, resulting in maximum penetration and in improved grain flow throughout the entire forging. Completed forgings are ejected manually or mechanically from the die cavities. It uses mechanical presses or hydraulic presses.

Drop forging process for the part shown. The various steps for drop forging of the part are shown in Fig. 14 (*c*).

1. Place the heated forging stock into the following impression 1 and reduce it by blows of the hammer with an increase in its length.
2. Carry out the edging operation in impression 2 to redistribute the metal. The cross-section is increased at certain places and reduced at others as required to fill the cavities of the die.
3. Perform the bending operation on the stock to give it proper shape, in bender 4.
4. Process the stock in the blocking (semifinish) impression 5. This reduces the wear of finishing operation
5. Forge the stock in the finish impression 3. This imparts the final shape to the forging.
6. The flash gutter " (to collect the excess metal) is provided only an. and the finish impression.

Fig. 14 (*b*) shows the bottom half of the die.

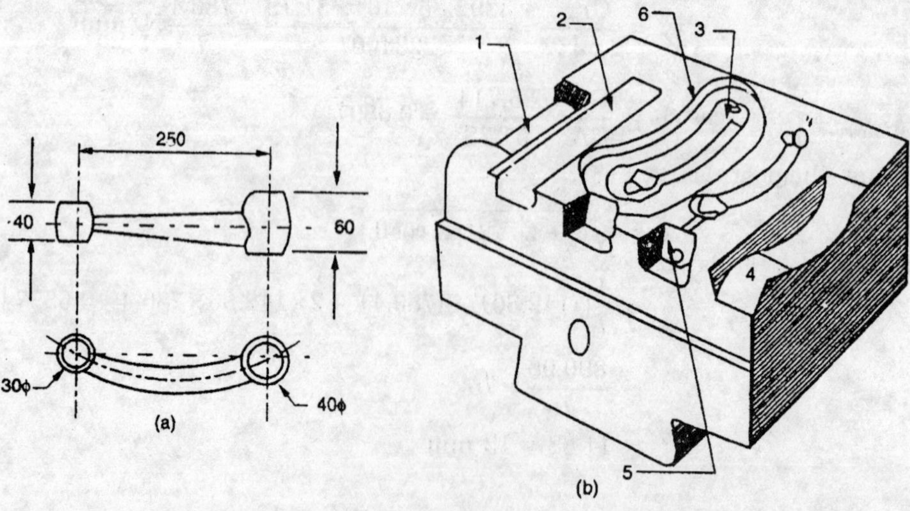

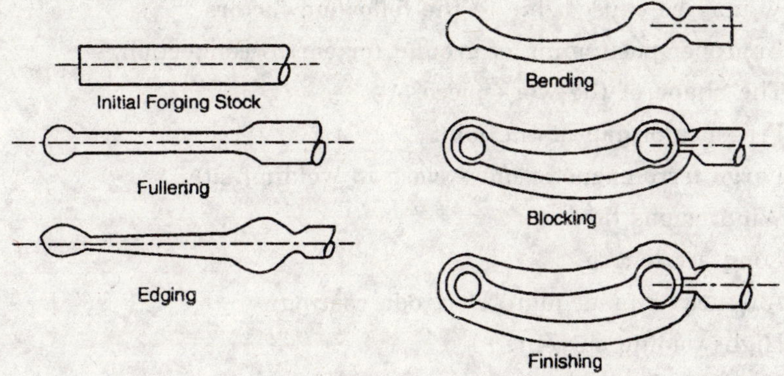

Fig. 14. Multi-empression forging die for the part.

(*b*) Calculate the size of a cylindrical riser, whose height and diameter are equal, to feed a steel slab casting 30 cm x 30 cm × 6 cm with a side riser. Casting is poured horizontally into the mould. Use modulus method. (15)

Sol. $h = d$ for the riser

Volume of casting. $V_c = 30 \times 30 \times 6 = 5400 \text{ cm}^3$

Surface area of casting,

$$A_c = 2 \times 30 \times 30 + 4 \times 30 \times 6 = 2520 \text{ cm}$$

Volume of rises, $V_r = \frac{\pi}{4}d^2h = \frac{\pi}{4}d^3$

Surface area of rises, $A_r = \pi\, dh + 2x\, \frac{3\pi}{2}\, d^2$

$$= \pi d^2 + \frac{\pi}{2}d^2 = \frac{3\pi}{2}d^2$$

$$\left(\frac{V}{A}\right)_r = \left(\frac{V}{A}\right)_c$$

$$\frac{\frac{\pi}{4}d^3}{\frac{3\pi}{2}d^2} = \frac{5400}{2520}$$

$$\frac{d}{6} = 2.14286$$

$$d = 12.86 \text{ cm}$$

$$h = 12.86 \text{ cm}$$

(c) What is an arc blow ? How is the arc blow problem taken care of in welding ? Calculate the melting efficiency in case of arc welding of steel with a potential of 20 V and current of 200 A. The travel speed is 5 mm/sec and the cross-sectional area of the joint is 20 mm. Heat required to melt steel may be taken as 10 J/mm^3 and the heat transfer efficiency as 0.85. (15)

Sol. Arc blow. Normally, the welding arc is oriented along the axis of the electrode rather than along the shortest distance between the electrode and the workpiece. However, an electric arc is an extremely flexible conductor which may be deflected from its intended path by a variety of factors. When such a phenomenon occurs during welding it is called an arc blow.

Arc blow may be caused due to the following factors :

(*i*) Improper positioning of ground (or earth) connection.

(*ii*) The shape of the work piece.

(*iii*) Presence of gap at the joint.

(*iv*) Large ferromegnetic mass close to welding site.

(*v*) Wind or gas flow.

(*vi*) Long arc length.

(*vii*) Absence or inadequate electrode coating.

(*viii*) High welding current.

(*ix*) Use of D.C. supply.

To eliminate or atleast to reduce the severity of arc blow, the following steps may be taken:

1. Use preferably A.C. for welding.
2. Place earth connection as far away as possible from the joint.
3. Point the electrode tip in the direction of arc blow.
4. Vary the angle between the electrode and the work to counteract the arc blow.
5. Use short arc length.
6. Use heavy coated electrodes in shielded metal arc welding.
7. Do not weld very close to heavy ferromagnetic mass.
8. Use low welding current consistent with good fusion.
9. Weld towards a heavy tack or a tab-out.
10. Use the sequence of back-step welding.
11. Wrap the ground cable around the workpiece so as to set up a magnetic field that would counteract the one producing arc blow.
12. Use appropriate ground connection for vertical welding.

$$V = 20 \text{ V}$$
$$I = 200 \text{ A}$$
$$V = 5 \text{ mm/s}$$
$$A = 20 \text{ mm}^2$$
$$Q = 10 \text{ J/mm}^3$$
$$h = 0.85$$
$$\eta = ?$$

Volume of fused metal = AV = 20 × 5 = 100 mm^3/s

Heat required for melting = QAV = 10 × 100 = 100 J/s

Heat generated = h VI = 0.85 × 20 × 200 = 3400 J/s

Efficiency of melting = $\dfrac{1000}{3400}$ = 0.294lor 29.41%.

Q.6. (*a*) Name the four types of statements in a complete APT part program. Prepare part program for geometry description of the contour shown in the Fig. 15 below : (15)

Sol.

The four types of statements in APT language are

(*i*) Geometry statements

(*ii*) Motion statements

(*iii*) Post-processor statements

(*iv*) Auxiliary statements

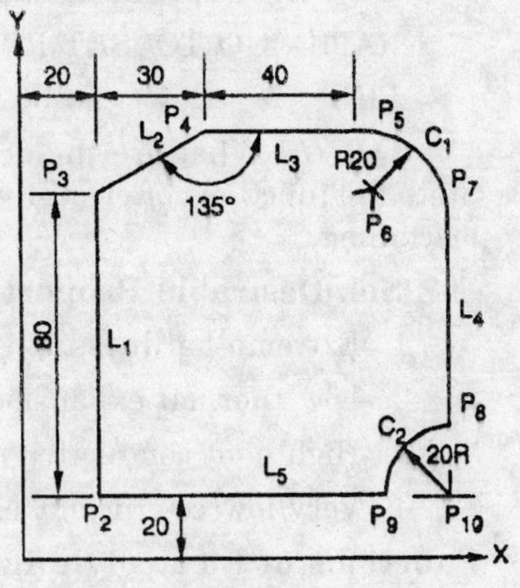

Fig. 15.

Part Program using APT Language (Fig. 16)

```
$ $ GEOMETRY
 SETPT = POINT/ – 20, – 20, – 25
    PI = POINT/90, 0, – 25
    P2 = POINT/0, 0, – 25
    P3 = POINT/0, 80, – 25
    P4 = POINT/30, 110 – 25
    P5 = POINT/70, 110, – 25
    P6 = POINT/70, 90, – 25
    P7 = POINT/90, 90, – 25
    P8 = POINT/90, 20, – 25
    P9 = POINT/70, 0, – 25
    Cl = CIRCLE/CENTRE, P6, RADIUS, 20
    C2 = CIRCLE/CENTRE, PI, RADIUS, 20
    LI = LINE/P2, P3
    L2 = LINE/P3, P4
    L3 = LINE/P4, P5
    L4 = LINE/P7, P8
    L5 = LINE/P9, P2
   PL1 = PLANE/P2, P3, P9
$ $ MOTION
 FROM/SETPT
 LOAD TL/1, ADJUST, 1
 CUTTER/20
 SPINDL/2000/CW
 FEDRAT/300, MMPM
 RAPID, GOLI PAST L5
 TLLFT, GO LFT/L1, PAST L2
 GO RGT/L3, TANTO, Cl
```

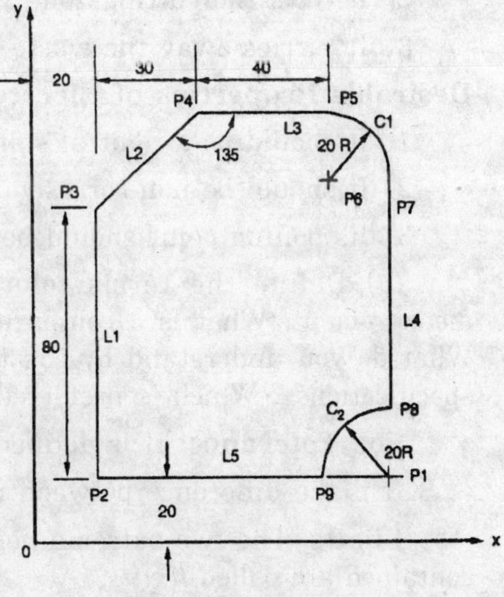

Fig. 16.

GOFWD/C1,PAST, L4

GO FWD/L4, TANTO, C2

GO FWD/C2, PAST, L5

RAPID, GO TO/ SET PT

FINI

Q.6. ***(b)*** What are the desirable properties that diamond cutting tool materials have ? State the functions of electrolyte and the properties it should have in case of electrochemical machining. (10)

Sol. Desirable Properties of Diamond cutting Tools Materials :

1. Extreme hardness.
2. Low thermal expansion.
3. High heat conductivity.
4. Very low co-efficient of friction.

Functions of Electrolyte in ECM:

1. It is essential for electrolytic process to work.
2. It cools the cutting zone which becomes hot due to the flow of high current.
3. It carries away the waste product.

Desirable Properties of Electrolyte in ECM:

1. It should be a neutral salt.
2. It should be non-corrosive
3. Its boiling point should be high.

(*c*) Define the terms tolerances, limits and fits with reference to dimensional measurement. What is a comparator ? How does it differ from a measuring instrument ? What do you understand by the terms hole basis and shaft basis in terms of assembly fit specifications ? Which is preferred and why is it preferred. (15)

Sol. Tolerance: It is defined as the total permissible variation of a size.

It is the difference between maximum limit and minimum limit of size.

Limits: The two extreme permissible sizes of a part between which the actual size is contained are called *limits.*

Fits: The fit between two mating parts is the relationship existing between them with respect to the amount of play or interference which is present when they are assembled together.

Comparator: A comparator is an indirect type of instrument with the help of which an unknown dimension of a workpiece is compared with a working standard (usually slip gauges).

A comparator gives the difference between the standard and the actual dimension of the workpiece, whereas a measuring instrument gives the dimension of the workpiece.

Hole basis system: In this system, the hole size is kept constant and the shaft diameter is varied to give the various types of fits, as shown in Fig. 17 *(a).* The basic size of the hole is taken as the low limit of size of the hole. The high limit of size of the hole and the two limit of size for the shaft are then selected to give the desired fit.

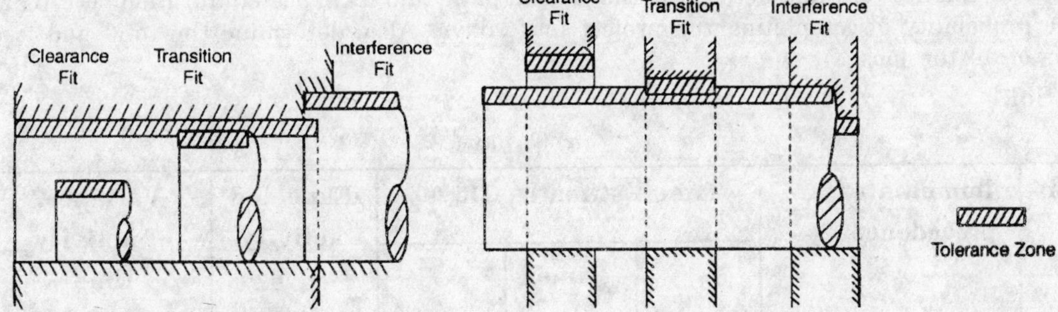

(a) Hole Basis System (b) Shaft Basis System

Shaft basis system: In this system, shaft size is kept constant and the size of hole is varied to give the various fits, as shown in Fig. 17 *(b)*. The basic size of shaft is taken as one of the limits of size (maximum limit) for the shaft. The other shaft limit of size and the two limits of size for the hole are then selected to give the desired fit.

Preference between Hole basis and Shaft basis: From manufacturing point of view, it is preferable to use the "Hole Basis" system, because it is economical. This is because a great many holes are produced by standard fixed size tools, such as, twist drills, reamers, core drills, taps, broaches, etc. The advantages of using fixed size tools is that the machine need not be set up to obtain the proper size of the hole. Subsequently, the shaft sizes are more readily available about the nominal size by means of turning or grinding operation. It is easier and more convenient to manufacture shafts of varying sizes than holes of varying sizes. The hole basis system is preferred, because it lessens the machining cost of drilling, boring and finishing holes of various sizes.

The shaft basis system is more advantageous in certain cases, *e.g.*, long shafts machined to the same size over their full lengths, and shafts ground on centreless grinding machines, etc.

Q.7. **(*a*)** A project consists of 7 jobs. Jobs A and F can be started and completed independently. Jobs B and C can start only after job A has been completed. Jobs D, B and G can start only after B, (C and D) and (E and F) are completed, respectively. Time estimates of all the jobs are given in the following table:

Job	Time estimates (Days)		
	Optimistic	**Pessimistic**	**Most likely**
A	3	7	5
B	7	11	9
C	4	18	14
D	4	12	8
E	4	8	6
F	5	19	12
G	2	6	4

Draw the network and determine the critical path, and its expected duration (T_e). What is the probability of completing the project in T_e days? Also, determine the total and free slacks of all the jobs. (15)

Sol.

(a) Table-1

Job	Immediate precedence	Time Estimates (Days) time			Expected activity $t_e = \frac{t_o + t_p + 4t_m}{6}$	Variance of activity time $\sigma^2 = [(t_p - t_o)/6]^2$
		i_0	t_p	t_m		
A	–	3	7	5	5	0.444
B	A	7	11	9	9	0.444
C	A	4	18	14	13	5.444
D	B	4	12	8	8	1.778
E	C,D	4	8	6	6	0.444
F	—	5	10	12	12	5.444
G	E, F	2	6	4	4	0.444

The net work has been drawn in the Figure 18 with t_e values from column 6 of above Table 1.

The critical path isA-B-D-E-G

Expected project duration, $T_e = 5 + 9 + 8 + 6 + 4 = 32$ days

Variance of project length $= 0.444 + 0.444 + 1.778 + 0.444 + 0.444 = 2.34$

Standard deviation, $\sigma_e = \sqrt{2.34} = 1.53$

Probability that the project will be completed in T_e days is given by:

Prob. $\left\{Z = \frac{T_e - T_e}{\sigma_e}\right\}$ = Prob {Z = 0}

From normal distribution table, Prob. {Z = 0} = 0.5 or 50%.

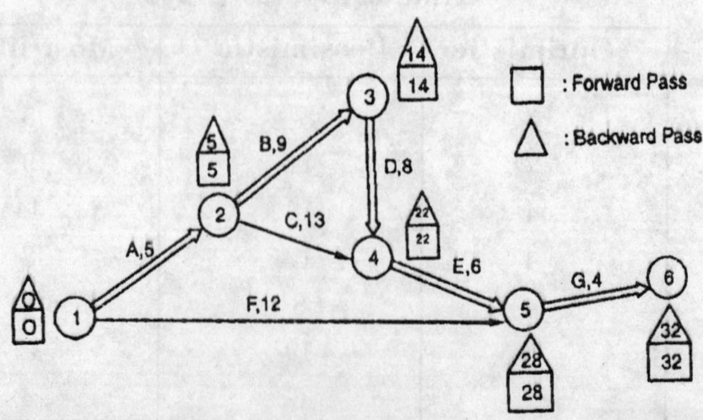

Fig. 18.

Total float (or slack): It is the difference between the time available to perform the activity (measured from the earliest start time to the latest finish time) and the expected completion time of the activity. For an activity (i, j),

Total float, $TF_{ij} = (Lj - E_i) - t_{ij} = LS_{ij} - ES_{ij} = LF_{ij} - EF_{ij}$

The total float has been calculated in column 7 of Table 2.

Free float (or slack): The time by which the completion of an activity can be delayed beyond its earliest finish time without affecting the earliest start time of a succeeding activity is called free float.

Free float, $FF_{ij} = (E_j - E_i) - t_{ij} = \min \{ES_{ij}$ for all immediate successors of $(i, j)\} - EF_{ij}$

The free float has been calculated in column 8 of Table 2.

Table-2

Activity	Duration	Earliest time		Latest time		Total float $(L_j - t_{ij}) - E_i$	Free float $(E_j - E_i) - t_{ij}$
(i, j)	t_{ij}	Start E_j	Finish $(E_i + t_{ij})$	Start $(L_j - t_{ij})$	Finish L_j		
A : 1 – 2	5	0	5	0	$^L J5$	0	0
B : 2 – 3	9	5	14	5, 9	14	0	0
C : 2 – 4	13	5	18	9	22	4	4
D : 3 – 4	8	14	22	14	22	0	0
E : 4 – 5	6	22, 18	28, 24	22	28	0	0
F : 1 – 5	12	0	12	16	28	16	16
G : 5 – 6	4	28, 24	32, 28	28	32	0	0

(*b*) Name the three costs involved in inventory control. A store procures and sells certain items. Information about an item is as follows :

Expected annual sales = 8000 units

Ordering cost = Rs 1,800 per order

Holding cost = 10% of average inventory value

The items can be purchased according to the following schedule :

Lot Size	Unit Price (Rs)
1—999	220
1000—1499	200
1500—1999	190
2000 and above	185

Determine the best order size. (10)

Sol. Cores involved in Inventory control:

1. Purchase cost.

2. Inventory carrying cost.

3. Ordering cost.

Annual sales, $D = 800$ units

Ordering Cost, C_0 = Rs. 1800 per order

Holding cost, C_c = 10% of average inventory value.

(*i*) Select the minimum unit price of Rs. 185.

$$C_c = 0.1 \times 185 = 18.5 \text{ Rs. unit}$$

$$\text{EOQ, } Q_4^* = \sqrt{\frac{2DC}{C_c}} = \sqrt{\frac{2 \times 8000 \times 1800}{18.5}} = 1248 \text{ units}$$

$$Q_4^* < 2000.$$

(*ii*)
$$C_c = 0.1 \times 190 = \text{Rs. } 19/\text{unit}$$

$$Q_3^* = \sqrt{\frac{2 \times 8000 \times 1800}{19}} = 1231 \text{ units}$$

$$< 1500$$

(*iii*)
$$C_c = 0.1 \times 200 = \text{Rs. } 20/\text{unit}$$

$$Q_2^* = \sqrt{\frac{2 \times 8000 \times 1800}{20}} = 1200 \text{ units}$$

$$Q_2^* < 1499 \text{ and } > 1000.$$

$$\therefore \quad EOQ = 1200 \text{ units.}$$

(c) Draw a flow chart and write a program in C to find out the diameter of a solid steel shaft to transmit 20 kW power at 200 r.p.m. The ultimate shear stress for steel is 360 MPa and the factor of safety is 8. (15)

(*c*)
```
# include < stdio. h>
# include < conio. h>
# include < math. h>
# define pi 3.1416 void main ()
float P, T, SU, FS, D;
clrscr (),
print / ("enter the value of power transmitted: ");
scan/("% f, & P);
print/( "enter the value of rpm: ");
scan/(" %l', & N);
print / ( "enter the value of shear stress :");
scan/( " % f, & SU);
print/( " enter the value of factor of safely :" );
scan/( " % f, & FS);
r=(60* 1000*P)/(2 *pi*N)
D = POW (( (16 * T* * 1000 * FS)/ (pi * SU)),
0.333333);
print/( " \ n D is : %l',£>);
getch ( ) ;
}
```

MECHANICAL ENGINEERING
PAPER II—2009

(Conventional)

SECTION-A

Q.1. ***(a)*** A balance mass of value $\frac{2}{3}$m is placed diametrically opposite to the crank at crank radius *r*. What is the expression for the unbalanced force along the line of stroke of a reciprocating engine for angular position θ of the crank with respect to the line of stroke?

Sol.

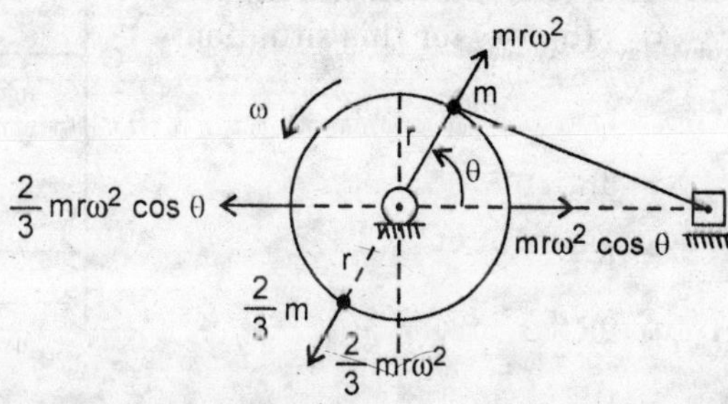

Fig. 1

Unbalanced force along the line of stroke

$$= mrw^2 \cos\theta - \frac{2}{3}mrw^2 \cos\theta$$

$$= \frac{1}{3}mrw^2 \cos\theta$$

(b) *(i)* What is an isochronous governor?

(ii) State the phenomenon of hunting in a spring controlled governor.

Sol.

(i) A governor with a range of speed zero is known as an isochronous governor. This means that for all positions of the sleeve or the balls, the governor has the same equilibrium speed.

(ii) It is a condition in which the speed of the engine controlled by the governor fluctuactes continuously above and below the mean speed.

(c) (*i*) Explain briefly vibration isolation and transmissibility.

(*ii*) The natural frequency of a system put on a foundation is ω_n. The operating frequency is ω. What is minimum value of $\left(\frac{\omega}{\omega_n}\right)$ beyond which the vibration isolation is possible or meaningful?

Sol.

(*i*) Vibration isolation is to minimise the transmission of unbalanced forces from the machine to the foundation.

Transmissibility is the ratio of the amplitude of force transmitted and the amplitude of the driving force.

(*ii*) For $\frac{\omega}{\omega_n} > \sqrt{2}$, vibration isolation is possible.

(d) The Mohr's circle for a plane stress is a circle of radius R with its origin at +2R on σ axis. Sketch the Mohr's circle and determine σ_{max}, σ_{min}, σ_{av}, $(\sigma_{xy})_{max}$ for this situation.

Sol. Refer to Fig. 2.

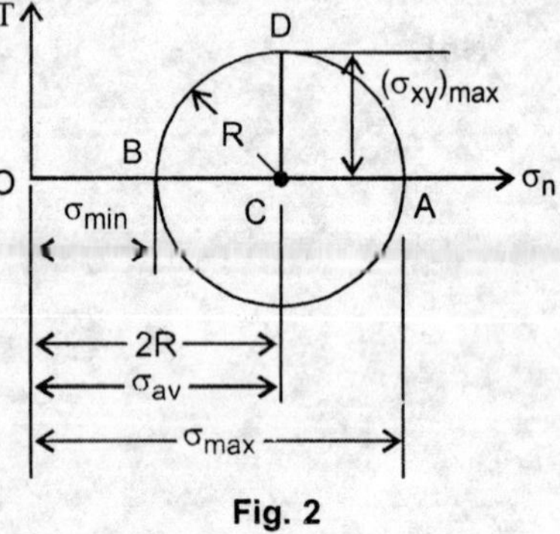

Fig. 2

$$\sigma_{max} = 3R$$

$$\sigma_{min} = R$$

$$\sigma_{av} = 2R$$

$$(\sigma_{xy})_{max} = R$$

(e) Briefly explain the role of idle gears in a simple gear train, with the help of neat sketches.

Sol. The role of idle gears is to provide same direction of rotation of the driven gear as the ariving gear. In the sketch, gears 2 and 4 are idle gears.

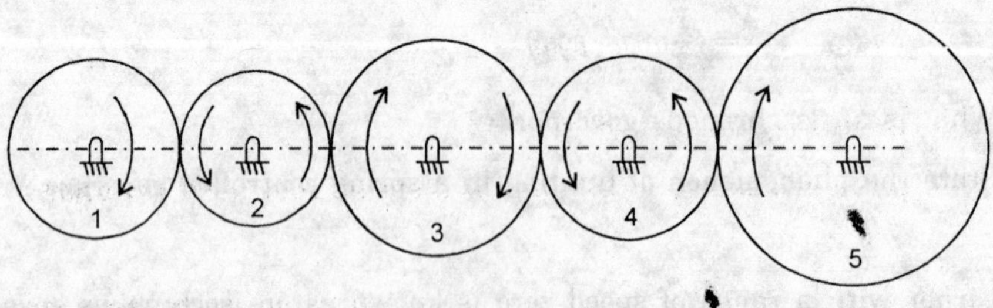

Fig. 3

(f) List two temporary fasteners and explain the working of one of these.

1. Screwed joint

2. Knuckle joint

Sol. Working of knuckle joint : It consists of two rods and a pin. The end of one of the rods is forged in the form of a fork while the end of the other rod has an eye, which can be inserted in the jaws of the fork. A cylindrical pin is driven through the holes in the forks and the eye. The pin is secured in position by a open pin or split pin. It is used when the axes of the rods are not in alignment but meet in a point. It allows small angular moment of one rod relative to the other rod.

(g) The state of stress at a point in a loaded machine member is given by the principal stresses.

σ_1 = 600 MPa, σ_2 = 0 and σ_3 = –600 MPa.

(*i*) What is the magnitude of the maximum shear stress?

(*ii*) What is the inclination of the plane on which the maximum shear stress acts which respect to the plane on which the maximum principal stress σ_1 acts?

Sol.

(*i*) Maximum shear stress,

$$\tau_{max} = \frac{1}{2}(\sigma_1 - \sigma_3) = \frac{1}{2}(600 + 600) = 600 \text{ MPa}$$

(*ii*)
$$\tan 2\theta = \frac{\sigma_1 - \sigma_3}{2\tau_{xy}} = \frac{600 + 600}{2 \times 0} = \infty$$

$$2\theta = 90°$$

$$\therefore \quad \theta = 45°$$

(h) Two long columns are made of identical lengths land flexural rigidities 'EI' Column 1 is hinged at both ends whereas for column 2 one end is fixed and the other end is free.

(*i*) Write the expression for Euler's buckling load for column 1.

(*ii*) What is the ratio of Euler buckling load of column, to that of column 2?

Sol.

(*i*)
$$P_{cr} = \frac{\pi^2 EI}{l^2}$$

(*ii*)
$$\frac{(P_{cr})_1}{(P_{cr})_2} = \frac{(\pi^2 EI/l^2)}{\left(\dfrac{\pi^2 EI}{4l^2}\right)} = 4$$

(i) A close coiled helical spring made of wire diameter *d* has mean coil radius R, number of turns *n* and modulus of rigidity G. The spring is subjected to an axial compression W.

1. Write the expression for the stiffness of the spring.

2. What is the magnitude of the maximum shear stress induced in the spring wire neglecting the curvature effect?

Sol.

1. Deflection, $\delta = \dfrac{64WR^3n}{Gd^4}$

   Stiffness, $k = \dfrac{w}{\delta} = \dfrac{Gd^4}{64R^3n}$

2. Maximum shear stress, $\tau_{max} = \dfrac{16\ WR}{\pi d^3}$

(*j*) List the conditions under which a rope drive is preferred to a belt drive.

Sol.

1. To transmit large amount of power.
2. The distance between the driving and driven shafts is very large.
3. Power may be taken off in any direction.
4. The shafts may not be strictly in alignment.

(***k***) What is achieved by using a metallic single crystal casting? Give one application of a single crystal casting made of wasp alloy.

Sol. When a single crystal casting solidifies gradually inwards, long column or crystals grow. This gives high strength to the casting.

(***l***) In investment casting process two types of ceramic slurries are used. Why do we use them and in what sequence are they applied?

Sol. First ceramic slurries are used for preparing the master mould. The second ceramic is used for making the wax pappern which is used for making the final mould in the same fashion as the conventional moulding process.

(***m***) What is the shape of a runner for making a sand mould? On what considerations is this shape selected?

Sol. Runners are normally made trapezoidal in cross-section. the runners are of larger cross-section and often stream lined to slow down and smooth out the flow. Runners are designed to provide approximately uniform flow rates to the various parts of the mould cavity.

(***n***) What does the following marking on a grinding wheel denote?

Sol.

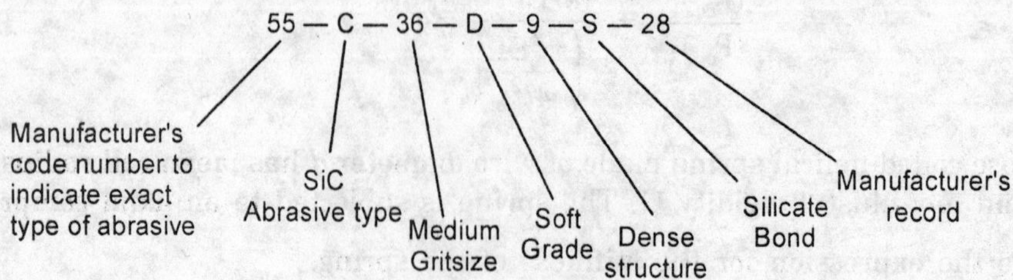

Alphanumeric system for grinding wheel specification.

(*o*) Show crater wear and flank wear on a single point cutting tool. State the factors responsible for wear on a turning tool.

Sol. The following factors are responsible for tool wear :

1. Forces
2. Temperature
3. Sliding action due to relative motion between tool and workpiece.

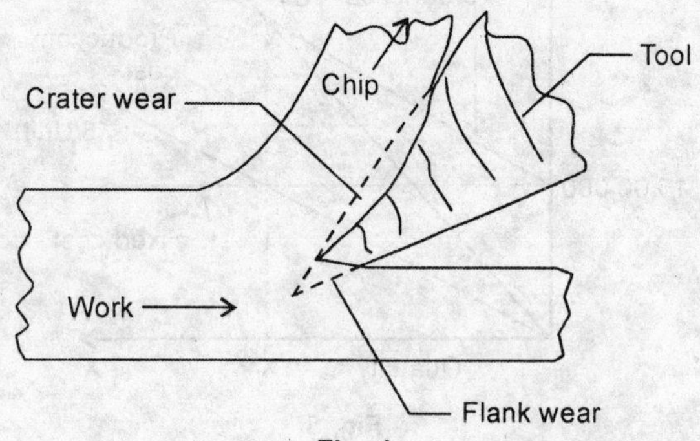

Fig. 4

(*p*) Compare the drilling and boring operations in relation to cutting action, accuracy and applications.

Parameter	Drilling	Boring
1. Cutting action	The cutting motion is provided to the two cutting edges (lips) by rotating the drill and the feeding is done by giving a rectilinear motion to the drill in axial direction	The cutting action is achieved by revolving the workpiece and giving feed to the tool parallel to the axis of the workpiece. The tool moves in already existing hole.
2. Accuracy	0.75 – 12.5 μm	0.05 – 25 μm
3. Applications	Used for drilling holes in a workpiece	Used for enlarging an already existing hole in a workpiece.

(*q*) What is the effect on order quantity when the demand increases by four-fold in basic order inventory system and other factors remain unchanged? Explain.

Sol.

$$\text{EOQ} = Q^* = \sqrt{\frac{2DC_0}{C_C}} \propto \sqrt{D}$$

$$\frac{Q_2^*}{Q_1^*} = \sqrt{\frac{4D}{D}} = 2$$

The order quantity is thus doubled.

(*r*) A company is faced with a situation where it can either produce some item by adding additional infrastructure which will cost them Rs. 15,00,000/- but unit cost of production will be Rs. 5/- each. Alternatively it can buy the same item from a vendor at a rate of Rs. 20/- each. When should the company add to its capacity in terms of demand of items per annum? Draw the diagram to show the BEP.

Sol.

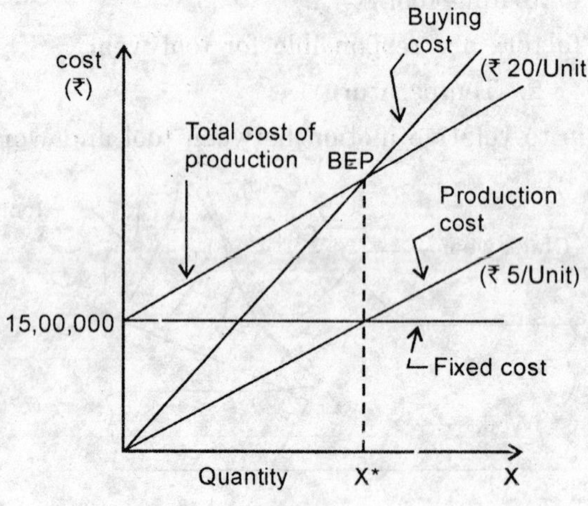

Fig. 5

Let X = number of items

At BEP,

$$20X = 5X + 15,00,000$$

$$X = 1,00,000 \text{ units}$$

(*s*) Define the term 'standard time' and list the common allowance given in work standard.

Sol. Standard time is the total time in which a job should be completed at standard performance. Standard time = observed time × Rating factor × (1 + % allowances). The common allowances are :

1. Contingency allowance for delay
2. Unoccupied time and interference allowance
3. Policy allowance

(*t*) Pointers are commonly used in 'C' for creating linked lists, Write a FORTRAN code to implement pointers, that is, to traverse the linked list.

Sol. The pointer operator is used to get the content of the address operator pointing to a particular memory element.

SECTION-B

Q.2. (*a*) A cantilever of circular solid cross-section is fixed at one end and carries a concentrated load P at the free end. The diameter at the free end is 200 mm and increases uniformly to 400 mm at the fixed end over a length of 2 m. At what distance from the free end will the bending stresses in the centilever be maximum? Also calculate the value of the maximum bending stress if the concentrated load P = 30 kN.

Sol. Let x = Distance from the free end at which bending stress in the cantilever is maximum (Fig. 6)

$$\frac{y-0.1}{x} = \frac{0.2-0.1}{2} = \frac{1}{20}$$

$$y = 0.1 + 0.05\,x$$

$$d = 2y = 0.2 + 0.1x$$

$$z_x = \frac{\pi d^3}{32} = \frac{\pi}{32}(0.2+0.1x)^3$$

$$\sigma_x = \frac{M_x}{z_x} = \frac{Px}{\frac{\pi}{32}(0.2+0.1x)^3}$$

$$= \frac{32\,Px}{\pi(0.2+0.1x)^3}$$

Fig. 6

For σ_x to be maximum,

$$\frac{d\sigma_x}{dx} = 0$$

or $$\frac{d}{dx}\left[\frac{x}{(0.2+0.1x)^3}\right] = 0$$

$$(0.2 + 0.1x)^3 \times 1 - x \times 3\,(0.2 + 0.1x)^2 \times 0.1 = 0$$

$$(0.2 + 0.1x)^2\,[0.2 + 0.1x - 0.3x] = 0$$

$$(0.2 + 0.1x)^2\,[0.2 - 0.1x] = 0$$

Either $0.2 + 0.1x = 0$, which gives, $x = -2m$

or $0.2 - 0.1x = 0$ which gives, $x = 2m$

∴ Maximum bending stress occurs at the fixed end of cantilever.

$$\sigma_{max} = \frac{32\times 30\times 10^3\times 2}{\pi\times(0.4)^3} = 9.55 \text{ MPa}$$

(***b***) In a torsion test, the specimen is a hollow shaft with 50 mm external and 30 mm internal diameter. An applied torque of 1.6 kN-m is found to produce an angular twist of 0.4° measured on a length of 0.2 m of the shaft. The Young's modulus of elasticity obtained from a tensile test has been found to be 200 GPa. Find the values of

(*i*) Modulus of rigidity,

(*ii*) Poisson's ratio.

Sol. Given : $d_0 = 50$ mm, $d_i = 30$ mm,
T = 1.6 kN·m, $\theta = 0.4°$, $l = 0.2$ m, E = 200 GPa

$$J = \frac{\pi}{32}(d_o^4 - d_i^4) = \frac{\pi}{32}(50^4 - 30^4)\times 10^{-12} = 5.34\times 10^{-7}\ \text{m}^4$$

(*i*) $$\frac{T}{J} = \frac{G\theta}{l}$$

$$\frac{1.6\times 10^3}{5.34\times 10^{-7}} = \frac{6\times\pi\times 0.4}{0.2\times 180}$$

$$G = 85.825 \text{ GPa}$$

(*ii*)
$$G = \frac{E}{2(1+\nu)}$$

$$1 + \nu = \frac{200}{2 \times 85.825} = 1.165$$

$$\nu = 0.165$$

(*c*) A cylindrical shell has the following dimensions :

Length = 3 cm

Inside diameter = 1 m

Thickness of metal = 10 mm

Internal pressure = 1.5 MPa

Calculate the change in dimensions of the shell and the maximum intensity of shear stress induced. Take E = 200 GPa and Poisson's ratio, ν = 0.3.

Sol. Given : l = 3 m, d_i = 1 m, t = 10 mm or 0.01 m, p = 1.5 MPa, E = 200 GPa, ν = 0.3

Hoop stress,
$$\sigma_\theta = \frac{pd_i}{2t} = \frac{1.5 \times 10^6 \times 1}{2 \times 0.01} = 75 \text{ MPa}$$

Longitudinal stress,
$$\sigma_l = \frac{pd_i}{4t} = \frac{1.5 \times 10^6 \times 1}{4 \times 0.01} = 37.5 \text{ MPa}$$

Hoop strain,
$$\varepsilon_\theta = \frac{1}{\varepsilon}(\sigma_\theta - \nu\sigma_l)$$

$$= \frac{1}{200 \times 10^3}(75 - 0.3 \times 37.5)$$

$$= 318.75 \times 10^{-6}$$

Longitudinal strain,
$$\varepsilon_l = \frac{1}{E}(\sigma_l - \nu\sigma_\theta)$$

$$= \frac{1}{200 \times 10^3}(37.5 - 0.3 \times 75)$$

$$= 75 \times 10^{-6}$$

$$\varepsilon_\theta = \frac{\delta d}{d}, \quad \delta d = 318.75 \times 10^{-6} \times 1 = 318.75 \times 10^{-6} \text{ m or } 0.318 \text{ mm}$$

$$\varepsilon_l = \frac{\delta l}{l}, \quad \delta l = 75 \times 10^{-6} \times 3 = 225 \times 10^{-6} \text{ m or } 0.225 \text{ mm}$$

Maximum shear stress,

$$\tau_{max} = \frac{1}{2}(\sigma_\theta - \sigma_l)$$

$$= \frac{1}{2}(75 - 37.5) = 18.75 \text{ MPa}$$

Q.3. (*a*) (*i*) Explain the term 'Interference' in involute gears and how this can be avoided.

(*ii*) Derive the expression for minimum number of teeth on a pinion for Involute Rack to avoid interference.

(*iii*) Find the minimum number of teeth on the pinion to avoid interference when the addendum for stup teeth is 0.84 module and the pressure angle f = $\cos^{-1} 0.95$, if (*a*) the gear ratio is 3 to 1 and (*b*) the pinion meshes with a rack.

Sol. (*i*) Interference in Involute Gears : An involute starts at the base circule and is generated outwards. It is therefore impossible to have an involute inside the base circle. The line of action is tangent to the two base circles of a pair of gears in mesh and these two points represent the extreme limits of the length of action. These two points are called *interference points*. If the teeth are of such proportion that the beginning of contact occurs before the interference points is met, then the involute portion of the driven gear will mate with a noninvolute portion of the driving gear, and involute interference is said to occur. This condition is shown in Fig. 7; E_1 and E_2 show the interference points that should limit the length of action. A shows the beginning of contact, and B shows the end of contact. It can be seen that the beginning of contact occurs before the interference point E_1 is met; therefore, interference is present. The tip of the driven tooth will gouge out or undercut the flank of the driving tooth as shown by the dotted line.

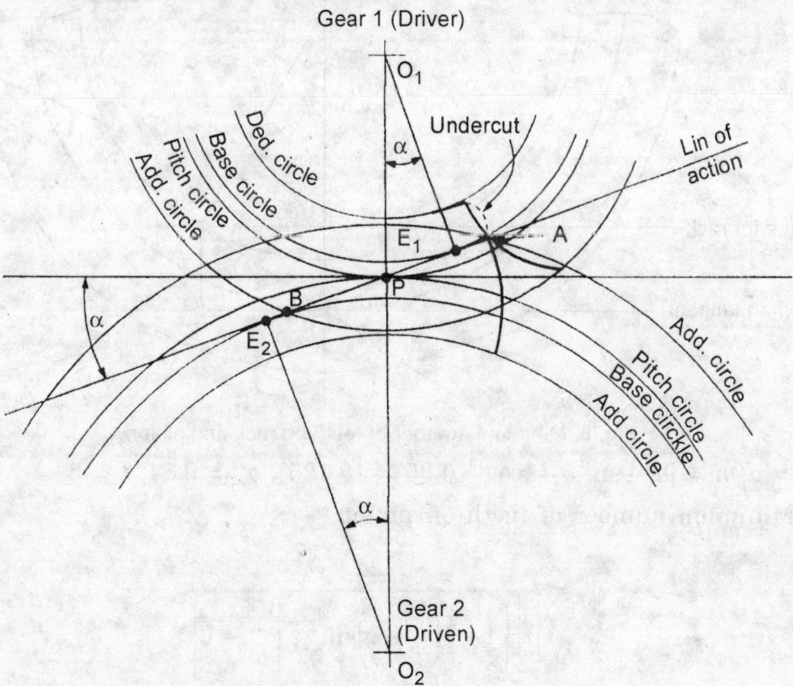

Fig. 7. Interference in rack and pinion.

Methods for avoiding interference : There are several ways of eliminating interference. Interference can be avoided by undercutting, making stub tooth increasing the pressure angle, and cutting the gears with long and short addendum gear teeth. The method of undercutting is to limit addendum of the driven gear so that it passes through the interference point E_1, thus giving a new beginning of contact. Interference and the resulting undercutting not only weaken the pinion tooth but may also remove a small portion of the involute adjacent to the base circle, which may cause a serious reduction in the length of action.

***(ii)* Minimum Number of Teeth on a Pinion for Involute Rack :** A rack is a gear of infinite pitch radius. Thus its pitch circle is a straight line, called the *pitch line*. the line of action is tangential to the base circle at infinity; hence the involute profile of the rack is a straight line and is perpendicular to the line of action. For the rack and pinion shown in Fig. 8, let

Addndum of rack, $h_r = a_r m$

$$h_r = AB = AP \sin\alpha = O_1P \sin\alpha \sin\alpha = O_1P \sin^2\alpha$$

$$= r_1 \sin^2\alpha$$

or $$a_r m = \left(\frac{mz_1}{2}\right)\sin^2\alpha$$

or $$a_r = \left(\frac{z_1}{2}\right)\sin^2\alpha$$

or $$z_1 = \frac{2a_r}{\sin^2\alpha}$$

Fig. 8. Minimum number of teeth on rack and pinion.

(iii) $h_a = a_p m = 0.84$ m, $\phi = \cos^{-1} 0.95 = 18.195°$, $a_p = 0.84$, $i = 3$

(*a*) Minimum number of teeth on pinion,

$$z_1 = \frac{2a_p}{i\left[\left[1+\frac{1}{i}\left(\frac{1}{i}+2\right)\sin^2\phi\right]^{1/2} - 1\right]}$$

$$= \frac{2\times 0.84}{3\left[\left\{1+\frac{1}{3}\left(\frac{1}{3}+2\right)\sin^2 18.195°\right\}^{1/2} - 1\right]} = \frac{1.68}{0.0372\times 3} = 15$$

(*b*) $$z_1 = \frac{2a_r}{\sin^2\phi} = \frac{2\times 0.84}{\sin^2 18.195°} = 17$$

(**b**) A vertical single cylinder is shown in Figure 9. The lower piston is connected to the centre crank. The stroke is 80 mm for the lower piston. The upper pistion operates the two outer cranks which are at 180° to the centre crank.

The reciproacting parts for the lower piston weigh 10 kg.

the reciproacting parts for the upper piston weigh 20 kg.

Find the stroke of the upper piston when the primary force is balanced.

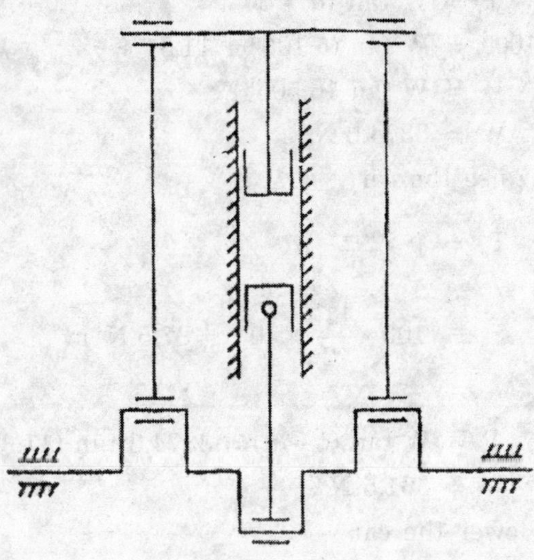

Fig. 9

Sol. Given : $m_1 = 10$ kg, $m_2 = 20$ kg, $L_1 = 80$ mm or $r_1 = 40$ mm

Total primary force,

$$F_p = m_1 r_1 \omega^2 \cos\theta + 2m_2 r_2 \omega^2 \cos(180° + \theta) = 0$$

$$10 \times 40 \times \omega^2 \cos\theta + 2 \times 20 \times r_2 \omega^2 (-\cos\theta) = 0$$

or $$400 - 40\, r_2 = 0$$

$$r_2 = 10 \text{ mm}$$

Stroke of upper piston, L_2 = 20 mm

Q.4. (**a**) The screw of a car lift has 150 mm dia. with square threads of 50 mm pitch and lead of 100 mm. Assume 1000 N is required for lifting of the car and thread friction coefficient is 0.20.

(*i*) Compute the torques required to rotate the screw to raise the car and lower the car. Also compute the overal efficiency of the screw.

(*ii*) What is the flaw in the above design?

Sol. (**i**) **Given :** $d_c = 150$ mm, $p = 50$ mm, L = 100 mm, P = 1000 N, $\mu = 0.20$

Depth of threads, $$h = \frac{p}{2} = \frac{50}{2} = 25 \text{ mm}$$

Mean diameter, $$d_m = d_c + h = 150 + 25 = 175 \text{ mm}$$

Outside diameter, $$d_0 = d_c + p = 150 + 50 = 200 \text{ mm}$$

Helix angle, $\alpha = \tan^{-1}\left(\frac{p}{\pi d_m}\right) = \tan^{-1}\left(\frac{50}{\pi \times 175}\right) = \tan^{-1} 0.09095 = 5.196°$

Angle of friction, $\phi = \tan^{-1} \mu$

$= \tan^{-1} 0.20 = 11.31°$

Raising of car :

$$P = W \tan (\alpha + \phi)$$

$$1000 = W \tan (5.196° + 11.31°)$$

$$= W \tan 16.506°$$

$$W = 3374.6 \text{ N}$$

Torque required to raise the car,

$$T_r = P \times \frac{d_m}{2}$$

$$= 100 \times \frac{175}{2} \times 10^{-3} = 87.5 \text{ N}\cdot\text{m}$$

Lowering of car :

$$P = W \tan (\phi - \alpha) = 3374.6 \tan (11.31 - 5.196°)$$

$$= 361.5 \text{ N}$$

Torque required to lower the car

$$T_l = P \times \frac{d_m}{2} = 361.5 \times \frac{175}{2} \times 10^{-3} = 31.63 \text{ N}\cdot\text{m}$$

Overall efficiency of the screw :

$$\eta = \frac{\tan \alpha}{\tan(\alpha + \phi)} \text{ for raising of load}$$

$$= \frac{0.09095}{0.29632} \times 100 = 30.7\%$$

$$= \frac{\tan(\phi - \alpha)}{\tan \alpha} \text{ for lowering of load}$$

$$= \frac{\tan 6.114°}{0.09095} \times 100 = 117.8\%$$

$\therefore$ Overall efficiency $= 30.7\%$

***(ii)* Flaw in the design :** The car is supported on a head which does not rotate along with the screw. The collar friction is not accounted for in this design. This is the major flaw in the design.

(*b*) A simple gear train consists of three gears with 18, 16 and 72 teeth. The 18 teeth pinion is connected to a 4 kW electric motor running at 2400 rpm. Design the shaft for 72 teeth gear if the shearing stress is not to exceed 60 MPa in the shaft.

Sol.

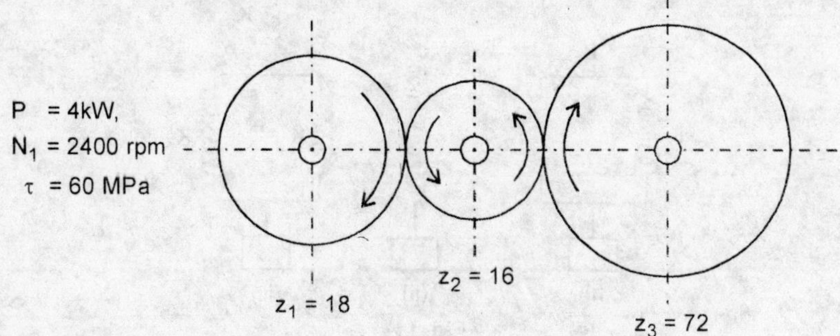

Fig. 10

Refer to Fig. 10

$$\frac{N_2}{N_1} = \frac{z_1}{z_2}$$

$$N_2 = \frac{18}{16} \times 2400 = 2700 \text{ rpm}$$

$$\frac{N_3}{N_2} = \frac{z_2}{z_3}$$

$$N_3 = 2700 \times \frac{16}{72} = 600 \text{ rpm}$$

Power transmitted, $$P = \frac{2\pi N_1 T_1}{10^3 \times 60}$$

$$4 = \frac{2\pi \times 2400 \times T_1}{10^3 \times 60}$$

$$T_1 = 15.915 \text{ N·m}$$

$$N_1 T_1 = N_3 T_3$$

$$T_3 = \frac{N_1}{N_3} \times T_1 = \frac{2400}{600} \times 15.915 = 63.66 \text{ N·m}$$

$$T_3 = \frac{\pi}{16} d^3 \tau$$

$$63.66 = \frac{\pi}{16} \times d^3 \times 60 \times 10^6$$

$$d = 0.0175 \text{ m or } 17.5 \text{ mm}$$

$$\simeq 18 \text{ mm}$$

(*c*) Find the equivalent spring constant if the below shown spring systems are replaced by a single spring.

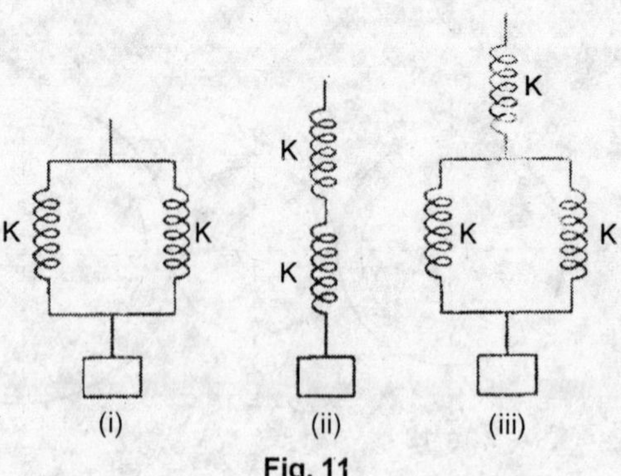

Fig. 11

Sol.

(*i*) The springs are in parallel. (Fig. 11)

Equivalent spring constant,

$$k_e = k + k = 2k$$

(*ii*) The springs are in series,

$$\frac{1}{k_e} = \frac{1}{k} + \frac{1}{k} = \frac{2}{k}$$

$$k_e = \frac{k}{2}$$

(*iii*)
$$\frac{1}{ke} = \frac{1}{k} + \frac{1}{2k} = \frac{3}{2k}$$

$$k_e = \frac{2k}{3}$$

(d) Most of the times the rotating disc-shafts have keys; under such conditions it cannot be assumed that the shaft has a uniform cross-section. For such a disc-shaft system how will you determine the limits of speed for instability?

Sol. For a shaft under torsion without keyway, transmitted torque,

$$T_1 = \frac{\pi}{16} d^3 \tau$$

and with a keyway,

$$T_2 = \frac{\pi}{16} d^3 \times (e\tau)$$

where
$$e = \frac{\text{Strength of shaft with keyway}}{\text{Strength of shaft without keyway}}$$

$$= 1.1 \frac{h}{d}$$

h = depth of keyway

d = shaft diameter

Let $\qquad P$ = power to be transmitted, kW

Then $\qquad P = \dfrac{2\pi N_1 T_1}{60 \times 10^3}$

$$N_1 = \frac{60 \times 10^3 \times P}{2\pi T_1} = \frac{60 \times 10^3 \times P}{2\pi \times \frac{\pi}{16} d^3 \tau} = \frac{48634 \times P}{d^3 \tau}$$

$$N_2 = \frac{60 \times 10^3 \times P}{2\pi \times \frac{\pi}{16} d^3 \times e\tau} = \frac{48634 \times P}{d^3 e\tau}$$

Thus limits of speed for instability can be determined.

SECTION-C

Q.5. (*a*) Explain the three types of oxy-acetylene flames. Indicate with the help of sketches the various zones, respective temperature ranges and applications of each type of flame.

Sol. Types of Oxy-Acetylene Flames : In an oxyacetylene flame, O_2 and C_2H_2 are mixed and burnt to release heat. The complete combustion of acetylene in an atmosphere of oxygen is represented by the following reaction :

$$C_2H_2 + 2.5O_2 = 2CO_2 + H_2O \text{ (vapour)} + 1284.57 \text{ kJ/mol}$$

An oxy-acetylene flame can be set to : neutral, carburising or oxidising condition depending upon the metal to be welded and the type of filler metal to be used. In a neutral flame, O_2 and C_2H_2 are mixed in equal amounts. In carburising or reducing flame, the percentage of C_2H_2 is more than that of O_2 and in the oxidising flame, the percentage of O_2 is more than of C_2H_2. These three types of flames are illustrated in Fig. 12.

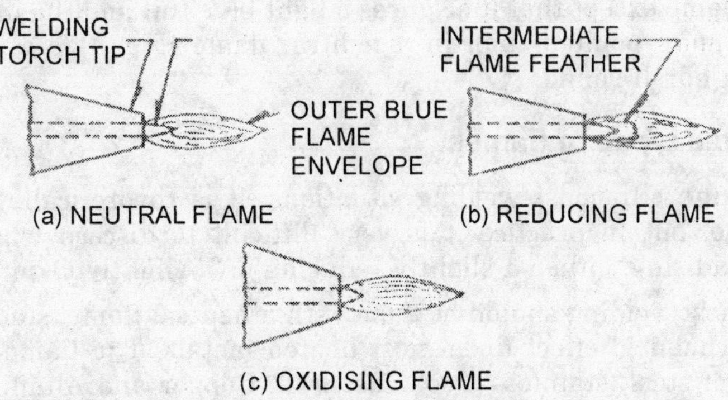

Fig. 12. OAW Flame Settings.

1. **Neutral Flame :** A neutral flame is obtained when equal amounts of O_2 and C_2H_2 are mixed and burnt in a torch. The flame is recognised by two sharply defined zones, the inner white cone flame and the outer blue flame evelope, (Fig. 12 (*a*)). The reaction at the inner cone for the neutral flame where equal volumes of cylinder oxygen and acetylene are used is,

$$C_2H_2 + O_2 \longrightarrow 2CO + H_2$$

This provides the most concentrated heat with the highest temperature for welding at a distance of 3 to 5 mm from the end of the inner cone. It is also apparent that the environment within the outer envelope consists of carbon monoxide and hydrogen and is relatively inert to materials that oxidise readily.

The reactions at the outer evelope are :

$$2CO + O_2 \longrightarrow 2CO$$

$$H_2 + \frac{1}{2}O_2 \longrightarrow H_2O \text{ (vapour)}$$

For these reactions, the oxygen is supplied from the surrounding air. During actual welding, the outer envelope spreads over the surface of the work material and serves as a protective shield from the ordinary atmosphere. Also, since the heat developed is not as concentrated, this outer envelope of flame contributes only to preheat the work material for welding.

2. **Carburising Flame :** A carburising or reducing flame is obtained when an excess of acetylene is supplied than which is theoretically required ($O_2 : C_2H_2 = 0.85$ to 0.95). A reducing flame is recognised by three distinct sections : the inner cone (which is not sharply defined) and an outer envelope as for the neutral flame. The third zone surrounds the inner cone and extends into the outer enveloping zone. It is whitish in colour and is called "excess acetylene feather". Its length is an indication of the amount of excess acetylene (Fig. 12 (*b*)).

   To obtain a neutral flame, first the reducing flame is obtained. Then the supply of oxygen is gradually increased until the intermediate feather disappears. The resulting flame will be a neutral flame.

3. **Oxidising Flame :** This flame has an excess of oxygen over that required for a neutral flame ($O_2 : C_2H_2 = 1.15$ to 1.50). To obtain an oxidising flame, the flame is first set to the neutral condition. Then the acetylene valve is turned down gradually to reduce the amount of acetylene, giving an excess of oxygen. The flame resembles the neutral flame except that it acquires a light blue tint and the inner cone is slightly shorter and more pointed than in a neutral flame (Fig. 12(*c*)). An oxidising flame burns with a harsh sound.

Applications of three types of flames

(*a*) **Neutral Flame :** In most welding situations, it is theoretically desirable to use a neutral flame, but in practice it is very difficult to discern whether the flame is neutral or oxidising, either a slightly reducing or a slightly Oxidising flame is used.

If possible, most welding should be done with a neutral flame, since such a flame has a minimum chemical effect upon most heated metals. The flame is widely used for the welding of steel, stainless steel, cast iron, Copper and Aluminium.

(*b*) **Reducing Flame :** The acetylene being excess in this flame, the available carbon is not completely consumed. With iron and steel, it will form iron carbide (hard and brittle). Therefore, metals that tend to absorb carbon shoud not be welded with a reducing flame. This flame is used for materials that oxidise rapidly like steel and aluminium. It is suitable for welding low alloy steels, low c-steels and for welding those metals (for example, non-ferrous) that do not tend to absorb carbon. Such flames are also used for welding Monel metal and Nickel and in hard surfacing with high speed steel and cemented carbides.

(*c*) **Oxidising Flame :** This flame has limited use because the excess oxygen tends to combine with many metals to form hard, brittle, low strength oxides. Also, excess of oxygen causes the weld bead and the surrounding area to have a skummy or dirty appearance. So, this flame is not used for welding steel. It is mainly used when welding materials which are not oxidised readily and which have a high solubility [of hydrogen in the molten state and low solubility] in the solid state, for example, brasses, bronzes and gold. For these metals the oxidising atmosphere creates a base metal oxide that protects the base metal. For example, in welding brass, Zinc has a tendency to separate and fume away. The formation of a covering copper oxide prevents Zinc from dissipating.

(***b***) Explain the processes of extrusion given below. Indicate one typical product made through each of these processes :

(*i*) Direct Extrusion

(*ii*) Indirect Extrusion

(*iii*) Hydro-static Extrusion

(*iv*) Impact Extrusion

Sol. (*i*) Direct Extrusion: The various direct extrusion processes are shown in Fig. 13(*a*).

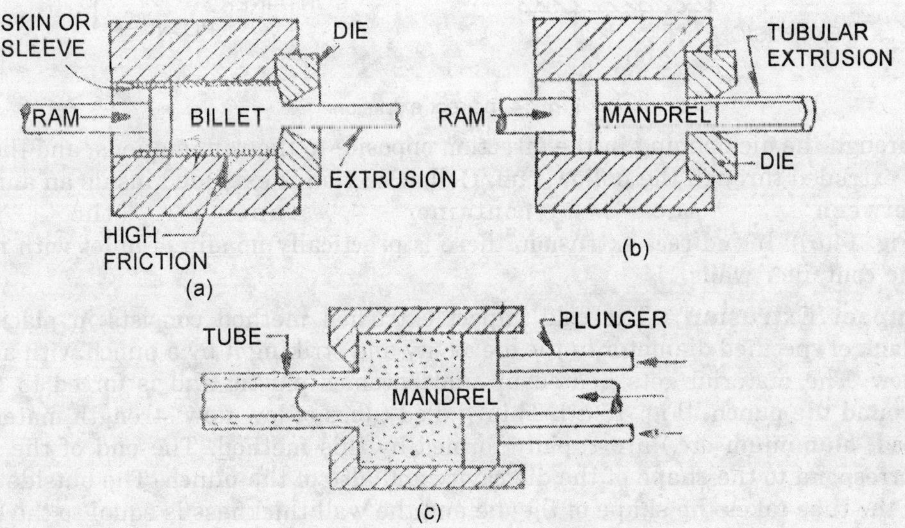

Fig. 13. Direct extrustion processes.

In the direct extrusion process (Fig. 13(*a*)), the heated billet is placed in the container. It is pushed by a ram towards the die. The metal is subjected to plastic deformation, slides along the walls of the container and is forced to flow through the die opening. At the end of the extruding operation a small piece of metal, called butt-end scrap, remains in the container and can not be extraded. To obtain a hollow section, a hollow billet is often used around a mandrel. In some instances, the billet is solid and is pierced by the mandrel or an axial hole is drilled in it by means of a drilling machine before the extrusion cycle begins. The mandrel may be fastened to the ram. As the ram moves in the direction towards the die, the metal is forced out through the annular clearance and form a tube, (Fig. 13(*b*)). With a modern extrusion press, the piercing mandrel is actuated by a separate hydraulic cylinder from the one which operates the ram, (Fig. 13(*c*)). The billet is first upset with the ram with the piercing mandrel withdrawn. Next the billet is pierced with the pointed mandrel ejecting a plug through the die. Then the ram advances and extrudes the billet over the mandrel to produce a tube. The outside diameter of the extruded tube equals the diameter of the die hole and the inside diameter equals that of the steel mandrel. This method is called as "Tubular extrusion".

(ii) **Indirect Extrusion :** For the production of a solid part, the die is mounted on the end of a hollow ram and enters the container, (Fig. 14(*a*)) the other end of container being closed by & closure plate. As the ram travels, the die applies pressure on the billet and the deformed metal flows

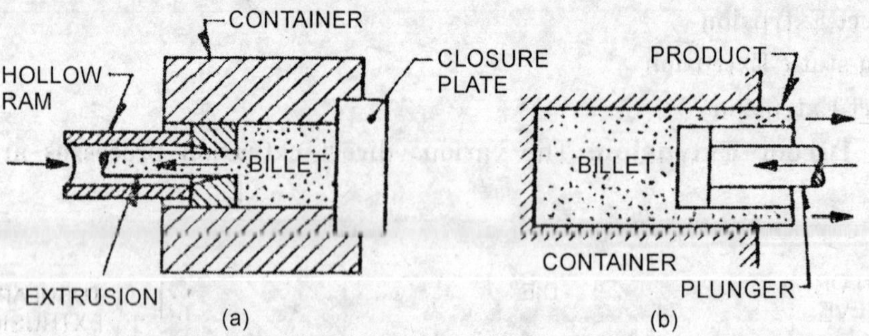

Fig. 14 Indirect extrusion.

through the die opening in the direction opposite to the ram motions, and the product is extruded through the hollow ram. Hollow section is extruded inside an annular gap between the container and the ram, (Fig. 14(*b*)). In indirect extrusion, there is practically no slip of billet with respect to the container walls.

(iii) **Impact Extrusion :** The cold impact extrusion method consists in placing a flat blank of specified diameter in the die cavity and striking it by a punch with a powerful blow. The material gets heated up and becomes plastic and is forced to squirt up around the punch, [Fig. 15(*a*)]. Thin walled tubes of low flow strength materials (tin, lead, aluminium etc.) are rapidly formed by this method. The end of the tube will correspond to the shape of the die cavity and also of the punch. The outside diameter of the tube takes the shape of the die and the wall thickness is equal to the clearance between the punch and the die. The operation is fully automatic and the production rate is as high as 50 tubes or more per minute. When the punch is on its upward stroke, the tube sticks to it. To effect release, either a stripper [See Fig. 15(*b*)] or compressed air is directed against the tube, thus stripping it from the punch. Threads may be formed at the end of these collapsible tubes by retractable die portions or by other methods. These collapsible tubes are used for cosmetics (cream, shaving cream), tooth paste, grease etc.

Other product applications include : Cans, fire extinguisher cases, radio shields, food containers, boxes for condensers and cigarette lighter cases (if symmetrical).

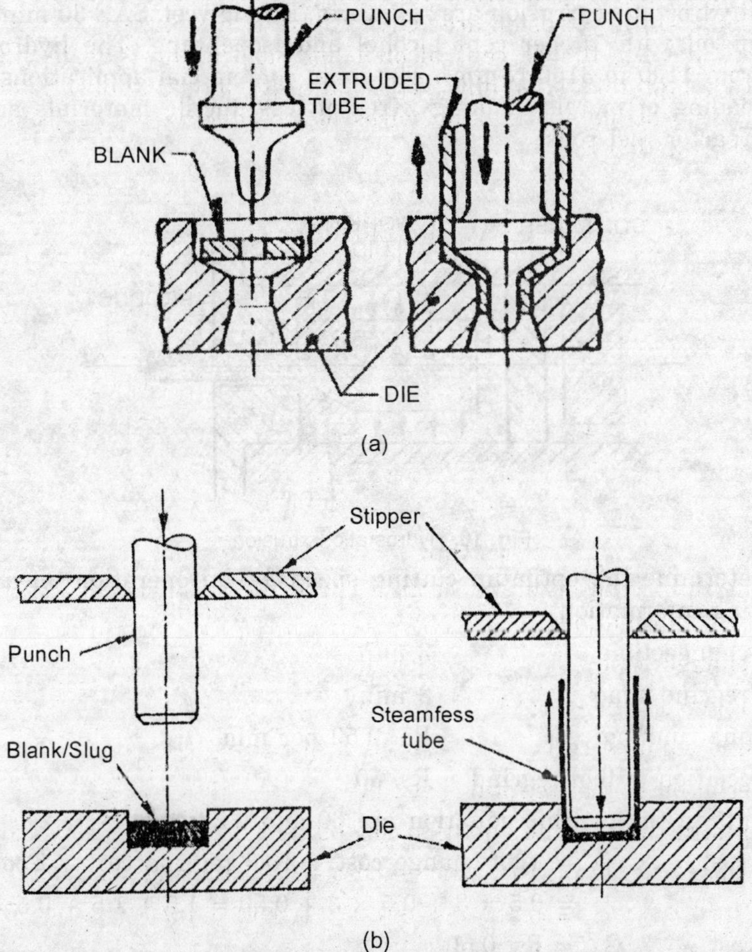

Fig. 15. Impact extrusion.

A variation of the method where a flat solid blank of slug (that is, without a hole) is used in the die cavity (the bottom of the die cavity is closed) is shown in Fig. 15(*b*). The method is also being employed to make thick walled products such as paint-spray-gun containers and grease-gun containers, by using large presses.

Impact extrusion speed is 100 to 350 mm/s.

(*iv*) **Hydrostatic Extrusion :** In hydrostatic extrusion, (Fig. 16) a high pressure liquid medium is used for the transmission of the force to the billet/blank. The liquid completely envelops the blank and enters die canals. Due to the hydrostatic pressure, the ductility of the material is increased. Even brittle materials like tungsten, cast iron and stainless steel etc. can be extruded. This also permits the extrusion of very long billets or even wire, accompanied by large reductions. There is no container friction and the pressurised fluid also acts as a lubricant and because of this, the extruded product has a good surface finish and dimensional accuracy. However, the absence of container friction combined with reduced die friction can increase the tendency to internal crack formation. The pressure transmitting fluids commonly used for hydrostatic extrusion are : Glycerin, Ethylglycol, SAE 30 mineral lubricating oil, castor oil with 10 per cent alcohol and isopentane. The hydrostatic pressure ranges from 1100 to 3150 N/mm^2. The main commercial applications of this process are : Cleading of metals, making wires of less ductile materials and extrusion of nuclear reactor fuel rods.

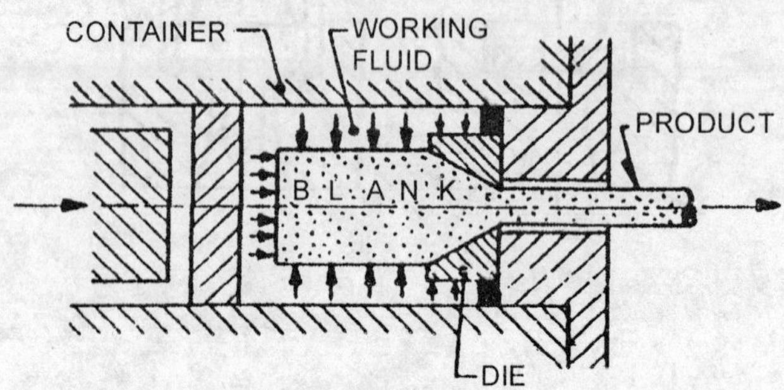

Fig. 16. Hydrostatic Extrusion

Q.6. (*a*) Determine the optimum cutting speed for an operation on a lathe machine using the following information :

Tool change time	3 min
Tool regrind time	3 min
Machine running cost	Re. 0.50 per min
Depreciation of tool regrind	Rs. 50

The constants in the tool life equation are 60 and 0.2.

Sol. Total cost, C_t = Tool change cost + Tool regrind cost + Tool depreciation

$$= 0.5 \times 3 + 0.5 \times 3 + 0.50 = 1.5 + 1.5 + 0.5 = \text{Rs. } 3.5$$

Machning cost, C_m = Rs. 0.50

$$C = 60, \; n = 0.2$$

Optimum cutting speed,

$$V_{opt} = C\left[\frac{C_m}{C_t} + \frac{n}{1-n}\right]^n$$

$$= 60\left[\frac{0.5}{3.5} + \frac{0.2}{1-0.2}\right]^{0.2}$$

$$= 49.8 \text{ m/min}$$

(*b*)

(*i*) Why are the alignment tests done on machine tools?

(*ii*) Give the list of common testing equipment needed for alignment tests on a general purpose lathe.

(*iii*) Write steps to check straightness of carriage in a horizontal plane with the help of sketches.

Sol. (*i*) Necessity of Alignment Tests

1. By alignment testing of a machine tool it can be put in a proper grade or precision class, thereby helping a purchaser to know what he can expect from the machine tool.
2. Testing helps in preventive maintenance of a machine tool.
3. Testing helps in determining the condition of the machine tool and hence its useful life.

The alignment tests are checks on the relationship of the various elements of the machine when it is idle and unloaded. These tests are performed by using ordinary measuring equipment together with a few special accessories. These, tests include the following :

(*a*) Checking the parallelism of the spindle and bed of a lathe.

(*b*) Squareness of the lathe movement of the spindle of a milling machine.

(*e*) Checking that the taper hole of a drilling machine is running true.

(*ii*) Testing Equipment Needed for Alignment Test

1. **Dial Gauges :** The graduation on the dial need not be finer than 0.01 mm. The initial measuring pressure should be from 40 to 100 gm; for very five measurements, a lower pressure as small as 20 gm is desirable.
2. **Test Mandrels :** They are used for checking the true running of the spindle. Their length varies from 100 mm to 300 mm.
3. **Straight edge and squares :** Straight edges of cast iron or steel should be heavy, well ribbed and reasoned. A square must have a wider bearing surface. The standard square should have a tolerance of ±0.01 mm and the precision square ±0.005 mm.
4. **Spirit levels :** They are used for high grade precision work. Spirit levels should have a sensitivity of about 0.04 to 0.06 mm per metre for each deflected division.
5. **Autocollimeter :** It is used for checking deflections of long beds in horizontal vertical or inclined planes.
6. **Waviness meter :** It is useful in recording and examining the surface waviness, 50 : 1 magnification is sufficient for them.

(*iii*) Checking Straightness of Carriage in Horizontal Plane

Carriage	Straightness of carriage in a horizontal plane	(*a*) Dial gauge and test mandrel (*b*) Taut wire and microscope	0.02 to 0.03 mm per 1000 mm

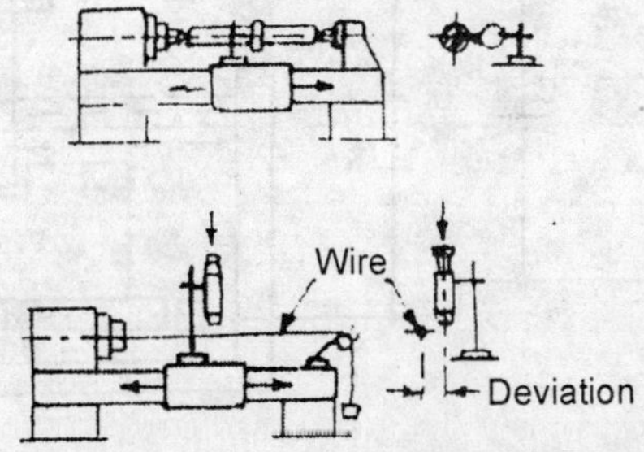

The straightness of carriage in horizontal plane by the following methods :

(*i*) *Straight edge method* : In this method, a precise straight edge applied on two gauge blocks of equal height is used. The distance under the straight edge to the ways is measured at various places by a set of gauge blocks and a thickness gauge. If, for example, at various points under the straight edge, thickness gauges of 0.10 mm, 0.05 mm and 0.08 mm will pass and the distance between the blocks supporting the straight-edge equals one metre, then the maximum error per metre will be 0.10 – 0.05 = 0.05 mm. Thus, the maximum error in straightness = Maximum size of thickness gauge – Minimum size of thickness gauge passing between the straight edge and the way.

It is evident that if thickness gauges of the same size pass under the straight edge, all along its length, the way will be as straight as the straight edge.

(*ii*) *Taut wire method* : The microscope is fixed to the saddle and travels along a tight wire. Deviations from straightness of the saddle slideways will be recorded, with the horizontal projection of the taut wire as the reference line, Fig. 17.

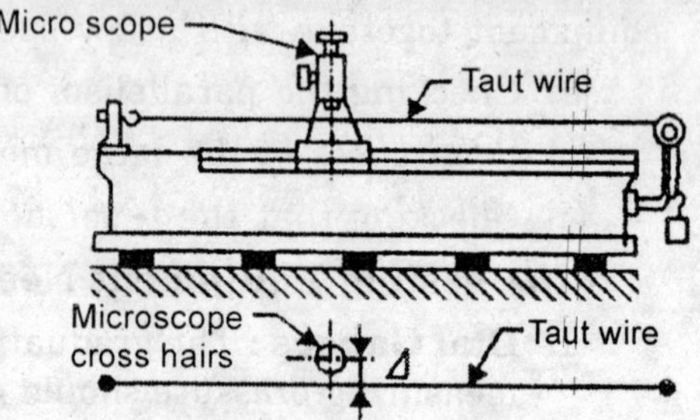

Fig. 17. Taut wire method.

(*c*)

(*i*) What is the principle of metal removal in EDM process?

(*ii*) Describe the process with the help of sketch.

(*iii*) List advantages and limitations of the system.

Sol. (*i*) Principle of Metal Removal in EDM : When a discharge takes place between two points of the anode and the cathode, the intense heat generated near the zone melts and evaporates the materials in the sparking zone. The workpiece is normally made the anode. A suitable gap, known as the spark gap, is maintained between the tool and the workpiece surfaces.

(*ii*) Description of Process :

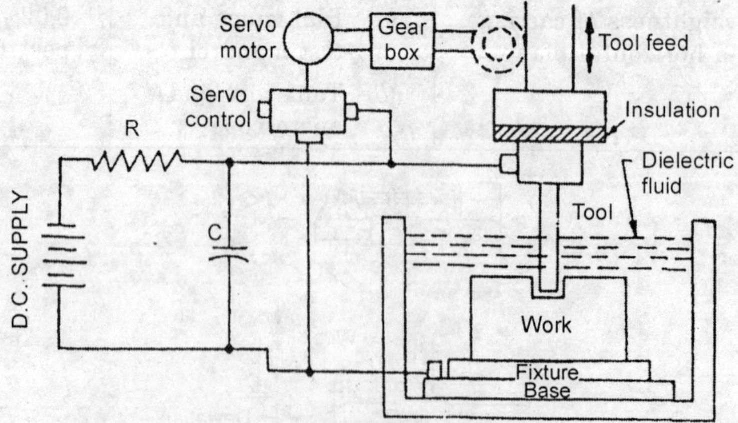

Fig. 18. Schematic Arrangement of EDM Method.

The schematic arrangement of EDM method is shown in Fig. 18. Power for generating the spark is fed from an A.C. source to a rectifier. The D.C. output is then fed to the spark generating circuit. The tool, work and tool slide servo-mechanism, are connected into the circuit. The tool and work re submerged in a fluid having poor electrical conductivity. The function of the servo- mechanism is to maintain a very small gap between the tool and the work. When the voltage across the gap becomes sufficiently large, the dielectric fluid becomes ionized and breaks down to produce an electrically conductive spark channel and the condensers discharge current across the channel in the form of a spark. When the voltage drops to about 12 V, the spark discharge extinguishes and the dielectric fluid once again becomes deionized. The condensers start to recharge and the process repeats itself. At each discharge, heat transfer from high temperature spark to both tool and work, melts partially vaporizes and partially ionizes the metal in a thin surface layer.

(*iii*) Limitations

1. Only electrically conductive materials can be machined.
2. High specific energy consumption.
3. Low material removal rate.
4. Gives rough surface for larger material removal rates.
5. Electrode wear and overcut are serious problems.
6. Perfectly square corners cannot be made.

Advantages

1. Metals, alloys and carbides can be machined.
2. Materials that are too hard or brittle can be machined.
3. No chips or burrs are left on the workpiece.
4. Cutting forces are virtually zero. So very delicate and fine work can be done.
5. Requires little operator's attention.

Q.7. (*i*) List common time-series forecasting models. Explain simple exponential smoothing method of forcasting demand. What are its limitations?

(*ii*) The monthly forecast and demand values of a firm are given below :

Month	Forecast units'	Demand units
Jan	100	97
Feb	100	93
Mar	100	110
Apr	100	98
May	102	130
Jun	104	133
Jul	106	129
Aug	108	138
Sep	110	136
Oct	112	124
Nov	114	139
Dec	116	125

Calculate Tracking Signal for each month. Comment on the forecast model.

(i) Time-Series Forecasting Models : These models use the part history of demand in making a forecast for the future. They perform well for short time forecast. Time series is a set of observations of some variable over a time period. The components of time series model are :

Trend (T), Cyclic variation (C), Seasonal variation (S), and Random variation (R).

Forecast, $Y = TCSR$

(a) Moving Average Method : The new moving average for each period is computed by taking the average demand for the most recent N (called the order of moving average) periods. It does not provide good forecast for seasonal components.

(b) Exponential Smoothing Methods : In this method, the weight assigned to a previous period's demand decreases exponentially as that date gets older. These methods work best for the forecasting of a large number of items.

If t is the current period, the forecast for the next period is :

$$F_{t+1} = \alpha D_t + (1 - \alpha) F_t$$

where α = Weightage given to the current period's observed demand

= 0 to 1

D_t = demand for the current month

F_t = forecast made for the current period

(iii)

Month	Forecast units	Demand units	Absolute error $\lvert e_t \rvert$	Squared error $(e_t)^2$
January	100	97	3	9
February	100	93	7	49
March	100	110	10	100
April	100	98	2	4
May	102	130	28	784
June	104	133	29	841
July	106	129	23	529
August	108	138	30	900
September	110	136	26	676
October	112	124	12	144
November	114	139	25	625
December	116	125	9	81
			$\Sigma\lvert e_t \rvert = 204$	$\Sigma(e_t)^2 = 4742$
			MAD $= \dfrac{204}{12} = 17$	MSE = 395.17

Mean absolute deviation, MAD = 17

Mean square error, MSE = 395.17

The absolute error for May, June, July, August, September and November is more than MAD. Therefore, the forecast units has to be revised.

(b) Four technicians are required to do four different jobs. Estimates of time to complete every job as provided by the technicians are as below:

Technician	Hours to Complete Job			
	Job 1	Job 2	Job 3	Job 4
A	20	36	31	27
B	24	34	45	22
C	22	45	38	18
D	37	40	35	28

Assign the jobs to technicians to minimize the total worktime. State the steps taken in the algorithm used.

Steps for Sequencing

Horo number of jobs = number of technicians ;

1st Step : Determine the minimum completion time for a job by a technician and allocate that job to that technician. Job 4 can be completed in minimum time of 18 hours by technician C. Place the associated job in the first available position in the sequence.

2nd step: Determine the next minimum time to complete a job. It is job 1 requiring 20 hours by technician A. Place this job in the last position. Eliminate these jobs and technicians for consideration in the next steps.

3rd step : Repeat the above process for jobs 2 and 3. Next minimum time is for job 2 requiring 34 hours by technician B.

Job sequence				Technician			
				A	B	C	D
4	—	—	—	—	—	18	—
4	—	—	1	20	—	—	—
4	2	—	1	—	34	—	—
4	2	3	1	—	—	—	35

The optimum sequence is : 4 – 2 – 3 – 1

(c) Prepare a flow chart and write a program in FORTRAN for the following problem:

The cutting tool life (T) for machining operation is given by Taylor's equation $vT^n = C$ where v is the cutting speed and C, n are constants. An MS workpiece of 50 mm dia, is turned with a carbide tool with n = 0.10 and C = 150. It is proposed to study the variation of tool life with the rpm of the workpiece at which turning is performed. Assume that the speed of spindle of lathe can be varied from 100 rpm to 2000 rpm in the steps of 100 rpm.

Sol. $vT^n = C$

$d = 50$ mm, $n = 0.10$, $c = 150$

Let s = speed in rpm

$$v = \frac{\pi d s}{10^3} \text{ m/min}$$

$$\therefore \quad \frac{\pi d s}{10^3} \times T^n = C$$

$$T = \left(\frac{10^3 \times C}{\pi d s}\right)^{1/n} = \left(\frac{10^3 \times C}{\pi d}\right)^{\frac{1}{n}} \times \frac{1}{s^{1/n}}$$

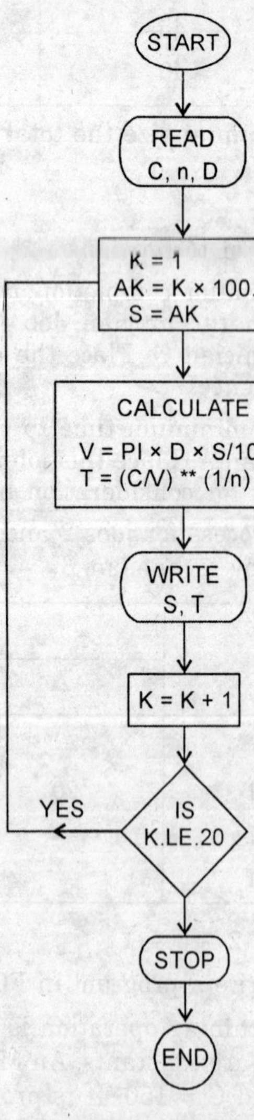

START

READ
C, n, D

K = 1
AK = K × 100.
S = AK

CALCULATE
V = PI × D × S/1000.
T = (C/V) ** (1/n)

WRITE
S, T

K = K + 1

IS
K.LE.20

YES

STOP

END

Fig. 19. Flow chart.

MECHANICAL ENGINEERING
PAPER II—2010

(Conventional)

SECTION A

Q.1. (*a*)

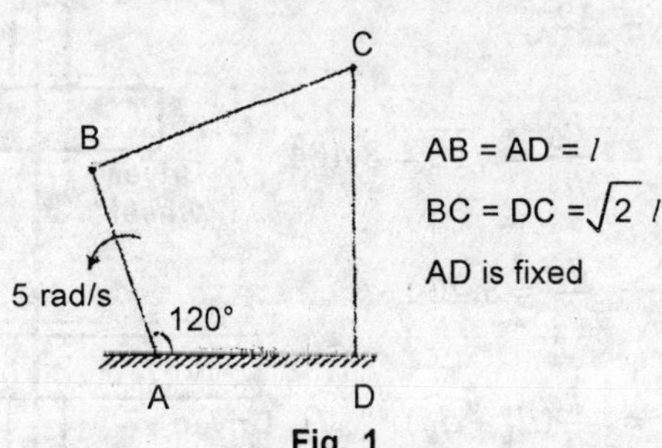

Fig. 1

The driver link AB of a four bar mechanism is rotated at 5.0 rad/s in counter clockwise direction as shown in the above figure. Locate and indicate the Instantaneous Centre (I.C.) of the coupler BC with respect to fixed link AD at an instant when ∠BAD = 180°. Find angular velocity of the coupler using I.C. method only.

Sol. Given : w_2 = 5 rad/s (CCW), n = 4

Number of instantaneous centres,

$$N = \frac{n(n-1)}{2} = \frac{4 \times 3}{2} = 6$$

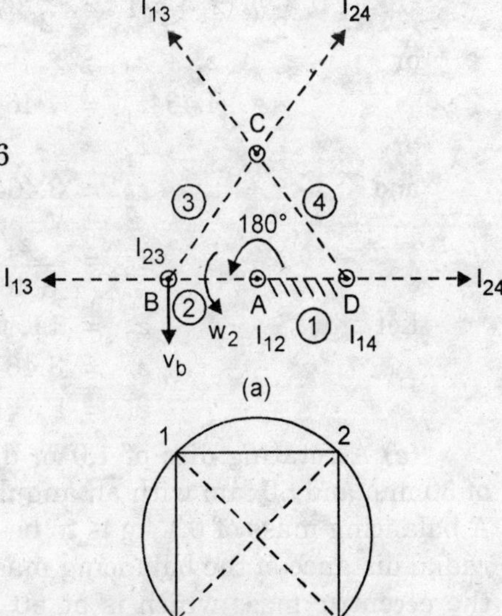

Fig 2. (b) Circle diagram

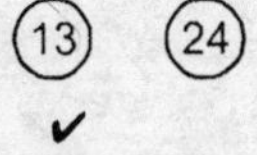

1	2	3	4
✔	✔	✔	
12	23	34	

(13) (24)

✔
14

(13) : 12, 23; 14, 34

(24) : 23, 34; 12, 14

$v_b = w_2 \times AB = 5 \times l = 5l$

The angular velocities of two links vary inverse only as the distances from their common instantaneous centre to their respective centres of rotation relative to the flame.

$$\frac{w_3}{w_2} = \frac{23-12}{23-13}$$

$$\frac{w_3}{5} = \frac{e}{\mathrm{I}_{13} - \mathrm{B}}, \quad w_3 = \frac{5l}{\mathrm{I}_{13} - \mathrm{B}}$$

(***b***) A suitable reverted gear train using four gear wheels is to be used for a clock, the minute hand of which is fixed to the driving spindle and the hour hand to a driven sleeve rotating freely on the same driving spindle axis. The modular pitch is to be kept same for all the wheels and each wheel should have least number of teeth but not 11 or less. Determine number of teeth on each wheels of the gear train.

Sol. Given : $m_1 = m_2 = m_3 = m_4 = m$

$$\frac{n_1}{n_4} = 12$$

$$\frac{n_1}{n_2} = \frac{n_3}{n_4} = \sqrt{12} = 3.464$$

$$\therefore \quad \frac{z_2}{z_1} = \frac{z_4}{z_3} = 3.464$$

Centre distance, $\mathrm{C} = \frac{1}{2}(d_1 + d_2)$

$$= \frac{1}{2}(d_3 + d_4)$$

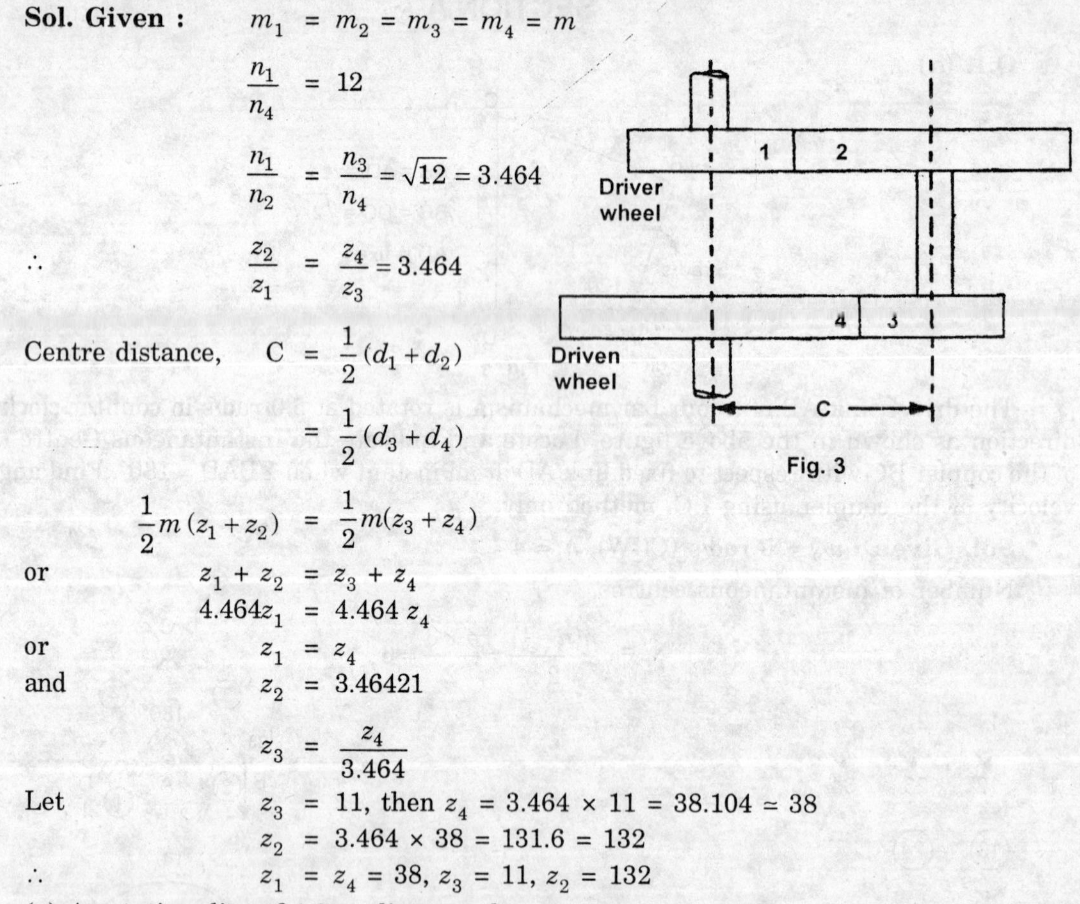

Fig. 3

$$\frac{1}{2}m(z_1 + z_2) = \frac{1}{2}m(z_3 + z_4)$$

or $\quad z_1 + z_2 = z_3 + z_4$

$$4.464z_1 = 4.464\, z_4$$

or $\quad z_1 = z_4$

and $\quad z_2 = 3.46421$

$$z_3 = \frac{z_4}{3.464}$$

Let $\quad z_3 = 11$, then $z_4 = 3.464 \times 11 = 38.104 \simeq 38$

$$z_2 = 3.464 \times 38 = 131.6 = 132$$

$$\therefore \quad z_1 = z_4 = 38,\ z_3 = 11,\ z_2 = 132$$

(***c***) A rotating disc of 1.0 m diameter has two eccentric masses of 0.5 kg each at radii of 50 mm and 60 mm with an angular interval between them as 150° on the plane of the disc. A balancing mass of 0.1 kg is to be attached in order to balance the disc. What should be the radial distance of the balancing mass from the centre of the disc and its angular position from the eccentric mass which is at 50 mm radius?

Sol. Given : $m_1 = m_2 = 0.5$ kg, $r_1 = 50$ mm, $r_2 = 60$ mm, $m_b = 0.1$ kg

Resolving the forces horizontally and vertically, we have

$0.5 \times 50 - 0.5 \times 60 \cos 30° - 0.1 \times r_b \cos \theta_b = 0$

$$25 - 25.98 - 0.1 \times r_b \cos \theta_b = 0$$

$$r_b \cos \theta_b = -9.8 \quad (1)$$

$$0.5 \times 60 \times \cos 60° = 0.1\, r_b \sin \theta_b$$

$$r_b \sin \theta_b = 150 \quad (2)$$

$$\tan \theta_b = -15.30612$$

$$\theta_b = -86.26°$$

$$\therefore \quad \angle m_1\, om_b = 180° - 86.26° = 93.74°$$

$$r_b = \sqrt{(150)^2 + (-9.8)^2} = 150.3 \text{ mm}$$

(*d*)

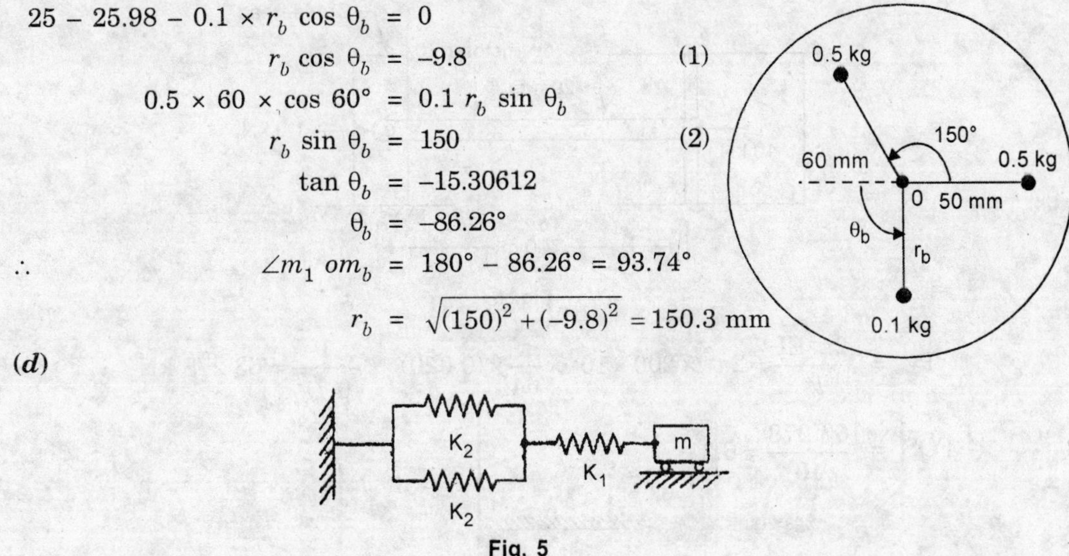

Fig. 5

Write down the differential equation of motion for the vibrating system shown in the above figure for free vibration, using D Alembert's principle and express its natural frequency in Hz, using system parameters as given.

Sol. Spring k_2 are in parallel. Their equivalent stiffness is $2k_2$. Now $2k_2$ and k_1 are in series. Their equivalent stiffness is,

$$\frac{1}{k_e} = \frac{1}{2k_2} + \frac{1}{k_1} = \frac{k_1 + 2k_2}{2k_1 k_2}$$

$$\therefore \quad k_e = \frac{2k_1 k_2}{k_1 + 2k_2}$$

Inertia force $= m\ddot{x} \longrightarrow$

Spring resisting force $= k_e x \longleftarrow$

Applying D' Alembert's principle, we have

$$m\ddot{x} + k_e x = 0$$

$$\ddot{x} + \frac{k_e}{m} x = 0$$

$$w_n = \sqrt{\frac{k_e}{m}} = \sqrt{\frac{2k_1 k_2}{m(k_1 + 2k_2)}} \text{ rad/s}$$

(*e*) The piston rod of diameter 20 mm and length 700 mm in a hydraulic cylinder is subjected to a compressive force of 10 kN due to internal pressure. The piston end of the rod is guided along the cylinder and the other end of the rod is hinged at the cross-head. The modulus of elasticity for piston rod material is 200 GPa. Estimate the factor of safety taken for the piston rod design.

Sol. The piston end of the rod is guided and may be taken as fixed end. The end of the rod in the cross head is pinned. Therefore, the piston rod may be taken as a column with one end fixed and the other hinged (see Fig. 6).

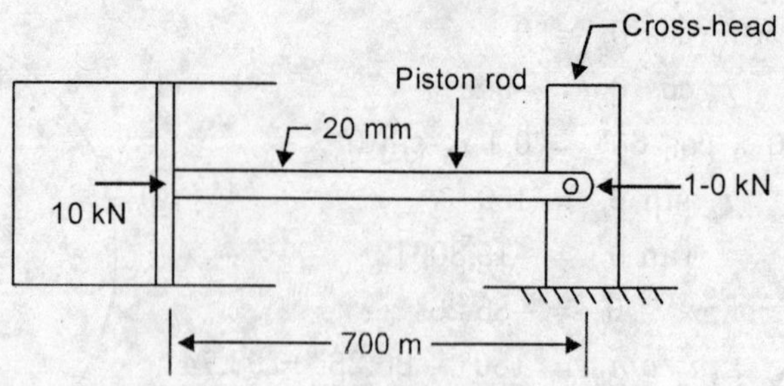

Fig. 6

$$\therefore \quad P_{cr} = \frac{2\pi^2 EI}{l^2} = 2\pi^2 \times 200 \times 10^9 \times \frac{\pi}{64} \times (0.020)^4 \times \frac{1}{(0.7)^2} = 63.278 \text{ kN}$$

$$\text{FOS} = \frac{63.278}{10} = 6.3$$

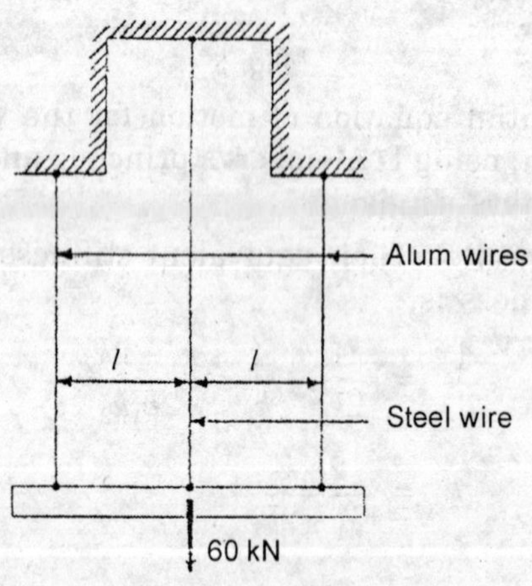

Fig. 7

(*f*) If a load of 60 kN is applied to a rigid bar suspended by 3 wires as shown in the above figure what force will be resisted by each wire?

The outside wires are of AL, corss-sectional area 300 mm^2 and length 4 m. The central wire is steel with area 200 mm^2 and length 8 m.

Initially there is no slack in the wires.

$$E = 2 \times 10^5 \text{ N/mm}^2 \text{ for Steel}$$

$$E = 0.667 \times 10^5 \text{ N/mm}^2 \text{ for Aluminium}$$

Sol. Given : P = 60 kN, A_{al} = 300 mm^2, L_{al} = 4 m, A_s = 200 mm^2, L_s = 8 m, $E_s = 2 \times 10^5$ N/mm^2, $E_{al} = 0.667 \times 10^5$ N/mm

Let P_{al}, P_s = load in each aluminium wire and steel wire respectively.

Then $$2P_{al} + P_s = P = 60 \qquad (1)$$

To the bar to remain horizontal,

$$\delta_{al} = \delta_s$$

$$\frac{P_{al}L_{al}}{A_{al}E_{al}} = \frac{P_sL_s}{A_sE_s}$$

$$\frac{P_{al} \times 4}{300 \times 6.667 \times 10^5} = \frac{P_s \times 8}{200 \times 2 \times 10^5}$$

$$P_{al} = 1.0005\ P_s \quad (2)$$

$$P_s(2 \times 1.0005 + 1) = 60$$

$$P_s = 19.993 \text{ kN}$$

$$P_{al} = 20.003 \text{ kN}$$

(g)

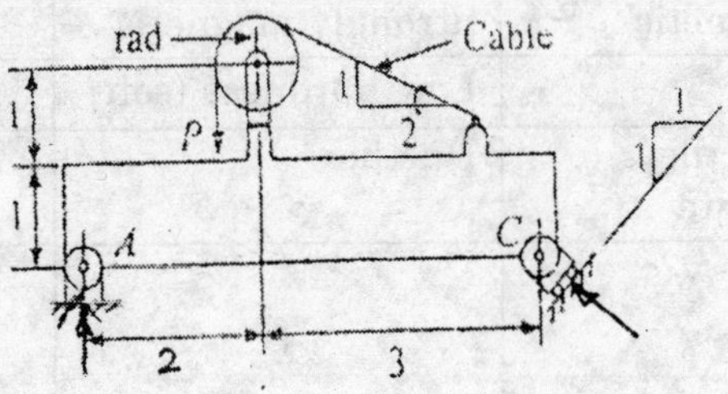

Fig. 8

For the planar structure shown in the above figure determine the reaction of all the reaction components.

All distaces shown are in metres. Radius of the pulley is 0.5 m. A load P = 10 kN is suspended from the cable.

Sol. Given : P = 10 kN, r_p = 0.5 m

Let R_A and R_C be the reactions at A and C respectively. For the equilibrium of the structure.

$$\Sigma F_V = 0 \text{ gives,}$$

$$R_A + R_C \cos 45° = P$$

or $$R_A + 0.707\ R_C = 10 \quad (1)$$

$$\Sigma M_A = 0 \text{ gives}$$

$$R_C \cos 45° \times 5 = P(2 - r_p)$$

$$3.535\ R_C = 10(2 - 0.5)$$

$$R_C = \frac{15}{3.535} = 4.24 \text{ kN} \quad (2)$$

$\therefore$ $$R_A = 10 - 0.707 \times 4.24 = 7 \text{ kN}$$

(h) Name 5 articles that are made of a thermoplastic material. Identify the name of the specific plastic used for any 3 of above applications.

Sol. 1. Toys, 2. Gears, 3. Knobs, 4. Gaskets, 5. Pipe fittings

Toys and knobs : Cellulose (Fibestos)

Gears : Polyamide (Nylon zytel)

Gaskets : TFE (Teflon)

(*i*) State clearly the similarities and differences among Austenite, Ferrite and cementite.

Sol.

Austenite : It is the solid solution of carbon in γ-iron.

Ferrite : It is the solid solution of carbon in α-iron.

Cementite : It is the chemical compound of iron and carbon (Fe_3C), C = 6.67% and has orthorhombic lattice.

Similarities and differences among austenite, ferrite and cementite

Property	Austenite	Ferrite	Cementite
1. Magnetic	Non-magnetic	Strongly magnetic	Ferro-magnetic below
2. Hardness	Soft	Low hardness (soft)	Extremly hard
3. Ductility	More ductile than ferrite	Ductile	No ductility and brittle
4. Tensile/ compressive Strength			Strong in compression
5. Density	Denser than ferrite		

(*j*) Name any 3 materials that can be classified as Ceramic. List 2 applications of each.

Sol. A ceramic may be defined as a solid formed by the combination of metallic and non-metallic elements.

Ceramic	Applications
1. Silica (SiO_2)	Abrasive
2. Tungsten carbide (WC)	Cutting tools
3. Boron nitride (BN)	Insulator

(*k*) What are effects of adding Silicon and Manganese to cast iron? Write in short about Zn-base die cast alloys.

Sol.

Alloying element to C.I.	Effects
1. Silicon	Increase softeness and machinability, It produces sound castings free from blow holes.
2. Manganese	Increases hardness of cast iron, helps to control the harmful effects of sulphur, increases strength and refines grains.

Zn-base die Cast Alloys

Zn is a corrosive resistance material and is used as a coating for the protection of iron and steel from corrosion. The method of applying zinc is called galvanizing. Zinc casts well and forms the base of various die-casting alloys.

(*l*) What is meant by interchangeable manufacture? Discuss a 'Go' gauge.

Sol. In mass production, work has to be done in a set competitive time, where great variations in dimensions of manufactured components will result. To control this certain variations in dimensions are allowed in the sizes of the machine elements or parts. This system of manufacture in which the dimensions of a part lie within some specified limits is called interchangeable manufacture. This implies that the parts which go into assembly may be selected at random from large number of produced parts.

Go gauge : It is a limit gauge which should pass through or over the part manufactured under its own weight. The part and the gauge must be at the same temperature.

(*m*) What are the advantages and disadvantages of a-c welding machine?

Sol. Advantages

1. Considerably less expensive equipment, light in weight, smaller in size and simpler in operation.
2. Lower maintenance and operating costs.
3. High efficiency.
4. Less efficiency.
4. Less electric energy consumption per kg of metal deposited as compared to D.C. welding.

Disadvantages

1. Low power factor.
2. Requires coated electrodes.
3. Unstable arc.
4. Unsuitbale for non-ferrous metals.

(*n*) Discuss the terms fineness and particle size distribution in powder metallurgy.

Sol. Fineness : The fineness of powder particles determines the size of grains and the distribution of grains of different sizes in the powder.

Particle size distribution : The particle size distribution refers to the quantity of each standard particle size in the powders.

(*o*) In GMAW process what is spray transfer and what gas mixtures are used?

Sol. Spray transfer : This is a technique in which the metal is transferred from the electrode to the weld zone in a fine spray. This method is used on the heavier butt welds or fillets.

Gas mixtures : Helium, Argon and CO_2 or their mixtures.

(*p*) Discuss a fixture, mentioning their materials, what are duplex fixtures?

Sol. Fixture : A fixture is a work holding device which only holds and positions the work, but does not in itself, locate or position the cutting tool. A fixture is bolted or damped to the machine table.

Materials for fixtures : Steel, case hardened steel, cast iron.

(*q*) Give the symbols activity names used in method study for charting of the processes.

Sol.

Symbol	Activity Name
○	Operation

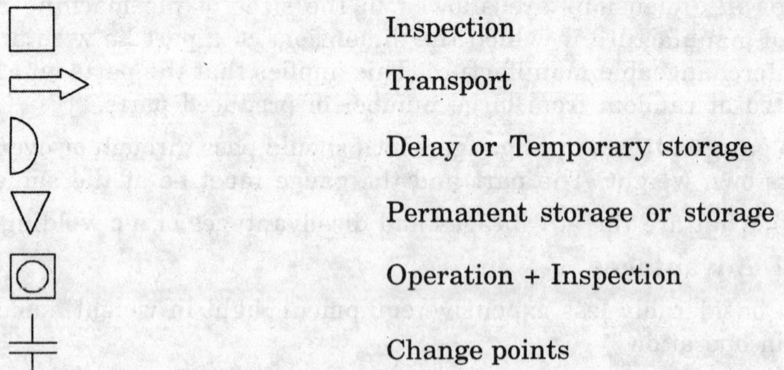

(*r*) What is float or slack and when does a subcritical path becomes critical?

Sol. Float : It is the length of time by which the duration of time of an activity can be increased without increasing the total time of the project.

Slack : It is the difference between the earliest time (TE) and latest time (TL) at which an event can take place without affecting the duration time of the project.

Critical path : A sub-critical path has positive float. When the float of all the activities along a path becomes zero it becomes a critical path.

(*s*) What is an ideal operating characteristics curve?

Sol. An ideal operating characteristics curve for a turbomachine is obtained by neglecting all the losses.

(*t*) Distinguish between material requirements planning and manufacturing resource planning.

Sol. Material requirements planning : It is the scientific way of determining the requirements of raw materials, components, spares and other items required for meeting production needs within the economic investment policies of a productive system.

Manufacturing resource planning : It is an integrated information system to synchronize all aspects of the business. It coordinates sales, purchasing, manufacturing, finance and engineering by adopting a local production plan and by using one data base to plan and update the activities of all systems of manufacturing.

Section B

Q.2. (*a*) In order to evaluate mass moment of inertia of a connecting rod, the following observations were made during experimentation:

Mass of connecting rod = 50 kg.

Distance between big end bearing and small end bearing centres = 1000 mm.

Diameter of big end bearing = 100 mm.

Diameter of small end bearing = 50 mm.

Time period of oscillation while connecting rod was suspended from big end = 1.75 second.

Time period of oscillation while connecting rod was suspended from small end = 2.0 second.

Evaluate radius of gyration of the connecting rod about an axis through the centre of mass perpendicular to the plane of oscillation as well as mass moment of inertia of the rod about the same axis.

Sol. Given : m = 50 kg, d_1 = 0.1 m, d_2 = 0.05 m, l = 1 m, T_1 = 1.75 s, T_2 = 2 s

Refer to Fig. 9.

For a simple pendulum, we have

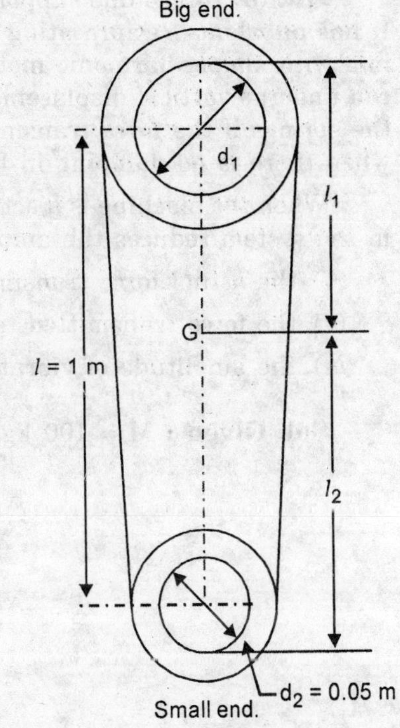

Fig. 9

$$T_1 = 2\pi\sqrt{\frac{l_{e1}}{g}}$$

$$1.75 = 2\pi\sqrt{\frac{l_{e1}}{9.81}}$$

$$l_{e1} = 0.761 \text{ m}$$

$$= \frac{K^2 + l_1^2}{l_1}$$

$$K^2 = 0.761\, l_1 - l_1^2 \quad (1)$$

Where k = radius of gyration of rod about G.

Fig

$$T_2 = 2\pi\sqrt{\frac{l_{e2}}{g}}$$

$$2 = 2\pi\sqrt{\frac{l_{e2}}{9.81}}$$

$$l_{e2} = 0.994 \text{ m}$$

$$= \frac{K^2 + l_2^2}{l_2}$$

$$K^2 = 0.994\, l_2 - l_2^2 \quad (2)$$

From eqs. (1) and (2), we get

$$0.761\, l_1 - l_1^2 = 0.994\, l_2 - l_2^2$$

Now

$$l_1 + l_2 = l + \frac{1}{2}(d_1 + d_2) = 1 + \frac{1}{2}(0.1 + 0.05) = 1.075 \text{ m}$$

$$l_2 = 1.075 - l_1$$

$$0.761\, l_1 - l_1^2 = 0.994\,(1.075 - l_1) - (1.075 - l_1^2)$$

$$= 1.06855 - 0.994\, l_1 - 1.155625 - l_1^2 + 2.15\, l_1$$

$$0.761\, l_1 = -0.087075 + 1.156\, l_1$$

$$0.395\, l_1 = 0.087075$$

$$l_1 = 0.220$$

$$l_2 = 1.075 - 0.220 = 0.855 \text{ m}$$
$$K^2 = 0.761 \times 0.220 - (0.220)^2 = 0.11902$$
$$K = 0.345 \text{ m}$$
$$I = mK^2 = 50 \times 0.11902 = 5.95 \text{ kg·m}^2$$

Q.2. (*b*) A machine supported symmetrically on four springs has a total mass of 100 kg. It has unbalance reciprocating parts of 2.0 kg which moves through a vertical stroke of 100 mm with simple harmonic motion. The machine is having only one degree of freedom and can undergo vertical displacement. Evaluate its natural frequency and combined stiffness of the springs if the force transmitted to the foundation is one-twentieth of the applied force, when there is no damping in the system. The machine is rotating at 800 rpm.

When the machine is a actually tested for vibration, it is found that the damping present in the system reduces the amplitude of successive free vibrations by 30%. Evaluate:

(*i*) the actual force transmitted to foundation at the running speed 800 rpm.

(*ii*) the force transmitted to foundation at reasonance, and

(*iii*) the amplitude of vibration of the machine due to unbalance mass, at resonance.

Sol. Given : M = 100 kg, m = 2 kg, L = 100 mm, $F_{tr} = \frac{1}{20}$ F, N = 800 rpm

$$\frac{X_{n+1}}{X_n} = 0.7$$

$$r = \frac{L}{2} = \frac{100}{2} = 50 \text{ mm}$$

$$\omega = \frac{2\pi N}{60} = \frac{2\pi \times 800}{60} = 83.776 \text{ rad/s}$$

$$r = \frac{1}{\beta^2 - 1}, \text{ when } \zeta = 0$$

$$0.05 = \frac{1}{\beta^2 - 1}$$

$$\beta^2 - 1 = 20$$

$$\beta = \sqrt{21} = 4.5826$$

$$\frac{w}{w_n} = 4.5826$$

$$w_n = \frac{83.776}{4.5826} = 18.28 \text{ rad/s}$$

$$w_n = \sqrt{\frac{k}{M}}$$

$$18.28 = \sqrt{\frac{k}{100}}$$

$$k = 33{,}416 \text{ N/m}$$

Lagarithmic decrement, $\delta = \ln\left(\frac{x_n}{x_{n+1}}\right) = \ln\left(\frac{1}{0.7}\right) = 0.35667$

$$= \frac{2\pi\zeta}{\sqrt{1-\zeta^2}}$$

$$4\pi^2\zeta^2 = 0.1272\ (1 - \zeta^2)$$

$$\zeta^2(4\pi^2 + 0.1272) = 0.1272$$

$$\zeta = 0.05667$$

(*i*)
$$\frac{F_{tr}}{F} = \frac{\sqrt{1+(2\zeta\beta)^2}}{\sqrt{(1-\beta^2)^2+(2\zeta\beta)^2}}$$

$$\frac{F_{tr}}{F} = \frac{\sqrt{1+(2\times 0.05667\times 4.5826)^2}}{\sqrt{\{1-(4.5826)^2\}^2+(2\times 0.05667\times 4.5826)^2}}$$

$$= \frac{1.12684}{20.007} = 0.0563$$

$$F = mrw^2 = 2 \times 0.050 \times (83.776)^2 = 701.84 \text{ N}$$

$$F_{tr} = 0.0563 \times 701.84 = 39.51 \text{ N}$$

(*ii*) At resonance, $\beta = 1$.

$$\frac{F_{tr}}{F} = \frac{\sqrt{1+4\zeta^2}}{2\zeta} = \frac{\sqrt{1+4(0.05667)^2}}{2\times 0.05667} = 8.8795$$

$$F = mrw_n^2 = 2\times 0.05(18.28)^2 = 33.41 \text{ N}$$

$$F_{tr} = 33.41 \times 8.8795 = 296.7 \text{ N}$$

(*iii*) Amplitude of vibration at resonance

$$= \frac{F_{tr} \text{ at } \beta = 1}{k}$$

$$= \frac{296.7\times 10^3}{33,416} = 8.87 \text{ mm}$$

(c) A semicircular steel ring of mean radius 300 mm is suspended vertically with the top end fixed as shown in the Fig. 10 and carries a vertical load of 200 N at the lowest point. Calculate the vertical deflection of the lower end if the ring is of retangular cross-section 20 mm thick and 30 mm wide.

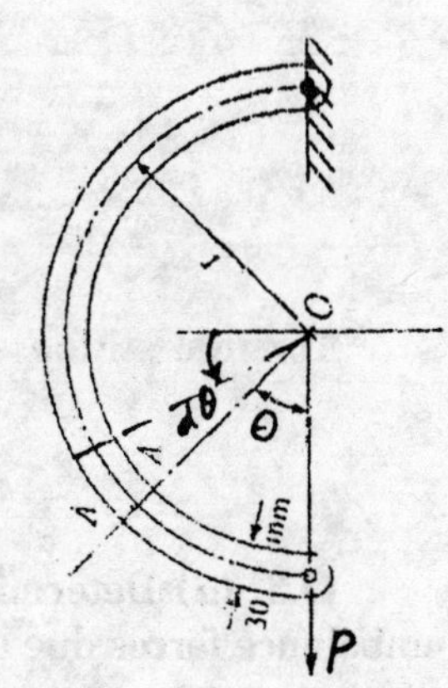

Fig. 10

Value of Elastic modulus is 2×10^5 N/mm^2.

Influence of circumferential and shearing forces may be neglected.

Sol. Given : $r = 0.3$ m, P = 200 N, $b = 30$ mm, $t = 20$ mm, E = 2×10^5 N/mm^2

$$I = \frac{bt^3}{12} = \frac{30\times 20^3}{12} = 2\times 10^4 \text{ mm}^4$$

Bending moment at section A – A , M = Pr sin θ

Stain energy, $$U = \int_0^{\pi} \frac{M^2 ds}{2EI} = \int_0^{\pi} \frac{P^2 r^2 \sin^2\theta \cdot r d\theta}{2EI}$$

$$= \frac{P^2 r^3}{2EI}\int_0^{\pi} \sin^2\theta \, d\theta = \frac{P^2 r^3}{2EI}\int_0^{\pi}\left(\frac{1-\cos 2\theta}{2}\right) d\theta$$

$$= \frac{P^2 r^3}{4EI}\left|\theta - \frac{\sin 2\theta}{2}\right|_0^{\pi} = \frac{P^2 r^3}{4EI}[\pi] = \frac{\pi}{4}\frac{P^2 r^3}{EI}$$

Vertical deflection of lower end of ring,

$$\delta_V = \frac{\partial U}{\partial P} = \frac{\pi P r^3}{2EI} = \frac{\pi \times 200 \times (0.3)^3}{2 \times 200 \times 10^9 \times 2 \times 10^{-8}}$$

$$= 2.12 \times 10^{-3} \text{ m} \quad \text{or} \quad 2.12 \text{ mm}$$

Q.2. (*d*) A hollow steel rod 200 mm long is to be used as torisonal spring. The ratio of inside to outside diameters is 1 : 2. The required stiffness of this spring is 100 N.m/degree.

Determine the outside diameter of rod.

Value of G is 8 × 10⁴ N/mm².

Sol. Given : l = 200 mm, $k = \frac{d_i}{d_0}$ = 0.5, k_t = 100 N.m/degree, G = 8 × 10^4 N/mm²

Polar moment of inertia of rod,

$$J = \frac{\pi}{32}(d_0^4 - d_i^4) = \frac{\pi d_0^4}{32}[1 - k^4]$$

$$= \frac{\pi}{32} d_0^4 \left[1 - \frac{1}{16}\right] = 0.09204\, d_0^4$$

$$\frac{T}{J} = \frac{G\theta}{l}$$

Torsinal stiffness of rod, $$k_t = \frac{T}{\theta} = \frac{GJ}{l} = \frac{8 \times 10^4 \times 0.09204\, d_0^4}{200} = 100 \times 10^3$$

$$d_0 = 7.2 \text{ mm}$$

$$d_i = 3.6 \text{ mm}$$

Q.3. (*a*) Determine the maximum and minimum values of the primary and secondary unbalance forces due to reciprocating masses for a 90° V-twin engine, whose details are given below:

Reciprocating mass per cylinder = 2.0 kg

Stroke for each cylinder = 12 cm

Connecting rod length of each cylinder = 20 cm

Crank shaft speed = 2000 rpm.

Indicate the crank positions corresponding to above maximum and minimum values of unbalance forces. Whether primary and secondary unbalance forces can be balanced completely or not?

Sol. Given : $R = 2$ kg, $L = 12$ cm,

$l = 20$ cm, $N = 2000$ rpm, $2\alpha = 90°$

$$\omega = \frac{2\pi N}{60} = \frac{2\pi \times 2000}{60} = 209.44 \text{ rod/s}$$

Crank radius, $$r = \frac{L}{2} = \frac{12}{2} = 6 \text{ cm}$$

$$n = \frac{l}{r} = \frac{20}{6} = \frac{10}{3}$$

Resultant primary force, $F_p = 2Rw^2r[(\cos^2\alpha \cos\theta)^2 + (\sin^2\alpha \sin\theta)^2]^{1/2}$

For $\alpha = 45°$, $F_p = Rw^2r = 2 \times (209.44)^2 \times 0.06 = 5263.8$ N

Resultant secondary force,

$$F_s = 2Rw^2\left(\frac{r}{n}\right)\left[(\cos\alpha \cos 2\alpha \cos 2\theta)^2 + (\sin\alpha \sin 2\alpha \sin 2\theta)^2\right]^{1/2}$$

For F_s to be maximum, $\sin 2\theta = \pm 1$, or $\theta = 45°$ or $135°$

$$(F_s)_{max} = \sqrt{2}\, Rw^2\left(\frac{r}{n}\right) = \sqrt{2} \times 2 \times (209.44)^2 \times \frac{0.06}{10} \times 3 = 2233.2 \text{ N}$$

$$(F_s)_{min} = 0 \text{ at } \theta = 0°$$

Q.3. (*b*) Two railway coaches are coupled with the help of two rods of a turn buckle with right and left handed threads having single-start square threads. Pitch and mean diameter of the threads are 8 mm and 30 mm respectively. How much energy will be spent in bringing two coaches closer through a distance of 320 mm against a steady load of 5.0 kN. Use coefficient of friction in screw thread and coupler nut as 0.12. Determine shear stress on the tie rod due to above torque.

Sol. Given : $p = 8$ mm, $d_m = 30$ mm, $s = 320$ mm, $W = 5$ kN, $\mu = 0.12$

Helix angle, $$\alpha = \tan^{-1}\left(\frac{p}{\pi d_m}\right) = \tan^{-1}\left(\frac{8}{\pi \times 30}\right) = 4.85°$$

Friction angle, $\phi = \tan^{-1}\mu = \tan^{-1} 0.12 = 6.84°$

Force required on the screw of turn buckle.,

$$P = W \tan(\alpha + \phi)$$
$$= 5000 \tan(4.85° + 6.84°) = 1034.54 \text{ N}$$

Torque on each rod, $$T = \frac{Pd_m}{2} = 1034.54 \times \frac{0.030}{2} = 15.518 \text{ N}\cdot\text{m}$$

Number of turns required to bring the coaches nearer by 320 mm,

$$N = \frac{s}{2p} = \frac{320}{2 \times 8} = 20$$

Energy spent $= 2\pi NT = 2\pi \times 20 \times 15.518 = 1950$ N·m

Shear stress in tie rod, $\tau = \dfrac{167}{\pi d_m^3} = \dfrac{16 \times 15.518 \times 10^3}{\pi \times 30^3} = 2.927 \text{ N/mm}^2$

Q.3. (*c*) The data obtained from a rectangular strain gauge rosette attached to a stressed steel member are

$$\in_{0°} = -220 \times 10^{-6}$$

$$\in_{45°} = 120 \times 10^{-6}$$

and $\in_{90°} = 220 \times 10^{-6}$

Given that value of $E = 2 \times 10^5 \text{ N/mm}^2$

and Poisson's Ratio, $\nu = 0.3$

Calculate the values of principal stresses acting at the point and their directions.

Sol. Given : $\in_{0°} = -220 \times 10^{-6}$, $\in_{45°} = 120 \times 10^{-6}$, $\in_{90°} = 220 \times 10^{-6}$, $E = 2 \times 10^5 \text{ N/mm}^2$, $\nu = 0.3$

$$\in_\theta = \frac{1}{2}(\in_x + \in_y) + \frac{1}{2}(\in_x - \in_y)\cos\theta + \frac{1}{2}\gamma_{xy}\sin 2\theta$$

$$\in_{0°} = \frac{1}{2}(\in_x + \in_y) + \frac{1}{2}(\in_x - \in_y) = \in_x$$

$\therefore$ $$\in_x = -220 \times 10^{-6}$$

$$\in_{45°} = \frac{1}{2}(\in_x + \in_y) + \frac{1}{2}\gamma_{xy}$$

$$120 \times 10^{-6} = \frac{1}{2}(-220 \times 10^{-6} + \in_y) + \frac{1}{2}\gamma_{xy}$$

$$240 \times 10^{-6} = -220 \times 10^{-6} + \in_y + \gamma_{xy}$$

$$\in_y + \gamma_{xy} = 460 \times 10^{-6} \quad (1)$$

$$\in_{90°} = \frac{1}{2}(\in_x + \in_y) - \frac{1}{2}(\in_x - \in_y)$$

$$220 \times 10^{-6} = \in_y$$

or $$\in_y = 220 \times 10^{-6}$$

From Eq. (1), we get

$$\gamma_{xy} = (460 - 220) \times 10^{-6} = 240 \times 10^{-6}$$

$$\in_{1,2} = \frac{1}{2}(\in_x + \in_y) \pm \frac{1}{2}\sqrt{(\in_x - \in_y)^2 + \gamma_{xy}^2}$$

$$= \left[\frac{1}{2}(-220 + 220) \pm \frac{1}{2}\sqrt{(-220 - 220)^2 + 240^2}\right] \times 10^{-6}$$

$$= \pm\frac{1}{2}(501.2) \times 10^{-6} = \pm 250.6 \times 10^{-6}$$

Principal stresses are :

$$\sigma_1 = \frac{E}{(1-\upsilon^2)}[\in_1 + \nu \in_2] = \frac{2\times10^5}{[1-(0.3)^2]}[250.6 - 0.3\times250.6]\times10^{-6}$$

$$= 38.55 \text{ N/mm}^2$$

$$\sigma_2 = \frac{E}{(1-\nu^2)}[\in_2 + \nu \in_1] = -38.55 \text{ N/mm}^2$$

Principal directions are given by :

$$\tan 2\theta = \frac{\gamma_{xy}}{\in_x - \in_y} = \frac{240\times10^{-6}}{-220\times10^{-6} - 220\times10^{-6}} = \frac{240}{-440} = 0.54545$$

$$2\theta = -28.61°$$

$$\theta_1 = -14.305°,\ \theta_2 = \theta_1 + 90° = 75.695°$$

Q.3. ***(d)*** A spherical shell of 150 mm internal diameter has to withstand an internal pressure of 30 M N/m^2. Calculate the thickness of the shell if the allowable stress is 80 M N/m^2.

Assume the stress distribution in the shell to follow the law :

$$\sigma_r = a - \frac{2b}{r^3} \text{ and } \sigma_\theta = a + \frac{b}{r^3}.$$

Sol. Given : r_i = 75 mm, $r_0 = r_i + t = (75 + t)$ mm, p_i = 30 N/mm^2, $(\sigma_\theta)_{\text{allow}}$ = 80 N/mm^2

$$\sigma_r = a - \frac{2b}{r^3},\quad \sigma_\theta = a + \frac{b}{r^3}$$

At $r = r_i$, $\sigma_r = -p_i = -30$ N/mm^2

$r = r_0$, $\sigma_r = 0$

$$-30 = a - \frac{2b}{75^3} \quad (1)$$

$$0 = a - \frac{2^b}{r_0^3} \quad (2)$$

Subtracting Eq. (2) from (1), we get

$$-3b = \frac{2b}{r_0^3} - \frac{2b}{75^3} = 2b\left(\frac{1}{r_0^3} - \frac{1}{75^3}\right) = 2b\left(\frac{75^3 - r_0^3}{75^3 r_0^3}\right)$$

$$2b = \frac{-30\times75^3 r_0^3}{(75^3 - r_0^3)},\ b = \frac{-15\times75^3\times r_0^3}{(75^3 - r_0^3)}$$

$$a = \frac{2b}{r_0^3} = \frac{-30\times75^3}{(75^3 - r_0^3)}$$

At $r = r_i$

$$\sigma_\theta = \frac{-30\times75^3}{(75^3 - r_0^3)} - \frac{(5\times75^3\times r_0^3)}{r_i^3(75^3 - r_0^3)}$$

$$= \frac{-30\times 75^3}{(75^3 - r_0^3)}\left[1+\frac{1}{2}\frac{r_0^3}{r_i^3}\right] = \sigma_{allow}$$

$$= 80$$

$$\frac{-30}{\left[1-\left(\frac{r_0}{75}\right)^3\right]}\left[1+\frac{1}{2}\left(\frac{r_0}{75}\right)^3\right] = 80$$

$$-30-15\left(\frac{r_0}{75}\right)^3 = 80-80\left(\frac{r_0}{75}\right)^3$$

$$65\left(\frac{r_0}{75}\right)^3 = 110$$

$$\frac{r_0}{75} = 1.1917$$

$$r_0 = 89.4 \text{ mm}$$

Thickness of shell, $t = r_0 - r_i = 89.4 - 75 = 14.9$ mm

Q.4. (a)

A stiff of negligible weight transfers a load P to a combination of three helical springs arranged in parallel as shown in the Fig. 11. The springs are made up of the same material and out of rods of equal diameters. They are of same free length before loading. The number of coils in those three springs are 10, 12 and 15 respectively, while the mean coil diameters are in ratio of 1 : 1.2 : 1.4 respectively. Find the distance 'x' as shown in figure, such that the stiff bar remains horizontal after the application of laod P.

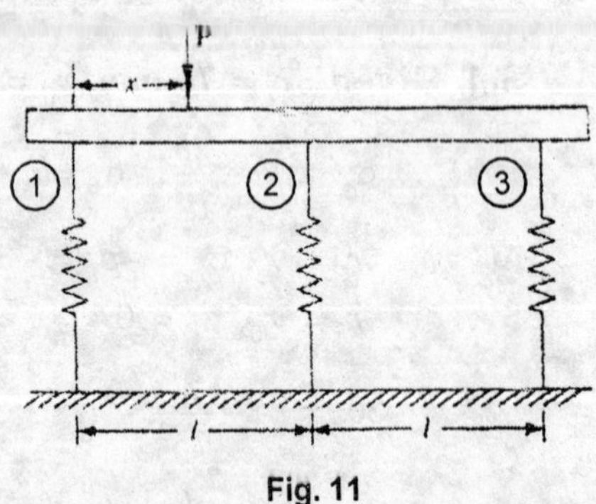

Fig. 11

Sol. Given : $\tau_1 = \tau_2 = \tau_3 = \tau$; $d_1 = d_2 = d_3 = d$; $l_{f1} = l_{f2} = l_{f3} = l_f$, $D_{m1} : D_{m2} : D_{m3}$ :: 1 : 1.2 : 1.4; $n_1 = 10$, $n_2 = 12$, $n_3 = 15$, $G_1 = G_2 = G_3 = G$, $\delta_1 = \delta_2 = \delta_3$ for the stiff bar to remain horizontal.

Let P_1, P_2, P_3 = load shared by springs 1, 2, 3 respectively

$$P_1 + P_2 + P_3 = P$$

$$\delta_1 = \frac{8P_1D_{m1}^3n_1}{Gd^4}, \delta_2 = \frac{8P_2D_{m2}^3n_2}{Gd^4}, \delta_3 = \frac{8P_3D_{m3}^3n_3}{Gd^4}$$

$$\therefore \quad P_1D_{m1}^3n_1 = P_2D_{m2}^3n_2 = P_3D_{m3}^3n_3$$

or

$$P_1D_{m1}^3\times 10 = P_2 \times (1.2\, D_{m1})^3 \times 12 = P_3 \times (1.4\, D_{m1})^3 \times 15$$

$$10P_1 = 20.736\, P_2 = 41.16\, P_3$$

$$P_2 = 0.48225\, P_1,\; P_3 = 0.24295\, P_1$$

$$\therefore \quad P_1(1 + 0.48225 + 0.24295) = P$$

$$P_1 = 0.57964\ P$$
$$P_2 = 0.27953\ P$$
$$P_3 = 0.14082\ P$$

Taking moments about the point of application of P, we have

$$P_1 \times x = P_2(l - x) + P_3(2l - x)$$
$$0.57964\ x = 0.27953\ (l - x) + 0.14082\ (2l - x)$$
$$= 0.27953\ l + 0.28164\ l - 0.42035\ x$$
$$x = \frac{0.56117\ l}{0.99999} = 0.56117\ l$$

Q.4. (*b*) A pair of standard spur gears has 16 and 18 teeth, module 12.5 mm and pressure angle 14.5°. Examine whether the pair will have interference. If so what should be the number of teeth in both the gears to avoid interference as well as the pair maintains the same speed ratio without modifying other parameters. Evaluate contact ratio for the new set.

Sol. Given : $z_1 = 16$, $z_2 = 18$, $m = 12.5$ mm, $\phi = 14.5°$

Minimum number of teeth on pinion to avoid interference,

$$(z_1)_{\min} = \frac{2a_p}{[\{1+i^2 \sin^2 \phi + 2i \sin^2 \phi\}^{1/2} - 1]}$$
$$i = z_2/z_1 = 18/16 = 9/8 = 7.125,\ a_p = h_{a1}/m = 1$$
$$(z_1)_{\min} = \frac{2\times 11}{[\{1+(1.125)^2 \sin^2 14.5° + 2\times 1.125 \times \sin^2 14.5°\}^{1/2} - 1]}$$
$$z_1' = (z_1)_{\min} = 19$$

Since $z_1 < (z_1)_{\min}$, therefore interference will occur.

$z_2' = 1.125 \times 19 = 21.375$, which gives fractional number of teeth.

We take $z_1' = 24$ so that $z_2' = 1.125 \times 24 = 27$

Contact ratio

$$r_1 = \frac{1}{2} m z_1' = \frac{1}{2} \times 12.5 \times 24 = 150 \text{ mm}$$
$$r_2 = \frac{1}{2} m z_2' = \frac{1}{2} \times 12.5 \times 27 = 168.75 \text{ mm}$$
$$h_{a1} = h_{a2} = m = 12.5 \text{ mm}$$
$$r_{a1} = r_1 + h_{a1} = 150 + 12.5 = 162.5 \text{ mm}$$
$$r_{a2} = r_2 + h_{a2} = 168.75 + 12.5 = 181.25 \text{ mm}$$
$$r_{b1} = r_1 \cos \phi = 150 \cos 14.5° = 145.2 \text{ mm}$$
$$r_{b2} = r_2 \cos \phi = 168.75 \cos 14.5° = 163.4 \text{ mm}$$

Length of path of recess,

$$L_r = \sqrt{r_{a1}^2 - r_{b1}^2} - r_1 \sin \phi$$

$$= \sqrt{(162.5)^2 - (145.2)^2} - 150 \sin 14.5°$$
$$= 35.4 \text{ mm}$$

Length of path of approach,

$$L_a = \sqrt{r_{a2}^2 - r_{b2}^2} - r_2 \sin \phi$$
$$= \sqrt{(181.25)^2 - (163.4)^2} - 168.75 \sin 14.5°$$
$$= 36.2 \text{ mm}$$

Length of path of contact,

$$L_p = L_r + L_a = 35.4 + 36.2 = 71.6 \text{ mm}$$

$$\text{Contact ratio} = \frac{L_p}{\pi m \cos\phi} = \frac{71.6}{\pi \times 12.5 \cos 14.5°} = 1.88 \simeq 2$$

Q.4. (*c*) A flat belt drive is required to transmit 10 kW from a motor running at 1000 rpm. The belt is 15 mm thick and has a mass density of 0.001 gm/mm³. Permissible tensile stress for the belt material is 2.5 N/mm². Diameter of the driving pulley is 250 mm, whereas the speed of the driven pulley is 367 rpm. Driving and driven shafts are 1.25 m apart. The coefficient of friction between belt and pulley may be taken as 0.25. Determine the width of the belt for safe working.

Sol. Given : $P = 10$ kW, $N_1 = 1000$ rpm, $t = 15$ mm, $\rho = 0.001$ gm/mm³,

$\sigma_t = 2.5$ N/mm², $D_1 = 250$ mm, $N_2 = 367$ rpm, $C = 1.25$m,

$\mu = 0.25$

$$D_2 = \frac{D_1 N_1}{N_2} = \frac{250 \times 1000}{367} = 681.2 \text{ mm}$$

Angle of arc of contact,

$$\theta = \pi - 2\sin^{-1}\left(\frac{D_2 - D_1}{2C}\right)$$
$$= \pi - 2\sin^{-1}\left(\frac{681.2 - 250}{2 \times 1250}\right)$$
$$= 160.13°$$

$$\frac{T_1}{T_2} = \exp(\mu\theta) = \exp\left(0.25 \times \frac{\pi}{180} \times 160.13\right) = 2.01$$

$$v = \frac{\pi D_1 N_1}{60 \times 10^3} = \frac{\pi \times 250 \times 1000}{60 \times 10^3} = 13.09 \text{ m/s}$$

$$P = \frac{(T_1 - T_2)v}{10^3}$$

$$T_1 - T_2 = \frac{10 \times 10^3}{13.09} = 763.94 \text{ N}$$

$$T_2(2.01 - 1) = 763.94$$

$$T_2 = 756.4 \text{ N}$$

$$T_1 = 2.01 \times 756.4 = 1520.3 \text{ N}$$

Mass of belt per metre length,

$$m = b \times t \times 10^{-6} \times 1 \times \rho$$

$$\rho = 1000 \text{ kg/m}^3,\ b = \text{width of belt in mm}$$

$$m = b \times 15 \times 10^{-6} \times 1000 = 0.015 \times b \text{ kg}$$

Centrifugal tension in belt,

$$T_c = mv^2 = 0.015 \times bx(13.09)^2$$

$$= 2.57 \times b \text{ N}$$

Maximum tension in belt,

$$T_{max} = \sigma_t \times b \times t = 2.5 \times b \times 15 = 37.5\ b \text{ N}$$

$$T_1 + T_C = T_{max}$$

$$1520.3 + 2.57 \times b = 37.5 \times b$$

$$b = 43.5 \simeq 44 \text{ mm}$$

Q.4. (*d*) The stress state at a point in a body is plane with $\sigma_1 = 60$ N/mm^2 and $\sigma_2 = -36$ N/mm^2. If the allowable stress for the material in simple tension or compression is 1000 N/mm^2 calculate the values of factor of safety with each of the following criteria for failure

(*i*) Max Stress Criteria

(*ii*) Max Shear Stress Criteria

(*iii*) Max Strain Criteria

(*iv*) Max distortion energy criteria

Sol. Given : $\sigma_1 = 60$ N/mm^2, $\sigma_2 = -36$ N/mm^2, $\sigma_0 = 100$ N/mm^2

(*i*)

$$\text{FOS} = \frac{\sigma_0}{\sigma_1} = \frac{100}{60} = 1.67$$

(*ii*)

$$\tau_{max} = \frac{\sigma_1 - \sigma_2}{2} = \frac{60+36}{2} = 48 \text{ N/mm}^2$$

$$\tau_0 = \frac{\sigma_0}{2} = \frac{100}{2} = 50 \text{ N/mm}^2$$

$$\text{FOS} = \frac{\tau_0}{\tau_{max}} = \frac{50}{48} = 1.04$$

(*iii*)

$$\varepsilon_{max} = \frac{1}{\varepsilon}(\sigma_1 - \nu\sigma_2)$$

Let

$$\nu = 0.3$$

$$\varepsilon_{max} = \frac{1}{E}[60 - 0.3 \times (-36)] = \frac{70.8}{E}$$

$$\varepsilon_0 = \frac{\sigma_0}{E} = \frac{100}{E}$$

$$FOS = \frac{\varepsilon_0}{\varepsilon_{max}} = \frac{100}{70.8} = 1.412$$

(*iv*)
$$\Sigma(\sigma_1 - \sigma_2)^2 = 2\left(\frac{\sigma_0}{FOS}\right)^2$$

$$(60 + 36)^2 + (-36)^2 + (-60)^2 = 2\left(\frac{100}{FOS}\right)^2$$

$$7056 = \left(\frac{100}{FOS}\right)^2$$

$$\frac{100}{FOS} = 84$$

$$F_{OS} = \frac{100}{84} = 1.19$$

SECTION C

Q.5. (*a*) (*i*) How do zirconium and thorium affect the tungsten electrode in the GTAW process? What is "weld decay" in Ni-Cr steels?

(*ii*) Is it possible to weld tantalum to steel, if yes, by which method? Explain the term hot cracks in welding and write four important causes.

(*iii*) Explain why the strength to weight ratio of die-cast parts increases with decreasing wall thickness. Explain the term stack molding.

(*iv*) Draw the typical configuration of Internal Centreless grinding mentioning main advantage and use. Draw tool life curves for cast alloy. High speed steel and ceramic tools.

Sol.

(*i*) In the tungsten electrode for GTAW, 1 to 2% thorium and zirconium are added to improve electron emission, arc stability, arc striking and current carrying capacity, and to increase the melting point of tungsten.

Weld decay : It is a defect caused in Ni-Cr steels on account of carbide precipitation during heat treatment.

(*ii*) Tantalum to steel can be welded by GMAW (Gas-Metal-Arc-Welding)

Heat cracks in welding. These cracks may occur in the weld during welding. The causes for these cracks are : structural stresses in the metal due to formation of martensite, heavy shrinkage, extra high amount of sulphur, phosphorus or carbon in the metal, excessively rigid clamping of the parts being welded or the presence of gases in the weld metal.

(*iii*) **Effect of wall thickness of die-cast parts on strength to weight ratio :** The strength-to-weight ratio of die-cast parts increases with decreasing wall thickness as the manufactured parts are more compact, the grains of metal get aligned according to the contour of the part. Decreasing wall thickness gives lesser weight and higher strength.

Stack moulding : It is a moulding process in which 10 to 12 flask sections are arranged one above the other, having a common sphere through which all the moulds are poured.

(*iv*) **Internal centreless grinding :** Centreless grinding is performed on workpieces which do not or cannot have centres. It can be done both for external and internal surfaces.

Fig. 12 shows the configuration for internal centreless grinding. The workpiece 1 is loaded into the grinding zone where it is held between three steel rolls-support 2, pressure 3 and regulating 4 -mounted in housing 5. The lever mounted pressure roll 3 exerts a preset force on the workpiece holding it against the support and regulating rolls. The regulating roll, linked to the workdrive, rotates the work piece at a peripheral speed of from about 0.7 to 1.0 m/s. An axial force acting on the workpiece, due to the fact that the regulating wheel axis is inclined to about 1/2°, holds the workpiece axially with one end face against a roller stop. The hole is ground with a wheel mounted on spindle 6. The rotation of the work on grinding is the result of the friction force between the work and the regulating roll. The workpiece is loaded and unloaded by moving the pressure roll outward.

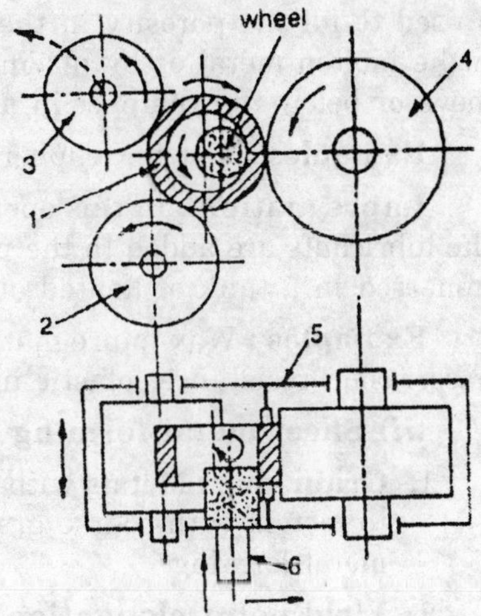

Fig. 12

Tool life curves : The tool life curves are shown in Fig.

$$VT^n = C$$

$$n = 0.1 - 0.2 \text{ for cost alloys}$$

$$= 0.2 - 0.4 \text{ for carbide}$$

$$= 0.4 - 0.6 \text{ for ceramics}$$

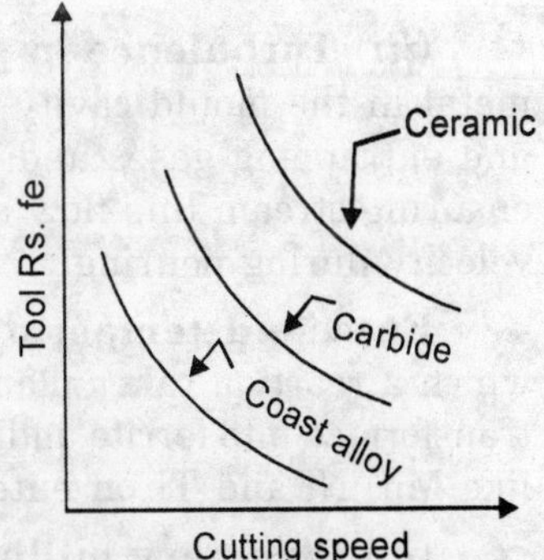

Fig. 13 Tool life curves.

Advantages of centreless grinding

1. High productivity
2. Less skilled operator needed.
3. The work is deflection free without tool vibrations.
4. Eassy control on size of the work.
5. Less grinding stock is required due to true floating condition prevailing during grinding.

Q.5. (*b*) (*i*) Explain the difference between infiltration and impregnation in powder metallurgy. Give at least two examples of each.

(*ii*) Why are the following characteristics of metal important in sheet metal forming:

1. Grain size and
2. Yield point elongation

Why are pure metals more easily cold worked than alloys?

(*iii*) In the design of the gating system what techniques are used for minimizing turbulence? Show with a figure the effect of alloying elements like Mn, Ni and Ti on the eutectoid temperature.

(*iv*) Explain with figures planetary milling. Indicate similarities between a vertical boring mill and a jig-boring machine.

Sol. (*i*) **Powder metallurgy : Infiltration :** It is the operation in which molten metal is used to fill the porosity in the sintered compact. It is carried out by immersing the part in the molten metal or by placing the infiltrant metal in the form of a sheet or slug either above or below the compact in a furnace. Capillary action fills the pores.

Examples : Copper, copper alloys and brass; silver infiltrating tungsten.

Impregnation : In this operation, the pores in a sintered compact are filled with an oil. The lubricants are added to the porous bearings, gears, and pump motors, etc. The parts are immersed in a tank of heated oil for a period of time to fill the pores.

Examples : Wax impregnation for lubrication; resin impregnation for strength, bonding or corrosion resistance; plastic impregnation to improve the outer surface for plating.

(*ii*) **Sheet metal forming :**

1. **Grain size :** Metals with fine grains have higher hardness and strength as compared to metals with coarse grains. Therefore, grain size plays on important role in sheet metal forming.
2. **Yield point elongation :** Higher yield point elongation ensures higher ductility of metal, which is more desirable in sheet metal forming. Pure metals are more easily cold worked than alloys due to their higher ductility and absence of dislocations.

(*iii*) **Turbulence in gating system :** A good gating design ensures distribution of the metal in the mould cavity at a proper rate without excessive temperature loss, turbulence, and entrapping gases and slags. The turbulence in the gating system can be minimized by ensuring stream line flow of liquid metal by avoiding sharp corners and optimizing the flow velocity during pouring.

Eutectoid temperature : It is the temperature (723°C) at which a reaction takes place in which austenite decomposes and is transformed into ferrite and cementite. The effect of alloying elements like Mn, Ni and Ti on eutectoid temperature is shown in Fig. 14.

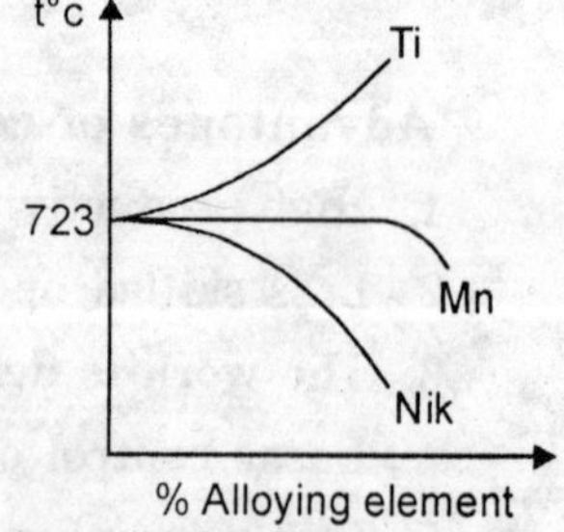

(*iv*) **Planetary milling :** In this type of milling, the work is held stationary, while revolving cutter moves in a planetary path to finish cylindrical surface on work, either internally, externally or simultaneously.

Similarities between a vertical boring mill and a jig-boring machine :

1. Both have heavy cast iron bed, which carries a horizontal circular table over it.
2. Both are vertical type boring machines.
3. The table can be moved precisely in two mutually perpendicular directions in a plane normal to the cylinder.

Q.6. (*a*) (*i*) Discuss short-circuiting metal transfer in GMAW mentioning its suitability. Also define the term "transition current", with figure.

(*ii*) Why is a unilateral tolerance preferred over bilateral tolerance? In surface roughness, discuss with a figure root-Mean-Square average method.

(*iii*) In metal casting define the terms chaplet and resin binder. Write the merits and demerits of shell moulding process.

(*iv*) What are a manipulator, wrist and end effector for a robot? Why has the wire EDM process become so widely accepted in industry?

Sol. (*i*) **Short-circuiting metal transfer in GMAW :** The short-circuiting metal transfer takes place with low current settings. Its main advantage is that it has low penetration, which can be effectively used for welding of thin sheets. It can also be used to weld difficult-to-reach positions and poorly fit joints.

Transition current : It is the value of the current at which the metal transfer changes from one type to another. For example, the current at which the metal transfer changes from globular to spray type is termed as the globular to spray transition current.

The transition current is shown in Fig. 15.

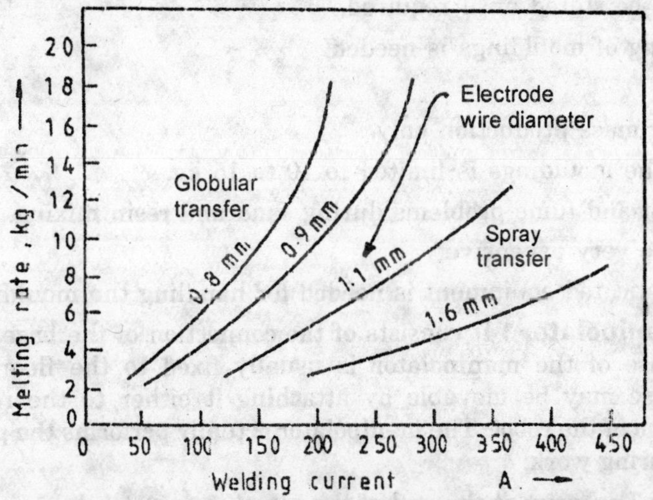

Fig. 15 Effect of welding current on melting rate for different sized carbon steel electrodes.

(*ii*) **Preference of unilateral tolerence over bilateral tolerence : In the unilateral** tolerence system, the variation in dimension from basic dimension is given in one direction only. The tolerence can be revised without affecting the allowance or clearance between mating parts *i.e.*, without altering the type of fit.

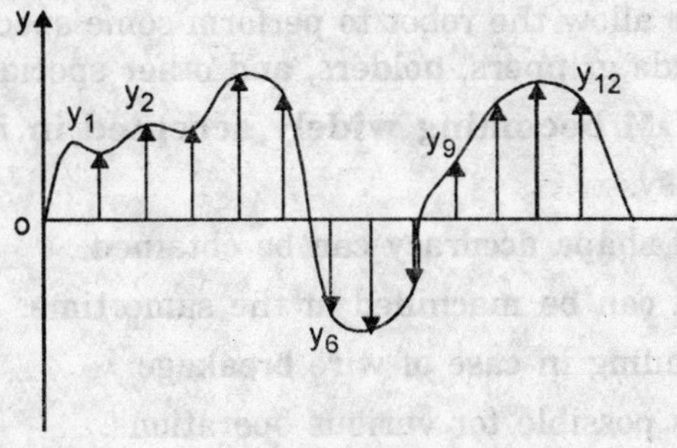

Fig. 16

RMS average method : The RMS value is obtained by setting many equidistant ordinates from the mean line, adding their squares and then taking the square root of their average value. (see Fig. 16).

$$\text{RMS value} = \sqrt{\frac{\sum_{i=1}^{n} y_i^2}{n}}$$

(*iii*) **Chaplet :** They are used to support the cores inside the mould cavity.

Resin binder : It is used in the core sand to provide proper strength to the core.

Shell Moulding

Merits

1. High dimensional accuracy of castings.
2. Smooth surface finish and reduced machining allowances possible.
3. Complex parts can be made.
4. Moulds can be stored until required.
5. Less cleaning of mouldings is needed.

Demerits

1. Suitable for mass production only.
2. Weight of the mouldings is limited to 10 to 15 kg.
3. Serious dust and fume problems during sand and resin mixing.
4. Patterns are very expensive.
5. More sophisticated equipment is needed for handling the mouldings.

(*iv*) **Robot Manipolator :** It consists of the connection of the base and the appendage of a robot. The base of the manipolator is usually fixed to the floor of the work area. Sometimes, the base may be movable by attaching it either to the rail or a track. The appendage is the arm of the robot. The manipolator actually performs the principal movements in doing manufacturing work.

Robot wrist : The wrist at the end of the robot's arm extends the mobility of the robot system and also increases the dimensions of the work envelope. The wrist flange allows the robot user to connect different end-of-arm tooling for different jobs.

End-effectors : The robot arm alone has no production capability, but the robot arm interfaced to production tooling becomes an effective production system. The tooling to perform the work task is attached to the tool plate at the end of the arm. Any special device or tool attached to the tool plate to allow the robot to perform some specilised task is called as end effector. End-effectors include grippers, holders, and other specialised tools.

Reasons for wire EDM becoming widely accepted in industry

1. The operation is easy.
2. Very high degree of shape accuracy can be obtained.
3. More than one part can be machined at the same time.
4. Automatic self-threading in case of wire breakage.
5. Computer control is possible for various operations.

Q.6. (*b*) (*i*) Explain why metal powders are blended. Describe what happens during sintering.

(*ii*) What is meant by Part Programming? Discuss point-to-point control, and its applications.

(*iii*) In FMS define the terms : Automatic Guided Vehicle (AGV) and DNC. Explain the terms chucking reamers and climb milling. What effects does the helix angle have on drill performance. Explain the terms combined cuts and multiple cuts.

Sol. (*i*) **Blending of Metal Powders :** Blending means the through intermixing of powders of different grain sizes of the same nominal composition. The reasons for blending of metal powders are :

1. To ensure filling up of the interstices between large particles, coarser fraction is blended with a fine fraction.
2. Binders are added to powder to develop adequate green strength.
3. Lubricants are added to the powders to reduce die wall friction and friction of the individual particles relative to one another. This increases the green compact density and flowability of powders.
4. To add volatilizing agents to give a desired amount of porosity.

Sintering : It can be defined as the bonding of adjacent surfaces of particles in a mass of metal powder, by heating. It consists of heating metal compacts in a furnace to a temperature below the melting point of the major constituent in an inert or reducing atmosphere under controlled conditions.

(*ii*) **Part programming :** The part programming is a step-by-step command to direct the action of the processing equipment.

Point-to-point control : This type of control moves the tool/machine only in straight lines. It is limited to drilling, reaming, boring, etc. and straight milling cuts parallel to a machine axis. Angles and arc segments are programmed as a series of straight lines.

(*iii*) **AGV :** An AGV is a computer controlled, driverless vehicle used for transporting materials from point-to-point in a manufacturing setting. They represent a major category of automated materials handling devices.

DNC : Direct numerical control (DNC) may be defined as a system connecting a set of numerically controlled machines to a common memory of part program or machine program storage with provision for on-demand distribution of data to machines. Thus DNC is a manufacturing system in which a number of machines are controlled by a computer through direct connection and in real time.

Chucking reamers : These reamers have ends of their teeth ground on a slight chamber for end cutting. In addition, cutting is also done from the side edges of the teeth along the circumference of the reamer. These reamers are considered precision machine reamers and can produce accurate holes.

Climb milling : In the climb or down milling, the direction of the worktable feed and rotation of the cutter is the same. the individual chips cut in the milling process are of variable thickness. The cutter tooth takes a chip of maximum thickness at the beginning of cut and of zero thickness at the end.

Effect of helix angle on drill performance : Fig. 17 shows two views of a twist drill. The normal rate angle α is approximately given by :

$$\alpha = \tan^{-1}\left[\frac{(2r/D)\tan\psi}{\sin\beta}\right]$$

where r = radius of the point on the cutting edge where normal rake is being evaluated.

D = nominal diameter of drill.

β = half point angle

ψ = helix angle

As ψ increases α also increases.

Cutting force F_c α cos (λ – α)

where λ = friction angle

As α increases, cos (λ – α) increases. Therefore F_c increases.

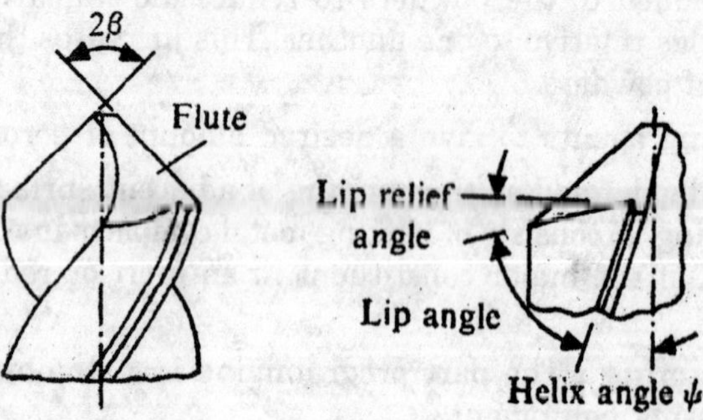

Fig. 17

Combined cuts : In combined cuts, two or more than two cutting operations like turning, thread cutting, drilling, etc. are performed simultaneously on the same workpiece. The various tools are mounted on the various slides of the machine tool.

Multiple cuts : In multiple cuts, the cutter is a multiple-edged rotary cutting tool having the shape of a solid of revolution with cutting teeth arranged either on the periphery or on end face or on both, as on a milling cutter. Usually, the cutter is held in a fixed (but rotating) position and the workpiece moves past the cutter during the machining operation.

Q.7. (*a*) Transportation costs from manufacturing plants to warehouses are given in table. They are in euros. Solve this problem to minimize the cost of transportation by stating the steps used in the algorithm.

Warehouse	P	L	A	NT
	A	B	C	D
1	10	8	10	8
2	10	7	9	10
3	11	9	8	7
4	12	14	13	10

Sol. Step I : Subtract the least element of each row from all elements of the respective row.

	A	B	C	D
1	2	0	2	0
2	3	0	2	3
3	4	2	1	0
4	2	4	3	0

Subtract the least element of each column for all the elements of the respective column.

	A	B	C	D
1	0	0	1	0
2	1	0	1	3
3	2	2	0	0
4	0	4	2	0

Step II : All the zeros can be covered by a minimum of 4 lines.

Since number of lines are 4 = 4 (rows and columns) is 4 = (n = 4). Hence optimal solution is reached.

Step III :

	A	B	C	D
1	0	0	1	0
2	1	0	1	3
3	2	2	0	0
4	0	4	2	0

In the first row there are three zeros, therefore, take the second row where there is only one zero at (2, B) element. Putting this element in a square, we have

	A	B	C	D
1	0	~~0~~	1	0
2	1	[0]	1	3
3	2	2	[0]	~~0~~
4	0	4	2	0

The third row has two zeros and fourth row has also two zeros. Now examine the columns. In first column there are two zeros. There is only one zero in third column at element (3, C). Put this element in a square.

Now there is a tie in each row and column. We make our choice and put the element (1, D) in square and cancel the other zeros in the first and fourth column. We get

	A	B	C	D
1	~~0~~	~~0~~	1	[0]
2	1	[0]	1	3
3	2	2	[0]	~~0~~
4	[0]	4	2	~~0~~

Now the first column contains only one zero at lement (4, A). Put it is square. This gives the optimal solution.

$$1 \longrightarrow D : 8$$
$$2 \longrightarrow B : 7$$
$$3 \longrightarrow C : 8$$
$$4 \longrightarrow A : 12$$

Optimum cost = 8 + 7 + 8 + 12 = 35.

Q.7. (*b*) What are moving average and exponential smoothing models for forecasting?

A dealership for Honda City cars sells a particular model of car in various months. Using the moving average method, find the exponential smoothing forecast for the month of October 2010. Take exponential smoothing constant as 0.2 :

Jan.	2010	80	cars
Feb.	2010	65	cars
March	2010	90	cars
April	2010	70	cars
May	2010	80	cars
June	2010	100	cars
July	2010	85	cars
Aug.	2010	60	cars
Sept.	2010	75	cars

Sol. Moving average method : In this method, the new moving average for each period in computed by taking the average demand for the most recent N periods.

Exponential smoothing method : In this method, the weightage of data diminishes exponentially as the data becomes older. The weightage of every previous data decreases by $(1 - \alpha)$ times, where α = exponential smoothing constant. It t is the current period, the forecast for the next period is :

$$F_{t+1} = \alpha D_t + (1 - \alpha) F_t$$

where D_t = demand for the current period

F_t = forecast for the current period

Given : $\alpha = 0.2$

Using 3 months period.

Month	Demand	Forecast
April	70	$\frac{1}{3}(80 + 65 + 90) = 78.33$
May	80	$\frac{1}{3}(65 + 90 + 70) = 75$
June	100	$\frac{1}{3}(90 + 70 + 80) = 80$
July	85	$\frac{1}{3}(70 + 80 + 100) = 83.3$
Aug.	60	$\frac{1}{3}(80 + 100 + 85) = 88.33$
Sep.	75	$\frac{1}{3}(100 + 85 + 60) = 81.66$

For the month of September,

$$F_t = 81.66, \; D_t = 75$$

Forecast for the month of October,

$$F_{t+1} = \alpha D_t + (1 - \alpha) F_t$$
$$= 0.2 \times 75 + (1 - 0.2) \times 81.66 = 80.33 \simeq 81 \text{ cars.}$$

Q.7. (*c*) Expand the following:

(*i*) ASCII (*ii*) BCD
(*iii*) MDI (*iv*) RAM
(*v*) ROM

Sol. ASCII : American Standard Code for Information Interchange
BCD : Binary Code Decimal
MDI : Multiple Document Interface
RAM : Random Access Memory
ROM : Read Only Memory

MECHANICAL ENGINEERING PAPER II—2011

(Conventional)

SECTION-A

Q.1. (*a*) (*i*) Define "Stability" and "Isochronism" for spring-controlled governors.

(*ii*) Draw the controlling force vs. radius of rotation of governor balls for unstable, stable and disochronous governors.

Sol.

(*i*) A governor is said to be stable, when for each speed within the working range, there is only one radius of the governor balls at which the governor is in equilibrium.

A governor is said to be isochronous, when the equilibrium speed is constant for all radii of rotation of the balls, within the working range.

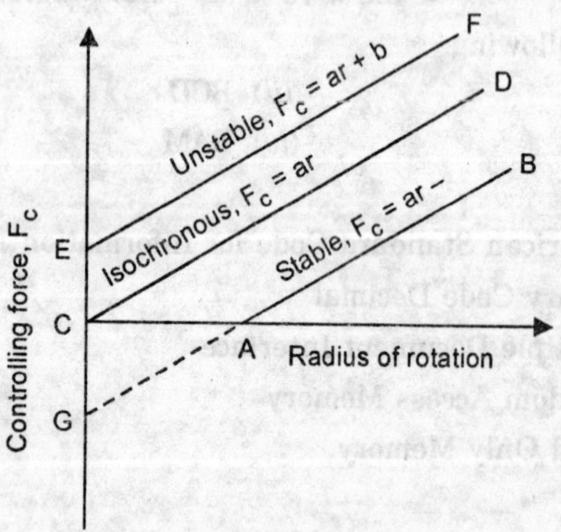

Fig. 1. Controlling force v's radius of rotation.

(*b*) A crank and slotted lever Quick Return Motion Mechanism used in a shaping machine has a distance of 200 mm between the centre of oscillation 'A' of the slotted lever and the centre of rotation 'B' of the crank. The radius of the crank BC is 100 mm.

(*i*) Show on a diagram the extreme positions of the lever AD during one complete rotation of the crank.

(*ii*) Find the ratio of the time of cutting to the time of the return stroke.

Sol. $\cos\alpha = \dfrac{BC_1}{AB} = \dfrac{100}{200} = \dfrac{1}{2}$, $\alpha = 60°$, $2\alpha = 120°$

$$\frac{\text{Time of cutting}}{\text{Time of return}} = \frac{360° - 2\alpha}{2\alpha} = \frac{360° - 120°}{120°} = \frac{240°}{120°} = 2$$

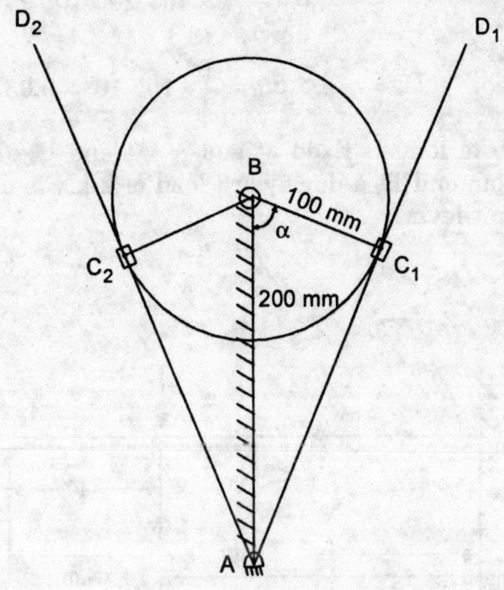

Fig. 2. Crank and slotted lever mechanism.

(*c*) **(*i*) What is critical damping for a spring, dashpot and mass vibrating system?**

(*ii*) Such a system has mass m = 10 kg, spring stiffness k = 4000 N/m and damping coefficient c = 40 N/m/s. Find the critical damping coefficient, and the damping factor ζ.

Sol.

(*i*) **For a critically damped system, ζ = 1, where**

$$\zeta = \text{damping ratio or damping factor}$$

$$= \frac{\text{damping coefficient}}{\text{critical damping coefficient}} = \frac{c}{c_c}$$

(*ii*) **Given :** m = 10 kg, k = 4000 N/m, c = 40 N/m/s

$$c_c = 2\sqrt{km} = 2\sqrt{4000 \times 10} = 400 \text{ N}\cdot\text{s/m}$$

$$\zeta = \frac{c}{c_c} = \frac{40}{400} = 0.1$$

(*d*) A tapered bar 200 mm long, tapers uniformly from a diameter of 40 mm to a diameter of 20 mm over its axial length. If an axial compressive force of 10 kN is applied on the bar, what is the strain energy absorbed in the bar? E = 100 kN/mm^2.

Sol. Given : l = 200 mm, d_1 = 40 mm, d_2 = 20 mm, P = 10 kN, E = 100 kN/mm^2

Compression of the bar, $\delta l = \dfrac{4Pl}{\pi d_1 d_2 E} = \dfrac{4 \times 10 \times 10^3 \times 200}{\pi \times 40 \times 20 \times 100 \times 10^3} = 0.0318$ mm

Strain energy absorbed, $U = \dfrac{1}{2} P \times \delta l = \dfrac{1}{2} \times 10 \times 10^3 \times 0.0318 = 0.159$ J

(*e*) A cantilever AB, 5 m long is fixed at end A. At end B, an upward load of 2 kN is applied. At point C, 2 m from end B, a downward load of 2 kN is applied. Draw the bending moment diagram of the cantilever.

Sol. B.M.D. (Fig. 3)

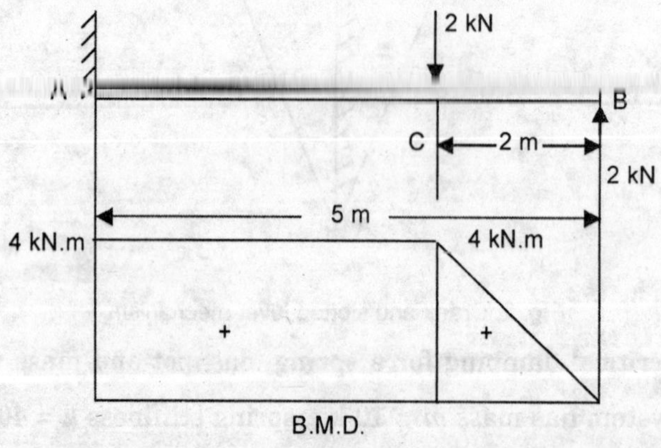

Fig. 3

Span BC :

$$M_x = +2x \text{ kN·m}$$
$$M_B = 0$$
$$M_C = 2 \times 2 = 4 \text{ kN·m}$$

Span CA :

$$M_x = 2x - 2(x - 2)$$
$$M_C = 2 \times 2 - 2\,(2 - 2) = 4 \text{ kN·m}$$
$$M_A = 2 \times 5 - 2\,(5 - 2) = 10 - 6 = 4 \text{ kN·m}$$

(*f*) A thin cylindrical shell of diameter 200 mm, wall thickness t is subjected to an internal pressure of 2 N/mm^2. The longitudinal joint efficiency of the shell is 80%. What should be the minimum wall thickness of shell, if allowable stress in shell is limited to 100 MPa?

Sol. Given : d = 200 mm, p = 2N/mm^2, η_l = 0.8, σ_a = 100 MPa

Longitudinal stress, $\sigma_l = \dfrac{pd}{4t\eta_l}$

$$100 = \frac{2 \times 200}{4 \times t \times 0.8}$$

$$t = 1.25 \text{ mm}$$

(g) Explain how grain refinement improves the strength of a polyerystalline material.

Sol. The strength of a polycrystalline material depends on the grain size of the material. Finer the grain size higher is the strength of the material. This is because the grain boundary acts as a barrier to the movement of dislocations, which gives poly crystalline materials higher strength with finer grains.

(h) Stub tooth spur gears are preferred for medium and heavy duty applications. Justify this statement in about twenty words.

Sol. Stub tooth spur gears have smaller moment arm for bending force which induces reduced bending stress. Hence they can transmit higher power and are preferred for medium and heavy duty applications.

(i) Which theory of failure will result in economic shaft design?

Sol. The maximum shear stress induced in the shaft according to the maximum distortion energy theory is lower than that of the maximum shear stress theory. Therefore, maximum distortion energy theory results in economic shaft design.

(j) Why are V-belts used in short centre drives?

Sol. There is no slip in a V-belt drive. It is a more positive drive and can transmit more power.

(k) What is meant by allotropism? What is its importance?

Sol. Allotropism is the property of some chemical elements to exist in two or more different forms at different temperatures. Allotropes have different structure but same chemical composition. They are very useful in getting materials of desired properties.

(l) List four undesirable affects of decarburization of steels after heat treatment.

Sol.

1. Lowered carbon content decreases the hardenability of steel.
2. Strength of steel is reduced.
3. Fatigue life is reduced.
4. Endurance limit is lowered.

(m) Name four fibre materials generally used in metal-matrix composites.

Sol. Fibre materials used in metal-matrix composites are :

Ni, Al, Cu, steel platelets

(n) Enumerate four defects caused due to residual stresses in welded joints.

Sol.

1. Elastic deformation.

2. Plastic deformation.
3. Angular distortion.
4. Cold cracking.

(*o*) What is isostatic pressing of metal powders? What are its advantages?

Sol. Isostatic pressing of metal powders : The blended mixture is compacted in moulds by the application of pressure to bring the metal powder into close proximity without permanent bonding.

Its advantages are :

1. The density, hardness and strength of powder increases.
2. Uniform grain structure is obtained.

(*p*) What is critical ratio in scheduling? What is the meaning of critical ratio scheduling value of 2?

Sol. Critical ratio (C.R.) in scheduling $= \dfrac{\text{Earliest due date (EDD)}}{\text{Processing time}}$

C.R. = 2 means that EDD is twice the processing time, *i.e.* the job can be completed in half of earliest due date.

(*q*) What are the steps involved in method study?

Sol. Steps involved in method study are :

1. Select
2. Record
3. Examine
4. Develop
5. Define
6. Install
7. Maintain

(*r*) What are the methods of finding initial solution in transportation problems?

Sol.

1. North-West corner method.
2. Least cost method.
3. Vogel's approximation method.

(*s*) A company purchases a product at ₹ 50 per unit and sells it at ₹ 90 per unit. The hiring charges for storing the unit are ₹ 20,000. Determine the number of units to be sold to achieve break-even. If the company sells 750 units, calculate the margin of safety.

Sol. Given : Fixed cost = ₹ 20,000

Variable cost = ₹ 50/unit

Selling price = ₹ 90/unit

$$\text{Break even quantity (BEQ)} = \frac{\text{Fixed cost}}{\text{Selling price-variable cost}}$$

$$= \frac{20,000}{90-50} = 500 \text{ units} = \text{argin of safety}$$

$$\text{Margin of sefety} = \text{Units sold} - \text{BEQ}$$

$$= 750 - 500 = 250 \text{ units}$$

(*t*) An array of 50 integer numbers is to be read by using Fortran program. Write the program lines for the same.

Sol. INTEGER A

```
DIMENSION A (50)
READ (7, *) A(I), I = 1,50
```

SECTION B

Q.2. (*a*) Three steel tubes of outer diameter 20 mm and inner diameter 16 mm each are welded together so that their centres form an equilateral triangle of side 20 mm. Three meter length of this composite tube is used as a simply supported beam at ends with a central point load. How much point load can be applied if the maximum stress in beam sections is not to exceed 100 MPa?

Sol.

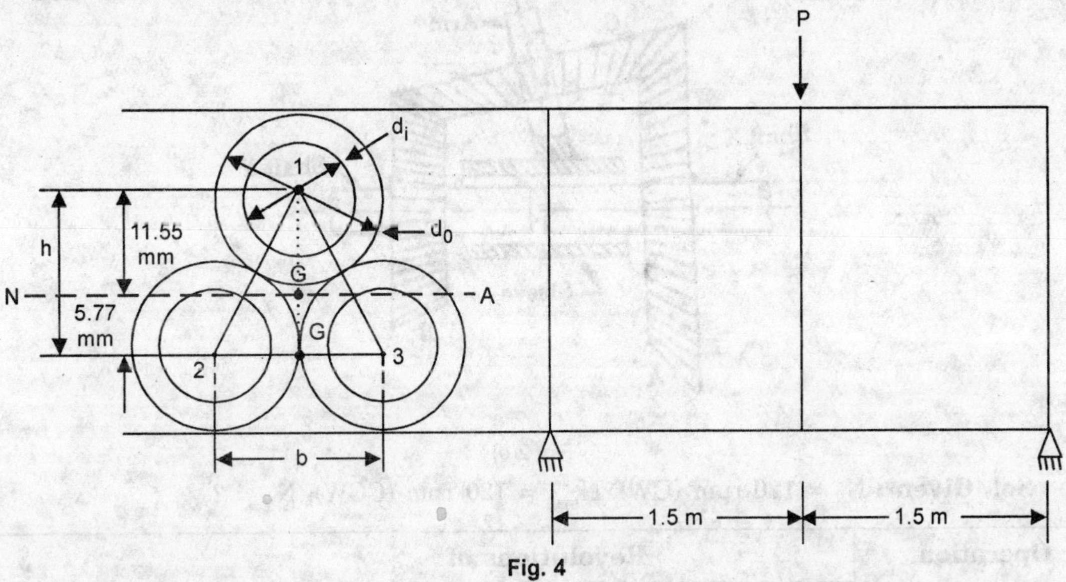

Fig. 4

Given : d_0 = 20 mm, d_i = 16 mm, l = 3 m, σ_{max} = 100 MPa, b = 20 mm $h = \frac{b}{2}$ tan 60° $= 10 \times \sqrt{3} = 17.32$ mm.

C.G. of equilateral triangle from the base $= \frac{h}{3} = \frac{17.32}{3} = 5.77$ mm

For composite section of three tubes,

$$I_{NA} = 2\left[\frac{\pi}{64}(20^4 - 16^4) + \frac{\pi}{4}(20^2 - 16^2) \times (5.77)^2\right]$$

$$+\left[\frac{\pi}{64}(20^4 - 16^4) + \frac{\pi}{a4}(20^2 - 16^2) \times (11.55)^2\right]$$

$$= 36{,}529 \text{ mm}^4$$

$$M_{max} = \frac{Pl}{4} = \frac{P \times 3 \times 10^3}{4} = 750 \times \text{ P N} \cdot \text{mm}$$

$$\sigma_{max} = \frac{M_{max} \times y_{max}}{I_{NA}}, \quad y_{max} = 11.55 + \frac{d_0}{2} = 11.55 + 10 = 21.55 \text{ mm}$$

$$100 = \frac{750 \times P \times 21.55}{36{,}529}$$

$$P = 226 \text{ N}$$

Q.2. (*b*) Two bevel gears A and B (having 60 teeth and 40 teeth, respectively) are rigidly mounted on two co-axial shafts X and y as shown in Fig. 5. A bevel gear C (having 50 teeth meshes with A and B and rotates freely on one end of an arm. At the other end of the arm is welded a sleeve and the sleeve is riding freely loose on the axes of the shafts X and Y. If the shaft X rotates at 120 r.p.m. clockwise and the arm rotates at 120 r.p.m. anticlockwise, find the speed of shaft Y.

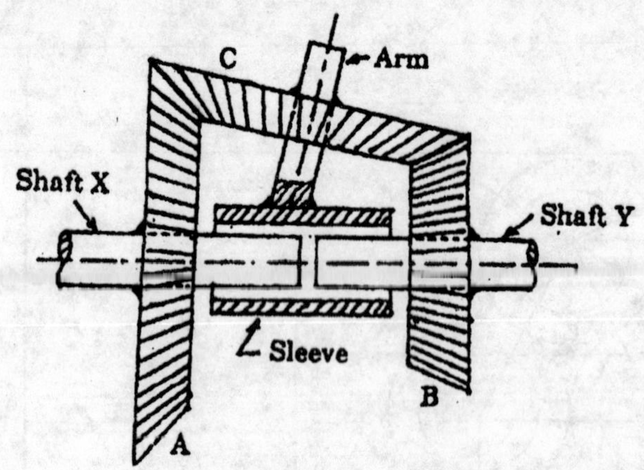

Fig. 2(*b*)

Sol. Given : $N_X = 120$ rpm (CW), $N_{arm} = 120$ rpm (CCW), $N_Y = ?$

Operation	Revolutions of			
	Arm	**Gear A** $Z_A = 60$	**Gear C** $Z_C = 50$	**Gear B** $Z_B = 40$
1. Arm fixed, gear A given 1 rev cw.	0	+1	$-\frac{Z_A}{Z_C} = -\frac{60}{50} = -\frac{6}{5}$	$\frac{Z_C}{Z_B} \times \frac{6}{5} = \frac{50}{40} \times \frac{6}{5} = \frac{3}{2}$
2. Multiply by x	0	$+x$	$-\frac{6}{5}x$	$\frac{3}{2}x$
3. Addy y	y	$x + y$	$-\frac{6}{5}x + y$	$\frac{3}{2}x + y$

$$y = -120 \text{ rpm}$$

$$N_A = N_X = 120 = x + y$$

$$\therefore \quad x = 120 - y = = 120 + 120 = 240$$

$$N_B = \frac{3}{2}x + y = \frac{3}{2} \times 240 - 120 = 240 \text{ rpm (cw)}$$

Q.2. (*c*) A hollow shaft whose ratio of internal diameter to external diameter (*k*) is 0.5, transmits 1.5 kW at 1400 rpm. At a certain section it is also subjected to a bending moment of 5 N.m. The shaft is to be made from chrome-vanadium steel having allowable shearing stress of 200 MPa. Design the shaft using the ASME code formula for transmission shafting as given below :

$$D_0 \text{ (outer diameter)} = \left[\frac{16}{\pi \times \tau_{all}}\left(C_m \cdot M^2 + C_t \cdot T^2\right) \times \frac{1}{1-k^4}\right]^{\frac{1}{3}}$$

Combined shock and fatigue factor for bending moment (C_m) is 1.5.

Combined shock and fatigue factor for torque (C_t) = 1.0.

Determine the internal and outer diameter of the shaft.

Sol. Given : $k = \frac{D_i}{D_0} = 0.5$, P = 1.5 kW, N = 1400 rpm, M = 5 N·m,

$\tau_{all} = 200$ MPa, $C_m = 1.5$, $C_t = 1.0$

$$P = \frac{2\pi NT}{60 \times 10^3}$$

$$T = \frac{1.5 \times 60 \times 10^3}{2\pi \times 1400} = 10.23 \text{ N} \cdot \text{m}$$

$$D_0 = \left[\frac{16}{\pi \times \tau_{all}}(C_m M^2 + C_t T^2) \times \frac{1}{(1-k^4)}\right]^{1/3}$$

$$= \left[\frac{16}{\pi \times 200}\left\{1.5 \times 5^2 + 1.0 \times (10.23)^2\right\} \times 10^6 \times \frac{1}{\{1-(0.5)^4\}}\right]^{1/3}$$

$$= 156.88 \text{ mm}$$

$$D_i = 78.44 \text{ mm}$$

Q.2. (*d*) What are supperalloys? Name 3 superalloys. Give composition of Nimonic alloy.

Sol. A superalloy is an alloy that exhibits excellent mechanical strength and creep resistance at high temperatures. They have good corrosion and oxidation resistance. Their base alloying element is usually nickel, cobalt, or nickel-iron. They have a matrix with an austenitic face-centered cubic crystal structure. Their typical applications are in the arrospace, industrial gas turbines, turbine blades for jet engines, etc.

Examples of superalloys are : Inconel, Nimonics, Incoloy, Hastealloy, Vitallium.

The common composition of Nimonic alloy is : Cr(15 to 18%), Co(15 to 18%), Mo(3 to 5%), Ti(1.2 to 4%), Al(1.2 to 5%) and the remainder Ni.

Q.3. (*a*) Design a close coiled helical spring to have spring index of 8. Axial deflection in spring is not to exceed 100 mm under an axial load of 2600 N and shear stress developed in spring is not to exceed 300 N/mm². Steel wires are available in diameters of 10, 11, 12, 13, 14 mm (in steps of 1 mm). Determine most suitable wire diameter, mean coil diameter and number of coils in spring. Given G = 84000 N/mm².

Sol. Given : $C = 8$, $\delta_{max} = 100$ mm,

$W = 2600$ N, $\tau_{max} = 300$ N/mm²,

$d = 10$ to 14 mm in steps of 1 mm,

$G = 84000$ N/mm².

Deflection, $$\delta = \frac{8\,WC^3 n}{Gd}$$

$$100 = \frac{8 \times 2600 \times 8^3 \times n}{84000 \times d}$$

$$\frac{n}{d} = 0.78876$$

Shear stress, $$\tau = K_w \times \frac{8WC}{\pi d^2}$$

Wall's correction factor,

$$K_W = \frac{4C-1}{4C-4} + \frac{0.615}{C}$$

$$= \frac{4 \times 8 - 1}{4 \times 8 - 4} + \frac{0.615}{8} = 1.184$$

$$300 = \frac{1.184 \times 8 \times 2600 \times 8}{\pi d^2}$$

$$d = 14.45 = 14 \text{ mm}$$

$$n = 0.78876 \times 14 = 11$$

$$D_m = 8 \times 14$$

$$= 112 \text{ mm}$$

Taking square and ground ends for wire,

total number of coils = 11 + 2 = 13

Q.3. (*b*) A two-cylinder uncoupled locomotive has inside cylinders 70 cm apart, the cranks are at right angles and are each 0.3 m long. The mass of the revolving parts per cylinder is 16 kg and the mass of the reciprocating parts per cylinder is 180 kg. The whole of the revolving and two-thirds of the reciprocating parts are to be balanced and the balance masses are to be placed in the planes of rotation of the driving wheels at a radius of 80 cm. The driving wheels are 2 m diameter and 1.5 m apart.

Find the magnitude and position of the balance masses. The driving crank speed is 300 r.p.m.

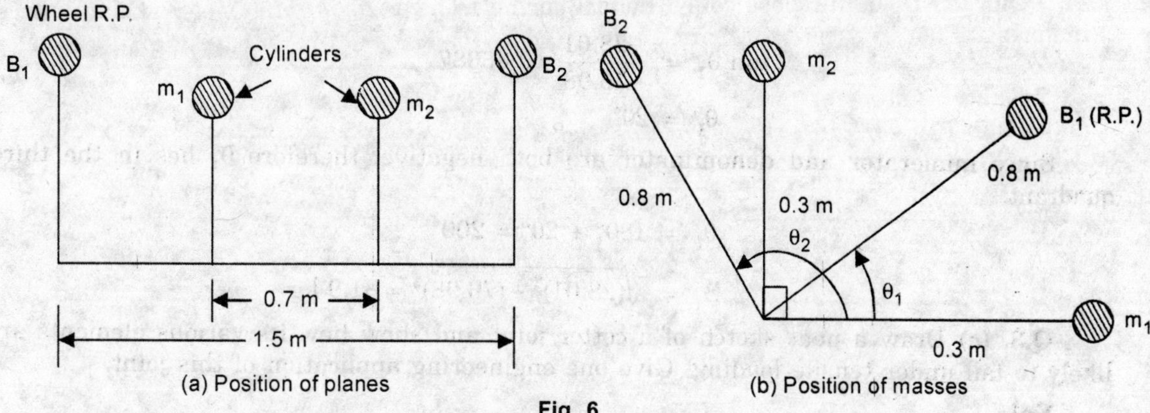

Fig. 6

Sol. Given : M = 160 kg, R = 180 kg, $c = \dfrac{2}{3}$

Total mass to be balanced, $m_1 = m_2 = M + cR = 160 + \dfrac{2}{3} \times 180 = 280$ kg

Mass m kg	**Radius r m**	**θ deg**	**Force/ω² mr kg·m**	**Distance from R.P. *l* m**	**Couple mr *l* kg·m²**
B_1	0.8	θ_1	0.8 B_1	0	0
$m_1 = 280$	0.3	0	84	0.4	33.6
$m_2 = 280$	0.3	90	84	1.1	92.4
B_2	0.8	θ_2	0.8 B_2	1.5	1.2 B_2

Couple balance :

$$1.2\ B_2 \cos\theta_2 + 33.6 + 0 = 0 \quad \text{or} \quad B_2 \cos\theta_2 = -28$$

$$1.2\ B_2 \sin\theta_2 + 0 + 92.4 = 0 \quad \text{or} \quad B_2 \sin\theta_2 = -77$$

$$\tan\theta_2 = \frac{-77}{-28} = 2.75, \quad \theta_2 = 70°$$

Since numerator and denominator are both negative, therefore θ_2 lies in the third quadrant.

$$\therefore \quad \theta_2 = 180° + 70° = 250°$$

$$B_2 = \sqrt{28^2 + 77^2} = 81.91 \text{ kg}$$

Force balance

$$0.8\ B_1 \cos\theta_1 + 84 + 0 + 81.9 \times 0.8 \cos 250° = 0$$

or $$B_1 \cos\theta_1 = -76.98$$

$$0.8\ B_1 \sin\theta_1 + 0 + 84 + 81.90 \times 0.8 \sin 250° = 0$$

$$B_1 \sin\theta_1 = -28.01$$

$$\tan \theta_1 = \frac{-28.01}{-76.98} = 0.36387$$

$$\theta_1 = 20°$$

Since numerator and denominator are both negative, therefore θ_1 lies in the third quadrant.

$$\therefore \quad \theta_1 = 180° + 20° = 200°$$

$$B_1 = \sqrt{(28.01)^2 + (76.98)^2} = 81.9 \text{ kg}$$

Q.3. (c) Draw a neat sketch of a cotter joint and show how its various elements are likely to fail under tensile loading. Give one engineering application of this joint.

Sol.

The gib and cotler joint in shown in Fig. 7.

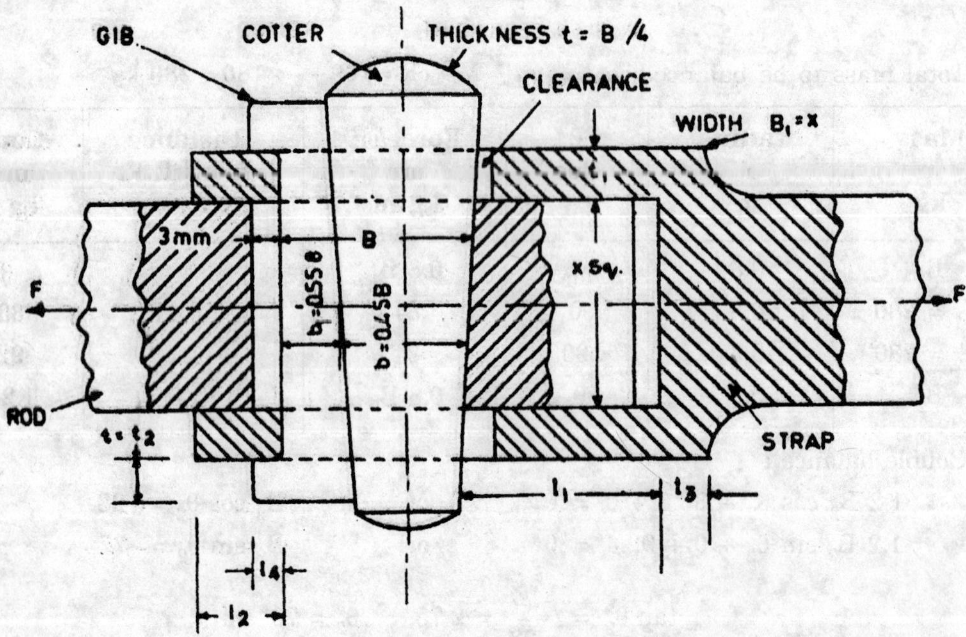

Fig. 7. Gib and cotter joint.

1. Size of rod : Let x be the one side of the square rod. Considering the failure of the rod in tension, we have

$$F = x^2 \sigma_t$$

From which x may be calculated.

Then, Width of strap, $B_1 = x$

Thickness of cotter, $t = \frac{B_1}{4} = \frac{x}{4}$

Thickness of gib $= t$

Height (t_2) and length (l_4) of gib head $= t$

3. Thickness of strap : Considering the failure of the strap end in tension at the location of gib and cotter (Fig. 2(a)), we have

$$F = 2(B_1 t_1 - t_1 t)\,\sigma_t$$
$$= 2(x t_1 - t_1 t)\,\sigma_t$$
$$= 2t_1(x - t)\,\sigma_t$$

From which t_1 may be calculated.

Fig. 8 (a)

To check the failure of strap or gib in crushing, we haved

$$F = 2t_1 t \sigma_c$$

σ_c should be less than the permissible crusshing strength.

4. Length of rod from cutter to end : Considering the double shear of the rod (Fig. (b)), we have

$$F = 2l_1\, x\tau$$

From which l_1 may be determined.

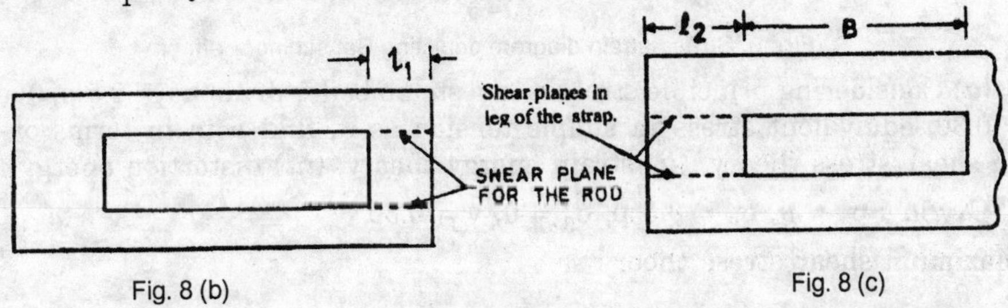

Fig. 8 (b) Fig. 8 (c)

Fig. 8

5. Length of the rod from gib to the end of strap : Considering the double shear of the strap (Fig. (c)), we have

$$F = 2 \times 2l_2\, t_1 \times \tau$$

From which l_2 may be calculated.

6. Other dimensions, $l_3 = \frac{2}{3}x$

Applications of cotter

1. Piston rod and cross-head connection of a steam engine.
2. Joining valve rod end and eccentric rod.
3. Foundation bolts.

Q.3. (*a*) What is Bauschinger's effect? Make a neat sketch of stress-strain diagram and explain how yield strength in compression is reduced than the yield strength in tension.

Sol. Bauschinger effect is the directionality of straining, *i.e.*, if strain hardening in a material takes place due to application of stress in a particular direction it will be easier to deform in the opposite direction. It can also be said that if a material has a yield point stress σ_{yt} in tension and is continued to deform from tension to compression in the same cycle it will show a yield point in compression σ_{yc}, where $\sigma_{yc} < \sigma_{yt}$. This effect is mainly due to the fact that lower stress is required to reverse the direction of slip on a certain slip plane than to continue slip in the original direction.

The stress-strain diagram-depicting Bauschinger effect is shown below.

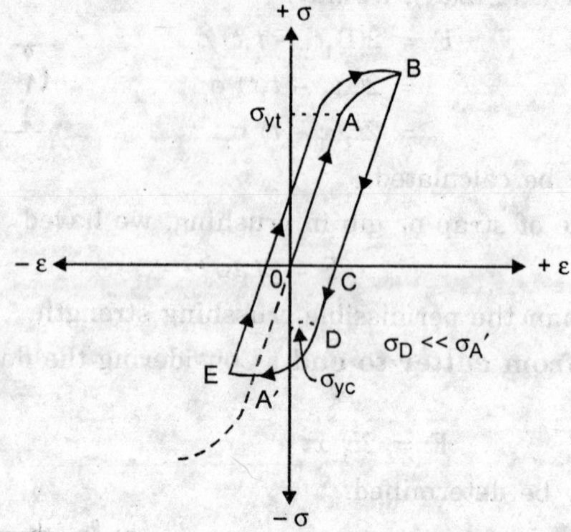

Fig. 9. Stress-strain diagram depicting Bauschinger effect.

Q.4. (*a*) Considering principle stresses in a steam boiler drum as p, 0.5 p, 0 and Poisson's ratio $\nu = 0.30$, equivalent stress in simple tension as σ, find p in in terms of σ due to (*i*) maximum shear stress theory, (*ii*) strain energy theory, (*iii*) distortion energy theory.

Sol. Given : $\sigma_1 = p$, $\sigma_2 = 0.5$ p, $\sigma_3 = 0$, $\nu = 0.30$

(*i*) Maximum shear stress theory

$$\tau_{max} = \frac{\sigma_1 - \sigma_3}{2} = \frac{p-0}{2} = \frac{p}{2} = \frac{\sigma_y}{2} = \frac{\sigma}{2}$$

$$\therefore \qquad p = \sigma$$

(*ii*) Strain energy theory.

$$\Sigma\sigma_1^2 - 2\nu\Sigma\sigma_1\sigma_2 = \sigma^2$$

$$p^2 + 0.25\,p^2 - 2\,\nu \times p \times 0.5\,p = \sigma^2$$

$$p^2 + 0.25\,p^2 - 0.3\,p^2 = \sigma^2$$

$$0.95\,p^2 = \sigma^2$$

$$p = 1.026\,\sigma$$

(*iii*)

$$\Sigma(\sigma_1 - \sigma_2)^2 = 2\sigma^2$$

$$[(0.5)^2 + (0.5)^2 + 1]\,p^2 = 2\sigma^2$$

$$p = 1.15\,\sigma$$

Q.4. (*b*) A steam engine develop 300 kW at 120 r.p.m. The coefficient of fluctuation of energy as found from the turning moment diagram is to be 0.1 and the fluctuation of speed is to be kept within ±1% of the mean speed Find the mass of the flywheel required, if the radius of gyration is 2 m.

Sol. Given : P = 300 kW, N = 120 rpm, $C_e = 0.1$, $C_s = \pm1\%$, K = 2 m

Work done per cycle, $$= \frac{P \times 60 \times 10^3}{N} = \frac{300 \times 10^3 \times 60}{120} = 150 \times 10^3 \text{ N}\cdot\text{m/cycle}$$

Coefficient of fluctuation of energy,

$$C_e = \frac{\text{Maximum fluctuation of energy}}{\text{Work done per cycle}}$$

$$0.1 = \frac{\Delta E_f}{150 \times 10^3}$$

$$\Delta E_f = 15 \times 10^3 \text{ N·m}$$

$$= I\omega_m^2 C_s = MK^2 \times \left(\frac{2\pi N}{60}\right)^2 \times C_s$$

$$15 \times 10^3 = M \times 2^2 \times \left(\frac{2\pi \times 120}{60}\right)^2 \times 0.02$$

$$M = 1187.36 \text{ kg}$$

Q.4. (*c*) (*i*) What are common modes of failure of rolling element bearings?

(*ii*) Write a brief comment on the holding torque acting on epicyclic gear train casings.

Sol. (*i*) **Modes of failure of rolling element bearings :** The bearings which are carefully installed, properly lubricated, kept free of foreign matter, and are not overload will fail by fatigue. The failure of the bearing appears as spalling or flaking of race-ways or rolling elements due to Hertzian contact stresses. Due to repeated contact the bearing will have a limited life. As pure rolling does not occur in the bearing and is always accompanied by sliding, small particles of metal are removed, which is called spalling failure. Failure of the bearing may also take place due to lubricant failure and overheating. The fatigue failure analysis data for bearings is statistical in nature.

(*ii*) **Holding torque acting on epicyclic gear train casings :** Consider the rotating parts of an epicyclic gear train shown in Fig. 10. If the rotating parts have no angular acceleration, then the gear train is kept in equilibrium by the following three externally applied toruqes:

1. Input torque (T_1) on the driving member.
2. Output torque (T_2) on the driven member.
3. Braking torque (T_3) on the fixed member.

The net torque applied on the gear train must be zero, i.e.,

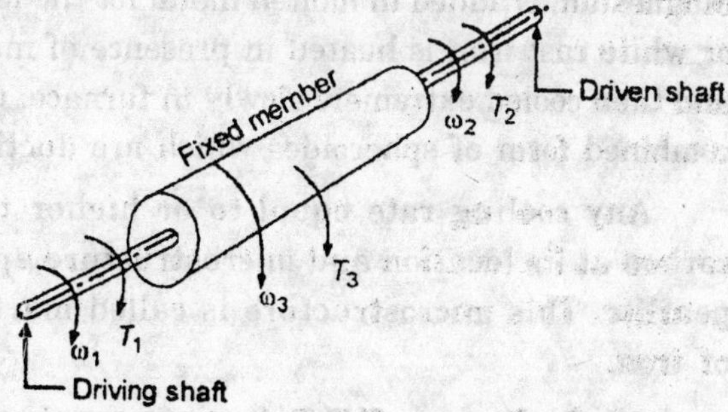

Fig. 10 Torque in epicyclic gear train.

$$T_1 + T_2 + T_3 = 0$$

If F_1, F_2 and F_3 are the externally applied forces at radii r_1, r_2 and r_3, then

$$F_1r_1 + F_2r_2 + F_3r_3 = 0$$

If ω_1, ω_2, and ω_3 are the angular velocities of the driving, driven, and fixed members respectively, and friction is neglected, then the net kinetic energy dissipated by the gear train must be zero.

$$T_1\omega_1 + T_2\omega_2 + T_3\omega_3 = 0$$

For a fixed member, $\omega_3 = 0$. Therefore

$$T_1\omega_1 + T_2\omega_2 = 0$$

$$T_2 = -\left(\frac{\omega_1}{\omega_2}\right)T_1$$

Holding torque,

$$T_3 = -(T_1 + T_2)$$

$$= T_1\left(\frac{\omega_1}{\omega_2} - 1\right)$$

$$= T_1\left(\frac{n_1}{n_2} - 1\right)$$

Q.4. (*d*) What is ductile SG iron? What is purpose of adding magnesium or cerium in molten metal? Why is its tensile strength more than the tensile strength of grey cast iron? When the metal is cooled fast from molten state, what type of microstructure is obtained? Make an approximate sketch of microstructure indicating the microconstituents.

Sol. Ductile SG iron is a spheroidal or nodular graphite iron. It contains graphite in the form of small spheres or nodules. It possesses high tensile strength and has good ductility as cast. This is known as ductile cast iron. It has good resistance to corrosion and wear. Magnesium is added in molten metal for the formation of spheroids, which are ductile. Chilled or white cast iron is heated in presence of magnesium and cerium to a certain temperature and then cooled extremely slowly in furnace. Carbon comes out of the structure and produces combined form of spheroides which are ductile in nature.

Any cooling rate equal to or higher than the critical cooling rate will freeze the carbon at its location and microstructure appears like colloidal solution of cementite and pearlite. This microstructure is called martensite which is most hard and brittle phase of iron.

All the lines on T-T-T diagram are that of decomposition of austenite into some other microstructure. Once austenite converts into some other phase, it never converts again. The fast cooling of molten state of a metal is shown in the T-T-T diagram, shown below in Fig. 11.

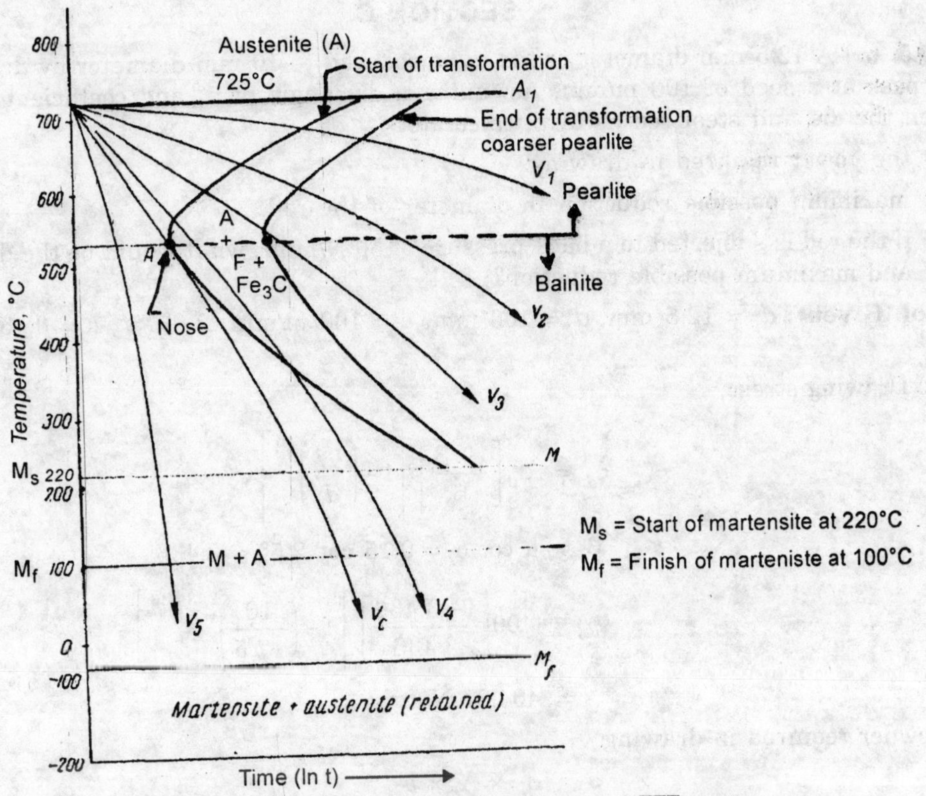

Fig. 11. Superimposition of cooling curves on a TTT curve (austenite decomposition). Cooling rates $V_1 < V_2 < V_3 < V_4 < V_5$

If the cooling rate is such that it just touches the nose of T-T-T diagram then it is called critical cooling rate (CCR), shown by curve V_c.

The microstructure of microconstituents is shown below in Fig. 12.

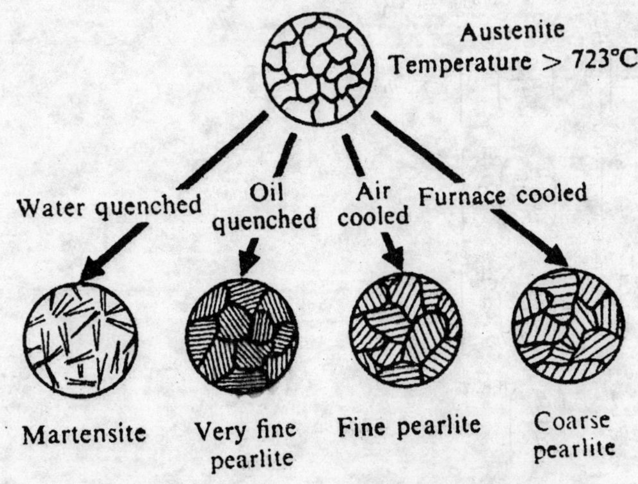

Fig. 12. Microstructure of microconstituents.

SECTION-C

Q.5. (*a*) A 12.5 mm diameter rod is to be reduced to 10 mm diameter by drawing in a single pass at a seed of 100 m/min. Assuming a die angle of 5° and coefficient of friction between the die and steel rod as 0.15, calculate :

(*i*) the power required in drawing.

(*ii*) maximum possible reduction in diameter of the rod.

(*iii*) if the rod is subjected to a back pressure of 50 N/mm², what would be the draw stress and maximum possible reduction?

Sol. Given : d_i = 12.5 mm, d_f = 100 mm, v = 100 m/min, 2α = 5°, μ = 0.15, σ_0 = 400 N/mm²

(*i*) Drawing stress,

$$\sigma_d = \sigma_0\left[\frac{1+B}{B}\right]\left[1-\left(\frac{d_f}{d_i}\right)^{2B}\right]$$

$$B = \mu \cot \alpha = 0.15 \cot 2.5° = 3.436$$

$$\sigma_d = 400\left[\frac{1+3.436}{3.436}\right]\left[1-\left(\frac{10}{12.5}\right)^{2\times 3.436}\right]$$

$$= 404.97 \text{ N/mm}^2$$

Powder required in drawing

$$= \sigma_d \times A_f \times v$$

$$= 404.97\times\frac{\pi}{4}(10)^2\times\frac{10}{60}\times 10^{-3}$$

$$= 53.01 \text{ kW}$$

(*ii*) For maximum reduction,

$$\sigma_d = \sigma_0$$

$$\sigma_0\left[\frac{1+B}{B}\right]\left[1+\left(\frac{d_f}{d_i}\right)^{2B}\right] = \sigma_0$$

$$\left(\frac{1+B}{B}\right)\left[1-\left(\frac{d_f}{d_i}\right)^{2B}\right] = 1$$

$$\left[\frac{1+3.436}{3.436}\right]\left[1-\left(\frac{d_f}{d_i}\right)^{2B}\right] = 1$$

$$1-\left(\frac{d_f}{d_i}\right)^{6.872} = 0.7746$$

$$\frac{d_f}{d_i} = 0.805$$

$$\frac{A_f}{A_i} = 0.648$$

Percentage reduction $= \left(1 - \frac{A_f}{A_i}\right) \times 100 = (1 - 0.648) \times 100 = 35.2\%$

(*iii*) Back pressure, $\sigma_b = 50 \text{ N/mm}^2$

$$\sigma_d = \sigma_0\left[\frac{1+B}{B}\right]\left[1-\left(\frac{d_f}{d_i}\right)^{2B}\right]+\sigma_b\left(\frac{d_f}{d_i}\right)^{2B}$$

$$= 404.97 + 50\left(\frac{10}{12.5}\right)^{2\times 3.436}$$

$$= 415.76 \text{ N/mm}^2$$

For maximum reduction, $\sigma_d = \sigma_0$

$$\left[\frac{1+3.436}{3.436}\right]\left[1-\left(\frac{d_f}{d_i}\right)^{6.872}\right]+\frac{50}{400}\left(\frac{d_f}{d_i}\right)^{6.872} = 1$$

$$1.291 - 1.291\left(\frac{d_f}{d_i}\right)^{6.872} + 0.125\left(\frac{d_f}{d_i}\right)^{6.872} = 1$$

$$1.166\left(\frac{d_f}{d_i}\right)^{6.872} = 0.291$$

$$\frac{d_f}{d_i} = 0.583$$

$$\frac{A_f}{A_i} = 0.34$$

Percentage reduction $= \left(1 - \frac{A_f}{A_i}\right) \times 100 = (1 - 0.34) \times 100 = 66\%$

Q.5. (*b*) A round casting is 20 mm in diameter and 50 mm in length. Another casting of the same metal is elliptical in cross-section, with a major to minor axis ratio of 2, and has the same length and cross-sectional area as the round casting. Both pieces are cast under the same conditions. What is the difference in the solidification times of the two castings?

Sol. Given : $d_r = 20$ mm, $l_r = 50$ mm; $\frac{a}{b} = 2$, $l_e = 50$ mm, $A_r = A_e$

Sodification time $t_s \; \alpha \left(\frac{V}{A_s}\right)^2$

Where V = Volume of casting, and A_s = surface area of casting.

r, e = subscripts for round and elliptical casting respectively.

Round casting :

$$V_r = \frac{\pi}{4} d_r^2 l_r = \frac{\pi}{4}(20)^2 \times 50 = 15708 \text{ mm}^3$$

$$A_{sr} = 2 \times \frac{\pi}{4} d_r^2 + \pi d_r l_r = 2 \times \frac{\pi}{4}(20)^2 + \pi \times 20 \times 50 = 1200 \times \pi \text{ mm}^2$$

Cross-sectional area,

$$A_r = \frac{\pi}{4} d^2 = \frac{\pi}{4} \times 20^2 = 100\pi \text{ mm}^2$$

Elliptical casting :

$$A_e = \pi ab = 2\pi b^2 = A_r = 100\,\pi$$

$$b = 7.07 \text{ mm and } a = 14.14 \text{ mm}$$

$$A_{se} = 2 \times \pi ab + \pi\sqrt{2}\sqrt{a^2 + b^2} \times l_e$$

$$= 2 \times 100\pi + \pi\sqrt{2}\sqrt{(14.14)^2 + (7.07)^2} \times 50$$

$$= 1317.8 \times \pi \text{ mm}^2$$

$$V_e = \pi ab \times l_e = 100\,\pi \times 50 = 15708 \text{ mm}^3$$

New $V_r = V_e$

$$\therefore \quad \frac{t_r}{t_e} = \left(\frac{A_{se}}{A_{sr}}\right)^2 = \left(\frac{1317.8\pi}{1200\pi}\right)^2 = 1.2$$

Q.5. (*c*) Discuss the process capalibilities and applications of Gas Metal Arc Welding, Gas Tungsten Arc Welding, and Diffusion Bonding processes.

Sol.

Welding process

1. Gas Tungsten Arc Welding (GTAW)

   It uses non-consumable tungsten electrode or TIG welding.

2. Gas Metal Arc Welding (GMAW)

   It uses consumable filler wire of small diameter and composition compatible with work material or MIG welding.

3. Diffusing bonding processes.

   A weld is achieved by the application of pressure, of the order of 5 to 75 MPa, while the pieces are held at about 70% of their melting point temperature.

Process Capabilities

1. Mild steel upto 3 mm thickness with argon as shielding gas.

   Stainless steel and high temperature alloys.

   Aluminium, Magnesium, copper and nickel, Titanium.

2. Aluminium alloys, carbon and low alloy steels and stainless steels with inlet gas shielding. Ferrous and non-ferrous metals with CO_2, N_2, argon and helium gas shielding.

3. Iron, titanium, steel, welding of steel to aluminium, tungsten, titanium, molybdenum, cermets, copper to titanium, titanium to platinum, etc.

Applications

1. Precision welding in aircraft, atomic energy and instruments industries, aircraft frame, jet engine casting and rocket motor casting, can sealing, tube-to-tube sheet welding of heat exchangers, die castings welding.
2. Production of carbodies, fabrication of structures, shipbuilding, pressure vessles, tanks, pipes, domestic equipment, aircraft engine manufacturing industries, manufacture of railway coaches, automobile industry, suitable for rotatic welding.
3. Tipping of heavy cutting tools with curbide tips or hard alloys, joining of vacuum tube components. In aerospace industry it is used for fabricating complex shaped components of titanium from simple structural shapes. Surfacing components to resist wear, corrosion or heat.

Q.6. (*a*) Discuss the effects of the following elements on the machinability of steels :

(*i*) Aluminium and Silicon

(*ii*) Sulphur and Selenium

(*iii*) Lead and Tin

(*iv*) Carbon and Manganese

(*v*) Molybdenum and Vanadium

Sol.

Elements	Effects on machinability of steels
1. Aluminium	Used as alloying element in nitriding steels decreases their machinability by producing aluminium oxides.
2. Silicon	Decreases machinability by producing silicates.
3. Sulphur	Improves machinability by producing sulphates in the presence of mangenese.
4. Selenium	Improves machinability.
5. Lead	Improves machinability by acting as an internal lubricent.
6. Tin	Increases machinability being soft.
7. Carbon	Increases hardenability and decreases machinability.
8. Manganese	Increases hardenability and decreases machinability. But in the presence of sulphur increases machinability by forming MnS.
9. Molybdenum	Increases hardenability and reduces machinability. But when alloyed with Ni, Cn and added to steel, gives better machining qualities.
10. Vanadium	It has very strong tendency to form carbides and reduces machinability.

Q.6. (*b*) Discuss the effects of inefficient dielectric and electrolyte circulation in the inter-electrode gap on the Electric Discharge Machining and Electro Chemical Machining processes respectively.

Sol. Effects of insufficient dielectric and electrolyte circulation in the inter-electrode gap

(*a*) Electric Discharge Machining

1. Causes overheating of the system due to insufficient cooling.
2. Causes accumulation of debris between electrodes and workpiece due to insufficient flushing. This decreases the life of electrode and increases machining time.
3. Dielectric fluid acts as an insulating medium. Its insufficiency slow down the potential.
4. Constant resistance across the gap cannot be maintained.

(*b*) Electrochemical Machining

1. The electrolytic process does not work efficiently.
2. The cutting zone is not cooled sufficiently and gets heated up due to the flow of high current.
3. The waste product of reaction is not flushed resulting in decreased efficiency of the system.

Q.6. (*c*) Why are tools coated? What are the common coating materials?

Sol. Tools are coated to improve their wear resistance, toughness, hot hardness and better thermal characteristics. This increases their tool life and can work at high cutting speed resulting in reduced machining time and operation cost.

Common coating materials are :

Titanium nitride, Titanium carbonitride, Aluminium oxide, Titanium carbide and Diamond.

Q.6. (*d*) What is creep feed grinding? Discuss its salient features, advantages, and application.

Sol. Creep Feed Grinding : In creep feed grinding, the entire depth of cut is completed in one pass only using very small in feed rates. The depth of cut is of the order of 1 to 30 mm with low work speeds of the order of 1 to 0.025 m/min. This reduces the idle time due to stopping of wheel and table reversal. The cutting forces and power consumed increases. This requires frequent dressing of wheel for efficient operation. The grinding fluids used are oil based in view of the low grinding speeds used. The creep feed grinding set up is shown in Fig.

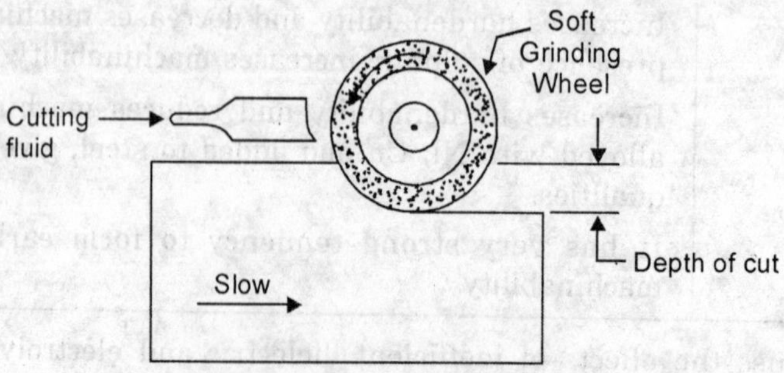

Fig. 13 (*a*). Creep feed grinding.

Q.6. (*e*) With the help of a neat sketch, explain the working of a diamond pin locator.

Sol. Diamond pin locator : Workpiece such as connecting rod or lever, which have two previously machined and finished holes at the two ends, may be located with the help of two pins projecting from the base surface of a jig or fixture, which will fit into the two holes in the workpiece, Fig. 14 (*a*). Assuming that the workpiece is effectively located on pin A, the only movement the workpiece can have is that of rotation about the pin A. Now, neither the workpiece nor the jig or fixture can be made to the exact dimensions. It means the centre distances between pins A and B and between holes A and B are subject to variation. When the centre distance dimensions for the pins and holes are at maximum and minimum conditions, a large allowance will result between the hole and pin at B in the Y direction. Due to this, the workpiece will have undesirable rotation about the pin A and the pin B becomes useless. Therefore, to locate the component completely, location faces opposed to this rotational movement should be provided at the hole B. This is achieved by relieving the pin B on two sides perpendicular to the X-axis.

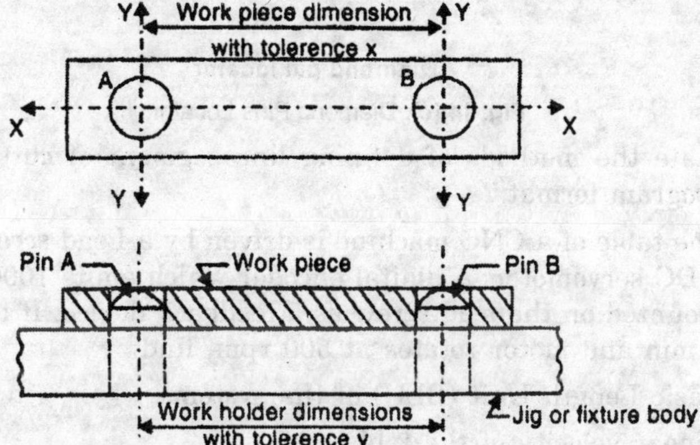

Fig. 14 (a). Radial Location.

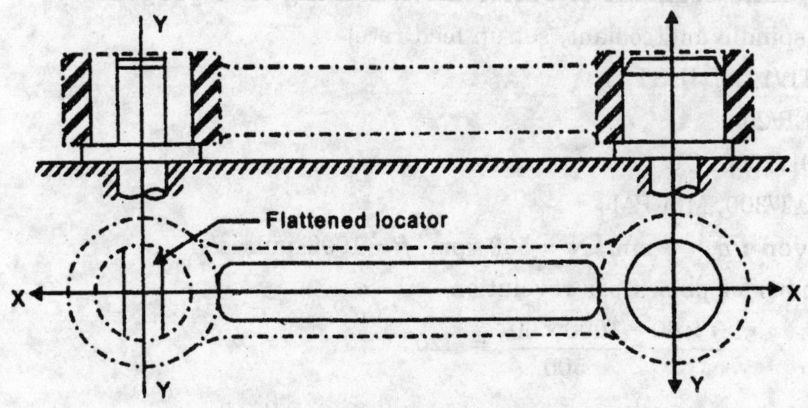

Fig. 14 (b). Flattened Pin Locator.

This will allow for variations in the X-direction but will provide cylindrical locating surfaces in the Y-direction. This will result in a flattened or diamond pin locator as shown in Fig. 14 (*b*) and Fig. 14 (*c*) respectively.

The important and accurate hole of the two holes should be used for principal cylindrical location with a full cylindrical pin. The diamond pin is used to constrain the pivoting of the workpiece around the principal location. The principal locator should be longer than the diamond pin so that the workpiece can be located and pivoted around it before engaging with the diamond pin. This simplifies and speeds up locating of the workpiece.

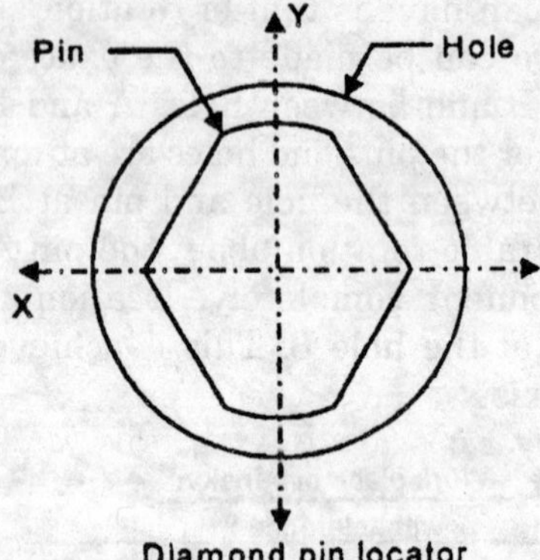

Fig. 14 (c). Diamond Pins Locator.

Q.6. (f) (*i*) State the methods of defining line segment of cutter motion using APT program format.

(*ii*) The table of a CNC machine is driven by a Lead screw which is rotated by a DC servomotor. A digital encoder which emits 1000 pulses per second is mounted on the lead screw as a feedback device. If the lead screw pitch is 6 mm and motor rotates at 500 rpm, find

1. Basic Length Unit (BLU) of the system.
2. Linear velocity of the table.
3. Frequency of pulses generated by the feedback device.

Sol. (*i*) **Line segment of cutter motion using APT program**

[Start spindle and coolant, set up feed rate]

Load TL/1, ADJUST, 1

CUTTER/20

SPINDL/2000, CW

FEDRAT/300, MM PM

(*ii*) **Given :** p = 6 mm, N = 500 rpm, f = 1000 pulses/s

1. Number of pulses per revolution

$$\frac{f \times 60}{\text{N}} = \frac{1000 \times 60}{500} = 120$$

$$\text{BLU} = \frac{p}{\text{pulses/rev}} = \frac{6}{120} = 0.05 \text{ mm}$$

2. Linear velocity of the table,

$$v = \text{BLU} \times \text{pulses/min} = 0.05 \times 1000 \times 60 \times 10^{-3}$$

$$= 3 \text{ m/min}$$

3. Frequency of pulses generated by the feedback device

$$= \frac{v}{\text{BLU}} = \frac{30 \times 10^3}{0.05} = 60{,}000 \text{ pulses/min}$$

Q.7. (*a*) A machine is used for turning operation and it takes 30 minutes to machine the component. Efficiency of the machine is 80% and scrap is 25%. The desired output is 1200 pieces per week. Considering 40 hours per week and 50 week in a year, determine the number of machines required in a year.

Sol. Given : t = 30 min, η_m = 0.8, scrap = 25%, Q = 1200 pieces/week,

Time available/year, T = 40 × 50 = 2000 hours/year

Number of components required per year = 1200 × 50 = 6 × 10^4

$$\text{Number of components required per hour} = \frac{6 \times 10^4}{2000} = 30$$

$$\text{Number of components produced per hour} = \frac{1}{t} = \frac{1}{0.5} = 2$$

Actual components produced per hour = 2 × η_m × (1 – scrap)

$$= 2 \times 0.8 \times 0.75 = 1.2$$

$$\text{Number of machines required} = \frac{30}{1.2} = 25$$

Q.7. (*b*) Draw the flow chart for solving assignment problems by Hungarian method.

Sol. Hungarian method for Assignment Problems : The flow chart is shown in Fig. 15.

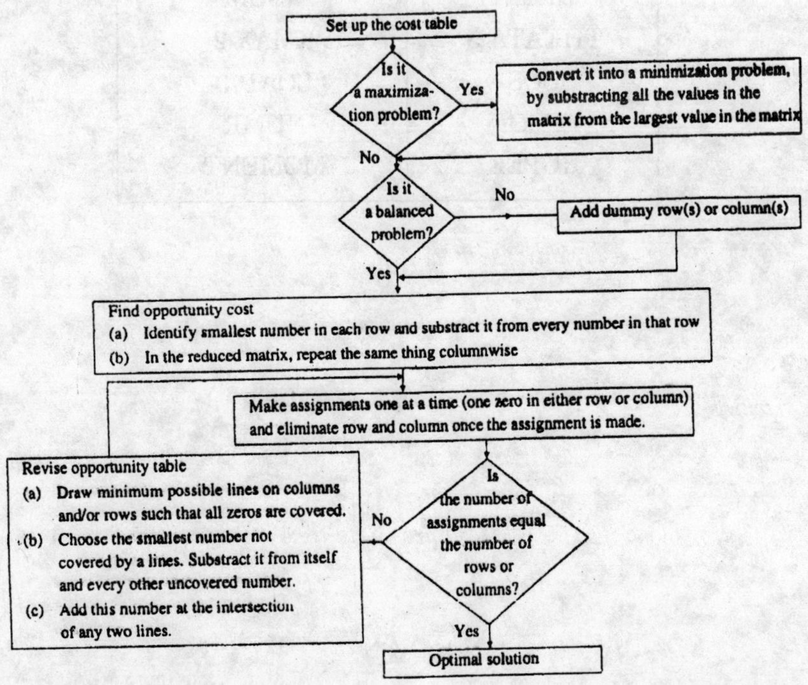

Fig. 15. Flow chart of steps in the Assignment Problems by Hungarian method.

Q.7. (*c*) Explain least square method of forecasting :

Sol. Least square Method of Forecasting : Simple regression analysis by the least square method enables the forecasted to predict the value of particular variable based on its relationship to another variable. This relationship is assumed to be linear as :

where n = number of observations

Coefficient of determination,

$$r^2 = 1 - \frac{\Sigma(Y - \hat{Y})^2}{\Sigma(Y - \bar{Y})^2}$$

$$= \frac{\alpha\Sigma Y + \beta\Sigma XY - n\bar{Y}^2}{\Sigma Y^2 - n\bar{Y}^2}$$

Q.7. (*d*) Group the following variables as integer and real variables in FORTRAN program language :

(*i*) NSUM
(*ii*) KMAX 2
(*iii*) DIA 24
(*iv*) LOAD 3
(*v*) THEATA 5
(*vi*) COS
(*vii*) EPSON
(*viii*) RUPEE
(*ix*) NUMR
(*x*) MOMENT

Sol.

Real variables	Integer variables
DIA 24	NSUM
THEATA 5	KMAX 2
COS	LOAD 3
EPSON	NUMR
RUPEE	MOMENT

MECHANICAL ENGINEERING PAPER II—2012

(Conventional)

SECTION-A

Q.1. **(*a*)** Determine the maximum and minimum transmission angle of the mechanism shown in Fig. 1 which is driven by member O_1A.

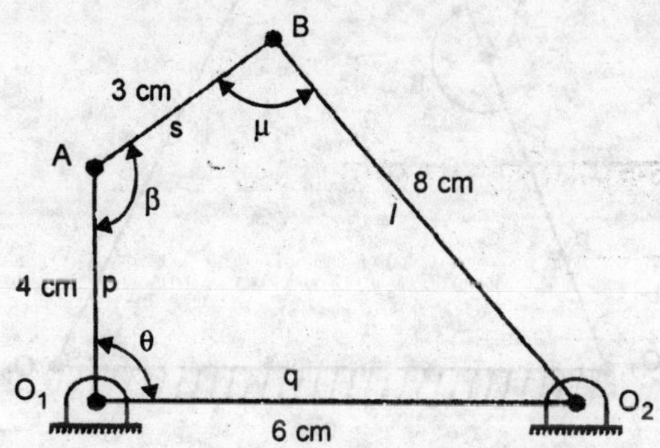

Fig. 1

Sol. Given : $l = 8$ cm, $s = 3$ cm, $p = 4$ cm, $q = 6$ cm

Since $(l + s) > (p + q)$, it belongs to class-II mechanism and is therefore a double-rocker mechanism.

(*i*) For transmission angle to be maximum, $\theta = 180°$ (Fig. 2a)

$$(p + q)^2 = l^2 + s^2 - 2ls \cos \mu$$

$$(4 + 6)^2 = 8^2 + 3^2 - 2 \times 8 \times 3 \times \cos \mu$$

$$100 = 64 + 9 - 48 \cos \mu$$

$$\cos \mu = -0.5625$$

$$\mu = -55.77° \quad \text{or} \quad 124.23°$$

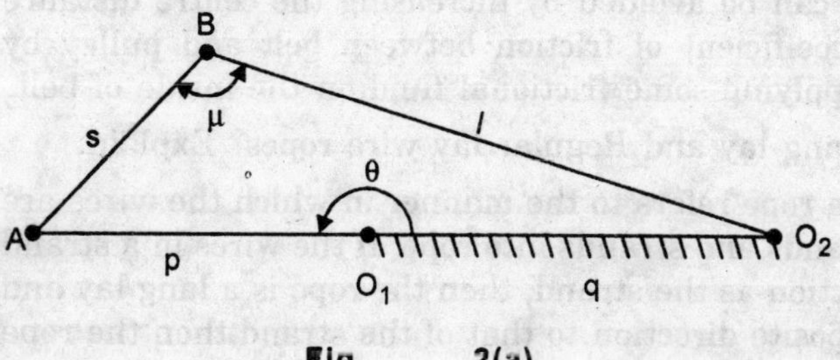

Fig. 2(a)

(*ii*) For transmission angle to be minimum, $\beta = 180°$ (Fig. 2b)

$$\therefore \quad q^2 = (p + s)^2 + l^2 - 2(p + s)\, l \cos \mu$$

$$6^2 = (4 + 3)^2 + 8^2 - 2\,(4 + 3) \times 8 \times \cos \mu$$

$$36 = 49 + 64 - 112 \cos \mu$$

$$\cos \mu = 0.6875$$

$$\mu = 46.57°$$

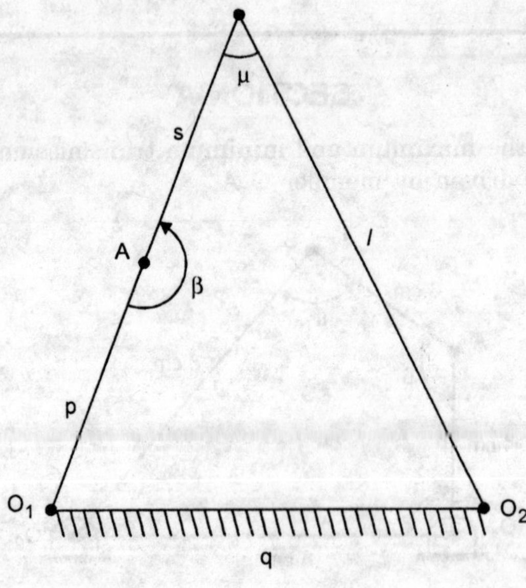

Fig. 2(b)

(*b*) For a cam whose acceleration is constant and is rotating with a constant speed, plot and show displacement diagram, velocity diagram, acceleration diagram and jerk diagram during rise.

Sol. When speed is constant then acceleration is zero. The acceleration and jerk are infinite at the beginning and end of rise. The various diagrams are shown in the Fig. 3.

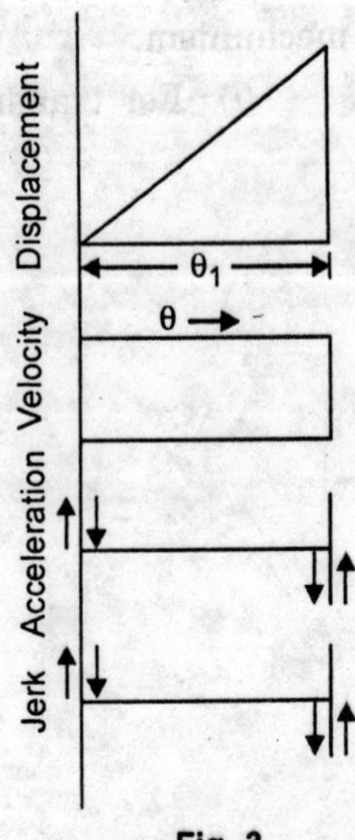

Fig. 3

(*c*) What is the effect of V belt over flat belt in transmission? How to avoid slip in flat belt?

Sol. The effect of V-belt over flat belt is that in case of slipping the V-belt goes down in the pulley groove and stop slipping.

Slip in flat belt can be avoided by increasing the centre distance and increasing the coefficient of friction between belt and pulley by sprinkling sand or applying some frictional fluid on the inside of belt.

(*d*) What are Lang lay and Regular lay wire ropes? Explain.

Sol. The lay of a rope refers to the manner in which the wires are helically laid into strands and strands into rope. If the wires in a strand coil in the same direction as the strand, then the rope is a lang lay and if they coil in the opposite direction to that of the strand then the rope is a regular lay.

(*e*) Fine pitch threads are more resistant to loosening due to vibrations than coarse pitch threads, why?

Sol. Fine pitch threads provide more area of contact as compared to coarse pitch threads. Hence they are more resistant to loosening due to vibrations.

(*f*) What is the difference between thick film lubrication and thin film lubrication?

1. **Thin film lubrication :** Thin film lubrication is defined as a condition of lubrication where the lurbicant film is relatively thin and there is partial metal-to-metal contact. This mode of lubrication is seen in door hinges and machine tool slides. It is also called boundary lubrication. The conditions resulting in thin film lubrication are excessive load, insufficient surface area or oil supply, low speed and misalignment.
2. **Thick film lubrication :** When the bearing and the journal is completely separated from each other by the lubricant film, then it is called thick film lurbication. Since there is no contact between the surfaces, the properties of surface, like surface finish, have little or no influence on the performance of the bearing. The resistance to relative motion between the journal and the bearing arises from the viscous resistance of the lurbicant. Therefore, the performance of the bearing is affected by the viscosity of the lubricant. Thick film lurbication could be hydrodynamic or hydrostatic. Thick film lubrication is provided in steam turbines, centrifugal compressors, crank shaft bearings, etc.

(*g*) Differentiate between basic static load rating and basic dynamic load rating of a ball bearing.

Sol. Basic Static Load Rating : The basic static load rating of radial ball bearing is defined as that radial load which corresponds to a total permanent deformation of element and race at the most heavily stressed contact of 0.0001 of the ball or roller diameter.

In single-row angular-contact ball bearings, the basic static load rating relates to the radial component of that load which causes a purely radial displacement of the bearings in relation to each other.

Basic Dynamic Load Rating : The basic dynamic load rating of radial ball bearings is that constant stationary radial load which a group of apparently identical ball bearings with stationary outer rings can endure for a rating life of one million revolutions of the inner rings.

In single-row angular-contact bearings, the dynamic load rating relates to the radial component of the load which results in a purely radial displacement of the bearing rings in relation to each other.

Load ratings, if given for specific speeds, are to be based on a rating life of 500 hours. This is equivalent to running the bearing for 500 hours at $33\frac{1}{3}$ rpm or a total of 10^6 revolutions.

(*h*) Differentiate clearly between Yield stress and 0.2% Proof stress for a ductile metal.

Sol. Yield stress : It is the stress at which a material exhibits a specific deviation from proportionality of stress and strain.

0.2% proof stress : It is the stress defined by drawing a line parallel to the stress-strain curve obtained in a simple tensile test at 0.2% of maximum strain.

(*i*) At a point in a stressed body, principal strain are e_1 and e_2. If E is the Young's modulus and ν is the Poisson's ratio, write down expressions for principal stresses in terms of e_1, e_2, E and ν.

Sol.

$$\sigma_1 = \frac{E}{1-\nu^2}[e_1 + \nu e_2]$$

$$\sigma_2 = \frac{E}{1-\nu^2}[e_2 + \nu e_1]$$

(*j*) A close coiled helical spring is subjected to axial load. Shear stress is developed in spring wire. Why the resultant shear stress at inner coil radius is more than the resultant shear stress at outer coil radius?

Sol. The maximum shear stress occurs at the inside of the spring coil. The reason for this is that the radius of curvature of the inside surface of the coil is smaller than the radius of curvature of the outer surface of the coil. Due to this the inside fibres are subjected to higher stress as compared to outer fibres.

(*k*) Mark the plane with Miller Indices $(1\ \bar{1}\ 0)$ in a cubic structure.

Reciprocals of miller indices are : (1, –1, ∞) plane has an intercept of 1 on x-axis, –1 on y-axis and is parallel to z-axis.

Plane $(1\ \bar{1}\ 0)$ is shown in Fig. 4.

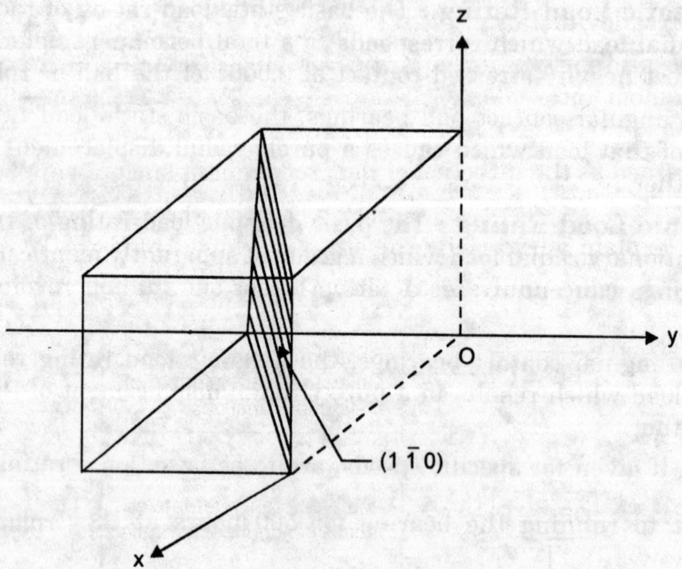

Fig. 4

(*l*) What is lead-tin eutectic alloy? What are special characteristics of a eutectic alloy? What is the composition of solder used in joining electrical connections?

Sol. Tin-lead eubectic alloy is formed by an eutectic reaction of lead and tin. Lead alloyed with tin forms solders which is used for joining electrical connections. Lead-tin eutectic alloy when alloyed with cadanium form bearing alloys, called white metal.

(***m***) What types of oxyacetylene flames you will use to weld (*i*) Copper base metals (*ii*) High carbon steel?

Why are all fuel lines threaded left handed?

Sol. Metal Type of oxyacetylene flame

(*i*) **Copper base metals :** Neutral flame

(*ii*) **Higher carbon steel :** Reducing flame

The regulators on fuel lines are to be operated by left hand and are therefore threaded left handed.

(***n***) Write the advantages and disadvantages of climb milling. In connection with grinding define grinding ratio.

Sol. In climb milling, the cutting and feed motion are in the same direction. It results in a better surface finish and longer tool life. The main disadvantage of climb milling is the tendency of the job being dragged into the cutter.

$$\text{Grinding ratio} = \frac{\text{Volume of work material removed}}{\text{Volume of wheel worn away}}$$

(***o***) In connection with surface texture define (*a*) waviness (*b*) flaws, and (*c*) lay. List three defects found on surfaces.

Sol. Waviness : It consists of those surface irregularities which are of greater spacing than roughness and it occurs in the form of waves. It is characterised by the ratio of their pitch to the peak of valley distance.

Flaws : These are surface irregularities or imperfections which occur at infrequent intervals and at random intervals. Such imperfections are : scratches, cracks, holes, pits, porosity, etc.

Lay : Lay is defined as the direction of the predominant surface pattern produced by tool marks.

(***p***) Name and explain very briefly the steps in Value Analysis.

Sol. The steps in value analysis are shown below :

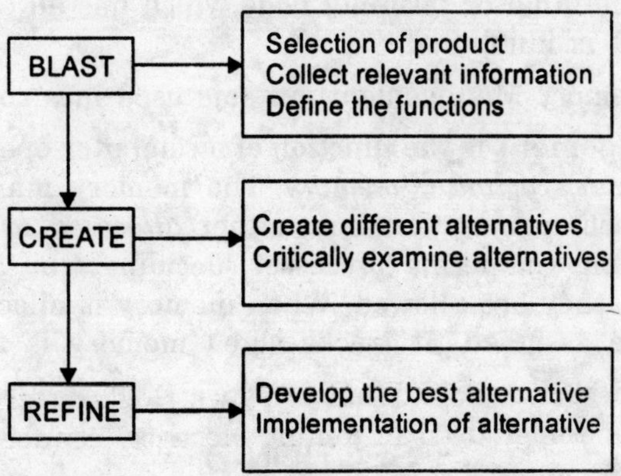

(***q***) Name and differentiate very briefly the principal classes of methods used in forcasting of demand.

1. **Qualitative or judgemental methods :** These methods rely on experts opinion in making a prediction for the future and are useful for long-range forecasting tasks.
2. **Extrapolative or time series method :** These methods use the past history of demand in making a forecast for the future and peform well for short time forecast.
3. **Causal or explanatory methdos :** Regression analysis.

(r) List the principal inputs and outputs of a Material Requirement Planning System.

Sol. Inputs

1. Master production plan
2. On hand inventories
3. Bill of materials
4. Purchased or manufactured order status by item
5. Replenishment rules by item

Outputs

1. **Primary outputs :** (*i*) Order release notice (*ii*) Planned orders to be released in future (*iii*) Rescheduling notices (*iv*) Cancellation notices (*v*) Inventory status.
2. **Secondary outputs :** (*i*) Performance reports of costs, item usage, lead times, etc. (*ii*) Inventory forecasts (*iii*) Exception reports.

(s) List the steps involved in numbering events and activities in a project so that CPM/ PERT can be used for scheduling it.

Sol. The steps involved in numbering the events as per the Fulkerson's rule are :

1. The initial event which has all outgoing arrows with no incoming arrow is numbered "1".
2. Delete all the arrows coming out from node "1". This will convert some more nodes (at least one) into initial events, number these events 2, 3, ... etc.
3. Delete all the arrows going out from these numbered events to create more initial events. Assign next numbers to these events.
4. Continue until the final or terminal node which has all arrows coming in, with no arrows going out is numbered.

(t) What are the Memory Management systems used in a computer?

Sol. Memory management is the function of a computer operating system responsible for managing the computer's *primary memory*. The memory management function keeps track of the status of each memory location, either *allocated* or *free*. It determines how memory is allocated among competing processes, deciding who gets memory, when they receive it, and how much they are allowed. When memory is allocated it determines which memory locations will be assigned. It tracks when memory is freed or *unallocated* and updates the status.

SECTION-B

Q.2. (a) The following particulars refer to a screw jack shown in Fig. 5 below :

Vertical load on the screw $W = 20{,}000$ N

Force applied at the lever and whose length $l = 80$ cm

Mean radius of the screw $r = 2.5$ cm

Pitch of the screw $p = 1$ cm

Mean radius of contact surface between swivel & head $r_m = 4$ cm

Coefficient of friction between screw thread and bearing $\mu' = 0.15$

Coefficient of friction between the swivel head and bearing $\mu = 0.1$

Determine :

(*i*) The force on the lever end required while raising and lowering the load.

(*ii*) The efficiency of the screw jack in both cases.

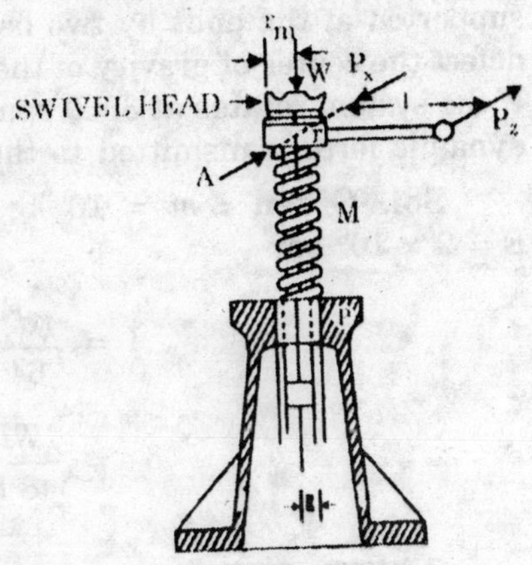

Fig. 5

Sol.

(*i*) Angle of helix of screw threads

$$\alpha = \tan^{-1}\left(\frac{p}{2\pi r}\right) = \tan^{-1}\left(\frac{1}{2\pi \times 2.5}\right)$$

$$= \tan^{-1}(0.06366) = 3.64°$$

$$\tan\phi = \mu' = 0.15, \quad \phi = 8.53°$$

Force required for raising the load,

$$P_1 = W\tan(\alpha + \phi)$$
$$= 20\tan(3.64° + 8.53°)$$
$$= 20\tan 12.11° = 4.313 \text{ kN}$$

Force required for lowering the load,

$$P_2 = W\tan(\phi - \alpha)$$
$$= 20\tan(8.53° - 3.64°)$$
$$= 20\tan 4.89° = 1.711 \text{ kN}$$

Total torque required for raising the load,

$$T_1 = P_1 \times r + \mu W r_m$$
$$= 4.313 \times 2.5 \times 10^{-2} + 0.1 \times 20 \times 4 \times 10^{-2}$$
$$= 0.1878 \text{ kN}\cdot\text{m}$$

Let F = Force applied at the end of the level

Then $T_1 = F_1 \times l$

$$F_1 = \frac{0.1878 \times 10^3}{0.8} = 234.8 \text{ N}$$

To lower the load,

$$T_2 = P_2 r + \mu W r_m$$
$$= 1.711 \times 2.5 \times 10^{-2} + 0.1 \times 20 \times 4 \times 10^{-2}$$
$$= 0.1228 \text{ kN}\cdot\text{m}$$
$$= F_2 \times l$$

$$F_2 = \frac{0.1228 \times 10^3}{0.8} = 153.5 \text{ N}$$

(*ii*) Efficiency of screw jack

$$\eta_1 = \frac{P_0 \times r}{P_1 r + \mu W r_m}, \text{ for raising the load}$$

where $P_0 = W \tan \alpha = 20 \tan 3.64° = 1.272$ kN

$$\eta_1 = \frac{1.272 \times 0.025}{0.1878} = 0.1694 \text{ or } 16.94\%$$

$$\eta_2 = \frac{P_0 \times r}{T_2} = \frac{1.272 \times 0.025}{0.1228} = 0.2589 \text{ or } 25.89\%$$

(*b*) A rotor of mass 10 kg is mounted midway on a 20 mm diameter horizontal shaft supported at the ends by two bearings. The bearing span is 0.8 m. Because of manufacture defect the center of gravity of the disc is 0.1 mm away from the geometric center of the rotor. If the system rotates at 3000 rpm, determine the amplitude of the steady state vibration and dynamic force transmitted to the bearing. Take E = 2 × 10⁶ bar. (10)

Sol. Given : m = 10 kg, d = 20 mm, l = 0.8 m, e = 0.1 mm, N = 3000 rpm, E = 2 × 10^6 bar

$$I = \frac{\pi d^4}{64} = \frac{\pi \times 20^4}{64} = 7854 \text{ mm}^4$$

$$\delta_{st} = \frac{Wl^3}{48 \text{ EI}} = \frac{10 \times 9.81 \times (0.8)^3}{48 \times 2 \times 10^{11} \times 7854 \times 10^{-12}} = 6.66 \times 10^{-4} \text{ m or } 0.666 \text{ mm}$$

Critical speed, $$w_n = \sqrt{\frac{g}{\delta_{st}}} = \sqrt{\frac{9.81}{6.66 \times 10^{-4}}} = 121.35 \text{ rad/s}$$

$$w = \frac{2\pi N}{60} = \frac{2\pi \times 3000}{60} = 314.16 \text{ rad/s}$$

$$\beta = \frac{w}{w_n} = \frac{314.16}{121.35} = 2.59$$

$$\frac{r}{e} = \frac{\beta^2}{\beta^2 - 1}$$

Amplitude of steady state vibration,

$$r = \frac{0.1 \times (2.59)^2}{(2.59)^2 - 1} = 0.1175 \text{ mm}$$

Dynamic load on bearing,

$$F_d = rk$$

$$k = m w_n^2 = 10 \times (121.35)^2 = 147{,}258 \text{ N/m}$$

$$F_d = 0.1175 \times 10^{-3} \times 147{,}258 = 17.3 \text{ N}$$

Dead weight on shaft,

$$W = mg = 10 \times 9.81 = 98.1 \text{ N}$$

Dynamic load on each bearing

$$= \frac{1}{2} F_d = \frac{1}{2} \times 17.3 = 8.65 \text{ N}$$

Total load on each bearing

$$= \frac{1}{2}(W + F_d) = \frac{1}{2}(98.1 + 17.3) = 57.7 \text{ N}$$

(c) Write short notes on the following :

(i) Epicyclic gear trains.

(ii) Interference and undercutting in involute gears.

Sol. Epicyclic gear trains : These are gear trains in which the axis of one or more gears move relative to the frame. The gear at the centre is called the sun, and the gears whose axes move are called the planets.

An epicyclic gear train is shown in Fig. 6. Arm 3 drives gear 1 about gear 2, which is a fixed external gear. Gear 1 rotates about its centre O_1 while this centre rotates about centre O_2 of the fixed gear. As gear 1 rolls on the outside of gear 2, a point on its surface will generate an epicycloid. If gear 2 happens to be internal gear and gear 1 rolls on the inside of gear 2, then a point on the surface of gear 1 will generate a hypocycloid. Because of the curves generated, it is called epicyclic gear train.

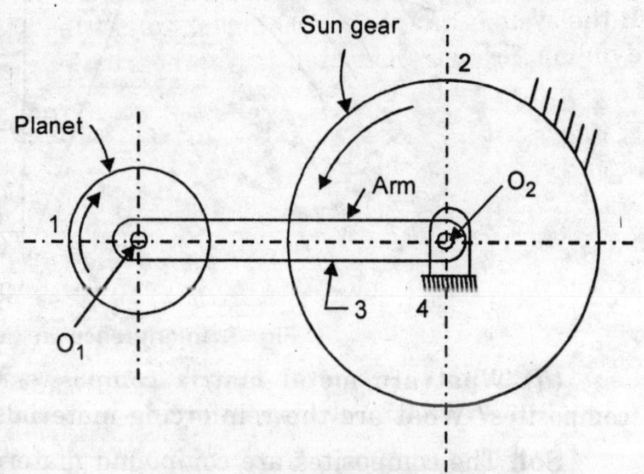

Fig. 6. Epicyclic gear train

***(ii)* Interference and undercutting in involute gears :** An involute starts at the base circle and is generated outwards. It is therefore impossible to have an involute inside he base circle. The line of action is tangent to the two base circles of a pair of gears in mesh, and these two points represent the extreme limits of the length of action. These two points are called *interference points*. If the teeth are of such proportion that the beginning of contact occurs before the interference point is met, then the involute portion of the driven gear will mate with a non-involute portion of the driving gear, and *involute interference* is said to occur This condition is shown in Fig. 7, E_1 and E_2 show the interference points that should limit the length of action. A shows the beginning of contact, and B shows the end of contact. It can be seen that the beginning of contact occurs before the interference point E_1 is met; therefore, interference is present. The tip of the driven tooth will gauge out or undercut the flank of the driving tooth as shown by the dotted line.

There are several ways of eliminating interference. Interference can be avoided by undercutting, making stub teeth, increasing the pressure angle, and cutting the gears with long and short addendum gear teeth. The method of undercutting is to limit addendum of the driven gear so that it passes through the interference point E_1, thus giving a new beginning of contact. Interference and the resulting undercutting not only weaken the pinion tooth but may also remove a small portion of the involute adjacent to the base circle, which may cause a serious reduction in the length of action.

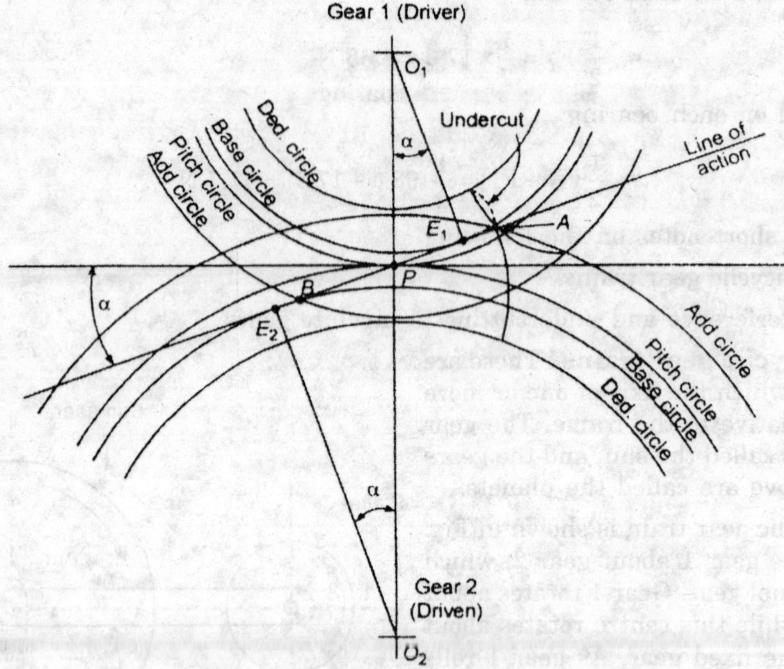

Fig. 7. Interference in gears and undercutting.

(*d*) What are metal matrix composites? What are special characteristics of these composites? What are the reinforcing materials? What are the applications of MMC? (10)

Sol. The composites are compound materials of cohesive structure made by physically combining two or more compatible materials, differing in composition and characteristics. The metal matrix composites are formed by reinforcing the liquid matrix by dispersing the metallic wires or fibres. The MMC have superior qualities as coompared to the parent constituents. The matrix may be of plastics or resins and the reinforcing metallic material may be steel wires, steel platelets, Ni, Al, Cu, Boron, Tungsten, Molybdenum, Al_2O_3, SiC, etc.

Special characteristics of MMC

1. High strength-to-weight ratio.
2. High stiffness-to-weight ratio.
3. Cost effective.
4. Long life.
5. No corrosion to reinforcing materials.
6. Suitable for high and low temperatures applications.

Applications

1. Structural parts.
2. Automotive parts.
3. Storage battery plates.
4. Bearings.

5. Aircraft and space craft structures.
6. Turbine and compressor blades.
7. Reflective, conductive and decorative coatings.

Q.3. (*a*) An open belt system is transmitting 10 kW power. Driving pulley is of diameter 200 mm and driven pulley is of diameter 480 mm. Centre distance between the pulleys is 2 m. Maximum stress in belt is not to exceed 2.2 MPa. Thickness of the bolt is 8 mm. Determine breadth and length of the belt.

Given density of belt material = 950 k/m³. Coefficient of friction between belt and pulley = 0.3. Driving pulley is rotating at 1000 rpm. Neglect bending stress in belt.

Sol. Given : p = 10 kw, d_1 = 200 mm, d_2 = 480 mm, C = 2 m, σ_{max} = 2.2 MPa, t = 8 mm, ρ = 950 kg/m³, μ = 0.3, n_1 = 1000 rpm

Linear speed of belt $v = \dfrac{\pi d_1 n_1}{10^3 \times 60} = \dfrac{\pi \times 200 \times 100}{10^3 \times 60} = 10.47$ m/s

Let b = belt width, mm

Mass of belt per metre length,

$$m = bt\rho = 6 \times 10^{-3} \times 8 \times 10^{-3} \times 950 = 7.6 \times 10^{-3} \times b \text{ kg}$$

Centrifugal tension, $T_c = mv^2 = 7.6 \times 10^{-3} \times b \times (10.47)^2 = 0.833 \times b$ N

$$\sin\alpha = \frac{d_2 - d_1}{2C} = \frac{480 - 200}{2 \times 2000} = \frac{280}{4000} = 0.07$$

$$\alpha = 4.014°$$

Angle of contact, $0 = 180° - 2 \times 4.014° = 171.972°$ or 3.0015 rad

Ratio of belt tensions, $\dfrac{T_1}{T_2} = \exp(\mu\theta) = \exp(0.3 \times 3.0015) = 2.96$

$$P = \frac{(T_1 - T_2)v}{10^3}$$

$$10 \times 10^3 = (T_1 - T_2) \times 10.47$$

$$T_1 - T_2 = 955.1 \text{ N}$$

$$T_2(2.46 - 1) = 955.1$$

$$T_2 = 654.2 \text{ N}, \quad T_1 = 1609.3 \text{ N}$$

$$T_1 + T_c = T_{max}$$

$$1609.3 + 0.833 \times b = b \times 8 \times 2.2$$

Width of belt, $b = 96$ mm

Length of open belt, $L_0 = 2C + \dfrac{(d_2 - d_1)^2}{4c} + \dfrac{\pi(d_1 + d_2)}{2}$

$$= 2 \times 2 + \frac{(480 - 200)^2 \times 10^{-6}}{4 \times 2} + \frac{\pi(200 + 480) \times 10^{-3}}{2}$$

$$= 4 + 0.0098 + 1.0681$$

$$= 5.0779 \text{ m}$$

(*b*) What are different types of threads used for power transmission? Make a simple sketch of each. Give special features of each thread and their applications.

Sol. Types of Screw Threads for Power Transmission : The following thread forms are used for power screws (Fig. 8) :

1. Square, 2. A4 cme, 3. ISO metric trapezoidal and 4. Buttress

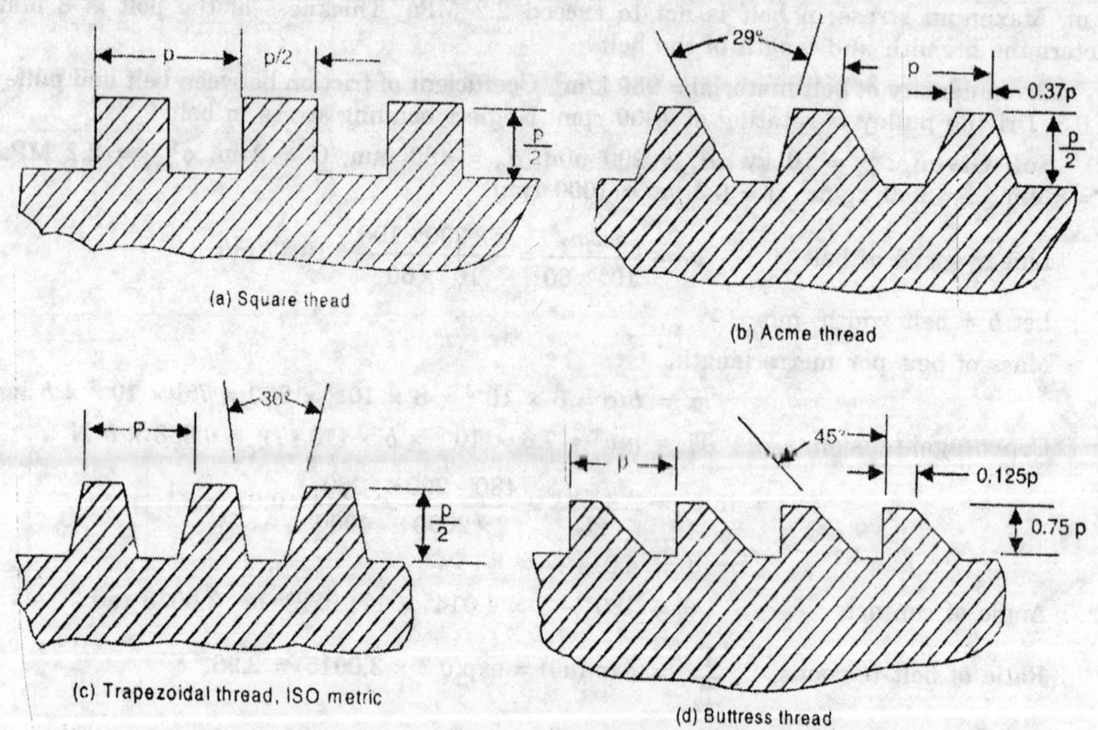

Fig. 8. Types of screw threads for power transmission.

Table 1. Speical festures and applications of power screws.

Type of screw threads	Speical features	Applications
(*a*) Square	• Can transmit power in either direction • Gives maximum efficiency • Produces minimum bursting pressure on the nut.	• Screw jacks • Presses • Clamping devices
(*b*) Acme	• Gives lower efficiency than square threads • Introduces some bursting pressure on the nut	• lathe lead screw • Power transmission in machine tools

Type of screw threads	Speical features	Applications
	• Stronger in shear than square threads • Can take up wear	
(*c*) Trapezoidal	• More stronger than square thread • Axial wear can be compensated	• Lathe lead screw • Power transmission machine tools
(*d*) Buttress	• Can transmit power in one direction only • Stronger than other threads • Limited use in power transmission.	• Used in vices

(***c***) Fatigue life of a 6 × 37 wire rope of nominal diameter of 12 mm is to be achieved to be 0.2 million bends. Tensile stress designation of wire rope is 1960 MPa. How much load can be lifted through a sheave of diameter 400 mm, through the arrangement shown in Fig. 9. Take ratio of bearing pressure between rope and sheave to ultimate strength of rope = 0.0024.

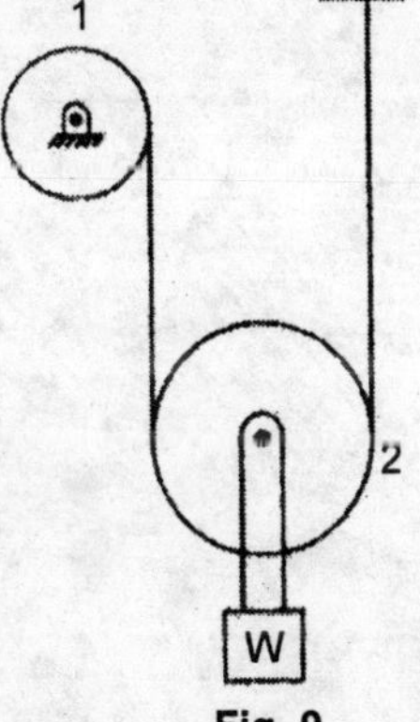

Fig. 9

Sol. Given : Wire rope = 6 × 37, d_r = 12 mm, number of bends = 0.2 million, tensile designation of wire rope, R_0 = 1960 MPa,

D = 400 mm, $\dfrac{p}{\sigma_{ut}} = 0.0024$

Number of bends :

Pulley 1 = 2, as the rope bends in opposite direction.

Pulley 2 = 1 for transition of rope from its straight position into bent one + 1 for bent position to straight one.

Total number of bends = 2 + 1 + 1 = 4

Rope life $\quad \alpha = \dfrac{1}{\text{number of bends}}$

For number of bends $= 4$, $D_{min}/d = 25$, $D_{min} = 25 \times 12 = 300$ mm < 400 mm.

Let $\quad$ W = load lifted

Load on each side of rope, $P = \dfrac{W}{2}$

$$p = \frac{2P}{d_r D} = \frac{W}{12 \times 400} = 2.083 \times 10^{-4} \times W$$

$$\frac{p}{\sigma_{ut}} = 0.0024, \; p = 1960 \times 0.0024 = 4.704 \text{ N/mm}^2$$

$$2.083 \times 10^{-4} \times W = 4.704$$

$$W = 22{,}583 \text{ N} \quad \text{or} \quad 22.583 \text{ kN}$$

Q.4. (*a*) A beam AB, 8 m long hinged at end A and roller supported at end B carries a point load of 8 kN at crank ED = 1 m, and a uniformly distributed load of 5 kN/m over CB = 4 m.

Determine :

(*i*) **Reactions at ends.**

(*ii*) **Bending moment at section where SF is zero.**

(*iii*) **Draw BM diagram only.**

Sol.

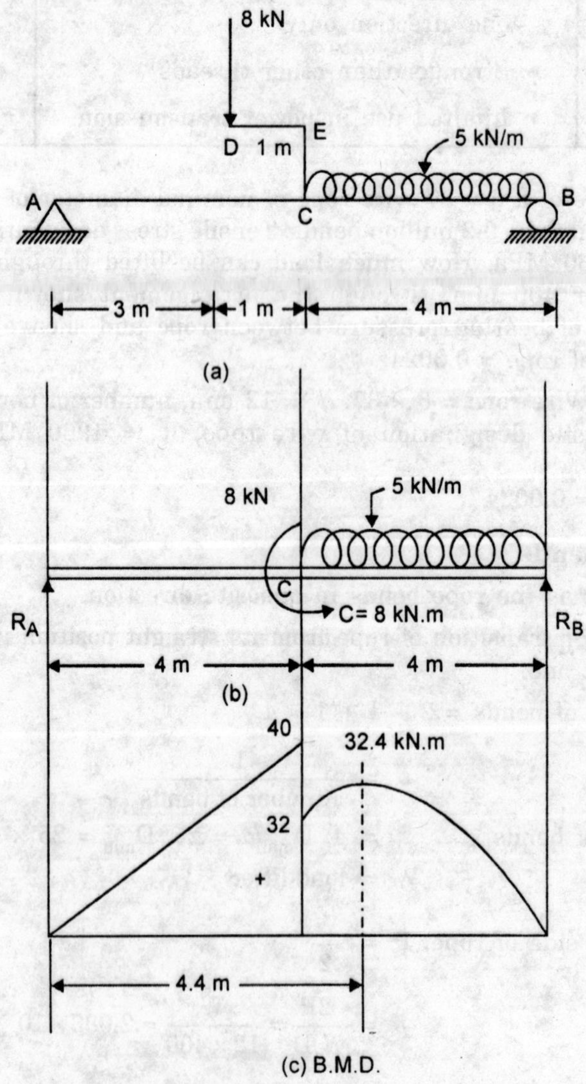

Fig. 10

(*i*) The load of 8 kN on the crank DE = 1m is equivalent to a couple C = 8 kN·m (ccw) at point C and a point load 8 kN at C, as shown in Fig. 10(*b*).

Let R_A and R_B be the reactions at the ends. Then taking moments about B, we have

$$8R_A = 8 \times 4 + 5 \times 4 \times 2 + 8$$
$$= 80$$
$$R_A = 10 \text{ kN}$$
$$R_B = 28 - 10 = 18 \text{ kN}$$

(*ii*) S.F. in span CB, taking x from A,

$$F_x = 10 - 8 - 5(x - 4) = 2 - 5(x - 4) = 0$$
$$x - 4 = 0.4,\ x = 4.4 \text{ m}$$

B.M.D.

Span AC :

$$M_x = 10x$$
$$M_A = 0,\ M_c = 10 \times 4 = 40 \text{ kN}\cdot\text{m}$$

Span CB :

$$M_x = 10 \times x - 8 - 8(x - 4) - \frac{5(x-4)^2}{2}$$
$$M_c = 10 \times 4 - 8 = 32 \text{ kN}\cdot\text{m}$$

At $x = 4.4$ m

$$M_x = 10 \times 4.4 - 8 - 8 \times 0.4 - 2.5 \times 0.4^2 = 32.4 \text{ kN}\cdot\text{m}$$
$$M_B = 10 \times 8 - 8 - 8 \times 4 - 2.5 \times 4^2 = 0$$

The bending moment diagram is shown in Fig. 10(*c*).

(*b*) A stepped shaft ABC, 1 m long made of steel is of diameter 20 mm from A to B = 0.5 m, 40 mm from B to C = 0.5 m. It is subjected to twisting moments 0.4 kNm (cw) at A, 0.8 kNm (ccw) at B and 0.4 kNm (cw) at C.

If G = 84 kN/mm² determine angular twist between A to C.

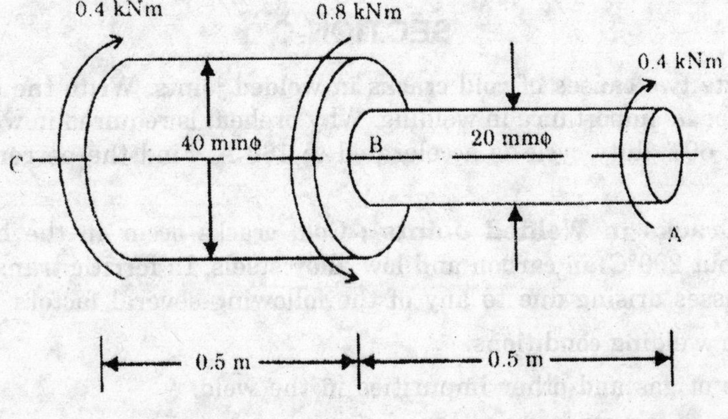

Fig. 11

Sol. As shown in Fig. 11,

$$T_{AB} = 0.4 \text{ kN}\cdot\text{m } (cw),\ T_{BC} = 0.8 - 0.4 = 0.4 \text{ kN}\cdot\text{m (ccw)}$$

$$\theta_{AB} = \frac{T_{AB}l_1}{GJ_1} = \frac{0.4\times10^6\times500\times32}{84\times10^3\times\pi\times20^4} = 0.15158 \text{ rad (ccw)}$$

$$\theta_{AB} = \frac{T_{BC}l_2}{GJ_2} = \frac{0.4\times10^6\times500\times32}{84\times10^3\times\pi\times40^4} = 0.00947 \text{ rad(ccw)}$$

Total angle of twist, $\theta = \theta_{AB} + \theta_{BC} = 0.15158 - 0.00947 = 0.1421 \text{ rad or } 8.142°\text{(cw)}$

(c) What do you understand by tempering? What are martempering and austempering process? What are their special features?

Sol. Tempering : It is the process of heating hardened steel to a temperature below 723°C and holding it there for the proper time, and then cooling slowly. Tempering is classified into three types according to the heating temperatures.

1. **Low-temperature tempering :** In this case, the work is heated to 150° to 250°C to relieve the internal stresses and to increase the ductility without changing the structure or hardness of steel. This process is applied to the heat treatment of carbon and low-alloy steel cutting tools, as well as measuring tools and parts that undergo surface hardening.
2. **Medium-temperature tempering :** This process involves heating the work to 350° to 450°C. This alters the structure of steel and matensite is transformed to troosite. The hardness get reduced and ductility increases.
3. **High-temperature tempering :** This process is performed in the range of 500° to 600°C to completely eliminate the internal stresses.

Martempering : It consists in heating steel to the hardening temperature, quenching in a medium having temperature of 150° to 300°C, holding to equalise the temperature and then cooled further to room temperature in air or oil. Austenite is converted into martensite. It results in less warping and quenching cracks.

Austempering : It is performed in the same manner as martempering but with a longer holding time at the hot bath temperature to ensure complete austenite decomposition into bainite and acicular troosite.

SECTION-C

Q.5. (a) Write two causes of cold cracks in welded joints. Write the carbon equivalent formula and decribe its importance in welding. Why preheat is required in welding? A machine rated at 100 A at 60% duty cycle is accelerated to 160 A. Find the percentage reduction of the output cycle.

Sol. Cold Cracks in Welded Joints : Cold cracks occur in the base metal at low temperatures (about 200°C) in carbon and low alloy steels. In ferritic transformable steels it is due to the stresses arising due to any of the following several factors :

1. Improper welding conditions.
2. Presence of gas and other impurities in the weld.
3. Wrong choice of filler rod.

4. Metallurgical factors such as excessive cooling rate resulting in the formation of martensite and formation of brittle phases in the weld when cold or the formation of phases which are brittle at high temperature.

Carbon Equivalent Formula

In welding, the carbon equivalent governs the hardenability of the parent metal. It is a rating of weldability related to carbon, manganese, chromium, molybdenum, vanadium, nickel and copper content. Carbon Equivalent, CE = C + Mn/6 + (Cr + Mo + V)/5 + (Cu + Ni)/15. The higher the carbon equivalent value, the faster the cooling rate, the higher the tendency for hard, brittle phases to form during cooling, and higher the tendency for cracking.

Preheating : 1. It is carried out to drive away moisture from weld area. preheating is performed by heating the surface of the material just above the boiling point of water. This will dry the plate surface and remove the undesirable contaminants that may otherwise cause porosity, hydrogen embrittlement, or cracking through the introduction of hydrogen during the welding process.

2. To lower the thermal gradient by reducing the temperature differential by preheating the base metal. This minimizes problems associated with distortion and excessive residual stress. Without perheating, rapid cooling will lead to formation of martensite and may cause cracking of metals with high hardenability.

Desired duty cycle, $D_d = ?$

Rated duty cycle, $D_r = 60\%$

Desired output current, $I_d = 160$ A

Rated output current, $I_r = 100$ A

$$\frac{D_d}{D_r} = \frac{(I_r)^2}{I_d}$$

$$D_r = 60\left(\frac{100}{160}\right)^2 = 23.437\%$$

$$\%\text{ reduction of output cycle} = \frac{60 - 23.437}{60} \times 100 = 60.94\%$$

(*b*) Name the various tests to be performed for moulding sand. Draw a top gate and write its advantages. What is a misrun and how is it caused? Write some advantages of cupola.

Sol. Various tests for moulding sand

1. Grain fineness test
2. Moisture content test
3. Clay content test
4. Permeability test
5. Strength tests
6. Hardeness test

Top Gate : In a top gate, the metal enters the mould cavity above the level of the parting line. The different designs of top gate are shown in the Fig. 12.

Advantages

1. It is conducive to favourable gradient for directional solidification towards the riser.
2. The pouring basin and gates themselves may serve as risers.
3. Suitable for larger castings made in moulds of erosion resistant material.

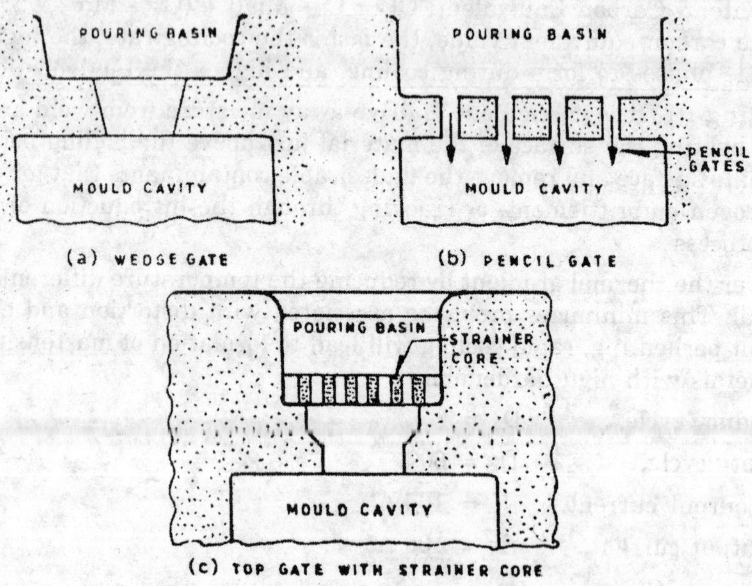

Fig. 12. Types of top gates.

Misrun : This is casting defect caused due to incomplete cavity filling. The causes for misrun are :

1. Inadequate metal supply.
2. Too low mould or melt temperature.
3. Improperly designed gates.
4. Too large length to thickness ratio of casting.

Advantages of Cupola

1. Gives continuous operation.
2. Low cost of operation.
3. Easy operation.
4. Composition of melt can be controlled.
5. Temperature of melt can be controlled.

(*c*) Define mistcooling and its effectiveness. Draw a Fig. showing relative severity of the following machining operations qualitatively.

Broaching (internal), milling, gear shaping and drilling, boring.

What happens to drilling if the lip relief angle is too small? Too large? In grinding, explain the meaning of the terms, loading and truing.

Sol. Mistcooling : The application of the cutting fluid as a fine atomized mist (combination of the carrier air and very fine drops of fluid) is called mistcooling. The size of the fluid drops is of the order of 10 to 25 μm. The mist is sprayed onto the cutting zone at high velocities of about 300 m/s or more and under pressure of the order of 0.28 to 0.42 MPa.

The effectiveness of mistcooling are :

1. Due to high velocity of fluid application the removal of heat by convection currents is intensified.
2. The surface area of the coolant is much greater in the form of mist than in flood application. This increases the cooling capacity of coolant.
3. Due to the expansion of mist in the issuing zone, its temperature falls down considerably.

Internal Broaching : In broaching operation, a special multipoint tool is used whose teeth remove the whole machining allowance in a single stroke. The only motion is the primary cutting motion of the tool. the feed is obtained by placing the teeth progressively deeper within the tool. The primary cutting motion is translational along the broach axis.

Milling : It is a machining process for removing excess material from a work piece with a rotating multiple-point cutting tool. Milling process involves simultaneous rotary cutter and linear feeding motion of the work.

Gear Shaping : The teeth are cut by the reciprocating motion of the cutter. The cutter can be either a rack-type cutter or a rotary pinion type cutter.

Drilling : It is an operation for making a hole with the help of a twist drill.

Boring : It is an operation for machining internal surface.

The average value of unevenness H_{av} for above machining operations are :

Process	Broaching	Milling	Shaping	Drilling	Boring
H_{av} (μm)	0.5–6.25	0.25–25	0.375–25	0.75–12.5	0.05–25

The relative severity of the machining processes is shown in Fig. 13.

Effect of lip relief angle in drilling : The heel of the drill point is backed off when ground to give relief behind the cutting lips. This will allow the cutting edges to cut without interference. If this angle is kept too small then the cutting edges will interfere during machining. It is kept 6° to 8°. If the relief angle is too large then the cutting force will increase.

Process	Roughness height µm										
	50	25	12.5	6.25	3.125	1.56	0.78	0.39	0.195	0.097	0.048
Sawing Planning, shaping Drilling											
Electric discharge machining Milling Broaching											
Reaming Boring, turning Electrolytic grinding											
Grinding Honing Polishing											
Lapping Superfinishing											

Key

—— Average application

The ranges shown above are typical of the processes listed.

Fig. 13

Grinding Operation

Loading : Grinding wheels are dressed to restore their cutting capacity which is lost when the voids between grains are loaded with the waste material of the grinding process dust (sintered metal). This defect is called loading.

Truing : It is an operation done on the grinding wheel to restore the correct geometric shape of the wheel, that is, it makes the face of the wheel concentric to its axis of rotation and its sides flat and parallel and normal to its axis of rotation. Truing is done by means of industrial diamonds.

(*d*) Write a FORTRAN program to find average height of boys in a class.

Sol.

```
C    AVERAGE HEIGHT OF BOYS IN A CLASS
C    N = NUMBER OF BOYS IN CLASS
C    H = HEIGHT OF A BOY
C    AH = AVERAGE HEIGHT OF BOYS
```

```
C    SUMH = SUM OF HEIGHTS
     READ (7, *) N, (H(J), J = 1, N)
     SUMH = 0
     Do 10 J = 1, N
     SUMH = SUMH + H(J)
10   CONTINUE
     AN = N
     AH = SUMH/AN
     WRITE (8, *) AH
     STOP
     END
```

Q.6. (*a*) Briefly describe some of the similarities between a robot and an NC machine. What are the five different types of robot systems? Discuss some important aspects of pendant programming in robotics.

Sol. Similarities between robot and NC machines

1. Power drive technology
2. Feedback system
3. Computer control

Robot Systems : The types of robot systems shown in Fig. 14, are described below.

1. **Cartesian coordinates :** A robot constructed around Cartesian coordinates configuration consists of three orthogonal slides [Fig. 14(*a*)]. the three slides are parallel to the *x*, *y* and *z* axes of the Cartesian coordinate system. By approrpiate movement of these slides, the robot is capable of moving its arm to any point within its three-dimensional rectangularly shaped workspace. Its work envelope is rectangular.

   A robot with Cartesian Coordinate axes is basically used for point-to-point applications. So it is a useful system for robots that are loading and unloading materials into machines. It is restricted to low-technology robotic systems.

2. **Cylindrical coordinates :** The cylindrical coordinate system (Fig. 14(*b*)) incorporates three degrees of freedom, or three axes : the rotational axis, θ; up-and-down axis, Z; and the reach or in-and-out axis, R. Its work envelope traces out a cylinder. It is found mostly in pick and place arms, for parts feeding and assembly.

3. **Polar (or spherical) coordinates :** It has three axes; the rotational axis, θ; the reach axis, R; and the bend axis, β. The work envelope traced out by the robot's manipulator is a sphere. This design is used mostly for machine loading, being well suited to a long straight reach into a press or moulding machine. The polar coordinates configuration is shown in Fig. 14(*c*).

4. **Revolute (or articulate) coordinates :** The revolute or jointed coordinate system (Fig. 14(*d*)) is identified by three axes : the θ axis (waist rotation); the β axis (shoulder rotation) and γ axis (elbow rotation). This configuration is similar in appearance to the human arm. The robot arm is mounted to a base which can be rotated to provide the robot with the capacity to work within a quasi-spherical space.

Table 2. Configuration and Work envelope of robots

Configuration	Work envelope
Cartesian	Rectangular
Cylindrical	Cylindrical
Polar	Spherical
Jointed-arm horizontal-axes	Spherical
Jointed-arm vertical-axes	Cylindrical
Pendulum Aym	Partial spherical
Multiple-joint arm	Spherical

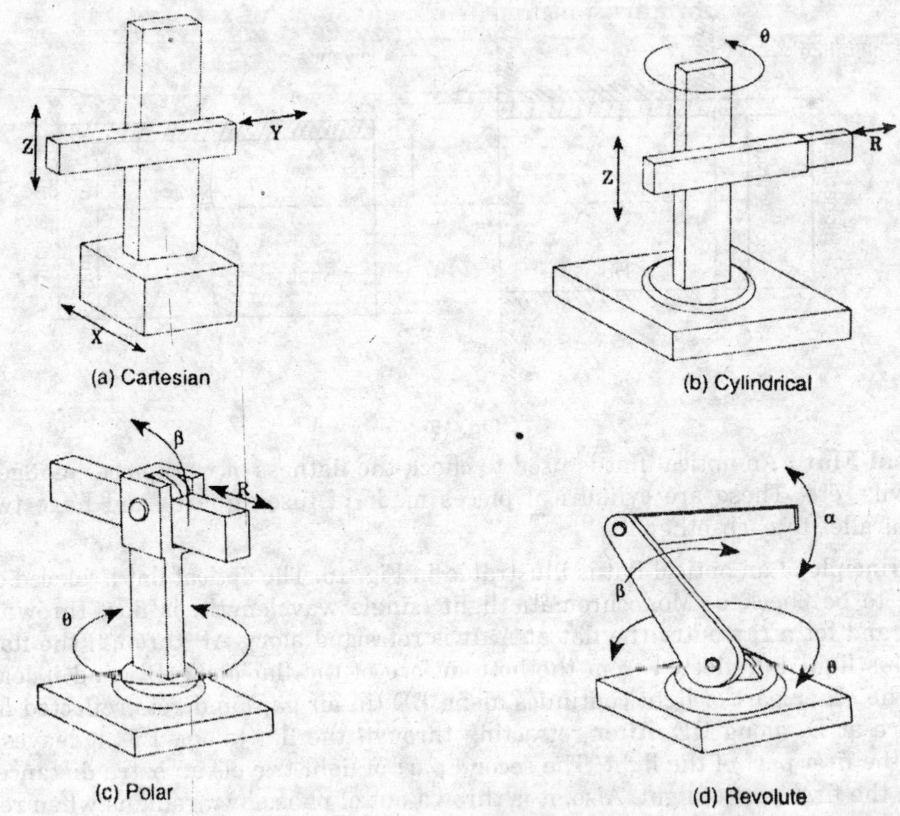

(a) Cartesian (b) Cylindrical

(c) Polar (d) Revolute

Fig. 14(a). Robot classification.

(***b***) Explain the difference between tolerance and allowance. Write in short about optical flat. Two fringe patterns are supplied for two completely different surfaces using optical flat, name the types of surfaces, and draw if required. Discuss with Fig., principle of working of a pneumatic gauge.

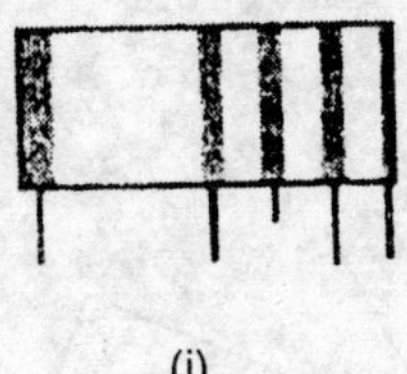

(i) (ii)

Fig. 14(b)

Fringe patterns for two completely different types of surfaces.

Sol. Tolerance : It is equal to the algebraic difference between the upper and lower deviations and has absolute value without sign.

Tolerance = Maximum limit of size – Minimum limit of size.

Allowance : It is an intentional difference between the maximum material limits of mating parts. For the shaft, the maximum material limit will be its high limit and for the hole, it will be its low limit. If the shaft is smaller than hole, the allowance is positive, but if the shaft is larger than the hole, it is negative. The above definitions are shown in Fig. 15.

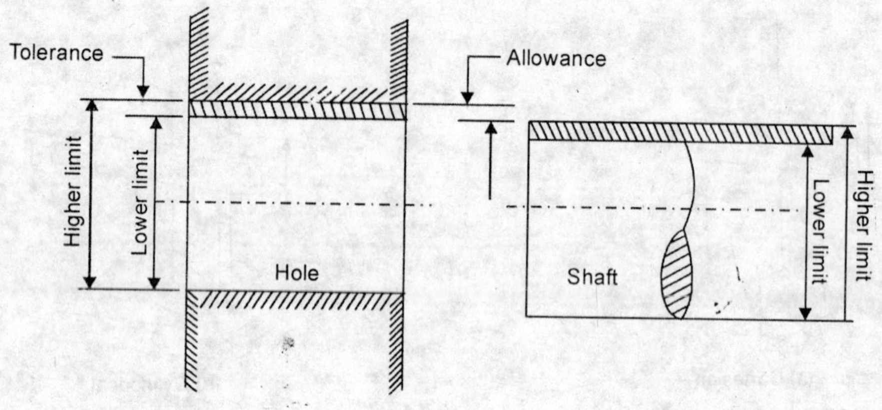

Fig. 15

Optical Flat : An optical flat is used to check the flatness of workpieces, gauge blocks, micron anvils, etc. These are cylindrical pieces made of fused quartz and have two faces perfectly parallel to each other.

The principle of an optical flat is illustrated in Fig. 16. The optical flat is placed over the work piece to be checked. Monochromatic light (single wavelength) is now thrown on the optical flat and let a ray enter the flat at A. It is refracted along AB through the flat. At B, a part of this light is reflected from the bottom face of the flat along BC and it leaves the flat at C. The other part of light continues along BD (in air gap) and gets reflected from the work surface at D, along DE. After refracting through the flat (along EF) it leaves it at F parallel to the first part of the light. The second part of light travels an extra distance of (BD + DE) than the first part of light. Also, it is thrown out of phase by π radians when reflecting from the work surface at D (when light travelling in air along BD is reflected at the work surface, which is optically more dense than air, there is a change in phase of π radians). Now, if the air gap at D is half the wavelength of light $\left(\frac{\lambda}{2}\right)$, the second part of light travels $\frac{\lambda}{2}$

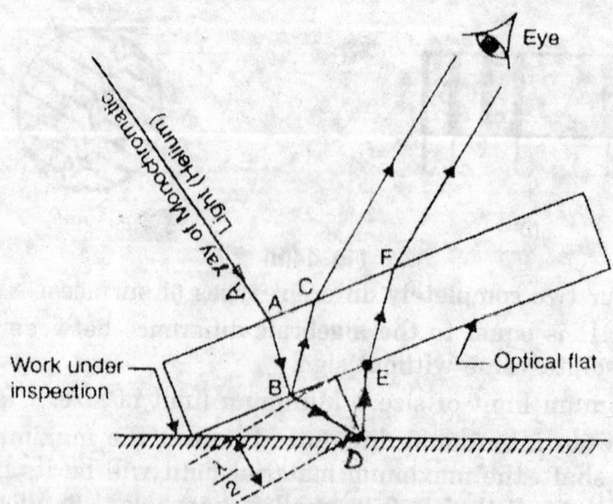

Fig. 16. Principle of Optical Flats.

down (BD) and $\frac{\lambda}{2}$ up (DE) in air. Thus, this part of light will be one full wavelength behind and out of phase by π radians relative to the first part of light. Thus, the two portions of light leaving the optical flat at C and F will cancel each other and a dark band will be seen by the observer. This will happen wherever the air gap is zero or product of any whole number times the half wave-length. On the ohter hand, if air gap is $\frac{1}{4}$ wavelength, $\frac{3}{4}$ wavelength and so on, the two parts of light will reinforce each other and we will get a bright band.

A trained inspector can tell the condition of the face of the work piece from the pattern of bright and dark bands. When these bands are straight, parallel and equally spaced, the surface is flat, [Fig. 17(*a*)]. If the bands curve towards the contact edge, the surface is concave [Fig. 17 (*b*)] and if the bands curve away from the contact edge, it is convex [Fig. 17(*c*)].

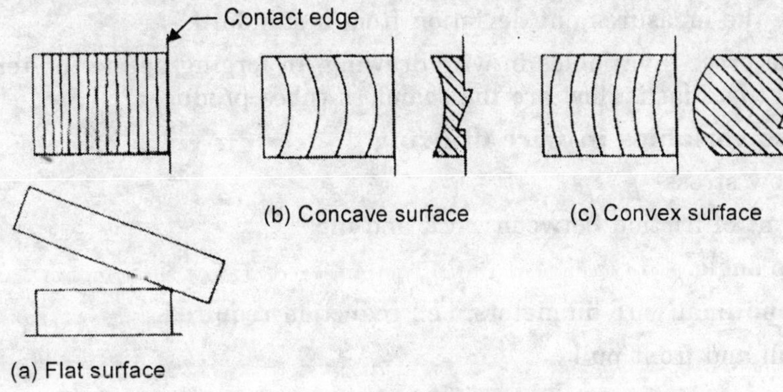

Fig. 17. Band Patterns

(*i*) is a flat surface.

(*ii*) is the gauge surface having a scratch

Pneumatic Gauge : Pneumatic gauge is a device for measuring the linear dimensions according to the flow rate of air escaping from a nozzle under pressure.

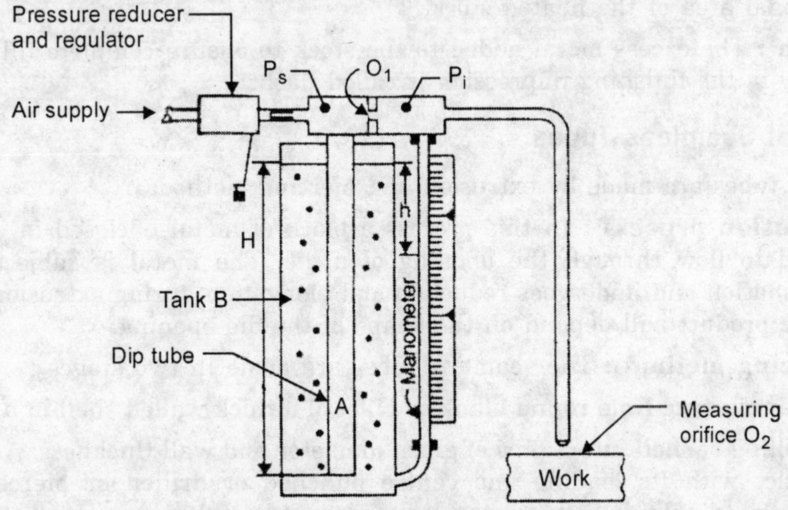

Fig. 18. Principle of Pneumatic gauge.

Fig. 18 shows the working principle of a pneumatic gauge. A pressure reducer and regulator reduces the high pressure air supply to a value of about 0.007 MPa and this pressure is maintained closely by inserting a restriction between the regulator and dip tube A. Initially, the levels of liquids in the tank B and the dip tube are the same. When the air under pressure is supplied to the system, the liquid in the dip tube will be pushed out into the tank. Any pressure greater than needed to just clear-the dip tube will escape through it into the tank as air bubbles. Thus, he pressure between the regulator and the control orifice O_1 will always be exactly the same, irrespectively of any variation in air pressure. Air after passing through the control orifice O_1, will reach the mouth of measuring orifice O_2. Back pressure will build up behind the measuring orifice if it is not able to pass the coming air flow. Due to this back pressure the level of the liquid in the manometer tube will change. Back pressure is created due to the restriction at the mouth of the orifice O_2 caused by the variations in the dimensions of the work being checked. Thus, the manometer tube can be calibrated to give the measurement deviation from a standard.

(c) Write the process variables in wire drawing. In forging define the terms *(i)* Edging *(ii)* Fullering and *(iii)* Flash. How are the seamless tubes produced?

Sol. Process variables in wire drawing

1. Mean flow stress
2. Coefficient of friction between work and die.
3. Semi-die angle.
4. Initial and final wire diameters, i.e., reduction required.
5. Back pull and front pull

Forging Process

(i) **Edging :** The function of edging is to distribute the metal longitudinally by moving metal from the portions where it is in excess to the portion which is deficient in metal.

(ii) **Fullering** : The fullering operating in forging may be defined to reduce the cross-sectional area of the heated stock.

(iii) **Flash :** The excess metal added to the stock to ensure complete filling of the die cavity in the finishing impression is called flash.

Preparation of Seamless Tubes

Seamless tubes are made by extrusion and piercing methods :

1. **Extrusion process :** In this process a block of metal enclosed in a container is forced to flow through the opening of a die. The metal is subjected to plastic deformation and undergoes reduction and elongation during extrusion. The section of the product will depend on the shape of the die opening.
2. **Piercing method :** The seamless tubes are made in two stages :

   (i) Manufacture from round blank or billet of a thick walled shell in a piercing mill.

   (ii) Rolling of shell into a pipe of given diameter and wall thickness. A heated round billet with its leading end centre punched or drilled or pierced, is pushed loingitudinally in between two large convex shaped (tapered) rolls that revolve in

the same direction, their axes being inclined at opposite angles of about 6° from the axis of the billet, Fig. 19(*a*). The clearance between the rolls is less than the diameter of the billet. As the billet is caught by the rolls and rotated, their inclination causes the billet to be drawn forward into them. The reduced clearance between the rolls forces the billet to deform into an elliptical shape. As the billet tries to rotate under high compressive forces created by the rolls, secondary tensile stress are set up at 90° to the compressive forces at the centre of the billet. The punched hole at the centre of the billet will tear open. The mandrel assists this action. As the billet rotates and feed towards the mandrel, the tearing action is propagated along the length of the billet forming a seamless shell (roughly formed tube). Upon completion of the operation, the mandrel is forced out of the shell. This piercing of the billet to form a rough tube is called "Mannesmann process".

The pierced shell is further processed in a "plug rolling mill", (Fig. 19(*b*)), which is a two-high reversing stand with a series of round passes in the rolls. A short mandrel (plug) is held in the centre of the pass by a long bar. The clearance between the mandrel and the pass determines the wall thickness of the rolled tube. This operation elongates the tube and reduces the wall thickness. The outside diameter of the tube is only approximate, however. The tube is still ina a rough state. Therefore, another operation, called "reeling" is performed between reelers and over a mandrel as shown in Fig. 19(*c*). the reeling operation improves the finish of the inside and outside surfaces, eliminates irregularities, scratches and out of roundness, and decreases the differences in wall thickness. Finally, a sizing operation is performed between sizing rolls without use of a mandrel, (Fig. 19 (*d*)). The final operations of reeling and sizing are often conducted on cooled tubes in order to improve their finish and size.

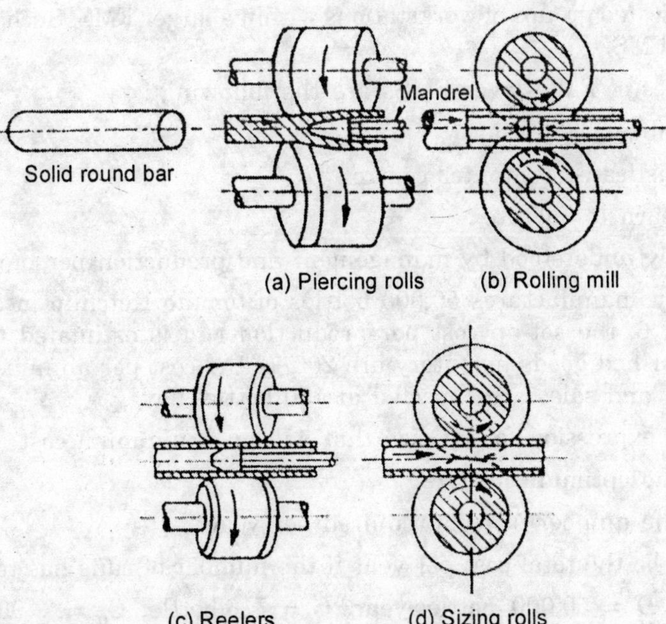

Fig. 19. Piercing methods for making seamless tubes.

(*d*) Explain with Fig. a drill-jig mentioning various provisions. What is the difference between FMS and FMC? Write some disadvantages of FMS.

Sol. Drilling-jig : A drilling jig for machining a hole in a casting is shown in Fig. 20.

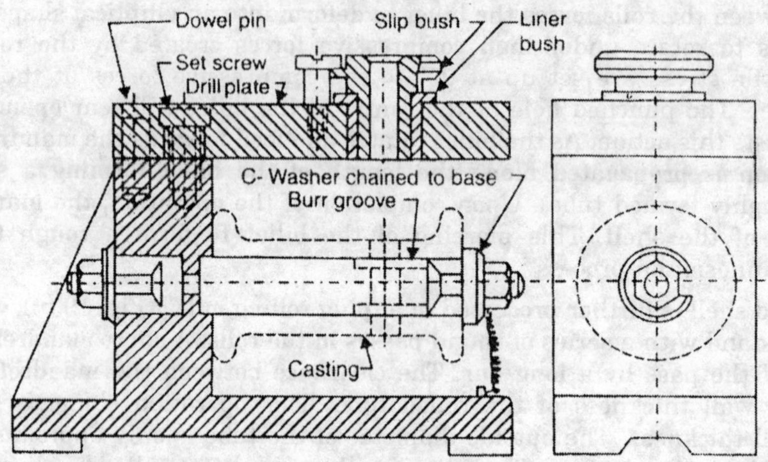

Fig. 20. A Drilling Jig,

FMC : A flexible manufacturing cell cosists of several CNC machine units served by a single robot.

FMS : A flexible manufacturing system is an individual machine or group of machines served by an automated materials handling system that is computer controlled and has a tool handling capability.

FMCs are dedicated groups of workstations within a larger FMS. Such cells are segregated components of an FMS.

The advantages of FMCs over FMSs are the following :

1. Less initial capital outlay.
2. Less sophisticated computer control.
3. Easy to learn to operate.
4. More easily understood by management and production personnel.

Q.7. (*a*) A unit manufactures 50,000 bottles of tomato Ketchup in a year. The factory cost per bottle is ₹ 6, the set-up cost per production run is estimated to be ₹ 90 and the carrying cost on finished goods inventory are 20% of the cost per annum. Production rate is 600 bottles per day and sales are expected at 150 bottles/day.

(*i*) Derive an expression for lot size that will give minimum cost (10)

(*ii*) What is the optimum lot size? (5)

(*iii*) What is the number of runs required per year? (5)

(*iv*) What will be the total cost per year if the number of runs has to be an integer? (5)

Sol. Given : D = 50,000 bottles/year, C = ₹ 6/bottle, C_0 = ₹ 90/set up, r = 20%, p = 600 bottles/day, d = 150 bottles/day.

(*i*) Total variable cost

PAPER II—2012

$$\text{TVC} = \left(1-\frac{d}{p}\right)\frac{QC_c}{2}+\frac{D}{Q}C_0$$

$$\frac{d(\text{TVC})}{dQ} = \left(1-\frac{d}{p}\right)\frac{C_c}{2}-\frac{D}{Q^2}C_0 = 0$$

$$\text{EOQ} = Q^* = \left[\frac{2DC_0}{C_c}\left(\frac{p}{p-d}\right)\right]^{1/2}$$

(*ii*) $$Q^* = \left[\frac{2\times 50{,}000\times 90}{0.2\times 6}\left(\frac{600}{600-150}\right)\right]^{1/2} = 3162.3 \simeq 3163 \text{ bottles}$$

(*iii*) Number of runs per year,

$$N^* = \frac{D}{Q^*} = \frac{50{,}000}{3163} = 15.8 \simeq 16$$

(*iv*) When $$N^* = 16,\ Q^* = \frac{50{,}000}{16} = 3125$$

$$\text{TVC}^* = \left(1-\frac{150}{600}\right)\times\frac{3125}{2}\times 0.2\times 6 + 16\times 90 = ₹\ 2846.25$$

(***b***) Ghosh and Company making pipes is considering investiment in an aggressive advertising campaign for the next year's sales. The data for the last year 2009-10 is :

Variable Expenses	
Direct Material	₹ 3.25 per piece
Direct Labour	₹ 8.00 per piece
Variable overhead	₹ 2.50 per piece
	₹ 13.75 per piece
Fixed Expenses :	
Manufacturing	₹ 25,000
Selling	₹ 40,000
Administrative	₹ 70,000
Total Fixed	₹ 1,35,000
Selling price per pipe	₹ 25
Sales 2009-10 (20,000 units)	₹ 5,00,000

The sales target for 2010-11 is ₹ 5,50,000 :

(*i*) What is the profit and break even point for the year 2009-10? (7)

(*ii*) How will the BEP change if the company spends ₹ 11,250 on advertising in 2010-11? (4)

(*iii*) What will be the profit if the company is able to sell 21,000 pipes only in 2010-11? (4)

Sol. Fixed cost, $F = ₹\ 1,35,000$

Variable cost, $V = ₹\ 13.75$ per piece

Selling price, $S = ₹\ 25$ per piece

(*i*) Break even point, BEP for 2009-10

Let N = number of peices

Total cost = Total revenue

$$F + V \times N = N \times S$$

$$N = \frac{F}{S-V} = \frac{1,35,000}{25-13.75} = 12,000 \text{ pieces}$$

Let R_B = break-even ₹

$$R_B = N \times S = 12,000 \times 25 = ₹\ 3,00,000$$

Profit = Revenue − Total cost

$$= 5,00,000 - (20,000 \times 13.75 + 1.35,000)$$

$$= ₹\ 90,000$$

(*ii*) Sales target for 2010 − 11 = ₹ 5,50,000

Fixed cost for 2010 − 11, F = 1,35,000 + 11,250 = ₹ 1,46,250

$$N = \frac{1,46,250}{25-13.75} = 13,000 \text{ pieces}$$

$$R_B = 13,000 \times 25 = ₹\ 3,25,000$$

(*iii*) Profit = 21,000 × 25 − (21,000 × 13.75 + 1,46,20)

$$= ₹\ 90,000$$

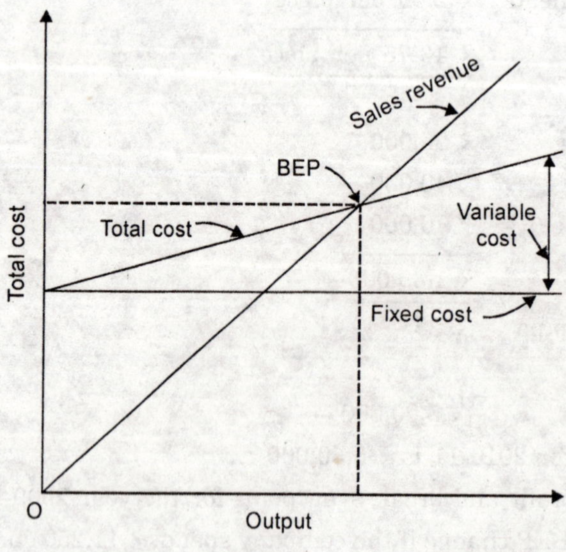

Fig. 21. Break even chart.

MECHANICAL ENGINEERING - 2013

PAPER - I

(Conventional)

Q.1(a) Liquefied Petroleum Gas (LPG) is sold in small cylinders for use as cooking gas. If one LPG cylinder contains 15 kg gas, express the quantity of gas contained in the cylinder in moles. Treat LPG as Butane (C_4H_{10}) only. 5

Sol. Total mass of LPG (C_4H_{10}) = 15000 grams

Molecular weight of $C_4H_{10} = 4 \times 12 + 10 \times 1 = 58$

Number of moles of $C_4H_{10} = \dfrac{\text{Weight in grams}}{\text{Molecular weight}}$

$= \dfrac{15000}{58} = 258.62$ Moles of LPG in cylinder.

Q.1(b) A heat engine operates between a source at 550°C and sink at 25°C. If the heat is supplied to the heat engine at a steady rate of 1200 kJ/min, determine the maximum power output of this heat engine in kW. 5

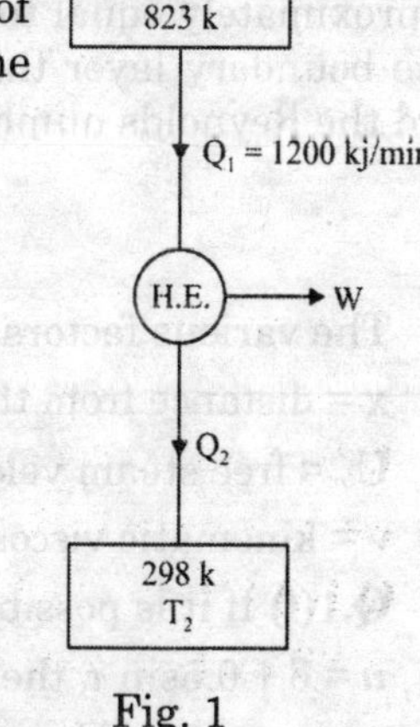

Fig. 1

Sol. Source temperature, $T_1 = 273 + 550 = 823$ K

Sink temperature, $T_2 = 273+25 = 298$ K

Heat supplied, $Q_1 = 1200$ kJ/min or 20 kJ/s

Maximum efficiency of heat engine (see Fig. 1.)

$$\eta = 1 - \frac{T_2}{T_1} = \frac{W_{max}}{Q_1}$$

$$1 - \frac{298}{823} = \frac{W_{max}}{20}$$

Maximum power output, $W_{max} = 20\left(1 - \dfrac{298}{823}\right) = 12.758\ kW$

Q.1(c) Explain the effect of combustion chamber shape and location of spark plug on detonation in an S.I engine. 5

Sol. ***Effect of combustion chamber shape on detonation in S.I. engine:*** The shape of combustion chamber should be such which promote turbulence to reduce detonation. Spherical shaped chambers minimizes knocking tendency. To prevent detonation the length of flame travel from the spark plug to the farthest point in the combustion space should be as short as possible.

Effect of spark plug location on detonation in S.I. engine: The location of spark plug should be such that the flame travel distance is less. As flame is initiated from the spark plug, a centrally located spark plug has least tendency for detonation. Also multiple spark plugs minimizes flame travel time and reduces knocking. To reduce the possibility of detonation the exhaust valve should be near the spark plug.

Q.1(d) Specify the conventional and alternative refrigerants for the following applications: **5**

(i) Domestic refrigerators and

(ii) Central Air-conditioning plants.

Sol.

Application	Conventional refrigerants	Alternative refrigerants
Domestic refrigerator	Freon- 12, R-12, R-22	R-134a, HC-270, HC-600a
Central air-conditioning plants	Freon- 11, R-22	HC-290, CO_2

Q.1(e) Discuss the various factors affecting boundary layer thickness.

Sol.

Boundary layer thicknesses – δ is defined as the distance from the boundary of a solid body measured in the y –direction to the point where the velocity of the fluid is approximately equal to 0.99 times the free stream velocity U_∞ of the fluid. Thus $u = 0.99\ U_\infty$. The boundary layer thickness is affected by the distance from the leading edge of the plate and tho Roynoldo number.

$$Re_x = \frac{U_\infty x}{v}$$

The various factors offecting boundary layer thickness are:

x = distance from the leading edge.

U_∞ = free steam velocity.

v = kinematic viscosity of fluid.

Q.1(f) If it is possible to describe a turbulent flow by the equation:

$u = \delta + 0.5\sin t$, then, calculate:

i. $\bar{u}'$ and $(\bar{u}'^2)^{1/2}$

Sol.

Given: $u = \delta + 0.5 \sin t$

At $t = \pi/2$, $u = 5.5$ and at $t = 3\pi/2$, $u = 4.5$

ii. Fluctuating component of velocity, (Fig.2)

$$\bar{u}' = \frac{0.5+(-0.5)}{2} = 0$$

iii.
$$(\bar{u}'^2)^{1/2} = \sqrt{\frac{(0.5)^2+(-0.5)^2}{2}} = \sqrt{0.25} = 0.5$$

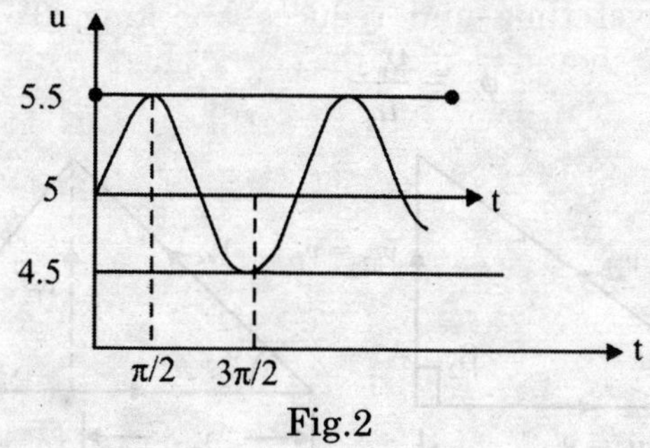

Fig.2

Q.1(g) Write expression for hydraulic efficiency of a water turbine. How does the efficiency vary with part load? 5

Sol. Hydraulic efficiency of water turbine η_{hyd} is defined as the ratio of power developed by the turbine runner to the power supplied by the jet at entrance to the turbine.

$$\eta_{hyd} = \frac{\text{power developed by the turbine runner}}{\text{Power supplied at the inlet of turnine}}$$

$$= \frac{(\upsilon_{w1} \pm \upsilon_{w2})u}{gH}$$

$$= \frac{\rho Q_a (\upsilon_{w1} \pm \upsilon_{w2})u}{\gamma Q_a H}$$

$$= \frac{H_r}{H}$$

Where $H_r = \frac{1}{g}(\upsilon_{w1} \pm \upsilon_{w2})u$ = Runner or Euler head.

Q_a = actual flow rate to turbine runner

$\upsilon_{w1}, \upsilon_{w2}$ = Velocities of whirl at inlet and outlet respectively

u = peripheral velocity of vanes

H = net head on the turbine runner

γ = specific weight of water

At part load efficiency reduces as turbine runner power decrease but power supplied at inlet to turbine remains same.

Q.1(h) Explain the terms slip and slip factor, as applied to centrifugal compressor. Write expressions for these terms. 5

Sol. In the ideal velocity diagram at outlet (Fig. 3) to compressor, it is assumed that $\upsilon_{w2} = u_2$. But this condition is not satisfied in actual practice due to secondary flow effects. Actually $\upsilon_{w2} < u_2$. The difference $(u_2 - \upsilon_{w2})$ is called the slip. Slip, $s = u_2 - \upsilon_{w2}$. Slip factor is defined as the ratio of actual whirl component to the ideal whirl component of velocities at outlet of the centrifugal compressor.

Slip factor, $\phi_s = \dfrac{v_{w2}}{u_2}$

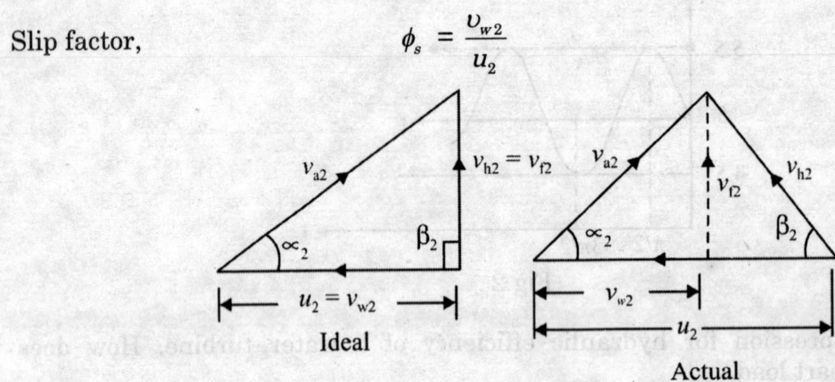

Fig. 3: Velocity diagram at outlet

Q.2(a) A spherical balloon of 1 m diameter contains a gas at 200 kPa and 300 K. The gas inside the balloon is heated until the pressure reaches 500 kPa. During the process of heating, the pressure is proportional to the diameter of the balloon. Determine the work done by the gas inside the balloon. 10

Sol. Given: $d = 1$ m, $p_1 = 200$ kPa, $T_1 = 300$ K, $p_2 = 500$ kPa, $p \propto d$

Volume of balloon, $V_1 = \frac{4}{3} \times \pi \left(\frac{d}{2}\right)^3 = \frac{4}{3} \times \pi \times \left(\frac{1}{2}\right)^3 = \frac{\pi}{6} m^3$

As $p \propto d$

$$\therefore \quad \frac{p_2}{p_1} = \frac{d_2}{d_1}$$

$$d_2 = 1 \times \frac{500}{200} = 2.5 m$$

$$V_2 = \frac{4}{3} \times \pi \left(\frac{2.5}{2}\right)^3 = 2.604\pi \ m^3$$

For a non-flow system,

Work done, $W = p_2 V_2 - p_1 V_1$

$$= (500 \times 2.604 - 200 \times \frac{1}{6}) \pi \times 10^3$$

$$= 3985.6 \text{ kJ}$$

Q.2(b) A gas turbine engine with regeneration operates with two stages of compression and two stages of expansion. The reassure ratio across each stage of compressor and turbine is 3.5. The air enters each stage of the compressor at 300 K and each stage of the turbine at 1200 K. The compressor and turbine efficiencies are 78 and 86 percent, respectively. The effectiveness of the regenerator is 72 percent. Determine the back work ratio and the thermal efficiency of the cycle, assuming constant specific heats for air at room temperature. 20

Sol. Given:

$$\frac{p_2}{p_1} = \frac{p_4}{p_3} = 3.5,\ p_3 = p_2 \text{ for compressor}$$

$$\frac{p_6}{p_7} = \frac{p_8}{p_9} = 3.5,\ p_7 = p_8 \text{ for turbine}$$

$$T_1 = T_3 = 300 \text{ K for compressor}$$

$$T_6 = T_8 = 1200 \text{ K for turbine}$$

$$\eta_{c1} = \eta_{c2} = 78\%,\ \eta_{t1} = \eta_{t2} = 86\%,\ \varepsilon_r = 72\%$$

The T-S diagram is shown in Figure. 4

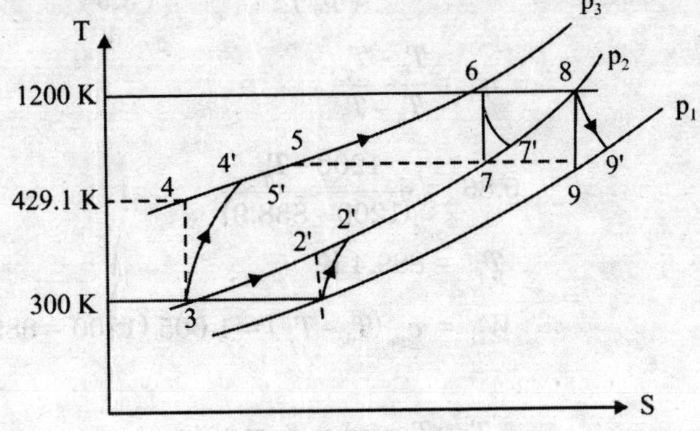

Fig. 4

Let $c_{pa} = 1.005 kJ/kg.K,\ \gamma = 1.4$

Compressor first stage:

$$T_2 = T_1\left(\frac{p_2}{p_1}\right)^{\frac{\gamma-1}{\gamma}} = 300(3.5)^{\frac{1.4-1}{1.4}} = 429.1\text{K}$$

$$\eta_{c1} = \frac{T_2 - T_1}{T_{2'} - T_1}$$

$$0.78 = \frac{429.1 - 300}{T_{2'} - 300},\ T_{2'} = 465.5 \text{ K}$$

Work done, $W_{c1} = C_{pa}(T_{2'} - T_1) - 1.005\ (465.5 - 300) = 166.32\ kJ/kg$

Compressor 2nd stage:

$$T_4 = T_3\left(\frac{p_3}{p_2}\right)^{\frac{\gamma-1}{\gamma}} = 300(3.5)^{\frac{1.4-1}{1.4}} = 429.1\text{K}$$

$$\eta_{c2} = \frac{T_4 - T_3}{T_{4'} - T_3}$$

$$0.78 = \frac{429.1 - 300}{T_{4'} - 300}$$

$$T_4 = 465.5K$$

$$W_{c2} - c_{pa}(T_{4'} - T_3) = 1.005(465.5 - 300) = 166.32 \text{ kJ/kg}$$

Total compressor work, $W_c = 2 \times 166.32 = 332.64 \ kJ/kg$

Turbine 1st stage:

$$T_7 = T_6 = \left(\frac{p_7}{p_6}\right)^{\frac{\gamma-1}{\gamma}} = 1200\left(\frac{1}{3.5}\right)^{\frac{1.4-1}{1.4}} = 838.9 \text{ K}$$

$$\eta_{t1} = \frac{T_6 - T_{7'}}{T_6 - T_7}$$

$$0.86 = \frac{1200 - T_{7'}}{(1200 - 838.9)}$$

$$T_{7'} = 889.4 \text{ K}$$

$$W_{t1} = c_{pa}(T_6 - T_{7'}) = 1.005\ (1200 - 889.4) = 312.15 \ kJ/kg$$

Turbine 2nd stage:

As $T_6 = T_8$ and $p_3/p_2 = p_2/p_1$

$\therefore$ $T_9 = T_7 = 838.9K,\ T_{9'} = 889.4K$

$$W_{t2} = W_{t1} = 312.15 \text{ kJ/kg}$$

Total turbine work, $W_t = 2 \times 312.15 = 624.30 \ kJ/kg$

$$W_{net} = W_t - W_c = 624.30 - 332.64 = 291.66 \ kJ/kg$$

Regenerator effectiveness, $\varepsilon_r = \dfrac{T_{5'} - T_{4'}}{T_{9'} - T_{4'}}$

$$0.72 = \frac{T_{5'} - 465.5}{889.4 - 465.5}$$

$$T_{5'} = 770.7K$$

Heat supplied, $q_s = c_{pa}\left[(T_6 - T_{5'}) + (T_8 - T_{7'})\right]$

$$= 1.005\ [(1200 - 770.7) + (1200 - 889.4)]$$

$$= 743.6 \ kJ/kg$$

Thermal efficiency, $\eta_{th} = \dfrac{W_{net}}{q_s} = \dfrac{291.66}{743.6} \times 100 = 39.22\%$

Back Work ratio $= \dfrac{W_{net}}{W_t} = \dfrac{291.66}{624.30} = 0.4672$

Q.2.(c) Derive expressions for $\Delta u, \Delta h$ and Δs for a gas that obeys the van der Waals equation of state for an isothermal process. 10

Sol. Van der Waal's Equation the volume of one mole of gas:

$$\left(p+\frac{a}{\bar{\upsilon}^2}\right)(\bar{\upsilon}-b)=\bar{R}T$$

where a, b = specific constants

$\bar{\upsilon}$ = molar volume of gas

$\bar{R}$ = Universal gas constant

p = gas pressure

T.ds = du + p.dυ → energy equation

T.ds = $c_\upsilon.dT + T(\frac{\delta p}{\delta T})_\upsilon d\upsilon$ → T.ds relation

$$\therefore du = c_\upsilon.dT + \left[T\left(\frac{\delta p}{\delta T}\right)_\upsilon - p\right]d\upsilon$$

For isothermal process, dT = 0

$$du = T\left[\left(\frac{\delta p}{\delta T}\right)_\upsilon - p\right]d\upsilon \quad \text{... (1)}$$

Differentiating Vander Wall's equation w.r.t. T, we have

Now $$p = \frac{\bar{R}T}{\bar{\upsilon}-b}-\frac{a}{\bar{\upsilon}^2}$$

$$\left(\frac{\delta p}{\delta T}\right)_\upsilon = \frac{R}{\bar{\upsilon}-b} \quad \text{... (2)}$$

From Eqs. (1) and (2), we have

$$du = T = \left[\frac{R}{\bar{\upsilon}-b}-\frac{\bar{R}T}{\bar{\upsilon}-b}+\frac{a}{\bar{\upsilon}^2}\right]$$

Integrating, We have

$$\int du = \left[T\left(\frac{\bar{R}}{\bar{\upsilon}-b}\right)-\frac{\bar{R}}{\bar{\upsilon}-b}+\frac{a}{\bar{\upsilon}^2}\right]dv$$

$$\int du = \int_{\bar{\upsilon}^1}^{\bar{\upsilon}^2}\frac{a}{\bar{\upsilon}^2}\,dv$$

$$\Delta u = \frac{a}{3}\left(\frac{1}{\bar{\upsilon}_2^3}-\frac{1}{\bar{\upsilon}_1^3}\right)$$

Similarly for Δs, assume s = f (υ , T)

$$ds = \left(\frac{\delta s}{\delta \upsilon}\right)_T d\upsilon + \left(\frac{\delta s}{\delta T}\right)_\upsilon dT$$

For isothermal process, dT = 0

$$ds = \left(\frac{\delta s}{\delta \upsilon}\right)_T d\upsilon$$

Maxwell's relation is, $\left(\frac{\delta p}{\delta T}\right)_\upsilon = \left(\frac{\delta s}{\delta \upsilon}\right)_T$

$$ds = \left(\frac{\delta p}{\delta T}\right)_\upsilon d\upsilon$$

From vander wall's equation, we have

$$\left(\frac{\delta p}{\delta T}\right)_\upsilon = \frac{\overline{R}}{\overline{\upsilon} - b}$$

$$\int ds = \int \left(\frac{\overline{R}}{\overline{\upsilon} - b}\right) d\upsilon$$

$$\Delta s = \overline{R}\ \ell n \left(\frac{\overline{\upsilon}_2 - b}{\overline{\upsilon}_1 - b}\right)$$

Q.3(a) An ammonia ice plant operates between a condenser temperature of 35°C and an evaporator temperature of – 15°C. It produces 10 tones of ice per day from water at 32°C to ice at – 3°C. Assuming simple saturation cycle, determine:

(i) the capacity of the refrigeration plant.

(ii) the mass flow rate of refrigerant.

(iii) the compressor cylinder diameter and stroke if its volumetric efficiency is 70%, rpm = 1250 and stroke to bore ratio is = 1.3.

(iv) the power of the compressor motor if the adiabatic efficiency of the compressor = 0.86 and mechanical efficiency = 0.94

(v) the theoretical and actual COP.

Assume latent heat of ice = 335 kJ/kg

Specific heat of water = 4.1868 kJ/kg.K

Specific heat of ice = 1.94 kJ/kg.K

The properties of NH_3 arc given blow:

T (°C)	$\upsilon_f \times 10^3$ (m³/kg)	υ_g (m³/kg)	h_f (kJ/kg)	h_g (kJ/kg)	s_f (kJ/kg-K)	s_g (kJ/kg-K)
–15	1.52	0.508	131.3	1443.9	07426	5.8223
35	1.7	0.096	366.1	1488.6	1.566	5.2086

Average specific heat of ammonia vapour is 2.2 kJ/kg-K. 20

Sol. Given: $t_1 = -15°C$, $t_3 = 35°C$, $t_w = 32°C$, $t_i = -3°C$, $\eta_{vc} = 70\%$, $N = 1250$ rpm, $L/D = 1.3$, $\eta_c = 0.86$, $\eta_{mech} = 0.94$, $L_i = 335$ kJ / kg, $c_{pw} = 4.1868$ kJ / kg – K, $c_{pi} = 1.94$ kJ / kg – K, Q =10 tones

$h_1 = 1443.9$ kJ /kg, $s_1 = 5.8223$ kJ /kg.K, $h_3 = 366.1$ kJ /kg, $h_4 = 131.3$ kJ /kg, $h_2 = 1488.6$ kJ /kg, $c_{p_v} = 2.2$ kJ /kg.K, $s_2 = 5.2086$ kJ /kg.K

The T – s diagram for simple saturation cycle is shown in Fig.5.

(i) Capacity of refrigeration plant,

$$RC = m \times RE = \frac{Q \times 10^3}{24 \times 3600}\left[c_{pw}(32-0) + L_i + c_{pi}(0+3)\right]$$

$$= \frac{10 \times 10^3}{24 \times 3600}\left[4.1868 \times 32 + 335 + 1.94 \times 3\right]$$

$$= 54.95 \text{ kW}$$

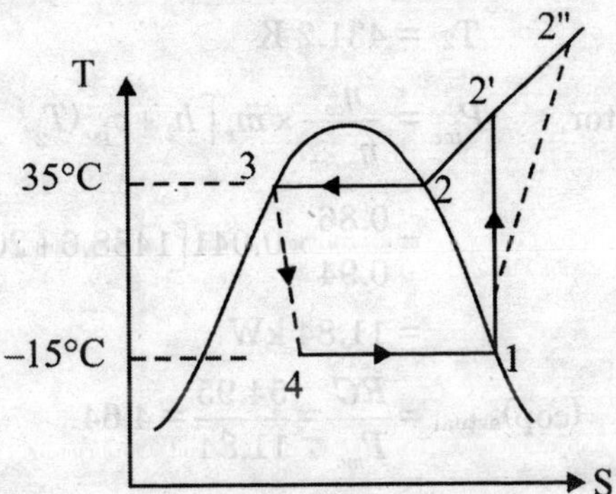

Fig.5: T- s diagram

(ii) $h_2' = h_2 + c_{pv}(T_{2'} - T_2)$

Also $s_1 = s_2' = s_2 + c_{pv} \ln\left(\frac{T_{2'}}{T_2}\right)$

$$5.8223 = 5.2086 + 2.2 \ln\left(\frac{T_{2'}}{308}\right)$$

$$T_2' = 407K$$

$$h_2' = 1488.6 + 2.2\,(407 - 308) = 1706.4 \text{ kJ/kg}$$

$$RC = \dot{m}_r\,(h_2' - h_3)$$

$$54.95 = \dot{m}_r\,(1706.4 - 366.1)$$

$$\dot{m}_r = 0.047 \ kg/s$$

Mass flow rate of refrigerant = 0.047 kg/s

(iii) Volumetric efficiency of compressor,

$$\eta_{vc} = \frac{V_{actual}}{V_{swept}} = \frac{\dot{m}_r \upsilon}{\frac{\pi}{4} D^2 LN} = \frac{\dot{m}_r \upsilon_{g1}}{\frac{\pi}{4} D^2 LN}$$

$$= \frac{0.041 \times 0.508 \times 4 \times 60}{\pi \times 1.3 D^3 \times 1250} = 0.7$$

$$D = 0.1118 \text{ m}, L = 1.3 D = 0.1454 \text{ m}$$

(iv) Adiabatic efficiency of compressor,

$$\eta_{ca} = \frac{T_{2'} - T_1}{T_{2''} - T_1}$$

$$0.86 = \frac{407 - 258}{T_{2''} - 258},$$

$$T_{2''} = 431.2 \text{ K}$$

Power of compressor motor,

$$P_{mc} = \frac{\eta_{ca}}{\eta_{mech}} \times \dot{m}_r \left[h_2 + c_{p\upsilon}(T_{2''} - T_2) - h_1 \right]$$

$$= \frac{0.86}{0.94} \times 0.041 \left[1488.0 + 202(431.2 - 308) - 1443.0 \right]$$

$$= 11.84 \text{ kW}$$

(v)

$$(\text{cop})_{\text{actual}} = \frac{RC}{P_{mc}} = \frac{54.95}{11.84} = 4.64$$

$$(\text{cop})_{\max} = \frac{T_1}{T_2 - T_1} = \frac{258}{308 - 258} = 5.16$$

Q.3.(b) In an oil engine, working on dual combustion cycle the temperature and pressure at the beginning of compression are 90^0 C and 1 bar. The compression ratio is 13:1. The heat supplied per kg of air is 1675 kJ, half of which is supplied at constant volume and half at constant pressure. Calculate:

(i) the maximum pressure in the cycle, and

(ii) the percentage of stroke at which cut-off occurs.

The γ for compression is 1.4, R = 0.287 kJ/kg.K and C_v for products of combustion is $0.71 + 20 \times 10^{-5} T$

where T is in K. 10

Sol. Given: $T_1 = 273 + 90 = 363$ K, $p_1 = 1$ bar, $r = 13$, $q_s = 1675$kJ/kg, $q_{sv} = 837.5$kJ/kg, $q_{sv} = 837.5$ kJ/kg, $\gamma = 1.4$, R = 0.287kJ/kg.K

$$C_{vg} = 0.71 + 20 \times 10^{-5} T$$

The p-v diagram for dual combustion cycle is shown in Fig.6

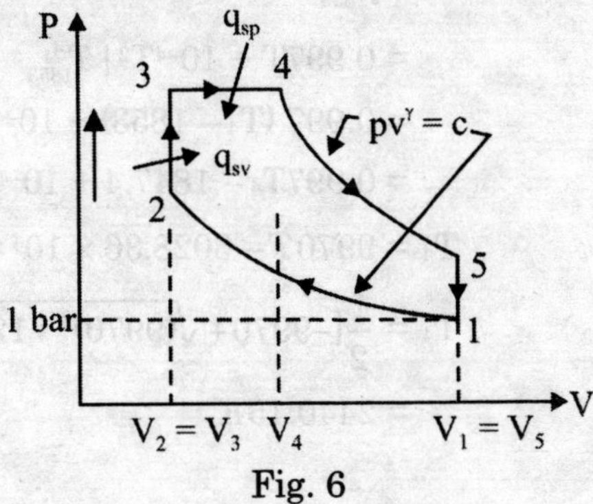

Fig. 6

Adiabatic compression-Process1-2 :

$$\frac{T_2}{T_1} = \left(\frac{p_2}{p_1}\right)^{\frac{\gamma-1}{\gamma}} = \left(\frac{v_1}{v_2}\right)^{\frac{\gamma-1}{\gamma}} = r^{\gamma-1}$$

$$T_2 = 363(13)^{(1.4-1)} = 1012.7K$$

$$p_2 = p_1\gamma_r = 9 \times 13^{1.4} = 32.267\ bar$$

Constant volume heat addition during process 2–3:

$$q_{sv} = \int_{T_2}^{T_3}\left(0.71 + 20\times10^{-5}T\right)dT$$

$$837.5 = 0.71T + 10^{-4}T^2 \Big|_{1012.7}^{T_3}$$

$$= 0.71(T_3 - 1012.7) + 10^{-4}[T_3^2 - (1012.7)^2]$$

$$= 0.71T_3 - 719.02 + 10^{-4}T_3^2 - 102.56$$

$$T_3^2 = 7100T_3 - 1659.08 \times 10^4 = 0$$

$$T_3 = \frac{1}{2}[-7100 \pm \sqrt{(7100)^2 + 4\times1659.08\times10^4}]$$

$$= \frac{1}{2}[-7100 + 10806]$$

$$T_3 = 1853K$$

For process 2-3, $\frac{p_2}{T_2} = \frac{p_3}{T_3}, p_3 = p_2\frac{T_3}{T_2} = 32.267\times\frac{1853}{1012.7} = 59.04\text{bar}$

(i) Maximum pressure in the cycle = 59.04 bar

(ii) Cut-off ratio at 4

$$q_{sp} = mc_p dT, c_p = R + c_v$$

$$837.5 = \int_{T_3}^{T_4}(0.287 + 0.71 + 20\times10^{-5}T)dT$$

$$= \int_{1853}^{T_4} (0.997 + 20\times10^{-5}T)dT$$

$$= 0.997T + 10^{-4}T^2 \mid T_{1853}^{4}$$

$$= 0.997\,(T_4 - 1853) + 10^{-4}\,[T_4^2 - (1853)^2]$$

$$= 0.997T_4 - 1847.4 + 10^{-4}\,T_4^2 - 343.36$$

$$T_4 = 9970T - 3028.36\times10^4 = 0$$

$$T_4 = \frac{1}{2}[-9970 + \sqrt{(9970)^2 + 12113.04\times10^4}\,]$$

$$= 2440.15K$$

For process 3-4:

$$\frac{V_3}{T_3} = \frac{V_4}{T_4}$$

$$\frac{V_4}{V_3} = \frac{T_4}{T_3} = \frac{2440.15}{1853} = 1.317 \qquad \text{...(1)}$$

Let V_4 occur at x fraction of stroke L, i.e. XL

$$V_1 - V_4 = \frac{\pi}{4}D^2 \times xL \qquad \text{...(2)}$$

Swept volume, $V_1 - V_2 = V_1 - V_3 = \frac{\pi}{4}D^2 L$ ...(3)

From Eqs.(1) to (3), we have

$$V_4 = V_1 - V_1 - \frac{\pi}{4}D^2 \times xL = V_3 + \frac{\pi}{4}D^2 L - \frac{\pi}{4}D^2 \times xL$$

$$= V_3(1-x)\,\frac{\pi}{4}D^2 L$$

$$= V_3 + (1 \quad x)\,(V_1 \quad V_3)$$

$$\frac{V_4}{V_3} = 1 + (1-x)\left(\frac{V_1}{V_3} - 1\right) = 1 + (1-x)\,(x-1)$$

$$= 1.317 = 1 + (1-x)\,(13-1)$$

$$x = 0.974$$

Hence cut-off occurs at 97.4% of stroke

Q.3.(c) Explain using psychrometric chart, the process of mixing of two air streams with condensation. 10

Sol. When a very cold and dry air mixes with very warm and high relative humidity air the resulting mixture condition may lie in two phases, (liquid and vapour). Hence, as a result there will be condensation of water vapour and some amount of water leaves the system as liquid. This results in increase in temperature of mixture with the release of latent heat of condensation.

The mixing of two air stream with condensation is shown in Fig.7.

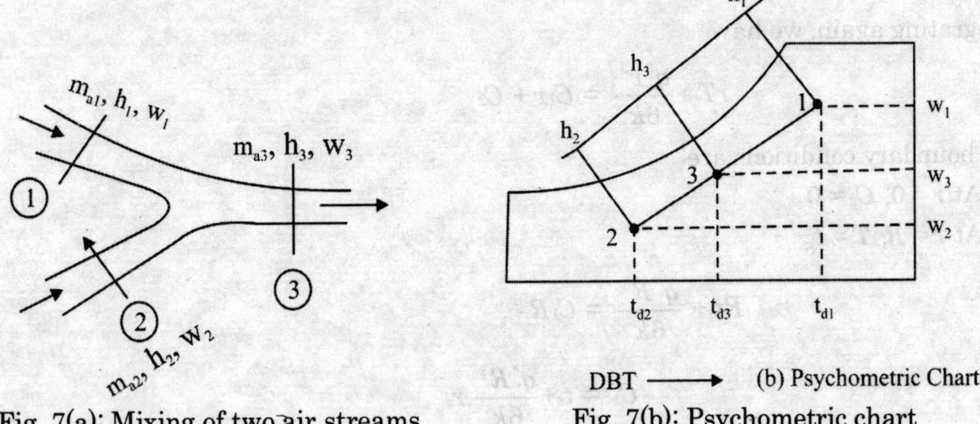

Fig. 7(a): Mixing of two air streams Fig. 7(b): Psychometric chart

Q.4(a) Consider a solid sphere of radius *R*, the surface of which is maintained at t_s. If the sphere has a uniform internal heat generation rate of q^* per unit volume, derive expression for the temperature distribution in the sphere and the temperature distribution in the sphere and the temperature at the sphere center. Take the thermal conductivity to be constant. 15

Sol. Consider a sphere of radius R having uniform heat generation q^* per unit volume, per unit time as shown in Fig.8.

Let k = constant thermal conductivity of sphere material

t_s = temperature of outer surface of sphere

Heat conduction equation in spherical coordinates is,

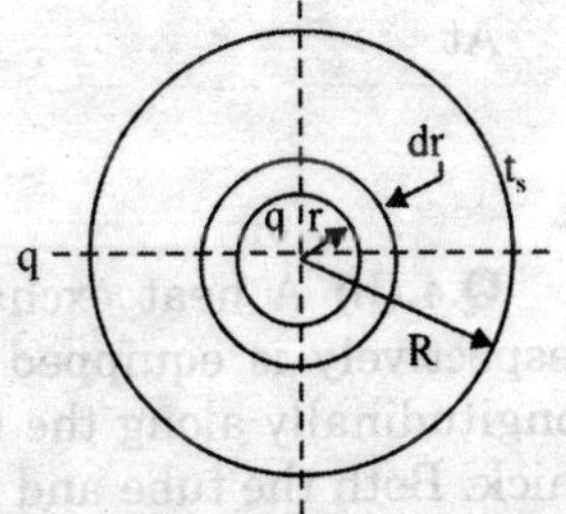

Fig. 8: Sphere with uniform heat generation

$$\frac{1}{r^2}\frac{\partial}{\partial r}\left(r^2\frac{\partial T}{\partial r}\right)+\frac{q^*}{k}=0$$

or $$\frac{1}{r^2}\left[r^2\frac{\partial^2 T}{\partial r^2}+2r\frac{\partial T}{\partial r}\right]+\frac{q^*}{k}=0$$

or $$\frac{d^2T}{dr^2}+\frac{2}{r}\frac{dT}{dr}+\frac{q^*}{k}=0$$

or $$r\frac{d^2T}{dr^2}+\frac{2dT}{dr}+\frac{q^*}{k}r=0$$

or $$\frac{d}{dr}\left[r\frac{dT}{dr}\right]+\frac{dT}{dr}+\frac{q^*r}{k}=0$$

Integrating, we have

$$r\frac{dT}{dr}+T+\frac{q^*r^2}{2k}=C_1$$

or $$\frac{d}{dr}(rT)+\frac{q^* r^2}{2k} = C_2$$

Integrating again, we have

$$rT+\frac{q^* r^3}{6k} = C_1 r + C_2$$

The boundary conditions are:

(i) At $r = 0$, $C_2 = 0$

(ii) At $r = R$, $T = t_S$

$$Rt_s+\frac{q^* R^3}{6k} = C_1 R$$

$$C_1 = t_s+\frac{q^* R^2}{6k}.r$$

$$\therefore \quad rT+\frac{q^* r^3}{6k} = t_s r+\frac{q^* R^2}{6k}.r$$

$$T+\frac{q^* r^2}{6k} = t_s+\frac{q^* R^2}{6k}$$

$$T = t_s+\frac{q^*}{6k}(R^2 - r^2)$$

At $r = 0$, $T = T_{max}$

$$T_{max} = t_s+\frac{q^*}{6k}R^2$$

Q.4.(b) A heat exchanger tube with inner and outer diameters 2.118cm and 2.54 cm respectively is equipped with 20 equally spaced straight fins of uniform thickness placed longitudinally along the tube. The fins are 2.5cm long in the radial direction and are 0.16cm thick. Both the tube and the fins are made of steel with k=45W/m. °C. The inside and outside heat transfer coefficients are 1130 and 255 $W/m^2.°C$, respectively. What is the overall heat transfer coefficient for the exchanger based on the outer exposed surface area? Compare this result with that for the same tube without fins, subjected to the same heat transfer coefficient.

Sol. Given: $d_1 = 0.02118m, d_2 = 0.0254m$, $n = 20$, $b = 0.025m$, $t = 0.0016m$, $k = 45 W/m.°C$, $h_1 = 1130 W/m^2.°C$, $h_2 = 2.55\ W/m^2.°C$ Length of fin = Length of tube = L = 1m Perimeter of fin, $P = 2(L+t) = 2(1+0.0016) = 2.0032m$ Area of cross section of fin, $A = Lt = 1.0\times 0.0016 = 0.0016\ m^2$.

Characteristic length of fin, $L_c = b+\frac{t}{2} = 0.025\frac{0.0016}{2}+0.0258m$

$$m = \sqrt{\frac{h_2 P}{kA}} = \sqrt{\frac{255\times 2.0032}{45\times 0.0016}} = 84.23$$

$$\sqrt{hPkA} = \sqrt{255 \times 2.0032 \times 45 \times 0.0016} = 6.065$$

For the fin, $\frac{1}{n} \times \frac{1}{\sqrt{hPkA}} = \tan h\,(m\,L_c) = \frac{1}{20} \times \frac{1}{6.065}\ \tanh(84.23 \times 0.0258) = \frac{1}{124.48}$

For the cylinder: $\frac{\ell n\left(\frac{r_2}{r_1}\right)}{2\pi kL} = \frac{\ell n\left(\frac{0.0254}{0.02118}\right)}{2\pi \times 4.5 \times 1} = 6.426 \times 10^{-4}$

For inner surface: $\frac{1}{h_1 A_1} = \frac{1}{1130 \times \pi \times 0.02118 \times 1} = \frac{1}{75.189}$

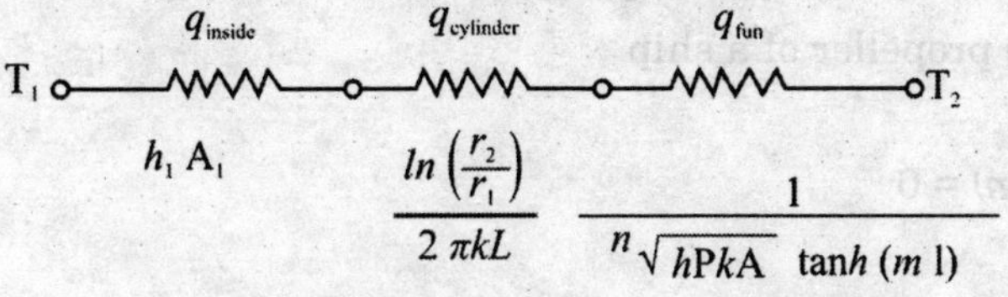

T_1 — q_{inside} ($h_1 A_1$) — $q_{cylinder}$ ($\frac{ln\left(\frac{r_2}{r_1}\right)}{2\pi kL}$) — q_{fun} ($\frac{1}{n\sqrt{hPkA}\ \tanh(m\,l)}$) — T_2

Fig. 9: Heat exchanger tube with fins

Overall heat transfer coefficient based on the outer exposed area (Fig. (9)

$$\frac{1}{UA_2} = \frac{1}{h_1 A_1} + \frac{\ell n(r_2 / r_1)}{2\pi kL} + \frac{1}{n\sqrt{hPkA}} = \tan(mL_c)$$

Outer Area, $A_2 = (tL + 2bL)\,n = (0.0016 \times 1 + 2 \times 0.025 \times 1)\,20 = 1.032 m^2$

$$\frac{1}{1.032U} = \frac{1}{75.189} + 6.426 \times 10^{-4} + \frac{1}{124.48}$$

$$U = 44.09\ W/m^{2\circ}C$$

Heat transfer coefficient without fins: Fig.(10)

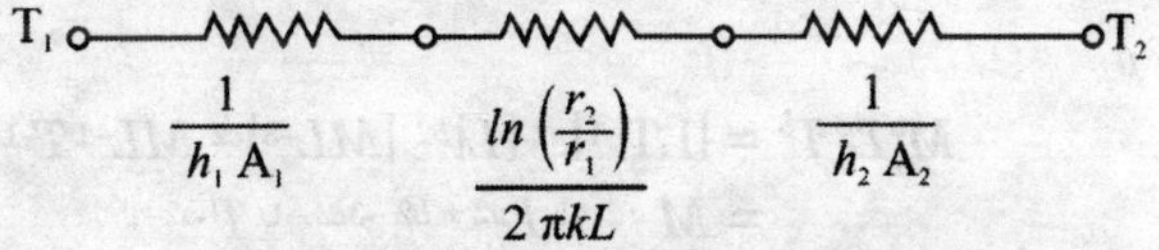

T_1 — $\frac{1}{h_1 A_1}$ — $\frac{ln\left(\frac{r_2}{r_1}\right)}{2\pi kL}$ — $\frac{1}{h_2 A_2}$ — T_2

Fig. 10: Heat exchanger tube without fains

$$\frac{1}{U_2 A_2} = \frac{1}{h_1 A_1} + \frac{\ell n(r_2 / r_1)}{2\pi kL} + \frac{1}{h_2 A_2}$$

$$A_2 = \pi d_2 L = \pi \times 0.0254 \times 1 = 0.0798\ m^2$$

$$\frac{1}{U_2} = \frac{0.0798}{75.189} + 0.0798 \times 6.426 \times 10^{-4} + \frac{1}{255}$$

$$U_2 = 198.64\ W/m^2.^\circ C$$

We find that heat transfer coefficient without fins is better than with fins. Fins should be used when h_2 is very low.

Q.4(c) Find the shape factor $F_{1\text{-}2}$ for the configuration shown in the Fig.11.

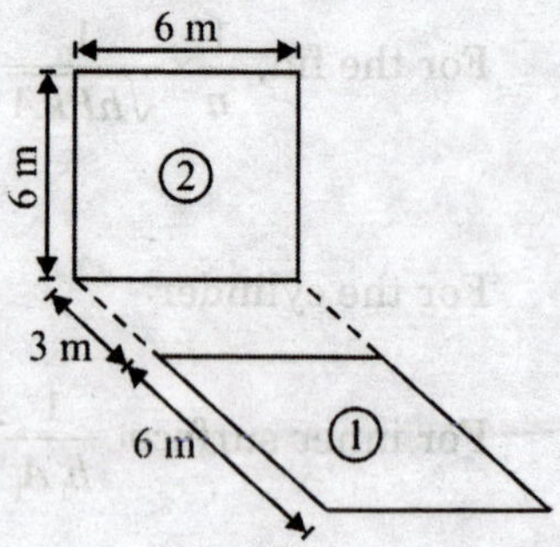

Fig.11

Sol. Given: $x = 6$ m, $y = 6$ m, $z = 6$m, $\frac{z}{x} = 1$, $\frac{y}{x} = 1$

For $\frac{z}{x} = 1$ and $\frac{y}{x} = 1$, shape factor, $F_{1\text{-}2} = 0.2$ from charts.

Q.5(a) Find the expression for the thrust F on the propeller of a ship. This thrust will be a function of mass density ρ and dynamic viscosity μ of the liquid and diameter d, speed of advance υ and rotational speed n of the propeller.

Sol. Thrust on the propeller of a ship

$F = f(\rho, \mu, d, \upsilon, n)$

Or $\phi(F, \rho, \mu, d, \upsilon, n) = 0$

Here n = 6, m = 3

Number of π - terms $= n - m = 6 - 3 = 3$

We choose υ, d, ρ as repeating variables.

1. $\pi_1 = \upsilon^{a1}\, d^{b1} \rho^{c1} F$

$$M^\circ L^\circ T^\circ = [LT^{-1}]^{a1}\,[L]^{b1}\,[ML^{-3}]^{c1}\,MLT^{-2}$$
$$= M^{c_1+1}\,L^{a1+b1-3c_1+1}\,T^{-a1-2}$$
$$-a_1 - 2 = 0,\ a_1 = -2$$
$$c_1 + 1 = 0, c_1 = -1$$
$$-2 + b_1 + 3 + 1 = 0,\ b_1 = -2$$
$$\pi_1 = \upsilon^{-2}\,d^{-2}\,\rho^{-1}\,F = \frac{F}{\upsilon^2 d^2 \rho}$$

2. $\pi_2 = \upsilon^{-a2}\, d^{\,b2}\, \rho^{c2}\, \mu$

$$M^\circ L^\circ T^\circ = [LT^{-1}]^{a2}\,[L]^{b2}\,[ML^{-3}]^{c2}\,ML^{-1}T^{-1}$$
$$= M^{c2+1}\,L^{a2+b2-3c2-1}\,T^{-a2-1}$$
$$-a_2 - 1 = 0,\ a_2 = -1$$
$$c_2 + 1 = 0,\ c_2 = -1$$
$$-1 + b_2 + 3 - 1 = 0,\ b_2 = -1$$
$$\pi_2 = \upsilon^{-1}\,d^{-1}\,\rho^{-1}\,\mu = \frac{\mu}{\upsilon d \rho}$$

3. $$\pi_3 = \upsilon^{a3}\,d^{\,b3}\,\rho^{c3}\,n$$
$$M^\circ L^\circ T^\circ = [LT^{-1}]^{a3}\,[L]^{b3}\,[ML^{-3}]^{c3}\,T^{-1}$$
$$= M^{c3}\,L^{a3+b3-3c_3}\,T^{-a3-1}$$
$$c_3 = 0,\ -a_3 - 1 = 0,\ a_3 = -1,\ -1 + b_3 = 0,\ b_3 = 1$$

$$\pi_3 = \upsilon^{-1} dn = \frac{dn}{\upsilon}$$

4. $$\frac{F}{\upsilon^2 d^2 \rho} = \phi\left[\frac{\mu}{\upsilon d\rho}, \frac{dn}{\upsilon}\right]$$

$$F = (\upsilon^2 d^2 \rho)\phi\left[\frac{\mu}{\upsilon d\rho}, \frac{dn}{\upsilon}\right]$$

$$= (\upsilon^2 d^2 \rho)\phi\left[\frac{1}{\text{Re}}, \frac{dn}{\upsilon}\right]$$

Q.5.(b) A smooth flat rectangular plate is placed edgewise in a stream of fluid. At what fraction of the length from the leading edge would drag force on the front portion to be equal to half of the total drag force? Assume the boundary layer to be laminar.

Sol. For laminar flow, local coefficient of drag for flow over a smooth flat rectangular plate is given by:

$$C_D = \frac{0.664}{\sqrt{\text{Re}_x}}$$

Average coefficient of drag, $$C_d = \frac{1.328}{\sqrt{\text{Re}_L}} = \frac{F_D}{\frac{1}{2}\rho A U_\infty^2} = \frac{\tau}{\frac{1}{2}\rho U_\infty^2}$$

Let the drag force become half at a fraction m of length L and B is the width of the plate. Then

$$\frac{F_D}{2} = \int_0^{mL} T_0 B dx = B\int_0^{mL} \frac{0.664}{\sqrt{\text{Re}_X}}\left(\frac{1}{2}\rho U_\infty^2\right)dx$$

now $$\sqrt{Re_x} = \sqrt{\frac{\rho U_\infty x}{\mu}}$$

$$\frac{F_D}{2} = B\int_0^{mL} \frac{0.664}{\sqrt{\frac{\rho U_\infty x}{\mu}}}\left(\frac{1}{2}\rho U_\infty^2\right)dx = 0.332B\int_0^{mL} \sqrt{\mu\ \rho U_\infty}.U_\infty.\frac{dx}{\sqrt{x}}$$

$$= 0.332B\sqrt{\mu\ \rho U\infty.U\infty}\int_o^{mL} \frac{dx}{\sqrt{x}}$$

$$= 0.332\ B\sqrt{upu_\infty.u_\infty}$$

$$\int_0^{ml} \frac{dx}{\sqrt{x}} = 0.332B\sqrt{\mu\ \rho U_\infty.U_\infty}\left|2\sqrt{x}\right|_0^{mL}$$

$$F_D = 1.328B\sqrt{\mu\ \rho U_\infty.U_\infty}\left(\sqrt{ml}\right) = 1.328B\sqrt{\mu\ \rho U_\infty.U_\infty}\sqrt{mL} \quad ...(1)$$

Also
$$F_D = \frac{1}{2} C_D \rho A U_\infty^2 = \frac{1.328}{\sqrt{\text{Re}_L}} . \frac{1}{2} \rho U_\infty^2 A$$

Now
$$A = L \times B, \sqrt{\text{Re}_L} = \sqrt{\frac{\rho U_\infty L}{\mu}}$$

$\therefore$
$$F_D = \frac{1.328}{2} \sqrt{\rho U_\infty mL} . BU_\infty = 0.664 \sqrt{\rho U_\infty mL} . BU_\infty \quad ...(2)$$

From Eqs. (1) and (2), we have

$$0.664\sqrt{m} = 0.332$$

$$m = 0.25$$

Hence at $\frac{1}{4}$ th of the length of plate the drag force will be half.

Q.5.(c) Derive an expression for the work done per second by water on the runner of Pelton turbine. Prove that for a given jet speed, volume flow rate, turning angle and wheel radius the maximum efficiency occurs when the turbine bucket move at half the jet speed.

Sol. Workdone and efficiency of a Pelton Turbine:

The velocity triangles for the split type buckets of a Pelton turbine are shown in Fig.12.

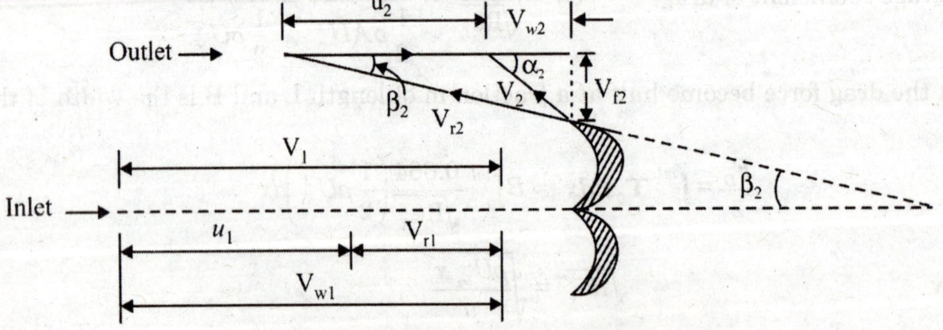

Fig.12: Velocity triangles for a Pelton turbine

Let
N = rpm of wheel

D = diameter of wheel, m

d = diameter of jet, m

u = mean peripheral velocity of runner ($u_1 = u_2 = u$)

$$= \frac{\pi DN}{60} m/s$$

v_1 = absolute velocity of water at inlet

v_{r1} = relative velocity of jet at inlet to bucket

α_1 = angle between the direction of jet and direction of motion of the bucket (called the guide angle)

β_1 = angle made by the υ_{r1} with the direction of motion of bucket at inlet (called vane angle at inlet)

υ_{w1}, υ_{f1} = components of υ_1 in the direction of motion of bucket whirl and perpendicular (flow) to it respectively.

υ_2 = absolute velocity of jet leaving the bucket

υ_{r2} = relative velocity of jet leaving the bucket

α_2 = angle made by υ_2 with υ_2

β_2 = angle made by υ_{r2} with u_2 (vane angle at outlet)

υ_{w2}, υ_{f2} = components of υ_2 in parallel (whirl) and perpendicular (flow) directions to u_2 respectively

Inlet: For $\alpha_1 = 0$ and $\beta_1 = 0$, the velocity triangle at inlet will be a straight line.

$$\therefore \quad \upsilon_{r1} = \upsilon_1 - u_1 = \upsilon_1 - u$$

$$\upsilon_{w1} = \upsilon_1$$

Outlet: Let $\upsilon_{r2} = k\upsilon_{r1}$

Where k = blade friction coefficient

= 0.85 – 0.95 = 1 for smooth blades

$\upsilon_{w2} = \upsilon_2 \cos \alpha_2 = \upsilon_{r2} \cos \beta_2 - u_2 = \upsilon_{r2} \cos \beta_2 - u$

Depending upon the magnitude of peripheral speed u, the turbine unit may have a slow, medium or fast runner.

(i) Slow runner: $\psi\alpha_2 > 90°$ and υ_{w2} is – ve, $\beta_2 < 90°$

(ii) Medium runner: $\alpha_2 > 90°$ and $\upsilon_{w2} = 0$, $\beta_2 < 90°$

(iii) Fast Runner: $\alpha_2 > 90°$ and υ_{w2} is + ve, $\beta_2 < 90°$

Force exerted by the jet water in the direction of motion,

$$F = \rho A\upsilon_1(\upsilon_{w1} \pm \upsilon_{w2})$$

Where $\quad A = \frac{\pi}{4}d^2$ = area of jet

$$\rho = \text{density of water}$$

Work done per second per unit weight of striking water

$$= \frac{\rho A\upsilon_1(\upsilon_{w1} + \upsilon_{w2})u}{\rho A\upsilon_1 g} = \frac{1}{g}(\upsilon_{w1} + \upsilon_{w2})u$$

Kinetic energy of jet at inlet $\quad = \frac{1}{2}m\upsilon_1^2 = \frac{1}{2}(\rho A\upsilon_1)\upsilon_1^2$

Hydraulic efficiency,

$$\eta_{hyd} = \frac{\text{work done by turbine per second}}{\text{K.E of jet per second}}$$

$$= \frac{\rho A \upsilon_1 (\upsilon_{w1} \pm \upsilon_{w2}) u}{\frac{1}{2} (\rho A \upsilon_1) \upsilon_1^2}$$

$$= \frac{2(\upsilon_{w1} \pm \upsilon_{w2}) u}{\upsilon_1^2}$$

For the velocity triangles shown in Fig. 12, we have

$$\upsilon_{w1} = \upsilon_1,\ \upsilon_{r1} = \upsilon_1 - u_1 = \upsilon_1 - u$$

$$\upsilon_{w2} = \upsilon_{r2} \cos \beta_2 - u_2 = k\ v_{r1} \cos \beta_2 - u$$

$$= k(\upsilon_1 - u) \cos \beta_2 - u$$

$$\eta_{hyd} = \frac{k(\upsilon_1 - u)\cos \beta_2 - u]u}{\upsilon_1^2}$$

$$= \frac{2[(\upsilon_1 - u)(1 + k \cos \beta_2)]u}{\upsilon_1^2}$$

The hydraulic efficiency will be maximum for a given value of υ_1, when

$$\frac{d\eta_{hyd}}{du} = 0$$

$$\frac{d}{du}[(\upsilon_1 - \upsilon)u] = 0$$

or $$(\upsilon_1 - u) \times 1 + u(-1) = 0$$

or $$\upsilon_1 - 2u = 0$$

$$u = \frac{\upsilon_1}{2}$$

Thus, the hydraulic efficiency of Pelton tribune shall be maximum when the velocity of wheel is half of the velocity of jet of water at inlet.

Maximum hydraulic efficiency becomes,

$$(\eta_{kyd})_{\max} = \frac{2[(\upsilon_1 - \frac{\upsilon_1}{2})(1 + k \cos \beta_2)] \frac{\upsilon_1}{2}}{\upsilon_1^2}$$

$$= \frac{1 + k \cos \beta_2}{2}$$

For no friction on the blades, $k = 1$, and

$$(\eta_{kyd})_{\max} = \frac{1 + \cos \beta_2}{2}$$

Q.6.(a) A pitot-static tube is inserted into an airstream of velocity U_0, pressure $1.02 \times 10^5 Nm^{-2}$ and temperature 28°C. IT is connected differentially to a mercury U-tube manometer. Calculate the difference of mercury levels in the two limbs of the manometer if the velocity U_o is:

(i) $50 ms^{-1}$ and

(ii) $420 ms^{-1}$.

Take the specific gravity of mercury as 13.6 and for air $\gamma = 1.4$, $R = 0.287\ kJkg^{-1} K^{-1}$

Sol. To measure the velocity of flow in pipe, the Pitot tube is inserted in the pipe as shown in Fig.13

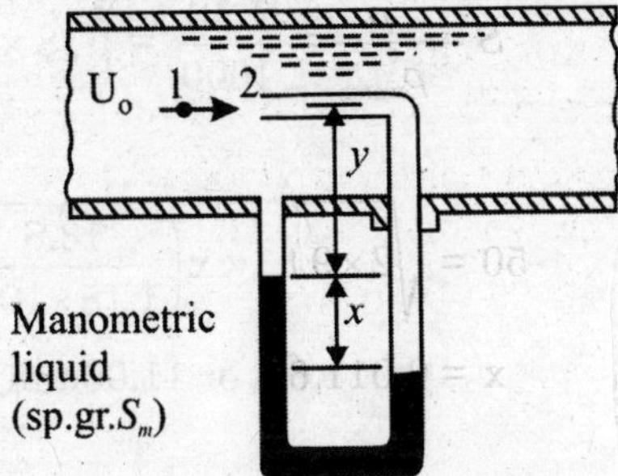

Fig. 13: Pitot Static Tube

This tube measures the stagnation pressure head (or total head) at its dipped end. The dynamic pressure head may be determined by connecting a differential manometer between the Pitot tube and the pressure tap.

Consider point 1 slightly upstream of the stagnation point 2 as shown in the Fig.13. Applying Bernoulli's equation between the points 1 and 2, we get

$$\frac{p_1}{\gamma} + \frac{\upsilon_1^2}{2g} = \frac{p_2}{\gamma} \quad \text{or} \quad \frac{p_2 - p_1}{\gamma} = \frac{\upsilon_1^2}{2g}$$

Where p_1, p_2 = pressure intensities at points 1 and 2.

$\upsilon_1 = \upsilon_0$ = velocity of airstream flow in the pipe at point 1.

γ = specific weight of fluid flowing through the pipe

Let S, S_m = specific gravitites of flowing fluid and the manometric liquid respectively

The equation for the flow through the manometer in meter of water may be written as:

$$\frac{p_1}{\gamma} S + yS + xS_m = (y + x)S + \frac{p_2}{\gamma} S$$

By simplifying, we get

$$\frac{p_2 - p_1}{\gamma} = x\left(\frac{S_m}{S} - 1\right)$$

$$= \frac{\upsilon_1^2}{2g}$$

$$v_1 = \sqrt{2gx\left(\frac{S_m}{S}-1\right)} = \sqrt{2gx\left(\frac{\rho_m}{\rho}-1\right)}$$

Given data for air:

$p = 1.02 \times 10^5\ N/m^2$, $T = 273 + 28 = 301K$, $R = 287$ kJ / $kg\ K$

$$p = \rho RT$$

Now
$$\rho = \frac{p}{RT} = \frac{1.02\times10^5}{287\times301} = 1.18 kg/m^3$$

Specific gravity of air,
$$S = \frac{\rho}{\rho_{water}} = \frac{1.18}{1000} = 1.18 \times 10^{-3}$$

i. $v_0 = v_1 = 50 m/s$

$$50 = \sqrt{2\times9.81\times x\left(\frac{13.6}{1.18\times10^{-3}}-1\right)}$$

$$x = 0.011.6\text{m or } 11.06\text{mm}$$

ii. $v_0 = v_1 = 420\ m/s$

$$420 = \sqrt{2\times9.81\times x\left(\frac{13.6}{1.18\times10^{-3}}-1\right)}$$

$$x = 0.780\ m\ or\ 780\ mm$$

Q.6.(b) It is required to convey 10 m^3/s of water at a mean velocity of 1.25 m/s. Calculate the dimension of the most efficient section of the following channels :

i. rectangular
ii. triangular
iii. trapezoidal and
iv. circular

Which of these has/have the least perimetere and the longest perimeter?

Sol. Given: $Q = 10m^3/s$, $v = 1.25m/s$

Now
$$Q = A \times v$$

$$A = \frac{10}{1.25} = 8m^2$$

For optimum section of a channel, its perimeter should be minimum.

i. Rectangular cross-section. (Fig.14)

$$A = by$$

$$P = b + 2y = \frac{A}{y} + 2y$$

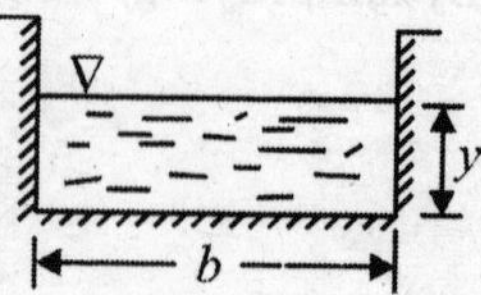

Fig.14: Rectangular Channel

For P to be minimum,

$$\frac{dP}{dy} = 0 \text{ or } -\frac{A}{y^2} + 2 = 0$$

$$A = 2y^2$$

$$b = \frac{A}{y} = 2y$$

$$y = \frac{b}{2} \text{ or } b = 2y$$

Width = twice the depth

$$\frac{d^2 p}{dy^2} = \frac{2A}{y^3}, \text{ which is positive}$$

$$8 = b \times \frac{b}{2} = \frac{b^2}{2}, b = \sqrt{16} = 4m, y = 2m$$

$$P = 4 + 2 \times 2 = 8m$$

ii. Triangular cross-section: (Fig.15)

Let y = depth of flow

θ = angle made bye the sides with the vertical

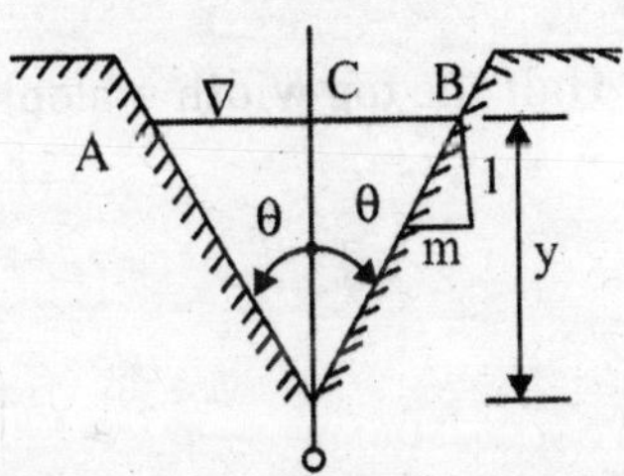

Fig.15: Triangular channel

In ΔOBC,

$$\tan \theta = \frac{BC}{OC} = \frac{BC}{Y}$$

$$BC = y \tan \theta$$

$$\text{Cos } \theta = \frac{OC}{OB} = \frac{y}{OB}$$

$$OB = y \sec \theta$$

Area across flow, $A = \frac{1}{2} AB \times OC = BC \times OC = y^2 \tan \theta, y = \sqrt{\frac{A}{\tan \theta}}$

Perimeter,

$$P = 2 \times OB = 2y \sec \theta$$

$$= 2\sqrt{\frac{A}{\tan \theta}} \times \sec \theta$$

For P to be minimum, $\frac{dP}{d\theta} = 0$

$$\frac{d}{d\theta}\left[2\sqrt{A} \times \frac{\sec \theta}{\sqrt{\tan \theta}}\right] = 0$$

Which gives, $\text{Sin } \theta = \frac{1}{\sqrt{2}}$ *or* $\theta = 45°$ *and* $m = 1$

$$8 = y^2 \text{ for } \theta = 45°$$

$$y = 2\sqrt{2}\ m$$

$$P = 2 \times \sqrt{2} \times \sec 45° = 8m$$

Trapezoidal cross-section (Fig.16).

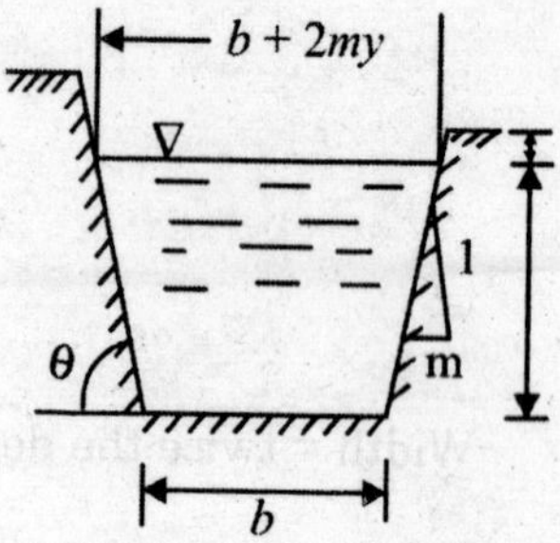

$$A = \frac{1}{2}(b + B)\,y = \frac{1}{2}(b + b + 2my)\,y$$

$$= (b + my)\,y$$

$$P = b + 2y\sqrt{1+m^2} = (\frac{A}{y} - my) + 2y\sqrt{1+m^2})$$

For P to be minimum,

$$\frac{dP}{dy} = \left(-\frac{A}{y^2} - m\right) + 2\sqrt{1+m^2} = 0$$

$$A = y^2(2\sqrt{1+m^2} - m) = y(b + my)$$

$$b = 2y(\sqrt{1+m^2} - m)$$

$$\frac{B}{2} = \frac{b + 2my}{2} = y\sqrt{1+m^2}$$

Bz

Fig.16: Trapezoidal Channel

Half the top width = sloping side of cross-section

$$\frac{dP}{dm} = -y + \frac{2my}{\sqrt{1+m^2}} = 0$$

$$\frac{2m}{\sqrt{1+m^2}} = 1$$

$$m = \frac{1}{\sqrt{3}}$$

$$\tan\theta = \frac{1}{m} = \sqrt{3}$$

$$or = 60°$$

$$b = 2y\left(\sqrt{1+\frac{1}{3}} - \frac{1}{\sqrt{3}}\right) = 1.1547y$$

$$B = 2y\sqrt{1+m^2} = 2y\sqrt{1+\frac{1}{3}} = 2.3094y$$

$$A = \frac{1}{2}(1.1547y + 2.3094y)\,y = 1.73205y^2 = 8$$

$$y = 2.15\text{m}$$

$$b = 1.1547 \times 2.15 = 2.48m$$

Now

$$B = 2.3094 \times 2.15 = 4.96m$$

$$P = 2.48 + 4.96 + 2 \times \frac{2.15}{\sin 60} = 9.92m$$

iv. Circular cross-section (Fig.17).

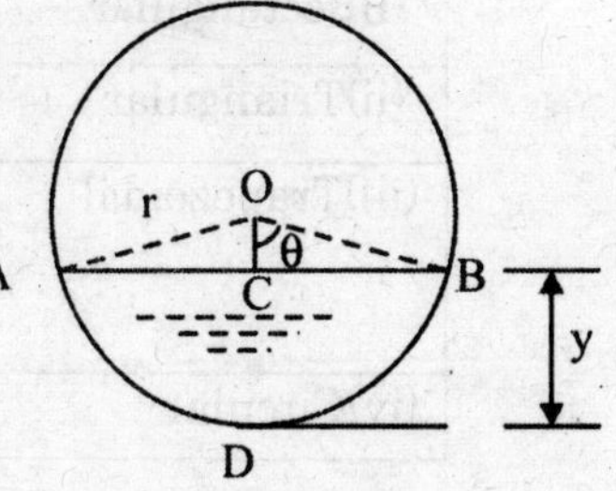

Fig.17: Circular channel

Area across flow, A = area $ADBA$ = area of sector OADBO-area of ΔOAB

$$= r^2\theta - \frac{1}{2} \times 2r\sin\theta \times r\cos\theta$$

$$= r^2\,\theta - r^2 \sin\theta\cos\theta$$

$$= r^2\left(\theta - \frac{\sin 2\theta}{2}\right)$$

Wetted perimeter, $P = 2r\theta$

Velocity of flow will be maximum when $\frac{A}{P}$ is minimum.

or $$\frac{d}{d\theta}\left(\frac{A}{P}\right) = 0$$

$$\frac{d}{d\theta}\left[\frac{r}{2}\left(1 - \frac{\sin 2\theta}{2\theta}\right)\right] = 0$$

$$\frac{r}{2}\left[0 - \frac{1}{2}\left\{\frac{\theta \times 2\cos 2\theta - \sin 2\theta \times 1}{\theta^2}\right\}\right] = 0$$

$$2\theta\cos 2\theta - \sin 2\theta = 0$$

$$\tan 2\theta = 2\theta$$

$$2\theta = 257.5° \quad \theta = 128.75°$$

Depth of flow, $$Y = OD - OC = r - r\cos\theta$$

$$= r\,(1 - \cos 128.75°) = 1.62r = 0.81d$$

$$8 = r^2\left(\frac{\pi}{180} \times 128.75^0 - \frac{1}{2}\sin 257.5^0\right)$$

$$2.735r^2 = 8,$$

$$r = 1.71m$$

$$P = 2 \times 1.71 \times \frac{\pi}{180} \times 128.75$$

$$= 7.68m$$

Cross-Section of channel	Dimensions		Perimeter , P, m
(i)Rectangular	$b = 4m$	$Y = 2m$	8
(ii)Triangular	$\theta = 45°$	$Y = 2\sqrt{2}\ m$	8
(iii)Trapezoidal	$\theta = 60°$	$Y = 2.15m$ $b = 2.48m$ $B = 4.96m$	9.92
(iv)Circular	$\theta = 128.75°$	$r = 1.71m$	7.68

Circular Cross-section has the least perimeter and trapezoidal section the maximum perimeter.

Q.6.(c) Derive an expression for area velocity relationship for a compressible fluid in the form :

$$\frac{dA}{A} = \frac{dV}{V}(M^2 - 1)$$

Further, explain the variation of velocity with change in area for:

i. Subsonic velocity

ii. Sonic velocity and

iii. Supersonic velocity

Sol. For a one-dimensional flow across an area of cross-section A with velocity v and constant density, ρ the continuity equation for steady flow is,

$$\rho Av = \text{const.}$$

Taking log of both sides, we have

$$\text{In } \rho + \text{In } A + \text{In } \upsilon = \text{In } const$$

Differentiating, we have

$$\frac{d\rho}{\rho} + \frac{dA}{A} + \frac{dV}{V} = 0 \qquad \text{...(1)}$$

If significant changes in cross-sectional are and, therefore, in v and ρ occur over only a short length of conduit, such as in a nozzle, frictional effects may be neglected. Euler's equation of motion for steady, frictionless flow neglecting the gravity and body forces may be written as:

$$VdV + \frac{dp}{p} = 0 \qquad \text{...(2)}$$

Now sonic velocity, $c = \sqrt{\dfrac{dp}{dp}}$ or $c^2 = \dfrac{dp}{dp}$

Eq.(2) may be written as:

$$VdV + \frac{d\rho}{\rho} \times c^2 \times \frac{d\rho}{dp} = 0$$

$$\frac{VdV}{c^2} + \frac{d\rho}{\rho} = 0 \qquad \text{...(3)}$$

From Eq.(1). We have

$$\frac{d\rho}{\rho} = -\frac{dA}{A} - \frac{dV}{V}$$

Eq. (3) becomes

$$\frac{dA}{A} + \frac{dV}{V} = \frac{VdV}{c^2}$$

or

$$\frac{dA}{A} = \frac{dV}{V}\left(\frac{v^2}{c^2} - 1\right) = \frac{dV}{V}(M^2 - 1) \qquad \text{...(4)}$$

Eq.(4) is known as the Hugoniot equation.

Also

$$\frac{dV}{V} = -\frac{dp}{\rho V^2} \qquad \text{from Eq. (2)}$$

Substituing in Eq. (1), we get

$$\frac{d\rho}{\rho} - \frac{dp}{\rho V^2} + \frac{dA}{A} = 0$$

or

$$\frac{dA}{A} = \frac{dp}{\rho V^2} - \frac{d\rho}{\rho} = \frac{dp}{\rho V^2}\left(1 - V^2\frac{d\rho}{dp}\right) = \frac{dp}{\rho V^2}\left(1 - \frac{V^2}{c^2}\right) = \frac{dp}{\rho V^2}\left(1 - M^2\right)$$

$$\therefore \quad dp = \frac{pV^2}{A}\left(\frac{1}{1-M^2}\right)dA = pV^2\left(\frac{1}{1-M^2}\right)\frac{dA}{A} \qquad \text{...(5)}$$

Flow through Ducts of Varying Area

The following conclusions of practical significance may be drawn from Eqs. (4) and (5)

1. **Subsonic flow: (M<1)**

i. For a convergent nozzle Fig. 18(a)

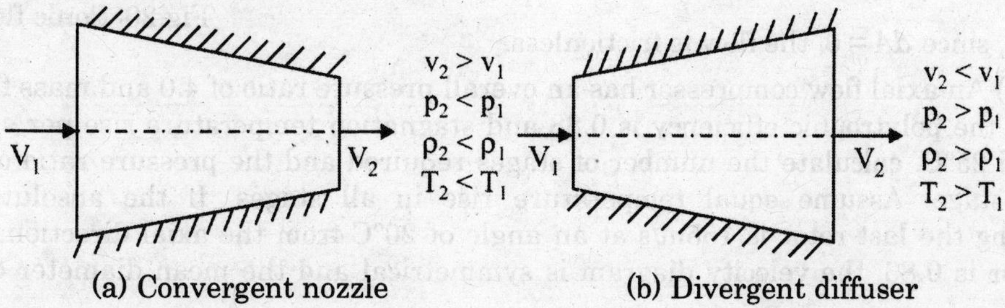

(a) Convergent nozzle (b) Divergent diffuser

Fig. 18: Subsonic flow (M<1)

$$\frac{dV}{V} > 0,\ \frac{dA}{A} < 0,\ dP < 0$$

ii. For a divergent diffuser, Fig. 18(b),

$$\frac{dv}{v} < 0,\ \frac{dA}{A} > 0,\ dp > 0$$

Eq.(4) shows that for $M < 1$ $(M^2 - 1)$ is negative. Therefore dA and dv must be of opposite sign.

In Eq. (5) likewise dp and dA must be of same sign.

2. Supersonic flow: (M > 1)

i. For a divergent mozzle, Fig.19 (a),

$$\frac{dV}{V} > 0, \frac{dA}{A} > 0, dp < 0$$

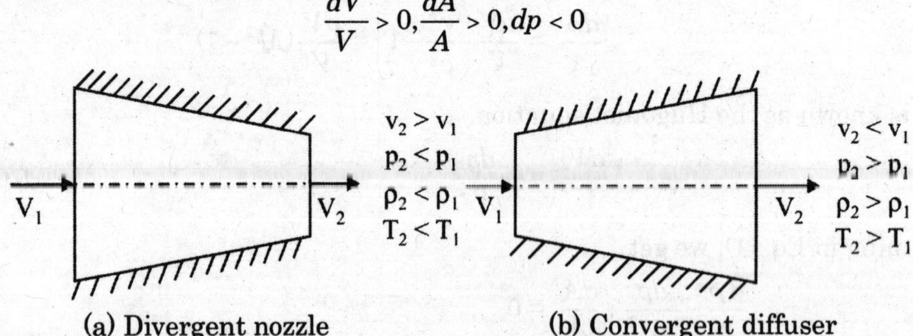

Fig.19: Supersonic flow (M > 1)

ii. For a convergent diffuser, Fig.19 (b),

$$\frac{dV}{V} < 0, \frac{dA}{A} < 0, dp > 0$$

For $M > 1$, dA and dv are of same sign but dp and dA are of opposite sign.

3. Sonic flow ($M = 1$) **Fig.20.**

$$\frac{dA}{A} = 0$$

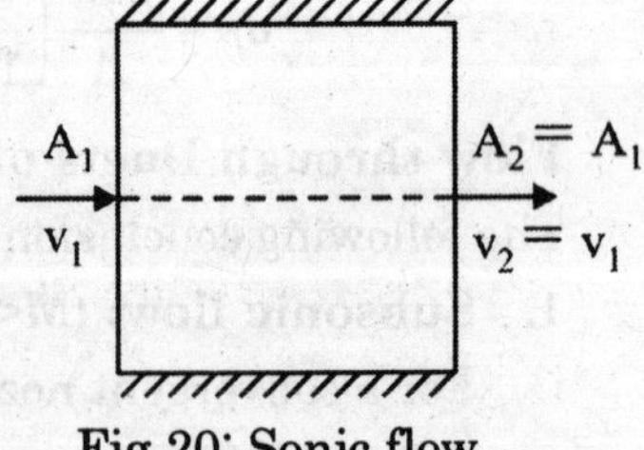

Fig.20: Sonic flow

$dp = 0$, since $dA = 0$, the flow is frictionless.

Q.7.(a) An axial flow compressor has an overall pressure ratio of 4.0 and mass flow of 160 kg/min. If the polytrophic efficiency is 0.88 and stagnation temperature rise per stage must not exceed 25°C, calculate the number of stages required and the pressure ratio of the first and last stage. Assume equal temperature rise in all stages. If the absolute velocity approaching the last rotor is 155m/s at an angle of 20°C from the axial direction, the work done factor is 0.83, the velocity diagram is symmetrical and the mean diameter of the last stage rotor is 180mm, compute the rotational speed and the length of the stage. Ambient conditions are 1.01 bar and 288 K.

Sol. Given: $r_p = 4, m_a = 160\ kg / \min, \eta_p$ 0.88, $(T_o)_{stage} \leq 25°C$, $Z = ?$, $\Delta p_z / \Delta p_1 = ?$, $(v_{a1})_z = 155$ m/s, $\phi\ w = 0.83$, $R_d = 0.5$ for symmetrical blading, $(d_m)_z = 180mm$, $p_1 = 1.01$bar, $T_1 = 288K$, $\alpha_1 = 20°$

Refer to Fig. (21).

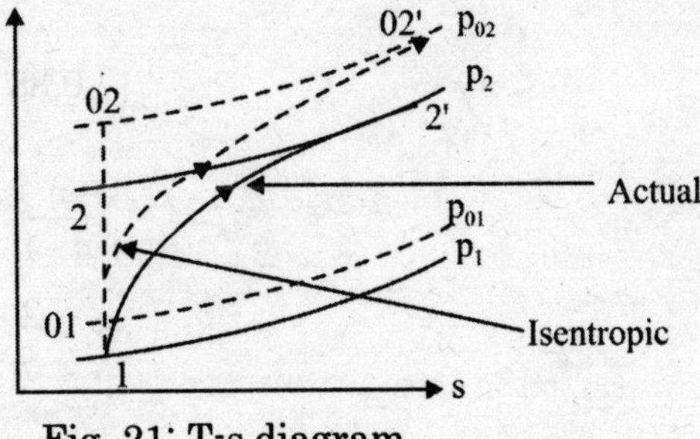

Fig. 21: T-s diagram

$$\frac{T_Z}{T_1} = r_p^{\left(\frac{\gamma-1}{\gamma}\right)}$$

$$T_2 = 288 \times 4^{\frac{1}{3.5}} = 428K$$

$$T_{OZ} = T_Z + \frac{(v_{a1})_z^2}{2c_p}$$

$$= 428 + \frac{(155)^2}{2 \times 1.005 \times 10^3} = 439.95K$$

$$\frac{p_{oz}}{p_1} = \left(\frac{T_{02}}{T_1}\right)^{\frac{\gamma}{\gamma-1}}$$

$$P_{oz} = 1.01\left(\frac{439.95}{288}\right)^{3.5} = 4.406 \text{ bar}$$

Now $\qquad (v_{a1})_1 = (v_{a1})_z = 155 m/s$

$$T_{01} = T_1 + \frac{(v_{a1})_1^2}{2c_p} = 288 + \frac{(155)^2}{2 \times 1.005 \times 10^3} = 299.95K$$

$$\frac{p_{01}}{p_{02}} = \left(\frac{T_{01}}{T_{02}}\right)^{\frac{\gamma}{\gamma-1}}$$

$$p_{01} = 4.406\left(\frac{299.95}{439.95}\right)^{3.5} = 1.153\text{bar}$$

Pressure ratio of first and last stage, $\frac{p_{oz}}{p_{01}} = \left(\frac{4.406}{1.153}\right) = 3.82$

Now $\qquad \frac{T_{oz'}}{T_{o1}} = \left(\frac{p_{oz}}{p_{o1}}\right)^{\frac{n-1}{n}}$

Polytropic efficiency, $\eta_p = \left(\frac{\gamma-1}{\gamma}\right)\left(\frac{n}{n-1}\right)$

$$0.88 = \left(\frac{1.4-1}{1.4}\right)\left(\frac{n}{n-1}\right)$$

$$\frac{n}{n-1} = 3.08$$

$$n = 3.08n - 3.08$$

$$n = 1.48$$

$$T_{oz'} = 299.95\left(\frac{4.406}{1.153}\right)^{\frac{0.48}{1.48}} = 463.3\text{K}$$

Number of stages, $Z = \frac{Toz - T_{0'}}{(dT_0')stage} = \frac{463.3 - 299.95}{25} = 6.53 \simeq 7$

Actual temperature rise per stage $= \frac{463.3 - 299.95}{7} = 23.3°C < 25°C$

Hence the actual temperature rise per stage is within the limit specified.

For symmetrical velocity diagram, $\alpha_1 = \beta_2$ as shown in Fig.22.

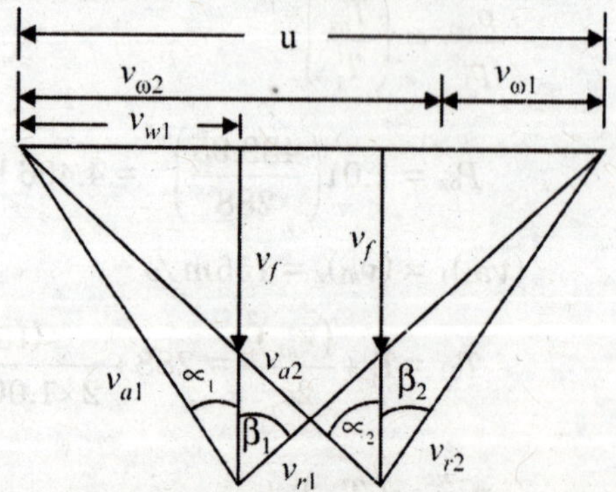

Fig. 22: Symmetrical Velocity triangles

Work done per stage, $w = c_p \times (dT_O')stage \times \phi w$

$$= 1.005 \times 23.3 \times 0.83$$

$$= 19.43 \text{ kJ/kg}$$

Also $w = (v_{w2} - v_{w1})u$

$$= (u - 2\, v_{w1})u$$

$$= (u - 2 \times va_1 \sin \alpha_1)u$$

$$= (u - 2 \times 155 \sin 20°)u$$

$$= (u - 106.03)u$$

$$u^2 - 106.03u - 19.43 \times 10^3 = 0$$

$$u = \frac{1}{2}\left[106.03 \pm \sqrt{(106.03)^2 + 4 \times 19.43 \times 10^3}\right]$$

$$= \frac{1}{2}\left[106.03 \pm 298.26\right] = 202.14\ m/s$$

Now $$u = \frac{\pi d_m N}{60}$$

$$N = \frac{202.14 \times 60}{\pi \times 0.18} = 21,448\ rpm$$

Area across flow, $$A_f = r d_m h$$

Density of air, $$\rho_1 = \frac{p_1}{RT_1} = \frac{1.01 \times 10^5}{287 \times 288} = 1.222 kg/m^3$$

Velocity of flow, $$v_{f1} = v_{a1} \cos \alpha_1 = 155 \cos 20° = 145.6 m/s$$

$$m_a^u = \rho_1 A_f v_{f1}$$

$$\frac{160}{60} = 1.222 \times \pi \times 0.18 \times h \times 145.6$$

$$h = 2.65 cm$$

Length of last stage rotor blade, $h = 2.65 cm$

Q.7.(b) What are the essentials of a good steam boiler? Enumerate the factors which should be considered while selecting a boiler.

Sol. Essentials of a Good Boiler

The essentials of a good steam boiler are:

1. It should be able to generate steam at the required pressure and quantity with minimum fuel consumption.
2. Its initial cost, installation, maintenance and operation cost should be as low as possible.
3. It should occupy small floor area and should be light in weight.
4. It should be able to meet the fluctuating demands of steam without fluctuations in steam pressure ... should be easily accessible for cleaning, inspection and ... have minimum of joints to avoid leakage of steam. ... erection times should be reasonable with minimum labour.

8. The water and flue gases velocities should be high for better heat transfer with minimum pressure drop through the system.

9. There should be no deposition of mud and foreign materials on the inside surfaces and soot deposits on the outside surface of heat transferring parts.
10. The boiler should conform to safety precautions as laid down in the Boiler Regulations.

Factors Affecting Boiler Selection

1. Power to be generated i.e. steam to be raised per hour.
2. Working pressure.
3. Initial capital.
4. Facilities available for erection.
5. Floor available.
6. Location of power house.
7. Availability of fuel and water.
8. Probable load factor.
9. Operating and maintenance cost.
10. Accessibility for inspection, cleaning and maintenance.

ENGINEERING SERVICES EXAMINATION (ES)-2013

MECHANICAL ENGINEERING

PAPER - II

(Conventional)

Section-A

Answer all the 10(ten) parts of the Questions No. 1. (Each part carries 4 marks)

Q.1.(a) With the help of a free body diagram, derive an expression for the effort P applied at the circumference of the screw to lift the load W. The coefficient of friction is μ and the helix angle is α.

Sol. The free body diagram for the screw to lift a load W is shown in Fig. 1. The forces acting on the screw are:

W = weight lifted by the screw

P = effort applied at the circumference of the screw

F = force of friction

R = normal reaction

R
P cos α
P
F
α
W cos α
α
W
P sin α

Fig. 1: Free body digram for screw

Consider the equilibrium of forces along the inclined plane, we have

$$W \sin \alpha + F = P \cos \alpha \quad ...(1)$$

For the equlibrium of forces perpendicular to the inclined plane, we have

$$W \cos \alpha + P \sin \alpha = R$$

Due to the coefficient of friction μ between the screw and body of screw jack,

$$F = \mu R$$

From eq.(1), $F = P \cos \alpha - W \sin \alpha$

From eq.(2), $\mu R = \mu \; W \cos \alpha + \mu \; P \sin \alpha$

$\therefore$ $P \cos \alpha - W \sin \alpha = \mu \; W \cos \alpha + \mu P \sin \alpha$

$$P(\cos \alpha - \mu \sin \alpha) = W(\mu \cos \alpha + \sin \alpha)$$

$$P = W\left(\frac{\mu \cos \alpha + \sin \alpha}{\cos \alpha - \mu \sin \alpha}\right)$$

Q.1.(b) Sketch and explain the working of centrifugal clutch. Give one application of it.

Sol. A centrifugal clutch as shown in Fig. 2 consists of spider keyed to the driving shaft weights. Two weights are attached to the spider. These weights are spring loaded and has friction lining on the outside surface. When the input shafts start rotating the weight moves radially out due to centrifugal force and presses against the rim attached to the driven shaft, thereby transmitting the torque. Assuming the angle subtended by the arc of contact of the weight with the rim to be small (<40°) the contact of the weight with the rim may be assumed as a point contact, so that the net radial force required to produce torque T is:

$$F_r = \frac{T}{\mu R n}$$

Where R = radius of rim and

n = number of weights,

μ = coefficient of friction

The net radial force, $F_r = F_a - F_s = mr\omega^2 - F_s$

Where F_c = centrifugal force, N

m = mass of rotating mass, kg

r = radius of gyration of mass, m

ω = angular speed of rotating shaft, rad/s

F_s = spring force, N

It is used in mopeds, a two wheeler for children.

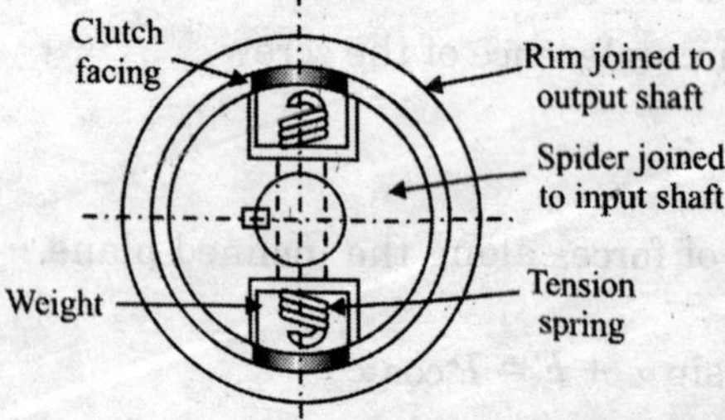

Fig. 2: Centrifugal clutch

Q.1.(c) A thin cylindrical shell with hemispherical ends is subjected to internal fluid pressure. For equal maximum stress to occur in both the cylindrical and the spherical portions, what would be the ratio of the thickness of the spherical portion to that of the cylindrical portion?

Sol. Cylindrical Shell with Hemispherical Ends

Cylindrical shell with hemispherical ends have the advantage over flat ends that the bending stresses are reduced. If t_1 and t_2 are the wall thickness of the cylindrical section and hemispherical ends respectively as shown in Fig. 3, then the circumferential strain in the cylinder is:

$$(\varepsilon)_c = \frac{pR}{Et_1}(1 - 0.5\nu)$$

Fig. 3: Cylindrical shell with hemispherical ends.

The circumferential strain in the hemispherical end is:

$$(\varepsilon)_s = \frac{pR}{2Et_2}(1-\nu)$$

These two strains must be equal so that the bending stresses are minimum at the junction and there is no distortion. Hence

$$\frac{pR}{Et_1}(1-0.5\nu) = \frac{pR}{2Et_2}(1-\nu)$$

or

$$\frac{t_1}{t_2} = 2\left(\frac{1-0.5\nu}{1-\nu}\right)$$

$$\frac{t_1}{t_2} = \frac{2-\nu}{1-\nu}$$

For equal maximum loop stress in cylinder and spherical ends,

$$\frac{pR}{t_1} = \frac{pR}{2t_2}$$

$$\therefore \quad \frac{t_1}{t_2} = 2$$

Q.1.(d) Two close coiled helical springs *A* and *B* made of the same wire show axial compressions of 8mm and 3mm, respectively, when they are subjected to the same axial load. The spring *A* has coils of mean diameter of 80 mm, while the spring *B* has 8 coils. Determine the mean coil diameter of the spring *B*.

Sol. Given: $G_A = G_D$, $\delta_A = 8mm$, $\delta_B = 3mm$, $W_A = W_B$, $n_A = 9$, $D_{mA} = 80mm$, $d_A = d_B$, $n_B = 8$, $D_{mB} = ?$

Deflection, $\delta = \frac{8WD_m^3 n}{Gd^4}$, shear Stress $\tau = \frac{8WD_m}{\pi d^3}$

$$\delta_A = \frac{8W_A D_{mA}^3 n_A}{G_A d_A^4}, \quad \delta_B = \frac{8W_B D_{mB}^3 N_B}{G_B d_B^4}$$

$$\therefore \quad \frac{\delta_A}{\delta_B} = \left(\frac{D_{mA}}{D_{mB}}\right)^3 \times \frac{n_A}{n_B}$$

$$\frac{8}{3} = \left(\frac{80}{D_{mB}}\right)^3 \times \frac{9}{8}$$

$$\left(\frac{80}{D_{mB}}\right)^3 = \frac{64}{27}$$

$$\frac{80}{D_{mB}} = \frac{4}{3}$$

$$D_{mB} = \frac{80 \times 3}{4} = 60mm$$

Q.1.(e) Explain the purpose of process annealing. How it is done? Discuss in short about bainite.

Sol. If the steel has been subjected to cold working to increase the tensile strength and yield point, it becomes work hardened and brittle. The metal can be softened by process annealing or recrystallization. Its purpose is to reduce the distortions of the crystal lattice produced by cold working.

Recrystallization annealing consists in heating the steel to the recrystallization temperature, i.e. to a temperature some what below the lower critical temperature, holding at this temperature for a prolonged period and slow cooling.

Recrystallization temperature, $T_{re} = 0.4 \times$ melting temperature of metal.

$= 600 - 700°$ C for steels

This process is frequently applied in the production of cold-rolled steel strip, in deep drawing and wire drawing operations as an intermediate process with the aim of increasing the plasticity of steel.

Bainite: An increased rate of cooling of steel, corresponding to an increase in the degree of super cooling of austenite leads to the formation of a finely dispersed pearlite, known as sorbite (at about 600°C). If the temperature is lowered from 550° C to 220° C, acicular (needle like) troosite is formed, which is known as bainite.

Q.1.(f) Name four materials used as shaped tool in Electrochemical machining. What types of materials are normally cut using high-pressure inert gas-assisted laser and why?

Sol. Materials used as shaped tool in ECM: copper, brass, bronze and aluminum. The materials which are normally cut using high-pressure inert gas assisted laser is diamond. Diamond is very hard and cannot be cut using conventional tool cutters which are less hard as compared to diamond. Also diamond is not a good electrical conductor, it cannot be cut using other non-conventional machining processes like ECM, EDM, and plasma arc, etc.

Q.1.(g) Write the reasons which result the metals to shrink during solidification and cooling in metal casting What are hot tears and cold shuts?

Sol. Shrinkage in metal casting during solidification and cooling: When the casting solidifies, it shrinks or contracts in size. This shrinkage occurs in three stages,

1. **Liquid contraction or shrinkage:** This shrinkage occurs when the metal is liquid. It occurs when the molten metal cools from the pouring temperature to the solidification temperature.
2. **Solidification shrinkage:** This occurs when the metal looses its latent heat, i.e. during the time the metal changes from the liquid state to the solid state.
3. **Solid shrinkage:** This occurs when the metal cools from the freezing temperature to the room temperature.

Hot tears: These are hot cracks which appear in castings in the form of irregular crevices with a dark oxidized fracture surface. They arise when the solidifying metal does not have sufficient strength to resist tensil forces produced during solidification. They occur mainly from an excessively high temperature of casting metal, increased metal contraction, incorrect design of gating system and non-uniform cooling. This defect can be avoided by improving the design of casting, having a mould of low hot strength and large hot deformation.

Cold Shut: This is an interface within a casting that is formed when two metal streams meet without complete fusion. Its causes are: inadequate metal supply, too low mould or melt temperature, improperly designed gates or length to thickness ratio of the casting is too large.

Q.1 (h) Explain the terms comminution and reduction used in powder metallurgy. Why lubricants are used to mix the metal powders?

Sol. Reduction: Pure metal is obtained by reducing its oxide with a suitable reducing gas at an elevated temperature but below the melting point. This pure metal is then crushed and milled to a powder. The resulting particles are of irregular shape and are quite porous. This process is cheap and a large amount of powder is made by this method.

Comminution: It is the process for reduction of particles of metal powders from one average size to a small average particle size through grinding or crushing using ball mills.

Lubricants are used to mix different metal powders so as to form a deposit layer over the particle surface which results in better interaction during mixing of powders.

Q.1 (i) Name 4 basic ways of establishing a time standard.

Sol. For basic ways to establish time standards are:

1. Time study by stop watch
2. Standard data system
3. Work Sampling
4. Predetermined motion time system (PMTS)

Q.1 (j) Consider the following formulas for statistics problems:

$$\text{Means} = \mu = \frac{1}{N}\sum_{i=1}^{n} di$$

$$\text{Standard devation} = \sqrt{\sigma}$$

$$\text{Variance} = \sigma^2 = \frac{1}{N}\sum_{i=1}^{n}(d_i - \mu)^2$$

write a *C* coding for the above.

Section - B

(Answer any two questions)

Q.2 (a) At a point in a loaded component the state of stress is given by

$$\sigma_x = 270\text{MPa},\ \sigma_y = 130\text{ MPa and}$$

$$\tau_{xy} = \pm 40\text{ MPa}.$$

Determine

(i) the maximum and minimum principal stresses and the planes on which they act.

(ii) the maximum shearing stress in magnitude and direction.

Sol. Given : $\sigma_x = 270\text{MPa}$, $\sigma_y = 130\text{MPa}$, $\tau_{xy} = \pm 40$ MPa

(i) Principal stresses are given by,

$$\sigma_{1,2} = \frac{1}{2}(\sigma_x + \sigma_y) \pm \frac{1}{2}\sqrt{(\sigma_x - \sigma_y)^2 + 4\tau_{xy}^2}$$

$$= \frac{1}{2}(270 + 130) \pm \frac{1}{2}\sqrt{(270 - 130)^2 + 4 \times 40^2}$$

$$= 200 \pm 80.62$$

Maximum principal stress, $\sigma_1 = 280.62$ MPa

Minimum principal stress, $\sigma_2 = 119.38$ MPa

Principal planes are given by,

$$\tan 2\theta = \frac{2\tau_{xy}}{\sigma_x - \sigma_y} = \frac{2 \times 40}{270 - 130} = \frac{80}{140} = 0.57143$$

$$2\theta = 29.74° \text{ Or } \theta = 14.87°$$

$$\theta_1 = 14.87°,\ \theta_2 = 90 + 14.87 = 104.87°$$

(ii) Maximum shearing stress, $\tau_{max} = \frac{1}{2}(\sigma_1 - \sigma_2)$

$$= \frac{1}{2}(280.62 - 119.38)$$

$$= 80.62 \text{ MPa}$$

Directions of maximum shearing stress are given by,

$$\tan 2\theta = \frac{\sigma_y - \sigma_x}{2t_{xy}} = \frac{130 - 270}{2 \times 40} = -1.75$$

$$2\theta = -60.26°,\ \theta = -30.13°$$

$$\theta_3 = -30.13°,\ \theta_4 = -30.13 + 90 = 59.87°$$

We observe that planes of principal stresses are inclined at 45°to the planes of maximum shear stress.

Q.2 (b) One of the turning pairs of a four-bar chain is replaced by a sliding pair, Draw the inversions by fixing different links. Give one application for each of the mechanism.

Sol. When one of the turning pairs of a four-bar chain is replaced by a sliding pair, it gives a single slider-crank mechanism, as shown in Fig.4:

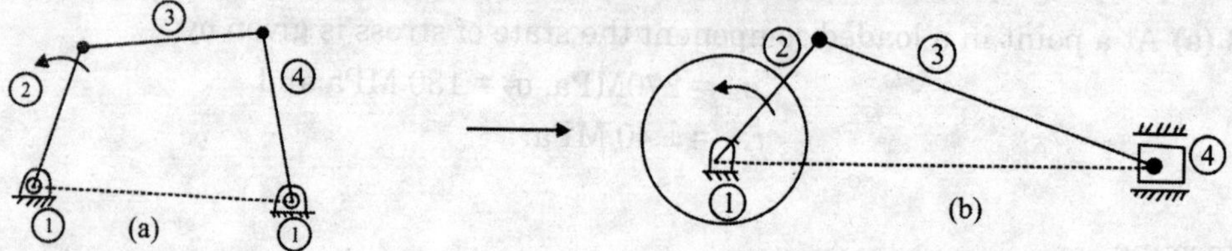

Fig. 4: (a) Four bar chain: all turning pairs (b) Single slider-crank chain

The inversions of single slider-crank mechanism are:

1. Pendulum pump
2. Oscillating cylinder engine
3. Rotary internal combustion engine
4. Crank and slotted lever quick-return motion mechanism
5. Whitworth quick-return motion mechanism.

These inversions have been shown in Fig. 5:

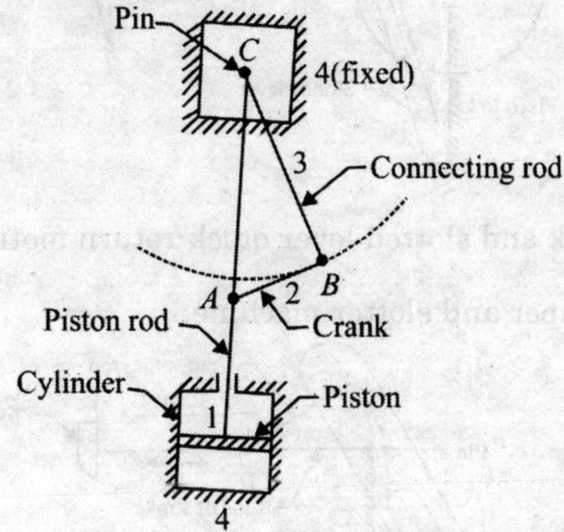

Fig. 5(a): Pendulum Pump

It is used to supply feed water to a boiler (hand pump).

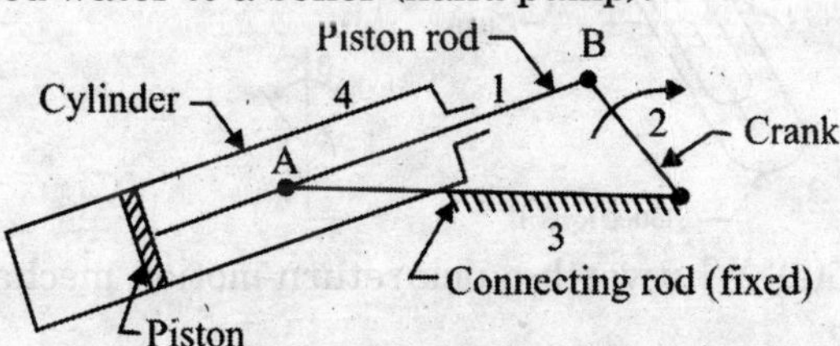

Fig. 5(b): Oscillating cylinder engine

It is used in hoisting engine mechanism, toys and supplying steam to the cylinder.

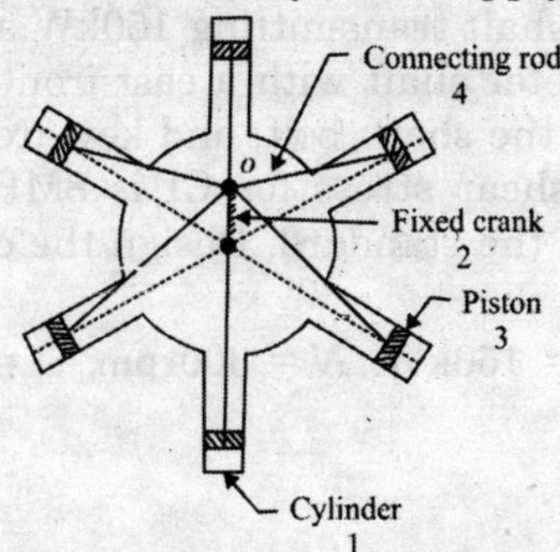

Fig. 5c: Rotary internal combustion engine

It was used earlier in aero-engines.

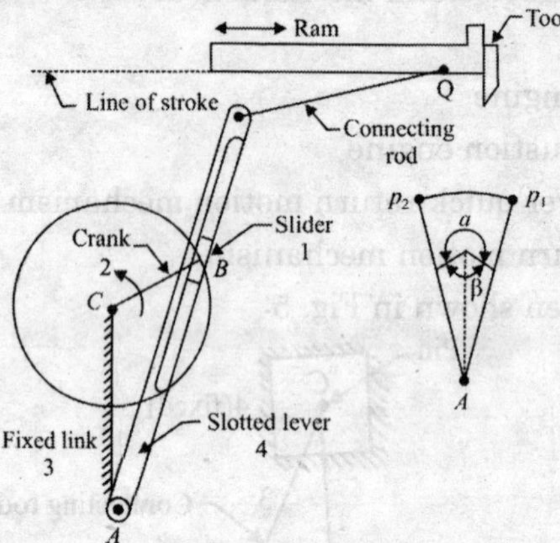

Fig. 5(d): Crank and slotted lever quick-return motion mechanism

It is mostly used in a shaper and slotter machine.

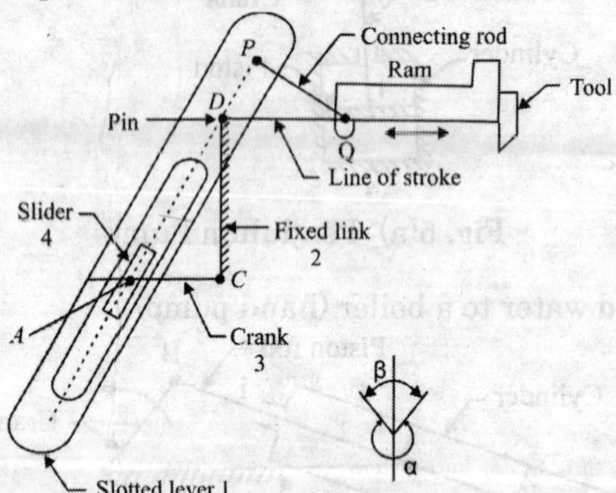

Fig. 5(e): Whitworth quick-return motion mechanism

It is used in shaper machine.

Fig. 5: Inversions of single slider-crank mechanism.

Q.2(c) A 100mm steel drive shaft transmitting 150kW at 300 r.p.m. has to be connected to a machine having same diameter shaft with a cast iron(CI) unprotected flange coupling. The permissible shear stress for the shaft. bolt, and key are 50 MPa. The bearing stress for bolt and key are 50 MPa and shear stress for CI is 8MPa. The basic dimensions of the coupling may be assumed as per the standard. Design the coupling and verify all the major dimensions for strength.

Sol. Given: d = 100mm, P = 150kW, N = 300rpm, $\tau_s = \tau_b = \tau_k$ = 50MPa, $(\sigma_b)_b = (\sigma_b)_k$ = 50MPa, τ_{ci} = 80MPa

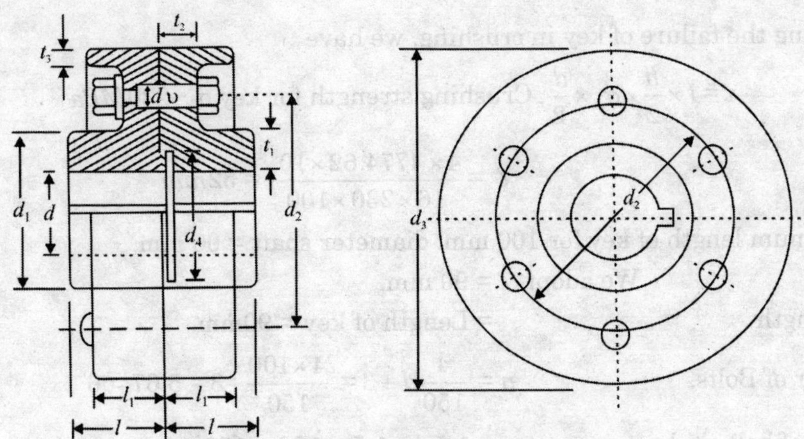

Fig. 6: Flange coupling-protected type

1. Power transmitted $\quad P = \dfrac{2\pi NT}{60\times10^3} kW$

$$\tau = \frac{150\times60\times10^3}{2\pi\times300} = 4774.65\text{N. m}$$

2. Diameter of hub,

The hub is designed as a follow shaft subjected to torsion.

let d_1= outer diameter of hub

Empirically, d_1= 1.75d + 6 mm = 1.75 × 100 + 6 = 181 mm

Checking the hub in shear, we have

$$\tau = \frac{\pi}{16}\left(\frac{d_1^4 - d^4}{d_1}\right)T$$

The hub is made of cast iron for which permissible shear stress, T_{ci} = 8MPa

$$\tau = \frac{16\times4774.65\times181\times10^3}{\pi(181^4-100^4)} = 4.52MPa.\text{ Hence safe.}$$

3. Key size,

For a shaft of diameter 100mm, the size of the key as per IS: 2048 is:

Width, b = 28 mm

Height, h = 16 mm

Considering the failure of key in shear, we have,

$$\tau = lb\frac{d}{2}\tau_k$$

$$4774.65\times10^3 = \ell\times28\times\frac{100}{2}\times50$$

$$\ell = 68.2 \simeq 70\text{ mm}$$

Considering the failure of key in crushing, we have

$$\tau = l \times \frac{h}{2} \times \sigma_c \times \frac{d}{2}, \text{ Crushing strength for key } \sigma_c \sim 230 MPa$$

$$\ell = \frac{4 \times 4774.62 \times 10^3}{16 \times 230 \times 100} = 52 mm$$

The minimum length of key for 100 mm. diameter shaft = 90 mm.

We adopt, $\ell = 90$ mm.

4. Hub length = Length of key = 90mm.

5. Number of Bolts, $n = \frac{4}{150} d + 3 = \frac{4 \times 100}{150} + 3 = 5.67 = 6$

6. Radius of bolt circle, $r_2 = 1.5d = 1.5 \times 100 = 150$mm.

7. Diameter of bolts,

Shearing force in each bolt, $F = \frac{T}{r_2 n} = \frac{4774.65 \times 10^3}{150 \times 6} = 5305N$

Let d_b = diameter of bolts.

$$\frac{\pi}{4} d_b^2 \tau_b = F$$

$$\frac{\pi}{4} d_b^2 \times 50 = 5305$$

$$d_b = 11.62 mm$$

The nearest standard size of hexagonal bolt is M12 as per IS: 1364.

8. Thickness of flange.

Empirically, thickness of flange, $t_2 = 0.5\, t_1 + 6$ mm.

Where $t_1 = \frac{1}{2}(d_1 - d)$

$$t_2 = \frac{1}{4}(181 - 100) + 6 = 26.25 \text{ mm}$$

Considering the bearing failure of flange, we have

$$F = n d_b t_2 \sigma_c$$

$$5305 = 6 \times 12 \times t_2 \times 230$$

$$t_2 = 0.32 \text{ mm}$$

we adopt, $t_2 = 26$ mm

9. Thickness of protecting flanges, $t_3 = 0.5\, t_2 = 13$ mm.

10. Hub overlap = 3mm.

The flange coupling is shown is Fig. 6.

Q.3(a) A steel beam of rectangular section has span of 8 m and simply supported at its ends. It is required to carry a total load of 60 kN uniformly distributed over the whole span. Find the minimum values of breadth and depth if the maximum bending stress is not to exceed 50 MPa. and the maximum deflection is limited to 10 mm E = 210 GPa.

Sol. Given : $L = 8\text{m}$, $wL = 60\ kN$, $b = ?$, $h = ?$, $\sigma_{max} = 50\text{MPa}$, $\delta_{max} = 10\text{mm}$, $E = 210\text{GPa}$, $B = ?$, $h = ?$

Let w = udl, N/m

Then $wL = 60$

$$w = \frac{60}{8} = 7.5kN/m$$

w per unit length

A B C $\frac{L}{2}$ $\frac{L}{2}$

Fig. 7

Maximum deflection at midspan C due to udl on *a* simply supported beam,

$$\delta_{max} = \frac{5}{384}\frac{wL^4}{E1}$$

$$10 \times \overline{10}^3 = \frac{5}{384} \times \frac{7.5 \times 10^3 \times 8^4}{210 \times 10^9 \times I}$$

$$I = 1.90476 \times \overline{10}^4\ m^4$$

$$M_{max} = \frac{wL^2}{8} = \frac{7.5 \times 8^2}{8} = 60kNm$$

$$\sigma_{max} = \frac{6M_{max}}{bh^2}$$

$$50 \times 10^6 = \frac{6 \times 60 \times 10^0}{bh^2}$$

$$bh^2 = 7.2 \times \overline{10}^3 \quad \ldots(1)$$

Now $$\frac{bh^3}{12} = 1.90476 \times \overline{10}^4$$

or $$bh^3 = 22.857 \times \overline{10}^4 \quad \ldots(2)$$

Dividing Eq. (2) by Eq. (1), we get

$$h = 0.31746 \text{ m.}$$

substituting in Eq. (1) the value of h, we get

$$\text{b} = \frac{7.2 \times \overline{10}^3}{(0.31746)^2} = 0.07144 \text{ m}$$

Q.3(b)(i) State the law of gearing to maintain the condition for constant velocity ratio between a pir of toothed wheels. Name two types of gear tooth profiles to satisfy these.

(ii) A pinion having 36 teeth drives a gear having 96 teeth. The profile of the gears are involute with 20° pressure angle, 10 mm. module and 10 mm. addendum. Find the length of path of contact, arc of contact and contact ratio.

Sol. i. Law of gearing

The fundamental law of gearing states that for constant angular velocity ratio of the two gears in contact the common normal at the point of contact must always intersect the line of centres at a fixed point (pitch point) and divide this line in the inverse ratio of the angular velocities of the two gears.

Involute and cycloidal profile gear tooth satisfy the law of gearing.

(i) Given $z_1 = 36$, $z_2 = 96$, $\alpha = 20°$, $m = 10mm$, $h_a = 10mm$.

Addendum radius of pinion, $r_{a1} = r_1 + h_{a1} = \dfrac{mz_1}{2} + h_{a1} = \dfrac{10 \times 36}{2} + 10 = 190mm$

Base circle radius of pinion, $r_{b1} = r_1 \cos\alpha = 180 \cos 20° = 169.14mm$

Addendum radious of gear, $r_{a2} = r_2 + h_{a2} = \dfrac{mz_2}{2} + h_{a2}\ \dfrac{10 \times 96}{2} + 10 = 490mm$

Base circle radius of gear, $r_{b2} = r_2 \cos\alpha = 490 \cos 20° = 460.45mm$

Length of path of recess, $L_r = \sqrt{r_{a1}^2 - r_{b1}^2} - r_1 \sin\alpha$

$$= \left[190^2 - (169.14)^2\right]^{1/2} - 180 \sin 20°$$

$$= 24.99 \text{ mm}$$

Length of path of approach, $L_a = \sqrt{r_{a2}^2 - r_{b2}^2} - r_2 \sin\alpha$

$$= \left[490^2 - (460.45)^2\right] - 480 \sin 20°$$

$$= 3.42 \text{ mm}$$

Length of path of contact, $L_p = L_r + L_a = 24.99 + 3.42 = 28.41mm$

Length of arc of contact, $L_c = \dfrac{L_p}{\cos\alpha} = \dfrac{28.41}{\cos 20°} = 30.23mm$

Contact ratio, $CR = \dfrac{L_p}{p_b} = \dfrac{L_p}{\pi m \cos\alpha} = \dfrac{28.41}{\pi \times 10 \times \cos 20°} = 0.962$

Q.3(c) In rotor system, three unbalanced masses $m_1 = 1.2$kg, $m_2 = 1.8$kg and $m_3 = 2.4$kg. The radius and angular positions of these masses from the horizontal plane are: $R_1 = 1.135m$ at 113.4°, $R_2 = 0.822\ m$ at 48.8° and $R_3 = 1.04m$ at 251.4°. Two balancing masses of m_A and m_B are on left and right ends of the rotor respectively. The distances of unbalanced masses and right side balance mass from the left end unbalance mass m_A are $l_1 = 0.854m$, $l_2 = 1.701m$, $l_3 = 2.396m$ and $l_b = 3.097m$. Find the mass-radius product and angular locations of balancing masses m_A and m_B for dynamic balance of the rotor system.

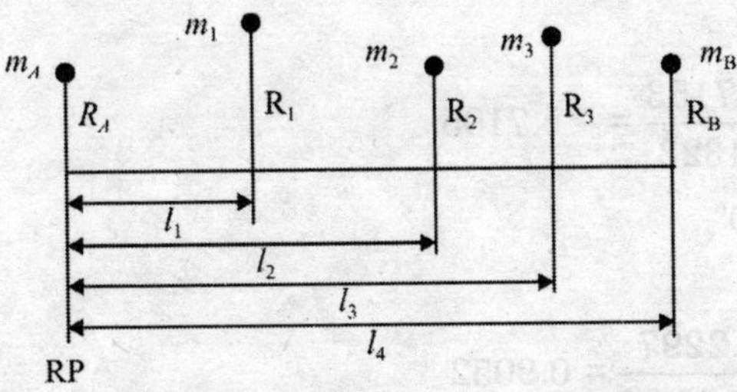

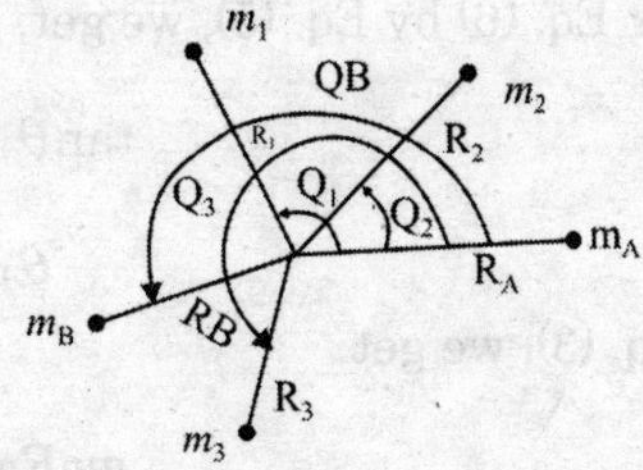

Sol. Fig. 8

Plane	Mass m, kg	Radius R	Angle θ deg	mR/w²	ℓ	mRℓ/w²
A (R.P)	m_A	R_A	$\theta_A = 0$	m_AR_A	0	0
1	1.2	1.135	113.4	1.362	0.854	1.163
2	1.8	0.822	48.8	1.4796	1.163	2.517
3	2.4	1.04	251.4	2.496	2.396	5.980
B	m_B	R_B	θ_B	m_BR_B	3.097	$m_BR_B \times 3.097$

$\Sigma F_x = 0$, *gives* : $m_AR_A \cos \theta_A + m_1 R_1 \cos \theta_1 + m_2 R_2 \cos \theta_2 + m_3 R_3 \cos \theta_3 + m_B R_B \cos \theta_B = 0$

$m_AR_A \cos \theta_A + m_B R_B \cos \theta_B = -1.362 \cos 113.4° - 1.4796 \cos 48.8° - 2.496 \cos 251.4° = 0.362$ ...(1)

$\Sigma F_y = 0$, *gives* : $m_AR_A \sin\theta_A + m_BR_B \sin\theta_B = -1.362 \sin 113.4° - 1.4796 \sin 48.8° - 2.496 \sin 251.4° = 0.00237$ ...(2)

Balancing couple: (A is the R.P.)

$\Sigma Cx = 0, m_AR_A\ell_A \cos \theta_A + m_1R_1\ell_1 \cos \theta_1 + m_2R_2\ell_2 \cos \theta_2 + m_3R_3\ell_3 \cos \theta_3 + m_BR_B\ell_B \cos \theta_B = 0$

$0 + m_BR_B \times 3.097 \times \cos\theta_B = -1.163 \cos 113.4° - 2.517 \cos 48.8° - 5.980 \cos 251.4 = 0.7113$

$m_BR_B \cos \theta_B = 0.2297$ ...(3)

$\Sigma C_y = 0, m_AR_Al_A \cos \theta_A + m_1R_1l_1 \sin\theta_1 + m_2R_2l_2 \cos \theta_2 + m_3R_3l_3 \sin\theta_3 + m_BR_Bl_4 \sin\theta_B = 0$

$0 + m_BR_B \times 3.097 \times \sin \theta_B = -1.163 \sin 113.4° - 2.517 \sin 48.8° - 5.980 \sin 251.4 = 2.7065$

$m_BR_B \sin \theta_B = 0.8739$ ...(4)

Dividing Eq. (4) by (3), are get

$$\tan \theta_B = \frac{0.8739}{0.2297} = 3.80452$$

$$\theta_B = 75.3°$$

From Eq. (1) and (2), we get

$$m_AR_A \cos \theta_A + 0.2297 = 0.362$$

$$m_AR_A \cos \theta_A = 0.1323 \quad ...(5)$$

From Eq. (2) and (4), we get

$$m_AR_A \sin \theta_A + 0.8739 = 0.00237$$

$$m_AR_A \sin \theta_A = -0.87153 \quad ...(6)$$

Dividing Eq. (6) by Eq. (5), we get,

$$\tan\theta_A = \frac{0.87153}{0.1323} = -5.7105$$

$$\theta_A = -80°$$

From Eq. (3), we get

$$m_B R_B = \frac{0.2297}{\cos 75.3°} = 0.9052$$

From Eq. (5), we have,

$$m_A R_A = \frac{0.1323}{\cos(-80°)} = 0.7619$$

Q.4(a) A hollow steel shaft 60 mm internal and 100 mm external diameter is to be replaced by a solid alloy shaft. If the polar modulus has the same value for both, calculate the diameter of the latter and the ratio of their torsional rigidities. Shear modulus G for steel = 2 times the G for the alloy.

Sol. Given: $d_i = 60$ mm, $d_o = 100$mm, $(Zp)_h = (Zp)_s$, $d = ?$, $G_h = 2G_s$

Let D = diameter of solid alloy shaft

$$Js = \frac{\pi d^4}{32};\ d_h = \frac{\pi}{32}\left(d_0^4 - d_i^4\right)$$

$$(Zp)_s = \frac{2J_s}{\alpha} = \frac{\pi}{16}d^3$$

$$(Zp)_h = \frac{2J_s}{d_0} = \frac{\pi}{16}d_0^3\left[1-\left(\frac{d_i}{d_0}\right)^4\right]$$

$$\frac{\pi}{16}d^3 = \frac{\pi}{16}d_0^3\left[1-k^4\right]$$

When $$k = \frac{d_i}{d_0}$$

$$d^3 = d_0^3(1 - k^4)$$

$$d^3 = 100^3[1 - (0.6)^4] = 0.8704 \times 100^3$$

$$d = 0.95478 \times 100 = 95.48 \text{ mm}$$

Torrional rigidity:

$$\frac{T}{J} = \frac{G\theta}{L},\ \frac{T}{\theta} = \frac{JG}{L} = \text{torrional rigidity.}$$

for same lengths of solid and hollow shafts,

Torrional rigidity, $k = \frac{T}{\theta}\alpha(JG)$

$$\frac{J_s G_s}{J_h G_h} = \frac{d_4 \times G_s}{(d_0^4 - d_i^4)G_h} = \frac{(95.48)^4}{(100^4 - 60^4) \times 2} = 0.4774$$

$$\frac{k_h}{k_s} = 2.09$$

Q.4(b) **A vibrating system is characterised by the following parameters:**

mass = m = 6kg

Stiffness of spring = k = 200N/m

Damping coefficient of the dashpot = c = 6 N-s/m

Determine

(i) the damping factor

(ii) the natural frequency of damped vibration

(iii) the number of cycle after which the original amplitude is reduced to 20 per cent.

Sol. Given m = 6kg, k = 200N/m, c = 6 $N.s/m$, $x_n = 0.2\ x_0$

Natural frequency, $\omega_n = \sqrt{\frac{k}{m}} = \sqrt{\frac{200}{6}} = 5.773 \text{ rad/s}$

Critical damping coefficient, $c_c = 2m\omega_n = 2 \times 6 \times 5.773 = 69.28\ N.s/m$

(i) Damping factor, $\zeta = \frac{c}{c_c} = \frac{6}{69.28} = 0.0866$

(ii) $\omega_d = \omega_n\sqrt{1-\zeta^2} = 5.773\sqrt{1-(0.0866)^2} = 5.75 \text{ rad/s}$

(iii) Logarithmic decrement, $\delta = \frac{2\pi\ \delta}{\sqrt{1-\delta^2}} = \frac{2\pi \times 0.0866}{\sqrt{1-(0.0866)^2}} = 0.546$

$$\delta = \frac{1}{n}\ell n\left(\frac{x_0}{x_n}\right)$$

$$0.546 = \frac{1}{n}\ell n\left(\frac{1}{0.2}\right)$$

$$n = 2.947 \sim 3$$

Q.4(c) **A full journal bearing has a journal diameter D of 25 mm, with a unilateral tolerance of -0.038 mm. The bushing bore has a diameter B of 25.038 mm and a unilateral tolerance of 0.075 mm. The D ratio is unit. The load is 1.1 kN and the journal runs at 18.33 r.p.s. The average viscosity is 55.2mPa-s.**

Minimum film thickness variable is 0.58 and coefficient of friction variable is 4.0. Find

(a) Sommerfeld number

(b) minimum film thickness

(c) frictional torque.

Sol. $D = 25\pm^{0.000}_{0.038}$ *mm*, $B = 25.038\pm^{0.075}_{0.000}$, $\frac{L}{D} = 1$, $W = 1.1kN$, $n = 18.33$rps, $\mu = 55.2mPas$,

$\bar{h} = \frac{h_0}{c} = 0.58$, $\bar{f} = \left(\frac{r}{c}\right) f = 4.0$

Minimum diameter of journal, $D_{min} = 25 - 0.038 = 24.962mm$

Maximum diameter of bushing bore, $B_{max} = 25.038 + 0.075 = 25.113mm$

Radial clearance, $c = \frac{1}{2}(B_{max} - D_{min}) = \frac{1}{2}(25.113 - 24.962) = 0.0755mm$

Journal pressure, $p = \frac{W}{LD} = \frac{11\times10^3}{2.5\times25} = 1.76$ n./mm^2

Sommerfeld number, **(i)** $S = \left(\frac{\mu n}{p}\right)\left(\frac{r}{c}\right)^2$

$$= \left(\frac{55.2\times10^{-9}}{1.76}\right)\left(\frac{12.5}{0.0755}\right)^2 = 0.01576$$

(ii) $\frac{L}{D} = 1$ and $S = 0.01576$

Coefficient of friction variable, $\bar{f} = f\left(\frac{r}{c}\right) = 4.0$

Coefficient of friction, $f = \frac{4\times0.0755}{12.5} = 0.02416$

Minimum film thickness variable, $\bar{h} = \frac{h_o}{c} = 0.58$

Minimum oil film thickness, $h_0 = 0.0755 \times 0.58 = 0.0438mm$

(iii) Frictional torque, $T = fWr = 0.02416 \times 1.1 \times 10^3 \times 12.5 \times 10^{-3} = 0.3322$ N.m

Q.4.d.(i) Why are fcc metals generally weak and ductile? Write the effect of adding chromium, nickel and molybdenum in steel?

(ii) Why plasticizers and lubricants are added to plastics? Name some important properties and applications of (a) nylon and (b) vinyls?

Sol.

(i) The ductile metals like steel, copper, aluminum etc. are having FCC crystal structure. The atomic packing factor (APF) for FCC crystal structure is 0.74 as compared to BCC, which has APF=0.68. The atoms in FCC metals are more tightly packed as compared to BCC metals .It is easier to deform ductile metals as compared to brittle metals. In the microstructural center, it is easier to slide the planes of atoms over one another in FCC metals. Therfore, FCC metals are generally weak and ductile.

Effects of adding alloying elements to steel are as follows:

Chromium (10 to 30%): Refines grain structure, increases strength, hardness and resistance to corrosion, increases wear resistance and heat resistance.

Nickel (0.25 to 5%): Refines grain structure, increases strength and toughness, increases resistance to corrosion.

Molybodenum (0.25 to 1.25%): Increase the strength and toughness. Increase high temperature strength.

(ii) Effects of adding plasticizers to plastics:

Plasticizers are liquids of high boiling point and low molecular weight, which are added to plastics to improve the plastic behavior. Its broad role is to separate the micro-molecules, thus decreasing the inter molecular forces and facilitating relative movement between molecules of the plastic, i.e. making deformation easier. They are essentially oily in nature .organic solvents, resins and even water are used as plasticizers.

(iii) Effects of adding lubricants to plastics:

Lubricants are added to plastics to reduce friction during processing, to prevent parts from sticking to mould walls, to prevent plastic films from sticking to each other and to impart an elegant finish to the final product. Commonly used libricants include: oils, soaps and waxes.

(ii)

Plastic	Important Properties	Applications
(a) Nylons	Low coefficient of friction. Dimensional stability Good electrical properties Lunbricants not required Abcasion resistant	Small moulded of gears and bushtubing. Brush bristle Fuel containers coating.
(b) Vinyls	Flame resistant. Resistant to chemicals, oils and solvents. Abrassion resistant. Sound and vibration deadening	Automotive panels. Parts for vaccum cleaners and refrigerators Floor covering. Pipe fittings Tamps, balls and floats.

Section - C

(Answer any two questions)

Q.5 a.(i) Write the advantages, applications, current and power input that may be required in flash welding. Why flashing is essential?

(ii) In metal casting write the purpose and types of muller. Why distortion allowance is provided on patterns?

Sol. (i)Flash Welding: In flash butt welding process, the parts to be welded approach towards each other and come into contact with the current switched on. After the parts are properly positioned and correct current, head speed and time are selected, a cycle start button is activated. This makes the movable head to approach the fixed head. As the abutting surfaces come very near to each other, extremely rapid heating tapes place when surface asperities first make contact. Molten metal is violently expelled and burning in air with considerable force. As this process continues for a few seconds, a very thin layer at the surface is melted. The current is then shut off and the two parts are rapidly pressed together causing a small upset. These sequences out liquid metal and oxides formed on the joint surfaces and the solid clean metel faces of the abutted parts are welded together.

Advantages:

1. No need of fillers metal.
2. Welded surfaces are very smooth.
3. Flashing action burns away the irregularities at the welded surface.

Applications:

1. Joining thin-walled tubes, rails, chainlinks, tools, press manufactured parts.
2. Dissimilar metals including refractory metals can be readily welded.
3. Joining the ends of wire and sheet coils to allow the continues operation of processing lines.
4. Joining of rims of wheels for automobiles and cycles, tubular parts etc.

Current input: current densities = 300 – 800 A/cm^2, welding voltage = 5 – 15V.

Power input: 1.5 to 12 kW.

Flashing is essential to burn away the irregularities at the welded surface.

(ii) Mullers are used to mix large quantity of send to get a uniform mixture. This sand is used for preparing moulds of casting. Mullers are rotary type mixing machines operated by hand through gear box or powered.

Distortion allowance is provided for castings of irregular shape which are distorted in the process of cooling because of metal shrinkage.

Q.5.b.(i) Name at least four methods by which high energy release rates are obtained. Why less springback be observed in HERF?

(ii) Write at least four factors on which the thrust force in drilling depends.

A hole is being drilled on a block aluminium alloy with 10 mm drill at a feed of 0.25 mm/rev.

The spindle is running at N= 850 r.p.m. Calculate the metal removal rate.

Sol. High Energy Rate Forming (HERF) Processes

1. Explosive forming
2. Electric spark forming
3. Electro- hydraulic forming
4. Electro-magnetic forming

The springback observed in HERF processed is very less because the energy needed for processing the material is released in a very short time, usually in milli-seconds or micro-seconds.

(i) Thrust force factors in drilling:

The total thrust force F in drilling can be expressed as:

$$F = 2F_t \sin \beta + F_{ch} + F_f$$

Where F_t = thrust components of cutting force

β = half the point angle of drill

F_{ch} = force from the chisel edge

F_f = friction force

The thrust force depends on the following:

1. Nominal diameter of drill and helix or shape angle.
2. Feed per revolution
3. Speed of drill bit
4. Work piece material

Given: D = 10mm, f = 0.25mm/rev., N = 850 rpm

Metal removal rate, $MRR = \left(\frac{\pi}{4} D^2\right) \times f \times N$

$$= \frac{\pi}{4} \times 10^2 \times 0.25 \times \frac{850}{60} = 278.16 mm^3 / s$$

Q.5.c(i) Why does titanium have poor machinability? Write at least four general characteristic that coatings for cutting tools and dies should possess.

You are asked to turn ductile cast iron with various microstructure and hardness as shown in the following table.

		Hardness(HB)	Ferrite	Pearlite
1.	Annealed	186	97 %	3%
2.	As cast	265	20%	80%
3.	Annealed	170	100%	–
4.	As cast	207	60%	40%

Draw a figure showing variation of tool life with cutting speed and the effect of workspaces hardness and microstructure.

(i) Schematically illustrate closed loop control system for a numerical control machine, mentioning the purpose of a feedback control. Write the advantage of CNC over conventional NC systems.

Sol. Machinability of Titanium

Titanium is a very hard material and difficult to machine. Hence, it has very poor machinability. Titanium is harder than most of the cutting tool materials. Characteristics of coatings for cutting tools and dies

1. The coating must be fine grained and free from binders and porosity.

2. It should be thick enough to enhance the tool life and thin enough to prevent brittleness.
3. It should have low coefficient of friction so that chips do not adhere to the shape face of the tool.
4. It should have high thermal conductivity so that the cutting heat generated is transferred quickly away from the cutting zone.

Taylor's tool life equation:

$$VT^n = C$$

Where V = cutting speed in m/min

T = actual cutting time between two resharpenings in minutes

C = a constant whose value depends upon other machine variables and work material variables.

Effect of microstructure of work piece:

In general, hard micro-constituents in the matrix of work piece result in poor tool life. If the grain size is larger, tool life is better. On the basis of tool life, the micro-constituents have been placed in the following descending order:

pearlite, sorbite, troosite, bainite and martensite

Effect of work piece hardness: The effect of work piece hardness is given by the equation,

$$V = \frac{\text{Constant}}{\text{BHN}^{1.68} \times \%\text{reduction}}$$

As the hardness increases the velocity decreases. Harder the material, lesser will be the tool life for the equivalent amount of material removal.

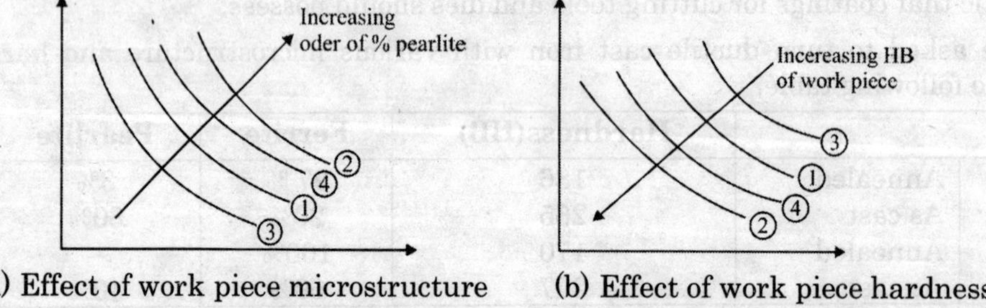

(a) Effect of work piece microstructure (b) Effect of work piece hardness

Fig. 9: Tool life V's cutting speed

(ii) Closed loop control system for a NC Machine

In the closed –loop(Fig. 10), the MCU generates a pulse signal until the signal returned from the feedback transducer agrees with the original number of pulses required to execute the movement . A comparator is used to compare the count of feedback pulses with the original value, and an error signal is output until the feedback pulse count agrees with the input value.

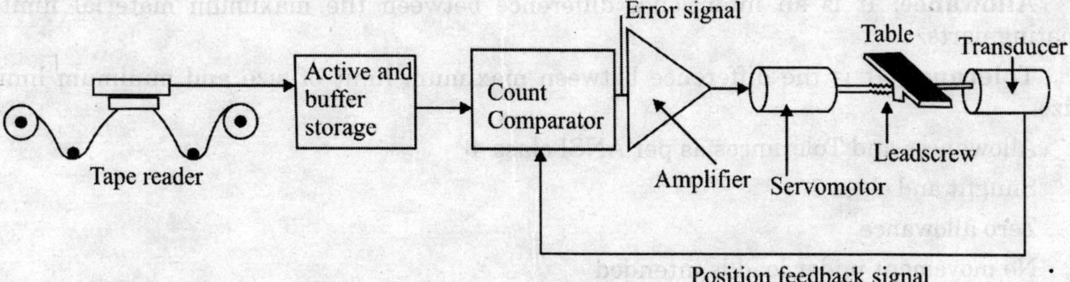

Fig. 10: Closed-loop feedback system

Feedback systems may be either digital or analogue. Digital systems generate pulses which are fed back to the control and count down linear to rotary motion in minimum moments on machine lead screws. Analogue systems sense and monitor variations in level of voltage.

The advantages of CNC are:

1. The part program tape and tape reader are used only once to enter the program into computer memory resulting in improved reliability.
2. Tape editing at the machine state
3. Tapes prepared in one units can be converted into another units.
4. Greater flexibility resulting into reduced obsolescence.
5. User-written programs can be called by MACRO.
6. CNC provides a total manufacturing system.

Q.6a.(i) Define the terms accuracy and precision. Name five main areas into the measuremont can be divided. Write the amount of allowance and tolerance that is permitted by the following classes of fit as per ANSI class 4: Snug fit and class 7: Medium force fit. Also mention applications.

(ii) Write two advantages of thread rolling and explain with figure two-die cylindrical machine.

Sol. (i)Accuracy: It refers to the fixed amount an instrument reading derivates from its known or calibrated input, regardless of how many times we make a measurement. Thus, an accuracy error is a fixed error.

Precision: It refers to how well an instrument can reproduce subsequent readings of an unchanged input. Thus, a precision error is not a fixed quantity.

Areas of measurement

The five main areas into which measurements can be divided are:

1. Metrology and precision measurements.
2. Motion measurements.
3. Pressure and temperature measurements.
4. Force, torque and power measurements.
5. Flow measurements.

Allowance: It is an intentional difference between the maximum material limits of mating parts.

Tolerance: It is the difference between maximum limit of size and minimum limit of size.

Allowances and Tolerances as per ANSI class 4:

Snugfit and class 7

Zero allowance

No movement under load is intended

Tightest fit that can be assembled by hand.

Application of snug fit: coupling, gear rings

Medium force fit

– an interference fit.

– requires considerable pressure

– assembled by heating and cooling

Application of medium force fit: Light drive fit for non ferrous parts, fastening wheels, cranks.

Fig. 11: Two die cylindrical thread rolling machine

(ii) Thread Rolling: Thread rolling is cold working process in which a blank of approximately equal to the pitch diameter of the required thread, is rolled between hardened steel rolling dies having the negative contour of the thread to be produced. As the thread shaped ridges on the dies penetrate the blank material, the material is displaced from the bottom of the thread crests.

In cylindrical die machine, the part to be threaded is rolled between rotating cylindrical dies. The machine can have two round ides located diametrically opposite each other (Fig. 11). Two dies machine has the main application of rolling the threads on taps.

Advantages of thread rolling

1. It is the fastest method of producing a thread.
2. There is no material wastage as the forming process is chipless.
3. The material is strained plastically during thread rolling and is work hardened. It becomes stronger against both tension and fatigue.

4. The rolled threads are stronger than machined threads as the grain fibres remain continuous and follow the contours of threaded surface.

Q.6.b. (i) Milling is an interrupted cutting process, show with figure convential face milling with cutting force diagram for Fe showing the above nature of the process.

(ii) Name the basic components of all robots. Write in short about welding robot.

Sol.(i).Face Milling: In face milling, the finished surface is at right angle to the cutter axis and it is obtained by the teeth on the periphery and the flat end of the cutter (Fig. a and b). It is a vertical milling operation. The scheme of chip formation during plain slab milling using a straight cutter as shown in Fig.(c).

Let D = cuter diameter

L = Depth of cut

The kinematics of chip formation in orthogonal milling with a straight edged cutter is shown in Fig. (d).

Let f = feed rate, mm/min

Then effective feed per tooth $= \dfrac{f}{NZ}$ mm.

Where N = cutter npm

Z = number of teeth or cutter.

MRR per unit width of job $= \dfrac{f}{d}$

If feed velocity is small as compared with the circumferential velocity of the cutter, then

AB $\simeq$ AC $\sin\beta$

Or $(t_1)_{max} \approx \dfrac{f}{NZ} \sin\beta$

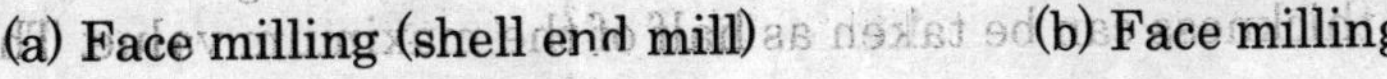

(a) Face milling (shell end mill) (b) Face milling

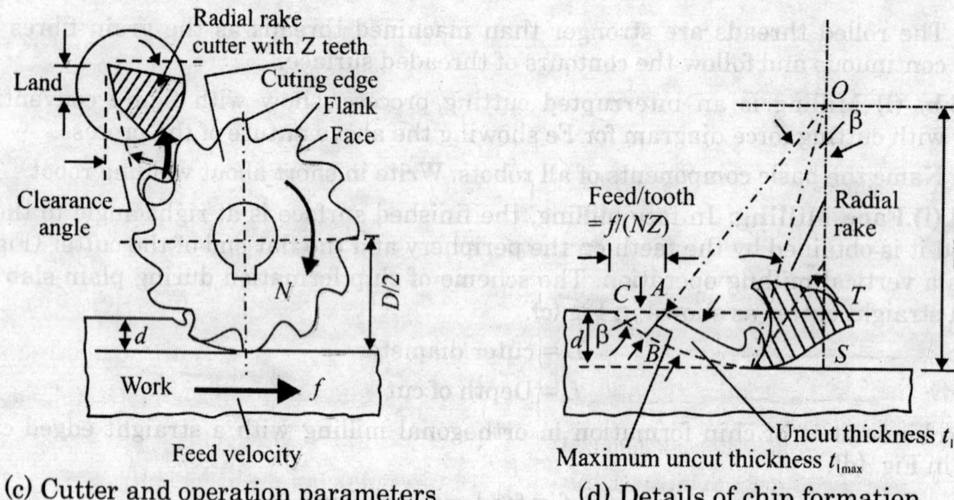

(c) Cutter and operation parameters (d) Details of chip formation

Fig. 12: Details of milling operation

Where β the angle included by the contact is are the cutter centre 0 in radians. Now, considering the triangle *OAT*. We have

$$\cos\beta \approx \frac{OT}{OA} = \left(\frac{D}{2} - d\right) / \frac{D}{2}.$$

Hence,

$$\sin\beta = \left[1 - \cos^2\beta\right]^{\frac{1}{2}} = \left[1 - \left(1 - \frac{2d}{D}\right)^2\right]^{1/2}$$

$$\approx 2\sqrt{\frac{d}{D}} \qquad \ldots(1)$$

neglecting the higher order terms in d/D as it is normally very small. Using this value of $\sin\beta$ in the expression of the maximum uncut thickness, we get

$$t_{1\max = \frac{2f}{NZ}\sqrt{\frac{d}{D}}} \qquad \ldots(2)$$

So, the cutting force components F_c and F_r (shown in Fig.13) not only change in direction but also in magnitude as cutting edge moves along the cut surface. It is obvious that when cutting with a straight cutter, there is no component of the cutting force along the cutter axis. The average uncut thickness can be taken as half of the maximum value. Thus,

$$t_{1\max} \approx \frac{f}{NZ}\sqrt{\frac{d}{D}} \qquad \ldots(3)$$

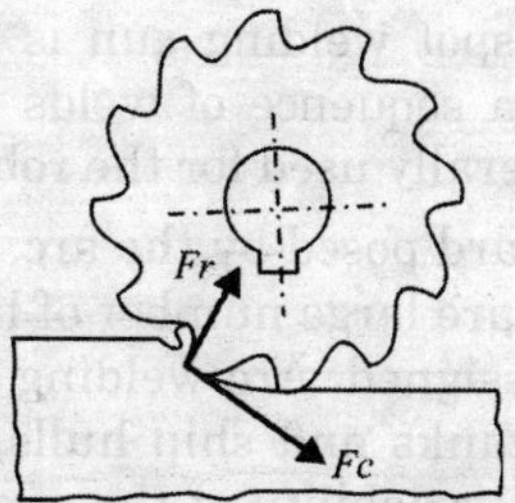

Fig. 13: Components of milling force

The average values of F_C and F_r can be approximately found out using this value of uncut thickness. Since F_r acts in the radial direction, it does not produce any torque and the arbor torque is due only to the component F_c. So, the torque M due to one cutting tooth is F_C D/2) and varies approximately as F_C.

(i) Basic components of a Robot

The basic component of a robot are:

1. Controller
2. Robot arm or manipulator
3. End of arm tools or end effectors
4. Power source

These are shown in Fig.14

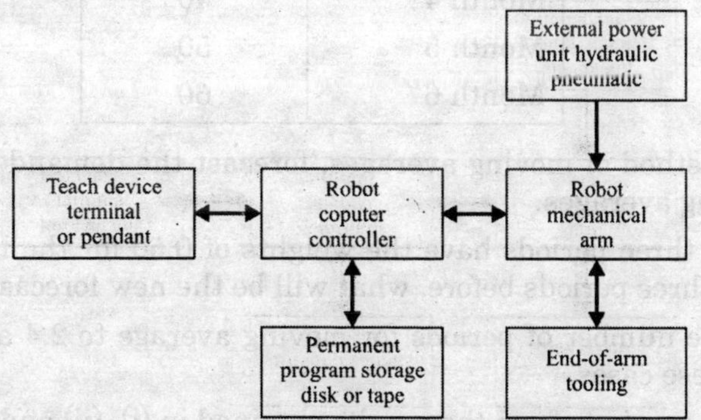

Fig. 14: Components of a robot system

Welding Robot

Welding is one of the first application where robots were used extensively Robots are used for spot welding as well as are welding.

Spot welding is the most widely used application particularly in the automobile body assembly lines. It is used to weld the frames and body panels together. Traditionally human operators were using the portable spot welders to make the spots at a number of locations. The operator needs to manipulate the position of the welding unit with all its associated cables, which are suspended from overhead hoist system. In view of the high speed of operation required, the quality and consistency of the weld is often a problem. It is also a hazardous operation. The heavy spot welding gun is attached so the end effector and the robot is programmed to perform a sequence of welds on the body as it is presented to the robot. Line tracking control is generally used for the robots used in assembly lines.

Because of the significant hazard posed by the arc, are welding is the process well suited for robotisation. However there is are large number of hurdles to be overcome before a robotic are welding operation could be designed. Arc welding is generally performed in in confined areas, such as the inside of the tanks and ship hulls, which are difficult to access for the robots. The other problems is the variability of the components to be welded, which is often difficult to control. The variability is in terms of the dimensions as well as the type of edges and surfaces presented to welding. This causes a variation in gap between the parts, which will have to be compensated by the welder. As a result, arc-welding robot will have to be provided with sensors to monitor these variations and the controller will have to compensate for that.

Q.6.(c) The demand for a certain raw material used in a manufacturing organization varies from month to month. The consumption pattern of the material during the last six months (in metric tons) is given below:

Table: Consumption pattern of raw material (metric tons)

Months	Demand
Month 1	20
Month 2	30
Month 3	20
Month 4	40
Month 5	50
Month 6	60

i. Using the method of moving averages, forecast the demand for month 7 using three period moving averages. 5

ii. Suppose the three periods have the weights of 0.50 for the immediate past and 0.25 for two and three periods before, what will be the new forecast?

iii. Changing the number of periods for moving average to 2.4 and 5 obtain the forecast in each of these cases.

iv. What do you conclude from the results obtained in (i), (ii) and (iii)? 5

Sol. (i) Using three months moving average, forecast for the 7th month

$$= \frac{\sum(\text{Demand of last 3 months})}{3} = \frac{(40+50+60)}{3} = \frac{150}{3} = 50$$

(ii) Forecast with weighted average moving method

Forecast for the 7th month $= 0.50 \times 60 + 0.25 \times 50 + 0.25 \times 40$

$= 30 + 12.5 + 10 = 52.5$

(iii) 2 peroids SMA $= \frac{60+50}{2} = 55$

Forecast for the 7th month = 55

$$\text{4 periods SMA} = \frac{60+50+40+20}{4} = \frac{170}{4} = 42.5$$

$$\text{5 periods SMA} = \frac{60+50+40+20+30}{4} = \frac{200}{4} = 50$$

(iv) (a) Forecast based on simple moving average is dependent upon the number of data points.

(b) Forecast varies randomly with number of data points for SMA.

(c) The weighted moving average gives forecast more closer to the demand of last month.

Q7. (a) A die making unit is planning to install a new CNC electric discharge machine in its job shop. Machine from two reputed manufacturers are available in the market. The relevant data about their product is

Costs and Life	Manufacturer A	Manufacturer B
Present Cost	₹ 1.0cr	₹ 1.50 cr
Annual Operating + maintenance	₹ 0.20 cr	₹ 0.12 cr
Salvage value at the end of useful life	₹ 0.05 cr	₹ 0.02 cr
Estimated useful life	10 yrs	10 yrs

Considering rate of return to be 12% per year, what is the best alternative?

(P_{wf} – series at 12% for 10 yrs = 5.65

P_{wf} - single payment at 12% for 10 yrs = 0.322

Sol.

Cost diagram for machine A

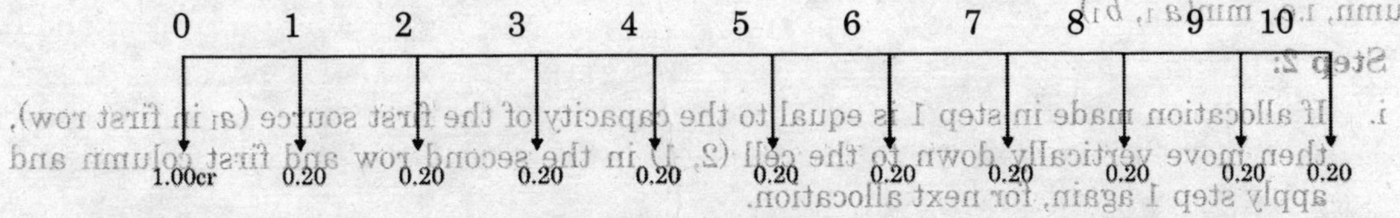

Present unit of the cost of machine A, on 10 year basis

$$P_{A10} = 1.00 + 0.20\,(P/A,\ 12\%,\ 10) + 0.05(P/F,\ 12\%,\ 10)$$

$$= 1.00 + 0.20 \times 5.65 - 0.05 \times 0.322$$

$$= 2.1139 \text{ cr}$$

Cost diagram for machine B

0 1 2 3 4 5 6 7 8 9 10

1.50cr 0.12 0.12 0.12 0.12 0.12 0.12 0.12 0.12 0.12 0.12

Present worth of the cost of machine B

$$P_{B10} = 1.5 + 0.12\ (P/A,\ 12\%,\ 10) - 0.02\ (P/F,\ 12\%,\ 10)$$
$$= 1.5 + 0.12 \times 5.65 - 0.02 \times 0.322$$
$$= 2.17156 \text{ cr}$$

Since $P_{A10} < P_{B10}$

Hence machine A is more economical.

Q.7.(b) A manufacture operates three factories from which items are shipped to four warehouses. Factory to warehouse shipping cost in Rs. quantities available at each of the factories *X, Y* and *Z* and requirements at each of the warehouses A, B, C and D are given in the table below:

Factory	Warehouses				Quantity Available
	A	B	C	D	
X	30	20	50	20	150
Y	20	10	40	40	240
Z	20	30	40	30	210
Quantity Required	130	120	160	190	600

Determine initial feasible solution by

i. North-West corner rule, and

ii. Lowest cost entry method

Sol. North-west corner method

It is an efficient method to obtain an initial solution. The method does not take into account the cost of transportation on any route of transportation.

Step 1: Start with the cell at the upper left (N-W) corner of the transportation matrix as allocate as much as possible equal to the minimum of the rim values for the first row and column, i.e., min(a_1, b_1)

Step 2:

i. If allocation made in step 1 is equal to the capacity of the first source (a_1 in first row), then move vertically down to the cell (2, 1) in the second row and first column and apply step 1 again, for next allocation.

ii. If allocation made in step 1 is equal to the requirement of the first destination (b_1, in first column), then move horizontally to the cell (1, 2) in the first row and second column and apply step 1 again for most allocation.

iii. If $a_1 = b_1$, allocate $x_{11} = a_1$ or b_1 and move diagonally to the cell (2, 2).

Step 3: Continue the procedure step by step till an allocation is made in the south-east (S-E) corner cell of the transportation table.

Remark: If during the process of making allocations for a particular cell, supply equals demand, then next allocation of magnitude zero can be made in cell either in the next row or column. This condition is known as degeneracy and may also occur in later steps.

The solution is feasible when $m + n - 1 =$ number of allocations.

DEMAND

Factory	Warehouses				Quantity
	A	B	C	D	Available
X	30 (130)	20 (20)	50	20	150 (20)(0)
Y	20	10	40(160)	40 (80)	240 (80)(0)
Z	20	30 (100)	40	30 (110)	210 (100)(0)
Quantity Required	130 (0)	120 (100)(0)	160 (0)	190 (110)(0)	600

SOURCE

$$m + n - 1 = 3 + 4 - 1 = 6$$

$$\text{Total cost} = 30 \times 130 + 20 \times 20 + 100 \times 30 + 160 \times 40 + 80 \times 40 + 110 \times 30$$

$$= 390 + 400 + 3000 + 6400 + 3200 + 3300$$

$$= \text{Rs } 16790$$

$$m + n - 1 = 3 + 4 - 1 = 6$$

Number of allocations = 6

Hence, the initial solution is feasible.

(ii) Lowest Cost Method

To minimize the total transportation cost, we must try to transport as much as possible through those routes (cells) where the unit transport cost is minimum.

Step 1: Select the cell with the smallest unit cost in the entire transportation table and allocate as much as possible to this cell and eliminate (line out) that row or column in which either supply or demand is exhausted. If both a row and column are satisfied simultaneously then only one may be crossed out.

In case the smallest unit cost cell is not unique, then select the cell where maximum allocation can be made.

Step 2: After adjusting the supply and demand for all uncrossed-out rows and columns repeat the procedure with the next smallest unit cost among the remaining all rows and columns of the transportation table and allocate as much as possible to this cell and eliminate (line out) that row and column in which either supply or demand is exhausted.

Step 3: Repeat the procedure until the entire available supply at various sources and demand at various destinations is satisfied.

Factory	Warehouses				Supply
	A	B	C	D	
X	30	20	50	20(150)	150 (0)
Y	20	10 (120)	40(120)	40	240 (1200)
Z	20 (130)	30	40	30 (40)	210 (1400)
Demand	130 (0)	120 (0)	160 (40)	190 (400)	600

$$\text{Total Cost} = 20 \times 130 + 10 \times 120 + 40 \times 120 + 20 \times 150 + 30 \times 40$$
$$= 2600 + 1200 + 4800 + 3000 + 1200$$
$$= \text{Rs. } 12800$$

Q.7.(c) A dual-card Kanban System needs to be designed for manufacturing process that has demand rate of 200 per hour of a certain item. The standard containers designed for the components hold 25 items. The conveyance time and production time for the components are 30 minutes and one hour, respectively. Assume a safety factor of 10%.

i. identify the number of C-Kanbans and P-Kanbans required in the system.

ii. If there is a rounding off involved in arriving at the number of Kanbans in the above case, compare the performance of the system when the number is rounded off to the immediate lower and next higher integer.

Sol. Given: Demand rate of items = 200 per hour

Capacity of standard containers = 25 items

Conveyance time for the components = 30 minutes

Production time for the components = 1 hour

Safety factor = 10 %

(i) Number of C-Kanbans or Conveyance Kanbans:

$$\text{Number of containers} = \frac{200}{25} = 8 \text{ per hour}$$

Total Conveyance time = 30 minutes for one container.

$$\text{Actual number of C-Kanbans} = \frac{8 \times 30}{60} = 4 \text{ per 30 minutes}$$

$$\text{Similarly number of P-Kanbans} = \frac{200}{25} \times \frac{1}{1} = 8 \text{ per hour}$$

(ii) With 10% factor of safety:

$$\text{Conveyance time} = 30 \times 1.1 = 33 \text{ minutes}$$
$$\text{Production time} = 1 \times 1.1 = 1.1 \text{ hour}$$
$$\text{Number of C-Kanbans} = \frac{8 \times 33}{60} = 4.4$$
$$= 4 \text{ for lower integer}$$
$$= 5 \text{ for higher integer}$$
$$\text{Number of P-Kanbans} = \frac{8 \times 1.1}{1} = 8.8$$
$$= 8 \text{ for lower integer}$$
$$= 9 \text{ for higher integer}$$

MECHANICAL ENGINEERING - 2014

PAPER - I

(Conventional)

Time Allowed: Three Hours　　　　　　　　　　　　　　　　　　　　　　Maximum Marks: 200

Instructions

Please read each of the following instructions carefully before attempting questions.

Candidates should attempt FIVE questions in all.

Question No. 1 is compulsory.

Out of the remaining SIX questions, attempt any FOUR questions.

The number of marks carried by a part of a question are indicated against it

Answers must be written in ENGLISH only.

Assume suitable data, if necessary, and indicate the same clearly.

For air R = 0.287 kJ/kg-K, Cp = 1.005 kJ/kg-K, γ = 1.4, M = 28.97 kg/kg-mole, Universal gas constant R = 8.314 kJ/kg mole-K.

Unless otherwise mentioned, symbols and notations have their usual standard meanings.

Neat sketches may be drawn, wherever required.

Attempts of questions shall be counted in chronological order. Unless struck off, attempt of a question shall be counted even if attempted partly.

Any page or portion of the page left blank in the answer book must be clearly struck off.

A psychrometric chart is attached to this question paper for necessary use by the candidate.

1. Answer all of the following:

(a) Give reasons why the Carnot cycle cannot be considered as the theoretical cycle for steam power plants even though its efficiency is the highest for the given heat source and sink temperatures. A Carnot cycle heat engine has an efficiency of 40%. If the high temperature is raised by 10%, what is the new efficiency keeping the same low temperature? 10

Sol. The Carnot cycle shown in Fig. 1 is an ideal cycle but cannot be considered as the theoretical cycle for steam power plants giving the maximum possible thermal efficiency for cycle operating on selected maximum and minimum temperature ranges. For the Carnot cycle 1-2-3-4, compressing a very wet stream at state 3 would require a compressor of size and cost comparable with a turbine. It would absorb work comparable to that developed by the turbine and its life would be short because of blade erosion and cavitation problems due to excessive moisture. For the cycle 1-2-5-6-1, the pump work, $W_p = (h_6 - h_5)$ is again very high and it is impossible to supply heat at infinite pressures and at constant temperature from state 6 to state 4. So, the Carnot cycle cannot be realised in practice, but it sets the upper limit to which the cycle efficiency of any thermal plant can be raised.

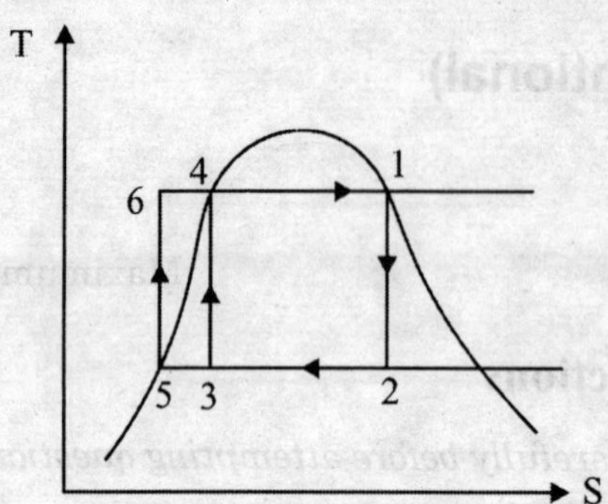

Fig.1: Carnot cycle.

Source
T_1
Q_1
E
W
Q_2
T_2
Sin k

Fig.2: Carnot engine.

Given: $\eta_E = 0.4, \quad T_1' = 1.1T_1, \quad \eta_E' = ?$

Refer to Fig. 2.

$$\eta_E = 1 - \frac{T_2}{T_1}$$

$$0.4 = 1 - \frac{T_2}{T_1}$$

or

$$\frac{T_2}{T_1} = 0.6$$

$$\eta_E' = 1 - \frac{T_2}{T_1'} = 1 - \frac{T_2}{1.1T_1}$$

$$= 1 - \frac{0.6}{1.1} = \frac{0.5}{1.1}$$

$$= 0.4545 \text{ or } 45.45\%$$

(b) What do you understand by the lumped capacity? What are the physical assumptions necessary for a lumped capacity unsteady state analysis to apply?

Determine time required in minutes for a 50 mm diameter steel sphere (ρ = 7800 *kg/m³*, c = 0.46 *kJ/kg-K*, k = 35 *W/m-K*) to cool from 600°C temperature to 100°C temperature if exposed to a cooling air at 30°C. The convection heat transfer coefficient is 10 *W/m²-K*.

Lumped capacity

All solids have finite thermal conductivity and there will be always a temperature gradient inside the solid whenever heat is added or removed. However, for solids of large thermal conductivity with surface areas that are large in proportion to their volume like plates and thin metallic wires, the internal resistance $\left(\dfrac{L}{hA}\right)$ can be assumed to be small or negligible in comparison with the convective resistance $\left(\dfrac{1}{hA}\right)$ at the surface. This type of heat flow occurs during heat treatment of metals and time response of thermocouples and thermometers.

The process in which the internal resistance is assumed negligible in comparison with its surface resistance is called the Newtonian heating or cooling process. The temperature in this process is considered to be uniform at a given time. Such an analysis is called *Lumped parameter analysis* because the whole solid, whose energy at any time is a function of its temperature and total heat capacity is treated as *one lump*.

Physical Assumptions for lumped capacity

1. Uniform temperature distribution throughout the body.
2. Thermal conductive resistance of the body is zero.
3. Convective heat transfer coefficient is constant and uniform.
4. Biot number $\left(\dfrac{hL_c}{k}\right)$, $Bi < 0.1$

where L_c = characteristic length

Given: $d = 50\ mm$, $\rho = 7800\ kg/m^3$, $c = 0.46 kJ/kg.K$, $k = 35 W/m.K$, $h = 10 W/m^2.K$, $t_i = 600°C$, $t = 100°C$, $t_\infty = 30°C$

For cooling of a sphere,

$$\frac{t - t_\infty}{t_i - t_\infty} = exp\left[-\frac{hA}{\rho Vc}t\right]$$

$$\frac{100-30}{600-30} = exp\left[-\frac{10\times\pi(50\times10^{-3})^2\times3\times t}{7800\times4\times\pi(25\times10^{-3})^3\times0.46\times10^3}\right]$$

$$0.1228 = exp\left[-3.3445\times10^{-4}t\right]$$

$$t = \frac{\ln 0.1228}{-3.3445\times10^{-4}} = \frac{-2.097198}{-3.3445\times10^{-4}} = 6270\,s$$

or 104.5 minutes

1(c) (i) From fundamental laws derive the Prandtl- Mayer relation for a normal shock in compressible fluid flow in a constant area duct for a perfect gas. 5

(ii) Show that the maximum efficiency for a jet striking at the center of a single symmetrically curved vane moving with a velocity u is given by $\dfrac{16}{27}\cos^2\dfrac{\theta}{2}$, where θ is the curvature of the vane. 5

Sol. (i) Prandtl: Mayer relation for a normal shock in compressible fluid flow in a constant area duct for a perfect gas.

Consider a stationary shock wave separating two uniform gaseous negions 1 and 2 shown in fig 3. The fundamental problem for a normal shock wave is to determine the pressure p_2, density p_2, velocity v_2 and temperature T_2 behind the shock wave when the corresponding values in front of the shock wave are known. The basic equation for one – dimensional flow are:

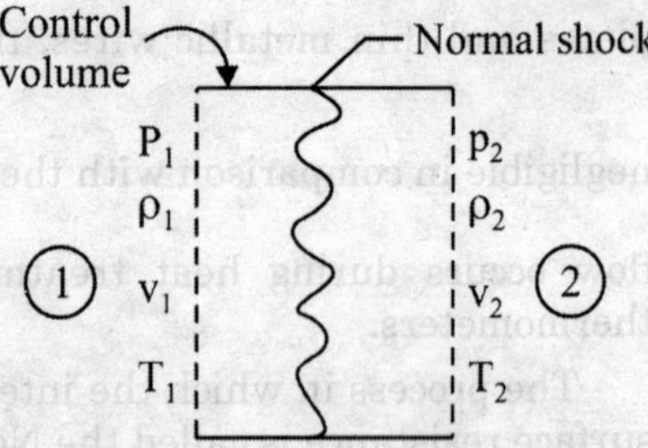

Fig. 3: Flow model for a normal shock

(i) Continuity equation: $\rho_1 v_1 = \rho_2 v_2 = \text{const} = m$ (say) ... (1)

(ii) Monentum equation: $p_1 + \rho_1 v_1^2 = \rho_2 v_2^2 + p_2$ ...(2a)

or $p_1 + v_1 m = p_2 + v_2 m$ ...(2b)

(iii) Energy equation: $\frac{v_1^2}{2} + \frac{\gamma}{\gamma - 1} \times \frac{p_1}{\rho_1} = \frac{v_2^2}{2} + \frac{\gamma}{\gamma - 1} \times \frac{p_2}{\rho_2} = \text{const.} = k(\text{say})$ ...(3)

(iv) For a perfect gas, equation of state: $p = \rho RT$ ...(4)

From Eq.(2*b*), $p_1 - p_2 = m(v_2 - v_1)$ ...(5)

From Eq.(1) we have $v_2 = \frac{m}{\rho_2}$ and $v_1 = \frac{m}{\rho_1}$ ...(6)

From Eqs.(5)and(6), we get

$$p_1 - p_2 = m^2 \left(\frac{1}{\rho_2} - \frac{1}{\rho_1} \right) \quad ...(7)$$

From Eq.(3), we get

$$\frac{\gamma}{\gamma - 1} \left(\frac{p_1}{\rho_1} - \frac{p_2}{\rho_2} \right) = \frac{1}{2} (v_2^2 - v_1^2)$$

$$= \frac{m^2}{2} \left(\frac{1}{\rho_2^2} - \frac{1}{\rho_1^2} \right)$$

$$= \frac{m^2}{2} \left(\frac{1}{\rho_2} - \frac{1}{\rho_1} \right) \left(\frac{1}{\rho_2} + \frac{1}{\rho_1} \right) \quad ...(8)$$

Dinviding Eq.(8) by (7), we have

$$\left(\frac{\gamma}{\gamma - 1} \right) \left[\frac{p_1 / \rho_1 - p_2 / \rho_2}{p_1 - p_2} \right] = \frac{1}{2} \left(\frac{1}{\rho_2} + \frac{1}{\rho_1} \right) \quad ...(9)$$

Rearranging Eq.(9), we have

$$\frac{\rho_2 - \rho_1(p_2/p_1)}{1-(p_2/p_1)} \times \frac{1}{\rho_1\rho_2} = \frac{\gamma-1}{2\gamma}\left(\frac{\rho_1+\rho_2}{\rho_1\rho_2}\right) \quad ...(10)$$

or

$$\frac{p_2}{p_1} = \frac{(\gamma-1)-(\gamma+1)(\rho_2/\rho_1)}{(\gamma-1)\rho_2/\rho_1-(\gamma+1)}$$

or

$$\frac{\rho_2}{\rho_1} = \frac{(\gamma-1)p_1+(\gamma+1)p_2}{(\gamma+1)p_1+(\gamma-1)p_2} \quad ...(11)$$

Local velocity of sound, $$c_1^2 = \frac{\gamma p_1}{s_1}, c_2^2 = \frac{\gamma p_2}{\rho_2} \quad ...(12)$$

Substituting Eq. (12) in Eq.(3), we have

$$\frac{v_1^2}{2}+\frac{c_1^2}{\gamma-1} = \frac{v_2^2}{2}+\frac{c_2}{\gamma-1} = k \quad ...(13)$$

Eq.(13) is exactly of the same from as energy equation for compressible isentropic flow when there is no shock wave. It follows that the critical speed v_c is the same on both sides of the shock.

When $\upsilon_1 = c_1 - \upsilon_c$ then Eq. (13) reduces to

$$\frac{\upsilon_c^2}{2}+\frac{\upsilon_c^2}{\gamma_1-1} = k$$

or

$$k = \frac{(\gamma-1)\upsilon_c^2}{2(\gamma-1)} \quad ...(14)$$

From Eq.(3) and (14), we get

$$\upsilon_1^2 + \frac{\gamma}{\gamma-1}\times\frac{p_1}{\rho_1} = \upsilon_2^2 + \frac{\gamma}{\gamma-1}\times\frac{p_2}{\rho_2} = \frac{(\gamma+1)\upsilon_c^2}{2(\gamma-1)} \quad ...(15)$$

or

$$\left\{\begin{array}{l} \dfrac{p_1}{\rho_1} = \dfrac{(\gamma+1)\upsilon_c^2}{2\gamma} - \dfrac{(\gamma-1)\upsilon_1^2}{2\gamma} \\ \dfrac{p_2}{\rho_2} = \dfrac{(\gamma+1)\upsilon_c^2}{2\gamma} - \dfrac{(\gamma-1)\upsilon_2^2}{2\gamma} \end{array}\right\} \quad ...(16)$$

From Eq.(2b), we get

$$\upsilon_2 - \upsilon_1 = \frac{1}{m}(p1-p2) = \frac{p_1}{\rho_1\upsilon_2} - \frac{p_2}{\rho_2\upsilon_2}$$

Substituting the values of p_1 and p_2 from Eq.(16), we get.

$$v_2 - v_1 = \frac{1}{v_1}\left[\frac{(\gamma+1)v_c^2}{2\gamma} - \frac{(\gamma-1)v_1^2}{2\gamma}\right] - \frac{1}{v_2}\left[\frac{(\gamma+1)v_c^2}{2\gamma} - \left(\frac{(\gamma-1)v_2^2}{2\gamma}\right)\right]$$

or
$$v_2 - v_1 = \frac{(\gamma+1)v_c^2}{2\gamma}\left[\frac{1}{v_1} - \frac{1}{v_2}\right] + \left(\frac{\gamma-1}{2\gamma}\right)(v_2 - v_1)$$

$$= \frac{(\gamma+1)v_c^2}{2\gamma v_1 v_2}(v_2 - v_1) + \left(\frac{\gamma-1}{2\gamma}\right)(v_2 - v_1)$$

or
$$1 = \frac{(\gamma+1)v_c^2}{2\gamma v_1 v_2} + \left(\frac{\gamma-1}{2\gamma}\right)$$

or
$$2\gamma v_1 v_2 = (\gamma+1)v_c^2 + (\gamma-1)v_1 v_2$$

or
$$v_1 v_2 = v_c^2 \qquad \text{...(17)}$$

Eq.(17) is known as Prandtl-Mayer relation.

1. (c –ii)

Given: v = velocity of jet

u = velocity of vane

Let m = mass flow rate of fluid

Refer to Fig. 4.

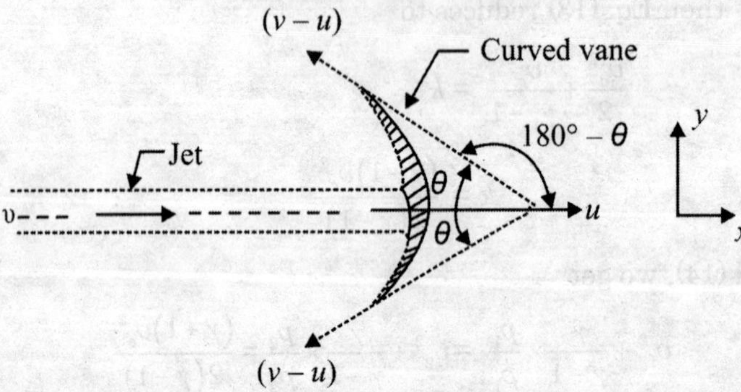

Fig. 4: Fluid jet striking a symmetrical curved vane

Force excerbed by jet, $F_x = \dot{m}\ [(v - u) - (v - u)\cos(180° - \theta)]$

$$= \dot{m}\ (v - u)(1 + \cos\theta)$$

$$Fx = \rho A(v-u)^2(1+\cos\theta)$$

Where A = area of cross-section of jet

ρ = density of fluid

Work done by the jet on the vane per second = $F_x \times u$

$$= \rho A\,(v - u)^2\, u(1 + \cos\theta)$$

Kinetic energy of jet plan per second = $\frac{1}{2}\rho A v^3$

$$\text{Efficiency of jet, } \eta = \frac{\text{Work done on vane by jet}}{\text{Kinetic energy of jet}}$$

$$= \frac{2u(v-u)^2(1+\cos\theta)}{v^3}$$

For maximum efficiency,

$$\frac{d\eta}{du} = 0$$

$$\frac{d\eta}{du} = \frac{2\left(v^2 3u^2 - 4uv\right)(1+\cos\theta)}{v^3} = 0$$

Since $2(1 + \cos\theta) / v^3 \neq 0$

$$v^2 + 3u^2 - 4uv = 0$$

or $(v - u) - (v - 3u) = 0$

or $v = u$ or $v = 3u$

For $v = u$, work done is zero, so not feasible.

For η_{max}, $v = 3u$

$$\eta_{\max} = \frac{2u \times 4u^2(1+\cos\theta)}{27u^3} = \frac{8}{27}(1+\cos\theta) = \frac{16}{27}\cos^2\frac{\theta}{2}$$

1(d) Define specific speed of a hydraulic turbine. Whore it is used? Show that the specific speed (N_s) can be expressed as: $N_s = \sqrt{\frac{N^2 P}{H^{5/2}}} = \frac{N\sqrt{P}}{H^{5/4}}$

N = Speed, P = Power developed or shaft power and H = Head under which the turbine is working.

What is the range of specific speed in r.p.m. for Pelton wheel? 10

Sol. Specific speed of a hydraulic turbine: It is the speed in rpm of a similar turbine working under unit head and delivering unit brake power.

Now $u = hDN$

unit speed, $u, \propto \sqrt{H}$

$\therefore$ $D \propto \frac{\sqrt{H}}{N}$

Also rake, $P_t \propto Q.H$

Where $Q \propto D^2\sqrt{H}$

$\therefore$
$$p_t \propto D^2 H^{\frac{3}{2}} \propto \frac{H}{N^2} \times H^{\frac{3}{2}} \propto \frac{H^{\frac{5}{2}}}{N^2}$$

or
$$N \times \left[\frac{H^{\frac{5}{2}}}{P_t}\right]^{\frac{1}{2}} \times H \frac{H^{\frac{5}{4}}}{\sqrt{P_t}}$$

or
$$N = N_s \frac{H^{\frac{5}{4}}}{\sqrt{P_t}}$$

or
$$N_s = \frac{N\sqrt{P_t}}{H^{\frac{5}{4}}}$$

Range of specific speed for Pelton wheel = 8.5 to 50(in SI Units)

1.(e) Calculate the volume rate of flow of water through a pipe of 100 mm dia. When measured by (a) An orifice plate of orifice size 50 mm. dia and (b) venturi tube of throat size 50 mm dia. The recorded D.P. 250 pa. The density of water is 1000 kg/m³. Assume orifice c_d equal to 0.6 and venture c_d equal to 0.9.

Sol. Given: $d_1 = 100 mm$, $d_2 = 50\ mm$, Depression pressure, $DP = 250Pa$, $\rho = 1000\ kg/m^3$ $c_{do} = 0.6$, $c_{dv} = 0.9$

$$Q = c_d \frac{A_1 A_2}{\sqrt{A_1^2 - A_2^2}} \sqrt{2gh}$$

$$A_1 = \frac{\pi}{4} \times d_1^2 = \frac{\pi}{4}(0.1)^2 = 7.854 \times 10^{-3} m^2$$

$$A_2 = \frac{\pi}{4} \times d_2^2 = \frac{\pi}{4}(0.05)^2 = 1.9635 \times 10^{-3} m^2$$

$$h = \frac{D.P}{\rho g} = \frac{250}{9.81 \times 10^{-3}} = 0.02548 m$$

$$\frac{A_1 A_2}{\sqrt{A_1^2 - A_2^2}} = \frac{7.854 \times 1.9635 \times 10^{-6}}{\sqrt{(7.854)^2 - (1.9635)^2} \times 10^{-3}} = 2.0279 \times 10^{-3}$$

$$\sqrt{2gh} = \sqrt{2 \times 9.81 \times 0.02548} = 0.707 m/s$$

$$Q_o = 0.6 \times 2.0279 \times 10^{-3} \times 0.707 = 8.602 \times 10^{-4} m^3/s$$

$$Q_v = 0.9 \times 2.0279 \times 10^{-3} \times 0.707 = 1.290 \times 10^{-3} m^3/s$$

Volume rate of follow of water throught the pipe = 1.290 litres/s

1.(f) A single acting reciprocating pump has a plunger 10 cm. diameter and a stroke of 200 mm. The centre of the pump is 4 m. above the water level in the sump and 14 m. below the level of water in a tank to which water is delivered by the pump. The diameter and length of suction pipe are 40mm. and 6m. respectively while those of delivery pipe are 30 mm. and

18 m. respectively. Determine the maximum speed at which the pump may be run without separation, if separation occurs at 7.848N/cm². Below the atmospheric pressure. Take atmospheric pressure head = 10.3 of water.

Sol. Given: $D = 100mm$, $L = 200mm$, $r = \frac{L}{2} = 100mm$, $H_s = 4m$, $H_d = 14m$, $d_s = 40mm$, $d_d = 30mm$, $L_s = 6m$, $L_d = 18m$, $\underline{H_{sep}} = 7.848N/cm^2$, $H_{atm} = 10.3$m of water.

(i) Speed of pump without separation during suction stroke: Pressure head in the cylinder at the beginning of suction stroke.

Acceleration head, $H_{as} = \frac{Ls}{g} \cdot \frac{A}{a_s} \cdot \omega^2 r$ at the beginning of suction stroke ($\theta = 0°$)

$$= \frac{6}{9.81} \times \left(\frac{100}{40}\right)^2 \times \omega^2 \times 0.1$$

$$= 0.38226\,\omega^2 m$$

Separation pressure, $H_{sep} = \frac{7.848 \times 10^4}{10^3 \times 9.81} = 8m$ (below atmospheric)

$$= 10.3 - 8 = 2.3m$$

$$H_{as} = H_{atm} - H_s - H_{sep}$$

$$0.3822\ \omega^2 = 10.3 - 4 - 2.3 = 4$$

$$\omega = 3.2348 \text{ red /s}$$

$$N = \frac{60 \times 3.2348}{2\pi} = 30.89 \text{ rpm}$$

(ii). Speed of pump without separation during delivery stroke. The possibility of separation is only at the end of the stroke.

Pressure head in the cylinder at the end of delivery stroke.

$\therefore$ $$H_{sep} = H_{atm} + H_d - H_{ad}$$

$$H_{ad} = 10.3 + 14 - 2.3 = 22 \text{ m}$$

$$H_{ad} = \frac{L_d}{g} \times \frac{A}{a_d} \times \omega^2 r$$

$$22 = \frac{18}{9.81} \times \left(\frac{100}{30}\right)^2 \times \omega^2 \times 0.1$$

$$\omega = 3.28496 \text{rad /s}$$

$$N = 31.369 \text{ rpm}$$

Maximum speed of the pump without separation is the minimum value of speed during suction and delivery strokes.

$\therefore$ $$N_{max} = 30.89 \text{ rpm}$$

1(g). Use the following mathematical relation for any three properties *x, y, z*

$$\left(\frac{\partial x}{\partial y}\right)_z \left(\frac{\partial y}{\partial z}\right)_x \left(\frac{\partial z}{\partial x}\right)_y = -1$$

to show the following relation for the joule - Thomson coefficient

$$\mu_j = \left(\frac{\partial T}{\partial P}\right)_h = \frac{T\left(\frac{\partial V}{\partial T}\right)_p - V}{c_p} = \frac{RT^2}{Pc_p}\left(\frac{\partial z}{\partial T}\right)_p$$

Z is the compressibility factor.

Sol.

$$dh = Tds + udp$$
$$Tds = c_p dT - ud_p$$
$$Tds = c_p dT - T\left(\frac{\partial \upsilon}{\partial T}\right)_p dp$$
$$dh = c_p dT - T\left(\frac{\partial \upsilon}{\partial T}\right)_p dp + \upsilon dp$$

If h = constant, then $dh = 0$, so that

$$c_p(dT)_h - \left[T\left(\frac{\partial \upsilon}{\partial T}\right) - \upsilon\right](dp)_h = 0$$

or
$$\left(\frac{\partial T}{\partial p}\right)_h = \frac{1}{c_p}\left[T\left(\frac{\partial \upsilon}{\partial T}\right)_p - \upsilon\right]$$

$$\mu_j = \left(\frac{\partial T}{\partial p}\right)_h = \frac{T^2}{c_p}\left[\frac{1}{T}\left(\frac{\partial \upsilon}{\partial T}\right)_p - \frac{\upsilon}{T_2}\right] = \frac{T^2}{c_p}\left[\frac{\partial}{\partial T}\left(\frac{\upsilon}{T}\right)\right]_p$$

$$pu = ZRT$$
$$\frac{\upsilon}{T} = \frac{ZR}{p}$$
$$\left[\frac{\partial}{\partial T}\left(\frac{\upsilon}{T}\right)\right]_p = \frac{R}{p}\left(\frac{\partial Z}{\partial T}\right)_p$$
$$\mu_j = \frac{RT^2}{pc_p}\left(\frac{\partial Z}{\partial T}\right)_p$$

1.(h) Write down the chemical formula of R-134a. What does 'a' signify in the nomenclature? Explain the effect of critical temperature and critical pressure of refrigerant on the COP of the cycle.

A straight-charged thermostatic Expansion Value (TEV) is designed to operate at an evaporator temperature of 7° C with a degree of superheat of 5° C R-134a is the refrigerant used in the refrigeration system as well as the bulb.

Find:

(i) The required spring pressure at the design condition,

(ii) Assuming the spring pressure to remain constant, find the degree of superheat, if the same TEV operates at an evaporator temperature of –23° C.

The saturation temperature and corresponding pressure of R-134a are given in the table below:

T sat,°C	–23	–13	–12	–11	2	7	12	13
Psat, kPa	116.39	117.92	185.24	192.8	314.62	374.63	443.01	457.76

Sol. Chemical formula of R134a: CH_2FCF_3 (Tetra-uoro floe ethane)

'a' signifies isomer.

Effect of critical temperature and critical pressure of refrigerant on the COP of the cycle:

For high COP, the critical temperature should be very high so that the condenser temperature line on the p-h diagram is far removed from the critical point. This ensures reasonable refrigerating effect which becomes very small if the state of the liquid before expansion is near the critical point.

The critical pressure should below so that the condensing pressure is low.

TEV

(i) At 7°C, $p_7 = 374.63$ kPa

At 2°C, $p_2 = 314.62$ kPa

At 5°C,

$$p_5 = p_2 + \frac{p_7 \quad p_2}{(7-2)} \times (5-2)$$

$$= 314.62 + \frac{(374.63 - 314.62)}{5} \times 3$$

$$= 350.626\text{kPa}$$

Spring pressure $p_7 - p_5 = 374.63 - 350.626 = 24.004$ kPa

(ii) Pressure at the evaporator inlet at – 23°C,

$$p - 23°C = 116.39 \text{ kPa}$$

For the same evaporator pressure, equilibrium pressure of charge,

$$p_{eqm} = 116.39 + 24.004 = 140.394 \text{ kPa}$$

The corresponding saturation temperature of the charge (R134a) would be in between –23°C, and –13°C.

$$t_{eqm} = -23 = \frac{(-13+23)}{p_{-13°C} - p_{-23°C}} \times \left(p_{eqm} - p_{-23°C}\right)$$

$$= -23 + \frac{10}{(177.92 - 116.39)} \times (140.394 - 116.39)$$

$$= -23 + 3.9 = -19.1°C$$

∴ At $p_{eqm} = 140.39 kPa,\ t = -19.1°C$

Degree of superheat = −19.1 + 23 = 3.9°C

2.(a) A cylinder having a piston restrained by a linear spring (of spring constant 15 kN/m) contains 0.5 kg of saturated vapour water at 120°C, as shown in the fig. 5. Heat is transferred to the water, causing the piston to rise. If the piston cross-sectional area is 0.05m², and the pressure varies linearly with volume until a final pressure of 500kPa is reached. Find the final temperature in the cylinder and the heat transfer for the process. The properties of water are given in the table below:

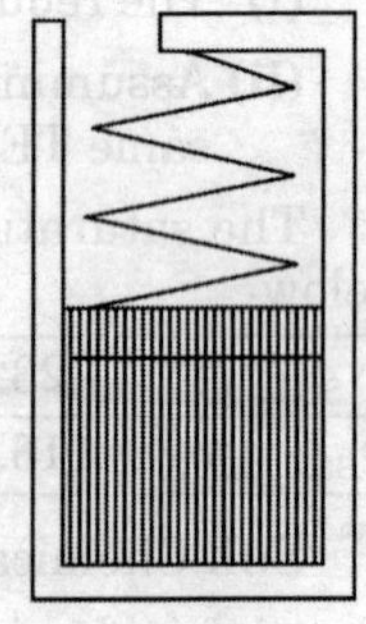

Fig. 5

t,°C	ρ, kPa	vg, m³/kg	ug, kJ/kg	hg, kJ/kg
120	198.50(p_{sat})	0.89186	2529.2	2705.9
151.83	500.00(p_{sat})	0.37477	2559.5	2746.6
801	500.00	0.99055	3664.2	4159.2
802	500.00	0.99147	3666.1	4161.6
803	500.00	0.99240	3668.0	4163.9
804	500.00	0.99333	3669.9	4166.3
805	500.00	0.99425	3671.8	4168.6

Sol. Given: $k = 15 kN/m$, $m_1 = 0.5 kg$, $t_1 = 120°C$, $A_p = 0.05 m_2$, $p \propto v$, $p_2 = 500 kPa$, $t_2 = ?$, $Q_{1\text{-}2} = ?$

Consider saturated water vapour as the control volume in the cylinder.

Continuity equation: $m_1 = m_2 = m$

Energy equation: $m(u_2 - u_1) = Q_{1-2} - W_{1-2}$

At $t_1 = 120°C$, from the given table of properties of water, we have

$$V_1 = 0.89186 m^3/kg,\ u_1 = 2529.2 kJ/kg$$

let x = compression of spring V/Ap

$$p_2 . A_p = p_1 . A_p + kx$$

Where p_1 = atmospheric pressure = 198.5 kPa

$$p_2 = p_1 + \frac{kx}{A_p}$$

$$p_2 = p_1 + \frac{k}{A_p^2} .m(\upsilon_2 - \upsilon_1)$$

$$\upsilon_2 = \upsilon_1 + \frac{A_p^2 (p_2 - p_1)}{km}$$

$$=0.89186+\frac{(0.05)^2(500-198.5)}{15\times0.5}$$

$= 0.99236 m^3/kg \simeq 0.99240 m^3/kg$

From the given table, temperature of water corresponding to 0.99236 $m^3 kg$. is 803°C

$\therefore \quad t_2 = 803°C$

$$W_{1-2} = \int p.dv = \frac{1}{2}(p_1 + p_2).m(\upsilon_2 - \upsilon_1)$$

$$=\frac{1}{2}(198.5+500)\times0.5(0.99240-0.89186)$$

$= 17.56 kJ$

$Q_{1-2} = m(u_2 - u_1) + W_{1-2}$

$= 0.5(3668.0 - 2529.2) + 17.56$

$= 586.96 kJ$

2(b) In an oil cooler the oil enters 10mm diameter tubes at 160°C and is cooled to 40°C. The mean velocity of oil in the tubes is 1.5m/s. Calculate the mean heat transfer coefficient. For turbulent flow of liquid being cooled take Nu = 0.0265 $Re^{0.8}$ $Pr^{0.3}$ and for laminar flow Nu = 3.65. Take all properties at bulk mean temperature and the properties are listed below:

t°C	P, kg/in³	Y, centi-stokes	K, W/m.K	C, kJ/kg.K
40	878	251.0	0.144	1.96
100	839	20.4	0.137	2.22
160	806	5.7	0.131	2.48

1 centi stoke = $10^{-6} m^2/s$.

Sol. Given: $d = 10mm$, $t_1 = 160°C$, $t_2 = 40°C$, $v = 1.5 m/s$, $h = ?$

Bulk mean temperature, $t_f = \frac{1}{2}(t_1 + t_2) = \frac{1}{2}(160 + 40) = 100°C$

At $t_f = 100°C$, $\upsilon = 20.4 cs = 20.4 \times 10^{-6} m^2/s$ and $k = 0.137 W/m.K$

Reynolds number, $$Re = \frac{\upsilon d}{\upsilon} = \frac{1.5\times10\times10^{-3}}{20.4\times10^{-3}} = 735.294$$

The flow is laminar as Re < 2300.

$\upsilon = 3.65$

$$\upsilon = \frac{\bar{h}d}{k} = \frac{\bar{h}\times10\times10^{-3}}{0.137} = 3.65, \ \bar{h} = 50 W/m^2.K$$

2(c) The cylinder bore diameter of a single acting reciprocating pump is 150 mm and stroke is 300 mm. The pump runs at 50 rpm and lifts water through a height of 25 m. The delivery pipe is 32 m. long and 100 mm diameter. Find the theoretical discharge and power required to run the pump. If the actual discharge is 4.2 lit/s find the percentage slip. Also determine the acceleration head at the beginning and middle of delivery stroke.

Sol. Given: $D = 150$ mm, $L = 300$ mm, $N = 50$ rpm, $H = 25$ m, $L_d = 32$ m, $d_d = 100$ mm, $Q_{th} = ?$, $P_{th} = ?$, $Q_{act} = 4.2$ lit/s, %slip=? $H_{ad} = ?$ at $\theta = 0°$ and 90°

Crank radius, $$r = \frac{L}{2} = \frac{300}{2} = 150 \text{ mm}$$

Theoretical discharge, $$Q_{th} = \frac{ALN}{60} = \frac{\frac{\pi}{4}(150\times10^{-3})^2 \times 300\times10^{-3}\times 50}{60}$$

$$= 4.41786 \times 10^{-3} \text{m}^3/\text{s}$$

$$= 4.41786 \text{ lit/s}$$

Theoretical power, $$P_{th} = \frac{rQH}{10^3} = \frac{9.81\times10^3\times4.41786\times10^{-3}\times25}{10^3} = 1.08348\text{kW}$$

$$\%\text{slip} = \left(\frac{Q_{th} - Q_{act}}{Q_{th}}\right)100 = \left[1 - \frac{4.2}{4.41786}\right]100 = 4.93\%$$

Acceleration head in delivery pipe:

$$H_{ad} = \frac{L_d}{g}.\frac{A}{a_d}\omega^2 r\cos\theta$$

At the beginning of delivery stock, $\theta = 0°$, so that $\cos\theta = 1$

$$H_{ad} = \frac{32}{9.81}\times\left(\frac{150}{100}\right)^2\times\left(\frac{2\pi\times50}{60}\right)^2\times0.15\times1$$

$$= 30.182 \text{ m}$$

At the middle of delivery stock, $\theta = 90°$, so that $\cos\theta = 0$

$$\therefore \quad H_{ad} = 0$$

3(a) In an air standard Otto cycle the maximum and minimum temperatures are 1400°C and 15°C. The heat supplied per kg of air is 800 kJ. Calculate the compression the ratio and cycle efficiency. Also calculate the maximum to minimum pressure ratio in the cycle

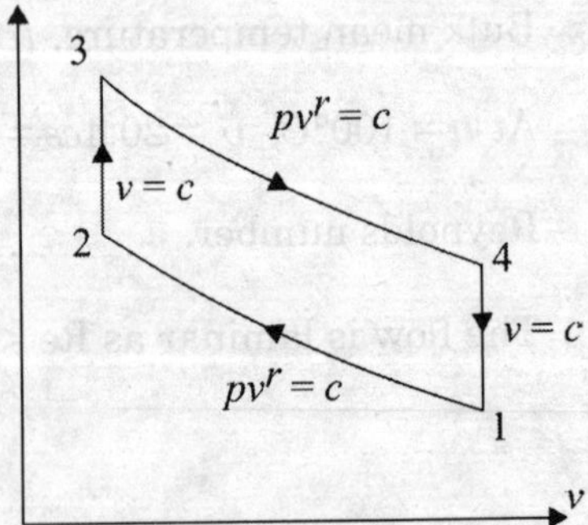

Fig. 6: p-v diagram for Otto Cycle

Given: $t_3 = 1400°C$, $t_1 = 15°C$, $q_s = 800$kJ/kg

The $p - v$ diagram for Otto cycle is shown in Fig.6.

$c_p - c_v = R$

$c_v = c_p - R = 1.005 - 0.287 = 0.718$kJ/kg.K

Heat supplied, $$q_s = c_v(T_3 - T_2)$$

$$800 = 0.718(1400 - t_2)$$

$$t_2 = 285.8°C. \text{ or } 558.8K$$

Adiabatic process 1 – 2:

$$T_1 v_1^{\gamma-1} = T_2 v_2^{\gamma-1}$$

Compression ratio, $$\gamma = \frac{v_1}{v_2} = \left(\frac{T_2}{T_1}\right)^{\frac{1}{\gamma-1}} = \left(\frac{558.8}{288}\right)^{\frac{1}{1.4-1}} = 5.24396$$

Cycle efficiency, $$\eta = 1 - \frac{1}{r^{\gamma-1}} = 1 - \frac{1}{(5.24396)^{0.4}} = 0.4846 \text{ or } 48.46\%$$

Constant volume process 2–3:

$$\frac{p_3}{T_3} = \frac{p_2}{T_2} \text{ and } p_2 = p_1\left(\frac{T_2}{T_1}\right)^{\frac{\gamma}{\gamma-1}}$$

$$\frac{p_{\max}}{p_{\min}} = \frac{p_3}{p_1} = \frac{T_3}{T_2} \times \left(\frac{T_2}{T_1}\right)^{\frac{r}{r-1}} = \frac{1673}{558.8} \times \left(\frac{558.8}{288}\right)^{\frac{1.4}{0.4}} = 30.462$$

3(b) An R717 based vapour compression refrigerating machine work between 38°C and –20°C temperature. The ammonia leaves the compressor dry and saturated Liquid ammonia is undercooled to 30°C temperature inside the condenser, before throttling. Find the theoretical COP of the machine. The c_p of saturated ammonia liquid is 4.91 kJ/kg-K. If net refrigeration required is 25.0 TR, find the mass flow rate of ammonia in kg/h, assuming relative COP 0.75. The saturation properties of R717 are given in the following table:

	Liquid		Vapour	
t,°C	h, kJ/kg	s, kJ/kg-K	h, kJ/kg	s, kJ/kg.K
38	380.78	1.6134	1489.36	5.1873
–20	108.55	0.6538	1437.68	5.8994

Sol. Given: h_1 = 1437.68 kJ/kg, h_2 = 1489.36 kJ/kg, h_3 = 380.78kJ/kg, $h_3 = h_4$, $c_{p\ell}$ = 4.91kJ/kg.K, RE =25TR, $\dot{m}_r$ = ?, $(COP)_{rel}$ = 0.75, T'_2 = 273 + 38 = 311K, c_{pv} = 1.145kJ/kg.K

The T– s diagram is shown in Fig. 7.

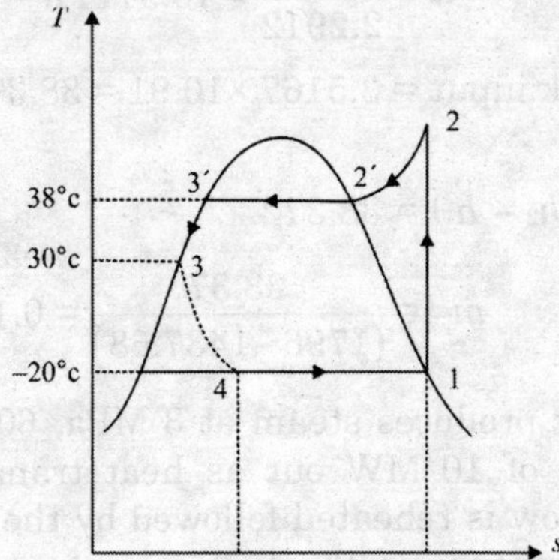

Fig. 7: T-S diagram for VCR m/c

Isentropic process 1-2:

$$s_1 = s_2 = 5.8994 \text{kJ/kg.K}$$

$$s_2 = 5.1873 \text{ kJ/kg.K}$$

$$s_2 - s_2{}^1, = c_{pv} \ln \frac{T_2}{T'_2}$$

$$5.8994 - 5.1873 = 1.145 \ln \frac{T_2}{311}$$

$$T_2 = 579.24 \text{K}$$

$$h_2 - h_2' = c_{pv} (T_2 - T'_2)$$

$$h_2 = 1489.36 + 1.145(579.24 - 311)$$

$$= 1796.5 \text{kJ/kg}$$

Now for under cooled liquid ammonia,

$$h_3' - h_3 = c_{pl} (T_3' - T_3)$$

$$380.78 - h_3 = 4.91(311 - 303)$$

$$h_3 = 341.5 kJ/kg$$

For throttling process 3 – 4

$$h_4 = h_3 = 341.5 kJ/kg$$

$$(\text{COP})_{\text{th}} = \frac{\text{Refrigerating effect (RE)}}{\text{Power input}} = \frac{h_1 \ h_4}{h_2 - h_1}$$

$$= \frac{1437.68 - 341.5}{1796.5 - 1437.68} = \frac{1096.18}{358.82} = 3.055$$

$$(\text{COP})_{\text{act}} = (\text{COP})_{\text{rel}} \times (\text{COP})_{\text{th}} = 0.75 \times 3.055 = 2.2912$$

$$(\text{COP})_{\text{act}} = \frac{RE}{W}$$

$$w = \frac{25}{2.2912} = 10.911 TR$$

$$\text{Work input} = 3.5167 \times 10.91 = 38.37 kW$$

Mass flow rate ammonia:

$$\dot{m}_r \ (h_2 - h_1) = 38.37$$

$$\dot{m}_r = \frac{38.37}{(1796 - 1437.68)} = 0.10693 \ hg/s \text{ or } 384.962 \ kg/h$$

3.(c) A smaller power plant produces steam at 3 MPa, 600°C in the boiler. It keeps the condenser at 45°C by transfer of 10 MW out as heat transfer. The first turbine section expands to 500 kPa and then flow is reheated followed by the expansion in the low pressure turbine. Find the reheat temperature so the turbine output is saturated vapour. For this reheat find the total turbine power output and the boiler heat transfer. Properties of water are given in the table below:

t,°C	p, kPa	υ_f, m³/kg	u_g, m³/kg	h_f kJ/kg	h_υ kJ/kg	s_f, kJ/kg.K	s_v, kJ/kg.K
45	9.59(Psat)	0.00101	15.252	188.42	2583.19	0.63861	8.1647
233.85	3000(Psat)	0.0012167	0.066664	1008.29	2803.99	2.6455	6.1870
600	3000	–	0.13245	–	3682.34	–	7.5084

Sol. The T- s diagram is shown in

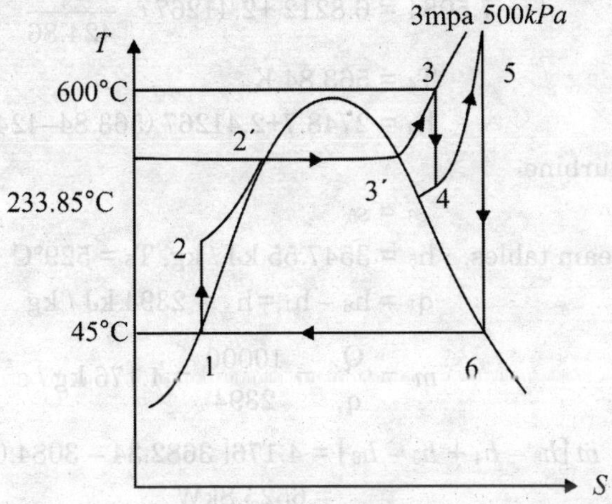

Fig. 8: T-s diagram

Given : $h_1 = h_{\ell 1} = 188.42$kJ /kg, $v_1 = \upsilon_{\ell 1} = 0.00101$m³/kg, $p_1 = 9.59$kPa, $t_1 = 45$°C,

$$h_3 = h_{\upsilon 3} = 3682.34 \text{kJ / kg}, \; p_3 = 3\text{MPa},$$

$$s_3 = s_{\upsilon 3} = 7.5084 \text{kJ / kg.K}, \; t_3 = 600°\text{C}$$

$$h_6 = h_{\upsilon 6} = 2583.19 \text{ kJ / kg},$$

$$s_6 = s_{\upsilon 6} = 8.1647 \text{ kJ / kg.K}$$

$$h_2 = h_1 + Wp = 188.42 + \upsilon_1 (p_2 - p_1)$$

$$= 188.42 + 0.00101 (3000 - 9.59)$$

$$= 191.44 \text{ kJ / kg}$$

Isentropic process 3 – 4, HP turbine:

$$s_3 = s_4$$

At 500kPa (state 4) : $h_{\ell 4} = 640.21$ kJ / kg, $h_{v4} = 2748.7$ kJ / kg,

$$s_{\ell 4} = 1.8606 \text{ kJ / kg.K}, \; s_{\upsilon 4} = 6.8212 \text{ kJ / kg.K},$$

$$c_p = 2.41267 \text{ kJ /kg.K}, \; t_s = 151.86°\text{C}$$

$$s_3 = 7.5084 = s_{\iota 4} + x_4 (s_{\upsilon 4} - s_{\iota 4})$$

$$= 1.8606 + x_4 (6.8212 - 1.8606)$$

$$x_4 = 1.138 > 1$$

Steam at state 4 is superheated.

$$h_4 = h_{\upsilon 4} + c_{p\upsilon}(t_{sup} - t_s)$$

and $$s_{sup} = s_g + c_p \ell_n \frac{T_{sup}}{T_s}$$

$$s_{sup} = s_3 = 7.5084 \text{ kJ / kg.K}$$

sg = $s_{\upsilon 4}$ = 6.8212 kJ / kg.K, Ts = 6.8212 kJ /kg.K, Ts = 273 +151.86 = 424.86 K

$$7.5084 = 6.8212 + 2.41267\, \ell_n \frac{T_4}{424.86}$$

$$T_4 = 563.84 \text{ K}$$

$$h_4 = 2748.7 + 2.41267\,(563.84 - 424.86) = 3084.01 \text{ kJ/ kg.K}$$

Process 5 – 6, L.P. turbine:

$$s_6 = s_5$$

At 500 kPa, from steam tables, $h_5 = 3547.55$ kJ / kg, $T_5 = 529°C$

For condenser, $q_1 = h_6 - h_1, = h_{fg} = 2394$ kJ / kg

Mass flow rate, $$\dot{m} = \frac{Q_\ell}{q_\ell} = \frac{10000}{2394} = 4.176 \text{ kg / s}$$

Turbine work, $W_t = \dot{m}\,[h_3 - h_4 + h_5 - h_6] = 4.176[\,3682.34 - 3084.01 + 3547.55 - 2583.19\,]$

$$= 6525.8 \text{kW}$$

Boiler heat, $Q_b = \dot{m}\,(h_3 - h_2 + h_5 - h_4) = 4.176(3682.34 - 191.44 + 3547.55 - 3084.01)$

$$= 16513.7 \text{kW}$$

4(a)

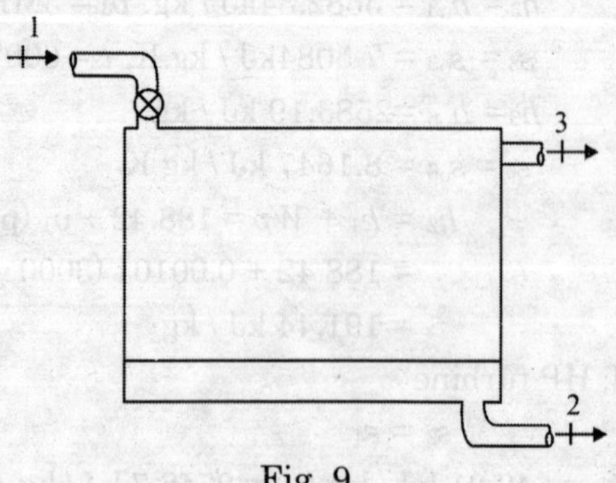

Fig. 9

A geothermal supply of hot water at 500 kPa, 150°C is fed to an insulated flash evaporator at the rate of 1.5 kg/s. A stream of saturated liquid at 200 kPa is drained from the bottom of the chamber and a stream of saturated vapour at 200 kPa is drawn from the top and fed to a turbine. Find the rate of entropy generation in the flash evaporator. Properties of water are given in the table below:

p, kPa	t°C	h_ℓ, kJ/kg	h_v, kJ/kg	s_ℓ, kJ/kg.K	s_v, kJ/kg.k
200	120.20(t_{sat})	504.68	2706.63	1.5300	7.1271
500	151.83(t_{sat})	640.08	2746.60	1.8603	6.8202
500	150.00	632.18	–	1.8417	–

Sol. Continuity equation $\dot{m}_1 = \dot{m}_2 + \dot{m}_3$

Energy equation $\dot{m}_1 h_1 = \dot{m}_2 h_2 + \dot{m}_3 h_3$

Entropy equation: $\dot{m}_1 s_1 + \dot{S}_{gen} + \int \frac{d\dot{Q}}{T} = \dot{m}_2 s_2 + \dot{m}_3 s_3$

Given: $h_1 = 632.18$ kJ / kg, $s_1 = 1.8417$ kJ /kg.K

$h_3 = 2706.63$ kJ / kg, $s_3 = 7.1271$ kJ / kg.K

$h_2 = 504.68$ kJ / kg, $s_2 = 1.53$ kJ / kg.K

From energy equation, $\dot{m}_3 = \frac{\dot{m}_1 (h_1 - h_2)}{(h_3 - h_2)} = \frac{1.5(632.18 - 504.68)}{(2706.63 - 504.68)} = 0.08685 kg / s$

$$\dot{m}_2 = \dot{m}_1 - \dot{m}_3 = 1.5 - 0.08685 = 1.4131.5 \text{ kg / s}$$

$$\dot{S}_{gen} = \dot{m}_2 s_2 + \dot{m}_3 s_3 - \dot{m}_1 s_1 \quad [\therefore Q = 0 \text{ for insulated chamber}]$$

$$= 1.41315 \times 1.53 + 0.08685 \times 7.1271 - 1.5 \times 1.8417$$

$$= 0.01856 \; W / K$$

4(b) In a chemical plant a solution of density 1100kg/m^3 and specific heat capacity 4.6 kJ/kg-K is to be heated from 65 to 100°C. The required flow rate of the solution is 11.8 kg/s. A tubular heat exchanger is used for this with the solution flowing at about 1.2 m/s in 25 mm bore iron tubes and being heated by wet steam at 115°C. The length of the tubes is not to exceed 3.5 m. Taking inside and outside heat transfer coefficient as 5 and 10 kW/m^2 K and neglecting the thermal resistance of the tube wall, estimate the number of tubes and the number of tube passes required.

Sol. Given: $\rho = 1100$ kg / m^3, $c_{pt} = 4.6$ kJ / kg.K, $t_{c2} = 65°C, t_{co} = 100°C$, $\dot{m}_c = 11.8$ kg / s, $\upsilon = 1.2 \; m / s$, $d = 25 mm$, $t_{hi} = t_{ho} = 115°C$, $\ell = 3.5 m$, $h_i = 5 \; kW / m2.K$, $h_o = 10 \; kW / m^2.K$

Refer to fig.10

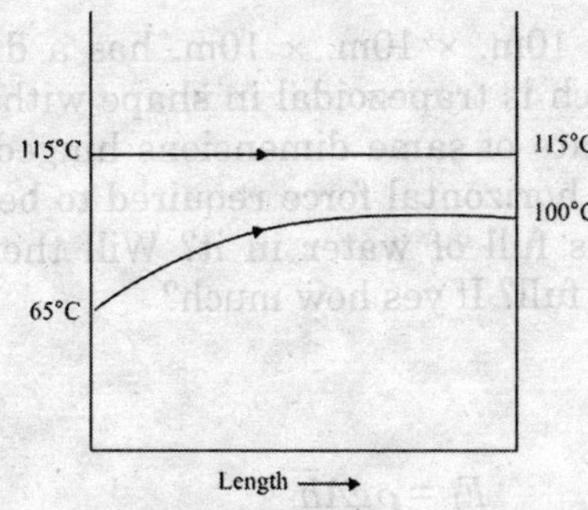

Fig. 10

Heat absorbed by cold solution,

$$\dot{Q} = \dot{m}_c c_{pc}\ (t_{co} - t_{ci})$$

$$= 11.8 \times 4.6(100 - 65) = 1899.8 \text{ kW}$$

$$\Delta t_1 = t_{hi} - t_{ci} = 115 - 65 = 50°C$$

$$\Delta t_2 = t_{ho} - t_{co} = 115 - 100 = 15°C$$

$$\text{LMTD} = \frac{\Delta t_1 - \Delta t_2}{\ell_n\left(\frac{\Delta t_1}{\Delta t_2}\right)} = \frac{50 - 15}{\left(\ell_n \frac{50}{15}\right)} = 29.07°C$$

Overall heat transfer coefficient:

$$\frac{1}{U} = \frac{1}{h_i} + \frac{\delta}{k} + \frac{1}{h} = \text{(Neglecting tube wall thickness resistance)}$$

$$U = \frac{h_i h_o}{h_i + h_o} = \frac{5 \times 10}{(5 + 10)} = 3.333\ kW/m^2.K$$

$$\dot{Q} = UA(LMTD)$$

$$A = \frac{\dot{Q}}{U \times LMTD} = \frac{1899.8}{3.333 \times 29.07} = 19.608\ m^2$$

Mass flow rate in tube, $m_c = \rho Q \upsilon$

Total area of tubes, $a = \frac{11.8}{1100 \times 1.2} 8.94 \times 10^{-3}\ m^2$

$$= n \times \frac{\pi}{4} d^2$$

Number of tubes, $n = \frac{4 \times 8.94 \times 10^{-3}}{\pi \times (0.025)^2} = 18.2 \cong 18$

Surface area of tubes per pass, $a_s = n \times d\ell = 18 \times \pi \times 0.025 \times 3.5 = 4.948 m^2$

Number of passes $= \frac{A}{a_s} = \frac{19.608}{4.948} = 3.963 \cong 4$

4(c) A water storage tank 10m. × 10m. × 10m. has a drainage opening on one of the vertical side at the bottom which is trapezoidal in shape with a width 2m at the bottom, 4m at the top and 1m height. A gate of same dimensions hinged along the top edge is used to close it. What is the minimum horizontal force required to be applied at the bottom to keep the gate closed if the tank has full of water in it? Will there be any change in the force required if the tank is only half full? If yes how much?

Sol. Tank full of water:

Refer to Fig. 11

Total force, $F_1 = \rho g A \bar{h}$

$$\bar{h} = h + \bar{y}$$

$$\bar{y} = \frac{d}{3}\left(\frac{2a+b}{a+b}\right)$$

$$= \frac{1}{3}\left[\frac{2\times4\times2}{4+2}\right] = \frac{1}{3}\times\frac{10}{6} = \frac{5}{9}m$$

$$\bar{h} = 9 + \frac{5}{9} = \frac{86}{9} = 9.556\text{m}$$

Area of tubezium $= \frac{1}{2}(a + b) \times d = \frac{1}{2} \times 6 \times 1 = 3\text{m}^2$

Force required to close the gate, $F_1 = 10 \times 9.81 \times 3 \times 9.556 = 281.22\ kN$

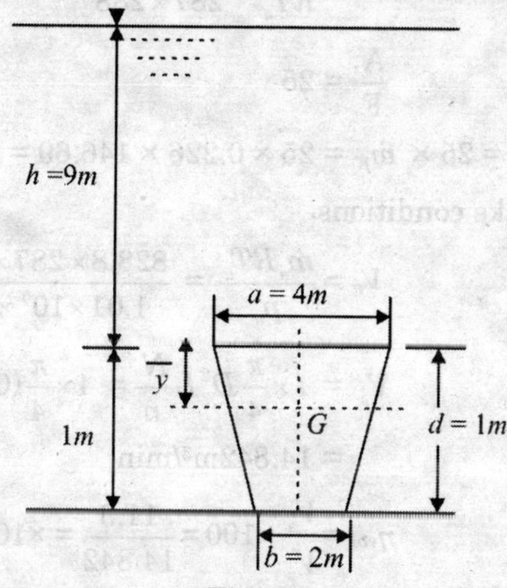

Fig. 11

(ii) Tank half full of water:

$$\bar{\text{h}} = 5 + \frac{5}{9} = 5.556m$$

$$F_2 = 10^3 \times 9.81 \times 3 \times 5.556 = 163.513\ kN$$

There is a change in the closing force.

Change in the closing force = $F_1 - F_2 = 281.22 - 163.513 = 117.707\ kN$

5.(a) A four cylinder 4 stroke diesel engine has a bore of 212 mm. and stroke 292 mm. At full loas at 720 rpm. the break mean effective pressure is 5.93 bar and specific fuel consumption is 0.226 kg/kWH. The air fuel ratio as determined by exhaust gas analysis is 25:1. Calculate the break thermal efficiency and volumetric efficiency of the engine. Atmospheric conditions are 1.01 bar and 15°C. The calorific value of fuel may be taken as 44200 kJ/kg.

Sol. Given: $i = 4$, $D = 212\ mm$, $L = 292\ mm$, $N = 720\ rpm$, $p_{bm} = 5.93$ bar, $\dot{m}_f = 0.226$ kg/kWh, $A/F = 25:1$, $(\eta_{th})_b = ?$, $\eta_{vol} = ?$, $P_a = 1.01$ bar, $t_a = 15°C$, $CV = 44200$ kJ/kg, $n = 2$

Brake power, $$\text{BP} = \frac{ip_{bm}\text{LAN}}{60\times10^3\times n} = \frac{4\times5.93\times10^5\times0.292\times\frac{\pi}{4}(0.212)^2\times720}{60\times10^3\times2}$$

$$= 146.69\ kW$$

Brake thermal efficiency, $$(\eta_{th})_b = \frac{\text{BP}}{\dot{m}_f\times\text{CV}} = \frac{146.69\times3600}{0.226\times146.99\times44200}$$

$$= 0.3604 \text{ or } 36.04\%$$

Density of atmospheric air, $$\rho a = \frac{p_a}{RT_a} = \frac{1.01\times10^5}{287\times288} = 1.2219 \text{ kg}/m^3$$

$$\frac{\text{A}}{\text{F}} = 25$$

$$\dot{m}_a = 25\times\dot{m}_f = 25\times0.226\times146.69 = 828.8\text{kg}/h$$

Volume flow rate at intake conditions:

$$V_a = \frac{\dot{m}_a RT_a}{p_a} = \frac{828.8\times287\times288}{1.01\times10^5\times60} = 11.3\text{m}^3/\text{min}$$

Stroke volume, $$\dot{V}_s = i\times\frac{\pi}{4}D^2L\frac{N}{n} = 4\times\frac{\pi}{4}(0.212)^2\times0.292\times\frac{720}{2}$$

$$= 14.842\text{m}^3/\text{min}$$

Volumetric efficiency, $$\eta_{vol} = \frac{\dot{V}_a}{\dot{V}_s}\times100 = \frac{11.3}{14.842} = \times100 = 76.13\%$$

5.(b) One kg of air at 35° C DBT and 60% RH is mixed with 2 kg of air at 20° C DBT and 13° C dew point temperature. Calculate the specific humidity, temperature and enthalpy of the mixture. Assume specific heat of steam as 1.88 kJ/kg.K and the following properties may be used:

Temperature, °C	Saturation pressure ps bar	Enthalpy of saturated steam, h_g kJ/kg
13	0.0150	2525.4
20	0.0234	2538.2
26	0.0336	2549.1
27	0.0356	2550.9
35	0.0563	2565.4

Sol. Given: $m_1 = 1$ kg, $t_{d1} = 35°$ C, $\phi_1 = 60\%$, $m_2 = 2$ kg, $t_{d2} = 20°C$, $t_{dp} = 13°$ C, $c_{ps} = 1.88$ kJ/kg.K

Air at 35° C DBT and 60 % RH:

From steam tables, at 35° C, $p_{vs} = 0.0563$ bar, $h_g = 2565.4$ kJ/kg

Relative humidity $\phi = \dfrac{p_v}{p_{vs}}$

$$p_v = 0.6 \times 0.0563 = 0.0338 \text{ bar}$$

Specific humidity, $w_1 = \dfrac{0.622\, p_v}{p - p_v} = \dfrac{0.622 \times 0.0338}{1.0132 - 0.0338} = 0.0214\text{kg/kg of d.a.}$

For $p_v = 0.0338$ bar, t_{dp} from the given table, by interpolation is,

$$t_{dp} = 26 + \left(\frac{0.0338 - 0.0336}{0.0356 - 0.0336}\right)(27 - 26) = 26.1°\text{C}$$

Enthalpy: $h_1 = c_p t_{d1} + w_1\, h_{vapour} = c_p t_{d1} + w_1\, [h_g + c_{ps}(t_{d1} - t_{dp})]$

$$= 1.005 \times 35 + 0.0214\, [2565.4 + 1.88\,(35 - 26.1)]$$

$$= 90.43 \text{ kJ/kg of d.a.}$$

Air at 20° C DBT and 13° C DPT:

At 13°C, $p_v = 0.015$ bar

Specific humidity, $w_2 = \dfrac{0.622\, p_v}{p - p_v} = \dfrac{0.622 \times 0.015}{1.0132 - 0.015} = 0.00935\text{kg/kg of d.a.}$

Enthalpy, $h_2 = c_p t_{d2} + w_2\, h_{vapour}$

$$= 1.005 \times 20 + 0.00935\, [2538.2 + 1.88\,(20 - 135)]$$

$$= 43.95 \text{ kJ/kg of d.a.}$$

Mass of vapour per kg of moist air $= \dfrac{1}{(m_1 + m_2)}\left[\dfrac{m_1 w_1}{1 + w_1} + \dfrac{m_2 w_2}{1 + w_2}\right]$

$$= \frac{1}{3}\left[\frac{1 \times 0.0214}{1.0214} + \frac{2 \times 0.00935}{1.00935}\right]$$

$$= 0.01316 \text{ kJ/kg of moist air}$$

$$w_{mixture} = \frac{0.01316}{1 - 0.01316} = 0.01333 \text{ kg / kg of d.a.}$$

$$h_{mixture} = \frac{1}{3}\left[\frac{90.43}{1.0214} + \frac{43.95 \times 2}{1.00935}\right]$$

$$= 58.54 \text{ kg / kg of moist air}$$

$$t_{mixture} = \frac{1}{3}\left[\frac{35}{1.0214} + \frac{20 \times 2}{1.00935}\right]$$

$$= 24.63° \text{ C}$$

5.(c) Air at 27° C temperature and 100 kPa pressure flows over a flat plate at a speed of 2.5 m/s. Calculate the boundary layer thickness at distance of 25 cm and 50 cm from the leading edge of the plate. Calculate the mass flow that enters the boundary layer between $x = 25$ cm and $x = 50$ cm from leading edge. The viscosity of air at 27° C is 30.8 μ Pa - m. Assume unit depth in z-direction.

Sol. Given: $T = 27 + 273 = 300\ K$, $p = 100\ kPa$, $v = 2.5$ m/s, $x_1 = 25$ cm, $x_2 = 50$ cm, $\dot{m}_a = ?$, $\mu = 30.8\ \mu$ Pa·m.

Density of air,
$$\rho = \frac{p}{RT} = \frac{100\times10^3}{287\times300} = 1.16144\ \text{kg / m}^3$$

At $x_1 = 25$cm,
$$\text{Re}_{x1} = \frac{vx_1\rho}{\mu} = \frac{2.5\times0.25\times1.16144}{30.8\times10^{-6}} = 23568.18$$

At
$$x_2 = 50\text{cm},\ \text{Re}_{x2} = \frac{vx_2\rho}{\mu} = \frac{2.5\times0.50\times1.16144}{30.8\times10^{-6}}$$
$$= 47136.36$$

Boundary layer thickness on a flat plase,

$$\delta = \frac{4.64\,x}{\sqrt{\text{Re}_x}}$$

$$\delta_{x1} = \frac{4.64\times0.25}{\sqrt{23568.18}} = 7.556\times10^{-3}\ \text{m or } 7.556\ \text{mm}$$

$$\delta_{x2} = \frac{4.64\times0.50}{\sqrt{47136.36}} = 0.01068\ \text{m or } 10.68\ \text{mm}$$

Mass flow rate in the boundary layer, $m = \int_0^\delta \rho u dy$

$$\frac{u}{U_\infty} = \frac{3}{2}\frac{y}{\delta} - \frac{1}{2}\left(\frac{y}{\delta}\right)^3$$

$$\dot{m} = \int_0^\delta \rho U_\infty\left[\frac{3}{2}\frac{y}{\delta} - \frac{1}{2}\left(\frac{y}{\delta}\right)^3\right]dy$$

$$= \rho U_\infty\left|\frac{3}{4}\frac{y^2}{\delta} - \frac{1}{8}\frac{y^4}{\delta^3}\right|_0^\delta$$

$$= \rho U_\infty\left[\frac{3}{4}\delta - \frac{1}{8}\delta\right] = \frac{\delta}{8}\rho U_\infty\delta$$

Mass flow entering the boundary layer between x_1 and x_2

$$= \frac{5}{8}pU_\infty(\delta_{x2} - \delta_{x1})$$

$$= \frac{5}{8} \times 1.16144 \times 2.5(10.68 - 7.556) \times 10^{-3}$$

$$= 5.669 \times 10^{-3} \text{kg / s}$$

6.(a) A 100 cc petrol Engine with a compression ratio of 6 compresses the air fuel mixture to 900 kPa and 375° C. At the end of compression the ignition is started and the pressure rises along a straight line and attains the highest value of 32.0 MPa after the piston has travelled 4% of the working stroke. The air fuel ratio is 15:1. Take R for mixture as 0.275 kJ/kg-K, calorific value of fuel = 44 MJ/kg and cv = 0.965 kJ/kg-K. Find the heat loss per kg of charge during explosion. Flue gases and air has some gas constant.

Sol. Given: $Vs = 100\ cc$, $r = 6$, $p_2 = 900$ kPa, $T_2 = 375 + 273 = 648\ K$, $p_s = 3$ MPa, $V_5 = 0.04 = V_s + V_2$, A/F = 15:1, $R = 0.275$ kJ/kg.K, $CV = 44$ MJ/kg, $C_V = 0.965$ kJ/kg.K, $R_g = R_a$

The p–v diagram for petrol engine is shown in Fig. 12. 1–2–3–4–1 is ideal cycle and 1–2–5–4–1 is actual cycle. During heat addition, process 2–3 shifts to 2–5.

$$Q_{2\to3} = \frac{cv}{m_f},\ m_f = \frac{m_a}{15}$$

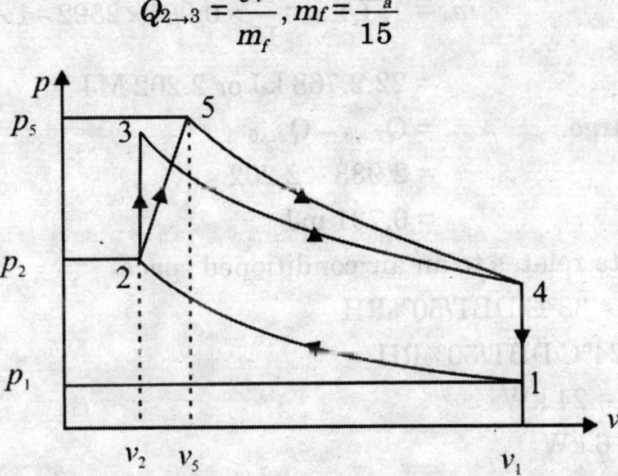

Fig. 12: p–v diagram for petrol engine.

For unit quality of air,

$$Q_{2\to3} = \frac{CV}{A/F} = \frac{44}{15} = 2.933 \text{ MJ}$$

$$V_1 - V_2 = V_s$$

$$V_2\left(\frac{V_1}{V_2} - 1\right) = V_s$$

$$V_2\ (r = 1) = V_s$$

$$V_2 = \frac{Vs}{r-1} = \frac{10}{6-1} = \frac{100}{5} = 20cc$$

$$V_5 = 0.04 \times 100 + 20 = 24\ cc$$

Process: 2–5:

$$\frac{p_2V_2}{T_2} = \frac{p_2V_5}{T_5}$$

$$\frac{T_5}{T_2} = \frac{p_5}{p_2} \times \frac{V_5}{V_2} = \frac{3 \times 10^6}{900 \times 10^3} \times \frac{24}{20} = 4$$

$$T_5 = 4 \times 648 = 2592 \text{ K}$$

$$Q_{2\to5} = m_5 \,.\, c_f \,.\, T_5 - m_2 \cdot c_a \, T_2$$

$$M_2 = m_a,\; m_5 = m_a \times m_f = m_a + \frac{m_a}{15} = \frac{16}{15} m_a$$

$$c_f = 0.965 \text{ kJ / kg.}K,\; c_a = 0.718 \text{ kJ / kg.K}$$

$$Q_{2\to5} = \frac{16\,m_a}{15} \times 0.965 \times 2592 - m_a \times 0.718 \times 648$$

for $m_a = 1,\; Q_{2\to5} = \frac{16}{15} \times 0.965 \times 2592 - 1 \times 0.718 \times 648$

$$= 22.2.768 \text{ kJ or } 2.202 \text{ MJ}$$

Heat loss per kg of charge $= Q_{2\to3} - Q_{2\to5}$

$$= 2.933 - 2.202$$

$$= 0.731 \text{ mJ}$$

6.(b) **The following data relates to an air-conditioned space:**

Outdoor condition = 38°C DBT/50%RH

Room condition = 24°C DBT/50%RH

Sensible heat load = 24 kW

Latent heat load = 6 kW

Bypass factor of the cooling coil = 0.16

If the ventilation requirement is such that on mass flow rate basis 20% fresh air is introduced and 80% supply air is recalculated, determine.

i. Supply air flow rate

ii. Outside air sensible heat

iii. Outside air latent heat

iv. Grand total heat

v. Effective room sensible heat factor

Sol. The schematic diagram of air-conditioning system is shown in Fig. 13(a) and the prosseces in Fig. 13(b).

Given: $t_{d1} = 38$°C, $\phi_1 = 50\%$, $t_{d2} = 24$°C, $\phi_2 = 50\%$, BPF = 0.16, RSH = 24 kW, RLH = 6 kW

1. First of all, mark the initial condition of air at 38°C DBT and 50% RH as points 1.

Now mark the room condition at 24°C DBT and 50% RH as point 2.

2. Calculate, $\text{RSHF} = \dfrac{RSH}{RSH + RLH} = \dfrac{24}{24+6} = \dfrac{24}{30} = 0.8$

3. Mark RSHR = 0.8 on the SHF scale as point 'a' and join with alignment circle point 'b' (26° C DBT and 50% RH).

4. From point 2, draw a line 2 – 4 parallel to line ab. The line 2 – 4 is called the RSHF line.

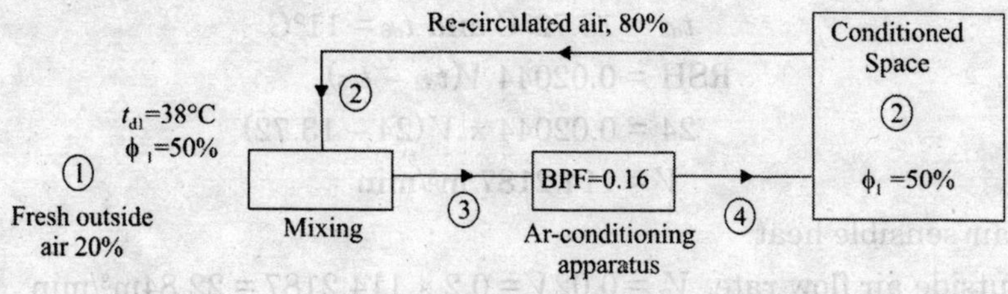

(a) Schematic diagram of A/c system

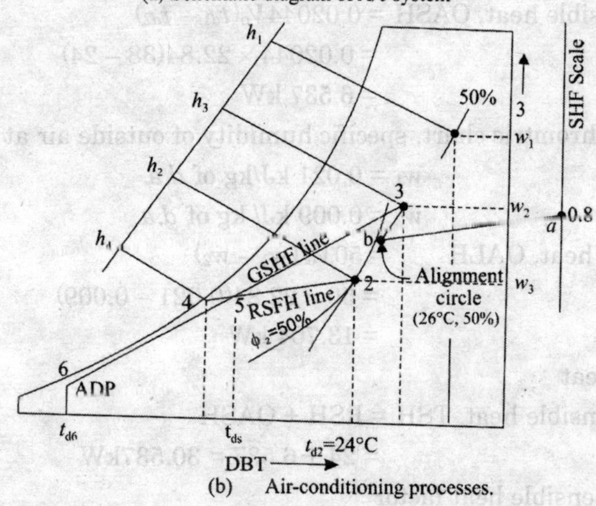

(b) Air-conditioning processes.

Fig.13

5. Since 20% fresh (or outside) air is mixed with 80% of supply air, therefore the condition of air entering the cooling coil after mixing process is marked on the line 1-2 by point 3, such that:

Length 2 – 3 = 0.2 × Length 1 – 2

6. Through point 3, draw a line 3-6 (GSHF line) intersecting the RSHF line at point 5 and the saturation curve at point 6, such that

$$\frac{\text{length } 5-6}{\text{length } 3-6} = \text{BPF} = 0.16$$

i. Supply air flow rate.

Let V = supply air flow rate, m³/min

t_{d5} = DBT of air leaving the cooling coil.

t_{d6} = ADP (Apparatus Dew Point)

From the psychrometric chart, are find that $t_{d3} = 28.5°C$

$$BPF = 0.16 = \frac{t_{d5} - ADP}{t_{d3} - ADP} = \frac{t_{d5} - t_{d6}}{t_{d3} - t_{d6}}$$

By hit and trial method, we find that

$$t_{d5} = 13.72°C \text{ and } t_{d6} = 11°C$$

$$RSH = 0.02044\ V(t_{d2} - t_{d5})$$

$$24 = 0.02044 \times V(24 - 13.72)$$

$$V = 114.2187 \text{ m}^3/\text{min}$$

ii. Outside air sensible heat

Outside air flow rate, $V_o = 0.02\,V = 0.2 \times 114.2187 = 22.84 \text{m}^3/\text{min}$

Outside air sensible heat, OASH $= 0.02044\,V_o(t_{d1} - t_{d2})$

$= 0.02044 \times 22.84(38 - 24)$

$= 6.537$ kW

iii. From the psychromtric chart, specific humidity of outside air at point 1 and 2,

$w_1 = 0.021$ kJ/kg of *d.a.*

$w_2 = 0.009$ kJ/ kg of *d.a.*

Outside air latent heat, OALH $= 50\,V_o\,(w_1 - w_2)$

$= 50 \times 22.84(0.021 - 0.009)$

$= 13.704$ kW

iv. Grand total heat

Total sensible heat, TSH = RSH + OASH

$= 24 + 6.537 = 30.537$kW

v. Effective room sensible heat factor

ERSH = RSH + OASH × BPF

$= 24 + 6.537 \times 0.16 = 25.046$ kW

ERLH = RLH + OALH × BPF

$= 6 + 13.704 \times 0.16 = 8.2$ kW

$$ERSHF = \frac{25.046}{25.046 + 8.2} = 0.753$$

6 (c)i. 1 m³ of air is heated at constant pressure from 15°C to 300°C and then cooled at constant volume back to its initial temperature. If the initial pressure is 1.03 bar calculate the net heat flow and overall change in entropy. Show the process on a T-S diagram.

ii. With simple schematic diagram explain the differences between fire tube and water tube boilers. Mention the merits and demerits of both.

Sol. i Given: $V_1 = 1m^3$, $p_1 = p_2$, $T_1 = 288k$, $T_2 = 573K$,

$p_1 = 1.03$ bar, $c_p = 1.005$ kJ/kg.K, $c_v = 0.718$kJ/kg.K,

$R = 0.287$ kJ/kg.K

$$\text{Mass of air, } m = \frac{p_1V_1}{RT_1} = \frac{1.03\times10^5\times1}{287\times288} = 1.246 \text{ kg}$$

Refer to Fig. 14

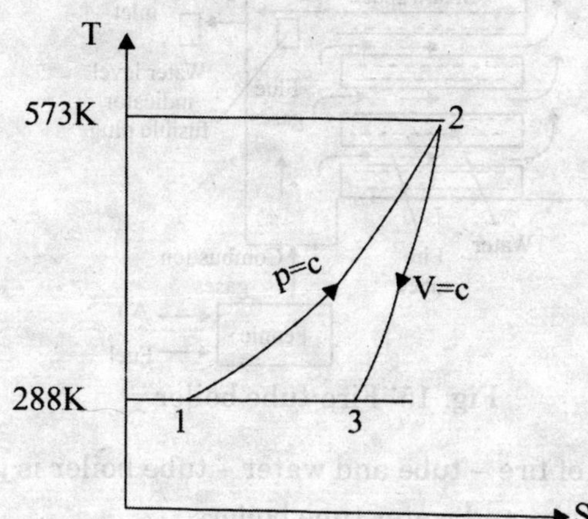

Fig. 14: T-s diagram

$$\begin{aligned} Q &= Q_{1-2} + Q_{2-3} \\ &= mc_p(T_2 - T_1) + mc_\upsilon(T_3 - T_2) \\ &= mc_p(T_2 - T_1) + mc_\upsilon(T_1 - T_2) \qquad [\therefore T_3 = T_1] \\ &= m(T_2 - T_1)(c_p - c_v) \\ &= 1.246\,(573 - 288)\,(1.005 - 0.718) \\ &= 101.916 \text{ kJ} \end{aligned}$$

$$S_2 - S_1 = mc_p\ell_\eta\left(\frac{T_2}{T_1}\right) = 1.246\times1.005\times\ell_n\left(\frac{573}{288}\right)$$

$$= 0.86144 \text{ kJ}/k$$

$$S_3 - S_2 = mc_\upsilon\ell_n\left(\frac{T_3}{T_2}\right) = mc_\upsilon\ell_n\left(\frac{T_1}{T_2}\right)$$

$$= 1.246\times0.718\ell_n\left(\frac{288}{573}\right) = -0.61544 \text{ kJ/K}$$

Overall change in entropy,

$$\Delta S = (S_2 - S_1) + (S_3 - S_2) = 0.86144 - 0.61544 = 0.246 \text{ kJ/K}$$

ii. In a fire-tube boiler, the products of combustion pass through its tube or flues, which are surrounded by water.

In a water – tube boiler, water flows through the tube and flue gases flow outside them. A schematative diagram of fire- tube boiler is shown in Fig. 15

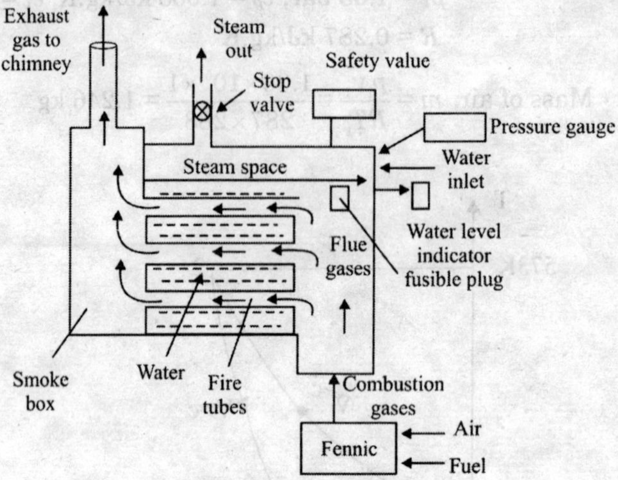

Fig. 15: Fire-tube boiler

The merits and demerits of fire – tube and water – tube boiler is given in Table 1

Table 1: Comparison of fire tube and water tube boilers

Sl. No.	Criteria	Fire tube Boilers	Water tube Boilers
1.	Raising of steam	Less rapid	More rapid
2.	Suitability for large power plants	Unsuitable because of lower rate of steam raised	Suitable because of higher rate of steam raised
3.	Operating pressure	≤ 25 bar	> 125 bar
4.	Floor area required per kg of steam	Less	More
5.	Reliability and cost	Greater reliability and low cost	Require careful attention and maintenance costs are higher
6.	Suitability for rapid changes in load	Suitable	Unsuitable
7.	Damage due to failure in feed water supply	No damage to boiler as it contains large quantity of water	The boiler shall be overloaded
8.	Risk of damage to property in case of bursting	Serious because of its large water capacity	Less serious
9.	Transportation	Very inconvenience because of large size	Can be easily transported and erected as various parts can be separated.

10.	Chances of bursting	Less	More
11.	Maintenance	Difficult for maintenance	All parts are easily accessible for cleaning, inspection and repairing.
12.	Treatment of feed water	Not essential	Essential

7(a). A sample of bituminous coal gave the following ultimate analysis by mass:

Carbon 81.9%, Hydrogen 4.9%, Oxygen 6%, Nitrogen 2.3% and Ash 4.9%.

Calculate

i. the stoichiometric air fuel ratio for complete combustion and

ii the volumetric analysis of wet and dry products of combustion when the air supplied is 25% in excess of that required for complete combustion.

Sol.

Constituent	Mass fraction	Mass of O_2 required for complete combustion, kg	Mass of products formed, kg	
			N_2	CO_2
C	0.819	$0.819 \times \frac{8}{3} = 2.184$		$0.819 \times \frac{11}{3} = 3.003$
H_2	0.049	$0.049 \times 8 = 0.392$		
Total		$\Sigma - 2.576$		$\Sigma = 3.003$
O_2	0.06	0.06		
N_2	0.023		0.023	

Total O_2 required for complete combustion = 2.576 – 0.06 = 2.516 kg/kg of fuel

$$\text{Stoichiometric air / fuel ratio required} = 2.516 \times \frac{100}{23} = 10.94 \text{ kg/kg of fuel}$$

For 25% excess air, total mass of air supplied = 1.25 × 10.94 = 13.675 kg/kg of fuel

Mass analysis of dry products of combustion with 25% excess air:

Mass of N_2 present in air = 0.77 × 13.675 = 10.53 kg

Mass of N_2 already present in fuel = 0.023 kg

Total mass of N_2 = 10.53 + 0.023 = 10.553 kg

Mass of free O_2 due to excess air = 10.94 × 0.25 × 0.23 = 0.629 kg

Constituent	Mass, kg	% by mass
CO_2	3.003	21.17
N_2	10.553	74.39
O_2	0.629	4.43
	$\Sigma = 14.185$	

Volumetric analysis

Constituent (1)	% by mass (2)	Molecular weight (3)	Proportional volume $(4) = \frac{(2)}{(3)}$	Percentage volume $(5) = \left[\frac{(4)}{\Sigma(4)}\right] \times 100$
CO_2	21.17	44	0.48141	14.68
N_2	74.39	28	2.6568	81.09
O_2	4.43	32	0.1384	4.23
			$\Sigma = 3.2763$	$\Sigma = 100$

7.(b) Three thin–walled long circular cylinders 1, 2 and 3 of diameters 150 mm, 250 mm and 350 mm respectively are arranged concentrically. Temperature of cylinder 1 is 80 K and that of cylinder 3 is 300 K. Emissivity of cylinder 1, 2 and 3 is 0.05, 0.1 and 0.2 respectively. Assuming that there is vacuum inside the annular spaces, determine the steady state temperature attained by cylinder 2. 10

Sol. Given: $d_1 = 0.15\ m$, $d_2 = 0.25\ m$, $d_3 = 0.35\ m$, $T_1 = 80k$, $T_3 = 300\ K$, $\varepsilon_1 = 0.05$, $\varepsilon_2 = 0.1$, $\varepsilon_3 = 0.2$, $T_2 = ?$

For steady state condition,

$$Q_{3\to2} = Q_{2\to1} \quad \ldots (1)$$

$$Q_{3\to2} = \frac{\left(T_3^4 - T_2^4\right)}{\frac{1-\varepsilon_3}{A_3\varepsilon_3} + \frac{1}{A_3F_{32}} + \frac{1-\varepsilon_2}{A_2\varepsilon_2}}$$

$$Q_{2\to1} = \frac{\sigma\left(T_2^4 - T_1^4\right)}{\frac{1-\varepsilon_2}{A_2\varepsilon_2} + \frac{1}{A_2F_{21}} + \frac{1-\varepsilon_1}{E_1\varepsilon_1}}$$

Eq.1 become,

$$\frac{\left(T_3^4 - T_2^4\right)}{\frac{1-\varepsilon_3}{A_3\varepsilon_3} + \frac{1}{A_3F_{32}} + \frac{1-\varepsilon_2}{A_2\varepsilon_2}} = \frac{\sigma\left(T_2^4 - T_1^4\right)}{\frac{1-\varepsilon_2}{A_2\varepsilon_2} + \frac{1}{A_2F_{21}} + \frac{1-\varepsilon_1}{A_1\varepsilon_1}} \quad \ldots (2)$$

$$\frac{T_3^4 - T_2^4}{\left(\frac{A_2}{A_3}\right)\left(\frac{1-\varepsilon_3}{\varepsilon_3}\right) + \left(\frac{A_2}{A_3}\right) \times \frac{1}{F_{32}} + \frac{1-\varepsilon_2}{\varepsilon_2}} = \frac{T_2^4 - T_1^4}{\frac{1-\varepsilon_2}{\varepsilon_2} + \frac{1}{F_{21}} + \left(\frac{A_2}{A_1}\right)\left(\frac{1-\varepsilon_1}{\varepsilon_1}\right)}$$

$$\frac{A_2}{A_3} = \left(\frac{0.25}{0.35}\right) = 0.7143, \quad \frac{A_2}{A_1} = \frac{0.25}{0.15} = 1.667$$

$$\frac{1-\varepsilon_1}{\varepsilon_1} = \frac{1-0.05}{0.05} = 19, \quad \frac{1-\varepsilon_2}{\varepsilon_2} = \frac{1-0.1}{0.1} = 9, \quad \frac{1-\varepsilon_3}{\varepsilon_3} = \frac{1-0.2}{0.2} = 4$$

$$F_{32}=\frac{d_2}{d_3}=\frac{0.25}{0.35}=0.7143,\ F_{12}=\frac{d_1}{d_2}=\frac{0.15}{0.25}=0.6$$

Substituting above values in Eq.(2), we have

$$\frac{300^4-T_2^4}{0.7143\times4+0.7143\times\dfrac{1}{0.7143}+9}=\frac{T_2^4-80^4}{9+\dfrac{1}{0.6}+1.667\times19}$$

$$\frac{300^4-T_2^4}{11.8572}=\frac{T_2^4-80^4}{42.34}$$

$$42.34\times300^4-42.34T_2^4=11.8572T_2^4-11.8572\times80^4$$

$$54.1972T_2^4=3.43439\times10^{11}$$

$$T_2=282.14k$$

7(c) **Find an expression for the drag force on smooth sphere of diameter *D*, moving with a uniform velocity *v* in a fluid of density ρ and dynamic viscosity μ.** 10

The drag force F_D is a function of, D, ν, ρ and μ.

$\therefore$ $$F_D=f(D,\nu,\rho,\mu)$$

or $$F_D=\phi(D^a,\nu^b,\rho^c,\mu^d)$$

where ϕ is a non-dimensional constant.

Using M–L–T system, we have

$$MLT^{-2}=\phi\,[L^a]\,[LT^{-1}]^b\,[ML^{-3}]^c\,[ML^{-1}T^{-1}]^d$$

$$[M]^1\,[L]^1\,[T]^{-2}=\phi\,[L]^{a+b-3c-d}\,[M]^{c+d}\,[T]^{-b-d}$$

For *M*: $c+d=1$ ... (1)

For *L*: $a+b-3c-d=1$ ... (2)

For *T*: $-a-b=-2$ ... (3)

There are four unknowns (*a, b, c, d*) and three equations. Therefore, it is not possible to find the values of *a, b, c* and *d* explicitly. Therefore, three of them can be expressed in terms of the fourth variable, which is most important. The role of viscosity μ is very important for drag force and hence *a, b, c* are expressed in terms of *d*.

$\therefore$ $$c=1-d$$

$$b=2-d$$

$$a=1-(2-d)+3(1-d)+d=2-d$$

For $$F_D=\phi\,[D]^{2-d}\,[v]^{2-d}\,[\rho]^{1-d}\,[\mu]^d$$

$$=\phi\,[D^2v^2\rho^2(D^{-d}v^{-d}\rho^{-d}\mu^d)$$

$$=\phi\left[\rho D^2\upsilon^2\left(\frac{\mu}{\rho\ \upsilon D}\right)^d\right]$$

$$=\rho D^2\upsilon^2\phi\left(\frac{\mu}{\rho\ \upsilon D}\right)$$

MECHANICAL ENGINEERING - 2014

PAPER - II

(Conventional)

Time Allowed: Three Hours   Maximum Marks: 200

Instructions

Please read each of the following instructions carefully before attempting questions.

Candidates should attempt FIVE questions in all.

Question No. 1 in section A is compulsory.

Out of the remaining attempt TWO from section–B and TWO from section–C .

All questions carry equal marks. The number of marks carried by a part of a question are indicated against it

Answers must be written in ENGLISH only.

Unless otherwise mentioned, symbols and notations have their usual standard meanings.

Neat sketches may be drawn, wherever required. All parts and sub – parts of a question are to be attempted together in the answer book.

Attempts of questions shall be counted in chronological order. Unless struck off, attempt of a question shall be counted even if attempted partly.

Any page or portion of the page left blank in the answer hook must be clearly struck off

Section - A

1. Answer all of the following (Each part carries 4 marks):

a. Draw a crank rocker mechanism and identify all instantaneous centres.

Sol. Consider a four – bar chain shown in Fig 1 (a) $O_2O_4 = \ell_1$ is the fixed link, $O_2A = \ell_2$ is the rank, $AB = \ell_3$ is the coupler and $O_4B = \ell_4$ is the lever or rocker. It becomes ac rank-rocker mechanism if any of the adjacent links of shortest link is fixed. For a four-bar mechanism to behave as a crank - rocker, the following conditions must be satisfied:

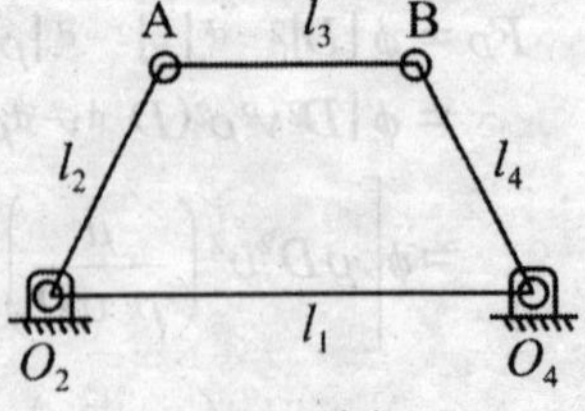

Fig. 1(a)

$$\ell_2 < \ell_1 \text{ or } \ell_3 \text{ or } \ell_4$$

and
$$\ell_1 < (\ell_3 + \ell_4 - \ell_2)$$

and $\ell_1 > (\ell_3 - \ell_4 + \ell_2)$

The instantaneous centers of crank-rocker mechanism are shown in Fig. 1(b)

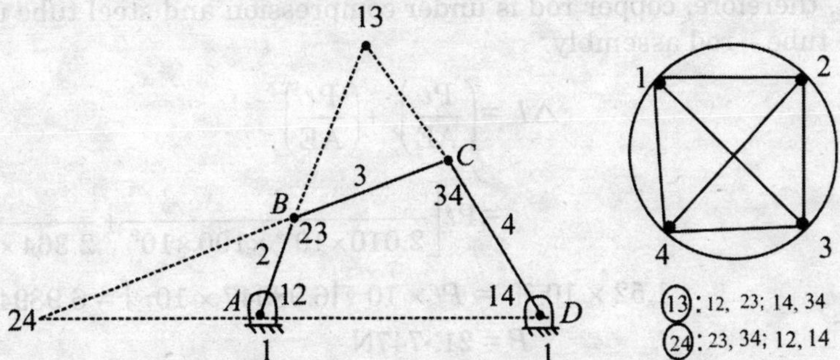

Fig. 1: Instantaneous centers of crank – rocker mechanism

1.(b) A steel tube 2.5 cm external diameter and 1.8 cm. internal diameter encloses a copper rod 1.6 cm. diameter to which it is rigidly joined at each end. If, at a temperature of 20°C there is no longitudinal stress, calculate the stresses in the rod and tube when the temperature is raised to 210°C.

Given: $E_s = 210\ GPa$ and $\alpha_s = 12\times10^{-6}/°C$

And $E_c = 100\ GPa$ and $\alpha_c = 20\times10^{-6}/°C$

Sol. Given: $d_{os} = 2.5$cm., $d_{is} = 1.8$cm., $d_c = 1.6$cm.,

$\Delta t - 210 - 20 = 190°C$, $E_s = 210\ GPa$, $\propto_s = 12 \times 10^{-6}/°C$

$E_c = 100GPa$, $\alpha_c = 20 \times 10^{-6}/°C$

Let ℓ = length of the tube rod assembly, m, (Fig.2)

$$A_s = \frac{\pi}{4}(d_{os}^2 - d_{is}^2) - \frac{\pi}{4}\left(2.52^2 - 1.8^2\right)$$

$= 2.364\text{cm}^2$ or $2.364 \times 10^{-4} m^2$

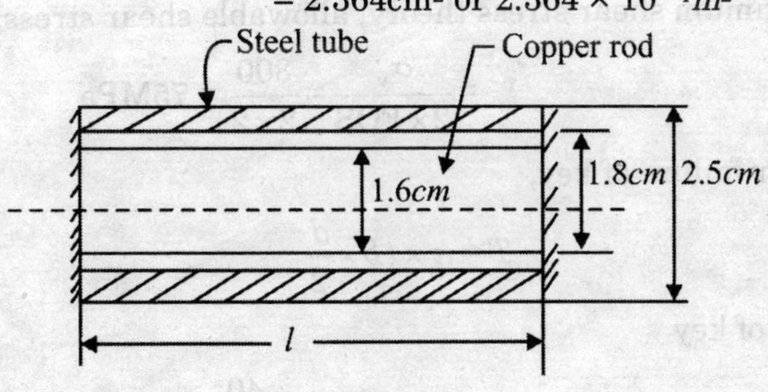

Fig.2

$$Ac = \frac{\pi}{4}d_c^2 = \frac{\pi}{4}\times1.6^2 = 2.010\text{cm}^2 \text{ or } 2.010\times10^{-4}\text{ m}^2$$

Difference in free thermal expansion of copper rod and steel tube,

$$\Delta\ell = \ell\,\alpha_c\,\Delta t - \ell\,\alpha_s\,\Delta t$$

$$= \ell\,(\alpha_c - \alpha_s).\ \Delta t$$

$= \ell\,(20 - 12) \times 10^{-6} \times 190$

$= 1.52{\times}10^{-3}\ \ell\ m$

$(\Delta\ell)_c > (\Delta\ell)_s$, therefore, copper rod is under compression and steel tube under tension. Let P = force in the tube – rod assembly.

$$\therefore \quad \Delta l = \left(\frac{P\ell}{AE}\right)_c + \left(\frac{P\ell}{AE}\right)_s$$

$$= P\ell\left[\frac{1}{2.010\times10^{-4}\times100\times10^{9}} + \frac{1}{2.364\times10^{-4}\times210\times10^{9}}\right]$$

$$1.52\times10^{-3}\ell = P\ell\times10^{-5}\,[6.98947\times10^{-3}] = 6.98947\times10^{-8}Pl$$

$$P = 21,\ 747\text{N}$$

Stress in steel tube, $\sigma_s = \dfrac{P}{A_s} = \dfrac{21{,}747}{2.364\times10^{-4}} = 92.184$ MPa (tensile)

Stress in copper rod, $\sigma_c = \dfrac{P}{A_c} = \dfrac{21.747}{2.010\times10^{-4}} = 108.194$ MPa (compressive)

1.(c) A drive shaft of 40 mm. diameter transmitting 25 kW at 300 rpm. is connected to a gear by a flat key of width 22 mm. and thickness 14 mm. It is made of steel having 300 MPa yield stress. Determine the length of the key to withstand shear. Use a factor of safety 2. 4

Sol. Given: d = 40mm., P = 25kW, N = 300rpm, b = 22mm., t = 14mm, σ_y = 300MPa, ℓ = ?, FOS= 2

$$P = \frac{2\pi NT}{60\times10^{3}}$$

$$T = \frac{25\times10^{3}\times60}{2\pi\times300} = 795.78\ \text{N.m}$$

According to maximum shear stress theory, allowable shear stress,

$$\tau = \frac{\sigma_y}{2\times\text{FOS}} = \frac{300}{2\times2} = 75\text{MPa}$$

Considering shear failure of key,

$$T = \tau\times\ell b\times\frac{d}{2}$$

Where ℓ = length of key

$$795.78\times10^{3} = 75\times\ell\times22\times\frac{40}{2}$$

$$\ell = 24mm$$

1.(d) A plate clutch has a single surface with an outside diameter of 250mm. and inside diameter of 100 mm. with a coefficient of friction 0.2. Find the required axial force to develop a maximum pressure of 0.65 MPa. Under this pressure, find the torque capacity of the clutch.

Sol. Given: $d_1 = 250$mm. or $r_1 = 125$mm, $d_2 = 100$mm. or $r_2 = 50$mm, $\mu = 0.2$, $F_a = ?$,

$p_{max} = 0.65$ MPa,

$T_f = ?$, $n = 1$

For uniform wear, $pr = \text{const}$

Axial force, $F_a = p_{max} \times 2\pi r_1 (r_1 - r_2)$

$= 0.65 \times 2\pi \times 50\ (125 - 50)$

$= 15315.26\ N$

Friction torque, $T_f = \mu\ F_a r_m$

Where $r_m = \frac{1}{2}(r_1 + r_2)$ for uniform wear

$= \frac{1}{2}\ (125+50) = 87.5$mm

$T_f = 0.2 \times 15315.26 \times 87.5 \times 10^{-3} = 268.02\ N.m$

1.(e) What are ceramics? Classify ceramics into four groups and describe their utilities and applications? 4

Sol. Definition: The ceramics may be defined as inorganic and non-metallic materials which are processed or used at high temperatures. According to the American Ceramic Society, ceramic products are those which are manufactured by the action of heat on raw materials.

Classification into Four Groups:

1. Natural or manufactured:
   - i. Natural ceramics – Silica, Silicates and clay materials
   - ii. Manufactured ceramics – Silicon carbide, (SiC), Aluminum Oxide (Al_2O_3), Silicon Nitride (Si_3N_4), etc.
2. Functional Classification:
   - i. Abrasives – Alumina, Carborundum
   - ii. Pure oxide Ceramics – MgO, Al_2O_3, SiO_2
   - iii. Fixed clay products – Bricks, Tiles, Porcelain, etc.
   - iv. Inorganic glasses – Window glass, lead glass, etc.
   - v. Cementing materials – Portland cement, Lime etc.
   - vi. Rocks – Granites, Sandstones etc.
   - vii. Minerals – Quartz, Calcite, etc.
   - viii. Refractories – Silica bricks, Magnesite, etc.
3. Structural Classification
   - i. Crystalline ceramics – MgO, Al_2O_3, (Abrasives)
   - ii. Non-crystalline ceramics – Window glass, bricks, floor tiles, etc.

iii. Glass bonded ceramics – Fixed clay products (refractories)

iv. Cements – Crystalline and non-crystalline phases, Portland cement, lime

Ceramic groups according to American ceramic society.

1. Whitewares
2. Glass
3. Refractories
4. Structural clay products
5. Enamels

Utilities and Applications of ceramics

1. Dinner wares, 2. Electrical and chemical porcelain, 3. Refractory bricks and tiles, 4. Glass, 5. Porcelain enamels, 6. Abrasives, 7. Cutting tools, 8. Cements and concretes, 9. Whitewares 10. Insulators.

1.(f) Describe the characteristics of tool materials? 4

Sol: Basic characteristics of tool materials:

1. **Hot hardness,** so that the tool does not loose its hardness and strength at the high temperatures developed during machining.
2. **Wear and abrasive resistance**, so that the tool retains its shape and cutting efficiency for a reasonably long time.
3. **Impact toughness,** so that the fine cutting edge of the tool does not break or chip, when the tool is suddenly loaded.

Additional properties:

1 Increased thermal conductivity.

2 Lower coefficient of thermal expansion.

3. Lower chemical and mechanical affinity for the work material.

4 High specific heat and low coefficient of friction between tool and work.

5. Easy to form, grind and sharpen to the desired tool geometry.

1.(g) Name the defects that may develop in arc welding of steel parts. 4

Sol. 1. Cracks, 2. Undercut, 3. Porosity or Voids, 4. Inclusions, 5. Lack of fusion, 6. Burning, 7. Internal residual stresses and distortion.

1.(h) Differentiate between orthogonal and oblique cutting. 4

Sol. Orthogonal cutting: It is a two – dimensional cutting in which the cutting edge of the tool is arranged perpendicular to the cutting velocity vector.

Oblique cutting: It is a three – dimensional cutting in which the cutting edge of tool is set at some angle other than 90° to the cutting velocity vector.

1.(i) What is ABC analysis?

Sol. ABC Analysis

An *ABC* analysis consists of separating the inventory items into three groups A, B, C, according to their annual cost volume consumption, *i.e.,* unit cost x annual consumption.

A. items – High Value: These are relatively few items whose value accounts for 60–70% of the total value of the inventory. These will usually be 10–15% of the total items.

B. item – Medium Value: A large number of items in the middle of the list, usually about 20–25% of the items whose total value accounts for about 20–25% of the total value.

C. item – Low Value: The bulk of items usually about 60–70% whose total inventory values are almost negligible. They account for 10–15% of the total value.

1.(j) Explain Queueing model and its applications. 4

Sol. Queueing Models

(M/D/l):(FCLS/∞/∞): Poisson arrival, constant service time (i.e. deterministic), number of serves = 1, queue discipline is FCLS, no limitations on the numbers in the system, and calling population is infinite.

(M/M/1) - Queue discipline is (FCFS/∞/∞).

(GI/E, /2): (∞/∞/FCFS): Inter arrival rate is based on General Independent probability distribution. Service times is based on Erlangian (Gamma) distribution with parameter k. Number of serves = 2.

(D/D/1) – Deterministic Model, constant inter – interval time and service time, FCFS/∞/∞ is implied. Service time is less than the time between arrivals.

(M/M/C):(FCFS/N) – Poisson arrival and exponential service. Multiple channel queue, C = number of servers, queue is FCFS, N = number in the system.

Applications

1. Retail shops
2. Air ports check-in
3. Traffic congestion
4. Automobile service centers
5. Railway wagon yard
6. Banking service

Section - B

Answer any TWO questions.

2.(a) A Power screw is made with Acme threads 34 mm-6 mm single start to lift and lower a load of 10 KN. The screw and nut are well lubricated Sliding friction is 0.15 and rolling friction is 0.02. Take semi-thread angle as 14.5° Determine:

i. raising torque
ii. eowering torque
iii. efficiency of the power screw

Sol. Given: $d = 34$ mm, $p = 6$ mm, $W = 10$ KN, $\mu_s = 0.15$, $\mu_r = 0.02$, $\theta = 14.5°$

Mean diameter of screw, $d_m = d - \frac{p}{2} = 34 - \frac{6}{2} = 31\text{mm}$

Collar diameter, $d_c = d - p = 34 - 6 = 28\text{mm}$

Helix angle, $\tan\alpha \frac{p}{\pi d_m} = \frac{6}{\pi \times 31} = 0.0616$

$\alpha = 3.525°$

(i) Raissing torque, $T_R = \frac{Wd_m}{2}\left[\frac{\mu_s \sec\theta + \tan\alpha}{1-\mu\sec\theta\tan\alpha}\right] + \frac{Wd_c}{2}\mu_r \sec\theta$

$$= \frac{10\times10^3\times31\times10^{-3}}{2}\left[\frac{0.15\sec14.5°+0.0616}{1-(0.15\sec14.5°)\times0.0616}\right] + \frac{10\times10^3\times31\times10^{-3}}{2}\times0.02\times\sec14.5°$$

$$= 155\left[\frac{0.21653}{0.99045}\right]2.1982 = 36.778 \text{ N.m}$$

ii. Lowering torque, $T_L = \frac{Wd_m}{2}\left[\frac{\mu_s\sec\theta - \tan\alpha}{1+\mu_s\sec\theta.\tan\alpha}\right] + \frac{Wd_c}{2}\mu_r\sec\theta$

$$= 155\left[\frac{0.15\sec14.5^o - 0.0616}{1+(0.15\sec14.5^o)\times0.0616}\right] + 2.89212$$

$$= 155\left[\frac{0.09333}{1.00954}\right] + 2.89212 = 17.2215 \text{ N.m}$$

iii. Efficiency, $\eta = \frac{\tan\alpha(1-\mu_s\sec\theta\tan\alpha)}{\mu_s\sec\theta+\tan\alpha}$

$$= \frac{0.0616(1-0.15\sec14.5^o\times0.0616)}{0.15\sec14.5^o+0.0616} = \frac{0.0610}{0.21653}$$

$$= 0.2817 \text{ or } 28.17\%$$

2.(b) Design a gip and cotter joint for two rods of square cross section transmitting an axial tensile load of 40 kN. The gib cotter and rods are made of same material having allowable tensile strength of 120 MPa and shear strength of 60 MPa and crushing strength of 160 Mpa. Assume the thickness of cotter as one- third of side of the square cross section.

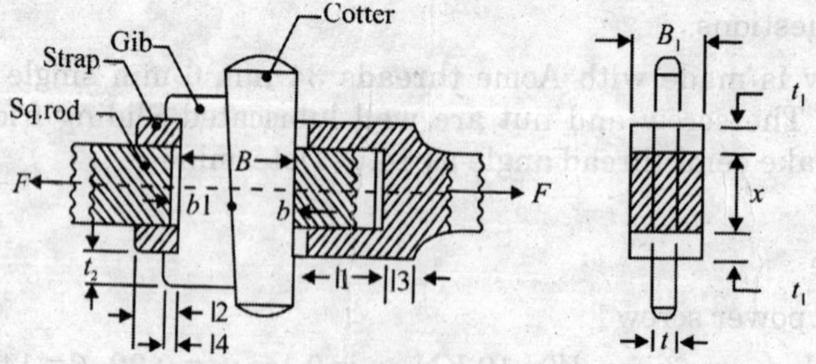

Fig. 3: Gib and cotter joint for square rods.

Given: $F = 40\ kN$, $\sigma_t = 120$MPa, $\tau = 60$MPa, $\sigma_c = 160$ MPa

The Gib and Cotter joint for square rods is shown in Fig.3.

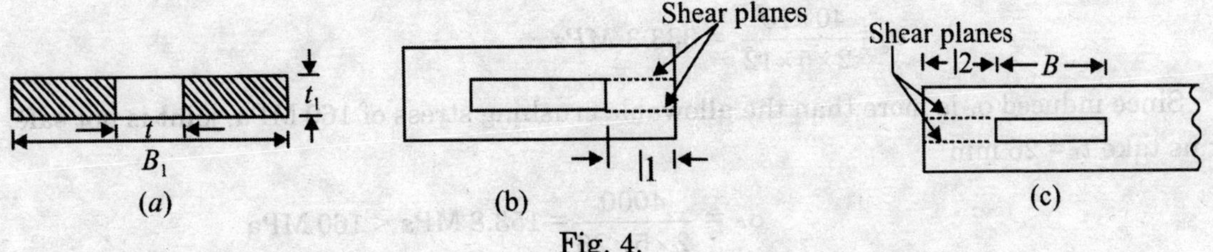

Fig. 4.

1. Size of square rod: Considering the failure of square rod of cross – section $x \times x$ under tensile face F, we have $\sigma_t . x^2 = F$

$$x = \sqrt{\frac{F}{\sigma_t}} = \sqrt{\frac{40\times10^3}{120}} = 18.257\ mm \simeq 20\ mm \text{ (say)}$$

The other dimensions are fixed as follows:

with of strap, $B_1 = x = 20\text{mm}$

Thickness of cotter, $t = \frac{B_1}{4} = \frac{20}{4} = 5\text{mm}$

Thickness of gib $\tau = 5\text{mm}$

Height of gib, $t_2 = t = 5\text{mm}$

Length of gib head, $\ell_4 = t = 5\text{mm}$

2. Width of gib and cotter

Considering the double shear of gib and cotter, we have

$$2Bt\tau = F$$

$$B = \frac{40\times10^3}{2\times5\times60} = 66.7\text{mm} \simeq 70\text{mm}$$

Width of gib, $b_1 = 0.55B$, for a single gib

$= 0.55\times70 = 38.5\text{mm} \simeq 40\text{mm}$

Width of cotter, $b = 0.45B$, for a single gib

$= 0.45 \times 70 = 31.5\text{mm} \simeq 32\text{mm}$

3. **Thickness of strap:** Considering the failure of strap end in tension at the location of gib and cotter, as shown in Fig 4(a), we have

$$2\,(B_1 - t)\,t_1\sigma_t = F$$

$$2\,(20 - 5)\,t_1 \times 120 = 40 \times 10^3$$

$$t_1 = \frac{40\times10^3}{2\times15\times120} = 11.1mm \simeq 12mm$$

4. Check for strap or gib in crushing

$$2tt_1\sigma_c = F$$

$$\sigma_c = \frac{40\times10^3}{2\times5\times12} = 333.3\ MPa$$

Since induced σ_c is more than the allowable crushing stress of 160 MPa, joint is not safe.

Let us take $t_1 = 26$ mm

$$\sigma_c = \frac{4000}{2\times5\times26} = 153.8\text{ MPa} < 160\text{ MPa}$$

5. Length of rod from cotter to end

Considering the double shear of rod, as shown in Fig. 4 (b), we have

$$2\ell_1 \times x \times T = F$$

$$\ell_1 = \frac{40\times10^3}{2\times20\times60} = 16.7\text{mm}$$

$$= 20\text{mm}$$

6. Length of rod from gib to the end of strap

Considering the double shear of strap, as shown in Fig .4(c), we have

$$2\times 2l_2t_1T = F$$

$$\ell_2 = \frac{40\times10^3}{4\times26\times60} = 6.4\text{mm} = 10\text{mm}$$

7. Other dimensions

Length of strap end, $\ell_3 = \frac{2}{3}x = \frac{2}{3}x = 20 = 13.33\text{mm} = 14\text{mm}$

Length of cotter = $4x = 4 \times 20 = 80$ mm

2.(c) Determine the diameter of a circular shaft subjected to a bending moment M = 13 kN·m and a torque T = 30 kN·m according to maximum shear stress theory Take σ_y = 700MPa and use a factor of safety of 2.6.

Sol. Given: M = 13 kN·m, T = 30 kN·m, σ_y = 700 MPa, FOS = 2.6

According to maximum shear stress theory, yield shear stress,

$$\tau_y = \frac{1}{2}\sigma_y = \frac{1}{2}\times 700 = 350\text{ MPa}$$

Allowable shear stress, $$\tau_a = \frac{\tau_y}{\text{FOS}} = \frac{350}{2.6} = 134.615\text{ MPa}$$

Equivalent torque, $$T_e = \sqrt{M^2+T^2} = \sqrt{13^2+30^2} = 32.695\text{ kN.m}$$

Let d = diameter of shaft, mm

$$T_e = \frac{\pi}{16} d^3 \tau_a$$

$$d = \left(\frac{32.695 \times 16 \times 10^6}{\pi \times 134.615} \right)^{1/3} = 107.35\text{mm} = 108\text{mm}$$

(a) **Compare the flexural strengths of the following three beams of equal weight and same material:**

i. **I-section 300 × 160mm with flanges 20mm. thick and web 16mm. thick**

ii. **Rectangular section having depth twice the width.**

iii. **Circular solid section.** **10**

Sol.

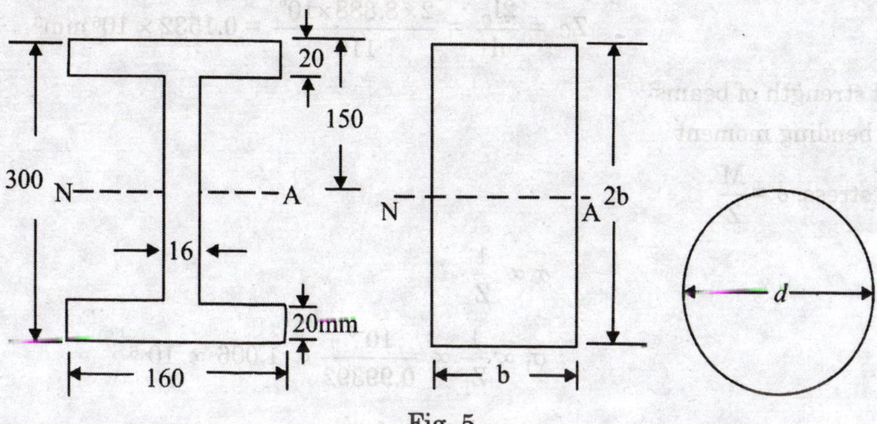

Fig .5

For the same material of three beams, weight is proportional to area of cross – section

$$A_I = 2 \times 160 \times 20 + 260 \times 16 = 10,560 \text{ mm}^2$$

$$A_R = b \times 2b = 2b^2$$

$$A_c = \frac{\pi}{4} d^2$$

For $A_R = A_I$

$$2b^2 = 10560, \ b = 72.66 \text{ mm} \sim 73 \text{ mm}$$

$$A_c = A_I$$

$$\frac{\pi}{4} d^2 = 10560, \ d = 115.95\text{mm} = 116\text{mm}$$

Moment of inertia of beams about N – A.

$$I_I = \frac{160 \times 300^3}{12} - \frac{144 \times 260^3}{12} = 149.088 \times 10^6 \text{ mm}^4$$

$$I_R = \frac{b \times (2b)^3}{12} = \frac{2}{3}6^4 = \frac{2}{3} \times 73^4 = 18.932 \times 10^6 \text{ mm}^4$$

$$I_C = \frac{\pi d^4}{64} = \frac{\pi}{64} \times 116^4 = 8.888 \times 10^6 \text{ mm}^4$$

Section modules of beams:

$$Z_I = \frac{I_I}{y_{max}} = \frac{149.088 \times 10^6}{150} = 0.99392 \times 10^6 \text{ mm}^3$$

$$Z_R = \frac{I_R}{b} = \frac{18.932 \times 10^6}{73} = 0.2593 \times 10^6 \text{ mm}^3$$

$$Z_C = \frac{2I_C}{d} = \frac{2 \times 8.888 \times 10^6}{116} = 0.1532 \times 10^6 \text{ mm}^3$$

Flexural strength of beams:

Let M = bending moment

bending stress, $\sigma = \frac{M}{Z}$

$$\therefore \quad \sigma \propto \frac{1}{Z}$$

$$\sigma_I \propto \frac{1}{Z_I} \propto \frac{10^{-6}}{0.99392} \propto 1.006 \propto 10^{-6}$$

$$\sigma_R \propto \frac{1}{Z_R} \propto \frac{10^{-6}}{0.25934} \propto 3.856 \propto 10^{-6}$$

$$\sigma_c \times \frac{1}{Z_c} \times \frac{10^{-6}}{0.15324} \times 6.526 \times 10^{-6}$$

$$\therefore \sigma_I : \sigma_R : \sigma_c :: 1.006 : 3.856 : 6.526 \quad \text{or} \quad 1:3.833:6.487$$

3.(b) The internal diameter of the cylinder of a hydraulic ram is 10 cm. Find the thickness required to withstand an internal pressure of 500 atm (1 atm = 98.07 kPa), if the yield point for the material (in tension as well as compression) is σ = 500 MPa, use a Factor of safety of 2. 10

Sol. Given: d = 10 cm, t = ?, p = 500 atm or 500 × 98.07 × 10^3 = 49.35 MPa, σ_y = 500 MPa, FOS = 2

Allowable tensile / compressive stress, $\sigma_a = \frac{\sigma_y}{FOS} = \frac{500}{2} = 250$ MPa

Hoop stress in cylinder, $\sigma_\theta = \frac{pd}{2t}$

Thickness, $t = \dfrac{pd}{2\sigma_a} = \dfrac{49.035 \times 100}{2 \times 250} = 9.807 \text{ mm} = 10\text{mm}$

3.(c) Draw the profile of a cam operating a knife-edged follower having a lift of 30 mm. The cam raises the follower with S.H.M. for 150° of its rotation followed by a period of dwell for 60°. The follower descends for the next 100° rotation of the cam with uniform velocity, again followed by a dwell period. The cam rotates at a uniform speed of 120 rpm and has a least radius of 20 mm.

What are the maximum velocity and acceleration of the follower during lift and return?

Sol. Given: $h = 30$ mm, $\theta_1 = 150°$, $\theta_2 = 60°$, $\theta_3 = 100°$, $\theta_4 = 50°$, $N = 120$ rpm, $r = 20$ mm

$$\omega = \frac{2\pi N}{60} = \frac{2\pi \times 120}{60} = 12.57 \text{rad/s}$$

During ascent with SHM for $\theta_1 = 150°$,

$$\upsilon_{max} = \frac{h}{2} \times \frac{\pi\,\omega}{\theta_1} = \frac{30}{2} \times \frac{\pi \times 12.57}{\left(\frac{\pi}{180} \times 150\right)} = 226.26 \text{ mm/s}^2$$

During descent with uniform velocity for $\theta_3 = 100°$

$$\upsilon_{max} = \frac{h\,\omega}{\theta_3} = \frac{30 \times 12.57}{\left(\frac{\pi}{180} \times 100\right)} = 216 \text{mm/s}$$

$$f_{max} = f = 0$$

Displacement diagram

Draw the displacement diagram as shown in Fig 6(a)

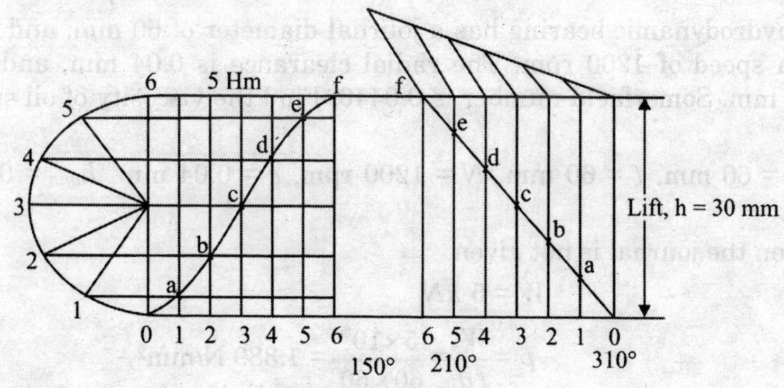

Fig. 6(a)

Cam profile

1. Draw a circle of radius equal to the bass circle radius of 20mm with center o.
2. Draw angles of ascent, dwell and descent of 150°, 60° and 100° respectively. Divide the angles 150° and 100° into 6 equal parts. Draw radial lines for these angles.

3. Match points 0 to 6 and 6′ to 0′ on the base circle in 150° and 100° respectively.
4. Measure distances 1a, 2b, ... from the displacement diagram from 150° and cut-off corresponding distances on the radial lines. Repeat the same process for 100°.
5. Join the points so obtained by a smooth curve to get the cam profile as shown in Fig. 6(b).

(b) Cam profile

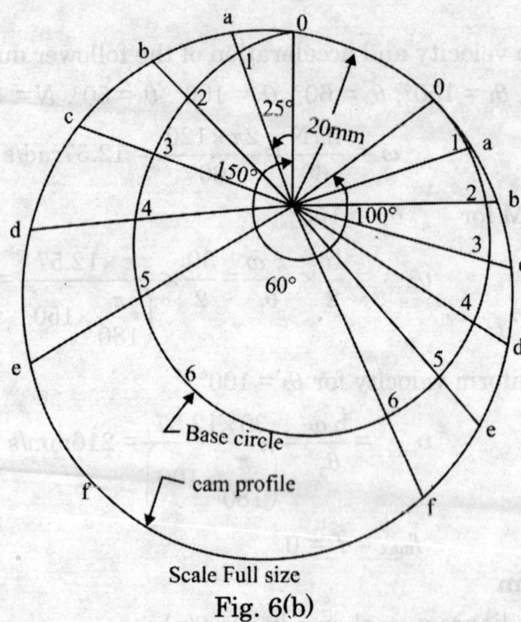

Fig. 6(b)

3 (d) A 360° hydrodynamic bearing has a journal diameter of 60 mm. and length 60mm. It is rotating at a speed of 1200 rpm. The radial clearance is 0.04 mm. and minimum oil thickness is 0.008 mm. Somerfaeld number is 0.0446. Find the viscosity of oil suitable for the bearing.

Sol. Given: d = 60 mm, ℓ = 60 mm, N = 1200 rpm, c = 0.04 mm, h_{min} = 0.008 mm, S = 0.0446, μ = ?

Radial load Won the journal is not given.

Let $$W = 5\ kN$$

$$p = \frac{W}{\ell d} = \frac{5 \times 10^3}{60 \times 60} = 1.389 \text{ N/mm}^2$$

Sommerfeld number, $$S = \left[\frac{\mu N}{60 p}\right]\left(\frac{\pi}{c}\right)^2 \times 10^{-6}$$

$$0.0446 = \left[\frac{\mu x 1200}{60 x 1.389}\right]\left(\frac{30}{0.04}\right)^2 \times 10^{-6}$$

$$= 8.1\ \mu$$

$$\mu = 5.5 \times 10^{-3} \text{Pa.s}$$

4.(a) The state of stress at a point in a structural component of elastic material are given as follows: A normal tensile stress of 160 MPa and a shearing stress of 120 MPa on one plane, a normal compressive stress of 110 MPa and the complementary shearing stress of 120 MPa on a second plane orthogonal to the first plane, and no stress on third plane which is orthogonal to the above two planes.

Determine:

i. the principal stresses and the positions of the planes on which they act, and

ii. the positions of planes on which there is no normal stress. 10

Sol. Given: $\sigma_x = 160$ MPa, $\sigma_y = -110$ MPa, $T_{xy} = 120$ MPa

$$\sigma_{1,2} = \frac{1}{2}(\sigma_x + \sigma_y) \pm \frac{1}{2}\sqrt{(\sigma_x + \sigma_y)^2 + 4T_{xy}^2}$$

$$= \frac{1}{2}(160 - 110) \pm \frac{1}{2}\sqrt{(160+110)^2 + 4x120^2}$$

$$= 25 \pm 180.624$$

$$\sigma_1 = 205.624 \text{ MPa}, \sigma_2 = -155.624 \text{ MPa}$$

i. $$\tan 2\theta = \frac{2\tau_{xy}}{\sigma_x - \sigma_y} = \frac{2 \times 120}{160 + 110} = \frac{240}{270} = 0.8889$$

$$2\theta = 41.63$$

$$\theta_1 = 20.816°$$

$$\theta_2 = \theta_1 + 90° = 20.816 + 90 = 110.816°$$

ii. The planes on which normal stress is zero, there shall be only shear stress whose value shall be maximum. These planes are given by.

$$\tan 2\theta_3 = \frac{\sigma_y - \sigma_x}{2\tau_{xy}} = \frac{-110 - 160}{2 \times 120} = \frac{-270}{240} = -1.125$$

$$2\theta_3 = -48.366°$$

$$\theta_3 = -24.183° \text{ and } \theta_4 = 90° - 24.183° = 65.816°$$

4.(b) Describe the methods of the following heat Treatment procedures of steels:

i. Nitrating

ii. Cyaniding.

Sol.

i. Nitrating: Nitrating is a surface hardening process in which the surface of steel is saturated with nitrogen. It consists in heating the part to a temperature of 480° to 650°C inside a chamber through which a stream of NH_3 is passed. Ammonia gets dissociated:

$$2NH_3 = 2N + 3H_2$$

The atomic nitrogen so formed diffuses into α-iron and saturates the metal. After nitrating, the work is cooled down to 200°C in a stream of ammonia.

The nitride parts acquire a very high surface hardness (730 to 1100 BHN). Nitrating increases the wear resistance, fatigue limit and corrosion resistance in air, water and water vapor.

Nitrating is usually applied to medium carbon and alloy steels containing Al, Cr, Mo and other elements capable of forming nitrides (nitralloys). Prior to nitrating, parts should be hardened, tempered and undergo the complete sequence of machining operations, including grinding. Only finish grinding or lapping is done after nitrating. The nitride case is usually from 0.2 to 0.4 mm. thick.

Nitrating is done at low temperatures as compared to hardening and carburizing, so it requires more time. But, since no quenching is necessary as the high hardness is obtained directly after the operation. This feature enables hardening defects to be avoided.

ii. Cyaniding: In this process of surface hardening, both carbon and nitrogen are added to the surface layer of steel. The process is based on the decomposition of cyanide compounds that easily release the cyan group (CN). Since the cyan group contains carbon and nitrogen atoms, iron is saturated with both at the same time. Cyaniding produces a thin (0.1 to 0.2 mm) but very hard case in a very short time. The process is most commonly applied to steels containing from 0.2 to 0.4% C. It is especially effective for medium and small parts, such as, screws, clamps, washer, gears, piston pins, small shafts, etc. Cyaniding involves heating the steal in a liquid or solid medium.

4.(c) Four rotating masses A, B, C and D rotating in four different parallel planes are completely balanced. Angular position of masses C and D are 90° and I95° respectively from B in anticlockwise direction looking from right hand side. The rotating masses and their radius of mass centers are given below:

$$m_b = 25.\text{kg},\ m_c = 40\text{ kg},\ m_d = 35.\text{kg}$$

$$r_b = 150\text{ mm},\ r_c = 100\text{ mm},\ r_d = 180\text{ mm}$$

Planes B and C arc 250 mm. apart and mass centre radius for mass A is 150 mm. Determine using graphical method:

i. The mass of A and its angular position with respect to B, and

ii. Position of planes A and D with respect to plane B. 10

Sol. Given: ∠ BOC = 90°, ∠ BOD = 195°, ∠ COD = 195° – 90° = 105° Reference plane (RP) B.(Fig.7)

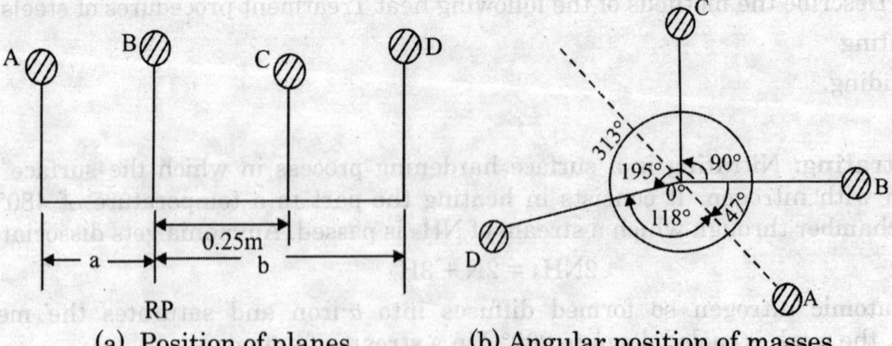

(a) Position of planes. (b) Angular position of masses.

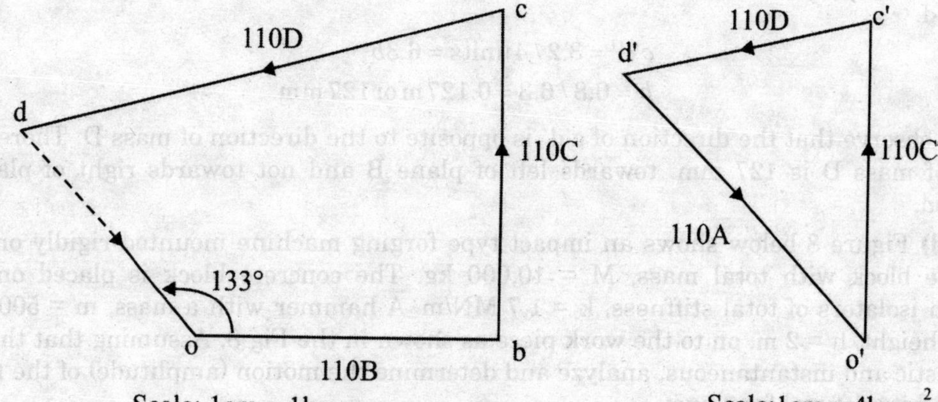

Fig.7. (d): Couple Polygon

Plane (1)	Mass m kg (2)	Radius r, m (3)	Centrifugal force mr/ω^2 kg·m (4)	Distance from RP ℓ, m (5)	Couple $mr\ell/\omega^2$ kg·m² (6)
A	m_a	0.15	$0.15m_a$	– a	$0.15m_a.a$
B(RP)	25	0.15	3.75	0	0
C	40	0.10	4.0	0.25	1.0
D	35	0.18	6.3	6	6.3b

Force Polygon: Draw the force polygon with data from column 4 as shown in (Fig.7c)

1. Draw ob = 3.75 units parallel to OB.
2. Draw bc = 4 units parallel to OC.
3 Draw cd = 6.3 units parallel to OD.
4. Join od and measure it

$$Od = 3.3 \text{ cm} = 0.15\ m_a.$$

$$m_a = 22 \text{ kg}$$

5. To locate the angular position of A, draw OA from O in Fig 7(b) parallel to od.

$$\angle AOB = 313° \text{ and } \angle AOD = 118°$$

Couple Polygon: Draw the couple polygon with the data from column 6 as shown in Fig.7(d).

1 Draw o'c' parallel to OC and equal to 1.0 units upwards

2. Draw a line from c' parallel to OD and another line from o' parallel to OA in Fig 7(b) to intersect at point d'.

3.

$$ó\,d' = 4.5/4 = -0.15\ m_a{\cdot}a$$

$$a = -\frac{(4.5/4)}{0.15 \times 22} = -0.34 \text{m or } -340 \text{ mm}$$

–ve sign indicates that plane A is towards right of plane B instead of on the left side as assumed.

$$c'd' = 3.2 / 4\,\text{units} = 6.3b$$

$$b = 0.8 / 6.3 = 0.127\,\text{m or } 127\,\text{mm}$$

We observe that the direction of c′d′ is opposite to the direction of mass D. Therefore, the plane of mass D is 127 mm. towards left of plane B and not towards right of plane B as assumed.

4.(d) Figure 8 below shows an impact type forging machine mounted rigidly on a large concrete block with total mass, M = 10,000 kg. The concrete block is placed on ground through isolators of total stiffness, k = 1.7 MN/m. A hammer with a mass, m = 500 kg falls from a height, h = 2 m. on to the work piece as shown in the Fig 8. Assuming that the impact is inelastic and instantaneous, analyze and determine the motion (amplitude) of the machine. Also find its natural frequency. 10

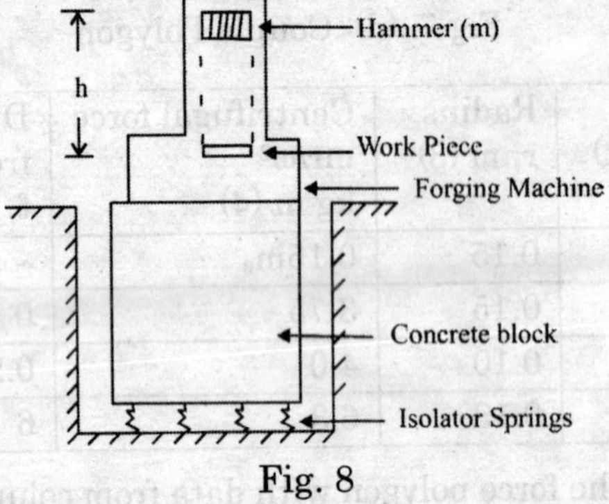

Fig. 8

Sol. Given: $m = 10,000$ kg, $k = 1.7$ MN/m, $m = 500$ kg, $h = 2$ m

Static deflection of springs, $\delta_{st} = \dfrac{mg}{k} = \dfrac{500 \times 9.81 \times 10^3}{1.7 \times 10^6} = 2.8853$ mm

$$\delta_{max} = \delta_{st}\left[1 + \sqrt{1 + \frac{2h}{\delta_{st}}}\right]$$

$$= 2.8853\left[1 + \sqrt{1 + \frac{2 \times 2000}{2.8853}}\right]$$

$$= 110.354 \text{ mm}$$

$$= \text{Amplitude of motion}$$

Natural frequency, $\omega_n = \sqrt{\dfrac{g}{\delta_{max}}} = \sqrt{\dfrac{9.81}{110.354 \times 10^{-3}}} = 9.428\,\text{rad/s}$

$$f_n = \frac{\omega_n}{2\pi} = \frac{9.428}{2\pi} = 1.5\,\text{Hz}$$

Section - C

Answer any TWO questions.

5.(a) Determine the dimensions of an optimum cylindrical riser attached to the side of a steel plate casting having dimensions 25 cm. × 12.5 cm. × 5cm. The volume shrinkage of steel during solidification is 3% and the volume of riser is 3 times that of dictated by the shrinkage consideration alone. 10

Sol. Volume of casting, $V_c = 25 \times 12.5 \times 5 = 1562.5 \text{ cm}^3$

For optimum cylindrical riser, $h = d$

Minimum volume of riser required to account for shrinkages,

$$V_r = \% \text{ shrinkages} \times V_c \times 3$$

$$= 0.03 \times 1562.5 \times 3$$

$$= 140.625 \text{ cm}^3$$

$$= \frac{\pi}{4}d^2h = \frac{\pi}{4}d^3$$

$$d = \left[\frac{4 \times 140.625}{\pi}\right]^{\frac{1}{3}} = 5.636 \text{ cm}$$

Dimensions of optimum riser: $h = d = 5.636$ cm

5. b (i) What kind of products are manufactured by wire drawing process?

(ii) How much force will approximately be required to draw a wire from 1.5mm. diameter steel wire to 1.0mm. diameter wire if the average yield strength of the work material is 300 MPa? 10

Sol.

i. The following kind of products are manufactured by wire drawing process:

1. Cables
2. Springs
3. Paper clips
4. Stringed musical instruments
5. Electric wiring
6. Spokes for wheels
7. Filaments for light bulbs
8. Stringed musical instruments Wire-frame structures
9. Nails
10. Screws
11. Bolts
12. Rivets

ii. Given: $d_i = 1.5$ mm, $d_f = 1.0$ mm, $(\sigma_y)_{avg} = 300\,\text{MPa}$, F = ?

Drawing force,
$$F = (\sigma_y)_{avg} A_f \ell n\left(\frac{A_i}{A_f}\right)$$

$$= 300 \times \frac{\pi}{4} \times 1^2 \times \ell n\left(\frac{1.5}{1.0}\right)^2$$

$$= 191.07 \text{ N}$$

5.(c) A bar of 70mm diameter is being cut orthogonally and is reduced to 68mm by a cutting tool. In case mean length of the chip is 68.9mm, find the cutting ratio. Determine shear angle also if the rake angle is 10°. 10

Sol. Given: $d_1 = 70$ mm, $d_2 = 68$ mm, $L_c = 68.9$ mm, $\phi = ?$, $\alpha = 10°$

Lenth of uncut chip,
$$L = \pi\left(\frac{d_1 + d_2}{2}\right) = \frac{\pi}{2}(70 + 68) = 216.77\text{mm.}$$

Cutting ratio,
$$r = \frac{t}{t_c} = \frac{\text{length of chip}}{\text{length of uncut chip}} = \frac{L_c}{L} = \frac{68.9}{216.77} = 0.318$$

$$\tan\phi = \frac{r\cos\alpha}{1 - r\sin\alpha} = \frac{0.318\cos 10°}{1 - 0.318\sin 10°} = 0.33147$$

$$\phi = 18.34°$$

5.(d) Name the machine tools by which teeth of straight toothed spur gears are produced in mass scale. Also mention what motions are imparted to the cutter and the blank in each of those machine tools. 10

Sol. The following machine tools are used for producing straight toothed spur gears on mass scale:

1. **Gear Hobbling:** It is a gear generating process in which the generating tool (cutter) is threaded and gashed, called the hob, and the process is called as gear hobbing. A hob resembles a worm, with gashes made parallel to its axis to provide cutting edges. Gear hobbing is a continuous cutting operation. The hob and the gear blank are connected by means of proper change gears. The ratio of hob and blank speeds is such that during one revolution of the hob, the blank turns through as many teeth as there are starts (threads) on the hob. The hob motion is rotary as well as translatory.
2. **Gear Shaping:** It is a profiling or form cutter method in which the spaces between the teeth are cut by means of a cutter conforming to the gear tooth form. The gear is driven by a gear train connected to the cutter drive. The cutter can be imparted translatory feed motion. The gear blanks are fixed on indexing spindle while the shaper tool oscillates as well as rotates as if in mesh with the work piece gear
3. **Gear planning:** It employs rack type cutters for generation of spur gears. Involube rack has straight edges and sharp corners. The cutter generates the teeth by its reciprocating planning action. The cutter oscillates as well as moves linearly as if in mesh with the workpiece gear.

6.(a) **Draw neat sketches of different Oxy-acetylene flames, label them and describe the two stages of combustion in gas welding. Show graphically the effects of current, pressure and time on the weld strength in resistance welding. What are the consequences of excessive or insufficient pressure during resistance welding?** 10

Sol. The different oxy-acetylene flames are shown in Fig. 9

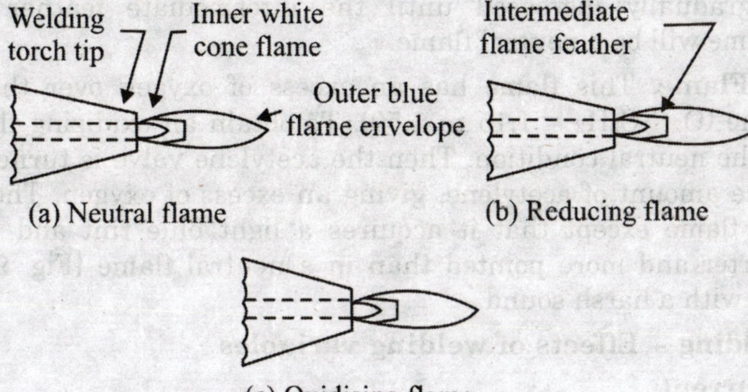

Fig. 9: OAW Flame Settings

1. **Neutral Flame.** A neutral flame is obtained when equal amounts of O_2 and C_2H_2 are mixed and burnt in a torch. The flame is recognized by two sharply defined zones, the inner white cone flame and the outer blue flame envelope, [Fig. 9 (a)]. The reaction at the inner cone for the neutral flame where equal volumes of cylinder oxygen and acetylene are used is,

First stage reaction:

$$C_2H_2 + O_2 \rightarrow 2CO + H_2 + 4.48 \text{ kJ/mol}$$

This provides the most concentrated heat with the highest temperature for welding at a distance of 3 to 5 mm. from the end of the inner cone. It is also apparent that the environment within the outer envelope consists of carbon monoxide and hydrogen and is relatively inert to materials that oxidise readily.

The reactions at the outer envelope are:

Second stage reaction:

$$2CO + O_2 \rightarrow 2CO_2$$

$$H_2 + 1/2\ O_2 \rightarrow H_2O \text{ (vapour)}$$

$$4CO + 2H_2O + 3O_2 \rightarrow 4CO_2 + 2H_2O + 812 \text{ kJ/mol}$$

For these reactions, the oxygen is supplied from the surrounding air. During actual welding, the outer envelope spreads over the surface of the work material and serves as a protective shield from the ordinary atmosphere. Also, since the heat developed is not as concentrated, this outer envelope of flame contributes only to preheat the work material for welding.

2. **Carburizing Flame.** A carburizing or reducing flame is obtained when an excess of acetylene is supplied than which is theoretically required ($O_2 : C_2H_2 = 0.85$ to 0.95). A reducing flame is recognized by three distinct sections: the inner cone (which is not

sharply defined) and an outer envelope as for the neutral flame. The third zone surrounds the inner cone and extends into the outer enveloping zone. It is whitish in color and is called "excess acetylene feather". Its length is an indication of the amount of excess acetylene [Fig. 9 (b)].

To obtain a neutral flame, first the reducing flame is obtained. Then the supply of oxygen is gradually increased until the intermediate feather disappears. The resulting flame will be a neutral flame.

3. **Oxidizing Flame.** This flame has an excess of oxygen over that required for a neutral flame ($O_2 : C_2H_2$ = 1.15 to 1.50). To obtain an oxidizing flame, the flame is first set to the neutral condition. Then the acetylene valve is turned down gradually to reduce the amount of acetylene, giving an excess of oxygen. The flame resembles the neutral flame except that it acquires a light blue tint and the inner cone is slightly shorter and more pointed than in a neutral flame [Fig. 9(c)]. An oxidising flame burns with a harsh sound.

Resistance welding – Effects of welding variables

1. Welding Current

The size of the weld nugget, and infact whether it will form or not depends upon the heat being generated faster than it is dissipated by conduction. Welding current is, thus, the most critical variable.

Both *a.c.* and *d.c.* are used to produce spot, seam, and projection welds. Most applications use single phase *a.c.* of mains frequency i.e. 50 hertz. However, *d.c.* is used for applications that need heavy current and the load for which can be balanced on a 3-phase power line. Also, with direct current machines the rate of current rise and fall can be programmed as per requirements. The current rise period or upslope and current decay period or down slope can be programmed with electronic control systems.

Control of upslope helps to avoid overheating and expulsion of molten metal at the beginning of the weld time as the interface resistance at that time is high. Down slope helps to control weld nugget solidification to avoid cracks in elements particularly in metals that are prone to quench-hardening and hot tearing

Accurate control of welding current is imperative for the success in resistance welding. Controls must, therefore, regulate the magnitude of cur-rent, its waveform, the timing, and the rest of the welding cycle. More accurately these parameters are controlled, better it is for the consistency of the welds.

2. Welding Time

The welding cycle time consists basically of four elements as explained below:

Squeeze Time: The time interval between the application of electrode pressure to the work and switching on the welding current is called the squeeze time. This time interval is provided to assure the contact between the electrode and the work and to initiate the application of force on it.

Weld Time: It is the time for which the welding current actually flows to melt the metal at the interface.

Hold Time: It is the time for which the electrodes are kept in position, after the welding current is switched off, to assure the application of pressure so as to consolidate the molten metal into a nugget which is then cooled by the dissipation of heat to the surrounding work material. If the applied force is excessive it may result in expulsion of molten metal from in-between the sheets.

Off Time: The time allowed to shift the work to the next location before the cycle is repeated is referred to as the off time. The electrodes are kept off the work during this time interval.

The time of current flow i.e. the weld time is controlled by electronic, mechanical, manual or pneumatic means. Timers may be synchronous or non-synchronous. The non-synchronous ones are those which start and stop the flow of welding current at any desired time with respect to the voltage waveform that is the opening and closing of the contactor is not necessarily synchronized with the waveform of the line-voltage. This may affect the a.c. frequency to the extent of ± 1 cycle. There are many non-critical applications where such a small deviation does not effect the quality of the weld to any considerable extent.

3. Pressure Control

Application of pressure through the electrode onto the workpieces ensures the completion of the electrical circuit. The force is applied by hydraulic, pneumatic, magnetic, or mechanical means. The pressure exerted depends upon the area of contact between the electrode and the work. The application of pressure serves a number of functions e.g.,

i. Brings the work pieces in close contact,

ii. Reduces the initial contact resistance at the interfaces,

iii. Suppresses metal expulsion between the workpieces, and

iv. Consolidates the molten metal into sound weld nugget.

The amount of pressure exerted depends upon the metal being welded. Soft metals may flatten under electrode pressure resulting in unsatisfactory weld or at least spoil the outer appearance of the work. Thus, apart from the welding current the clamping and squeezing pressures must be on the basis of the parent material, its thickness and the type of welding current used.

Most ferrous metals are welded with constant pressure but better results are obtained by the application of variable pressure for high conductivity, low resistivity metals. During the weld (or heating) time, a higher forging pressure may have to be exerted in order to obtain a sound rather than a superficial weld. To avoid metal expulsion it is essential for the work pieces to be forced together under high pressure after the required zone has reached the fusion temperature.

The effect of welding variables is shown in Fig.10

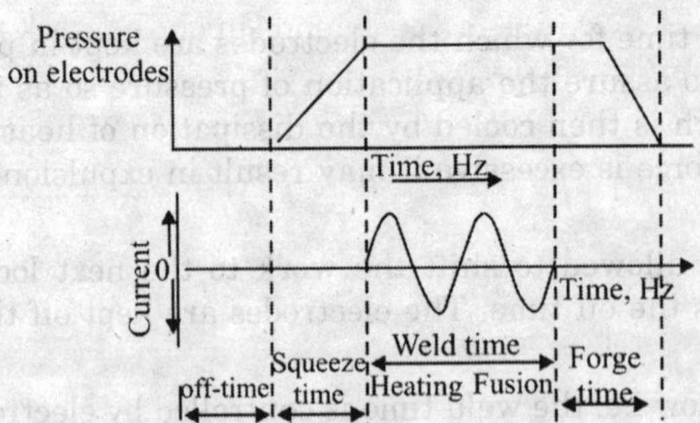

Fig.10: Effects of welding variables on weld strength in resistance welding.

6.(b) Show schematically the Merchant's force circle in orthogonal cutting. Derive the equations for shear and friction forces in terms of the material properties and cutting process parameters. State also the assumptions made while arriving at the final equations.

Sol. Merchant's Force cycle in Orthogonal cutting

The Orthogonal machining mechanism is shown in Fig.11

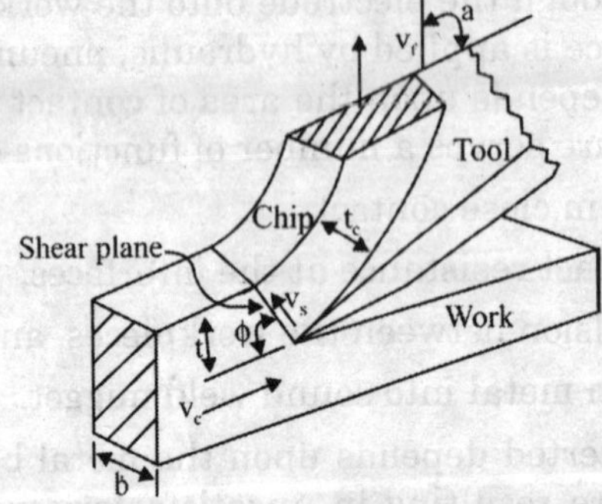

Fig. 11: Orthogonal machining mechanism.

The Merchant circle for mechanics of chip formation is shown in Fig. 12.

Angle of friction, $\beta = \tan^{-1} \mu = \tan^{-1}$ (F/N) where μ = coefficient of friction between tool and chip

F = force of friction, and N = normal reaction.

$$2\phi + \beta - \alpha = C$$

where C = machining constant for the work material.

The shear angle relations are given in Table below.

Table: Shear Angle Relations

Source	Result
Ernst and Merchant	$2\phi + \beta - \alpha = \pi/2$
Merchant's second solution	$2\phi + \beta - \pi = C_m$
Lee and Shaffer	$\phi + \beta - \alpha = \pi/4$
Stabler	$\phi + \beta - \alpha/2 = \pi/4$

The machining constant (C_m) in degrees depends upon the work material. It is around 70° for steel, 80° to 90° for stainless steel.

$$F = F_c \sin\alpha + F_t \cos\alpha$$
$$N = F_c \cos\alpha - F_t \sin\alpha$$

Resultant force, $R = [F_c^2 + F_t^2]^{\frac{1}{2}}$

Cutting force, $F_c = F_s \cos\phi + F_n \sin\phi$

Thrust force, $F = F_n \cos\phi - F_s \sin\phi$

Force parallel to the shear plane, $F_s = F_c \cos\phi - F_t \sin\phi$

Force perpendicular to the shear plane, $F_n = F_c \sin\phi + F_t \cos\phi$

Shear stress on the shear plane, $\tau_s = F_s/A_s$

Where

$A_s = b\, t/\sin\phi$, is the area of the shear plane.

$\tau_s = (F_c \sin\phi \cos\phi - F_t \sin^2\phi)/(b\, t)$

$F_c = F_s \cos(\beta - \alpha) / \cos(\phi + \beta - \alpha)$.

$$F_s = \frac{b_t T_s}{\sin\phi}$$

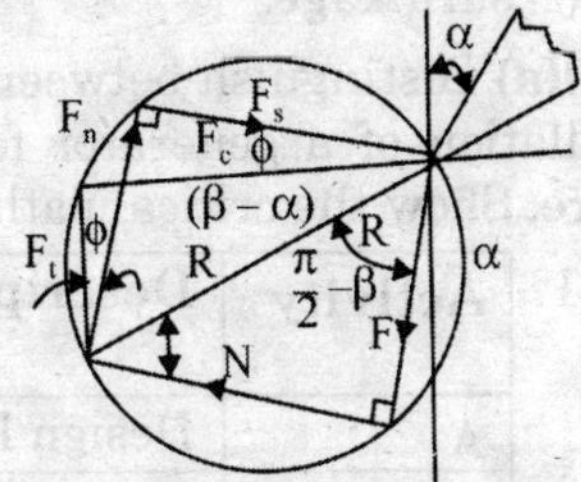

Fig. 12: Merchant's force circle for orthogonal cutting.

Assumptions

1. The work material is rigid, perfectly plastic.
2. The distribution of normal and shear stresses on the shear plane is uniform.
3. The cutting process is two – dimensional.
4. The work material is deformed across the shear plane.
5. The tool edge is sharp.

6.(c) Discuss the effect of welding speed on grain structure and properties of weld metals.

Sol. A large amount of deposition of the filler metal is caused due to too low a travel speed. On the other hand too high a speed does not give enough time for melting the base metal and consequently results in less deposition. When welding speed is increased, the amount of heat input, and corresponding filler metal per unit length of the joint decreases and also the joint reinforcement. The opposite would be time when the welding speed is decreased.

The grain structure is dependent on the temperature reached by the base metal during welding and the cooling rate.

6.(d) State the principles of Riser Design and discuss the factor affecting riser efficiency. (10)

Sol. Principles of Riser Design

1. To be effective, a riser must be the last part of the casting to solidify, such that all the shrinkages that are likely to occur should be in the riser.

2. The volume of the riser must be sufficient to compensate for metal shrinkage within the casting.
3. The riser must cover completely the casting section that requires feeding.
4. The fluidity of the metal inside the riser must be maintained so that the metal can flow from it and penetrate to the last contraction cavity.
5. The riser should be effective in establishing a pronounced temperature gradient with the casting so that the casting will solidify directionally towards the riser.

Factors affecting Riser efficiency

1. Materials for feeding aids should be exothermic or insulators.
2. Solidification time should be longer.
3. Exothermic material should be kept at the top of riser to eliminate centre line shrinkage.

7.(a) Distinguish between CPM and PERT. The table given below is for manufacture and installation of a generator for a small scale industry to generate power in case of supply failure. Show the critical path and calculate the project duration time.

Activity	Description of Activity	Duration of Activity	Immediate Predecessor
A	Design Plant Layout	12 days	—
B	Select Site	8 days	A
C	Select Vendors	4 days	A
D	Select Personnel	3 days	A
E	Prepare Site	12 days	B
F	Manufacture /Assemble Generator	18 days	C
G	Prepare Operation Manual	5 days	C
H	Install Generator	4 days	E,F
I	Train Opertors	9 days	D,G
J	Obtain Licence	6 days	H,I

Sol. Compesison of CPM and PERT.

CPM		PERT	
1.	Uses activity network.	1.	Uses event oriented network.
2.	Used extensively in construction projects.	2.	Used mostly in R & D projects.
3.	Uses deterministic concept.	3.	Uses probabilistic model concept.
4.	Can control both time and cost when planning.	4.	Is basically a tool for planning.
5.	Duration of activity may be estimated with a fair degree of accuracy.	5.	Estimate of time for activities are not so accurate and definite.

The network for the given data is shown in Fig.13.

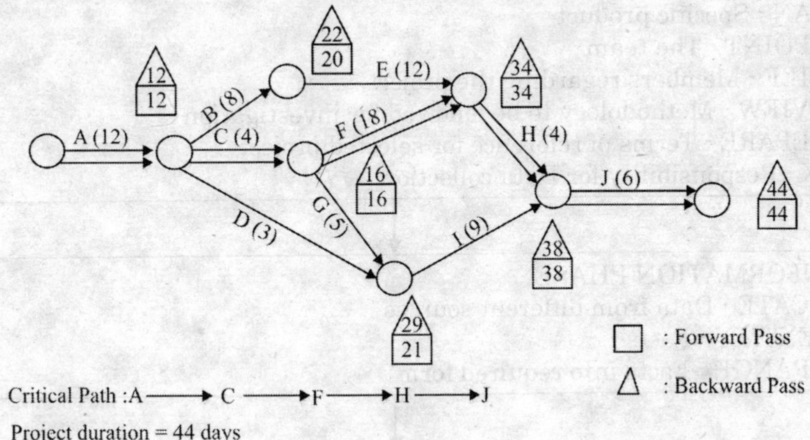

Fig. 13: Network

7.(b) There is a proposal to purchase a machine for Rs. 30,000/-. Manufacturing cost of a component from this machine is estimated at Rs. 6.0. The component can be sold in the market for Rs. 9.0. The life of the machine is estimated as 18 years. An attractive rate of return is reckoned as 12% and estimated average insurance and taxes as 3%. What quantity should be produced in year to make this proposal profitable?

Sol. Net rate of return, a = Actual rate of return – insurance and taxes = 12 – 3 = 9%

$$\text{Present worth of machine} = P\left[\frac{(1+r)^n - 1}{r(1+r)^n}\right]$$

$$= 30{,}000\left[\frac{(1+0.09)^{18} - 1}{0.09(1+0.09)^{18}}\right]$$

$$= \text{Rs.}262{,}668.75$$

At the break even point,

Fixed cost + variable cost × B.E. quantity = selling price × B.E. quantity

$$\text{F.C.} + \text{V.C.} \times \text{Q} = \text{S.P.} \times \text{Q}$$

$$\text{Q} = \frac{F.C.}{S.P.-V.C.} = \frac{262{,}668.75}{9-6} = 87{,}556.25 \sim 87{,}557$$

$\therefore$ Break even quantity = 87,557

7(c) (i) Describe seven phase job plan of value engineering.

(ii) Enlist the advantages of DNC over CNC.

PHASES AND CONSTITUENT ELEMENTS OF VALUE ENGINEERING

1. ORIENTATION PHASE
• SELECT - Project to be handled by V.E. Team
• ESTABLISH - Priorities
• PLAN - Specific product
• APPOINT - The team
• BRIEF - Members regarding the project
• REVIEW - Methodology to be followed for investigation
• PREPARE - Terms of reference for selected project
• FIX - Responsibility for Data collection

↓

2. INFORMATION PHASE
• LOCATE - Data from different sources
• IDENTIFY - Facts
• ARRANGE - Facts into required form

↓

3. FUNCTIONAL ANALYSIS PHASE
• LIST - The components or parts
• PREPARE - Verb-Noun description of functions
• ESTABLISH - Cost of BASIC FUNCTIONS
• ESTIMATE - Worth of each Basic function
• DETERMINE - Value improvement potential

↓

4. SPECULATION OR CREATIVE PHASE
• CONDUCT - Creative problem-solving
• GENERATE - Ideas, Combine and/or rearrange them to provide different ways to accomplish basic functions

↓

5. EVALUTION PHASE
• SHORT LIST - Creative ideas
• DEVELOP - Ideas functionally
• EVALUATE - Ideas according to the prefixed criteria
• RANK - Alternatives according to the cost reduction potential
• SELECT - One or more alternatives for development

↓

6. DEVELOPMENT PHASE
• ANALYSE - The alternatives for technical requirement
• CONFIRM - The economic feasibility
• ARRANGE - The necessary test runs
• SELECT - One or two alternatives for recommendations

↓

7 RECOMMENDATION PHASE
- PREPARE - The reports of V.A. proposals
- MAKE - Presentation to the decision-making authority
- SUBMIT - A definite plan for action
- SECURE - Approval for implementation

↓

8. IMPLEMENTATION PHASE
- OBTAIN - Approval to the proposal
- FIX - Responsibility for implementation
- AUTHORISE - Change to convert recommendations into actions
- EXPEDITE - Implementation
- RESOLVE - Problem identified during implementation
- TRAIN - Workmen
- RETIME - Jobs

↓

9. FOLLOW-UP PHASE
- AUDIT - Savings and compare results with original expectations
- SUBMIT - Cost savings achievement reports to the management
- EVALUATE - The effectiveness
- RECOMMEND - Changes/corrective actions and awards for contributors to value engineering proposals

(ii) Advantages of DNC over CNC

The major advantages of DNC system over CNC are:

i. Elimination of punched tapes and tape readers.

ii. Greater computational capability and flexibility.

iii. Convenient storage of NC part program in computer files.

iv. Programs stored as CLFILE, i.e., cutter path data rather than postprocessed programs for specific machine tools.

v. Reporting of shop performance.

vi. Establishes the frame work for the evolution of the fiture computer automated factory.

OTHER BOOKS BY THE SAME AUTHOR

Strength of Materials

STRENGTH OF MATERIALS

About the Book

Strength of Materials is a Textbook for the B.E. / B.Tech. students of UP Technical University, Lucknow and Uttrakhand Technical University, Dehradun. The book has been written as per the new syllabus of these Universities. The subject matter has been explained in the simplest possible way for easy assimilation by the students. This has been reinforced by a large number of solved examples.

Salient Features

- A large number of solved examples.
- Short answer type questions chapter wise.
- Unsolved end - of chapter exercises.

Edition : Latest ISBN : 978-93-81068-80-9 Pages (Approx.) : 600

FLUID MECHANICS

About the Book

Fluid Mechanics is a textbook for the B.E. / B.Tech. students of all Indian Universities and institutions. The subject matter has been explained in the simplest possible way for easy assimilation by the students. This has been reinforced by a large number of solved examples.

Salient Features

- A large number of solved examples,
- Short answer type questions chapter wise.
- Unsolved end - of chapter exercises.
- Multi - choice questions from ESE/ CSE/ GATE

Edition : Latest ISBN : 978-93-81068-65-6 Pages (Approx.) : 680

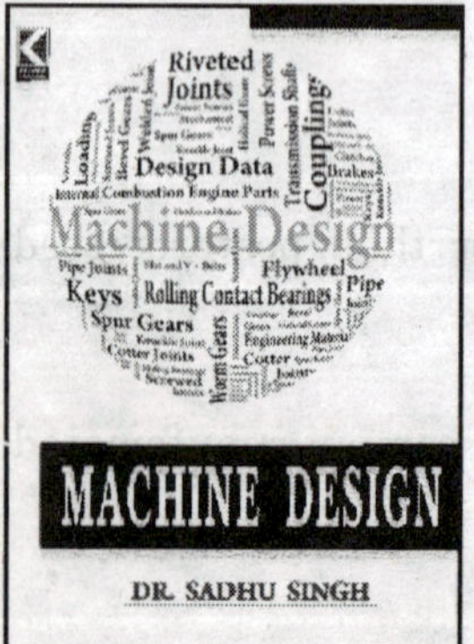

MACHINE DESIGN

About the Book

This book is useful for the students of B.Tech/ B.E. for all Indian Universities and Institutions. This book covers all the topics related to Machine Design Components and Machines. The matter contains data as per BIS. A lot many problems have been solved taking data from BIS. Utilizing the latest standards and codes, it is expected that the book shall be appreciated by the students as well as teachers

Edition : Latest ISBN : 978-93-82609-57-5 Pages (Approx.) : 832

MACHINE DESIGN DATA BOOK

Topics Covered

Basic Design Data ♦ Engineering Materials ♦ Design for Static Loading ♦ Design for Variable Loading ♦ Riveted Joints ♦ Welded Joints ♦ Screwed Joints ♦ Power Screws ♦ Transmission Shafts ♦ Keys and Couplings ♦ Cotter Joints ♦ Knuckle Joint ♦ Pipe Joints ♦ Mechanical Springs ♦ Flat and V – Belts ♦ Chains and Ropes ♦ Spur Gears ♦ Helical Gears ♦ Bevel Gears ♦ Worm Gears ♦ Sliding Bearings ♦ Rolling Contact Bearings ♦ Flywheel ♦ Clutches and Brakes ♦ Internal Combustion Engine Parts

Edition : Latest ISBN : 978-93-82609-51-3 Pages (Approx.) : 300